D1577035

The ENCYCLOPEDIA *of* MINERALOGY

ENCYCLOPEDIA OF EARTH SCIENCES SERIES

Volume

- I THE ENCYCLOPEDIA OF OCEANOGRAPHY/*Rhodes W. Fairbridge*
- II THE ENCYCLOPEDIA OF ATMOSPHERIC SCIENCES AND ASTROGEOLOGY/ *Rhodes W. Fairbridge*
- III THE ENCYCLOPEDIA OF GEOMORPHOLOGY/*Rhodes W. Fairbridge*
- IVA THE ENCYCLOPEDIA OF GEOCHEMISTRY AND ENVIRONMENTAL SCIENCES/ *Rhodes W. Fairbridge*
- IVB THE ENCYCLOPEDIA OF MINERALOGY/*Keith Frye*
- VI THE ENCYCLOPEDIA OF SEDIMENTOLOGY/*Rhodes W. Fairbridge and Joanne Bourgeois*
- VII THE ENCYCLOPEDIA OF PALEONTOLOGY/*Rhodes W. Fairbridge and David Jablonski*
- VIII THE ENCYCLOPEDIA OF WORLD REGIONAL GEOLOGY, PART 1: Western Hemisphere (Including Antarctica and Australia)/*Rhodes W. Fairbridge*
- XII THE ENCYCLOPEDIA OF SOIL SCIENCE, PART 1: Physics, Chemistry, Biology, Fertility, and Technology/*Rhodes W. Fairbridge and Charles W. Finkl, Jnr.*
- XV THE ENCYCLOPEDIA OF BEACHES AND COASTAL ENVIRONMENTS/*Maurice L. Schwartz*

ENCYCLOPEDIA OF EARTH SCIENCES, VOLUME IVB

The ENCYCLOPEDIA *of* MINERALOGY

EDITED BY

Keith Frye
Old Dominion University

Hutchinson Ross Publishing Company
Stroudsburg, Pennsylvania

Dedicated to the memory of Nancy Ferguson Frye (1943–1975)

Library of Congress Catalog Number: 81–982
ISBN: 0–87933–184–4

85 84 83 82 81 5 4 3 2 1
Manufactured in the United States of America.

Library of Congress Cataloging in Publication Data

Main entry under title:
The Encyclopedia of minealogy.
(Encyclopedia of earth sciences; v. 4B)
Includes bibliographies and indexes.
1. Mineralogy–Dictionaries. I. Frye, Keith, 1935– . II. Series.
QE355.E49 549′.03′21 81–982
ISBN 0–87933–184–4 AACR2

Distributed world wide by Academic Press,
a subsidiary of Harcourt Brace Jovanovich,
Publishers.

PREFACE

Information about certain minerals and their uses was already part of human knowledge by the time of the first hominids, as seen by the artifacts they left behind. One can only speculate on when the first observations on the relative merits of one kind of stone over another were made and transmitted. The oldest-known mineral artifacts reveal selectivity on the part of a collector; as is evident from the tools found at Olduvai Gorge, Tanzania, *Homo habilis* preferred chalcedony for choppers 1,750,000 years ago. The dietary use of the mineral *halite* (common table salt) also reveals knowledge about its properties and is of great antiquity (see *Salt Economy*).

Other early human activities requiring knowledge of mineral properties emerged much later and not before the late Paleolithic. They include selection of the best clays for making pots, identification of ore minerals that could be reduced to yield metals (see *Metallic Minerals; Metallurgy*), evaluation of certain minerals for their decorative qualities (see *Gemology*), as well as selection of minerals on the basis of hardness (see *Mohs Scale of Hardness*) and toughness (see *Jade*) for chopping and cutting tools.

Although tabulations of minerals and their properties, both physical and fanciful, were set forth before the 16th century, mineralogy as a science began only in the 1660s with the works of Robert Hooke and Niels Stensen (Nicolaus Steno) on the inorganic growth and consequent geometric shapes of crystals (see *Crystallography: History*). After three centuries of ceaseless study our science continues to expand, and to yield new secrets.

Coverage and Scope

The interests of mineralogists and the methods of their science transcend the study of minerals *sensu stricto*. Many of these broader interests and mineralogical activities are treated in articles such as *Abrasive Materials; Gemology; Glass: Theory of Crystallization; Invertebrate and Plant Mineralogy; Human and Vertebrate Minerals; Metallurgy; Portland Cement Mineralogy; Resin and Amber;* and *Synthetic Minerals.*

In addition to articles by practicing mineralogists about the many aspects of their science, this volume contains a mineral glossary and two indexes to assist users in finding mineralogic information. Topics not immediately found among the alphabetically listed article titles and cross-references may be searched for in the subject index. The mineral glossary contains nearly 3000 entries, approximately 2400 of which are considered to be valid mineral species. All mineral species are defined by chemical composition, structural group, and crystal system, with one or two references given. In addition, 530 of the more important minerals in terms of abundance, distribution, or economic importance are described in greater detail. Each mineral entry is accompanied by cross-references to articles in which it is mentioned. The second index is to authors cited in the references following the main entries.

In selecting and soliciting entries for this volume, the editor was faced with the question: Who are the likely readers? The advanced professionals certainly possess extensive libraries and would have little use for our volume. Mainly we are writing for the nonmineralogists. Since all levels of scientific and technical sophistication are covered in this category, we expect different levels of background among the users of the volume. Hence, each article has its own level of assumed background knowledge based on the judgments of the authors and the editor as to who might want to know about a particular aspect of the science.

Each article is accompanied by references considered to be sufficient to lead the user into more advanced and detailed treatment of each topic. Following them is a list of cross-references that includes related articles. In addition, further information concerning minerals mentioned in each entry may be found in the mineral glossary. Thus, for the user of this volume, each entry may be thought of as a starting point for study of the topics covered.

Since the various aspects of mineralogy may be divided and grouped together in many different ways, decisions regarding which entry should discuss a particular topic in detail were arbitrary in the final analysis. The policy of the editor was to let the experts define their own topics in their own ways. One result of this policy has been that there is an overlap of coverage in some cases, with different authors expounding the same material in different contexts. A second result has been the creation of gaps in coverage that the editor has attempted to fill with short entries of his own. But by means of the cross-references, users should have no difficulty in finding the particular mineralogic information for which they are searching.

As in any science that is active and growing, there are differences of opinion on certain topics. The editor has made no attempt to reconcile these differences—when one finds disagreement among experts, one finds disagreement among experts.

The editor has strong prejudice about the use of some words: "Parameter" refers to lattice parameters only (conic sections are not considered in this volume). "Lattice" is used in the strict crystallographic sense only, which does not admit to defects or vibrations. Rocks are "felsic" or "mafic," not "acid" or "basic." "Hydroxyl" refers to (OH) and "hydrous" to H_2O in chemical formulas except where otherwise designated as a mineral name—for example, *hydroxyapatite,* which was priority.

Minerals and Mineral Names

The purpose of this volume is to set forth the science of minerals as it exists today. In the strict sense, as adopted by mineralogists, a mineral is first of all crystalline and secondly a natural geochemical or biochemical product. Naturally formed materials that are noncrystalline are "mineraloids" (see *Opal; Resin and Amber*); crystalline materials of human manufacture are "synthetics" (see *Synthetic Minerals*). Mineraloids, however, may become minerals when, over periods of time, their constituent atoms become rearranged into an orderly crystalline pattern; synthetic materials may yield minerals when they react with their environments to form new crystalline compounds (an example of the latter occurs when a long-buried Roman coin becomes encrusted with *azurite*).

Each mineral has a name. Some names, such as *gold,* have roots lost in antiquity, while others are introduced year by year for newly discovered mineral species (see *Naming of Minerals*). Unlike names of compounds in organic and inorganic chemistry, which usually state systematically the composition of the compound, mineral names are trivial. Often they are simply identification labels not necessarily containing any reference to the chemistry, structure, or physical properties of any mineral species. In some cases they do, but many new minerals are named for their place of discovery or in honor of someone.

A nonmineralogist may see little net gain in calling natural NaCl *halite* rather than sodium chloride, but for the compound $KAl_2(OH)_2[AlSi_3O_{10}]$ a systematic name like potassium dialuminum dihydroxyl aluminotrisilicate is not likely to displace *muscovite* as the name of the mineral. Indeed, complex chemical compounds are rarely called by their full systematic names; they are trivialized, reduced to acronyms, or called by initial letters.

In addition to valid names for mineral species—that is, older names agreed upon by custom or newer ones approved by the International Mineralogical Association Commission on New Minerals and Mineral Names—many mineral-like names are used by both specialists and nonspecialists. They include varietal names such as amethyst for violet *quartz* and ruby for red *corundum.* Then there are series names, such as ***olivine*** for the complete crystal solution series between *forsterite* (Mg_2SiO_4) and *fayalite* (Fe_2SiO_4) (see *Olivine Group*), group names, such as ***mica*** for all the minerals with the muscovite structure, and synonyms, such as octahedrite for *anatase.* Moreover, a single name may refer to species (e.g., *pyrite*) and also to a group of minerals having the same structure (e.g., ***pyrite*** group).

In order to assist readers in recognizing the kind of mineralogic entity referred to by a mineral-like name, these names are encoded by typeface throughout this volume. All names of valid mineral species (e.g., *albite*) are set in lower-case italic type. All names of series or groups (e.g., ***plagioclase*** or ***feldspar***) are set in boldface italic letters. Varietal names, synonyms, and discredited names (i.e., a mineral named and later discovered not to be a valid species) are not distinguished by special typeface. Questions of the correct level of a mineral name are resolved in the mineral glossary of this volume.

Italic type is used not only for the names of valid mineral species but also for foreign words and phrases and for the titles of books and journals. Furthermore, when a mineral name is used in a sense that does not refer to a mineral species or group proper, no special designating typeface is utilized—e.g., *spinel* ($MgAl_2O_4$), a mineral species assigned to the ***spinel*** group because of its spinel structure; quartz syenite, the name of a rock different from *quartz*-bearing syenite.

Although many synonyms and discredited names are included in the mineral glossary, it is not intended as a compendium of ancient mineral names. For these names, the user is referred to Dana's *System of Mineralogy,* Hey's *Chemical Index of Minerals,* or Lazarenko and Vynar's *Mineralogical Dictionary.* In the non-English-speaking countries, mineral names are usually identical to those adopted here, with certain minor linguistic modifications (e.g., German forms drop the final "e" in *ite*; Spanish, French, and Italian add accents to help with pronunciation). While most mineral names (except for the oldest) conventionally end with *-ite,* it should be appreciated that many rock and fossil terms also have this suffix (see *The Encyclopedia of Petrology* and *The Encyclopedia of Paleontology*).

Abbreviations, Units, and Symbols

Although modern metric units (*Système International*) have been employed by the majority of the authors in this volume, two exceptions to the SI units are used by mineralogists and are retained in this volume. Unit-cell edges and other linear dimensions are given in angstrom units (10 Å = 1 nm) and pressures are given in bars and kilobars (1 bar = 100 kPa). Some SI equivalents are given in Table 1, exponential prefixes for units in Table 2, and mineralogic and crystallographic symbols in Table 3.

Literature

If mineralogy were an isolated discipline with no application of mineralogic information or techniques in other branches of science, and if the scientists of other disciplines had no interest whatsoever in minerals, no more than a handful of periodical publications would suffice to report the findings of all the mineralogists in the world. Such is not the case.

From characteristics of chemical bonding and crystal structures of *Abrasive Materials,* through cement chemistry (see *Portland Cement Mineralogy*), to the ion-exchange properties of *Zeolites,* mineral characteristics are of interest to scientists of many other disciplines. Hence, mineralogic data are published in a wide variety of periodicals, many of which have as their main focus of interest a topic or discipline that is not intrinsically mineralogic. In addition, some groups of minerals—an example would be clay minerals—are so complex and important that they are studied by specialists who even have their separate publications.

Both circumstances—interest from other disciplines and internal special interests—lead to a large number of publications in which mineral data and mineralogic techniques are the subject in part, at least, of scientific articles. Indeed, mineralogists themselves are hard pressed to keep abreast of all published developments that may be of interest. Part of this difficulty is linguistic; few mineralogists are proficient in all languages in which mineralogic information is published. Another part, previously mentioned, is the diversity of publications.

One publication that brings together mineralogic data for English-speaking readers is *Mineralogical Abstracts,* published jointly by the Mineralogical Society of Great Britain and the Mineralogical Society of America. In 1976, 44 abstractors from 10 different countries organized abstracts from 21 contributors from national mineralogical society members of the International Mineralogical Association.

The accompanying list of periodicals includes those that are currently publishing or have recently published mineralogic information. They range in character from periodicals presenting predominantly mineralogical information to those publishing it only incidentally. The decision to include or exclude

TABLE 1. American Equivalents to SI Units

Length	
millimeter (mm)	0.039 inch
centimeter (cm)	0.394 inch
meter (m)	3.281 feet
kilometer (km)	0.621 mile
Area	
centimeter^2 (cm^2)	0.155 inch^2
meter^2 (m^2)	10.764 feet^2
kilometer^2 (km^2)	0.386 mile^2
Volume	
centimeter^3 (cm^3)	0.061 inch^3
liter (l)	1.057 quarts
meter^3 (m^3)	35.316 feet^3
kilometer^3 (km^3)	0.240 mile^3
Mass	
gram (g)	0.035 ounce avoirdupois
kilogram (kg)	2.205 pounds avoirdupois
Mass per unit volume (density)	
grams/centimeter^3 (g/cm^3)	0.574 ounce per inch^3
Pressure	
kilopascal (kPa)	0.145 pound per inch^2

TABLE 2. Exponential Prefixes for Units

Exponent	Prefix	Symbol
12	tera	T
9	giga	G
6	mega	M
3	kilo	k
2	hecto	h
1	deca	da
-1	deci	d
-2	centi	c
-3	milli	m
-6	micro	μ
-9	nano	n
-12	pico	p
-15	fento	f
-18	atto	a

10^{-9} m = 1 nm = 10 Å

TABLE 3. Crystallographic Symbols Used in This Volume

Symbol	Meaning
a, b, c	Lengths of unit-cell edges.
a, b, c	Unit-cell vectors.
α, β, γ	Interaxial angles $y \wedge z$, $z \wedge x$, $x \wedge y$.
x-, y-, z-, (u-)	Directions of crystallographic axes of coordinates.
x, y, z	Coordinates of any point (equivalent position) in the unit cell, expressed in terms of a, b, c.
u, v, w	Coordinates of any lattice point, expressed in terms of a, b, c as units.
$[u\ v\ w]$	Indices of a direction in the direct (real-space) lattice (zone axis).
$\langle u\ v\ w \rangle$	Indices of a "form" of zone axes (related by symmetry).
$(h\ k\ l)$	Indices of a crystal face, or of a single plane, or of a set of parallel planes.
$h\ k\ l$	Indices of the reflection from a set of parallel planes; coordinates of a reciprocal lattice point.
$\{h\ k\ l\}$	Indices of a form of planes, or of the reflections from a form of sets of parallel planes.
$\{h\ k\ i\ l\}$	Ditto for the hexagonal coordinate axes x-, y-, u-, z-.
a^*, b^*, c^*	Lengths of reciprocal-lattice unit-cell edges.
a* b* c*	Unit-cell vectors in reciprocal space.
$\alpha^*, \beta^*, \gamma^*$	Interaxial angles in reciprocal space.
d_{hkl}	Interplanar spacing of planes (hkl).
V or V_c; V^*	Unit-cell volume in direct and in reciprocal space.
$p, c; \rho$	Primitive and centred two-dimensional lattices; one-dimensional lattice.
P, R, A, B, C, F, I, H	Lattice symbols in three dimensions.
1, 2, 3, 4, 6	Rotation-axis symbols (X) (or rotation-point symbols in two dimensions).
$\bar{1}, \bar{2}, \bar{3}, \bar{4}, \bar{6}$	Inversion-axis symbols ($\bar{X}$).
$2_1, 3_1, 3_2, 4_1, 4_2, 4_3, 6_1, 6_2, 6_3, 6_4, 6_5$	Screw-axis symbols.
m	Mirror-reflection plane in three dimensions, mirror-reflection line in two dimensions, or mirror-reflection point in one dimension.
g	Glide-reflection line in two dimensions.
a, b, c, n, d	Glide-reflection planes in three dimensions.
n, m	Any integers.
H	Mohs hardness
G	Specific gravity (or density in g/cm^3)

periodicals at the latter end of the spectrum was somewhat arbitrary.

Besides the title of the periodical, most of the entries include data of first publication, any sponsoring organization, and the publisher. Publications of special interest to mineral collectors appear with the entry *Collecting Minerals.*

Periodicals

Academiae Scientiarum Finnicae, *Ann.* Series A.III, 1909–, Akateeminen Kirjakauppa, Helsinki 10, Finland.

Académie des sciences, *Comptes rendus,* Ser. D, Sciences Naturelles, 1835–, Gautier-Villars, Paris, France.

Academy of Sciences of the USSR, *Doklady,* Earth Sciences sections (Translated from *Doklady Akademiia Nauk SSSR*), American Geological Institute, Scripta Publishing Company, Washington, DC 20005, USA.

Acta Chemica Scandinavica, 1947–, Munksgaard, Noerre Soegade 35, DK-1370 Copenhagen K, Denmark.

Acta Crystallographica, Section A: Crystal Physics, Diffraction, Theoretical and General Crystallography; Section B: Structural Crystallography and Crystal Chemistry, 1948–, Munksgaard, Noerre Soegade 35, DK-1370 Copenhagen K, Denmark.

Acta Geologica, 1952–, Magyar Tudományos Akademia, H 1363 Budapest, Hungary.

Acta Geológica Hispánica, 1966–, Instituto Nacional de Geologia, Avenida José Antonia 585, Barcelona, Spain.

Acta Geologica Polonica, 1950–, Polska Akademia Nauk, Komitet Nauk Geologicznych, Warszawa, Poland.

Acta Geologica Sinica, Plenum Publishing Corporation, New York and London.

Acta Geologica Taiwanica, 1947–, Science Reports of the National Taiwan University, Taipei, Taiwan.

Akademiya Nauk SSSR, *Izvestiya,* 1936–; *Doklady,* Seriya Geologicheskaya, Moscow, USSR.

American Ceramic Society, *Journal,* 1905–; *Bulletin,* 1922–, 65 Ceramic Drive, Columbus, Ohio 43214, USA.

American Chemical Society, *Journal,* 1879–, Washington, DC 20036, USA.

American Crystallographic Association, *Transactions,* 1965–, Polycrystal Book Service, Pittsburgh, Pennsylvania, USA.

American Journal of Science, 1818–, Yale University, New Haven, Connecticut 06520, USA.

of America, 1909 K Street NW, Washington, DC 20006, USA.

Arkiv för Mineralogi och Geologi, 1949–, Almqvist & Wilsell, Stockholm, Sweden.

Der Aufschluss–Zeitschrift für die Freunde der Mineralogie und Geologie (VFMG), Dantestrasse 50, Heidelberg, GFR.

Australasian Institute of Mining and Metallurgy, *Proceedings,* 1898–, Clunis Ross House, 191 Royal Parade, Parkville, Victoria 3052, Australia.

Australian Gemmologist, 1958–, Gemmological Association of Australia, Box 149, GPO Sydney, N.S.W. 2001, Australia.

Australian Geology and Geophysics, See *B M R Journal.*

Australian Mineralogist, 1976– (in *Australian Gems and Crafts Magazine*), GPO Box 1071J, Melbourne, 3001 Victoria, Australia.

Beiträge zur Mineralogie und Petrographie, see *Contributions to Mineralogy and Petrology.*

B M R Journal of Australian Geology and Geophysics, 1976–, Bureau of Mineral Resources, Geology and Geophysics, Box 378, Canberra City, A.C.T. 2601, Australia.

Boletin Geológico y Minero. Instituto Geológica y Minero de España, Ríos Rosas, 23, Madrid, 3, Spain.

British Ceramic Society, *Transactions and Journal,* 1902–, Shelton House, Stoke Road, Shelton, Stoke-on-Trent, UK.

British Museum (Natural History), *Bulletin, Mineralogy,* 1952–, Cromwell Road, London SW7, UK.

Bulletin de Minéralogie (Société français de Minéralogie et de Cristallographie), 1878–, Masson et Cie., 120 Bd. Saint-Germain, 75280 Paris Cedex 06, France. (Publication title changed vol. **101**, 1978, see Société français.)

Canadian Journal of Earth Sciences, 1964–, National Research Council of Canada, Ottawa, Ontario, Canada K1A 0R6.

Canadian Mineralogist, 1957–, Mineralogical Association of Canada, Royal Ontario Museum, Department of Mineralogy, Toronto, Ontario, Canada M5S 2C6.

Carnegie Institution Yearbook, 1902–, Annual Report to the Director, The Geophysical Laboratory, 2801 Upton Street NW, Washington, DC 20008, USA.

Časopis pro Mineralogii a Geologii, 1956–, Societas Mineralogica et Geologica Bohemoslovaka, Vodickova 40, 112 29 Prague 1, Czechoslovakia.

Chemical Abstracts, 1907–, American Chemical Society, The Ohio State University, Columbus, Ohio 43210, USA.

Chemical Geology, 1966–, An International Journal, Elsevier Scientific Publishing Company, P.O. Box 211, Amsterdam, The Netherlands.

Chemical Society, *Journal,* 1847–; *Dalton Transactions; Faraday Transactions,* Burlington House, London W1V OBN, UK.

Chemie der Erde, 1914–, Zeitschrift für chemische Mineralogie, Petrographie, Bodenkunde, Geochemie und Meteoritenkunde, Villengang 2, Postfach 176, 69 Jena, East Germany.

Clay Minerals, 1947–, Journal of the Clay Minerals Group of the Mineralogical Society, Oxford 0X2 0E1, England, UK.

Clay Science (Nendo Kagaku), 1960–, Nippon Nendo Gakkai, Department of Mineral Industry, School of Science and Engineering, Waseda University, 4–170 Nishi Okubu, Shinjuku-ku, Tokyo 160, Japan.

Clays and Clay Minerals, 1952–, Journal of the Clay Minerals Society, Headington Hill Hall, Oxford, England, UK.

Compass, 1920–, Sigma Gamma Epsilon, University of Oklahoma, Norman, OK 73019, USA.

Contributions to Mineralogy and Petrology, Beiträge zur Mineralogie and Petrologie, 1947–, International Mineralogical Association, Springer-Verlag, Heidelberger Platz 3, D-1000 Berlin 33, GFR, and 175 Fifth Avenue, New York, NY 10010, USA.

Crystallography (Soviet Physics), Translation of *Kristallografiya,* American Institute of Physics, 335 East 45th Street, New York, NY 10017, USA.

Crystal Structure Communications, 1971–, Università di Parma, Via M. d'Azeglio 85, 43100 Parma, Italy.

CSIRO Minerals Research Laboratories, Annual Reports; Investigation Report, East Melbourne, Australia.

Cuadernos Geología 1970–, Universidad de Granada, Secretariado de Publicaciones, Granada, Spain.

Current Science, India, 1932–, Current Science Association, S. R. S. Sastry, Bangalore 560006, India.

Dansk Videnskabernes Skrifter, 1928–, Det Kongelige Danske Videnskabernes Selskab, Dantes Plads 5, København V, Denmark.

Denmark, Geological Survey, *Bulletin,* Noerre Soegade 35, DK-1370, København K, Denmark.

Deutschen Gemmologische Gesellschaft, *Zeitschrift,* 1951–, Gewerbehalle, Postfach 2717, 6580 Idar-Oberstein 2, GFR.

Doklady–Earth Science Sections (translation of *Doklady Akademii Nauk SSSR*), 1959–, Scripta Technica, Inc., 1511 K Street NW, Washington, DC 20005.

Earth and Planetary Science Letters, 1966–, Elsevier Scientific Publishing Company, P.O. Box 211, Amsterdam, The Netherlands.

Earth Science Reviews, 1965–, Elsevier Scientific Publishing Company, Box 211, Amsterdam, The Netherlands.

Economic Geology, 1906–, The Bulletin of the Society of Economic Geologists, New Haven, Connecticut, 06520, USA.

Finland, Geological Survey, *Bulletin,* 1895–, Helsinki, Finland.

Fortschritte der Mineralogie, 1911–, Deutsche Mineralogische Gesellschaft, E. Schweizerbart'sche Verlagsbuchhandlung, Johannesstrasse 3 a, 7000 Stuttgart, West Germany.

Gemmological Society of Japan, *Journal (Hoseki Gakkaishi),* 1974–, Tohoku University, Sendai 980, Japan.

Gems and Gemology, 1934–, Gemological Institute of America, 1660 Stewart Street, Santa Monica, CA 90404, USA.

Geochemistry International (replaced *Geochemistry,* 1964), Translation of *Geokhimiya Akademii Nauk SSSR,* American Geological Institute and the American Geophysical Union, Scripta Publishing Company, Washington, DC 20005, USA.

Geochimica (China), in translation: Plenum Publishing Corporation, New York and London.

Geochimica et Cosmochimica Acta, 1963–, The Geochemical Society and the Meteorological Society, Pergamon Press, Oxford OX3 OBW, UK.

Geoderma, 1967–, Elsevier Publishing Company, Box 211, Amsterdam, The Netherlands.

Geokhimiya, Akademiya Nauk SSSR, 1956–, Institut Geokhimii i Analitichnoi Khimmi im. V. I. Vernadskogo, Vorobevskoe Shosse, 47a, Moscow, USSR.

Geological Association of Canada, *Proceedings,* 1947–, Toronto, Ontario, Canada.

Geological Institute, *Bulletin,* 1892–, University of Uppsala, Uppsala, Sweden.

Geological Journal, 1964–, Seel House, Seel Street, Liverpool L1 4AY, England.

Geological Magazine, 1864–, Cambridge University Press, London, NW1 2DB, UK.

Geological Society, *Journal,* 1845–, Scottish Academic Press, 33 Montgomery Street, Edinburgh EH7 5JX, Scotland.

Geological Society of America, *Bulletin,* 1888–; *Abstracts with Programs,* 1969–, 3300 Penrose Place, Boulder, Colorado 80301, USA.

Geological Society of Australia, *Journal,* 1953–, Adelaide, NSW, Australia.

Geological Society of Denmark, *Bulletin,* 1894–, 35 Noerre Soegade, DK-1370 Copenhagen K, Denmark.

Geological Society of Finland (Suomen Geologinen Seura), *Bulletin,* 1929–, Helsinki, Finland.

Geological Society of India, *Journal,* 1959–, 16 Ali Asker Road, Bangalore 560052, India.

Geological Society of Japan (Nihon Chishitsu Gakkai), *Journal,* 1893–, University of Tokyo, 7-3-1 Hongo, Bunkyo-ku, Tokyo 113, Japan.

Geological Society of South Africa, *Transactions,* 1896–, Marshalltown, Transvaal, South Africa.

Geological Survey of Canada, *Annual Report,* 1843–; *Memoirs,* 1910–; *Papers,* 1935–, Department of Mines, Ottawa, Ontario, Canada.

Geological Survey, Finland, *Bulletin,* 1950–, Geologinen Tutkimuslaitos, Otaniemi, Helsinki, Finland.

Geological Survey, Great Britain, *Bulletin,* 1939–, London.

Geological Survey of Japan, *Bulletin,* 1950–, 135 Hisamato, Takatsu-ku, Kawasaki 213, Japan.

Geologicky Zbornik, 1950–, Slovenska Akademia Vied, Klemensova 19, 895 30 Bratislava, Czechoslovakia.

Geologiska Föreningens i Stockholm, *Fërhandlingar,* 1872–, 104 05 Stockholm, Sweden.

Geologists' Association, *Proceedings,* 1859–, Chelsea College, 171-3 King Street, London W6 9LZ, England, UK.

Geologiya i Geofizika, 1960–, Institut Geologii i Geofiziki, Novosibirsk, Akademgorodok, USSR (see *Soviet Geology*).

Geology, 1973–, Geological Society of America, 3300 Penrose Place, Boulder, Colorado 80301, USA.

Geoscience Canada, 1974–, Geological Association of Canada, 111 Peter Street, Toronto 1, Canada.

Geoscience Magazine (Chigaku Kenkyu), Kamigyo-ku Kyoto, Japan.

Groupe français des Argiles, *Bulletin,* Centre national de la recherche scientifique, Paris, France.

Grønlands Geologiske Undersøgelse, *Bulletin,* 1948–, København, Denmark.

High Temperatures–High Pressures, 1969–, Pion Ltd., 207 Bronesbury Park, London NW2 5JN, UK.

Indian Journal of Earth Sciences, 1974–, Indian Society of Earth Sciences, Presidency College, Calcutta 700073, India.

Indian Mineralogist, 1959–, Journal of the Mineralogical Society of India, Manasagangotri, Mysore 6, India.

Industrial Diamond Review, 1940–, N. A. G. Press, London, UK.

Institute of Geological Sciences, *Annual Report,* London SW7 2DE, UK.

Institution of Mining and Metallurgy, *Transactions* (Section B: Applied Earth Science), 1892–, 44 Portland Place, London W1N 4BR, UK.

International Geological Congress, *Proceedings,* Published in cities of meetings, usually every four years.

Japanese Association of Mineralogists, Petrologists and Economic Geologists, *Journal* (Ganseki Kobutsu Kosho Gakkaishi), 1959–, Tohoku University Faculty of Science, Aobayama, Sendai-shi 980, Japan.

Journal of Applied Crystallography, 1968–, International Union of Crystallography, Copenhagen, Denmark.

Journal of Chemical Physics, 1933–, American Institute of Physics, 335 East 45th Street, New York, NY 10017, USA.

Journal of Crystal Growth, 1967–, North-Holland Publishing Company, P. O. Box 211, Amsterdam, The Netherlands.

Journal of Gemmology, 1947–, Gemmological Association of Great Britain, Proceedings, London, UK.

Journal of Geochemical Exploration, 1972–, Association of Exploration Geochemists, Elsevier Publishing Company, Box 211, Amsterdam, Netherlands.

Journal of Geology, 1893–, University of Chicago Press, Chicago, Illinois 60657, USA.

Journal of Geophysical Research, 1896–, American Geophysical Union, Washington, DC 20006, USA.

Journal of Materials Science, 1966–, Chapman and Hall, London EC4P 4EE, UK.

Journal of Petrology, 1960–, The Clarendon Press, Oxford, UK.

Journal of Physical Chemistry, 1896–, American Chemical Society, 1155 16th Street NW, Washington, DC 20036, USA.

Journal of Physics C: Solid State Physics, 1968–, Institute of Physics, Techno House, Redcliffe Way, Bristol BS1 6NX, UK.

Journal of Sedimentary Petrology, 1931–, Society of Economic Paleontologists and Mineralogists, Tulsa, Oklahoma; USA.

Journal of Solid State Chemistry, 1969–, Academic Press, 111 Fifth Avenue, New York, NY 10003, USA.

Korean Institute of Mining, *Journal (Kwangsan Hakhoe Chi),* 1968–, 172 Kangnung-Dong, Dobang-Ku, Seoul 130–02, South Korea.

Kristallografiya, 1956–, Akademiya Nauk SSSR, Leninskii, 14, Moscow V-71, USSR (see *Crystallography*).

Kwartalnik Geologiczny, 1957–, Instytut Geologiczny, Rakowiecka 4, Warsaw, Poland.

Lapidary Journal, 1945–, P. O. Box 2369, San Diego, California, USA.

Lapis, 1976–, Christian Weiss Verlag, 8 München 2, Oberanger 6, GFR.

Lithos, 1968–, National Councils for Scientific Research in Denmark, Finland, Norway, and Sweden, Universitetsforlaget, Box 307, Oslo 3, Norway.

Materials Research Bulletin, 1966–, An international journal reporting on research in crystal growth and materials preparation and characterization, Pergamon Press, Oxford OX3 OBW, UK, and Elmsford, New York 10523, USA.

Meddelelser om Grønland, 1879–, Copenhagen, Denmark.

Metallurgical Transactions A, Physical Metallurgy and Materials Science, 1970–, American Society for Metals, Metals Park, OH 44073, USA.

Mineralia Slovaca, 1969–, Leningradska 11, 896 26 Bratislava, Czechoslovakia.

Mineralium Deposita, 1966–, Society for Geology Applied to Mineral Deposits, Springer-Verlag, Berlin-Heidelberg-New York.

Mineralogia Polonica, 1970–, Polskie Towarzystwo Mineralogiczne, Al. Michiewicza, Krakow, Poland.

Mineralogica et petrographica acta (Acta Geologica alpina), 1948–, Contributions from the Istituto di mineralogie e petrografia, Università di Bologna, Bologna, Italy.

Mineralogical Abstracts, 1920–, (separate 1959–), The Mineralogical Society of Great Britain and the Mineralogical Society of America, 41 Queen's Gate, London SW7 5HR, UK.

Mineralogical Journal, 1953–, Mineralogical Society of Japan (Nihon Köbutsugakki), Tokyo, Japan.

Mineralogical Magazine, 1876–, Journal of the Mineralogical Society of Great Britain, 41 Queen's Gate, London SW7 5HR, UK.

Mineralogical Record, 1970–, Tucson, AZ 85740, USA.

Mineralogicheskii Sbornik (Lvov), 1947–, L'vovskii Gosudarstvennyi Universitet, Lvov, USSR.

Minerals Science and Engineering, 1969–, National Institute for Metallurgy, Private Bag 7, Private Bag X3015, South Africa.

Natural Resources Forum, 1976–, United Nations, D. Reidel Publishing Company, Dordrecht, Holland.

Nature, 1869–, Macmillan Journals Ltd., London WC2R 3LF, UK, and Washington, DC 20045, USA.

Die Naturwissenchaften, 1913–, Springer-Verlag, D-6900 Heidelberg 1, GFR.

Nendo Kagaku (see *Clay Science*).

Neues Jahrbuch für Mineralogie, 1830–, Monatshefte, Abhandlungen, Abteilungen, E. Schweizerbart'sche Verlags buchhandlung, Stuttgart, GFR.

New Zealand Journal of Geology and Geophysics, 1958–, Department of Scientific and Industrial Research, Wellington, New Zealand.

New Zealand Journal of Sciences, 1958–, Department of Scientific and Industrial Research, Box 9741, Wellington, New Zealand.

Norges Geologiske Undersoekelse, Bulletin, 1972–, Universitetsforlaget, Oslo, Norway.

Norsk Geologisk Tidsskrift, 1905–, Norsk Geologisk Forening, Universitetsforlaget, Box 307, Blindern, Oslo 3, Norway.

Österreichische Akademie der Wissenschaften, Mathematischnaturwissenschaftliche Klasse, Anzeiger, Springer, Vienna, Austria.

Periodico di Mineralogica, 1930–, Rome, Italy.

Petrologie, 1975–, Doin Editeurs, 8, Place de l'Odeon, F75006 Paris, France.

Philosophical Magazine, 1798–, Taylor and Francis Ltd., 10-14 Macklin Street, London WC2 B5NF, UK.

Philosophical Transactions, Series A, 1665–, Royal Society, 6 Carlton House Terrace, London SW1Y 5AG, UK.

Physics and Chemistry of the Earth, 1956–, Pergamon Press, Elmsford, NY 10523, USA.

Physics and Chemistry of Minerals, 1977–, International Mineralogical Association, Springer-Verlag, Postfach 105280, D-6900 Heidelberg 1, GFR.

Polska Akademia Nauk, Prace Mineralogiczne, 1965–; *Archiwum Mineralogiczne,* 1925–, ul Zwirki i Wigury 93, Warsaw 22, Poland.

Precambrian Research, 1974–, Elsevier Scientific Publishing Company, Box 211, Amsterdam, Netherlands.

Progress in Crystal Growth and Characterization, 1977–, Pergamon Press, Maxwell House, Fairview Park, Elmsford, NY 10523, USA.

Progress in Materials Science, 1949–, Pergamon Press, Elmsford, NY 10523, USA.

Rocks and Minerals, P.O. Box 29, Peekskill, New York, 10566, USA.

Rumania, Comitetul de Stat al Geologiei, Institutul Geologic, *Studii Technice si Economice, Seria I, Mineralogie-Petrographie,* Bucharest, Rumania.

Schweizerische mineralogische und petrographische Mitteilungen, 1920–, Zurich, Switzerland.

Science, 1883–, American Association for the Advancement of Science, 1515 Massachusetts Avenue, NW, Washington, DC 20005, USA.

Sciences de la Terre, Nancy, Université École Nationale Supérieure de Geologie Appliquée et de Prospection Minière, France.

Scientia Geologica Sinica, 1975–, Plenum China Program, 227 West 17th Street, New York, NY 10001, USA.

Scientific Society of Warsaw, *Mineralogic Archives,* 1925–, Warszawa, Poland.

Scottish Journal of Geology, 1965–, Scottish Academic Press, 33 Montgomery Street, Edinburgh EH7 5 JX, Scotland.

Scripta Geologia, 1971–, Universita J. E. Purkyne, Brno, Czechoslovakia.

Sedimentary Geology, 1965–, Elsevier Scientific Publishing Company, Box 211, Amsterdam, Netherlands.

Silicáty/Silicates, 1957–, Ceskoslovenska Akademie Ved, Vodicova 40, 112 29 Prague 1, Czechoslovakia.

Société français de Minéralogie et de Cristallographie, *Bulletin,* 1878–1977 (name changed to *Bulletin de Minéralogie* with vol. **101**, 1978), Masson et Cie, Paris, France.

Société Géologique de Belgique, *Annales,* 1877–, Universite de Liége, 7 Place du Vingt-Août, 4000 Liége, Belgium.

Sociedade Geologica de Portugal, *Boletim,* Falculdade de Ciencias, Rua da Escola Politecnica, 58, Lisbon, 2, Portugal.

Società Italiana de Mineralogia e Petrologia, *Rendiconti,* Musea Civico di Stoua Naturale, Corso Venezia, 55, Milan, Italy.

Société Royale des Sciences de Liége, *Bulletin,* 1837–, 15 Avenue des Tilleuls, B-4000 Liége, Belgium.

Soil Science Society of America, *Journal,* 1936–, Madison, Wisconsin 53715, USA.

Soil Science, 1916–, Williams and Wilkins, 428 E. Preston St., Baltimore, MD 21202, USA.
Sovetskaya Geologiya, 1958–, Ministerstvo geologii i okhrany nedr, Moscow, USSR.
Soviet Geology and Geophysics (English translation of *Geologiya i Geofizika*), Allerton Press, 150 Fifth Ave., New York, NY 10011, USA.
Soviet Physics–Crystallography (translated *from Kristallografiya,* see *Crystallography*).
Soviet Physics-Doklady, 1961–, (translated from *Doklady Akademiya Nauk SSSR*), American Institute of Physics, 335 East 45th Street, New York, NY 10017, USA.
Structure Reports, International Union of Crystallography, Oosthoek, Utrecht, The Netherlands.
Sveriges Geologiska Undersoekning, Årsbok, 1907–, Svenska Reproduktions AB, Fack, S-162 10 Vaellingby 1, Sweden.
Thermochimica Acta, 1970–, Elsevier Scientific Publishing Co., Box 211, Amsterdam, Netherlands.
Tschermak's Mineralogische und Petrographische Mitteilungen, 1878–, Springer, Vienna, Austria.
United States Bureau of Mines, *Reports of Investigations, Minerals Yearbook,* annual, Department of the Interior, Washington, DC 20402, USA.
United States Geological Survey, *Professional Papers,* 1902–; *Bulletin,* 1883–; *Circular,* Washington, DC 20402, USA.
Ussher Society, *Proceedings,* British Museum (Natural History), Cromwell Road, London SW7, UK.
Vignana Bharathi, 1975–, University of Bangalore, Bangalore 560056, India.
Západne Karpathy, Ložiska Séria Geologia, 1976–, Seria Mineralogia, Petrografia, Geochémia, 1975–, Geologicky Ustav Dionyza Stura, Mlynska Dolina 1, 809 40 Bratislava, Czechoslovakia.
Zapiski, Vsesayuznyi Mineralogicheskogo Obshchestva, 1933–, Leningrad, USSR.
Zeitschrift für Kristallographie, Kristallgeometrie, Kristallphysik, Kristallchemie, 1977–, Akademische Verlagsgesellschaft, 6 Frankfurt am Main, GFR.
Zentralblatt für Mineralogie, 1807–, Schweizerbart'sche Verlagsbuchhandlung, Johannesstrasse 3a, 7000 Stuttgart 1, GFR.

In addition to journals, reference is made to compilations of mineral data and information and the user of this volume will find additional mineral data in the following reference books.

Mineral Reference Books

Battey, M. H., and Tomkeieff, S. I., 1964. *Aspects of Theoretical Mineralogy in the U.S.S.R.* New York: Macmillan, 505p.
Bragg, L.; Claringbull, G. F.; and Taylor, W. H., 1965. *Crystal Structures of Minerals.* Ithaca, N.Y.: Cornell, 409p.
Clark, S. P., Jr., ed., 1966. *Handbook of Physical Constants,* rev. ed. Boulder, Colorado: The Geological Society of America, 587p.
Deer, W. A.; Howie, R. A.; and Zussman, J., 1962–1963. *Rock-Forming Minerals:* vol. 1, *Ortho-/and Ring Silicates;* vol. 2, *Chain Silicates;* vol. 2A, *Single-Chain Silicates;* vol. 3, *Sheet Silicates;* vol. 4, *Framework Silicates;* vol. 5, *Non-Silicates.* London: Longmans.
Deer, W. A.; Howie, R. A.; and Zussman, J., 1964. *An Introduction to the Rock-Forming Minerals.* London: Longmans, 528p.
Dietrich, R. V., 1969. *Mineral Tables.* New York: McGraw-Hill, 237p.
Donnay, J. D. H., and Ondik, H. M., eds., 1973. *Crystal Data,* 3rd ed. Washington: U.S. Department of Commerce, National Bureau of Standards and the Joint Committee on Powder Diffraction Standards.
Fleischer, M., 1980. *Glossary of Mineral Species,* 3rd ed. Tucson, Arizona: Mineralogical Record.
Frondel, C., 1962. *Dana's System of Mineralogy,* vol. III, 7th ed., *Silica Minerals.* New York/London: Wiley, 334p.
Hey, M. H., 1962. *An Index of Mineral Species & Varieties Arranged Chemically.* London: British Museum (Natural History), 728p. (2nd ed. 1963, Appendix I, 1963; II, 1974.)
Kostov, I., 1968. *Mineralogy.* London: Oliver and Boyd, 587p.
Lazarenko, E. K., and Vynar, O. M., 1975. *Mineralogical Dictionary, Ukranian-Russian-English.* Kiev: Ukranian Academy of Sciences.
O'Donoghue, M., ed., 1976. *The Encyclopedia of Minerals and Gemstones.* London: Orbis, 304p.
Palache, C.; Berman, H.; and Frondel, C., 1944, 1951. *Dana's System of Mineralogy,* vol. I, *Elements, Sulfides, Sulfosalts, Oxides;* vol. II, *Halides, Nitrates, Borates, Carbonates, Sulfates, Phosphates, Arsenates, Tungstates, Molybdates, etc.* New York/London: Wiley.
Phillips, W. R., and Griffin, D. T., 1981. *Optical Mineralogy: The Nonopaque Minerals.* San Francisco: W. H. Freeman, 677p.
Roberts, W. L.; Rapp, G. R., Jr.; and Weber, J., 1974. *Encyclopedia of Minerals.* New York: Van Nostrand Reinhold, 693p.
Strunz, H., 1957. *Mineralogische Tabellen,* 3rd ed. Leipzig: Akadem. Verlag.
Trögger, W. E., 1959. *Optische Bestimmung der gesteinbilden Minerale,* Teil 1, *Bestimmungstabellen,* 3rd ed. Stuttgart: Schweizerbart'sche, 147p.
Wells, A. F., 1962. *Structural Inorganic Chemistry,* 3rd ed. London: Oxford, 1055p.
Winchell, A. N., and Winchell, H., 1951. *Elements of Optical Mineralogy,* Part II, *Descriptions of Minerals,* 4th ed. New York/London: Wiley, 551p.

For the reader wishing to pursue the subject of mineralogy in a systematic fashion, several textbooks on general mineralogy are listed. Textbooks specifically treating crystallography are given at the ends of the entries *Directions and Planes, Lattice, Point Groups,* and *Symmetry.*

Textbooks on Mineralogy

Battey, M. H., 1972. *Mineralogy for Students.* Edinburgh; Oliver and Boyd, 323p.
Berry, L. G., and Mason, B., 1959. *Mineralogy: Concepts, Descriptions, Determinations.* San Francisco: Freeman, 630p.
Correns, C. W., 1968. *Einführing in die Mineralogie*

(Kristallographie und Petrologie), 2nd ed. Heidelberg: Springer-Verlag, 458p.
Correns, C. W., 1969. *Introduction to Mineralogy, Crystallography and Petrology*, 2nd ed. (translated by William D. Johns). Heidelberg/New York: Springer-Verlag, 484p.
Cox, K. G.; Price, N. B.; and Harte, B., 1967. *An Introduction to the Practical Study of Crystals, Minerals, and Rocks.'* London/New York: McGraw-Hill, 233p.
Dent Glasser, L. S., 1977. *Crystallography and Its Applications.* London: Van Nostrand Reinhold, 224p.
Ernst, W. G., 1969. *Earth Materials.* Englewood Cliffs, N.J.: Prentice-Hall, 150p.
Ford, W. E., 1932. *Dana's Textbook of Mineralogy*, 4th ed. New York: Wiley, 851p.
Frye, K., 1974. *Modern Mineralogy.* Englewood Cliffs, N.J.: Prentice-Hall, 325p.
Hurlbut, C. S., Jr., and Klein, C., 1977. *Manual of Mineralogy* (after James D. Dana), 19th ed. New York/London: Wiley, 532p.
Mason, B., and Berry, L. G., 1968. *Elements of Mineralogy.* San Francisco: Freeman, 550p.
Morimoto, N.; Sunagawa, I.; and Miyashiro, A., 1975. *Kobutsugaku* [*Modern Mineralogy*]. Tokyo: Iwanami Shoten Publ., 654p.
Nickel, E., 1971–1975. *Grundwissen in Mineralogie:* Teil 1, 1971, 207p.; Teil 2, 1973, 301p.; Teil 3, 1975, 269p. Thun: Ott Verlag.
Parker, R. L., and Bambauer, H. U., 1975. *Mineralienkunde*, 5th ed., Thun: Ott Verlag, 368p.
Ramdohr, P., and Strunz, H., 1978. *Klockmann's Lehrbuch der Mineralogie.* Stuttgart: F. Enke Verlag, 876p.
Read, H. H., 1970. *Rutley's Elements of Mineralogy.* 26th ed. London: Thomas Murby, 560p.
Shelley, D., 1975. *Manual of Optical Mineralogy.* Amsterdam: Elsevier, 239p.
Sinkankas, J., 1966. *Mineralogy: A First Course*, 2nd ed. Princeton, N.J.: Van Nostrand, 587p.

Acknowledgments

Contributors to this volume are listed alphabetically following the list of Main Entries. Several of these contributors also acted as reviewers concerning content and level of articles submitted, as sources of names of potential contributors, and as reviewers for the entries that the editor was forced to prepare during the final stages of preparation of the volume. The editor wishes especially to acknowledge the valuable advice of F. Donald Bloss (Virginia Polytechnic Institute), William T. Holser (Oregon), and the late Marjorie Hooker.

Articles written by the volume editor were reviewed and improved by Rhodes W. Fairbridge (Columbia), William T. Holser (Oregon), Robert A. Howie (London), Robert E. Newnham (Penn State), Paul H. Ribbe (Virginia Tech), Steven Scott (Toronto), Gene C. Ulmer (Temple), Eugene White (Penn State), William B. White (Penn State), and William S. Wise (Santa Barbara). Mineral descriptions prepared by the volume editor were reviewed by Jan Kutina (American University, Washington, D.C.) and by John S. White, Jr. (Smithsonian), while the entire mineral glossary was greatly improved by a heroic overview by Paul Brian Moore (Chicago). The writing style of the volume editor also benefited greatly from the editing of it by Franklin P. Ferguson (emeritus, Penn State). Gail Nicula (science librarian, Old Dominion University) and Pat McAndrew (reference librarian, University of Virginia) gave invaluable assistance in locating needed reference materials. Production of the encyclopedia was greatly facilitated and the end product enhanced by the meticulous proofreading of the final pages by John Frye, my father.

My sincere gratitude is extended to all those individuals named and unnamed who helped to bring this volume to light.

KEITH FRYE
Tyro, Virginia

MAIN ENTRIES

CONTRIBUTORS

ALLEN M. ALPER, GTE Sylvania Incorporated, Towanda, Pennsylvania 18848. *Mineral Properties; Refractory Minerals; Synthetic Minerals.*

STURGIS W. BAILEY, Dept. of Geology and Geophysics, University of Wisconsin, Madison, Wisconsin 53706. *Chlorite Group.*

THOMAS F. BATES, Mail Stop 701, U.S. Geological Survey, Denver, Colorado 80225. *Electron Microscopy (Transmission).*

M. HUGH BATTEY, Dept. of Geology, The University of Newcastle, Newcastle upon Tyne, NE1 7RU, England. *Density Measurements; Goniometry.*

GEORGE H. BEALL, Research and Development Laboratories, Corning Glass Works, Corning, New York 14830 *Glass, Devitrification of Volcanic.*

LEONARD G. BERRY, Dept. of Geological Sciences, Queen's University, Kingston, Ontario, Canada K7L 3N6. *Polymorphism.*

M. R. BLOCH, The Negev Institute for Arid Zone Research, Beer Sheba, Israel. *Salt Economy.*

F. DONALD BLOSS, Dept. of Geological Sciences, Virginia Polytechnic Institute and State University, Blacksburg, Virginia 24061. *Dispersion, Optical; Optical Orientation.*

WALTER L. BOND, deceased. *Piezoelectricity.*

EMMY BOOY, Dept. of Geological Engineering, Colorado School of Mines, Golden, Colorado 80401. *Smectite Group;* mineral descriptions.

H. P. BOVENKERK, Special Materials Department, General Electric Company, Worthington, Ohio 43085. *Abrasive Materials.*

OTTO BRAITSCH, deceased. *Saline Minerals (Soluble Evaporites).*

GEORGE W. BRINDLEY, Materials Research Laboratory, The Pennsylvania State University, University Park, Pennsylvania 16802. *Clays, Clay Minerals; Phyllosilicates.*

C. ERVIN BROWN, Eastern Mineral Resources Branch, U.S. Department of the Interior, Geological Survey, Reston, Virginia 22092. *Pigments and Fillers.*

JOHN G. BURKE, Dept. of History, University of California, Los Angeles, California 90024. *Crystallography: History.*

R. W. CAMERON, Dept. of Geology, Columbia University, New York, New York 10027. Mineral descriptions.

KEITH E. CHAVE, Dept. of Oceanography, University of Hawaii, Honolulu, Hawaii 96822. *Invertebrate and Plant Mineralogy.*

ROBERT CROWNINGSHIELD, Gemological Institute of America, 580 Fifth Avenue, New York, New York 10036. *Gemology.*

EDGAR F. CRUFT, Nord Resources Corporation, 2300 Candelaria Road, N.E., Albuquerque, New Mexico 87107. *Anhydrite and Gypsum;* mineral descriptions.

J. BARRY DAWSON, Dept. of Geology, The University of Sheffield, Sheffield, S1 3JD, England. *Diamond;* mineral descriptions.

PAUL S. DeCARLI, Stanford Research Institute, Menlo Park, California 94025. *Coesite and Stishovite.*

ROBERT C. DOMAN, Ceramic Product Development, Corning Glass Works, Corning, New York 14830. *Mineral Properties; Refractory Minerals.*

COLIN H. DONALDSON, Dept. of Geology, University of St. Andrews, Fife, Scotland, KY16 9ST. *Olivine Group.*

J. D. H. DONNAY, Emeritus Professor, The Johns Hopkins University; Present address: Department of Geological Sciences, McGill University, 3450 University Street, Montreal, PQ, Canada H3A 2A7. *Epitaxy; Monotaxy; Polycrystal; Syntaxy.*

DAVID A. DUKE, Research and Development Laboratories, Corning Glass Works, Corning, New York 14830 *Glass: Theory of Crystallization.*

KINGSLEY DUNHAM, Charleycroft, Quarryheads Lane, Durham DH1 3DY, England. *Mineral and Ore Deposits.*

ARTHUR J. EHLMANN, Dept. of Geology, Texas Christian University, Fort Worth, Texas 76129. Mineral descriptions.

W. GARY ERNST, Dept. of Earth and Space Sciences, University of California, Los Angeles, California 90024. *Asbestos.*

SAMUEL F. ETRIS, American Society for Testing and Materials, Philadelphia, Pennsylvania 19103. *American Society for Testing and Materials.*

ERNEST E. FAIRBANKS, Old Orchard Beach, Maine 04064. *Gangue Minerals.*

RHODES W. FAIRBRIDGE, Dept. of Geology, Columbia University, New York, New York 10027. *Allotropy; Authigenic Minerals; Diagenetic Minerals; Enhydro, Enhydrite, or Waterstone; Mica Group;* mineral descriptions.

CHARLES W. FINKL, JNR., Ocean Science Center, Nova University, 8000 North Ocean Drive, Dania, Florida 33004. *Soil Mineralogy.*

FRANKLIN F. FOIT, JR., Dept. of Geology, Washington State University, Pullman, Washington 99163. *Tourmaline Group;* mineral descriptions.

KURT FREDRIKSSON, Division of Meteorites, Smithsonian Institution, U.S. National Museum, Washington, D.C. 20560. *Electron Probe Microanalysis.*

C. G. I. FRIEDLAENDER, R.R. 1, P.O. Box 325, Westport, Ontario, Canada K0G 1X0. *Quartz;* mineral descriptions.

MICHAEL J. FROST, Dept. of Geography and Geology, The Polytechnic of North London, Holloway, London N7 8DB, England. *Garnet Group; Metallic Minerals; Mohs Scale of Hardness; Refractive Index; Twinning; Widmanstätten Figures.*

KEITH FRYE, Dept. of Geophysical Sciences, Old Dominion University, Norfolk, Virginia 23508. *Alunite Group, Beudantite Group, Crandallite Group; Aragonite Group; Astrophyllite Group; Autunite (Torbernite) Group, Meta-Autunite (Metatorbernite) Group; Barometry, Geologic; Biopyribole; Clinopyroxenes; Directions and Planes; Epidote Group. Feldspar Group; Feldspathoid Group; Frondelite Group; Halotrichite Group; Humite Group; Kaolinite-Serpentine Group; Lattice; Linnaeite Group; Minerals, Uniaxial and Biaxial; Native Elements and Alloys; Osumilite Group; Point Groups; Polysomatism; Pyrite Group; Pyrochlore Group, Stibiconite Group; Pyroxenoids; Rock-Forming Minerals; Rupture in Mineral Matter; Rutile Group; Sodalite Group; Space Group Symbol; Symmetry;* mineral descriptions.

LOUIS H. FUCHS, Chemistry Division, Argonne National Laboratories, Argonne, Illinois 60439. Mineral descriptions.

G. DONALD GARLICK, Dept. of Geology, Humbolt State University, Arcata, California 95521. *Mantle Mineralogy.*

DONALD E. GARRETT, President, Saline Processors, Inc., 911 Bryant Place, Ojai, California 93023. *Saline Minerals (Soluble Evaporites).*

JOSEPH I. GOLDSTEIN, Dept. of Metallurgy, Lehigh University, Bethlehem, Pennsylvania 18015. *Metallurgy.*

J. A. GILBERT, Dept. of Geology, Columbia University, New York, New York 10027. Mineral descriptions.

VIVIEN GORNITZ, Institute for Space Studies, National Aeronautics and Space Administration, New York, New York 10025. *Phantom Crystals; Skeletal Crystals.*

T. J. GRAY, Director, Atlantic Industrial Research Institute, Nova Scotia Technical College, Halifax, Nova Scotia, Canada. *Crystals, Defects in.*

MICHAEL W. GRUTZECK, Materials Research Laboratory, Pennsylvania State University, University Park, Pennsylvania 16802. *Portland Cement Mineralogy.*

WALTER E. HILL, C. F. and I. Steel Corporation, Pueblo, Colorado 81002. Mineral descriptions.

RALPH J. HOLMES, deceased. *Gemology.*

MARJORIE HOOKER, deceased. *International Mineralogical Association.*

R. A. HOWIE, Dept. of Geology, King's College, Strand, London WC2R 2LS, England. *Orthopyroxenes;* mineral descriptions.

AZUMA IIJIMA, Geological Institute, Faculty of Science, University of Tokyo, Tokyo, Japan. Mineral descriptions.

JOHN T. JENKINS, Dept. of Geology, Concordia University, Montréal, Québec, Canada H4B 1R6. *Mineral Classification: Principles.*

J. B. JONES, Dept. of Geology and Mineralogy, University of Adelaide, Adelaide, South Australia 5001, Australia. *Opal.*

KLAUS KEIL, Director, Institute of Meteoritics, University of New Mexico, Albuquerque, New Mexico 87106. *Electron Proble Microanalysis.*

WILLIAM C. KELLY, Dept. of Geology, University of Michigan, Ann Arbor, Michigan 48104. *Gossan.*

MARTIN KLEIN, JR., Technical Staffs Division, Corning Glass Works, Corning, New York 14830. *Sands: Glass and Building.*

GUNNAR KULLERUD, Dept. of Geosciences, Purdue University, Lafayette, Indiana 47907. *Thermometry, Geologic.*

JAN KUTINA, Director, Laboratory of Global Tectonics and Metallogeny, The American University, Washington, D.C. 20016. *Vein Minerals.*

RALPH L. LANGENHEIM, Dept. of Geology, University of Illinois, Urbana, Illinois 61801. *Resin and Amber.*

THOMAS G. LANGTON, U.S. Department of the Interior, Bureau of Mines, Washington, D.C. 20241. *Mineral Industries.*

L. J. LAWRENCE, 15 Japonica Road, Epping, New South Wales 2121, Australia. *Ore Microscopy (Mineragraphy).*

W. LAYTON, Layton and Associates, Pty., Ltd., Brisbane, Queensland 4000, Australia. *Amphibole Group; Pyroxene Group.*

CHENG K. LY, Dept. of Geology, University of Newcastle, Newcastle, New South Wales 2308, Australia. *Electron Microscopy (Scanning) and Clastic Mineral Surface Textures.*

JAMES W. McCAULEY, Ceramics Research Division, Army Materials and Mechanics Research Center, Watertown, Massachusetts 02172. *Calcite Group.*

DUNCAN McCONNELL, Professor Emeritus, The Ohio State University, Columbus, Ohio 43210. *Apatite Group; Human and Vertebrate Minerals.*

DAVID J. McDOUGALL, Dept. of Geology, Concordia University, Loyola Campus, Montréal, Québec, Canada H4B 1R6. *Etch Pits; Thermoluminescence.*

ROBERT N. McNALLY, Manager, Ceramic Research Division, Sullivan Science Park, Corning, New York 14830. *Mineral Properties; Synthetic Minerals.*

BRYANT MATHER, President, American Society for Testing and Materials; Chief Concrete Laboratory, U.S. Army Engineer Waterways Experiment Station, Vicksburg, Mississippi 39180. *American Society for Testing and Materials.*

CHARLES MILTON, Dept. of Geology, George Washington University, Washington, D.C. 20052. *Green River Mineralogy.*

JAMES K. MITCHELL, Dept. of Civil Engineering, University of California, Berkeley, California 94720. *Thixotropy.*

GERARD C. MOERSCHELL, Dept. of Geology, Columbia University, New York, New York 10027. Mineral descriptions.

PAUL BRIAN MOORE, Dept. of Geophysical Sciences, University of Chicago, Chicago, Illinois 60637. *Pegmatite Minerals.*

DEREK J. MORRISON-SMITH, Cranleigh School, Cranleigh, Surrey GU6 8QQ, England. *Plastic Flow in Minerals.*

ROBERT F. MUELLER, Route 1, Box 250, Staunton, Virginia 24401. *Order-Disorder.*

KURT NASSAU, Bell Laboratories, Murray Hill, New Jersey 07974. *Crystal Growth.*

ROBERT E. NEWNHAM, Materials Research Laboratory, The Pennsylvania State University, University Park, Pennsylvania 16802. *Jade; Magnetic Minerals.*

EDWARD OLSEN, Curator of Mineralogy, Field Museum of Natural History, Chicago, Illinois 60605. *Meteoritic Minerals.*

ADOLF PABST, Dept. of Geology and Geophysics, University of California, Berkeley, California 94720. *Barker Index of Crystals; Metamict State; Topotaxy;* mineral descriptions.

OLE V. PETERSEN, Curator, Mineralogical Museum, 1350 Copenhagen, Denmark. *Museums, Mineralogical.*

E. PICCIOTTO, Service de Géologie et Géochimie Nucleaires, Université Libre de Bruxelles, Brussels 5, Belgium. *Pleochroic Halos.*

CHARLES T. PREWITT, Dept. of Earth and Space Sciences, State University of New York, Stony Brook, New York 11794. *International Mineralogical Association.*

REX T. PRIDER, Dept. of Geology, University of Western Australia, Nedlands, Western Australia, Australia. *Blowpipe Analysis of Minerals; Crystallography: Morphological.*

JOHN D. RIDGE, 1402 N.W. 18th Street, Gainesville, Florida 32605. *Mineral Deposits: Classification.*

ROMANO RINALDI, Institute of Mineralogy and Petrology, University of Modena, 1-41100 Modena, Italy. *Zeolites.*

PHILIP E. ROSENBERG, Dept. of Geology, Washington State University, Pullman, Washington 99163. *Tourmaline Group;* mineral descriptions.

D. ROSTOKER, Assistant Director, Research and Development, Materials Division, Norton Company, Worcester, Massachusetts 01606. *Sands: Glass and Building.*

DELLA M. ROY, Materials Research Laboratory, The Pennsylvania State University, University Park, Pennsylvania 16802. *Portland Cement Mineralogy.*

CARLETON N. SAVAGE, deceased. *Blacksand Minerals and Metals; Placer Deposits.*

CECIL J. SCHNEER, Dept. of Geology, University of New Hampshire, Durham, New Hampshire 03824. *Crystal Habit.*

E. R. SEGNIT, C.S.I.R.O., Division of Mineral Chemistry, Port Melbourne, Victoria 3207, Australia. *Opal.*

R. KARL SMITH, Technical Staffs Division, Corning Glass Works, Corning, New York 14830. *Synthetic Minerals.*

DAVID H. SPEIDEL, Dept. of Geology, Queens College of the City University of New York, Flushing, New York 11367. *Mineralogical Phase Rule;* mineral descriptions.

LLOYD W. STAPLES, Professor Emeritus, University of Oregon, Eugene, Oregon 97403. *Mineral Classification: History; Naming of Minerals; Nitrate Minerals.*

JOHN STARKEY, Dept. of Geology, University of Western Ontario, London, Ontario, Canada N6A 3K7. *Plagioclase Feldspar.*

IAN M. STEELE, Dept. of Geophysical Sciences, University of Chicago, Chicago, Illinois 60637. *Lunar Mineralogy.*

HAROLD B. STONEHOUSE, Dept. of Geology, Michigan State University, East Lansing, Michigan 48823. *Pseudomorphism.*

T. P. THAYER, U.S. Department of the Interior, Geological Survey, Washington, D.C. 20242. Mineral descriptions.

GENE C. ULMER, Dept. of Geology, Temple University, Philadelphia, Pennsylvania 19122. *Spinel Group.*

MINORU UTADA, Geological Institute, Faculty of Science, University of Tokyo, Tokyo, Japan. Mineral descriptions.

CHARLES J. VITALIANO, Dept. of Geology, Indiana University, Bloomington, Indiana

47401. *Color in Minerals; Luminescence; Optical Mineralogy.*

DOROTHY B. VITALIANO, Bloomington, Indiana 47401. *Color in Minerals; Luminescence; Optical Mineralogy.*

ERNEST E. WAHLSTROM, 2522 174th Avenue, N.E., Redmond, Washington 98052. *Anisotropism; Isotropism; Polarization and Polarizing Microscope.*

S. ST. J. WARNE, Dept. of Geology, University of Newcastle, Newcastle, New South Wales 2308, Australia. *Electron Microscopy (Scanning) and Clastic Mineral Surface Textures; Staining Techniques.*

JAMES F. WESCOTT, A. P. Green Refractories Company, Mexico, Missouri 65265. *Fire Clays.*

JOHN SAMPSON WHITE, Division of Mineralogy, National Museum of Natural History, Washington, D.C. 20560. *Collecting Minerals; Wolframite Group;* mineral descriptions.

W. ARTHUR WHITE, Earth Materials Technology Section, Illinois State Geological Survey, Urbana, Illinois 61801. Mineral descriptions.

WILLIAM B. WHITE, Materials Research Laboratory, The Pennsylvania State University, University Park, Pennsylvania 16802. *Cave Minerals.*

E. J. W. WHITTAKER, Dept. of Geology and Mineralogy, Oxford University, Oxford OX1 3PR, England. *Isomorphism.*

LEE WILSON, Dept. of Geology, Columbia University, New York, New York 10027. Mineral descriptions.

M. M. WILSON, Dept. of Geology and Mineralogy, University of Queensland, Brisbane, Queensland 4007, Australia. *Amphibole Group; Pyroxene Group.*

PHILLIP N. YASNOWSKY, U.S. Department of the Interior, Bureau of Mines, Washington, D.C. 20241. *Mineral Industries.*

DAVIS A. YOUNG, Department of Physics, Calvin College, Grand Rapids, Michigan 49506. *Alkali Feldspars.*

ABRASIVE MATERIALS

People worked with tools before recorded history. In shaping tools, they must have soon discovered that certain substances are harder than others and not only would make more durable tools but could be used to abrade or form tools into more desirable shapes than those occurring naturally. This discovery led to the purposeful selection of hard materials useful as abrasives and, eventually, to a classification of minerals into a rough scale of hardness. Abrasives are those hard materials used to shape, to surface, and to sharpen tools and other things. In general, abrasives are used to clean, to form, and to dimension workpieces.

Later, as industrial needs expanded, mining and processing of natural abrasives developed, as did the fabrication of tools and machinery for their applications. These developments were followed by the manufacture of abrasives, now a key and important industry. The annual world consumption of abrasives and abrasive tools has a value in excess of one billion dollars.

About 1820, Mohs, a German mineralogist, introduced a scale of hardness that employed 10 minerals as standards. In the *Mohs scale,* quantitative differences were based on scratching or abrading ability (see *Mohs Scale of Hardness*). For example, the mineral *topaz,* 8 on the Mohs scale, could be abraded by *corundum,* which was 9; and *topaz* could abrade *quartz,* which was 7. This was a convenient scale for practical application and was useful for mineral identification. The hardness of some common abrasives is shown in Table 1. Minerals with a hardness of about 6 or higher on the Mohs scale could be considered naturally occurring abrasives.

TABLE 1. Mohs Hardness of Common Abrasives

Abrasive	Hardness
Tripoli	5–6
Quartz	7
Garnet	8
Corundum, alumina, silicon carbide	9
Cubic boron nitride	9+
Diamond	10

Naturally occurring abrasives in use include sand (*quartz*), ***garnet,*** emery, *corundum,* and *diamond.* Manufactured abrasives include alumina, silicon carbide, boron carbide, cubic boron nitride, and *diamond.* Today, most industrial abrasives are manufactured to ensure control of properties and improve consistency (see *Synthetic Minerals*).

Consumption or production of the major abrasives is tabulated in Table 2 from data in *Minerals Yearbook, 1972.* Consumption of abrasive *diamond* in the USA for 1977 is estimated to have been about 30 million carats (6 metric tons); world consumption is estimated to have been about 85 million carats (17 metric tons). Uses other than abrasive are included in the consumption figures. Tripoli, for example, is extensively used as a filler; silicon carbide and silica also are used as fillers (see *Pigments and Fillers*); and silicon carbide and alumina are used as refractory materials (fire brick) (see *Refractory Minerals*).

Tripoli is a naturally occurring abrasive that is, as found in North Africa, basically an amorphous silica or diatomaceous earth. It is relatively weak and soft compared to other abrasives; hence it is used as a finishing material on relatively soft workpieces.

Silica stones were used in large pieces as grinding stones and wear surfaces, but have now been replaced almost entirely by manufactured abrasives.

Garnet mined for abrasive use is usually a mixed aluminum, magnesium, iron silicate. Its chief uses are as coated abrasives for working wood, glass, and some metallic materials and as a grit blasting agent.

TABLE 2. Production/Consumption of Abrasives, 1972 (metric tons)

Abrasive	United States	World
Natural		
Tripoli	80,000	n.a.
Silica stones	3,000	n.a.
Garnet	17,000	19,000
Emergy and *corundum*	2,600	120,000[a]
Artificial		
Silicon carbide[a]	150,000[b]	460,000
Aluminum oxide	168,000[b]	630,000
Metallic abrasives	215,000[b]	n.a.

[a]Includes material for refractories and other nonabrasive uses.
[b]Production in the United States and Canada 1972.

Emery is a type of natural *corundum* containing *magnetite* and *hematite.* Its uses include nonslip fillers for surfacing and as a coated or tumbling abrasive.

Corundum, an impure form of naturally occurring aluminum oxide, is used for grinding and finishing glass and has been replaced almost entirely by manufactured aluminum oxide. World production of natural *corundum* in 1972 has been estimated as 7900 metric tons.

The processing of natural abrasives includes separation, crushing, grinding, screening, and classification for specific sizes and shapes of the grit particles.

Silicon carbide is made by reacting silica, petroleum coke, salt, and sawdust in an electric furnace using the process developed in 1891 by Edward Acheson. Two types of grit are produced, one colored green and the other black. The green grit is generally the purer form and is more brittle and friable. It is used for grinding hard brittle materials such as cemented carbides. The black grit is tougher and is used more widely for shaping and cutting nonmetallic workpieces such as glass, ceramics, natural stones, masonry, and concrete aggregate.

Aluminum oxide or alumina is made by melting bauxite (an impure natural aluminum oxide) in an electric furnace. Coke and iron are added to remove impurities. Use of various additives and variations in the process make it possible to produce a wide range of grit strengths or friabilities. Titanium is the principal additive used to control grit strength. The major application of alumina is in grinding all types of ferrous-based and similar metallic materials.

Metallic abrasives—steel and iron shot—are used to clean and to work harden surfaces of metallic materials by air blasting and shot peening.

Industrial *diamond* abrasives are both supplied from natural deposits and produced synthetically. Mined *diamond* is crushed and sorted by shape and size. Manufactured *diamond* is grown from *graphite* and certain catalytic agents under conditions of extreme pressure and temperature. For the most part, manufactured *diamond* is grown directly to the sizes and shapes to be used. Great variation in grit strength and other properties are possible through control of the manufacturing process, that *diamond* grits are produced with properties matched to the bonding system and to specific use requirements. Abrasive *diamond* is used principally in the sawing, grinding, and polishing of nonmetallic materials such as glass, stone, ceramics, cermets, and masonry materials.

The newer synthesized abrasives include cubic boron nitride, which is increasingly important in the finish grinding of hard alloy steels, and zirconia-alumina, which is used primarily in coated abrasives.

Abrasives are used as bonded abrasives, as coated abrasives, or in the loose form. Typical of bonded abrasives are abrasive tools wherein the abrasive is held by resinous, metallic, or vitreous matrices and formed into grinding wheels, saws, discs, and burrs used in powered tools. Other forms of bonded abrasives include sticks for honing, stones for hand sharpening tools, and bricks for finishing castings. Coated abrasives are fixed on paper, cloth, fiber, or composite backings, generally with organic binders. They are used as belts, discs, and sheets for power-tool or hand use. Loose abrasives are generally carried as a slurry with water- or oil-based carriers, sometimes with the addition of lubricants or inhibitors. Abrasive slurry is circulated between a rotating plate or oscillating lap and a workpiece. Generally, loose abrasives are used for finer finishing or polishing operations or where abrasive cost is very low, such as for *quartz*-based abrasive. However, even *diamond* is used in this fashion in the polishing of hard materials.

H. P. BOVENKERK

References

Coes, L., Jr., 1971. *Abrasives.* New York: Springer-Verlag.

Jensen, M. L., and Bateman, A. M., 1979. *Economic Mineral Deposits,* 3rd ed. New York: Wiley, 571–580.

Minerals Yearbook, published annually by the Bureau of Mines, Department of the Interior. Washington, D.C.: U.S. Government Printing Office.

Cross-references: *Mineral Industries; Mohs Scale of Hardness; Sands—Glass and Building; Synthetic Minerals.*

ALKALI FELDSPARS

Alkali ***feldspars*** rank with *quartz* and ***plagioclase*** as the most common minerals in the earth's crust. Chemically they are represented by the formula $(K,Na)AlSi_3O_8$. The chemical end member $KAlSi_3O_8$ (*Or*) includes the minerals *orthoclase, microcline,* and K-rich *sanidine;* and the end member $NaAlSi_3O_8$ (*Ab*) is the mineral *albite.* In most natural specimens, there is also a small amount of the anorthite component $CaAl_2Si_2O_8$. Traces of Rb, Tl, Pb, Ba, Sr, Fe^{3+}, Ga, and Ge may also be present. Alkali ***feldspars*** occur extensively in a variety of geologic environments, and, owing to their compositional and structural complexity, are especially useful as petrologic indicators.

Most natural specimens are whitish, gray, or pinkish. They have two well-developed cleavages at or nearly at right angles to one another; relatively low density (2.55–2.63 gm/cm^3); and monoclinic or triclinic crystal symmetry. In thin sections, alkali *feldspars* are colorless and can be identified by their cleavage, low relief, and low birefringence. A great variety of twin types such as Carlsbad, albite, Pericline, or polysynthetic tartan twinning also are useful in their identification.

Crystal Structure

Alkali *feldspars* are framework aluminosilicates composed of SiO_4 and AlO_4 tetrahedra, each of which shares all four of its apical oxygen atoms with neighboring tetrahedra. The tetrahedra are thus linked in all three dimensions, and form four-membered ringlike groups, which are further joined to one another in somewhat distorted chainlike structures elongated parallel to the *a* crystallographic axis. Adjoining chainlike groups are cross-linked (see vol. IVA: *Mineral Classes: Silicates*) by shared apical oxygens and not by cations as in the true chain silicates. Between the chainlike groupings are large sites in which are situated potassium ions, coordinated by nine oxygen ions, or sodium ions, coordinated by six or seven oxygen ions. The gross structure is, therefore, a three-dimensional tetrahedral network of composition $AlSi_3O_8^{-1}$ electrostatically balanced by univalent K and Na cations.

Order-Disorder

The feldspar structure contains an average of one aluminum (Al) for every three silicon (Si) ions; but, in different *feldspars* (q.v.), the Al shows varying degrees of preference for a particular tetrahedral site. Where the Al ions are randomly distributed throughout the tetrahedral sites, the structure is disordered; if the Al ions are situated in a preferred tetrahedral site, the *feldspar* has an ordered structure (see *Order–Disorder*). There are, therefore, numerous possible polymorphic modifications, or structural states, of alkali *feldspars*, depending upon the extent of ordering of Al ions throughout the structure. A continuum of structures seems to exist between completely disordered and completely ordered end members, although most natural alkali *feldspars* tend to occur in a few selective structural states.

The Al–Si ordering is chiefly a function of the rate and temperature of annealing. High-temperature polymorphs such as high *sanidine* and high *albite* have disordered structures, whereas, lower-temperature phases, such as *microcline* and low *albite,* are well ordered.

Alkali-Feldspar Series

It is convenient to arrange alkali *feldspars* into series in which each member of the series, no matter what its composition, has the same structural state. The high-*sanidine*–high-*albite* series consists of *feldspars* formed at elevated temperatures and possessing very little Al–Si ordering. This series includes high *sanidine* (K-rich), *anorthoclase,* and high *albite* (Na-rich). Complete solid solution exists between the potassium (K) and sodium (Na) end members of the series. High *albite,* however, displays triclinic symmetry, whereas *sanidine* is monoclinic, the monoclinic-triclinic inversion being located approximately at composition $Or_{40}Ab_{60}$ (40% $KAlSi_3O_8$).

The most completely ordered alkali *feldspars* are maximum *microcline* and low *albite,* a pair of low-temperature minerals common in granites and in pegmatites. Solid solution is limited and intermediate bulk compositions are stable as two coexisting phases, one K rich and the other Na rich. Maximum *microline* and low *albite* are triclinic.

Crystal Symmetry

All sodium *feldspars* are triclinic at room temperature; on the other hand, potassium *feldspar* can be monoclinic or triclinic. High-temperature, nearly disordered phases such as high and low *sanidine* are monoclinic. *Orthoclase,* a partially ordered *feldspar,* has overall monoclinic symmetry and morphology, although some specimens contain extremely small twin-related domains of triclinic symmetry. All *microclines* are triclinic; indeed, they show varying degrees of triclinicity or obliquity depending upon the degree of ordering. Therefore, maximum *microcline,* the most ordered K-*feldspar,* is also the most oblique; that is, its crystallographic angles α and γ show greater deviation from 90° than in less ordered *microclines.* The separation between the 131 and $1\bar{3}1$ X-ray diffraction peaks can be used to estimate the obliquity of a *microcline.* Adularia is a K-rich *feldspar* that may be monoclinic like *orthoclase* or triclinic like *microcline.* It has a distinctive {110} habit. Adularia is an authigenic or a very low-temperature hydrothermal variety and, presumably, ought to be highly ordered. It has been suggested that the rate of growth may be too fast or that crystallization temperature may be too high to permit extensive Al–Si ordering.

Cell Dimensions, Structure, and Composition

The unit-cell dimensions of alkali feldspars

$a = 8.139\text{–}8.578$ Å

$b = 12.789\text{–}13.030$ Å

$c = 7.106\text{–}7.220$ Å

are a function both of composition and of structural state. Fig. 1 shows in simplified form the general relationship between chemical composition, structural state, and computer-refined *b* and *c* unit-cell dimensions of alkali ***feldspar.*** Such a *b-c* diagram may be used to estimate the chemical composition and degree of Al–Si ordering in natural alkali ***feldspars*** if lattice parameters have been determined. If refined unit-cell dimensions are unavailable, the position of the $\bar{2}01$ X-ray diffraction peak may provide an estimate of the composition of heated, low-Ca alkali ***feldspars*** (see Fig. 2).

Perthite

With decreasing temperature, alkali ***feldspars*** undergo increased ordering of the alkali atoms as well as Al and Si. At elevated temperatures there is complete solid solution between the $NaAlSi_3O_8$ and the $KAlSi_3O_8$ end members; below 660°C at about 1 kilobar pressure (and approximately 810°C at 10 kilobars pressure), however, ordering of K and Na begins to take place. Indeed, the ordering is of such a nature that segregations of relatively Na-rich ***feldspar*** develop within a relatively K-rich phase (or vice versa, depending on bulk composition). If equilibrium cooling proceeds far below the critical temperature, the segregations become progressively more Na-rich and the host more K-rich. This exsolving process of Na-***feldspar*** from K-***feldspar*** produces the group of alkali ***feldspars*** known as "perthite." Exsolution texture visible to the unaided eye is called macroperthite and that visible only by means of the optical microscope is called microperthite. Moonstone is a variety of cryptoperthite detectable only by X-ray diffraction or by electron microscopy.

The perthitic segregations typically develop along crystallographic planes such as ($\bar{6}01$), producing intergrowths of subparallel blebs, rods, strings, or lamellae of exsolved material in an optically continuous host. Although most perthites probably form by exsolution, replacement and simultaneous crystallization have also been advanced as modes of perthite development.

Phase Relationships

Melting of $KAlSi_3O_8$. It is often difficult to obtain equilibrium in high-temperature experiments with alkali ***feldspars*** because of their sluggish reaction rates; but in spite of this fact, a good deal is known about their phase relationships. The melting point of $KAlSi_3O_8$

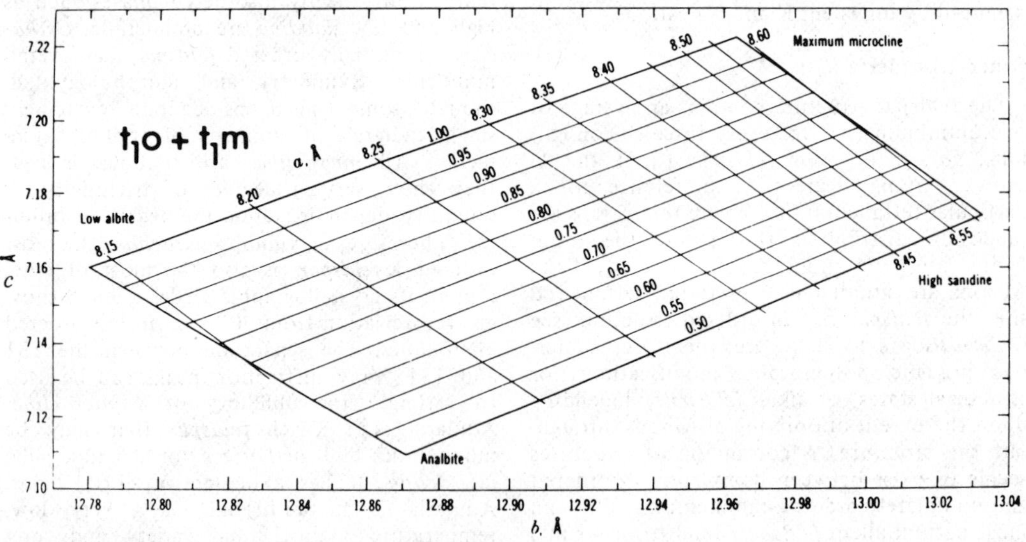

FIGURE 1. Plot of *b* vs. *c* unit-cell dimensions for alkali ***feldspars*** based on the end members low *albite,* maximum *microcline,* high *albite* (analbite), and high *sanidine.* Lines of equal Al content in T_1 sites are based on complete disorder in the high-*albite*–high-*sanidine* series and on complete order in the low-*albite*–maximum-*microcline* series. The mineral *orthoclase* has its *a* unit-cell dimension bewteen *sanidine* and *microcline* (from Ribbe, 1975, after Stewart and Wright, 1974).

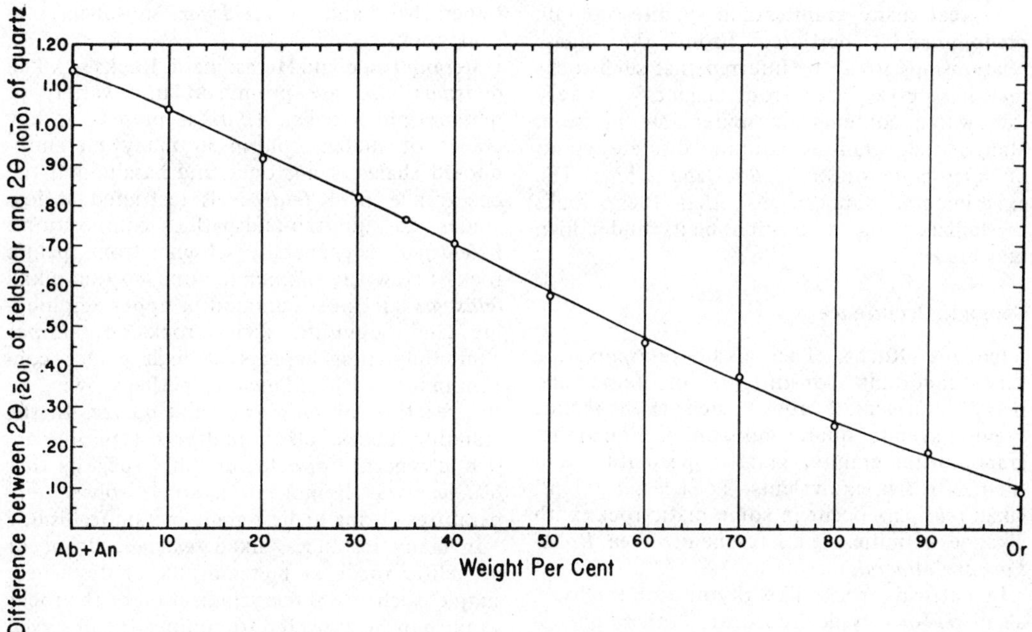

FIGURE 2. Position of the $\bar{2}01$ X-ray diffraction peak as a function of composition for heat-treated, low-Ca, alkali ***feldspar*** (from Tuttle and Bowen, 1958).

at 1 atm (ca. 1 bar) pressure is about 1150°C, at which temperature high *sanidine* melts incongruently to *leucite* plus liquid. In the presence of H_2O under pressure, the melting point is reduced considerably, and high *sanidine* melts congruently at water pressures as low as 2.6 kilobars.

Melting of $NaAlSi_3O_8$. In contrast to $KalSi_3O_8$ (high *sanidine*), $NaAlSi_3O_8$ (high *albite*) melts congruently at all pressures, whether or not H_2O is present in the system. The melting point at one atmosphere is about 1118°C. Increasing water pressure, however, drastically lowers the melting point so that at 5 kilobars P_{H_2O} *albite* melts at only 748°C. At every high pressure and low temperature, *albite* breaks down to *jadeite* plus *quartz*.

$KAlSi_3O_8$-$NaAlSi_3O_8$-H_2O. The melting and subsolidus relationships of the system $KAlSi_3O_8$-$NaAlSi_3O_8$-H_2O are particularly useful in leading to an understanding of the genesis and history of alkali-***feldspar***–bearing igneous and metamorphic rocks. Some of the phase relationships for different values of P_{H_2O} are summarized in Fig. 3. During equilibrium cooling from liquid, all compositions within the system produce one homogeneous feldspar solid-solution phase if P_{H_2O} is less than about 4.5 kilobars. The temperature of initial formation of that phase is considerably lowered at higher values of P_{H_2O}. During further slow cooling, most compositions pass below the solvus, beneath which an initially homogeneous ***feldspar*** undergoes exsolution to produce perthite.

At water pressures in excess of about 4.5 kilobars, two ***feldspars*** crystallize simultaneously at solidus temperatures for intermediate compositions in the system so that K-rich and Na-rich ***feldspar*** coexist as free phases rather than as a perthitic intergrowth. Upon continued cooling, however, the K-rich feldspar may become slightly perthitic if disequilibrium prevails.

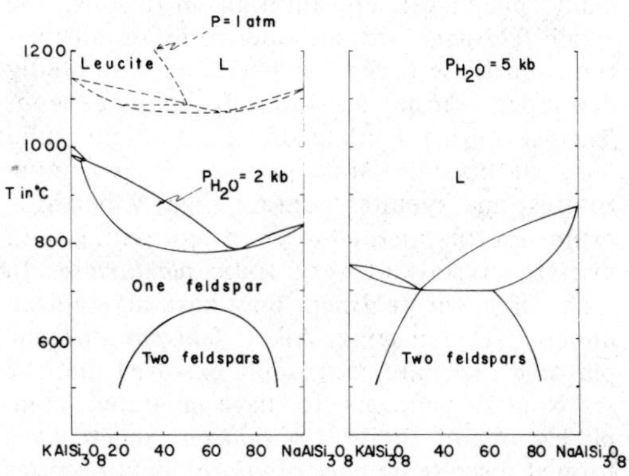

FIGURE 3. Equilibrium relationships of phases in the system $KAlSi_3O_8$-$NaAlSi_3O_8$-H_2O at 1 atm and at 2 and 5 kilobars pressure.

A great many granites and syenites contain predominantly perthite. From the phase relationships it can be inferred that such rocks may have crystallized from magmas with very low water contents or under low pressure. Many other granites contain discrete grains of *microcline* or *orthoclase* and *albite*. The experimental data suggest that these rocks crystallized from water-rich melts under high pressure.

Geologic Occurrence

Igneous Rocks. The alkali ***feldspars*** are very important constituents of felsic and many intermediate rocks such as nepheline syenite, syenite, quartz monzonite, monzonite, granodiorite, granite, granitic pegmatite, and their extrusive equivalents. To a lesser extent, alkali ***feldspars*** occur in some mafic rocks with alkaline affinities, e.g., teschenite (see *Rock-Forming Minerals*).

In extrusive rocks like rhyolite or trachyte, alkali ***feldspar*** typically occurs both as phenocrysts and in the fine-grained groundmass. Owing to rapid cooling during extrusion, high-temperature solid-solution phases like *sanidine* and *anorthoclase* commonly are quenched so that more ordered alkali ***feldspars*** are relatively rare in volcanic rocks. Quenching also prevents perthitic texture from developing on greater than a submicroscopic scale. In general, more K-rich *sanidines* tend to occur in rhyolites and other felsic volcanics whereas sodic phases such as *anorthoclase* or sodic *sanidine* are predominant in strongly alkaline extrusives like phonolite or pantellerite.

Felsic and intermediate plutonic rocks contain lower-temperature ***feldspars.*** Many granites and syenites contain strongly perthitic *orthoclase* or *microcline;* all the *albite* is present as material exsolving from the K-***feldspar.*** In many deep-level Precambrian intrusions, the alkali ***feldspars*** are mesoperthitic or antiperthitic. In these rocks, the perthite undoubtedly developed from an initially homogeneous ***feldspar*** during cooling of an essentially solid mass of crystals. Most granites, quartz monzonites, and syenites contain slightly perthitic *orthoclase* or *microcline* grains coexisting with discrete crystals of very sodic ***plagioclase.*** In such cases, the ***feldspars*** may have crystallized directly from magma. Alkali ***feldspar*** in some plutonic rocks are so strongly exsolved that the sodic phase appears to have migrated completely out of the host K-***feldspar*** and to have formed discrete patches of *albite*. Such textures may, in some instances, represent intermediate stages of development between one-***feldspar*** hypersolvus and two-***feldspar*** subsolvus plutonic rocks.

Metamorphic and Metasomatic Rocks. Alkali ***feldspars*** also are prominent in a variety of metamorphic rocks. *Albite* is present at low grades of metamorphism in many metamorphosed shales, sandstones, and basalts; but the occurrence of K-***feldspar*** is restricted at low grades to quartzo-feldspathic compositions. K-***feldspar*** is generally absent from pelitic rocks below the *sillimanite* zone, so that alkali ***feldspars*** are most common in upper amphibolite and granulite facies rocks. K-***feldspar*** commonly first appears at high grades concomitantly with *sillimanite,* perhaps owing to the reaction of *muscovite* and *quartz*. In the granulite facies, alkali ***feldspars*** typically are strongly perthitic, a feature that suggests that ***feldspar*** was homogenized above solvus temperatures during highest-grade metamorphism.

In many instances, alkali ***feldspars*** do occur in pelitic rocks at metamorphic grades where simple isochemical recrystallization of the rocks would not be expected to produce the observed ***feldspar.*** The ***feldspar*** commonly occurs as large crystals called "porphyroblasts," typically bearing inclusions of other minerals. It is possible that such alkali ***feldspar*** has been produced metasomatically, that is, through addition of alkalis to the rock by means of a permeating intergranular fluid phase in the rocks. This interpretation is enhanced by the fact that such ***feldspar*** porphyroblasts are quite common in the vicinity of large felsic plutonic bodies like granite that could provide a source for the migrating alkalis. Recent experimental studies have further demonstrated the high mobility of K and Na in fluids.

Metasomatic development of alkali ***feldspars*** also occurs in some igneous rocks such as granite, granodiorite, and quartz keratophyre, where a Na-rich feldspar replaces K-rich ***feldspar,*** or vice versa.

Sedimentary Rocks. *Orthoclase, microcline,* and *albite* are rather abundant constituents of clastic sediments, particularly arkosic sandstones. Although most sedimentary K-***feldspar*** is detrital, authigenically grown crystals are not uncommon (see *Authigenic Minerals*). Typically, such ***feldspar*** is very low in Na content, is never perthetic, and never shows polysynthetic tartan twinning. The structural state of authigenic K-***feldspar,*** commonly called adularia, is dependent upon the structure of the detrital grain around which the new material grows, but generally resembles that of *orthoclase*. Most authigenic ***feldspar*** crystallizes simultaneously with sedimentation, although later replacement of fossils during dolomitization of calcareous sediments has been documented.

DAVIS A. YOUNG

References

Deer, W. A.; Howie, R. A.; and Zussman, J., 1963. *Rock-Forming Minerals,* vol. 4, *Framework Silicates.* New York: Wiley, 435p.

Feldspar, 1962. *Norsk Geol. Tidsskr.*, **42**(2).

Hovis, G. L., 1980. Angular relations of alkali feldspar series and the triclinic-monoclinic displacive transformation. *Amer. Mineral,* **65**, 770–778.

Instituto "Lucas Mallada," 1961. *Cursillos y Conferencias del,* **8.**

Luth, W. C., 1974. Analysis of experimental data on alkali feldspars: unit cell parameters and solvi, in W. S. MacKenzie, and J. Zussman, 249–296.

MacKenzie, W. S., and Zussman, J., eds., 1974. *The Feldspars.* New York: Crane, Russak, 717p.

Ribbe, P. H., ed., 1975. *Feldspar Mineralogy,* vol. 2, Mineral. Soc. Am. Short Course Notes.

Sipling, P. J., and Yund, R. A., 1976. Experimental determination of the coherent solvus for sanidine-high albite. *Amer. Mineral,* **61**, 897–906.

Smith, J. V., 1974. *Feldspar Minerals,* vol. 1, *Crystal Structure and Physical Properties;* vol. 2, *Chemical and Textural Properties.* New York: Springer-Verlag.

Stewart, D. B., and Wright, T. L., 1974. Al/Si order and symmetry of natural alkali feldspars, and the relationship of strained cell parameters to bulk composition. *Bull. Soc. fr. Minéral. Cristallogr.*, **97**, 356–377.

Tuttle, O. F., and Bowen, N. L., 1958. Origin of granite in the light of experimental studies in the system $NaAlSi_3O_8$-$KAlSi_3O_8$-SiO_2-H_2O, *Geol. Soc. Am. Mem.*, **74**, 153p.

Yund, R. A., 1974. Coherent exsolution in the alkali feldspars, in A. W. Hofmann et al., eds., *Geochemical Transport and Kinetics.* Washington: Carnegie Institution of Washington, 173–183.

Cross-references: *Authigenic Minerals; Feldspar Group; Order-Disorder; Plagioclase Feldspars; Polymorphism; Rock-Forming Minerals; Staining Techniques; Twinning.* Vol. IVA: *Mineral Classes: Silicates; Phase Equilibria; Solid Solution.*

ALLOTROPY

Allotropy (so named by J. J. Berzelius in 1841) refers to the existence of a chemical element in two or more distinct forms having different crystalline structures and/or physical properties. Allotropes may differ with respect to density, melting point, molar volume, color, and other physical properties. In some cases, there is a reversible, in others an irreversible transition from one allotrope to another.

Examples of allotropy include:

- carbon: *chaoite, graphite,* and *lonsdaleite* (hexagonal); *diamond* (isometric).
- *sulfur:* native or *α-sulfur* (orthorhombic); γ-sulfur or *rosickyite* (monoclinic).
- phosphorus: white/yellow (two forms: cubic and orthorhombic), violet and black (thus, four allotropes of contrasting properties); red phosphorus is a mixture.
- *tin:* white (tetrahedral); gray (cubic).
- *iron:* α-iron or *kamacite* (body-centered cubic and magnetic); γ-iron or *taenite* (face-centered cubic and nonmagnetic); δ-iron or ferrite (body-centered cubic and stable only above 1400°C).
- oxygen and ozone: oxygen, O_2 (density 1.429 g/liter); ozone, O_3 (density 2.144 g/liter).

RHODES W. FAIRBRIDGE

Reference

Weaver, E. C., 1966. Allotropes, in G. L. Clark, ed., *Encyclopedia of Chemistry,* 2nd ed. New York: Reinhold.

Cross-reference: *Polymorphism.*

ALLOYS—*See* METALLURGY; NATIVE ELEMENTS AND ALLOYS

ALUNITE GROUP, BEUDANTITE GROUP, CRANDALLITE GROUP

Alunite Group	*Beudantite Group*	*Crandallite Group*
Alunite	*Beudantite*	*Crandallite*
Ammoniojarostie	*Corkite*	*Dussertite*
Argentojarosite	*Hidalgoite*	*Eylettersite*
Beaverite	*Hinsdalite*	*Florencite*
Hydronium jarosite	*Kemmlitzite*	*Gorceixite*
Jarosite	*Schlossmacherite*	*Goyazite*
Natroalunite	*Svanbergite*	*Lusungite*
Natrojarosite	*Weilerite*	*Plumbogummite*
Osarizawaite	*Woodhouseite*	*Waylandite*
Plumbojarosite		*Zairite*

These isostructural groups of trigonal minerals have the general formula

$$AB_3(XO_4)(OH)_6$$

The X site is tetrahedral; the B site is octahedral; and the A site has a coordination number of 12. ***Alunites*** are sulfates (X = S) where A = K,Na,(NH_4),Ag,(H_3O),Pb; ***Crandallites*** are phosphates with one (OH) replaced by (H_2O) and A = Ca,Sr,Ba,Pb,Bi,Ce,Th; and ***Beudantites*** are arsenates or phosphate sulfates where A = Ca,Sr,Ba,Pb. The B site in all three groups contains (Al,Fe^{3+}) with the possibility of Cu substitution in the *Alunites.*

Most members of these groups have basal cleavage. Crystals of ***Alunites*** are rare; but ***Beudantites*** and, to a lesser extent, ***Crandallites*** display rhombohedral pseudocubes.

Alunites occur in rocks altered by sulfuric acid liberated by the oxidation of *pyrite* and by hydrothermal action. ***Beudantites*** are typically found in vugs in *quartz* veins. ***Crandallites*** are secondary minerals, but are also found in detrital sands.

KEITH FRYE

Reference

Botinelly, T., 1976. A review of the minerals of the alunite-jarosite, beudantite, and plumbogummite groups, *J. Res. U.S. Geol. Survey*, **4**, 213–216.

Cross-references: See glossary.

AMERICAN SOCIETY FOR TESTING AND MATERIALS

ASTM is an independent, nonprofit corporation founded in 1898 and incorporated in 1902 by the engineering community of the USA. Its purpose was to bring together highly knowledgeable people to make technical judgments on technical matters. This 25,000-member society is organized into committees that work within specific technical areas to reach a national consensus on specifications criteria for products and natural and commercial materials and on procedures for characterizing and evaluating such products and materials. These specifications, test methods, definitions, etc., collectively known as *standards* and originally used as convenient references for purchase contracts, have attained value as supportive documents for national and local codes and in litigation, and have become avenues for working with regulatory agencies.

The standards-writing committees are composed of representatives from producer, consumer, academia, and government; companies, organizations, and agencies pertinent to the scope of the committee, must all participate in order to reach a truly national consensus on any given judgment.

The original work of ASTM covered steel used for railroad rail, boiler vessels, buildings, bridges, etc. Work expanded after the turn of the century to cover iron, cement, wood, petroleum, coal, paint, clay pipe, electrical wire, road and paving materials, etc. Research programs were also established on corrosion of metals, deterioration of paint, fire testing of materials, etc.

In recent years the value of consensus standards for such matters as performance of commercial products, safety of consumer goods, measurement of pollutants, biocompatibility of materials, and technical aspects of products liability litigation has considerably enlarged the activities of ASTM.

The ASTM committees having pertinent projects in the field of mineralogy are given in Table 1. Numerous special subcommittee activities and standard developments are specifically related to matters concerned with mineralogy. For example, Committee C-18 on Natural Building Stones has subcommittees on test methods; on durability, wear, and weathering; on material specifications; on composites; on color, coatings, and finishes; and on nomenclature and definitions. Committee D-4 on Road and Paving Materials has subcommittees on aggregate tests and on aggregate specifications. Committee D-18 on Soil and Rock for Engineering Purposes has subcomittees concerned with the identification and classification of soils, with physico-chemical properties of soils and rocks, with dynamic properties of soils, and with rock mechanics. Committee E-4 on Metallography deals with microhardness, X-ray methods, methods of thermal analysis, electron microscopy and

TABLE 1. ASTM Committee Activities Relevant to Minerology

Materials or Test Techniques	Committee Number
Material Class	
Aggregates, pozzolans	C-9
Bitumins	D-4
Cement	C-1
Clays	C-21
Coal	C-5
Lime	C-7
Metal Bearing Ores	E-16
Natural gas	D-3
Paint Pigments	D-1
Petroleum	D-2
Stones for building	C-18
Soil and rock for engineering	D-18
Test Techniques	
Chemical analysis of metals	E-3
Chromatography	E-19
Color and photography	E-12
Emission spectroscopy	E-2
Metallography, all techniques	E-4
Microscopy, all techniques	E-25
Molecular spectroscopy	E-13
Petrography	C-9
Thermal analysis	E-37

diffraction, and scanning microscopy and microprobe analysis. Committee C-9 on Concrete and Concrete Aggregates has a subcommittee (C-09.02.06) on the petrography of concrete and concrete aggregates, which has under its jurisdiction the ASTM Standard Recommended Practice for Petrographic Examination of Aggregate for Concrete, Designation: C295; and the Standard Descriptive Nomenclature of Constituents of Natural Mineral Aggregates, Designation: C294. Committee C-9 also has a subcommittee (C-09.02.08) on concrete for radiation shielding, which has under its jurisdiction the Standard Descriptive Nomenclature of Constituents of Aggregates for Radiation-Shielding Concrete, Designation: C638. In this latter designation, there is a table of the commercially important boron minerals that are used in radiation-shielding concrete.

The ASTM Committee on Terminology is responsible for publication of the *Compilation of ASTM Standard Definitions.* This glossary of terms compiles, in one book, the standard definitions developed by all the committees. In this volume are found many standard definitions useful in many aspects of mineralogy. For example, in the 1981 Glossary, Committees C-18 on Natural Building Stones, C-7 on Lime, and C-14 on Glass and Glass Products each have defined limestone in slightly different ways. Committee C-18 has a scientific definition of granite in Standard C119, Committee C-14 a standard of quartz glass, Committee C-18 a standard definition of quartzitic sandstone, and C-8 on refractories has a standard definition of quartzite (ganister).

The ASTM publishes an *Annual Book of ASTM Standards,* the *Journal of Testing and Evaluation,* the monthly *Standardization News,* and a series of special technical publications. Many papers dealing with matters of interest to mineralogists will be found in these latter, more technical volumes. ASTM STP169 on the *Significance of Tests and Properties of Concrete and Concrete Aggregates,* first issued in 1955, has always included two chapters dealing with the petrographic examination of hardened concrete and the petrographic examination of concrete aggregates. The third edition of this publication, entitled *Significance of Tests and Properties of Concrete and Concrete-Making Materials* (ASTM STP 169-B), was published in 1978. ASTM STP83, published in 1948, included the proceedings of a symposium on mineral aggregates that was sponsored jointly by Committees C-9 on Concrete and Concrete Aggregates and D-4 on Paving Materials. It is still one of the outstanding compilations of review papers in the field of aggregates. The former Committee E-1 on Methods of Testing sponsored a symposium on light microscopy in 1952. The proceedings of this symposium, published in ASTM STP143, included a discussion of the application of light microscopy to research on concrete. This paper was recognized as the most outstanding paper on concrete and concrete aggregates published by ASTM that year. Since 1979, ASTM has published *Cement, Concrete, and Aggregates,* which deals in part with the application of mineralogical examination techniques, both standard and novel, to these synthetic materials.

Membership on ASTM technical committees is open to people interested in the fields covered, and inquiries as to membership, publications, and standards issued may be obtained by writing to: Information Center, ASTM, 1916 Race St., Philadelphia, Pa. 19103, USA. (Telephone: 215 299-5475.)

BRYANT MATHER
S.F. ETRIS

Cross-references: *Abrasive Materials; Asbestos; Clays, Clay Minerals; Electron Microscopy: Scanning; Electron Microscopy: Transmission; Electron Probe Microanalysis; Fire Clays; Mineral Industries; Mineral Properties; Optical Mineralogy; Ore Microscopy; Pigments and Fillers; Portland Cement Mineralogy; Refractory Minerals; Sands–Glass and Building; Soil Mineralogy; Zeolites.* Vol. IVA: *X-ray Diffraction Analysis; X-ray Spectroscopy.*

AMPHIBOLE GROUP

The term *amphibole,* derived from the Greek *amphibolos* meaning "ambiguous," was first used by Haüy in 1801. The complex nature of amphibole compositions is easily appreciated when it is remembered that all the eight elements that, together, make up 98% of the earth's crust, are present in common ***amphiboles.*** As a result of this range of composition, the group has a multiplicity of names (see Leake, 1978).

The system of nomenclature used here is given in Fig. 1.

Structure

The ***amphiboles*** have either orthorhombic or monoclinic symmetry. They are classified as inosilicates and have a structural arrangement of continuous double chains of alternately cross-linked silicon–oxygen tetrahedra. The relationship of this structure, with a 4:11 silicon–oxygen ratio, to crystal form and cleavage is shown in Fig. 2. The double chains repeat at intervals of 5.3 Å approximately.

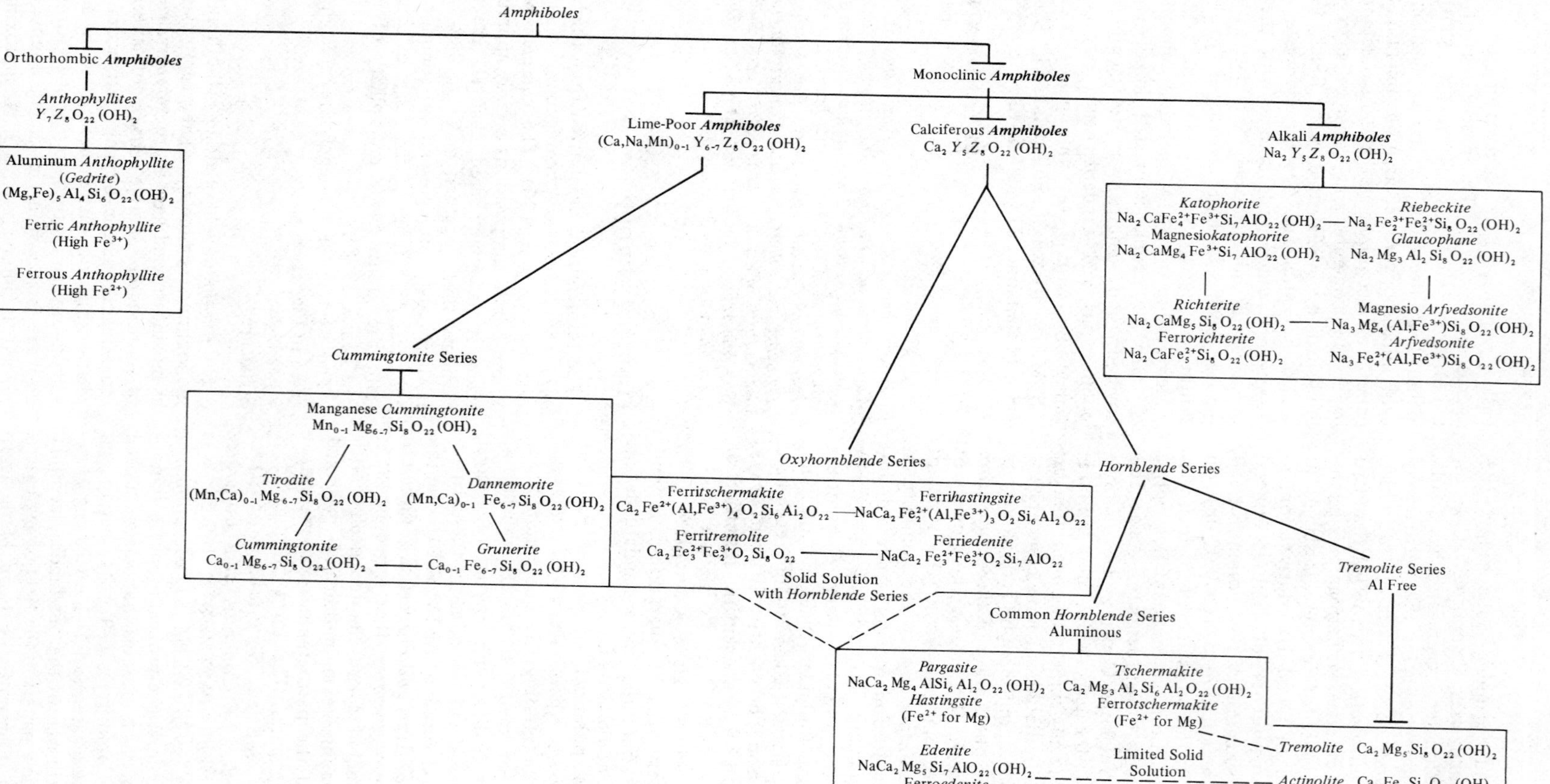

FIGURE 1. Classification of the ***amphibole*** group.

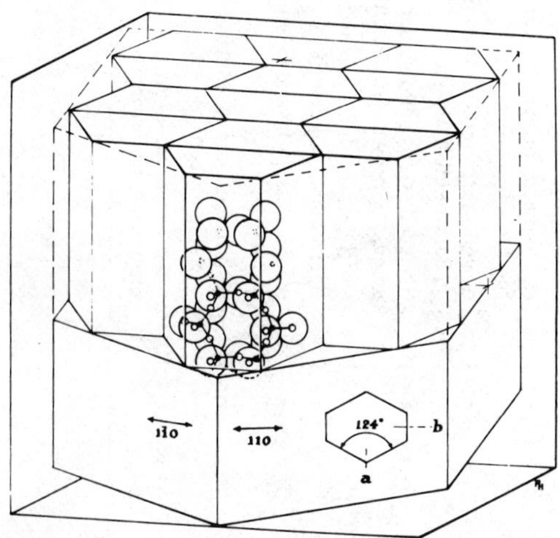

FIGURE 2. The relationship between atomic structure and cleavage in the ***amphiboles.*** The planes of weakness are shown by heavy lines, and resulting cleavage directions {110} by broken lines (from *Mineralogy, Concepts, Descriptions, Determinations* by L. G. Berry and Brian Mason, copyright © 1959 by W. H. Freeman and Company and used by permission).

Fig. 3 shows that this defines the *c* parameter of the unit cell.

Cations of various sizes occupy the M_1, M_2, M_3, and M_4 sites (Fig. 3) and act as spacers between the chains. Broadly, the alkali and calciferous varieties have cations larger than manganese in M_4 sites. The nature of the spacers, the presence of OH groups, and the substitution of aluminum for divalent and tetravalent ions with concomitant filling of the vacant *A* sites in the structure produces a range of cell parameters within the group (Table 1).

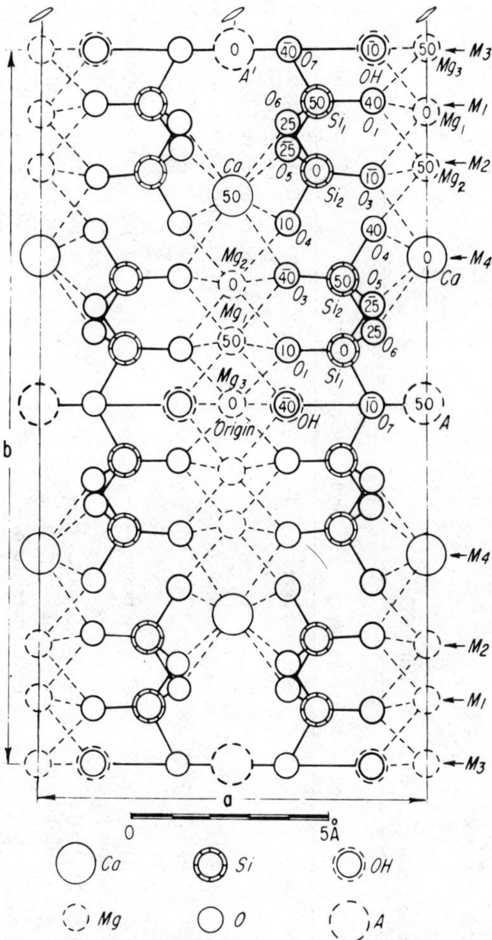

FIGURE 3. Structure of *tremolite* ***amphiboles*** (after Warren, 1930).

Chemistry

Since the group exists in two different classes of symmetry, it is convenient to consider the chemistry similarly. All ***amphiboles*** can be fitted into two general compositional expressions:

- monoclinic: $X_{2-3}Y_5Z_8O_{22}(O,OH,F,Cl)_2$
- orthorhombic: $Y_7Z_8O_{22}(O,OH,F,Cl)_2$

where X = Ca,Na,K
Y = Mg,Cr,Li,Fe^{2+},Fe^{3+},Mn,Al
Z = Si,Al,Ti, and sometimes B, in part

The species description of an ***amphibole*** is based on the end-member concept (Phillips and Layton, 1964; Leake, 1978). This system was originally developed by means of the partial triangle and the partial prism (Hallimond, 1943). It is based on the types of substitution that can take place in the *tremolite* structure (see Fig. 4). However, dark colored (Fe-bearing) calcic ***amphiboles,*** which, for lack of chemical data, cannot be confidently identifiable as near to an end member are called *hornblende.* Most *hornblendes* are actinolitic, tschermakitic, edenitic, pargasitic, or hastingsitic ***amphiboles*** (Leake, 1978).

Substitution. Substitition of Al^{3+} for Si^{4+} occurs in the Si-O chain and immediately leads to a positive charge deficiency in the structure. It is compensated by the introduction of Na into the vacant *A* sites in the monoclinic structure (Fig. 3). *Edenite, pargasite,* and their iron analogs are formed in this way. Alternatively, *Y*-group magnesium or ferrous iron may be replaced by trivalent aluminum or ferric iron. *Tschermakites, anthophyllites,* and,

TABLE 1. Physical Properties of the Amphibole Group

Mineral Series	Symmetry	D	Hardness	Cleavage	Twinning	Color	Sign	Optic orient.	Disp.	Pleochroism
Anthophyllite	O	2.85–3.57	5.5–6	$(210):(2\bar{1}0)$ 54½°	none	white, gray, green, clove brown	±	$\gamma=z$	$r \lesseqgtr V$	$\gamma=\beta>\alpha$; $\gamma>\beta=\alpha$; α, yellows, pale brown; β, yellows; γ, greens, blues, browns
Cummingtonite	M	3.10–3.60	5–6	$(110):(1\bar{1}0)$ $\cong 55°$	{100} simple lamellar	dark green brown	+	$\beta=y$	$r \lesseqgtr V$	magnesian-rich nonpleochroic; others $\alpha=\beta$, colorless; γ, pale green; γ, pale brown
Grunerite	M	3.10–3.60	5–6	$(110):(1\bar{1}0)$ $\cong 55°$	{100} simple lamellar	dark green brown	-	$\beta=y$	$r \lesseqgtr V$	$\alpha=\beta$, pale yellow or brown
Tremolite	M	3.02–3.44	5–6	$(110):(1\bar{1}0)$ $\cong 56°$	{100} simple lamellar	colorless to dark green with Fe	-	$\beta=y$	$r<V$	α, pale yellow; γ, green to blue; β, pale yellow green
Common *Hornblende*	M	3.02–3.45	5–6	$(110):(1\bar{1}0)$ $\cong 56°$	{100} simple lamellar	green to dark green		$\beta=y$	$r \lesseqgtr V$	variable green, yellow green, bluish green, brown; $\gamma \geqslant \beta > \alpha$; $\beta>\gamma>\alpha$
Oxyhornblende	M	3.19–3.30	5–6	$(110):(1\bar{1}0)$ $\cong 56°$	{100} simple lamellar	brown to black	±	$\beta=y$	$r<V$	α, yellow; β, brown; γ, dark red brown
Kaersutite	M	3.20–3.28	5–6	$(110):(1\bar{1}0)$ $\cong 56°$	{100} simple lamellar	reddish brown to brown	-	$\beta=y$	$r>V$	α, yellows; β, reddish brown; γ, dark reddish brown
Richterite	M	2.97–3.45	5–6	$(110):(1\bar{1}0)$ $\cong 56°$	{100} simple lamellar	brown, yellow, or greens	-	$\beta=y$	$r<V$	$\beta>\gamma>\alpha$; α, β, yellow, lilac, greens; γ, brown, greens, blue
Arfvedsonite-Eckermannite	M	3.00–3.50	5–6	$(110):(1\bar{1}0)$ $\cong 56°$	{100} simple lamellar	bluish green, black	-	$\beta=y$	$r<V$ $r \lesseqgtr V$	$\gamma<\beta<\alpha$; α, bluish greenish, β, lavender; γ, greenish
Glaucophane	M	3.08–3.30	6	$(110):(1\bar{1}0)$ $\cong 56°$	{100} simple lamellar	gray, lavender blue	-	$\beta=y$	$r<V$	α, colorless; β, lavender blue; γ, blue
Riebeckite	M	3.02–3.42	5	$(110):(1\bar{1}0)$ $\cong 56°$	{100} simple lamellar	dark blue to black	±	$\gamma=y$	$r \lesseqgtr V$	α, prussian blue; β, indigo blue; γ, yellow green
Katophorite	M	3.20–3.50	5	$(110):(1\bar{1}0)$ $\cong 56°$	{100}	rose red, dark red, brown black	-	$\gamma=y$ $\beta=y$	$r \lesseqgtr V$	$\gamma>\beta>\alpha$; *kataphorite* $\gamma<\beta>\alpha$; magnesio*katophorite*

	α	β	γ	$2V\alpha$	$\gamma : Z$	δ	Å a_0	Å b_0	Å c_0	β	Z	Space group
Anthophyllite	1.596–1.694	1.605–1.710	1.615–1.722	78°–111°	0	0.013–0.028	18.5–.6	17.7–18.1	5.27–5.32	–	4	*Pnma*
Cummingtonite	1.635–1.665	1.644–1.675	1.655–1.698	65°–90°	15°–21°	0.020–0.030	≅ 9.6	≅ 18.3	≅ 5.3	≅ 101°50′	2	*C*2/*m*
Grunerite	1.665–1.696	1.675–1.709	1.698–1.727	90°–96°	10°–15°	0.030–0.045	≅ 9.6	≅ 18.3	≅ 5.3	≅ 101°50′	2	*C*2/*m*
Tremolite	1.559–1.688	1.612–1.697	1.622–1.705	86°–65°	21°–10°	0.027–0.017	≅ 9.85	≅ 18.1	≅ 5.3	≅ 105°50′	2	*C*2/*m*
Common *Hornblende*	1.613–1.705	1.618–1.714	1.632–1.730	120°–10°	13°–34°	0.014–0.028	≅ 9.9	≅ 18.0	≅ 5.3	≅ 105°30′	2	*C*2/*m*
Oxyhornblende	1.662–1.690	1.672–1.730	1.680–1.760	60°–82°	0°–18°	0.018–0.070	≅ 10.0	≅ 18.1	≅ 5.3	≅ 106°	2	*C*2/*m*
Kaersutite	1.670–1.689	1.690–1.741	1.700–1.772	66°–82°	0°–19°	0.019–0.083	9.9	18.21	5.4	106°	2	*C*2/*m*
Richterite	1.605–1.685	1.618–1.700	1.627–1.712	66°–87°	15°–40°	0.015–0.029	9.82	17.96	5.27	104°20′	2	*C*2/*m*
Arfvedsonite-Eckermannite	1.612–1.700	1.625–1.709	1.630–1.710	80°–0°	$\alpha : Z$ 0°–53°	0.005–0.020	≅ 9.7	≅ 17.7	≅ 5.3	≅ 104°	2	*C*2/*m*
Glaucophane	1.606–1.661	1.622–1.667	1.627–1.670	50°–0°	4°–15°	0.008–0.022	≅ 9.7	≅ 17.7	≅ 5.3	104°	2	*C*2/*m*
Riebeckite	1.654–1.701	1.662–1.711	1.668–1.717	40°–90°	$\alpha : Z$ 3°–21°	0.006–0.016	≅ 9.75	≅ 18	≅ 5.3	103°	2	*C*2/*m*
Katophorite	1.639–1.681	1.658–1.688	1.660–1.690	0°–50°	$\alpha : Z$ 36°–70°	0.007–0.021						

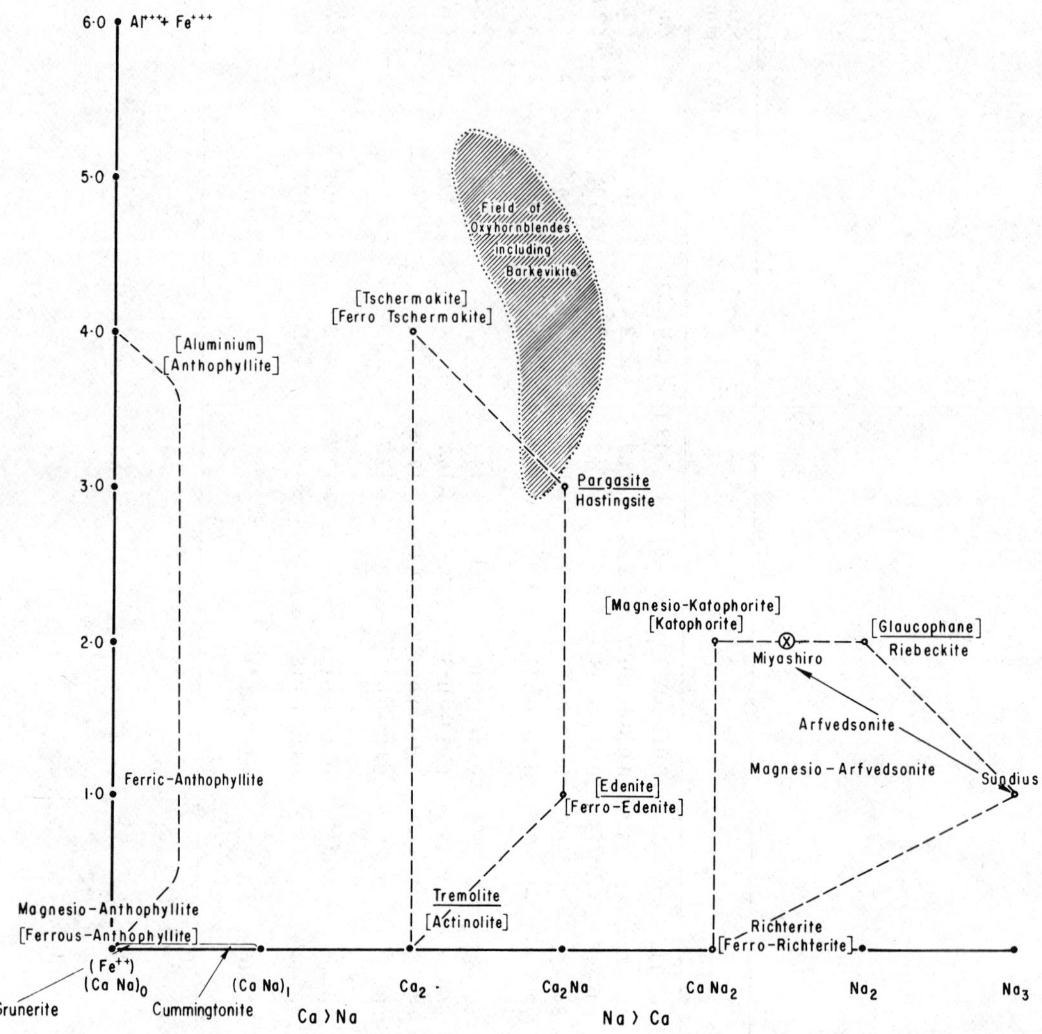

FIGURE 4. Classification of ***amphiboles*** by the end-member method. Fe^{2+} is normal to the plane of the paper. All quantities are expressed in atomic proportions.

in part, *pargasites,* have this type of compensation.

Aluminum does not normally substitute for more than one-fourth of the silicon atoms in the chain structure. Technically, the compositions produced by the substitution described here are not true end members. Substitution may continue further but rarely does.

Sodium Substitution. The basic substitutions required to form the alkali ***amphibole*** subgroup are Na for Ca and 2Na for Ca. The Na for Ca substitution in M_4 sites leads to a positive charge deficiency. This is balanced by $(Al,Fe^{3+})_2$ substituting for $(Mg,Fe)_2$ in M_2 sites, thus forming *glaucophane-riebeckite.*

Substitution of 2Na for Ca does not affect the charge balance and gives *richterite.* This is not a true end member since further similar substitution produces *eckermannite-arfvedsonite.* The charge deficiency caused by this further substitution is balanced by a *Y*-group change of $(Mg,Fe)^{2+}$ for $(Al,Fe)^{3+}$.

Iron Substitution. A very common substitution in the ***amphibole*** group is that of Fe for Mg. In the orthorhombic ***amphiboles,*** all four sites, M_1, M_2, M_3, and M_4, are filled by Mg^{2+} and Fe^{2+} in varying proportions. In the monoclinic amphiboles, M_4 is occupied by the larger cations Na, Ca, and, to a lesser extent, by other cations. The Mg^{2+} and Fe^{2+} ions enter the M_1, M_2, and M_3 sites in the monoclinic varieties.

However, as a result of the differing electronic configurations of Fe^{2+} and Mg^{2+}, crystal-field phenomena (Burns, 1968) have a significant effect upon the ranges of substitution of the ions in the different sites M_1, M_2, M_3, and M_4 in the ***amphiboles.*** The symmetry configurations of the sites, either octahedral, tetragonal, or monoclinic, affect the energy separations between the *d* electrons; and the energy differences figure in the ordering of Fe^{2+} and Mg^{2+} in the sites. Generally, the entry of Fe^{2+} into the distorted sites is favored by the crystal-field effects. The relative enrichment of Fe^{2+} in *actinolite* is $M_1 > M_3 > M_2$ (Burns and Greaves 1971); for *cummingtonite,* it is $M_4 > M_1, M_3 > M_2$.

Minor Element Substitution. Trace-element (Zn,Co,Ni,Cu,Cr) substitutions of a range of base-metal elements occur in the ***amphiboles.*** These substitutions occur predominantly as a result of similarities in ionic size and charge between the trace components and the major elements. The principal substitutions are for Fe^{2+} and Mg^{2+}. Crystal-field effects are important in trace-element substitutions and site ordering in the ***amphiboles.***

Role of Water

In 1890, Penfield showed water to be an essential part of the ***amphibole*** structure. W. T. Schaller, in 1916, derived the correct chemical formula for *tremolite* minerals and included the OH groups. The structural positions of these are shown in Fig. 3.

Apart from ***oxyhornblendes,*** most ***amphiboles*** contain approximately two OH groups per formula unit. The hydroxyl radical is difficult to determine chemically. There is considerable debate in the literature on the position, number, and effect of OH groups in the ***amphibole*** structure (Phillips, 1963). It is for these reasons that many analyses have been recalculated on the basis of 23 oxygen atoms per formula unit, thus omitting water from the main part of the calculation. This is unsatisfactory since, in the ***oxyhornblendes,*** excess charge is compensated by oxidation of ferrous ions. This transformation was described by V. E. Barnes in 1930; and Phillips (1963) suggests a number of checks to test the credibility of analyses on this basis.

The hydroxyl-deficient ***amphiboles*** ($OH + F + Cl < 1.00$) constitute a separate mineral subgroup known as ***oxyhornblendes*** (basaltic *hornblendes*). ***Oxyhornblende*** includes some *kaersutite, pargasite,* and *hastingsite.*

The OH group is partly interchangeable with fluorine. W. Eitel, in 1954, and J. A. Kohn and J. E. Comeforo, in 1955, synthesized fluorine and boron-bearing (in two positions) *edenite.* However, only about one-third of the OH sites can be occupied with F (Leake, 1968). It has been suggested that the limit of substitution is related to the Fe-F bonding between the O_3 site (Fig. 3) and M_1-M_2-M_1, and to the geometry of the structures. The F-rich ***amphiboles*** are generally found in marbles.

Solid-Solution Relationships

The extensive substitution effects in the ***amphibole*** group do not imply continuous solid solution either within the major subgroups or between them. A. N. Winchell, in 1945, and N. Sundius, in 1946, demonstrated the existence of important miscibility gaps—for example, the gap between the aluminous *hornblendes.* There are nonsequences between the calciferous and alkali ***amphiboles.*** Continuous substitution of Na for Ca does not take place. There are marked solution breaks between *hornblendes* and *anthophyllites;* nor can complete compositional gradation be detected between *cummingtonites* and the other series.

Exsolution phenomena are thus common in the ***amphiboles.*** Ross, Papike, and Shaw (1969) have studied these relationships in *cummingtonite-actinolite* and *hornblende-anthophyllite-gedrite* from a wide variety of natural occurrences. Experimental evidence (Cameron, 1975) confirms the natural observations. The compositions of the unmixed phases may be used as a geothermometer. The unmixing is usually of a Ca-poor phase oriented on ($\bar{1}01$) and (100) of a Ca-rich host. However, exsolution of *cummingtonite* from *actinolite* and *actinolite* from *cummingtonite* may occur.

Other complex intergrowths, interpreted as stable coexistence of two ***amphiboles,*** include associations of *glaucophane-barroisite* and *cummingtonite-anthophyllite.*

The Cummingtonite Problem

This subgroup, with a general formula close to that of the *anthophyllites,* has a monclinic structure similar to *tremolite.* It is a bridging series between two structural states of the ***amphibole*** group. Investigation of this problem has shown the following criteria to be important:

1. Manganese is a common constituent, particularly in iron-poor *cummingtonites.* It is not important and rarely present in quantity in *anthophyllites.* It does occur in other monoclinic minerals such as *richterite* and *tremolite.*
2. Aluminum is important in *anthophyllites* but uncommon in *cummingtonites.*

Analyses by Klein (1964) show that iron-rich *cummingtonite* (*grunerite*) has little Ca or Mn, but that magnesian varieties have these elements in quantity. The ionic radius of Fe is such that $Fe^{2+} \ll Mn^{2+} \ll Ca^{2+}$ but Mn^{2+} and Fe^{2+} are not widely different in size. It would then appear that Mn and Ca can stabilize the monoclinic structure when the smaller magnesium ion is present, but when Fe^{2+} is present, this ion is large enough to maintain monoclinic symmetry without assistance.

It is significant in this respect that the *anthophyllite* structure (Fig. 5) is never filled out to $Fe_7Si_8O_{22}(OH)_2$. J. C. Rabbitt, in 1948, observed that high iron content is always accompanied by high aluminum content in this series. The orthorhombic structure seems unable to tolerate a full complement of ions the size of Fe^{2+} without some relief of strain by smaller-sized Al^{3+} (Layton, 1964). The Fe_7Si_8 ***amphibole*** configuration is therefore monoclinic.

F. R. Boyd's attempts to synthesize minerals of the general formula $(Mg,Fe)_7Si_8O_{22}(OH)_2$ during 1956–1957 showed that the orthorhombic varieties richer in iron than Mg50Fe50 could not be made. Monoclinic forms were obtained only in the composition range Mg85Fe15 to Mg70Fe30. Layton and Phillips (1960) considered these to represent clino-*anthophyllites* and not *cummingtonites.* M. G. Bown of Cambridge (personal communication) has since identified a natural clino-*anthophyllite.*

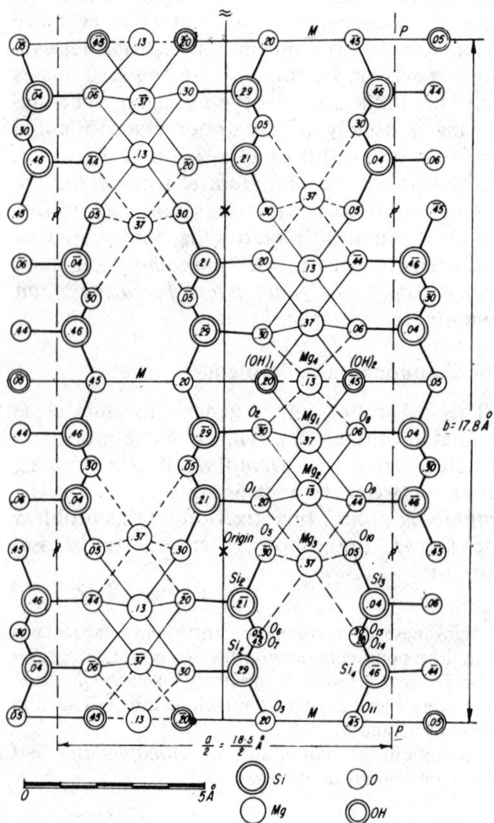

FIGURE 5. Structure of the *anthophyllites.*

A review of the evidence on *cummingtonite* suggests that the magnesium varieties with essential calcium, manganese, or both are close to *tremolite* in character. The iron-rich varieties, although monoclinic because of the large quantity of iron in the structure, have considerable affinities to *anthophyllite.*

Comparisons with Pyroxene

Amphiboles and ***pyroxenes*** both have inosilicate structures. The latter are characterized by a single-chain arrangement where each silicon-oxygen tetrahedron is linked to the next by a common oxygen. This results in an (Si_2O_6) grouping as compared with the previously described ***amphibole*** configuration of (Si_4O_{11}).

In both groups, the chains are aligned along the *c* direction in the crystals; and in both cases, larger cations such as, Ca, Mg, and Fe act as spacers. B. E. Warren, in his 1930 structural analysis of *tremolite* and *diopside,* demonstrated that there were whole blocks of the structure of *tremolite* that were similar to that of *diopside.* In addition, if the *diopside* structure is sheared so that the lower part is displaced by $(a/2) + (c/2)$, a structure essentially the same as that of *tremolite* is produced. In terms of paragenesis, ***pyroxenes*** could be expected to convert to ***amphiboles*** under stress conditions–and they do.

A comparison of the unit-cell data for *tremolite* and *diopside* shows that they are identical except for a doubling along the *b* direction in ***amphiboles:***

	a	*b*	*c*	*β*
Tremolite	9.74 Å	17.80 Å	5.26 Å	105°14′
Diopside	9.73 Å	8.91 Å	5.25 Å	105°50′

In view of the parity of structure and composition, the color, luster, hardness, and density of the two groups are similar. The doubling of the chains affects the form and cleavage of the crystals (see *Pyroxene Group,* Figs. 1 and 2). The cleavage angle thus becomes an important distinguishing property.

It is of some interest to compare the *cummingtonite* and *anthophyllite* minerals with the ***pyroxene*** group. The calcic ***pyroxenes*** are monoclinic, as are the high-temperature forms of the lime-free ***pyroxenes***. There is an obvious analogy between the lime-poor ***pyroxene*** *pigeonite* [(Ca,Mg,Fe)(Mg,Fe)Si_2O_6] and *cummingtonite.* There are similar parallels between the alkali members of both groups.

Distinguishing Features

There is considerable overlap of properties among subgroups of ***amphiboles.*** The various physical properties are listed in Table 1. In general, the orthorhombic ***amphiboles*** are distinguished from their monoclinic relatives by parallel extinction in all [001] zone sections. But care should be exercised when using this property since Rabbitt has drawn attention to several classic cases of misidentification. The difference in *d* spacing for the highest intensities is more reliable:

	d	*I*	*d*	*I*	*d*	*I*
Anthophyllite	3.030 Å	10	3.235 Å	9	2.531 Å	8
Tremolite	3.142 Å	10	2.705 Å	9	2.522 Å	8
Cummingtonite	2.784 Å	10	1.406 Å	9	2.137 Å	8

N. Sundius demonstrated in 1946 that *anthophyllite* and *cummingtonite* could be distinguished from all other ***amphiboles*** by their smaller cleavage angle.

Occurrence

The ***amphibole*** group is comprised of common rock-forming minerals (q.v.). In total, the group covers almost the entire range of igneous rocks. Some member of the group is often an essential mineral of metamorphic rocks of low-to-moderate regional grade and thermal varieties. ***Amphiboles*** convert easily at higher temperatures to ***pyroxenes.*** They are useful indicators of pressure–temperature conditions of formation—e.g., *glaucophanes* indicate relatively high pressure and low temperatures of formation; ***oxyhornblendes*** occur in high-temperature, relatively dry, igneous rocks. *Actinolites* in greenschist facies metamorphic rocks transform to *hornblende* in higher-grade amphibolite and, in part, granulite facies rocks. Changes in pressure–temperature conditions of formation are inevitably reflected in changes in either Ca for Na, Mg for Fe, Si for Fe, or O for OH substitution in the structure.

The paragenesis of the ***amphibole*** group is briefly summarized in Table 2 and illustrated in part in Fig. 6.

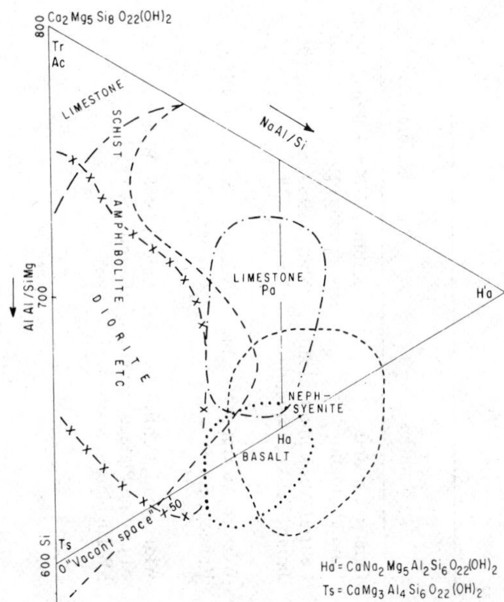

FIGURE 6. Occurrence and composition of the calciferous ***amphiboles*** (after Hallimond, 1943).

Uses

Fibrous minerals of the ***amphibole*** group, particularly *riebeckite* (crocidolite), amosite (*cummingtonite*), *anthophyllite, tremolite,* and *actinolite,* are characterized by the length, strength, flexibility, and heat resistance of their fibers. With the ***serpentine*** mineral *chrysotile,* they comprise *asbestos* (q.v.).

Crocidolite or blue asbestos comes principally from Cape Province, Transvall, South Africa, where it occurs usually in cross-fiber seams. Amosite is usually grayish green in color with a considerable length of fiber, sometimes up to 12 in (30 cm). It has good flexibility and better heat resistance than crocidolite. The other ***amphibole*** minerals are of minor importance and none has the same silkiness, length, or strength of fiber as *chrysotile* (see *Mineral Industries*).

The principal uses of *amphiboles* are in asbestos cement products, asbestos paper- and millboard, filters, insulators, and thermoplastics. The majority of the world's supply of ***amphibole*** asbestos comes from South Africa, Australia, France, and Finland.

W. LAYTON
M. M. WILSON

References

Berry, L. G., and Mason, B., 1959. *Mineralogy.* San Francisco: Freeman, 612 p.

Burns, R. G., 1968. Crystal-field phenomena and iron enrichments in pyroxenes and amphiboles, *Internat. Mineral. Assoc. Proc.,* 5th General Meeting, 170–183.

TABLE 2. Paragenesis of the Amphibole Group

Mineral Series	Rock Types	Associated Minerals	Localities	References[a]
Anthophyllites	Magnesian-rich metamorphic rocks of medium grade; hydrothermally altered ultrabasic rocks.	*talc, cordierite*	Orijavi, Finland; Kamiah, Idaho	Rabbitt, 1948; Eskola, 1914; Anderson, 1931
Cummingtonite	Calcium-poor, iron-rich, medium-grade regional metamorphics; also in igneous hypersthene dactites, diorites.	*hornblende*, ***pyroxene***, *fayalite, hedenbergite, almandine*	Central Abukuma Plateau, Japan; Teisko, Finland	Shido, 1958; Seitsaari, 1952; Kuno, 1938
	Grunerites occur in metamorphosed iron-rich siliceous sediments, under a wide range of conditions with some stress component.		Western Australia; Mesabi Range, US	Miles, 1946; James, 1955; Gruner, 1922
Hornblende	Widespread distribution in acid, alkali, and intermediate plutonic rocks; also in some ultrabasics. Common in low- and medium-grade regional metamorphics as well as in early-stage thermal metamorphics, e.g., impure siliceous dolomites (see Fig. 6). *Tremolites* can remain stable to 850°C (Boyd, 1954[a]) and occur as tremolite amphibolites. With deficient silica, they convert to *pargasite* and increase in intensity of color. At higher temperatures ***pyroxene*** is the dominant phase.	*epidote*, ***chlorite***, ***pyroxene***, ***feldspars***, ***garnet***, *kyanite*	Widespread; Central Abukuma Plateau, Japan; SW Highlands, Scotland	Shido, 1958; Wiseman, 1934; Compton, 1958
Oxyhornblendes	Generally occur in basaltic rocks, basalts to trachytes; *kaersutite* has wider paragenesis and is found in trachyandesites, alkali rhyolites, and camptonite dykes. *Ferro-hornblende* has been recorded in essexites, nepheline and sodalite syenites, foyaites, and camptonite dykes. Sequence of crystallization may be: *Ferro-hornblende* → *kaersutite* → *hastingsite* → *arfvedsonite*	***pyroxenes***, ***feldspars***, *ilmenite, magnetite*	San Juan, Colo.; Oki Islands, Japan; Alno, Sweden	Larsen, 1937; Tomita, 1934; Eckermann, 1948; Yagi, 1953
Glaucophanes	*Glaucophane* rocks are usually found in folded geosynclinal regions but the mineral has been found in nepheline syenites in SE Greenland. *Glaucophane* is generally regarded as a high-pressure mineral, particularly in association with *jadeite*. *Riebeckite* has igneous paragenesis and occurs in syenites, granites, rhyolites, and trachytes.	*lawsonite, pumpellyite, epidote*, ***chlorite***, *muscovite, stilpnomelane, jadeite, almandine*	California; Kaito, Japan; Northern Nigeria; Cape Province, S Africa	Ernst, 1957–58; Miyashiro, 1957; Brothers, 1954; Seki, 1958; Jacobson et al., 1958; Peacock, 1928; Hall, 1930

Richterites	Occur in metamorphosed limestones and have similar paragenesis to *pargasite* and *tremolite*. In alkaline rocks they are associated with late-stage hydrothermal activity.	***pyroxenes***, *tremolite*, *pargasite*, manganese ores	Iron Hill, Colo.; Western Australia; Chikla, India	Larsen, 1942; Prider, 1939; Bilgrami, 1955
Katophorites	Occurs in monzonites and theralites. They also appear in some phonolites and trachytes.	*arfvedsonite*, *hornblende*, **garnet**, *ilmenite*, *magnetite*	Tanganyika; Kenya; Oslo, Norway	Harkin, 1960; Smith, 1931; Oftedahl, 1953
Arfvedsonites	This group is characteristic of plutonic alkali igneous rocks and pegmatites. *Arfvedsonite* occurs in *quartz*-bearing syenites and nepheline syenites. *Arfvedsonite* is associated with hastingsitic ***amphiboles*** in alkaline rocks.	*aegerine*, *aegerine-augite*, *hastingsite*	Sakhalin, Japan; Norra Kärr, Sweden	Iwao, 1938; Adamson, 1942–44

[a] References may be found in Deer, Howie and Zussman, 1962.

Burns, R. G., and Greaves, C., 1971. Correlations of infrared and mossbauer site population measurement of actinolites, *Am. Mineralogist,* **56,** 2010–2033.

Cameron, K. L., 1975. An experimental study of actinolite-cummingtonite phase relations with notes on the synthesis of Fe-rich anthophyllite, *Am. Mineralogist,* **60,** 375–390.

Deer, W. A.; Howie, R. A.; and Zussman, J., 1962. *Rock-Forming Minerals,* vol. 2, Chain Silicates. London: Longmans, 203–364.

Ernst, W. G., 1968. *Amphiboles: Crystal Chemistry, Phase Relations and Occurrence.* New York: Springer-Verlag, 125p.

Hallimond, A. F., 1943. On the graphical representation of the calciferous amphiboles, *Am. Mineralogist,* **28,** 65–89.

Klein, J., C., 1964. Cummingtonite-grunerite series: A chemical, optical and X-ray study, *Am. Mineralogist,* **49**(7 and 8), 963–983.

Layton, W., 1964. Factors governing the natural and synthetic occurrence of members of the amphibole group of minerals, *Neues Jahrb. Mineral., Monatsh.,* **5,** 135–147.

Layton, W., and Phillips, R., 1960. The Cummingtonite problem, *Mineralog. Mag.,* **32**(251), 659–663.

Leake, B. E., 1968. A catalog of analyzed calciferous and sub-calciferous amphiboles, together with their nomenclature and associated minerals, *Geol. Soc. Am. Spec. Paper, 98,* 1–210.

Leake, B. E. (compiler), 1978. Nomenclature of amphiboles, *Am. Mineralogist,* **63,** 1023–1052.

Phillips, R., 1963. The recalculation of amphibole analyses, *Mineralogical Mag.,* **33**(263), 625–674.

Phillips, R., and Layton, W., 1964. The calciferous and alkali amphiboles, *Mineralog. Mag.,* **33**(267), 1097–1109.

Ross, M.; Papike, J. J.; and Shaw, K. W., 1969. Exsolution textures in amphiboles as indicators of sub-solidus thermal histories, *Mineral. Soc. Am. Spec. Paper 2,* 275–299.

Troll, G., and Gilbert, M. C., 1972. Fluorine-hydroxyl substitution in tremolite, *Am. Mineralogist,* **57,** 1386–1403.

Warren, B. E., 1930, The crystal structures and chemical composition of the monoclinic amphiboles, **Zeitschr. Krist., 72,** 493–517.

Cross-references: *Asbestos; Mantle Mineralogy; Mineral Classification: Principles; Naming of Minerals; Order-Disorder; Plastic Flow in Minerals; Pleochroic Halos; Polysomatism; Pyroxene Group; Rock-Forming Minerals; Soil Mineralogy.* Vol. IVA: *Inosilicates; Mineral Classes: Silicates; Solid Solution.*

ANHYDRITE AND GYPSUM

The calcium sulfate in evaporites sometimes occurs as *gypsum*, $CaSO_4 \cdot 2H_2O$, sometimes as *anhydrite*, $CaSO_4$, and sometimes as both minerals together. Near-surface material is almost always *gypsum* because of the ease of weathering and hydration of $CaSO_4$, and deep-seated subsurface material is always *anhydrite* because of dehydration effects. Numerous examples of replacement of one of these minerals by another are known (Murray, 1964; Stewart, 1953; Borchert and Baier, 1953; Ogniben, 1955; Sund, 1959).

The water solubilities of *gypsum* and *anhydrite* have been investigated by Posnjak (1938), Bock (1961), Marshall and Slusher (1966), and others; and MacDonald (1953) has made thermochemical calculations. For many years, it was believed that *gypsum* crystallized out of pure water at a temperature less than about 40°C, that *anhydrite* was the stable phase above this temperature, and that increasing salinity lowered the transition point (see Fig. 1).

However, laboratory evidence for the synthesis of *anhydrite* directly from NaCl-rich waters has shown that *gypsum* is always the stable phase (Zen, 1965).

Hardie (1967) converted *gypsum* to *anhydrite* in solutions containing high concentrations of H_2SO_4, and Cruft and Chao (1969) showed that under conditions of very high supersaturation, i.e., high Ca^{2+} or high $SO_4{}^{2-}$, *anhydrite* would form and be stable as a primary phase. Their conclusion is that the true thermodynamic transition curve between *gypsum* and *anhydrite* is close to that defined by MacDonald (1953) and others, but that kinetic considerations do not permit *anhydrite* to nucleate except under high concentrations of

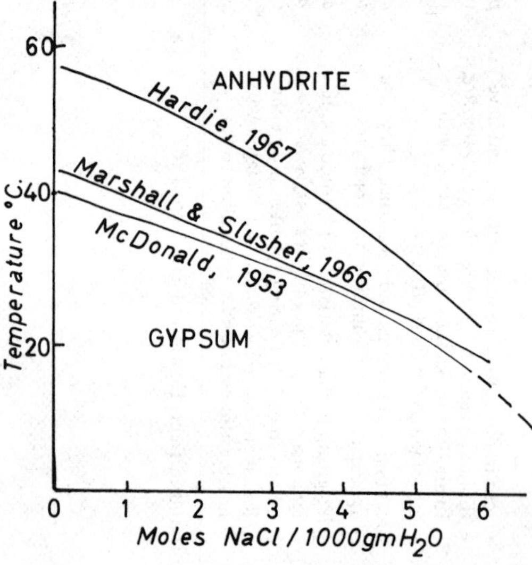

FIGURE 1. Transition temperature of *gypsum-anhydrite* at 1 atm pressure as a function of NaCl in solution according to various workers (after Cruft, 1969).

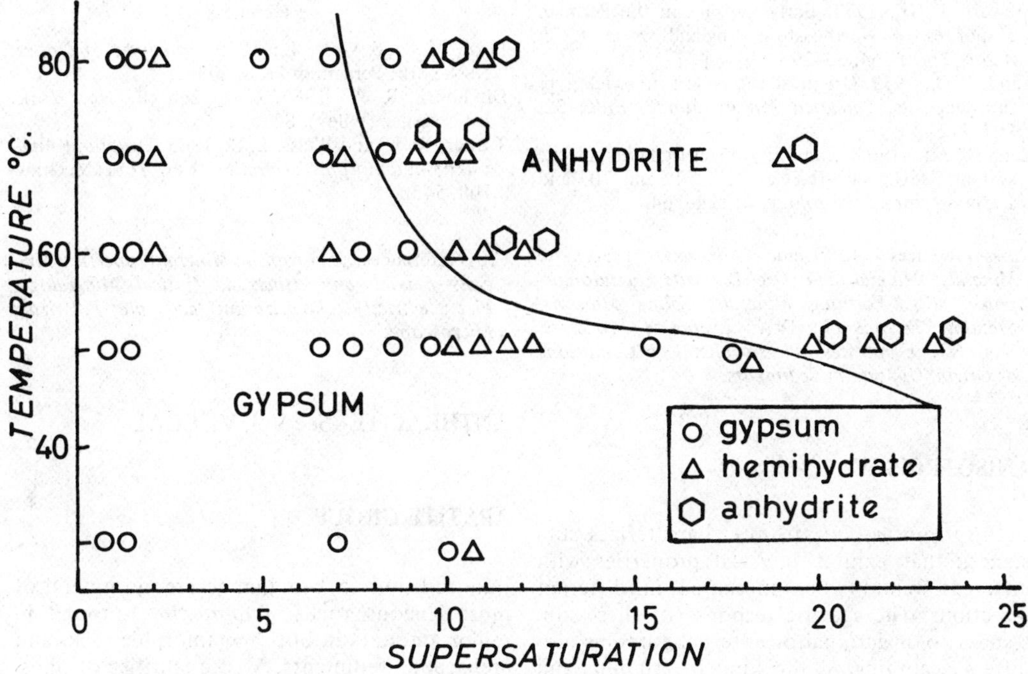

FIGURE 2. Dependence of transition temperatures on supersaturation.

sulfate or calcium (see Fig. 2). Some nonmarine solutions, such as saline groundwaters, can attain these supersaturations, particularly in arid climates; but it is very unlikely marine solutions would ever do so. A possible mechanism for the formation of marine *anhydrite* is reaction of early-precipitated *gypsum* with deeper-bottom brine pools of abnormal composition, as in the Red Sea, or by reaction of sedimentary pore waters with *gypsum*. This could produce *anhydrite* formed by diagenetic changes immediately after precipitation as *gypsum* in the marine environment.

That some extensive marine *anhydrite* beds were not formed by dehydration of the whole unit on deep burial is evident from the fine individual laminae (1-3 mm) of *anhydrite* that have been described by Anderson and Kirkland (1966, and personal communication) and are correlated over 65 miles in the Castile Formation of the Permian of Texas and New Mexico. Extensive dehydration by burial would surely have destroyed these fine laminae extending over such great distances, and their formation could well be an example of early-formed *gypsum* sinking to the bottom of a basin, well below wave-base or turbulent currents, and reacting in a deep brine pool to form *anhydrite*.

E. F. CRUFT

References

Anderson, R. Y., and Kirkland, D. W., 1966. Intrabasin varve correlation, *Geol. Soc. Am., Bull.*, 77, 241-256.

Bock, E., 1961. On the solubility of anhydrous calcium sulfate and of gypsum in concentrated solutions of sodium chloride at 25°C, 30°C and 50°C, *Canadian J. Chem.*, 39, 1746-1751.

Borchert, H., and Baier, E., 1953. Zur metamorphose ozeaner Gipsablagerungun, *Neues Jahrb. Mineral. Abh.*, 86, 103-154.

Cruft, E. F., and Chao, P. C., 1969. Kinetic considerations of the gypsum-anhydrite transition, *3rd Internat. Salt Symp.*, Cleveland, Ohio.

Davidson, C. F., 1965. A possible mode or origin of stratabound copper ores, *Econ. Geol.*, 60, 942-954.

Hardie, L. A., 1967. The gypsum-anhydrite equilibrium at one atmosphere pressure, *Am. Mineralogist*, 52, 171-200.

MacDonald, G. J. F., 1953. Anhydrite-gypsum equilibrium relationships, *Am. J. Sci.*, 251, 884-898.

Marshall, W. C., and Slusher, R., 1966. Thermodynamics of calcium sulfate dihydrate in aqueous sodium chloride solutions, 0-110°, *J. Phys. Chem.*, 70, 4015-4028.

Murray, R. C., 1964. Origin and diagenesis of gypsum and anhydrite, *J. Sed. Petrology*, 34, 512-523.

Ogniben, L., 1955. Inverse graded bedding in primary gypsum of chemical deposition, *J. Sed. Petrology*, 25, 273-281.

Posnjak, E., 1938. The system $CaSO_4$-H_2O, *Am. J. Sci.*, 238, 559-568.

Stewart, F. H., 1953. Early gypsum in the Permian evaporites of Northeastern England, *Proc. Geol. Assoc. Yorks.*, **64**, 33–39.
Sund, J. O., 1959. Origin of the New Brunswick gypsum deposits, *Canadian Mining Metall. Bull.*, **52**, 707–712.
Zen, E. An., 1965. Solubility measurements in the system $CaSO_4$-NaCl-H_2O at 35°, 50° and 70° and 1 atm pressure, *J. Petrology*, **6**, 124–164.

Cross-references: *Authigenic Minerals; Diagenetic Minerals; Mineral and Ore Deposits; Pseudomorphism; Rock-Forming Minerals; Saline Minerals; Staining Techniques.* IVA: *Evaporite Processes.* Vol. VI: *Evaporites; Physicochemical Conditions of Origin; Gypsum in Sediments.*

ANISOTROPISM

Anisotropism (*anisotropy*) characterizes substances that exhibit physical properties with different values when measured in different directions. The specific response of these substances to an internal or external stimulus also differs according to the kind of stimulus and the manner and direction or directions of its application. Anisotropism to some kinds of stimuli is observed in all crystals, in many strained, noncrystalline substances, and in aggregates and accumulations of discrete particles or larger bodies in which there exists at least some degree of parallelism of a linear or planar fabric or of structural elements. *Optical anisotropism* is exhibited by light-transmitting substances through which light of a particular frequency travels with different velocities for different directions of transmission. Such substances doubly refract transmitted light and are described as *birefringent*. Optically, anisotropic opaque substances reflect or absorb light in varying degrees depending on the angle and state of polarization of the incident light.

Most rock bodies are anisotropic and microscopically or megascopically display linear or planar elements of fabric. Familiar examples are stratified sedimentary rocks and metamorphic rocks such as schist and gneiss.

Anisotropism may or may not be characterized by elements of axial or planar symmetry, so that a distinction may be made between *triclinic anisotropism* (no symmetry) and *symmetrical anisotropism*. Examples of symmetrical anistropism are observed in optically uniaxial and biaxial crystals and in anisotropic rock bodies that respond symmetrically to stresses applied symmetrically with respect to linear or planar elements of fabric.

ERNEST E. WAHLSTROM

References

Born, M., and Wolf, E., 1959. *Principles of Optics.* New York: Pergamon Press, 803p.
Ditchburn, R. W., 1963. *Light*, 2nd ed., New York: Interscience (Wiley), 833p.
Turner, R. J., and Weiss, L. E., 1963. *Structural Analysis of Metamorphic Tectonites.* New York: McGraw-Hill, 545p.

Cross-references: *Isotropism; Mineral Properties; Minerals–Biaxial and Uniaxial; Optical Mineralogy; Piezoelectricity; Polarization and the Polarizing Microscope.*

ANTHRACITE–*See* Vol. VI, COAL

APATITE GROUP

In addition to being an accessory mineral of most igneous rocks, *fluorapatite* is found in minor amounts in both metamorphic rocks and arenaceous sediments. Vast quantities of phosphatic rocks, called phosphorites, are found in various parts of the world, and these are predominately *francolite* (carbonate fluorapatite). As a biomineral, *dahllite* (carbonate hydroxyapatite) is the mineral component of bones and teeth of all living vertebrates (see *Human and Vertebrate Mineralogy*).

Chemistry

Besides these phosphatic species, several structurally related minerals, including arsenates, vanadates, and silicates, have the apatite structure; these minerals are:

- *fluorapatite* $Ca_{10}F_2(PO_4)_6$
- *hydroxyapatite* $Ca_{10}(OH)_2(PO_4)_6$
- *chlorapatite* $Ca_{10}Cl_2(PO_4)_6$
- *francolite* $Ca_{10}F_2(PO_4,CO_3)_6$
- *dahlite* $Ca_{10}(OH)_2(PO_4,CO_3)_6$
- *belovite* $(Sr,Ce,Na,Ca)_{10}(OH)_2(PO_4)_6$
- *fermorite* $(Ca,Sr)_{10}(OH)_2(AsO_4,PO_4)_6$
- *hedyphane* $(Ca,Pb)_{10}Cl_2(AsO_4)_6$
- *wilkeite* $Ca_{10}(O,OH,F)_2(SiO_4,PO_4,SO_4)_6$
- *britholite* $(Ce,Ca)_{10}(OH,F)_2(SiO_4,PO_4)_6$
- *britholite-(Y)* $(Y,Ca)_{10}(OH,F)_2(SiO_4,PO_4)_6$
- *strontiapatite* $(Sr,Ca)_{10}(F)_2(PO_4)_6$

Synthetics. Numerous artificial compounds have the following isomorphic substitutions in the general formula

$$A_{10}Z_2(XO_4)_6$$

where A = Ca,Sr,Ba,Pb,Na,K,Cd,Eu^{2+},Sn^{2+}
Z = F,Cl,OH,Br,I
X = P,As,V,Cr,S,Si

Not all combinations are possible, however; A. G. Cockbain, in 1968, and Kreidler and Hummel (1970) related compounds that have the apatite structure to the ratio of the radii of the A to the X cations. The latter authors, however, separate apatitic structures into two series, depending upon whether the principal Z anion is F or Cl.

One of the more interesting ***apatites*** is prepared by heating the rare pegmatitic mineral, *morinite*. The composition of the resulting ***apatite*** is $(Ca_{4.5}Na_{2.3}Al_{3.2})F_2(P_{4.5}Al_{1.5})O_{24}$ indicating statistical substitution of Al in both Ca and P positions (Fisher and McConnell, 1969). A small amount of Mn can also substitute for P, although it usually substitutes for Ca in natural material.

Other cations introduced into synthetic apatites, in amounts up to nearly half the ten Ca positions, include Zn, Cu, Ni, Mg, and Co as *chlorapatites*. Substitutions of B and W in P positions appear to be possible, although these compounds have not been thoroughly investigated.

Through heating of the metamict mineral spencite (*tritomite*-Y), Jaffe and Molinski, in 1962, obtained a glass and an apatitic phase they believed to have the composition $[Y_3(Ce,Pr,Th)Ca]O(Si_2B)O_{12}$. In this compound, one O^{2-} substitutes for one F^- and raises the question of *voelckerite* (oxyapatite): Is this mineral $A_{10}Z_2(XO_4)_6$ or is it $A_{10}Z\square(XO_4)_6$? Putting the question slightly differently: Is one of the two F positions in the structure vacant (as represented by □ in the formula)? There are excellent theoretical reasons for assuming that such an anion would not be missing in oxygenated structures of this type, and the data that indicate such stoichiometry are unsatisfactory from an analytical standpoint. Nevertheless, this matter—along with numerous other crystallochemical topics of a similar nature—remains to be resolved. The problem was considered by S. Mehta and D. R. Simpson in 1975. They found that a "monofluorophosphate" contains 12% F and corresponds to the formula $Ca_6Na_4(PO_3F)_6O_2$, the Ca-bonded oxygens (Z oxygens) having a six-fold coordination.

Carbonate Apatites. ***Carbonate apatites*** are the most perplexing members of the ***apatite*** group from the standpoint of both chemistry and structure. Not only do they have CO_3 groups substituting for PO_4 groups, but they usually show an excess of water above F+OH=2. It has been postulated that this excess water enters the structure in any or all of the following ways: (1) simple substitution of H_2O for OH (or F); (2) substitution of H_3O^+ or H_2O for Ca; and (3) substitution of $(OH)_4$ for PO_4—analagous to the substitution of $(OH)_4$ for SiO_4 in ***hydrogarnets*** and in several other minerals and compounds.

Francolite is the F-bearing mineral of the ***carbonate apatites.*** On analysis, it often shows an excess of fluorine in addition to significant amounts of water. *Dahllite*, comparatively free from fluorine, may contain as much as 6%, by weight, CO_2 and three (or more) times the theoretical amount of water. The importance of *francolite* as the essential component of phosphorite and the significance of *dahllite* as the mineral component of teeth and bones have already been mentioned.

Apatites with Complex Compositions. *Ellestadite* is the end member $Ca_{10}(Cl,F,OH)_2(SiO_4)_3(SO_4)_3$; *wilkeite* is intermediate in composition between *ellestadite* and *fluorapatite*. *Strontiapatite* is $(Sr,Ca)_{10}F_2(PO_4)_6$. *Manganapatite* contains MnO; *fermorite* contains Sr and As; *saamite* contains 3 to 5% rare earths in addition to Sr; *britholite* contains Ce as well as variable amounts of Al and Si replacing P and differs from abukumalite (*britholite*-Y) only in that the latter contains Y and Th. Beckelite and lessingite contain Si, whereas sulfatapatite contains more than 1% SO_3. Dehrnite and lewistonite contain 7.11 and 4.34 percent Na_2O as well as more than 1% K_2O; in addition, dehrnite contains 1.49% CO_2. *Belovite* is a strontium-containing *hydroxyapatite* with replacement of Sr by Ce, Na, and Ca. (Note: Only those names shown in italics are considered valid species. ***Mineral groups*** and ***series*** are in bold italics and varietal names are in roman.)

A *fluor-chlor-hydroxyapatite* has been described from a serpentinized marble in Ceylon by P. G. Cooray, in 1970, the ratios F:Cl:OH of which are 1.0:1.17:1.36. These ratios support the supposition of complete isomorphous mixing of the three end members, *fluorapatite*, *chlorapatite*, and *hydroxyapatite*. Incidentally, this ***apatite*** showed 1.18% CO_2 on analysis.

Pyromorphite seems to form a complete series with *mimetite*, the former being a lead chlorphosphate and the latter a lead chlorarsenate. There is Ca substitution for Pb, but very little substitution of F or OH for Cl. Slightly more than 4% V_2O_5, smaller amounts of Cr and Fe, as well as traces of Sr, Ba, Mn, and rare earths have been reported for this series. When the Ca:Pb ratio exceeds 1:1, the variant becomes *hedyphane* [$(Ca,Pb)_{10}Cl_2(AsO_4)_6$], but the variants between *pyromorphite* and *chlorapatite* have not been studied adequately by synthetic methods to indicate whether or not this series is complete. *Svabite* [$Ca_{10}F_2(AsO_4)_6$] at Franklin, New Jersey, contains 1.54% ZnO; at Jakobsberg, Sweden, it contains 4.52% PbO.

The hydroxy equivalent of *svabite, johnbaumite* [$Ca_{10}(OH)_2(AsO_4)_6$], was reported at Franklin, New Jersey, by P. J. Dunn, D. R. Peacor, and N. Newberry in 1980. Dunn and R. C. Rouse reported morelandite [$Ba_{10}Cl_2(AsO_4)_6$] from near Nordmark, Sweden, in 1978.

Vanadinite [$Pb_{10}Cl_2(VO_4)_6$] may occur with As:V=1:1, in which case this isomorphic variant is called *endlichite*. Considering the small differences in the ionic radii (V^{5+}=0.355 Å and As^{5+}=0.335 Å for four-fold coordination) a complete series would be expected with *mimetite*; Ca, Zn, and Cu may substitute for Pb. *Bellite* [$(Pb,Ag)_{10}Cl_2(Cr,As,Si)_6O_{24}$] has essential amounts of three tetrahedral cations and resembles *wilkeite* [$Ca_{10}F_2(P,S,Si)_6O_{24}$] in this respect.

Minor Constituents and Impurities. Antimony is quite rare as a constituent of any of the ***apatite*** minerals, although it has been reported (Sb_2O_5 0.55%) in *svabite* at Långban, Sweden. Iron is sometimes reported in small amounts, but it may represent admixture of another phase–*hematite* in the case of *francolite*–when it exceeds more than a few tenths of one percent. Magnesium, with its small radius, occurs to a limited extent in members of the ***apatite*** group (about 0.5% being the usual maximum).

Physical Properties

Crystallography. *Apatite* crystallizes in the hexagonal dipyramidal class, showing considerable variation in habit from prismatic (Fig. 1) through equant to tabular (Fig. 2). Some varieties are pseudohexagonal; microcrystalline *dahllite* and *francolite* are probably triclinic; *chlorapatite* and *hydroxyapatite* are sometimes monoclinic. *Mimetite* is sometimes monoclinic–as probably are other members of the ***pyromorphite*** subseries–but all are at least pseudohexagonal and may occur as prismatic crystals.

Crystal Structure. The space group of ***apatite*** was determined as $P6_3/m$ in 1923, and the structure has been known since 1930. *Fluorapatite* is shown in Fig. 3. The monoclinic structures involve a doubling of one of the lateral hexagonal axes, such that $a=b/2$ with the angle $\gamma \approx 120°$ and the space group becoming $P2_1/b$.

With the introduction of CO_3 groups substituting for PO_4 groups–actually $3PO_4 \rightarrow 4CO_3$–the symmetry probably decreases still further,

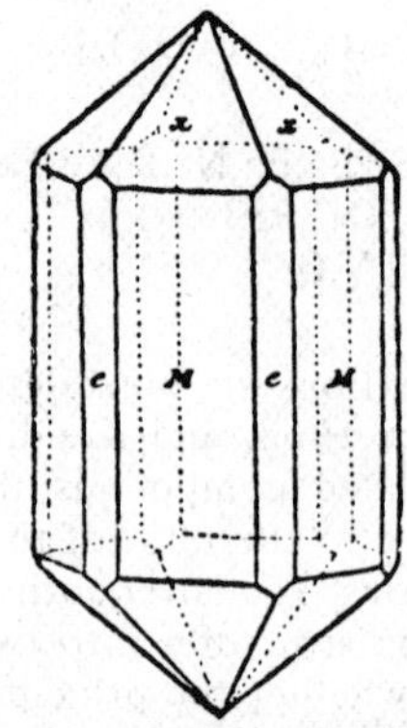

FIGURE 1. Typical crystal of ***apatite*** showing the forms: *M* (10.0), *e* (11.0), and *x* (10.1).

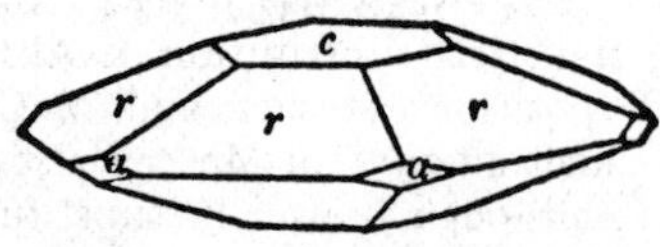

FIGURE 2. Tabular habit of crystals that occur at Hebron, Maine. Forms present are: *a* (11.0), *r* (10.2), and *c* (00.1).

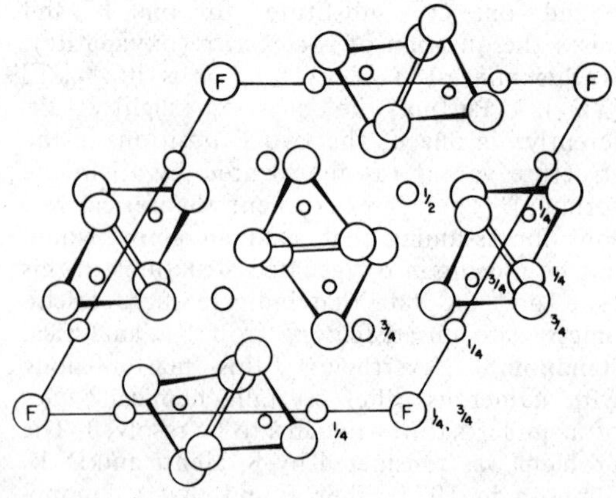

FIGURE 3. Atomic arrangement in *fluorapatite*, $Ca_{10}(PO_4)_6F_2$. Largest circles are oxygens forming tetrahedral groups (two oxygens at c = 3/4 and two others at 0.57 and 0.93c). Phosphorus atoms (smallest circles) are at 0.25 and 0.75c. Calcium atoms are at 0.25 and 0.75c for those surrounding the 6_3 axes; others are at zero and 0.50c. F indicates fluorine on the symmetry planes 0.25 and 0.75c.

becoming triclinic. However, this reduction in symmetry is not detectable from powder diffraction diagrams of these microcrystalline substances, but has become evident from optical properties. The explanation of the structure of the ***carbonate apatites*** is shown as Figs. 4 and 5. Unit-cell dimensions are given in Table 1.

Optical Properties. *Fluorapatite* is uniaxial and negative ($\omega > \epsilon$). However, some varieties are biaxial and ***carbonate apatites*** may show small extinction angles with respect to twins having the composition plane (10•0). Dispersion is high, so measurements should be made using monochromatic light–usually 589 nm. Refractive indices have been reported for artificial compounds having the compositions indicated (except *svabite, johnbaumite*, and *morelandite*) in Table 2. Colored ***apatites*** may show pleochroism with absorption greater for epsilon (ϵ) than omega (ω); dispersion is greater for ω. Although *pyromorphite* and *mimetite* are similar to ***apatite***, the absorption of *vanadinite* is greater for ω.

Coloration of ***apatite*** is common: greenish yellow, pale pink, lavender, and orange. Blue ***apatite*** from near Keystone, South Dakota, apparently owes its color to the presence of

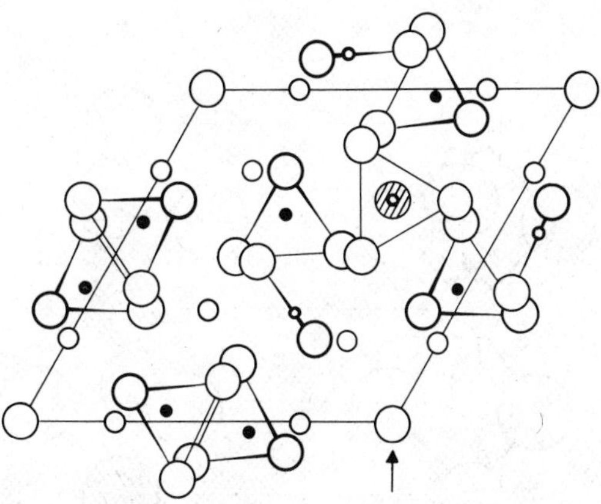

FIGURE 4. Diagrammatic representation of $3PO_4 \rightarrow 4CO_3$ (from McConnell, 1952). Small open circles are C atoms and large shaded circles are H_3O+ at (0, ½*c*). The arrow indicates the origin, which is the viewpoint for Fig. 5.

$MnO_4{}^{3-}$ ions and can be bleached within a few minutes by heating at 600°C. *Pyromorphite* is green, yellow, or brown; *mimetite* is usually colorless to pale yellow; *vanadinite* is yellow, red or brown. All these colors are very pale in thin sections.

Fluorescence and luminescence (q.v.) among natural ***apatites*** seem to be attributable to the presence of rare earths, yttrium, and manganese. Among artificial phosphors (called halophosphates), color can be varied by control of the ratio of Cl:F as well as by the Mn and Sb concentrations. Experimentation with other activators continues, including Al, Zr, Ce, La, and Pb (in addition to Mn and Sb).

Other Properties. Calculated densities are given in Table 1. The melting points are *fluorapatite*, 1615–1622°C; *hydroxyapatite*, 1614°C; *chlorapatite*, 1612°C, and seem to decline somewhat for the *pyromorphite* subseries, as does the hardness. ***Apatite*** is 5 in the

TABLE 1. The Dimensions of Apatite End Members

	a (Å)	*c* (Å)	Density (calc.) (g/cm³)
Fluorapatite	9.37	6.88	3.20
Hydroxyapatite	9.42	6.88	3.15
Chlorapatite	9.63	6.78	3.18
Ellestadite (hydroxy)	9.48	6.93	3.08
Johnbaumite	9.70	6.93	3.73
Svabite	9.75	6.92	3.71
Morelandite	10.17	7.315	5.33
Pyromorphite	9.97	7.32	7.15
Mimetite	10.26	7.44	7.28
Vanadinite	10.33	7.34	6.93

TABLE 2. Optical Properties of Apatite End Members

	Composition	$\omega(\gamma)$	$\epsilon(\alpha \approx \beta)$	Δ
Fluorapatite	$Ca_{10}F_2(PO_4)_6$	1.632	1.629	0.003
Hydroxyapatite	$Ca_{10}(OH)_2(PO_4)_6$	1.651	1.647	0.004
Chlorapatite	$Ca_{10}Cl_2(PO_4)_6$	1.668	1.667	0.001
Ellestadite	$Ca_{10}(OH)_2(SiO_4)_3(SO_4)_3$	1.654	1.650	0.004
Johnbaumite	$Ca_{10}(OH)_2(AsO_4)_6$	1.687	1.684	0.003
Svabite	(natural)	1.706	1.698	0.008
Morelandite	$Ba_{10}Cl_2(AsO_4)_6$	1.880	1.884	–0.004
Pyromorphite	$Pb_{10}Cl_2(PO_4)_6$	2.058	2.048	0.010
Mimetite	$Pb_{10}Cl_2(AsO_4)_6$	2.147	2.128	0.019
Vanadinite	$Pb_{10}Cl_2(VO_4)_6$	2.416	2.350	0.066

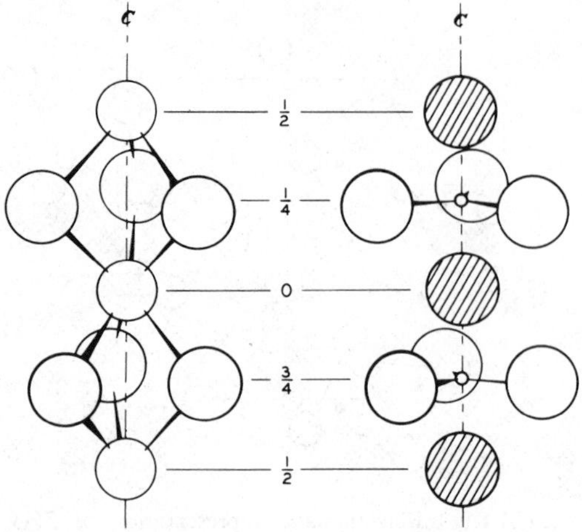

FIGURE 5. Arrangement of oxygen atoms adjacent to threefold axis (from McConnell, 1952). Left: Ca atoms at (0, ½ c) and O atoms at (¼, ¾). Right: Displaced Ca atoms have in their sites H_3O^+ ions (shaded) and C atoms form CO_3 groups.

Mohs scale, whereas *pyromorphite* and *mimetite* are 3.5–4 and *vanadinite* is 3.

Cleavage is distinct in ***apatite***, one parallel with the basal pinacoid and another perpendicular to it. *Pyromorphite* is described as "traces of (10•1) cleavage," *mimetite* as "very poor prismatic cleavage," and *vanadinite* merely as having "uneven to conchoidal fracture."

Inasmuch as magnetic susceptibility and the dielectric constant are markedly affected by impurities, they are of little interest with respect to ***apatite*** except to note that they are very low to low. Mechanical properties are of little practical interest except in conjunction with teeth and bones, where they are difficult to measure because of textural difference of the microcrystals.

Occurrences

Although *fluorapatite* commonly occurs as an accessory mineral in most igneous rocks, it may reach significant proportions in certain nepheline syenites (Kola Peninsula) and in dikes with titanium oxides (nelsonites, Nelson County, Virginia), which represent segregations. Crystals of *fluorapatite*—to a lesser extent *hydroxyapatite* and *chlorapatite*—are encountered in vugs within phosphate-bearing pegmatites (see *Pegmatite Minerals*) in many localities, where they may be associated with lithium-containing minerals, *hornblende, barite, calcite, allanite*, etc. Hydrothermal veins, contact metamorphic zones, and carbonatites frequently contain ***apatites*** (in the first case frequently *francolite*). Indeed, when phosphorus is present, as it usually is, ***apatite*** is the usual mode of crystallization for this element.

Manganiferous varieties are essentially confined to pegmatites, as at Branchville, Connecticut, and Varuträsk, Sweden. Strontium-containing ***apatite*** is known from a syenite dike cutting an ***apatite*** pyroxenite, near Libby, Montana, from the Khibiny region (Kola Peninsula), and from an alkaline-syenite pegmatite associated with the Burpala intrusion, where it is a voelckerite also high in rare earths (RE_2O_3 ca. 13%). *Strontiapatite* (ca. 46% SrO) has a complex paragenesis.

Francolite occurs in carbonatites in southern and eastern Africa and in hydrothermal veins at numerous localities, including Magnet Cove, Arkansas; Estremadura, Spain; Staffel, Germany; and Tonopah, Nevada. It may occur as an overgrowth on earlier ***apatite*** crystals.

Chlorapatite is found in the vicinity of Balmle, Norway, and in stony meteorites (see *Meteoritic Minerals*). *Hydroxyapatite* seems to occur in metamorphic rocks where there was a deficiency of fluorine and (or) chlorine. It has been reported from talc schists in the Hospenthal, Uri, Switzerland, and at Holly Springs, Georgia.

Some lunar *fluorapatites* contain moderate amounts of Y_2O_3 and RE_2O_3 (1.2 and 2.5%, respectively), but others contain much smaller amounts (<0.1 and 0.3%) (see *Lunar Minerals*).

Pyromorphite and *mimetite* occur as secondary minerals associated with *smithsonite, hemimorphite, anglesite, cerussite*, and others. *Pyromorphite* is found in Chester County, Pennsylvania; Cherokee County, Georgia; Hampden County, Massachusetts; and Shoshone County, Idaho. *Mimetite* is less frequently encountered. It occurs in Pennsylvania; Pinal and Yavapai Counties, Arizona; Inyo County, California; the Tintic district, Utah; and the Eureka district, Nevada.

Vanadinite is a secondary mineral of lead deposits in several mining districts of the United States—chiefly in New Mexico, California, and Arizona. Arsenical specimens are found in Sierra County, New Mexico.

Pyromorphite, mimetite, and *vanadinite* are very seldom encountered in these mining districts today. For other occurrences—chiefly foreign—see Palache, Berman, and Frondel (1951).

DUNCAN McCONNELL

References

Deer, W. A., Howie, R. A.; and Zussman, J., 1962. *Rock-Forming Minerals*, vol. 5. London: Longmans, 323–338.

McConnell, D., 1952. The problem of the carbonate apatites. IV. Structural substitutions involving CO_3 and OH. *Bull. Soc. Min. Crist.*, 75, 428–445.
McConnell, D., 1973. *Apatite: Its Crystal Chemistry, Mineralogy, Utilization and Geologic and Biologic Occurrences.* New York/Wien: Springer-Verlag, 111p.
Palache, C.; Berman, H.; and Frondel, C.; 1951. *The System of Mineralogy*, 7th ed. New York: Wiley, 877–906.
Strunz, H., 1970. *Mineralogische Tabellen*, 5th ed. Leipzig: Geest & Portig K.-G., 326–329.
Winchell, A. N., and Winchell, H., 1964. *The Microscopic Character of Artificial Inorganic Solid Substances.* New York/London: Academic Press, 203–205.

Cross-references: *Blowpipe Analysis; Human and Vertebrate Mineralogy; Mohs Scale of Hardness; Pegmatite Minerals; Pleochroic Halos; Refractory Minerals; Rock-Forming Minerals; Staining Techniques; Vein Minerals.* Vol. VI: *Phosphate in Sediments.*

ARAGONITE GROUP

The ***aragonite*** group and the ***calcite*** group (q.v.) of carbonate minerals share the formula ACO_3 and together include most of the common carbonate minerals. Cations with radii larger than Ca^{2+} crystallize with the aragonite structure, those with smaller radii crystallize with the calcite structure. Minerals of the ***aragonite*** group include

aragonite: $CaCO_3$	*strontianite:* $SrCO_3$
cerussite: $PbCO_3$	*witherite:* $BaCO_3$

They are orthorhombic ($Pmcn$) in contrast to the trigonal ($R\bar{3}c$) structure of the ***calcite*** group.

Ordinarily, calcium carbonate ($CaCO_3$) crystallizes as *calcite* under normal atmospheric pressure and as *aragonite* at high pressure. But under special conditions and above 20°C, *aragonite* may precipitate at normal atmospheric pressure, for instance: (1) in some hot springs [e.g., Karlovy Vary (Carlsbad), Czechoslovakia] and cave deposits (see *Cave Minerals*); (2) in some recent shallow marine oolites; (3) as an unstable form in modern seas either biogenetically or as an epitactic overgrowth on nuclei having the aragonite structure; (4) in the oxidized zone of ore deposits; (5) in veins in altered mafic igneous rocks; (6) in some sulfur deposits; and (7) as skeletal material in living organisms and in recent fossils (see *Human and Vertebrate Mineralogy; Invertebrate and Plant Mineralogy*).

Witherite occurs with *barite* in hydrothermal veins. *Strontianite* is a low-temperature hydrothermal mineral, but also occurs in geodes and concretions in limestone and clay. *Cerussite* is typically found as a secondary mineral with oxides and other carbonates in the oxidized portion of ore deposits.

KEITH FRYE

References

Deer, W. A.; Howie, R. A., and Zussman, J., 1966. *An Introduction to the Rock-Forming Minerals.* London: Longmans, 497–503.
DeVilliers, J. P. R., 1971. Crystal structures of aragonite, strontianite, and witherite. *Am. Mineralogist*, 56, 758–767.

Cross-references: *Calcite Group; Cave Minerals; Epitaxy; Human Mineralogy; Invertebrate and Plant Mineralogy; Vein Minerals;* see also glossary.

ASBESTOS

Asbestos is a term applied to any of several varieties of silky or finely fibrous, flexible, relatively refractory minerals. Asbestos occurs principally in veins transecting ultramafic igneous rocks; minor amounts occur in siliceous dolomitic marbles and low-grade schists and meta-ironstones. The fibers may be aligned normal or somewhat inclined to the walls (cross fiber) or parallel to the walls (slip fiber).

Chrysotile ***serpentine*** is the chief source of asbestos, constituting approximately 95% of the world production (Bowles, 1959); but several species of ***amphibole*** asbestos occur. These latter include amosite (a fibrous member of the *cummingtonite-grunerite* series), *tremolite-actinolite*, crocidolite (a fibrous member of the *riebeckite-magnesioriebeckite* series), and, rarely, *anthophyllite* and *eckermannite*.

Chrysotile, like other ***serpentine*** minerals, is a phyllosilicate (q.v.); the structure consists of two basic units, a pseudohexagonal sheet composed of Si–O tetrahedra (each silicon is surrounded by four anions), which is cross-linked to a "brucite layer" composed of Mg–OH octahedra (each magnesium is surrounded by six anions). The distance between successive composite sheets is about 7 Å. Fibrosity results from the fact that the sheet is curved into the shape of tubes with an outer diameter of about 260 Å and an inner diameter of about 110 Å—as deduced from X-ray diffraction and electron microscopy (Whittaker, 1957).

The amphibole structure consists of paired Si–O tetrahedral chains of indefinite length, cross-linked to each other and to cations of sixfold or greater coordination. The cation "layers" may be imprecisely described as dis-

TABLE 1. Chemical Analyses of Chrysotile

	1	2	3	4	5
SiO_2	41.80	41.97	41.83	42.02	42.54
TiO_2	0.05	–	0.02	0.00	–
Al_2O_3	0.11	0.10	0.30	0.52	5.68
Fe_2O_3	0.68	0.38	1.29	0.19	1.06
FeO	0.05	1.57	0.08	0.11	0.74
MnO	0.04	–	0.04	0.03	–
MgO	42.82	42.50	41.39	41.44	35.57
CaO	0.10	–	trace	0.00	0.13
Na_2O	0.03	–	–	–	–
K_2O	0.01	0.08	–	–	–
H_2O+	14.04	13.56	13.66	14.04	13.26
H_2O-	0.28	–	1.57	1.64	0.38
Other	0.10	0.10	–	–	–
Total	100.11	100.26	100.18	99.99	99.36

From W. A. Deer, R. A. Howie, and J. Zussman, *Rock-Forming Minerals,* Vol. III, 1962, pp. 566–567. Reprinted with permission of Longman Group Limited, London.

continuous segments of brucite sheets. ***Amphibole*** varieties of asbestos owe their fibrosity to the pronounced growth of single crystals parallel to extension of the Si–O tetrahedral double chains (Whittaker, 1949).

Composition

The structural formula for ***serpentine*** is $Mg_6Si_4O_{10}(OH)_8$. Chemical analyses, presented in Table 1, show that *chrysotile* closely conforms to the theoretical end member. Apparently a slight deficiency of Si^{4+} is compensated by an equivalent excess of OH^-, thus reducing the amount of O^{2-} per formula unit. For a detailed description of the physical and chemical properties of ***serpentine*** group minerals, see Faust and Fahey (1962).

Amphibole asbestos has the following structural formulas:

- amosite; *anthophyllite:* $(Mg,Fe^{2+})_7Si_8O_{22}(OH)_2$
- *tremolite; actinolite:* $Ca_2(Mg,Fe^{2+})_5Si_8O_{22}(OH)_2$
- crocidolite: $Na_2(Mg,Fe^{2+})_3Fe_2^{3+}Si_8O_{22}(OH)_2$
- *eckermannite:* $Na_{2.5}Ca_{0.5}(Mg,Fe^{2+},Fe^{3+},Li,Al)_5Si_{7.5}Al_{0.5}O_{22}(OH)_2$

Representative chemical analyses are presented in Table 2. These types of asbestos, similar to

TABLE 2. Chemical Analyses of Amphibole Asbestos

	1	2	3	4	5	6	7
SiO_2	49.47	47.35	47.04	51.94	50.66	55.44	56.48
TiO_2	0.25	trace	trace	–	–	0.04	0.30
Al_2O_3	0.63	4.20	7.02	0.20	0.04	0.22	1.22
Fe_2O_3	4.15	3.34	2.43	18.64	22.64	16.77	8.38
FeO	35.63	36.60	26.10	19.39	17.05	5.23	2.67
MnO	0.61	0.28	0.15	–	–	0.09	–
MgO	6.57	5.80	4.96	1.37	1.99	12.30	17.40
CaO	0.52	0.77	10.84	0.19	0.01	2.17	2.70
Na_2O	0.02	trace	trace	6.07	5.15	6.76	8.09
K_2O	0.20	trace	trace	0.04	0.09	0.15	1.82
H_2O+	2.33	1.25	1.05	2.58	2.62	0.55	0.87
H_2O-	0.07	0.35	0.45	0.31	0.15	0.05	–
Other	–	0.09	0.15	–	–	0.26	–
Total	100.45	100.03	100.19	100.73	100.40	100.03	99.93

1. Amosite (Deer, Howie, and Zussman, 1963).
2. Amosite (Peacock, 1928).
3. Aluminous *ferrotremolite* (Peacock, 1928).

4,5. Ferrous *crocidolite* (Peacock, 1928).

6. Magnesium *crocidolite* (Ernst, 1960).
7. *Eckermannite* (Deer, Howie, and Zussman, 1963).

amphiboles in general, show wide ranges of composition due to their structural complexity: cation sites of different sizes are available for the accommodation of appreciable quantities of nearly all cations present in the earth's crust.

Genesis

Chrysotile most commonly occurs as veins in ultramafic rocks. Here its presence apparently results from low-temperature hydrothermal alteration of the host rock along fissures and solution channelways (Cooke, 1937). Bowen and Tuttle (1949) studied the stability relationships of minerals in the system $MgO–SiO_2–H_2O$ and concluded that the maximum thermal stability of ***serpentine***, in a rock of its own bulk composition, is about 500°C under most crustal conditions (see Fig. 1; for a recent review, see Evans, 1977). Where pure dunite undergoes hydration, the high-temperature stability limit of ***serpentine*** is reduced. Siliceous *dolomite* apparently may undergo serpentinization according to the reaction: *dolomite*+*quartz*+water=***serpentine***+*calcite*+carbon dioxide. This reaction must occur at temperatures considerably below the degradation of pure ***serpentine***, inasmuch as presence of CO_2 in the fluid phase lowers the chemical potential of H_2O and, thus, the thermal range of the hydrous silicate. Furthermore, in the presence of excess SiO_2, the ***serpentine***+*calcite* assemblage gives way to a higher-rank *tremolite*-bearing assemblage, also below the breakdown curve for pure ***serpentine***. *Chrysotile* deposits are, therefore, confined to low-grade metamorphic or retrograded environments.

Anthophyllite has a field of stability in magnesian silicate rocks at moderately elevated temperatures and fluid ($\approx H_2O$) pressures, according to Greenwood (1963). At 2000 bars fluid pressure, the low-temperature assemblage *talc*+*forsterite* reacts to form *anthophyllite* at about 670°C (see Fig. 1). In environments where the condensed bulk composition departs from that of *anthophyllite* itself, or where the chemical potential of H_2O is low (i.e., the fluid is "contaminated" or is not present as a separate phase), the stability field of *anthophyllite* declines to lower temperatures. The occurrence of *anthophyllite* asbestos is, thus, a complex function of rock bulk composition, temperature, and chemical potential of water.

Stability relations of amosite have not been determined; but judging from compositions presented in Table 2, its stability range may be intermediate between those of *anthophyllite* and *tremolite-actinolite*. Some samples described as amosite appear to have compositions approaching that of *ferrotremolite* (the iron end member of the *tremolite-actinolite* series).

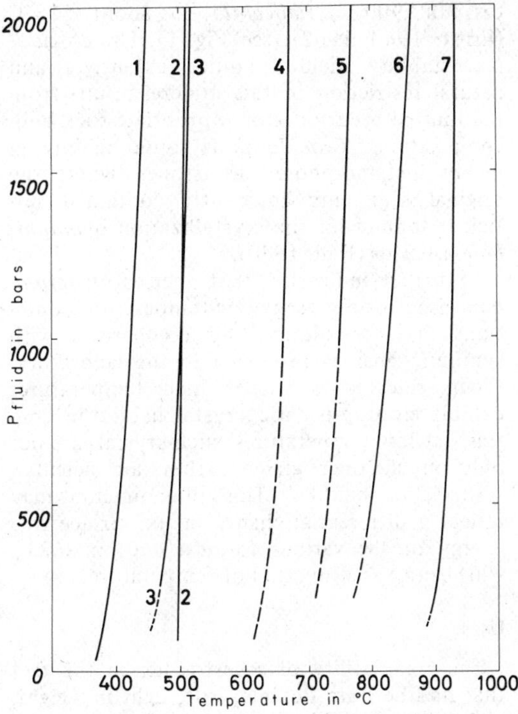

Figure 1. Experimentally determined stability ranges for asbestos minerals under conditions where fluid pressure approximates total pressure (after Ernst, 1960). Reactions involved are: (1) ***serpentine*** + *brucite* = *forsterite* + fluid (Bowen and Tuttle, 1949); (2) ***serpentine*** = *forsterite* + *talc* + fluid (Bowen and Tuttle, 1949); (3) *riebeckite* = *hematite* + *magnetite* + *acmite* + *quartz* + fluid (Ernst, 1962); (4) *talc* + *forsterite* = *anthophyllite* + *fluid* (Greenwood, 1963); (5) *anthophyllite* = *enstatite* + *quartz* + fluid (Greenwood, 1963); (6) *tremolite* = *diopside* + *enstatite* + *quartz* + fluid (Boyd, 1959); and (7) *magnesioriebeckite* = *hematite* + *magnesioferrite* + ***olivine*** + *hypersthene* + liquid + fluid.

The high-temperature stability limit of *tremolite* has been determined by Boyd (1959), and lies near 875°C at 2000 bars fluid ($\approx H_2O$) pressure (see Fig. 1). Substitution of ferrous iron for magnesium lowers the thermal range of this ***amphibole*** drastically (see Ernst, 1966), as does reduced chemical potential of H_2O and departure of rock bulk composition from that of the ***amphibole***. Actinolitic amosite in nature is, therefore, restricted to relatively low-grade metamorphic environments where vein fluid plus reacting country rock have produced the requisite chemical environment.

At 2000 bars fluid ($\approx H_2O$) pressure, *magnesioriebeckite* is stable to temperatures in excess of 900°C, *riebeckite* to about 515°C (Ernst, 1960, 1962) (see Fig. 1). The *crocidolite* stability field is quite extensive, and natural restriction of this mineral results from the limited occurrence of appropriate rock bulk compositions. *Crocidolite* is found mainly in feebly metamorphosed ironstones, where the original sediments apparently contained sufficient sodium for the crystallization of ***amphibole*** asbestos (Hall, 1930).

A further indication that species of ***amphibole*** asbestos represent low-temperature conditions may be adduced by comparison with synthetic analogs produced in the laboratory. Chain silicates grown at high temperatures exhibit stout prismatic crystal habits; in contrast, at low temperatures, such crystals are not only much finer grained, they are acicular, fibrous, or hairlike. This phenomenon may reflect a differential change in the surface free energy of the various forms ($\{hk0\}$ vs $\{hk1\}$, $\{001\}$, etc.) as a function of temperature.

Uses

The great utility of asbestos lies in the fact that its fibers are flexible, soft, light in weight, and refractory. Of greatest value are the spinning fibers, long-fiber grades of *chrysotile* and *crocidolite*. Such material is woven into fire-resistant tape, conveyor belts, ropes, clothing, gloves, and other insulating fabrics. Nontextile uses include fabrication as packings, gaskets, brake linings, asbestos paper, and noncorrosive electrochemical filters.

Nonspinning fibers include shorter-fiber varieties of *chrysotile* and *crocidolite*, as well as other types of ***amphibole*** asbestos. Such materials are employed in the manufacture of a great variety of nonflammable products including clutch-plate facings, gaskets, brake shoes, roofing shingles, wallboard, floor tile, asbestos cement, fillers, lightweight insulation aggregate, electrical panels, putties, plastics, and filters.

Recently, some of these applications have been severely curtailed due to the appreciation that certain fibrous varieties may be responsible for high incidences of pulmonary cancer.

World Distribution

Production of asbestos reflects fairly adequately the world distribution. Canada leads with 45% of the estimated world production, followed by the Soviet Union, 34%; Union of South Africa, 7%; Southern Rhodesia, 6%; United States and Italy, 2% each; Swaziland, 1%; and the rest of the world, 3%. Percentages are based on the 1957 output (Jenkins, 1960). Estimated world reserves appear to be sufficient for 50–100 years' production at the present rate. Canada leads in estimated reserves, followed by the Soviet Union, Southern Rhodesia, Republic of South Africa, Australia, and the Peoples Republic of China (see *Mineral Industries*).

For a comprehensive discussion of world production, reserve tonnages, mining and milling methods, physical properties of the varieties of asbestos, and geology, see Jenkins (1960).

W. G. ERNST

References

Bowen, N. L., and Tuttle, O. F., 1949. The system $MgO-SiO_2-H_2O$, *Geol. Soc. Am. Bull.*, **60**, 439-460.

Bowles, O., 1959. Asbestos, a materials survey, *U.S. Bur. Mines Inf. Circ. 7880*, 94p.

Boyd, F. R., 1959. Hydrothermal investigations of amphiboles, pp. 377-396, in P. H. Abelson, ed. *Researches in Geochemistry*. New York: Wiley, 511p.

Cooke, H. C., 1937. Thetford, Disraeli and eastern half of Warwick map areas, Quebec, *Geol. Surv. Canada, Mem.*, **211**, 160p.

Deer, W. A.; Howie, R. A.; and Zussman, J., 1962. *Rock-Forming Minerals*, vol. 3, *Sheet Silicates*. New York: Wiley (London: Longmans), 270p.

Deer, W. A.; Howie, R. A.; and Zussman, J., 1963. *Rock-Forming Minerals*, vol. 2, *Chain Silicates*. New York: Wiley (London: Longmans), 379p.

Ernst, W. G., 1960. The Stability relations of magnesioriebeckite, *Geochim. Cosmochim. Acta*, **19**, 10–40.

Ernst, W. G., 1962. Synthesis, stability relations and occurrence of riebeckite and riebeckite-arfvedsonite solid solutions, *J. Geol.*, **70**, 689–736.

Ernst, W. G., 1966. Synthesis and stability relations of ferrotremolite, *Am. J. Sci.*, **264**, 37–65.

Evans, B. W., 1977. Metamorphism of alpine peridotite and serpentinite, *Ann. Rev. Earth Planet. Sci.*, **5**, 397-447.

Faust, G. T., and Fahey, J. J., 1962. The serpentine-group minerals, *U.S. Geol. Surv. Prof. Paper 384A*, 92p.

Greenwood, H. J., 1963. The synthesis and stability of anthophyllite, *J. Petrology*, **4**, 317–351.

Hall, A. L., 1930. Asbestos in the Union of South Africa, *Geol. Surv. Union So. Africa, Mem.*, **12**, 329p.

Jenkins, G. F., 1960. Asbestos, pp. 23–53 in *Industrial Minerals and Rocks*, Am. Inst. Mining, Met., and Petrol. Engrs., 934p.

Peacock, M. A., 1928. The nature and origin of the amphibole asbestos of South Africa, *Am. Mineralogist*, **13**, 241–285.

Whittaker, E. J. W., 1949. The structure of Bolivian crocidolite, *Acta Crystallogr.*, **2**, 312–317.

Whittaker, E. J. W., 1957. The structure of chrysotile. V. Diffuse reflexions and fibre texture, *Acta Crystallogr.*, **10**, 149–156.

Cross-references: *Amphibole Group; Electron Microscopy (Transmission); Human and Vertebrate*

Mineralogy; Mineral Industries; Museums, Mineralogical; Pigments and Fillers; Refractory Minerals.

ASTERISM–*See* GEMOLOGY

ASTROPHYLLITE GROUP

Members of the ***astrophyllite*** group are chain silicates that do not belong to either the ***pyroxene*** or the ***amphibole*** group. They have the general formula

$$A_3B_7C_2Si_8O_{24}(O,OH)_7$$

where A = K,Na,Cs
B = Fe,Mn
C = Ti,Nb,Zr

Members of the group are as follows:

- *Astrophyllite:* $(K,Na)_3(Fe,Mn)_7Ti_2Si_8O_{24}(O,OH)_7$
- *Cesium-kupletskite:* $(Cs,K,Na)_3(Mn,Fe)_7(Ti,Nb)_2Si_8O_{24}(O,OH,F)_7$
- *Kupletskite:* $(K,Na)_3(Mn,Fe)_7(Ti,Nb)_2Si_8O_{24}(O,OH)_7$
- *Niobophyllite:* $(K,Na)_3(Fe,Mn)_6(Nb,Ti)_2Si_8(O,OH,F)_{31}$
- *Zircophyllite:* $(K,Na,Ca)_3(Mn,Fe)_7(Zr,Nb)_2Si_8O_{27}(OH,F)_4$

A solid-solution series exists between *astrophyllite* (Fe>Mn) and *kupletskite* (Mn>Fe) and between *kupletskite* and *cesium-kupletskite* (Cs>K,Na). *Niobophyllite* has Nb>Ti and *zircophyllite* has Zr>Ti in the octahedral sites.

The structure consists of chains of silica tetrahedra cross-linked into sheets by *C*-ion octahedra, giving these minerals perfect basal {001} cleavage. Unlike the ***micas***, the cleavage flakes are brittle. Triclinic crystals are tabular. Occurrence is rare; the minerals are associated with volcanic alkalic rocks.

KEITH FRYE

Reference

Phillips, W. R., and Griffen, D. T., 1981. *Optical Mineralogy, The Nonopaque Minerals.* San Francisco: Freeman, p. 530.

AUTHIGENIC MINERALS

Authigenesis is any process involving growth in situ, i.e., on the spot. Its antonym is "allogenic". Introduced by Kalkowsky (1880), the term describes the origin of any mineral that is formed subsequent to the origin of its matrix or surroundings but is not a product of transformation or recrystallization; it is customarily reserved for low-temperature sedimentary situations. Tester and Atwater (1934) have pointed out that the term should be reserved for discrete crystallographic units of "exotic" nature rather than be used for the rock-forming minerals. The term is not normally used for overgrowths and metasomatic replacement; thus, massive dolomitization is excluded but discrete *dolomite* crystals formed, for example, during the early burial phase are clearly authigenic. Many minerals formerly considered exclusive "high-temperature" igneous or hydrothermal indicators (e.g., ***feldspars***, and the metal sulfides such as *galena* and *chalcopyrite*) are now frequently recognized as authigenic, particularly under the bacterially induced ph/Eh environments of syndiagenesis.

A useful review of authigenic minerals in marine sediments is that by Bonatti (1972). Many striking features of sedimentary rocks are due to authigenic minerals–e.g., the *hematite* in many red beds (Walker, 1967, 1974; Baskin, 1956; Van Houten, 1972); the ***zeolites*** in tuffs (Walton, 1975); the ***feldspars*** in many graywackes and sandstones (Topkaya, 1950; Buyce and Friedman, 1975) and in shales (Weiss, 1954); the *muscovite* or *biotite* in certain sandstones (Dapples, 1967); or *epidote* in others (Fediuk, 1962); *phillipsite* and *glauconite* in deep-sea deposits (Bonatti, 1972); the *pyrite* or *marcasite* in coal or shale (Newhouse, 1927; Edwards and Baker, 1951). Authigenic minerals may occur diffused in interstitial spaces, in discrete beds, or in concretions. Those formed under soft, early burial conditions are commonly euhedral and under the microscope can be seen to have pushed their way into the matrix or penetrated adjacent grains.

Some authorities prefer to restrict the term authigenesis to minerals formed before burial and consolidation (i.e., exclusively in the syndiagenetic stage); but, in general usage (Teodorovich, 1961; Fairbridge, 1967), the term recognizes new "exotic" minerals formed also during anadiagenesis and epidiagenesis (see Table 1 and *Diagenetic Minerals*). Because of such ambiguities, the term *neoformation* (or neogenesis) is sometimes used to distinguish true authigenesis from diagenetic alteration or *transformation*, e.g., in Europe (Millot, 1970; Müller, 1967).

RHODES W. FAIRBRIDGE

References

Baskin, Y., 1956. A study of authigenic feldspars, *J. Geol.*, **64**, 132–155.

Bonatti, E., 1972. Authigenic Minerals–Marine, in

TABLE 1. More Common Authigenic Minerals (Excluding Halides)

Mineral	Formula	Frequency R= Rare C= Common	Usual Development: Syn-dia-genetic	Ana-dia-genetic	Epi-dia-genetic
Anatase	TiO_2 (tetr.)	R		x	x
Anhydrite	$CaSO_4$	C		x	
Ankerite	$Ca(Mg,Fe)(CO_3)_2$	R		x	x
Apatite		C	x		x
Aragonite	$CaCO_3$ (orth.)	C	x	x	x
Azurite	$Cu_3(CO_3)_2(OH)_2$	R		x	x
Barite	$BaSO_4$	C	x		x
Bornite	Cu_5FeS_4	R	x		x
Brookite	TiO_2 (orth.)	R		x	
Calcite	$CaCO_3$ (trig.)	C	x	x	x
Celestite	$SrSO_4$	C	x		x
Cerussite	$PbCO_3$	R		x	
Chalcedony	SiO_2	C	x	x	x
Chalcopyrite	$CuFeS_2$	R	x		x
Chamosite	$(Fe,Mg)_5Al(Si_3Al)O_{10}(OH,O)_8$	R	x	x	
Chlorite group	–	C	x	x	
Dolomite	$CaMg(CO_3)_2$	C	x	x	x
Galena	PbS	R	x		x
Glauconite	$(K,Na)(Al,Fe,Mg)_2(Al,Si)_4O_{10}(OH)_2$	C	x	x	x
Greenalite	$Fe_6Si_4O_{10}(OH)_8$	R		x	
Gypsum	$CaSO_4 \cdot 2H_2O$	C	x		x
Halite	NaCl	C	x		x
Hematite	Fe_2O_3	C			x
Hydromagnesite	$Mg_5(CO_3)_4(OH)_2 \cdot 4H_2O$	R			x
Illite group	–	C	x		
Kaolinite	$Al_2Si_2O_5(OH)_4$	C			x
K-***Feldspars***	$KAlSi_3O_8$	C		x	x
Leucoxene	Ti oxides	C		x	x
Limonite	Fe oxides	C			x
Magnesite	$MgCO_3$	R			x
Malachite	$Cu_2CO_3(OH)_2$	R		x	x
Marcasite	FeS_2 (orth.)	C	x		x
Montmorillonite	$(Na,Ca)_{0.33}(Al,Mg)_2Si_4O_{10}(OH)_2 \cdot nH_2O$	C	x		x
Muscovite	$KAl_3Si_3O_{10}(OH)_2$	C			
Natron	$Na_2CO_3 \cdot 10H_2O$	R			x
Nesquehonite	$Mg(HCO_3)(OH) \cdot 2H_2O$	R	x		x
Opal	$SiO_2 \cdot nH_2O$	C	x		x
Phillipsite	$(Ca,Ba,K,Na)_6Al_8(Al,Si)_2Si_{10}O_{40} \cdot 15\text{–}20H_2O$	R	x		
Plagioclase	$(Ca,Na)(Al,Si)AlSi_2O_8$	C		x	x
Psilomelane (Wad)	Mn oxides	C	x		
Pyrite	FeS_2 (iso.)	C	x		x
Pyrolusite	MnO_2 (tetr.)	C	x		
Quartz	SiO_2	C		x	x
Rhodochrosite	$MnCO_3$		x	x	
Rutile	TiO_2 (tetr.)	R		x	x
Siderite	$FeCO_3$	C	x	x	
Sphalerite	ZnS (iso.)	R	x		x
Strontianite	$SrCO_3$	R		x	
Sulphur	S (orth.)	R	x		x
Tourmaline group	–	R		x	
Witherite	$BaCO_3$	R		x	x
Zeolites (*Phillipsite, Heulandite, Laumonite, Chabazite, Natrolite, Analcime*)		C	x		x
Zircon	$ZrSiO_4$	R		x	

From R. W. Fairbridge, Phases of Diagenesis and Authigenesis, in *Diagenesis in Sediments*, 1967, G. Larsen and G. V. Chilingar, eds. Reprinted with permission from Elsevier.

Rhodes W. Fairbridge, ed. *Encyclopedia of Geochemistry and Environmental Sciences.* New York: Van Nostrand Reinhold, 48–55.
Buyce, M. R., and Friedman, G. M., 1975. Significance of authigenic K-feldspars in Cambrian-Ordovician carbonate rocks of the Proto-Atlantic shelf in North America, *J. Sed. Petrology*, **45**, 808–821.
Cronan, D. S., 1974. Authigenic minerals in deep-sea sediments, in E. D. Goldberg, ed. *The Sea*, vol. 5. New York, 491–545.
Dapples, E. C., 1967. Silica as an agent in diagenesis, in G. Larsen and G. V. Chilingar, eds., *Diagenesis in Sediments*. Amsterdam: Elsevier, 323–342.
Divis, A. F., and McKenzie, J. A., 1975. Experimental authigenesis of phyllosilicates from feldspathic sands, *Sedimentology*, **22**, 147–155.
Edwards, A. B., and Baker, G., 1951. Some occurrences of supergene iron sulphides in relation to their environment of deposition, *J. Sed. Petrology*, **21**, 34–46.
Fairbridge, R. W., 1967. Phases of diagenesis and authigenesis, in, G. Larsen and G. V. Chilingar, eds., *Diagenesis in Sediments*. Amsterdam: Elsevier, 19–89.
Fediuk, F., 1962. Zur Entstehung des Epidotes in den Kambrischen Sedimenten des Pribramer Reviers (Zentralboehmen), *Casopis Mineral. Geol.*, **7**, 89–99.
Funk, H., 1975. The origin of authigenic quartz in the Helvetic Siliceous Limestone (Helvetischer Kieselkalk), Switzerland, *Sedimentology*, **22**, 299–306.
Kalkowsky, E., 1880. Über die Erforschung der archaeischen Formationen, *Neues Jahrb. Min. Geol. Palaeont., Monatsh.*, **1**, 1–29.
Middleton, G. V., 1972. Albite of secondary origin in Charny Sandstones, Quebec, *J. Sed. Petrology*, **42**, 341–349.
Millot, G., 1970. *Geology of Clays*. New York: Springer, 429p. (French edition, 1964. *Géologie des Argiles*. Paris: Masson, 499p.)
Müller, G., 1967. Diagenesis in argillaceous sediments, in G. Larsen and G. V. Chilinger, eds., *Diagenesis in Sediments*. Amsterdam: Elsevier, 127–177.
Newhouse, W. H., 1927. Some forms of iron sulphide occurring in coal and other sedimentary rocks, *J. Geol.*, **35**, 73–83.
Stewart, R. J., 1974. Zeolite facies metamorphism of sandstone in the western Olympic Peninsula, Washington, *Geol. Soc. Am. Bull.*, **85**, 1139–1142.
Teodorovich, G. I., 1961. *Authigenic Minerals in Sedimentary Rocks*. New York: Consultants Bureau, 120p.
Tester, A. C., and Atwater, G. I., 1934. The occurrence of authigenic feldspars in sediments, *J. Sed. Petrology*, **4**, 23–31.
Topkaya, M., 1950. Recherches sur les silicates authigènes dans les roches sedimentaires, *Bull. Lab. Géol. Minéral. Geophys. Musée Géol. Univ. Lausanne*, **97**, 132p.
Van Houten, F. B., 1972. Iron and clay in tropical savanna alluvium, northern Colombia: a contribution to the origin of red beds, *Geol. Soc. Am. Bull.*, **83**, 2761–2772.
Walker, T. R., 1967. Formation of red beds in modern and ancient deserts, *Geol. Soc. Am. Bull.*, **78**, 353–368.
Walker, T. R., 1974. Formation of red beds in moist tropical climates: a hypothesis, *Geol. Soc. Am. Bull.*, **85**, 633–638.
Walton, A. W., 1975. Zeolite diagenesis in Oligocene volcanic sediments, Trans-Pecos, *Geol. Soc. Am. Bull.*, **86**, 615–624.
Weiss, M. P., 1954. Feldspathized shales from Minnesota, *J. Sed. Petrology*, **24**, 270–274.
Woodward, H. H., 1972. Syngenetic sanidine beds from Middle Ordovician Saint Peter Sandstone, Wisconsin, *J. Geol.*, **80**, 323–332.

Cross-references: *Diagenetic Minerals.*

AUTUNITE (TORBERNITE) GROUP, META-AUTUNITE (METATORBERNITE) GROUP

Autunite Group

autunite	*novacekite*	*torbernite*
fritzscheite	*sabugalite*	*troegerite*
heinrichite	*saleeite*	*uranocircite*
kahlerite	*sodium autunite*	*uranospinite*
		zeunerite

Meta-autunite Group

abernathyite	*metanovacekite*
bassetite	*metatorbernite*
meta-ankoleite	*meta-uranocircite*
meta-autunite	*meta-uranospinite*
metaheinrichite	*metazeunerite*
metakahlerite	*sodium uranospinite*
metakirchheimerite	*uramphite*
metalodevite	

The autunite structure is characterized by layers of $(P,As,V)O_4$ tetrahedra and highly distorted UO_6 octahedra with A ions and H_2O molecules occupying cavities between the layers.

Minerals of the ***autunite*** and ***meta-autunite*** groups share the general structural formula

$$A(UO_2)_2(XO_4)_2 \cdot nH_2O$$

where, in the ***autunites***,

$A = Ca,Na_2,Ba,Mg,Fe^{2+},Cu,Mn$
$X = P,As,V$
$n = 8\text{–}12$

and, in the ***meta-autunites***,

$A = K_2,(NH_4)_2,Ca,Ba,Mg,Fe^{2+},Co,Cu,Zn$
$X = P,As$
$n = 4\text{–}8$

The two groups differ in water content and in the polytypic stacking of the layers. The transition between the two groups occurs near or somewhat above room temperature, depending upon ambient humidity and mineral composition.

The layers are identical in each group, the difference being in the stacking. Layers in ***autunites*** are off-set by ½ unit-cell edge from ***meta-autunites***, producing a more closely packed structure in the former.

Minerals of these groups are found associated with one another and with other secondary uranium minerals as oxidation products of *uraninite* and other uranium-bearing primary minerals.

KEITH FRYE

References

Nininger, R. D., 1954. *Minerals for Atomic Energy*. New York: Van Nostrand, 32-34.

Phillips, W. R., and Griffin, D. T., 1981. *Optical Mineralogy*. San Francisco: Freeman, p. 424, 444, 448.

Cross-references: See mineral glossary.

AXIS OF SYMMETRY–*See* SYMMETRY

B

BARKER INDEX OF CRYSTALS

Throughout the nineteenth century and the early decades of the twentieth century the description of the morphology of both natural and artificial crystals on the basis of goniometric measurement was an important activity of most mineralogists and of some chemists. By far the most comprehensive compilation of the data so produced is Paul Groth's *Chemische Krystallographie*, Vols. I–V (1906–1919, Engelmann, Leipzig), which covers over 7000 substances. The arrangement is chemical and the crystallographic data are not presented in a form suitable for determinative purposes.

E. von Fedorow, the first to derive the 230 space groups and the inventor of the two-circle goniometer and the universal stage, attempted to devise a scheme for the identification of crystalline substances on the basis of morphology. In theory, such a scheme should be applicable to all but isometric crystals, since the interfacial angles of all other crystals are, in general, characteristic for each substance. The work of Fedorow and his collaborators did not appear until after his death. It was published by the Russian Academy of Sciences in Petrograd (now Leningrad) in 1920 under the German title *Das Krystallreich. Tabellen zur Krystallochemischen Analyse.* The work was not successful and, possibly because of the place and time of its publication, did not become widely known.

A first problem in carrying out such a scheme as that of Fedorow is the establishment of a basis for arrangement of the data that is suitable for determinative purposes. T. V. Barker, a young Englishman who had worked with Fedorow, applied himself to this task and published his recommendations in a small book entitled *Systematic Crystallography* (Murby, London, 1930). He described a set of simple rules for arriving at the "classification angles" for crystals. Barker died in 1931 before he could put his plans into effect. During the next fifteen years crystallographers in Russia, Britain, the Netherlands, Belgium, and the United States continued the unfinished work of Fedorow and Barker; and some tables of limited scope were published, for instance those of J. D. H. Donnay and J. Melon, *Crystallo-Chemical Tables for the Determination of Tetragonal Substances* (1934, Johns Hopkins University, Studies in Geology, No. 11, pp. 305–388). However, it was not until 1945 that a group of crystallographers in Britain undertook to complete the work of Barker.

The *Barker Index of Crystals* consists of three volumes, bound in seven parts: Volume I (1951), *Tetragonal, Hexagonal, Trigonal and Orthorhombic Systems:* Part 1, "Introduction and Tables"; Part 2, "Crystal Descriptions"; Volume II (1956), *Monoclinic System:* Part 1, "Introduction and Tables," Parts 2 and 3, "Crystal Descriptions"; Volume III (1964), *Anorthic System:* Part 1, "Introduction and Tables," Part 2, "Crystal Descriptions, Atlas of Configurations" (the publisher is W. Heffer & Sons Ltd., Cambridge). The principal compilers were Mary W. Porter and R. C. Spiller (first two volumes) and Mary W. Porter and L. W. Codd (last volumes), but several others, especially M. H. Hey, wrote important parts of the Index and many contributed to it.

The Barker Index now covers 7394 substances. For each substance, classification angles are tabulated, a single angle for tetragonal and hexagonal crystals, two or more for crystals of lower symmetry. In general, it should be possible by following the Barker rules to arrive at a classification angle or angles from goniometric measurements of a crystal. If the substance is included in the Index it will be listed in the tables by classification angle. Identification can be checked by reference to the descriptions in the latter parts of each volume.

The Barker rules for anorthic (triclinic) crystals are rather complicated; their statement covers six pages and is difficult to apply. For the other systems the rules can be learned and applied rather easily, though difficulties may arise in some cases. The Barker Index is based largely on the data compiled by Groth before 1919. Data for only a few hundred substances are based on later work. Nearly all of the data on minerals are old and triclinic minerals are omitted.

A. PABST

References

Pabst, A., 1965. The Barker index of crystals, 1951–1964, *Science*, **149**, 45-46.

Terpstra, P., and Codd, L. W., 1961. *Crystallometry*. New York: Academic Press, 420p.

Cross-references: *Axis of Symmetry; Crystal Habits; Crystallography: History; Crystallography: Morphological; Directions and Planes; Lattice; Symmetry.*

BAROMETRY, GEOLOGIC

One goal of earth scientists is to determine the temperature and pressure of formation of rocks that, by one way or another, are brought to the surface of the earth for study. Minerals in rocks from outcrops, mines, and drill cores may have crystallized at temperatures and pressures considerably above those prevailing when they are studied. Certain physical and chemical characteristics of these minerals may be examined to determine what conditions prevailed during crystallization or during recrystallization. The resulting estimates of preexisting temperatures and pressures are known as *geologic thermometry* and as *geologic barometry*, respectively (also *geothermometry* and *geobarometry;* see *Thermometry, Geologic*).

If, from experimental studies of phase equilibria (see vol. IVA), a certain mineral or mineral assemblage is determined to be in equilibrium only within a defined pressure range, the presence of that mineral or assemblage in a rock indicates that it was once subjected to pressures within that range.

Precise pressures of mineral formation are elusive, however, since stability ranges may be quite large and dependent upon temperature as well. Mineral assemblages generally yield more precise pressure estimates than do individual mineral species because they commonly have a narrower stability range than do species.

Pressures at the time of equilibration may also be estimated if crystals contain trapped liquid carbon dioxide inclusions. Such inclusions, highly compressed and nearly pure, have been found in ***olivine, pyroxene,*** and ***plagioclase*** crystals in ***olivine***-bearing xenoliths in basalts (Roedder, 1965). These inclusions are generally rounded to spherical, although some show faceted negative crystals. They rarely exceed 30 μm (micrometers) and most are smaller than 5 μm in diameter. Density of the CO_2 inclusions can be calculated from data that are available on CO_2. From the density and an independent estimate of temperature, pressure of equilibration may be estimated. The estimate is based upon the assumption that no leakage has occurred. Results of such studies indicate that the nodular xenoliths in basalts crystallized at pressures of 2500–5000 bars—equivalent to depths of 8–16 km in the lithosphere.

The presence of a solvus representing variation in the amount of crystal solution between two or more end-member mineral species may permit interpretation of pressures from mineral compositions (see vol. IVA: *Exsolution, Phase Equilibria*). Higher temperatures tend to allow greater amounts of ionic substitution in crystal structures than do lower temperatures. Increased pressure, on the other hand, tends to make crystals less tolerant of foreign ions in their structures.

The ideal geologic barometer would be a solvus that is pressure dependent while insensitive to variations in temperature. Since this relationship of pressure, temperature, and mineral composition occurs rarely, if ever, geologic temperatures must be estimated independently to utilize solvus compositions to estimate geologic pressures.

The solvus between *enstatite* ($Mg_2Si_2O_6$) crystal solutions and *diopside* ($CaMgSi_2O_6$) crystal solutions has been used as a geologic barometer. The *diopside* limb of the solvus is insensitive to variations in pressure, but the amount of Ca tolerated by *enstatite* coexisting with *diopside* varies measureably with pressure. For pure end-member compositions at 1200°C, the ratio Ca/(Ca+Mg) in *enstatite* ranges from >0.035 at 5 kbar to <0.025 at 30 kilobar (Nehru, p. 580, in Boettcher, 1976). However, the presence of Fe increases substantially the solubility of Ca in *enstatite*. The use of this geologic barometer is further complicated by the presence of Al, Ti, and Cr in *enstatite* in some assemblages.

Temperature and pressure effects on distribution coefficients of ion pairs in coexisting minerals at equilibrium have also been studied. For example, xenolithic nodules associated with kimberlites have been analyzed by electron microprobe in an attempt to check geothermal gradient estimates (Presnall, pp. 582–588, and MacGregor and Basu, pp. 715-724, in Boettcher, 1976).

The dependence of the solubility of Fe in *sphalerite* (ZnS) in the presence of *pyrite* (FeS_2) and hexagonal *pyrrhotite* ($Fe_{1-x}S$) on pressure of equilibration is another geologic barometer for some rocks. The mole percentage of Fe in *sphalerite* drops from 20% at 0.5 kilobar to 10% at 10 kilobar throughout the temperature range 300–700°C (Scott, p. 660, in Boettcher, 1976; Lusk and Ford, 1978).

There still remain, however, uncertainties in defining appropriate chemical and mineralogic reactions and determining their temperature and pressure dependence.

KEITH FRYE

References

Boettcher, A., ed., 1976. Geothermometry-Geobarometry, a collection of papers presented at the International Conference on Geothermometry and Geobarometry convened at the Pennsylvania State University, October 5–10, 1975, *Am. Mineralogist*, **61**, 549–816.

Lusk, J., and Ford, C. E., 1978. Experimental extension of the sphalerite geobarometer to 10 kbar, *Am. Mineralogist*, **63**, 516–519.

Roedder, E., 1965. Liquid CO_2 inclusions in olivine-bearing nodules and phenocrysts from basalts, *Am. Mineralogist*, **50**, 1746–1782.

Cross-references: *Garnet Group; Mantle Mineralogy; Mineralogical Phase Rule; Pyroxene Group; Thermometry, Geologic;* Vol. IVA: *Exsolution; Fluid Inclusions; Phase Equilibria.*

BEAD TESTS—*See* BLOWPIPE ANALYSIS

BECKE LINE—*See* OPTICAL MINERALOGY

BEUDANTITE GROUP—*See* ALUNITE GROUP

BIAXIAL MINERALS—*See* MINERALS, UNIAXIAL AND BIAXIAL

BIOPYRIBOLE

The mineral terms "biopyribole" and "pyribole" were coined by A. Johannsen from parts of the names *(bio)tite, **(pyr)oxene**,* and ***amph-i(bole)*** as labels to be used in the field when more precise identification was not immediately possible. These terms were revived by J. B. Thompson, Jr., for crystallochemical reasons to illustrate the kinship among ***micas, pyroxenes***, and ***amphiboles***.

By dividing the structures of these mineral groups into modules (see *Polysomatism*), the amphibole structure and composition may be conceived as a mineralogical hybred composed of alternating modules of pyroxene and mica structures and compositions in the ratio 1:1 (see Table 1).

Other combinations of these structural modules also belong to the biopyriboles. The structure of *jimthompsonite*, a triple-chain silicate, may be constructed from mica and pyroxene modules in the ratio 2:1. Similarly, *chesterite*, containing both double and triple chains of silica tetrahedra in its idealized structure, contains these modules in the ratio 3:2.

As is the case with the ***phyllosilicates***, modules may be stacked in different ordered arrays, e.g., orthorhombic *jimthompsonite* vs. mono-

TABLE 1. Polysomatic Series in the Biopyriboles

mica		***pyroxene***		***amphibole***
talc	+	*diopside*	=	*tremolite*
$Mg_3Si_4O_{10}(OH)_2$		$Ca_2Mg_2Si_4O_{12}$		$Ca_2Mg_5Si_8O_{22}(OH)_2$
talc	+	*enstatite*	=	*anthophyllite*
$Mg_3Si_4O_{10}(OH)_2$		$Mg_4Si_4O_{12}$		$Mg_7Si_8O_{22}(OH)_2$
talc	+	*jadeite*	=	*glaucophane*
$Mg_3Si_4O_{10}(OH)_2$		$Na_2Al_2Si_4O_{12}$		$Na_2Mg_3Al_2Si_8O_{22}(OH)_2$
Fe-talc	+	*acmite*	=	*riebeckite*
$Fe_3Si_4O_{10}(OH)_2$		$Na_2Fe_2Si_4O_{12}$		$Na_2Fe_5Si_8O_{22}(OH)_2$
phlogopite	+	*diopside*	=	K-edenite
$KMg_3AlSi_3O_{10}(OH)_2$		$Ca_2Mg_2Si_4O_{12}$		$KCa_2Mg_5AlSi_7O_{22}(OH)_2$
wonesite	+	*diopside*	=	*edenite*
$NaMg_3AlSi_3O_{10}(OH)_2$		$Ca_2Mg_2Si_4O_{12}$		$NaCa_2Mg_5AlSi_7O_{22}(OH)_2$
talc	+	CATS	=	*tschermakite*
$Mg_3Si_4O_{10}(OH)_2$		$Ca_2Al_4Si_2O_{12}$		$Ca_2Mg_3Al_4Si_6O_{22}(OH)_2$
mica		***pyroxene***		multichain silicates
2 *talc*	+	1 *enstatite*	=	1 *jimthompsonite*
$2[Mg_3Si_4O_{10}(OH)_2]$		$1[Mg_4Si_4O_{12}]$		$1[Mg_{10}Si_{12}O_{32}(OH)_4]$
3 *talc*	+	2 *enstatite*	=	1 *chesterite*
$3[Mg_3Si_4O_{10}(OH)_2]$		$2[Mg_4Si_4O_{12}]$		$1[Mg_{17}Si_{20}O_{54}(OH)_6]$

After Thompson, 1978.

clinic *clinojimthompsonite*, or they may be disordered in their stacking.

KEITH FRYE

References

Thompson, J. B., Jr., 1978. Biopyriboles and polysomatic series, *Am. Mineralogist*, **63**, 239–249.

Veblen, D. R.; Buseck, P. R.; and Burnham, W. C., 1977. Asbestiform chain silicates: New minerals and structure groups, *Science*, **198**, 359–365.

Cross-references: *Amphibole Group; Phyllosilicates; Pyroxene Group*. Vol. IVA: *Mineral Classes: Silicates.*

BIREFRINGENCE—*See* OPTICAL MINERALOGY

BISECTRIX—*See* OPTICAL MINERALOGY

BLACKSAND MINERALS AND METALS

Long before history was being recorded, men must have observed the concentration of heavy, black minerals on beaches, along streams, and in deserts. Lightweight minerals in these environments are naturally winnowed from heavier, often dark or black detrital mineral components by wave, current, stream, and wind action. These natural process systems produce concentrations of minerals with different densities because of continually changing levels of energy within, and related to, different environments (Daily, 1973). Locally, the resultant heavy-mineral concentrations (Fig. 1) may be called placers (see *Placer Deposit*). As early as 1742, Shellrocke's *Voyages to California* told of abundant beach deposits of black minerals discovered along the west coast of America. Later it was ascertained that these black mineral deposits contained *platinum, magnetite, chromite, garnet*, and *rutile*, among other minerals. Black minerals, such as *magnetite, ilmenite,* and *hematite,* are particularly common in almost any placer deposit because these minerals are (1) plentiful in the earth's crust and (2) relatively stable when subjected to weathering and transportation.

Gold prospectors of the middle and late 1800s were familiar with accumulations of black minerals. They knew that placer *gold* commonly occurred in darker sands and yet they often tried to avoid the heaviest concentrations because the black minerals clogged up their sluice boxes and *gold* pans and made it difficult to extract the desired *gold* particles. A cubic yard of gravel at some localities has been known to yield over 100 pounds of black minerals.

The convenient term "blacksand" has been used by many investigators to designate any concentration of sand-sized minerals that are conspicuously dark or black; in other words, a type of sand deposit containing certain groups of minerals. However, many potentially valuable mineral commodities that are not black in color may be present in such sand deposits. Thus, these deposits may yield multimineral coproducts or by-products, thereby increasing their potential value.

Some authorities tend to restrict the words "black sand" for beach sand deposits that are rich in the black mineral *magnetite*, whereas others use the words for dark sand of volcanic origin. It would seem logical to accept the increasingly common usage of the term "blacksand" for accumulations of dark, detrital minerals (and their lighter-colored associates) occurring in placer-type deposits (Table 1; Fig. 1). The combined form, "blacksand placer deposit" is convenient and meaningful, and there is much to commend its general use.

A wide variety of minerals may be found in a blacksand deposit, but the heavier, more-resistant minerals are more common in such concentrations of sand. Heavy minerals are those with specific gravity from 2.85 to 22.0 (Table 2). It is interesting to note that the hard minerals, for example, lightweight, light-colored *zircons* and relatively heavy, generally dark-colored *garnet*, are unusually common in blacksands (Table 2). These minerals may travel long distances from their bedrock sources. Because of the brittleness of many hard-to-very-hard heavy minerals, they are likely to be found in deposits nearer to their source rocks. Soft, heavy minerals are likely to accumulate very close to their source. Some very heavy minerals, such as *gold* and *platinum*, are very resistant to chemical attack. Though soft, they may be carried long distances by transporting agents; they are likely to be very fine by the time they are deposited. *Gold*, for example, may be carried great distances in flour form because in flour size the grains tend to be protected from further attrition, probably because of the cushioning effect of surrounding water or air of the environment in which the mineral grains are being transported. The same factor—fine size—may apply to other minerals, resulting in longer distances of transportation.

Included in blacksand minerals are oxides, multiple oxides, and silicates. Important are the so-called transition elements, including the

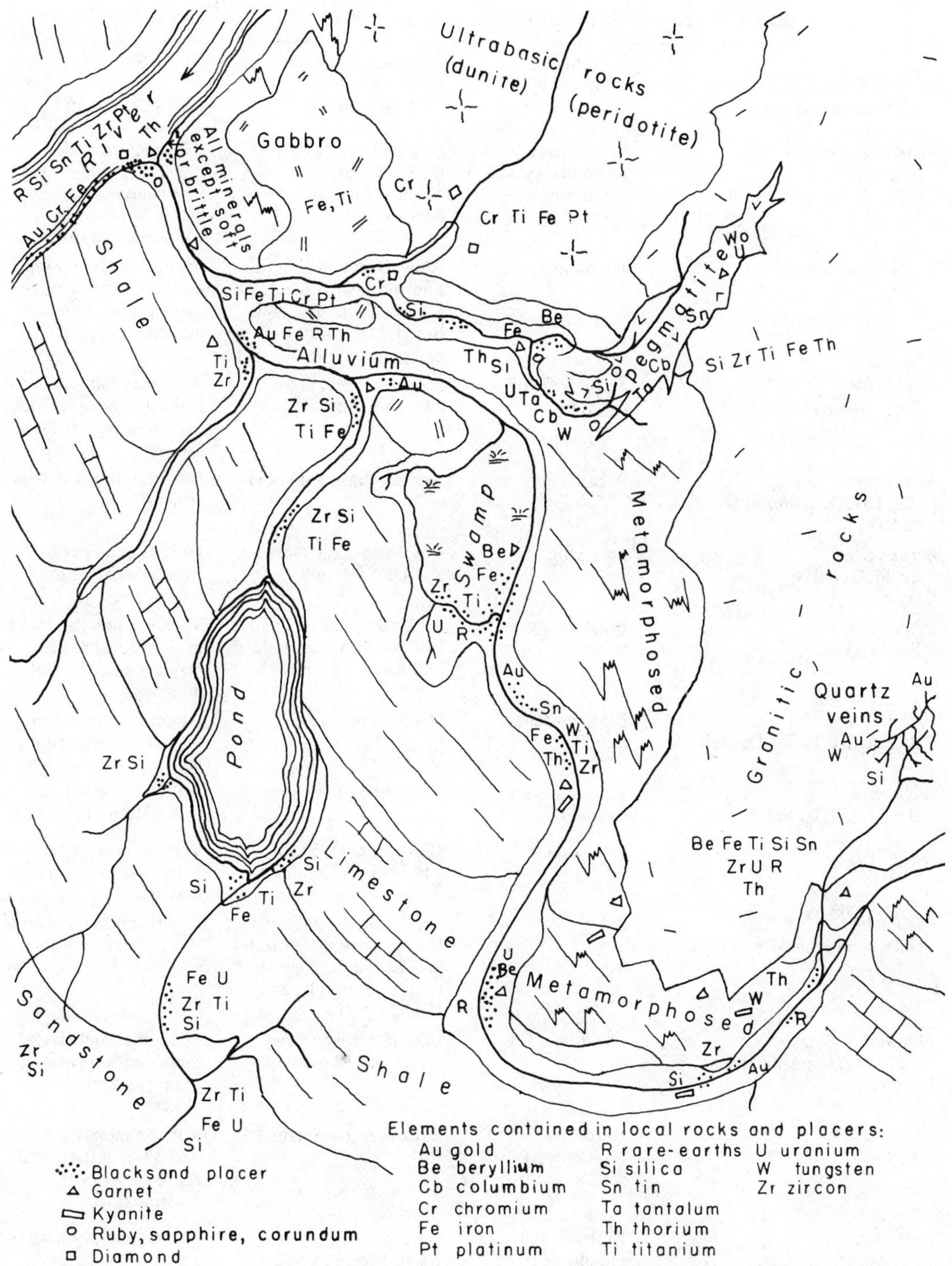

FIGURE 1. Idealized geologic map of blacks and sources.

lanthanides (rare earths) and the actinites (transuranium series). The rare-earth metals are becoming of increasing interest because of their physical properties—yttrium and the lanthanides, for example (Brobst and Pratt, 1973). Studies of placer deposits indicate that they consist of more than fifty kinds of heavy minerals, containing among other elements, niobium (columbium), tantalum, yttrium, zirconium, hafnium, uranium, vanadium, thorium,

TABLE 1. Selected Blacksand Minerals with Actual or Potential Commercial Value

Mineral and Formula	Actual or Potential(?) Commodity	Selected General Areas of Occurrence	Comments
Allanite (orthite) $(Ca,Ce,La)_2(Al,Fe,Mg)_3(SiO_4)_3(OH)$ with U,TH, and Be	Rare earths, radioactive blacks, and beryllium(?)	E Greenland, N Italy, USA (Idaho), and Arctic submarine deposits	Like *epidote;* accessory mineral in deep-seated rocks and pegmatites, metaigneous rocks, volcanics, and iron ores
Anatase (octahedrite) TiO_2	Titanium(?)	Sandstones and grits of England, Switzerland, France, Sri Lanka, and Brazil; blacksands in many areas	From igneous and metamorphic rocks; alteration of other titanium minerals
Baddelyite ZrO_2 (some Fe and HfO_2)	Zirconium and hafnium	Sri Lanka, Brazil, Sweden, and Zaire	Accessory mineral in magnetite-pyroxenites; may be with detrital minerals
Bastnaesite $(Ce, La)CO_3F$ (minor U and Th)	Rare earths	Russia (Ural), gold sands	Not common as detrital mineral
Bertrandite $Be_4Si_2O_7(OH)_2$	Beryllium	Czechoslovakia, Norway, and USA (Maine)	Not common as detrital; in pegmatite terrain
Beryl $Be_3Al_2(SiO_3)_6$	Beryllium, gem stones	Sri Lanka, Brazil, Colombia, South Africa, Ireland, and Russia (S Urals)	Source in granitic rocks including pegmatites; may occur as detrital mineral
Betafite $(U,Ca)_2(Nb,Ta,Ti)_2O_6\cdot(OH)$	Niobium-tantalum(?)	Madagascar	Source in pegmatites; may occur as detrital mineral
Brannerite $(U, Ca, Ce)(Ti, Fe)_2O_6$	Rare earths and thorium(?)	USA (Idaho)	Source metamict mineral in detrital materials
Brookite TiO_2	Titanium(?)	USSR (Urals), Brazil, and USA (N Carolina)	Veins in gneiss and schist and in placer deposits
Cassiterite SnO_2, sometimes Ta and Nb	Tin	Brazil, Malaysia, Thailand, Burma, New South Wales and Tasmania, Nigeria, Bolivia, Tanzania, and Uganda	Granite and pegmatites; also placer deposits
Chromite $FeCr_2O_4$, some Mg and Al	Chromium	USA (Calif. and Oreg.), Japan and Rhodesia	Peridotites and related rocks and serpentines; stream and beach placers
Cinnabar HgS	Mercury (quicksilver)	Bavaria and Surinam	Not common, but may occur as a detrital mineral near rocks of volcanic origin
Columbite $(Fe,Mn)(Nb,Ta)_2O_6$ *Tantalite* $(Fe,Mn)Ta_2O_6$	Niobium and tantalum	USA (Idaho), W Australia, Nigeria, USSR (S Urals), Zaire, and SW Africa	Granite and pegmatites grains and pebbles in detrital minerals
Corundum Al_2O_3	Abrasive *corundum;* gemstones (ruby and sapphires)	Sri Lanka, Upper Burma, Rhodesia, Malagasy Republic, USA (Mont.), Canada (Quebec), Colombia, and Brazil	Igneous and metamorphic rocks, and in detritals

TABLE 1. (Continued)

Mineral and Formula	Actual or Potential(?) Commodity	Selected General Areas of Occurrence	Comments
Cyrtolite (*zircon*) $ZrSiO_4$, with U, Th, and Y	Zirconium and rare earths(?)	Sri Lanka and Madagascar	Igneous rocks, limestones, and metamorphic rocks; not unusual in placers
Diamond C	Abrasive and cutting materials; gemstones	SE, S, SC, and W Africa (Tanzania, South Africa, Rhodesia, Angola, Nambia, Congo, Ghana, Guinea, Ivory Coast, Liberia, and Sierra Leone), Brazil (Guyana), India; Indonesia; Venezuela; and USSR.	Ultramafic rock; placers
Eschynite (*aeschynite*) $(Ce,Ca,Fe,Th)(Ti,Nb)_2O_6$ Priorite (*aeschnite*-(Y)) $(Y,Er,Ca,Fe,Th)(Ti,Nb)_2O_6$	Rare earths, thorium and niobium(?)	USSR and Africa (Swaziland)	Nepheline syenite, pegmatites and granite; in detritals
Euxenite $(Y,Ca,Ce,U,Th)(Nb,Ta,Ti)_2O_6$ *Polycrase* (similar to *euxenite*)	Rare earths, uranium, niobium, and tantalum	USA (Idaho and Montana), Norway, Greenland, and Madagascar	Associated with pegmatites and granitic rocks; placers
Feldspars K,Na,Ca,Al silicates	Ceramic materials	May be a by-product of silica sand production	Common in most rocks and placers
Fergusonite $(Y,Er,Ce,Fe)(Nb,Ta,Ti)O_4$ *Formanite* $(U,Zr,Th,Ca)(Ta,Nb,Ti)O_4$	Rare earths, thorium, and niobium-tantalum(?)	Sri Lanka, Africa (Swaziland), SW South Africa, Rhodesia, W Australia, and USA (N Carolina)	Occurs as detritals in granitic and pegmatitic rock areas
Gadolinite $Be_2FeY_2Si_2O_{10}$	Rare earths	Sweden	Granitic rocks and pegmatites; in some detritals
Garnets Fe,Mg,Mn,Ca,Cr,Al silicates (minor germanium, Idaho)	Abrasive; gemstones	S Africa, Italy, Finland, Sri Lanka, India, Brazil, S Greenland, USSR (Urals), and USA (Idaho and Ariz.); actually very widespread occurrences	Wide occurrence in igneous and metamorphic rocks; in detrital minerals
Gold Au	Gold	Ghana, Nigeria, Australia, Urals, Brazil, Colombia, and USA (Alaska, Calif., Colo., Mont., Wyo., and Idaho)	Placers in regions of granitic rock and *quartz* veins
Hematite Fe_2O_3	Iron(?)	Occurrence widespread	As specularite or nodules
Hjelmite (*yttromicrolite*) $(Y,Ca)(Ta,Nb)_2O_6(OH)$	Lanthanides, actinides, niobates, and tantalates(?)		Environment similar to *columbite* and *tantalite*
Ilmenite $FeTiO_3$ with Mg and Mn	Titanium	Widespread occurrences; Sri Lanka, Colombia, Nigeria, New Zealand, Queensland, and USA	Common in igneous and metamorphic rocks and in placers
Iridosmine (including *osmiridium*) Ir, Os	Platinum group minerals	Australia (New South Wales), USA (N Calif. and southern Oreg.), and USSR (Urals)	Occurs as flakes or flattened grains in alluvial deposits; from mafic rocks
Kyanite Al_2SiO_5	High-alumina refractory; gemstone	Sri Lanka, Brazil, and USA (Va. and S Carolina)	Accessory mineral in gneiss and schist; beach-sand detritals

TABLE 1. (Continued)

Mineral and Formula	Actual or Potential(?) Commodity	Selected General Areas of Occurrence	Comments
Magnetite Fe_3O_4	Iron		Abundant and widespread; occurring detrital mineral
Micas Li,Mg,Fe,K,Na,Ca,Al silicates	Scrap *mica*(?)		Abundant and widespread occurrence; not a heavy detrital mineral
Monazite $(Ce,La,Y,Th)PO_4$	Rare earths and thorium	Australia, India, Taiwan, Brazil, Colombia, Sri Lanka, Travancore, Nigeria, England, Indonesia, and USA (Idaho, Mont., Ala., Va., N and S Carolina)	Common as placer mineral
Palladium Pd usually alloyed with Pt and Ir	Platinum group minerals	Guyana, Alaska, Colombia, Ethiopia, and Urals	Detrital sands from ultramafic rocks
Phenakite Be_2SiO_4	Beryllium(?)	Brazil and USSR (Urals)	Resembles *quartz;* found in areas underlain by granitic rocks or schist
Platinum Pt with other platinum minerals and Fe	Platinum group minerals	USSR (Urals), South Africa, Colombia, USA (Oreg., Calif., and Alaska), Japan, Australia (New South Wales and Tasmania), Indonesia, and New Guinea	Mafic igneous rocks are source of detritals
Pyrochlore $(Na,Ca)Nb_2O_6F$ *Microlite* $(Na,Ca)_2Ta_2O_6(O,OH,F)$	Niobium-tantalum(?)	France; probably in other areas where rock is suitable	Occurs in pegmatites of granitic rocks
Quartz SiO_2	Silica; gemstone	Widespread occurrences	Silica rich rocks yield detrital deposits
Rutile TiO_2	Titanium; gemstone	Austria, Italy, Switzerland, France, Norway, S Australia, Brazil, India, Senegal, South Africa, and USA	Occurs as accessory mineral in igneous rocks; common detrital mineral in beach sand
Samarskite $(Y,Er,Ce,U,Ca,Fe,Pb,Th)(Nb,Ta,Ti,Sn)_2O_6$	Lanthanides, actinides, niobates, and tantalates(?)		Granite pegmatite, a common source of detrital grains
Scheelite $CaWO_4$	Tungsten	Canada (NW Ter.) and New Zealand	Source of detritals in crystalline igneous rocks; often near contact zones
Sillimanite Al_2SiO_5	High-alumina refractory; gemstone	(see *kyanite*)	(see *kyanite*)
Silver Ag	Silver	Occasionally with *gold* in *gold* placers	(see *gold*)
Sphene (*titanite*) $CaTiSiO_5$	Titanium(?)	Many areas	Occurs widespread as accessory minerals in igneous rocks; common detrital mineral

TABLE 1. (Continued)

Mineral and Formula	Actual or Potential(?) Commodity	Selected General Areas of Occurrence	Comments
Tapiolite $FeTa_2O_6$ Mossite $Fe(Nb,Ta)_2O_6$	Niobium and tantalum(?)		Occurs as detrital mineral in areas underlain by granite pegmatite
Thorianite ThO_2 with U	Thorium and uranium(?)	Sri Lanka and Malagasy Republic	Granite pegmatite source
Thorite $ThSiO_4$	Thorium(?)	USA (Idaho)	Like *zircon* in structure; occurs as detrital mineral in regions of *thorite* mineralization
Topaz $Al_2SiO_4(F,OH)_2$	Gemstone	USA (Calif. and Texas), USSR (Urals), Japan, Brazil, Sri Lanka, Australia, and Ghana	Source from highly siliceous rocks
Uraninite UO_2	Uranium		Occurs as detrital mineral in areas underlain by granite pegmatite
Wolframite $(Fe,Mn)WO_4$	Tungsten	Malaysia, Bolivia, and Nigeria	Alluvial deposits derived from *quartz* vein deposits
Xenotime YPO_4 and rare earths	Yttrium(?) and rare earths	USA (Georgia and N. Carolina) and Brazil	Detrital derived from granite, pegmatite, and gneiss
Zircon $ZrSiO_4$ with HfO_2	Zirconium; hafnium; gemstone	USA (Fla. and Ga.), Australia, Brazil, Thailand, and Malaysia. Widespread occurrance as detrital mineral (see *baddelyite*)	Common accessory mineral in igneous and metamorphic rocks
Zirkelite $(Ca,Ce,Th,U)Zr(Ti,Nb)_2O_7$ and rare earths	Actinides, lanthanides, and zirconium (?)	Sri Lanka and Brazil	

titanium, gold, iron, silica, and the rare earths (Savage, 1961a). In some parts of the world blacksands yield *platinum, chromite*, tungsten, beryllium, *tin*, and *diamonds*. All the minerals found in blacksands are too numerous to list here, but a few are enumerated in Table 1. One Russian publication illustrates more than 145 varieties of placer minerals (Trushkova and Kukharenko, 1961); however, these include many less dense and also soft minerals that are not normally common in blacksands.

Some blacksands contain *metamicts*, which are minerals that are radioactive amorphous pseudomorphs of original crystalline minerals (see *Metamict State*). Metasomatic processes may be partly responsible for the metamict state, a condition in which the individual minerals have a disordered crystalline structure caused by radiation bombardment. Such minerals are usually very complex and their identification is most difficult; hence, metamicts are often referred to simply as "radioactive blacks." The principal method for the identification of some of the metamict black minerals involves the use of X-ray powder patterns. It is necessary to restore the original crystal structures by heating the mineral material before the patterns are produced. Complex intergrowths of radioactive blacks and nonradioactive minerals and rare earths, for example, like some of the phosphates (*monazite* and *xenotime*) are not unusual, adding to the general problems of academic research and economic exploitation of such minerals.

Milner et al. (1962), Gardner (1955), Kukharenko (1961), and Hutton (1950) have published some of the most complete and useful information on the blacksand minerals, based upon their broad experience with deposits all over the world. More than 8000 mineral varie-

TABLE 2. Relative Hardness (Mohs Scale) and Weight of Commercial or Potentially Commercial Detrital Minerals

Sp. Gr.	Soft (2–4.5)	Soft to Hard (3–6)	Hard (5–6)	Hard to Very Hard (5–7.5)	Very Hard (6.5–7.5)	Hardest (8–10)
Light (2.00–2.85)	*mica*[a]		*feldspar*		*quartz*[a] *bertrandite*	*beryl*
Heavy (2.86–5.8)	*bastnaesite*[a,b]	cyrtolite[b] *betafite*[a,c] *brannerite*[a,c] *xenotime*[a,c] *thorite*[a,c]	*anatase*[a] sphene[a,b] *chromite*[a] *aeschynite*[a,c] *samarskite*[a,c] *allanite*[a,b] *monazite*[a,c] *brookite*[a] *ilmenite*[a] *zirkelite*[a,c] priorite[a,c] hjelmite[a,c] *columbite*[a,d] *microlite*[a,b] *pyrochlore*[a,b]	*fergusonite*[a,c] *formanite*[a,c] *euxenite*[a,c] *polycrase*[a,c] *kyanite* *sillmanite* *magnetite*[a] *hematite*[a] *rutile*[a]	*gadolinite*[b] *zircon* *baddelyite*[a,d] *garnet*[a]	*phenacite* *topaz* *corundum* *diamond*[a]
Very heavy (5.9–7.5)		*scheelite*	*wolframite*[a]	*cassiterite*[a]		
Heaviest (7.6–22.0)	*cinnabar*[a] *silver* *gold* *platinum*	*palladium*	*uraninite*[a,c]	*tapiolite*[a] mossite[a] *tantalite*[a,d] *iridosmine*	*thorianite*[a,c]	

[a] Always or sometimes black or dark, producing blacksand in placer deposits.
[b] May be radioactive or not depending upon whether some actinide elements are present.
[c] Generally radioactive because actinides are present in major proportions.
[d] Radioactive only when containing intergrowths of radioactive minerals.

ties are known, but only about fifty are common in blacksands (Table 1). The potentially valuable minerals are those that are (a) of importance as sources of metals; (b) gemstones and other industrial-use-type minerals in placers; (c) rare-earth minerals; (d) minerals of potential importance as technologies are improved; and (e) minerals of no commercial value, but important as associates and, therefore, as indicators of the possible presence of valuable minerals (Brobst and Pratt, 1973).

Very valuable metal commodities, including *gold* and *platinum*, are available from blacksands. Several of these metals are being exploited from deposits at various places in the world (Tables 1, 2, and 3). Newly discovered sources and uses result in a rapidly changing focus of geographic availabilities. In the past, some publications have referred to several of these blacksand metals as: miracle, rare, newer, future, highly reactive, elusive, and less-familiar metals. To some extent all these descriptions are correct, but several of the terms may be misleading; for example, the "rare metals" are not particularly rare in comparison to the amount of other metals in the earth's crust (Table 3). The term "elusive metal" is good, because many of the metals available in blacksands are difficult to produce in their pure metallic state; the technological or metallurgical problems related to their production are numerous (Hampel et al., 1961). Whether or not any of these metals merit the name "miracle metal" still remains to be demonstrated. Most important, however, is the fact that some of the metals are fulfilling the need for versatile materials of unusual physical, structural, and nuclear properties (see *Metallurgy*). In the space age, some of these metals have properties that are very desirable, such as light weight and strength; toughness and durability; and resistance to high temperatures, corrosive chemicals, and radiation damage (Savage, 1961b). Many of these qualities, and the use of such metals in space exploration, have resulted in an increasing market for newly fabricated items

TABLE 3. Properties of Selected Elements in Blacksand Minerals

Element	Density at 68° F (20° C)[a] (lb/in.³)	(g/cm³)	Melting Point[a] (°F)	(°C)	Grams per ton in Crustal Rocks[b]	Cost[c] ($)
Beryllium (Be)	0.067	1.85	2345	1284	2	103.00 m lb
Cerium (Ce)	0.241	6.66	1463	795	46	108.00 m kg
Chromium (Cr)	0.259	7.19	3407	1875	200	110.00 mc mt
Gadolinum (Gd)	0.285	7.90	2394	1312	6	430.00 m kg
Gold (Au)	0.698	19.32	1945	1063	0.005	500.00 m oz
Hafnium (Hf)	0.480	13.29	3902	2150	5	90.00 m lb
Iron (Fe)	0.284	7.87	2795	1513	50,000	203.00 m st
Lanthanum (La)	0.224	6.17	1688	920	18	108.00 m kg
Lutetium (Lu)	0.356	9.84	3008	1652	0.8	13200.00 m kg
Mercury (Hg)	0.489	13.55	−37.96	−38.87	0.5	370.00 m 76 lb
Neodymium (Nd)	0.253	7.00	1875	1024	24	250.00 m kg
Niobium (Nb)	0.309	8.57	4474	2468	24	35.75 m lb
Palladium (Pd)	0.434	12.02	2826	1552	0.01	75.00 m oz
Platinum (Pt)	0.772	21.40	3216	1769	0.005	340.00 m oz
Praseodymium (Pr)	0.245	6.78	1715	935	6	290.00 m kg
Promethium (Pm)	–	–	1895	1035	–	
Samarium (Sm)	0.273	7.54	1962	1072	7	280.00 m kg
Scandium (Sc)	0.108	2.99	2802	1539	5	9000.00 m kg
Silicon (Si)	0.084	2.33	2570	1410	277,200	0.565 m lb
Silver (Ag)	0.379	10.49	1761	960.5	0.1	21.79 m oz
Tantalum (Ta)	0.600	16.6	5425	2996	2	200.00 m lb
Thorium (Th)	0.421	11.66	3182	1750	10	15.00 m lb
Tin (Sn)	0.263	7.3	449	232	3	8.28 m lb
Titanium Ti)	0.164	4.54	3035	1668	4,400	10.73 m lb
Tungsten (W)	0.697	19.3	6170	3410	1	15.50 m lb
Uranium (U)	0.688	19.07	2070	1132	2	–
						320.00 m kg
Yttrium (Y)	0.162	6.98	2748	1509	40	35.00 m lb
Zirconium (Zr)	0.233	6.45	3366	1852	160	—

[a]C. A. Hampel, and others, 1961.
[b]Mason, 1958.
[c]U.S. Bureau of Mines, 1980.
m = metal
mc = metal concentrate
st = short ton
mt = metric ton
lb = pound
kg = kilogram
oz = ounce (Troy)
m 76 lb = 76 lb flask

for use on earth. This fact has been used as an argument to encourage financial support for more space research.

Industry-supported studies at the facilities of Battelle Memorial Institute, Stanford Research Institute, and many state, federal, and company-owned laboratories have emphasized and continued to demonstrate the importance of blacksand minerals (Savage 1961b). Frequent articles appearing in the *Battelle Technical Review*, concerning the metallurgy of blacksand metals, include accounts by V. D. Barth, J. A. DeMastry, B. W. Gonser, R. C. Himes, and L. W. Coffer.

More than fifty major alloys of chromium, silicon, vanadium, tungsten, zirconium, niobium, tantalum, boron, calcium, and titanium are produced for the metals industry (see *Mineral Industries*). Developments in powder metallurgy demonstrate how far we have come since the first electrosmelting work in 1892 at Spray, North Carolina.

Methods for processing the less-common metals by powder metallurgy have special significance in nuclear engineering. Beryllium as a neutron reflector is an example. The powder technique in the construction of rocket-propelled missiles is very important. It is possible to use heat-resistant metals of the chromium-cobalt-nickel class in environments up to 1500–

1800°F (815–980°C). Other materials, some of which are obtainable from blacksands, give promise of suitable performance at temperatures of 2000–5000°F (1095–2760°C) (Clark, 1963), e.g., niobium, tantalum, and tungsten. Powder metallurgy makes possible the production of materials that in later stages may be forged to desirable shapes with high density and still with fine and uniform grain size.

Ceramic materials and metal combinations, the "cermets," now include carbides, nitrides, borides, diborides, and silicides–substances with high chemical stability, resistance to oxidation (including high temperatures), high strength, and low density. These materials are necessary in a world demanding increased energy and fabricated goods, in both developed and underdeveloped countries; continuing exploration and space travel; and a rapidly developing, sophisticated electronics industry.

Several metals that have been long known, but little used up to recent time, have, and will continue to fulfill more and move major roles in the future. Among these metals, derived from minerals of blacksand origin, are zirconium, hafnium, niobium, tantalum, yttrium, and chromium. They may be taken from blacksands in economically valuable amounts as coproducts or by-products.

Niobium, tantalum, and zirconium carbides and diborides produce structural materials useful at some of the high temperatures mentioned above. New uses are being found for oxidation- and corrosion-resistant zirconium alloys; welded zirconium tubing; niobium-tantalum alloys for aircraft, spacecraft, electronics, and chemical engineering industries; alloys of niobium and numerous metals for high ductility; corrosion-resistant tantalum alloys; and numerous other applications. Even metals with poor structural qualities are being put to good use, for example, as semiconductors.

Metallurgists are working on new frontiers with many alloys that are used to coat or clad "older" (familiar) structural materials. Their rate of investigation will influence the future use and production of blacksand minerals. This in turn should lower prices (Table 3), provided production increases.

Of considerable importance to the future is the fact that, by comparison to the familiar metals, some of the recently developed metals are plentiful in the earth's crust (Table 3). This fact might result in an increase in the tempo of research into new uses for these metals. Probably there will be a continued future demand for the abrasive materials (q.v.), gemstones (see *Gemology*), and precious metals that are recoverable from some blacksands. Other factors that will influence development of blacksand deposits in the future include the favorable or unfavorable impact of recycling, development of technologies that permit recovery of widely dispersed minerals in "low-grade ore rock," and the constraints of environmental laws on surface mining.

Most future ventures into the blacksand market will probably involve only economical, large-scale production of the raw materials from a single deposit for more than one commodity and coproduct or by-product recovery operations. Newly developed upgrading, sorting, and metallurgical techniques will make possible production for what could be a broad and expanding future market. However, in order to fulfill the many requirements of regulatory measures imposed by environmental laws, development of many known or potential blacksand deposits in the developed countries will be either curtailed or much more expensive operations. Underdeveloped countries, on the other hand, may not be restricted by strong environmental constraints for some years into the future. Underground mining and recovery of placer deposits at depth may, in some localities, solve the environmental problems of surface disturbance and any required restoration costs (Popov, 1971).

In summary, major mineral commodities are still being obtained from blacksand placers (Brobst and Pratt, 1973; Council on International Economic Policy, 1974; and Daily, 1973). *Gold* placers have been most important as a source of *gold* through time but such sources appear to be approaching exhaustion. *Diamonds* recovered from both land and ocean placers supply the bulk of world requirements. The world's most important tin source, *cassiterite,* is also derived from both terrestrial and submarine placers located principally in Malaysian territory and the Strait of Molluca. In recent years marine placers extending from west of Thailand and the southern extension of Burma, south and southeast to the Java Sea, have shown great promise. Dredging and hydraulic-pumping recovery methods are proving successful in recovering such deposits (Popov, 1971).

Titanium in the form of *rutile* (TiO_2) and *ilmenite* ($FeTiO_3$) and also *monazite* [(Ce,La,Y,Th)PO_4] and *zircon* ($ZrSiO_4$) are being found in commercial amounts in widespread beach blacksand accumulations. The same is true of titaniferous iron minerals. Recovery of *platinum*, once largely from placer deposits, appears to be slowing down.

C. N. SAVAGE

References

Brobst, D. A., and Pratt, W. P., ed., 1973. United States Mineral Resources, *U.S. Geol. Survey Prof. Paper 820*, 722p.

Clark, F., 1963. *Advanced Techniques in Powder Metallurgy*. An AEC monograph prepared under direction of Am. Soc. for Metals. New York: Rowman and Littlefield, 180p.

Council on International Economic Policy, 1974. *Special report–Critical Imported Materials (Mineral Commodities)*. Washington, D.C.: U.S. Government Printing Office, 49p. Commodity analysis (Appendix), 61p.

Daily, A. F., 1973. Placer mining, in *Mining Engineering Handbook*, vol. 2. New York: Society of Mining Engineers, AIME, 17-151 to 17-180.

Gardner, D. E., 1955. Black-sand, heavy-mineral deposits of eastern Australia, *Australian Bur. Min. Res. Geol. Geophys. Bull.*, **28**, 103p.

*Hampel, C. A., et al., 1961. *Rare Metals Handbook*, 2nd ed. New York: Reinhold, 715p.

*Hutton, C. O., 1950. Studies of heavy detrital minerals, *Geol. Soc. Am. Bull.*, **61**, 635-710.

Kukharenko, A. A., 1961. *Placer Mineralogy*. Moscow: USSR Geological Research Inst., State Publ. House of Sci. and Tech., Literature on Geology and Conservation of Mineral Resources, 317p.

Mason, B., 1958. *Principles of Geochemistry*, 2nd ed. New York: Wiley.

Milner, H. B., et al., 1962. Diagnostic properties of sedimentary rock minerals, in *Sedimentary Petrography*: vol. 2, 4th rev. ed. London: George, Allen and Unwin, 15-207.

Popov, G., 1971. *Placer Deposits in the Working of Mineral Deposits*. Moscow, USSR: MIR Publishers (in English), 567-597.

Savage, C. N., 1961a. Economic geology of central Idaho blacksand placers, *Idaho Bur. Mines Geology Bull.*, **17**, 160p.

Savage, C. N., 1961b. Metals from blacksands, *Idaho Bur. Mines Geology Circ.*, **10**, 34p.

Trushkova, N. N., and Kukharenko, A. A., 1961. *Atlas of Placer Minerals*. Moscow: USSR Geol. Res. Inst., State Publ. House of Sci. and Technol., Literature on Geology and Conservation of Mineral Resources, 536p.

US Bureau of Mines, 1975. *Commodity Data Summaries (to 1974)*. Washington: Bureau of Mines and Geological Survey, US Department of the Interior, 193p.

Wells, J. H., 1973. *Placer Examination–Principles and Practices*. Washington: US Dept. of Interior, Bur. of Land Management, Tech. Bull. 4 (Field edition), 155p.

*Additional references may be found in this work.

Cross-references: *Abrasive Materials; Diamond; Garnet Group; Gemology; Metallurgy; Metamict State; Mineral Industries; Native Elements and Alloys; Order-Disorder; Placer Deposits; Pseudomorphism; Rutile Group;* also, see mineral glossary.

BLOWPIPE ANALYSIS OF MINERALS

Analysis is the determination of the chemical elements in a compound. If only the presence or absence of the elements is ascertained, the analysis is called "qualitative"; if their amounts are determined it is called "quantitative." A few elementary methods of qualitative analysis will be considered here.

Minerals are, in general, insoluble in water and many of them are insoluble in acids. Their examination by the "wet" methods of chemical analysis usually require a preliminary fusion and solution to obtain solutions that can be used for the normal "wet-way" analysis. With the blowpipe, analytical tests are applied directly to the mineral and are quickly carried out with very little material and with very simple, easily portable apparatus and reagents. Blowpipe analysis is sometimes called "dry-way" analysis. It can be carried out easily in the field by a prospector.

Technique

Apparatus. A simple blowpipe is a tube bent at right angles, one extremity having a mouthpiece, the other being terminated by a finely perforated jet. By blowing through this pipe, keeping the cheeks distended, a steady blast can be directed into a flame and can modify its character. A gas flame is very convenient for blowpipe experiments; but the blowpipe can be used with the flame of a candle, a spirit lamp, or an oil lamp. When a lamp or candle is used, the wick should be bent in the direction in which the flame is blown.

For the analyses outlined in this article, the following apparatus and reagents are also needed: two or three blocks of charcoal at least 18 cm long, 3 cm thick, and 3 cm wide; a piece of platinum wire (about 26 gauge) 5 cm long, mounted in a short length of glass rod; a pair of forceps; about 30 g (1 oz.) each (preferably in wide-mouthed, glass-stoppered bottles) of borax, salt of phosphorus (microcosmic salt), and sodium carbonate; small, glass-stoppered bottles of pure hydrochloric and sulfuric acid, cobalt nitrate, silver nitrate, barium chloride, and hydrogen peroxide; some zinc shavings, tinfoil, magnesium ribbon, and cupric oxide (CuO); a piece of dark blue glass (an 8-cm square will be enough); a dozen or more pieces of "hard" glass tubing about 1 cm in diameter

This entry has been extracted largely from a chapter in *Elementary Practical Geology*, E. Clarke, R. T. Prider, and C. Teicher, University of Western Australia Press, Perth, Australia, 1946.

and about 10 cm long; two or three test tubes about 15 cm long.

Flames. Two processes of great importance in blowpipe analysis are oxidation and reduction. A chemical change by which an element or compound loses one or more electrons (gaining oxygen) is called oxidation. For example, if metallic copper is heated in contact with air, it is changed to a black oxide of copper:

$$2Cu + O_2 = 2CuO$$

Two types of flame can be produced by means of the blowpipe, one in which oxidation of the substance under examination is brought about and the other in which reduction of the substance takes place.

Oxidizing Flame. Introducing the nozzle of the blowpipe into the flame at about one-third the breadth of the flame (Fig. 1A) will produce an oxidizing flame. This flame is blue and feebly illuminating; in it the air from the blowpipe is well mixed with the gases from the flame and complete combustion takes place. There are two positions in this flame at which operations useful to the mineralogist are performed. The hottest part is just outside the inner blue cone, and is called the "point of fusion" (Fig. 1A). The best point for oxidation, the "point of oxidation" (Fig. 1A), is just beyond the visible part; at this point, the substance being heated absorbs oxygen from the air and oxidation takes place.

Reducing Flame. Holding the nozzle of the blowpipe some little distance from the flame (Fig. 1B) will produce a reducing flame, which is bright yellow, luminous, ragged, and noisy. In this flame, the stream of air from the blowpipe drives the whole flame rather feebly before it, and there is little mixture of air with the gases of the flame. The result is that these gases are not completely burned and readily combine with the oxygen of any substance introduced into their midst; in consequence, the substance undergoes reduction. The assay must be completely surrounded by the reducing flame (as at *R* in Fig. 1B) otherwise it will absorb oxygen from the air and will become oxidized and not reduced.

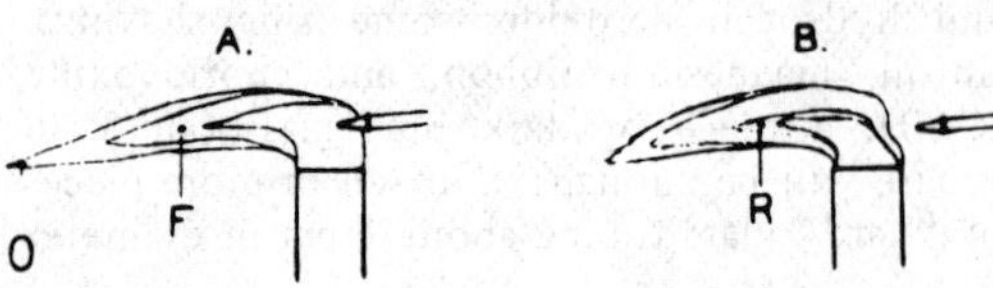

FIGURE 1. Production of (A) the oxidizing and (B) the reducing flames by means of the blowpipe. In (A), the point O is the point of oxidation and F is the point of fusion. In (B), the point R is the point of reduction.

When using a blowpipe with a Bunsen burner, the best results are obtained by shutting off the air supply and reducing the flame to a length of about 3 cm. When this is done, the reducing and oxidizing flames can be obtained with little effort.

Supports. When heating assays with a blowpipe, it is necessary to support the material in some way. The most important supports are platinum wire and charcoal.

Platinum wire is expensive, so note carefully the precautions mentioned below.

Charcoal forms a good support by virtue of its infusibility, low capacity for conducting heat, and reducing action. The carbon of which it is formed readily combines with the oxygen the assay material may contain and so the material becomes reduced; many minerals may be reduced to metals by heating in the reducing flame on the charcoal block, e.g., *galena* may be reduced to metallic lead.

In using the charcoal block, the assay should be placed in a small hollow scraped in the charcoal and there should be a large area of cool charcoal beyond the hollow to provide the best conditions for the deposition of any encrustation (a deposit from the volatile constituents that are driven off by the heat). Should the assay crackle or fly about (decrepitate) another should be made by powdering the substance and mixing it into a thick paste with water.

When the assay is heated, it may emit a characteristic odor or form an encrustation on the charcoal: the odor emitted, the color of the encrustation and its distance from the assay, or the nature of the residue will indicate the presence of some element. These results are described later, in the sections on charcoal tests.

Fluxes. Certain substances are added to the assay to reduce the temperature of melting. They are also very useful when the constituents of the assay form a characteristic colored compound with these substances. These substances are called fluxes. The three most important fluxes are discussed here.

Borax. Hydrous sodium borate, $Na_2B_4O_7 \cdot 10H_2O$ (*borax*), serves principally to reduce substances to the form of oxides. From the color and other properties that these oxides impart to a borax bead, we are able to ascertain to a certain extent the nature of the substance under examination. *Borax* also serves to reduce metallic ores in such a way that a minute globule of the metal may be separated from the other elements with which it is combined. The method of testing is to make a small borax bead on the end of the platinum wire and introduce into it a small amount of the mineral being tested. This bead is first heated in the reducing flame and the color noted both when the bead is hot and when it is cold: the colors are those imparted by the low-oxygen compound of the metal since heating in the reducing flame tends to remove some of the oxygen and form the lower oxide. The bead is then heated in the oxidizing flame and the colors of the hot and cold bead are noted. These colors are due to presence of the higher oxide, which contains more oxygen since heating in an oxidizing flame has encouraged absorption of oxygen from the atmosphere. For example, after heating any manganese salt in the reducing flame, the borax bead is colorless (due to the presence of the lower oxide MnO); whereas, after heating in the oxidizing flame, the color of the same bead is reddish-violet (due to the presence of the higher oxide Mn_2O_3). After heating copper compounds in the reducing flame, the color of the bead is opaque red (due to metallic copper and the lower oxide Cu_2O), whereas, after heating in the oxidizing flame the color is blue (due to the higher oxide CuO).

Salt of Phosphorus. Microcosmic salt is hydrated sodium ammonium phosphate. The action of this flux is similar to that of *borax*; it reduces the oxides of the metals to complex phosphates. The characteristic colors imparted to the bead when hot and cold often differ according to whether the flame used is oxidizing or reducing.

Sodium Carbonate. When mixed with finely powdered oxides or sulfides of metals together with powdered charcoal and then heated on charcoal, $Na_2CO_3 \cdot 10H_2O$ serves to reduce these materials to the metallic state.

Qualitative Blowpipe Tests

Flame Tests. The substance is held on platinum wire. Certain sulfides and arsenides reduce readily to the metal, which alloys with platinum. The metal, therefore, should be changed to a less easily reducible state by roasting before being heated in contact with platinum. Hold near the blue tip in the oxidizing flame and note coloration of the outer flame. If no result is obtained, moisten the substance with HCl or H_2SO_4. Always note the ease or otherwise with which a result is obtained—this may be characteristic of the mineral. After each test, clean the wire by dipping it in HCl and holding in the flame until no further coloration of the flame is noticeable. Flame tests may be made by holding small slivers or chips in the flame with the forceps. Conclusions to be drawn from flame tests are given in Table 1.

Borax Bead Tests. Heat the end of the platinum wire (which should not be bent) to redness and plunge it into powdered *borax*, some of which will adhere to the wire. When the adherent *borax* is heated in the blowpipe flame, the powder froths up or intumesces, and gradually fuses to a clear transparent bead, which should not exeed 1.5 mm in diameter. Allow some of the substance to be examined to touch the hot bead so that a small amount sticks to it. Some minerals should be added in very small amounts; otherwise the reaction will be masked and the bead will either be opaque (dark gray or black) or its color will be difficult to determine.

If the assay is metallic in luster or high in specific gravity, roast it thoroughly on the charcoal block before adding it to the bead; otherwise the platinum wire may be destroyed. This is because metallic arsenides or sulfides will alloy with the platinum, making it brittle.

Heat the bead first in the reducing flame and then in the oxidizing flame, noting, in both cases, the color of the bead both when hot (immediately after the red heat has disappeared) and when cold (Table 2).

Salt of Phosphorus or Microcosmic Bead Tests. The salt of phosphorus bead is made in the same manner as the borax bead but, owing to the greater fluidity of salt of phosphorus when first heated, it is more difficult to manipulate. This is a good means of testing for silicates because the bases are dissolved in the bead but the silica remains as an insoluble skeleton (see Table 3).

Charcoal Tests Without Flux. Place small chips or powder of the assay in a shallow cavity near one end of the charcoal. Hold the charcoal almost horizontal and heat the assay directly with a good oxidizing flame so as to traverse the length of the block. The object of this is to blow any volatile oxides, formed during the heating of the assay, onto the cooler parts of the charcoal where they may be deposited as an encrustation.

Raise the temperature gradually, interrupting the operation occasionally to examine the assay for results. If fusion has taken place, crush and roast carefully. If the residue is black or colored, use it for borax bead tests. As heating proceeds note the following.

- Change of condition–
  - a. Fusion: easily (e.g., *stibnite*) or with difficulty (e.g., *kaolinite*).
  - b. Decrepitation: crackling and flying into fragments (e.g., a chip of *sphalerite*).
  - c. Intumescence (e.g., *borax*).
  - d. If fusible: note whether the residue glows brightly in the oxidizing flame (e.g., *magnesite*).
- Evolution of odors–A sulfurous odor indicates the presence of a sulfide (e.g., *pyrite*); an odor of garlic indicates the presence of arsenic (e.g., *arsenopyrite*).

TABLE 1. Flame Test Results

Color	Indicates	Remarks	Suitable Minerals[a]
Brick-red	Ca	Dingy green through indigo glass	*Calcite* ($CaCO_3$) or *gypsum* ($CaSO_4 \cdot 2H_2O$)
Crimson	Sr and Li		*Strontianite* ($SrCO_3$)
Emerald-green	Cu	With HCl gives blue flame of $CuCl_2$; not good with *chalcopyrite*	*Chalcopyrite* ($CuFeS_2$)[b], or *malachite* (hydrous copper carbonate)
Bluish green	P	Flash only with H_2SO_4	*Apatite* (phosphate and fluoride of calcium)
Grass-green	B	Requires H_2SO_4	*Borax* (hydrated sodium borate)
Yellowish green	Ba	Long heating of *barite* is necessary to eliminate sodium flame; better result with powder	*Barite* ($BaSO_4$)
Yellowish green	Zn		*Sphalerite* (ZnS)[b]
Azure blue	$CuCl_2$	Other Cu compounds with HCl	*Malachite* (hydrous copper carbonate)
Lavender with pale green border	Sb		*Stibnite* (Sb_2S_3)[b]
Whitish blue	As		*Arsenopyrite* (FeAsS)[b]
Whitish blue	Pb		*Cerussite* ($PbCO_3$)
Lavender	K	Usually masked by Na, but visible through blue glass	*Sylvite* (KCl)
Yellow	Na		*Halite* (NaCl)

[a]Minerals suitable for illustrating the various tests are given under this heading in this and succeeding tables.
[b]Roast on charcoal before using.

TABLE 2. Diagnosis of Borax Bead Test Results

	Oxidizing Flame		Reducing Flame		
Element	Hot	Cold	Hot	Cold	Suitable Minerals
Co	Blue	Blue	Blue	Blue	*Smaltite* [$(Co,Ni)As_2$][a]
Fe	Yellow to brown	Colorless to yellow	Green to yellowish green	Bottle green	*Hematite* (Fe_2O_3) or limonite ($2Fe_2O_3 \cdot 3H_2O$)
Cr	Brown	Yellowish green	Emerald	Emerald	*Chromite* ($FeO \cdot Cr_2O_3$)
Cu	Green	Blue	Colorless	Opaque red	*Cuprite* (Cu_2O) or *malachite* [$CuCO_3 \cdot Cu(OH)_2$]
Mn	Violet	Violet	Colorless	Colorless	Psilomelane (impure hydrous manganese dioxide)
Ni	Violet	Yellow to brown	Gray	Metallic Gray	*Garnierite* (hydrous nickel-magnesium silicate)
Colors resulting from mixtures of coloring oxides					
Fe and Cu as in roasted *chalcopyrite*	Green	Green	Green	Red	*Chalcopyrite* ($CuFeS_2$)[a]
Fe and Mn as in *wolframite*	Violet	Brown	Green	Bottle green	*Wolframite* [$(Fe,Mn)WO_4$]

[a]Roast on charcoal before testing.

TABLE 3. Bead Tests with Salt of Phosphorus

Element	Oxidizing Flame	Reducing Flame	Suitable Minerals
W	Colorless	Gray-violet or blue-green	*Scheelite* ($CaWO_4$)[b]
Mo	Bright green	Green	*Molybdenite* (MoS_2)[b]
Ti	Colorless[a]	Violet	*Rutile* (TiO_2)
Si	(Remains undissolved)		*Orthoclase* ($KAlSi_3O_8$)

[a]Dissolves very slowly.
[b]Must be saturated.

- Change in the color–Note the color of the assay both when hot and when cold (Table 4).
- Encrustation–If the oxides formed during this heating are volatile, they will be carried from the assay and deposited on the cooler parts of the block. Test the volatility of the encrustation by touching the deposit with the tip of the oxidizing flame and noting the ease with which it disappears. Note the colors of the encrustation both near and away from the assay (Table 5).

All white residues and encrustations, when thoroughly cold, should be moistened with cobalt nitrate solution, heated strongly in an oxidizing flame, and the color noted. Satisfactory results are difficult to obtain and the tests must be interpreted with caution. (see Table 6).

Charcoal Tests with Flux. Finely powder the assay material and mix with two or three times its volume of sodium carbonate. Moisten the mixture with water, and pack it into a hollow on the charcoal block. Heat strongly in the reducing flame, and note whether reduction to metal takes place without encrustation (Table 7) or with encrustation (Table 8). Treat any white encrustation with cobalt nitrate as directed under Charcoal Tests without Flux.

Remove some of fused mass from the charcoal and, when cold, place on a silver coin moistened with a drop of water. Let it stand for a few minutes. A black stain on the coin indicates the presence of S. Apply this test to *pyrite* (FeS_2) and *gypsum* ($CaSO_4 \cdot 2H_2O$) as well as other sulfur compounds. This test is very useful for detecting sulfur in sulfates from which SO_2 fumes are not evolved on heating.

Other Simple Qualitative Tests

Closed-Tube Tests. Introduce a small quantity of the assay into a small glass tube, closed at one end, and heat. In some cases, a deposit

TABLE 4. Assay Color Changes

Original Color	Color after Heating		Indicates	Suitable Materials
	Hot	Cold		
Blue or green	Black	Black	Cu salt	*Malachite* [$CuCO_3 \cdot Cu(OH)_2$]
Red	Black	Red	Fe_2O_3	*Hematite* (Fe_2O_3)
Yellow or brown	Black	Black	$FeCO_3$	*Siderite* ($FeCO_3$)
White or yellow	Yellow	White	Sn and Zn salts	*Cassiterite* (SnO_2) and *sphalerite* (ZnS)

TABLE 5. Color of Encrustation on Charcoal

Near Assay		Away from Assay		Indicates	Suitable Minerals
Hot	Cold	Hot	Cold		
White (volatile)	White	Blusih white	Bluish white	Sb	*Stibnite* (Sb_2S_3)
White (very volatile)	White	Grayish white	Grayish white	As	*Realgar* (AsS)
Yellow	Yellow	Bluish white	Bluish white	Pb	*Galena* (PbS)
Yellow	White	White	White	Zn	*Sphalerite* (ZnS)
Pale yellow	White	White (difficult to obtain)	White (difficult to obtain)	Sn	*Cassiterite* (SnO_2)
Yellow (purple black if there is slightest suggestion of reducing flame but can be changed to yellow and white by a touch of oxidizing flame)	White	Bluish	White	Mo	*Molybdenite* (MoS_2)

TABLE 6. Color of Residues on Reheating

After Heating	Indicates	Suitable Minerals
Gray	Ca, Ba, Sr	*Calcite* ($CaCO_3$), *barite* ($BaSO_4$), *strontianite* ($SrCO_3$)
Green	Zn	*Sphalerite* (ZnS)
Pink	Mg	*Magnesite* ($MgCO_3$)
Earthy blue	Al	*Kaolinite* ($Al_2O_3 \cdot 2H_2O \cdot 2SiO_2$)
Glazed blue	Fusible silicates (and certain borates and phosphates)	*Borax* (hydrated sodium borate)
Yellowish green	Certain borates and phosphates	***Apatite*** (calcium fluoride and phosphate or calcium chloride and phosphate)
Bluish green	Sn	*Cassiterite* (SnO_2)
Dirty dark green	Sb	*Stibnite* (Sb_2S_3)

(called a *sublimate*) is formed on the cooler parts of the tube and the color and nature of this sublimate may indicate one or more of the elements present in the assay. The assay itself may be converted into a new compound with characteristic color and properties. Thus, brown limonite (hydrous ferric oxide) may be converted into black magnetic oxide and, at the same time, water will be expelled from the mineral and will collect on the cool upper part of the tube. Before making the test, heat the tube gently to ensure that it is perfectly dry and then allow it to cool; then introduce the assay, and, holding the tube at an angle of about 45° to the horizontal, apply heat to the lower, closed end. Look for the following effects.

a. Glowing, phosphorescence, decrepitation, fusion, etc.
b. Gas is given off. This indicates the presence of O from oxides, CO_2 from carbonates, SO_2 from sulfides, or F from fluorides. It is impracticable to distinguish among these gases (except SO_2, which is recognizable by its odor) in such small tubes.
c. Drops of water condense in the top of tube. This indicates that the mineral is a hydrous compound (e.g., *gypsum*).
d. The color of the assay changes as on the charcoal block (see Table 4).
e. A sublimate is formed. The diagnosis is given in Table 9.

Open-Tube Tests. Heat the assay in a tube 8–10 cm long and open at both ends. The assay should be placed about 3 cm from one end of the tube and the tube held in an inclined position with the assay in the lower part. When the assay is heated, a current of air passes over the heated substance, which may then absorb oxy-

TABLE 7. Reduction Products with Sodium Carbonate

Globules	Indicates	Suitable Minerals
Brilliant white metal, malleable	Sn, Ag	*Cassiterite* (SnO_2)[a] *Argentite* (Ag_2S)
Yellow metal, malleable	Au	
Red scales or globule, malleable	Cu	*Malachite* [$Cu(OH)_2 \cdot CuCO_3$], *chalcopyrite* ($CuFeS_2$)
Gray powder, strongly magnetic	Fe	Limonite (ferric hydroxide)
Gray powder, feebly magnetic	Co, Ni	*Smaltite* [$(Co, Ni)As_2$]

Note: To test the magnetic properties of the residue, powder it, place it on paper, and apply a magnet underneath the paper. If the powder is not moved Fe must be absent; now touch the powder with the magnet; if Co and Ni are present they, being feebly magnetic, will be attracted.

[a]The Sn bead is obtained with difficulty; the *cassiterite* must be ground to a very fine powder.

TABLE 8. Encrustation Formed by Reduction with Socium Carbonate

Globule	Encrustation	Indicates	Suitable Mineral
White, malleable, does not mark paper	Very slight	Sn[a]	*Cassiterite* (SnO_2)
White, malleable, marks paper	Yellow (hot and cold)	Pb	*Galena* (PbS) or *cerussite* ($PbCO_3$)
White, brittle	White, close to substance	Sb	*Stibnite* (Sb_2S_3)
None	Yellow (hot); white (cold)	Zn	*Sphalerite* (ZnS)
None	White, white fumes, and smell of garlic	As	*Arsenopyrite* (FeAsS) or *realgar* (AsS)

[a]Obtained only with difficulty.

TABLE 9. Closed-Tube Sublimates

Sublimate		Indicates	Suitable Mineral
Hot	Cold		
Orange	Orange-yellow (opaque)	S	*Pyrite* (FeS_2)
Deep red	Reddish yellow	AsS and As_2S_3	*Realgar* (AsS)
Black	Brownish red	Sb_2S_3	*Stibnite* (Sb_2S_3)
Black	Black	HgS	*Cinnabar* (HgS)
Black	Metallic mirror	As	*Arsenopyrite* (FeAsS)

gen and release volatile oxides that may form a sublimate on the cool upper part of the tube. Diagnosis of open-tube results is given in Table 10.

Tests for the Acid Radical. *Carbonate.* Carbonates effervesce with HCl. Note that some carbonates, such as *magnesite* and *dolomite*, require slight heating before the reaction will commence.

Chloride. Mix the mineral powder with CuO and moisten with H_2SO_4 to form a paste. Place on charcoal and heat with the oxidizing flame. Note flame coloration, first greenish blue, then brilliant blue. Test *halite* (NaCl). A useful wet-way test for chloride (which may be readily applied since many chlorides are water-soluble) is the addition of silver nitrate solution. The result will be the precipitation of white silver chloride if dissolved chloride is present. (Distilled water must be used for making the test solution since undistilled water often contains dissolved chlorides.)

Fluoride. Fluorides are soluble in concentrated H_2SO_4 with evolution of HF, which etches glass. To test a fluoride, place the material (e.g., *fluorite*, CaF_2), in a test tube, add a little concentrated H_2SO_4, and warm gently. A moistened slip of glass held near the mouth of the test tube will be etched by the HF fumes evolved. The glass must be dried to detect the etching.

Phosphate. Place the powdered mineral with about 10 mm of magnesium ribbon on charcoal and ignite by touching with the oxidizing flame. Allow the mass to cool and then moisten with water. Phosphuretted hydrogen, recognized by its odor of bad fish, is liberated. Test ***apatite*** (phosphate and fluoride of calcium or phosphate and chloride of calcium, but generally both fluoride and chloride are present).

TABLE 10. Open-Tube Sublimates

Odor	Sublimate	Given by	Indicates	Suitable Mineral
Garlic	White volatile crystals far from assay	As_2O_3	As	*Smaltite* $(Co,Ni)As_2$
Suffocating	None	SO_2	S	*Pyrite* (FeS_2)
	White, dense fumes	Sb_2O_3	Sb	*Stibnite* (Sb_2S_3)
	White; the assay fuses to yellow drops, which are white when cold	$PbSO_4$	PbS	*Galena* (PbS)
	Metallic mirror	Hg	Hg	*Cinnabar* (HgS)
	Yellow when hot; white when cold; crystalline near assay	MoO_3	Mo	*Molybdenite* (MoS_2)

Silicate. Silicates leave a residual skeleton of undissolved SiO_2 in the salt of phosphorus (microcosmic salt) bead. Test *orthoclase* ($KAlSi_3O_8$).

Sulphate. The sulfur of the sulfate compounds cannot usually be so easily detected as the sulfur of the sulfide minerals, from which SO_2 is evolved when the sulfide is heated on the charcoal block or in the closed tube. Sulfur in sulfates is best detected by the silver-coin test. A very useful wet-way test is to heat the assay as above on a charcoal block with sodium carbonate, dissolve it in dilute HCl, and add barium chloride solution. A white precipitate of barium sulfate indicates the presence of sulfur.

Sulphide. The sulfur of sulfides can usually be detected by the odor of SO_2 when the substance is heated either on charcoal or in the closed or open tube. It may also be detected by the method described above for sulfates. H_2S is evolved from some sulfides on addition of HCl.

Titanate (or titanium as the metallic radical). Fuse the substance with sodium carbonate on charcoal. Dissolve the residue with dilute H_2SO_4. Add hydrogen peroxide. An amber color will indicate the presence of titanium.

Tungstate. Saturate a salt of phosphorus (microcosmic salt) bead with the substance. Dissolve the bead in dilute HCl (one part of HCl to three or four of water), add a small piece of tinfoil, and boil. A deep blue solution indicates that tungsten is present.

Another useful test for tungsten is to fuse with sodium carbonate, dissolve the fused mass in dilute HCl as above, add a few zinc shavings, and boil. Blue coloration indicates the presence of tungsten.

Test for *Cassiterite.* Place grains or chips of the assay in a watch glass together with some zinc shavings and HCl. If, after some time, the mineral grains become silvery in color, this indicates that the mineral is *cassiterite.* Note that sufficient zinc shavings should be added to neutralize the acid; otherwise any coating on the grains will be immediately removed by the acid.

Hints for Determining Minerals

Blowpipe tests should be regarded as affording confirmation of the determination of a mineral by its physical properties, rather than as being a means of determining the chemical constituents of an unknown powder. Its various physical properties indicate the nature of a mineral and narrow down the possibilities to comparatively few species. Suitable blowpipe tests may then be applied to determine which of the minerals is indicated by the elements present. For example, earthy *hematite* and earthy *cuprite* may appear very similar (although *hematite* is harder than *cuprite*) since they are alike in color, streak, and specific gravity. A flame test will readily determine to which of these species the mineral belongs, since *cuprite* will yield the deep green flame of copper whereas the *hematite* will not.

The light-colored minerals, consisting of silicates, carbonates, sulfates, oxides, or phosphates of the light metals, rarely yield positive results with the borax bead or charcoal block, but many yield some color in the flame. Dark-colored minerals with dark-colored streaks are predominantly oxides, sulfides, or arsenides of the heavy metals, which generally give positive

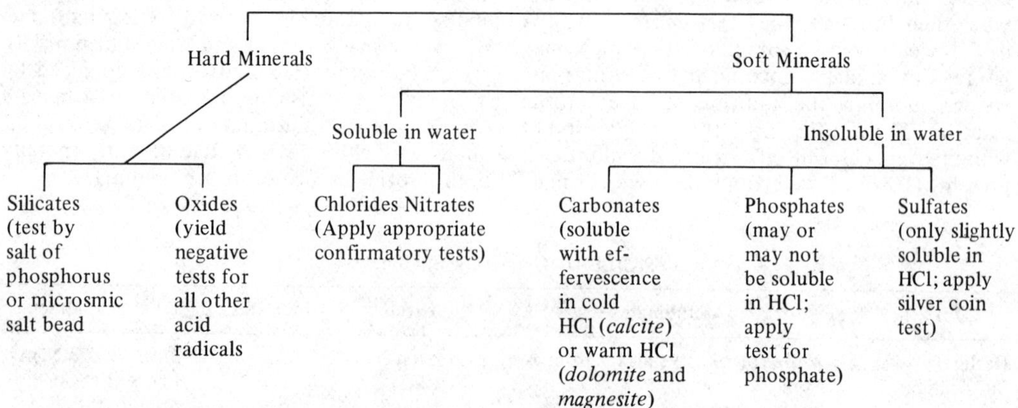

NOTE 1. Among the common silicates the micaceous minerals, *talc,* ***chlorite,*** and the ***micas,*** are soft.
2. Many of the light-colored minerals are hydrous compounds and, in all cases, a closed-tube test should be made to determine the presence or absence of water in the mineral.

FIGURE 2. Light-colored minerals.

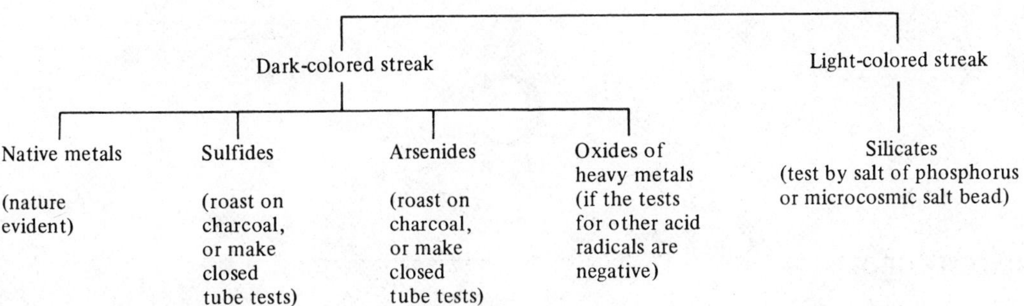

NOTE: Among the common sulfides, *sphalerite* has a light-colored streak.

FIGURE 3. Dark-colored minerals.

results in the borax bead or on the charcoal block. Therefore, in applying tests for a metallic radical, always apply the flame test as preliminary to the examination of a light-colored mineral, or apply a borax bead or charcoal test in the case of the dark-colored heavy minerals. For determination of an acid radical the scheme set out in Figs. 2 and 3 will be found useful.

In some instances the less common acid radicals such as tungstate, titanate, etc., may be present in both the light- and dark-colored groups. These may be indicated by negative tests for the other acid radicals.

REX T. PRIDER

References

Berzelius, J. J., 1845. *The Use of the Blowpipe in Chemistry and Mineralogy*, (transl. by J. D. Whitney). Boston: W. Ticknor and Co., 237p.

Butler, G. M., 1910. *Pocket Handbook of Blowpipe Analysis*, 1st ed. New York: Wiley 80p.

Clarke, E.; Prider, R. T.; and Teicher, C., 1946. *Elementary Practical Geology*. Nedlands, Australia: University of Western Australia, 170p.

Davidson, E. H., 1940. *Field Tests for Minerals*, 2nd ed. London: Butler & Tanner, 60p.

Elderhorst, W., 1874. In H. B. Nason and C. F. Chandler, eds. *Manual of Qualitative Blow-Pipe Analysis and Determinative Mineralogy*, 6th ed. New York: Zell, 312p.

Fletcher, E. F., 1894. *Practical Instructions in Quantitative Assaying with the Blowpipe*. New York: Wiley, 142p.

Galbraith, F. W., 1963. Chemical tests for the elements, *Geotimes*, 7(8), 35-36; 8(1), 35-36 (AGI Data Sheets 42a and 42b).

Cross-references: Vol. IVA: *Geochemistry: Testing for Elements; Mineralogy; Oxidation and Reduction.*

BONE MINERALS—*See* HUMAN AND VERTEBRATE MINERALOGY

BRAVAIS LATTICE—*See* LATTICE

C

CALCITE GROUP

Calcium carbonate ($CaCO_3$)–including the three naturally occuring polymorphs *calcite, aragonite,* and *vaterite*–seems to be almost ubiquitous in nature. Although the overall abundance of this group is not as great as some other minerals, its members have the most varied occurrence of any mineral family. They are found in all rock types, in ore deposits, in caves, in geyser and hot springs deposits, etc., and as structural and nonstructural components in living organisms (see *Cave Minerals; Human and Vertebrate Mineralogy; Invertebrate and Plant Mineralogy*). Many aquatic organisms–including shellfish, sea urchins, and seal coral–secrete $CaCO_3$ as an integral part of their body. Inoué and Okazaki (1977) have recently summarized biocrystal formation in sea urchins and have beautifully demonstrated the relationship of *calcite* crystal growth to the spicules. In humans, $CaCO_3$ has been studied in relation to the formation of carries in teeth and in the calculi, or stones, that form in the gall bladder, kidney, and urinary bladder. Calcium carbonates have extensive ornamental use as marble and mother of pearl, practical use as chalk, and have been used as money (wampum), by North American Indians. In the form of limestone and marble they are very important as structural building materials. Moreover, $CaCO_3$ can be seen as stalactite-like structures hanging from the ceilings of man-made concrete tunnels, as troublesome deposits in air conditioners, and in many other places where water and air can react with Ca-bearing material.

The name *calcite* is derived from the latin word for lime (*Calx, calcis*) and the Greek word that described burnt lime; calcspar is an older name no longer used. Occasionally it is referred to as Iceland spar. Other substances that consist primarily of $CaCO_3$ include, travertine, stalactites, stalagmites, and onyx. The most extensive treatments of this mineral family can be found in Palache, Berman, and Frondel (1951); Deer, Howie, and Zussman (1963); and Lippmann (1973).

Phase Relationships

In order to appreciate the complexity and understand the semantics of calcium carbonate, one must be first aware of its temperature-pressure phase-stability diagram. A slightly schematic diagram compiled from various sources is presented in Fig. 1. It shows that $CaCO_3$ can exist in at least five different modifications of crystal structure, each of which is thermodynamically stable only for a certain temperature and pressure range. What this diagram does not show, however, are the intriguing environmentally controlled kinetic effects that can occur during aqueous precipitation of $CaCO_3$, drastically altering the interpretation of the equilibrium picture. Lippman (1973) has given extensive treatment to these effects (see also McCauley and Roy, 1974). According to equilibrium (reversible) thermodynamics, *aragonite* and *calcites* II and III are high-pressure phases, whereas *calcite* is stable at normal and high temperature, and *vaterite*, apparently, is stable at low temperature. With the exceptions of *vaterite* and *calcites* II and III, these experimental observations have been in general confirmed by natural occurrences. One problem in interpretation of natural occurrences is that observations are made on specimens at normal temperature and pressure, where *calcite* is always stable, and only extrapolated to conditions at the time of formation. Many factors can contribute to a change in the crystal structure of $CaCO_3$ and, hence, lead to erroneous deductions concerning initial phases.

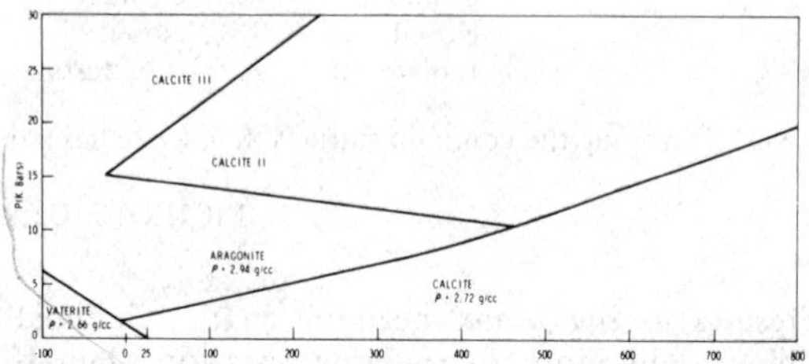

FIGURE 1. Phase diagram for $CaCO_3$ (adapted from Bridgman, 1939; Boettcher and Wyllie, 1968; Johannes and Puhan, 1971; and Albright, 1971).

Calcite Crystal Structure

Because of the early availability of good crystals and its very perfect rhombohedral cleavage, *calcite* was one of the first crystal structures to be determined by means of X-ray diffraction (Bragg, 1914). The accepted space group for *calcite* is $R\bar{3}c$ ($D^6{}_{3d}$, No. 167). In descriptive fashion, its crystal structure is directly analogous to a face-centered cubic NaCl structure that has been compressed along any cube diagonal to 76.66% of its original length. For comparative purposes, Na and Cl are replaced with Ca and CO_3, respectively. Since Cl is roughly spherical in shape, whereas the carbonate group is triangular to a first approximation, distortions from the simple NaCl-type of structure occur in $CaCO_3$, resulting in more complex analytical ways of describing the unit cell for X-ray or for morphologic investigations. Describing $CaCO_3$ is a compressed NaCl structure results in a cell with 4 formula units corresponding directly to the common cleavage or morphologic rhombohedron. If this cell is used for X-ray crystallographic calculations, X-ray diffraction theory demands that one must double the vertical body parameter, resulting in a cell containing 32 formula units and making calculations extremely difficult. Hence, X-ray crystallographers have chosen a simpler unit cell, which contains only two formula units. This cell is also a rhombohedron, but the rhombohedral angle is actue (about 46°) and not obtuse (about 101°) as in the cleavage cell. This is clearly shown in Fig. 2. However, as with all rhombohedral type structures pictoral and descriptive presentations are extremely difficult, so that hexagonal geometry is generally used to describe the structure. Fig. 3 shows this relationship the number of formula units in the unit cell must be tripled to match the observed density to the calculated X-ray density, since the hexagonal unit cell has three times the volume of the rhombohedral one. Table 1 summarizes these relationships. The following formulas are useful for relating rhombohedral to hexagonal unit cells.

HKL = hexagonal and
hkl = rhombohedral indices

$$\begin{aligned} \text{where } H &= h - k \\ K &= k - l \\ L &= h + k + l \\ h &= 1/3(2H + K + L) \\ k &= 1/3(-H + K + L) \\ l &= 1/3(-H - 2K + L) \end{aligned}$$

and, if 4 hexagonal indices are used, $I = -(H + K)$.

The lattice parameters are

$$a_R = \frac{1}{3}\,(3a_H^2 + c^2)^{\frac{1}{2}}$$

$$\sin\frac{\alpha}{2} = \frac{3}{2[3+(c/a_H)^2]^{\frac{1}{2}}}$$

Of particular note is that the d-spacing of the cleavage plane $(10\bar{1}4)$ was used as a standard for defining the fundamental X-ray wavelength. The d-spacing of $(10\bar{1}4)$ was set at 3.029 kX units (1 kX=1.00202 Å).

Comparison to Aragonite. The crystal structure of *aragonite* is very similar to that of *calcite*. In both structures, the cation is surrounded by six anion groups. For *calcite*, this results in six oxygen ions (from the carbonate groups) being close to calcium. However, in the transition from *calcite* to *aragonite*, there is

TABLE 1. Calcite Unit Cells

		X-ray Smallest Cell	Cleavage Rhomb Pseudo-cell	Cleavage Rhomb True Cell
Rhombohedral axes	a_{rh}(Å)	6.37	6.42	12.85
	α_{rh}	46°05′	101°55′	101°55′
	Z_{rh}	2	4	32
	Cleavage rhomb indices	{211}	{100}	{100}
Hexagonal axes	a_{hex}(Å)	≅ 5	≅ 10	≅ 20
	c_{hex}(Å)	≅ 17	≅ 8.5	≅ 17
	Z_{hex}	6	12	96
	Cleavage rhomb indices	$\{10\bar{1}4\}$	$\{10\bar{1}1\}$	$\{10\bar{1}1\}$

From W. A. Deer, R. A. Howie, and J. Zussman, 1963, *Rock-Forming Minerals*, vol. V. Reprinted with permission from Longman Group Limited, London, p. 230.

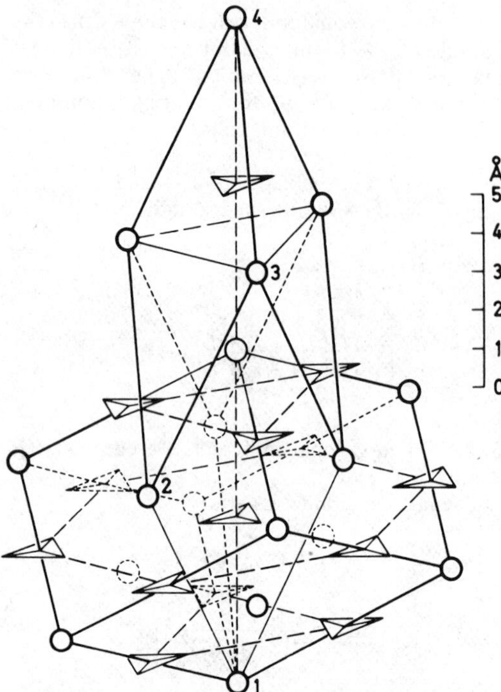

FIGURE 2. Cleavage cell or morphological rhombohedron of *calcite* (lower part of drawing), containing $4CaCO_3$. The center of symmetry, coinciding with the Ca at the apex of the cleavage rhombohedron, imposes opposite orientations on the two CO_3 groups which lie above and below, along the vertical body diagonal (broken line connecting 1 and 4). Hence, the 4-formula cleavage cell cannot be a true unit cell of *calcite*, because its vertical body diagonal represents only half the identity period. The full period, however, is contained in the acute cell with $2CaCO_3$. A true unit cell, shaped like the cleavage rhombohedron, must have a vertical body diagonal of doubled length. Consequently, all of the linear dimensions of the small cleavage cell here presented ought to be doubled. Such a cell is composed of 8 small cells and will thus contain a total of $32CaCO_3$. It is self-evident that such a cell would be rather unwieldy in crystallographic computations. Therefore, the acute cell containing only $2CaCO_3$ is generally preferred in modern crystallography. Of the Ca atoms (open circles) those marked 1, 2, 3, 4 show the interrelation with the hexagonal unit cell in Fig. 3. The vertical scale applied to *calcite*. Triangles: CO_3 groups. (from Lippman, 1973).

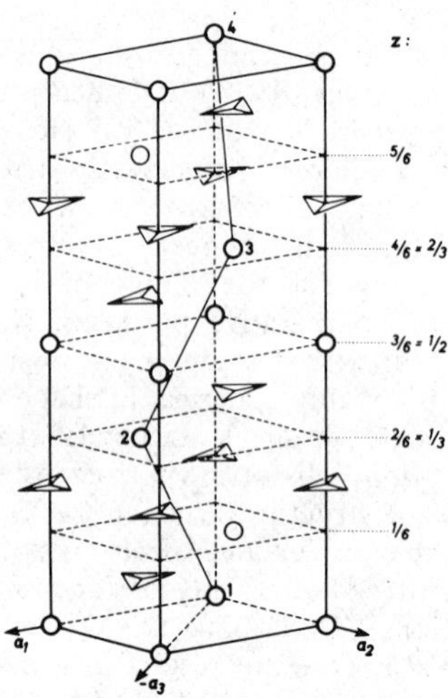

FIGURE 3. Hexagonal unit cell of *calcite*, height: c = 17.06 Å, containing $6CaCO_3$ showing alternating levels of Ca (open circles) and of CO_3 groups (triangles). The z coordinates, in fractions of c, to the right apply to the Ca levels. CO_3 levels are exactly halfway with multiples of 1/12 as z coordinates. The Ca atoms marked 1, 2, 3, 4 correspond to those in the rhombohedral cell (Fig. 2) marked the same way. 2 has the coordinates 2/3, 1/3, 1/3; 3 has 1/3, 2/3, 2/3. These sites are generated from the primitive points 1 or 4 by rhombohedral centering, which makes the hexagonal cell triply primitive (from Lippman, 1973).

a rotation of the carbonate groups by approximately 30° around an axis perpendicular to the CO_3 plane, so that three more oxygen ions are brought into a tight coordination cage around the Ca ion. This structure is a tighter, more dense one and is, therefore, more stable at high pressures than is *calcite*. Hence, the natural occurrence of *aragonite* in environments that are unambiguously known to be in the temperature-pressure range for *calcite* have been a puzzle to geologists for years. McCauley and Roy (1974) have presented evidence showing that the major kinetic interactions in the formation of *aragonite* outside of its thermodynamic stability region are related to entrapment of HCO_3^- groups in early formed $CaCO_3$ nuclei and to the epitaxial overgrowth of *aragonite* on $SrCO_3$, $PbCO_3$ and $MgCO_3 \cdot 3H_2O$ nuclei.

Cleavage of Calcite. The cleavability of *calcite* can be explained by its structure. The rhombohedral atomic planes are held together by the electrostatic interaction of oppositely charged ions perpendicular to these planes. By an applied external physical force, a plane of atoms can shift position, bringing indentically charged ions opposite each other, which results in repulsion and, thence, cleavage of the material.

Crystal Morphology. An extensive description of the known morphologic habits (and their accepted nomenclature) of *calcite* is given in Palache, Berman, and Frondel (1951). More than 700 forms have been reported (see *Crystal Habits*).

Other Members of the Group

In the *calcite* group are all those carbonate

TABLE 2. Crystallographic Data for Calcite Group Minerals

Name	Chemistry	Unit Cell (Å) a	b	c	ρ_o(g/cm³)	C.I.R.[a]
Aragonite	$CaCO_3$	4.961	7.967	5.740	2.94	1.18
Calcite	$CaCO_3$	4.990		17.061	2.72	1.00
Otavite	$CdCO_3$	4.920		16.298	4.96	0.95
Rhodochrosite	$MnCO_3$	4.771		15.664	3.70	0.82
Siderite	$FeCO_3$	4.689		15.373	3.96	0.77
Sphaerocobaltite	$CoCO_3$	4.658		14.958	4.13	0.95
Smithsonite	$ZnCO_3$	4.653		15.025	4.43	0.75
Magnesite	$MgCO_3$	4.633		15.016	3.00	0.72
	$NiCO_3$	4.598		14.723		0.70
Vaterite	$CaCO_3$	7.15		16.94	2.66	
Hydrocalcite	$CaCO_3 \cdot H_2O$	6.09		7.53	2.42	
Nitratine[b]	$NaNO_3$	5.06		16.81		
Niter[b]	KNO_3	5.41	9.17	6.42		

[a]C.I.R. = Cation Ionic Radius in Å.
[b]Included for comparison.

minerals that have crystal structures identical to that of *calcite*. In crystal chemical terms, members of the ***calcite*** group consist of a divalent carbonate radical with a positive bivalent cation, the size of which is about the same as or smaller than Ca. Cations larger than Ca result in the formation of carbonates with the aragonite structure. Table 2 summarizes key crystallographic data for minerals of the ***calcite*** group. Data for *aragonite, vaterite,* and $CaCO_3 \cdot H_2O$ are included for comparison. Note that with the exception of $CoCO_3$ the lattice parameters depend on the ionic size of the divalent cations. Furthermore, Na and K nitrates also are commonly associated with the ***calcite*** and ***aragonite*** groups since they are isostructural with *calcite* and *aragonite*, respectively. Their lattice parameters are also listed in Table 2. The sizes and shapes of the ions are very similar to those of the carbonate group, but the charges are half as large, predicting weaker bonding and less stable structures. This is indeed the case, and the transformations of these substances being at lower temperatures and pressures make their investigation easier.

Optical Properties

The refractive indices for minerals of the ***calcite*** group are listed in Table 3. The exceptionally large birefringence for the ***calcite*** group of minerals results in two distinct images (double refraction) when small, clear cleavage rhombs are placed over any printed material. Furthermore, before the invention of polarizing plastic sheets, *calcite* was almost exclusively used as a polarizer because specially cut material could eliminate completely the second ray of light and, thus, only pass one polarized ray of light. Used in microscopes, these specially prepared *calcite* (Iceland Spar) prisms of optical quality are called *nicols* (see *Polarization and Polarizing Microscope*).

TABLE 3. Refractive Indices for Calcite Group Minerals

Carbonate	ϵ[a]		ω	$\epsilon-\omega$
Calcite	1.486		1.658	-0.172
$MgCO_3$	1.509		1.700	-0.191
$MnCO_3$	1.597		1.816	-0.219
$FeCO_3$	1.635		1.875	-0.240
$CoCO_3$	1.60		1.855	-0.255
$ZnCO_3$	1.621		1.848	-0.227
Vaterite	1.650		1.550	+0.100
$CaCO_3 \cdot H_2O$	1.543		1.590	+0.047
	α	β	γ	$\alpha-\gamma$
Aragonite	1.530	1.682	1.686	-0.156

[a]Refractive indices for monochronatic sodium D light.

The exceptionally high birefringence of the carbonates results from a very strong interaction of the planar carbonate groups with one of the rays of light. The other ray has a much smaller interaction with the large carbonate groups, since its polarization direction is perpendicular rather than parallel to the plane. In minerals of both the ***calcite*** and ***aragonite*** groups the plane of the carbonate groups are perpendicular to the *c* axis. The carbonate groups are parallel to the *c* axis in *vaterite*, which is clearly reflected in its positive birefringence. In $CaCO_3 \cdot H_2O$, the plane of the groups is inclined to the *c* axis, resulting in much smaller positive birefringence.

The variation in refractive index with chemistry can also be seen in Table 3. Because of the similarity of all the structures, there is much replacement (contamination) of the pure end-member carbonates by other cations of the ***calcite*** group. These substitutions can significantly change the various refractive indices, depending on the amount of replacement.

Chemistry

Impurities. There can be much contamination of these carbonates by other ***calcite*** group cations, and also other ***aragonite*** group (Pb,Sr,Ba) cations. Analyses show that there is much more substitution in the former case, than in the latter. Much analytical data are available with excellent summaries in Deer, Howie, and Zussman (1963) and in Graf (1960). A great deal of laboratory work has been carried out to explain impurities in natural carbonates and to deduce the temperature-pressure-environmental composition conditions of their formation. The largest efforts have been directed toward the magnesium carbonate-calcium carbonate system, because of the puzzles surrounding the low-temperature occurrence of *dolomite* [$CaMg(CO_3)_2$] and the formation of magnesian *calcite* in certain marine organisms. Laboratory duplication of assumed natural formation conditions of *dolomite* and magnesian *calcites* have been successfully carried out only with great difficulty using circuitous routes. A beautiful review of these problems can be found in Lippmann (1973).

Thermal-Pressure Characteristics. Diminishing natural supplies of *calcite* has stimulated much work on the synthesis of *calcite* for use as polarizers in precise optical devices. However, an economically feasible technique has still eluded discovery. It seems that the gel approach (McCauley and Roy, 1974) may still be the best. The reasons for the great difficulty encountered in these synthesizing efforts is the low solubility of *calcite* in water ($k=10^{-8.35}$, 25°C, 1 atm) and its disassociation to CaO and CO_2 at high temperature (895°C, 1 atm). For *calcite* to melt into a liquid of identical composition (congruently), a CO_2 pressure of 1025 atm is needed at 1339°C. Wyllie and Tuttle (1960), in a classic paper, investigated these phenomena as related to the formation of carbonatite igneous rocks. Fig. 4 schematically depicts the melting-decomposition relationships of $CaCO_3$ (*calcite*) as a function of temperature and pressure. Note the line for congruent melting of calcite (*CC=L*) extending from the point *M* at about 1339°C and 1025 atm.

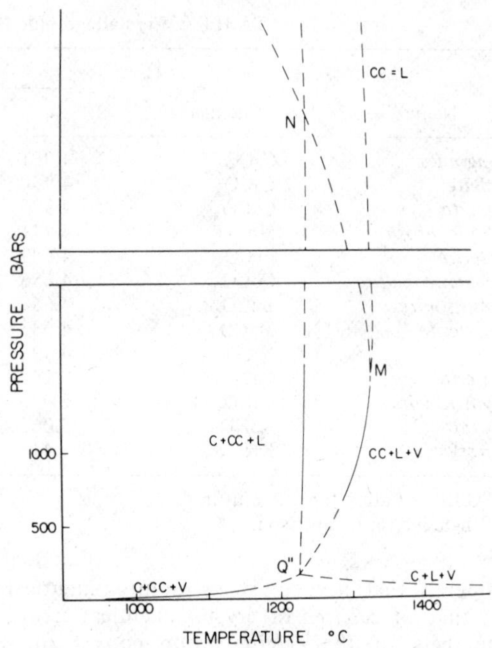

FIGURE 4. Schematic *PT* projection of univariant equilibria in the system $CaO–CO_2$ (from Wyllie and Tuttle, 1960). *C* = CaO, *CC* = $CaCO_3$, *L* = liquid, *V* = vapour. Four univariant curves meet at the invariant point *Q"*.

The solubility of *calcite* in water is essentially controlled by the temperature, CO_2 pressure, and pH of the aqueous system. Fig. 5 illustrates (slightly idealized) these relationships at 1 atm total pressure and 25°C. Also plotted on this figure are the relative concentration of various ionic carbonate species in water as a function of pH. *Calcite's* solubility as a function of temperature and pressure has been determined by Sharp and Kennedy (1965). Their experimental data show that 1) solubility of *calcite* increases with total pressure, 2) solubility at constant total pressure decreases with increasing temperature, and 3) at constant total pressure and temperature, the solubility passes through a maximum as a function of CO_2 concentration.

Miscellaneous Properties

Calcite is also noted for its thermal and ultraviolet induced *luminescence* (q.v.). Recent work (Sommer, 1972) also shows that *calcite* luminesces from the bombardment of electrons or other charged particles. This technique can be used to distinguish the various polymorphs of $CaCO_3$ because of the different colors of

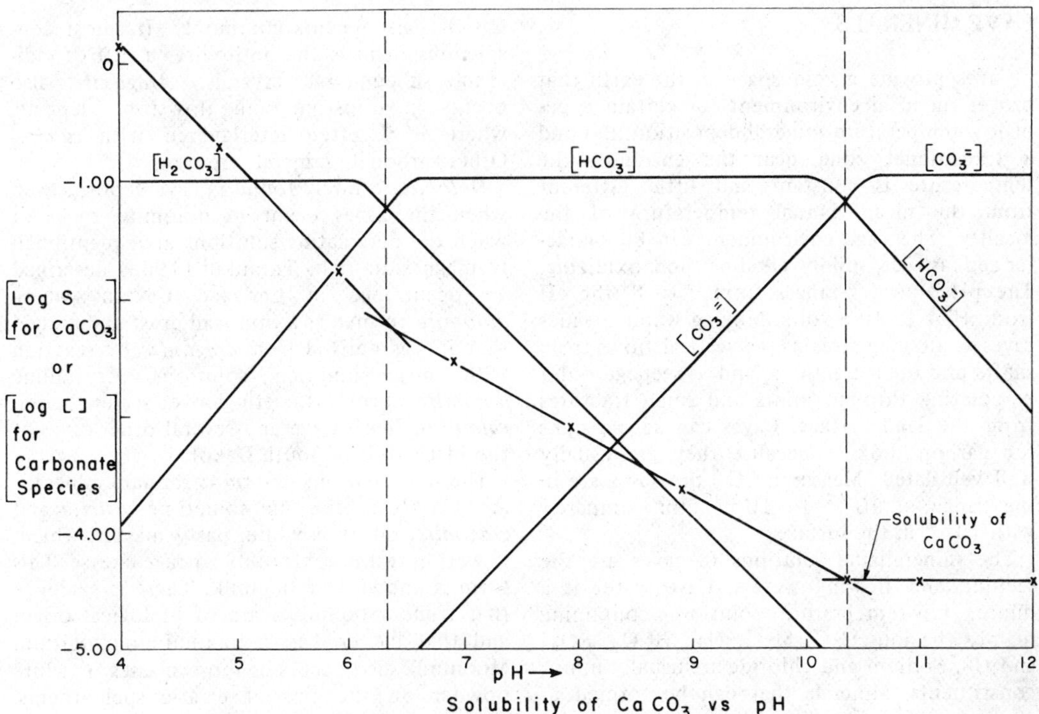

FIGURE 5. Solubility of $CaCO_3$ vs. pH.

luminescence. Manganese is primarily responsible for the ultraviolet luminescence.

JAMES W. McCAULEY

References

Albright, J. N., 1971. Vaterite stability, *Am. Mineralogist*, **56**, 620–624.

Boettcher, A. L., and Wyllie, P. J., 1968. The calcite-aragonite transition measured in the system CaO-CO_2-H_2O, *J. Geol.*, **76**, 314–330.

Bragg, W. L., 1914. The analysis of crystals by the X-ray spectrometer, *Proc. Roy. Soc. (London)*, ser. A, **89**, 468–489.

Bragg, L.; Claringbull, G. F.; and Taylor, H. H., 1965. *The Crystalline State*, vol. IV, *Crystal Structure of Minerals.* Ithaca, N.Y.: Cornell University Press, 490p.

Bridgman, P. W., 1939. The high pressure behavior of miscellaneous minerals, *Am. J. Sci.*, **237**, 7–18.

Deer, W. A.; Howie, R. A.; and Zussman, J., 1963. *Rockforming Minerals*, vol. 5, *Non-Silicates*. London: Longmans, 371p.

Graf, D. L., 1960. Geochemistry of Carbonate Sediments and Sedimentary Carbonate Rocks, parts 1-5. Urbana: Illinois State Geological Survey.

Inoué, S., and Okazaki, K., 1977. Biocrystals, *Scientific American*, **263,** 83–92.

Johannes, W., and Puhan, D., 1971. The calcite-aragonite transition, reinvestigated, *Contr. Mineralogy Petrology*, **31**, 28–38.

Kerr, P. F., 1977. *Optical Mineralogy*, 4th ed. New York: McGraw-Hill, 492p.

*Lippman, F., 1973. *Sedimentary Carbonate Minerals.* Berlin: Springer-Verlag, 228p.

*McCauley, J. W., and Roy, R., 1974. Controlled nucleation and crystal growth of various $CaCO_3$ phases by the silica gel technique, *Am. Mineralogist*, **59**, 947–963.

Palache, C.; Berman, H.; and Frondel, C., 1951. *The System of Mineralogy*, vol. II. New York: Willey, 1124p.

Sharp, W. E., and Kennedy, G. C., 1965. The system CaO-CO_2-H_2O in the two phase region calcite and aqueous solution, *J. Geol.*, **73**, 391–403.

Sommer, S. E., 1972. Cathodoluminescense of carbonates, 1. Characterization of cathodoluminescence from carbonate solid solutions, *Chem. Geol.*, **9**, 257–273.

Strunz, H., 1970. *Mineralogische Tabellen*. Leipzig: Geest and Portig, 621p.

Wyllie, P. J., and Tuttle, O. F., 1960. The system CaO-CO_2-H_2O and the origin of carbonatites, *J. Petrology,* **1**(1), 1–46.

*Extensive reference lists in these works.

Cross-references: *Cave Minerals; Crystal Habits; Human and Vertebrate Mineralogy; Invertebrate and Plant Mineralogy; Luminescence; Polarization and Polarizing Microscope.*

CAVE MINERALS

Caves provide a void space in the earth that proves an ideal environment for certain types of low-temperature mineral deposition. Beyond a transitional zone near the entrance, the temperature is constant and little different from the mean annual temperature of the locality. The cave environment can be characterized as wet, mildly alkaline, and oxidizing. The pH of water ranges from 7 to 8, the eH from +0.4 to +0.6 volts. Moving water invades caves as flowing streams, as vertical flows from shafts and open fractures, and as seepage water percolating through joints and small fractures from the land surface. Caves can act as sinks for carbon dioxide because they are usually well ventilated. Measured CO_2 pressures are in the range of $10^{-2.2}$ to $10^{-2.9}$ atm compared with $10^{-3.5}$ at the surface.

The mineralizing solutions in caves are the seeping and flowing waters. Cave water is a dilute, low-temperature solution, containing mainly the ions Ca^{2+}, Mg^{2+}, Na^+, HCO_3^-, SO_4^{2-}, and Cl^-. Sodium and chloride are usually minor constituents. Minerals that can be formed in caves are, therefore, limited to phases containing these six components. The main suites are carbonates and sulfates and these are treated in the sections that follow. Isolated examples occur in which a specific cave is invaded by ore-forming solutions, by geothermal waters, or by waters with some other exotic chemistry. In such cases, other minerals are formed and the complete list of all minerals observed in caves totals 80 to 90 species (Moore, 1970). This article is restricted to those minerals that form from ordinary carbonate groundwaters and those that result from organic processes in caves.

Carbonate Minerals

Carbonate minerals of calcium and magnesium that occur in caves are listed in Table 1. Most cave $CaCO_3$ is low-magnesium *calcite*. *Aragonite*, the high-pressure polymorph of $CaCO_3$, also occurs commonly. Its most conspicuous form is the anthodite, a tuft of radiating subeuhedral crystals. *Aragonite* also occurs in dripstone and flowstone deposits where it is often interlayered with *calcite*. Other carbonate minerals are rare.

Dolomite is rarely found in cave deposits even when the caves occur in dolomite rocks or when the percolating solutions are exceptionally magnesium-rich. Thrailkill (1968) described an occurrence in Carlsbad Caverns where *dolomite* formed as a thin wall crust and argued that it was derived from *aragonite* by reaction with magnesium-rich solutions. Crystalline *dolomite* occurs as coatings over scalenohedral *calcite* in Jewel Cave and several other caves of the Black Hills of South Dakota.

The hydrated magnesium carbonate minerals, along with *huntite* and sometimes *calcite* and *aragonite*, occur as white, pasty masses, which, if wet, have the texture of cottage cheese. This form is known as moonmilk. There is evidence that some moonmilks are of biological origin and that bacteria assist in their precipitation. Moonmilk also occurs as loose masses of white powder on the tips of erratic speleothems, where it appears to be the magnesium-rich residue left when the solution evaporated (Fischbeck and Müller, 1971). Other hydrated magnesium carbonate minerals, *artinite, lansfordite*, and *dypingite*, have not been reported from caves. Most analyzed cave samples have been collected with few precautions to prevent water loss or to preserve low-temperature forms. Other hydrated magnesium carbonates and hydrated calcium carbonates could be sought in cave deposits with refined collecting techniques.

The mechanism by which *calcite* and *aragonite* are deposited in caves was established by Holland et al. (1964), and most later work (for example Thrailkill, 1971) has confirmed the model. Rainwater falling on the earth has a carbon dioxide pressure roughly in equilibrium with the CO_2 in the atmosphere. In the soil, where it becomes infiltration water and picks

TABLE 1. Carbonate Minerals Found Commonly in Limestone Caves

Mineral	Composition	System and Space Group	Common Forms
Aragonite	$CaCO_3$	Orthorhombic *Pmcn*	Acicular crystals, dripstone
Calcite	$CaCO_3$	Trigonal $R\bar{3}c$	Dripstone, flowstone, erratic forms
Dolomite	$CaCO_3 \cdot MgCO_3$	Trigonal $R\bar{3}$	Rare decomposition product
Huntite	$CaCO_3 \cdot 3MgCO_3$	Trigonal $R32$	Moonmilk
Hydromagnesite	$4MgCO_3 \cdot Mg(OH)_2 \cdot 4H_2O$	Orthorhombic $C222_1$	Moonmilk
Magnesite	$MgCO_3$	Trigonal $R\bar{3}c$	Moonmilk
Nesquehonite	$MgCO_3 \cdot 3H_2O$	Monoclinic $P2_1/n$	Moonmilk

up additional CO_2 generated by organic processes, the pressure may reach 0.1 atm or higher. The details are complex, depending on soil type and thickness, degree of moisture saturation, type of plant cover, and time in the growing season. Most of the reaction between highly acidic soil water and limestone or dolomite bedrock takes place in the contact zone at the base of the soil. If the water comes into complete equilibrium with *calcite*, the dissolved $CaCO_3$ may reach a level of 400 to 500 ppm. There is little additional reaction when the *calcite*-rich water moves downward through tight joints and fractures, because there is no possibility of CO_2 exchange. When the water seeps through the ceilings of cave passages, however, it is highly supersaturated with respect to the ambient CO_2 pressure of the cave atmosphere. CO_2 is outgassed and carbonate minerals are precipitated.

If the deposition rate is slow, only small supersaturations will be achieved and only *calcite* will be precipitated. If, however, outgassing of CO_2 occurs faster than *calcite* deposition, or if the nucleation of *calcite* is inhibited (for example by Mg^{2+}, Sr^{2+}, or SO_4^{2-} in the water), the necessary supersaturation for deposition of *aragonite* can occur and this normally metastable mineral can be precipitated (Curl, 1962).

Crystal growth of *calcite* in the cave environment appears to be a very slow process and requires conditions that remain constant for long periods of time. Speleothems (cave deposits) with the largest and most perfect crystals are typically found in caves without entrances. In caves that are subject to seasonal changes in temperature or moisture or to flooding, speleothems are composed of small crystals often with sand, silt, and other clastic materials trapped on grain boundaries.

The *c*-axis direction in *calcite* is the fast-growth direction. Speleothem shapes are often controlled by it. Straw stalactites may be mosaic single crystals with the optic axis oriented along the straw axis. Larger stalactites may have a core surrounding the central canal with a *c*-axis orientation along the stalactite axis while the bulk of the stalactite consists of wedge-shaped grains with axes perpendicular to the stalactite axis (Moore, 1962a).

The shapes of speleothems are guided by trade-offs between the forces of crystal growth and hydraulic forces from flowing or dripping water (Andrieux, 1962, 1965; Curl, 1972, 1973). Speleothems in their natural environment are some of the most esthetically pleasing objects in the mineral world and they have accordingly picked up a great variety of names. A somewhat simplified classification is listed in Table 2. Speleothems are described in considerable detail in a recently published monograph by Hill (1975).

Sulfate Minerals

The sulfate minerals that have been identified in cave deposits are listed in Table 3. The compositions are as expected from Ca^{2+}, Mg^{2+}, and Na^{+} as the common cations in groundwater. *Gypsum* is the most common of these and is found in many caves of the Appalachian Mountains, the Interior Lowlands, and Southwest USA. *Gypsum* is only slightly water soluble; the other minerals listed in Table 3 are highly soluble and require very dry caves for precipitation and preservation.At nominal ground temperatures of 10°C and at high relative humidities, *epsomite* is the stable magnesium sulfate salt. There is evidence that the critical water-vapor pressure for the decomposition of *epsomite* to *hexahydrite* lies within the range spanned by the cave environment and, as a result, the mineral form changes from *epsomite* to *hexahydite* and back again with the changing seasons. Both *epsomite* and *mirabilite* occur widely in caves. The double salts have been found only in the Flint Mammoth cave system of Kentucky. The anhydrous sulfates *barite* and *celestite* occur sparsely as authigenic crystals within the clastic sediments of the cave floor. *Celestite* has been observed as crusts on cave walls.

Most transport and deposition of sulfate minerals within the cave is apparently a simple matter of solution of the material in percolating water followed by deposition when the solutions evaporate. The sulfates are present in ground water at lower concentrations, and the sulfate minerals are more soluble, than the carbonate minerals. Sulfates growing on the outside of *calcite* speleothems are known, but mixed sulfate-carbonate assemblages have not been observed.

The ultimate origin of the sulfate ion is a difficult question. The following mechanisms for sulfate mobilization and transport have been identified in specific localities:

1. Solution and transport by percolating ground water of evaporites that occur as part of the sedimentary sequence.
2. Hydration of *anhydrite* that occurs within the carbonate rocks.
3. Oxidation of sulfides, mainly *pyrite*, that occurs locally within the cavernous limestone.
4. Oxidation of sulfides from elsewhere in the stratigraphic section and transport of sulfate into the cave by percolating water.

The sulfide oxidation mechanism may be promoted by the action of iron- and sulfur-

TABLE 2. Common Forms of Speleothems

Form	Size Range	Comments
A. Dripstone and flowstone forms (Subaerial, hydraulic control)		
1. Stalactite[a,b,c,d]	cm to meters	Pendulant. Flow through central canal. Result of water dripping from ceilings.
2. Stalagmite[e]	cm to ten meters	Columnar or mound-like. Layered structure. No canal. Result of water dripping onto floor.
3. Drapery	cm to meters	Furled sheets. No canal. Formed by dripping and trickling water from and under ceilings and ledges.
4. Flowstone	Meters to tens of meters	Layered deposit formed from flowing sheets of water.
B. Erratic forms (subaerial, crystal growth control)		
1. Shield[f]	10 cm to 5 meters	Disc-like objects. Medial crack acts as canal. Result of water flow under pressure. *Calcite.*
2. Helictite[g,h]	1 to 20 cm	Curved stalactite-like forms. Central canal. *Calcite* and sometimes *aragonite* mineralogy.
3. Botryoidal Forms[i]	few cm	Bead-, knob- or coral-like objects. Project from walls. Layered structures build on small projections. No canal. *Calcite* or *aragonite* mineralogy.
4. Anthodite	1 to 20 cm	Radiating acicular crystals. Subhedral to euhedral. Sometimes dendritic. *Aragonite* mineralogy.
5. Oulopholite	1 to 50 cm	Curved, radiating bundles of bladed or fibrous crystals. Flower-shaped. *Gypsum, epsomite* or *mirabilite* mineralogy.
6. Moonmilk[i,j,k]	–	Loose powder, or wet, pasty mass. *Calcite, aragonite,* or hydrated carbonate minerals.
C. Subaqueous forms		
1. Rimstone dams	meters to tens of meters	Natural dams of travertine in cave streams and pools. Mainly *calcite.*
2. Concretions[l]	mm to cm	Concentrically layered, unattached structures. Mainly *calcite.* Also called cave pearls.
3. Crystal linings	cm to many meters	Crystal coatings in pools and entire caves deposited from standing water. Scalenohedral *calcite* the most common crystal.

For more detail, particularly on mechanisms of development see following references:

[a]Hicks (1950) [b]Moore (1962a) [c]Curl (1972) [d]Andrieux (1962)
[e]Curl (1973) [f]Moore (1962b) [g]Moore (1954) [h]Andrieux (1965)
[i]Thrailkill (1971) [j]Fischbeck and Müller (1971) [k]Mélon and Bourguignon (1962) [l]Baker and Frostick (1947,1951)

consuming bacteria, notably *Thiobacillus thiooxidans* and *Ferrobacillus ferrooxidans*. The iron from the sulfate is immobilized as insoluble iron hydroxide but the sulfate can be transported as sulfuric acid. The reaction of sulfuric acid with limestone is dependent on carbon dioxide pressure. Build-up of CO_2 can stop the reaction and, for this reason, the sulfate-bearing solutions can percolate relatively long distances through the limestone without being consumed. The open cave passages act as a sink for carbon dioxide, permitting the reaction to proceed and the sulfate minerals to be deposited by direct replacement of *calcite* by *gypsum* in situ.

Gypsum and the other sulfates are monoclinic and their crystallization habit leads to forms of speleothems different from the carbonate minerals. Sulfate-bearing solutions percolate slowly and the volume of fluid is small. As a result, the form of the deposit is dominated by crystal-growth forces; hydraulic controls are rarely in evidence. Stalactites of *gypsum* do occur, usually as very porous masses.

TABLE 3. Sulfate Minerals Found Commonly in Limestone Caves

Mineral	Composition	System and Space Group	Common Forms
Barite	$BaSO_4$	Orthorhombic *Pnma*	Authigenic, in cave soils
Celestite	$SrSO_4$	Orthorhombic *Pnma*	Authigenic, in cave soils, wall crusts
Thenardite	Na_2SO_4	Orthorhombic *Fddd*	Decomposition product of *mirabilite*
Gypsum	$CaSO_4 \cdot 2H_2O$	Monoclinic *A2/a*	Crusts, massive crystals, flowers
Epsomite	$MgSO_4 \cdot 7H_2O$	Orthorhombic $P2_1 2_1 2_1$	Stalactite, flowers, effluorescences
Hexahydrite	$MgSO_4 \cdot 6H_2O$	Monoclinic *C2/c*	Effluorescenses
Mirabilite	$Na_2SO_4 \cdot 10H_2O$	Monoclinic $P2_1/a$	Stalactites, flowers
Bloedite	$MgSO_4 \cdot Na_2SO_4 \cdot 4H_2O$	Monoclinic $P2_1/a$	Crusts
"Labile salt"	$CaSO_4 \cdot 2Na_2SO_4 \cdot 2H_2O$	Unknown	Minor phase in *mirabilite* stalactites

Stalactites of *epsomite* and *mirabilite* are often transparent and as water-clear as icicles. The most common form of all sulfate minerals is wall crusts of granular crystals. *Gypsum* occurs as long bladed crystals that may reach lengths of a meter or more and may range in size down to thin fibers fractions of a millimeter in cross section. *Gypsum* blades and needles are usually twinned, the butterfly twin being most common. *Gypsum, epsomite*, and *mirabilite* occur as masses of curved, petal-like crystals known as "*gypsum* flowers."

Other Minerals

Among the halide minerals, *halite* has been confirmed as a cave deposit in the Mullamulang Cave in arid Western Australia where it is found as curved petals and stalactite forms.

Quartz occurs as crystal linings in a few caves that have been invaded by hydrothermal waters, and as an authigenic mineral in some cave fills. Perennial *ice* occurs in high-altitude and high-latitude caves and should be regarded as a cave mineral. Most common among the oxide minerals are iron and manganese oxide hydrates. Stalactites of limonite occur as well as minor limonite and *goethite* crusts. The increase in alkalinity when surface streams sink into limestone caves causes Mn^{2+}, in solution as a trace constituent, to oxidize with the precipitation of poorly crystallized and poorly characterized black coatings on the beds of many cave streams. Those analyzed appear to be Ca- or Ba-rich so that they would be classified as *todorokite*, psilomelane, or *hollandite*, but their crystalinity is so poor that they are amorphous to X-rays and the crystal structures have not been identified.

Nitrate minerals, either *nitrocalcite, niter* or *nitratine* occur in the soils of dry caves where they are often mixed with water-soluble sulfate minerals. The extraction of cave saltpeter (*niter*) for the manufacture of gun powder was an important operation in the War of 1812 and in the Civil War. Despite arguments about organic sources (guano), and bacteriological fixation of nitrogen, the mechanism of formation of these minerals remains uncertain.

TABLE 4. Phosphate and Related Organic Minerals Found in Caves

Mineral	Composition	System and Space Group	
Whitlockite	$Ca_3(PO_4)_2$	Trigonal	$R\bar{3}c$
Monetite	$CaHPO_4$	Triclinic	$P\bar{1}$
Brushite	$CaHPO_4 \cdot 2H_2O$	Monoclinic	*I2/a*
Hydroxyapatite	$Ca_5(PO_4)_3(OH)$	Hexagonal	$C6_3/m$
Newberyite	$MgHPO_4 \cdot 3H_2O$	Orthorhombic	*Pbca*
Struvite	$NH_4MgPO_4 \cdot 6H_2O$	Orthorhombic	$Pm2_1n$
Phosphammite	$(NH_4)_2HPO_4$	Monoclinic(?)	
Biphosphammite	$NH_4H_2PO_4$	Tetragonal	$I\bar{4}2d$
Crandallite	$CaAl_3(PO_4)_2(OH)_5 \cdot H_2O$	Trigonal	
Taranakite	$3(K,Na,NH_4,Ca)_2O \cdot 5(Al,Fe)_2O_3 \cdot 7P_2O_5 \cdot 43H_2O$	Trigonal	
Weddellite	$CaC_2O_4 \cdot 2H_2O$	Tetragonal	*I4/m*
Urea	$CO(NH_2)_2$	Tetragonal	$P42_1m$
Uricite	–	–	–
Guanine	$C_5H_3(NH_2)N_4O$	–	–

The guano excreted by bats and birds that use caves as roosting and nesting places in tropical regions provides a source of organic materials for complex mineralization processes. Crystallization from water and urine in the guano itself produces a number of organic minerals. Reaction of leach solutions percolating through the guano with the underlying wall rock produces a complex suite of calcium, magnesium, and ammonium phosphate minerals. There is also a reaction between the leach solutions and the clay minerals present in the clastic fills of the cave, which produces several complex aluminum phosphate minerals. A partial listing is given in Table 4. Few detailed studies of guano mineralogy have been made. The best known are the caves of Western Australia (Bridge, 1973a,b; 1974), the guano deposits of Mona Island caves (Kaye, 1959) and Pig Hole Cave, Virginia (Murray and Dietrich, 1956).

WILLIAM B. WHITE

References

Andrieux, C., 1962. Etude cristallographique des édifices stalactitiques, *Bull. Soc. fr. Minéral. Cristallogr.*, **85**, 67–76.

Andriex, C., 1965. Morphogenese des hélicites monocristallines, *Bull Soc. fr. Minéral. Cristallogr.*, **88**, 163–171.

Baker, George, and Frostick, A. C., 1947. Pisoliths and ooliths from some Australian caves and mines, *J. Sed. Petrology*, **17**, 39–67.

Baker, George, and Frostick, A. C., 1951. Pisoliths, ooliths and calcareous growths in limestone caves at Port Campbell, Victoria, Australia, *J. Sed. Petrology*, **21**, 85–104.

Bridge, P. J., 1973a. Urea, a new mineral, and neotype phosphammite from Western Australia, *Mineralog. Mag.*, **39**, 346–348.

Bridge, P. J., 1973b. Guano minerals from Murra-elelevyn Cave, Western Australia, *Mineralog. Mag.*, **39**, 467–469.

Bridge, P. J., 1974. Guanine and uricite, two new organic minerals from Peru and Western Australia, *Mineralog. Mag.*, **39**, 889–890.

Curl, Rane L., 1962. The aragonite-calcite problem, *Bull. Natl. Speleol. Soc.*, **24**, 57–73.

Curl, Rane L., 1972. Minimum diameter stalactites, *Bull. Natl. Speleol. Soc.*, **34**, 129–136.

Curl, Rane L., 1973. Minimum diameter stalagmites, *Bull. Natl. Speleol. Soc.*, **35**, 1–9.

Fischbeck, Reinhard, and Müller, German, 1971. Monohydrocalcite, hydromagnesite, nesquehonite, dolomite, aragonite, and calcite in speleothems of the Frankische Schweiz, Western Germany, *Contr. Mineralogy Petrology*, **33**, 87–92.

Hicks, Forrest L., 1950. Formation and mineralogy of stalactites and stalagmites, *Bull. Natl. Speleol. Soc.*, **12**, 63–72.

Hill, Carol A., 1976. *Cave Minerals*: Huntsville, Alabama: National Speleological Society, 137p.

Holland, Heinrich D., et al., 1964. On some aspects of the chemical evolution of cave waters, *J. Geol.*, 72, 36–67.

Kaye, Clifford A., 1959. Geology of Isla Mona, Puerto Rico, and notes on the age of Mona Passage, *U.S. Geol. Surv. Prof. Paper 317-C*, 141–178.

Mélon, J., and Bourguignon, P., 1962. Etude du Mondmilch de quelque grottes de Belgigue, *Bull. Soc. fr. Mineral. Cristallogr.*, **85**, 234–241.

Moore, George W., 1954. The origin of helictites, *Natl. Speleol. Soc. Occ. Paper No. 1*, 16p.

Moore, George W., 1962a. The growth of stalactites. *Bull. Natl. Speleol. Soc.*, **24**, 95–106.

Moore, George W., 1962b. Role of earth tides in the formation of disc-shaped cave deposits, *Proc. 2nd Internat. Cong. Speleol.*, **1**, 500–506.

Moore, George W., 1970. Checklist of cave materials, *Natl. Speleol. Soc. News*, **28**, 9–10.

Murray, John W., and Dietrich, Richard V., 1956. Brushite and taranakite from Pig Hole Cave, Giles County, Virginia, *Am. Mineralogist*, **41**, 616–626.

Thrailkill, John, 1968. Dolomite cave deposits from Carlsbad Caverns, *J. Sed. Petrology*, **38**, 141–145.

Thrailkill, John, 1971. Carbonate deposition in Carlsbad Caverns, *J. Geol.*, **79**, 683–695.

CHAIN SILICATES—*See* AMPHIBOLE GROUP; PYROXENE GROUP. Vol. IVA: MINERAL CLASSES: SILICATES

CHLORITE GROUP

The ***chlorite*** group of minerals derives its name from the common green color of most varieties, although specimens of many diverse colors are known to exist. The morphology ranges from fine-grained earthy masses to spherules and rosettes to scaly flakes and pseudohexagonal platelets. There is a perfect basal cleavage parallel to (001). The cleavage flakes are flexible but inelastic, with a luster varying from pearly or vitreous to dull and earthy. The hardness on the cleavage is about 2½. The specific gravity varies between 2.6 and 3.3 as a function of composition.

Chlorite is a common accessory mineral in low- to medium-grade regional metamorphic rocks, and may be the most abundant mineral in metamorphic rocks of the chlorite zone. It is an occasional constituent of igneous rocks, in most cases probably forming secondarily by deuteric or hydrothermal alteration of primary ferromagnesian minerals. ***Chlorite*** is found in pegmatites and fessure vein deposits. It is a common constituent of altered basic rocks and of hydrothermal alteration zones around ore bodies. In sedimentary rocks, ***chlorite*** is a common, but usually minor, component.

Structure

The crystal structure of ***chlorite***, determined by Pauling (1930), consists of a negatively charged 2:1 layer, similar to that found in most *phyllosilicates* (q.v.), which is compensated electrostatically by a positively charged hydroxide interlayer sheet (Fig. 1). The 2:1 layer is composed of two pseudohexagonal sheets of interlinked tetrahedra that are inverted relative to one another and joined together so that the tetrahedral apical oxygens plus hydroxyl groups at the center of each pseudohexagonal ring provide octahedral coordination around medium-sized cations at the median plane of the layer. The interlayer sheet consists of medium-sized cations that are octahedrally coordinated between two planes of hydroxyl groups. This interlayer sheet must be positioned so that the surface OH groups are linked by long hydrogen bonds of approximate length 3.0 Å to tetrahedral basal oxygens on the surfaces of the 2:1 layers above and below. The interlayer sheet is bonded to the 2:1 layers in addition by electrostatic bonds resulting from appreciable cation substitutions within both the 2:1 layer and the interlayer.

Composition

Most ***chlorites*** are trioctahedral and contain close to 3.0 cations per formula unit in both the octahedral sheet within the 2:1 layer and in the octahedral interlayer sheet. The composition of the 2:1 layer can be expressed as

$$(R^{2+}_{3-a-b-c}R^{1+}_{a}R^{3+}_{b}\square_{c})(Si_{4-x}R^{3+}_{x})O_{10}(OH)_2$$

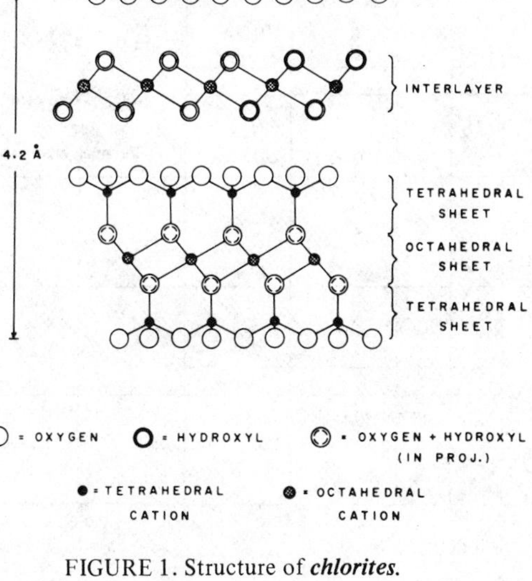

FIGURE 1. Structure of ***chlorites***.

and that of the interlayer as

$$(R^{2+}_{3-d-e-f}R^{1+}_{d}R^{3+}_{e}\square_{f})(OH)_6$$

Tetrahedral substitutional of R^{3+} for Si is primarily by Al^{3+}, occasionally by Fe^{3+} or Cr^{3+}. The main constituents of the 2:1 octahedral sheet and of the interlayer octahedral sheet are Mg, Fe^{2+}, Fe^{3+}, and Al, but with important substitutions of Cr, Ni, Mn, V, Cu, or Li in certain species. The negative charge (x) resulting from substitutions in the tetrahedral sheets can be increased by substitution of R^{1+} or of vacancies (□) for R^{2+}, but is more likely to be decreased by substitution of R^{3+} for R^{2+}. Most of the negative charge (x), however, is neutralized by a positive charge on the interlayer sheet that results from substitution of R^{3+} for R^{2+}. The composition of the entire structural unit can be expressed as

$$(R^{2+}_{6-w-y-z}R^{1+}_{w}R^{3+}_{y}\square_{z})(Si_{4-x}R^{3+}_{x})O_{10}(OH)_8$$

where $x = y - w - 2z$.

Foster (1962), in a study of 150 chlorite analyses, found a complete substitutional series between Mg and Fe^{2+}. In most cases, octahedral Al is low and other trivalent cations, such as Fe^{3+} or Cr^{3+}, are present to proxy for Al in balancing the negative charges on the tetrahedral sheets. If the total number of trivalent octahedral cations is approximately equal to tetrahedral R^{3+}, the octahedral occupancy is close to 6.0 atoms. If the total number of trivalent octahedral cations is greater than tetrahedral R^{3+}, Foster showed that total octahedral occupancy then is less than 6.0 cations by an amount equal to one half the excess of octahedral R^{3+} over tetrahedral R^{3+}. Thus, the excess octahedral trivalent cations replace divalent cations in the ideal formula in the ratio 2:3. The range of tetrahedral substitution found in Foster's study was 0.55 to 1.66 R^{3+} atoms per four tetrahedral sites.

The amount of Fe^{3+} present bears no necessary relation to the amount of Fe^{2+} in a particular specimen, but there is a general trend for Fe^{2+}-rich ***chlorites*** to contain more Fe^{3+} than do the Fe^{2+}-poor chlorites. In most cases, this Fe^{3+} must be considered primary unless there are definite indications of oxidation. Foster (1964) found many Fe-rich ***chlorites*** to be (OH)-deficient according to analysis. There is little correlation between ferric iron and excess oxygen in these analyses, however, and internal oxidation and dehydration will explain (OH) deficiency in only some of the Fe-***chlorites***.

Stacking

The combination of a 2:1 layer plus an interlayer measures about 14.2 Å in thickness. The tetrahedral and octahedral networks have pseudohexagonal symmetry and usually are described by an orthohexagonal lattice of dimensions a=5.1–5.3 Å and $b=a\sqrt{3}$=8.9–9.2 Å. The stacking sequence of adjacent layers reduces the resultant symmetry to monoclinic or triclinic in most cases, even if the unit cell has orthorhombic shape. In clay mineral analyses, ***chlorite*** is differentiated from ***smectite*** and ***vermiculite*** of similar layer-plus-interlayer thickness by lack of interlayer collapse on heating, lack of interlayer expansion upon solvation with glycerol or ethylene glycol, and a severalfold enhancement of intensity of the 14 Å basal X-ray reflection on heating to 450–550°C.

Bailey and Brown (1962) have shown that the main structural variations in ***chlorite*** involve the position of the interlayer sheet relative to the 2:1 layer. Firstly, the interlayer octahedra can be oriented either parallel to the octahedra in the 2:1 layer (orientation I), or antiparallel (orientation II) as shown in Fig. 1. Secondly, for either orientation, the interlayer can be positioned in two different ways to achieve hydrogen bonds with the 2:1 layer below—position *a*, in which interlayer cations superimpose directly over tetrahedral cations in the sheet immediately below, and position *b*, in which these cations do not superimpose. Thus, there are four structural types of 14 Å units that can be termed I*a*, I*b*, II*a*, and II*b*. The next 2:1 layer then can be positioned in six different ways above the interlayer to achieve optimum hydrogen bonding.

Single-crystal X-ray studies have shown that most ***chlorites*** either have regular one-layer stacking sequences or irregular sequences in which successive layers adopt at random any of three superpositions separated by $\pm b/3$ that provide hydrogen bonds with the interlayer. X-ray patterns of the latter are characterized by sharp $k=3n$ reflections and continuous or partial streaking of $k \neq 3n$ reflections. Bailey and Brown have shown that 12 regular one-layer structures and 6 irregular or semirandom sequences theoretically are possible. The latter can be identified by X-ray powder patterns, and 4 of the 6 possible semirandom sequences have been recognized to date. These structural types cut across the arbitrary compositional boundaries that have been used to define ***chlorite*** species in the past. The II*b* structural type is especially common in nature and is believed to be the most stable arrangement. It is predominant in chlorite-grade metamorphism and in geologic environments in which sufficient thermal energy is available to produce the stable form. The I*a*, I*b* (β=90°), and I*b* (β=97°) structures are believed to crystallize metastably at lower temperatures.

Classification

Foster's (1962) classification scheme for trioctahedral ***chlorites*** is illustrated in Fig. 2, where the main compositional species are differentiated by tetrahedral Si/Al and octahedral Mg/Fe ratios. Because composition has been

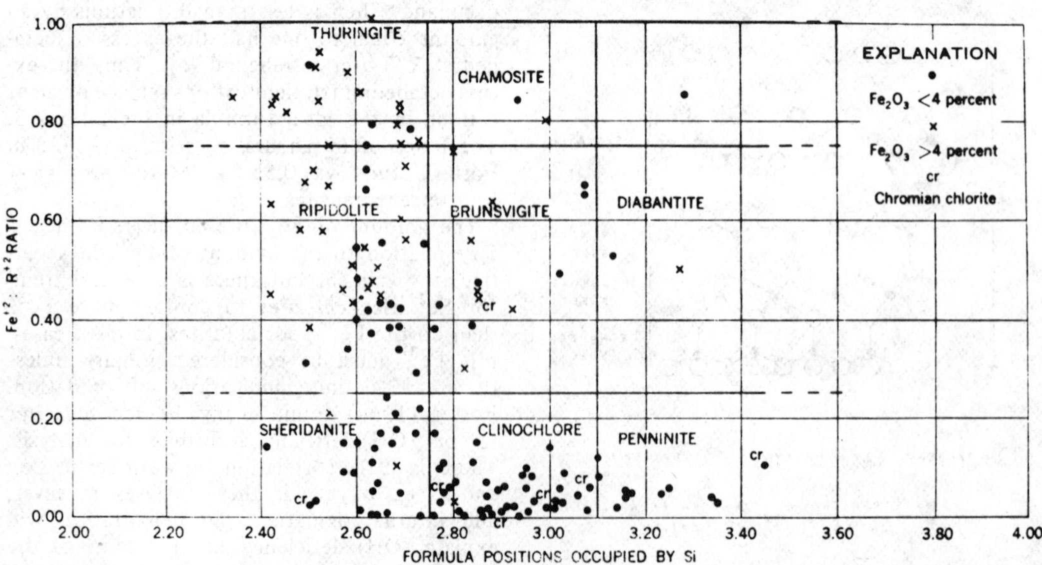

FIGURE 2. Classification of ***chlorites*** (from Foster, 1962).

shown to have little influence on the structural arrangement of layers, Bayliss (1975) has recommended that most of the species names shown in Fig. 2 be discarded. There now is international agreement that the species names for trioctahedral ***chlorites*** be determined by the dominant divalent cation present. Approved names are *clinochlore* (Mg dominant), *chamosite* (Fe^{2+} dominant), *pennantite* (Mn^{2+} dominant), and *nimite* (Ni dominant). Significant cationic substitutions can be indicated by adjectival modifers, e.g. ferrian *clinochlore* or aluminian *chamosite*. A few ***chlorites*** are known in which one or both octahedral sheets are dioctahedral, i.e., have close to 2.0 rather than 3.0 octahedral cations per formula unit. Such ***chlorites*** are highly aluminous. Examples include *cookeite*, with a dioctahedral 2:1 layer and a trioctahedral Al-Li interlayer; *sudoite*, with a dioctahedral 2:1 layer and a trioctahedral Al-Mg interlayer; and *donbassite* with two dioctahedral sheets. In the latter case, the positive charge on the interlayer arises primarily from Al cations in excess of 2.0 in accord with the formula $Al_{2+x/3}(OH)_6$ for layer charge (x).

S. W. BAILEY

References

Bailey, S. W., and Brown, B. E., 1962. Chlorite polytypism: I. Regular and semi-random one-layer structures, Am. *Mineralogist*, **47**, 819–850.

Bayliss, P., 1975. Nomenclature of the trioctahedral chlorites, *Can. Mineralogist*, **13**, 178–180.

Foster, Margaret D., 1962. Interpretation of the composition and a classification of the chlorites, *U.S. Geol. Surv. Prof. Paper 414-A*, 33p.

Foster, Margaret D., 1964. Water content of micas and chlorites, *U.S. Geol. Surv. Prof. Paper 474-F*, 15p.

Pauling, L., 1930. The structure of the chlorites, *Proc. Natl. Acad. Sci.*, **16**, 578–582.

Cross-references: *Clays, Clay Minerals; Diagenetic Minerals; Green River Mineralogy; Mica Group; Phyllosilicates; Pleochroic Halos; Topotaxy;* also, see mineral glossary.

CLASSIFICATION OF MINERALS—*See* MINERAL CLASSIFICATION: HISTORY; MINERAL CLASSIFICATION: PRINCIPLES

CLAYS, CLAY MINERALS

The term *clay* has no single and universally accepted definition. Clays occur as rock-forming materials and in soils, and may constitute an entire member of a rock formation or amount to no more than a small fraction filling cracks or acting as a cement between larger particles. Clays are characterized primarily by their small particle size, which usually is taken as less than 2 μm. Coarse, medium, and fine clays have size ranges about 2–0.5, 0.5–0.2, and below 0.2 μm, respectively. The effective size is determined by the rate of settling in water or by direct electron-microscope measurements. On this basis, any material ground to less than a 2-μm particle size becomes a clay.

In ceramic practice, emphasis is placed also on the plastic properties developed when clays are suitably crushed and mixed with water. Most, but not all, clays are composed of platy particles (Fig. 1) with a large ratio of surface area to mass (ranging from about 10 m^2/g to several hundred m^2/g). Water absorbed on these surfaces causes the particles to adhere and also to slide over one another when a clay mass is deformed. Not all ceramic clays have or develop these plastic properties; a flint clay, for example, is a hard, flint-like material in which fine platy crystals are firmly cemented so that plastic properties are not readily developed.

Clay Minerals

A wide variety of minerals, characterized by chemical composition and crystal structure, occur in clays. Some are similar to or even identical with macroscopically developed minerals while others occur mainly as clay-size particles. They are studied by a wide variety of physical and chemical methods and with modern methods of analysis they are known with the same degree of detail as larger-size minerals.

Generally, the first step in studying the mineralogy of a clay material is the separation of the clay fraction from coarser material either by sedimentation in water or by centrifugation after the material has been disaggregated. A spherical particle of density 2.65 g/cm^3 and diameter D = 2 μm settles 5 cm in water at 20°C in about 4 hr. Since the settling time for a given distance varies as $1/D^2$, smaller particles take correspondingly longer times and vice versa. Often the main problem in separating the clay fraction is disaggregating the material. Physical methods such as grinding and sieving, vigorous agitation in water by ultrasonic and other methods, and various chemical procedures are used (see Jackson, 1958).

Clay minerals are principally silicates of aluminum, iron, and magnesium and belong to the phyllosilicate (or layer silicate) family of minerals. The crystallinity of the minerals, as indicated by their form seen in the electron microscope and by their crystal (atomic) struc-

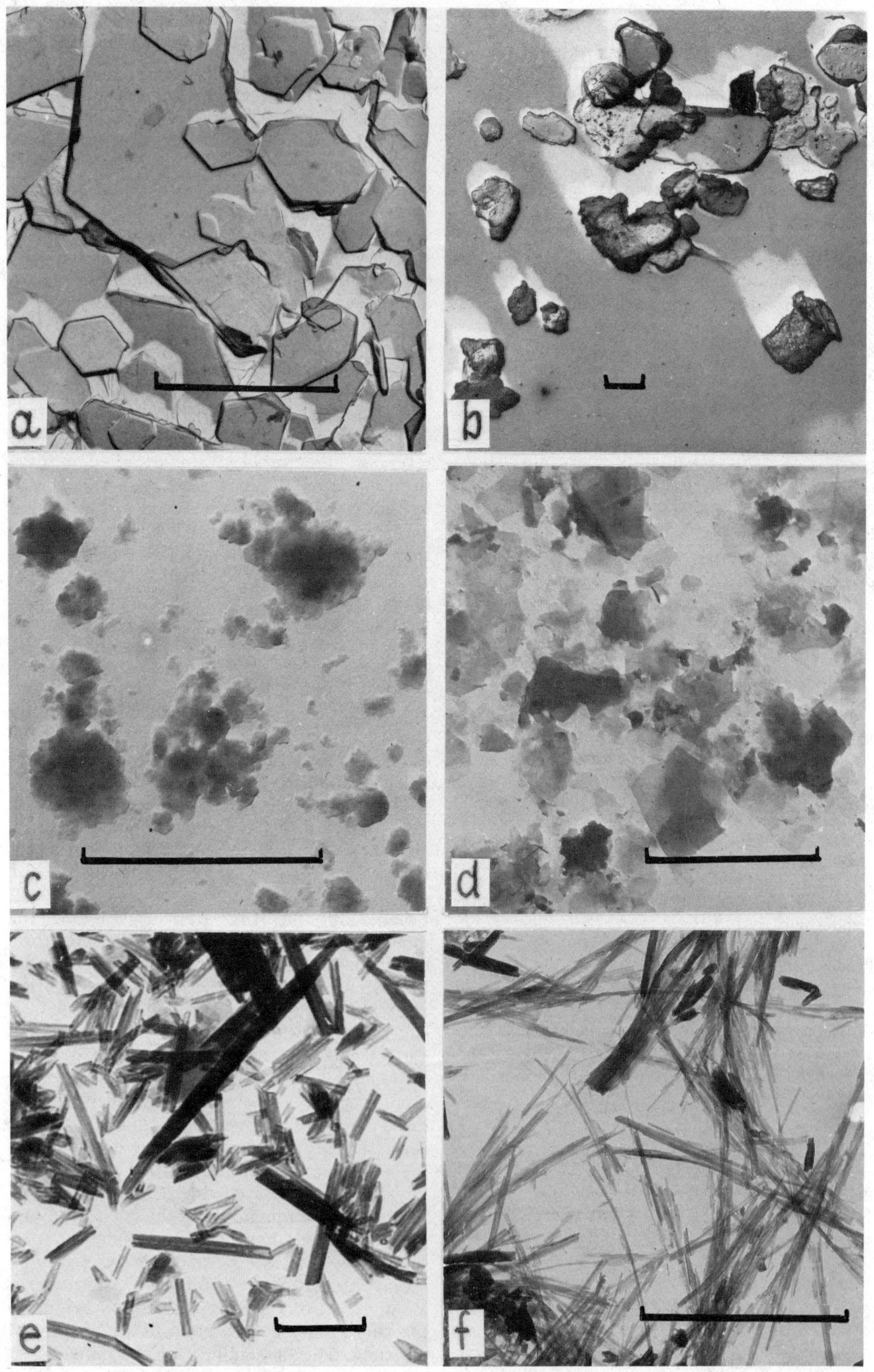

FIGURE 1. Electron micrographs of some clay minerals. (a) *Kaolinite,* of a type used for paper coating; Pt-shadowed, carbon replica. (b) *Kaolinite,* particles with blocky texture; Pt-shadowed, carbon replica. (c) *Illite.* (d) *Montmorillonite.* (e) *Halloysite.* (f) *Sepiolite.* One-μm length marked on each photograph. (Photographs courtesy of Mineral Constitution Laboratory, Pennsylvania State University.)

ture indicated by X-ray diffraction analysis, vary greatly from well-crystallized forms to poorly crystallized and structurally disordered forms and to amorphous materials. Clay minerals are classified in groups according to crystal structure (determined by X-ray methods) and to chemical composition.

The principal clay minerals are as follows: The ***kaolinite-serpentine*** group includes the ***kaolinite*** minerals of which *kaolinite* and *halloysite* are the most commonly occurring and the ***serpentine*** minerals, which include *lizardite* and the fibrous mineral *chrysotile*. The clay ***micas***, commonly called ***illites***, are similar to *muscovite* and *biotite* in structure and composition. The ***smectites***, of which *montmorillonite* is the most important, are extremely fine grained and have important swelling properties. ***Vermiculites*** and ***chlorites*** also occur with clay-grade particle size. A wide range of ordered and disordered interstratified minerals are found in which layers of different kinds are combined. *Allophane* is an amorphous or nearly amorphous clay and *imogolite*, a more recently recognized clay mineral, may be an early stage of crystallization of *allophane*.

The individual particles of these minerals are mainly of platy form and this morphology is clearly related to their layer-type crystal structures. The plates may develop, however, to form elongated laths and ribbonlike particles. The layers may curl and develop fibrous, rodlike, or even tubular forms and spheroidal forms. Under appropriate conditions, ***smectites*** and ***vermiculites*** can expand in water and salt solutions so that the individual structural layers become widely separated and the material passes progressively from a microcrystalline to a gel state.

Methods of Studying Clays

The crystal structures of clay minerals are studied by X-ray powder diffraction and by electron-diffraction methods. The structures are now, for the most part, well understood. Bailey (1980) has given a detailed survey of the ordered structures and Brindley (1980) of the disordered structures. X-ray diffraction is the principal method for identification of clay minerals, together with relevant thermal and chemical tests, and also for their quantitative mineralogical analysis (Brindley and Brown, 1980).

Clays are ideally suited for electron microscope examination (Fig. 1). Compilations of electron micrographs have been given by Beutelspacher and van der Marel (1968) and by Gard (1971). Combined electron-microscope and electron-diffraction data enable external form and internal structure to be correlated. Electron diffraction analysis of clay minerals has been treated in detail by Zvyagin (1967). The development of electron imaging techniques can be expected to lead to more detailed understanding of structural imperfections in clays. Important thermal effects occur when clays are heated; these can be observed by thermal methods of analysis, especially thermogravimetric and differential thermal analysis (Mackenzie, 1957). Infrared spectrometry is particularly useful for studying the hydroxyl groups in clay minerals and the organic complexes formed by ***smectites*** and ***vermiculites*** (Farmer, 1974).

Other methods include chemical procedures for purification of clay minerals; for determinations of their ion-exchange capacities, particularly their cation-exchange capacities; and for the measurement of surface areas by gas-adsorption and liquid-adsorption techniques (Jackson, 1958).

In many applications of clay mineralogy (Grim, 1962), the colloidal properties of clay-water and clay-fluid systems are of very great importance (Van Olphen, 1963).

Kaolinite Group Minerals

Minerals of the ***kaolinite*** group are 1:1 type phyllosilicates. Their chemical compositions are close to $Al_2Si_2O_5(OH)_4$ and the three principal polytypes are *kaolinite-1Tc, dickite-2M*, and nacrite-*2M* (see *Phyllosilicates* for structures). The thickness of the unit layer, basal spacing, is 7.15 Å. *Kaolinite* is by far the commonest of these minerals and is the major component of "china clay," "ball clay," "fire-clay," etc. It varies widely in degree of crystal-structure perfection and in regularity of particle form (Fig. 1). Many *kaolinites*, particularly those formed hydrothermally, exhibit a more or less hexagonal crystal form and have well-crystallized structures, giving a sharp and many-lined X-ray powder diagram. Other *kaolinites* show poorly outlined crystals and have less perfect internal structure, giving broadened and weakened X-ray diffractions. The results have been interpreted in terms of random displacements of the component layers, and of misfitting domain structures.

Halloysite, a mineral of the ***kaolinite*** group, exists in two forms. One, with composition $Al_2Si_2O_5(OH)_4$, has a basal spacing of about 7.2 Å, in agreement with the other ***kaolinite***-group minerals. A second, more hydrous form with composition $Al_2Si_2O_5(OH)_4 \cdot 2H_2O$, has an expanded spacing of 10.1 Å due to a single sheet of water molecules between the 1:1 aluminosilicate layers, (see *Phyllosilicates*, Fig.

3e). The "extra" water is removed, more or less irreversibly, at temperatures as low as about 75°C. X-ray diffraction data indicate that *halloysites* have a high degree of layer disorder. Electron micrographs show fiber-like forms (Fig. 1e), many of which seem to be tubular; rolled forms showing the edge of the rolled structure have been recorded. Electron-diffraction data on single particles of *halloysite*, by several investigators, have indicated a higher degree of crystalline order and have shown that some *halloysites* possess a characteristic two-layer structure different from the structures of *dickite* and *nacrite* and that the particles appear to be prism shaped rather than tubular (Honjo et al., 1954; Chukhrov and Zvyagin, 1966).

For many years, *halloysite* was the only ***kaolinite*** group mineral recognized as also existing in a hydrous, expanded form. The expansion could be varied by introducing polar organic molecules such as methanol, ethylene glycol, acetone, etc., either in place of, or in addition to, the water molecules. Complexes have been formed also with many inorganic salts such as halides and acetates of potassium (K), ammonium (NH_4), cesium (Cs), etc., with basal spacings up to about 14 Å. This led to the discovery that *kaolinite, dickite*, and *nacrite* also can be expanded to about 14.1 Å by treatment with concentrated potassium acetate solutions. Thorough washing with water removes the salt, leaving a hydrated complex which varies with the particular ***kaolinite*** mineral. It is not yet fully understood how these complexes are formed but hydrogen bonding seems to be an important feature. Other molecules capable of forming hydrogen bonds, such as urea, H_2NCONH_2, formamide, $HCONH_2$, and hydrazine, N_2H_4, will penetrate between the layers of *kaolinite* (Weiss et al., 1966).

Smectites

The ***smectite*** group of minerals is also named the ***montmorillonite-saponite*** group after the principal di- and trioctahedral minerals involved. The shorter name is generally preferred. The basic structure consists of layers of the 2:1 type similar to those in ***micas*** but with smaller negative charges, about 0.3–0.6 per formula unit as compared with 1.0 for the ideal mica composition (see *Phyllosilicates*). The number of interlayer cations is correspondingly smaller as is the ionic bonding of the layers. These features give rise to the two most important properties of ***smectites***, namely (a) their intracrystalline swelling in water and in many organic media and (b) their high cation-exchange capacities. Structurally, the ***smectites*** are almost always turbostratic, i.e., the layers are stacked in a disordered way so that no three-dimensional X-ray diffractions are obtained, apart from the special case of the basal, 00*l* reflections. The minerals *beidellite* and *saponite*, however, exhibit a tendency towards three-dimensional regularity.

For pure ***smectites***, the cation exchange capacity is related directly to the interlayer cations (apart from any "edge" effects) and can be expressed in monovalent cations per unit cell, or per formula unit, but for impure materials (and most clay materials are impure) it is expressed in milliquivalents per 100 g dry clay. Various methods are used for measuring the exchange capacity: Sodium-saturated material is calcium-saturated by treatment with aqueous solutions of $CaCl_2$. After careful washing to remove excess $CaCl_2$, the calcium in the clay is exchanged for sodium with *N* NaCl solution and the displaced Ca is measured by a titration procedure. A *montmorillonite* with 0.33 monovalent cation per formula unit (see later) has a calculated exchange capacity of 92 meq/100 g dry clay. Experimental values for ***smectites*** are commonly in the range 80–120 meq/100 g clay.

The compositions of the principal ***smectites*** are given in Table 1. The di- and trioctahedral compositions can be regarded as derived from *pyrophyllite* and *talc*, respectively. In these formulas, M^+ signifies monovalent cations such as Li^+, Na^+, K^+, NH_4^+, and, in some circumstances, ions such as $[Al(OH)_2]^+$, or organic cations such as RNH_3^+. Also M^+ may be replaced by one half M^{2+}, where M^{2+} is Ca^{2+}, Mg^{2+}, etc. The amount of substitution in the basic formulas is indicated by y, which is commonly in the range 0.3–0.5; but, in *saponite*, two kinds of substitution are shown for which $x \approx 0.85$, $y \approx 0.50$, and $x - y \approx 0.35$. For many ***montmorillonites***, y is close to 0.33 and this is often taken as an "ideal" value. Of these minerals, *montmorillonite* is by far the most common. To each formula, nH_2O is added to indicate the interlayer water, which varies with the interlayer cations and the humidity conditions. The variable basal spacings are listed in Table 2; expansion from a minimum spacing of about 9.5 Å (obtained after heating to slightly elevated temperatures) to about 12.5 Å corresponds to the introduction of single sheets of water molecules, and expansion to 15.5 Å corresponds to double sheets of water molecules. Still further expansions may occur on immersion in water.

Polar organic molecules also enter between the silicate layers. The effective thickness Δ of interlayer material is approximately $d_{001} - 9.5$ Å. Values of Δ for a wide range of organic materials have been obtained and often correspond

TABLE 1. Structural Formulas of Smectites and of the Prototype Minerals, Pyrophyllite and Talc

Dioctahedral	Trioctahedral
$Al_2Si_4O_{10}(OH)_2$ *pyrophyllite*	$Mg_3Si_4O_{10}(OH)_2$ *talc*
$M_y^+[(Al_{2-y}Mg_y)Si_4O_{10}(OH)_2]^{-y} \cdot nH_2O$ *montmorillonite*	$M_y^+[(Mg_{3-y}Li_y)Si_4O_{10}(OH)_2]^{-y} \cdot nH_2O$ *hectorite*
$M_y^+[Al_2(Si_{4-y}Al_y)O_{10}(OH)_2]^{-y} \cdot nH_2O$ *beidellite*	$M_{x-y}^+[(Mg_{3-y}Al_y)(Si_{4-x}Al_x)O_{10}(OH)_2]^{-x+y} \cdot nH_2O$ *saponite*
$M_y^+[Fe_2(Si_{4-y}Al_y)O_{10}(OH)_2]^{-y} \cdot nH_2O$ *nontronite*	$(x > y)$

to single or double sheets of oriented molecules. The probable orientation can often be deduced from molecular dimensions but precise interpretations are difficult because questions arise about the packing of the molecules against the silicate surfaces and among themselves.

The variable basal spacing of ***smectites*** is an important diagnostic feature in their X-ray identification, but it is generally better to form either an ethylene glycol or a glycerol complex by simply adding an excess of these liquids to the clay; or by exposing the clay to the organic vapor at a slightly elevated temperature, such as 60°C. The resulting complexes have basal spacings around 16.9 and 17.8 Å, respectively, and are identified easily by X-ray diffraction without confusion from other clay minerals that may be present; they are independent of the precise humidity conditions.

Hydration expansion of Na-*montmorillonite* in electrolytes such as NaCl and Na_2SO_4 solutions varies with the concentration, C, of the electrolyte, the interlayer spacing ranging from around 20 Å (3–4 sheets of water molecules) in strong concentrations to values around 150–200 Å in very dilute concentration. Approximately, d_{001} is proportional to $1/C^{1/2}$. These large expansions are no longer intracrystalline since the separation of the layers is comparable to their lateral dimensions. They reflect the transition from the crystalline state to a colloi-

TABLE 2. Basal Spacings, *d* (in Å) of Some Variously Substituted ***Montmorillonites*** and ***Vermiculities*** Under Various Conditions

	Na^+		Mg^{2+}	Ca^{2+}		Ba^{2+}	
	R.H.	*d*	*d*	*R.H.*	*d*	*R.H.*	*d*
Montmorillonites							
In water vapor	0–0.2	9.8		0	10.2	0	10.2
	0.2–0.5	12.5		0.3–0.8	15.5	0.1–0.5	12.5
	>0.7	15.6				>0.7	15.5
Immersed in water		19.0–200[a]	19.2		18.9		15.5
With excess ethylene glycol		16.9–17.1	16.9–17.1		16.9–17.1		16.9–17.1
With excess glycerol		17.6–17.8	17.6–17.8		17.6–17.8		17.7–17.8
Vermiculites							
Immersed in water		14.8	14.8		15.4		15.4
With excess ethylene glycol	mainly	16.1–16.3	14.3, 16.3[b]	mainly	16.1–16.3		16.0–16.2
With excess glycerol		14.3–14.8	14.3		14.3, 17.6[b]		14.3, 17.6[b]

[a]The spacing of Na-montmorillonite in water varies with the electrolyte concentration and has been followed to <about 200 Å at the lowest concentrations. Li-montmorillonite behaves similarly.
[b]Spacing varies with the layer change.

dal gel state. ***Montmorillonites*** show similar effects when saturated with Li^+ and with K^+ ions but not when saturated with divalent cations.

Combined electron-microscope and electron-diffraction studies of ***smectites*** have shown that the ultimate particles are often very small; they have strong tendencies to aggregate in edge-to-edge associations and to stack together in irregular face-to-face associations (Mering and Oberlin, 1967; Gard, 1971). Most ***montmorillonites*** appear to have an ultimate particle size of about 300 Å; by Wyoming ***montmorillonites*** show larger particles, i.e., from 2000–5000 Å (0.2–0.5 μm). The particles are platy, usually with irregular outlines. *Nontronite, hectorite,* and *saponite* form elongated, lath-like particles; the *hectorite* particles show strong tendencies for edge-to-edge associations.

Vermiculites

Vermiculites are similar to ***smectites*** in many aspects. Their layer charge is about twice that of ***smectites***. They swell readily in water and in many organic liquids. Their cation-exchange capacities are about twice those of ***smectites*** and lie mainly in the range 120–200 meq/100 g. ***Vermiculites*** exist as macrocrystalline mica-like sheets and also as clay-grade minerals. ***Vermiculites*** are principally trioctahedral minerals and the following is a typical formula:

$$Mg_{.3}[(Mg_{2.3}Al_{.2}Fe^{3+}_{.5})(Si_{2.7}Al_{1.3})O_{10}(OH)_2]^{-.6}\bullet nH_2O$$

Two processes have been suggested for their development from trioctahedral ***micas***, particularly *biotites*: (a) replacement of interlayer K^+ ions by H_2O with concomitant oxidation of Fe^{2+} to Fe^{3+} ions and (b) replacement of $2K^+$ by Mg^{2+} and H_2O.

Dioctahedral ***vermiculites*** are probably widely dispersed in soils, particularly acid soils, where they may have formed by alteration of fine-grained dioctahedral ***micas***. Their identification and characterization are difficult because they occur in complex mixtures with other clay minerals.

Because of their higher layer charge, ***vermiculites*** show a less ready expansion than ***smectites*** when immersed in water and in organic liquids. This applies to the diagnostic tests with ethylene glycol and glycerol. The preferred technique for distinguishing fine-grained ***vermiculites*** and ***smectites*** is to saturate the clay with Mg ions and then to form the glycerol complex; ***smectites*** give a basal spacing of around 17.6–17.8 Å and ***vermiculites*** a spacing around 14.3 Å. With cations other than Mg present, the distinction is sometimes less certain.

Macrocrystalline ***vermiculites*** provide excellent material for detailed structural studies of hydration and organic complexes of clay minerals. Several such studies of hydrated ***vermiculites*** have been made and the results have shown how the water molecules are grouped in hydration complexes around the interlayer Mg^{2+} ions. Under normal humidities, Mg-***vermiculites*** expand rather exactly to 14.35$\pm$ 0.02 Å. In water, they expand further to 14.8 Å. At lower humidities, spacings of 13.8 and 11.6 Å are obtained. Each of these spacings is a different hydration state and much detailed work is still required to analyze these complexes and to carry out similar work with other interlayer cations.

Clay-grade micas, Illites, Glauconites

Micas and mica-like materials are among the most prevalent clay-grade minerals, but their detailed study has encountered many difficulties. Although many clay minerals can be found in nearly monomineralic conditions, clay-grade ***micas*** are seldom found in this state. Macrocrystalline ***micas*** (see *Phyllosilicates*) have been studied in detail and many of the results are applicable to the clay-grade ***micas***. If the X-ray reflection indexed as 060 can be recognized unambiguously, then the *b* unit-cell parameter can be found and the di- or trioctahedral character of the ***mica*** established. Most clay-grade ***micas*** appear to be dioctahedral. If certain characteristic features of the X-ray diffraction pattern can be ascertained, the polytypic nature of the ***mica*** may be determined; *1Md*, *1M*, and $2M_1$ are the polytypes most often encountered. Chemical analyses show an important K_2O content. These characteristics indicate that many clay-grade ***micas*** are related to *muscovite*. Other clay ***micas***, however, are trioctahedral and these may be related to *biotite*.

The detailed study of many clay-grade ***micas*** is made difficult by their poor crystallinity and by the complex mixture of minerals in which they occur. Many disordered clay ***micas*** belong to the category of interstratified layer minerals, and are probably ***mica-montmorillonites***. This name implies a proportion of expandable (hydratable) montmorillonite-like layers interspersed randomly with nonexpandable mica-like layers. The evidence for this interstratification is that the basal X-ray reflections do not fit the Bragg reflection equation, $n\lambda = 2d\sin\theta$ (λ=wavelength of X-rays, θ=reflection angle, d=spacing of the basal planes, n=an integer,

1,2,3,...), where *n* has integral values for a fixed spacing *d*. The reflections tend to be broad and asymmetrical. Internal inhomogeneity of this kind makes interpretation of all analytical data difficult and uncertain.

Certain chemical characteristics of clay-grade ***micas*** appear to be general or at least widespread, particularly low K_2O and high H_2O contents as compared with data for macrocrystalline *muscovites*. One suggestion is that the interlayer cations, wholly or largely K^+ in *muscovite*, are K^+ and $(H_3O)^+$ in clay-grade ***micas***; the analyses of a few rather well-purified clay ***micas*** have been interpreted along these lines. However, it must be noted that when macrocrystalline *muscovite* is ground to a 1-μm particle size, the H_2O content is appreciably increased, from about 4.5% to about 6.5% in one set of experiments, and the water is removed over a wide range of temperatures in much the same way as for clay ***micas***. Therefore, surface hydration may contribute to the high water content of many clay-grade ***micas***. The low K_2O content might be due to a leaching of the outer layers of the fine particles. If clay-grade ***micas*** are interstratified systems containing more hydrous layers, this also will cause high H_2O and low K_2O contents.

The uncertainty of using the names of macroscopic ***micas***, such as *muscovite* and *biotite*, for ill-defined clay-grade ***micas*** led to the name "***illite***" (Grim, 1953, p. 35), not as a specific mineral name, but as a general term for the clay mineral constituent of argillaceous sediments belonging to the ***mica*** group. It has become a very useful, general term; attempts to define ***illite*** more specifically are contrary to the original intention. Detailed studies in favorable cases may enable more specific names to be used, but it should not be supposed that the detailed characteristics are those of *all* ***illites***. The main usefulness of the term is that it is nonspecific for clay-grade ***micas***.

X-ray analyses of interstratified ***mica-montmorillonite*** clays have shown (Hower and Mowatt, 1966) an approximately linear relationship between the nonexchangeable K and Na ions, which presumably belong to the ***mica*** layers, and the percentage of nonexpandable layers. This relation extrapolates to about 0.75 (K+Na) fixed ions for 100% nonexpandable layers, and about 90% expandable layers when fixed (K+Na) ions are zero. On this basis, when the number of fixed (K+Na) ions is about 0.6–0.7 per formula unit, which corresponds to the average composition of many ***illites***, about 12% of expandable layers may be present. The fact that the first diffraction peak from an ***illite*** may correspond to a spacing not greatly in excess of 10 Å, the value for *muscovite*, may suggest that little interstratification is present. However, the apparent spacings of interstratified minerals do not vary linearly with the proportions of the kinds of spacings present, and probably 10% of expanded spacings could be interstratified without a corresponding increase in the apparent spacing of the mineral.

Glauconite is similar to the ***illites*** in being fine grained and poorly crystalline, often with high contents of H_2O, which have been attributed to $(H_3O)^+$ ions, and with interstratified hydrated layers. Analyses generally show important contents of ferric and ferrous iron. It is generally dioctahedral, with the 1*Md* or 1*M* type of structure.

Chlorites and Swelling Chlorites

Chlorites (q.v.; and see *Phyllosilicates*) are principally trioctahedral minerals which, like ***micas*** and ***vermiculites***, occur both as macroscopic and as clay-grade materials. Chemical and structural analyses have been confined mainly to well-crystallized, macroscopic forms but the results are applicable to clay-grade materials. The simpler polytypic forms of ***chlorites*** are recognizable from X-ray powder diffraction data, provided the diffraction records are not confused by the presence of other clay minerals (Bailey and Brown, 1962). Many ***chlorites***, including well-crystallized macroscopic materials, are structurally disordered by random layer displacements of $nb/3$, which is a feature also of many ***kaolinite*** group minerals. The bonding principle in ***chlorites*** is similar to that in ***kaolinites***, namely OH–O bonds between oxygen and hydroxyl sheets, but ionic forces between negatively charged 2:1 layers and positively charged hydroxide interlayers also are important.

Chlorites exhibit a wide variety of isomorphous substitutions in both the tetrahedral and octahedral sheets forming the layer structure, and varieties are named principally on the basis of chemical composition. A partial identification by X-ray methods depends on certain features of the diffraction data that are related to composition. A general structural formula for ***chlorites*** can be written

$$(Mg_{6-x-y}Fe^{2+}_{y}Al_x)(Si_{4-x}Al_x)O_{10}(OH)_8$$

The basal spacing d_{001} varies almost linearly with x, i.e., $d=14.55-0.29x$; the b-parameter varies almost linearly with y, i.e., $b=9.21+my$, where $m\approx0.032$–0.037. The value of y can be estimated also from the basal intensities. These methods are not very precise but may give an indication of chemical composition when other methods cannot be applied.

When ***chlorites*** are heated at progressively rising temperatures, the hydroxide layers of the structure dehydroxylate at around 500–600°C and the 2:1 mica-like layers at around 700–800°C; precise temperatures depend on chemical composition, crystal size and perfection, rate of temperature increase, etc. Partial dehydroxylation at around 600°C produces marked changes in the X-ray diffraction patterns. Clay-grade ***chlorites***, often poorly crystallized, dehydroxylate at lower temperatures, around 400°C, with little or no differentiation between the reactions of the component layers.

Chlorites normally do not show swelling behavior in water or organic liquids and have basal spacings in the range 14.0–14.4 Å. Among clay materials, so-called "swelling chlorites" have been found that expand to about 17.8 Å in glycerol, and, therefore, might be mistaken for ***smectites***. But these show little collapse when heated to 500°C, as compared with ***smectites***, which collapse to a spacing of about 10 Å when heated at about 150°C. It is thought that swelling ***chlorites*** have imperfect hydroxide layers that consist partly of hydroxide material and partly of hydrated Mg ions (as in ***vermiculites***); the interlayer bonding is weakened sufficiently to permit the expansion, but the hydroxide material is sufficient to prevent structural collapse on heating. Chlorite-like layers capable of swelling but not of collapse are found also as components of interstratified minerals and possibly occur more frequently in this form than as individual minerals.

Dioctahedral ***chlorites*** occur principally, perhaps entirely, among soil clays and were recognized first as components of interstratified dioctahedral ***chlorite-montmorillonite*** materials (see *Soil Mineralogy*). A nearly pure mineral has been isolated with the composition

$$Ca_{0.11}(Al_{3.02}Fe^{3+}_{0.35}Mg_{1.18})(Si_{3.26}Al_{0.74})O_{10}(OH)_8$$

containing 4.55 octahedral cations as compared with 4 for a strictly dioctahedral ***chlorite*** and 6 for a trioctahedral ***chlorite*** (Sudo and Sato, 1966). The mineral *cookeite*, with composition approximately $(LiAl)_4(Si_3Al)O_{10}(OH)_8$, has Al_2 in the dioctahedral sheet of the 2:1 layer and $LiAl_2$ in the hydroxide sheet, and can be called a di-, trioctahedral ***chlorite***.

Interstratified Structures

Many kinds of interstratified phyllosilicates are found among clay minerals, with all degrees of regularity and irregularity in the layer sequence. A few minerals have a highly regular alternation of two kinds of layers. *Rectorite* is such a mineral, in which beidellite-like layers, with small charge and capable of swelling, alternate with mica-like layers, with higher charge and nonswelling. *Corrensite* has a regular alternation of nonswelling and swelling chlorite layers. ***Chlorite-smectite*** interstratifications of various kinds have been described. ***Chlorites*** themselves are regular interstratifications of mica-like and brucite-like layers. Regular interstratifications are revealed by their large basal spacings which equal the sum of the component spacings. Thus, *rectorite* normally shows a spacing of 24.6 Å, compounded of about 10.0 Å for the nonswelling component and 14.6 Å for the expanded layer. These minerals are usually not well crystallized, by comparison with ***micas*** or ***chlorites***, and the interstratifications are probably not completely regular.

Minerals with highly irregular sequences, in which the ratios of the components may take almost any value, are much more common, particularly among soil clays. They are characterized by broad and asymmetric X-ray scattering maxima which do not conform to integral orders in the simple Bragg equation. Many ***illites*** appear to be of this kind. Detailed studies of these irregular structures are difficult, but one general and simple procedure follows from the fact that most of them contain one nonvariable kind of layer (mica-like or chlorite-like) and one variable kind of layer (smectite-like or swelling chlorite-like). Variations of the X-ray characteristics observed after various heating and swelling procedures usually provide a good clue to the nature of the interstratified components. Other methods, analogous to one-dimensional Patterson syntheses in X-ray diffraction analysis, permit (with certain simplifications and assumptions) a direct evaluation of the kinds of layers that are present and of the probable sequence of layers. Thus, if layers of types A and B are present, the analysis indicates the probability of A being followed by another A layer or by a B layer. These methods have been developed particularly by MacEwan and his collaborators (Amil, Garcia, and MacEwan, 1968). For more recent work by Reynolds, see Brindley and Brown (1980).

Palygorskite (Attapulgite) and Sepiolite

Both *palygorskite* and *sepiolite* are basically fibrous, though macroscopically they appear in various forms, e.g., earthy, compact, spongy, fibrous, or paper-like. In electron micrographs these minerals appear as fibers, sometimes many microns long and small fractions of a micron in cross section (see Fig. 1f). The structures are composed of ribbons of 2:1-type

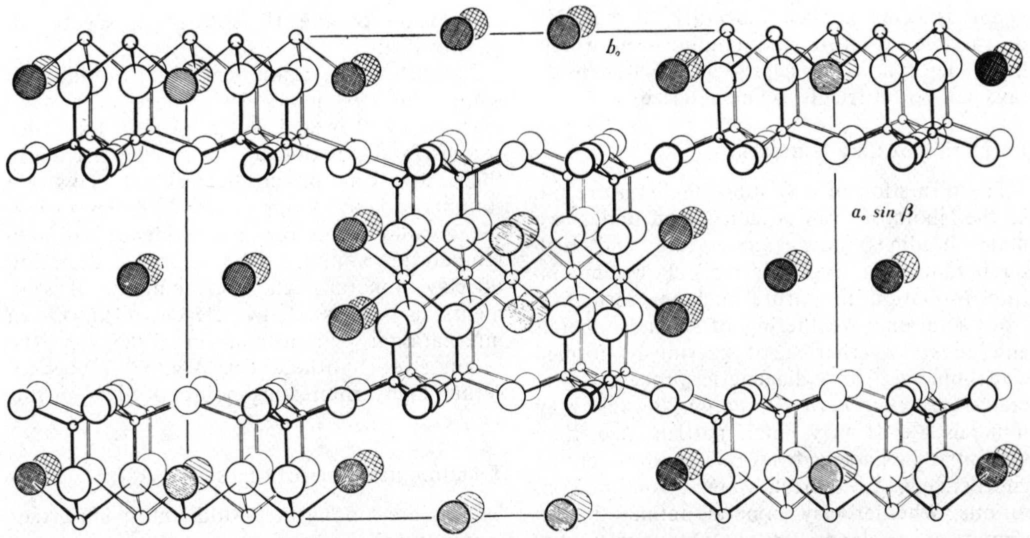

FIGURE 2. Drawing of the structure of *palygorskite* (from Brindley and MacEwan, 1953). Large open circles, oxygen; large shaded circles, hydroxyls; large cross-shaded circles, water molecules; medium-small circles, octahedrally coordinated cations, mainly Mg; small circles, tetrahedrally coordinated cations, mainly Si. Observe the channels through the structure, normal to the page, containing water molecules.

structure alternating with channels containing water, as shown in Fig. 2 for *palygorskite*, for which the *b* parameter is about 18 Å and corresponds to twice the *b* parameter of the mica structure. *Sepiolite* has a similar block structure with $b \approx 27$ Å, equivalent to three units of mica structure. The tetrahedral sheets are continuous across the structure but point in alternate directions in passing from block to block. The octahedral sheets have continuity only along the fiber axis (*c* axis) normal to the plane of the figure. The *c* parameter, about 5.25 Å, corresponds to *a* of the mica structure. There are structural similarities with the chain silicates (see *Amphibole Group*) and some mineralogists place *palygorskite* and *sepiolite* in an intermediate class between layer and chain silicates. Others, including the writer, regard them as a group within the layer silicate class; other layer silicates can be cited in which the silicon-oxygen tetrahedra point in opposite directions within a single structural sheet, so that *palygorskite* and *sepiolite* are not unique in this respect.

The ideal compositions are as follows:

- *palygorskite:* $Mg_5Si_8O_{20}(OH)_2 \cdot (OH_2)_4 \cdot 4H_2O$
- *sepiolite:* $Mg_8Si_{12}O_{30}(OH)_4 \cdot (OH_2)_4 \cdot 8H_2O$

There may be a small partial replacement of Si by Al and the electrical charges are balanced by exchangeable cations in the channels. The exchange capacities lie approximately in the range 10–40 meq/100 g. In the formulas, (OH_2) represents water molecules attached to edges of the structural ribbons and held more firmly than H_2O molecules in the channels.

Allophane

Allophane is a gel-like, clayey material that appears to be amorphous or nearly amorphous to X-rays. It is principally a hydrous aluminosilicate, which may also contain or be associated with iron oxide(s) and P_2O_5. Field evidence suggests that it may sometimes be an intermediate stage between aluminosilicate gels and *halloysite* or *kaolinite*, or between *halloysite* and *gibbsite*. It appears particularly as a weathering product of volcanic ash (Bates, 1962).

The Al_2O_3/SiO_2 mole ratio varies between about 0.5 and 1, but larger values also are possible. Electron-microscope data are variable; particles with rounded, shell-like structure, thin films, and also thin, hair-like, branching fibers have been described. It has been stated (Wada, 1967) that data for "well-characterized soil *allophane*" indicate two possible end-member compositions,

$$Al_2O_3 \cdot 2SiO_2 \cdot 3H_2O \quad \text{and}$$
$$Al_2O_3 \cdot SiO_2 \cdot 2H_2O$$

Chains of corner-linked SiO_4 tetrahedra linked to chains of $Al(O,OH,H_2O)$ octahedra by oxygen sharing are compatible with these compositions. It is very probable that the term *allo-*

phane is being applied currently to materials of variable structure and composition which, being amorphous or nearly amorphous to X-rays, are very difficult to characterize.

Formation of Clay Minerals

The formation of clay minerals in nature and in the laboratory can be considered under two main headings, low temperature and high temperature (or hydrothermal). Low-temperature formation in nature includes formation from solutions, weathering of silicate minerals and rocks, weathering of previously formed clay minerals, and diagenetic processes generally (see *Diagenetic Minerals*). Because clay minerals are of very small particle size, they may alter in passing from one environment to another, e.g., from fresh water to marine conditions. Whether clay minerals reflect the environment of deposition, or the source area, has been much discussed; generally it is better to consider the extent to which a particular clay reflects one or the other aspect, since both are important (Millot, 1964, 1970). Low-temperature laboratory studies have aimed at elucidating the physicochemical conditions under which clay minerals can have formed or altered in nature. Since low temperatures give rise to slow reactions and often to very low solubilities, these studies commonly involve long times and small yields; reduction of the time factor by using higher temperatures and more concentrated solutions means a departure from natural conditions.

High-temperature or hydrothermal formation of clay minerals by reactions of high-temperature fluids with rock materials is more easily duplicated in the laboratory. Phase-equilibrium studies of two, three, and more component systems involving the common rock-forming oxides, SiO_2, Al_2O_3, Fe_2O_3, MgO, FeO, CaO, K_2O, Na_2O, and especially H_2O, under controlled conditions of temperature, pressure, and composition, have resulted in the synthesis, not only of the nonswelling phyllosilicates (*kaolinite, serpentine,* ***micas, chlorites,*** *talc*, and *pyrophyllite*), but also of the swelling minerals (***smectites, vermiculites***, and various interstratified structures). Many of the clay minerals are synthesized easily and with a high degree of crystalline perfection, often equal to or better than that found in natural clays. The chemistry of clay mineralogy has been enlarged by synthetic processes that include many ions not commonly found in natural clays, such as Ge in place of Si, Ga and In in place of Al, Ni and other divalent ions in place of Mg. These synthetic clays have been particularly useful in the elucidation of the IR absorption spectra of phyllosilicates.

Hydrothermal equipment and equilibrium studies up to about 1956 have been summarized by Roy and Tuttle (1956). Extensive studies in the period up to 1967, particularly those involving phyllosilicates and clays, are summarized by Wones (1967). Keller (1964, 1967) summarizes the field evidence and low-temperature studies of the origin and alteration of clay minerals. The geochemistry of sediments is discussed by Degens (1965). An international symposium in 1965 on the "Genèse et Synthèse des Argiles" brings together many interesting points of view on this subject.

Technological Uses of Clays

Clays are among the world's most important raw materials because of their diverse uses, vast amounts, and widespread distribution (Grim, 1962). Traditionally, clays are used for making, for example, earthenware, fine china, stoneware, bricks, tiles, pipes. *Kaolinite* is the most important clay mineral in these applications. The properties of the clay-water system are fundamental in the initial forming operations and the chemical composition of the clay and associated minerals is basically important in relation to properties during and after firing.

In foundry molding sands, the clay content plays a very important part; *montmorillonite* is used most commonly, but fine-grained *kaolinites* are also used (see *Refractory Minerals*).

Clays have great technological importance in relation to soil mechanics—the colloidal properties of clay-water systems especially are involved. The thixotropic properties of these systems are of exceptional importance in connection with so-called *quick clays*, which, if disturbed, lose their strength and become liquid-like; massive landslides and enormous damage result from the movement of such clays (see *Soil Mineralogy*).

Clays have major importance in relation to many aspects of petroleum technology. Recognition of particular clay horizons plays a part in geological mapping. Permeability of rock formations to water and petroleum depends on their clay content, amount, and kind. Drilling fluids (muds) contain considerable proportions (up to about 30%) of clay minerals. Their thixotropic behavior and wall-building properties are important (see *Thixotropy*). ***Montmorillonites*** and, to a lesser extent, *attapulgites* are employed. Clays are used as cracking catalysts for heavy petroleum fractions and are prepared principally from *montmorillonite, halloysite*, and *kaolinite* by various acid and thermal treatments. The exact nature of the catalytic process is not easily explained (see Grim, 1962). Clays are used also to decolorize oils—both filtration and contact processes are involved. So-called "fuller's earths" have been used for this purpose.

Clays are used in large quantities in paper making, both as fillers and for surface coating. Pure *kaolinites* giving good orientation of the platy particles are especially useful. The colloidal properties of the clay-water system are very important. *Attapulgite* is used in the manufacture of a special paper that produces copies under normal writing and typing pressures without use of carbon paper (see *Pigments and Fillers*).

Other areas of industry and technology in which clays are utilized are the following (see Grim 1962, for more details): adhesives; animal bedding; atomic (radioactive) waste disposal; cement, mortor, and aggregates (see *Portland Cement Mineralogy*), clarification of wines, beer, etc.; seed coating; desiccants, absorbents, and molecular sieves; emulsifying agents; fabrics; fertilizers; food; greases; ink; leather; medicines, pharmaceuticals, and cosmetics; paint; pesticides; plastics; rubber; soaps and polishing compounds; water clarification.

G. W. BRINDLEY

References

Amil, A. Ruiz; Garcia, A. Ramirez; and MacEwan, D. M. C., 1968. *X-ray Diffraction Curves for Analysis of Interstratified Structures.* Madrid: Volturna Press, 38p. + 140p. of figures.

Bailey, S. W., 1980. Structures of layer silicates, p. 1–123 in G. W. Brindley and G. Brown, eds., *Crystal Structures of Clay minerals and Their X-ray Identification.* London: The Mineralogical Society, 495p.

Bailey, S. W., and Brown, B. E., 1962. Chlorite polytypism: I. Regular and Semi-Random One-Layer Structures, *Am. Mineralogist*, **47**, 819–850.

Bates, T. F., 1958. *Selected Electron Micrographs of Clays and Other Fine-Grained Minerals*, publication no. 51, College of Mineral Industries, Pennsylvania State Univ., University Park, Pa.

Bates, T. F., 1962. Halloysite and gibbsite formation in Hawaii, *Clays, Clay Minerals*, **9**, 315–328.

Beutelspacher, H., and Van der Marel, H. W., 1968. *Atlas of Electron Microscopy of Clay Minerals and Their Mixtures.* New York: Elsevier. 333p.

Brindley, G. W., 1980. Order-disorder in clay mineral structures, p. 125–195, in G. W. Brindley and G. Brown, eds., *Crystal Structures of Clay Minerals and Their X-ray Identification.* London: The Mineralogical Society, 495p.

Brindley, G. W., and Brown, G. (eds.), 1980. *Crystal Structures of Clay Minerals and Their X-ray Identification.* London: The Mineralogical Society, 495p.

Brindley, G. W., and MacEwan, D. M. C., 1953, in A. T. Green and G. H. Stewart, eds., *Ceramics, A Symposium* 15–93. England: British Ceramic Society.

Chukhrov, F. V., and Zvyagin, B. B., 1966. Halloysite, a crystallochemically and mineralogically distinct species, *Proc. Intern. Clay Conf., Jerusalem, Israel*, **1**, 11–25.

Degens, E. T., 1965. *Geochemistry of Sediments.* Englewood Cliffs, N.J.: Prentice-Hall, 382p.

Farmer, V. C., ed., 1974. *The Infrared Spectra of Minerals.* London: Mineralogical Society, 539p.

Gard, J. A., ed., 1971. *The Electron-optical Investigation of Clays.* London: Mineralogical Society, 383p.

Grim, R. E., 1962. *Applied Clay Mineralogy.* New York: McGraw-Hill, 422p.

Grim, R. E., 1968. *Clay Mineralogy*, 2nd ed. New York: McGraw-Hill. 596p.

Honjo, G.; Kitamura, N.; and Mihama, K., 1954. A study of clay minerals by electron-diffraction diagrams due to individual crystallites, *Acta Crystallogr.*, **7**, 511–513.

Hower, J., and Mowatt, T. C., 1966, Mineralogy of illites and mixed-layer illite-montmorillonites, *Am. Mineralogist*, **51**, 825–854.

Jackson, M. L., 1958. *Soil Chemical Analysis.* Englewood Cliffs, N.J.: Prentice-Hall, 498p.

Keller, W. D., 1964. Processes of Origin and Alteration of Clay Minerals, pp. 3–76 in C. I. Rich and G. W. Kunze, eds., *Soil Clay Mineralogy.* Chapel Hill, N.C.: Univ. of N. Carolina Press.

Keller, W. D., 1967, Geologic occurrence of the clay-mineral layer silicates, in *Layer Silicates. A.G.I. Short Course Notes.* Washington, D.C.: American Geological Inst., p. WK 1–105, WKR 1–25.

Mackenzie, R. C., ed., 1957. *Differential Thermal Investigation of Clays.* London: Mineralogical Society, 456p.

Mering, J., and Oberlin, A., 1967, Electron-optical study of smectites, *Clays Clay Minerals*, **15**, 3–15.

Millot, G., 1964. *Geologie des Argiles.* Paris: Masson et Cie, 499p.

Millot, G., 1970. *Geology of Clays*; Weathering, Sedimentology, Geochemistry, W. R. Farrand and H. Paquet, trans. New York: Springer, 429p.

Roy, R., and Tuttle, O. F., 1956. Investigations under hydrothermal conditions, in L. H. Ahrens et al., eds., *Physics and Chemistry of the Earth*, Vol. 1. New York: Pergamon Press, 138–180.

Sudo, T., and Sato, M., 1966. Dioctahedral chlorite, *Intern. Clay Conf. Proc.*, Jerusalem, Israel, **1**, 33–40.

Van Olphen, H., 1963. *Introduction to Clay Colloid Chemistry*, New York: Wiley. 318p.

Wada, K., 1967, Structural scheme of soil allophane, *Am. Mineralogist*, **52**, 690–708.

Weaver, C. E., and Pollard, L. D., 1973. *The Chemistry of Clay Minerals.* Amsterdam/New York: Elsevier, 213p.

Weiss, A.; Thielepape, W.; and Orth, H., 1966, Neue Kaolinit Einlagerungsverbindungen, *Internat. Clay Conf. Proc.*, Jerusalem, Israel, **1**, 277–294.

Wones, D., 1967, Compositional variations and phase equilibria of some layer silicates, in *Layer Silicates.*

A.G.I. Short Course Notes. Washington, D.C.: American Geological Inst. p. DRW 1-142.

Zvyagin, B. B., 1967. *Electron Diffraction Analysis of Clay Mineral Structures.* New York: Plenum Press.

Cross references: *Crystallography: Morphological; Electron Microscopy (Transmission); Fire Clays; Isomorphism; Mica Group; Mineral Industries; Order-Disorder; Phyllosilicates; Polysomatism; Portland Cement Mineralogy; Soil Mineralogy; Straining Techniques; Rock-Forming Minerals.*

CLEAVAGE—*See* RUPTURE IN MINERAL MATTER

CLINOPYROXENES

The *clinopyroxenes* and the *orthopyroxenes* (q.v.) are the two divisions of the *pyroxene* group (q.v.). Both divisions are characterized by single chains of SiO_4 tetrahedra sharing two of their four apical O^{2-} ions so as to form an endless tetrahedral chain.

The *clinopyroxenes,* the larger of the two divisions, differ from the *orthopyroxenes* in that the stacking of the tetrahedral chains is staggered in the *z*-axis direction (Fig. 1). This stagger results in a monoclinic unit cell in contrast to the orthorhombic unit cell of the *orthopyroxenes.* A second difference between the two divisions is that, whereas, the *b* and *c* unit-cell edges are approximately the same, the *a* unit-cell edge of the *orthopyroxenes* is approximately twice that of the *a* unit cell of the *clinopyroxenes.*

Composition

The *clinopyroxenes* vary widely in chemical composition, but they may be represented by the structural formula

$$(M_2)^{VI-VIII}_{1-x}(M_1)^{VI}_{1+x}Si_2O_6 \qquad 0<x<1$$

where M_2 = Ca,Mg,Na
M_1 = Mg,Fe^{2+},Mn,Li,Ni,Al,Fe^{3+},Cr,Ti

There is some substitution of Al for Si in the tetrahedral site in *fassaite* and *augite*. The M_2 site, which may be occupied or vacant, changes coordination number in response to the size of its cation. In *clinoenstatite*, the M_2 ion is Mg with a coordination of 6 O^{2-} ions. By contrast, the M_2 occupant is Ca in *diopside* and the coordination is 8, 6 of the O^{2-} ions contributing 1/3 electron each to satisfy the 2+ charge on Ca^{2+}, and 2 O^{2-} ions contributing no charge to the site by reason of being shared by 2 tetrahedra in the chain.

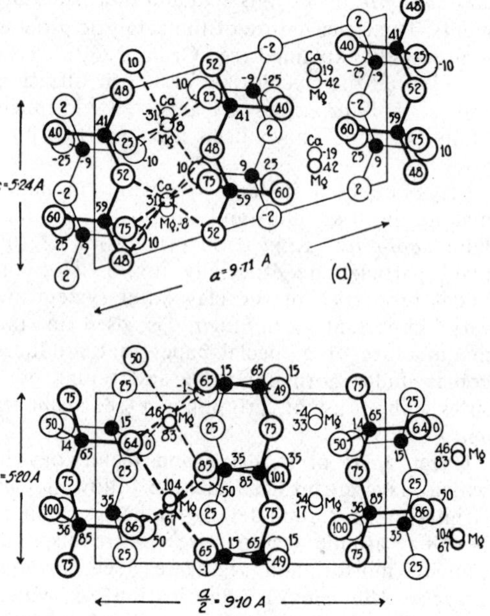

FIGURE 1. Comparison of the structures of ***ortho-pyroxene*** and ***clinopyroxene*** (reprinted from Sir Lawrence Bragg, G. F. Claringbull, and W. H. Taylor: *Crystal Structures of Minerals.* © 1965 by G. Bell & Sons, Ltd. and used by permission of Cornell University Press). Numbers indicate the percentage of the distance of the ion in the unit-cell *b* direction. (a) *Diopside.* (b) *Enstatite.* Note that only $\frac{1}{2}a$ is shown for *enstatite.*

The *clinopyroxenes* (see Table 1) may be divided into several solid-solution series:

- *clinoenstatite–clinoferrosilite,* where M_2=Mg and M_1=Mg,Fe^{2+}
- *diopside–hedenbergite–johannsenite,* where M_2=Ca and M_1=Mg,Fe^{2+}, and Mn, respectively
- *augite–omphacite–fassaite,* where M_2=Ca,Na and M_1=Mg,Fe^{2+},Mn,Fe^{3+},Al,Ti, and where Si is partially replaced by Al
- *pigeonite,* where M_2<0.2 Ca and M_1=Mg,Fe^{2+}
- *augite-acmite,* where M_2=Na and M_1=Fe^{3+}

Acmite is the end-member mineral of a solid-solution series extending toward *augite* in composition, i.e., substitution of Na by Ca with concomitant replacement of Fe^{3+} by Fe^{2+} and Mg. The name "aegirine" has been applied to members of this series with between 0.7 and 1.0 Fe^{3+} ions per unit cell. The term "aegirine-augite" refers to members of the series with Fe^{3+} between 0.1 and 0.7 ions per unit cell. Use of the term "*acmite*" follows Fleischer (1980). See Deer, Howie, and Zuss-

TABLE 1. Clinopyroxenes

Acmite	$NaFeSi_2O_6$
Augite	$(Ca,Na)(Mg,Fe,Al,Ti)(Si,Al)_2O_6$
Clinoenstatite	$Mg_2Si_2O_6$
Clinoferrosilite	$(Fe,Mg)_2Si_2O_6$
Clinohypersthene	$(Mg,Fe)_2Si_2O_6$
Diopside	$CaMgSi_2O_6$
Fassaite	$Ca(Mg,Fe^{3+},Al)(Si,Al)_2O_6$
Ferrosilite	$Fe_2Si_2O_6$
Hedenbergite	$CaFeSi_2O_6$
Jadeite	$NaAlSi_2O_6$
Johannsenite	$CaMnSi_2O_6$
Kanoite	$(Mn,Mg)_2Si_2O_6$
Omphacite	$(Ca,Na)(Mg,Fe^{2+},Fe^{3+},Al)Si_2O_6$
Pigeonite	$(Mg,Fe,Ca)(Mg,Fe)Si_2O_6$
Spodumene	$LiAlSi_2O_6$
Ureyite (cosmochlore)	$NaCrSi_2O_6$

man (1978) vol. 2A, p. 483 for a review of the nomenclature for this series.

Additional clinopyroxenes, in which ionic substitution is rather restricted, include

- *spodumene*, where M_2 = Li and M_1 = Al
- *jadeite*, where M_2 = Na and M_1 = Al
- *ureyite* (cosmochlore), where M_2 = Na and M_1 = Cr

Occurrence

Clinoenstatite and *clinoferrosilite* are rare in terrestrial rocks, but are found in some meteorites (see *Meteoritic Minerals*). *Diopside* is characteristic of thermally metamorphosed calccareous sedimentary rocks and is commonly associated with *calcite* or *calcite* plus *forsterite*. It is also found with *hedenbergite* in skarns and is the ***pyroxene*** in some extrusive mafic rocks, e.g., basalt and picrite.

Hedenbergite occurs in metamorphosed iron-rich sediments, in *quartz* syenites, and in some *fayalite* granites. *Johannsenite*, associated with *bustamite* and *rhodonite*, is found in some limestones metasomatized by nearby felsic-to-intermediate igneous rocks.

Augite is the most common of the ***clinopyroxenes***, being the essential ferromagnesian mineral in gabbros, diabases (dolerites), and basalts. It may occur with an ***orthopyroxene*** or with *pigeonite*. *Omphacite* is an essential constituent, along with *pyrope-almandine*, in eclogites (see *Mantle Mineralogy*) and in granulites. In the former, Al concentrates in the octahedral sites, whereas in the latter, it replaces Si in the tetrahedral sites. *Fassaite* is found in *quartz*-free environments, commonly in metamorphosed limestones.

Pigeonite is restricted to the ground mass of volcanic rocks of intermediate composition. If not quenched, *pigeonite* inverts to an intergrowth of *augite* plus an ***orthopyroxene***.

Acmite (aegirine) is a product of the crystalization of alkaline magmas where it is associated with *arfvedsonite* in syenites and *riebeckite* in some alkali granites.

Clinopyroxenes intermediate in composition between *acmite* and *augite* (aegirine-augite) are found associated with *glaucophane*, *crossite*, and *riebeckite* in some schists.

Spodumene occurs as megacrysts up to several meters long in Li-rich granite pegmatites (see *Pegmatite Minerals*), where it is associated with *lepidolite*, *beryl*, and members of the ***tourmaline*** group.

Jadeite (see *Jade*) is always associated with *albite* in metamorphic rocks formed under high pressures and low temperatures. *Ureyite* is a meteoritic mineral (q.v.).

KEITH FRYE

References

Deer, W. A.; Howie, R.A.; and Zussman, J., 1978. *Rock-Forming Minerals*, vol. 2A, *Single-Chain Silicates*. London: Longmans, 668p.

Fleischer, M., 1980. *Glossary of Mineral Species*. Tucson, Arizona; Mineralogical Record, Inc., 192p.

Papike, J. J., et al., eds., 1969. Pyroxenes and amphiboles: Crystal chemistry and phase petrology, *Mineral. Soc. America Spec. Paper #2*, 314p.

Cross-references: *Glass, Devitrification of Volcanic; Jade; Orthopyroxenes; Plastic Flow in Minerals; Pyroxene Group; Rock-Forming Minerals.*

CLOSE-PACKED STRUCTURE—*See* Vol. IVA: CRYSTAL CHEMISTRY

COAL—*See* Vol. VI: COAL

COESITE AND STISHOVITE

These minerals, high-pressure, metastable polymorphs of silicon dioxide, first became known as a result of laboratory syntheses. *Coesite* was synthesized in 1953 and *stishovite* in 1961. They are named by Sosman (1954) and by Chao et al. (1962) for their discoverers, L. Coes and S. M. Stishov, respectively. The unit cell of *coesite* has 6 and *stishovite* has 2 SiO_2 formula units.

Physical Properties

Coesite is similar to *quartz* and to other low-density forms of silicon atom tetrahedrally bonded to four oxygen atoms, each of which is shared by a silicon atom in an adjoining tetrahedron. The 9.1% greater density of *coesite* (2.911 g/cc) relative to *quartz* is due to a more efficient packing of the silicon-oxygen tetrahedra. *Coesite* is monoclinic, optically positive, and has refractive indices of 1.599 and 1.604. The low solubility of *coesite* in dilute hydrofluoric acid permits concentration of *coesite* from mixtures with lower-density silica minerals.

Stishovite (4.287 g/cc) is 47.3% denser than *coesite* and is structurally unlike the other silica minerals. The rutile-like *stishovite* structure consists of silicon atoms in sixfold coordination with oxygen; each oxygen atom is shared by three silicon atoms. *Stishovite* has refractive indices of 1.799 and 1.826. *Stishovite* is quite resistant to attack by cold concentrated hydrofluoric acid, although it can be decomposed by prolonged treatment in a mixture of hot dilute hydrofluoric acid and a small amount of nitric acid.

Occurrence

Naturally occurring *coesite* and *stishovite* were first discovered in the impact metamorphosed Coconino Sandstone at Meteor Crater, Arizona, in 1960 and 1962, respectively. They have since been found in the vicinity of other craters suspected to be of meteorite-impact origin. *Coesite* has also been recovered from partially fused alluvium near a crater produced by the detonation of 500 tons of TNT, and *stishovite* has been synthesized by subjecting quartzose rocks to explosively produced pressure pulses of less than ten microseconds duration.

In the earth's mantle (see *Mantle Mineralogy*), *coesite* should be the stable form of silica at depths between approximately 100 and 300 km. At greater depths, *stishovite* should be the stable form. It is possible, but not likely, that *coesite* formed at depth could survive transport to the surface of the earth. *Coesite* inverts rapidly to lower-density forms of silica at temperatures above 350°C and pressures below 1 atm. *Stishovite* begins to invert to lower-density forms at temperatures of 1000°C and a pressure of 30,000 atm; hence, it is virtually impossible that *stishovite* could survive transport to the surface.

The presence of *coesite* in a metamorphosed rock strongly suggests either meteorite impact or large-scale conventional or nuclear explosion as the probable cause of the metamorphism. The additional presence of *stishovite* in the rock is very good evidence of impact or explosively produced metamorphism.

P. S. DeCARLI

References

Boyd, F. R., and England, J. L., 1960. On coesite, *J. Geophys. Res.*, **65**, 749–756.

Cailleux, A.; Guillemaut, A.; and Pomerol, C., 1964. Présence de coesite, indice de hautes pressions, dans l'accident circulaire des Richât (adran Mauritanien); *Comptes Rendus Acad. Sci.*, **258**, 5488–5490.

Chao, E. C. T., et. al., 1962. Discovery of stishovite in nature, *J. Geophys. Res.*, **67**, 419–421.

Dachille, F.; Zeto, R. J.; and Roy, R., 1963. Thermal stability of stishovite and coesite, *Science*, **140**, 991–993.

De Carli, P. S., and Milton, D. J., 1965. Stishovite: synthesis by shock waves, *Science*, **147**, 144.

Frondel, C., 1962. *The System of Mineralogy*, vol. III. New York and London: Wiley, 310–318.

Sosman, R. B., 1954. New high-pressure phases of silica, *Science*, **119**, 738–739.

Cross-references: *Mantle Mineralogy; Polymorphism; Quartz.*

COLLECTING MINERALS

Mineral collecting comprises at least two distinctly different forms of the collecting. These forms have in common the goal of gathering a suite of minerals, but this is where the similarity ends. Traditionally, we think of mineral collecting as putting together a collection of minerals by going into the field and personally removing specimens from the earth. Mineral collecting, in this sense, is not only building a collection, but is also the act of looking for specimens in mines, quarries, roadcuts, and any other place where they may be exposed.

Today, many collectors have never donned old clothes and boots and dirtied their hands by venturing into the field after mineral specimens. Their efforts have been confined entirely to purchasing specimens from dealers, or using the "silver pick," as this method has come to be known. There really is nothing new about this, for many of Europe's wealthy royalty were known to assemble cabinets of natural curiosities centuries ago (see *Museums, Mineralogical*). Predictably, the products of these two forms of collecting, personal versus buying, usually differ significantly. Collectors of the former kind tend to find satisfaction from less spectacular material, largely representative of what one might expect from local sources. The pride one derives from self-discovery provides great personal pleasure even though the specimens may never approach in quality those available through purchase. The other type of collector is really in pursuit of natural objects of art. He is after minerals that are, as a rule, colorful, sculptural, and free of readily discernible damage. This type of collector is only interested in minerals that are breathtakingly beautiful and of great intrinsic value.

Naturally, there is much overlap. Numerous examples can be found of collectors who combine these aims in varying proportions. In many cases, a collector may begin as one type and gradually slip over into the other. There is a tendency for the field-oriented collectors to become far more knowledgeable of general mineralogy and it is not at all unusual for such collectors to evolve an understanding of the science, on their own, that is equivalent to having taken several undergraduate courses in mineralogy and geology. The emphasis of this article is placed upon the generalist who devotes a substantial portion of his effort to building a systematic collection of all available species—a collector who owns both self-collected specimens and specimens purchased from dealers.

A mineral specimen is the objective of the collector. Usually it consists of a visible mass of a given mineral, nearly always attached to or embedded in some of the rock with which it was found. Ideally, it is a moderately large mass of the mineral and is more or less pure. The more highly prized specimens are those in which a particular mineral exists in well-defined crystals that are essentially complete and free of damage (see *Crystal Habit*).

A distinction is made between study and exhibit specimens. The former may be complete individual crystals of moderate size; more often they are simply fragments of larger crystals or amorphous pieces of polycrystalline material. They may be handled, measured, and tested in a variety of ways without their importance as study specimens being diminished. In terms of market values, fine study pieces of common minerals are practically worthless. In the case of the more rare species, the market value tends to increase relative to the rarity even when the specimens are not attractive.

Where Are Collectors Found?

The greatest concentration of collectors in the world is in the western part of the United States, specifically California and Arizona. The reason for this is obvious; there is a high density of good collecting localities in these states. A century ago, the concentration was in the New York City to Philadelphia region where many good localities combined with several active natural science academies and museums to foster intense interest in mineral collecting. In general, concentrations of collectors occur where there are concentrations of prolific localities. Exceptions may be found where economic conditions preclude such activities; Mexico and Brazil have been exceptional examples in contemporary history. Outstanding private and public collections simply do not exist in these countries even though they have provided a disproportionate share of fine mineral specimens for a long time. Ironically, this fact too is related to economics. The more primitive a mining operation is, the more likely it is that mineral specimens will be produced. Mechanization and large-volume mining are detrimental to the recovery of specimens, as nearly all rock is moved in huge quantities without discrimination. Many highly industrialized countries were great sources of specimens when the mining was primarily by hand methods but are no longer important producers.

Major mining companies, in fact, try to discourage mineral collecting on their properties and they are not pleased when miners encounter good specimens since they may be tempted to interrupt their mining activities to try to salvage the specimens. Furthermore, mine owners are liable for injuries to collectors while on their land and their failure to prohibit collecting may mean loss of insurance coverage.

Building a Collection

Specimen Acquisition. Most of the more serious collectors begin by trying to find good specimens in rock exposures to which they have ready access. Increasing sophistication usually leads to a desire to expand the collection beyond the limitations of self-collecting. The collector then may participate in either buying from dealers, exchanging with other collectors, or both. There are dealers liberally distributed around the world. Most operate from fixed locations but also lease space at mineral shows where large numbers of buyers may gather. Some are primarily wholesalers and others concentrate on retailing, although many function in either capacity as the situation dictates. The main sources of their stock are miners, collectors living near good localities, and old collections they manage to acquire.

Exchanging has attracted a growing number of participants in recent years, particularly in view of the increased mobility of present-day collectors. Exchanging by mail has long been practiced but not always with satisfaction since the parties involved had little or no opportunity to select the materials that were traded. Today, collectors and dealers travel around the world with specimens for trade in their possession. They visit museums, other collectors, and mineral shows in search of minerals equal in value to what they are offering. The results of this kind of exchange include greater mutual satisfaction.

A collector who is fortunate enough to live near a prolific, or at least important, source of exceptional specimens can build a first-class collection by exchanging. His proximity to such a place permits frequent visits, greatly increasing the odds of being there when the best specimens are encountered, either in the course of the legitimate mining, through his own personal mining, or thorough searching. Of course, the collector may amass a large quantity of personally collected specimens and sell them in bulk to a major mineral merchant, so that he may purchase a wider variety of specimens.

Classification. Mineral collectors tend to succumb to the urge to classify. Minerals may be arranged in a variety of ways, but a system based upon chemical composition is almost universally adopted. In this system, the major categories are native elements, sulfides, sulfosalts, oxides, etc., ending with the silicates. The silicates are commonly subdivided according to the manner in which the basic building block, the SiO_4 tetrahedron, is arranged in the structure of the compound. Silicates are approximately equal in number to all of the members of the other major chemical groups combined (see Vol. IVA: *Mineral Classes: Nonsilicates; Mineral Classes: Silicates*).

Collection Maintenance. Cataloging, the recording of pertinent data relating to the specifics of a specimen's point of origin and method and date of acquisition (cost, if purchased), is generally believed to be a necessity in building a mineral collection. Nevertheless, many collections, large and small, exist for which a catalog has not been prepared. The locality information, at the very least, is an essential ingredient of the individual label. Specimens lose their value when their source can no longer be determined. Certain minerals from certain localities have such distinctive characteristics that experienced collectors can tell precisely where they came from, but this is the exception rather than the rule and examples are all too rare.

Catalogs and labels are the traditional means of assuring the preservation of important data about the specimens. These are customarily linked together through the use of numbers sequentially assigned to the minerals and entered in the catalog. The catalog may be in book form, on index cards, or even on loose sheets of paper; it makes little difference. Ideally, the catalog numbers are painted on the specimens, or are printed on paper glued to the specimen, on the underside so that they do not detract from the appearance of the specimen. Labels are often prepared with the collector's name on them. Most allow room for the specimen name and locality, and the catalog number should also be entered. The three elements—catalog, specimen number, and specimen label—should always be employed. Some collectors combine two of them by preparing very small labels that can be glued directly onto the specimens.

Most collections are housed in one of two widely adopted methods. They are either arranged on shelves within a glass case or are set in open trays or boxes, which in turn are kept in broad shallow drawers in cabinets. Obviously, the former method is preferred for attractive, exhibit-type specimens, many of which tend to be delicate, for they can be viewed without handling and are not subjected to rolling induced by opening and closing drawers.

It is fortunate that most minerals can be washed, even scrubbed, in detergent with stiff brushes and suffer no damage. However, there are minerals that cannot be cleaned in this manner and there are others that are not even stable in the atmosphere; these require special treatment. Many specimens can be improved in appearance by exposing certain crystals at the expense of others in which they are embedded. The removal of unwanted material may be accomplished mechanically with chisels and scraping tools, or by dissolving it with certain solvents. Cleaning and "improving" mineral specimens has become an art and many collectors are masterful practitioners.

Specialization. There are those who elect to specialize in lieu of trying to build a comprehensive collection. The reasons for specializing are many. The specialist may be stimulated by the challenge of a specialty, or despair over the futility of trying to obtain specimens of every known mineral. If the limits of the specialty are narrow enough, one may look ahead to the day when the collection is as complete as it possibly can be. Specializing can also be the result of a genuine interest in some single aspect of collecting, such as the minerals of a specific locality or those of a particular group or family. In the northern New Jersey–New York City area, there is intense interest among local collectors in the ***zeolite*** minerals of the traprock quarries so prevalent in the area. There are probably more collectors of copper minerals than in any other specialty, because the colors of many copper minerals are so spectacular. Some collect only uranium minerals or minerals from one state, *quartz*-family minerals, single crystals, twinned crystals, carbonates, on and on. There are also those who collect only one mineral. For a mineral to be a good candidate it must be known to occur in a wide variety of forms, colors, and associations. Chief among these is *calcite*, and *calcite* specialists abound.

Most of the specialists have a general collection, in addition to their specialty, for the beauty and appeal of many minerals is difficult to resist. While nearly all collectors seek attractive crystals and crystal groupings, there are specialists who ignore the more obvious attributes to collect rare minerals, which seldom are found in showy specimens. The goal is obtaining a sample of every known mineral. Such collectors are known as "Dana" collectors, because Dana devised the most widely adopted system of mineralogy.

Another category of specialization is based upon specimen size. One may collect only miniatures, thumbnails, micromounts, or cabinet-sized specimens. Miniatures are specimens small enough to fit within a cube two inches on an edge but large enough that they may be readily appreciated at a moderate distance, say, several feet. Thumbnails are somewhat smaller but not so small that magnification is required to see them well. Micromounts, the smallest class, are so small that they must be viewed under a microscope to be fully appreciated. Anything larger than a miniature (no upper limit) is regarded as a cabinet specimen.

There are many other specialties not mentioned here; but it should be obvious that, in mineral collecting as in other hobbies, the varieties of specialization are endless.

Organizations, Activities, and Services

Clubs and Federations. In North America, mineral hobbyists have developed a loosely structured, three-tiered organization. On the local level, there are *mineral clubs*, most of which are incorporated as nonprofit entities. The clubs within specific regional divisions are affiliated with their respective *regional federations*. These, in turn, are all part of the *American Federation of Mineralogical Societies*. The regional federations actively manage a variety of projects, the most important of which include sponsorship of Federation shows, management of the competitive displays by their members at these shows, and conducting an active and generous scholarship program for college students in the earth sciences.

The practice of exhibiting collections in competition at mineral shows has grown to the point where it is a major activity consuming major amounts of the time and effort of Federation officers, show committees, and club members. An elaborate set of rules has evolved. Various categories of competing have been established (beginning, novice, advanced) for a wide variety of specialization, such as size (thumbnail, miniature, cabinet), minerals from one locality, minerals having one element in common, pseudomorphs, and minerals belonging to one group (***garnets, zeolites, micas***). There are also categories for mineral materials that have been fashioned into gemstones and jewelry. Trophies, ribbons, and certificates are awarded to the winners.

The clubs, of course, incorporate into their membership, and the federations into their activities, many people whose principle interest is either lapidary (cutting and polishing) or fossil collecting. In fact, an overwhelming majority of the 500,000 or so members of clubs that are affiliated with federations are primarily oriented toward lapidary activities.

The mineral clubs are highly variable insofar as what they offer to their members. The stronger clubs, in terms of good programs and solid accomplishments, tend to be found near concentrations of highly productive mineral localities or in large cities where active mineral collecting was good long ago and where museums and science academies are located. Club meetings may offer lectures, demonstrations, auctions, swaps, and classes on mineralogy. Micromounters usually have work sessions. Members bring their own microscopes and actually work on specimen material at the meetings.

Increasing in popularity are symposia where specific subjects are discussed over the course of several days with invited speakers who are knowledgeable on the topic for the meeting. The great success of micromount symposia in recent years has lead to the extension of the concept to other phases of the hobby. It is expected that the number and diversity of symposia for mineral collectors will increase for some time to come.

The *Friends of Mineralogy* is a relatively new international organization comprised of mineral collectors with a somewhat more academic interest in mineralogy, although this certainly is not a requirement for membership. Its members include many amateur mineralogists and a great number of specimen-oriented professional mineralogists, most of whom are museum curators and professors of or researchers in mineralogy. Here, too, exists a regional breakdown with various officers at different levels. The regions may act in virtual autonomy so long as their programs are not in conflict with the basic objectives of the group. The purposes of the Friends of Mineralogy are diverse but, in general, the unifying theme is a dedication to promoting better mineral appreciation, education, and preservation.

Identification Help. One of the greatest drawbacks to the development of a young mineral collector is not knowing where to go for help in learning to identify minerals. Too often, help is available but not utilized. One should, of course, begin on the most local level available. In most cases, this would be someone in the science faculty at the neighborhood high school or a member of the nearest mineral club. A college or university with a geology program will likely have someone competent to identify most of the common minerals. If there is a local one, this is very fortunate. Otherwise, the unknowns will have to be taken to the state university. Most states also have geological surveys and (or) mines departments where help in identification can be found. Ingenious use of existing facilities can usually solve the problem of getting minerals identified.

Mineral Shows. Mineral shows are usually small productions, commonly in school gymnasiums and most often promoted by local mineral clubs. Every year, however, there is a large American Federation "national" show. Also, each regional federation usually puts on an annual show. These shows are moved to a different city each year within the confines of the appropriate federation. In addition to the rather large federation shows, there are several club shows that have become so popular and well attended that they easily rival the best federation efforts.

The size and importance of a show is reflected in the quality of the special exhibits and the type of mineral dealer allowed to lease space there. Exhibits include those entered in competition, special educational displays, and displays from well-known collectors, museums, and universities.

The burden of producing a show falls upon members of the mineral club or clubs in the city where it is held. The mineral dealers bring a great variety of merchandise–from expensive crystal specimens to fancy jewelry, tumbled stones, fossils, and shells. While some may emphasize one kind of material, they all tend to carry goods in many categories.

If competitive exhibits are included, the judging occurs in the morning of opening day. As exhibitors win on one level of competition at one show, they advance to other categories and keener competition in bigger shows. "National" trophies, for example, can be won only at national shows.

Many shows still provide space for swappers. Here people with self-collected specimens are provided with space where they can spread their minerals out to try to promote an exchange for those of another collector.

Publications. A few of the periodical publications available to the collector are listed as *References* below, along with their mailing addresses.

JOHN SAMPSON WHITE

References

der Aufschluss, Blumenthalstrasse 40, 6900 Heidelberg, West Germany.

Earth Science Digest, P.O. Box 1815, Colorado Springs, Colorado 80901, USA.

Lapidary Journal, P.O. Box 80937, San Diego, California 92138, USA.

Lapis, 8000 München 2, Oberanger 6, West Germany.

Mineralienfreund, 8807 Breienbach, Switzerland.

Mineralogical Record, P.O. Box 35565, Tucson, Arizona 85740, USA.

Le Monde et les Mineraux, 4 Av. de la Porte de Villiers, Paris 75017, France.

Notizie del Gruppo Mineralogico Lombardo, Via Ozanam 3, 20129 Milano, Italy.

Rocks and Minerals, 4000 Albemarle Street, N.W., Washington, D.C. 20016, USA.

Cross-references: *Crystal Habit; Gemology; Mineral Classification: Principles; Mineral Properties; Museums, Mineralogical; Naming of Minerals; Pegmatite Minerals; Vein Minerals;* also, see glossary.

COLOR IN MINERALS

The phenomenon of color in minerals, as in other substances, is due to the absorption of light. When white light is passed through a mineral or reflected from its surface, a certain amount is absorbed and converted to heat energy. If the light is absorbed completely the resulting color is black; if all wavelengths are absorbed equally but not completely, the result is gray; and if absorption is selective, the mineral shows a color that depends on the wavelength(s) transmitted or reflected.

In addition to being a useful aid in identification, the color of minerals may be of commercial importance, as in the case of the ochres and other mineral pigments, and, most notably, the gems and semiprecious stones. Mineral color may be described in terms of hue (the portion of the spectrum represented) and tone (relative brightness); for instance, "*Malachite* is emerald-green to grass-green; *azurite*, light azure to deep blue" (Kraus and Slawson, 1939, p. 207).

The color of a mineral may be due directly to its chemical composition, in which case it will be fairly constant and characteristic; the yellow of *sulfur* is an often-quoted example. Such minerals are called *idiochromatic*. In other cases, the color may be due to the presence of foreign ions in the crystal structure, or to tiny inclusions of pigmenting impurities. When the color is variable, the minerals are called *allochromatic*.

Impurities that impart color to minerals may be uniformly distributed, as the ferric oxide responsible for the red color of carnelian, or unevenly distributed, illustrated by the banding of agate or the blotchiness of bloodstone. Impurities may be submicroscopic in size, or large enough to be identified with a lens. Color can be produced artificially by soaking porous minerals in pigmenting solutions, or by inducing defects by irradiation with X-rays or ultraviolet light (see *Defects in Crystals*).

The visible spectrum extends from 400–700 μm. Color results from absorption bands due mainly to electronic transitions. If an absorption band in a mineral removes a particular color from a beam of white light, the complementary color is seen. The Fe^{2+} in minerals of the ***olivine*** series absorbs red light, giving these minerals their typical green color.

Electronic transitions causing absorption in the visible spectrum are mainly of four types: internal transitions, charge transfer, defects, and band-gap transitions.

Internal transitions occur in ions with incomplete electron shells such as the partially filled *d* orbitals of transition metals and *f* orbitals of rare-earth elements. In ionically bonded minerals, the absorption bands and the resulting complementary color depend primarily upon the electronic structure of the cation and secondarily upon the strength and the symmetry of the crystal field (see Vol. IVA: *Crystal Field Theory*).

Chrysoberyl, $(Al,Fe^{3+})BeO_4$, and members of the ***olivine*** series, $(Mg,Fe^{2+})SiO_4$, have very similar structures, but Fe^{3+} ions absorb violet and Fe^{2+} ions absorb red wavelengths. These minerals typically show the complementary colors yellow and green, respectively.

Crystal field strength influences the absorption color of Cr^{3+} ions in different minerals. The substitution of Cr^{3+} for Al^{3+} in emerald (*beryl*, $Be_3Al_2Si_6O_{18}$) imparts a green color, whereas in ruby (*corundum*, Al_2O_3) the substitution produces red. The bond distances are shorter in ruby than in emerald, creating a larger crystal field that shifts the absorption band.

Minerals of the ***tourmaline*** group (q.v.) contain both Fe^{2+} and Fe^{3+} in octahedral coordination. The octahedra share edges in the *a*-axis plane. These minerals absorb light strongly for electric-vector vibration directions perpendicular to the *c* axis, but weakly for vibration directions parallel to *c*. The strong absorption results from charge transfer of electrons from Fe^{2+} to Fe^{3+} ions in the layers of edge-sharing octahedra, but not between them. ***Tourmalines*** are, thus, strongly dichroic, as are some other minerals with tetragonal, hexagonal, or trigonal symmetry. Minerals crystallizing in the orthorhombic, monoclinic, and triclinic systems may be trichroic.

Yellow *halite* (NaCl) and magenta *sylvite* (KCl) derive their colors from defects in their structures. Light is absorbed by single electrons trapped in anion vacancies. This type of chromophore is called an *F-center* (*Farbzentrum*).

Greenockite (CdS) is a yellow semiconductor because blue and violet light have sufficient energy to promote electrons into the conduction band, whereas light of longer wavelengths does not.

The *streak*, which is the color of the powdered mineral (usually obtained by scratching the mineral on an unglazed porcelain streak plate), is a more reliable indication of the intrinsic color of a mineral than is the general appearance, being less influenced by impurities than by the inherent color of the mineral itself.

When color is imparted by the play of light rather than true pigmentation, a mineral is said to be *pseudochromatic*. *Opalescence* and *iridescence* are pseudochromatic phenomena. Opalescence is the result of reflection and refraction of light from layers of different refractive index, as in opal (q.v.), or in *labradorite* ***feldspar*** (see *Feldspar Group*). Iridescence is shown by surface films that develop as a result of tarnishing; for example, fresh surfaces of *bornite* (Cu_5FeS_4) are bronzy, but upon exposure quickly oxidize and acquire a purple-violet iridescence that has earned this mineral its popular name "peacock copper ore."

In some cases the presence of foreign ions alone is insufficient to produce color unless the ions are stimulated by light or some other form of radiation. A common example is the purple color induced in bottle glass by the action of sunlight. Color may be artificially induced in certain gems, including *diamonds, corundum*, and amethyst (*quartz*), by exposure to X-rays or to radium. The effect of radioactivity on color is also illustrated by the *pleochroic halos* (q.v.) seen under the microscope surrounding tiny radioactive mineral inclusions in certain minerals, usually of *zircon* in *biotite*.

CHARLES J. VITALIANO
DOROTHY B. VITALIANO

References

Berry, L. G., and Mason, B. H., 1959. *Mineralogy*. San Francisco: Freeman, 612p.

Kraus, E. H., and Slawson, C. B., 1939. *Gems and Gem Materials*. New York: McGraw-Hill, 287p.

Loeffler, B. M., and Burns, R. G., 1976. Shedding light on the color of gems and minerals, *Am. Scientist*, **64**, 636–647.

Nassau, K., 1978. The origins of color in minerals, *Am. Mineralogist*, **63**, 219–229.

Nassau, K., 1980. The causes of color, *Scientific American* (Oct.), 124–152.

Newnham, R. E., 1975. *Structure-Property Relations.* New York: Springer-Verlag, 234p.

Przibram, K., 1956. *Irradiation Colours and Luminescence.* London: Pergamon, 332p.

Walton, J., 1952. *Physical Gemology.* London: Pitman, 304p.

Winchell, A. N., 1937. *Elements of Optical Mineralogy*, Part 1, *Principles and Methods.* New York: Wiley, 263p.

Cross-references: *Gemology; Luminescence; Mineral Properties; Optical Mineralogy.* Vol. IVA: *Crystal Field Theory.*

COORDINATION POLYHEDRA—*See* Vol. IVA: CRYSTAL CHEMISTRY

CRANDALLITE GROUP—*See* ALUNITE GROUP

CRUDE OIL—*See* Vol. VI: CRUDE OIL COMPOSITION AND MIGRATION; Vol. XIV

CRYSTAL—*See* CRYSTAL HABITS

CRYSTAL DEFECTS—*See* CRYSTALS, DEFECTS IN; ORDER-DISORDER

CRYSTAL DRAWING—*See* CRYSTALLOGRAPHY: MORPHOLOGICAL

CRYSTAL FACE—*See* CRYSTAL HABIT; CRYSTALLOGRAPHY: MORPHOLOGICAL

CRYSTAL FORMS—*See* CRYSTAL HABIT; CRYSTALLOGRAPHY: MORPHOLOGICAL

CRYSTAL GROWTH

Mineralogy books often define a crystal in terms of the external crystal faces; a better definition uses the concept of an ordered internal arrangement of atoms or ions. A hypothetical perfect crystal is then a solid in which perfect three-dimensional order exists. All real crystals however contain many types of defects (e.g., dislocations, vacancies, interstitials, inclusions, low-angle grain boundaries, etc.; see *Crystals, Defects in*), so that a rigorous definition becomes difficult. The dividing line between a "single crystal" and a "polycrystalline mass" may even depend on the end use. For jewelry uses, quite large defects may be tolerated; whereas, for nonlinear optical purposes, even a low-angle grain boundary of 0.1° may make a crystal behave as if it consisted of two separate crystals.

General Considerations

The aim in crystal growth is to permit the formation of one or only a few crystal nuclei, and to provide controlled conditions so that the continuing growth from the nucleus to the desired size occurs without the nucleation of additional crystals and the formation of a polycrystalline mass. Any specified level of purity, required impurities, and desired degree of perfection will impose further limitations on the crystal grower.

In some techniques, nucleation is controlled by permitting it to occur in only a small space and in a large temperature gradient, e.g., Bridgman-Stockbarger growth (see Melt Techniques, below), or a seed crystal may be used to avoid the necessity for nucleation as in Czochralski crystal pulling. In epitaxial growth, a thin layer of crystal is grown on a substrate (see *Epitaxy*). This substrate may be a crystal of the same material, or a different material if some atomic planes have approximately the same spacing and can, therefore, supply nucleation (see section on Vapor-Phase Techniques).

Which of the many crystal-growth techniques to try first for a given crystal is a difficult choice even for an experienced crystal grower. The melting point, solubility, stability, and other physical and chemical properties of the material, as well as the requirements on size, purity, and perfection (freedom from solid or gaseous inclusions, low-angle grain boundaries, strains, etc.), will all affect the decision. In solution growth, for example, it is necessary to know the phase diagram of the solvent–solute system, and since few of these are available, it is usually the first task of the crystal grower to determine the phase diagram (see Vol. IVA: *Phase Equilibria*). Considerable judgment is involved, particularly when the needed data is incomplete, leading to the conclusion that crystal growth is as much an art as it is a science.

Why Grow Crystals?

Single crystals are grown for a wide variety of reasons. In the field of technology, the best known crystal devices are oscillators, transducers, and filters made from *quartz* crystals (Fig. 1) and semiconductor devices made from crystals of silicon and germanium. Among more recently discovered uses may be included microwave devices utilizing iron-containing ***garnets*** such as yttrium iron garnet ($Y_3Fe_5O_{12}$) and laser crystals based on yttrium aluminum garnet ($Y_3Al_5O_{12}$), *corundum* (Al_2O_3), *scheelite* ($CaWO_4$), *fluorite* (CaF_2), and gallium arsenide (GaAs).

Synthetic crystals are widely used in the jewelry field. Here "synthetics"—which duplicate natural gemstones, such as ruby (Al_2O_3 + Cr)—are distinguished from "artificial" or "imitation" stones—which merely show a resemblance in some respect. Synthetic crystals used to imitate *diamond* have included the soft but brilliant strontium titanate ($SrTiO_3$) and *rutile* (TiO_2) and the harder but somewhat dull rare-earth ***garnet***-type crystals: YAG-for yttrium aluminum garnet ($Y_3Al_5O_{12}$) and GGG-for gadolinium gallium garnet ($Gd_3Ga_5O_{12}$). More recently, synthetic cubic zirconia (ZrO_2 containing about 10% Y_2O_3) has been found to provide an imitation distinguishable from *diamond* only by the trained gemologist.

In the area of research, synthetic crystals are used extensively to study the physical properties of pure substances, of materials containing controlled amounts of impurities, and of isomorphous substitution series, as well as to study the stability fields of crystal phases and the possible conditions under which natural minerals may have formed. Since minerals often contain a variety of impurities, the distribution of individual impurities during crystal synthesis is studied to provide information on conditions of mineral formation. This is also the most direct way to establish which of the various impurities present in a mineral is the cause of the color or other physical characteristics.

Classification of Crystal-Growth Techniques

Crystal-growth techniques may be conveniently classified by the number of components present in the growth medium and by the state of matter from which the crystal forms. In "one-component growth," the composition of the growth medium is essentially the same as the composition of the growing crystal. Thus, the controlled solidification of pure water (H_2O) yields crystalline *ice* (H_2O). In "multicomponent growth," additional chemical species are present—most commonly as a solvent, as in the growth of salt crystals (*halite*, NaCl) from the brine ($NaCl-H_2O$) system.

In multicomponent growth, extra time is

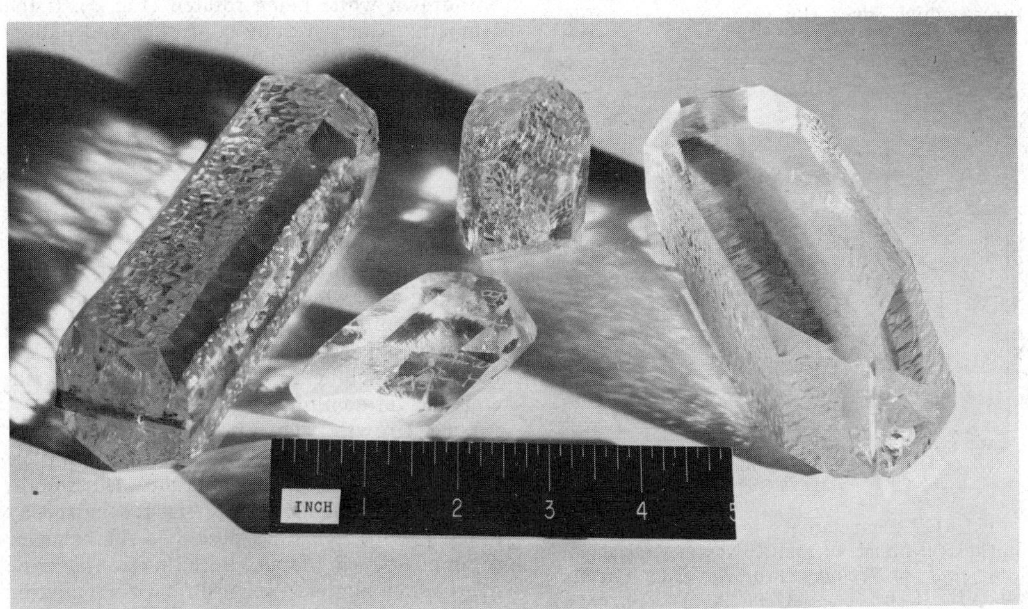

FIGURE 1. Hydrothermally grown *quartz* crystals (from K. Nassau, Crystal growth techniques, in *Techniques of Inorganic Chemistry*, vol. VII, H. B. Jonassen and A. Weissberger, eds., Interscience, New York, 1968; reprinted by permission of John Wiley & Sons, Inc. Photograph courtesy of Bell Telephone Laboratories).

needed for the unwanted components (e.g., the solvent) to diffuse away from the growing crystal. Thus, "solution techniques" (multicomponent) give growth rates in millimeters per day or week as against millimeters per hour in the case of "melt techniques" (one-component). However, many materials of interest melt at inconveniently high temperatures, decompose at or below the melting point, or form glasses rather than crystals from the melt, so that lower-temperature multicomponent techniques must be used.

General treatments on crystal growth have been given by Laudise (1970), Smakula (1961), and Wilke (1973). Gemstone synthesis has been covered by Nassau (1980). Articles on current research can be found in the following journals: *Journal of Crystal Growth; Journal of Applied Physics; Journal of the American Ceramic Society; American Mineralogist;* and *Mineralogical Magazine.*

Melt Techniques

Solidification in a Crucible. A container or crucible is filled with essentially pure material, which is completely melted and then solidified in a carefully controlled fashion to yield a single crystal. The most common arrangement consists of a crucible with a pointed lower end, which is lowered from the center of a vertical tube furnace as shown in Fig. 2b. A better arrangement with a sharper temperature gradient uses two furnaces set at different temperatures, which gives the temperature profile of

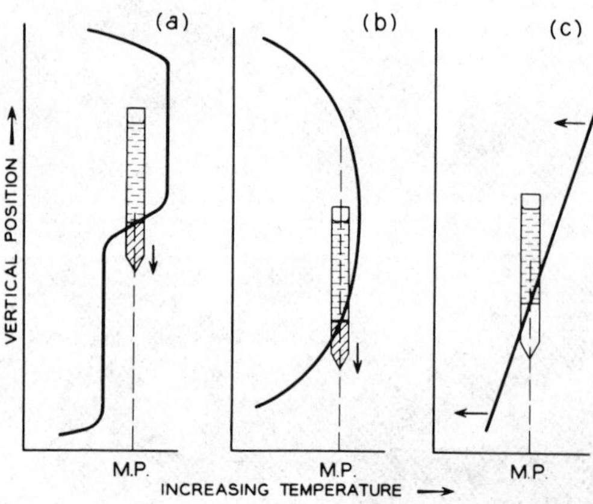

FIGURE 2. Temperature profiles in Bridgman-Stockbarger furnaces (from K. Nassau, Crystal growth techniques, in *Techniques of Inorganic Chemistry,* vol. VII, H. B. Jonassen and A. Weissberger, eds., Interscience, New York, 1968; reprinted by permission of John Wiley & Sons, Inc.). (a) Two-heater winding arrangements. (b) Single-furnace arrangement. (c) Graded-winding arrangement.

Fig. 2a. If a graded furnace winding is used, the temperature can be lowered slowly with the crucible being held still as in Fig. 2c.

Because the small region at the pointed tip has the lowest temperature in the melt, only one nucleus usually results, and at the completion of the experiment the solidified crucible contents consist of one single crystal. This technique is frequently known as the Bridgman-Stockbarger technique after two early experimenters. Many variations exist, including one in which a horizontal boat with a pointed end is moved horizontally through a furnace. Crystals of *fluorite, halite, sylvite,* and the like up to one foot in diameter have been grown in this way.

With very high-melting (refractory) materials such as cubic zirconia (M.P. = 2750°C) a related technique called skull-melting is used. Here a series of closely spaced water-cooled copper fingers form a cup. Energy in the form of radio-frequency waves penetrates and melts the contents, which are contained in a "skull" of powder prevented from melting by the cold fingers. On slowly cooling the container, directional solidification produces a cluster of columnar crystals up to several centimeters in size.

Pulling from the Melt. This technique is often designated Czochralski growth after its originator. A crucible filled with molten material is again used, but this time a seed crystal is inserted into the top of the melt and slowly withdrawn while being rotated (Fig. 3). If the temperature is carefully controlled and pulling is slow and steady (about 1 cm per hour) crystal boules many centimeters in length and several centimeters in diameter can be readily pulled.

For high-melting materials, use of platinum, iridium, or rhodium crucibles may be necessary. Many different sources of heat can be used (radio frequency, resistance, gas) and close temperature control is necessary. Convection currents, which can become quite complex in nature, must be carefully controlled. Crystals that can be readily grown by the Czochralski technique include ruby, *scheelite, fluorite,* silicon, and germanium.

Zone Growth. This technique is primarily used to purify materials, but can produce single crystals at the same time. The importance of this technique was first recognized by Pfann (1966), who has described the behavior of impurities in detail. Both horizontal zone melting in a boat and vertical float-zone melting are widely used to purify and crystallize silicon and germanium for the electronics industry.

For float zoning (Fig. 4), a solid rod of material is supported vertically between two

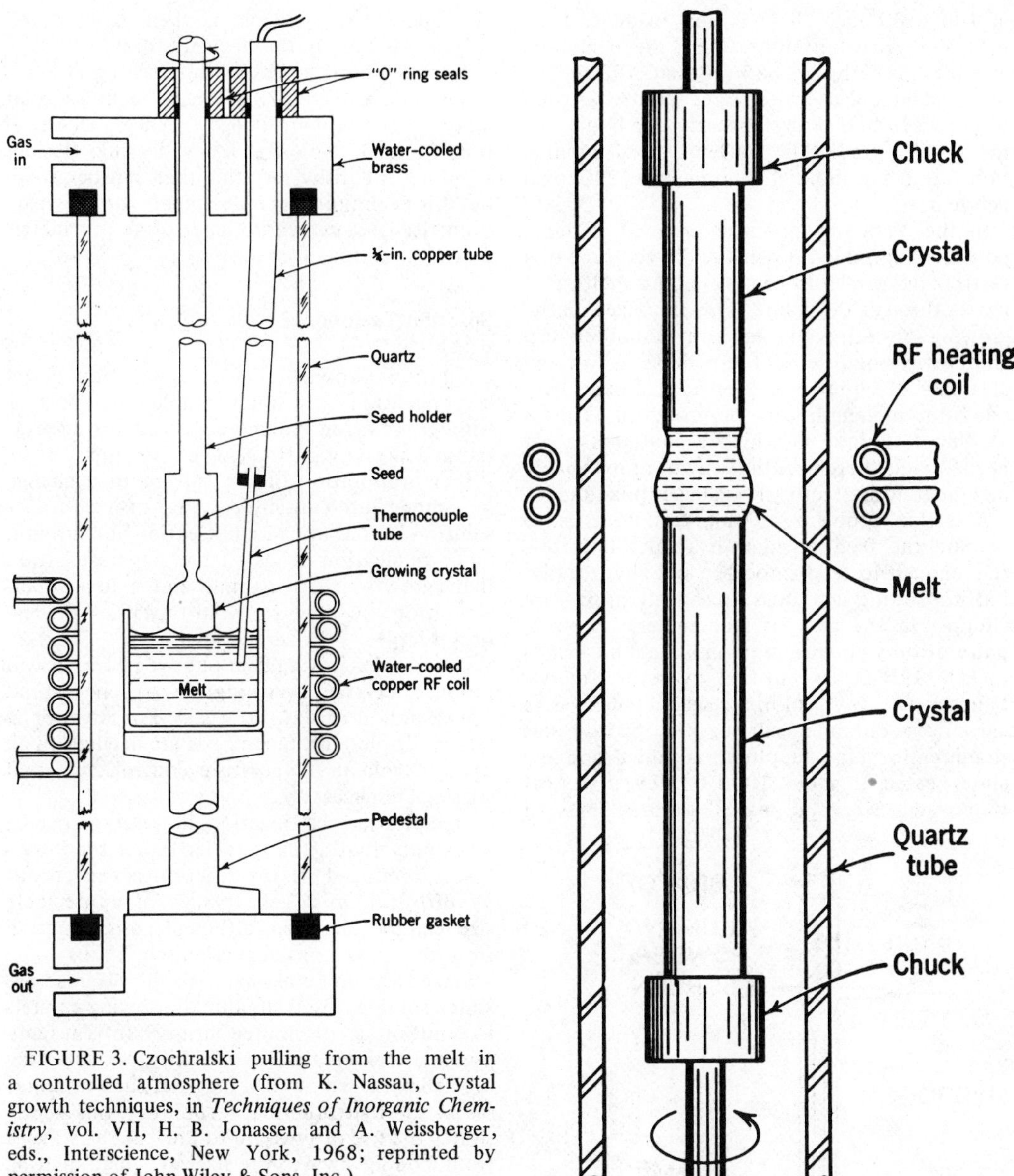

FIGURE 3. Czochralski pulling from the melt in a controlled atmosphere (from K. Nassau, Crystal growth techniques, in *Techniques of Inorganic Chemistry*, vol. VII, H. B. Jonassen and A. Weissberger, eds., Interscience, New York, 1968; reprinted by permission of John Wiley & Sons, Inc.).

FIGURE 4. Floating-zone apparatus (from K. Nassau, Crystal growth techniques, in *Techniques of Inorganic Chemistry*, vol. VII, H. B. Jonassen and A. Weissberger, eds., Interscience, New York, 1968; reprinted by permission of John Wiley & Sons, Inc.).

chucks and rotated. By the use of a radio-frequency coil or other heating device a narrow zone of the rod is melted and slowly moved up the rod (perhaps 1 cm per hour). Material melts from the upper part of the rod and solidifies onto the lower part as a single crystal if conditions are suitable. Surface tension supports the floating zone; this is another "crucibleless" technique.

Depending on the distribution coefficient (see the section on Theory of Crystal Growth), any given impurity may move with or against the motion of the zone, and many repeat passes of the zone may be made to obtain a desired degree of purification.

Flame-Fusion Growth. Flame-fusion crystal growth was first described by A. V. L. Verneuil in 1902, although there is now evidence that, in a somewhat different form, the technique was used as early as 1885 (Nassau, 1980). The most celebrated use of this technique has been to grow synthetic ruby and sapphire boules for the jewelry trade, the current world production rate being about 1 billion carats (200 tons) per year.

In the Verneuil apparatus (Fig. 5), alumina powder is sprinkled down the oxygen tube of a vertical oxygen-hydrogen torch and melts as it passes through the flame. This impinges on the growing crystal boule to form a molten cap and slowly solidifies as the pedestal is lowered at a rate of about 1 cm per hour. Boules over one foot in length can be produced. This is another technique in which a container is not needed and, therefore, the growth of extremely high-melting (refractory) crystals is possible.

A major problem of this technique is to prepare the feed powder in a suitable state—fine enough to feed smoothly and melt readily, but not so fine as to dust excessively or to form clumps. In the case of feed material for sapphire or ruby growth, ammonium alum [$NH_4Al(SO_4)_2 \cdot 12H_2O$] is purified by several recrystallizations. Any coloring agent (such as a chromium-containing salt for ruby or iron plus titanium for blue sapphire) is added and the alum heated at about 1000°C. This drives off water and SO_3 and leaves a fluffy mass of alumina, Al_2O_3, which is then ground and screened for use in the Verneuil torch.

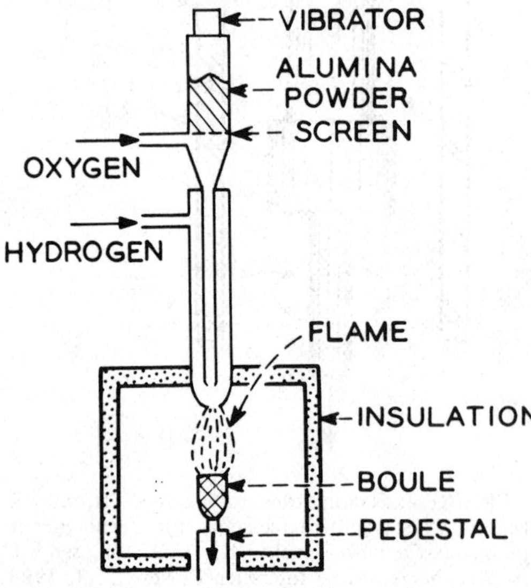

FIGURE 5. Verneuil apparatus for flame-fusion growth (from K. Nassau, Crystal growth techniques, in *Techniques of Inorganic Chemistry*, vol. VII, H. B. Jonassen and A. Weissberger, eds., Interscience, New York, 1968; reprinted by permission of John Wiley & Sons, Inc.).

Several modifications of this technique exist, using alternate forms of heat, such as solar image or carbon-arc image furnaces, electrical resistance, or radio-frequency heating. Besides sapphire and ruby, crystals that can be grown by this technique include *spinel, rutile*, strontium titanate, *magnetite*, and refractory metals such as tungsten.

Solution Techniques

Solution Growth. The crystallization or recrystallization of water-soluble materials is widely used for commercial products such as salt and sugar (Van Hook, 1961; Mullins, 1961). Either evaporation of the solvent or a change in temperature (usually cooling) of a saturated solution is used to cause nucleation and growth. Since nucleation needs a larger supersaturation than growth, seed crystals and a low supersaturation are used to avoid unwanted nucleation if large crystals are desired.

The Holden crystallizer shown in Fig. 6 will produce crystals weighing up to one pound over a period of days to weeks. For larger crystals, multiple-tank techniques are needed. In all cases, excellent temperature control and good stirring are necessary.

Precipitation by reaction in solution occurs so rapidly that a fine powder rather than crystals is produced. If the reaction is carried out by diffusion in a gel, crystals of appreciable size can be obtained, although contamination from the gel is a problem (Henisch, 1970).

Since few mineral materials of interest are water soluble, the technique of growing crystals in solution is of limited utility. It is usually possible to increase the solubility by increasing the temperature. With a material such as water (as in the hydrothermal technique), this necessitates the use of pressure to prevent the solvent from boiling away. Alternatively, high-melting nonvolatile solvents can be used (as in the flux technique).

Hydrothermal Growth. Originated by H. de Senarmount in 1851, hydrothermal growth is the crystal growing technique closest to the manner in which most natural minerals were formed. Hydrothermally grown *quartz* crystals (Fig. 1) are currently in commercial production. Many materials besides *quartz* have been grown by this technique, including emerald, sapphire and ruby, *calcite, zincite,* ***tourmaline***, etc., but generally only in small crystals. Extended studies are needed to optimize conditions for the growth of large crystals, and this is usually only economically feasible if at least

FIGURE 6. The Holden crystallizer with crystals of ammonium dihydrogen phosphate growing from water solution (from K. Nassau, Crystal growth techniques, in *Techniques of Inorganic Chemistry,* vol. VII, H. B. Jonassen and A. Weissberger, eds., Interscience, New York, 1968; reprinted by permission of John Wiley & Sons, Inc. Photograph courtesy of Bell Telephone Laboratories).

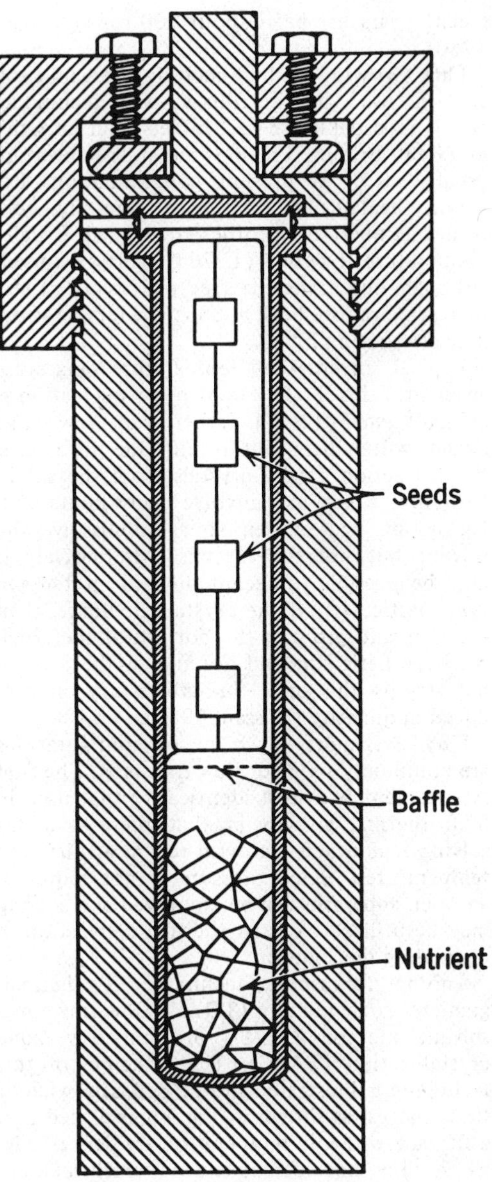

FIGURE 7. Laboratory hydrothermal bomb (from K. Nassau, Crystal growth techniques, in *Techniques of Inorganic Chemistry,* vol. VII, H. B. Jonassen and A. Weissberger, eds., Interscience, New York, 1968; reprinted by permission of John Wiley & Sons, Inc.).

medium-scale industrial production is anticipated.

The pressure vessels or autoclaves used for hydrothermal growth are made of special steel to withstand the high temperature and pressure. Platinum, silver, or gold linings may be needed to prevent corrosion, and the autoclave closures become quite elaborate. Since explosions occur occasionally, the vessels are also known as "bombs."

Laboratory bombs (Fig. 7) may range from a few inches to over a foot in length. Feed material (crushed natural *quartz*, for example) is placed in the lower part of the bomb and thin seed plates in the upper part. The bomb is partially filled with solution and the closure tightened. As the temperature passes about 100°C the solution boils and from this point on the bomb is full of fluid.

A commercial bomb is typically 12 ft. (3.7 m) long and 14 in. (36 cm) in diameter with walls 4 in. (10 cm) thick. The bomb may be filled to 80% with 4% sodium hydroxide solution; the temperature may be about 360°C at the top and 400°C at the bottom. This produces a pressure of about 25,000 psi. Such a bomb yields about 40 crystals weighing over 1 lb. (0.5 kg) each in less than three weeks. The world annual *quartz* production capacity in recent years has been about 700 tons (Nassau, 1980).

Flux Growth. A high-melting inorganic solvent is used in the flux-growth technique, which is also known as "fluxed melt growth" in Great Britain. Precious-metal crucibles are usually used to prevent contamination. Resistance-heated furnaces with carefully controlled temperature profiles are needed. Usually temperature lowering (1°C/hr. or less) is used, although flux transport (see next section) and flux evaporation, both at constant temperature, may also be employed.

Typical solvents and temperature ranges are given in Table 1. The control of nucleation is difficult, many small rather than few large crystals often being the result. One problem is the extraction of the crystals from the solidified flux, which may involve many hours boiling in an acid chosen so as to dissolve the solvent but not the crystals. Alternatively it may be possible to decant the still molten solvent, particularly if the crystals are denser than the flux and grow at the bottom of the crucible. Synthetic emerald for use in the jewelry industry is the only flux-grown crystal produced in quantity (Nassau, 1980).

Flux Transport. Two major configurations are commonly used for flux transport. The first arrangement is almost identical to that used in hydrothermal growth in that there is a dissolving region and a growth region at different temperatures. Since pressure is not required, an open apparatus can be employed and stirring may be utilized to assist thermal convection in circulating the solution.

Another transport technique is thermal-gradient zone melting (TGZM), the travelling-solvent method (TSM), or thin-alloy zone crystallization (TAZC). The principle of this technique is illustrated in Fig. 8. A sandwich of three pieces, as shown in Fig. 8b, is placed into a furnace with the temperature gradient of Fig. 8a. A thin sheet of solvent A, which melts at the temperature used, is located between a crystal seed B and the feed material C.

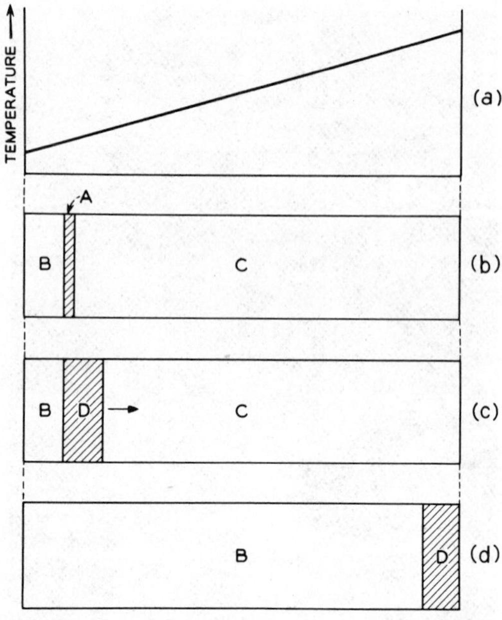

FIGURE 8. Thermal-gradient zone melting (from K. Nassau, Crystal growth techniques, in *Techniques of Inorganic Chemistry,* vol. VII, H. B. Jonassen and A. Weissberger, eds., Interscience, New York, 1968; reprinted by permission of John Wiley & Sons, Inc.). (a) Temperature profile; (b) starting configuration; (c) growth configuration; and (d) final configuration.

Under the driving force of the temperature gradient, the feed material C dissolves in the solvent A to form the saturated solution D of Fig. 8c, while crystallization occurs onto the seed B. This produces motion of the flux zone to the right and, after a period of time, all the feed material has crystallized, resulting in the state shown in Fig. 8d.

Various configurations are possible, and modifications have also employed an electric field instead of a thermal gradient. The TGZM technique has been used for the growth of crystals of silicon carbide, silicon, gallium arsenide, etc.

TABLE 1. Typical Conditions for Flux Growth

Flux	Temp. Range °C	Crystals Grown
PbF_2, PbO, and mixtures	1300–900	Rare-earth ***garnets,*** emerald, ruby, MgO, ZnO, BeO, ferrites
MoO_3:Li_2O, WO_3:Na_2O, etc.	1300–700	$CaWO_4$, $SrMoO_4$, $ZrSiO_4$, TiO_2, BeO, emerald
B_2O_3:Na_2O,B_2O_3:BaO, etc.	1200–900	***Garnets,*** ferrites, CeO_2
KF, KCl, etc.	1100-400	$BaTiO_3$, $KNbO_3$, ZnS, CuCl

Solid-Solid Transformations

Small crystals, for X-ray diffraction studies for example, can often be obtained by "sintering," i.e., heating a compacted powder close to the melting point (or a transition temperature if present).

In the case of readily deformable materials such as metals, the strain-anneal technique is useful. A sheet, bar, or wire is deformed by rolling, bending, or drawing and then annealed. The process is repeated a number of times. Some crystal grains grow at the expense of others and, under favorable conditions, single crystals as large as one inch across can be obtained, with aluminum, for example.

At high pressure (above 1 million psi) and high temperatures (above 2000°C), carbon will convert to *diamond* by a solid-solid transformation. This process can be speeded up by the presence of metals, such as nickel, but growth of the *diamond* crystals then occurs from nickel solution essentially by the TGZM method. The apparatus needed to contain these high pressures and temperatures is very complex, but growth times are short, on the order of seconds to minutes only for small crystals (grit); crystals up to one carat have been grown over week-long periods (Nassau, 1980).

Vapor Phase Techniques

Condensation Growth. Under appropriate conditions many materials will condense from the vapor directly into the crystalline state without passing through a liquid phase. If a liquid is not formed during the evaporation step, then this process is sublimation.

Open and closed systems are illustrated in Figs. 9a and 9b, the temperature gradient of Fig. 9c being suitable in both cases. The feed material evaporates at the higher temperature at A and deposits on a seed crystal at B at the lower temperature (or spontaneously nucleates at C). In the open system of Fig. 9a a stream of inert gas is used; in the closed system of Fig. 9b, either a vacuum or an inert gas may be employed.

Growth rates are quite slow and nucleation is difficult to control in this technique. Crystals grown by condensation include zinc oxide and cadmium sulfide.

Reaction Growth. Vapor-phase reaction growth is usually carried out in an apparatus similar to the open condensation system of Fig. 9a. Separate inert-gas streams may carry the components of the compound to be grown, and active gases may also be involved. Reaction may occur where the gas streams come together or, on a heated substrate, where epitaxial growth may occur. Many variations are possible. The reaction may involve combination of elements in an inert gas stream, for example:

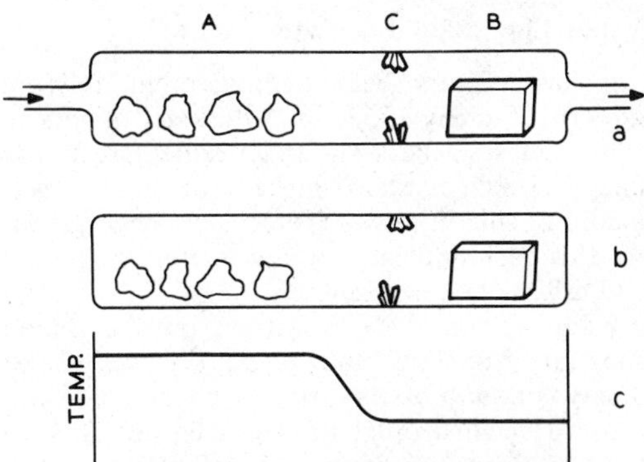

FIGURE 9. Growth by condensation (from K. Nassau, Crystal growth techniques, in *Techniques of Inorganic Chemistry*, vol. VII, H. B. Jonassen and A. Weissberger, eds., Interscience, New York, 1968; reprinted by permission of John Wiley & Sons, Inc.). (a) Growth in a gas stream; (b) growth in a vacuum; and (c) temperature profile.

$$Cd + S \rightarrow CdS$$

decomposition in an inert or active gas stream:

$$CH_3SiCl_3 \rightarrow SiC + 3HCl \quad (\text{in } H_2)$$

or reaction with an active gas stream:

$$7SiCl_4 + C_7H_8 + 10H_2 \rightarrow 7SiC + 28HCl$$

Transport Growth. Vapor-phase transport growth is often carried out in a sealed tube similar to the condensation system of Fig. 9b. In addition to the feed material and a possible seed crystal or substrate, the tube contains an active carrier gas.

In the growth of germanium by iodine transport, a small amount of iodine is added. This reacts with the germanium to form GeI_2. Growth occurs at the higher temperature (500–700°C) by the disproportionation reaction:

$$2GeI_2 \rightarrow GeI_4 + Ge$$

The GeI_4 vapor travels to the cooler (400°C) end of the tube, where fresh germanium is picked up by the reverse reaction:

$$GeI_4 + Ge \rightarrow GeI_2$$

With about 5 mg I_2 per cm^3, growth rates of about 10 μm/hr. are obtained.

Transport can also be performed in an open flow system (Fig. 9a) if the transporting agent is present in the gas stream.

Vapor-Liquid-Solid Growth

It has recently been demonstrated that the growth of many types of "whiskers" from the vapor phase occurs via an intermediate liquid phase. Single-crystal whiskers of silicon, germanium, silicon carbide, etc., have been grown by this technique up to centimeters in length and millimeters in diameter.

When silicon is being grown onto a silicon substrate from the vapor phase, the presence of a tiny amount of impurity, such as gold, will cause a small droplet of Au-Si liquid alloy to form. Silicon is taken up by this alloy droplet from the vapor phase, and crystallization of silicon occurs onto the substrate. This results in the growth of a silicon pillar, the alloy droplet riding on the pillar as shown in Fig. 10. Growth terminates when the gold, which has a small solubility in the growing silicon, is completely used up.

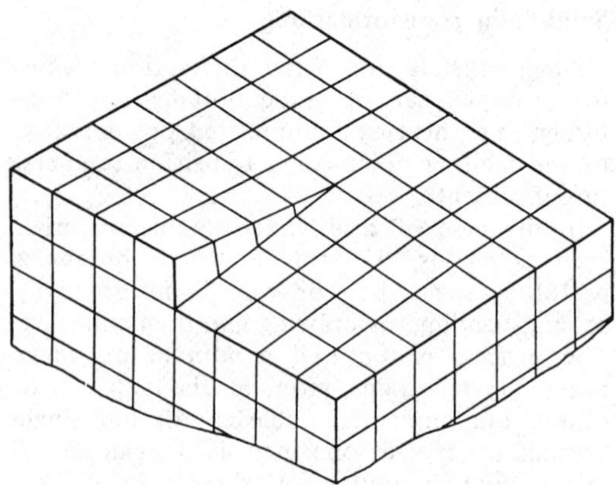

FIGURE 11. Schematic representation of a screw dislocation (from K. Nassau, Crystal growth techniques, in *Techniques of Inorganic Chemistry,* vol. VII, H. B. Jonassen and A. Weissberger, eds., Interscience, New York, 1968; reprinted by permission of John Wiley & Sons, Inc.).

Theory of Crystal Growth

Three aspects of crystal growth are particularly interesting from a theoretical point of view: nucleation, growth, and impurity distribution.

Nucleation may be homogeneous–i.e., within the bulk of a phase–or heterogeneous–i.e., on a surface, which may be the wall of a container, an epitaxial substrate, or an impurity particle. In the theory of homogeneous nucleation, the formation of an "embryo" is the critical stage. This arises from the random formation of a small ordered cluster of atoms, which may grow to become a nucleus, i.e., a seed crystal, or may be dispersed again by the same random processes that formed it.

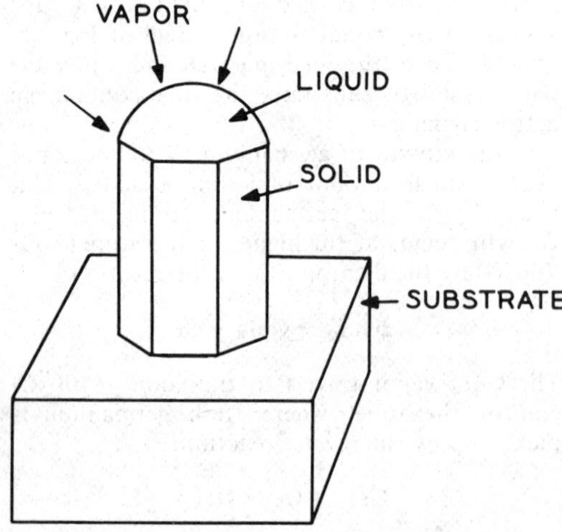

FIGURE 10. Vapor-liquid-solid growth (from K. Nassau, Crystal growth techniques, in *Techniques of Inorganic Chemistry,* vol. VII, H. B. Jonassen and A. Weissberger, eds., Interscience, New York, 1968; reprinted by permission of John Wiley & Sons, Inc.).

The rate of formation of nuclei rises very rapidly with the degree of supersaturation (or supercooling). If the temperature is lowered too much, the nucleation rate decreases again due to kinetic effects.

Several growth mechanisms have been postulated. The simplest, that of the addition of building units to a partially completed crystal layer, is probably unimportant, since on completion of the layer, nucleation of a new layer needs more supercooling or supersaturation than is commonly observed. The "screw dislocation" mechanism avoids this difficulty because, however many units are added to the step of a screw dislocation (Fig. 11), the dislocation always remains present. In metals, the crystal faces usually remain rough on the atomic level, so that nucleation of a new layer does not arise. The mechanism of "vapor-liquid-solid" growth has already been described above.

The distribution of an impurity is described by the distribution coefficient, $k = C_s/C_l$, where C_s is the concentration of the impurity in the solid phase (crystal) and C_l is the concentration in the liquid phase (melt). The derivation of k from a phase diagram is shown in Fig. 12; k is a constant in dilute solution during slow (equilibrium) growth.

If a quantity of melt is solidified by "normal freezing" (Bridgman-Stockbarger or Czochralski growth, as against "zone freezing" in zone or Verneuil growth) the distribution of an im-

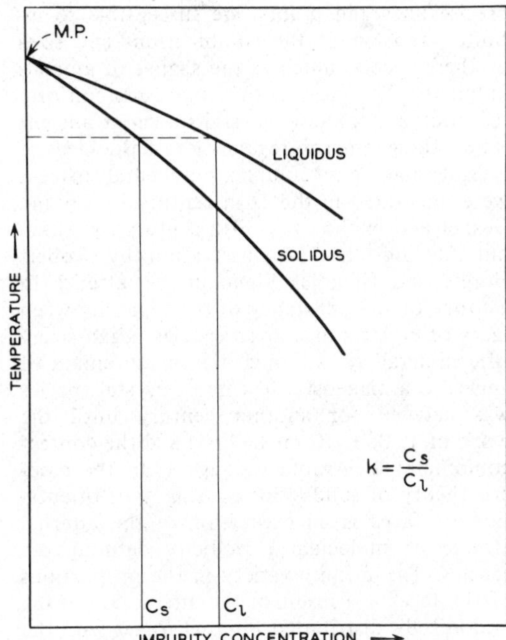

FIGURE 12. Phase diagram illustrating the depression of melting point by an impurity and the distribution coefficient k (from K. Nassau, Crystal growth techniques, in *Techniques of Inorganic Chemistry*, vol. VII, H. B. Jonassen and A. Weissberger, eds., Interscience, New York, 1968; reprinted by permission of John Wiley & Sons, Inc.).

purity with starting concentration C_0 is given to a first approximation by

$$C/C_0 = k(1-g)^{k-1}$$

where the concentration is C after a fraction g of the melt has solidified and ideal stirring is assumed. This gives curve (a) of Fig. 13. In the absence of stirring, the excess impurity (C_0-kC_0) must diffuse away from the solidification interface, and the result is the distribution of curve (b) of Fig. 13.

Different considerations apply to zone freezing, where the distribution after one pass of a zone of length l is given by

$$C/C_0 = 1-(1-k)e^{-kg/l}$$

This produces a curve somewhat similar to (b) in Fig. 13. Equations and curves for many different cases have been given by Pfann (1966).

Crystal Growth in Nature

Igneous rocks are formed during the solidification of magmas, which consist predominately of the oxides of Si, Al, Ca, Mg, Fe, Na, and K.

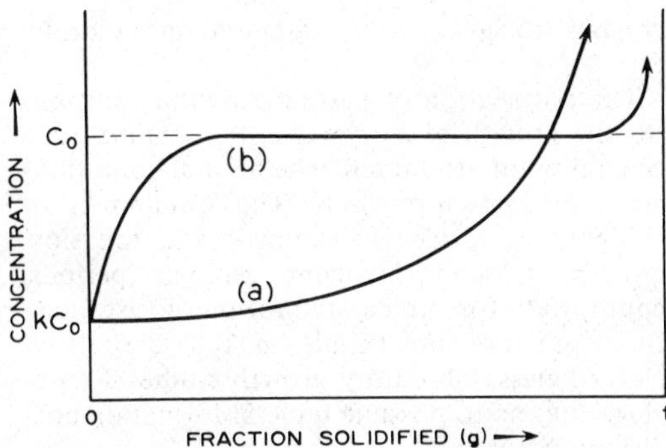

FIGURE 13. Impurity distribution for $k < 1$ (from K. Nassau, Crystal growth techniques, in *Techniques of Inorganic Chemistry*, vol. VII, H. B. Jonassen and A. Weissberger, eds., Interscience, New York, 1968; reprinted by permission of John Wiley & Sons, Inc.). (a) After normal freezing with ideal stirring and (b) with limited stirring.

Depending on the exact composition and conditions, the crystallization of magmas produces rocks containing crystals of *quartz*, ***feldspar***, ***feldspathoids***, ***pyroxenes***, *hornblende*, *biotite*, and ***olivine*** (see *Rock-Forming Minerals*). The process is intermediate between the laboratory melt and flux and hydrothermal growths.

Those components rejected during this primary crystallization of magma form the less viscous pegmatite and hydrothermal vein fluids (see *Pegmatite Minerals* and *Vein Minerals*). These fluids contain large amounts of volatile material, such as H_2O, CO_2, HCl, HF, and sulfur compounds. Their crystallization is thus close to the laboratory hydrothermal growth, but with quite large mineralizer concentrations. This leads to the spectacular pegmatite gem and mineral crystals such as, for example, *quartz*, ***feldspars***, ***micas***, *beryl*, and ***garnet***. Diluted with groundwater, such solutions continue hydrothermal growth in hot-spring areas. Study of the fluid inclusions trapped during crystal growth yields information on the growth temperature and fluid composition (Roedder, 1962; see also *Barometry, Geologic; Thermometry, Geologic*).

The growth of crystals such as calcium carbonate, sodium chloride, sulfates, and borates in evaporites corresponds to the laboratory water-solution growth (see *Evaporite Minerals; Saline Minerals*). The equivalent of solid-state transformation is seen in a very complex manner in contact and regional metamorphism, with the possibilities of high pressure, localized melting, and the presence of volatile material being further complications. Vapor-phase condensation and reaction growth occurs in the fumaroles of active volcanic regions, leading to crystals such as *sulfur*, ammonium chloride, etc.

The duplication of natural growth conditions for the growth of a crystal in the laboratory is generally not attempted. The natural conditions are never known precisely. The containment of a highly corrosive environment and the slow growth, possibly involving geologic periods, appropriate for the crystallization of vast quantities are not practicable in the laboratory. Nevertheless, laboratory growth studies do provide clues as to possible mineral-formation conditions in nature.

K. NASSAU

References

Henish, H. K., 1970. *Crystal Growth in Gels*, University Park, Pa.: Pennsylvania State University Press, 111p.

Laudise, R. A., 1970. *The Growth of Single Crystals*, Engelwood Cliffs, N.J.: Prentice-Hall, 352p.

Mullin, J. W., 1961. *Crystallization*. London: Butterworths, 480p.

Nassau, K., 1980. *Gems made by Man*. Radnor, Pa.: Chilton Book Co., 364p.

Pfann, W. G., 1966. *Zone Melting*. 2nd ed. New York: Wiley, 310p.

Roedder, E., 1962. Ancient Fluids in Crystals. *Sci. Am., 207*, 38.

Smakula, A., 1961. *Einkristalle*, Berlin: Springer-Verlag, 431p.

Van Hook, A., 1961. *Crystallization*, New York: Reinhold, 325p.

Wilke, K.-Th., 1973. *Kristallzüchtung*. Berlin: VEB Deutscher Verlag der Wissenschaften, 923p. plus 177-page reference booklet.

Cross-references: *Abrasive Materials; Barometry, Geologic; Crystallography: Morphological; Crystals, Defects in; Diamond; Epitaxy; Gemology; Glass: Theory of Crystallization; Magnetic Minerals; Metallurgy; Quartz; Synthetic Minerals; Thermometry, Geologic;* Vol. IVA: *Crystal Chemistry; Fluid Inclusions; Hydrothermal Solutions; Phase Equilibria; Silicates.*

CRYSTAL HABIT

The general shape of a crystal and the proportions of its bounding sides—such as elongate, tabular, bladed, or other—is known as its habit, from the Latin *habitus*. The term habit as a nontechnical characterization of appearance first came into mineralogical usage from biology early in the 20th century. The surfaces of minerals, though commonly consisting of well-defined plane facets meeting in geometrically precise lines and points, are susceptible to infinite variation in the combinations and sizes of these facets, much as the shapes of animals or plants. The faces at the same time conform to strict and precise laws. Known in ancient times, the geometric regularities of the faces of crystals making up complex polyhedral surfaces were illustrated in the 16th century by Conrad Gesner and in the early 17th century by Anselmus de Boodt. The observation by Robert Hooke and Nicholas Steno in the latter 17th century of the constancy of the angles between faces of crystals of a given species began scientific mineralogy, although the measurement of angles as a diagnostic feature of crystal species was delayed for another century—until the work of J. B. L. Romé de l'Isle and the contact goniometer of Arnold Carangeot. In the modern theory of solids, this constancy of orientation of faces is an expression of the internal atomic or molecular periodicity defining the species. The infinite variety in the proportions of the faces is a result of the interaction of the periodically distributed internal forces with the environment during crystal growth (q.v.). The term "form," ordinarily synonymous with shape, is reserved by crystallographers for the ensemble of faces generated by the operations of "symmetry" (q.v.).

Description of Habits

The habits of single specimens are described by any convenient term seldom requiring explanation, as tabular (***micas***), acicular or needle-like (*rutile*), prismatic, columnar, bladed, blocky, stubby (*zircon*), dendritic or arborescent (snowflake), wiry (*silver*), leafy, sparry. Terms may be geometric, as pyramidal, orthogonal; or crystallographic terms may be used to characterize crystals (octahedral, rhombic, scalenohedral). A single species may occur in many very different habits (see Figs. 1 & 2). Victor Goldschmidt, between 1913 and 1923, published the 18-volume *Atlas der Krystallformen* correlating the morphological measurements and illustrations of the entire mineralogical literature. For minerals such as *barite* ($BaSO_4$), more than 700 different specimens were illustrated, while for *pyrite* or *calcite*, the numbers reached into the thousands. Morphologists measured the angles of each face of a specimen with respect to three intersecting axes selected according to the principles of crystallography. Faces could be indexed and grouped by symmetry into forms. The assemblage of forms present on a specimen constituted a combination of forms which, with the habit, characterized the specimen morphologically.

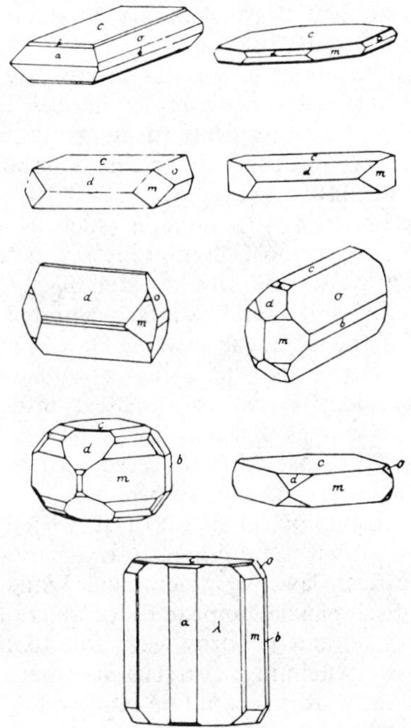

FIGURE 1. Nine distinct habits of *barite* (from Schneer, 1970).

Similar general terms of appearance characterizing groups of crystals are also employed, sometimes blurring the distinctions between single crystals, aggregates of crystals, twinned (butterfly, iron-cross) crystals, and even cleavage (fibrous or asbestiform ***serpentine***), and rocks (oolitic limestone and oolitic *hematite*). *Hematite* may also occur in specular or mirror-like habit, or with completely distinct appearance and properties such as earthy. Still other habit terms in the literature for groups of roughly spherical aggregates are botryoidal (globular like grapes) and reniform (kidney-like). Mammillary or botryoidal *malachite* is semiprecious (see *Gemology*). Most bauxite is pisolitic (pea-like) or, as at Les Baux (France), oolitic. Native metals, such as Michigan *copper*, may be arborescent. *Gypsum*, *barite*, and *calcite*, among others, occur as radiating groups or rosettes called "desert roses."

Surface textures due to conditions of growth are also considered as habits. "Striated cubic" describes the characteristic surface appearance of common *pyrite*, which may also be framboidal (with bumps like a raspberry) or massive (fine-grained). The superficial appearance of intergrowths of more than one species are often

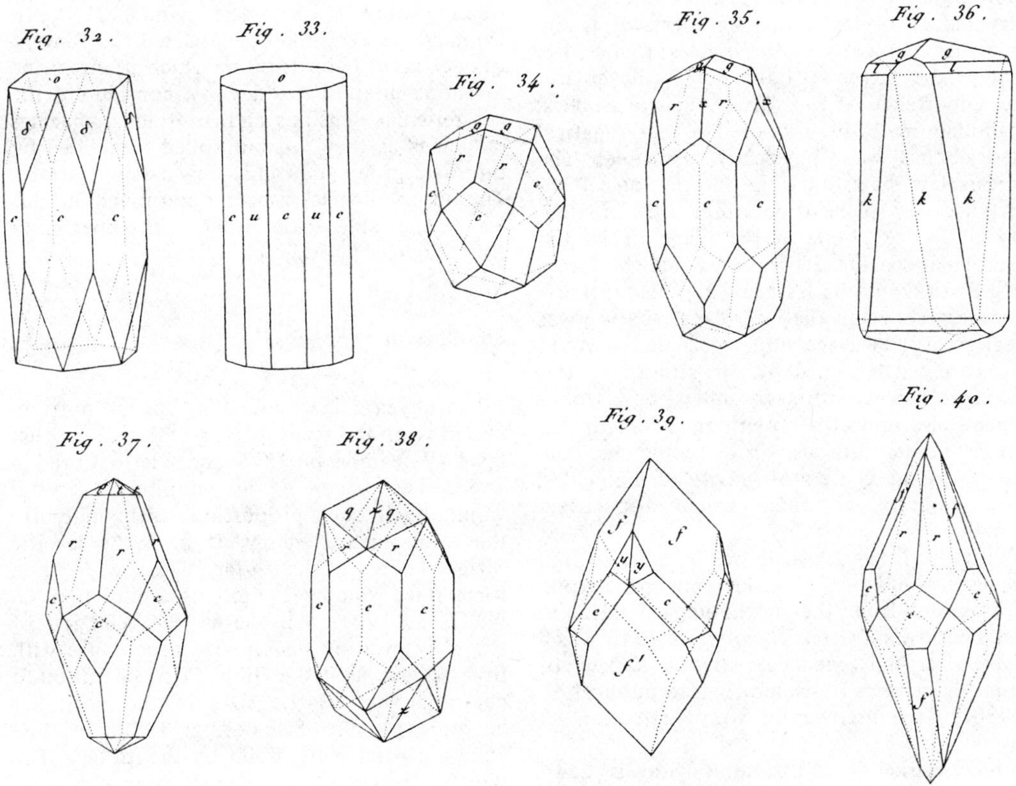

FIGURE 2. Nine habits of *calcite* (from Haüy, 1801).

characteristic (rutilated *quartz*, graphic granite). Mode of occurrence, such as crusty or drusy (a surface growth of fine crystals) or vuggy (lining a cavity) are other terms for habits. "Dog-tooth spar" identifies pointed scalenohedra of *calcite* approaching each other from the opposite walls of a cavity. In ordinary usage the imagination and good sense of the mineralogist are the only limitations on the terminology of crystal habits.

Origin of Habit

In the strict sense, the habit of a crystal consists of the faces present and their proportionate sizes. Although Kepler, Descartes, Hooke, Huyghens, and others in the 17th century had accounted for the plane polyhedral regularities of crystals by postulating an internal periodic molecular structure, the relationship of this structure to the actual forms present on the specimen is usually attributed to René Just Haüy, who observed that the tangents of angles made by surface planes with the crystallographic axes were in simple integer proportions. These integers, suitably transformed, became the Miller indices (see *Directions and Planes*) of today. Haüy's rule states that the order of morphological importance (size and frequency of occurrence) of the forms of a crystal is the order of complexity of the indices; that is, the simpler the integers of the index, the greater the importance of the form. Auguste Bravais, beginning in 1849, proposed a mathematical theory that was physically plausible. The Bravais law states that the morphological importance of a plane is determined by the density of lattice points in the plane or by the interplanar spacing. He reasoned that the forces of growth would be least and the plane, therefore, largest where the interplanar spacing was greatest. The cleavage along such planes would be entirely determined by this structure; morphology, however, Bravais understood would depend also upon the growth environment. His lattice points, Bravais wrote, could be considered as the centers of gravity of molecules, the symmetry of which would also affect growth.

With the Bravais law, it was possible to find the lattice and often considerable space-group information from the morphological data. E. von Fedorov's *Crystal Kingdom* of 1913 was a massive crystal chemical atlas—a survey of structure rendered obsolete before publication (1920) by the discovery of X-ray diffraction.

Donnay-Harker Generalization of Bravais' Law

E. Mallard, G. Junghan, H. Baumhauer, L. Sohncke, P. Groth, and C. Peacock, among others, applied Bravais' theory to mineralogy. According to Haüy's rule, the order of importance of forms of all species should have been from simplest to most complex indices. For the case of isometric crystals (using modern notation), forms in order of decreasing importance would be {100}, {110}, {111}, {210}, {211}, etc. By contrast, if a mineral (such as *magnetite*) occurred most commonly as octahedra {111}, Bravais' rule would refer *magnetite* to the "octahedral mode" or face-centered cubic lattice of the isometric system. The (111) plane of this lattice is the plane of greatest net reticular density and of greatest interplanar spacing. Georges Friedel, by showing in detailed studies that entire observed sequences of forms in order of their morphological importance (morphological aspect or M.A.) fell in the required order, established Bravais' theory as an empirical law of mineralogy. While occasional discrepancies appeared for single forms, both Friedel and Fedorov were able to demonstrate overwhelming correlation where complete series of forms could be observed.

With the advent of structure determination by X-ray diffraction, many of the observed discrepancies were considered by P. Niggli and others to result from the simplification in treating only ideal lattice points. J. D. H. Donnay and D. Harker generalized the Bravais-Friedel law to the form in which it enters the textbook literature today by considering the net reticular densities and interplanar spacings of planes as they are multiplied or divided by the space-group glide and screw-axis operations. The morphological aspect determined in this way is the same as the order of the lines in an X-ray powder photograph.

Equilibrium

The physical laws governing the proportionate sizes of the faces of a crystal were written by J. W. Gibbs in 1878 and Pierre Curie in 1888. The faces of the equilibrium crystal would be in such proportions as to minimize the total surface energy. If σ_i is the specific surface energy of the ith face with area F_i, then for a fixed volume $\Sigma_i\, \sigma_i F_i$ is a minimum. G. Wulff, in 1901, considering the selection of forms as given by structure in accordance with Bravais' law, used the Gibbs-Curie condition to determine the relative sizes of the faces. Let h_i represent the face-normal (Wulff vector) from a central Wulff point on the ith face. The Wulff vectors represent the velocities of growth. Wulff's theorem (von Laue, 1943) states for the equilibrium crystal that

$$h_i$$
$$\sigma_i = h_j$$
$$\sigma_j = h_k$$
$$\sigma_k = ... = C$$

The surface of terminus of all possible vectors from the Wulff point is called the γ-plot and planes perpendicular to the vectors at cusps of the γ-plot enclose the equilibrium or Wulff solids (Herring, 1951).

Wulff recognized that σ_i was a function of the environment as well as the structure and, with Gibbs and Curie, discussed limitations on the application of equilibrium theory to nature. R. Parker in 1923, F. Braun in 1932, and P. Niggli in 1941 devised procedures and made statistical studies of the persistence of forms, their sizes, and their provenance in order to cancel out accidental effects. Assuming that the morphological distribution throughout geologic time and space represented by these persistences (P_i) approached a statistical equilibrium, specific surface energies (σ_i) were calculated with the function

$$\sigma_{hkl} = kT \ln P_{hkl} \text{ (Schneer, 1970).}$$

Kinetic-Molecular Theories

Dislocation theory led W. K. Burton, N. Cabrera and F. C. Frank to the spiral theory of crystal growth and the introduction of the concepts of flat, stepped, and kinked faces. Hartman and Perdok (1955) calculated energies of the strongest vectors in a crystal which they called periodic bond chains or *PBC* vectors. An *F* or flat face was defined as a face containing at least two nonparallel *PBC* vectors; an *S* or stepped face contains one; and a *K* or kinked face, none. *F* faces are the most stable, *S* next, and *K* faces the fastest growing, therefore, least stable. Calculations of electrostatic attachment energies for slices parallel to crystal faces have been made using these (and comparable, cf. Dowty, 1977) methods. Because of the importance in these calculations attached to the thickness of the slice d_{hkl}, results approach those of the Bravais-Donnay-Harker law. These are structure-determinant accounts of crystal habit. By replacing the geometric points of the original Donnay-Harker law with physical centers of force, such discrepancies as the predominance of the cube face in rocksalt when the octahedron is the plane of greatest net reticular density, are explained. Computer simulations of crystal growth based on kinetic theory (Gilmer, 1980) demonstrate the significance for growth rates of the structure.

Principle of Equivalence

The striking correlation between morphological distributions and X-ray diffraction, particularly Debye-Scherrer patterns, was noted by Niggli, Wells, Buerger, and Donnay and Donnay. D. McLachlan Jr., in 1952, derived equations for the prediction of the prominence of a form–equations essentially equivalent to the structure-factor equations. Schneer compared the statistically derived specific surface energies for minerals of the ***barite*** group (morphological distribution) with calculated structure factors. Using the statistical magnitudes as coefficients, both Fourier and Patterson transforms have been computed locating the periodically distributed centers of forces of growth in the heaviest ions (Fig. 3).

Variation in Occurrence of Forms

The subject has been studied by R. Kern and discussed extensively by Sunagawa. Although the sequence of importance of forms of a

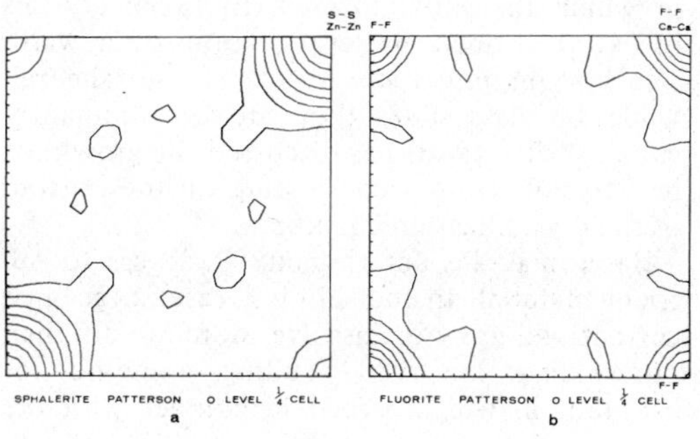

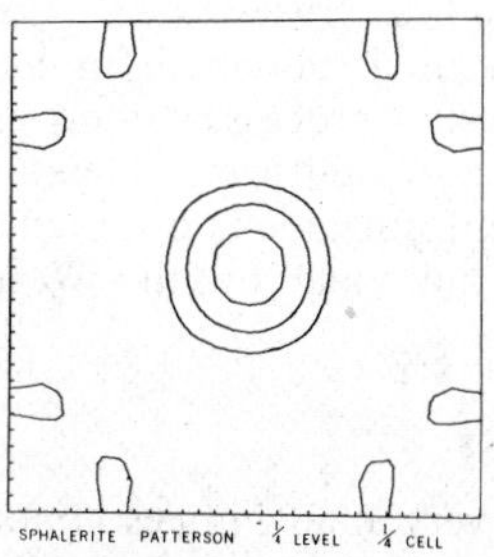

FIGURE 3. Fourier (Patterson) transforms of the observed morphological persistences of (a and c) *sphalerite* and (b) *fluorite* (a and b from Schneer, 1970).

species and their statistical distribution are determined primarily by the structure, the growth of any single specimen is an interaction of the structure-determined forces with the environment. The growth of a crystal is a process in time during which the conditions of growth and, therefore, the changes in habit are subject to variation. The sequence of forms for crystal individuals has been modified by selective adsorption of impurities (Frondel, 1935; Buckley, 1951) and by chemical variation in the environment, effects also believed significant in the natural occurrence of minerals. Temperature is believed to alter habit, since it alters growth rates. High temperature makes for simple morphology. Persistent minerals such as *pyrite* may display distinct habits in zoning of ore deposits. Grain size is believed to affect habit; also the stages of crystallization revealed by zoning within the crystal are often evidence of changing habit. It is probable that pressure also influences habit as, for example, in the growth of euhedral crystals during metamorphism, or flat or needle habits (LeChatelier's principle).

Variation in Proportion of Sizes of Forms

In the experimental synthesis of snow, Nakaya (1954) obtained three distinct habits, needles, plates, and flakes, as a function of the temperature of the water vs. the temperature of the air in which the snow formed. Epitaxial crystals of NH_4I formed by evaporation of a water solution on *muscovite* flakes are tetrahedral, while on glass slides they are predominantly cubic. Wells (1946) has discussed the growth of half-crystals from seeds resting on the bottom of the crystallization chamber.

Habit may be determined by access to nutrient material. In the simplest case, a face in a current will grow or dissolve more rapidly than other faces. In crystal pulling, a method of synthesis in which a seed in contact with the surface of a melt or solution is raised slowly enough to permit crystallization, it is possible to obtain endless crystals (see *Crystal Growth*). The same result is obtained for *corundum* with the flame-fusion process, namely endless single-crystal rods of uniform diameter. *Quartz* is grown in uniform blocks oriented for optimum processing and yield by the Western Electric Company.

Vicinal Faces

Faces of very high indices almost parallel to major faces of a crystal are called "vicinal." Niggli explained their appearance as a growth phenomenon determined by the depth of a surface layer of unsaturated bonds.

Ideal Habits

I. Kostov in Bulgaria has been engaged in the compilation of an atlas of normative crystal habits (genotypes) for comparison with observed habits (phenotypes). Wulff had observed that the Bravais law required that a crystal take the form of the Dirichlet region of a reciprocal lattice point, i.e., a polyhedron in which each inner element is closer to the given lattice point than any other point of the lattice. There are 24 such Wulff ideal habits (Leonyuk et al., 1980), which correspond to the first Brillouin zones of the solid-state physicists. The polyhedra complementary to the 97 morphological aspects given by the Donnay-Harker law have been designated as Donnay-Harker zones, and polyhedra complementary to the full crystal structures obtained from the Wulff or γ-plots of Herring (1951) are designated Wulff zones (Schneer, 1978).

Conclusion

The habit of a crystal is the end result of a process in time and a record of the equilibrium stages of the forces to which it has been subjected. The infinite variety of habits of crystals of a single species fall, nevertheless, within definite constraints fixed by the internal structure common to all specimens and expressed by statistically determined sequences of importance of forms and morphological persistence distributions. ". . . the electron distribution is probably a basic criterion describing the crystal face-formation as a function of its structure and composition" (Leonyuk et al., 1980). The variation of individual habits from the structure-determined norm is a measure of the forces which have acted upon the crystal during growth and a record of its history.

CECIL J. SCHNEER

References

Boetius de Boodt, Anselmus, 1636. *Gemmarum et Lapidum, Historia*. Leyden: Joannis Maire, 576p.

Bravais, Auguste, 1866. *Études Cristallographiques*. Paris: Gauthier-Villars, 176p.

Braun, F., 1932. Morphologische, genetische und paragenetische Trachtstudien an Baryt. *Neues Jahrb. Mineral.*, **65**A, 173–222.

Buckley, H. E., 1951. *Crystal Growth*. New York: Wiley, 571p.

Burke, John G. 1966. *Origins of the Science of Crystals*. Berkeley: Univ. Calif. Press, 198p.

Burton, W. K., and Cabrera, N., 1949. Crystal growth and surface structure, *Discuss. Faraday Soc., No. 5*, 33–48.

Curie, P., 1970. On the formation of crystals and the capillary constants of their different faces (transl. C. Schneer), *J. Chem. Edu.*, **47**, 636–637.

Donnay, J. D. H., and Harker, D., 1937. A new law of crystal morphology extending the law of Bravais, *Am. Mineralogist*, **22**, 446-467.
Dowty, E., 1977. Crystal structure and crystal growth: I. The influence of internal structure on morphology: a reply. *Am. Mineralogist*, **62**, 1063-1064.
Fedorov, E. von, 1920. *Das Krystallreich*, Petrograd: Acad. des Sciences de Russie, VIII ser., vol. XXXVI, 1050p. + 213p.
Frank, F. C., 1949. The influence of dislocations on crystal growth, *Discuss. Faraday Soc., No. 5*, 48-54.
Friedel, Georges, 1907. Études sur la loi de Bravais, *Bull. Soc. fr. Mineral.*, **30**, 326-455.
Gesner, C., 1565. *De Rerum Fossilium*. Zurich: Gesner, 169p. num. (338p.).
Gibbs, J. W., 1875-1878. On the equilibrium of heterogeneous substances, *Trans. Conn. Acad. Sci.*, **3**, 108-248, 343-524.
Goldschmidt, Victor, 1913-1923. *Atlas der Kristallformen*. Heidelberg: Winter.
Hartman, P., and Perdok, W. G., 1955. On the relation between structure and morphology of crystals, *Acta Crystallog.*, **8**, 49-52, 521-524, 525-529.
Haüy, René Just, 1801. *Traite de Mineralogie*. Paris: Conseil des Mines, Atlas pl. XXXV.
Herring, C., 1951. Some theorems on the free energies of crystal surfaces. *Phys. Rev.*, **82**, 87-93.
Kepler, Johann, 1966. *The Six-Cornered Snowflake* (transl. C. Hardie). Oxford: Clarendon Press, 77p.
Leonyuk, N. I., Gaulin, R. V., Al'shinskaya, L. I., and Delone, B. N., 1980. Practical determination of perfect habits of crystals. *Zeitschr. fur Kryst.*, **151**, 263-269.
McLachlan, D., Jr., 1952. Some factors in the growth of crystals: 1, Extensions of the Donnay-Harker law, *Bull. Univ. Utah Expt. Sta.*, **57**, 1-12.
Metzger, Hélène, 1969. *La Genèse de la Science des Cristaux*. Paris: Blanchard, 248p.
Nakaya, U., 1954. *Snow Crystals*. Cambridge: Harvard, 510p.
Niggli, P., 1941. *Lehrbuch der Mineralogie und Kristallchemie*, 3rd ed. Berlin: Borntraeger, 688p.
Schneer, C. J., 1970. Morphological basis for the reticular hypothesis, *Am. Mineralogist*, **55**, 1466-1488.
Schneer, C. J., 1978. Morphological complements of crystal structures. *Can. Mineral.*, **16**, 465-470.
Sunagawa, I., 1957. Variation in crystal habit of pyrite, *Rep. Geol. Soc. Japan*, **175**, 42p. + 10pl.
Sunagawa, I., and Endo, Y., 1971. Morphology of nucleus of galena and phlogopite, *Mineral. Soc. Japan, Spec. Paper 1*, 25-29.
von Laue, M., 1943. Der Wulffsche Satz für die Gleichgewichtsform von Kristallen, *Zeit. Kristallogr.*, **105**, 124-133.
Wells, A. F., 1946. Crystal habit and internal structure, *Phil. Mag.*, **37**, 184-199, 217-236.
Wulff, G., 1901. Zur Frage der Geschwindigheit des Wachstums und der Auflosung der Kristallflachen, *Zeit. Kristallogr.*, **34**, 449-530.

Cross-references: *Barker Index of Crystals; Crystallography: History; Crystallography: Morphological; Directions and Planes; Lattice; Point Group; Polycrystal; Skeletal Crystals; Space-Group Symbol; Symmetry; Twinning.*

CRYSTAL SOLUTION–*See* Vol. IVA: **SOLID SOLUTION**

CRYSTAL STRUCTURE–*See* glossary. Vol. IVA: **CRYSTAL CHEMISTRY**

CRYSTALLIZATION–*See* **CRYSTAL GROWTH; GLASS: THEORY OF CRYSTALLIZATION; SKELETAL CRYSTALS; SYNTHETIC MINERALS; VEIN MINERALS**

CRYSTALLOGRAPHY: HISTORY

The word crystal derives from the Greek word κρύσταλλος, meaning clear *ice*, reflecting the belief that persisted until the 17th century that transparent *quartz*, rock crystal, was permanently frozen water. The Latin *crystallus* denoted any geometrically shaped precious gemstone, and the designation *crystal* was extended in the 17th century to describe all regularly formed chemical salts. Robert Hooke, in his *Micrographia* (London, 1665), was the first to attempt to account for the polyhedral shapes of crystals, postulating that their composing particles were spherical and showing that regular geometric forms resulted from the stacking of contiguous spheres.

The most extensive 17th-century treatise about crystals was Nicolaus Steno's *De solido intra solidum naturaliter contento dissertationis prodromus* (Florence, 1669). Steno dismissed the current belief that mineral crystals in the earth grew in the same manner as vegetables and asserted that their growth takes place by the addition of particles which came directly from an external fluid and were superimposed on existing external faces. He recognized that crystal growth is directional in character and understood that the eventual shape of a crystal depends entirely upon the growth rate in various directions.

Steno's emphasis on solution conditions was important in later thought, as was his remark that the modifications of the crystal form as it grows can be compared to a process in which the original form is truncated in certain ways at its solid angles and along its edges. This statement provided a rationale for future crystal classification.

Two other important 17th-century essays were Erasmus Bartholinus' *Experimenta crystalli Islandici disdiaclasti quibus mira & insolita refractio detegitur* (Copenhagen, 1669), and

Christiaan Huygens' *Traité de la lumière* (Leyden, 1690). Bartholinus discovered and described the double refraction of light passing through Iceland crystal (*calcite*), and Huygens attempted brilliantly though unsuccessfully to give a scientific explanation of the phenomenon. In the same period, such men as Antony van Leeuwenhoek extended the microscopic study of crystals.

Eighteenth-century scientists undertook the task of crystal classification. The first major work was Maurice Cappeller's *Prodromus crystallographiae, de crystallis impropriae sic dictis commentarium* (Lucerne, 1723), in which 40 crystalline substances were classified according to nine different forms. Linnaeus, in his *Systema naturae* (Stockholm, 1768), classified mineral and saline crystals under four generic shapes–hexagonal prism, cube, tetragonal prism, and octahedron.

The major classificatory work of the 18th century was that of J. B. L. Romé de l'Isle. His initial effort, *Essai de cristallographie* (Paris, 1772), was followed by a four-volume masterpiece, *Cristallographie, ou description des formes propres à tous les corps du règne minéral* (Paris, 1783). He classified about 450 crystals under six primitive forms–regular tetrahedron, cube, regular octahedron, rhombohedron, rhombic octahedron, and dodecahedron with triangular faces. He attributed the variety of forms that a substance might assume to a process of truncation, which operated on the primitive form and the cause of which, he confessed, was unknown. But his second work was most important because of the application of *goniometry* (q.v.). The constancy of the interfacial angles of a crystalline substance had been recognized in *quartz* by Steno and must certainly have been known to such men as Huygens and Domenico Guglielmini. This law, however, was first clearly stated in Romé de l'Isle's work and was drawn to his attention by his student, Arnold Carangeot, the inventor of the contact goniometer.

Abbé René Just Haüy, beginning in 1781, discovered the mathematical relationships between the various forms of a crystalline substance and published his initial treatise, *Essai d'une théorie sur la structure des crystaux*, in 1784. Guglielmini had suggested in 1705 that cleavage fragments represented the primitive polyhedra from which crystals were constructed; Christian F. Westfeld and Johann G. Gahn expressed the opinion in 1767 that *calcite* crystals were formed of rhombohedral molecules; and Torbern Bergman had derived the *calcite* scalenohedron from the cleavage rhombohedron in 1773. Bergman, however, failed to derive the hexagonal prism of *calcite*, a problem that Haüy solved.

Haüy's method consisted of carefully cleaving a crystal until it appeared that the resulting shape would suffer no change by additional cleavage. Cleavage of a variety of substances showed him that the number of different nuclei so obtained could be reduced to six diverse shapes, which he called primitive forms. These, in turn, could be mathematically subdivided into three shapes which Haüy termed "integrant molecules": the tetrahedron, the triangular prism, and the parallelepipedon, which were the building blocks of crystals.

Haüy named the mathematical relationships (now subsumed under the law of rational indices) "laws of decrement," each of which either individually or in combination with one another served to explain the appearance of any crystal form. Every crystal face could be considered as having been built up by the progressive addition of lamellae formed of contiguous integrant molecules, one molecule thick. Assuming that the nucleus of a crystal is a cube and that the added integrant molecules are also cubes, if one row of cubic molecules is subtracted from each successively applied layer, the cube will eventually take the form of a dodecahedron with rhombic faces (Fig. 1). Decrement might also occur at the angles as well as the edges of the lamellae and entails no more than six rows of molecules. Beginning with the analysis of ***garnet*** and *calcite*, Haüy extended his studies over the next forty years to hundreds of crystalline substances.

The fact that many substances do not satisfactorily yield to cleavage caused mistakes in Haüy's analyses, and the perfection of the reflecting goniometer by William Hyde Wollaston in 1809 uncovered other errors. Further, Haüy had identified the integrant molecule and the chemical molecule of the substance, asserting

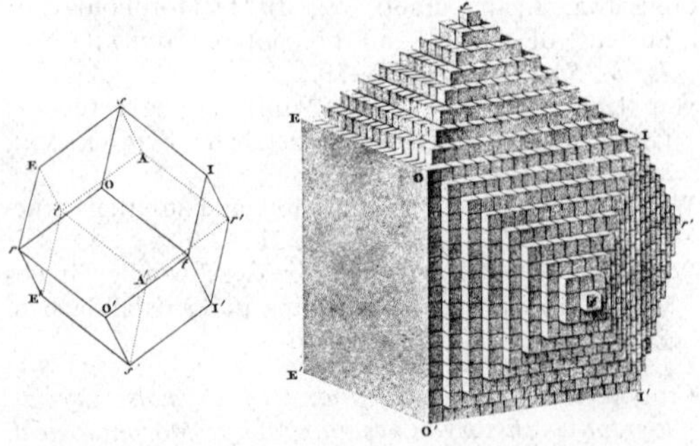

FIGURE 1. Dodecahedron constructed from cubic integrant molecules (from Haüy, 1801).

the priority of crystallography in the determination of a mineral species. The discovery of isomorphism and polymorphism by Eilhard Mitscherlich in 1819 and 1822 overthrew this contention, whereas the recognition of a definite relationship between optical properties and crystal form and the establishment of the six crystal systems led to the gradual abandonment of Haüy's method.

Jean B. Biot and David Brewster independently discovered optical biaxiality in 1812, and shortly thereafter both determined that uniaxial crystals might be optically either positive or negative. Both recognized that the optical axes corresponded to axes about which the crystal was symmetrical; and, over a period of years, Brewster classified almost 300 crystalline substances according to whether they were uniaxial, biaxial, or optically isotropic (see *Minerals, Uniaxial and Biaxial*).

In 1808, Christian S. Weiss published his *De Indagando formarum cristallinarum caractere geometrico principali*, which delineated his idea that all crystals had at least one axis that was a unique, principal, and dominating direction and that crystal forms might be described in terms of the inclinations of the faces to these axes. In 1815, Weiss distinguished the present isometric, tetragonal, orthorhombic, and hexagonal crystal systems; and in 1816 he published a notational system to describe crystal forms. The present Miller indices, introduced by W. H. Miller in 1839, are the reciprocals of Weiss's symbols (see *Directions and Planes*). Friedrich Mohs' work paralleled that of Weiss. In 1822 he published the first volume of his *Grund-Riss der Mineralogie*, in which he classified crystals into systems as Weiss had done, but advanced one step further in recognizing the possibility of systems in which the axes were not all mutually perpendicular. Two years later, Mohs in the second volume of his work, and independently Karl F. Naumann, identified the present triclinic and monoclinic crystal systems. In 1830, Johann F. C. Hessel contributed to the analysis of external crystal symmetry by demonstrating that, from a geometrical point of view, there can exist only 32 point groups, that is, combinations of crystallographic symmetry elements. These are the present crystal classes. Hessel's work was overlooked, and the 32 classes were defined independently by Auguste Bravais in 1848 (see *Point Groups*).

In the early 19th century, many attempts were made to construct models of the interior structure of crystals from considerations of their cleavage, external symmetry, and directional physical properties. In 1824 Ludwig Seeber suggested the concept of an internal crystal lattice, proposing to substitute minute spherical atoms in the center of Haüy's integrant molecules, which would in effect trace a space lattice. At the same time, two French mathematicians, Claude Navier and Augustin Cauchy, investigating problems of elasticity, proposed the idea of conceiving solids as arrays of mathematical points extended symmetrically throughout space. In Germany, a mathematical error prevented Moritz Frankenheim from successfully deriving the correct number of crystal lattices in his book, *Die Lehre von der Cohäsion* (Breslau, 1835). Gabriel Delafosse, in 1840, asserted that Haüy's integrant molecule, rather than being the chemical molecule of the substance, was instead the representation of the small intermolecular spaces or the outline of the nodes of a crystal lattice. All of these studies prefaced the work of Bravais.

Beginning in 1848, Bravais wrote a series of papers in which he initially treated the types of geometric figures formed by points distributed regularly in space. He then applied these considerations to crystals, with the points viewed as the centers of gravity of the chemical molecule or as poles of forces. With this approach, Bravais was able to explain the cleavage and external symmetry of crystals as a function of the reticular density. Most important, Bravais demonstrated that there was a maximum of 14 space lattices, differing by symmetry and geometry, whose translational repetition in space maintained the symmetrical arrangement of the points of a unit cell. These space lattices could be subsumed under the previously recognized six crystal systems (see *Lattice*).

Just how the chemical atoms or molecules were arranged in the unit cells forming the space lattice remained a matter of speculation. In 1879, Leonard Sohncke was able to arrive at 65 different arrangements of points by considering the points from an internal as well as external orientation. E. S. Federov, in the 1880s, demonstrated that there were only 230 possible space groups; but because his work was in Russian, it did not become generally known until after the publication in 1891 of the work of Artur Schoenflies, who arrived at the same result by considering the problem as one of geometrical group theory. The result incorporated the introduction of two new symmetry elements, the *screw axis* and the *glide plane*.

But even these important mathematical advances did not reveal the actual structure of atoms, ions, or molecules in crystals. The answer to this problem could only have been determined, as it finally was, by the X-ray diffraction technique discovered by Max von Laue in 1912. The perfection and extension of this method, primarily by William H. Bragg and William Lawrence Bragg, ushered in the modern period of crystallography.

JOHN G. BURKE

References

Burke, J. G., 1966. *Origins of the Science of Crystals.* Berkeley: Univ. of California Press, 196p.

Groth, P., 1926. *Entwicklungsgeschichte der Mineralogischen Wissenschaften.* Berlin: Springer-Verlag, 261p.

Metzger, H., 1918. *La Genèse de la Science des Cristaux.* Paris: Alcan, 248p. (Reprint, 1969, Paris: Blanchard).

Cross-references: *Barker Index of Crystals; Crystal Habits; Crystallography: Morphological; Directions and Planes; Lattice; Optical Orientation; Point Groups; Space-Group Symbol; Symmetry; Twinning.*

CRYSTALLOGRAPHY: MORPHOLOGICAL

A substance possessing an orderly crystal structure or arrangement of atoms or ions is said to be a crystal if it has developed a definite geometric external form. It is described as crystalline if it lacks a regular external form, but possesses an orderly internal structure. The term *cryptocrystalline* is applied to an aggregate of interlocking, irregularly shaped crystalline grains, visible only with the aid of a microscope.

When actual crystal faces are developed, they are very helpful in identifying the mineral because each mineral is characterized by constancy of interfacial *angles*. In crystallography, the interfacial angle is the angle between the normals to the two faces concerned (see *Barker Index of Crystals*). Crystals grow by addition of material to existing faces, this material being deposited parallel to surfaces already existing (see *Crystal Growth*). Consequently, if more material is added to one face than to another, the faces become unlike in size and shape, but the interfacial angles remain the same. The important thing in crystallography, therefore, is not the size or shape of the faces, but the angles between the faces (see *Goniometer*).

Definitions

Faces. The plane surfaces of crystals are known as faces, although they are sometimes slightly curved as in *dolomite, diamond*, etc. Like faces have the same properties; unlike faces have different properties.

Cleavage. Many minerals tend to break with smooth surfaces, or cleave, in certain crystallographic directions, i.e., parallel to certain crystal faces or to possible crystal faces. For example, *galena*, which commonly crystallizes in cubes, has a cubic cleavage, whereas *fluorite*, which also crystallizes in cubes, has an octahedral cleavage. The cleavage is defined according to the direction it takes in the crystal and is referred to the crystallographic axes in exactly the same way as are the crystal faces (see *Rupture in Mineral Matter*).

Form. A form in crystallography includes all the faces that have a like position relative to the crystallographic axes. A form is open if its faces do not completely enclose a space. The faces of a crystal may, therefore, belong entirely to one form, if that form is closed; but there must be faces of at least two forms on a perfect crystal if those forms are open. A crystal in which all the faces are alike is termed a simple form. For example, a cube is a simple form consisting of six similar square faces, and an octahedron is also a simple form of eight triangular faces. A crystal that shows two or more such simple forms developed is called a combination. The general appearance of a crystal, which depends on the forms developed, is termed the habit (see *Crystal Habit*).

Symmetry. In a crystal, there is a regularity of position of like faces, edges, etc., and the degree of this regularity constitutes the degree of symmetry. The degree of symmetry depends on the number of planes and axes of symmetry in the crystal and on whether or not it has a center of symmetry (see *Symmetry*).

A plane of symmetry is a plane that divides it into halves, each of which is the mirror image of the order (Fig. 1).

If a crystal, when rotated, occupies the same position in space more than once in a complete rotation, the axis about which it has been rotated is an axis of symmetry (Fig. 1). An axis of symmetry is of twofold, threefold, fourfold, or sixfold symmetry, depending on whether the crystal occupies the same position two, three, four, or six times in a complete rotation (see *Symmetry*).

When like faces of a crystal are arranged in pairs in .corresponding positions on opposite sides of a central point, that point is known as the center of symmetry (Fig. 1).

Classification of Crystals. Every crystal can be classified according to the degree of symmetry it exhibits. Crystals may be primarily classified into 32 symmetry classes, not all of

This entry has been extracted largely from a chapter in *Elementary Practical Geology*, E. Clarke, R. T. Prider, and C. Teicher, University of Western Australia, Nedlands, Australia, 1946.

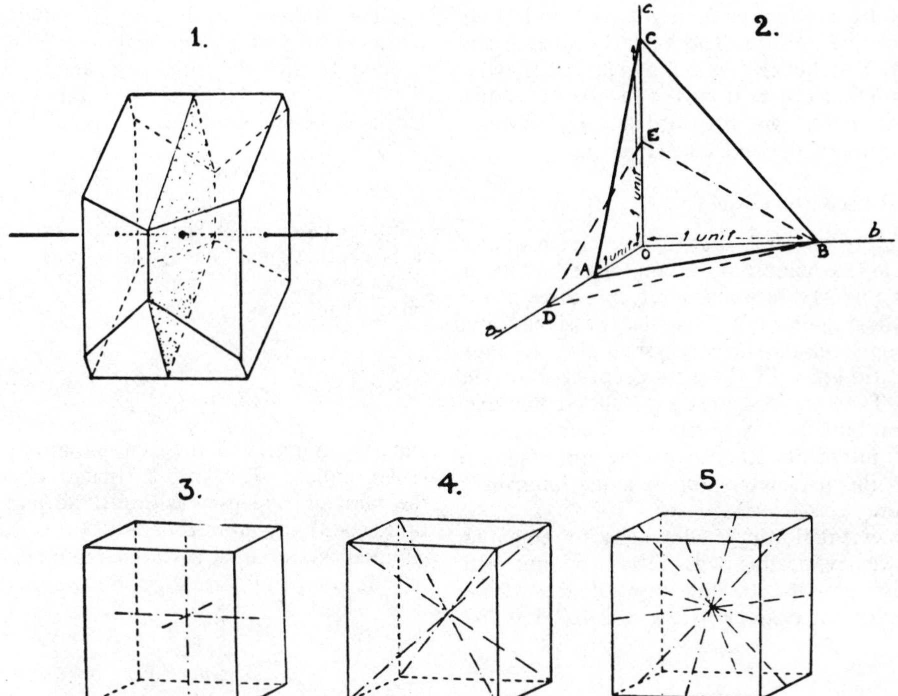

FIGURE 1. Symmetry, parameters, and intercepts. (1) *Orthoclase,* illustrating plane of symmetry (shaded), axis of symmetry, and center of symmetry. (2) Illustrating parameters and intercepts (see text). (3) Three axes of fourfold symmetry (= crystallographic axes). (4) Four axes of threefold symmetry. (5) Six axes of twofold symmetry.

which are important in mineralogy. The symmetry classes are usually grouped into six crystal systems according to the nature of the crystallographic axes to which they can be referred (see Table 1).

Crystallographic Axes

In describing a crystal, it is necessary to have some method of describing the faces that bound it. In crystallography, reference axes are not necessarily at right angles to one another nor are the units of measurement necessarily the same on the different axes. These reference axes are termed the crystallographic axes. Some crystals, e.g., the simple cube, have thirteen symmetry axes (Fig. 1), but only three, or at the most four, crystallographic axes need be assumed in any crystal because every face can be defined by naming its intercepts on those axes. Whenever a suitable number of axes of symmetry exist, they are chosen as the crystallographic axes.

By convention, the axis that is placed vertical is *c*, that running from right to left is *b*, and that running from front to back is *a*. Measurements on the *a* axis are positive when measured toward the observer; on the *b* axis, measure-

TABLE 1. Criteria for Crystal Symmetry

System	Criteria
Triclinic	*No* axes or mirror planes; may have center of symmetry
Monoclinic	*One* twofold axis and(or) one mirror plane
Orthorhombic	*Three* mutually perpendicular twofold axes; can have any combination of three axes and(or) mirror planes
Tetragonal	*One* fourfold axis or fourfold inversion axis with any combination of twofold axes and mirror planes
Hexagonal	*One* threefold or sixfold axis or a threefold inversion axis with any combination of twofold axes and mirror planes
Isometric	*Four* threefold axes or inversion axes; can have various combinations of fourfold axes and inversion axes, twofold axes, and mirror planes

ments are positive to the right; and on c, they are positive upward. The angle between a and b is γ, that between a and c is β, and that between b and c is α. If two of the axes are equal they are both a and the third is c, or if all three axes are equal they are all a (Fig. 2).

Crystal Faces (or Forms)

In every mineral that forms crystals, it is necessary to choose as a "unit face" (ABC in (2) of Fig. 1) a face that intercepts three of the crystallographic axes. This face is assigned unit intercepts on the three crystallographic axes. The ratio $a{:}b{:}c$ of these intercepts is the axial ratio. The lengths of the crystallographic axes are constant for any particular mineral species. These intercepts a, b, c of the unit face are called the parameters (*para*, along; *metron*, a measure; see *Lattice*).

It is essential in crystallography to remember that we are dealing with ratios and not with definite lengths. In the case of the ***olivine*** *forsterite*, for example, $a{:}b{:}c$ is 0.467:1:0.587.

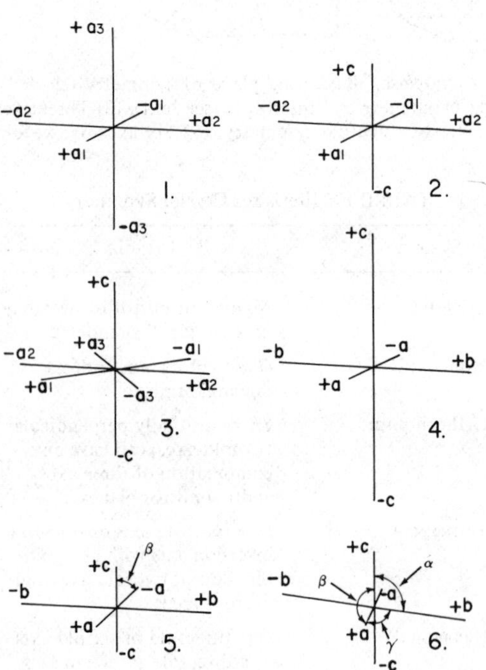

FIGURE 2. Clinographic projections of the crystallographic axes of the six systems: (1) Cubic system. (2) Tetragonal system: axes for *zircon* in which c = 0.64. (3) Hexagonal system: axes for *calcite* in which c = 0.85. (4) Orthorhombic system: axes for *barite* which has $a{:}b{:}c$ = 0.81:1:1.31. (5) Monoclinic system: axes for *gypsum* with axial ratio 0.69:1:0.41. $\beta = 80°42'$. (6) Triclinic system: axes for *plagioclase* with axial ratio 0.63:1:0.56, $\alpha = 94°3'$, $\beta = 116°29'$, $\gamma = 88°9'$.

Miller Indices. Each face is described by numbers by which the parameters must be divided to give the intercepts of that face on the crystallographic axes. The intercepts of a face may be represented by:

$$na{:}mb{:}pc \quad \text{or} \quad \frac{a}{h}{:}\frac{b}{k}{:}\frac{c}{l}$$

For instance, if $2a{:}3b{:}4c$ are the intercepts of a face, they may be divided by 12 without making any real difference since they are only ratios. They then become

$$\frac{a}{6}{:}\frac{b}{4}{:}\frac{c}{3}$$

and the numbers in the denominators are the Miller indices. They are sufficient to identify the face and are used without the parameters thus, (643) or, more generally, (hkl). The Miller indices are equal to the parameters divided by the intercepts (see *Directions and Planes*):

$$h{:}k{:}l = \frac{a}{na}{:}\frac{b}{mb}{:}\frac{c}{pc}$$

Crystallographic "Laws"

From the study of crystals, the following generalizations have been made:

- *The Law of Constancy of Symmetry.* All crystals of the same mineral species have the same symmetry.
- *The Law of Constancy of Interfacial Angles.* The angle between two like faces of a given form is constant for all crystals of a given mineral (measured at the same temperature).
- *The Law of Rational Intercepts.* All crystal faces make intercepts on the crystallographic axes that are either infinity or small, rational multiples of the intercepts made by the unit form.

Crystal Aggregates and Twin Crystals

Composite and Twin Crystals. In nature, crystals seldom occur loose and isolated. A mass is an "irregular aggregate" if there is no fixed relationship between the orientation of any two individual crystals; it is a "parallel growth" if all the individual crystals have the same crystallographic orientation.

Crystals in which the orientation is partially the same are "twin crystals," in which one or two axes of one indivual are parallel to the corresponding axes in the other individual. In twin crystals of the simple or contact type, one part is in the reverse position to the other part and the twin may be thought of as produced by the rotation of one half of the crystal through an angle. The plane of rotation is called the twin plane and the axis about which rotation has taken place is the twin axis and is perpendicular to the twin plane. The plane along which the two sides of the twin are joined is called the composition plane; it may or may not be the twin plane.

1. Penetration twins are those in which two or more complete crystals interpenetrate and appear to be crossing through one another. Normally, the crystals have a common center of symmetry (an example is *staurolite*).
2. Polysynthetic, repeated, or multiple twins are those in which twinning according to the same law is repeated in the crystal. Some multiple twins are made up of many individuals so narrow that they appear merely as striations (e.g., the **plagioclase feldspars**).

Twin features are referred to the crystallographic axes in the same way as are crystal faces (see *Twinning*).

Crystallographic Systems

There are six crystal systems, each characterized by the relative lengths and inclinations of their axes (see Fig. 2). Within each crystal system there is a holohedral, or normal, class, which possesses the greatest degree of symmetry that is possible within each system, and other classes of lesser symmetry. Not all of the 32 crystal classes are important in terms of mineral species having the symmetry of the class. The symmetry and forms of the holohedral and other mineralogically important classes are described here in detail and common minerals crystallizing in each class are listed. The first class symbol is the Hermann-Mauguin symbol and the second, in parentheses, is the Schoenflies symbol for class symmetry (see *Point Groups*).

A. The Isometric (or Cubic) System. This system has 3 equal axes at right angles; they are interchangeable and all are denoted by *a*. There are 5 symmetry classes.

Class $\frac{4}{m}\bar{3}\frac{2}{m}$ (O_h). Hexoctahedral, Holohedral, Ditesseral, or Normal Class. Symmetry: the highest possible. Planes: 9 (3 axial, 6 diagonal). Axes: 13 (3 of fourfold, 4 of threefold, 6 of twofold symmetry) (see *3, 4, 5* in Fig. 1). A center of symmetry. The crystallographic axes are the axes of fourfold symmetry.

This class has the following forms:

(a) Cube: A 6-faced solid formed by faces that cut one axis and are parallel to the other two. The indices are therefore $\{100\}$.

(b) Rhombic dodecahedron: A solid having 12 faces, each face cutting two axes at equal distances and parallel to the third. The indices are $\{110\}$.

(c) Octahedron: A solid having 8 equilateral triangular faces, each face cutting all three axes at the same distance and, therefore, having the indices $\{111\}$.

(d) Tetrahexahedron: A solid having 24 faces, each face an isosceles triangle, and formed by a 4-faced pyramid on each face of the cube. Each face is parallel to one axis and cuts the other axes at unequal lengths. The general indices are $\{hk0\}$ and a common form is $\{210\}$.

(e) Trisoctahedron: A solid having 24 faces, each face an isosceles triangle, appearing as a 3-faced pyramid on each face of the octahedron. Each face cuts two axes at equal distances and the third axis at a greater distance. The general indices are $\{hhl\}$ and a common form is $\{221\}$.

(f) Trapezohedron: A solid having 24 faces, each face a quadrilateral. Each face cuts two axes at an equal length and the third axis at a smaller length. The general indices are $\{hll\}$ and a common form is $\{211\}$.

(g) Hexoctahedron: A solid having 48 faces, each face a scalene triangle. Each face cuts the three axes at unequal lengths. The general indices are $\{hkl\}$ and a common form is $\{321\}$.

Minerals: *gold, copper, pentlandite, galena, halite, sylvite, fluorite, periclase, spinel, magnetite, chromite,* ***garnet*** group.

Class 432 (*O*). Pentagonal-icositetrahedral, Plagiohedral-hemihedral, Gyroidal, or Tesseral-holoaxial Class.

Minerals: None important.

Class $\bar{4}3m$ (T_d). Hextetrahedral, Tetrahedral-hemihedral, or Ditesseral-polar Class. Symmetry: Planes: 6 diagonal. Axes: 7 (3 of twofold, 4 of threefold symmetry). No center of symmetry. The crystallographic axes are the axes of twofold symmetry.

This class has the following forms:

(a) Tetrahedron: A 4-faced solid, each face meeting the axes at equal distances and, therefore, having the indices $\{111\}$ similar to the octahedron, from which it is produced by the development of the alternate faces.

(b) Other forms are developed in this class but will not be considered here.

Minerals: *sphalerite, lazurite, tetrahedrite, diamond.*

Class $\frac{2}{m}\bar{3}$ (T_h). Pyritohedral, Dyakisdodecahedral, Pentagonal-hemihedral, Diploidal, or Tesseral-central Class. Symmetry: Planes: 3 axial. Axes: 7 (3 of twofold, 4 of threefold symmetry). A center of symmetry. The crystallographic axes are axes of twofold symmetry.

This class has the following forms:

(a) Pyritohedron: A solid bounded by 12 pentagonal faces. Each face cuts two axes at unequal lengths and is parallel to the third, thus having indices $\{210\}$ similar to the tetrahexahedron $\{210\}$ from which it is produced by the development of the alternate faces.

(b) Diploid: A solid bounded by 24 faces, each face a quadrilateral. Each face cuts all three axes at unequal lengths and a common form is $\{321\}$ with the same indices as the hexoctahedron from which it is produced by the development of the alternate faces at the expense of the others.

(c) Other forms are similar geometrically to those of the holohedral class but have lower crystallographic symmetry because of the unlike character of the faces, e.g., in a cube of *pyrite* it will be noted that the faces are striated and that the striations on the three pairs of faces lie in three directions at right angles, parallel to the crystallographic axes. These striations result from an oscillatory combination, both the cube and pyritohedron having endeavored to exert their form during crystallization.

Minerals: *pyrite.*

Class 23 (*T*). Tetratoidal, Tetrahedral-pentagonal-dodecahedral, or Tesseral-polar Class.

Minerals: None important.

B. The Hexagonal System. This system has 3 equal lateral axes (a_1, a_2, a_3), making angles of 120° with each other and a fourth vertical axis, *c*, of greater or less length at right angles to the plane of the lateral axes. There are 12 possible symmetry classes.

Class $\frac{6}{m}\frac{2}{m}\frac{2}{m}$ (D_{6h}). Dihexagonal-dipyramidal, Holohedral, Dihexagonal-equatorial, or Normal Class. Symmetry: Planes 7 (4 axial, 3 diagonal). Axes: 7 (6 of twofold, 1 of sixfold symmetry). A center of symmetry. Crystallographic axes: *c* is the axis of sixfold symmetry; a_1, a_2, a_3 are axes of twofold symmetry.

This class has the following forms:

(a) Basal Pinacoid or Plane: An open form of 2 faces cutting the vertical axis and parallel to the plane of the lateral axes. The indices are therefore $\{0001\}$.

(b) Hexagonal Prism of the First Order: Six faces parallel to the vertical axis and one lateral axis and cutting the other two lateral axes at unit distances. The indices are $\{1\bar{1}00\}$ (see *4* in Fig. 3).

(c) Hexagonal Prism of the Second Order: Six faces parallel to the vertical axis, meeting two lateral axes at unit distance and the third at one-half this distance. The indices are $\{11\bar{2}0\}$ (see *5* in Fig. 3).

(d) Dihexagonal Prism: Twelve faces each parallel to the vertical axis and cutting the lateral axes at unequal distances. The indices are $\{21\bar{3}0\}$ (see *6* in Fig. 3).

At first sight, this form seems to have 12 vertical planes of symmetry; but, as with the ditetragonal prism and pyramid, adjacent angles are unequal and, therefore, there are only 6 vertical planes of symmetry.

(e) Hexagonal Pyramid of the First Order: Twelve faces, each cutting the vertical axis, two of the lateral axes at unit distance and parallel to the third. The indices of the unit form are $\{10\bar{1}1\}$.

(f) Hexagonal Pyramid of the Second Order: Twelve faces, each cutting the vertical axis, two of the lateral axes at unit distance, and the third at one-half this distance. The indices of a common form are $\{11\bar{2}1\}$.

(g) Dihexagonal Dipyramid: Twenty-four faces, each cutting the three lateral axes at unequal distances and also cutting the vertical axis. The indices of a common form are $\{21\bar{3}1\}$. This form has the same appearance of greater symmetry as has the dihexagonal prism.

Minerals: *graphite, pyrrhotite, covelite, molybdenite, beryl.*

Class 622 (D_6). Hexagonal-trapezohedral, Trapezohedral-hemihedral, or Hexagonal-holoaxial Class.

Minerals: high *quartz*.

Class 6*mm* (C_{6v}). Dihexagonal-pyramidal, Holohedral-hemimorphic, or Dihexagonal-polar Class.

Minerals: *wurtzite, zincite, ice.*

Class $\bar{6}m2$ (D_{3h}). Ditrigonal-dipyramidal, Trigonal-hemihedral, Trigonotype, or Ditrigonal-equatorial Class.

Minerals: *benitoite.*

Class $\frac{6}{m}$ (C_{6h}). Hexagonal-dipyramidal, Tripyra-

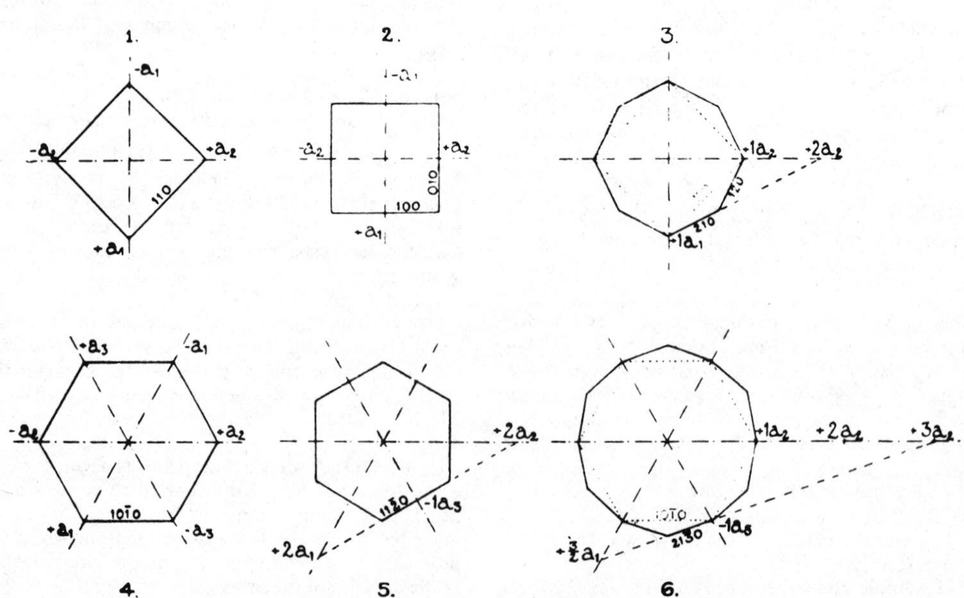

FIGURE 3. Plans showing the relations of the prisms in the tetragonal and hexagonal systems in which the lateral axes are interchangeable. (1) Tetragonal prism of 1st order (110). (2) Tetragonal prism of 2nd order (100). (3) Ditetragonal prism (210). (4) Hexagonal prism of 1st order (10$\bar{1}$0). (5) Hexagonal prism of 2nd order (11$\bar{2}$0). (6) Dihexagonal prism (21$\bar{3}$0).

midal, Pyramidal-hemihedral, or Hexagonal-equatorial Class.

Minerals: ***apatite*** group.

Class 6 (C_6). Hexagonal-pyramidal, Pyramidal-hemimorphic, Pyramidal-hemihedral-hemimorphic, or Hexagonal-polar Class.

Minerals: None important.

Class $\bar{6}$ (C_{3h}). Trigonal-dipyramidal, Trigonal-tetartohedral, or Trigonal-equatorial Class.

Minerals: None important.

Class $\bar{3}\ \frac{2}{m}$ (D_{3d}). Hexagonal-scalenohedral, Ditrigonal-scalenohedral, Rhombohedral-hemihedral, or Dihexagonal-alternating Class. Symmetry: Planes: 3 vertical diagonal. Axes: 4 (1 of threehold, 3 of twofold symmetry). A center of symmetry. Crystallographic axes: c is the axis of threefold symmetry; a_1, a_2, a_3 are the axes of twofold symmetry.

This class has the following forms:

(a) Rhombohedron: A solid bounded by 6 rhomb-shaped faces. Each face cuts the vertical axis and two of the lateral axes at unit distance and is parallel to the third. A common form is $\{10\bar{1}\bar{1}\}$, similar to the hexagonal pyramid from which it is produced by the development of the alternate faces of the upper part of the hexagonal pyramid and the other alternate set of faces of the lower part.

(b) Scalenohedron: A solid bounded by 12 faces, each face cutting all four axes at unequal distances. A common form is $\{2131\}$ and it is developed in a similar fashion to the rhombohedron from the dihexagonal pyramid, just as the rhombohedron is developed from the pyramid.

(c) Other forms are similar to those of the normal class, but they have lower crystallographic symmetry.

Minerals: *corundum, hematite, brucite, magnesite, siderite, smithsonite, calcite, rhodochrosite.*

Class 32 (D_3). Trigonal-trapezohedral, Trapezohedral-tetartohedral, or Trigonal-holoaxial Class.

Minerals: low *quartz, cinnabar.*

Class 3*m* (C_{3v}). Ditrigonal-pyramidal, Rhombohedral-hemimorphic, Trigonal-hemihedral-hemimorphic, or Ditrigonal-polar Class.

Minerals: ***tourmaline*** group.

Class $\bar{3}$ (C_{3i}). Rhombohedral, Trirhombohedral, Trigonal-rhombohedral, Rhombohedral-tetartohedral, or Hexagonal-alternating Class.

Minerals: *ilmenite, dolomite.*

Class 3 (C_3). Trigonal-pyramidal, Trigonal-tetartohedral-hemimorphic, or Trigonal-polar Class.

Minerals: None important.

C. The Tetragonal System. This system has 2 equal lateral axes, (a_1, a_2)–which are interchangeable–and one vertical axis, c, of greater or lesser length; all the axes are at mutual right angles.

There are 7 classes, but the normal class alone is well populated with minerals.

Class $\frac{4}{m}\ \frac{2}{m}\ \frac{2}{m}$ (D_{4h}). Ditetragonal-dipyramidal, Holohedral, Ditetragonal-equatorial, or Normal Class. Symmetry: Planes: 5 (3 axial, 2 diagonal). Axes: 5 (4 of twofold, 1 of fourfold symmetry). A center of symmetry. Crystallographic axes: c is the axis of fourfold symmetry; a_1, a_2 are axes of twofold symmetry.

This class has the following forms:

(a) Basal Pinacoid or Basal Plane: This form consists of 2 faces, which are parallel to the plane of the lateral axes and, therefore, have the indices $\{001\}$. This is an open form since it does not enclose a space; it, therefore, cannot occur alone on a crystal.

(b) Tetragonal Prism of the First Order: An open form of 4 faces, each cutting the two lateral axes at equal distances and parallel to the vertical axis (the lateral axes emerge at the centers of the vertical edges) (see *1* in Fig. 3). The indices are $\{110\}$.

(c) Tetragonal Prism of the Second Order: An open form of 4 faces, each cutting one of the lateral axes and parallel to the vertical axis and the other lateral axis (the axes emerge from the centers of the faces; see *2* in Fig. 3). The indices are $\{100\}$.

(d) Ditetragonal Prism: This is an open form of 8 faces, each face is parallel to the vertical axis and cuts the lateral axes at unequal distances. A common form is $\{210\}$, the general indices being $\{hk0\}$ (see *3* in Fig. 3). At first sight, this form seems to have 8 vertical planes of symmetry, but measurement with a goniometer shows that adjacent angles are not equal and, therefore, that there are only 4 vertical planes of symmetry.

(e) Tetragonal Dipyramid of the First Order: A form of 8 faces, each face cutting the vertical axis and cutting the lateral axes at equal distances. A common form is $\{111\}$.

(f) Tetragonal Dipyramid of the Second Order: A form of 8 faces, each face cutting the vertical axis and one lateral axis and parallel to the other lateral axis. A common form is $\{101\}$.

(g) Ditetragonal Dipyramid: A form of 16 faces, each face cutting the vertical axis and the two lateral axes at unequal distances. It has the same appearance of greater symmetry as has the ditetragonal prism. The general indices are $\{hkl\}$ and a common form is $\{211\}$.

Minerals: *rutile, cassiterite, zircon.*

Class 422 (D_4). Tetragonal-trapezohedral, Trapezohedral-hemihedral, or Tetragonal-holoaxial Class.

Minerals: *cristobalite.*

Class 4*mm* (C_{4v}). Ditetragonal-pyramidal, Holohedral-hemimorphic, or Ditetragonal-polar Class.

Minerals: None important.

Class $\bar{4}2m$ (D_{2d}). Tetragonal-scalenohedral, Tetragonal-sphenoidal, Sphenoidal-hemihedral, Didigonal-scalenohedral, or Ditetragonal-alternating Class.

Minerals: *bornite, chalcopyrite.*

Class $\frac{4}{m}$ (C_{4h}). Tetragonal-dipyramidal, Tripyramidal, Pyramidal-hemihedral, or Tetragonal-equatorial Class.

Minerals: *leucite.*

Class 4 (C_4). Tetragonal-pyramidal, Pyramidal-hemimorphic, Hemihedral-hemimorphic, or Tetragonal-polar Class.

Minerals: None important.

Class $\bar{4}$ (S_4). Tetragonal-disphenoidal, Sphenoidal-tetartohedral, or Tetragonal-alternating Class.

Minerals: None important.

D. The Orthorhombic System. This system has 3 unequal axes (a, b, c), all at right angles. The a axis is shorter than the b axis and is termed the *brachy-axis*, the b axis is termed the *macro-axis*. There are 3 classes but the normal class is the only important one.

Class $\frac{2}{m}\ \frac{2}{m}\ \frac{2}{m}$ (D_{2h}). Rhombic-dipyramidal, Orthorhombic-dipyramical, Holohedral, Didigonal-equatorial, or Normal Class. Symmetry: Planes: 3 axial. Axes: 3

(of twofold symmetry). A center of symmetry. The crystallographic axes are the axes of symmetry.

The forms for this class are listed below. In this and, to some extent, in the monoclinic and triclinic systems the following terms are used: Dome—a form that intersects the *c* axis and one lateral axis and is parallel to the other lateral axis; Pinacoid—a form that intersects one axis and is parallel to the other two.

(a) Basal Pinacoid: An open form of two faces that are parallel to the plane of the lateral axes and, therefore, have the indices {001}.

(b) Macropinacoid: An open form of 2 faces that are parallel to the vertical and macro-axes. The indices are {100}.

(c) Brachypinacoid: An open form of 2 faces that are parallel to the vertical and brachy-axes. The indices are {010}.

(d) Prism: An open form of 4 faces, each face parallel to the vertical axis and cutting the two lateral axes. The indices of the unit prism are {110}. Note, however, that other prisms such as {210} or {130} may be developed.

(e) Macrodone: An open form of 4 faces, each face parallel to the macro-axis but cutting the other two axes. The general indices are {*h*01}. The unit form is {101}.

(f) Brachydome: An open form of 4 faces, each face parallel to the brachy-axis but cutting the other two axes. The general indices are {0*kl*}. The unit form is {011}.

(g) Dipyramid: A closed form of 8 faces, each face cutting all three axes. The unit dipyramid has the indices {111} and the general indices for the form are {*hkl*}.

Minerals: alpha *sulfur, stibnite, marcasite, chrysoberyl, columbite, tantalite, goethite, aragonite, witherite, cerussite, anhydrite, celestite, barite, anglesite,* ***olivine****, sillimanite, andalusite, topaz, zoisite,* ***orthopyroxenes, orthoamphiboles.***

Class 222 (D_2). Rhombic-disphenoidal, Orthorhombic-disphenoidal, or Digonal-holoaxial Class.

Minerals: None important.

Class *mm*2 (C_{2v}). Rhombic-pyramidal, Orthorhombic-pyramidal, or Didigonal-polar Class.

Minerals: *chalcocite.*

E. The Monoclinic System. This system has three unequal axes, the *c* axis vertical, the *b* axis at right angles to *c*, and the *a* axis making an oblique angle with the plane containing *b* and *c*. The *b* axis is termed the *ortho-axis* and the *a* axis the *clino-axis*. There are 3 classes of which only the normal class is important in mineralogy.

Class $\frac{2}{m}$ (C_{2h}). Prismatic, Holohedral, Normal, or Digonal-equatorial Class. Symmetry: Planes: 1. Axes: 1 (of twofold symmetry). A center of symmetry. Crystallographic axes: *b* is the axis of twofold symmetry; *a* and *c* lie in the plane of symmetry.

The following forms occur:

(a) Basal Pinacoid: An open form {001} of 2 faces parallel to the clino-axis and ortho-axis.

(b) Clinopinacoid: An open form {010} of 2 faces parallel to the vertical axis and clino-axis.

(c) Orthopinacoid: An open form {100} of 2 faces parallel to the vertical axis and ortho-axis.

(d) Prism: An open form of 4 faces, parallel to the vertical axis and cutting both the lateral axes. The indices of the unit prism are {110}, the general indices are {*hk*0}.

(e) Clinodome: An open form of 4 faces, each parallel to the clino-axis and cutting both the ortho-axis and the vertical axis. The unit form is {011}.

(f) Hemi-orthodomes: No form that is parallel to the ortho-axis and cuts the other axes can have 4 faces, since the ortho-pinacoid is not a plane of symmetry. Only hemi-orthodomes are therefore possible. These are 2 in number and each has 2 faces, each face being parallel to the ortho-axis and cutting both the other axes. The unit form is {101}.

(g) Hemi-Bipyramids: These are open forms of 4 faces that cut all 3 axes. The general indices are {*hkl*}. The unit is {111}.

Minerals: *arsenopyrite, realgar, orpiment, azurite, malachite, borax, gypsum, monazite, lazulite, sphene, clinozoisite, epidote,* ***clinopyroxenes, clinoamphiboles****, talc, muscovite, biotite,* ***chlorite*** group, *kaolinite, chrysotile, sanidine, orthoclase, tridymite.*

Class 2 (C_2). Sphenoidal, Hemimorphic, or Digonal-polar class.

Minerals: None important.

Class *m* (C_s). Domatic, Clinohedral, Hemihedral, or Planar class.

Minerals: None important.

F. The Triclinic System. This system has 3 unequal axes, none of them at right angles. The *a* axis is shorter and is termed the *brachy-axis*, the *b* axis is longer and is termed the *macro-axis*. There are 2 classes, of which only the normal class is of importance in mineralogy.

Class $\bar{1}$ (C_i). Pinacoidal, Holohedral, Central, or Normal Class. Symmetry: Planes: none. Axes: none. A center of symmetry. Since there is only a center of symmetry, the presence of any one face necessitates only the presence of an opposite parallel face; each form, therefore, consists of 2 faces only.

The forms are analogous to those of the orthorhombic system, but each form is limited to 2 faces. Further description, therefore, will not be given here, but common forms are: (a) Basal Pinacoid {001}; (b) Macropinacoid {100}; (c) Brachypinacoid {010}; (d) Hemi-prisms {110}; (e) Hemi-macrodomes {101}; (f) Hemi-brachydomes {011}; and (g) Quarter Bipyramid {111}.

Minerals: *Turquois, kyanite, microcline,* ***plagioclases.***

Class 1 (C_1). Pedial, Hemihedral, or Assymetric Class.

Minerals: None important.

Crystallographic Projections

The relationship between angles and planes is best represented by the projection of face poles onto the surface of a surrounding sphere (see Fig. 4). A spherical projection is three dimensional, and cannot be transposed to the two-dimensional surface of a page without distortion of either angles or areas. Since crystallographers are primarily interested in an angle-true projection, part of this problem may be circumvented through the utilization of either a stereographic or a gnomonic projection, both

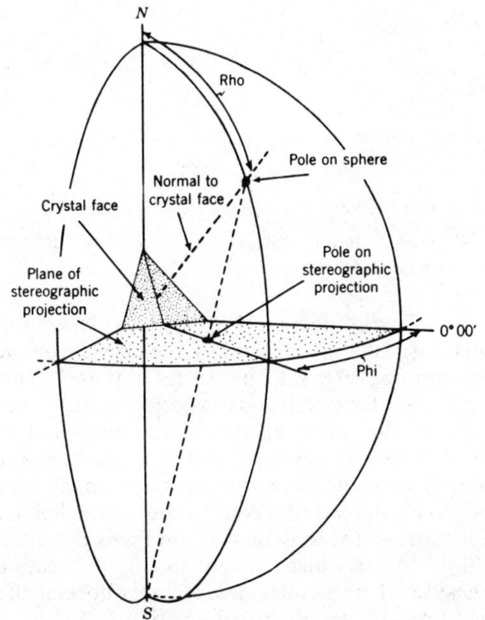

FIGURE 4. Diagram showing relationship between spherical projection and stereographic projection of the pole of a crystal face (from Wahlstrom, 1955).

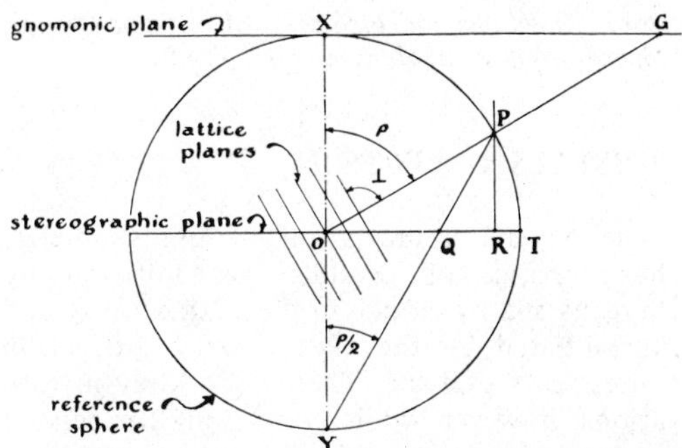

FIGURE 5. The relation between the stereographic, orthographic, gnomonic, and spherical projections in a section including an axis of the sphere, *XOY*, and the perpendicular *OP* to a set of lattice planes. (from *Mineralogy, Concepts, Descriptions, Determinations* by L. G. Berry and Brian Mason; copyright © 1959 by W. H. Freeman and Company, and used by permission). *P* is the pole point, the intersection of the perpendicular *OP* with the sphere; *Q* is the corresponding stereographic projection pole; *R* is the orthographic projection pole; and *G* is the gnomic projection pole representing the lattice planes. The plane of the orthographic projection is shown coinciding with the stereographic plane.

of which are derived from the extension of the spherical poles (of an imaginary spherical projection) onto precisely positioned planes (see Fig. 5). The pattern formed of points of intersection of the poles with the equatorial plane constitutes a stereographic projection.

A gnomonic projection is formed if a plane tangential to a predetermined point, usually the north pole, on the surface of a sphere is used. In this projection, imaginary lines are extended from the center of the reference sphere normal to the crystal faces until they intersect the plane of projection. The stereographic and the gnomonic projections are both angle true, but not area true; the reasons for this are more fully developed in conjunction with problems found in mapping the earth.

Inasmuch as stereographic and gnomonic projections record angular relationships among crystal faces, they, therefore, also record clearly the symmetrical patterns of crystal forms, in many cases more obviously than do crystals themselves. Additionally, accurate drawings from any perspective may be constructed from a stereographic projection of a crystal.

REX T. PRIDER

References

Berry, L. G., and Mason, B., 1959. *Mineralogy*, San Francisco: Freeman, 612p.

Bragg, W. H., and Bragg, W. L., 1924. *X-Rays and Crystal Structure*. London: G. Bell, 322p.

Clarke, E.; Prider, R. T.; and Teichert, C., 1946. *Elementary Practical Geology*. Perth: Univ. of West Australia, 170p.

Donnay, J. D. H.; Donnay, G.; Cox, E. G.; Kennard, O.; and King, M. V., 1963. *Crystal Data–Determinative Tables*. Washington, D.C.: Am. Crystallogr. Assoc., 1302p.

Goldschmidt, V., 1913. *Atlas der Krystallformen*. Heidelberg: Carl Winters Universitätsbuch,. 9 vols.

International Union of Crystallography, 1952–1962. *International Tables for X-Ray Crystallography*. Birmingham, England: Kynach Press, 3 vols.

Porter, M. W., and Spiller, R. G., 1952–1964. *The Barker Index of Crystals*. Cambridge, England: W. Heffer, 3 vols.

Ramachandran, G. N. (ed.), 1964. *Advanced Methods of Crystallography*. Berlin: Lange and Springer, 279p.

Shubnikov, A. V., 1960. *Principles of Optical Crystallography*. New York: Consultants Bur., 186p.

Smits, D. W. (ed.), 1960. *World Directory of Crystallographers and of Other Scientists Employing Crystallographic Methods*. Utrecht: Internat. Union of Crystallography, 134p.

Wahlstrom, E. E., 1955. *Petrographic Mineralogy*. New York: Wiley, 408p.

Wood, E. A., 1964. *Crystals and Light*. Princeton, N.J.: Van Nostrand, 160p.

Wyckoff, R. W. G., 1963. *Crystal Structures*, 2nd ed. New York: Interscience, 5 vols.

Cross-references: *Axis of Symmetry; Barker Index of Crystals; Crystal Habits; Crystallography: His-*

tory; Directions and Planes; Lattice; Point Groups; Space-Group Symbol; Symmetry; Twinning.

CRYSTALS, DEFECTS IN

The general appreciation of the geometric characteristics of crystals was initiated by Huygens and by Hooke in the 17th century and substantiated by the classic X-ray diffraction experiments of Laue. These very X-ray investigations, however, while confirming the general geometric postulates, also brought the realization that although certain minute units might themselves be perfect, disalignment or other imperfections existed; this led Darwin to propose a mosaic type of structure. Since that time, a wealth of information has been accumulated regarding the nature and extent of "defects" in crystals and a continually developing appreciation for their significance in chemical and physical properties.

Defects in a crystalline solid can best be appreciated by subdivision into four broad classes. Of these, the first are the electronic defects, due to photon interaction, which will always be present even in theoretically pure and geometrically perfect assemblies of ions. These defects can most readily be visualized as a temporary change in ionization state resulting in the transitory freeing of an electron and generation of an electron absence or positive hole. Recombination occurs at some site but may be only partial, giving rise to an "exciton."

The second general grouping is concerned with structural defects, which, again, will always exist in any system, no matter how pure, at any temperature above absolute zero. These point imperfections arise in accordance with thermodynamic requirements for a system to minimize its free energy. When an ion leaves a normal site to take up an abnormal position, such as an interstitial site, there is a free energy of formation involved. This, however, is offset by a decrease in configurational entropy and it can readily be shown (after Schottky) that at any real temperature there will be a net decrease in free energy when defects are created by thermal agitation. It is possible on this basis to calculate the equilibrium number of defects for any particular temperature; however, it should be clearly appreciated that below the Tammann temperature (approximately 0.51 T melting point), the crystal structure is essentially frozen and a larger number of defects will persist in crystals grown at a higher temperature.

The two most common types of point defects are illustrated in Fig. 1.

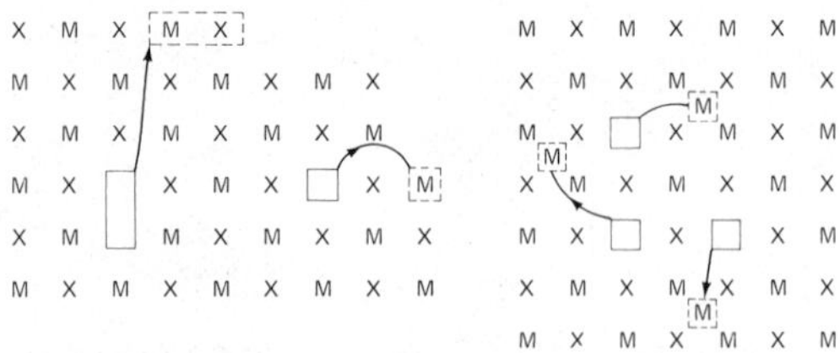

FIGURE 1. Point defects. (a) Schottky defects. (b) Frenkel defects.

In Schottky defects (Fig. 1a), ions leave normal positions in the bulk of the crystal to take up normal sites at the surface, thus leaving vacancies. Originally, anion-cation pairs, in the case of the alkali halides, were considered to migrate but the pairing is not implicit in the overall concept. The second predominant type of point defect observed in the silver halides and in zinc oxide is that of the Frenkel defects (Fig. 1b), in which an ion, usually the cation because of its smaller size, leaves a normal site and takes up an interstitial position.

Other point defects are possible, such as a mixture of interstitial ions of both types or exchange of position; but, at least in the case of strongly ionic solids, these are less probable.

These defects are observed for all materials, even those of ultimate purity and exact stoichiometry. In the case of impure materials and those deviating even very moderately from stoichiometry, the number of point defects may exceed the thermodynamic equilibrium value by many orders of magnitude.

The fourth general group of defects can be classified as impurity ions and can be exemplified in the technological application of crystals to transistors. Typically, if a tetrahedrally coordinated atom such as silicon is considered in a state of theoretical purity, the crystal should be nearly insulating. However, the presence of an altervalent ion in a tetrahedral site modifies the electron cloud distribution so that, effectively, there may be an excess or deficiency of electronic charge in the vicinity of the impurity (Fig. 2). Thus, a normally trivalent ion such as boron will occasion a net positive hole (electron deficiency) in its vicinity, whereas a pentavalent ion such as arsenic will donate a quasi-free electron to the structure. At any real temperature, a number of these states will be ionized, constituting the characteristic semiconductivity of the "doped" material. Similarly, in oxide, sulfide, and similar systems, valence control operates to moderate semiconducting or insulating character.

The final general group of defects is associated with dislocations. Under this category can

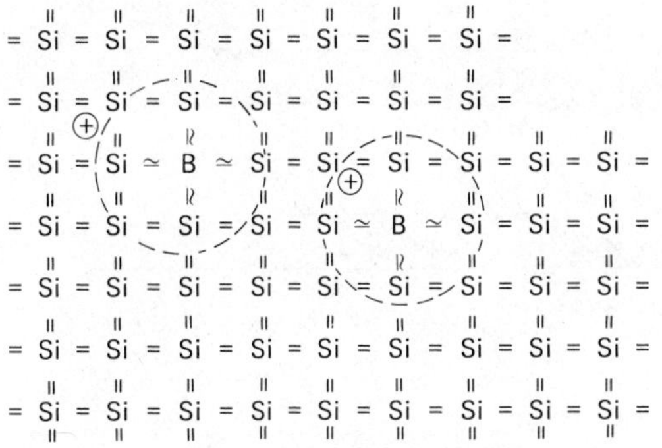

```
  ||     ||     ||     ||     ||     ||     ||
= Si  =  Si  =  Si  =  Si  =  Si  =  Si  =  Si  =
  ||     ||     ||     ||     ||     ||     ||
= Si  =  Si  =  Si  =  Si  =  Si  =  Si  =  Si  =
  ||  (+)||     ≀≀     ||     ||     ||     ||     ||     ||
= Si  =  Si  ≃  B   ≃  Si  =  Si  =  Si  =  Si  =  Si  =  Si  =
  ||     ||     ≀≀     ||     ||(+)  ≀≀     ||     ||     ||
= Si  =  Si  =  Si  =  Si  =  Si  ≃  B   ≃  Si  =  Si  =  Si  =
  ||     ||     ||     ||     ||     ≀≀     ||     ||     ||
= Si  =  Si  =  Si  =  Si  =  Si  =  Si  =  Si  =  Si  =  Si  =
  ||     ||     ||     ||     ||     ||     ||     ||     ||
= Si  =  Si  =  Si  =  Si  =  Si  =  Si  =  Si  =  Si  =  Si  =
  ||     ||     ||     ||     ||     ||     ||     ||     ||
= Si  =  Si  =  Si  =  Si  =  Si  =  Si  =  Si  =  Si  =  Si  =
  ||     ||     ||     ||     ||     ||     ||     ||     ||
```

FIGURE 2. Boron-doped silicon.

be considered not only dislocations themselves but extended and associated dislocations, grain and mosaic boundaries, and all the characteristics of crystal growth (q.v.). Dislocations differ from all preceding groups, which are essentially point defects, in that they are one-dimensional phenomena. They may be characterized under the associated *Burgers vector*, which is the closure error in traversing a full circuit in a crystal surrounding a dislocation in the process. This indicates the nature and strength of the structural singularity. A dislocation is classified as an edge dislocation if the Burgers vector is perpendicular to the line of dislocation, or as a screw dislocation if parallel to it. Typical dislocation types are illustrated in Fig. 3.

Dislocations can move in a crystal, and have been correlated with plastic flow in crystals (see *Plastic Flow in Minerals*). Under stress, dislocations can multiply as described by F. C. Frank and W. T. Read, Jr. They also migrate under thermal stress. Dislocations are extremely important in providing the sink and source for other defects and, in addition, they function as internal surfaces and as nuclei in crystallization phenomena. Stress fields exist across dislocations, resulting in significant contributions to both physical and chemical properties.

It may be stated as a generalization that physical and chemical properties of solids are intimately related at all times to their defect constitution.

T. J. GRAY

References

Cotrell, A. H., 1953. *Dislocations and Plastic Flow in Crystals*. Oxford: Clarendon Press, 223p.

Gray, T. J., 1957. *Defect Solid State*. New York: Interscience Pub., 511p.

Wert, C. A., and Thomson, R. M., 1964. *Physics of Solids*. New York: McGraw-Hill, 436p.

Cross-references: *Crystal Growth; Crystallography: Morphological; Etch Pits; Order-Disorder; Plastic Flow in Minerals.*

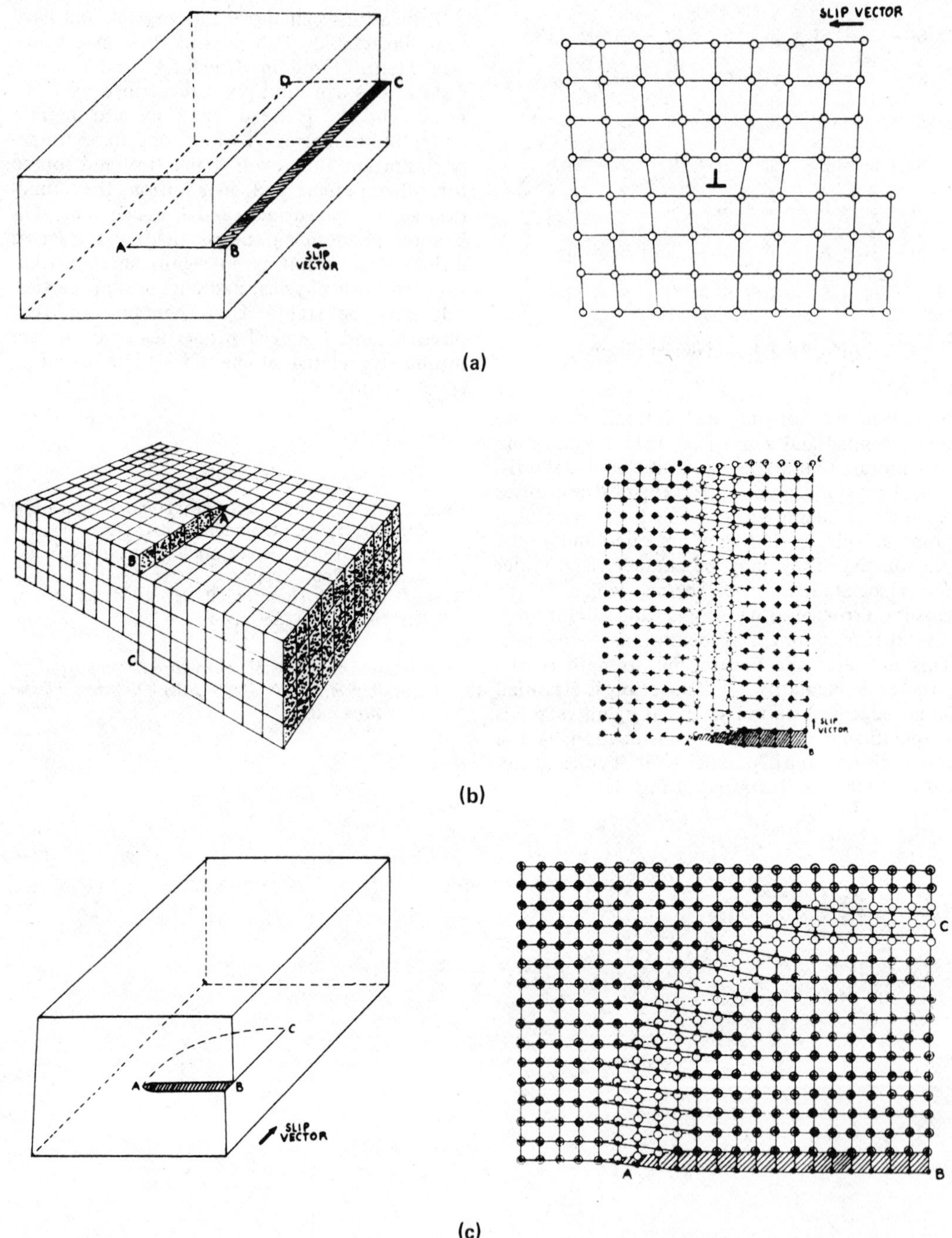

FIGURE 3. (a) An edge dislocation *AD* (⊥) normal to the slip vector produced by slip over the area *ABCD.* (b) A screw dislocation at *A* parallel to *BC* and to the slip vector results in a spiral arrangement of atoms around the dislocation. Open circles are atoms in the plane above the slip plane and solid circles those in the plane below. (c) A curved dislocation *AC* arises from unit slip in the slip plane *ABC.* Open and solid circles illustrate the arrangement of atoms above and below the slip plane.

D

DENDRITE—*See* SKELETAL CRYSTALS

DENSITY MEASUREMENTS

The mass of unit volume of a substance is its density. Its dimensions are M/L^3 and the units of measurement must be stated. Since most substances change in volume with temperature the latter should be stated; but, except at extremes of temperature, this correction may be neglected in mineralogy, as is the correction for the weight of air displaced in weighing.

Density is usually stated in g/cm^3 and symbolized by D or ρ or sometimes by G in mineralogy. (G, an abbreviation of an abbreviation, comes from $S.G.$ the specific gravity or relative density, which is the ratio of the weight of the substance to that of an equal volume of water and is simply a number without units. Hence G for density is undesirable.)

Usually the volume of a substance is measured by displacement of a fluid (generally water) so that a correction for the temperature of the fluid must be applied. The correction for water temperature may be important and the water temperature at the time of measurement should be stated as a subscript, or the density value should be reduced to that of water at 4°C, symbolized D or $D_{4°C}$ respectively. For measurement at extremes of temperature, the temperature of the mineral is added as a superscript, D^{tm}_{tw} or $D^{tm}_{4°C}$.

Sample Preparation

Two problems of density measurement are the selection of a pure sample free from contaminating grains, especially included grains of other minerals or of fluids, and avoiding trapped air in the sample between grains or along flaws during immersion in a displacement fluid.

Translucent substances may be examined microscopically for inclusions or contaminating grains. Although large, pure fragments of minerals are sometimes available, in practice small grains often have to be used and larger pieces may have to be crushed for purification. Opaque substances are difficult to check for purity and a sample of the grains may have to be polished and examined under the reflected-light microscope to ensure homogeneity.

After a pure sample is obtained, it is weighed in air. Its volume must then be found by immersion in a displacement fluid since direct measurement is virtually impracticable. At this stage, air must be removed from around the grains and from any cracks. With large fragments, this may be done by inspection and touching off air bubbles with a small brush. A drop of a wetting agent in the water, which should not materially affect its density, may help the displacement of air.

For small fragments and piles of grains, it is better to put the sample immersed in water into an evacuable desiccator from which the air may then be exhausted with a water pump.

A fluid of known density and lower viscosity than water, e.g., carbon tetrachloride, may be used as the displacement fluid and will more readily displace the air. The logical extension of this idea is to use air itself as the displacement fluid (see below).

With these principles in mind the available methods of measurement may be described.

Direct Weighing in Air and in Liquid

If the sample can be held in a nylon loop, it may be hung from the hook of a chemical balance, weighed first in air, and then suspended in liquid (Fig. 1). The uncorrected density is given by

$$D = \frac{W_1}{W_1 - W_2}$$

W_1 and W_2 being the weights in air and water. If a liquid other than water is used, the uncorrected density is given by

$$D = \frac{W_1 \times D_f}{W_1 - W_2}$$

where W_2 is the weight in this fluid and D_f is the density of the fluid.

For small particles, a glass fiber with two small pans along its length may be hung from the arm of a torsion balance (Fig. 2). The mineral grains are transferred from the upper

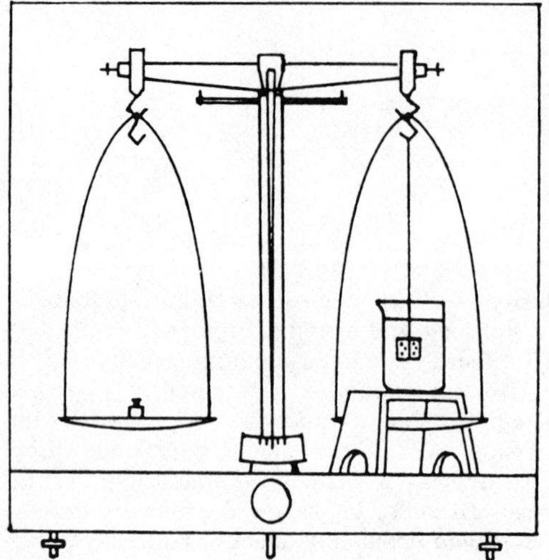

FIGURE 1. Arrangement for direct weighing in air and in water (from M. H. Battey, *Mineralogy for Students,* Longman Group Ltd, London, 1972).

to the lower pan between weighings and the level of the liquid adjusted so that the length of glass fiber immersed is the same during each weighing.

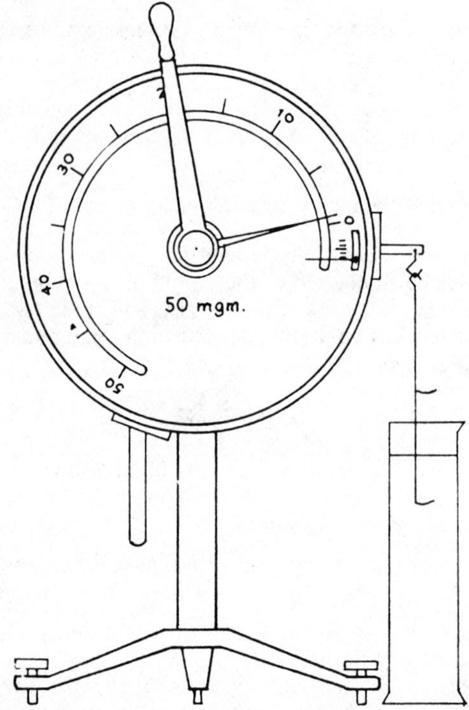

FIGURE 2. Torsion balance (from M. H. Battey, *Mineralogy for Students,* Longman Group Ltd, London, 1972).

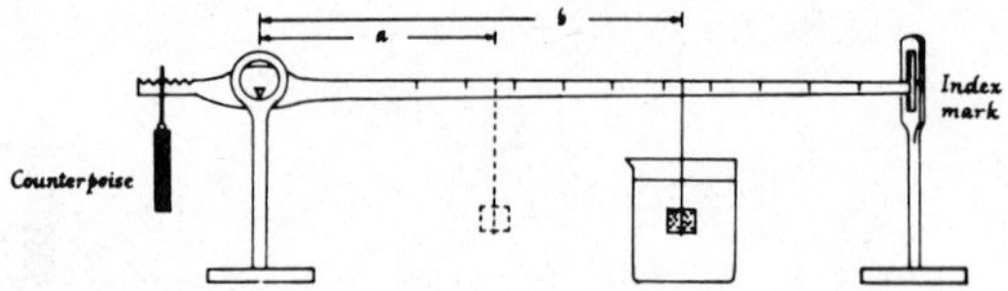

FIGURE 3. Walker steelyard balance (from M. H. Battey, *Mineralogy for Students,* Longman Group Ltd, London, 1972).

Relative Weighing in Air and Liquid

A counterpoised beam balance (Fig. 3) or a spring balance (Fig. 4) can give the ratio of weights in air and in water without direct weighing.

On the beam balance, the sample balances the counterpoise at distance a from the fulcrum in air and at distance b when the sample is immersed. Using water as the fluid,

$$D_{tw}^{tm} = \frac{1/a}{(1/a)-(1/b)} = \frac{b}{b-a}$$

Since a relatively large mineral fragment is required, homogeneity may be difficult to ensure.

The spring balance has a weak coiled spring supporting two pans, the lower of which is immersed throughout the measurement. The extension of the spring is measured by observing the position of a bead on the spring relative to a fixed scale engraved on a mirror. Coincidence of the bead and its reflection during reading eliminates parallax error. The balance is read unloaded (a), with the mineral on the upper pan (b) and with the mineral on the lower pan (c). The differences $b-a$ and $c-a$ are

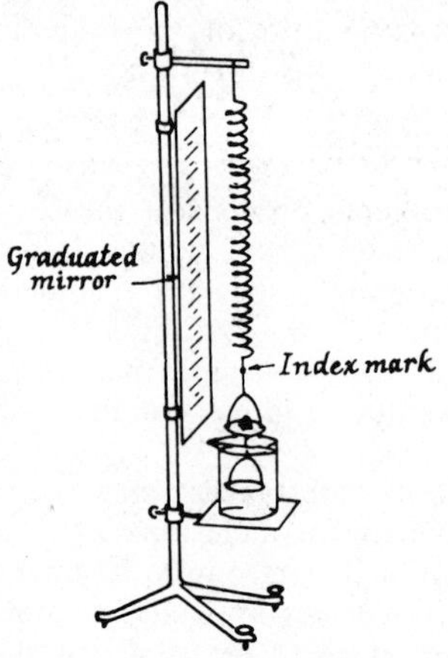

FIGURE 4. Jolly balance (from M. H. Battey, *Mineralogy for Students,* Longman Group Ltd, London, 1972).

directly proportional to the weights in air and liquid. Using water

$$D_{tw}^{tm} = \frac{b-a}{(b-a)-(c-a)} = \frac{b-a}{b-c}$$

Pyknometer or Specific Gravity Bottle

This is a small flask (10–50 ml capacity) with a long stopper fitting accurately in a ground-glass joint at the neck (Fig. 5). The stopper has a capillary hole down its axis. When the flask is filled to the brim and the stopper is inserted, the stopper should always seat at the same level and the liquid should rise through the capillary. When surplus liquid has been wiped off, the volume of liquid should be the same on every occasion.

To determine the density of a solid, weighings are made in the following order to avoid transfer and possible loss of mineral powder: (*a*) pyknometer, dry and empty; (*b*) pyknometer + mineral (usually as crushed grains); (*c*) pyknometer + mineral + liquid; and (*d*) pyknometer + liquid alone (this should be constant for a given liquid at a given temperature). Then, $b-a$ gives the weight of the mineral in air and $c-d$ the weight of the mineral less the weight of an equal volume of water (or other liquid).

Using water,

$$D_{tw}^{tm} = \frac{b-a}{(b-a)-(c-d)}$$

Errors may arise from,

i. variation in the seating of the stopper;
ii. evaporation from the capillary during weighing;
iii. variations in temperature.

The Method of Floating Equilibrium

The central problems of homogeneity and exclusion of air are largely avoided by the indirect method of placing a small number of grains of a mineral in a liquid of high density and adjusting the concentration of the liquid until the grains remain suspended. The density of the liquid is then determined.

This procedure demands only a small amount of mineral; the grains can be inspected individually for purity; and a sufficient amount of liquid can be used to allow its density to be determined, without serious error, using a large pyknometer. Alternatively, the Westphal balance (see below) may be used, or for more approximate work, interpolation between the densities of two grains of known density that respectively sink and float in the liquid.

The Westphal balance (Fig. 6), a modification of the steelyard balance, has a glass sinker combined with a thermometer hung at the end of the graduated arm, and a counterpose trimmed so that a pointer attached to it is opposite an index mark when the sinker is immersed in water. When the sinker is hung in a liquid denser than water, the addition of calibrated riders to the graduated arm to bring the pointer back to the index mark makes possible a direct reading of the density of the liquid.

A set of sinkers of varying densities can be made by sealing small pieces of steel inside short lengths of glass tubing; these are useful for approximate measurements. A set of mineral grains of known densities is useful in adjusting liquids to roughly the density required for a particular determination.

Because of its accuracy, flexibility, and convenience, floating equilibrium is here regarded as the best method for routine determinations of mineral densities.

Heavy Liquids. The following are the heavy liquids commonly used for the method of floating equilibrium, and for separating minerals of different densities from one another.

Bromoform, $CHBr_3$, with a maximum density of 2.90 g/cm^3, may be diluted with acetone. It

FIGURE 5. Pyknometer (pycnometer) (from M. H. Battey, *Mineralogy for Students*, Longman Group Ltd, London, 1972).

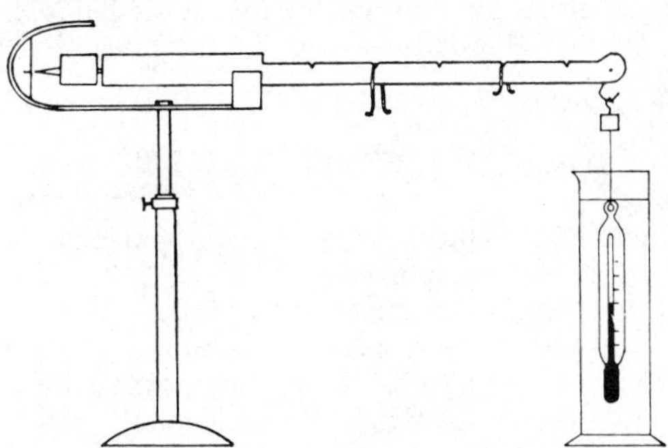

FIGURE 6. Westphal balance (from M. H. Battey, *Mineralogy for Students*, Longman Group Ltd, London, 1972).

may be recovered by shaking the bromoform-acetone mixture in a large volume of distilled water, allowing the bromoform to sink, and drawing it off. The vapor of bromoform is toxic; work should be done near an extract fan, and vessels should be kept covered. It is decomposed by light.

Thoulet's solution, with a density of 3.19 g/cm^3, is an aqueous solution of potassium mercuric iodide. It may be diluted with water and reconcentrated by gentle evaporation on a water bath. It is prepared by dissolving 270 g HgI_2 and 230 g KI in 80 ml of cold, distilled water, and then gently evaporating until a crystal of *fluorite* floats. A small excess of KI will not matter, but HgI_2 should not be in excess. The solution is poisonous and tends to corrode the skin.

Methylene iodide (di-iodomethane), CH_2I_2, with a density of 3.325 g/cm^3, may be diluted with chloroform and recovered by gentle volatilization of the diluent. It is decomposed by exposure to light.

Clerici's solution, with a density of 4.4 g/cm^3, is an aqueous solution of thallium formate malonate. To prepare it, neutralize two equal parts of Tl_2CO_3 with equivalent parts of malonic and formic acids, separately, and then mix the two solutions and concentrate by gentle evaporation. Clerici's solution corrodes the skin, but, because of its high density and the fact that it can be diluted with water and readily reconcentrated, it is one of the most useful liquids for flotation.

Density of a Mineral Heavier Than a Liquid. This case will quite often arise. Make a small buoy a few mm across by enclosing a small piece of heavy metal in a ball of wax, so that its density is between 1 and 2 g/cm^3. Weigh this, and weigh some small grains of the mineral to be determined. Lightly press the mineral grains into the wax and determine the density of buoy + mineral by flotation. Remove the mineral grains and determine the density of the buoy alone by flotation in the diluted liquid.

To find the density, Dm, of the mineral we use,

$$Dbm = \frac{Wb + Wm}{Vb + Vm} = \frac{Wb + Wm}{Wb/Db + Wm/Dm}$$

where Dbm = the density of buoy and mineral combined
Wb = the weight of the buoy
Db = the density of the buoy
Vb = Wb/Db, the volume of the buoy
Wm = the weight of the mineral
Vm = Wm/Dm, the volume of the mineral

and
$$Dm = \frac{Dbm\,Wm}{Wb + Wm - Dbm\,Wb/Db}$$

Instead of a wax buoy, a horseshoe loop of glass, the springiness of which allows it to grip the mineral grain, may be a more satisfactory and permanent float.

The Volumenometer

The use of air as a displacement fluid is limited, because of the difficulty of measuring small volumes of air accurately. Its use is required in two cases: (i) for fine powders; and (ii) for substances soluble in, or reacted upon by displacement liquids.

The principle of the apparatus is shown in Fig. 7. The volume of air, v_1, in the test chamber is compressed with the manometer to volume v_2, by raising the level of the liquid in the chamber from the lower to the upper of two fixed marks. The head, h_1, required to do this is measured.

The liquid level is lowered to the bottom mark, and a weighed amount of the substance whose density is to be determined is placed in the specimen cup on the gauze support.

The chamber is closed, the liquid brought again to the upper mark, and the head, h_2, required is noted. The volume, v_3, of the introduced substance is to be found.

At an atmospheric pressure of P mm mercury, with h_1 and h_2 also measured in mm of mercury, $P_1 = P + h_1$ and $P_2 = P + h_2$. Then, at any given temperature

$$P_1v_2 = k = P_2(v_2 - v_3)$$
$$P_1v_2 = P_2v_2 - P_2v_3$$
$$P_2v_3 = P_2v_2 - P_1v_2$$
$$v_3 = v_2(P_2 - P_1)/P_2$$

Calculation of Density

The density of a mineral is directly related to the volume of the unit cell of the structure

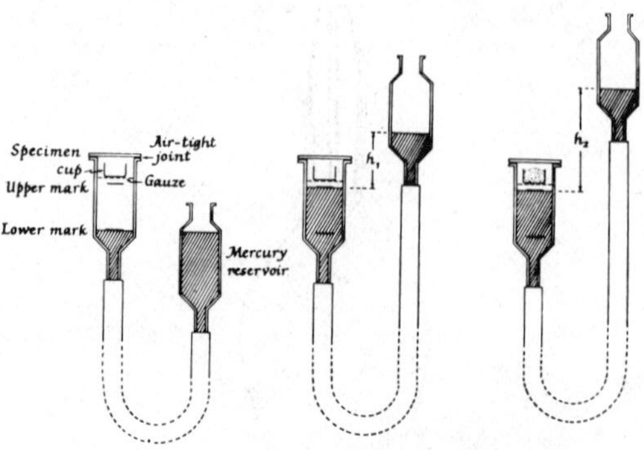

FIGURE 7. Principle of a volumenometer (from M. H. Battey, *Mineralogy for Students,* Longman Group Ltd, London, 1972).

and the atomic weights of the atoms in the cell. The relationship is

$$D = \frac{AW \times (1.6602 \times 10^{-24})}{V \times 10^{-21}}$$

where D is the density in g/cm^3, AW is the sum of the atomic weights of the atoms in the unit cell, and V is the cell volume in nm^3. The constant (1.6602×10^{-24}) is the unit of atomic weight in grams (the reciprocal of Avogadro's Number), while the cell volume in nm^3 must be multiplied by 10^{-21} to convert it to cm^3.

Such a calculation is often of value in cross-checking the results of a chemical analysis of a mineral against measurements of its density and cell size.

Because of atomic substitution, most minerals do not have exactly the density and cell size to be expected from the formula of a hypothetically pure compound. Where sufficient data on the effect at atomic substitution have been collected, the measurement of density or cell size may be a useful guide to composition. For example, there is a close relationship between density and composition in many of the series of ferromagnesian silicates such as the ***garnet*** group or the ***olivine*** group.

M. H. BATTEY

Reference

Battey, M. H., 1972. *Mineralogy for Students.* London: Longmans, 323p.

Cross-references: *Crystals, Defects in; Gemology; Mineral Properties.*

DEVITRIFICATION—*See* GLASS, DEVITRIFICATION OF VOLCANIC; GLASS: THEORY OF CRYSTALLIZATION

DIAGENETIC MINERALS

The process of diagenesis involves all the physical and chemical changes that may occur in a sediment from the moment of deposition to a stage where it is either exposed to destruction by weathering or passes into the realm of metamorphism (Von Gümbel, 1868; somewhat modified by Walther, 1894; Fairbridge, 1967). The metamorphic boundary varies according to the stability characteristics of any given mineral, but is usually taken to be in the range of 200–300°C, with pressure (not critical) in the order of 1000 bars. Thus, in sedimentary basins under a geothermal gradient of 30°C/km, the boundary is commonly at approximately 10,000 m (Müller, 1967). Comparable transformations, such as of snow to glacier *ice* or of peat to coal (Teichmüller and Teichmüller, 1967), are also "diagenetic." Subsurface waters likewise undergo a diagenetic evolution (Degens and Chilingar, 1967).

Diagenesis may involve a variety of physicochemical and biochemical systems leading to lithification, such as compaction, cementation, recrystallization, and replacement. (The Greek prefix means "through" or "thorough", implying *all* systems leading to lithification.) New minerals that are created primarily from interstitial solutions during diagenesis, a process sometimes called "neoformation," are known as authigenic minerals (q.v.). Other minerals, resulting from in situ alteration of the parent material, are distinguished as due to "transformation." Both categories, thus all minerals formed after deposition of a sediment and prior to its distintegration by weathering or alteration by metamorphism, are diagenetic minerals.

Some authorities recognize a preburial stage of mineral particle modification effected during transportation to the site of deposition (e.g., Müller, 1967); it is questionable whether this halmeic stage ("halmyrolysis" or "aquatalysis") should be called diagenetic in view of the definition of diagenesis.

The writer distinguishes three evolutionary stages of diagenesis related to the history of any given sedimentary basin (Fig. 1). These phases ideally may be sequential but, because of eustatic oscillations or tectonic events, they may be terminated abruptly, bypassed, or recycled (Fairbridge, 1967). The three phases can be defined as follows.

Syndiagenesis is often marked by two substages: near-surface—which is usually oxidizing—and early burial at a few centimeters to a few meters depth—which is reducing and marked by a low pH. This phase is mainly influenced by the nature of the sea floor or lake floor and is affected by diffusion with bottom waters; where the latter are euxinic, the first substage is omitted.

Anadiagenesis, the deep burial condition, is influenced by progressive compaction and by the upwards or lateral motion of trapped connate waters and other fluids such as hydrocarbons. The temperature and pressure are progressively more elevated with depth, as also are usually the ionic concentrations and pH controls. At maximal burial (over 10,000 m), there will be a gradational transition, mineral by mineral, to the metamorphic suites (beginning with the Zeolite stage). Anadiagenesis

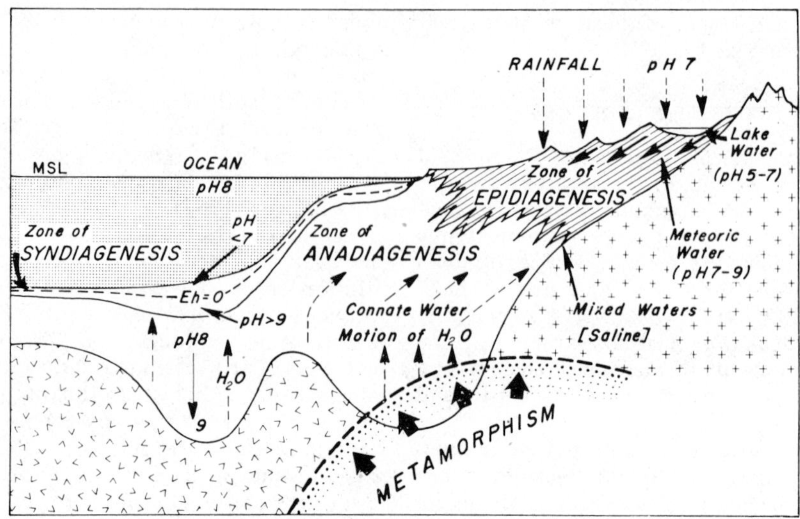

FIGURE 1. Idealized profile through a continental margin, showing the sites of contemporary marine sedimentation and the three phases of diagenesis. (from F. W. Fairbridge, Phases of diagenesis and authigenesis, in *Diagenesis in Sediments,* G. Larsen and G. V. Chilingar, eds., Elsevier, Amsterdam, 1967). Note the (1) diffusion potential during syndiagenesis; (2) upward liquid motion in anadiagenesis; and (3) downward motion in epidiagenesis.

may be interrupted at any time, but in some basins may continue for as long as 10^9 years.

Epidiagenesis, the emergence phase, may follow anadiagenesis or occur immediately after syndiagenesis. It is marked by descending meteoric waters that are initially of low pH and low ionic concentrations. Weathering and dissolutions occur at the surface and below this there is epigene authigenesis (Perelman, 1967).

Diagenetic Minerals in Sedimentary Rocks

Sedimentary rocks are variously distinguished by distinctive diagenetic minerals. Some important categories are discussed below:

Quartz* Sandstones.** Cements are introduced at several stages. These include opal, evolving to chalcedony, or *quartz*, often as overgrowths, sometimes *calcite* (known as "Fontainebleau Sandstone type" if in crystal continuity), sometimes *hematite* or other minerals (Dapples, 1967). Authigenic ***feldspars, *epidote,* ***micas,*** and *kaolinite* may be added, particularly in the early anadiagenetic stage (100–2000 m).

Limestones. In the syndiagenetic stage, metastable minerals such as *aragonite* and magnesium-rich *calcite* usually invert to stable *calcite* with low impurities; under minor loading (over 1000 m) or tectonic dynamic pressures, there is recrystallization in the plane of maximum stress; or under anadiagenesis, "saline reflux" solutions rich in magnesium tend to replace the calcium ion by ion to create vast metasomatic deposits of *dolomite* (Fairbridge, 1957; Adams and Rhodes, 1960; Friedman and Sanders, 1967; Bathurst, 1975). Dolomitic crusts or near-surface enrichment may appear during epidiagenesis and, less commonly, dedolomitization or calcitization may take place (Fairbridge, 1967, p. 69).

Clay-Shale Rocks. A mixture of clastics and newly formed authigenic minerals, produced by halmyrolysis during transportation and under early burial, may include *kaolinite,* ***illites,*** *palygorskite, sepiolite,* ***chlorites, smectites***, and ***zeolites*** (the last two if volcanic glass is being deposited). At deeper, anadiagenetic levels, below about 1000 m, ***chlorites*** and ***illites*** are also formed. From initially deposited clays there are also massive transformations, notably to ***illites*** and ***chlorites,*** increasing greatly with depth. *Montmorillonite* and *kaolinite* transformations develop during syndiagenesis, more so during early anadiagenesis at depths of 100–2000 m, but decrease at greater depths (Müller, 1967).

Evaporite Deposits. *Gypsum* formed under surface conditions tends to dehydrate to *anhydrite* at depths over 100 m (early anadiagenesis), but hydrates once more in the epidiagenetic phase (Murray, 1964; West, 1964). The halides, sulfates, and carbonates of evaporite sequences undergo a complex sequence of transformations and recrystallizations with progressively deeper burial under what has been called "geothermal metamorphism"–

actually quite low-temperature anadiagenetic diagenesis (Borchert and Muir, 1964; Braitsch, 1962; Stewart, 1965).

Sedimentary Ore Deposits. Thanks to the experimental work of Baas Becking and others (1960) it is now established that many metallic sulfides and other ore deposits are initiated under conditions of bacterial syndiagenesis. Aspects of their development have been reviewed by Kutina (1963), Amstutz and Bubenicek (1967), and Stanton (1972).

Plate tectonic studies, especially of deep seafloor spreading centers, confirm the theories suggesting exhalation of metallic ore solutions and the idea of near-surface hydrothermal systems (Rona and Lowell, 1980).

RHODES W. FAIRBRIDGE

References

Adams, J. E., and Rhodes, M. L., 1960. Dolomization by seepage refluxion, *Bull. Am. Assoc. Petrol. Geologists*, **44**, 1912–1920.

Amstutz, G. C., and Bubenicek, L., 1967. Diagenesis in sedimentary mineral deposits, in Larsen and Chilingar, 417–475.

Baas Becking, L. G. M., et al., 1960. Limits of the natural environment in terms of pH and oxidation-reduction potentials, *J. Geol.*, **68**, 243–284.

Bathurst, R. G. C., 1975. *Carbonate Sediments and Their Diagenesis*, 2nd ed. Amsterdam: Elsevier, 658p.

Borchert, H., and Muir, R. O., 1964. *Salt Deposits.* London: Van Nostrand, 300p.

Braitsch, O., 1962. Entstehung und Stoffbestand der Salzlagerstätten, *Mineral. Petrog. Mitt.*, **3**, 232p.

Chilingar, G. V.; Bissell, H. J.; and Fairbridge, R. W., eds., 1967. *Carbonate Rocks.* Amsterdam: Elsevier, 471p.

Dapples, E. C., 1967. Diagenesis of sandstones, in Larsen and Chilingar, 91–126.

Degens, E. T., and Chilingar, G. V., 1967. Diagenesis of subsurface waters, in Larsen and Chilingar, 477–502.

Fairbridge, R. W., 1957. The dolomite question, in R. J. LeBlanc, and J. G. Breeding, eds. Regional Aspects of Carbonate Deposition. *Soc. Econ. Paleont. Mineral. Spec. Publ. 5*, 125–178.

Fairbridge, R. W., 1967. Phases of diagenesis and authigenesis, in Larsen and Chilingar, 19–89.

Friedman, G. M., and Sanders, J. E., 1967. Origin and occurrence of dolostones, in Chilingar, Bissell, and Fairbridge, 267–349.

Kirkland, D. W., and Evans, R., eds., 1973. *Marine Evaporites.* Stroudsburg, Pa.: Dowden, Hutchinson & Ross, 426p.

Kossovskaja, A. G., and Shutov, V. D., 1965. Facies of regional epi- and metagenesis [*Isvesta A.N. SSR, ser. Geol.*, 1963, 7, 3–18], in English, *Intern. Geol. Rev.*, 7, 1157–1167.

Kutina, J., ed., 1963. Symposium Problems of Post-magmatic Ore Deposition, *Geol. Surv. Czechoslovakia, Prague*, **1**.

Larsen, G., and Chilingar, G. V., eds., 1967. *Diagenesis in Sediments*, Amsterdam: Elsevier, 551p (see also 1980 revision).

Müller, G., 1967. Diagenesis in argillaceous sediments, in Larsen and Chilingar, 127–178.

Murray, R. C., 1964. Origin and diagenesis of gypsum and anhydrite, *J. Sed. Petrology*, **34**, 512–523.

Perelman, A. I., 1967. *Geochemistry of Epigenesis.* New York: Plenum (transl. from Russian), 266p.

Price, N. B., 1973. Chemical diagenesis in sediments, *Woods Hole Oceanog. Inst., Mass., Tech. Rept.*, Oct. 1972–Feb. 1973, 95p.

Rona, P. A., and Lowell, R. P., eds., 1980. *Seafloor Spreading Centers: Hydrothermal Systems.* Stroudsburg: Dowden, Hutchinson & Ross, 424p.

Stanton, R. L., 1972. Sulfides in sediments, in R. W. Fairbridge, ed., *Encyclopedia of Geochemistry and Environmental Sciences.* New York: Van Nostrand Reinhold, 1134–1140.

Stewart, F. H., 1965. The mineralogy of the British Permian evaporites, *Mineralog. Mag.*, **34**, 460–470.

Teichmüller, M., and Teichmüller, R., 1967. Diagenesis of coal (coalification), in Larsen and Chilingar, 391–415.

Von Gümbel, C. W., 1868. *Geognostische Beschreibung des ostbayrischen Grenzgebirges.* Kassel, 700p.

Walther, J., 1894. *Einleitung in die Geologie als historische Wissenschaft.* Jena: Fischer, 1055p.

West, I. M., 1964. Evaporite diagenesis in the lower Purbeck beds of Dorset, *Proc. Yorkshire Geol. Soc.*, **34**, 315-326.

Cross-references: *Authigenic Minerals; Rock-Forming Minerals.* Vol. VI: *Authigenesis; Diagenesis.*

DIAMOND

Diamond is a high-pressure polymorph of carbon, C, named by Pliny (77 A.D.) from the corruption of the Greek word meaning "the invincible."

Physical Properties

Diamond belongs to the isometric crystal system. The crystal habit is dominantly octahedral, less commonly dodecahedral and cubic, and rarely tetrahedral. The faces are often curved due to growth and solution facets, and are frequently striated or etched (see Fig. 1). It is the hardest known natural mineral, being 10 in the Mohs scale of hardness. It is not attacked by acids, and has a greasy touch. The structural cell, $a_0 = 3.5595$ Å, contains eight carbon atoms and is face-centered cubic. Twinning is common on {111} and on {001}. The refractive index n, ranges from 2.417 to 2.421 (Na light). Dispersion is strong, and the characteristic luster of the polished stones is due to a combination of the high refractive index and the extreme hardness. The optical and physical properties depend mainly upon the presence or absence of nitrogen impurities, either in dispersed or plate-

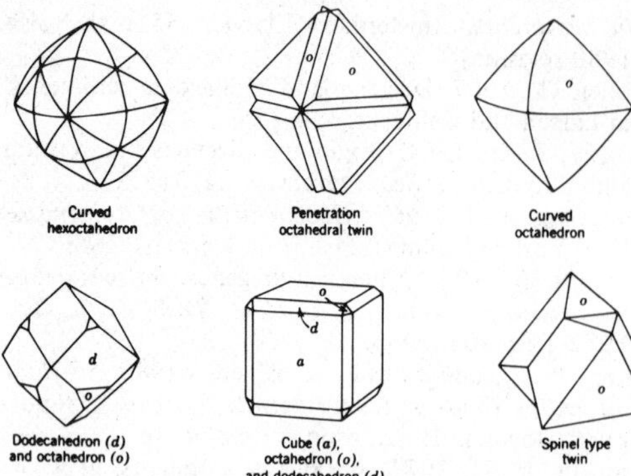

FIGURE 1. Examples of the commoner crystallographic forms of *diamond* (from Vanders and Kerr, 1967; by permission of John Wiley & Sons). High pressure synthesis experiments have shown that the cubic *diamond* is the crystallographic type formed at relatively low temperature, whereas the octahedral *diamonds* are very high-temperature forms. Although mixtures of the other crystallographic types are known in most kimberlite diatremes and alluvial *diamond* deposits, cubic *diamonds* are abundant only in the alluvial *diamond* fields of Kasai (Congo) and Brazil.

let form, which cause *diamond* to be divided into Type I (containing nitrogen) or Type II types (Berman, 1965). The various differences for the two Types are listed in Table 1. The internal growth features of *diamonds* indicate that they had a complex and turbulent growth history (Frank, 1969). The color is highly variable; clear, blue, red, pink, orange, yellow, brown, black, gray, and green stones having been recorded. The yellow, brown, and green varieties may be due to small amounts of iron, aluminum, and magnesium (Fesq et al., 1975), whereas gray or black stones owe their color to inclusions of carbonaceous material. The inferior varieties of *diamond* known as "bort" and "carbonado" are granular or cryptocrystalline *diamonds* that owe their dark color to impurities and inclusions. Irradiation can cause coloration of *diamonds*, neutron bombardment causing a green color and electrons producing a blue color. The green *diamonds*, when heated, pass through various stages of yellow but never become entirely clear again. Green *diamonds* have been found in the gold-producing conglomerates of the Witwatersrand, South Africa, and their color is attributed to irradiation from associated uranium minerals. The purest *diamond* has a specific gravity of 3.511, compared with 2.23 for the low-pressure polymorph of carbon, *graphite*. *Diamond* burns in air at 850°C to give CO_2, and in vacuum is partially converted to graphite at $>1500°C$.

Occurrence

The primary source of *diamond* is the mica-peridotite, kimberlite (q.v., in Vol. V), which occurs in small diatremes (pipes), sills, and dikes. Diamondiferous kimberlites, many of which are Jurassic or Cretaceous in age, are known in the USA, the USSR (Yakutia and the Don Basin), India, Borneo, Mali, Sierra Leone, Zaire, Tanzania, Angola, Rhodesia, and many localities in the Republic of South Africa (Dawson, 1980). *Diamond* has also been found in eclogite nodules in kimberlite pipes in South Africa (Rickwood and Mathias, 1970), in eclogite xenoliths from the Mir and Udachnaya kimberlites, Yakutia (Sobolev et al., 1972; Ponomarenko et al., 1973), in rare garnet serpentinite xenoliths from the Aykhal pipe, Yakutia (Sobolev et al., 1969), and in a mica-garnet lherzolite xenolith from the Mothae pipe, Lesotho (Dawson and Smith, 1975).

Recent alluvial *diamond* deposits related to known kimberlite intrusions are known in Borneo, W Africa, Zaire, Angola, and S Africa; and it is possible that kimberlites may be found in the neighborhood of the alluvial deposits in Brazil, Guyana, China, Australia, and the Hoggar area of the Sahara where no primary source of *diamond* is known as yet. Alluvial *diamonds* have been found in Czechoslovakia near basaltic breccia intrusions that have yet to be proved diamondiferous. Several *diamonds*

TABLE 1. Basic Physical Properties of the Two Main Types of Diamond

Physical Property	Type I *Diamond*	Type II *Diamond*
Infrared absorption	Show absorption between 8 and 10μm	No absorption between 8 and 10 μm
Ultraviolet absorption	Complete beyond 3300 Å	Transparent down to 2200 Å
Photoconductivity	Weak	Strong
Birefringence	Usually strong	Weak or absent
Thermal conductivity	Very good	Extremely good
Electrical conductivity	None	Type IIb are semiconductors
X-ray diffraction	Additional spots and streaks	Normal
Morphology	Well-formed single crystals	Poorly developed
Cleavage	Relatively uneven	Relatively perfect

are reported as having been obtained from Precambrian phyllites in Ghana (Junner, 1943). *Diamonds* are known from ancient placer deposits (q.v.) in Brazil, Russia, S Africa, Rhodesia, India, and Australia. In recent years, much of the *diamond* produced in SW Africa has been won from raised beach deposits to the north and south of the mouth of the Orange River, and *diamonds* are now being recovered from submarine sediments being dredged up by off-shore vessels in the same area. In the USA, several small *diamonds* have been found in the Quaternary glacial deposits in Wisconsin and Ohio, and kimberlites have recently been found in the Ontario-Quebec area of Canada across which the glaciers traveled (Lee and Lawrence, 1968).

Small *diamonds* have been found in several meteorites. Some recent experimental work (Lipschutz and Anders, 1961) favors the theory that these *diamonds* have formed by a direct shock transformation of *graphite* to *diamond* upon impact of the meteorite with the earth, rather than the *diamonds* having been produced under high gravitational pressure in a meteorite parent body (see *Meteoritic Minerals*). However, synthesis conditions of a hexagonal polymorph of *diamond* with the wurtzite structure, named *lonsdaleite* (Frondel and Marvin, 1967), which has recently been found in some meteorites, suggest that the *diamonds* in some meteorites were produced by intense shock pressures acting on crystalline *graphite* inclusions present within the meteorite before impact (Hanneman et al., 1967).

Although most older reports of *diamonds* being found in Alpine-type peridotites and dunites have been regarded as open to question, the recent report of *diamonds* in peridotite and in adjacent alluvial deposits in the Sevan ophiolite belt, Armenia, USSR, (Pavlenko et al., 1974) reopens the whole question of *diamonds* in the ultramafic rocks of the fold-mountain belts (Dawson, 1979). The recent report of *diamond* in basalt from Kamchatka (Kutyev and Kutyeva, 1975) requires confirmation.

Inclusions in Diamond

Many *diamonds* contain small minerals enclosed within the *diamond* crystals during their growth. The most abundant inclusions are magnesian *chromite, forsterite,* ***garnet, clinopyroxene***, and *graphite*; the order of abundance varies from locality to locality. Other less common inclusions are ***orthopyroxene***, *pyrrhotite, pentlandite, rutile, coesite, diamond, ilmenite, magnetite, kyanite, phlogopite*, and *biotite* (Meyer and Boyd, 1972; Sobolev, 1977; Giardini et al., 1974; Prinz et al., 1975). Although most inclusions within a *diamond* are either single crystals or several inclusions of a single mineral species, Sobolev (1977) has recognized inclusions of the following combinations of mineral species within individual *diamonds:* ***garnet*** *+****olivine****±enstatite±chromite* (dunite/harzburgite paragenesis); ***garnet*** *+****olivine****+enstatite+diopside±chromite* (lherzolite paragenesis); ***garnet*** *+****olivine****+diopside* (wehrlite paragenesis); and ***garnet*** *+omphacite±rutile* (eclogite paragenesis). Most inclusions are the site of stresses within the *diamonds*, identified by birefringence or by dark-colored microfractures (Harris, 1972). In addition to recognizable crystalline inclusions, some Type I *diamonds* contain white clouds of minute particles $<3\ \mu m$ in diameter—probably of gas and (or) liquid. In addition to nitrogen, Melton and Giardini (1974) have detected water, hydrogen, carbon dioxide, argon, ethylene, ethyl alcohol, butane, and oxygen gas in *diamonds* from Africa and Brazil. Fesq et al. (1975), on the basis of high-resolution neutron activation analyses of inclusion-free *diamonds*, have proposed the presence (in some *diamonds*) of micro-droplets of liquid of ***garnet*** *+diopside±****olivine*** chemistry, together with high amounts of sulphur and carbon dioxide.

Synthesis and Paragenesis

In the latter part of the 19th century and at the beginning of this century, several scientists attempted to synthesize *diamond*. The first attempts were by J. B. Hannay in 1880 and minute crystals produced during one of his experiments, when lithium, bone oil, and paraffin were heated together under pressure, have been authenticated as *diamond* by modern X-ray techniques. Later syntheses claimed by F. H. Moissan, Sir William Crookes, and Sir Charles Parsons are questionable.

The first repeatable syntheses of *diamond* were made in 1953 by the Swedish company Allmana Avenska Elektriska Aktiebolaget; since then, more syntheses have been carried out in several laboratories in the USA and the USSR; in the De Beers Adamant Research Laboratory, Johannesburg; and at the Australian National University, Canberra. A report in 1959 (Bovenkerk, et al., 1959) showed that in a pure carbon system the direct transition of *graphite* to *diamond* could not take place even at 120,000 atm pressure, but certain metals added to the system acted as catalysts enabling *diamond* to be formed in the temperature range 1,200–2,400°C and in the pressure range 55,000–100,000 atm. The catalysts used were tantalum, chromium, iron, cobalt, nickel, ruthenium, palladium, indium, and platinum. *Diamond* has also been synthesized at subatmospheric pressures by deposition on *diamond* powder from methane gas at 1050°C and 0.3 torr pressure (Angus et al., 1968).

Many scientists believe that natural *diamond* originates within the earth's mantle (see *Mantle Mineralogy*) because (i) the principal parent rock of *diamond* (kimberlite) is ultramafic in composition like the mantle; (ii) the syngenetic inclusions resemble those found in high-pressure, upper-mantle paragenesis; and (iii) the extremes of pressure and temperature necessary for *diamond* formation will normally only be achieved within the mantle, a fact confirmed by the experiments in synthesizing *diamond*. Most geologists and physical chemists believe that natural *diamond* can only form above the *graphite/diamond* equilibrium line in the presence of molten silicate material, i.e., above the peridotite solidus.

One estimate of the minimum conditions for the formation of natural *diamond*, arrived at by the intersection of the *graphite/diamond* equilibrium curve and the slope of the dry solidus of natural garnet peridotite, is 1800°C and 68 kbrs, approximating to 200 kms depth (Kennedy and Nordlie, 1968). However, if the peridotite solidus can be lowered due to the presence of *phlogopite*, it is possible that the minimum conditions for *diamond* formation may be as low as 1200°C and 45 kilobars (Dawson, 1972). The importance of abundant volatile material for the formation of *diamond* is testified to by the high amounts of water and CO_2 in the host kimberlite (Dawson, 1980), and it has been suggested that partial pressure of carbon dioxide essentially equal to the confining pressure is necessary in order that *diamond* crystals may be stable in an environment that contains substantial ferrous and ferric iron (Kennedy and Nordlie, 1968). The possibility of the formation of seed nuclei in the upper mantle, later overgrown by metastable *diamond* during kimberlite ascent, has been proposed by Mitchell and Crockett (1971). The discovery of *diamond* in garnet lherzolite (a major constituent of the upper mantle) raises the possibility that the *diamond* seeds for kimberlite *diamonds* may be derived from garnet lherzolite (Dawson and Smith, 1975).

Uses

Because of its luster and hardness, *diamond* is without equal as a gemstone (see *Gemology*). The finest stones are clear or blue-white, though some magnificent colored stones have been found. The names of some of the more famous uncut (or raw) *diamonds*, with their weight in carats (1 carat=0.2 g) and their country of origin are: Cullinan/3106, Excelsior/995, Jonker/726, Jubilee/650, De Beers/440, and the yellow Colenso/133 (all from S Africa); Pitt/410 (India); Darcy Vargas/460, Star of the South/262, and the pink Cross of the South/118 (Brazil). Famous cut stones are the Koh-i-Nur/108, Orlov/199, and the Shah/88. Apart from its everyday uses in glass cutting, wire-drawing, and borehole drilling, *diamond* is a strategic mineral used in heavy industry for machine tooling, drilling, and grinding (see *Abrasive Materials*). Its high thermal conductivity is leading to increased use for heat sinks in the electronic industry and its low coefficient of friction makes it an attractive (though costly) alternative to traditional metals in the manufacture of microtomes and surgical scalpels (Caveney, 1979).

J. B. DAWSON

References

Angus, J. C.; Will, H. A.; and Stanko, W. S., 1968. Growth of diamond seed crystals by vapor deposition, *J. Appl. Physics*, **39**, 2915–2922.

Berman, R., 1965. *Physical Properties of Diamond*. Oxford: Clarendon Press, 443p.

Bovenkerk, H. P., et al., 1959. Preparation of diamond, *Nature*, **184**, 1094–1098.

Bruton, E., 1978. Diamonds. 2nd ed. London: National Association of Gemology Press, 532p.

Caveney, R. J., 1979. Non-abrasive industrial uses of diamond, in J. E. Field, ed. *The Properties of Diamond*. London: Academic Press, 619–639.

Dawson, J. B., 1979. New aspects of diamond geology, in J. E. Fields, ed. *The Properties of Diamond*. London: Academic Press, 539–554.

Dawson, J. B., 1980. *Kimberlites and their Xenoliths*. Berlin: Springer-Verlag, 252p.

Dawson, J. B., 1972. Kimberlite and its relationship to the upper mantle, *Phil. Trans. Roy. Soc. London. A.*, **271**, 297–311.

Dawson, J. B., and Smith, J. V., 1975. Occurrence of diamond in a mica-garnet lherzolite xenolith from kimberlite, *Nature*, **254**, 580–581.

Fesq, H. W., et al., 1975. A comparative trace-element study of diamonds from Premier, Finsch and Jagersfontein Mines, South Africa, *Phys. Chem. Earth*, **9**, 817–836.

Frank, F. C., 1969. Diamonds and deep fluids in the upper mantle, in S. K. Runcorn, ed. *The Application of Modern Physics to the Earth and Planetary Interiors*. New York: Wiley/Interscience, 247–250.

Frondel, C., and Marvin, U. B., 1967. Lonsdaleite, a hexagonal polymorph of diamond, *Nature*, **214**, 587–589.

Giardini, A. A., et al., 1974. Biotite as a primary inclusion in diamond: Its nature and significance, *Am. Mineralogist*, **59**, 783–789.

Hannemann, H. E.; Strong, H. M.; and Bundy, F. P., 1967. Hexagonal diamonds in meteorites: Implications, *Science*, **155**, 995–996.

Harris, J. W., 1972. Black material on mineral inclusions and in internal fracture planes in diamond, *Contrib. Mineral. Petrol.*, **35**, 22-33.
Junner, N. R., 1943. The diamond deposits of the Gold Coast, *Bull. Geol. Surv. Gold Coast, No. 12.*, 23p.
Kennedy, G. C., and Nordlie, B. E., 1968. The genesis of diamond deposits, *Econ. Geol.*, **63**, 495-503.
Kutyev, F. Sh., and Kutyeva, G. V., 1975. Diamonds in the basaltoids of Kamchatka, *Doklady Akad. Nauk S.S.S.R.*, **221**, 183-186 (in Russian).
Lee, H. A., and Lawrence, D. E., 1968. A new occurrence of kimberlite in Gauthier Township, Ontario, *Geol. Surv. Canada, Paper 68-22*, 16p.
Lipschutz, M. E., and Anders, E., 1961. The record in the meteorites: Part 4, *Geochim. Cosmochim. Acta*, **24**, 83-105.
Mitchell, R. W., and Crockett, J. H., 1971. Diamond genesis–a symposium of opposing views. *Mineral. Deposita*, **6**, 392-403.
Melton, C. E., and Giardini, A. A., 1974. The composition and significance of gas released from natural diamonds from Africa and Brazil, *Am. Mineralogist*, **59**, 775-782.
Meyer, H. O. A., and Boyd, F. R., 1972. Composition and origin of crystalline inclusions in natural diamonds, *Geochim. Cosmochim. Acta*, **36**, 1255-1273.
Pavlenko, A. S., et al., 1974. On the diamonds in the ultramafic belts of Armenia, *Geochemistry Internat.*, **11**, 282-294.
Ponomarenko, A. I.; Serenko, V. P.; and Lazko, E. E., 1973. First finds of diamond-bearing eclogite in the kimberlite of the pipe Udachnaya, *Doklady Akad. Nauk S.S.S.R.*, **209**, 188-189 (in Russian).
Prinz, M., et al., 1975. Inclusions in diamonds: garnet lherzolite and eclogite assemblages, *Phys. Chem. Earth*, **9**, 797-816.
Rickwood, P. C., and Mathias, M., 1970. Diamondiferous eclogite xenoliths in kimberlite, *Lithos*, **3**, 223-235.
Smith, G. F. H., 1958. *Gemstones*, 13th ed. London, Methuen, 560p.
Sobolev, N. V., 1977. *Deep-seated Inclusions in Kimberlites and the Problem of the Composition of the Earth's Mantle*, Washington, DC: American Geophysical Union, 279p.
Sobolev, V. S., et al., 1969. Xenoliths of diamond-bearing pyrope serpentinite from the 'Aykhal' pipe, Yakutia, *Doklady Acad. Sci. U.S.S.R.*, **188**, 168-170 (American Geological Institute translation).
Sobolev, V. S.; Sobolev, N. V.; and Lavrentyev, Yu. G., 1972. Inclusions in diamond from diamondiferous eclogite, *Dokl. Acad. Sci. U.S.S.R., Earth Sci. Sec.*, **207**, 121-123.
Vanders, I., and Kerr, P. F., 1967. *Mineral Recognition*. New York: Wiley, 316p.

Cross-references: *Abrasive Materials; Blacksand Minerals and Metals; Gemology; Mantle Mineralogy; Meteoritic Minerals; Mineral Industries; Mohs Scale of Hardness; Placer Deposits; Synthetic Minerals.*

DIATOMITE—*See* Vol. VI

DIRECTIONS AND PLANES

Crystals are bounded by planar faces because they are made up of layers of atoms or ions. Planes are characteristic of crystal structures, of lattices (q.v.), of twinning (q.v.), and of mineral cleavage (see *Rupture of Mineral Matter*), as well as of faces.

A pole, the line perpendicular to a plane, has direction in space. The intersection of two crystal faces defines a line, the zone direction. Physical properties of minerals, such as hardness, piezoelectricity, magnetism, etc., vary systematically with crystallographic direction.

Lengths of unit-cell edges of lattices and the angles between these edges are determined by X-ray diffraction (q.v., Vol. IVA). The lengths a, b, and c along the crystallographic axes x, y, and z and the interaxial angles α between y and z, β between x and z, and γ between x and y constitute the parameters of a lattice, but symmetry (q.v.) may reduce the number of parameters needed to characterize a lattice.

Planar directions are given by Miller indices, which are reciprocals of axial intercepts of crystal planes (Fig. 1). Thus, if the intercepts are proportional to $2a$, $2b$, and $1c$, the reciprocals, after dividing through by $1a$, $1b$, and $1c$, are 1/2, 1/2, and 1/1, which, cleared of fractions are the Miller indices (112) (read, "one, one, two," not "one hundred twelve").

For planes lying parallel to a crystallographic axis, an intercept is at infinity (∞) and the reciprocal is zero. A plane lying parallel to a and to b but having an intercept with c would be noted as $1/\infty a$, $1/\infty b$, and $1c$, which yield the Miller indices (001) ("zero, zero, one" or "oh, oh, one"). Planes with negative intercepts are indicated with a bar over the negative index number. Thus, the plane with intercepts proportional to $1a$, $1b$, and $-1c$ has Miller indices $(11\bar{1})$ ("one, one, bar one"). Since Miller indices are used to designate slopes of planes, they are also cleared of any common factor; e.g., there is no face (333).

Prior to the analysis of crystal structures by X-ray diffraction, axes were selected so as to be parallel to axes of morphological symmetry and (or) to important edges between crystal faces (zone axes). Unit length along each axis was defined at that time with reference to a parametral plane with intercepts on all three axes selected to give the simplest total intercepts for all crystal faces.

All symmetrically related faces constitute a form (see *Crystal Habits*) and are designated $\{khl\}$. Thus, the six faces of a cube have the Miller indices (100), (010), (001), $(\bar{1}00)$, $(0\bar{1}0)$, and $(00\bar{1})$, which are represented by {100}.

All of the planes in a set, the intersections of

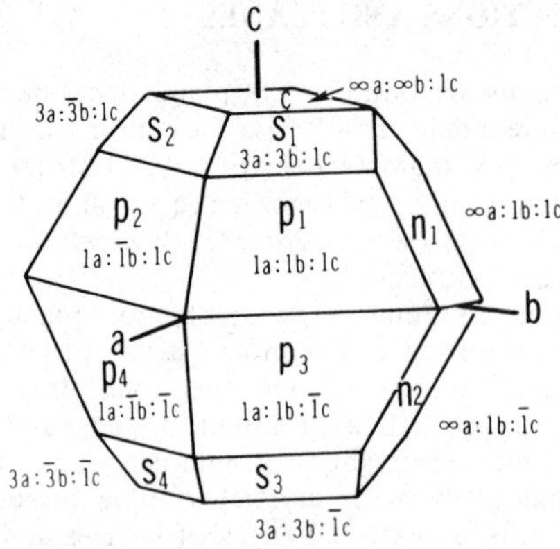

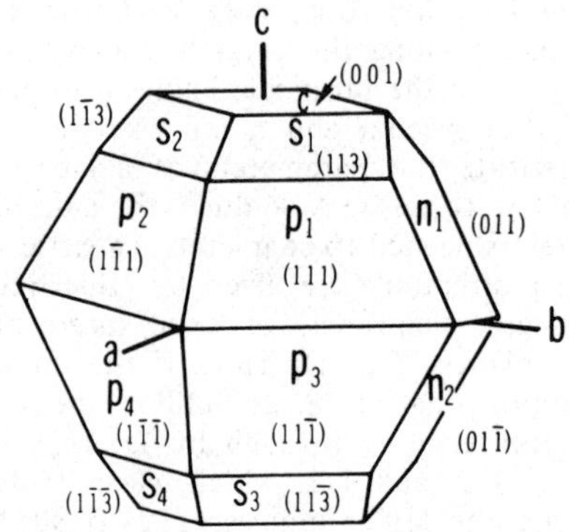

FIGURE 1. Orthorhombic crystal (from *Crystallography and Crystal Chemistry: An Introduction* by F. Donald Bloss. Copyright © 1971 by Holt, Rinehart and Winston. Reprinted by permission of Holt, Rinehart and Winston). (a) Crystal faces with Weiss parameters. (b) Crystal faces with Miller indices.

which are parallel, constitute a zone and the common direction is the zone axis, designated $[uvw]$, where u, v, and w are the coordinates of a point on a line from the origin. A plane (hkl) is in a zone $[uvw]$ if $hu+kv+lw=0$. A form of symmetrically related zone axes is designated $\langle uvw \rangle$. These symbols also are used to designate planes and directions of deformation in crystals (see *Plastic Flow in Minerals*).

The slope of any rational plane or the direction of any rational line in a crystal may be designated by its three indices. For crystals belonging to the hexagonal or to the trigonal system, i.e., having one sixfold or one threefold axis of symmetry, addition of a redundant fourth index number i is convenient. In these two systems, the crystallographic axes are three coplanar a axes of equal length at mutual angles of 120° and a c axis at right angles to the a plane. Thus, any plane not parallel to the a plane will intercept at least two of the three a axes. The reciprocal of the intercept with the a_3 axis is designated i, resulting in the Miller-Bravais indices $(hkil)$. The planes $(10\bar{1}0)$ and $(11\bar{2}0)$ are, therefore, vertical at 30° to one another. The third index i is redundant since $h+k+i=0$. It may be omitted, in which case the indices are given as $(hk \cdot l)$.

In analysis of crystal structures by X-ray diffraction, both planar slope (hkl) and interplanar spacing d are determined. Families of planes having the same slope but separated in space are designated by Bragg indices: hkl or d_{hkl} (no parentheses, braces, or brackets enclose Bragg indices). Unlike Miller indices, Bragg indices may contain a common factor and 111 represents twice the interplanar spacing of 222, thrice 333, etc. The common factor in Bragg indices represents the order of diffraction: first, second, or third, etc. (see Vol. IVA: *X-ray Diffraction*).

KEITH FRYE

References

Bloss, F. D., 1971. *Crystallography and Crystal Chemistry*. New York: Holt, Rinehart & Winston, 545p.

Buerger, M. J., 1963. *Elementary Crystallography*. New York and London: Wiley, 528p.

Megaw, H. D., 1973. *Crystal Structures: A Working Approach*. Philadelphia: Saunders, 563p.

Cross-references: *Axis of Symmetry; Barker Index of Crystals; Crystal Habits; Crystallography: History; Crystallography: Morphological; Lattice; Point Groups; Space-Group Symbol; Symmetry; Twinning.*

DISPERSION, OPTICAL

The angular divergence and consequent resolution into spectral colors of polychromatic or white light upon obliquely crossing an interface between two transparent media (Fig. 1A) constitutes "optical dispersion." The divergence of the different wavelengths (λ_1,λ_2,λ_3, etc.) increases to the extent that the dispersion curves for the two media—that is, the lines that show for each how refractive index n changes with wavelength λ—differ in slope and (or) intercepts on the axis along which n is plotted (Fig. 1B). If the relation between n and λ is plotted on Hartmann dispersion net paper, particularly for media with refractive indices under 1.65, the dispersion curves usually become straight lines (Fig. 1C).

For light of wavelengths λ_1, λ_2, and λ_3 (with angles of refraction r_1, r_2, and r_3 in Fig. 1A), let n_{11}, n_{12}, and n_{13}, respectively, represent the refractive indices for medium 1 and n_{21}, n_{22}, and n_{23} those for medium 2. If i represents the angle of incidence of the nondispersed

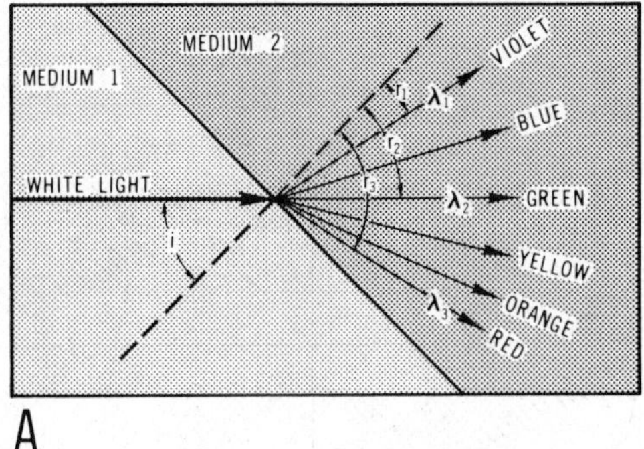

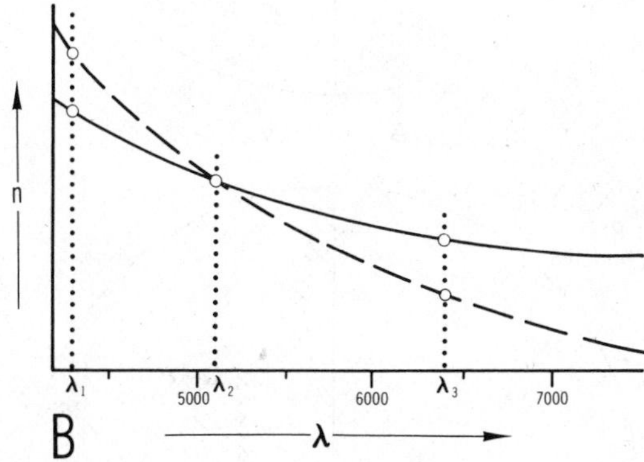

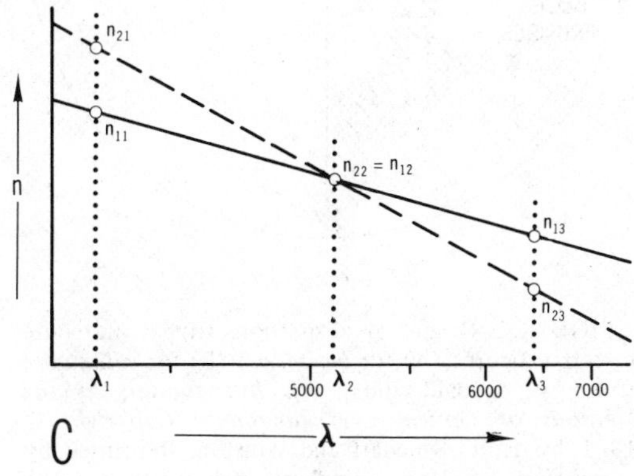

FIGURE 1. (A) Dispersion of a ray of white light into colored rays upon obliquely crossing an interface between two transparent media. (B) The change in refractive index (n) versus wavelength of light—that is, dispersion curves—for medium 1 (solid line) and medium 2 (dashed line) as plotted on ordinary graph paper. (C) This same data, plotted on Hartmann Dispersion Net paper is seen to become linear. The refractive indices for medium 1 and medium 2 at wavelengths λ_1, λ_2, and λ_3 are singled out. See text for further discussion.

beam, then from Snell's law

$$n_{11} \sin i = n_{21} \sin r_1$$
$$n_{12} \sin i = n_{22} \sin r_2$$
$$n_{13} \sin i = n_{23} \sin r_3$$

If, as shown in Fig. 1C, $n_{11} < n_{21}$, $n_{12} = n_{22}$, and $n_{13} > n_{23}$, then from the above equations, $r_1 < r_2 = i < r_3$, as illustrated in Fig. 1A.

Dispersion Equations

Various formulas have been proposed to describe the change of refractive index (n) with respect to wavelength (λ) in isotropic materials. Cauchy determined the relationship to be

$$n = C_1 + \frac{C_2}{\lambda^2} + \frac{C_3}{\lambda^4} + \ldots$$

where C_1, C_2, C_3, ... are Cauchy constants characteristic of the material. For a substance with refractive index below 1.65, the Hartmann dispersion equation

$$n = n_0 + \left(\frac{C}{\lambda} - \lambda_0\right)$$

may pertain, C being a constant and n_0 representing the material's refractive index at a particular wavelength λ_0. In this case, the dispersion curve plots precisely as a straight line on Hartmann dispersion net paper, which is based on Hartmann's equation. For materials that are deeply colored because they strongly absorb certain wavelengths of the visible spectrum, if λ_0 now represents the wavelength at which absorption is maximum, the relationship between n and λ may be better approximated by Sellmeier's formula

$$n^2 = 1 + \frac{A\lambda^2}{\lambda^2 - {\lambda_0}^2}$$

where A is the Sellmeier constant, which is characteristic of the substance, and λ_0 represents the wavelength at which absorption is greatest. For such strongly absorbent materials, n may no longer decrease steadily with increase in λ as in what is called "normal dispersion" (Fig. 1C). Instead n may increase as λ increases, such dispersion being sometimes called "anomalous dispersion."

Using Fraunhofer's symbolism for specific visible wavelengths—that is, A, 7594 Å; B, 6870 Å; C, 6563 Å; D, 5893 Å; E, 5269 Å; F, 4861 Å; G, 4308 Å—an isotropic substance's refractive indices for these particular wavelengths may be symbolized as n_A, n_B, n_C, n_D, ..., n_G. Using these symbols, the dispersive ability of a given isotropic material may be expressed as its "coefficient of dispersion," which is defined as $n_F - n_C$. A superior expression, however, is its "dispersive power," which is defined as $(n_F - n_C)/(n_D - 1)$. Occasionally, particularly for liquids, dispersive power is stated as the reciprocal of the foregoing value, that is, as $(n_D - 1)/(n_F - n_C)$.

In general, the dispersive power is greater, that is, the dispersion curve slopes more steeply, (i) for liquids as compared to solids, (ii) for materials with high refractive indices as compared to those with low, and (iii) for materials containing transition elements such as iron and titanium.

Dispersion in Uniaxial Crystals

The dispersion curves for ϵ and for ω in uniaxial crystals are usually not parallel. In rare instances, for example, in *metatorbernite* (Bloss, 1961, p. 141), the ϵ and ω dispersion curves may so differ in slope that they intersect. The crystal is then isotropic for λ_I, the wavelength of intersection, whereas its optic sign for $\lambda < \lambda_I$ is reversed from that for $\lambda > \lambda_I$.

Dispersion in Biaxial Crystals

For biaxial crystals, the principal refractive indices α, β, and γ change value with wavelength so that a dispersion curve exists for each. This dispersion of α, β, and γ causes the optic axial angle, $2V$, to change with wavelength. In other words, the optic axes occupy different positions for different wavelengths of light, that is, dispersion of the optic axes exists. On the basis of such dispersion, biaxial crystals are divisible into two types: (*1*) those for which $2V$ for red light exceeds $2V$ for the shorter violet or blue wavelengths, this being expressed as $2V_r > 2V_v$ or simply as $r > v$, and (*2*) those for which $r < v$. If there is pronounced difference in value of $2V$ for red compared to $2V$ for blue light, the crystal's interference figure for white light will display bluish and reddish fringes at the edges of the isogyres, particularly in the region of the melatopes (Fig. 2). The reddish fringe locates the melatope and a portion of the isogyre for blue light; the bluish fringe that for red light. Thus, if blue light were used to illuminate the crystal, the reddish fringes would become black areas; if red light were the illuminant, the bluish fringes would be sites of blackness.

For orthorhombic crystals the principal axes of the indicatrix, X, Y, and Z, coincide with the crystallographic axes for all wavelengths of light (see *Optical Orientation*). Normally, therefore, orthorhombic crystals only show dispersion of the optic axes. *Brookite*, however, is unusual because its dispersion curve for the index of light vibrating parallel to its *a* axis intersects the curve for light vibrating parallel to its *c* axis (Fig. 3). Thus, for the 5550 Å wavelength, it is

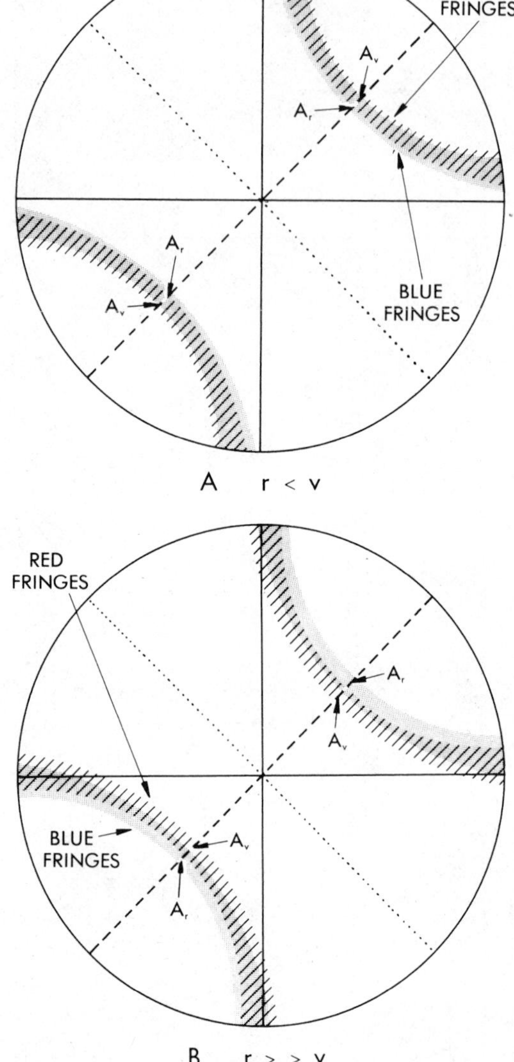

FIGURE 2. Origin of dispersion fringes in acute bisectrix figures (A) for *barite* and (B) for *sillimanite* (from F. Donald Bloss, *An Introduction to the Methods of Optical Crystallography*. *Copyright* © 1961 by Holt, Rinehart and Winston. Reprinted by permission of Holt, Rinehart and Winston). The isogyre locations for red light are shown as shaded areas; for blue light, as ruled areas. Red light emerging along the red (shaded) isogyre is extinguished at the analyzer whereas blue light is transmitted. Blue light emerging along the blue (ruled) isogyre is extinguished whereas red light is transmitted. Where both isogyres are superimposed, all wavelengths are extinguished; these areas thus appear black. Where they are not superimposed, a blue coloration is developed along the red isogyre and melatope whereas a red coloration is developed along the blue isogyre and melatope. The dotted and the dashed lines represent the traces of principal planes.

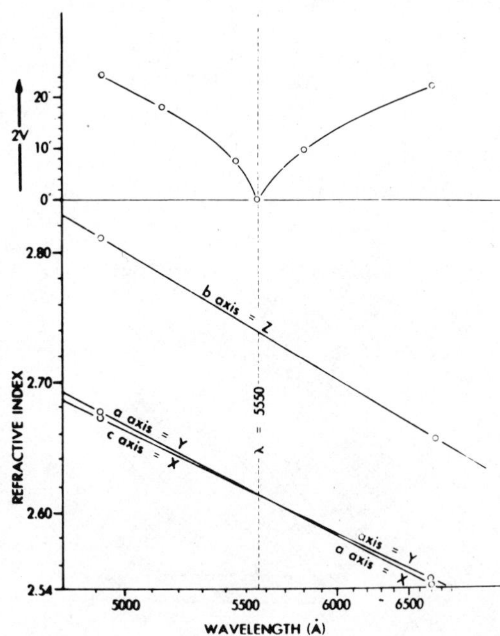

FIGURE 3. Variation of refractive indices with respect to wavelength for light vibrating parallel to the *b* axis, *a* axis, and *c* axis in *brookite* (from F. Donald Bloss, *An Introduction to the Methods of Optical Crystallography.* Copyright © 1961 by Holt, Rinehart and Winston. Reprinted by permission of Holt, Rinehart and Winston). Top scale: variation of the optic angle $2V$ for different wavelengths of light.

uniaxial (+), whereas it is biaxial (+) and its XZ or optic plane is parallel to (100) for $\lambda <$ 5550 Å, but to (001) for $\lambda >$ 5550 Å. This is called "crossed-axial plane dispersion."

In monoclinic crystals, one of the three principal axes of the indicatrix coincides in direction with the single symmetry axis (2-fold and (or) $\bar{2}$-fold) for all wavelengths of light (see *Optical Orientation*). The remaining two principal axes change position as wavelength changes, this constituting "dispersion of the principal axes." For monoclinic crystals the two principal axes undergoing dispersion always remain (i) in the plane normal to the single symmetry axis and (ii) at 90° to each other as wavelength is varied. An acute-bisectrix-centered figure of a monoclinic crystal for which dispersion of two of its principal axes is pronounced will display

- *inclined dispersion* (see Bloss, 1961, Fig. 9–28), if the invariant, symmetry-coincident, principal axis is the optic normal Y;
- *parallel or horizontal dispersion* (see Bloss, 1961, Fig. 9–29), if the invariant axis is the obtuse bisectrix of the indicatrix; or

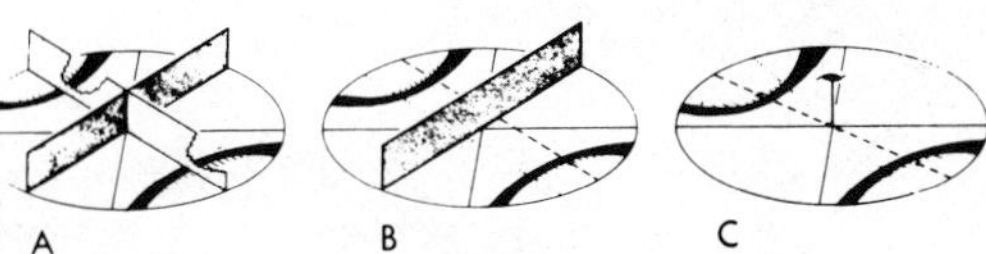

FIGURE 4. Differentiation of orthorhombic and monoclinic crystals from the symmetry of the dispersional color fringes in their acute bisectrix-centered figures (from F. Donald Bloss, *An Introduction to the Methods of Optical Crystallography.* Copyright © 1961 by Holt, Rinehart and Winston. Reprinted by permission of Holt, Rinehart and Winston). Orthorhombic crystals can always be considered to have two planes of symmetry perpendicular to the interference figure (A). Monoclinic interference figures are symmetrical only with respect to one such plane (B) or to a twofold axis perpendicular to the plane of the interference figure (C). The *b* axis of monoclinic crystals either is the direction perpendicular to this plane (B) or it coincides with the twofold axis (C).

- *crossed dispersion* (see Bloss, 1961, Fig. 9–30), if the invariant axis is the acute bisectrix.

Much more complicated interference figures might result if, in a situation like that for *brookite*, a monoclinic crystal were discovered for which the dispersion curves for α, β, and γ intersected.

For triclinic crystals, all three principal axes of the indicatrix are subject to dispersion. Dispersional fringes in interference figures are thus more difficult to interpret. In general, the dispersional color fringes in the acute-bisectrix-centered figures of triclinic crystals are asymmetric, lacking either a plane of symmetry or a 2-fold axis normal to the plane of the interference figure, whereas a monoclinic crystal may possess one such plane or 2-fold (Fig. 4B, C) and an orthorhombic crystal more than one (Fig. 4A).

F. DONALD BLOSS

References

Bloss, F. D., 1961. *An Introduction to the Methods of Optical Crystallography*. New York: Holt, Rinehart & Winston, 294p.

Emmons, R. C., 1943. The universal stage, *Geol. Soc. Am., Mem. 8.*

Wahlstrom, E. E., 1960. *Optical Crystallography*, 3rd ed. New York: Wiley, 356p.

Winchell, A. N., 1937. *Elements of Optical Mineralogy*, 5th ed. New York: Wiley, pt. I.

Cross-references: *Color in Minerals; Isotropism; Minerals: Uniaxial and Biaxial; Optical Mineralogy; Optical Orientation; Refractive Index*; also, see glossary.

DOUBLE REFRACTION—*See* OPTICAL MINERALOGY

E

ELECTRON MICROSCOPY (SCANNING) AND CLASTIC MINERAL SURFACE TEXTURES

Scanning Electron Microscope

Three years after the suggestions of Knoll in 1935, the first experimental scanning electron microscope (SEM) was built by Ardenne. An improved unit with resolution of about 500 Å was produced by Zworkykin, Hillier, and Snyder in 1942, but it was not until 1965 that the first commercial unit based on the work of Oatley was produced by Cambridge Scientific Instruments.

Instrumentation. In a way similar to the transmitted electron microscope (TEM), the SEM, under vacuum, uses a hot tungsten-filament electron gun and a series of electromagnetic lenses to bombard the sample with a fine beam of electrons. The acceleration potentials involved range from 1 to 30 kV.

When the electron beam is made to scan and bombard the surface of the subject with primary electrons, in a similar (but slower) way to a television raster, a number of phenomena result (Fig. 1). These are the reflection of primary electrons (back-scattered electrons), low-energy (secondary) electron emission, cathodoluminescence (as visible light of various wave-lengths), X-ray excitation-beam-induced conductivity (electromotive force), electron transmission (if the specimen is thin enough), and radiation damage.

The secondary electron-emission mode is used to produce the SEM picture by detecting variations over the sample surface. The detectors in turn synthesize these data and display them on a cathode-ray tube by another synchronously scanning raster (see Long, 1967; Willard, 1968; and Krinsley and Doornkamp, 1973).

Field of Application. With the development of the SEM, a whole new field of microscopy was opened up, one where, over a very large range of magnifications (e.g. 20X to 100,000X), the surface features of materials, both organic and inorganic, could be examined nondestructively without involving the thin-section or replication techniques required for TEM examinations or the depth of focus and upper magnification limitations of the optical light microscope.

INCIDENT ELECTRON PROBE

Electron detector
Backscattered electrons
Photon detector
Cathodoluminescence
Electron detector
Secondary electrons
Electromotive force
Absorbed electrons
Electron detector
Transmitted electrons

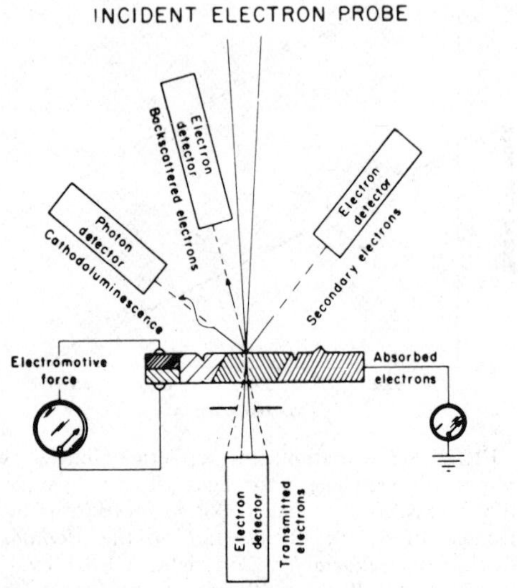

FIGURE 1. Phenomena produced and detected when a specimen is bombarded with a narrow beam of primary electrons (from Willard, 1968).

Almost without exception in mineralogy, indirect replica examinations were used for TEM and the direct examination of the original material for SEM.

The facility to examine entire specimens of the actual subject material over an extremely wide range of magnifications and unhampered by depth-of-focus limitations and replica distortions and artifacts has led to the wide and rapidly increasing use of scanning electron microscopy.

A comparison of the important features exhibited by SEM, the optical light microscope (OLM), and the conventional transmission electron microscope (TEM) is made in Table 1.

Sample Preparation. The sample surfaces are cleaned chemically, generally by boiling in concentrated hydrochloric acid, stannous chloride solution to remove iron oxide, or a solution of potassium dichromate and potassium permanganate dissolved in concentrated sulphuric acid to remove organic particles and coatings. Due

TABLE 1. Comparison of Important Features Exhibited by Optical Light (OLM), Transmission Electron (TEM), and Scanning Electron (SEM) Microscopes

	OLM	TEM	SEM
Lens system	Mostly glass lenses	Electromagnetic lenses	Electromagnetic lenses
Image	Displayed on eye retina or viewing screen	Displayed on fluorescent viewing plate	Line scan display on cathode-ray tube for viewing
Energy	White or monochromatic light	Accelerated electrons, 50–1000 kv	Accelerated electrons 5–25 kv
Wave lengths	4000–7000 Å	Generally less than 1 Å	Less than 1 Å
Resolution	Down to 2000 Å	7–50 Å	Down to 250 Å
Depth of focus	Very poor at high magnifications	Good at all magnifications	Exceptionally good at all magnifications
Magnification range	Up to 2,500X	600X–250,000X	45X–100,000X
Sample requirements	Standard or irregular size thin sections, polished or rough surfaces. Thickness can be important.	Thickness critical: 6 microns or less, and a few mm diameter for transmission. Replicas are very thin films, generally carbon, for irregular surfaces.	Thickness not critical: a dimension less than about 25 mm. Surface can be rough; must be vacuum coated with Au or Pt.

From Willard, 1968.

to induced surface damage, ultrasonic cleaning is not recommended.

Samples are mounted with double-sided sticky tape or silver paint on metal plugs in configurations designed to aid in the later identification of specific grains. The plug is then positioned on a small rotating table in a vacuum evaporator. Here, by an electrical evaporation process, it is thinly coated, over a period of about 20 min., with gold, platinum-palladium, or carbon. Of these, gold appears to give marginally the best results. The coated specimens are then ready for insertion into the SEM for viewing.

Clastic Mineral Grain Surface Textures and Features

Early work describing the surface textures and features of *quartz* sand grains in relation to environmental discrimination and diagenesis was carried out by the TEM technique. Subsequently, the SEM technique has been extensively used to study the surface textures of mineral grains for

1. determing depositional sedimentary environments of recent sediments,
2. determining the source and provenance of sediments,
3. study of different stages of diagenesis and the nature and origin of silica overgrowths on quartz grains, and
4. determining the origin of grain frosting in the desert environment.

A little work has been carried out on rock fragments, not as a tool for environmental discrimination, but to further the understanding of diagenesis. The SEM has also been used to study the surface textures of heavy mineral grains.

Various minerals have been examined by the SEM, but the many different surface morphological features exhibited by clastic grains of the mineral *quartz* have attracted by far the greatest interest, attention, and investigation.

Quartz. Most authors have concluded that surface textures can be used in the study of the depositional environmental of sands. Each environment appears to be characterized by processes and energy levels that are not only particular to each specific environment, but that result in the production of features and textures diagnostic of it. The origin of these features may be mechanical or chemical.

Mechanical features

1. V-shaped pits
2. Straight or curved grooves
3. Conchoidal breakage patterns
4. Imbricated breakage patterns
5. Upturned plates
6. Flat cleavage surfaces
7. Disk-shaped concavities
8. Graded arcs
9. Meandering ridges

Etching and diagenetic features

10. Solution and (or) precipitation
11. Oriented V-forms
12. Smooth precipitation surfaces
13. Crystal growths
14. Surface cracks

These features may be described as follows:

1. *Mechanical V-shaped pits* are the most diagnostic features of high-energy beach environments. They are usually best seen at 3000X to 5000X magnifications (Fig. 2), although they may sometimes be observed at magnifications as low as 1000X. These pits are triangular in shape and are scattered randomly over the surface of the grains. The size of these pits varies from approximately 0.2 μm to 3 μm; the most common size is around 0.5 μm. The presence of these pits on *quartz* grains from beach environments has led to the general belief that they are due to impact between grains in this turbulent, aqueous, high-energy environment. The exact mechanism is, however, still not clear, as Margolis and Krinsley (1974) have invoked the cleavage of *quartz* as a significant factor.

2. *Straight or curved grooves* can be seen best at magnifications between 1000X and 2000X (Fig. 3) and occur scattered irregularly over the grain's surface. The size ranges from a few μm to as much as 15 μm in length and up to 0.4 μm wide and 0.5 μm deep (Krinsley and Doornkamp, 1973). They are found associated with the mechanical V-shaped pits on grains from beach environments and are considered to be of similar origin.

3. *Conchoidal breakage patterns*, although variable in size, are usually small (several μm) on grains from beach and dune environments and large (on the order of tens of μm) on material from glacial environments (see plates 29 to 42 in Krinsley and Doornkamp, 1973). They consist of curved, irregular depressions or convex elevations, which are typically stepped in a series of shell-like ridges.

4. *Imbricated breakage patterns* are observed with the relatively large (up to 100 μm) conchoidal breakage patterns on grains of glacial environmental origin.

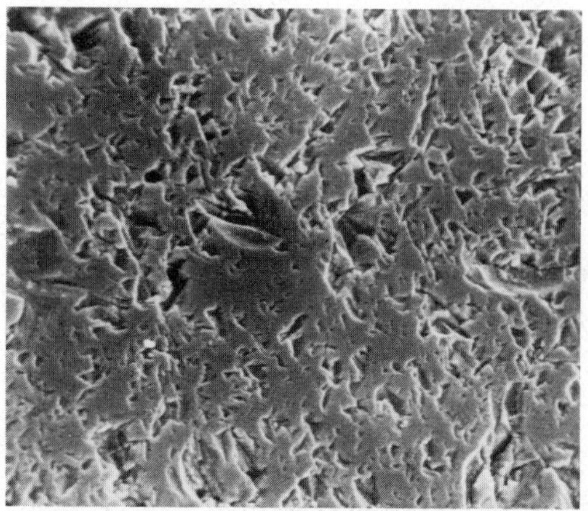

FIGURE 2. Mechanical V-shaped pits seen on *quartz* grains from a modern beach (X2080).

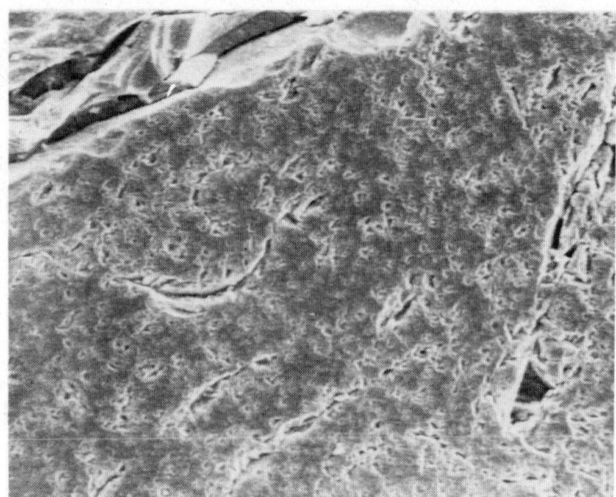

FIGURE 3. Mechanical curved grooves seen on *quartz* grains from a modern beach (X666).

5. The *mechanical upturned plates* of Krinsley and Doornkamp (1973) are exhibited on grains of any size from most natural and some experimental environments. They are composed of a series of parallel plates, perfectly oriented in one direction and separated by depressions. *Precipitated* upturned silica plates are also sometimes observed in weathered source materials and in aeolian and diagenetic environments. In tropical desert sands, the upturned plates are often modified by the precipitation of silica in depressions between the plates (Krinsley and Doornkamp, 1973).

6. *Flat cleavage surfaces* are most commonly found on smaller *quartz* grains (<200 μm) and mainly occur on the grains from weathered source material and glacial environments.

7. *Disk-shaped concavities* occur on grains from hot and cold aeolian desert environments where they are probably the product of strong abrasion (Krinsley and Doornkamp, 1973, plates 85 and 86). The surfaces of the disk-shaped concavities from hot deserts usually show solution-precipitation features of silica, while some of those from cold deserts are smooth.

8, 9. *Graded arcs and meandering ridges* are described as mechanical features observed on grains from aeolian environments. The meandering ridges result from the intersections between two or more disk-shaped concavities or conchoidal breakage patterns produced on grains from aeolian environments by grain-to-grain collision.

10. *Solution- and (or) precipitation chemical-etching features* may be seen on grains from most environments. Chemical etching may produce holes of various shapes and sizes (Fig. 4).

TABLE 2. Relative Abundance of Mechanical and Chemical Features Seen on Quartz Grains from Various Environments[a]

Environment	Grain Morphology	Mechanical Features[b]									Chemical Features[c]				
		1	2	3	4	5	6	7	8	9	10	11	12	13	14
High-energy beach	Well rounded to very well rounded	Very abundant	Present	Rare	Very rare	Very rare	Very rare	–	–	–	Rare	–	–	–	–
Low-energy beach	subrounded to rounded	Present	Very rare	–	–	Very rare	Very rare	–	–	–	Abundant	Common	Present	–	–
Nearshore marine	Subangular to rounded	Present	Very rare	Very rare	Very rare	Very rare	Very rare	–	–	–	Abundant	Present	Present	–	–
Coastal dune	Well rounded to very well rounded	Common	Present	Rare	Very rare	Very rare	Very rare	Rare	Present	Rare	Very common	Rare	Present	–	–
Desert dune	Subangular to well rounded	–	–	Present	Rare	Present	Very rare	Common	Rare	Common	Common	–	Present	–	Common
Glacial	Very angular with jagged edges	–	–	Very common	Common	Present	Rare	–	–	–	Present	–	Present	–	–
Source material	Very angular to angular	–	–	Very common	Common	Present	Present	–	–	–	Very rare	–	–	–	–
Fluvial	Angular to subangular	Very rare	–	Common	Present	Present	Rare	–	–	–	Present	–	–	–	–
Estuarine, coastal lagoon	Angular to very well rounded	Present	Rare	Present	Rare	Rare	Present	Very rare	Very rare	–	Common	Rare	Present	–	–

[a]Approximate relative abundances: very abundant, > 80%; abundant, 80–60%; very common, 60–40%; common, 40–20%; present, 20–5%; rare, 5–1%; very rare, < 1%.
[b]Mechanical features: (1) V-shaped pits, (2) straight or curved grooves, (3) conchoidal breakage patterns, (4) imbricated breakage patterns, (5) upturned plates, (6) flat cleavage surfaces, (7) disk-shaped concavities, (8) graded arcs, (9) meandering ridges.
[c]Chemical (etching and diagenetic) features: (10) solution and(or) precipitation, (11) oriented V-forms, (12) smooth precipitation surfaces, (13) crystal growth, (14) surface cracks.

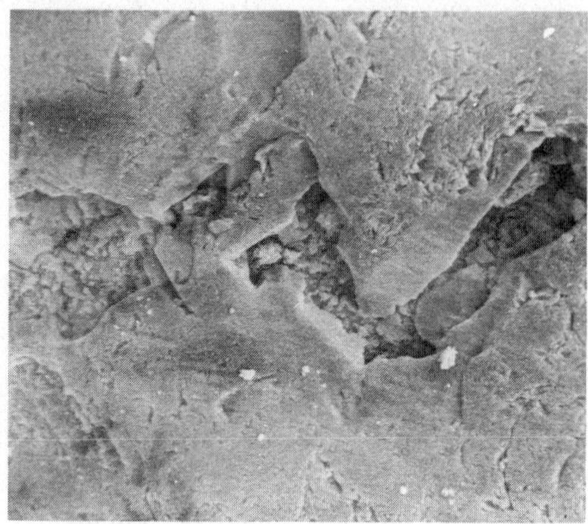

FIGURE 4. Solution and precipitation features seen on a *quartz* grain from a Pleistocene dune deposit (X425).

The precipitation of secondary silica tends to produce the "rolling topography" of Krinsley and Doornkamp (1973). Thus, grains become increasingly rounded due to the processes of solution and precipitation.

11. *Oriented etched V forms* differ from mechanical V-shaped pits by their preferred orientation (see plates 99 to 102 of Krinsley and Doornkamp, 1973). They are believed to be produced by etching in sea water, and similar features have been obtained by artificial etching with hydrofluoric acid and sodium hydroxide. Their size varies from <1 μm to 1600 μm (Krinsley and Doornkamp, 1973).

12. *Smooth precipitation surfaces* are found on grains from diagenetic environments. Very thin precipitates of silica of various origins appear as smooth surfaces in small depressions on the grains or they may nearly cover the whole grain (Krinsley and Doornkamp, 1973).

13. *Quartz crystal growths*, typified by the presence of recognizable crystal faces, have been reported in "early diagenetic" and high-energy chemical environments. They occur in small surface depressions and are not necessarily in optical continuity with the host grain.

14. *Surface cracks* found on desert grains have been reported by Krinsley and Doornkamp (1973), who concluded that these surface cracks only occur on small grains. These cracks of arcuate, circular, and polygonal shape are only seen with high magnifications, usually >1000X. They occur in depressions on the grain surfaces and have been attributed to the action of dew.

A number of the features described above have been attributed by many authors to specific depositional environments. However, some recent workers have concluded that, as several features may be found on grains from individual environments, a statistical evaluation of the most common features obtained from the examination of a number of grains is of greater diagnostic value.

In order to stress the application and importance of the presence of various surface features on *quartz* grains to the identification and interpretation of their environments of deposition, these features and their distribution have been summarized in Table 2, while their application to late Quaternary deposits has been elucidated by Ly (1978).

S. ST. J. WARNE
CHENG K. LY

References

Barbaroux, L., et al., 1972. Examen au microscope électronique a balayage de grains de sable de diverses origines - Essai de topologie, signification environnementale. *B-R-G-M Bull. No. 4*, 3-31.

Krinsley, D. H., and Doornkamp, J., 1973. *Atlas of Quartz Sand Surface Textures.* New York: Cambridge University Press, 91p.

Krinsley, D. H., and Margolis, S. V., 1971. Quartz sand grain surface textures, in R. E. Carver, ed., *Procedures in Sedimentary Petrology.* New York: John Wiley and Sons.

Krinsley, D. H.; Biscaye, P. E.; and Turekian, K. K., 1973. Argentine basin sediment sources as indicated by quartz surface textures, *J. Sed. Petrology*, **43**, 251-257.

Long, J. V. P., 1967. Electron Probe Microanalysis, in J. Zussman, ed., *Physical Methods in Determinative Mineralogy.* London: Academic Press, 514p.

Ly, C. K., 1978. Grain surface features in environmental determination of late Quaternary deposits of New South Wales. *J. Sed. Petrology*, **48**, 1219-1226.

Margolis, S. V., and Krinsley, D. H., 1974. Processes of formation and environmental occurrence of microfeatures on detrital quartz grains. *Am. J. Sci.*, **274**, 449-464.

Setlow, L. W., and Karpovich, R. P., 1972. "Glacial" microtextures on quartz and heavy mineral sand grains from the littoral environment, *J. Sed. Petrology*, **42**, 864-875.

Whalley, W. B., and Krinsley, D. H., 1974. A scanning electron microscope study of surface textures of quartz grains from glacial environments, *Sedimentology*, **21**, 87-105.

Willard, R. J., 1968. Scanning election microscope, *Geotimes*, **13**, 16-18.

ELECTRON MICROSCOPY (TRANSMISSION)

The world of the geologist extends from the most minute particles of clay to the galaxies in the universe, and knowledge of both extremes is necessary to understand the origin and nature of the earth on which we live. Just as the telescope brings to our view the vast bodies scattered throughout the cosmos, the electron microscope enables us to see the minutiae of the microscopic world all about us. Fig. 1, for example, is an electron microscope picture of a sheet of paper like that on which the picture is printed. The total area shown is no bigger than the period at the end of this sentence, yet many flat, hexagonal-shaped crystals of the clay mineral *kaolinite*—invisible to the naked eye—are revealed in sufficient detail to permit critical measurement of shape, orientation, packing, and other factors of vital importance to paper manufacturer, clay producer, and scientist. These particles provide the gloss required on paper used in "slick sheet" magazines and books in which photographs of high quality are reproduced. The rough surface on some of the particles is starch, which is also needed to coat some types of paper.

The electron microscope reveals features not visible in the light microscope because the ability to resolve detail depends primarily on the wavelength of the medium used, and the wavelength of the electron beam is approximately 1/84,000 that of light. However, because of diffraction effects and lens aberrations, the limit of resolution of electron microscopy at present is approximately 4 angstrom units compared to the generally accepted limit of 2000 Å for light microscopy. Differences in design of electron and light microscopes exist because of the need, in the former, for a high vacuum, electromagnetic rather than glass lenses, and a fluorescent screen and photographic plate on which to view and record the image. (For details the reader is referred to books on electron microscopy such as Zworykin, et al., 1948; Siegel, 1964).

Techniques

Ability to apply the electron microscope to the study of geological subjects depends upon the successful use of various techniques. In the simplest of these, the transmission method (TEM), many types of geologically significant particulate matter can be placed in the beam of the instrument and their images studied by direct observation either on a fluorescent screen or in photographs. This is the case in Fig. 2 showing flakes and shreds of the clay mineral *montmorillonite*. Some measure of the thickness of the particles is obtained by the density of the image as determined by the ability of the electron beam to penetrate the material. The thin film of collodion on which the clay rests in the specimen holder is easily penetrated and therefore appears light whereas the thicker portions of the particles have not

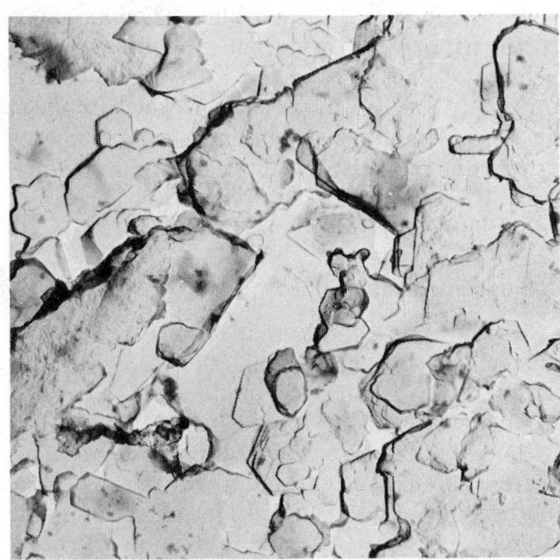

FIGURE 1. Replica of the surface of a sheet of magazine paper (X5600). The crystals are the clay mineral *kaolinite* and the rough areas are due to starch that was not completely removed from the paper surface (from *Encyclopedia of Microscopy*, George L. Clark, ed. © 1961 by Litton Educational Publishing, Inc. Reprinted by permission of Van Nostrand Reinhold Company).

FIGURE 2. A sample of the clay mineral *montmorillonite* from Clay Spur, Wyoming. The material was dispersed by ultrasonics and mounted on the electron microscope screen using the freeze-dry technique (X4550) (micrograph taken by Joseph J. Comer, electron microscopist, formerly of The College of Earth and Mineral Sciences, The Pennsylvania State University).

permitted the passage of electrons and appear dark in the final image.

The low penetrating power of the electron beam in most electron microscopes is such that only objects less than about 0.1 μm (.000004 in.) in thickness appear translucent or transparent. However, a great variety of fine-grained minerals fall in this size range–e.g., clays (see *Clays, Clay Minerals*), varieties of asbestos (q.v.), hydrated calcium silicates important in cement (see *Portland Cement Mineralogy*), as well as the fine particles produced by the chemical or physical breakdown of materials commonly found in coarser sizes. Also, techniques have been developed that permit the manufacture of rock and mineral sections thin enough to allow penetration of the electron beam and, thus, to be studied by the transmission method (see Fig. 3).

More complex in character and execution, but more useful for a wide diversity of applications and materials, is the so-called "replica technique." First devised to permit the electron microscope study of surface characteristics of metals, the method involves making a replica of the features to be studied in the form

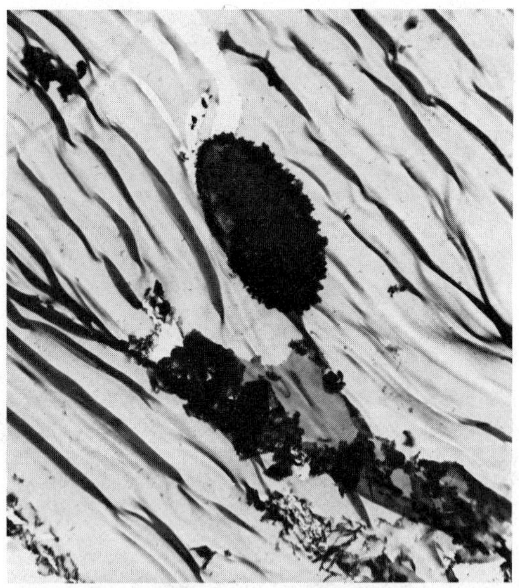

FIGURE 3. Section of a metallic zinc spherule (X450). This micrograph illustrates an extension of the conventional biological sectioning technique to materials of geologic interest. In an effort to gain electron diffraction data on metallic spheres recovered from dull cores an individual spherule was embedded in epoxy and sectioned using an ultra microtome and diamond knife. Slight distortion of the metallic slice is evident as is shearing of the epoxy (courtesy of A. J. Tousimis, Biodynamics Research Corporation).

of a film thin enough to be successfully penetrated by the electron microscope beam. The film may consist simply of plastic, such as collodion or formvar, which has been spread on the specimen, allowed to harden, and then stripped off, or it may consist of a material of low density, such as beryllium, carbon, or silicon, which has been vacuum evaporated onto the specimen and then separated from it by one of several procedures. In either case, surface features on the original specimen are precisely duplicated in the replica.

To make irregularities and boundaries more visible, a metal shadowing technique is commonly used either on replicas or where the actual particles to be studied lie upon the plastic substrate. A dense metal, such as gold, chromium, or platinum, is evaporated in a vacuum at a low angle to the plane of the specimen so that a ridge on the replica, for example, or a particle on the substrate receives a coating of metal on the "windward" side toward the metal source, but none on the "lee" side. When the replica is then observed and photographed in the electron microscope, the electron beam penetrates the metal-coated side of the ridge with more difficulty than the metal-free side and the image of the ridge or particle is accented. In Fig. 2 and 4 for example, the shadows (which appear white on the print since they are dark on the negative) are easily observed. Shadows may also be used as a measure of particle thickness if the shadowing angle is known. An excellent diagrammatic treatment of replicating and metal shadowing techniques can be found in Kay (1965, ch. 5). For many of the purposes for which replicas are used, use of the scanning electron microscope (SEM) is also effective and may involve less difficulty in sample preparation [see *Electron Microscopy (Scanning)*].

Geological Applications

Clays. Some of the applications of electron microscopy to geology are illustrated in the pictures. Clays and other minerals that usually occur in crystals too small to be studied in sufficient detail in the light microscope have received much attention from the electron microscopist. Figs. 1 and 4 picture the mineral *kaolinite*, a clay with a multitude of uses, among them the making of fine china and whiteware bodies and producing the gloss on high-grade paper. Fig. 1 illustrates the usefulness of the replica technique in evaluating such characteristics as the orientation and packing of the pseudo-hexagonal clay particles and aggregates on the paper surface. Fig. 4 shows how a replica of particles dispersed on a glass slide

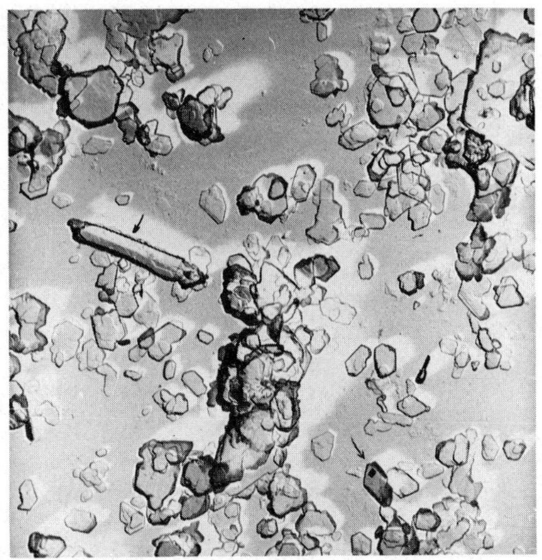

FIGURE 4. Crystals of *kaolinite* clay from a deposit in Ione, California (X3800). The replica was prepared by (1) dispersing the crystals onto a glass slide; (2) evaporating platinum at a low angle onto the slide to create the shadow effect; (3) evaporating carbon over crystals, platinum, and all to produce a thin, stable film; (4) stripping the platinum-carbon film and crystals from the glass slide; (5) dissolving the clay crystals in hydrofluoric acid; and (6) mounting the platinum-carbon replica on a collodion substrate for viewing with the electron microscope (micrograph taken by Joseph J. Comer, electron microscopist, formerly of The College of Earth and Mineral Sciences, The Pennsylvania State University).

FIGURE 5. A platinum-carbon replica of a fracture surface of a sample of the clay mineral *halloysite*. This material consists of laths and tiny hollow tubes that are chemically similar to *kaolinite*. A few hexagonal flakes of the latter mineral can also be seen (X2775); (micrograph taken by Joseph J. Comer, electron microscopist, formerly of The College of Earth and Mineral Sciences, The Pennsylvania State University).

reveals details of size and shape as well as the surface features of individual crystals and cleavage fragments. Fig. 4 also demonstrates the usefulness of the electron microscope in evaluating sample purity. Arrows indicate the presence of a small opaque crystal (lower right) having the form and size common to *anatase* (TiO_2), and an elongate particle of *halloysite* (upper left) such as those observed in Fig. 5. This latter clay mineral has a composition almost identical to that of *kaolinite*, yet it occurs in the form of hollow cylindrical tubes and laths such as those shown here. As might be expected, it also is used for high-quality ceramic materials, but because of its shape cannot provide the smooth surface necessary for a successful paper-coating clay. Fig. 5 also illustrates how easily the smooth, hexagonal *kaolinite* particles can be identified even though present in only minor amount in the *halloysite* sample.

Fig. 2 illustrates another clay, *montmorillonite*, that is obviously very different in appearance and character from *kaolinite* and *halloysite*. Individual flakes and sheets of this mineral approach the unit cell thickness of 10 to 30 Å when the material is dispersed and, thus, provide tremendous areas of surface per unit weight of clay. This property, plus the hydration and ion-exchange characteristics that relate to it, makes this clay mineral important in industry as a filler and as a carrier for insecticides, dyes, and other materials (see *Pigments and Fillers*). It is also the major constituent of the bentonites, deposits important to the oil industry as a source of drilling mud (see *Thixotropy*).

Fig. 2 also illustrates the usefulness of another technique, the freeze-dry method, for preparing samples for electron microscope study. Here, a drop of an aqueous clay suspension was frozen and the ice subsequently removed by sublimation in a vacuum. In this way, the clay particles settle onto the substrate retaining the convoluted forms they had while in suspension. Metal shadowing then is used to produce the heightened contrast and three-dimensional appearance noted in the micrograph.

Other Minerals. Although the electron microscope is particularly useful in mineralogy for studying varieties of clay, asbestos, and other materials having individual crystals that rarely attain "light microscope" dimensions, many of the more coarsely crystalline mineral

FIGURE 6. Replica of rhombohedral crystals of the mineral *dolomite* (X12,000); (courtesy of H. P. Studer, Shell Development Company).

FIGURE 7. Replica of *quartz* crystals occurring on the fracture surface of a sample of novaculite-type chert from Caddo Gap, Arkansas (X3600); (from *Encyclopedia of Microscopy,* George L. Clark, ed. © 1961 by Litton Educational Publishing, Inc. Reprinted by permission of Van Nostrand Reinhold Company).

varieties are also found in particles and crystals a few microns or less in size. Such is the case for the *dolomite* rhombs (Fig. 6) from a sediment rich in this carbonate mineral and in *montmorillonite*. The *quartz* crystals (Fig. 7) with prism and rhombohedron faces were observed in replicas made of fracture surfaces of novaculite, a variety of chert from Arkansas. They apparently occur along minute fissures where there was room for the crystal faces to develop.

The texture more typical of novaculite and of other cherts and flints is illustrated in Fig. 8. Here, with one or two exceptions, the surfaces bounding the polyhedral shapes do not appear to be crystal faces, but simply those planes at which adjacent crystalline units met, possibly during growth from a gel. Other chert types are characterized by fracture surfaces possessing a spongy texture.

Electron microscope replicas are useful in studying cleavage and fracture patterns produced in minerals as well as in a fine-grained rock such as chert. Fig. 9 was taken at the National Bureau of Standards as part of a study of the piezoelectric properties of *quartz*. Details not otherwise visible were produced by etching the surface with hydrofluoric acid prior to the making of the replica. Fig. 10 illustrates the excellent cleavage produced by the fracturing of a

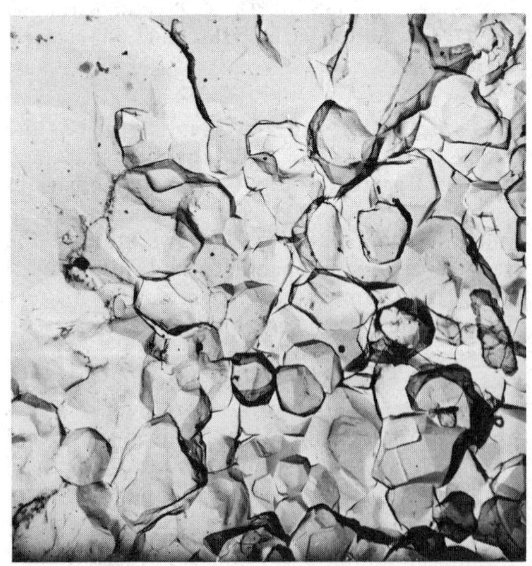

FIGURE 8. The polyhedral units revealed on this fracture surface of chert from Williamsburg, Pennsylvania, do not appear to be crystallographically controlled. Their surfaces may simply represent the boundaries at which adjacent units met during growth (X2470); (micrograph taken by Joseph J. Comer, electron microscopist, formerly of The College of Earth and Mineral Sciences, The Pennsylvania State University).

FIGURE 9. Replica of a quartz fracture surface after etching with hydrofluoric acid (X2550); (courtesy of A. van Valkenburg and Elisabeth Mitchell, formerly of the Constitution and Microstructure Section, Mineral Products Division, National Bureau of Standards).

crystal of synthetic *sylvite* (KCl). Here the cubic cleavage contrasts markedly with atypical curved surfaces.

Crystal faces that appear smooth to the naked eye often reveal a complexity of features under electron microscope inspection. Such is the case for the clear crystal of Iceland Spar (*calcite*) shown in Fig. 11. Although the magnification is relatively low for electron microscope work, the area of the crystal face represented in the micrograph is only 1200 μm^2, considerably smaller than the point of a pin.

Materials of organic origin, such as microfossils, are also of geological significance. Diatoms have always been a favorite with microscopists and Fig. 12 shows several found in a lake-bottom mud. Whereas diatoms are siliceous in nature, coccoliths, such as that shown in Fig. 13, are tiny calcareous bodies produced by marine organisms known as coccolithophores. Very common in present-day phytoplankton, they have also been found in great abundance in marine sediments throughout the world dating back to the Mesozoic. Because they have evolved steadily with time, these characteristics place coccoliths among the best index fossils we know for correlation of geologic strata over long distances. In contrast

FIGURE 10. Synthetic *sylvite* (KCl). The curved surfaces are in sharp contrast to the typical cubic cleavage (X4650); (from *Encyclopedia of Microscopy*, George L. Clark, ed. © 1961 by Litton Educational Publishing, Inc. Reprinted by permission of Van Nostrand Reinhold Company).

FIGURE 11. Incipient cleavage lines are shown in this replica of the surface of a clear calcite crystal (X600); (courtesy of Electron Microscope Laboratory, University of Texas, and of R. Shoji and R. L. Folk for whom the micrograph was taken).

FIGURE 12. Diatoms and aragonite needles from a recent lake bottom mud. An X-ray powder diffraction pattern established the identity of the mineral as aragonite. Electron micrographs revealed the presence of several species of diatoms (X2220); (courtesy of the U.S. Geological Survey).

to coccolithophores and diatoms, some organisms make use of a variety of mineral substances. Such is the case of the protozoan having the shell illustrated in Fig. 14.

The figures used herein are merely illustrative of a much larger variety of materials of geological importance that have been effectively studied with the electron microscope. In the 30 years since its lenses were first focused on test specimens of clay and diatoms, this remarkable instrument has taken its place among the many other tools now in use for the study of geology. The electron microscope performs most effectively in that portion of the size spectrum between 10 and 10,000 millimicrons (0.010 to 10 μm) and thus helps to bridge the gap between the fields of the X-ray crystallographer and the light microscopist. When used in conjunction with electron diffraction and electron microprobe work (see *Electron Probe Microanalysis*), it provides the necessary morphological data to accompany the structural and chemical information revealed by the other techniques.

FIGURE 13. Coccolith found in deep sea mud taken from a depth of 1500 m in the Bahamas (X8300); (courtesy of the Electron Microscope Laboratory, University of Texas).

FIGURE 14. The shell of the microscopic foraminifera *Haplogragmoides canariensis* (Protozoa) is composed of an agglutination of mineral grains selected by the organism from the ocean floor. In the electron microscope, the grains of quartz and feldspar are seen to be beautifully fitted together and tightly packed in a subordinate cement composed of organic substances and colloidal iron oxide, (X1500); (courtesy of K. M. Towe, Smithsonian Institution).

THOMAS F. BATES

References

Andersen, C. A., ed., 1973. *Microprobe Analysis.* New York: Wiley/Interscience, 571p.

Bates, T. F., 1949. The electron microscope applied to geological research, *New York Acad. Sci. Trans.*, ser. II, **11**, 100-108.

Birks, L. S., 1971. *Electron Probe Microanalysis*, 2nd ed. New York: Wiley/Interscience, 190p.

Comer, J. J., 1959. The electron microscope in the study of minerals and ceramics, *Am. Soc. Testing Materials, Spec. Tech. Publ. 257*, 94-120.

Dwornik, E. J., 1966. Use of the scanning electron microscope in geological studies, *U.S. Geol. Surv. Prof. Paper 550-D*, D209-D213.

Dwornik, E. J., and Ross, M., 1955. Application of electron microscope to mineralogic studies, *Am. Mineralogist*, **40**, 261-274.

Fahn, R., 1956. Applications of the electron microscope for the investigation of rocks and minerals, *Tonind. Ztg. Keram. Rundsch.*, **80**, 171-180.

Glauert, A. M., ed., 1972. *Practical Methods in Electron Microscopy*, Amsterdam and London: North Holland, vols. 1 and 2.

Goldstein, J. I., et al., 1975. *Practical Scanning Electron Microscopy*. New York: Plenum, 582p.

Heinrich, K. F. J., ed., 1968. *Quantitative Electron Probe Microanalysis.* Washington, D.C.: NBS Special Publ. 298, 299p.

Jenkins, R., 1974. *An Introduction to X-ray Spectrometry*. London: Heyden, 163p.

Kay, D. H., ed., 1965. *Techniques for Electron Microscopy*, 2nd ed. Philadelphia: Davis.

McKee, T. R., and Brown, J. L., 1977. Preparation of specimens for electron microscope examination, p. 809-846 in J. B. Dixon and S. B. Week, *Minerals in soil environments*, Madison, Wisconsin: Soil Science Society of America.

Proceedings of the Annual Meetings of Microbeam Analysis Society (formerly Electron Microprobe Society of America, 5 ring-bound volumes).

Proceedings of Annual IITRI SEM meetings (8 bound volumes).

Proceedings of Annual Electron Microscope Society of America meetings.

Siegel, B. M., 1964. *Modern developments in electron microscopy*. New York: Academic Press.

Thomas, G., 1962. *Transmission Electron Microscopy of Metals*. New York: Wiley, 299p.

Wells, O. C., 1974. *Scanning Electron Microscopy.* New York: McGraw-Hill, 421p.

Wenk, H. R., ed., 1976. *Electron Microscopy in Mineralogy*. New York: Springer-Verlag, 564p.

Woldseth, R., 1973. *X-ray Energy Spectrometry*. Burlington, Calif.: Kevex Corp., 150p.

Zworykin, V. K., et al., 1948. *Electron Optics and the Electron Microscope*. New York: Wiley, 766p.

Cross-references: *Clays and Clay Minerals; Crystallography: Morphological; Electron Microscopy (Scanning); Electron Probe Microanalysis; Optical Mineralogy.* Vol. I: *Phytoplankton; Marine Sediments.* Vol. VI: *Chert and Flint; Novaculite.*

ELECTRON PROBE MICROANALYSIS

Nondestructive qualitative and quantitative analysis of solid samples as small as a few cubic microns in volume, either singular grains or localized areas in polished or thin sections, may be done by electron probe microanalysis. Thus, the sample weight can be of the order of 10^{-11} g and the detection limit for most elements, ranges from ≈50 ppm to 0.1%, depending on the element itself and on the elements with which it is combined in the mineral. However, microprobe analysis is most suitable for determination of relatively high concentrations (>0.1%) in small samples, particularly in intimately mixed multiphase systems.

The introduction of the electron microprobe to mineralogy, petrology, meteoritics, etc., has revolutionized science. Its impact on earth sciences can only be compared to that of the introduction of the polarizing microscope (petrographic microscope, see *Polarization and Polarizing Microscope*) to petrography more than 125 years ago (Keil, 1967, 1973). Now, most major geology department, government, and industrial laboratories make extensive use of the electron microprobe in research and in teaching. The importance of the instrument is illustrated by the hundreds of publications per year dealing with its application to research problems in the earth sciences, as well as by the large number of new minerals that have been discovered and described with the aid of this tool (some 500 as of February 1981).

Instrumentation

The first electron microprobe was developed by Castaing (1951), although an American patent (no. 2,418,029), based on similar ideas, was issued to J. Hillier in 1947. Fig. 1 illustrates schematically the essential parts of presently marketed machines, although the designs vary widely among different makes. The working principle is that each element in the sample emits a characteristic X-ray spectrum when bombarded with electrons of sufficient energy. These spectra allow the identification of any particular element. The intensity of a particular characteristic spectral line of an element emitted from an unknown mineral, when compared to the intensity of the same line emitted under the same excitation conditions from a standard of known composition, makes it possible to calculate the concentration of the element in the unknown sample.

The instrument operates under vacuum conditions and constitutes basically an X-ray tube with the specimen as the target, in principle similar to some instruments used by de Hevesy and others in the early days of X-ray emission spectroscopy. However, in electron microprobes, the beam can be focused to a spot size approximately 0.1 μm, the high voltage beam current are highly stabilized, and more or less

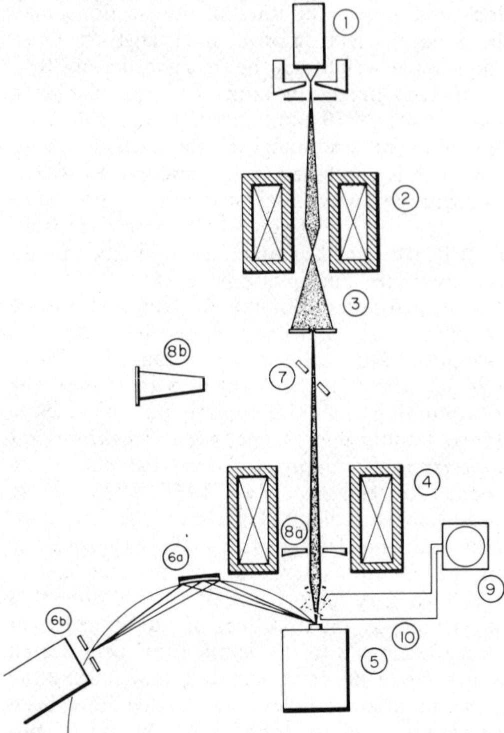

FIGURE 1. Electron Microprobe X-ray analyzer. Schematics of main parts: 1. Electron gun. 2. Electromagnetic lens ("condenser"). 3. Electron beam (shaded). 4. Electromagnetic lens ("objective"). 5. Sample stage with flat polished sample (on top). 6. X-ray spectrometers (usually 2 to 4, *a.* curved, ground crystal, *b.* X-ray detector). 7. 45-degree mirror with aperture. 8. Optical viewing system, *a.* reflecting objective with central aperture, *b.* eyepiece and specimen illuminator. 9. Oscilloscope; *x* and *y* sweep synchronously with electron beam via 10. 10. Sweep amplifier and deflection plates. Not shown are electronics for the electron gun, electromagnetic lenses, and the X-ray detectors. The whole system except 8*b*, is under vacuum, $\approx 10^{-5}$ torr. (Drawing by B. Swope, University of California, La Jolla.)

complicated sample stages as well as light-optics systems are provided so that the desired sample area can be brought under the beam and observed during the analysis. Furthermore, most commercial instruments have two or more focusing curved-crystal spectrometers with a Bragg-angle (θ) coverage of about 10 to 75°. By using different crystals (e.g., *quartz*, LiF, ADP, KAP, lead stearate decanoate), a wavelength range from about 0.5–100 Å can be covered. The X-ray detectors are usually sealed or flow proportional counters coupled to conventional ratemeters, pulse-height analyzers, and scalers similar to the systems used in X-ray fluorescence analysis (see vol. IVA: X-ray Spectroscopy; for comprehensive discussions of electron microprobe analysis, see Birks, 1971; Malissa, 1966; Keil, 1967, 1973; Zussman, 1967; Obst et al., 1972; Andersen, 1973; Reed, 1975).

Analysis

The sample is usually a metallographic polished section or a polished petrographic thin section (no cover glass). For accurate quantitative work, it is essential that the sample surface be flat, polished, and clean. Loose grains are best embedded in some metal or plastic and then ground and polished. Electrically nonconductive samples must be coated with a thin film of vacuum-evaporated metal ($\approx$50 Å) or carbon ($\approx$200 Å) to avoid charging of the sample, which would deflect the electron beam.

A qualitative analysis of a sample for elements between uranium (Z=92) and boron (Z=5) can be performed within 30–60 minutes by automatically scanning the spectrometers through their respective wavelength ranges and recording the ratemeter output. In this mode, the detectability is of the order of 0.1 to 1 weight percent depending on allowable sample current (excessive heating must be avoided), element, matrix, etc. The recorded spectrum also allows semiquantitative estimates of amounts present, even in a complex sample, if the line intensities are known for a couple of pure elements under identical analytical conditions.

Recently, solid-state energy-dispersion spectrometers, consisting of a liquid-nitrogen-cooled, lithium-drifted silicon solid-state detector, a multichannel pulse-height analyzer, and associated electronics, have been applied successfully to qualitative and quantitative spectral analysis with the electron microprobe (Fitzgerald, et al., 1968; Russ, 1971; Reed and Ware, 1973; Ware and Reed, 1973; Reed, 1972; Woldseth, 1973). The energy resolution of the best detectors is better than 150 eV and serves to analyze the X-ray spectrum on the basis of characteristic energies of X-rays. The advantage of the energy-dispersion spectrometer over the wavelength-dispersion spectrometer is that with the former, qualitative and quantitative analysis can be carried out in a few minutes.

The basis for quantitative analysis is that the intensity of the electron-excited characteristic radiation, in a first approximation, is proportional to the concentration (in weight percent) of the element. This relation may be expressed as follows (for a pure element standard; Castaing, 1951):

$$C_A = \frac{I_A}{I_{(A)}} \times 100$$

where C_A is the concentration of element A in percent in the sample; I_A is the true intensity of a characteristic line of the desired element A generated from the compound of unknown composition; and $I_{(A)}$ is the true intensity of the same line generated under the same conditions from a standard of known composition. The observed radiation, however, is attenuated by absorption or increased by secondary fluorescence, generally to a degree different in the sample and in the standard. These differences, as well as differences in back-scattering and electron retardation, must be corrected. The following corrections of the observed intensities are usually considered (Wittry, 1964; Keil, 1967; Andersen, 1973; Reed, 1975):

1. *Drift*, mostly due to warping of the filament, results in variations in the beam current and, hence, counting rates during the course of analysis. It can be corrected for by interpolating between count rates obtained from a standard at certain time intervals, or by accurately monitoring the beam current.
2. *Wavelength Shift*–Particularly in the soft region ($\lambda \geqslant 4$Å, i.e., K-lines of elements with $Z<19$, potassium; or L-lines of elements with $Z<48$, cadmium), the characteristic lines shift in wavelength with the valence state, coordination, etc., of the element. This shift may cause in error of 20–30% if the spectrometer has high resolution. It may be corrected by resetting the spectrometer on sample and standard or by using an experimentally determined correction factor. Wavelength shifts are undesirable in quantitative analysis, but are useful in the determination of valence state and coordination number (e.g., White and Gibbs, 1967). Shifts in wavelength are also connected with changes in intensity ratios between certain lines of various compounds (Andersen, 1967a). Since changes in intensity can be measured more easily and accurately than actual wavelength shifts, the former method is preferred.
3. *Dead Time* of the X-ray detectors (and electronics)–The dead-time is the time interval after a pulse in which the system cannot respond to another pulse (1 to 2 μsec. for proportional counters, up to 10 μsec. for associated electronics).
4. *Background* is mainly due to the continuous X-ray spectrum generated by the electrons in the sample, as well as to small contributions from cosmic rays, scattered radiation, and radioactivity in the sample. It can be the major source of errors in the analysis of minor or of trace elements.
5. *Mass Absorption*–The X-rays are generated at certain depths within the sample. Consequently, the emerging radiation is absorbed, usually to a different degree in sample and standard. This is frequently the major sample-dependent correction; it can, however, be estimated reasonably well (see, e.g., Philibert, 1963; Duncumb and Shields, 1966).
6. *Secondary Fluorescence*–X radiation (characteristic lines or continuum) of shorter wavelength than the absorption edge of an element will excite characteristic lines of the same element. This X radiation has to be subtracted from the recorded intensity readings. For K-lines excited by K-lines, the correction can be carried out reasonably accurately (Wittry, 1964; Reed, 1965); for other combinations and for fluorescence by the continuum, the situation is not as favorable (Henoc et al., 1964).
7. *Atomic Number*–After the above corrections (1-6), Castaing's first approximation may be applied. If sample and standard vary widely in average atomic number ($\bar{Z}$), an additional correction has to be made for differences in electron stopping power and for back-scattering of electrons. This is the atomic number or α-correction (Castaing, 1951; 1960; Thomas, 1964).

Many minerals are composed of relatively light elements (e.g., $Z<26$), where the magnitude of the theoretical corrections is often great when pure-element standards are used. In order to minimize these corrections and, hence, minimize the effects of uncertainties in the theoretical corrections on the analytical data, mineral standards of compositions close to the minerals to be analyzed are used almost exclusively in mineralogical applications. The raw X-ray-intensity data can then be corrected by the above-named corrections, or some other correction procedure may be applied. The method proposed by Bence and Albee (1968), which applies empirical correction factors based on pure oxides or simple minerals, has become accepted in mineral analysis as a convenient and simple correction procedure, which requires no more than an electronic desk computer for data reduction. Finally, corrections except those for drift, can be eliminated altogether when calibration curves are used. However, this requires the availability of a series of standards of variable compositions and is easily applicable only to simple systems where essentially only two elements are substituted for each other (e.g. ***olivines***, $Mg_2SiO_4-Fe_2SiO_4$; ***orthopyroxenes***, $MgSiO_3-FeSiO_3$; ***plagioclases***, $NaAlSi_3O_8-CaAl_2Si_2O_8$; Fredriksson, 1965).

Scanning

Most commercial electron microprobes have a scanning capability that allows qualitative and semiquantitative estimation of the spatial elemental distribution in a section (Duncumb, 1959; Wittry and Fitzgerald, 1961). The finely focused electron beam is scanned over a sample area usually a few to several hundred microns square. The scan is synchronous with that of the cathode ray on one or several oscilloscopes of which the intensities (Z-axis) are modulated by the output of an X-ray detector, a sample-current meter, or a back-scattered or secondary-electron detector. Thus, a light spot is obtained on the oscilloscope when the electron beam excites a spectral line of an element to which the spectrometer is tuned. In this manner, the distribution of an element can be photographed on the oscilloscope screen with a magnification of < 20,000 times and a resolution of <1 μm.

Similarly, the distribution of back-scattered electrons, secondary electrons, or current collected by the sample can be imaged, providing information on the average atomic number variations as well as the "topography" of the sample. The resolution is better than for X-rays, and magnifications of as much as 100,000X are possible for secondary electrons. Fig. 2 shows an application of the beam-scanning technique.

Applications

In the last few years, electron probe analysis has been applied to a wide variety of problems, predominantly in metallurgy and mineralogy (Keil, 1967, 1973). Microprobe techniques have also been applied successfully to biological and medical specimens (Andersen, 1967b), pigments, ceramics, meteorites, lunar rocks, and many other natural and artificial materials. Whenever essentially nondestructive, quantitative, in-situ, elemental analysis of micron-sized material is required, the electron microprobe X-ray analyzer is unsurpassed in effectiveness, accuracy, and speed. It should also be noted that the electron microprobe can be used to determine the bulk composition of heterogeneous materials, such as rocks, by using a broad electron beam (approximately 100–200 μm) or by analyzing pellets or beads prepared from the rocks (Keil, 1967; 1973).

Much of the present work in electron microprobe analysis (other than applications) is directed toward a better theoretical understanding of electron and X-ray interactions with matter and, hence, better theoretical correction procedures; preparation of homogeneous mineral and alloy standards of well-known composition; automation of analytical procedures including raw-data correction by high-speed computer; and application of new devices such as solid-state energy dispersion spectrometers to quantitative analysis. Most electron microprobes are now computerized and automated, which allows X-ray scanners and, occasionally, the sample stage to be positioned automatically according to programmed instructions. These automated electron microprobes not only offer greater analytical speed but, more importantly, allow all desired elements to be determined in one run of a few minutes. Hence, a complete analysis of all elements is available after only a few minutes. Also, "one run" eliminates the analytical errors due to the need for repositioning the electron beam when many elements are to be measured with an electron microprobe, particularly in the case of minerals with variable compositions.

In recent years, a related technique, namely ion microprobe analysis, has been developed and promises to become a major analytical tool of great potential in mineralogical applications. The instrument uses a primary focused-ion beam to sputter secondary ions from the unknown mineral in a polished thin section; the secondary ions are then analyzed in a mass spectrometer. Thus, isotopic and trace-element analysis of micron-sized material is possible (e.g., Andersen, 1973).

KLAUS KEIL
KURT FREDRIKSSON

References

Andersen, C. A., 1967a. The quality of X-ray microanalysis in the ultra-soft X-ray region, *Brit. J. Appl. Phys.*, **18**, 1033–1043.

Andersen, C. A., 1967b. An introduction to the electron probe micro-analyzer and its application in biochemistry, in D. Glick, ed., *Methods of Biochem. Anal.*, **15**, 147–270.

Andersen, C. A., ed., 1973. *Microprobe Analysis*. New York: Wiley, 571p.

Bence, A. E., and Albee, A. L., 1968. Empirical correction factors for the electron microanalysis of silicates and oxides, *J. Geol.*, **76**, 382–403.

Birks, L. S., 1971. *Electron Probe Microanalysis*, 2nd ed. New York: Wiley/Interscience, 190p.

Castaing, R., 1951. Application des sondes électroniques à une methode d'analyse ponctuelle chimique et cristallographique. Thesis, University of Paris.

Castaing, R., 1960. Electron probe microanalysis, *Adv. Electron. Electron Phys.*, **13**, 317–386.

Duncumb, P., 1959. The X-ray scanning microanalyser, *Brit. J. Appl. Phys.*, **10**, 420–427.

Duncumb, P., and Shields, P. K., 1966. Effects of critical excitation potential on the absorption correction in X-ray microanalysis. *Proc. Symp. Electron Microprobe* (Washington, D.C., 1964), T. D. McKinley, K. F. J. Heinrich, and D. B. Wittry, eds. New York: Wiley, 284–295.

Fitzgerald, R.; Keil, K.; and Heinrich, K. F. J., 1968. Solid-state energy dispersion spectrometer for electron microprobe X-ray analysis, *Science,* 159, 529–530.

Fredriksson, K., 1965. Standards and correction procedures for microprobe analysis of minerals, in IVe Congrès International sur l'optique des Rayons X et la Microanalysis, Orsay, September 1965. Paris: Hermann, p. 305–309.

Heinrich, K. F. J., 1967. Quantitative electron probe microanalysis, *Natl. Bur. Stand. Spec. Publ. 298.* Washington D.C.: U.S. Government Printing Office, 299p.

Henoc, M. J.; Maurice, F.; and Kirianenko, A., 1964. Microanalyseur à sonde électronique: Étude de la correction de fluorescence due au spectre continu, *Rapport de Commiss. à l'énergie atomique*, CEA-R 2421.

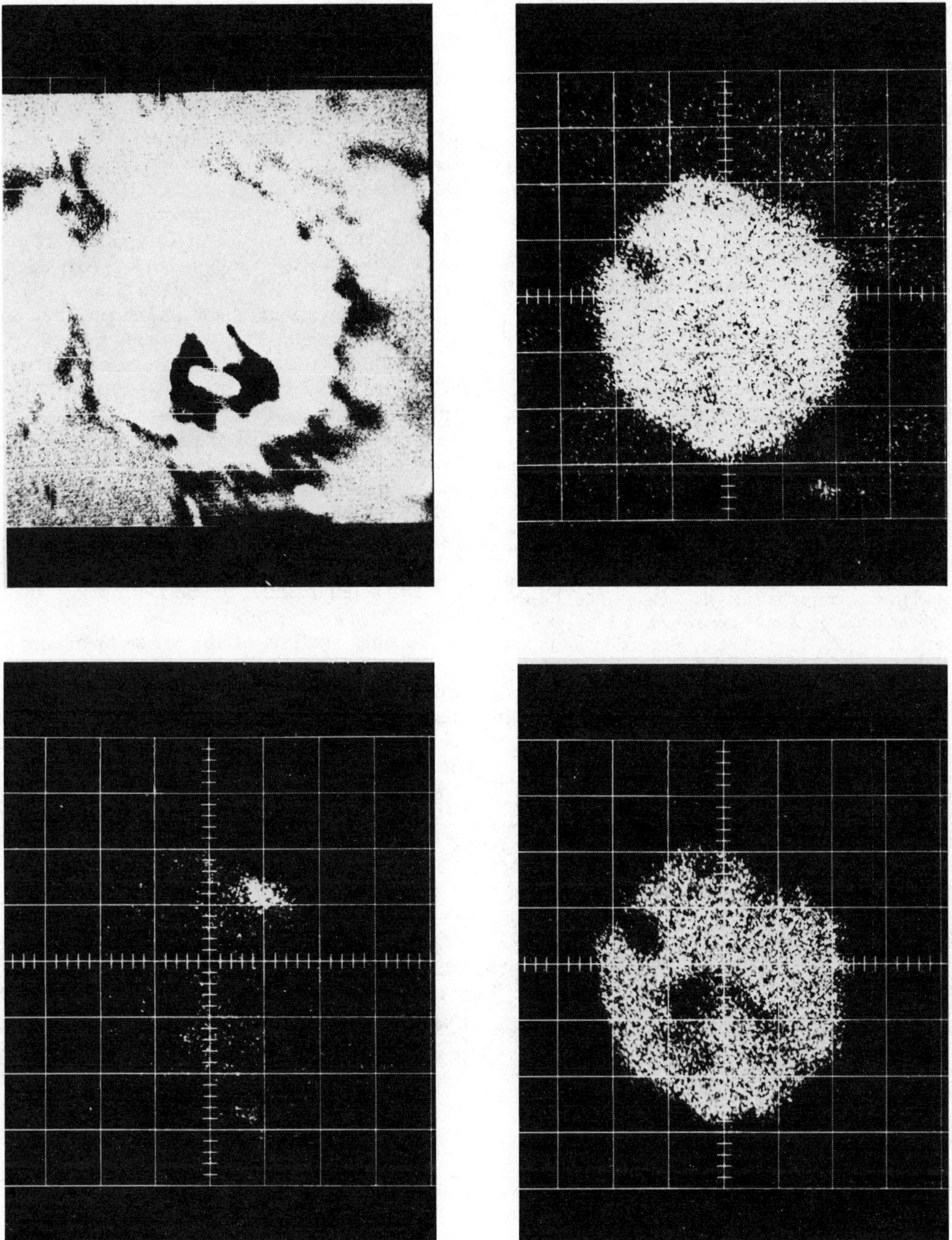

FIGURE 2. Electron beam-scanning pictures of *lawrencite* (Fe,Ni)Cl_2 in a chondrule of the "Bjurböle" chondrite. The hexagonal crystal is surrounded by magnesium-iron silicates. *Upper left:* BSE (back scattered electron) distribution, illustrating the higher average atomic number ($\bar{Z}$) of *lawrencite* (bright) and the "topography"; a hole in the *lawrencite* shows up dark. *Upper right:* Fe distribution, ≈40%, in *lawrencite*, ≈17% in silicate. *Lower left:* Ni distribution; a few percent in *lawrencite* and no detectable amounts in matrix. *Lower right:* Cl distribution; ≈50% in *lawrencite.* Scanned area is 50 × 50 μm.

Keil, K., 1967. The electron microprobe X-ray analyzer and its application in mineralogy, *Fortschr. Mineral.*, **44**, 4–66.

Keil, K., 1973. Applications of the electron microprobe in Geology, in Andersen, 1973, 189–239.

Malissa, H., 1966. *Elektronenstrahl-Mikroanalyse.* Wien: Springer-Verlag, 154p.

Obst, K. H.; Münchberg, W.; and Malissa, H., 1972. *Elektronenstrahl-Mikroanalyse (ESMA) zur Untersuchung basischer feuerfester Stoffe.* Wien: Springer-Verlag, 125p.

Philibert, J., 1963. A method for calculating the absorption correction in electron probe micro-analysis, *Proc. 3rd Internat. Symp. X-ray Optics X-ray Microanal.* (Stanford, 1962), H. H. Pattee, V. E. Cosslett, and A. Engström, eds. New York: Academic Press, 379–392.

Reed, S. J. B., 1965. Characteristic fluorescence correction in electron-probe microanalysis, *Brit. J. Appl. Phys.*, **16**, 913–926.

Reed, S. J. B., 1972. Dead time correction for X-ray intensity measurements with a Si (Li) detector, *J. Phys. E (Sci. Instrum.)*, **5**, 994–996.

Reed, S. J. B., 1975. *Electron Microprobe Analysis.* Cambridge: Cambridge University Press, 400p.

Reed, S. J. B., and Ware, N. G., 1973. Quantitative electron microprobe analysis using a lithium drifted silicon detector, *X-ray Spectrometry*, **2**, 69–74.

Russ, J. C., ed., 1971. Energy dispersion X-ray analysis: X-ray and electron probe analysis, *ASTM Spec. Publ.*, **485**, 285p.

Thomas, P. M., 1964. A method for correcting atomic number effects in electron probe microanalysis, *Atomic Energy Res. Establ.*, Harwell Publ. No. R4593.

Ware, N. G., and Reed, S. J. B., 1973. Background corrections for quantitative electron microprobe analysis using a lithium drifted silicon X-ray detector, *J. Phys. E (Sci. Instrum.)*, **6**, 286–288.

White, E. W., and Gibbs, G. V., 1967. Structural and chemical effects on the SiKβ X-ray line for silicates, *Am. Mineralogist*, **52**, 985–993.

Wittry, D. B., 1964. Methods of quantitative electron probe analysis, *Proc. 12th Annu. Conf. Appl. X-Ray Anal.* (Denver, 1963), vol. 7, W. M. Mueller, G. Mallett, and M. Fay, eds., New York: Plenum Press, 395–418.

Wittry, D. B., and Fitzgerald, R., 1961. Equipment for beam scanning and step scanning in electron-probe analysis, *Adv. X-Ray Anal.*, **5**, 538–553.

Woldseth, R., 1973. *X-ray Energy Spectrometry.* Burlingame, Calif.: Kevex Corp.

Zussman, J., ed., 1967. *Physical Methods in Determinative Mineralogy.* New York: Academic Press, 514p.

ENHYDRO, ENHYDRITE, OR WATER-STONE

Popularly known as "water-stones," or "water agates," enhydrites or endhydros are water-filled geodes or "nodules of chalcedony containing water, sometimes in large amount" (Dana, 1920, 6th ed.). The water is gradually lost by diffusion and the interior is then partly or totally filled with *quartz*, as an inward-directed layer.

Well-known examples are found in Australia, at Beechworth, Victoria, but only one specimen (at the Geological Museum in Melbourne) still retains its water (Dunn, 1872, 1913; Boutakoff and Whitehead, 1952). Other occurrences have been reported from Gredell Co., North Carolina, and Silver City, Idaho (Lindgren, 1900); Little Namaqualand, S Africa, and Waihai Gold Mine, New Zealand (Dunn, 1913).

Specimens vary in shape and form; they are about 1–10 cm in size, with walls from 0.1 to 5 mm thick (Fig. 1). They are usually honey yellow to brown in color. The chalcedony is laminated along flat planes parallel to the bounding surfaces (of the surrounding formation). It is fibrous, parallel to the surface near the contact in three sets at 60°, but inside the fibers are radial. There are often layers of clay between the laminae and in the interior. In some there are hexagonal cavities. The fibers are elongated parallel to the fast optical direction.

The only analysis of the water on record is by Foord (1872); it is clear, richly mineralized with chlorides and sulfates of sodium, magnesium, calcium, and dissolved silica. The clay was reported by Lindgren to be *kaolinite* with or without *sericite*.

Environment. At the Victorian occurrence,

FIGURE 1. Enhydros (from Boutakoff and Whitehead, 1952). (a) Filled with fluid with an air bubble. (b) Filled with *quartz*. (c) Thick walls showing fibrous structure.

a weathered pegmatite dike intrudes an Ordovician sandstone, associated with fault breccia, the affected zone being largely reduced to clay and chalcedony. The latter is often in an open boxwork or in scales resembling books of ***mica***.

Origin. The enhydrites are evidently epidiagenetic; they are not pseudomorphs, but cavity-filling between the preexisting faces of pegmatite ***feldspars***, *calcite*, and other minerals, now removed by deep weathering.

RHODES W. FAIRBRIDGE

References

Boutakoff, N., and Whitehead, S., 1952. Enhydros or water-stones, *Mining and Geol. J.* (Melbourne), **4** (5), 14-18.

Cooksey, T., 1892. Some suggestions regarding the formation of "Enhydros" or water-stones, *Rec. Aust. Mus.* (Sydney), **2**, 92-94.

Dunn, E. J., 1872. Notes on enhydros found at Beechworth, *Roy. Soc. Vict.*, **10**, 71-76.

Dunn, E. J., 1913. The Woolshed Valley, Beechworth, *Bull. Geol. Surv. Vict.*, no. 25, 1-20.

Foord, J., 1872. Notes on enhydros found at Beechworth, *Roy. Soc. Vict.*, **10**, 71-76.

Frondel, C., 1962. *The System of Mineralogy, III*, 7th ed. New York and London: Wiley, 230-231.

Lindgren, W., 1900. The gold and silver veins of Silver City, De Lamar, and other mining districts of Idaho, *20th Annu. Rept., U.S. Geol. Surv.*, pt. 3, 75-256.

Cross-references: *Pegmatite Minerals, Pseudomorphism*. Vol. VI: *Geodes*.

EPIDOTE GROUP

Members of the ***epidote*** group of orthosilicates have the general formula

$$A_2B_3\,(SiO_4)\,(Si_2O_7)\,O\,(OH)$$

where A = Ca,Sr,Pb,Ce,Y
B = Al,Fe,Mn,V

The structure of the group consists of chains, oriented parallel to the *y* axis, consisting of ions in octahedral coordination cross-linked by independent SiO_4 and Si_2O_7 tetrahedral units. The group includes the following:

- *allanite:* $(Ce,Ca,Y)_2\,(Al,Fe)_3\,(SiO_4)\,(Si_2O_7)\,O\,(OH)$
- *clinozoisite:* $Ca_2Al_3\,(SiO_4)\,(Si_2O_7)\,O\,(OH)$
- *epidote:* $Ca_2\,(Al,Fe)_3\,(SiO_4)\,(Si_2O_7)\,O\,(OH)$
- *hancockite:* $(Pb,Ca,Sr)_2(Al,Fe)_3(SiO_4)(Si_2O_7)O(OH)$
- *mukhinite:* $Ca_2\,(Al_2V)\,(SiO_4)\,(Si_2O_7)\,O\,(OH)$
- *piemontite:* $Ca_2\,(Al,Mn,Fe)_3\,(SiO_4)\,(Si_2O_7)\,O\,(OH)$
- *zoisite:* $Ca_2Al_3\,(SiO_4)\,(Si_2O_7)\,O\,(OH)$

Zoisite is orthorhombic and has little substitution of Fe^{3+} for Al in the octahedral sites. *Clinozoisite* and *epidote*, both monoclinic, permit up to one third of the octahedral sites to be occupied by Fe^{3+}. In *piemontite* two thirds of the Al may be replaced by Mn and Fe. *Allanite* (orthite) permits the coupled substitution of trivalent rare-earth elements for Ca and Fe^{2+} for Fe^{3+} to maintain electrostatic neutrality. *Mukhinite* has one third of the Al replaced by V and *hancockite* has Pb replacing Ca. The latter two minerals are quite rare.

Allanite occurs mainly as an accessory mineral in felsic and some alkaline plutonic rocks and associated pegmatites. It is commonly in a metamict state (q.v.). *Piemontite* is found in low-grade schists, in some manganese ore deposits, and as a late-stage mineral in some felsic and intermediate volcanic rocks. *Epidote* is a characteristic mineral in regional metamorphic rocks of the greenschist facies where it occurs with *albite* and ***chlorite***. It is also a common break-down product of ***plagioclase*** during retrograde metamorphism of mafic igneous rocks. *Epidote* and *clinozoisite* may be the products of late-stage crystallization of magmatic rocks. Saussuritization is the hydrothermal production of the assemblage *albite, zoisite, epidote, clinozoisite*, and *calcite* from ***plagioclase feldspar***. *Zoisite* is common in regionally metamorphosed argillaceous calcareous sandstones of medium grade.

KEITH FRYE

References

Deer, W. A., Howie, R. A., and Zussman, J., 1966. *An Introduction to the Rock-Forming Minerals*. London, Longmans: 61-69.

Phillips, W. R., and Griffen, D. T., 1981. *Optical Mineralogy: The Nonopaque Minerals*. San Francisco: Freeman, 145-154.

EPITAXY

Royer (1928) introduced the term "epitaxy" to designate the mutual orientation of crystals of different species, when a crystal that grows *on* another crystal is oriented by it and with respect to it (from the Greek *epi* = on + *taxis* = arrangement). An example of epitactic overgrowth is alkali halide growing on a cleavage plane of ***mica***. This is a case of two-dimensional structural control: the mesh of the (111) net of the halide crystal nearly coincides in shape and size with a mesh of the pseudo-hexagonal net (001) of the ***mica*** substrate, a geometrical condition that is necessary but not sufficient.

"Epitaxy" has also been used to refer to three-dimensional control, where "syntaxy" (q.v.) would be the proper term; for example,

Trillat and Laloeuf (1949) state that the making of artificial snow by means of silver iodide is due to epitaxy because hexagonal AgI and ice have similar cell dimensions. The correct adjectival form (Schneider, 1963) is "epitactic" (Greek derivation) or "epitaxic" (taken from the French).

J. D. H. DONNAY

References

Royer, L., 1928. Recherches expérimentales sur l'épitaxie ou orientation mutuelle de cristaux d'espèces différentes, *Bull. Soc. franc. Minéral.*, **51**, 7-156.

Schneider, H. G., 1963. Bemerkung zur Terminologie des Begriffes Epitaxie, *Acta Cryst.*, **16**, 1261-1262.

Trillat, J. J., and Laloeuf, A., 1949. Étude, par diffraction électronique, des vapeurs d'iodure d'argent, *Compt. Rend.*, **228**, 81-83.

Cross-references: *Monotaxy; Syntaxy; Topotaxy.*

ETCH PITS

Several characteristics of crystals and crystalline aggregates can be investigated by etching natural and artificial surfaces. This technique has found application in crystallography, petrology, mineralography, paleontology, metallography, and other fields of study. Chemical etching agents are commonly used, but electrolytic, thermal, and ion-bombardment methods are also employed. Etching of crystals and aggregates may result either in uniform removal of material over the entire exposed surface, or in preferential attack and removal of material (a) at grain boundaries, (b) at specific crystal faces or axes, or (c) at a number of small discrete areas within the boundaries of individual grains. When preferential etching attacks small, discrete areas, the depressions formed are called "etch pits" (or etch figures). Controlled etching will often produce etch pits with well-defined shapes and faceted sides that can be related to external or internal characteristics of the crystalline material.

History

The earliest use of an etching technique appears to have been by Widmanstätten in 1808; by about 1820 several other investigators had applied etching methods to crystallographic studies. Following 1850, very intensive investigations were made of the effects of etching and the formation of etch pits, particularly as applied to differential solution and the crystallographic symmetry of minerals (see *Crystallography: Morphological*). Virtually all of this work was done in Germany, and much of it has been summarized by Baumhauer in *Die Resultate der Aetz-Methode in der Krystallographischen Forschung* published in 1894 (see Honess, 1927).

With the advent of widespread use of X-ray crystallography in the 1920s, the use of etch pits as an aid in the analysis of crystal symmetry fell into disuse (except for investigations of *quartz*). However, X-ray studies provided considerable amounts of data demonstrating that the internal structure of crystals is usually far from perfect due to defects such as vacancies, impurity ions, and dislocations (see *Crystals, Defects in*). The subsequent development (1950s) of the dislocation theory and its application to the deformation of metals led to renewed interest in the use of etch pits as a means of observing the distribution of dislocations in a variety of crystalline materials. Much of this recent work was initiated by metallurgists (see *Metallurgy*) and their techniques have been adapted to material characterization studies of natural and artificial nonmetallic materials.

Dislocations

Screw and edge dislocations are generally considered to be created during deformation of crystalline materials. However, both types of dislocation may also be formed during the process of crystallization or during subsequent stages of heating and cooling. Regardless of the method of formation, the dislocations may have "free energy" and the resultant enhanced chemical reactivity causes a selective removal of material at the locus of the intersection of the dislocation and the etch surface. However, the type of etching, time, and temperature will affect the number, size, shape, and perfection of facets and the orientation of etch pits. In some minerals, quite different etching procedures may be required to produce etch pits on adjacent crystal faces.

Dislocation concentration in single crystals may range from as low as $10^3/cm^2$ (annealed, undeformed condition) to $10^{12}/cm^2$ (highly deformed condition); but good resolution of etch pits is difficult if the dislocation concentration is in excess of $10^6/cm^2$. Observations of dislocation planes defined by etch pits usually requires either optical- or electron-microscope examination, whereas etch pits on crystal faces may be etched to sufficient size to be seen with either the naked eye or low magnification.

Symmetry

The use of etch pits as an aid in determining the symmetries of crystals that are doubtful on

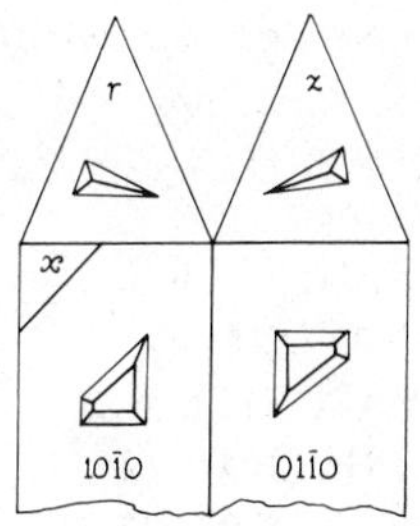

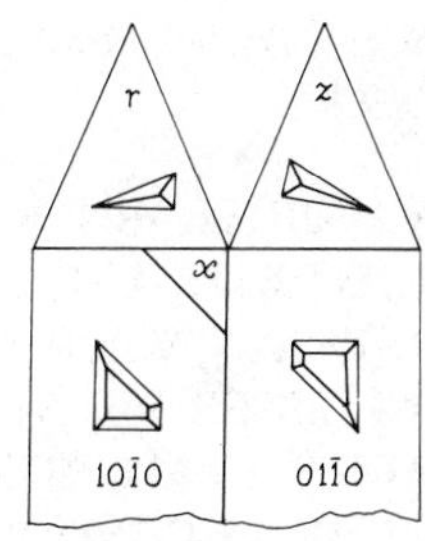

FIGURE 1. Etch pits on *quartz* display enantiomorphism. Left-handed crystals have the left positive trapezohedron (x) to the left of the prism $(10\bar{1}0)$ below the dominant positive rhombohedron (r); right-handed crystals have the right positive trapezohedron to the right. Etch pits reveal handedness in the absence of diagnostic crystal faces.

other grounds appears to have met with conspicuous success in some cases and ratther less success in others. The unsuspected effects due to deformation undoubtedly contributed to some of the problems encountered in attempted applications. *Quartz* (q.v.) provides one of the better examples of etch pits indicating the crystallographic symmetry, since they can show the trigonal trapezohedral symmetry, the right- or left-handedness of the crystal, and the difference between positive and negative forms (Frondel, 1962). Examples of the shape and distribution of etch pits on *quartz* crystal faces are shown in Fig. 1.

Deformation

When etch-pit techniques are used in studies of deformation, it is usually possible to distinguish edge dislocations by straight-line arrays of etch pits; and, in special cases, positive and negative edge dislocations may be seen (Hull, 1965). In addition, screw dislocations are characterized by the asymmetry of their etch pits (Gilman and Johnston, 1957). Descriptions of "solute canals" and "beaks" (Honess, 1927) correspond closely to descriptions of etch pits

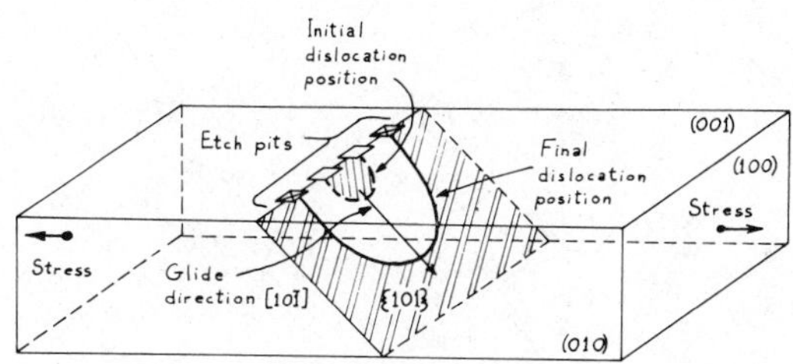

FIGURE 2. Etch pits mark an edge dislocation in a LiF crystal (see text and *Plastic Flow in Minerals*).

due, respectively, to edge and screw dislocations. The relationship of etch pits to a dislocation plane in lithium fluoride is shown in Fig. 2. The diagram shows an edge dislocation on a $\{101\}$ glide plane expanding in a $[10\bar{1}]$ direction (Gilman and Johnston, 1957). In *calcite*, various alignments of etch pits can be expected on $\{1011\}$ cleavage faces as follows (Thomas and Renshaw, 1967):

1. Twin gliding on $\{01\bar{1}2\}$ with a slip direction $[0\bar{1}11]$ results in linear arrays of "close-spaced" etch pits in a $\langle 2\bar{1}\bar{1}0\rangle$ direction.
2. Normal translational gliding on $\{10\bar{1}1\}$ with a slip direction $[\bar{1}012]$ results in alignments of etch pits along $\langle 10\bar{1}1\rangle$ (see *Plastic Flow in Minerals*).
3. The less common translational gliding on $\{0\bar{2}21\}$ with a slip direction $[\bar{1}012]$ results in alignment of etch pits along $\langle 10\bar{1}1\rangle$ and $\langle 2\bar{1}\bar{1}0\rangle$.
4. Translational gliding at elevated temperatures (above 500°C) on $\{10\bar{1}0\}$ with a slip direction $[0\bar{2}10]$ results in an alignment of etch pits along $\langle\bar{1}213\rangle$.

A preliminary study of etch-pit concentrations on cleavage surfaces in a white marble showed statistically that grain growth in *calcite* is comparable to grain growth in strained and annealed metals. Large cleavage surfaces ($300mm^2$ and larger) have a small number of pits ($6 \times 10^2 cm^{-2}$); medium range cleavage surfaces (1 to 2 mm^2) approximate the overall average pit density ($4 \times 10^3 cm^{-2}$); and small cleavage surfaces (about $0.2mm^2$) have a large number of pits ($3 \times 10^4 cm^{-2}$). The low density in the large grains of *calcite* can be attributed to grain growth and the higher density in the smallest grains probably reflects a level of strain prior to a thermal event (McDougall 1970).

The formation of etch pits can be observed in the following two simple experiments:

1. Immerse a small, clear *quartz* crystal in concentrated hydrofluoric acid (HF) and heat gently. (Etching with HF must be done in a fume hood and acid must not be spilled on the skin.) Etch a crystal for 1 hr, rinse in tap water, dry, and examine under a binocular microscope. The shape and (or) orientation of the etch pits will only be similar on alternate faces, indicating the threefold symmetry through the *c* axis.

2. Prepare a fresh cleavage surface of Iceland spar (*calcite*). (Avoid touching the surface with the finger tips.) Flood the cleavage surface with concentrated formic acid for 15 sec, then rinse with tap water and with alcohol. Allow to air dry and examine under reflected light with about X250 magnification. Arrays of etch pits should be clearly visible.

DAVID J. McDOUGALL

References

Frondel, C., 1962. *Dana's System of Mineralogy*, III, *Silica Minerals*. New York: Wiley, 158–162.

Gilman, J. J., and Johnston, W. G., 1957. The origin of glide bands in lithium fluoride crystal, in J. C. Fisher, et al., eds., *Dislocation and Mechanical Properties of Crystals*. New York: Wiley.

Honess, A. P., 1927. *The Nature, Origin and Interpretation of Etch Figures in Crystals*. New York: Wiley, 176p.

Hull, D., 1965. *Introduction to Dislocations*. Oxford: Pergamon, 259p.

McDougall, D. J., 1970. Technique to determine the relative concentration of crystallographic dislocations in marble, Abstracts, GAC-MAC meeting, Winnipeg, Manitoba.

Thomas, J. M., and Renshaw, G. D., 1967. Influence of dislocation on the thermal decomposition of calcium carbonate, *J. Chem. Soc.* (A), 2058–2061.

Cross-references: *Crystallography: Morphological; Crystals, Defects in; Electron Microscopy; Metallurgy; Plastic Flow in Minerals; Widmanstätten Figures.*

EVAPORITE MINERALS—*See* ANHYDRITE AND GYPSUM; SALINE MINERALS. Vol. IVA, EVAPORITE PROCESSES. Vol. VI: EVAPORITE FACIES; EVAPORITES: PHYSICO-CHEMICAL CONDITIONS OF ORIGINS

EXSOLUTION—*See* AMPHIBOLE GROUP; SPINEL GROUP; WIDMANSTÄTTEN FIGURES

EXTINCTION ANGLE—*See* OPTICAL MINERALOGY

F

FACE, CRYSTAL–*See* CRYSTAL HABITS; CRYSTALLOGRAPHY: MORPHOLOGICAL

FELDSPAR GROUP

Feldspars are a group of aluminosilicate minerals in which the silicon and aluminum tetrahedra share all four apical oxygen ions with neighboring tetrahedra in such a way that the structure is a three-dimensional framework (see Vol. IVA: *Mineral Classes: Silicates*). If the tetrahedra in a framework structure are entirely occupied by silicon ions (Si^{4+}) coordinated by oxygen ions (O^{2-}), the structure is electrostatically neutral and has a composition of SiO_2 (see *Quartz; Coesite and Stishovite*).

One fourth to one half of the tetrahedra are occupied by aluminum ions (Al^{3+}) in the ***feldspars***, causing the framework to have a net negative charge, +1 for each aluminum tetrahedron. Electrostatic neutrality is achieved in the structure by the inclusion of sodium (Na^+), potassium (K^+), or calcium (Ca^{2+}) ions in cavities within the framework (Fig. 1). In addition, there are rare boron (B^{3+}), barium (Ba^{2+}), and ammonium (NH_4^+) ***feldspar*** minerals. The group includes the following minerals:

- *albite:* (***Plagioclase***, An^0–An^{10}) $NaAlSi_3O_8$
- *andesine:* (***Plagioclase***, An^{30}–An^{50})
- *anorthite:* (***Plagioclase***, An^{90}–An^{100}) $CaAl_2Si_2O_8$
- *anorthoclase:* (Na,K) $AlSi_3O_8$
- *buddingtonite:* (NH_4) $AlSi_3O_8$
- *bytownite:* (***Plagioclase***, An^{70}–An^{90})
- *celsian:* $BaAl_2Si_2O_8$
- *hyalophane:* (K,Ba) Al $(Al,Si)_3O_8$
- *labradorite:* (***Plagioclase***, An^{50}–An^{70})
- *microline:* $KAlSi_3O_8$
- *oligoclase:* (***Plagioclase***, An^{10}–An^{30})
- *orthoclase:* $KAlSi_3O_8$
- *paracelsian:* $BaAl_2Si_2O_8$
- *plagioclase:* (Na,Ca) Al (Al,Si) Si_2O_8
- *reedmergnerite:* $NaBSi_3O_8$
- *sanidine:* (K,Na) $AlSi_3O_8$

In addition to their importance in industry and commerce, the ***feldspars*** are the major rock-forming minerals (q.v.), constituting 50% by volume of earth's crust. They also are a major constituent of the earth's upper mantle (see *Mantle Mineralogy*), of the lunar crust (see *Lunar Mineralogy*), and of some meteorites (see *Meteoritic Minerals*).

Because ***feldspar*** structures are very complex in detail, conditions at the time of their formation and of their subsequent history may be deduced from precise analyses of their composition and structure (see *Barometry, Geologic; Thermometry, Geologic*). Physical and chemical histories of individual rock formations, hence the earth and other solid bodies of the solar system, may be gleaned from such deductions.

Composition

Feldspar chemistry may be characterized by the general formula MT_4O_8, where M represents large cations that occupy cavities within the framework structure and T represents the small cations in tetrahedral coordination that make up the framework proper. In common rock-forming ***feldspars***, M sites are occupied predominantly by K^+, Na^+, and Ca^+, with minor to trace amounts of Ba^{2+}, Sr^{2+}, Rb^+, Cs^+, Pb^{2+}, and rare-earth elements. Additionally, Fe^{2+}, Mg^{2+}, or H_3O^+ (hydronium) may be present in the M sites.

The T sites are occupied predominantly by

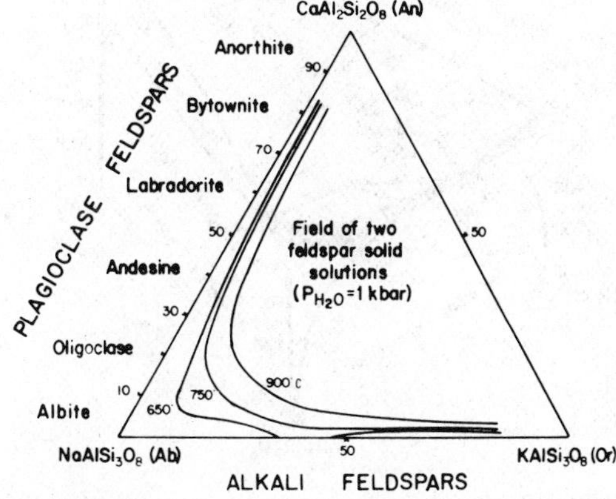

FIGURE 1. Solid-solution limits in the system *albite-orthoclase-anorthite* as a function of temperature at a water-vapor pressure of one kilobar. Decreasing pressure increases the area of two feldspars (from Ribbe, 1975, p. R-2).

Al^{3+} and Si^{4+}, with minor to trace amounts of Fe^{2+}, Fe^{3+}, P^{5+}, and Ti^{4+}.

The common ***feldspars*** are divided into two series:

1. the alkali ***feldspars*** (q.v.), where M is Na^+ and K^+, and
2. the ***plagioclase feldspars*** (q.v.), where M is Na^+ and Ca^{2+}.

Since, at high temperatures, there is a complete solid solution (q.v., Vol. IVA) within, but not between, each of the two series, mineral compositions are conveniently represented as mole percentages of the end-member compositions *Or* (*orthoclase, microcline*, or *sanidine*), *Ab* (*albite*), and *An* (*anorthite*). The names of the ***plagioclases***, and their compositional ranges, are displayed graphically in Fig. 1.

The extent of solid solutions among the ***feldspars*** varies with temperature of formation, being more restricted at lower than at higher temperatures (Fig. 1). Recrystallization at temperatures below the solidus restricts the extent of solid solution to an even greater degree (Fig. 2; see also *Alkali Feldspars; Plagioclase Feldspars*).

Structure

Feldspar structures best may be visualized by building from components. (Note: Although the terms "ring" and "chain" are used in conceptualizing the structure, the ***feldspar*** structure is not a ring nor a chain silicate; see Vol. IVA: *Mineral Classes: Silicates*).

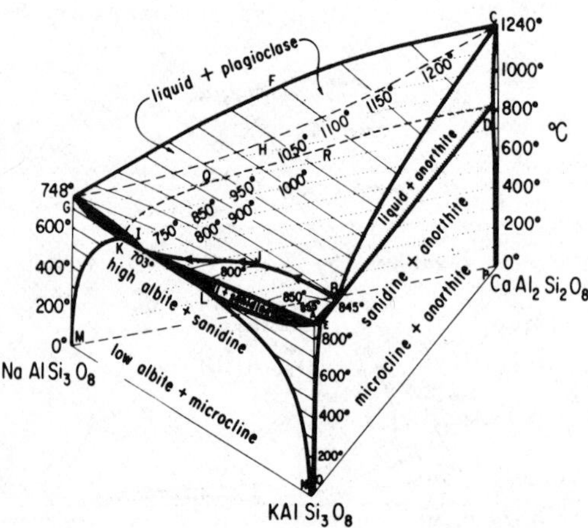

FIGURE 2. Schematic temperature-composition diagram for the system *albite-orthoclase-anorthite* at approximately 5 kilobars water-vapor pressure. Note that the subsolidus region has been simplified (from Smith, 1974, vol. 1, p. 10, reprinted with permission of Springer-Verlag).

The primary unit, as in all silicates except those formed under extremely high pressure (see *Coesite and Stishovite; Mantle Mineralogy*), is the TO_4 tetrahedron wherein each T ion is surrounded by 4 O^{2-} ions. The second level of construction is the linking of these tetrahedra by the sharing of apical O^{2-} ions between tetrahedra in such a way as to form a 4-membered ring of tetrahedra (Fig. 3a). Next, alternate pairs of tetrahedra in the 4-membered rings are cross-linked to give the framework structure of ***feldspar*** (Fig. 3b,c).

The structural state of ***feldspar*** may be characterized by the degree of ordering of the T ions that make up the framework and of the M ions that fill the cavities within it (see *Order-Disorder*). In the alkali ***feldspars,*** K^+ and Na^+ occupy the M sites. Because Na^+ (r=0.10 nm) is smaller than K^+ (r=0.13 nm; 1 nm=10Å), the cavity in the framework collapses from a coordination of 9 O^{2-} ions about K^+ to 6 or 7 about Na^+. Above 800°C, thermal vibration of M-site ions is sufficient to permit Na^+ and K^+ to be interchangeable in the structure and M-site occupancy is disordered (i.e., each ion being randomly distributed). Annealing at temperatures lower than 800°C permits segregation between expanded and collapsed M sites and the formation of K-rich and Na-rich regions in the structure. These regions range in size from submicroscopic lamellae (cryptoperthite) through lamellae several mm across (macroperthite) to discrete, homogeneous grains, depending upon duration of recrystallization and the presence of mineralizers.

Because of the similarity of size of Na^+ and Ca^{2+} ions, M-site ordering in the ***plagioclases*** is of less primary significance than in the alkali ***feldspars.***

Ordering of Si^{4+} and Al^{3+} in the T sites with falling temperature affects the symmetry and twinning of alkali ***feldspars.*** T-site occupancy is completely disordered in the high-temperature alkali ***feldspars*** (*sanidine*, high *albite*). This disorder gives *albite* above 1000°C and *sanidine* monoclinic symmetry ($C2/m$). Disorder requires that the T-site occupancy be statistically random, that the probability of finding an Al^{3+} ion in any particular 4-membered ring of tetrahedra be 1.0, and that the probability of finding an Al^{3+} ion in any particular tetrahedron within a ring be 0.25. However, the T sites are not identical, the oxygens around T_1 being more closely bonded to an M ion than those around T_2. During annealing, the Al^{3+} ions migrate preferentially into T_1 as does Si^{4+} into T_2 sites. This degree of T-site ordering is characteristic of *orthoclase* and of low-temperature *sanidine*, both of which are monoclinic, and of intermediate *albite*, which is triclinic.

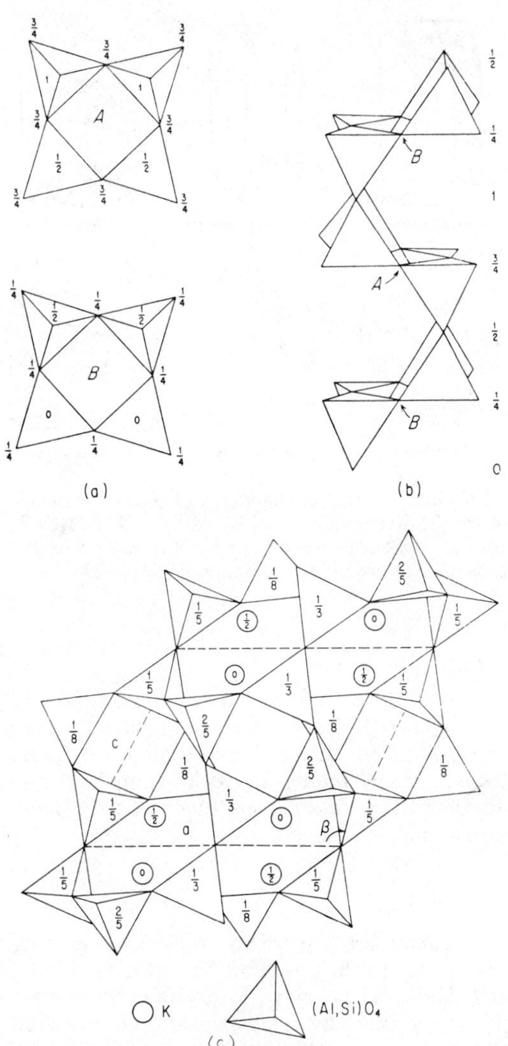

FIGURE 3. Feldspar structure (from Keith Frye, *Modern Mineralogy,* © 1974, pp. 88, 89; reprinted by permission of Prentice-Hall, Inc., Englewood Cliffs, N.J.). (a) Four-membered "ring" of tetrahedra; numbers show elevations of the apices of the tetrahedra in two rings. (b) "Rings" of tetrahedra linked into "chains"; numbers and letters as in (a). (c) ***Feldspar*** structure projected onto the *a-c* crystallographic plane with numbers showing elevations of the tetrahedral centers in the *b* direction.

Further differentiation of the T_1 and the T_2 sites permits the Al^{3+} ions to collect in just one of two T_1 sites leaving Si^{4+} ions to the two T_2 sites and the other T_1 site. This degree of ordering of the T sites characterizes low-temperature *albite* and maximum *microcline*, both of which are triclinic in symmetry. Intermediate states of disorder exist in particular *albite* and *microcline* specimens.

A different constraint on T-site occupancy affects the ***feldspars*** where the Si:Al ratio is 1:1 (*anorthite, celsian*). Cation electrostatic repulsion tends to distribute the Si^{4+} ions as far as possible from one another. This repulsion results in Si^{4+} ions alternating with Al^{3+} ions in the T sites and a doubling of the size of the unit cell in the c-axis direction. Another consequence of this repulsion-generated order is the discrimination of 16 unique T sites (see Ribbe, 1975).

Displacive transformations upon cooling lead to slight but significant symmetry changes in the structure. During cooling and annealing, decreasing disorder and resulting symmetry changes nucleate (see *Crystal Growth*) simultaneously at a multitude of locations within the structure, with the possibility of different orientations of the T_1 and T_2 sites. Each of these nucleation centers may grow into a region that is microscopically or macroscopically visible and related to adjacent regions by a twin law (see *Twinning*). Lack of twinning is, thus, an indicator of a low temperature of crystallization (see *Authigenetic Minerals*).

At near-melting temperatures, disorder in the T sites permits solid solution between *albite* (*Ab*) and *anorthite* (*An*) in the ***Plagioclases.*** However, the T-site order of the two end members of the series is completely different and they are totally incompatible with one another at low temperatures. Thus, ordered ***plagioclases*** of intermediate composition are not possible between approximately An_2 and An_{95} in the series. This structural incompatibility leads to a variety of exsolution phenomena. Since Al^{3+} and Si^{4+} diffuse between T sites extremely slowly, homogeneous regions are generally submicroscopically small.

In the bulk composition range of approximately An_2 to An_{16}, T-site ordering of domains of nearly pure *albite* forces the exsolution of a disordered phase near An_{25} in composition. The resulting submicroscopic, two-phase intergrowths are known as "peristerites" (compare "perthite," which results from M-site ordering). Similarly, at the *anorthite*-rich end of the series, "Huttenlocher intergrowths" are composed of an ordered An_{85-95} phase and an An_{60-70} phase. The irridescence of some *labradorite* specimens (labradorescence) results from "Bøggild intergrowths" of phases with compositions on opposite sides of a miscibility gap between An_{40-45} and An_{55-60}. Those specimens, in which the lamellae are spaced at distances corresponding to wavelengths of visible light, act as optical diffraction gratings to produce the characteristic iridescence. At around 500°C, the miscibility gap in ***plagioclase*** extends from An_{39} to An_{88}.

Melting Relationships

Melting and crystallization in the system *Ab-Or* under different water-vapor pressures are summarized in *Alkali Feldspars* (q.v.). Those for *Ab-An* are in *Plagioclase Feldspars* (q.v.). As can be seen in Fig. 2, at 5 kilobars pressure of water vapor, the system *Or-An* is characterized by a eutectic at 845°C, owing to limited solid solution between the two silicate end members.

The liquidus surface in the system *Ab-Or-An* at 5 kilobars water-vapor pressure is characterized by a large field of primary crystallization for ***plagioclase*** separated from a small field for *sanidine* by a thermal valley descending from the *Or-An* eutectic at 845°C toward the minimum in the system *Or-Ab* at 703°C. (see Vol. IVA: *Phase Equilibria* for definitions).

Physical Properties

Feldspars are monoclinic ($C2/m$) or triclinic ($C\bar{1}$) depending upon the size of the *M*-site ion and the structural state (see above). Some typical habits are illustrated in Fig. 4; common twins are illustrated in Fig. 5. Twin laws are tabulated in Table 1. Cleavage is perfect prismatic at right angles in monoclinic species and at near right angles in triclinic species. Pure ***feldspars*** are white or colorless, but impurities and inclusions provide a variety of shades (e.g., amazonite, green; aventurine, reddish; sunstone, golden). H=6-6½; G=2.55-2.63. Optically transparent; α=1.514-1.529, β=1.518-1.533; γ=1.521-1.539; $2V_x$=18-103° (see individual minerals).

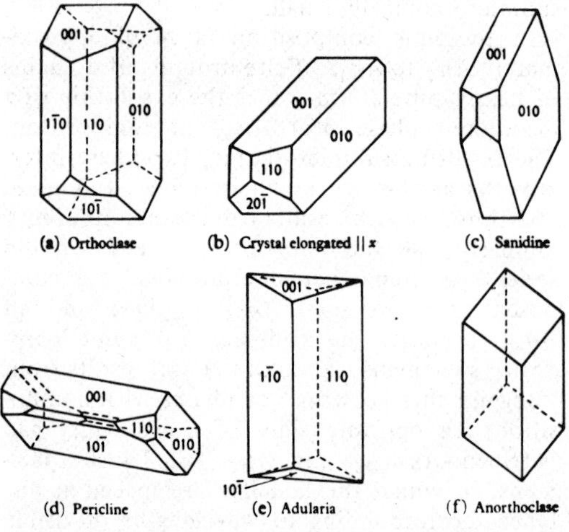

FIGURE 4. Some habits of minerals of the ***feldspar*** group (from W. A. Deer, R. A. Howie, and J. Zussman, *Rock-Forming Minerals* vol. IV. Longman Group Limited, London, 1963, p. 21).

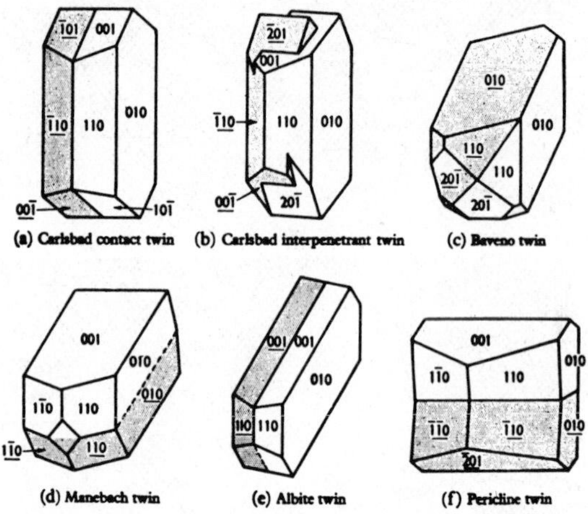

FIGURE 5. Some common twin forms of minerals of the ***feldspar*** group (from W. A. Deer, R. A. Howie, and J. Zussman, *Rock-Forming Minerals,* vol. IV. Longman Group Limited, London, 1963, p. 23).

Uses

The ***feldspars*** are important in manufacture of porcelains, ceramic glazes, and glass. Colored and iridescent varieties are semiprecious gem stones (see *Gemology*). Rocks composed predominantly of ***feldspar*** are quarried for dimension stone.

Occurrence

Feldspars are the major rock-forming minerals (q.v.) in the crusts of the earth, the moon, and Mars. They may be crystallized from a magma, formed by recrystallization in the solid state in metamorphic rocks, deposited from hydrothermal solution in veins, or crystallized as authigenic minerals from interstitial waters in sedimentary rocks (see *Alkali Feldspars; Plagioclase Feldspars*; individual minerals for details of occurrence).

KEITH FRYE

References

Christie, O. H. J., ed., 1962. Feldspar volume, *Norsk Geolog. Tidsskrift*, **42**, 606p.

Deer, W. A.; Howie, R. A.; and Zussman, J., 1963. *Rock-Forming Minerals*, vol. 4. London: Longmans, 435p.

Mackenzie, W. S., and Zussman, J., eds., 1974. *The Feldspars.* Manchester, England: Manchester University Press, 717p.

Ribbe, P. H., ed., 1975. *Feldspar Mineralogy.* Washington, D.C.: Mineral. Soc. Am. Short Course Notes, vol. 2.

Smith, J. V., 1974. *Feldspar Minerals*, vols. 1-2. New York: Springer-Verlag, 1317p.

TABLE 1. Feldspar Twin Laws

Name	Twin Axis	Composition Plane	Remarks
Normal twins			
Albite	⊥ (010)	(010)	Repeated; triclinic only
Manebach	⊥ (001)	(001)	Simple
Baveno (right)	⊥ (021)	(021)	Simple, rare in plagioclases
Baveno (left)	⊥ (0$\bar{2}$1)	(0$\bar{2}$1)	
X	⊥ (100)	(100)	
Prism (right)	⊥ (110)	(110)	
Prism (left)	⊥ (1$\bar{1}$0)	(1$\bar{1}$0)	
Parallel twins			
Carlsbad	[001] (*z* axis)	(*hk*0), usually (010)	Simple
Pericline	[010] (*y* axis)	(*h*0*l*), "rhombic section" parallel to *y*	Repeated; triclinic only
Acline A	[010] (*y* axis)	(001)	
Acline B	[010] (*y* axis)	(100)	
Estérel	[100] (*x* axis)	(0*kl*) "rhombic section" parallel to *x*	Repeated
Ala A	[100] (*x* axis)	(001)	
Ala B	[100] (*x* axis)	(010)	
Complex twins			
Albite–Carlsbad (Roc Tourné)	⊥ *z*	(010)	Repeated
Albite–Ala B	⊥ *x*	(010)	
Manebach–Acline A (Scopie)	⊥ *y*	(001)	
Manebach–Ala A	⊥ *x*	(001)	
X–Carlsbad	⊥ *z*	(100)	
X–Acline B	⊥ *y*	(100)	

From W. A. Deer, R. A. Howie, and J. Zussman, 1963, *Rock-Forming Minerals,* vol. IV. Reprinted with permission from Longman Group Limited, London, p. 22.

Spear, F. S., 1977. Evidence for a miscibility gap in plagioclase feldspar in the composition range An_{39}–An_{88}, Annual Report, Director Geophysical Laboratory, *Carnegie Inst. Wash. Yr. Bk. 76,* 619–621.

Tuttle, O. F., and Bowen, N. L., 1958. Origin of granite in the light of experimental studies in the system $NaAlSi_3O_8$–$KAlSi_3O_8$–SiO_2–H_2O, *Geol. Soc. Amer. Mem.*, **74**, 153p.

Cross-references: *Blacksand Minerals; Crystal Growth; Diagenetic Minerals; Electron Microscopy (Transmission); Fire Clays; Mineral Classification: Principles; Mineral Industries; Mineral and Ore Deposits; Museums, Mineralogical; Optical Orientation; Order-Disorder; Pegmatite Minerals; Rock-Forming Minerals; Sands, Glass and Building; Soil Mineralogy; Staining Techniques; Thermoluminescence; Twinning.* Vol. IVA: *Mineral Classes: Silicates.*

FELDSPATHOID GROUP

The minerals *nepheline, kalsilite*, and *leucite* comprise the ***feldspathoid*** group. They have framework structures of silicon (SiO_4) and aluminum (AlO_4) tetrahedra and crystallize in silica-poor environments.

- *kalsilite:* $KAlSiO_4$
- *leucite:* $KAlSi_2O_6$
- *nepheline:* $Na_3KAl_4Si_4O_{16}$

Nepheline and *kalsilite* are not isostructural, but both have tridymite-type structures of cross-linked, six-membered rings, half the tetrahedral Si being replaced by Al. Electrostatic neutrality is maintained by the presence of alkali ions in cavities in the framework. The framework is distorted, resulting in alkali sites of two different sizes. One-fourth of the sites are coordinated by nine oxygen ions and, ideally, are occupied by potassium. Three-fourths of the sites have eight-fold coordination and, ideally, are occupied by sodium. Most *nepheline* samples deviate from the ideal 3Na:K ratio, indicating $Na \rightleftarrows K$ substitution. In addition, more than half the tetrahedra may contain Si, resulting in vacant cavities that maintain neutrality. *Nepheline* and *kalsilite* are hexagonal.

The leucite structure is composed of four- and six-membered rings that are cross-linked by sharing corners to build up a framework of tetrahedra. One-third of the tetrahedral sites

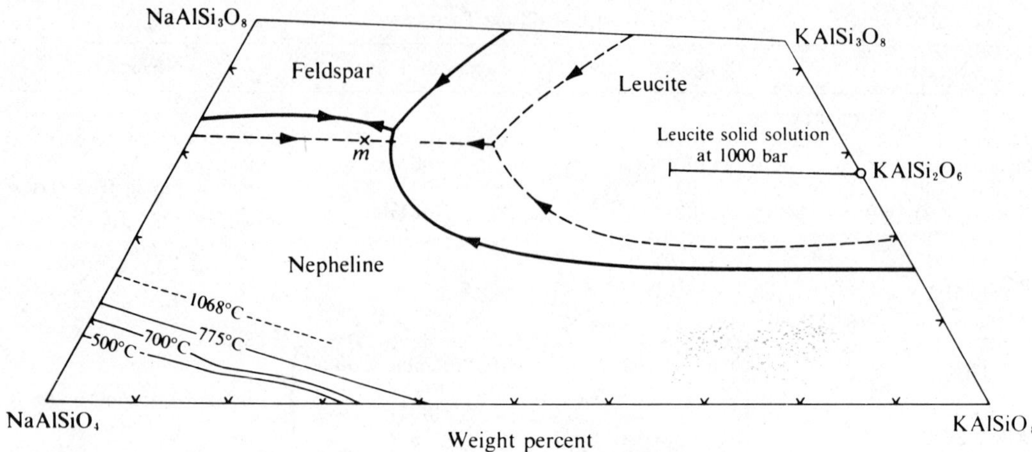

FIGURE 1. Liquidus surface of a portion of the system $NaAlSiO_4$-$KAlSiO_4$-SiO_2-H_2O showing the primary fields of *nepheline, leucite,* and *alkali* ***feldspar,*** field boundaries, and directions of falling temperatures (from I. S. Carmichael, F. J. Turner, and Jean Verhoogen, *Igneous Petrology,* McGraw-Hill Book Co., New York, 1974; reprinted with permission). Field boundaries for 1 bar pressure are solid; field boundaries for 1 kilobar pressure are dashed.

are occupied by Al and two-thirds by Si. Electrostatic neutrality is maintained by the presence of K in cavities within the framework. Although the framework has cubic symmetry, at low temperatures the K ion is too small to fill its cavity; and, below about 625°C, the structure collapses partially to tetragonal symmetry.

The liquidus surface of the system $KAlSiO_4$-$NaAlSiO_4$-SiO_2-H_2O at 1 kilobar (Fig. 1) shows a minimum on the alkali ***feldspar*** tie line ($NaAlSi_3O_8$-$KAlSi_3O_8$) with temperatures falling away from the minimum both toward and away from the SiO_2 apex. With crystallization, liquid compositions migrate away from the tie line toward SiO_2 (silica-saturated liquids) or away from SiO_2 (silica-unsaturated liquids). (See vol. IVA: *Phase Equilibria* for definitions.)

Nepheline occurs in *quartz*-free, felsic plutonic and volcanic rocks such as nepheline syenites and phonolites. It also forms by reaction between magmas and calcareous host rocks. In syenite pegmatites, it is associated with alkali ***feldspars,*** with minerals of the ***sodalite*** group, and with other silica-undersaturated minerals.

Leucite occurs in K-rich mafic and ultramafic volcanic and plutonic rocks. In some cases, it is replaced by an intergrowth of alkali Feldspar and *nepheline* known as pseudoleucite.

KEITH FRYE

Reference

Deer, W. A.; Howie, R. A.; and Zussman, J.; 1963. *Rock-Forming Minerals*, vol. 4. London: Longmans, 435p.

Cross-references: *Crystal Growth; Mineral Classification: Principles*. Vol. IVA: *Phase Equilibria.*

FERRIMAGNETISM; FERROMAGNETISM—*See* MAGNETIC MINERALS

FILLERS—*See* PIGMENTS AND FILLERS

FIRE CLAYS

Fire clay is defined in the *AGI Glossary of Geology* as follows:

> A siliceous clay rich in hydrous aluminum silicates, capable of withstanding high temperatures without deforming and useful for the manufacture of refractory ceramic products. It is deficient in iron, calcium, and alkalies, and approaches *kaolin* in composition, the better grades containing at least 35% alumina when fired.

In the 1979 ASTM Standards, fire clay is defined as follows:

> An earthy or stony mineral aggregate which has as the essential constituent hydrous silicates of aluminum with or without free silica, plastic when sufficiently pulverized and wetted, rigid when subsequently dried, and of suitable refractoriness for use in commercial refractory products.

Ries (1927) uses the term "fire-clay" and expresses the opinion that no clay should be classed as a fire-clay unless the fusion is higher than cone 27. Norton refers to "fireclay" as including nearly all clays that have a fusion point above approximately 1600°C. (≈2900°F) and are white burning. His definition would include the high alumina clays but would eliminate the white-burning *kaolins*.

The following standards are those set forth in the 1979 ASTM Standards for fire clay and high-alumina refractory brick:

(a) Fire clay brick are divided into five different classes:

	Min. PCE
Super duty	33
High duty	31½
Semi-silica	–
Medium duty	29
Low duty	15

The super-duty and high-duty classes are divided further into three types under each class.

(b) High-alumina brick are divided into seven different classes:

	Min. PCE
50 percent alumina	34
60 percent alumina	35
70 percent alumina	36
80 percent alumina	37
85 percent alumina	–
90 percent alumina	–
99 percent alumina	–

Usage as set forth by the ASTM standards precludes the high-alumina clays from being included as fire clays. There is a gradation from fire clays into the high-alumina clays, especially in the flint-to-burley–flint clays of Pennsylvania and Missouri.

The term "fire clay" was originally applied to those underclays of coal beds that were heat resistant and were used in the manufacture of fire brick. This application has been abused in that the term was applied to all underclays, the majority of which are not refractory enough for the manufacture of fire brick. Many fire clays are not underclays at all and have no association with coal.

Fire clays are characterized by refractoriness, which is the most important property, and vary greatly in plasticity, density, shrinkage, strength, and color. Fire clays are primarily hydrous alumina silicates with varying amounts of impurities such as iron oxides, lime, magnesia, alkali, and free silica. If the percentage of impurities is too large, the clay will fuse at a comparatively low temperature and, therefore, cannot be used as a refractory material. The chemical analyses of some typical fire clays is shown in Table 1.

Fire clays are composed predominantly of the mineral *kaolinite* $Al_2(Si_2O_5)(OH)_4$ but may have other clay minerals of the ***kaolinite*** group present. In addition, there may be minor amounts of ***illite*** or ***montmorillonite*** present. A considerable amount of research has been accomplished in the last twenty years regarding the true nature of clay. Grim, Keller, and Paul F. Kerr have added much knowledge to the understanding of the clay minerals. The use of differential thermal analyses, X-ray diffraction, and SEM have been of great help to the clay mineralogist. This research has been carried on by universities and colleges, by research foundations, and by corporate research laboratories.

The term "fire clay" is now considered to include, not only the refractory underclays, but also the flint clays, kaolins, and kaolinitic clays. Norton (1968) refers to fire clay as those clays that are buff burning and kaolins as those clays that are white burning. Fire clays range in hardness from the very soft kaolins to the hard Missouri and Pennsylvania flint clays. The plasticity or workability varies greatly among fire clays from different localities. The fire clays referred to as plastic clays present good plasticity, although some of them require finer grinding than others. The hard flint clays on the other hand show little or no plasticity even when finely ground. The plastic and semi-plastic fire clays slake down in water, while some flint clays do not break down at all.

Fire clays vary in color depending upon the amount of organic material and iron minerals present. Some of the kaolinitic clays range from white to black within the same deposit, depending upon the amount of lignitic material present. Plastic and semi-plastic clays vary from a light gray to greenish to dark gray, depending on minute particles of organic material. Flint clays may vary from white to dark gray or may be red to purple in color. Some irony flint clays present a mottled appearance. Plant remains are found in some fire clays, such as the carbon imprints in the Mercer clays of Pennsylvania and the lignitic rocks and roots found in the kaolinitic clays of the bauxite deposits in Arkansas.

All fire clays contain minerals of a nonclay type, and these may greatly influence the refractoriness and the economic value of the fire clay. Some of the minerals commonly found in fire clays are as follows: *quartz* (sand grains), ***feldspars*** (small fragments *orthoclase* and *plagioclase*), ***micas*** (usually small flakes), iron minerals (*hematite, magnetite*, limonite, *pyrite, siderite*), titanium minerals (*rutile, anatase*), limestone (*calcite, dolomite*), *magnesite, gypsum*, ***garnets***, and ***tourmalines***. The iron min-

TABLE 1. Chemical Analyses of Typical Fire Clays

	L. Kittanning Clay (New Brighton, Pa.)	Bolivar Flint Fire Clay (Salina, Pa.)	No. 1 Fire Clay (Woodbridge, N.J.)	Coal Measures Clay (Boone Furnace, Ky.)	Plastic Fire Clay (St. Louis, Mo.)	Mt. Savage Clay (Piedmont, W. Va.)	Athens (Texas)	Golden (Colorado)	Dry Branch (Ga.)
Silica (SiO_2)	62.89	51.92	51.56	46.56	59.36	61.44	73.00	50.35	40.28
Alumina (Al_2O_3)	21.49	31.64	33.13	37.47	23.26	26.18	15.79	33.64	34.72
Ferric oxide (Fe_3O_3)	–	–	.78	trace	3.06	.30	.63	.75	.84
Ferrous oxide (FeO)	1.81	1.13	–	–	–	.36	–	–	–
Lime (CaO)	.38	.03	trace	.112	.65	.12	1.29	–	.05
Magnesia (MgO)	.56	.44	trace	trace	.42	–	1.53	trace	.04
Potash (K_2O)	2.52	.40	trace	.28	.63	–	.10	.49	trace
Soda (Na_2O)			trace	.28		.02	.16	.09	–
Sulphur trioxide (SO_3)	–	–	–	–	.35	–	–	–	trace
Titania (TiO_2)	1.82	1.16	1.91	–	1.01	1.39	.43	.80	1.15
Water (H_2O)	7.58	13.49	12.50	13.03	10.20	9.07	5.76	11.75	12.39
Moisture	1.16	–			2.74	.77	–	2.13	10.72
Total	100.21	100.21	99.88	97.732	101.68	99.65	98.69	100.00	100.19

Data from Reis, 1927, p. 336.

erals, *gypsum*, and free silica are the most common accessory minerals that affect the refractory value of the deposit and may make mining a difficult selection process. *Siderite* ($FeCO_3$) is abundant in Pennsylvania and Kentucky fire clays. It is also found in the kaolinitic clay of Arkansas and Texas. *Pyrite* (FeS_2) is common in the Missouri, Ohio, and Pennsylvania fire clays and is also found in some of the Georgia, Alabama, and Texas kaolinitic clays. *Quartz* in the form of sand grains is present to some extent in almost all types of deposits.

Fire clays may be residual, such as the kaolins of North Carolina, or transported, such as the sedimentary fire clays of Pennsylvania and the kaolinitic clays of Georgia and Texas. The majority of the fire-clay production in the United States is from sedimentary deposits. These deposits are sediments that have been transported and deposited in lagoons, swamps, shallow lakes, or seas under freshwater conditions. These sediments were altered during transportation and deposition by the leaching of the Na, Ca, Mg, K, and Si with the resultant enrichment of the alumina. The minerals of the ***kaolinite*** group are readily formed from Na- and K-rich rocks in a hot climate with a high rainfall and favorable topographic conditions. The degree of alteration at particular time periods results in the presence of different clays within the same deposit. An example is the presence of both flint clay and plastic clay underlying coal seams in Pennsylvania and Kentucky. A high degree of leaching results in the formation of high-alumina clays.

There are three general types of fire-clay deposits. The first is the layered (bedded) deposit, which is represented by the fire clays occurring as underclays in Pennsylvania, Kentucky, Ohio, northern Alabama, and Illinois. The second type is the basin deposit formed either as a result of the leaching of limestones and dolomites or the erosion channels formed by stream action. These are represented by the filled sink deposits of Missouri, the kaolinitic deposits in Arkansas and Idaho, and the channel deposits in Colorado. The third type of deposit is lenticular and occurs primarily in the kaolins and kaolinitic clays of South Carolina, Alabama, Georgia, Texas, and California. There are, of course, variations of these three types occurring within each area.

In New Jersey, fire clays occur as a series of plastic clays, spar sand, sandy clays, along with lignitic sands and clays that are not suitable for refractories. The fire-clay deposits occurring in Pennsylvania, West Virginia, Ohio, Kentucky, Illinois, and northern Alabama are bedded deposits that thicken or thin and may vary laterally from good refractory material to clays too sandy and too irony to be used. The fire clays of Missouri probably represent the greatest variation, ranging from low-heat-duty clays to super-duty clays to high-alumina clays. There are fire clays, which have been fully reported in the literature of the state geological agencies, located in most of the states. Many of these deposits have been worked extensively, but many have not been used commercially because of their poor quality, small size, and location.

Fire clays also occur in Canada, primarily in the provinces of Saskatchewan and British Columbia. Fire clays are mined and used in Mexico, although the majority of the known deposits are of hydrothermal origin. Fire clays are mined and used for refractories throughout the various countries of South America. England has extensive deposits of commercial fire clays, which have been mined for many years and are still in active production. France and Germany also have extensive deposits of fire clays that are used for refractory production.

Fire clays are primarily mined and used for the production of refractories, which includes not only the categories set forth under the 1979 ASTM standards, but also specialty products, mortars, and foundry clays. Fire clays are also used in minor amounts in the manufacture of face brick and tile and sewer pipe. They are also used to make retorts for smelting various ores.

Fire clays are located and prospected by a variety of methods. In addition to the location and sampling of outcrops, trenching and dozing slots along suspected clay horizons are also used, primarily on bedded deposits. In most cases, geological field work followed by wildcatting with power drills, is used to locate and determine the quality of the deposits. Auger drills, churn drills, and core drills are used extensively in the United States. Complete drilling and testing of the fire clays in a deposit is necessary as a guide for quality control purposes during mining of the deposit.

Most fire-clay mining, now carried on in the United States, is done by open pit methods. Until recent years, a considerable amount of fire clay was mined from underground operations. But, this type of operation has now been largely abandoned in favor of open-pit mining because of safety and economic factors.

Open pit mines are stripped by many different methods. Drag lines and large stripping shovels are used in many areas; scrapers and self-propelled loaders are used on large stripping operations where the overburden is not too hard and consolidated. Mining in open pits ranges from hand separation of high-alumina clays in the Missouri diaspores and flint pits to loading of large trucks by shovel, front-end loader, or drag line. In all cases, mining is controlled by the results of the drilling and testing done prior to the actual mining operation. Quality control of the raw materials is imperative in order to supply the refractory industry with the type and quality of fire brick demanded today.

JAMES WESTCOTT

References

Bolger, R. C., and Weitz, J. H., 1952. Mineralogy and origin of the Mercer fire clay of North-Central Pennsylvania, *Problems of Clay and Laterite Genesis*, Symposium AIME, 81-94.

Grim, R. E., 1953. *Clay Mineralogy*. New York: McGraw-Hill, 384p.

Keller, W. D.; Westcott, J. F.; and Bledsoe, A. O., 1954. The origin of Missouri fire clays, in Clays and clay minerals, *Natl. Acad. Sci.-Natl. Res. Council, Pub. 327*, 7-46.

Norton, F. H., 1968. *Refractories*. New York: McGraw-Hill, 61-78.

Reis, H., 1927. *Clays: Occurrence, Properties and Uses*, 3rd ed. New York: Wiley, 333-334.

Searle, A. B., 1924. *Refractory Materials*. London: Griffin, 13-18.

Cross-references: *Clays and Clay Minerals; Mineral Industries; Refractory Minerals; Rock-Forming Minerals.*

FLUORESCENCE—*See* LUMINESCENCE

FLUORESCENCE ANALYSIS—*See* ELECTRON PROBE MICROANALYSIS. Vol. IVA: X-RAY SPECTROSCOPY

FORM, CRYSTAL—*See* CRYSTAL HABITS; CRYSTALLOGRAPHY: MORPHOLOGICAL

FRAMEWORK SILICATES—*See* individual mineral groups. Vol. IVA: CRYSTAL CHEMISTRY

FRONDELITE GROUP

Minerals of the ***frondelite*** group are hydroxyl phosphates of iron and manganese; a member is *frondelite*, if manganese exceeds iron and *rockbridgeite*, if iron exceeds manganese. The names originated with Lindberg (1949), for Clifford Frondel, American mineralogist, and Frondel (1949), for the occurrence at Rockbridge County, Virginia, USA. The group composition is

$$(Mn^{2+},Fe^{2+})Fe_4^{3+}(PO_4)_3(OH)_5$$

frondelite, $Mn^{2+} > Fe^{2+}$ and *rockbridgeite*, $Fe^{2+} > Mn^{2+}$.

These minerals are orthorhombic, in botryoidal masses and crusts, with H=4½; G=3.3-3.5. They are brittle, vitreous to dull green to greenish black, becoming brownish to reddish on oxidation, and subtranslucent.

Frondelites are found as an alteration product of iron-manganese phosphates in pegmatites. *Frondelite* occurs at Conselheira Pena, Minas Geraes, Brazil, and at North Groton, New Hampshire, USA. *Rockbridgeite* occurs in Rockbridge County, Virginia, and Greenbelt, Maryland, USA; at Herdorf, Ullersreuth, Hagendorf, and Kreuzberg, West Germany; and at Chanteloube, France.

KEITH FRYE

Reference

Moore, P. B., 1970. Crystal chemistry of the basic iron phosphates, *Am. Mineralogist*, 55, 135-169.

FUELS, MINERAL—*See* Vols. XIV and XVI

G

GANGUE MINERALS

Ore, as mined, contains ore minerals, gangue minerals, and country rock. A mass of mineral matter is not considered *ore* unless one or more of its ore minerals can be recovered on a profitable basis. As originally used, ore minerals were metal bearing (see *Metallic Minerals*). More recently, nonmetallic minerals of commercial importance are considered ore minerals (see *Mineral Industries*).

The name, *ore*, appears to have been derived from the Middle English, *oor*, but its similarity to earlier names for copper ore and brass suggest a derivation. *Gangue* is from the Greek, *gang*, meaning "vein of metal," the form gangue being of French derivation.

As generally used, gangue minerals have no commercial importance in a particular period of time, possibly becoming ore minerals at a later date. They are commonly silicates, carbonates, or fluorides, more rarely sulfides. The gangue minerals of ore deposits formed at high temperatures differ from those deposited at lower temperatures. The principal ones are as follows:

- hydrothermal: *apatite*, ***garnet***, ***mica***, *quartz*, *topaz*, ***tourmaline;***
- mesothermal: *barite*, carbonates, *quartz*;
- epithermal: adularia, ***alunite***, *calcite*, chalcedony, *fluorite*, opal, *quartz*.

Ore in huge tonnages is mined and treated, especially when the ore mineral sought has sufficient value. The Rand, South Africa, mines produce 400 tons of gold a year, treating over 60,000,000 tons of ore to produce this amount (gold in the earth's crust has been estimated at 0.005%; the minimum amount required as ore is 6 ppm).

E. E. FAIRBANKS

Cross-reference: *Mineral and Ore Deposits.*

GARNET GROUP

The commonest ***garnets*** are essentially iron or calcium aluminum silicates. They are commonly red or brown and transparent with a vitreous luster. They have a hardness of 6–7½, no cleavage, often appear as dodecahedra or trapezohedra, and occur mainly in schists, gneisses, and marbles.

Because of its hardness and angular fracture, ***garnet***, mainly the iron-aluminum variety, is used as a mild abrasive (see *Abrasive Materials*). A number of varieties of ***garnets*** are also cut as gemstones (see *Gemology*).

The name ***garnet*** comes indirectly from the Latin *granatus*, meaning "like a grain," and directly from pomegranate, in reference to the small red seeds of this fruit.

Chemistry and Structure

Garnet has the general formula $A_3B_2(SiO_4)_3$, where A may be calcium, magnesium, iron, or manganese, and B may be aluminum, iron, or (less commonly) chromium, vanadium, or zirconium. The silicon may be partially replaced by aluminum and (SiO_4) groups by $(OH)_4$. Other replacements can occur; and ***garnets*** containing substantial quantities of titanium, phosphate, and even yttrium have been recorded. Most ***garnets*** show extensive solid solution.

Structurally, ***garnet*** is an orthosilicate (or nesosilicate) consisting of independent silicon-oxygen tetrahedra linked to trivalent and divalent ions. Natural ***garnets*** are solid solutions of several end members and are named after the predominant component. The equations given are for the pure end members.

The principal members of the ***garnet*** group are subdivisions that represent end-members of isomorphous series. In addition to the minerals listed in Table 1, the ***garnet*** group also includes *henritermierite, hydrougraudite, majorite*, and *schorlomite*.

Pyrope: $Mg_3Al_2(SiO_4)_3$; G=3.58; named by A. G. Werner in 1803, from the Greek meaning "firelike"; not recorded in a pure natural state, but as a component with *almandine* and *grossular*; color is vermillion, ranging from a crimson to a purplish shade with additional almandine component present. Occurs most commonly in peridotites or kimberlites associated with *diamond*, ***serpentine***, *spinel*, ***pyroxenes***, and ***olivines***; transparent blood-red

TABLE 1. Garnet Group

		n	D	a
Almandine	$Fe_3^{2+}Al_2Si_3O_{12}$	1.830	4.318	11.526
Andradite	$Ca_3Fe_2Si_3O_{12}$	1.887	3.859	12.048
Goldmanite[a]	$Ca_3V_2Si_3O_{12}$	1.821	3.74±	12.011
Grossular	$Ca_3Al_2Si_3O_{12}$	1.734	3.594	11.851
Hydrogrossular	$Ca_3Al_2Si_2(OH)_4O_8$	1.675±	3.13±	12.16±
Kimzeyite[a]	$Ca_3Zr_2(Al_2Si)O_{12}$	1.94	4.0	12.46
Knorringite[a]	$Mg_3Cr_2Si_3O_{12}$	1.803	3.756	11.659
Pyrope	$Mg_3Al_2Si_3O_{12}$	1.714	3.582	11.459
Spessartine	$Mn_3Al_2Si_3O_{12}$	1.800	4.190	11.621
Uvarovite	$Ca_3Cr_2Si_3O_{12}$	1.86±	3.90±	12.00

In part, after Deer, Howie, and Zussman, 1962.
[a]Data on natural material showing considerable solid solution.
n = index of refraction
D = specific gravity
a = length of unit-cell edge, in Å

varieties are used as gemstones, "cape rubies" (from South Africa) or "Arizona rubies."

Almandine (almandite): $Fe_3Al_2(SiO_4)_3$; G=4.32; name was probably derived by Agricola (1546) from Alabanda, Turkey, where the alabandic carbuncles described by Pliny (77 AD) were cut; color is deep red to brownish black. It is the commonest species of ***garnet***, characteristically occurring in mica schists and gneisses; used as a gemstone when transparent or exhibiting asterism; the Gore Mountain, New York, abrasive deposits are principally *almandine.*

Spessartine (spessartite): $Mn_3Al_2(SiO_4)_3$; G=4.19; named by F. S. Beudant in 1832, from a locality at Spessart, W Germany; color is usually a rich shade of brown or orange; a less common ***garnet*** found in granite pegmatites, rhyolites, and Mn-rich assemblages, in association with *rhodonite, quartz*, and *sphalerite*. Large orange crystals of gem quality are found in Minas Gerais, Brazil, and the Malagasy Republic.

Grossular (grossularite): $Ca_3Al_2(SiO_4)_3$; G=3.59; named by Werner in 1811 from the resemblance of the pale green varieties to the gooseberry, *R. grossularia*; lightest colored of the garnets, usually yellow-green, white, pink, cinnamon, or light brown. Its chief occurrences are in contact and regionally metamorphosed limestones associated with *wollastonite, calcite, vesuvianite* (idocrase), *diopside*, and *epidote*; important localities are Mondos, Mexico, and the Transvaal, South Africa.

Andradite: $Ca_3Fe_2(SiO_4)_3$; G=3.85; named by Dana in 1868 in honor of J. B. d'Andrada, the Portuguese mineralogist, who first described this species; color is highly variable, ranging from brown to red to green to black to honey-yellow. Occurs typically in contact metamorphosed calcareous sediments at Franklin, New Jersey; in serpentine and chlorite schists of the Urals, USSR, and San Benito County, California; the transparent apple-green variety, "demantoid," and the honey-yellow "topazolite" are used as gemstones.

Uvarovite: $Ca_3Cr_2(SiO_4)_3$; g=3.90; named by G. H. Hess in 1832 in honor of Count S. S. Uvarov, formerly the president of St. Petersburg (Leningrad) Academy; color is a distinct bright, deep emerald green. It is found in serpentinites, closely associated with *chromite* in chromium deposits at Outokumpu, Finland; Bisersk, Urals, USSR; Röros, Norway; and Pic Posets, France.

Physical Properties

Garnets are commonly various shades of red or brown, but ***garnets*** rich in titanium may be almost black, the calcium-rich ***garnets*** with chromium or vanadium are green, and the pure calcium-aluminum ***garnets*** may be colorless or white. ***Garnet*** belongs to the isometric crystal system and is usually distinctly crystallized as rhombic dodecahedra or trapezohedra, either alone, in combination, or with hexoctahedral modification, but it may be found in rounded grains or granular masses. It has no cleavage but may show a rhombododecahedral parting. The calcium- and manganese-rich ***garnets*** are commonly weakly birefringent, probably due to internal strain, and may show complex twinning. Variations of refractive index and density are given in Fig. 1.

Many properties of ***garnets*** vary approximately linearly with end-member composition. The properties of any ***garnet*** of known composition may, therefore, be estimated from the properties of the pure end members. The refractive

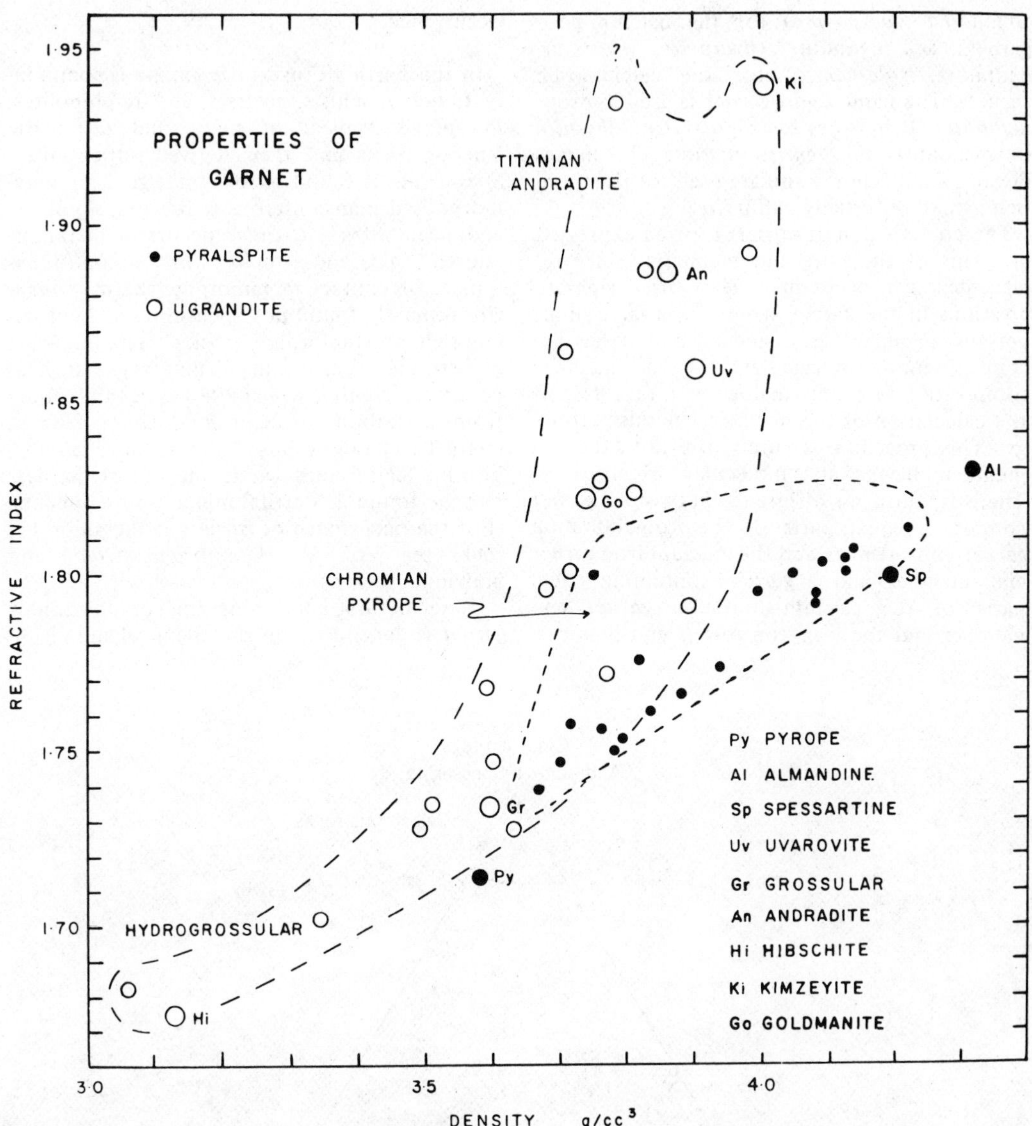

FIGURE 1. Refractive index of ***garnets*** related to density.

indices, specific gravities, and lengths of the unit-cell edge of the common end members are given in Table 1.

Composition may be estimated from properties if data for at least three independent properties are available. Besides the properties already mentioned, magnetic susceptibility and colorometrically determined total iron and manganese have been used for this purpose.

Nomenclature

Because of the wide variation in composition, the nomenclature of ***garnet*** presents many problems. Most natural ***garnets*** belong to one of nine groups, according to their dominant components. These groups, with their ideal compositions and properties, are given in Table 1. The same names are used for the theoretical end members, with the addition of calderite for the manganese-iron end member, blythite for the manganese-manganese end member, skiagite for the iron-iron end member, and hanleite for the magnesium-chromium end member. Although there is extensive solid solution throughout the ***garnets,*** most tend to be either distinctly calcium rich or calcium poor. It is, therefore, convenient to use the name ***pyralspite*** (*pyrope,* *almandine, spessartine*) for the calcium-poor garnets and ***ugrandite*** (*uvarovite, grossular, andradite, goldmanite*) for the calcium-rich garnets. The name ***hydrogarnet*** is used hydrous ***ugrandite***. It includes *hydrogrossular, hibschite* (= plazolite) and *henritermierite*. The names melanite and schorlomite are used for titanium-rich ***ugrandite*** (usually *andradite*).

The composition of ***garnets*** is often expressed in terms of the pure end members. Since replacement can occur in at least three separate positions in the garnet structure, a large number of end members is needed and it has become common to calculate the end member composition in a way analogous to that used in the calculation of the norm in igneous petrology. This procedure is purely arbitrary, the end members having no "molecular" significance. There is, thus, no difference between a ***garnet*** containing equal parts of the iron-aluminum ***garnet*** end member and the calcium-iron ***garnet*** end member and a ***garnet*** containing equal parts of the calcium-aluminum ***garnet*** end member and the iron-iron ***garnet*** end member.

Occurrence

In the Earth's Crust. *Almandine* is commonly found in schists, gneisses, and amphibolites. *Pyrope* is typical of mafic and ultramafic igneous rocks and their derived serpentinites. *Spessartine* is found in low-grade schists, metamorphosed manganiferous sediments, rhyolites, and pegmatites. *Grossular* occurs in metamorphosed calcareous rocks while *andradite* is typical of contact metamorphic skarns. *Uvarovite* is mainly found in serpentinites and chromium-rich metamorphic rocks. Titanium-rich ***garnets*** are common in alkaline rocks such as nepheline syenite. *Kimzeyite* has been recorded from a carbonatite and *goldmanite* from a contact metamorphosed uranium-vanadium–bearing calcareous sandstone. Most ***garnets*** may be found as detrital minerals in sediments. (For the occurrence of ***garnets*** in metamorphic rocks see Vol. V, *Metamorphic Grade* and individual rock names.)

Under natural conditions, the composition of ***garnet*** is dependent on the chemical and physi-

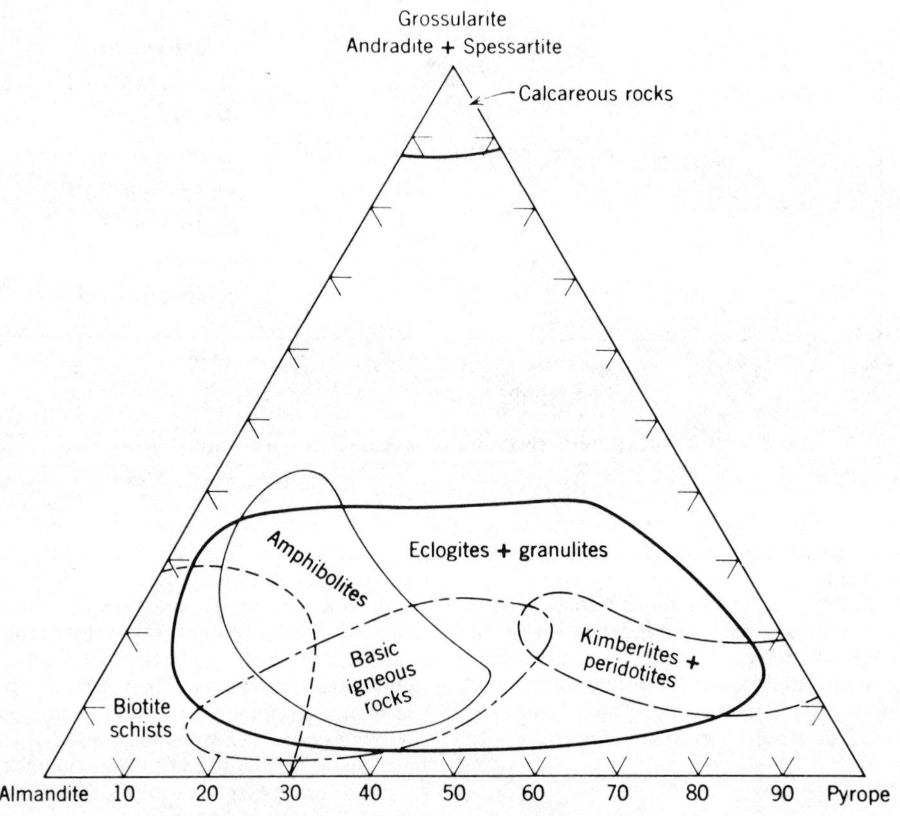

FIGURE 2. Petrographic affinities of ***garnets*** (from T. F. W. Barth, *Theoretical Petrology;* copyright © 1962, John Wiley & Sons, Inc.; reprinted by permission).

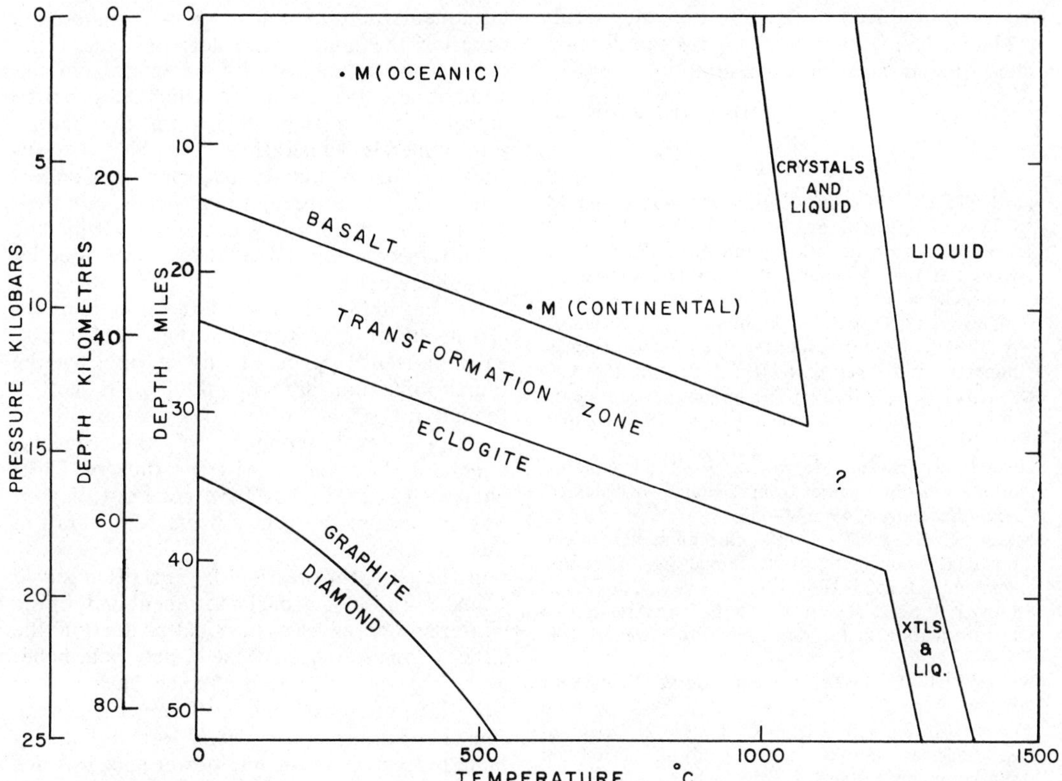

FIGURE 3. Pressure-temperature diagram for basalt/eclogite compared with carbon (*graphite/diamond*) (after Yoder and Tilley, 1962).

cal conditions of its growth. The gross variation with environment is indicated in Fig. 2. In sequences of regionally metamorphosed pelitic sediments, it is usual for the ***garnets*** formed at the lowest grade of metamorphism to be relatively rich in calcium or manganese with a systematic decrease in these constituents with increasing grade of metamorphism (see *Thermometry, Geologic; Rock-Forming Minerals*).

In the Mantle. Basalt reacts under high pressure to give eclogite. The reaction is of the following type:

$CaMgSi_2O_6$ + $NaAlSi_3O_8$ + $CaAl_2Si_2O_8$ +
diopside *labradorite*

Mg_2SiO_4 →
forsterite

$CaMgSi_2O_6$ + $NaAlSi_2O_6$ + $CaMg_2Al_2Si_3O_{12}$
omphacite **garnet**

+ SiO_2
quartz

The stability of basalt, and that of carbon for comparison, is given in Fig. 3. Regardless of the changes responsible for the Mohorovičić (M) discontinuity, the mantle material is probably garnet rich (see *Mantle Mineralogy*). Since primary basalt magma is believed to originate within the mantle, the petrology of these rocks is of great importance. Yoder and Tilley (1962) suggest that they are mainly garnet peridotites. Work on the possible equilibria in the lower crust and mantle has been summarized by Wyllie (1971).

This basalt→eclogite reaction results in a volume change of roughly 10% and is probably exothermic; and, along with postulated changes (e.g., ***olivine→spinel***) that occur at still greater depths, it is playing an increasing part in geotectonic thought. Kennedy (1959) has, for instance, pointed out that the volume change induced by pressure is capable of greatly extending the epeirogenic and orogenic readjustments previously explained by isostasy and that the temperature increase during this change may later reverse the reaction and cause emergence of a previously sinking trough.

It has also been suggested that deep-seated temperature changes, by their effect on this reaction, may cause epeirogenic uplift or depression. The higher pressure-phase changes have been suggested (and denied) as a cause of certain deep-focus earthquakes, and evidence has been produced to indicate that they would probably help rather than hinder the development of convection in the mantle.

MICHAEL J. FROST

References

Barth, T. W., 1962. *Theoretical Petrology*, 2nd ed. New York: Wiley, 416p.

*Deer, W. A.; Howie, R. A.; and Zussman, J., 1962. Rock-Forming Minerals, vol. 1. London: Longmans, Green, 77-112.

Gaudefroy, C.; Orliac, M.; Permingeat, F.; Parfenoff, A.; 1969. L'henritermierite, une nouvelle espèce minerale. *Bull. Soc. franc. Min. Crist.*, **92**, 185-190.

Kennedy, G. C., 1959. The origin of continents, mountain ranges and ocean basins, *Am. Scientist*, **47**, 491-504.

Moench, R. H., and Meyrowitz, R., 1964. Goldmanite, a vanadium garnet from Laguana, New Mexico, *Am. Mineralogist*, **49**, 644-655.

Nixon, P. H., 1968. A new chromium garnet end member, knorringite, from kimberlite, *Am. Mineralogist*, **53**, 1833-1840.

Smith, J. V., and Mason, B., 1970. Pyroxene–garnet transformation in Coorara meteorite, *Science*, **168**, 832-833.

Wyllie, P. J., 1971. *The Dynamic Earth*. New York: Wiley, 392p.

*Yoder, H. S., and Tilley, C. E., 1962. Origin of basalt magmas: an experimental study of natural and synthetic rock systems, *J. Petrology*, **3**, 342-532.

*Additional bibliographic references may be found in this work.

Cross-references: *Abrasive Materials; Blacksand Minerals; Density Measurements; Gemology; Mantle Mineralogy; Placer Deposits; Refractory Minerals; Synthetic Minerals*; also, see glossary.

GEMOLOGY

A gem or gemstone, in the strict sense, is a fashioned mineral whose beauty and durability are such that it is employed by man for personal adornment. In a broader sense it also includes fashioned minerals not durable enough to wear but beautiful objects for collectors. In view of this, gemology (Br. gemmology; Latin *gemma*—a gem) is often regarded merely as the aspect of mineralogy concerned with gem minerals. Such a definition is too limited, for, although gemology is primarily concerned with the composition, structure, properties, occurrence, origin, fashioning, and means of identification of gem minerals, it embraces much more than this. It deals with aspects of minerals of little or no interest to the mineralogist, and with many other materials, both natural and artificial, totally unrelated to the field of mineralogy. The gemologist must deal with cut or fashioned stones as well as "rough" material, a situation the mineralogist does not face. Gemology is also concerned with man-made gem substitutes, and the means by which they can be detected and distinguished from the natural gem minerals which they resemble. It deals with organic materials of plant and animal origin, such as amber, jet, and pearls, and their substitutes, none of which falls within the definition of a mineral, as that term is used by the mineralogist.

The beginnings of man's interest in gems are lost in the mists of time, but they were already an important aspect of his culture by the fourth millenium BC, some 6000 years ago. Although the science of gemology is usually regarded as a development of this century, mineralogists from at least the time of Theophrastus (372–287 BC) have gone out of their way to discuss the gem aspects of minerals in their treatises.

In the past, the wearing of gems often served a dual purpose—personal adornment and, in the wearer's mind at least, personal protection, the latter a consequence of the widely held belief in the supernatural power of gems. Such powers were attributed to gems at a very early date; and man has never completely lost his belief in their protective talismanic power and, in some cases, capacity for evil, even in sophisticated cultures, the United States being no exception. Opals (q.v.) are still viewed by many Americans with awe and fear. Gems have also, from the earliest times, been associated with authority. The assignment of certain types of jade (q.v.) to particular ranks in Imperial China, and especially the widespread and ancient practice of using engraved stones of the recessed type (intaglios and cylinder seals) to imprint wax in the sealing of documents and letters, are aspects of this.

The Development of Modern Gemology

The modern approach to the study of gem materials dates from the latter part of the 19th century but the intensive development of modern gemology is largely a product of the last fifty years. The published works of P. Groth (1887) and C. Doelter (1893) in Germany and George F. Kunz (1890 and later) in America was followed by the appearance of Max Bauer's *Edelsteinkunde* in 1896. This immediately became and remained the authoritative reference on gems. Revised by K. Schlossmacher in 1932, it is still a mine of gemological information. The English translation by L. J. Spencer (1904) has recently been reproduced (1969). A number of excellent books have appeared in English. Those of B. W. Anderson,

G. F. Herbert Smith, and Robert Webster in England are noteworthy. For many years the most widely used American work on gemology was that of E. H. Kraus and E. F. Holden, later Kraus and C. B. Slawson. Following the rise of the gemological movement in America, the publications of R. M. Shipley, R. T. Liddicoat, G. R. Crowningshield, and others connected with the Gemological Institute of America have added greatly to the technical literature, especially in the field of modern methods of gem identification. Capt. J. Sinkankas has written several notable books on gemstones and localities.

In the period prior to 1931, courses in gemology were already offered as part of the mineralogy program in several American colleges and universities, beginning with one given by G. M. Butler, at the Colorado School of Mines from 1909 to 1913. A. J. Moses established a course at Columbia University in 1913 and E. H. Kraus introduced a similar course at the University of Michigan in 1916. It remained, however, for Robert M. Shipley, Sr., to establish gemology as a recognized essential of the jewelry industry.

As early as 1908, a resolution was adopted at the annual conference of the National Association of Goldsmiths of Great Britain advocating the teaching of courses and the conduct of examinations in gemology. The first examinations were held in 1913. The scheme became dormant from 1915 to 1922; it was revived after World War I and has remained active ever since. Almost from the beginning the influence of G. F. Herbert Smith is evident. His designing of the portable pocket refractometer in 1906 and the publication of the first of the 12 editions of his book *Gemstones* in 1912 did much to encourage the study of gemology.

Robert M. Shipley, Sr., a former retail jeweler of Wichita, Kansas, on completing his gemological studies in London, returned to America in 1929; and, in 1931, he founded the Gemological Institute of America at Los Angeles. With the backing of a few farsighted jewelers and the encouragement and support of a number of professional mineralogists, Shipley, a man of remarkable perseverance, established the scientific study of gemology on a sound footing in America. From its initial offering of a single correspondence course designed for retail jewelers, the Institute has developed into a major organization, which, in addition to its correspondence courses, operates residence courses in Santa Monica, New York, and, at intervals, in other cities of the United States, Canada, and Japan. The Institute maintains elaborate gem-testing and diamond-grading laboratories at its headquarters in Santa Monica as well as in Los Angeles and in New York—the latter are known as the "G.I.A.—Gem Trade Laboratory Inc." Since its founding, more than 6000 students, including many foreigners, have completed Institute courses and its gem-testing and diamond-grading laboratories have become world renowned.

Shipley's concern was not limited to technical aspects of gemology. From his earlier experience as a practicing jeweler, he realized the need for improvement in the business ethics of the industry. To accomplish this, he founded the American Gem Society in 1934 to promote gemological education and to raise and maintain at a high level the ethical standards of the jewelry trade. As part of its program in carrying out these purposes, the American Gem Society grants the titles of "Registered Jeweler" and "Certified Gemologist." Within its membership of 1611, the American Gem Society counts many of the major firms and leaders in the manufacturing, supplier, and retail jewelry industry of the United States and Canada. Both the G.I.A. and the A.G.S. have achieved positions of international recognition in the jewelry trade.

Although it had existed with a less formal organizational setup as far back as 1908, the Gemmological Association of Great Britain assumed full status as a Branch of the National Association of Goldsmiths of Great Britain also in 1931, In more recent yeras, similar organizations have appeared in Australia, Canada, and several European countries.

Gem Minerals

True gems are all minerals but they constitute only one of the categories covered by the term "gem materials," which also includes cuttable but not wearable minerals and such naturally occurring nonminerals as the mineraloids, amber and jet, natural pearls, and a multitude of man-made gem substitutes. The latter include synthetics, imitations, assembled stones, cultured and imitation pearls, etc., as well as heat-treated, irradiated, coated, dyed, and otherwise altered stones (see *Synthetic Minerals*). Many of these are of recent invention, but glass imitations have been found along with some of the earliest known real gems. Man learned to fake gems almost as early as he developed an interest in the real ones. Pliny the Elder (23–79 AD) notes in his *Natural History* that "there is no fraud or deceit in the world which yields greater profit than that of counterfeiting gems."

If gemologists had to deal only with naturally occurring gem minerals, the problems of identification would be relatively simple, since only a small percentage of the known minerals have ever been employed for personal adornment. It is not possible to arrive at an exact figure for the total number of known minerals, because of differences of opinion among mineralogists as to what constitutes a valid species. However, the figure is undoubtedly in the vicinity of 2500 to which about 50 new ones are added each year. Occasionally a new gem mineral or variety is found, a spectacular recent addition being tanzanite, a highly colored purplish blue and strongly pleochroic *zoisite* from Tanzania, formerly Tanganyika. Another is Tsavorite, an emerald green variety of *grossular* from Tanzania and Kenya.

The nomenclature of gemology was firmly established long before man had any reliable means of identifying minerals. Names were often applied on the basis of color, equating color difference with mineral species. Later scientific methods of identification showed that, in some cases, two or more gems, named in this way, were color varieties of a single mineral species. Ruby and sapphire are now known to be merely color variations of the mineral *corundum*. Most gems bear either the mineral name—*diamond, topaz*, etc.—or a name applied to a color or structural variety of a single mineral—sapphire, ruby, emerald, onyx, carnelian, etc. No fewer than 25 varieties of the mineral *quartz* have received special gem names—rock crystal, amethyst, rose quartz, chalcedony, sard, sardonyx, carnelian, tiger's eye, etc. Jade (q.v.) is an exception among gems in that the name is applied to two entirely unrelated minerals—*jadeite*, a ***pyroxene*** and *nephrite*, an ***amphibole***.

No more than 100 of the more than 2000 recognized minerals are known to have ever been cut for gem purposes. Scarcely more than 25 of these have been widely used and fewer than 20 can be regarded as important. The more important gem minerals and some of the principal gems obtained from them are: *Diamond*—diamond; *corundum*—ruby, sapphire of many colors (also star rubies and star sapphires); *beryl*—emerald, aquamarine, golden beryl, morganite; *chrysoberyl*—chrysoberyl, alexandrite, cat's eye; *spinel*—various colors especially red spinel; *jadeite* and *nephrite*—jade; *topaz*—topaz; ***olivine***—sold as peridot; ***tourmaline***—rubellite, etc.; opal—precious opal, fire opal; *zircon*—zircon (red, brown, blue, and white; the last two are products of heating.); *zoisite*—tanzanite; *turquoise*—turquoise; *haüyne* and *lazurite*—lapis lazuli (lapis, actually a rock consisting largely of *haüyne* with *pyrite*, etc.); ***garnet*** group—*spessartine* (brownish red), *pyrope* (red, Bohemian garnet), rhodolite (light red purple), *almandine* (purple red, almandine), *andradite* (green, demantoid), *grossular* (hessonite, Tsavorite); ***feldspar*** group—*albite* and *orthoclase* (moonstone), *labradorite* (labradorite), *microcline* (Amazon stone, amazonite); and *quartz*—(a) coarsely crystalline *quartz*—rock crystal, amethyst, rose quartz, smoky quartz (cairngorm), citrine ("topaz quartz") and (b) cryptocrystalline (chalcedonic or microcrystalline *quartz*)—sard, carnelian, sardonyx, jasper, agate, moss agate, bloodstone, chrysoprase, etc.

Prerequisites of a Gem Mineral

The major prerequisites of a gem mineral are beauty, durability, rarity, and demand (vogue or fashion).

Beauty (Splendor). Beauty or splendor, based on color, brilliance, luster, transparency, and phenomena, is the primary requirement of any gem mineral.

Durability. Since a wearable gem must not only be beautiful but must remain so, properties that insure durability, such as hardness (resistance to abrasion) and tenacity or toughness (resistance to cleaving or fracturing), are essential. These two properties have no necessary relationship to each other, but the terms are often confused. *Diamond*, with its easy cleavage in four directions, is more brittle than many other gem minerals, although at the same time it is the hardest natural substance known (see *Mohs Scale of Hardness*). Both jade minerals, on the other hand, although ranking only 6.5 to 7 in hardness, possess far greater tenacity or toughness than any other gem (see *Jade*). To rank as a major gem, a mineral should have a hardness of 7 or greater, since the dust in the air, which is largely quartz (hardness 7), settles onto the surfaces of clothing and by constant abrasion destroys the polish of materials of inferior hardness.

Rarity. Rarity, an essential basis of demand, hence of value, reflects the universal human weakness to crave what others cannot have. Amethyst, the beautiful but plentiful purple variety of *quartz* is, in consequence, one of the less-desirable and relatively inexpensive stones. There is, however, a limit to the favorable influence of rarity, since interest wanes, regardless of beauty, if a mineral is so rare that comparatively few know if its existence. The striking blue mineral *benitoite* exemplifies this as does tanzanite the availability of which has been curtailed by political factors in Tanzania.

Demand (Vogue or Fashion). A fourth factor of importance is demand, also referred to as vogue or fashion. The demand for some gems has varied greatly over the years. The red ***garnet*** *pyrope* (Bohemian garnet), so popular in the last century, is today a gem of minor importance and value. *Turquoise* has risen and declined in popularity several times in this century. Some of the vagaries of demand can be accounted for, others not. There is evidence that opals brought higher prices in Roman times than now, and one of the reasons for this was Roman faith in the stone's talismanic nature, especially as a protection for men in battle. This is in strange contrast to the later and still current distrust of opal, dating back to the publication of Sir Walter Scott's *Anne of Geierstein* in the early part of the 19th century. In this novel, evil and tragedy are associated with an opal. Scott's works were so avidly read and widely translated that, by the middle of the 19th century, the market for opals is said to have virtually disappeared. There are many people today, even in America, for whom possession of an opal is unthinkable.

Perfection. Surface and internal imperfections, or inclusions, especially the position of internal flaws in relation to the surface, have a marked influence on the value of cut stones. Detection and the accurate location of such defects in a stone are important aspects of gemological examination. The influence of defects on value varies among the different gemstones. Imperfections that would not be tolerated in a *diamond* might cause little concern in an emerald, since the latter is recognized as usually imperfect. Some highly flawed emeralds of excellent color are often elaborately carved to conceal the numerous flaws, yet still bring high prices.

Size. It is almost axiomatic among minerals that the larger the crystal the less perfect it will be, hence small crystals of gem quality are likely to be more common than large ones. Diamond is an exception. Many of the largest stones graded by the GIA Gem Trade Laboratory, Inc. have been flawless or nearly so. Since, among the major precious stones (*diamonds,* corundums, emeralds), larger stones are avidly sought after and the demand far exceeds the supply, the price per carat of large specimens of a particular gem bears little relation to the price of small ones of the same quality. Since larger stones are avidly sought after and the demand far exceeds the supply, the price per carat of large specimens of a particular gem bears little relation to the price of small ones of the same quality. However, stones such as very large *quartz* gems and aquamarines are less costly per carat than more wearable sizes.

Fashioning. The term "fashioning," not to be confused with "fashion" in the sense of vogue, refers to the style and perfection of cutting and polishing to which a gem has been subjected. Few natural stones have ever been found that were of such quality, shape, and size that they could be mounted and worn in their natural state (Fig. 1). As in other things, styles in gem fashioning change over the years. There are also great differences in the skill of various cutters, with a consequent wide range in the quality of the product. The gemologist, in evaluating cut stones, is concerned not only with identification but with the stone's present worth, which involves, among other things, its perfection and the style and quality of the fashioning. This always raises the question of the need for recutting. This may be called for because of the poor quality of the present fashioning or because of an outmoded style. Since recutting always involves an appreciable loss of weight, a stone requiring it, suffers a loss in per-carat value, unless the recutting enhances, materially, some other factor.

Portability. A characteristic possessed by all gems to some degree and by the more valuable ones in particular is that they represent great

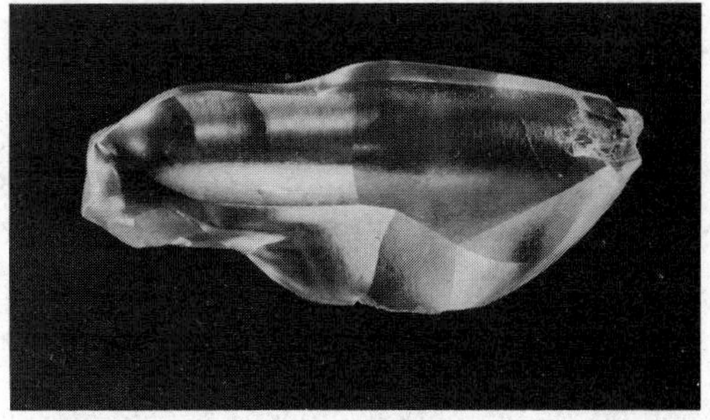

A

B

FIGURE 1. The Star of Arkansas, one of the largest *diamonds* found in America. A. The "rough," as found. B. The faceted gem. The "rough" is a good example of the unusual natural shapes in which *diamond* sometimes occurs. The apparent irregular complexity of the faceted stone is an illusion, due to multiple internal reflections. This splendid stone was found in 1955 at the only primary American *diamond* desposit, near Murfreesboro, Arkansas. The 15.31 carat flat, fish-shaped rough stone was cut into a beautiful 8.28 carat marquise, an unusually high recovery. Faceted gem is 3 cm long.

concentrations of value in a small object—the factor of portability. This is one of the major reasons for man's interest in gems, aside from his desire for personal adornment. It underlies the practice of purchasing gems (especially *diamonds*) as a form of investment; they are easily concealed and readily portable objects, the value of which is great and far more stable than many currencies. The accumulation of gems for this purpose is a practice of long standing in many parts of the world, although Americans until the 1970's tended to belittle its importance because of the long-continued stability of their own currency and political system. Their attention was focused on gem investment before and during World War II, when refugees often arrived with one resource—gems—which they had been able to conceal and escape with because of their small size relative to value—portability. Spurred by the oil crisis in the early 1970s and aided by GIA-GTL, Inc. quality reports on diamonds, Americans begun to buy them as a monetary hedge. Their liquidity has yet to be tested (1981).

The Optical Basis of Beauty in Gems

The ultimate basis of beauty in a gem is to be found in its chemical composition and crystal structure, but the immediate one is optical. Five optical properties contribute in a major way to the appearance of gems: color, brilliance, luster, transparency, and phenomena.

Color. Color is without doubt the most important of the optical properties (see *Color in Minerals*). With the exception of *diamond*, colorless varieties of gem minerals are of little value beyond the cost of cutting. *Diamond* owes its appeal, despite the absence of color, to its remarkable brilliance, in consequence of its high refractive index (2.43), and to its unusually strong dispersion (fire) (see *Refractive Index; Dispersion, Optical*). These place *diamond* in a class by itself among natural gemstones.

Selective Absorption. The basis of true color, the green of emerald and the red of ruby, is the selective absorption of white light, an optical property to be distinguished from the flashes of dispersive color of *diamond* and the somewhat similar-appearing interference colors exhibited by precious opal. Colored gem materials are either idiochromatic or allochromatic, mostly the latter. Idiochromatic materials occur in one color only and this is a fundamental property of the substance. *Malachite*, either when found in nature or synthesized, is always a deep green. In contrast, the majority of gem minerals such as *diamond* and *quartz*, as well as the bulk of gem substitutes, are normally colorless and said to be allochromatic. What color they possess is due to the chance presence of impurities or to defects in the crystal structure referred to as "color centers" (see *Defects in Crystals*). Since the number of chance factors is large, such minerals may exhibit a great range of colors. The color range in some allochromatic minerals may be so extreme that the relation of the several color types was at first unrecognized, each color variety being regarded as a distinct and unrelated mineral. The mineral *corundum* (Al_2O_3), colorless when pure, is deep red if traces of chromium are present. This is the gem ruby. The beautiful blue variety of this mineral, sapphire, owes its color to traces of iron and titanium. *Quartz* (SiO_2), another striking allochromatic material, provides gems of such widely different appearances as colorless rock crystal, purple amethyst, pink rose quartz, yellow-brown citrine ("topaz quartz"), and brown to black smoky quartz.

In some minerals, there is a difference in selective absorption (hence color) along different directions in the crystal. This optical peculiarity, known as "pleochroism," occurs only in minerals belonging to crystal systems other than the cubic (isometric). The pleochroism of ruby, sapphire, and ***tourmaline*** is so strong that it was recognized by gem cutters centuries ago and they paid careful attention to the orientation of stones cut from crystals of these minerals in order to secure the most pleasing colors in the finished gem.

Dispersion and Interference. Color of two types, other than that due to selective absorption, is observed in gem materials. The flashes of spectral colors (the jewelers' "fire") shown especially well by a properly cut *diamond*, are due to the dispersion of white light, the faceted stone acting as a complex prism. Dispersion is present in all nonopaque gems, but is apparent only in extreme cases such as *diamond* and *titanite* (sphene). Precious opal provides a somewhat similar effect (see *Opal*). The play of colors is due to interference. The most recent explanation is based on electron microscope data indicating that precious opal, unlike common opal, has an orderly arrangement of rows of submicroscopic silica spheres with a periodic spacing, a condition favoring interference. Unlike selective absorption, dispersion and interference effects are not true colors, the mineral in both cases being virtually colorless.

Brilliance. Following colors, the closely related optical effects brilliance and luster are of the greatest importance. Brilliance is a measure of the light thrown back to the eye by a gem, compared with the amount of incident light falling on it. Luster, a less exact term, refers more to the quality than the quantity of the light thrown back. The maximum potential brilliance of any transparent gem is determined by the size of the critical angle, which in turn bears an inverse relationship to the refractive index of the stone. The critical angle, which in three dimensions can be visualized as a cone whose half angle is the critical angle, determines how much of the incident light entering a cut gem will be lost by "light leakage" at the back facets. The higher the index of refraction, the smaller the critical angle cone; the smaller the cone, the less light leakage, since only light rays that strike a back facet, within the critical angle cone, will be "leaked out." The less light lost in this way, the more remains to emerge from the crown of the stone to be thrown back to the eye of the observer, which is the measure of brilliance. This explains why a piece of glass or *quartz* can never equal the brilliance of *diamond*, regardless of the proportioning of the stone or the perfection of its polish. The index of refraction of *diamond* (2.43), more than half again that of *quartz* (1.55), provides it with an exceptionally small critical angle (24°), only a little more than half as great as that of *quartz* (41°), with a consequent reduction in light leakage (Fig. 2).

Luster. Luster refers more to the quality than to the quantity of reflected light, although some terms applied to luster, such as adamantine (diamond-like) and vitreous (glassy), actually are descriptive of brilliance as well. The luster described as "pearly" is exhibited by materials possessing a microscopically fine platy structure in which the plates more or less parallel the surface. A natural pearl consists of a multitude of essentially but not perfectly concentric spheroidal, onion-like shells from the center outward, which accounts for its fine pearly luster. Such a structure results in diffraction of light as well as simple reflection. Similarly, materials with a fibrous structure, especially those in which the fibers are parallel, exhibit a silky luster, for example, the satin spar variety of *gypsum*. This effect is related to the phenomena of chatoyancy and asterism.

Transparency. In the gem trade, today, there is a premium on transparency; gems such as *diamond* and ruby are held in highest esteem. Four transparent stones, *diamond*, ruby, sapphire, and emerald, are often singled out and described as "precious," all others being regarded as "semiprecious." Application of the term "semiprecious" to such superb gems as alexandrite, fine jade, and black opal, which often sell at prices comparable with those obtained for the four "precious" stones, is inconsistent and there is a growing tendency to abandon these terms and use *gemstone* for all.

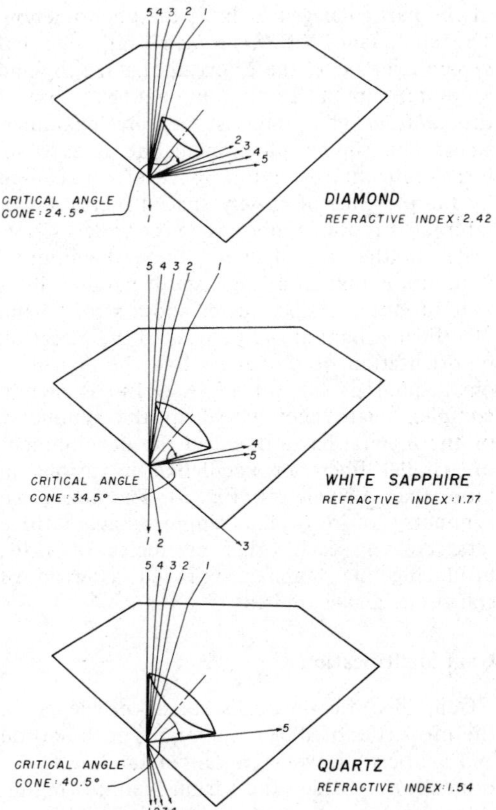

FIGURE 2. Diagrammatic cross sections of three different brilliant-cut stones illustrating the inverse relation between refractive index and critical angle, as well as the direct relation between the critical-angle cone (the half angle of which is the critical angle) and light loss.

Until man learned to facet stones, the effective beauty of transparent gems could not be fully appreciated. For this reason in ancient times, prior to the development of faceting, transparent gems did not enjoy the popularity they do now, and the translucent to opaque ones, such as lapis lazuli and *turquoise*, and the numerous color and structural varieties of chalcedonic *quartz*, such as carnelian, sard, sardonyx, etc., were highly regarded. There has since been a virtual reversal and today such opaque and translucent materials are considered inferior, or at least relatively inexpensive.

Phenomena. Unusual optical effects are exhibited by some specimens of certain gem materials. The term "phenomena" is applied to these effects and the stones themselves are said to be "phenomenal." There are many types, and in no case is the effect a fundamental optical property, since only occasional specimens of the particular gem exhibit the phenomenon. The chatoyancy of *chrysoberyl* cat's eye and *quartz* tiger's eye, the asterism of star ruby and sapphire, and the shimmering adularescense of the ***feldspar*** gem, moonstone, are examples. These and similar phenomena are due to internal reflection or interference effects caused by the presence of closely spaced inclusions or internal structures present on a microscopic scale. In the case of chatoyancy, the internal structure is that of a single set of parallel fibers that produces a silky luster and a strong band of reflected light at right angles to the direction of orientation of the fibers like the high light on a spool of silk thread. Asterism is merely complex chatoyancy in which the symmetry of the mineral has influenced the development of parallel fibers or needlelike inclusions in more than a single set (Fig. 3). The hexagonal symmetry of ruby and sapphire favors three sets, crossing each other at angles of 120°, producing the familiar six-rayed asterism of star rubies and sapphires.

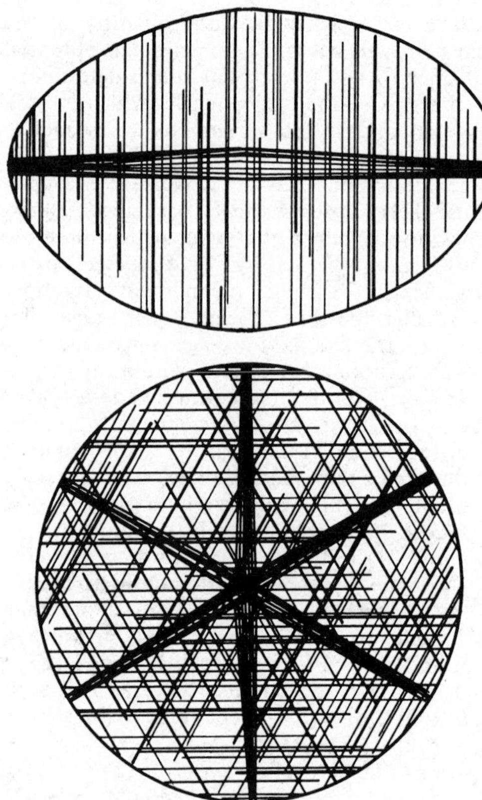

FIGURE 3. Upper diagram shows orientation of a reflected narrow band of light (chatoyancy) in relation to the single set of fibrous or needlelike inclusions or structures present in such stones as *chrysoberyl,* cat's eye and tiger's eye. In lower diagram this is compared with the effect produced by the triple set of needlelike inclusions present in star stones such as star ruby and star sapphire. The asterism of six-rayed star stones is due to the intersecting reflections from three sets of elongated inclusions oriented at angles of 120°, a sort of triple chatoyancy, the orientation of which is determined by the hexagonal symmetry of the mineral.

Gem Identification

Gem identification calls for a knowledge of the properties of all gem materials, both natural and artificial, as well as an understanding of the principles on which the various gem-testing instruments operate, their use, and their limitations. The gemologist must be prepared to deal with both rough and cut stones, the former being handled as one would other minerals except for the exercise of suitable concern for the value of the material. Cut stones, however, make up the bulk of the items a gemologist is called upon to identify.

The identification of cut gems differs from identification of ordinary minerals or gem roughs in two ways. In the first place, the value of the material and the risk of damage limit the determinative methods to those that are essentially nondestructive. This restriction increases the difficulty of identification. It is further increased if the stone is cut "en cabochon" instead of being faceted or carved with a matte finish, since this makes it impossible to secure precise refractive-index data on a refractometer, one of the keystones of gem identification. Identification is also made difficult when the stone is mounted, as it often is, since specific gravity determinations are then impossible. In addition, unlike the mineralogist, the gemologist is always confronted with the possibility that the material may be artificial and, even if he establishes that it is natural, it may still have been altered or tampered with in any of a large number of ways. It is clear that a knowledge of the properties of the numerous gem substitutes and the many ways in which natural material can be altered and "upgraded" is as essential as a knowledge of the characteristics of gem minerals themselves.

The identification of a gem involves

1. differentiation between various gem minerals of similar appearance;
2. differentiation between natural gem minerals and man-made gem substitutes of all kinds; and
3. detection, in the case of natural gem materials, of any of the known methods of alteration or tampering.

Gem Testing Instruments

The identification of gems and their substitutes depends, in part on the determination of fundamental optical and physical properties by means of special instruments, and in part on the recognition of inclusions and internal structure that can be observed only under considerable magnification. Microscopic observations are especially valuable in differentiating synthetic and natural stones.

The most frequently used instrument is the refractometer, by means of which indices of refraction are determined by measuring the critical angle (angle of total reflection) (Fig. 4). The refractive index (R.I. or *n*) is the most important single optical property of a gem material. The flat, polished surfaces of faceted stones are ideal for refractometer determinations, whereas precise measurements cannot be made on the curved surfaces of cabochons. Conventional gem refractometers are limited to readings of 1.80 and below. The introduction of recent diamond substitutes has led to the development of instruments to measure surface reflectivity and thus indicate identity of stones with R.I. above 1.80.

The polariscope, a device permitting observation of gems in polarized light, consists of two polaroid plates between which the stone is placed. It enables one to determine whether a stone is singly or doubly refracting, although this can also be done for most stones of R.I. 1.80 and below on the refractometer. One of its principal uses is the detection of such optical features as anomalous double refraction, although this can also be done on the refractometer. One of the principal uses of the polariscope is the detection of such optical features as anomalous double refraction, a characteristic of synthetic spinel, one of the most widely used gem substitutes.

Interference figures can also be observed in a polariscope by adding a low-power lens to its optical system. Cabochon stones act as lenses themselves. Interference figures establish the optic character of the stone and make it possible to assign it to one of five major crystallographic categories. This is a powerful aid in identification, since it limits the possible crystal system of an unknown. Amorphous materials and isometric crystals do not produce interference figures. Uniaxial figures, either positive or negative, are formed only by tetragonal or hexagonal crystals, whereas orthorhombic, monoclinic, and triclinic crystals produce either positive or negative biaxial figures (see *Minerals, Uniaxial and Biaxial*).

The dichroscope, essentially a device for detecting pleochroism (dichroism and trichroism), consists of a prism of *calcite* mounted in a tube with a small port or window at one end. If a transparent mineral is placed between the instrument and a light source, the doubly refracting *calcite* separates the pleochroic colors so that one observes a double image of the window, in which the different colors of even a weakly pleochroic material are plainly visible and can be compared side by side. In the case of nonpleochroic materials, both images will be the same color.

Both petrographic polarizing microscopes (see *Polarization and Polarizing Microscope*) and binocular stereoscopic microscopes are essential in a gem-testing laboratory, the latter being the more generally useful for work with cut stones. The Gemolite is a specially designed dark-field binocular microscope produced by the Gemological Institute of America. It differs from an ordinary binocular chiefly in the excellence of its lighting arrangements. Lighting is especially important in observing faceted gems, the highly reflective surfaces of which tend to obscure the interiors of cut stones.

Specific gravity, the most important of the physical properties, may be determined precisely by means of a specific gravity balance or approximately, which is often sufficient, by employing the sink-float technique with liquids of known density (see *Density Measurements*).

Since different materials of apparently the same or closely similar colors often provide distinctive and characteristic absorption spectra, a hand spectroscope attached to a microscope—or mounted independently with a light source—is an extremely valuable tool, especially in the detection of artificially irradiated colored *diamonds*.

Fluorescence under ultraviolet light is also highly diagnostic of some gem materials; it is especially useful in the detection of synthetic emerald and natural colored black pearls (see *Luminescence*).

Pearls pose a special problem, first solved by A. E. Alexander in the early 1930s. His technique is a modification of that used by the medical profession, namely, roentgenography, otherwise known as X radiography or X-ray shadow photography. It has proven to be the only really satisfactory method of distinguishing, with certainty, between natural and the best quality cultured pearls.

Although of secondary importance in pearl testing, X-ray diffraction (q.v., vol. IVA), especially the powder methods, are as important in gemology as in mineralogy in the identification of rough gem material. X-ray diffraction methods can also be used to identify cut stones when other methods fail and the stone is of sufficient size and interest to justify the sacrifice of a small amount of material and the cost of repolishing or recutting.

An X-ray diffraction apparatus, designed by R. J. Holmes and G. S. Switzer in 1947, pro-

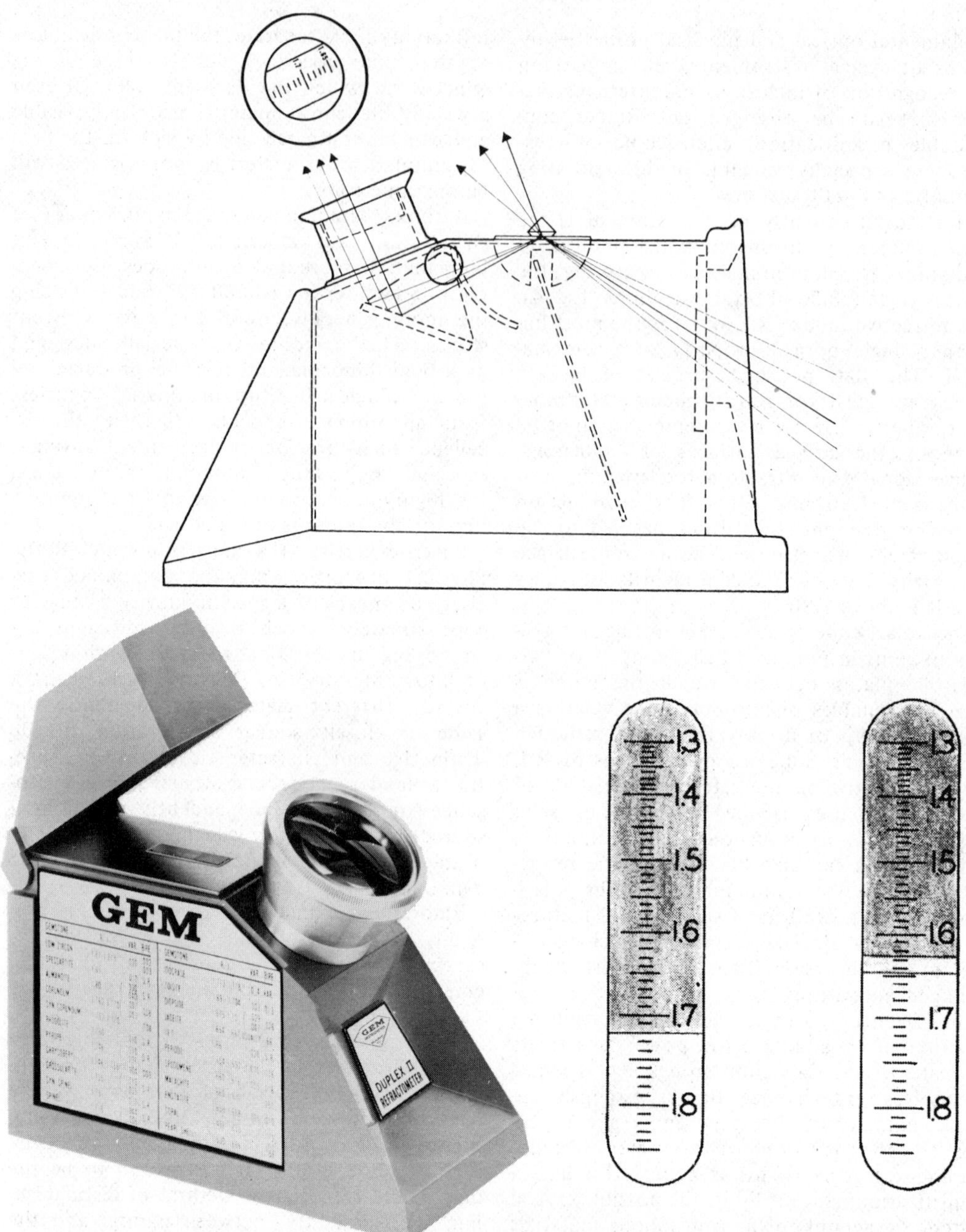

FIGURE 4. A modern gem refractometer (total reflectometer) used in gem identification (lower left). The diagrammatic cross section (above) shows the optical system of the instrument. Two examples of the effect observed in the eyepiece scale are shown, that on the left for a singly refracting material (*spinel*) with a refractive index (n = 1.72) compared with the double reading obtained from ***tourmaline***, a strongly doubly refracting stone with two indices, 1.625 and 1.648. (Reprinted with permission of the Gemological Institute of America.)

vides a means of securing fairly satisfactory powder-type patterns from single-crystal gems without damaging the stone. It is essentially a Laue-type camera in which a motor driven goniometer head has a partial universal motion. As gemology has become more complex, research instruments such as spectrophotometers and scanning electron microscopes with energy dispersive units and cryogenic environments are in regular use.

Gem Materials and Problems of Identification

Natural Gem Minerals. Differentiating between gem minerals of the same color usually presents no serious difficulty. The number of possibilities is limited and the fundamental physical and optical properties of the minerals involved are usually so different that, in spite of similarity in color, routine tests to determine basic optical and physical properties readily establish identity.

Nonmineral Natural-Gem Materials. Two types of materials are covered by the term "nonmineral natural-gem materials":

1. organic substances, such as amber and jet, found within the crust of the earth but not regarded as minerals by mineralogists and usually assigned by them to the category of "mineraloids" (see *Resin and Amber*); and
2. the products of living organisms—for example, natural pearls and precious coral—which are in no sense mineral.

Neither amber nor jet is in much demand at the present. Amber, however, has an ancient and honorable history, being one of the earliest as well as one of the most widely and continuously used of all gems. It can be distinguished from its imitations, usually plastics, by its extremely low specific gravity (it floats in sea water) and from other gums by the latter's reaction to organic solvents. Natural pearls rank with the finest gems in desirability and costliness. However, the once important natural pearl industry has suffered severely in this century from the inroads of Japanese pearl culture. The cultured pearl, a "semiartificial" object, appears to have been first produced in a spherical form, using deliberately planned experiments, by Tokichi Nishikawa, son-in-law of Kokichi Mikimoto. A specimen was exhibited to the Emperor at the graduation exercises at Tokyo University in July 1909. Tatsuhei Mise, without using scientific methods, had also produced a spherical pearl as early as 1904. Although Mikimoto, who is regarded as the father of the Japanese culture-pearl industry, had obtained a patent in 1896, it was for a method of producing blister pearls grown on the shell, not spherical ones grown in a pearl sac. He did not apply for a patent for spherical pearls until 1914. Mikimoto was chiefly responsible for the spectacular success of the Japanese culture-pearl industry. The problem of differentiating between natural and culture pearls calls for special apparatus and techniques, largely X-ray.

Imitations. Imitations are man-made products that superficially resemble natural gemstones but differ from them in chemical composition, internal structure, or both. Numerous materials, mostly amorphous rather than crystalline, such as glass (paste), plastics, and ceramics, have been used; glass has been the most common. Imitations usually present little difficulty in identification, since the difference in chemical composition and structure insure that their basic optical and physical properties will be distinct from those of the natural gems they imitate. Glass imitations rank among the oldest of all gem substitutes, glass beads being known from Mesopotamian sites that date back to at least 3000 BC.

Imitation pearls are solid or wax-filled glass beads coated with "essence d'orient," an aqueous solution of "guanine," which is a colorless organic substance ($C_5H_5N_5O$) consisting of microscopic rhombic plates and derived from the scales of certain fish. Imitation pearls, despite the remarkable similarity in appearance, are easily distinguished from both natural and cultured pearls, by their smooth surface, in contrast to the rough "sandy" feel of natural and cultured pearls, when touched to the teeth.

Reconstructed Gems. The only reconstructed gems worthy of note are "reconstructed ruby" and pressed amber. The former has often been described as having been produced in the 1880s by sintering or incipient fusion of tiny fragments of natural ruby. Since the particles were never really molten, the product would be a sintered aggregate of randomly interlocked fragments. It has been successfully established by K. Nassau and R. Crowningshield that material of this type was not produced by the methods claimed and that the material so described represents early attempts at synthesis. "Reconstructed" has also been incorrectly used as a synonym for synthetic corundum, synthetic spinel, and even for glass imitations. These improper usages are deliberate attempts to upgrade synthetics and imitations. Pressed amber is still produced by subjecting small fragments to controlled heat and pressure in the presence of a solvent.

Synthetics. A synthetic is an artificial gem substitute that has, not only a superficial resemblance to, but essentially the same chemical composition and crystal structure as the gem mineral it represents (Fig. 5). Since chemical composition and crystal structure determine the fundamental physical and optical properties of any crystalline substance, it follows that a natural gem and its synthetic equivalent have essentially identical properties, hence the determination of basic properties usually does not provide a means of differentiating the two. Synthetics, therefore, provide the gemologist with the most frequent and most difficult identification problems. The essential identity in basic properties of a synthetic and its equivalent natural gem renders many of the more routine gem-testing methods of little value for differentiating between them. In fact, it would

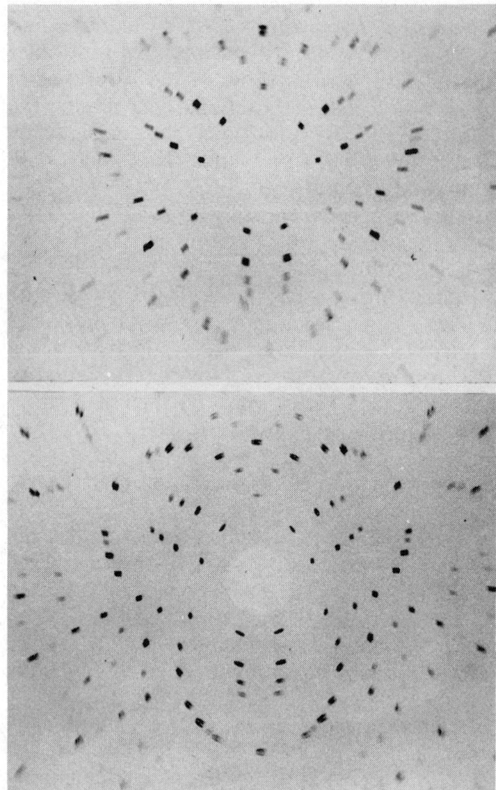

FIGURE 5. Laue X-ray diffraction patterns of synthetic ruby (above) and natural ruby (below), showing essential similarity of position and relative intensity of the diffraction spots.

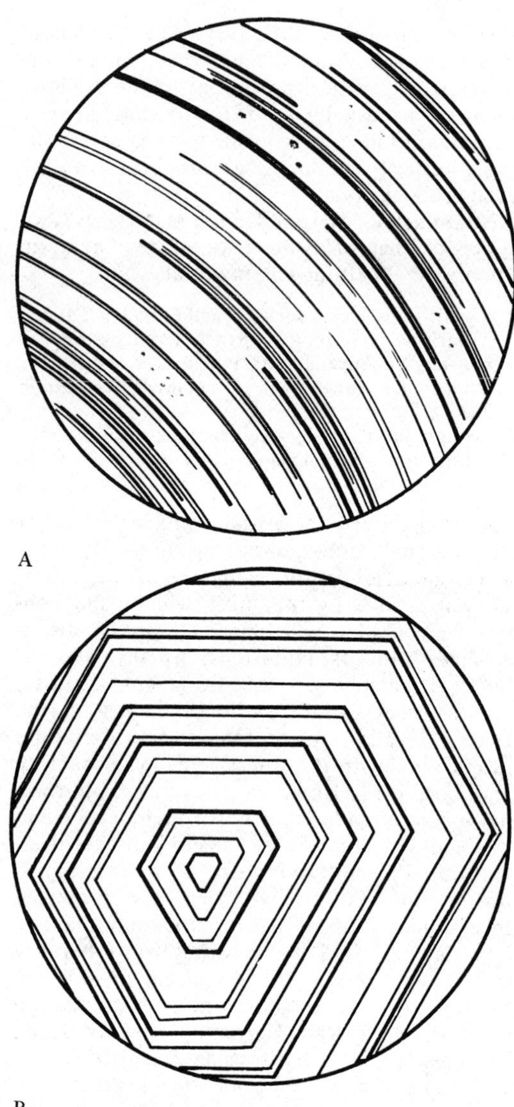

FIGURE 6. Zonal color distribution in (A) synthetic sapphire and (B) natural sapphire. Curved color bands and occasional round bubbles are characteristic of synthetic *corundum* produced by the Verneuil flame fusion process, since the boules formed by this process are somewhat rounded. Color variation in natural *corundum* (sapphire) usually exhibits a distinct hexagonal pattern in which the bands are straight or angular, not curved.

appear at first glance that it would be impossible to detect a synthetic. Fortunately, the methods employed in the production of synthetics involve conditions of growth markedly different from those under which the equivalent natural mineral crystallizes. In consequence, synthetics usually exhibit certain gross features—largely in the form of inclusions, internal structures, color zoning, etc.—that differ from those observed in natural gems (Fig. 6). These features enable gemologists to detect synthetics. Such determinations, however, are often difficult, call for careful microscopic examination under ideal conditions, and may require additional special techniques. Extensive experience in dealing with synthetics is also essential, since many of the critical differences are not obvious or even apparent to the inexperienced. The growing number and increasing perfection of synthetics render their detection even more difficult. The problem of detecting synthetics is especially difficult in the case of those that are flawless. The discovery by W. Plato that synthetic *corundum*, produced by the flame fusion method, exhibits a strain structure consisting of two sets of lines intersecting at angles of 60° makes it possible to detect such synthetics even though they are free of all inclusions, color bands, striae, etc. This structure can be seen if the stone is observed in the optic-axis direction in polarized light when immersed in a liquid of essentially the same refractive index as the stone. The chevron-like strain structure does not appear in natural *corundum*.

Beginning with the synthetic ruby and colorless white sapphire of Verneuil at the turn of the century, the list of synthetics has slowly grown. Many gem minerals, including *diamond*, have been synthesized, but not in sizes or of a quality suitable for gem purposes. The following synthetics suitable for gem use have been produced up to the present:

- *corundum*–ruby, and sapphire of many colors, including star rubies and star sapphires. Flux-grown synthetic ruby in recent years has been a great source of potential fraud and misuse as the gem investment craze has grown. Jewelers trained to recognize Verneuil synthetics by their curved growth lines have misidentified flux corundums with their natural appearing inclusions.
- *spinel*–many colors resembling those of various gems; often sold improperly under the name of the gem mineral whose color is imitated, as "synthetic aquamarine" or "synthetic zircon"
- emerald–now produced by various methods and available from several sources. The original German synthetic emerald (Igmerald) of the 1930s was apparently never commercially marketed. Chatham's synthetic emeralds are sold under the name "Chatham-created" emeralds. Synthetic emeralds were also produced by Linde and are currently produced in France by Gilson, and in Japan by the Kyocera Company.
- *rutile*–a true synthetic, but does not resemble natural *rutile* in color. Formerly it sold on the basis of its resemblance to *diamond* under such names as Titania, Kenya Stone, and many others, until better substitutes appeared.
- *quartz*–both synthetic citrine and amethyst have been commercially available since 1970. If flawless, they may be unidentifiable by the practicing gemologist.
- *chrysoberyl*–synthetic alexandrite appeared on the market in the 1970s, manufactured in California and Japan.

Some synthetic crystalline gem materials are not the equivalent of any known mineral. Strontium titanate sold under the name of Fabulite is an example. Its success, like that of synthetic *rutile*, is based largely on its superficial resemblance to *diamond* in both brilliance and fire. Unfortunately its hardness, which is between 5 and 6, makes it an unsatisfactory gem material. Another synthetic material that has no natural counterpart is an yttrium compound ($Y_3Al_5O_{12}$) with a garnet structure–an yttrium garnet known in the trade as "YAG." This is a colorless transparent material with a hardness of approximately 8 and a brilliance and dispersion that render it a passable imitation of *diamond*. Since 1975 synthetic cubic zirconia known in nature but unnamed to date, has outrivalled all other *diamond* imitations and has led to the introduction of thermal conductivity meters. *Diamond*'s uniquely high conductivity distinguishes it from all transparent substitutes.

The "Synthetic Emerald," produced by Johann Lechleitner of Austria and formerly distributed by Linde, represents a type of gem substitute that challenges classification. It consists of completely shaped and faceted pieces of colorless or very pale common *beryl* or aquamarine that have been given a uniform overgrowth of synthetic emerald by means of a hydrothermal process. The surfaces are then polished. The bulk of the stone is a natural mineral, the surface is synthetic!

Assembled or Composite Stones. Assembled stones consist of two or more shaped pieces cemented or bonded together (Fig. 7). There are many types and they were formerly produced in great quantities but, except for opal doublets, they have been largely replaced by synthetics. Some types are of great antiquity. Pliny states that sardonyx doublets, consisting of two pieces of onyx of different colors and cemented together, were well known in Roman times.

Doublets have been produced both to improve cheap glass imitations so that they would stand up better under severe wearing conditions and deliberately to deceive the buyer. In the first group are the well known "garnet-top" doublets, by far the most common of all assembled stones and still encountered although no longer manufactured. These consist of a thin slab of a red *garnet* "fused" to a colored glass

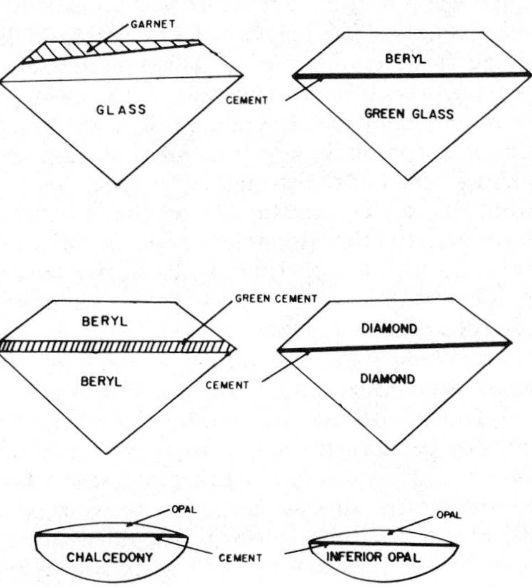

FIGURE 7. Several types of assembled or composite gem stones (see text).

base that makes up the bulk of the stone. The ***garnet***, which is much harder and less brittle than glass, covers the crown of the stone including the edges of the table facet, the most vulnerable part of a faceted stone. Strangely, when viewed from above, the red color of the ***garnet*** is not visible, being completely masked by the colored glass base. These composites are easily detected since the difference in luster of the ***garnet*** and the glass renders the boundary between the two parts of the stone clearly visible, even under a hand lens. They were made in many colors to resemble all the major gems, and so carelessly that it is clear they were not designed to deceive.

In the category of deceptive assembled stones, no gem was more widely imitated than emerald. "Emerald" doublets and triplets of several types, in addition to the usual garnet-tops, were used. These were carefully made in two pieces with flat surfaces joined at the girdle plane. Most often they consist of two pieces of *quartz* or two pieces of colorless synthetic *spinel*, joined with a green cement. These present no particular difficulty in identification. A more dangerous type consists of two pieces of colorless *beryl* joined at the girdle with green cement. In this case, tests on either the crown or pavilion would check out for emerald, since that gem is merely a color variety of *beryl* and has the same basic properties. The difficulty is compounded by use of gypsy-type mount, in which the girdle is completely hidden and, therefore, the junction between the pieces is concealed. It should be noted that even a thin green film at the girdle plane gives the optical illusion that the entire stone is uniformly green when viewed from above. If you look across such doublets, parallel with the table, especially if the stone is immersed in a liquid, the colorless portions become apparent. Also, most fail to appear red under an emerald filter. Although this observation is not conclusive since a few natural emeralds also fail to appear red, it would be a clue that further testing is needed.

At present, the assembled stones most often encountered are opal doublets or triplets in which very thin slabs of fine opal are cemented to thicker pieces of either chalcedony or poorer-grade opal or ironstone matrix, the latter being especially dangerous. The existence of quantities of fine opal in veinlets too thin for use as a gem has encouraged this practice. Many have a third portion consisting of a cabochon of rock crystal or glass on top of the thin opal. Because of the prevalence of assembled stones, one should always view with suspicion gypsy mountings and, even more so, gems mounted with a concealed back that renders impossible the testing of the pavilion of the stone. Fossilized ammonite polished and assembled with a top of quartz or glass resembles fine black opal.

The foil back, also a type of assembled stone, exists in many types and serves several purposes. A colored metallic foil, either attached to the pavilion facets or placed in a covered mounting behind a pale stone of similar color, deceptively enhances the color of the stone. The silvery metallic foil coating the pavilion facets of rhinestones acts as a mirror, preventing light leakage and falsifies their brilliance, giving these pieces of glass an undeserved diamondlike appearance. The use of foil backs, bearing a starlike design, can give the illusion of asterism where none exists. Blue foil, or even blue paint on the back, can give color to a pale star sapphire or lead one to think that a piece of rose *quartz* showing weak asterism is a star sapphire. Any pale star stones that have been so enhanced can be readily recognized by looking across the stone, parallel to the base. In this direction they have a colorless or very pale and transparent appearance.

Cultured Pearls. Cultured pearls are unlike any other gem substitutes. They are produced by placing round mother of pearl beads, as a large seed, within the body of the living pearl "oyster"—which is not a true oyster, but a mollusc of the genus *pinctada* (*meleagrina*). The animal coats the "seed" with a thin veneer of pearly nacre identical in composition and structure with that in a natural pearl. This nacreous coating becomes sufficiently thick in about three years to provide a fine pearly luster and the cultured pearl is then removed.

The distinction between natural and cultured pearls is not difficult, even if the pearls have not been drilled for stringing. In any case, X-ray methods are the only really reliable means of making the distinction. X radiography or shadow photography is the most useful of the X-ray techniques, especially since a large number of pearls can be examined at one time. Since the mother of pearl bead, which has an entirely different internal structure from the concentric shell arrangement of a natural pearl, makes up all but a thin veneer of the cultured pearl, the X radiographs of the two types are different, providing striking visual evidence of the type of pearl present (Fig. 8).

Altered Natural Gemstone. The number of ways in which natural gem materials can be altered are legion and not all can be detected by available methods. Such tampering has many purposes, which include improving, removing, or changing the color of a stone; concealing imperfections; strengthening the material; improving its general appearance; or making larger more usable-sized pieces out of smaller ones.

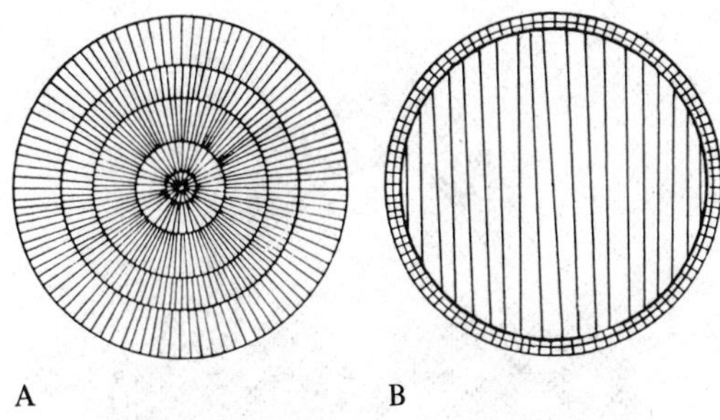

FIGURE 8. Cross sections of (A) natural and (B) cultured pearls. Only a thin outer veneer of pearly nacre with the characteristic radiating needlelike development of *aragonite* crystals is present in the latter, whereas such a radiating needlelike structure is found throughout a natural pearl. The structure of the mother of pearl bead used as a seed in the production of cultured pearls has a subparallel layered structure.

The methods involve the application of heat or heat and pressure plus "powder baths" to diffuse color into pale sapphires; subjecting the stone to radiation; or impregnating with wax, oil, dyes, etc. The following are some of the more frequently encountered types.

Dyeing or staining has been applied mostly to porous materials, especially the cryptocrystalline forms of *quartz*, but *turquoise* and even jade have been treated in this way. The dyeing of chalcedonic *quartz*, which is highly porous, was practiced by the Romans and has been a major activity at the famous gem center of Idar-Oberstein in West Germany since 1819. Unlike crystal-aggregate materials, single crystal gems can be stained only if the crystal is badly fractured. Badly cracked emeralds are frequently soaked in oil to conceal the cracks, and dyes are sometimes added to this oil to improve the color.

Color changes in single-crystal gems have usually been achieved by heating or formerly by cyclotron treatment and gamma radiation. Yellow citrine *quartz* or "topaz *quartz*" has been produced from poor-quality amethyst or smoky *quartz* by heating. The beautiful blue *zircons* as well as the colorless *zircons* are produced from the normal red-brown stones by intense heating. The same method is used in the "pinking" of *topaz*. The striking, recently discovered gem, tanzanite, a purplish blue *zoisite* from Tanzania, is much improved by heating—which gives it a more pleasing deep blue color. Most specimens of Brazilian aquamarine are believed to have had their color improved by heating, but it is not possible to prove that a particular stone has been so treated.

With the virtual loss of Burma as a source of fine rubies and sapphires, the gem market has gradually realized that many blue sapphires available since about 1975 were originally cloudy gray Sri Lankan rough stones. Heated in crucibles in Bangkok, their impurities are rearranged to give the stones an attractive blue color and in the process the stones become transparent. Not all such heated stones can be identified and the trade has come to accept them without demerit much as it has heat-treated aquamarines.

More insidious and potentially fraudulent is a heating process whereby faceted clear sapphires lacking the necessary impurities are heated in a powder bath of the requisite elements—aluminum, iron, and titanium to produce a skin-deep coloration. Other colors than blue are similarly produced using different powder in accordance with a 1975 United States patent (U.S. Patent 3,897,529 assigned to Union Carbide Corporation).

Moderate heat and pressure, in the presence of a solvent, has long been used at the amber centers of East Prussia to produce blocks of pressed amber from small pieces—producing the so-called reconstructed amber.

Porous stones such as *turquoise* are often impregnated with wax or liquid plastic to make friable "chalky" specimens more compact so that they will take a better polish, resulting in an improved appearance or even a better color. To accomplish the latter, dyes may be added.

In recent years large numbers of *diamonds* and other gems have been subjected to bombardment by high-speed particles, first in cyclotrons and similar accelerators, later in an atomic pile and with various radioactive isotopes such as americium 241 oxide. Interesting, apparently permanent, color changes have resulted from this treatment. Yellowish or brownish off-color *diamonds* can be enhanced in this way by converting them into definite colored "fancies," enhancing their beauty. The natural appearing colors produced are many tones of yellow, brown, orange-brown, blue-green, and black. Long before the cyclotron was invented, it was found that *diamonds* turned green by long immersion in a solution of radium bromide. Unfortunately, stones so treated are always highly radioactive and are potentially dangerous to wear. Cyclotron-, atomic pile-, and isotope-colored *diamonds* can be detected by characteristic lines in their spectra, while the more common yellow and brown treated diamonds yield their identity to the hand spectroscope or recording spectrophotometer.

Since even slight deviations from white toward yellow or brown in the color of *diamonds* materially affect the per-carat price, various fraudulent devices have been used to conceal the off-color nature of such stones.

One of these is to coat the pavilion, in whole or in part, with a slightly bluish plastic, enamel, or even indelible ink to offset the yellowish tinge of an inferior stone. Some treatments of this type have proved difficult to detect.

Many gemstones have been altered in color in a gamma cell. Perhaps the most commercially successful are the permanent blue given to certain colorless *topaz* and the gray color imparted to off-color cultured pearls.

Gemstone Fashioning (Cutting and Polishing)

A major concern of the gemologist in the examination of cut stones is the recognition and evaluation of the type and quality of fashioning. This requires a knowledge of the purposes of fashioning and the principles involved in selection of the cutting style best suited to a particular material.

Man's first gems were undoubtedly colorful stream pebbles that he used as found. Although he soon learned that they were improved by polishing, he made no effort for a long time to change the original shape or size (Fig. 9). The Imperial Crown of Charlemagne (800 AD) contains a large number of crudely rounded asymmetric cabochons of various sizes and shapes—essentially polished pebbles (Fig. 10). Later, cabochons were ground to symmetrical shapes, round or oval, as they are today. At present, cabochon cuts are employed largely for nontransparent gems such as *turquoise* or opal, or for phenomenal stones, since chatoyancy and asterism are brought out most effectively in a stone with a curved surface.

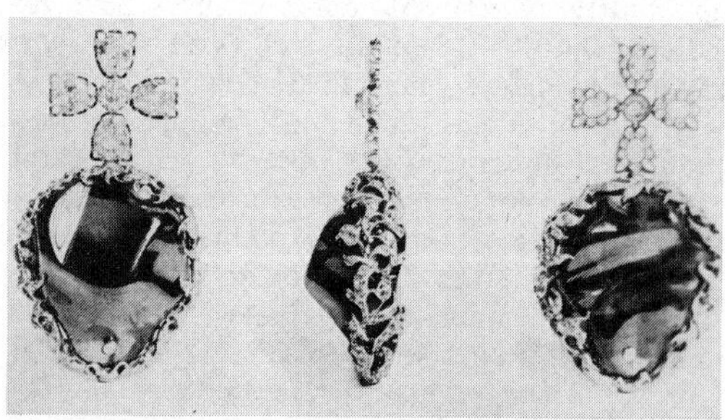

FIGURE 9. The great "Ruby," supporting the cross pâtée of the Russian Imperial Crown (from USSR, The People's Commissariat of Finances, 1925, "Russia's Treasure of Diamonds and Precious Stones"). Actually a magnificent red *spinel* surrounded by small *diamonds*, it is a superb example of the earliest type of gem fashioning. Essentially, it is a polished pebble of irregular shape; no attempt to shape it appears to have been made. Note the small hole drilled through the lower end to permit its being tied to clothing by a strong thread, a common practice before it was the fashion to mount stones in metal.

FIGURE 10. The Crown of Charlemagne, created about 800 A.D. (Top) A photo of the crown, preserved at Nuremburg. (Bottom) A sketch made by Albrecht Dürer (1471–1528). A modest attempt was made to make some of the stones symmetrical but most of them are merely pebbles polished as found.

At a still later date, sometime prior to 1400 AD, it was found that the potential beauty of transparent stones could be brought out better by faceting. At first, faceted stones were also asymetric, the irregular surface of the rough being replaced by a minimal number of randomly placed, flat facets. The magnificent Shah *diamond* in the Russian Crown Jewels is a fine example (Fig. 11). Later faceted stones were cut, not only in symmetrical form, but with special attention to proportions and angles, particularly in the case of *diamond*.

Through the centuries, a multitude of styles of faceting has been developed. These vary in the number, shape, distribution, and angular relationships of the facets as well as in the

FIGURE 11. The Shah, one of the world's great *diamonds* (after Fersman, 1926). The Shah is an 88-carat stone of Indian origin. Formerly part of the Persian regalia and later one of the two principal *diamonds* in the Russian crown jewels (the other being the Orloff), it is a splendid example of primitive asymmetric faceting in which an irregularly shaped stone was polished to achieve a minimum of loss of material. The groove that encircles the small end of the stone facilitates an early method of attaching jewels to clothing by means of a strong thread. The Shah is also remarkable as an engraved *diamond* and bears the names of several of the Persian Shahs.

general proportions of the stone. In the case of colored stones, the shape, especially the depth, is adapted to the depth of color of the material, to provide the most pleasing effect.

The critical importance of *diamond* fashioning places it in a class by itself. Since *diamond* is essentially colorless, interest in it as a gemstone depends entirely on its remarkable brilliance and fire (dispersion). These are purely optical effects and are highly sensitive to and dependent on correct proportioning and faceting. Long experience has shown that the unusual potential brilliance and fire of the diamond are brought out best by the so-called "brilliant" style of cutting (Fig. 12).

The general shape and proportions of this cut are based on the octahedron, the common geometric form of *diamond* crystals as well as its cleavage form. The brilliant cut originated in the mid-15th century as the simple "table cut." In this, the upper and lower portions of an octahedral crystal were ground off to form a large table above and a small parallel culet below. This provided two facets of a ten-faceted stone with a square outline; the remaining eight facets were merely the slightly modified faces of the octahedral crystal. From this modest beginning, the brilliant cut has evolved, by stages, into the modern standard 58-faceted

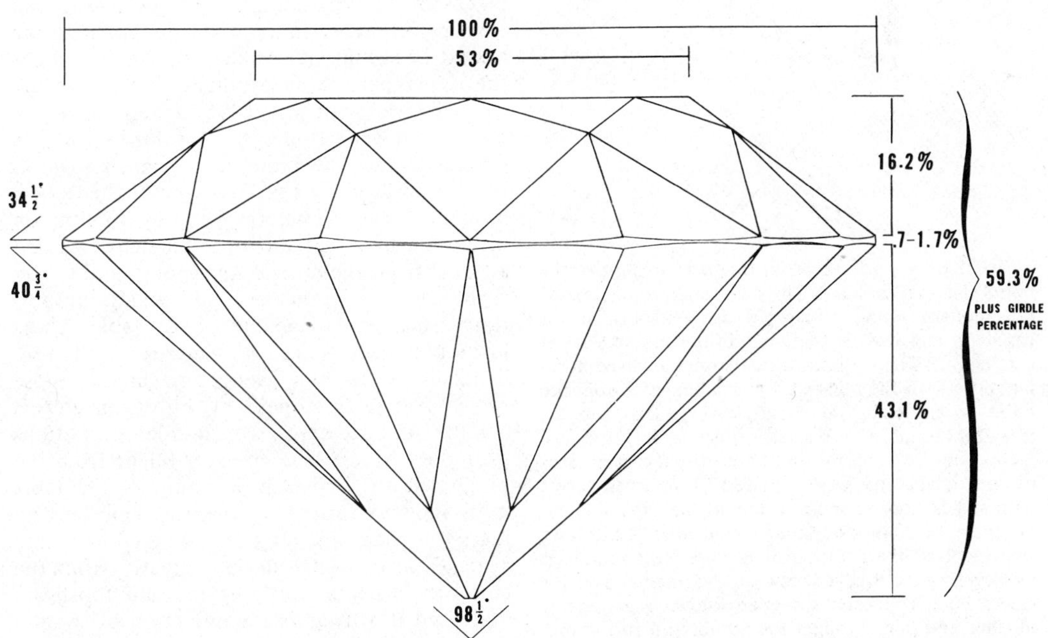

FIGURE 12. The American cut, which is theoretical, is based on the designs of M. Tolkowsky (1919). Well-cut *diamonds* closely approximating the angles and proportions indicated are more costly to produce than spread stones commonly cut today. (Reprinted with permission of the Gemological Institute of America).

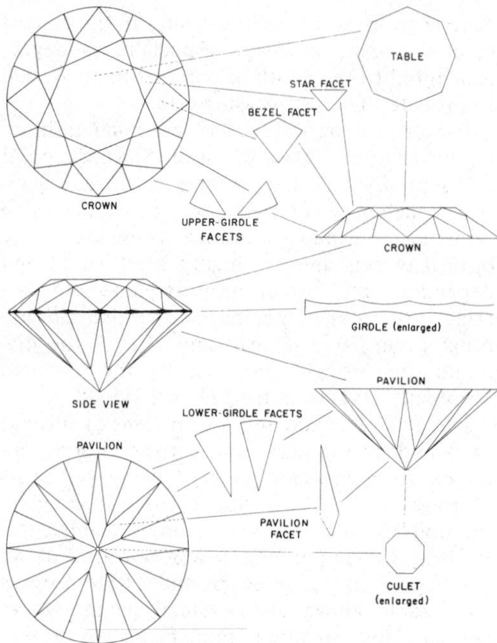

FIGURE 13. A breakdown of the American cut showing the shape, angular relationships, number, and names of the 7 types of facets. (Reprinted with permission of the Gemological Institute of America.)

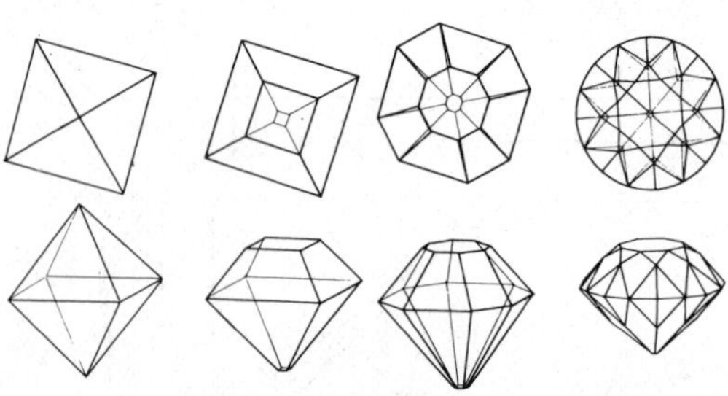

FIGURE 14. Early stages in the development of the round *diamond* brilliant. The shape and proportion of the modern round 58-faceted *diamond* brilliant is based on and had its beginning in the simple 8-faced octahedron. The earliest faceted stones involved nothing more than the removal, by grinding, of 2 opposite points of an octahedral crystal, leaving a large top surface (table) and a very small surface (culet), for protection only, at the bottom. Later, the 8 crystal faces of the octahedron were replaced by polished facets with angles close to those of the original crystal faces, producing a 10-faceted stone (table cut). A still later step involved the addition of 8 more faces produced by beveling the 8 edges between the original 8 major facets. Such 18-faceted stones are known as single cut. Further additions brought the number up to the present 58 of the standard brilliant. On the far right is the typical European brilliant–round, but with a very high crown compared with the American cut in Fig. 13.

round brilliant (33 in the crown and 25 in the pavilion; see Fig. 13).

The modern brilliant is perfectly round and has a low crown but this has been so only in this century. Formerly, *diamond* brilliants were more obviously influenced by the shape of the octahedral crystal from which they have evolved, being square in outline with very high crowns (old-mine cut). Although the square outline gave way later to a circular one, the crown remained high (European cut) (Fig. 14). It was known for some time that these high-crowned stones lacked maximum brilliance, but economic factors prevailed. Both the square outline and the high crown of the old-mine cut were arrived at long ago as a compromise between so-called ideal proportions and economic realism. The cut represented an effort to reduce to a minimum *diamond* loss during cutting, under the conditions then prevailing, and still produce a satisfactory stone (Fig. 15). At that time the entire shaping operation had to be carried out by grinding, the more recent techniques of cleaving and sawing being unknown. Whatever *diamond* was removed in the operation was reduced to virtually worthless powder so that reluctance to lower the height of the table, any more than was absolutely necessary, is understandable.

The discovery, in the 1870s, that a *diamond* crystal could be sawed in the three mutually perpendicular directions, paralleling the axial planes of an octahedral crystal, eventually led to a major change in *diamond* design. With the advent of sawing, it became possible to remove the upper part of an octahedral crystal at any desired level with virtually no loss, since the piece removed could be fashioned into a smaller stone. The improved design, known as the "American" cut or "Tolkowski theoretical brilliant," was characterized (1) by a reduction in the height of the crown to a height regarded as ideal from the optical standpoint, and (2) by slight changes in angles and general proportions, such as broadening of the table. These lower-crowned, so-called American cut brilliants were at first characterized as "those spread American stones" by European cutters (Fig. 16). However, economic conditions prompted by a desire to save weight from the rough led to further broadening of the table until modern brilliants cut with table percentages less than 60% are rare with 64% being the average. Stones that deviate greatly from the currently accepted proportions and finish are evaluated downward as a reflection of the undue weight saved in reaching the finished weight.

The standard 58-faceted brilliant exists in a

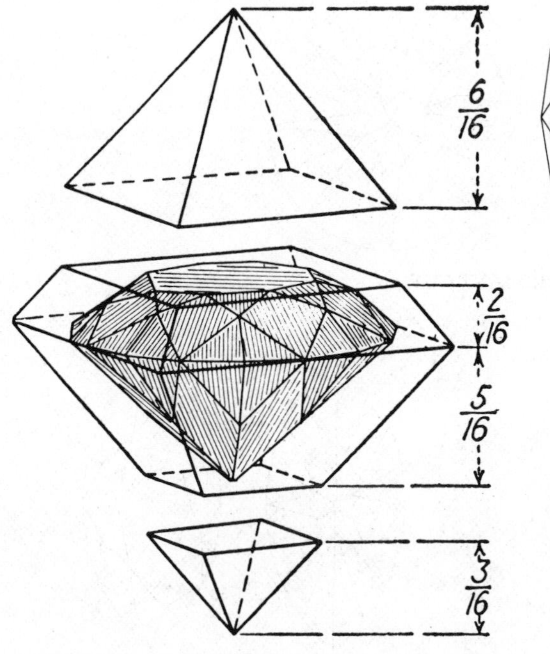

FIGURE 15. Relationship of a round *diamond* brilliant to the 8-faced octahedral crystal rough (from E. H. Kraus and C. B. Slawson, *Gems and Gem Materials*, 5th ed., McGraw-Hill, New York, 1947; reprinted with permission of the McGraw-Hill Book Company). To obtain a cut stone of the correct proportions it is necessary to remove two relatively large portions of an octahedral *diamond* crystal. Today, the two tips of the octahedron, the large upper and small lower, can be sawed off with little loss of weight and used to make smaller stones. Prior to the discovery, in the latter part of the nineteenth century, that *diamonds* could be sawed in the three axial-plane directions, these two tips had to be removed by grinding, a very wasteful process since the removed material was reduced to essentially worthless *diamond* dust.

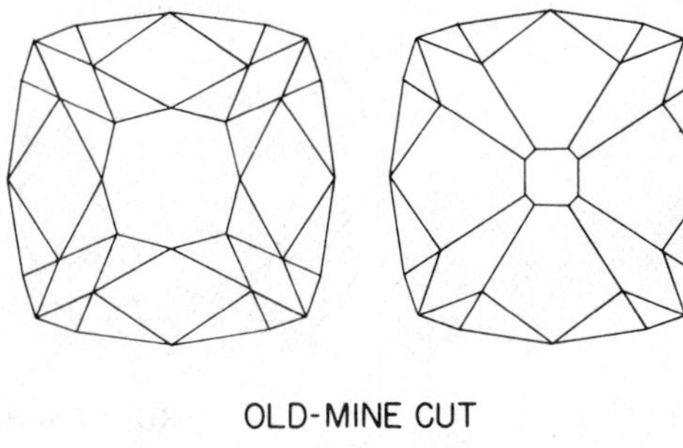

OLD-MINE CUT

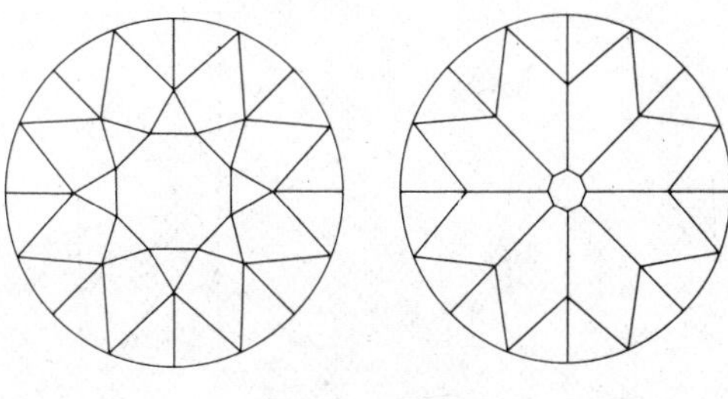

OLD EUROPEAN CUT

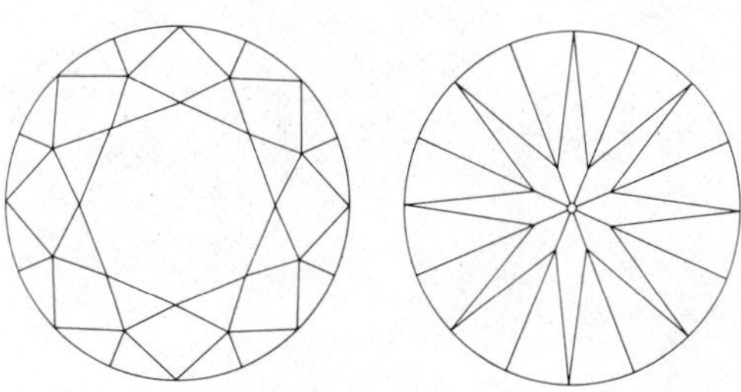

AMERICAN CUT

FIGURE 16. Three important late stages in the development of the modern standard round *diamond* brilliant. Both the old-mine and the later European cut had high crowns and consequently small tables. The change from the square old-mine to the round European cut involved a certain sacrifice of weight since the natural diamond crystal is often octahedral with a square cross section. The lower crown and resulting broader table facet, as well as the changes in the shape of the lower pavilion facets, of the modern American cut produced a more pleasing stone and one of greater brilliance. All three types have 58 facets. (Reprinted with permission of the Gemological Institute of America.)

modified form in other modern cuts, such as the oval, marquise, pendaloque, and heart-shaped stones (Fig. 17). Careful observation reveals that these are all basically the same as the round brilliant. They differ only in the outline of the stone and in slight differences in facet shape and interfacet angle and, in some cases, in the number of facets. Other styles of *diamond* fashioning, such as the emerald cut or rose cut, differ fundamentally from the brilliant cut and are known to provide stones of inferior brilliance, regardless of the perfection with which they have been cut. Recently diamonds with cushion octagon outlines and modified brilliant cut faceting have been promoted as improvements in brilliance for this shape.

Closely related to crown height is the overall proportion or ratio of depth to diameter of the stone, which determines the angular relationship of the back facets to the table. This is critical, as it influences light leakage at the back

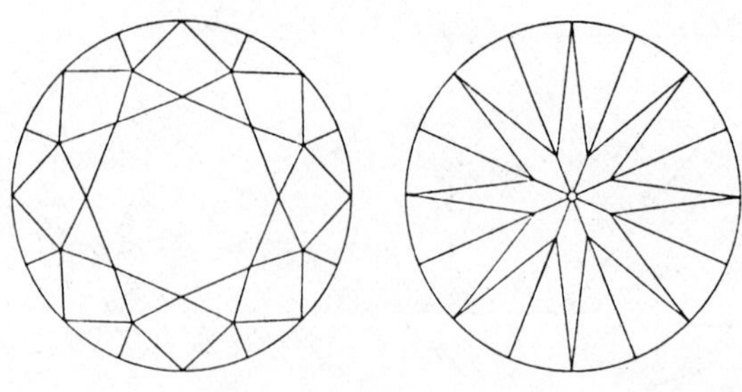

ROUND BRILLIANT CUT

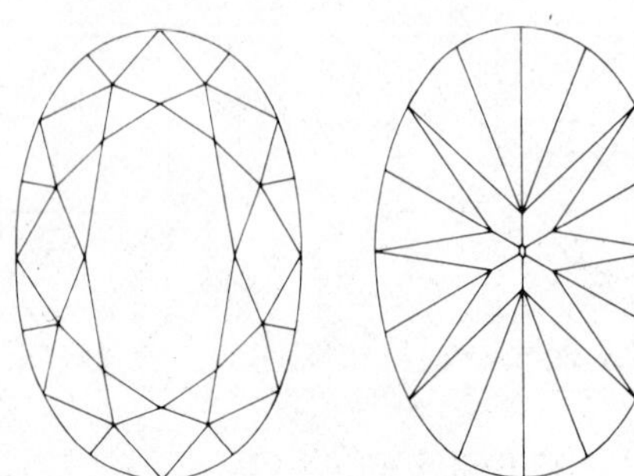

OVAL BRILLIANT

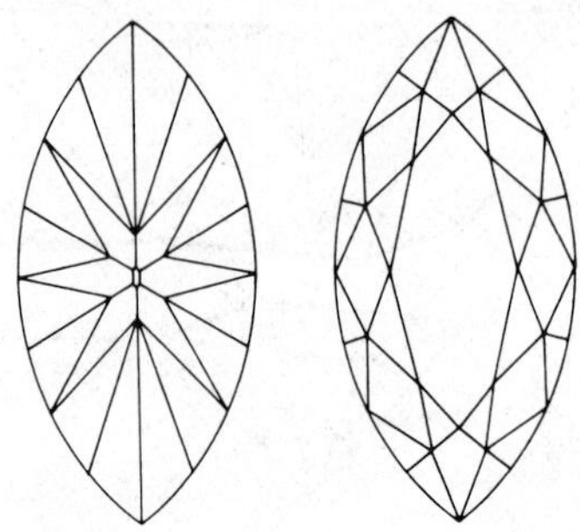

MARQUISE or NAVETTE BRILLIANT

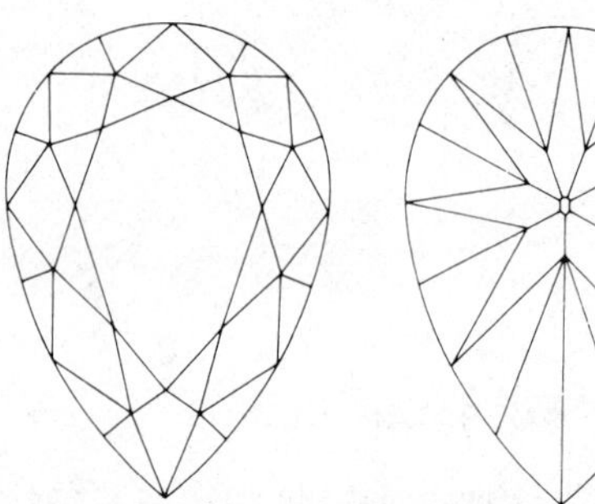

PEAR-SHAPE BRILLIANT

HEART-SHAPE BRILLIANT

FIGURE 17. Modifications of the standard round brilliant. Four less common modern *diamond* cuts, although quite different in overall appearance, are merely modifications of the standard round brilliant. There is essential agreement in the number, shape, and relationship of the facets in all 5 types. (Reprinted with permission of the Gemological Institute of America.)

of the stone and materially affects its brilliance (Fig. 18).

It is clear that several major factors influence the grading or evaluation of cut gems, especially *diamonds*. These factors are often referred to as the "four Cs"—color, clarity, carat weight, and cutting. In the case of *diamond*, color usually refers more to absence of color or deviation from colorless. Clarity covers transparency and the degree to which the material lacks imperfections. Carat weight refers to the size of the stone, and cutting to the quality and perfection of the fashioning. All are of great importance in arriving at an evaluation of a cut stone's worth.

RALPH J. HOLMES
Revised by
Robert Crowningshield

References

Anderson, B. W., 1948. *Gemtesting*. New York: Emerson, 377p.

Bauer, Max, 1896, 1909. *Edelsteinkunde*. Rev. 1932 by K. Schlossmacher.

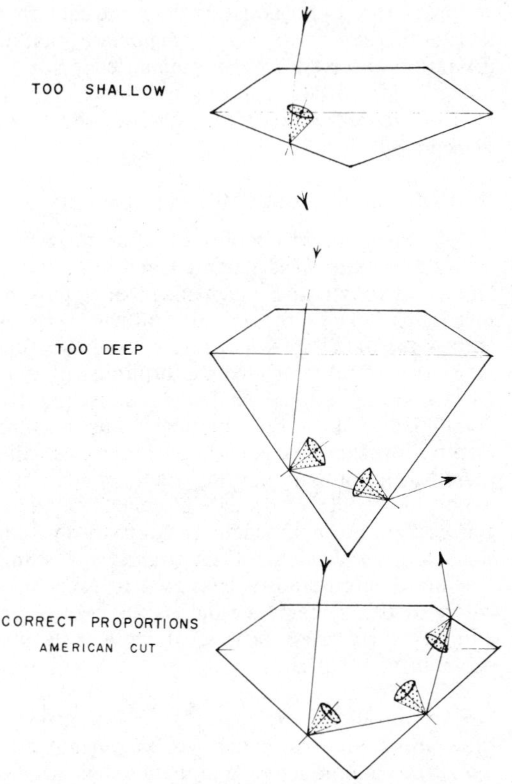

FIGURE 18. Correct proportioning of a *diamond* brilliant. Maximum brilliance in a round *diamond* brilliant is achieved by correct proportioning, the principal aim of which is to reduce to a minimum the light loss at the back facets of the stone. A beam of light that has entered a gem must be totally reflected twice from the back facets on opposite sides of the stone if it is to emerge from the crown and contribute to the gem's brilliance. To achieve this, it is essential that the slopes of the pavilion (back) facets be such as to insure that the light will strike them at angles exceeding the critical angle. Diagrams show how light is lost at the "first back facet" if the stone is cut too shallow and at the "second back facet" if the stone is too deep. The ideal proportions represented by the American cut insure minimum light loss at both back facets and consequent maximum brilliance.

Bauer, Max, 1969. *Precious Stones*. L. J. Spencer, trans. Rutland, Vt.: Charles E. Tuttle, 647p.

Gill, J. O., 1978. *Gill's Index to Journals, Articles and Books Relating to Gems and Jewelry*. Santa Monica: Gemological Institute of America, 420p.

Kraus, E. H., and Slawson, C. B., 1947. *Gems and Gem Materials*, 5th ed. New York: McGraw-Hill, 332p.

Liddicoat, Richard T., Jr., 1969. *Handbook of Gem Identification*, 5th ed. Los Angeles: Gemological Institute of America, 430p.

Nassau, K., 1980. *Gems Made by Man*. Radnor, Pa.: Chilton, 364p.

Shipley, Robert M., 1951. *Dictionary of Gems and Gemology*. Los Angeles: Gemological Inst. America, 261p.

Sinkankas, John, 1959. *Gemstones of North America*. Princeton, N.J.: Van Nostrand, 675p.

Smith, G. F. Herbert, 1958. *Gemstones*, 13th ed., revised by F. C. Phillips. London: Methuen, 560p.

Webster, Robert, 1962. *Gems: Their Sources, Descriptions and Identification*. London: Butterworths, 792p.

GEOLOGIC BAROMETRY—*See* BAROMETRY, GEOLOGIC

GEOLOGIC THERMOMETRY—*See* THERMOMETRY, GEOLOGIC

GLASS, DEVITRIFICATION OF VOLCANIC

Composition and Structure of Glassy Rocks

Volcanic glasses have been classified descriptively as follows: obsidians, those glasses with a vitreous luster; pitchstones, those with a resinous luster; and perlites, those that contain a multitude of irregular tension fractures. Chemically, all three types may range in composition from basalt through andesite, to rhyolite or trachyte. Compositions that do not fall on the main lines of liquid development during fractional crystallization of basalt are not represented. The chemical distribution as a function of silica content shows a dominant peak at about 73% of silica, corresponding to rhyolite glass. In addition, a distinct maximum at about 52% corresponds to basaltic glasses. No volcanic glasses are known with less than 35% or more than 80% of silica. Rhyolite glass seems about 3 times more abundant than basalt or andesite glass, despite the general 2:1 preponderance in nature of crystalline basalt over rhyolite. The greater stability of viscous rhyolite (of major normative components *orthoclase, albite*, and *quartz*) as compared to the more fluid basaltic glass (of major normative components ***pyroxene, plagioclase***, and ***olivine***), is undeniable.

Most volcanic glasses contain a few large phenocrysts and (or) a myriad of small crystallites. The orientation and distribution of these crystals frequently show flow structure, indicating that the original liquid had partly crystallized prior to cooling. Other glasses show individual microlites or spherulites crossing flow lines; this is indicative of partial devitrification (crystallization from a rigid supercooled condition).

Crystalline extrusives, as well as minor intrusives of Tertiary age, frequently show small amounts of residual glass, as do, on occasion, certain chondritic meteorites. Most terrestrial glass occurs in rocks younger than the Miocene age ($<$25 m yr), but lunar and meteorite glass can be as old as 4 b yr.

Devitrification Through Spontaneous Nucleation

According to the classical theory of spontaneous nucleation and growth of a crystal species in a supercooled liquid, the maximum rate of growth occurs at a higher temperature than the maximum rate of nucleation. Partial or complete devitrification occurring during supercooling of volcanic flows would, therefore, be expected to result in a crystal-size variation from fine near the borders to coarser toward the interior and more slowly cooling portion of the body. If the layer of molten material was thin enough to be entirely quenched to glass by its cooler surroundings, devitrification upon subsequent regional reheating would not result in such a distribution. Since most thick flows and dikes are not cooled rapidly enough to be quenched to a uniform glass, it is not surprising that the gradational grain-size distribution is most common. Typical size variation in rhyolitic obsidians, where crystal growth rates are limited by high viscosities, is from 1 to 100 μm. Crystals are usually somewhat coarser in the more fluid andesites and basalts.

Rates of nucleation and growth as a function of temperature would be expected to vary for the different phases separating from a glass. Thus, in a supercooled basalt supersaturated with respect to both ***pyroxene*** and ***plagioclase***, the respective rates of nucleation and growth at the temperature of devitrification would control the resulting texture, providing that no liquid phase separation occurred prior to crystallization. If the growth rate for ***plagioclase*** were much greater, this mineral could crystallize almost completely before the ***pyroxene*** substantially developed. Such a phenomenon could not develop under equilibrium conditions where cotectic (simultaneous) crystallization of both phases would occur at a very early stage.

The ophitic texture in diabase (tholeiitic basalt) flows and dikes in which ***plagioclase*** occurs as euhedral laths in association with anhedral ***pyroxene*** could represent such nonequilibrium crystallization from a supercooled melt. The observation that this texture is normally absent in coarser gabbros of the same composition that have cooled very slowly supports this hypothesis, as does the fact that it is ***pyroxene*** rather than ***plagioclase*** that is generally observed to have separated prior to cotectic precipitation during fractional crystallization of differentiated tholeiitic sills (e.g., Palisade Sill, N.Y.).

Devitrification Through Microimmiscibility

Controlled devitrification of glass to a fine crystalline material is accomplished in industry largely through the mechanism of microimmiscibility. The composition of the glass is adjusted so that it will separate upon heating into two finely divided amorphous phases. One or both of these phases will subsequently crystallize and a fine homogeneous texture results. In the case of rhyolitic glasses, the most common in nature, such phenomena would not likely occur. Chemically, these glasses can be included in the ternary system silica-*albite-orthoclase*, the "granite" system. No liquid immiscibility is known to exist anywhere in this system; anomalous flat regions on liquidus surfaces, indicative of metastable immiscibility, are absent.

Basic low-silica glasses, as found in nature, are chemically more complex. Normatively, they are composed largely of ***plagioclase*** and ***pyroxene*** components, and hence fall in the system SiO_2-Al_2O_3-CaO-MgO-FeO-Na_2O. Although no liquid immiscibility is known on the liquidus of any multicomponent compositions in this system, there is strong evidence for metastable immiscibility, particularly with compositions containing considerable FeO. In the systems *fayalite* (Fe_2SiO_4)-*anorthite*-silica, *fayalite-albite*-silica, and *fayalite*-alumina-silica, the *fayalite* liquidus surface displays the anomalous flat character indicative of metastable immiscibility. Moreover, in the *fayalite-leucite* ($KAlSi_2O_6$)-silica system, stable immiscibility exists on the liquidus. In this case, a potassium aluminosilicate-rich liquid coexists with a liquid rich in ferrous iron and silica. Evidently, there is a general tendency for ferrous silicate glass to separate from various aluminosilicate glasses.

In the case of natural basalt, experiments performed at Corning Glass Works upon vitrified Holyoke diabase have shown that this type of immiscibility can be induced by simple heating. An iron-rich silicate glass, probably containing some magnesia and lime, separates from the remaining calcium-sodium aluminosilicate (***plagioclase***) glass.

The nature of this liquid-liquid phase separation depends on the melting conditions. An extremely fine ($<$0.1 μm), homogeneous emulsion forms if the glass is oxidized, while larger, widely scattered blebs appear in the reduced glass. Upon further heating, the iron-rich glass develops randomly oriented *magnetite* crystals (100–500 Å), which act as nucleii for subsequent crystallization and growth of ***pyroxene***. Coarser crystallization occurs in less oxidized glass due to fewer crystals of *magnetite*, which has a ferric to ferrous iron ratio of 2:1. The bulk of the ***plagioclase*** glass is stable and does not crystallize below the solidus (≈1050°C) in a matter of a few hours.

Since basaltic glasses occur in nature, it is likely that semicrystalline ***pyroxene-"plagioclase*** glass" composites exist. They are extremely fine grained (≈1 μm), and very homogeneous except where FeO predominates over Fe_2O_3, which results in a reduced tendency toward immiscibility and a lack of abundance of nucleation sites.

It should be noted that this sort of devitrification can only occur if basaltic lava is supercooled below the solvus of metastable immiscibility (probably well below the solidus) prior to crystallization. The sequence of devitrification is likely: glass → iron-rich glass + aluminosilicate glass emulsion → *magnetite* + glass composite → *magnetite* + ***clinopyroxene*** (major crystalline phase) + glass composite → very fine crystalline basalt. The "two-glass" and "*magnetite* + glass" stages may be short in duration due to the relative ease of crystallization of silicate glasses rich in iron.

Time Factor in Devitrification

The activation energies for thermal reconstruction of volcanic glasses are high, roughly 50–100 Kcal/mole. Temperatures of several hundred degrees centigrade are required to cause appreciable devitrification even within the entire space of geologic (or meteoric) time (≈5 b yr). Such temperatures occur near the surface of the earth only in areas of unusual geothermal gradient.

Observations of the degree of devitrification versus water content of related natural perlites, and experimental studies of surface devitrification of perlites under hydrothermal conditions, have shown that the activation energy for devitrification is reduced by a factor of two by the presence of water. Hydroxyl-ion breakage of the Si–O–Si bridges presumably accounts for the reduction in glass stability. Thus, within a given time, volcanic glass in the presence of water vapor will devitrify to a given extent at temperatures some 200°C lower than will glass under dry conditions. Since water vapor is abundant in most geologic environments, it is, therefore, not surprising that few volcanic glasses predate the Tertiary period. The cool and dry meteoritic and lunar environments, on the other hand, may account for the persistence of glass in lunar rocks and meteorites.

GEORGE H. BEALL

References

Bowen, N. L., 1928. *The Evolution of the Igneous Rocks*. New York: Dover.

Marshall, R. R., 1961. Devitrification of natural glass, *Geol. Soc. Am. Bull.*, **72**, 1493–1520.

Cross-references: *Clinopyroxenes; Crystal Growth; Feldspar Group; Glass: Theory of Crystallization; Lunar Minerals; Meteoritic Minerals; Rock-Forming Minerals; Synthetic Minerals.* Vol. II: *Meteorites.* Vol. V: *Andesite; Basalt; Glass, Natural Glasses, and Volcanic Glass; Rhyolite; Trachyte.*

GLASS: THEORY OF CRYSTALLIZATION

Any crystallization process consists of three fundamental steps: (1) attainment of metastability, (2) the formation of nuclei, and (3) crystal growth. Each of these steps is represented in the crystallization of a glass, with some modifications due to the unique properties of glass.

Glass has been defined in many ways. A general definition is that a glass is a supercooled liquid with a viscosity greater than $10^{14.5}$ poise. A restricted definition, of more use to the earth scientist, is that a glass is an inorganic product of fusion, which has cooled to a rigid state without crystallizing. This implies that the amorphous structure is statistically random but does not preclude short-range order.

A typical glass-forming silicate melt is unique among solutions in its ability to dissolve nearly all of the chemical elements. Such a melt is, therefore, a solution of oxides in thermodynamic equilibrium. Another unusual property of a glass-forming melt is its enormous change in viscosity with temperature. This change results in high-temperature equilibrium states being maintained as the melt is cooled. Solutes present in the melt become supersaturated in the glass and high-temperature oxidation states are preserved. This nonequilibrium, as the result of the sharp viscosity increase upon cooling, satisfies the first step in the crystallization process.

Nucleation

Adjustment of any metastable state, such as a glass, to a stable configuration requires overcoming a free-energy barrier and subsequent passage to a lower free-energy state. This initial free-energy barrier is the nucleation barrier and it can be overcome by either homogeneous (spontaneous) or heterogeneous (induced) nucleation.

Homogeneous Nucleation. The formation of a liquid droplet or a crystalline particle within a homogeneous melt requires the expenditure of some energy to create the liquid or solid surface. This is the origin of the energy barrier. The free energy required to form a stable nucleus (ΔF) is, therefore, the sum of the free energy required to form the surface (F_s) and the free energy of forming the bulk of the material (F_v). This latter term is negative at temperatures below the melting point so that

$$\Delta F = F_s - F_v$$

Assuming the particle being formed is spherical,

$$\Delta F = 4\pi r^2 \Delta f_s - \frac{4}{3}\pi r^3 \Delta f_v \qquad (1)$$

where Δf_s is the free energy per unit area of surface between the two phases and Δf_v is the magnitude of the free-energy change per unit volume in transforming from one phase to the other. For any value of Δf_s and Δf_v, there is a maximum ΔF for some critical radius r^*. Spherical regions smaller than r^*, called embryos, are continually forming and dissolving since they require an increase in free energy to grow. Regions larger than r^* can grow with a decrease in the free energy and are, therefore, stable nuclei. Embryos have an enhanced vapor pressure or solubility when first formed because of their small size. This explains qualitatively why they dissolve rather than grow. The Kelvin equation relates the supersolubility of the droplet to its size.

$$\ln \frac{P_r}{P_\infty} = \ln \alpha = \frac{2M\sigma}{RT\rho r} \qquad (2)$$

where P_r = vapor pressure of droplet of radius r
P_∞ = vapor pressure of a plane surface
α = measure of supersolubility as either a pressure or concentration ratio
σ = surface tension
M = molecular weight of liquid forming droplet
ρ = density of droplet

Using the critical radius r^* and solving for ΔF from Eq. (1),

$$\Delta F^* = \frac{16\pi M^2 \sigma^3}{3\rho^2 R^2 T^2 \ln^2 \alpha} \qquad (3)$$

which gives a measure of the nucleation barrier in terms of the degree of supersolubility. The nucleation rate, I, is proportional to the probability of overcoming this nucleation barrier. The absolute-reaction-rate theory gives

$$r = \frac{NkT}{h} \exp\left[-\frac{(\Delta F^* + Q)}{kT}\right] \qquad (4)$$

where N = number of atoms per unit volume
ΔF^* = free energy at r^*
Q = activation energy of diffusion

When ΔF^* is comparable with Q, the nucleation rate abruptly reaches a maximum value.

Heterogeneous Nucleation. One means of lowering ΔF^* is the formation of a liquid-liquid emulsion in the glass. The energy per unit area of surface between two glasses, Δf_s, is small so that emulsions will nucleate easily in unstable regions. The energy barrier is also lowered considerably by the presence of foreign particles that can act as nucleation sites. The theory for heterogeneous nucleation follows that of homogeneous nucleation except that allowance is made for the interfacial energy between the impurity and the glass. Volmer (Chemical Engineering Symposium, 1952) showed that the free energy is now

$$\Delta F' = \Delta F\left[\frac{(2 + \cos\theta)(1 - \cos^2\phi)}{4}\right]$$

where θ is a measure of the registry of the crystal structure of the impurity and the crystallizing phase. This registry corresponds to the angle of wetting in a liquid–solid system. ΔF is the homogeneous nucleation barrier previously discussed.

For the case of a complete nonaffinity between the crystalline solid and the foreign surface ($\phi = 180^\circ$), $\Delta F' = \Delta F$. For a partial affinity ($0 < \phi < 180^\circ$) the nucleation can occur at a lower supersaturation since $\Delta F' < \Delta F$. This is the most common case. When $\phi = 0$ there is a complete affinity, the nucleation barrier is zero, and the foreign particle is an effective seed.

Each glass will nucleate by the mechanism requiring the least energy. If the glass is one that will phase-separate, this will be the probable mechanism because of the low energy required. If several crystalline phases are capable of forming, the first to nucleate will be the one with the lowest free-energy nucleation barrier. This is generally the phase with a structure most like the glass and is usually a crystalline phase of low stability. This early crystalline phase can then act as a nucleation site for other crystalline phases. Epitaxial growth is then possible if the two structures are compatible (see *Epitaxy*). The crystallization path in a many-component system can involve the successive formation of many phases or groups of crystalline phases depending on the mechanism responsible for the nucleation. The equilibrium phase assemblage may never be reached within a reasonable period of time.

Growth

Once the free-energy barrier to nucleation has been overcome by one of the discussed mechanisms, the passage to the lower energy state by crystal growth can proceed. Growth may be limited in the glass by (1) diffusion of material to the growing crystal, and (2) adsorption and orientation at the glass–crystal interface. Desorption and counterdiffusion will be considered only as analogous to 1 and 2. Usually the phase crystallizing is not of the same composition as the bulk glass, and material must be transported an appreciable distance. If the activation energy for diffusion, Q, is larger than the activation energy at the interface, Q', the reaction will be diffusion controlled. However, if $Q<Q'$, the rate-controlling step will be due to adsorption and orientation at the glass–crystal interface. The smaller the particle and the smaller the degree of supersaturation, the more likely Q' will be greater than Q. However, in the few cases where growth rates in glasses have been measured, the time dependence was based on a diffusion-controlled rate as follows:

$$U = A \exp\left[-\frac{CT_1}{H(T_1 - T)kT} - \frac{Q}{kT}\right] \qquad (5)$$

where A, C = constants
T_1 = the liquidus temperature
T = the temperature at the glass–crystal interface
H = the heat of two dimensional condensation
Q = the activation energy for diffusion

It has been found empirically that the activation energy for viscosity can be used so that Eq. (5) becomes

$$U = \frac{A}{\eta} \exp - \left[\frac{ZT_1}{T_1(T_1 - T)}\right] \qquad (6)$$

where η is the viscosity and Z is a constant. This implies that the species whose movement controls crystal growth also determines the viscosity.

Many theories have been proposed to account for the details of growth, but these are beyond the scope of this discussion.

DAVID A. DUKE

References

Chemical Engineering Symposium, 1952. Nucleation phenomena, *Industrial and Engineering Chemistry*, **44**, pt. 1, 1269–1339.

Hench, L. L., and Freiman, S. W., 1971. Advances in Nucleation and crystallization in glasses, *Am. Ceram. Soc. Spec. Publ. 5*.

Hinz, W., 1977. Nucleation and Crystal growth, *Survey Papers of the XIth International Congress on Glass*, vol. **1**, *Glass Science*, 217–259.

McMillen, P. W., 1979. *Glass-Ceramics*, 2nd ed. New York: Academic Press, 285p.

Stookey, S. D., and Maurer, R. D., 1961. Catalyzed crystallization of glass in theory and practice, *Progr. Ceram. Sci.*, **2**, 77–101.

Van Hook, A., 1961. *Crystallization: Theory and Practice*. New York: Reinhold, 325p.

Cross-references: *Crystal Growth; Epitaxy; Glass, Devitrification of Volcanic.*

GLASS SANDS—*See* SANDS, GLASS AND BUILDING

GONIOMETRY

A goniometer (Greek: *gonia*, angle and *metron*, measure) is an instrument for measuring angles, particularly angles in crystals. Goniometry is the theory and practice of measuring angles.

The measurement of angles between planes in a crystal is the first step in determining its symmetry and the ratio of the edge lengths of one unit of its atomic pattern (i.e., of its "unit cell"). Classification by symmetry, coupled with other data, leads by successive stages to identification of the substance and a knowledge of its complete atomic arrangement. The crystallographic planes in question are fundamentally planes of atoms in regular array, but these atomic planes govern the growth, under favourable circumstances, of visible plane faces bounding the crystal. The recognition that the angles between corresponding faces of crystals of the same substance are always the same is attributed to Nicolaus Steno (1638–1686), and the rule that common crystal faces are planes of greatest reticular density of atoms is attributed to Auguste Bravais (1811–1863).

Contact Goniometer

The simplest method of measurement of interfacial angles of crystals is with the contact goniometer, which is a protractor with a pivoting arm (Fig. 1). The figure sufficiently shows the use of the instrument: the only precaution to be observed is that the plane of the protractor must be held normal to the line of intersection of the two faces whose interfacial angle is to be measured. The angle recorded is, by convention, the *normal interfacial angle* indicated in Fig. 1 (i.e., the angle between the normals to the two faces). This is the most

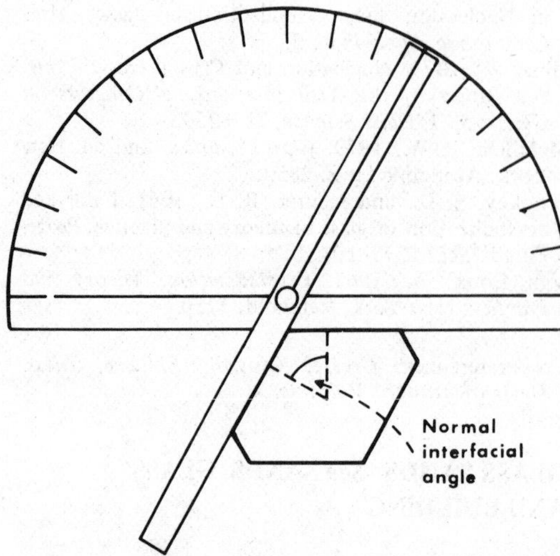

FIGURE 1. Contact goniometer (from M. H. Battey, *Mineralogy for Students*, Longman Group Ltd., London, 1972, reprinted with permission).

convenient angle for subsequent graphical procedures. The contact goniometer is not suitable for small crystals.

Optical Goniometer

Greater accuracy is achieved by the use of an optical goniometer, of which Wollaston's

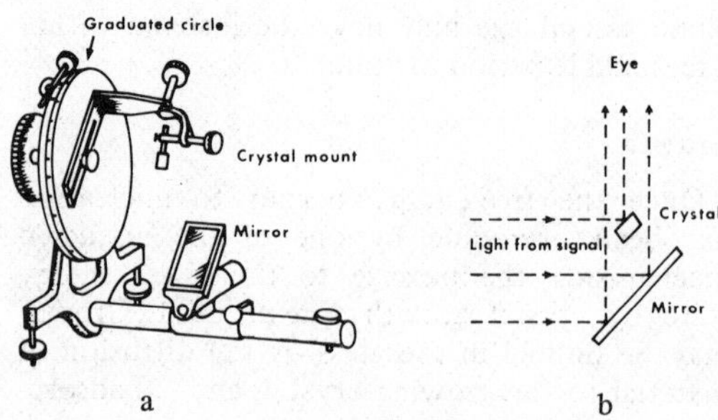

FIGURE 2. (a) Wollaston's single-circle optical goniometer. (b) Light path through goniometer.

(1809) instrument (Fig. 2a) was the earliest. With Wollaston's instrument, the image of a distant light source reflected by the crystal faces in turn is brought into coincidence with the reflection of the same source in a fixed mirror below the crystal (Fig. 2b). Two types are now in common use, the single-circle goniometer and the more sophisticated two-circle type.

Single-Circle Goniometer. For use on the single-circle goniometer (Fig. 3), the crystal, which must have perfect faces for good results and will usually be small (2–5 mm), is first sketched and the faces identified by shape and natural blemishes or artificial marks. It is then set in plasticene on the table of the instrument, which is supported on two lateral adjusting

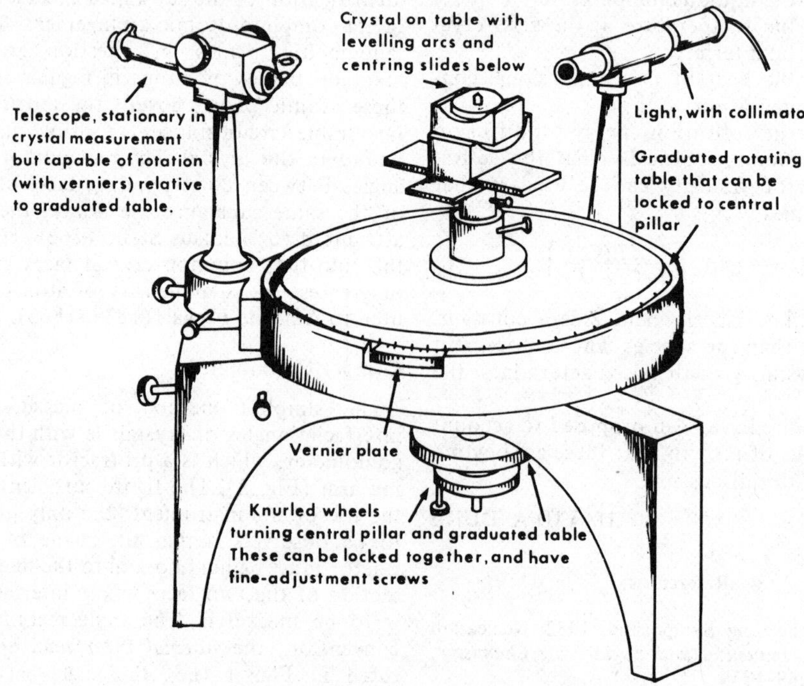

FIGURE 3. Single-circle goniometer.

slides and two arcs, and has a range of movement along the axis of the central pillar. The crystal is placed with a zone axis (the common direction of intersection of a prominent set of faces) as nearly vertical as can be done by eye and one of the faces should be roughly parallel to the plane of one of the arcs.

The telescope is set approximately at right angles to the light source and collimator and locked in that position. The central pillar is then turned until the reflection of the collimator light beam (generally an hour-glass shaped signal) by the crystal face set parallel to an arc is seen in the telescope. This light signal is adjusted, by moving the arc normal to the face, so that it is symmetrical about the horizontal cross-wire of the telescope. Then the central pillar is turned until a second good signal is seen in the telescope, preferably from a face at a fairly high angle to the first one. This signal is placed symmetrically on the horizontal cross-wire by movement of the arcs. A return to the first face for adjustment should result in the zone axis being set vertical.

The pillar is then locked to the graduated circle and, these two being turned together, each successive reflection from faces in the zone is placed on the vertical cross-wire of the telescope and its angular position read with the verniers. The differences between successive readings give the normal interfacial angles.

The crystal is then dismounted and a new zone set vertical and measured, starting from an identified face in the first zone. This procedure is repeated until all the faces on the crystal have been included in one or more zones.

Two-Circle Goniometer. The two-circle goniometer (Fig. 4) is designed to allow all zones to be measured without dismounting the crystal. The spindle, slides, and arcs that carry the crystal are here set horizontally, and the horizontal graduated circle of the single-circle instrument becomes the vertical circle. The whole of this assembly is capable of being rotated around a vertical column, its rotation being measured on a horizontal graduated circle. Optical arrangements are made to enable the graduations of both circles to be read

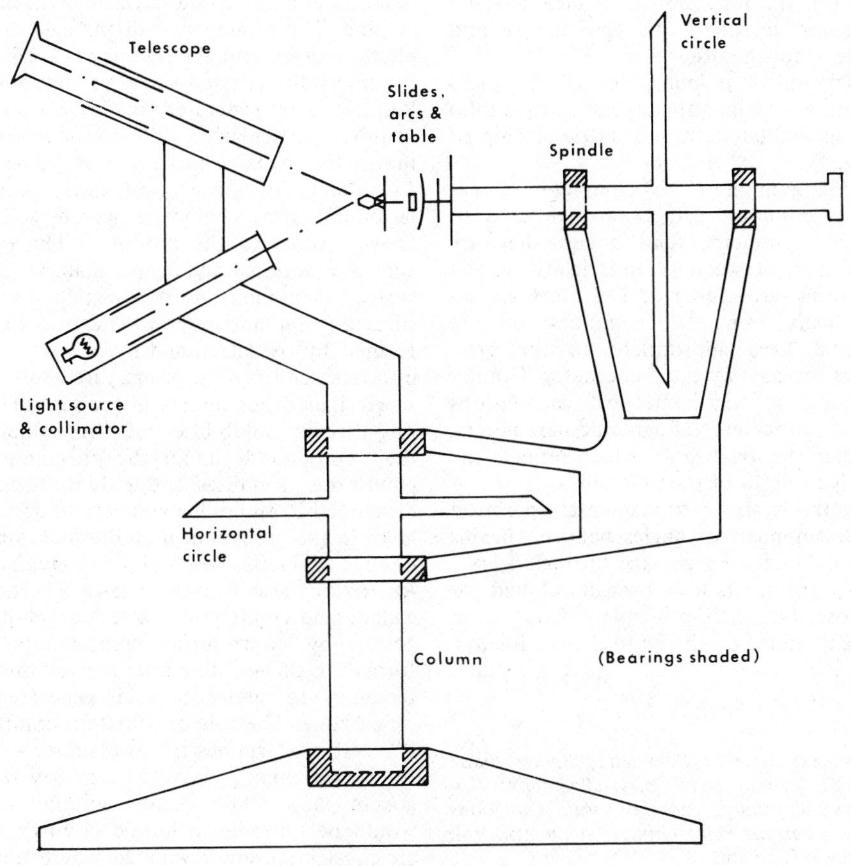

FIGURE 4. Two-circled goniometer.

through one low-power microscope (not shown in Fig. 4).

The light source with its collimator and the telescope for observing reflections are mounted in a vertical plane and carried on an arm that is an integral part of the central column. The angle between collimator and telescope is fixed. For convenience, the whole vertical column may be turned in its base.

The crystal is set with one face roughly normal to the spindle of the crystal table: this face is called the *pole face*. The spindle is then set in the plane of the collimator and telescope, by observing the reflection from the pole face as the spindle is rotated. When the reflection describes a circle around the center of the telescope field, the spindle is in the required plane. The size of the circle described is then reduced to zero by adjustment of the arcs, after which the pole face will be normal to the spindle. The assembly carrying the vertical circle is then clamped to the central column and the horizontal circle is read. This reading becomes the zero for polar angles, ρ, read on the horizontal circle. The angles, ν, read on the vertical circle are referred to an arbitrary V_0, the reflection of a face selected as the *prime meridian*, to give the second coordinate, ϕ for the other faces.

Measurements of ρ and ϕ for all the faces (except those attaching the crystal to the table) are then accumulated from a single setting of the crystal.

When the goniometry is complete, stereographic or gnomonic projections are used to exhibit the symmetry; and angles that are closely similar, between symmetrically related pairs of faces, are averaged. The faces are indexed (usually using Miller indices) on the conventional axes appropriate to the symmetry class revealed by the goniometry. Graphical methods or trigonometrical calculations then supply other crystallographic parameters, in particular the *axial ratio*, which reflects the relative edge lengths of the unit cell.

Goniometry is also a fundamental operation in the measurement of angles between beams of X-rays diffracted by passage through a crystal. Many instruments have been developed for this purpose, but their principles of operation are basically those of the optical instruments.

M. H. BATTEY

Cross-references: *Axis of Symmetry; Barker Index of Crystals; Crystal Habit; Crystallography: History; Crystallography: Morphological; Directions and Planes; Lattice; Point Groups; Symmetry*. Vol. IVA: X-ray Diffraction.

GOSSAN

The Cornish term *gossan* is applied to chemically weathered, iron-stained outcrops of sulfide mineral deposits. Synonyms include the French *chapeau de fer*, the German *eiserner Hut*, and the Spanish *colorados*. The term *pacos ore* is used in Bolivia for gossans over tin veins that are residually enriched in *cassiterite*. Most workers reserve the term "capping" for gossans over disseminated sulfide deposits of the porphyry copper type.

Significance

Gossans provide many clues to the character of ore once present in the plane of the topographic surface and, therefore, aid in exploration for deposits concealed at depth. Also, some gossans are residually enriched in the less soluble metals like gold, silver, and tin to the extent that the gossan itself constitutes ore.

Origin

Gossans are the normal end products of oxidation and leaching of iron-bearing sulfide deposits exposed to the surface environment by erosion. The principal oxidizing agent is atmospheric oxygen and the leaching is accomplished by descending groundwaters of meteoric origin. Sulfuric acid generated by oxidation of the sulfides, particularly *pyrite* and *marcasite*, promotes the leaching process. Many gossans consist chiefly of *quartz* and ferric oxides, the latter imparting characteristic red, yellow, and brown colors to the outcrops. The *quartz* is typically residual hypogene material although surface accumulations of iron-stained supergene silica are not uncommon. The iron oxides are formed by oxidation and hydrolysis of ferrous iron released from the primary iron sulfides. The survival of other metals in the gossan depends on intrinsic solubilities of those metals and their compounds under the prevailing surface conditions. Acidities and oxidation potentials developed in the oxide zone are important variables in the control of mobility and mineralogy of specific metals (see Vol. IVA: *Oxidation and Reduction*; also Garrels, 1954). The hypogene gangue and country rock affect development of gossan by contributing components for the formation of insoluble salts and by their varied tendency to neutralize acids generated in the oxide zone. The role of climate is complex, but it clearly determines the availability of solvent for the leaching process. In very dry regions, a gossan may retain highly soluble salts that would be unstable in humid country. Gossans are most maturely developed where the rate of erosion has been slow relative to the rate of oxidation and leaching.

Mineralogy

The minerals found in gossan are either inert hypogene species like *quartz*, which survive weathering, or newly formed oxidation products that are relatively stable under surface conditions. Iron normally occurs as *goethite, hematite*, or *jarosite; goethite* is the chief constituent of gossan "limonite." Zinc is easily leached but may remain, commonly as *smithsonite, hemimorphite*, or *hydrozincite*. Copper may be completely leached or persist in supergene minerals such as *malachite, azurite*, and *chrysocolla*. In arid terrane, a complex variety of copper halides, sulfates, and carbonates may occur in the gossan. Silver is less mobile than either zinc or copper, particularly in the presence of halide ions. It commonly occurs in the outcrop as native *silver, chlorargyrite, iodargyrite*, and *bromargyrite*. Lead strongly resists solution by the formation of not readily soluble *anglesite, cerussite, wulfenite*, or a number of other less-common oxysalts. *Gold* is quite inert and tends to be residually enriched in the gossan. Under special conditions of very high chloride-ion activity, low pH, and high oxidation potential, it may be leached from the gossan. Tin usually remains in the outcrop in the primary form of *cassiterite*. Molybdenum may persist in the outcrop as relict *molybdenite* or in a variety of molybdenum oxides or ferric molybdates.

Interpretation

The interpretation of gossans involves reconstruction of the mineralogy, grade, and extent of ore from its weathered surface remains and the selection of favorable ground where ore may lie at depth (Locke, 1926; Blanchard, 1939, 1968). It is a sophisticated skill calling for extensive field experience and thorough knowledge of the chemistry of the oxide zone. Clues to the mineralogy of the original ore may include relict hypogene minerals; supergene oxide mineralogy; trace elements in the gossan; and subtle variations in the color, mineralogy, and textures of the iron precipitates. Limonite "boxworks" are diagnostic of specific primary sulfides and are always sought in examination of leached outcrops (Blanchard, 1968). After mental reconstruction of the hypogene mineralogy, the grade is usually estimated visually. During oxidation, iron oxides may migrate beyond the limits of their sulfide source and stain the surrounding country rock. For this reason, estimates of the size, shape, and location of ore once present requires recognition and, in some cases, careful mapping of indigenous versus transported iron oxides.

WILLIAM C. KELLY

References

Blanchard, R., 1939. Interpretation of leached outcrops, *J. Chem. Met. Min. Soc. South Africa*, **39**, 344–373.

Blanchard, R., 1968. Interpretation of leached outcrops, *Nevada Bur. Mines Bull.*, **66**, 196p.

Garrels, R. M., 1954. Mineral species as functions of pH and oxidation-reduction potentials, *Geochim. Cosmochim. Acta*, **5**, 153–168.

Locke, A., 1926. *Leached Outcrops as Guides to Copper Ore*. Baltimore: Williams and Wilkins, 175p.

Cross-references: *Gangue Minerals; Metallic Minerals; Mineral Deposits Classification; Mineral and Ore Deposits; Vein Minerals;* see also mineral glossary. Vol. IVA: *Oxidation and Reduction.*

GREEN RIVER MINERALOGY

The Green River formation, a series of dolomitic marlstones and oil shales, extends over 16,000 square miles in Utah, Colorado, and Wyoming, USA (Fig. 1). The formation is over 7,000 ft thick, and includes many altered tuff beds, but no known igneous intrusives. Nevertheless, the mineralogy of the formation is

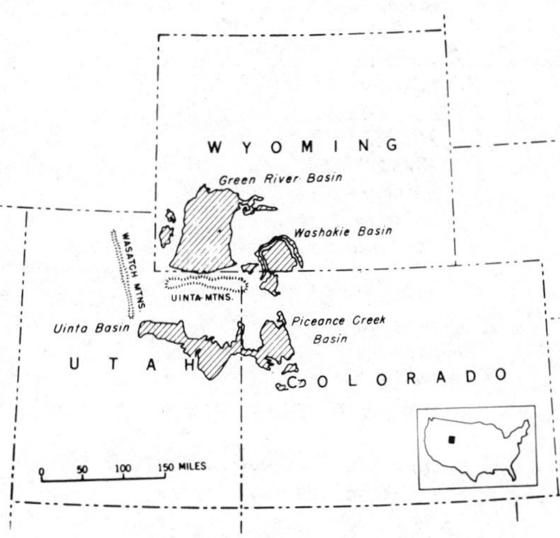

FIGURE 1. Basins of Green River deposition. Index map showing major structural basins containing lacustrine strata of early and middle Eocene ages in NW Colorado, NE Utah, and part of SW Wyoming, USA. Three basins each have a distinctive and differing type of Green River mineralogy—the Green River (Bridger) Basin in Wyoming, the Uinta Basin in Utah, and the Piceance Creek Basin in Colorado.

TABLE 1. Green River Authigenic Minerals

Elements	
sulfur	S
Sulfides	
marcasite	FeS_2
pyrite	FeS_2
pyrrhotite	$Fe_{1-x}S$
wurtzite	ZnS
Halides	
cryolite	Na_3AlF_6
fluorite	CaF_2
halite	NaCl
NEIGHBORITE	$NaMgF_3$
Oxides	
quartz	SiO_2
chalcedony	$SiO_2 \cdot H_2O$
anatase	TiO_2
Hydroxides	
nordstrandite	$Al(OH)_3$
Phosphates	
fluorapatite	$Ca_{10}(PO_4)_6F_2$
collophane	$Ca_{10}(PO_4)_6CO_3 \cdot H_2O$
Carbonates (Simple)	
nahcolite	$NaHCO_3$
trona	$Na_2CO_3 \cdot NaHCO_3 \cdot H_2O$
WEGSCHEIDERITE	$Na_2CO_3 \cdot 3NaHCO_3$
thermonatrite	$Na_2CO_3 \cdot H_2O$
natron	$Na_2CO_3 \cdot 10H_2O$
calcite	$CaCO_3$
magnesite	$MgCO_3$
strontianite	$SrCO_3$
witherite	$BaCO_3$
siderite	$FeCO_3$
aragonite	$CaCO_3$
MCKELVEYITE	$(Na_2Ca_1)(Ba_4RE_{1.7}U_{0.3})(CO_3)_9$
EWALDITE	$Ba(RE, Na,Ca)(CO_3)_2$
burbankite	$Na_2(Ca, Sr, Ba, Ce)_4(CO_3)_5$
dolomite	$CaCO_3 \cdot MgCO_3$
dawsonite	$NaAlCO_3(OH)_2$
EITELITE	$Na_2CO_3 \cdot MgCO_3$
SHORTITE	$Na_2CO_3 \cdot 2CaCO_3$
pirssonite	$Na_2CO_3 \cdot CaCO_3 \cdot 2H_2O$
gaylussite	$Na_2CO_3 \cdot CaCO_3 \cdot 5H_2O$
barytocalcite	$CaCO_3 \cdot BaCO_3$
NORSETHITE	$MgCO_3 \cdot BaCO_3$
Carbonates (Compound with Other Anions)	
BRADLEYITE	$MgNa_3CO_3PO_4$
northupite	$Na_2Mg(CO_3)_2 \cdot NaCl$
FERROAN northupite	$Na_2(Mg, Fe)(CO_3)_2 \cdot NaCl$
tychite	$Na_2Mg(CO_3)_2 \cdot Na_2SO_4$
burkeite	$Na_2CO_3 \cdot 2Na_2SO_4$
Sulfates	
barite	$BaSO_4$
celestite	$SrSO_4$

TABLE 1. (Continued)

siderotile (rozenite)	$FeSO_4 \cdot 4H_2O$
gypsum	$CaSO_4 \cdot 2H_2O$
szomolnokite	$FeSO_4 \cdot H_2O$
starkeyite	$MgSO_4 \cdot 4H_2O$
bloedite	$MgSO_4 \cdot Na_2SO_4 \cdot 4H_2O$
(also, *burkeite* and *tychite*)	
Silicates	
	(Clay Minerals)
kaolinite	$Al_2Si_2O_5(OH)_4$
montmorillonite	hydrous Mg-Fe-Al silicate
trioctahedral *smectite*	hydrous Mg-Fe-Al silicate
nontronite	hydrous Fe-Al silicate
illite	*muscovite-montmorillonite*
	(Magnesium Silicates)
LOUGHLINITE	$Na_2Mg_3Si_6O_{16} \cdot 8H_2O$
sepiolite	$H_2Mg_2Si_3O_{10}$
stevensite	hydrous magnesium silicate
talc	$Mg_3(OH)_2Si_4O_{10}$
chlorite	hydrous Fe-Mg-Al silicate
(Micas)	
hydrobiotite?	hydrous Na-Fe-Mg-Al silicate
biotite?	hydrous Na-Fe-Mg-Al silicate
	(Feldspars)
orthoclase	$KAlSi_3O_8$
albite	$NaAlSi_3O_8$
REEDMERGNERITE	$NaBSi_3O_8$
	(Borosilicates)
searlesite	$NaBSi_2O_6 \cdot H_2O$
GARRELSITE	$NaBa_3Si_2B_7O_{16}(OH)_4$
leucosphenite	$CaBaNa_3BTi_3Si_9O_{29}$
	(Titanosilicates)
labuntsovite	$(K,Ba,Na,Ca,Mn)(Ti,Nb)(Si,Al)_2(O,OH)_7 \cdot H_2O$
vinogradovite	$Na_5Ti_4AlSi_6O_{24} \cdot 3H_2O$
(Pyroxene)	
acmite	$NaFeSi_2O_6$
(Amphibole)	
magnesioarfvedsonite	$(Na,K)_{0.76}(Mg,Fe^{2+},Fe^{3+},Al,Ti)_{4.82}$ $(Na,Ca)_{2.00}Si_{8.06}O_{22}(OH,F_2)_{1.46}$
	(Zirconosilicates)
elpidite	$Na_2ZrSi_6O_{15} \cdot 3H_2O$
catapleite	$Na_2ZrSi_3O_9 \cdot 2H_2O$
(Zeolites)	
analcime	$NaAlSi_2O_6 \cdot H_2O$
harmotome-wellsite	$(Ba,Ca,K_2)Al_2Si_6O_{10} \cdot 6H_2O$
natrolite	$Na_2Al_2Si_3O_{10} \cdot 2H_2O$
clinoptilolite-mordenite	$(Ca,Na_2,K_2)(AlSi_5O_{12})_2 \cdot 6H_2O$
apophyllite(?)	$KCa_4FSi_8O_{20} \cdot 8H_2O$
Hydrocarbons	
gilsonite, uintahite, utahite, tabbyite, ozokerite, ingramite, albertite, kerogen	
ABELSONITE (nickel porphyrin)	$C_{31}H_{36}N_4Ni$

characterized by a unique assemblage of minerals known elsewhere in igneous or pegmatitic paragenesis. Because of the economic importance of vast deposits of *trona* now mined and potential reserves of hydrocarbons (oil shale) and *dawsonite* (alumina), the mineralogy of the formation has long been intensively studied, first by Bradley, in 1929. Some 80 authigenic species have been identified, of which 13 were first found in the formation, 7 so far nowhere else. Table 1 lists the minerals, with the 13 in capitals.

As Table 1 shows, beside the usual rock-forming elements constituting sedimentary rocks generally and oil shales especially, the Green River sedimentary mineralogy includes unusual concentrations of such elements as rare earths, niobium, uranium, titanium, boron, and zirconium—all of "cold-water," nonigneous, geochemical origin.

A comprehensive review of the mineralogy with extensive bibliography appeared in the *Mineralogical Record* (Milton, 1977).

Minerals

Sulfur is rarely seen, *gypsum* being an alteration product of sulfide minerals in weathered outcrops. *Pyrite* and *pyrrhotite* are abundant and widespread, always in discrete crystals or crystal aggregates, not in beds or massive aggregates. *Marcasite* is very rare and confined to weathered outcrops. *Wurtzite* is widespread, but always in microscopic, typically hemimorphic, orange-red crystals. Of the halides, only *halite* is found in massive beds, in the deeper areas of the formation. *Fluorite* occurs in disseminated crystals, sometimes associated with thin beds of *cryolite* (in Colorado); *neighborite* is rare, in Utah and Colorado. *Quartz* is ubiquitous, as chalcedony and crystal *quartz*, the latter in simple microscopic combinations of ±rhombohedra and prisms; no twinning has been observed. *Nordstrandite* is extremely rare, and is known only as microscopic fissure coatings in Colorado. *Fluorapatite* is scanty but widespread, except in Wyoming, where extensive beds, slightly uraniferous, are found. These may be the only known nonmarine phosphate bedded deposits. Collophane occurs as disseminated fossil remains.

The Green River (simple) carbonates include many very rare or unique species, a few with notable rare earths or uranium. Compound carbonates include carbonate-phosphate, sulfates, and chlorides; only *northupite* is widespread and abundant.

Sulfates comprise two groups—one found by weathering processes on sulfide-bearing outcrops; these are all hydrous sulfates. *Barite* and *celestite* are found, rarely and sparsely, in deep cores. *Anhydrite* is not known in the formation. Hydrocarbons include a rather poorly defined variety of organic substances, of generally complex composition (mixtures), apparently uncrystallized, with the exception of *abelsonite*, which probably has a relatively simple composition and occurs in macroscopic crystals. Silicates, the most abundant group, include various clay minerals, ***feldspars, amphiboles, pyroxenes***, and a number of complex boron-titanium silicates, as well as zirconosilicates. ***Zeolites*** are uncommon, with the exception of ubiquitous *analcime*. However, some altered tuff beds consist largely of *clinoptilolite-mordenite*.

Conditions of Origin

The Green River formation includes clastic beds, such as siltstones and sandstones, fossiliferous limestones, and ash beds, with an extensive series of saline beds (mostly carbonate, but also chloride and phosphate). These represent evaporites and chemical precipitates; and diagenesis has reconstituted the ionic constituents of the bedded deposits, forming silicates such as *searlesite, leucosphenite, acmite*, and *labuntsovite*. An obscure, but conceivably significant role in the mineral formation may have been the existence for long periods of time—hundreds of thousands if not millions of years—of fresh oxygenated waters, teeming with life, animal and vegetable, overlying an anaerobic stagnant hypolimnion. The role of microscopic living organisms in concentrating trace elements has become known in recent years through studies of atomic fallout. Such organisms may, over great periods of time, concentrate uranium, zirconium, niobium, and zinc, among other elements. In the ecology of the ancient lakes, further concentrations would be effected as these organisms became food for higher forms of life, with eventual deposition in the tissues of fishes, which form characteristic fossils of many Green River beds. Following this biogenic concentration of rare elements and sedimentation of the nitrogeneous organic debris in the hypolimnion, fugitive substances, such as methane, hydrogen sulfide, and ammonia, must have been generated in the bottom muds and had a role in the special mineralization of the Green River formation. Indicative of generally reducing conditions governing Green River mineral formations is the virtually complete fixation of sulfur as sulfide rather than sulfate. Sulfates (*gypsum, celestite*, and *siderotil*) are only known as very minor recent weathering products; *barite,* however, may be diagenetic. Of course, the oil shale itself was formed under reducing conditions. There have been reports, apparently reliable, of ammonia and hydrogen being found in some quantity in certain wells in Wyoming and Colorado. The existence of an ammonium feldspar, *buddingtonite*, in the geologically somewhat similar Searles Lake beds in California suggests that unstable ammonia compounds, with ammonia now replaced by alkalis, may have once existed.

There is a definite variation in authigenic mineral content or in morphology of individual minerals from basin to basin in the Green River, presumably reflecting variations in conditions of sedimentation or, more probably, diagenesis.

Studies have been made of the physical chemistry of carbonate mineral formation in the Green River, notably by Hans Eugster (Milton and Eugster, 1959). In Green River petrogeny, temperatures could hardly have exceeded 250°C, with moderate pressures of at most a few thousand feet of overlying sediments.

Following discovery of the boron-*albite reedmergnerite*, it, with boron-*biotite* and other silicates with boron replacing aluminum, has been synthesized. Many studies have been made of chemical systems related to Green River mineralogy, particularly in relation to soda ash production by the Solvay process; thus, *wegscheiderite* was known as Wegscheider's salt for fifty years prior to its discovery in Wyoming and Colorado. Although the respective component systems had been investigated, it seemed exhaustively, neither $Na_2Ca_2(CO_3)_3$, *shortite*, nor $NaBSi_3O_8$, *reedmergnerite*, were known prior to their discovery in the Green River.

CHARLES MILTON

References

Bradley, W. H., 1964. Geology of Green River formation and associated Eocene rocks in southwestern Wyoming and adjacent parts of Colorado and Utah, *U.S. Geol. Surv. Prof. Paper 496-A*, 86p.

Bradley, W. H., and Eugster, H. P., 1969. Geochemistry and paleolimnology of the trona deposits and associated authigenic minerals of the Green River formation of Wyoming, *U.S. Geol. Surv. Prof. Paper 496-B*, 71p.

Brobst, D. A., and Tucker, J. D., 1973. X-ray mineralogy of the Parachute Creek member, Green River formation, in the Northern Piceance Creek Basin, Colorado, *U.S. Geol. Surv. Prof. Paper 803*, 1-53.

Eugster, H. P., and Surdam, R. C., 1973. Depositional environment of the Green River formation of Wyoming: a preliminary report, *Geol. Soc. Am. Bull.*, **84**, 1115-1120.

Eugster, H. P., and Surdam, R. C., 1974. Depositional environment of the Green River formation of Wyoming: Reply, *Geol. Soc. Am. Bull.*, **85**, 1192.

Fahey, Joseph J., 1962. Saline minerals of the Green River formation with a section on X-ray powder data for saline minerals of the Green River formation by Mary E. Mrose, *U.S. Geol. Surv. Prof. Paper 405*.

Goodwin, Jonathan H., and Surdam, R. C., 1967. Zeolitization of tuffaceous rocks of the Green River formation, Wyoming, *Science*, **157**, 307-308.

Hay, R. L., 1966. Zeolites and zeolitic reactions in sedimentary rocks, *Geol. Soc. Am. Spec. Paper 85*, 130p.

Heinrich, E. W., 1966. *Geology of carbonatites*. Chicago: Rand McNally, 126-128.

Lundell, Leslie L., and Surdam, R. C., 1975. Playa-lake deposition: Green River formation, Piceance Creek Basin, Colorado, *Geology*, **3**, 493-497.

Milton, C., 1957. Authigenic minerals of the Green River formation of the Uinta Basin, Utah, *Intermountain Assoc. Pet. Geol., Guidebook to the Uinta Basin*, 8th Field Conf., 136-143.

Milton, C., 1971. Authigenic minerals of the Green River formation, *University of Wyoming, Contributions to Geology–Trona Issue*, **10**, 57-63.

Milton, C., 1977. Green River mineralogy, *Mineral. Rec.*, **8**, 368-379.

Milton, C., and Eugster, H. P., 1959. Mineral Assemblages of the Green River formation, in P. H. Abelson, ed., *Researches in Geochemistry*. New York: Wiley, 118-150.

Milton, C., and Fahey, J. J., 1960a. Classification and association of the carbonate minerals of the Green River formation, *Am. J. Sci.*, (Bradley Vol.), **258-A**, 242-246.

Milton, C., and Fahey, J. J., 1960b. Green River Mineralogy–a historical account, *Wyoming Geol. Assoc. Guidebook 15th Annu. Field Conf.*, 159-162.

Milton, C., et al., Silicate mineralogy of the Green River formation of Wyoming, Utah, and Colorado, *Internat. Geol. Cong.*, 21st Session, Norden, pt. 21, 171-184.

Surdam, R. C., and Parker, R. B., 1972. Authigenic aluminosilicate minerals in the tuffaceous rocks of the Green River formation, Wyoming, *Bull. Geol. Soc. Am.*, **83**, 689-700.

Surdam, R. C., and Wolfbauer, C. A., 1975. The Green River formation–A playa-lake complex, *Geol. Soc. Am. Bull.*, **86**, 335-345.

H

HALOTRICHITE GROUP

This group of monoclinic, hydrous sulfate minerals has the general formula

$$AB_2(SO_4)_4 \cdot 22H_2O$$

where $A = Fe^{2+}$,Mg,Mn,Zn
B = Al,Fe^{3+},Cr^{3+}

The group includes the following minerals:

- *apjohnite* $MnAl_2(SO_4)_4 \cdot 22H_2O$
- *bilinite* $Fe^{2+}Fe^{3+}_2(SO_4)_4 \cdot 22H_2O$
- *dietrichite* $(Zn,Fe,Mn)Al_2(SO_4)_4 \cdot 22H_2O$
- *halotrichite* $Fe^{2+}Al_2(SO_4)_4 \cdot 22H_2O$
- *pickeringite* $MgAl_2(SO_4)_4 \cdot 22H_2O$
- *redingtonite* $(Fe,Mn,Ni)(Cr,Al)_2(SO_4)_4 \cdot 22H_2O$

The name is from the Latin *halotrichium* based on the older German *Haarsalz*, hair salt, in reference to the clusters and mats of hair-like fibrous crystals found as efflorescences in old mine workings and other sheltered places and currently occurring with other sulfates such as *epsomite, melanterite, slavikite*, etc., in places where there is weathering of *pyrite*-bearing shales (*vitriolschiefer*). The minerals are water soluble with an astringent taste.

KEITH FRYE

Reference

Menchetti, S., and Sabelli, C., 1976. The halotrichite group: the crystal structure of apjohnite, *Mineral. Mag.*, **40**, 599-608.

Cross-references: See mineral glossary.

HOPPER CRYSTALS–*See* SKELETAL CRYSTALS

HUMAN AND VERTEBRATE MINERALS

The normal skeletal and dental tissues of humans can be disposed of quickly; they consist of the ***apatite***, *dahllite* (carbonate hydroxy-apatite), with one exception–a portion of the inner ear, which is calcium carbonate. (See below under *aragonite, calcite, vaterite*, and *monohydrocalcite*.)

The pathogenic precipitates that occur in humans are very complex and include numerous biominerals, chiefly phosphates and oxalates. Again, *dahllite* is well represented among pathogenic mineralization (calcification) products, and possibly has been precipitated by bacterial action. Although intracellular *dahllite* is produced by, for example, *Bacterionema matruchotii*, normal bone growth takes place in animals bred under germ-free conditions. Thus, although pathogenic mineralizations normally imply the existence of microorganisms, in most cases, these pathogens have not been identified specifically with respect to their contributed biominerals. In fact, the nature of the pathogen may not be as important as its metabolites–carbonic anhydrase has been demonstrated to produce *dahllite* crystallization in vitro; lithofellic acid ($C_{20}H_{36}O_4$) seems to be associated with intestinal or stomach stones (bezoars) and may be the principal component of some.

While these remarks will be confined essentially to human biomineralogy, on a few occasions it seems essential to extend them beyond *Homo* on the assumption that concretionary bodies that occur in lower primates, for example, can also occur in man (see Table 1).

By far the best methods for studying biominerals are the use of immersion liquids in conjunction with a polarizing microscope and the use of X-ray powder diffraction patterns. Since some uroliths, for example, may contain two or more phases (minerals), it is essential to know whether calcium carbonate is actually present before reporting it, despite effervescence obtained by a chemical test, because *dahllite* contains detectable amounts of CO_2. Chemical tests fail to distinguish, furthermore, between the two common calcium oxalates, the carbonates and calcium phosphates. Many of the earlier reports–prior to use of X-ray diffraction methods–were subject to error in interpretation of the mineral components.

The occurrences are mentioned under the mineral names.

Apatite. $Ca_{10}F_2(PO_4)_6$ does not form under physiological conditions, although F is presumed to replace

TABLE 1. Minerals Occurring in Primates[a]

Phosphates	Carbonates	Sulfates
dahllite (end.)	*aragonite* (end.)	*barite* (exo.)
bobierrite (end.)	*calcite* (end.)	*gypsum* (end.)
brushite (end.)	*monohydrocalcite* (?)	*epsomite* (?)
dittmarite (?)	*vaterite* (end.)	*hexahydrite* (?)
hannayite (?)	**Silicates**	**Oxalates**
hopeite (end.)	*actinolite*, asbestiform (exo.)	*weddellite* (end.)
monetite (end.)	*antigorite*, asbestiform (exo.)	*whewellite* (end.)
newberyite (end.)	*tremolite*, asbestiform (exo.)	**Others**
"octacalcium phosphate" (end.)	*kaolinite* (exo.)	*thorianite* (exo.)
phosphorroesslerite (?)	opaline silica (end.)	*uricite* (end.)
stercorite (end.)	*quartz* (exo.)	metals (exo.)
struvite (end.)	*vermiculite* (exo.)	*cinnabar* exo.)
whitlockite (end.)	*talc* (exo.)	

[a]Endogenous (end.) and exogenous (exo.) origins are indicated; (?) indicates uncertainty of the mineral's occurrence.

the OH of *dahllite* in bone and dental enamel to an extent of about 0.1% by weight of the entire inorganic portion. Of 87 cases of urinary calculi, Herman, Mason and Light, in 1958, found 11 to contain more than 1000 ppm of F. (The theoretical amount of F in $Ca_{10}F_2(PO_4)_6$ is 3.77% by weight.)

Aragonite. Although $CaCO_3$ (as *aragonite*) is stated in many textbooks to occur as otoliths of vertebrates, the statoconia of man are *calcite*, and various other biominerals occur among vertebrates. *Aragonite* has been found in gallstones and, very rarely, as a urolith. Because of the possible inversion, *aragonite* to *calcite*, fresh material must be studied, rather than material that has been preserved in formalin.

Asbestos. Ca, Mg, Fe silicates (*tremolite* and *actinolite*) of the ***amphibole*** group and hydrous magnesium silicate (*antigorite*) of the ***serpentine*** group occur as exogenous minerals in cases of pneumoconiosis (see vol. IVA: *Mineral Particles and Human Disease*).

Barite. $BaSO_4$ has been aspirated from the lungs of a foundry worker (*Parsons*, 1963).

Bobierrite. $Mg_3(PO_4)_2 \cdot 8H_2O$ is said to occur rarely as a urinary deposit of man, but was the essential component of a mammalian enterolith (intestinal stone) studied by Hutton in 1942. This iron-bearing *bobierrite* contained *vivianite*, 36%, and $Mn_3(PO_4)_2 \cdot 8H_2O$, 7%, in isomorphic substitution.

Brushite. $CaHPO_4 \cdot 2H_2O$ occurs, probably as a transitional endogenous mineral phase, in both oral and urinary calculi where the more stable component is *dahllite*, particularly when the pH exceeds 6.4 (see also *monetite* below).

Calcite. This polymorph of $CaCO_3$ occurs normally as statoconia within the labyrinth of the human ear. It also occurs as gallstones, in one case in combination with *aragonite* and *vaterite*. *Calcite* is frequently associated with cholesterol in gallstones; it has been reported in a stone from the pancreas (Gibson, 1974). Since *calcite* is the stable form of $CaCO_3$, it should be searched for in any situation where *aragonite* or *vaterite* is found.

Cinnabar. HgS has been reported from a tumor formed over a skin tattoo (Parsons, 1963).

Coal. Dust of coal may produce in the lungs anthracosis (a form of penumoconiosis).

Dahllite. A carbonate *hydroxyapatite* in which CO_3 groups replace PO_4 groups is the sole inorganic component of normal teeth and bones of humans and vertebrates in general. Supposedly isomorphic replacement of Ca takes place to a limited extent, Na, Mg, and K being present in amounts of about 1.4, 1.0, and 0.1 respectively as percentages of the oxides. The carbonate content is about 4–5%. *Dahllite* is the predominant inorganic constituent of oral calculus (supragingival, subgingival, and glandular stones) and many urinary calculi of the phosphate type (McConnell, 1963). Although *hydroxyapatite* is said to occur, all such reported occurrences probably are based on inadequate testing for CO_2, because *hydroxyapatite* (without CO_2) is very rare in nature and seems to be impossible to synthesize under simulated physiological conditions. Pathogenic occurrences, in addition to urinary and oral calculus, are in the lungs (in connection with histoplasmosis), the spleen, prostate gland, appendix, testes, and the walls of the bronchi. Induced calcification of the aorta of cattle and the induced corneal calcification of rabbits are apatitic. Periarticular calcifications are *dahllite*. In addition, the "sand" of the pineal gland is apatitic, but this may be normal rather than pathologic; the quantity seems to bear a rough correlation with age; and its function, if any, is unknown.

Dittmarite. $NH_4MgPO_4 \cdot H_2O$ possibly occurs with *newberyite* and *hannayite* (see below).

Gypsum. $CaSO_4 \cdot 2H_2O$ is a rare component of uroliths, particularly those composed essentially of uric acid.

Epsomite. $MgSO_4 \cdot 7H_2O$ has been reported as a possible component of uroliths.

Halite. NaCl is a questionable endogenous mineral of urinary calculi; it may represent the evaporation product from storage in saline solutions.

Hannayite. $Mg_3(NH_4)_2H_4(PO_4)_4 \cdot 8H_2O$ is a possible component of uroliths that also contain *struvite* and *newberyite* in addition to *dahllite*. Five cases are reported by Gibson (1974).

Hexahydrite. $MgSO_4 \cdot 6H_2O$ was reported as a component for one of 14,500 uroliths (Gibson, 1974), where it may have represented the partial dehydration of *epsomite* ($MgSO_4 \cdot 7H_2O$).

Hopeite. $Zn_3(PO_4)_2 \cdot 4H_2O$ has been reported as a most unusual component of two uroliths (Parsons, 1963).

Hydroxyapatite. See *dahllite* above.

Kaolinite. $Al_2Si_2O_5(OH)_4$ is an exogenous mineral sometimes associated with *quartz* in pneumoconiosis.

Metals. Metallic objects introduced orally are usually passed within a few days; those introduced as puncture wounds may require surgical removal.

Mica. See *vermiculite* below.

Monetite. $CaHPO_4$ is a rarely encountered component of urinary calculi where the pH is about 5.1 or lower; it is ordinarily associated with oxalates (*weddellite* and *whewellite*). It has been reported also in dental calculus.

Monohydrocalcite. $CaCO_3 \cdot H_2O$ occurs, insofar as present knowledge is concerned, only as statoconia of the tiger shark, where it was found in association with *aragonite* and *calcite*.

Newberyite. $MgHPO_4 \cdot 3H_2O$ occasionally is found in uroliths of the *dahllite-struvite* type, where it may form from decomposition of *struvite*.

Octacalcium phosphate. $Ca_8H_2(PO_4)_6 \cdot 5H_2O$ has been reported as a minor component of some dental calculus samples. This substance is not particularly stable and probably is converted to *dahllite*; it is unknown as a mineral—other than a biomineral—and consequently no mineral name is applied.

Opaline silica. Hydrous SiO_2 has been found in association with *calcite* and *weddellite* as a component of urinary calculi of cattle; in humans it possibly forms as a result of silica pneumoconiosis.

Phosphorroesslerite. $MgHPO_4 \cdot 7H_2O$ has never been reported but like *newberyite* might result from decomposition of *struvite* or *hannayite*.

Quartz. SiO_2 occurs as exogenous nodules within the lungs of patients suffering from pneumoconiosis produced by this mineral. *Quartz* also occurs as a minor constituent, again exogenous, in stomach stones of mammals.

Stercorite. $NaNH_4HPO_4$ (microcosmic salt) occurs in human urine.

Struvite. $MgNH_4PO_4 \cdot 6H_2O$ is often a principal phosphatic component of uroliths. It is one of a few biominerals that may be associated with a particular bacterium, *Proteus mirabilis*, an ammonia producer.

Talc. $Mg_3(OH)_2Si_4O_{10}$ has been reported as an exogenous component of the pericardium where it produced death through formation of a granuloma (Parsons, 1963).

Thorianite. ThO_2 has been reported as crystals within the spleen and liver of a person who had received injections of thorotrast (Parsons, 1963).

Uricite. $C_5H_4N_4O_3$. Mineral name for the monoclinic anhydrous form of uric acid.

Vaterite. $CaCO_3$ has been reported in a gallstone that also contained *aragonite* and *calcite*.

Vermiculite. A micaceous mineral, *vermiculite* has been found in lung tissue of horses following its use as an absorbent on stall floors.

Vivianite. See *bobierrite* above.

Weddellite. $Ca(COO)_2 \cdot 2H_2O$ is a principal component of many urinary calculi, where it occurs in association with *whewellite* (Lonsdale et al., 1968).

Whewellite. $Ca(COO)_2 \cdot H_2O$, with *weddellite*, is a principal component of many urinary calculi of the uric acid type.

Whitlockite. $\beta Ca_3(PO_4)_2$, particularly that containing small amounts of MgO and occasionally small amounts of CO_2, is a component of dental calculus where *dahllite* is the principal component. Chronic renal failure has produced calcified visceral tissues that are either *whitlockite* or an immediate precursor of this biomineral. It has been reported as a testicular calcification, as prostatic calculi, and in urinary calculi. In the last case it may be pseudomorphous after *brushite*.

Substances introduced into the body as restorative or prosthetic devices have not been considered here because they are neither normal nor pathogenic.

Numerous organic substances that are not recognized as minerals may occur in uroliths: uric acid, uric acid dihydrate, ammonium acid urate, sodium acid urate monohydrate, l-cystine, and xanthine (?). Gallstones may be primarily cholesterol, while stomach and intestinal stones (mostly from animals) may involve numerous organic compounds, such as keratin (hair), ellagic acid, and lithofellic acid.

Finally, it can be said that, among phosphates, any of the 71 biominerals listed by McConnell (1973), except *francolite*, might occur. And to this list now must be added *biphosphammite*, $NH_4H_2PO_4$, and *phosphammite*, $(NH_4)_2HPO_4$. However, those minerals high in aluminum or iron would not be expected to occur unless the pathologic conditions were most unusual.

DUNCAN McCONNELL

References

Gibson, R. J., 1974. Descriptive human pathological mineralogy, *Am. Mineralogist*, **59**, 1177–1182.

Lonsdale, K.; Sutor, D. J.; and Wooley, S., 1968. Composition of urinary calculi by X-ray diffraction. Collected data from various localities, *Brit. J. Urology*, **40**, 33–36, 402–411.

McConnell, D., 1963. Concretions, pathological, including urinary "stones," oral calculus, and gallstones: diffraction analysis, in G. L. Clark, ed., *Encyclopedia of X-rays and Gamma Rays*. New York: Reinhold, 178–179.

McConnell, D., 1973. Biomineralogy of phosphates and physiological mineralization, in E. J. Griffith et al., eds., *Environmental Phosphorus Handbook*. New York: Wiley, 425–442.

Parsons, J., 1963. Medical problems: Applications of X-ray diffraction anaysis, in G. L. Clark, ed., *Encyclopedia of X-rays and Gamma Rays*. New York: Reinhold, 588–590.

Cross-references: *Apatite Group; Asbestos; Calcite Group; Invertebrate and Plant Mineralogy;* see also mineral glossary. Vol. IVA: *Medical Geology; Mineral Particles and Human Disease.*

HUMITE GROUP

Humites are magnesium or manganese orthosilicates similar in structure to the ***olivines*** (see *Olivine Group*). The humite structure is a slightly distorted hexagonal, close-packed array of oxygen, fluoride, or hydroxyl ions. Silicon occurs in isolated tetrahedra and magnesium or manganese occupies octahedra that share edges to form zigzag chains in the structure.

Humite compositions may be idealized by the general formula

$$nM_2SiO_4 \cdot M(OH,F)_2$$

where $n=1,2,3,4$. The group includes *norbergite*, where M=Mg, $n=1$; *chondrodite*, where M=Mg, $n=2$; *alleghenyite*, where M=Mn, $n=2$; *humite*, where M=Mg, $n=3$; *manganhumite*, where M=Mn, $n=3$; *clinohumite*, where M=Mg, $n=4$; and *sonolite*, where M=Mn, $n=4$, thus:

- *alleghanyite:* $Mn_5(SIO_4)_2(OH)_2$
- *chondrodite:* $(Mg,Fe)_5(SiO_4)_2(F,OH)_2$
- *clinohumite:* $(Mg,Fe)_9(SiO_4)_4(F,OH)_2$
- *humite:* $(Mg,Fe)_7(SiO_4)_3(F,OH)_2$
- *manganhumite:* $Mn_7(SiO_4)_3(OH)_2$
- *norbergite:* $Mg_3(SiO_4)(F,OH)_2$
- *sonolite:* $Mn_9(SiO_4)_4(F,OH)_2$

Substitution of F for OH is extensive for most members of this group, although *clinohumite* can occur without F and *norbergite* may be very low in OH. At any particular locality, however, the F:OH ratio is constant. Substitution of Fe, Ca, or Ti for Mg or Mn is limited; but there is complete solid solution between the Mg and the Mn members of the group—except for *norbergite*.

Manganhumite, *norbergite*, and *humite* are orthorhombic; the other members of this group are monoclinic. The magnesium ***humites*** are colorless, yellow, or brown, whereas the manganese ***humites*** tend to be of reddish shades.

Paragenesis is restricted. Magnesium ***humites*** occur in metamorphosed and metasomatized calcareous rocks and in skarns at contacts with felsic plutonic rocks. Manganese ***humites*** occur singly or together with other Mn-bearing minerals in hydrothermal ore deposits.

KEITH FRYE

References

Deer, W. A.; Howie, R. A.; and Zussman, J., 1962. *Rock-Forming Minerals*, vol. 1. London: Longman, 47–58.

Francis, C. A., and Ribbe, P. H., 1978. Crystal structures of the humite minerals: Magnesian manganhumite. *Am. Mineralogist*, **63**, 874–877.

Jones, N. W.; Ribbe, P. H.; and Gibbs, G. V., 1969. Crystal chemistry of the humite minerals, *Am. Mineralogist*, **54**, 391–411.

Robinson, K., Gibbs, G. V., and Ribbe, P. H., 1973. The crystal structures of the humite minerals. IV. Clinohumite and titanoclinohumite. *Am. Mineralogist*, **58**, 43–49.

Cross-references: *Polysomatism;* see also mineral glossary.

I

INCLUSIONS, FLUID–*See* **BAROMETRY, GEOLOGIC.** Vol. IVA: **FLUID INCLUSIONS**

INDEX OF REFRACTION–*See* **OPTICAL MINERALOGY; REFRACTIVE INDEX**

INDUSTRIAL MINERAL DEPOSITS –*See* **MINERAL INDUSTRIES; PLACER DEPOSITS**

INTERNATIONAL MINERALOGICAL ASSOCIATION

The International Mineralogical Association (IMA) is an organization of national mineralogical societies or groups representing individual countries and limited to one member from each country. IMA was founded to promote and implement cooperation among these societies in aspects of mineralogy of international interest and concern.

International cooperation is accomplished through day-to-day work of nine commissions to which each national society may appoint a member (Table 1). Each commission reports periodically on activities, progress, and accomplishments.

The IMA published a *World Directory of Mineralogists* in 1962 and 1970, a volume containing the Proceedings of the meetings held between 1958 and 1972, and a *World Directory of Mineral Collections* prepared by the Commission on Museums in 1974 with a second edition in 1977. From time to time, the IMA issues *IMA NEWS* containing news of the national societies and the commissions.

Officers and councilors are elected for a term of four years. Voting power at the biennial meetings ranges from one to five, determined by the number of resident members of a member organization.

The make-up of the IMA Councils in previous

TABLE 1. IMA Commissions

Commission	Office	Name
1. Abstracts	Chairman:	T. Hügi (Switzerland)
	Secretary:	R. A. Howie (Great Britain)
2. Cosmic mineralogy	Chairman:	D. P. Grigoriev (USSR)
	Secretary:	K. Keil (USA)
3. Crystal growth of minerals	Chairman:	I. Sunagawa (Japan)
	Secretary:	A. Baronnet (France)
4. History and teaching	Chairman:	H. E. Wenden (USA)
	Secretary:	P. Paulitsch (Federal Republic of Germany)
5. Mineral data and classification	Chairman:	A. Preisinger (Austria)
	Secretary:	P. B. Moore (USA)
6. Museums	Chairman:	P. C. Zwaan (Netherlands)
	Secretary:	O. V. Petersen (Denmark)
7. New minerals and mineral names	Chairman:	A. Kato (Japan)
	Vice-Chairman:	M. H. Hey (Great Britain)
	Secretary:	G. Gottardi (Italy)
8. Physics of minerals	Chairman:	S. Hafner (Federal Republic of Germany)
	Vice-Chairman:	A. S. Marfunin (USSR)
	Secretary:	C. Prewitt (USA)
9. Ore microscopy	Chairman:	G. A. Desborough (USA)
	Vice-Chairman:	C. Levy (France)
	Secretary:	E. F. Stumpfl (Austria)

TABLE 2. IMA Council (1958–1978)

Council	1958–1960	1960–1964	1964–1970	1970–1974	1974–1978
President	*Parker SWZ	Fisher USA	*Tilley UK	Strunz FRG	Sobolev USSR
Past President	+Claringbull UK	*Parker SWZ	Fisher USA	*Tilley UK	Strunz FRG
1st Vice President	Wickman SWD	*Tilley UK	Strunz FRG	Watanabe JPN	Guillemin FR
2nd Vice President	Grigoriev USSR	Barsanov USSR	Korzhinskii USSR	Guillemin FR	Fornaseri IT
Secretary	Amorós SPN	Amorós SPN	Preisinger AUS	*Hooker USA	*Hooker USA
Treasurer	Fisher USA	Berry CAN	Berry CAN	Berry CAN	Berry CAN
Councillors	Orcel FR	*Naidu IND	*Barth NOR	Fornaseri IT	Coombs NZ
	*Onorato IT	Sahama FIN	Kutina CSR	Hügi SWZ	Font-Altaba SPN
	Ito JPN	Winkler FRG	Watanabe JPN	Kostov BUL	Howie UK
					Sunagawa JPN
					Zemann AUS

+Prefounding adviser
*Deceased

TABLE 3. IMA Officers and Councilors (1978–1982)

President	C. Guillemin (France)
Vice-Presidents	I. Kostov (Bulgaria)
	P. J. Wyllie (USA)
Secretary	C. Tennyson (Federal Republic of Germany)
Treasurer	L. G. Berry (Canada)
Councilors	D. S. Combs (New Zealand)
	M. Font-Altaba (Spain)
	R. A. Howie (UK)
	H. Sorensen (Denmark)
	Ichiro Sunagawa (Japan)
Past President	V. S. Sobolev (USSR)

years and the officers and councilors for 1978–1982 are shown in Tables 2 and 3.

Data provided by
MARJORIE HOOKER
CHARLES T. PREWITT

INVERTEBRATE AND PLANT MINERALOGY

The vast majority of invertebrate and plant skeletal parts are composed of calcium carbonates and opal. However, a variety of other minerals occur occasionally. Table 1 lists the minerals found as skeletal parts of invertebrates and plants. In the table, references are provided for only the exotic occurrences.

In addition to the minerals listed in Table 1, various amorphous hardparts have been reported. It is almost certain that, with further work, additional minerals will be found as skeletal parts of invertebrates and plants.

The mineralogy of skeletal parts of organisms appears to be influenced by environmental factors, notably temperature. Chave (1954) showed that in many marine invertebrates that deposit Mg-*calcite* hardparts, the amount of magnesium in the *calcite* increased with increasing temperature. Furthermore, the rate of magnesium increase with increasing temperature appeared to be phylogenetically controlled, with higher forms (ostracods and barnacles) showing the least effects of temperature.

Lowenstam (1954) showed that in three groups of bimineralic carbonate-secreting marine invertebrates (bryozoans, serpulid worms, and molluscs), the ratio of *aragonite* to *calcite* increased with increasing temperature.

Most of the common mineral phases used by organisms as skeletal parts have been shown to be unstable at near-surface conditions of pressure and temperature (Jamieson, 1953; Clarke, 1957; Harker and Tuttle, 1955). Thus, during diagenesis it is quite likely that much of the fossil record is lost or modified. Prediction of the original mineralogy of many fossil forms is risky.

KEITH E. CHAVE

References

Chave, K. E., 1954. Aspects of the biogeochemistry of magnesium 1. Calcareous marine organisms, *J. Geol.*, **62**, 266–283.

Clarke, S. P., 1957. A note on calcite-aragonite equilibrium, *Am. Mineralogist*, **42**, 564–566.

Harker, R. I., and Tuttle, O. F., 1955. Studies in the system CaO-MgO-CO_2. II. Limits of solid solution along the binary join $CaCO_3$-$MgCO_3$, *Am. J. Sci.*, **253**, 274–282.

Jamieson, J. C., 1953. Phase equilibrium in the system calcite-aragonite, *J. Chem. Phy.*, **21**, 1385–1390.

Lowenstam, H. A., 1954. Factors affecting the aragonite:calcite ratios in carbonate-secreting marine organisms, *J. Geol.*, **62**, 284–322.

Lowenstam, H. A., 1962. Goethite in radular teeth of Recent marine gastropods, *Science*, **137**, 279–280.

Lowenstam, H. A., 1967. Lepidocrocite, an apatite mineral, and magnetite in teeth of chitons (Polyplacophora), *Science*, **156**, 1373–1375.

Lowenstam, H. A., 1968. Weddellite in a marine gastropod and in Antarctic sediments, *Science*, **162**, 1129–1130.

TABLE 1. Mineralogy of Mineralized Invertebrates and Plants

Mineral	Composition	Organism
calcite	$CaCo_3$	All coccolithophores All planktonic Foraminifera Parts of some molluscs
aragonite	$CaCO_3$	All Chlorophyta All Phaeophyta Some Rhodophyta Some Phanerogams Some benthic Foraminifera All or parts of some Sclerospongia All Hydrozoa Some Anthozoa All or parts of Bryozoa All or parts of Amphineura All Scaphopoda All or parts of Gastropoda All or parts of Pelecypoda Some Cephalopoda All or parts of Polychaeta
Mg-*calcite*	$(Ca_{1-x}Mg_x)CO_3$ (x = trace to ≈ 0.30)	Some Rhodophyta Some benthic Foraminifera All calcareous Porifera Some or parts of Anthozoa Some or parts of Bryozoa All articulate Brachiopoda Some Cephalopoda All or parts of Polychaeta All Echinodermata All Ostracoda All Cirripedia All Malacostraca
opal	$SiO_2 \cdot nH_2O$	All diatoms All silicoflagellates Some Phanerogams Some Radiolaria All Hexactinellida All Demospongia Parts of some Sclerospongia Parts of some Gastropoda
dahllite	$Ca_{10}(PO_4)_6(CO_3)(OH)$ (With many ionic substitutions– < 1% fluorine)	Parts of some Pelecypoda (Watabe, 1956)
francolite	$Ca_{10}(PO_4)_6(CO_3)F$ (With many ionic substitutions– >1% fluorine)	All inarticulate Brachiopoda (McConnell, 1963)
celestite	$SrSO_4$	Some Radiolaria (Odum, 1951)
fluorite	CaF_2	Parts of some Gastropoda (Lowenstam and McConnell, 1968) Parts of some Malacostraca (Lowenstam and McConnell, 1968)
weddellite	$CaC_2O_4 \cdot 2H_2O$	Parts of some Gastropoda (Lowenstam, 1968)
lepidocrocite	$\gamma FeO(OH)$	Parts of some Amphineura (Lowenstam, 1967)
goethite	$\alpha FeO(OH)$	Parts of some Gastropoda (Lowenstam, 1962)
magnetite	Fe_3O_4	Parts of some Amphineura (Lowenstam, 1967)
vaterite	$CaCO_3$	Some Tunicata (Lowenstam and Abbott, 1975)

Lowenstam, H. A., and Abbott, D. P., 1975. Vaterite: A mineralization product of the hard tissues of a marine organism (Ascidiacea), *Science*, **188**, 363-365.

Lowenstam, H. A., and McConnell, D., 1968. Biologic precipitation of fluorite, *Science*, **162**, 1496-1498.

McConnell, D., 1963. Inorganic constituents in the shell of the living brachiopod *Lingula, Geol. Soc. Am. Bull.*, 74, 363-369.

Odum, H. T., 1951. Notes on the strontium content of sea water, celestite radiolaria, and strontianite snail shells, *Science*, **114**, 211-213.

Watabe, N., 1956. Dahllite identified as a constituent of prodissoconch I of *Pinctada martensii, Science*, **124**, 630-630.

Cross-references: *Calcite Group; Human and Vertebrate Mineralogy; Opal; Resin and Amber;* see also mineral glossary.

ISOMORPHISM

The concept of isomorphism played an important role in chemistry, crystallography, and mineralogy from its enunciation by Mitscherlich (1819) until the second quarter of the twentieth century. However, with the growth of our understanding of the structure of crystals, resulting from the development of X-ray crystallography, it has become evident, that, in the ultimate analysis, the concept of isomorphism does not correspond precisely to any single category of phenomena; it corresponds in part to the concepts of isostructural crystals, solid-solution series, and epitaxy, but not to the whole of any one of them. Nevertheless, because of its historical importance, isomorphism remains a widely used, and useful, concept.

Etymologically, isomorphism means similarity of crystal form. To be regarded as isomorphous, two substances must form crystals of identical symmetry, the interfacial angles of which are approximately equal. However, the degree of latitude to be allowed in this approximation has never been defined nor has the mere identity of interfacial angles ever been regarded as a sufficient criterion to decide that substances are isomorphous. This is perhaps surprising until it is recalled that all substances that crystallize in the cubic system necessarily have identical interfacial angles, and to treat them as isomorphous on this basis alone would not be profitable.

Mitscherlich originated the concept of isomorphism when he found that certain pairs of salts (e.g., phosphates and arsenates) that developed closely similar crystalline forms also possessed very similar chemical formulas, differing only by the replacement of one element by another, atom for atom. The importance of the concept of isomorphism to nineteenth-century chemistry lay in this relationship because, if corresponding compounds of two elements could be shown to be isomorphous, this provided evidence for such similarity in their formulas at a time when the atomic weights of many elements were unknown or in dispute. If the formula of a compound A could be assigned on the basis of its isomorphism with another compound B of known formula, then the atomic weight of any element in compound A could be determined by quantitative analysis of that compound. Because the development of chemistry was vitally dependent on the determination of correct atomic weights, it was important to avoid the possibility of a false attribution of similar formulas to a pair of compounds on the basis of a merely accidental similarity of crystal form. Hence, there arose the idea of "tests of isomorphism" over and above similarity of crystal form, and so it came about that the concept of the isomorphism of two substances acquired the meaning of *a proved association of similarity of crystal form and chemical formula.*

Tests for Isomorphism

Three tests have been used to confirm isomorphism between two compounds:

1. Formation of solid solutions between the two compounds (see vol. IVA: *Solid Solution*). Such solid solutions may extend over the whole range of composition from one compound to the other, or they may occur only over restricted ranges of composition near those of the two compounds.
2. It may be impracticable to prepare solid solutions of two compounds because of great differences in their melting points, solubilities, and chemical stabilities. In this case, an alternative test is the formation of overgrowths of one compound around a crystal of the other—that is, a crystal of one compound will continue to grow by forming a zone of the second compound around it, the two parts of the crystal being in crystallographic continuity.
3. A much less extreme form of this phenomenon is the formation of oriented overgrowths of small crystals of one compound on a crystal face of the other (see *Epitaxy*), as for example of sodium nitrate on crystals of *calcite*.

Inadequacies in the Concept of Isomorphism

The basis of isomorphism is fundamentally one of similarity of crystal structure, not merely of chemical composition. Obviously, substances of closely similar formula may form crystals of quite different structure just as a single substance of one given composition may form different structures (see *Polymorphism*).

In modern usage, isomorphism is in fact generally regarded as involving close similarity of structure. Such similarity strictly requires that the two structures should have identity of space group (see *Space-Group Symbol*), a much more rigorous requirement than the identity of point group (see *Point Groups*) that is demanded by similarity of crystal form.

In order to satisfy the tests for isomorphism, still further restrictions are imposed. The formation of solid solutions between substances of similar structure normally requires similarity in size of the structural units (ions, atoms, or molecules) and in the nature of the forces between them (e.g., similar degrees of ionic, covalent, or metallic bonding). The formation of oriented overgrowths also requires a match in the size of the repeating units; but as this match need only be in particular plane sections of the structure, it is possible for structures that are widely different to satisfy these tests, and alone they do not, therefore, provide very strong evidence of isomorphism.

Simple examples of these difficulties, which could not have been appreciated by nineteenth-century chemists, are provided by sodium chloride (NaCl), potassium chloride (KCl) and lead sulfide (PbS). All three have exactly the same structure type and have the same space group. However, because of the difference in size of the Na^+ and K^+ ions, the extent of solid solution between NaCl and KCl is very limited, although it is sufficient for them to be regarded as isomorphous. On the other hand, the bonding in PbS is much less ionic than that in the alkali halides, and as a result solid solution does not occur between them.

Thus, the concept of isomorphism involves not only similarity of formula and structure, but also an undefined degree of chemical similarity. In this respect, it differs from the modern concept of isostructural crystals, which is based purely on the geometrical arrangement of interatomic links in the crystal, without regard to the chemical nature or even the identity of the linked atoms. It is, therefore, possible for isostructural crystals to differ from one another in space group, or even in point group. Thus *sphalerite* (ZnS) and *diamond* are isostructural, although the symmetry of *sphalerite* is lower than that of *diamond*, because of the nonequivalence of alternate atoms, and they would certainly not be regarded as isomorphous. On the other hand, structural similarity can in some cases take precedence over strict similarity of formula even in substances regarded as isomorphous. Thus, the ***amphiboles*** *tremolite* $[Ca_2Mg_5Si_8O_{22}(OH)_2]$ and *richterite* $[Na_2CaMg_5Si_8O_{22}(OH)_2]$ are regarded as isomorphous in spite of their containing different numbers of atoms in their formulas. This possibility arises from the existence of a relatively rigid structural framework within which certain sites may be either filled or left empty. One is thus led to the concept that isomorphous substitution in an isomorphous series may involve substitution not only of one atom for another, but also of an atom for a vacancy and vice versa.

E. J. W. WHITTAKER

Reference

Mitscherlich, E., 1819 (1820). Sur la relation qui existe entre la forme cristalline et les proportion chimiques, *Ann. Chim. Phys.*, **14**, 172.

Cross-references: *Crystal Growth; Crystallography: Morphological Epitaxy; Polymorphism.*

ISOTOPIC VARIATIONS IN MINERAL DEPOSITS–*See* Vol. IVA

ISOTROPISM

Isotropism (isotropy) is a property whereby a substance exhibits an equal or uniform interaction in all directions or along all possible reference axes between it and an internal or external stimulus. Optical isotropism characterizes substances such as unstrained gases, liquids, and isometric (cubic) crystals, which transmit visible light waves of a specified frequency with equal velocity in all directions. Isotropism in volumes larger than single atoms depends on the nature of the stimulus and the volume under consideration, i.e., a crystal optically isotropic to visible radiation is not isotropic in its interaction with X-radiation. Also, a sufficiently large volume of rock consisting of randomly oriented anisotropic crystals (see *Anisotropism*) may behave isotropically in a stress field. Crystal or fragment accumulations in volumes large enough to behave isotropically in stress fields sometimes are described as being "massive."

ERNEST E. WAHLSTROM

References

Born, M., and Wolf, E., 1959. *Principles of Optics.* New York: Pergamon, 803p.

Ditchburn, R. W., 1963. *Light*, 2nd ed. New York: Interscience, 833p.

Turner, R. J., and Weiss, L. E., 1963. *Structural Analysis of Metamorphic Tectonites*, New York: McGraw-Hill, 545p.

Cross-references: *Anisotropism; Dispersion, Optical; Optical Mineralogy.*

J

JADE

Jade carving has been practiced for centuries by Chinese craftsmen, who have given the world many exquisite jade statues, much jewelry, and other articles shaped to extremely delicate forms. Jade is found in the Sinkiang province of China, and also in Taiwan, Burma, Japan, Mexico, Guatemala, New Zealand, Australia, Rhodesia, as well as the states of Alaska, Wyoming, and California. Ornamental jades include two different minerals, *jadeite* and *nephrite*. *Nephrite* is an ***amphibole*** in the *tremolite-actinolite* series, $Ca_2(Mg,Fe)_5(Si_4O_{11})_2(OH)_2$, derived from alpine-type peridotite-dunite intrusives. The rarer *jadeite* is a ***pyroxene*** of composition $NaAlSi_2O_6$, often found as stream-worn boulders. *Jadeite* is the harder of the two minerals, 7 on the Mohs scale compared to 6½, although *nephrite*, "the axe stone," is generally considered to be tougher and more resistant to fracture. *Jadeite* is 10% denser than *nephrite*, and sinks in methyl iodide. It also has an "icy" appearance caused by reflections from individual crystallites. Green, white, black, and yellow-brown are common colors for both jades, but vivid emerald-greens, tomato-reds, and lavenders are found only in *jadeite* (Zara, 1969).

Beauty and durability are the essential attributes of a gemstone (see *Gemology*). Durability requires both hardness and toughness to provide resistance to abrasion and impact. Toughness is the outstanding property of jade.

To the mineralogist, hardness means resistance to scratching and is empirically defined by the Mohs scale of hardness (q.v.). Toughness is defined as resistance to breakage. Among mineralogists, there is no widely accepted measure of toughness, although some minerals such as *nephrite* are regarded as much tougher than others.

The hardest minerals are not always the toughest; prospectors who have "tested" *diamonds* with hammers have learned this the hard way. To a gem cutter, toughness means resistance to fracture or cleavage during the normal cutting, grinding, and polishing procedures. The toughness scale given in Table 1 is an approximate ranking based on lapidary experience.

TABLE 1. Gemstones Arranged in Order of Decreasing Toughness and Hardness

Toughness	Hardness
carbonado	*diamond*
nephrite	carbonado
jadeite	*corundum*
chrysoberyl	*chrysoberyl*
corundum	*topaz*
quartz	*beryl*
diamond	***garnet***
peridot	*zircon*
beryl	*jadeite*
garnet	*quartz*
zircon	*spodumene*
topaz	*nephrite*
orthoclase	peridot
spodumene	*orthoclase*

After Parsons, 1969.

Dense polycrystalline minerals tend to be tougher than single crystals, and single crystals with easy cleavage are weakest of all. The chemical composition and crystal structure of *jadeite* is very similar to that of *spodumene* ($LiAlSi_2O_6$), yet *jadeite* is far tougher because it occurs as a dense polycrystalline material rather than as single crystals.

Several engineering parameters have been devised to describe toughness. Quantitatively, it can be represented by work-to-fracture, the amount of energy required to create a unit area of fracture surface, or by fracture strength, the stress required to break a specimen under specified conditions.

Fracture strengths of several varieties of *nephrite* and *jadeite* have been measured (Bradt et al., 1973), using an Instron testing machine. The fracture strength of *nephrite* is about 200 MN/m^2; that of *jadeite* is about 100. The strengths of the two jades exceed most commercially available ceramics (about 70), but are less than the hot-pressed oxides and nitrides used for cutting tools and turbine vanes, which usually have strengths of 500 MN/m^2.

Jade is even more outstanding in work-to-fracture measurements. Fracture energy is determined by breaking the specimen at a slow, controlled rate while measuring the displacement and applied forces. Integration of the

FIGURE 1. Fracture surfaces of *jadeite*. Note the presence of numerous cleavage steps, and the directional nature of the cleavage pattern, parallel to the ***pyroxene*** chains. Scanning electron micrograph (X730).

FIGURE 2. Fracture surfaces of *nephrite*. Note the fibrous texture of the fracture, the random orientation of the fiber bundles or colonies, and the very rough fracture surface, indicative of a tortuous crack path. Scanning electron micrograph (X700).

force-displacement curve gives the work, and fracture energy is then calculated by dividing the work by twice the cross-sectional area produced.

The average fracture energy for a number of Russian and Alaskan *nephrites* is 225 J/m^2, an exceptionally large value. Burmese *jadeite* specimens have fracture energies 120 J/m^2. Jade is about an order of magnitude tougher than most ceramic materials. Fracture energies for dense oxide ceramics seldom exceed 50, and single crystal values are usually about 1 J/m^2.

The high fracture energy and fracture strength of jade are not directly related to atomic bonding per se, but to the microstructure of the jade and the restrictions this microstructure imposes on crack propagation.

A scanning electron micrograph of the fracture surface of *jadeite* is shown in Fig. 1. *Jadeite* fracture surfaces characteristically show a very high percentage of transgranular cleavage fractures. When a crack propagates through *jadeite*, it does not proceed in an intergranular fashion, following the boundaries between the grains; rather, it proceeds directly through the grains in a transgranular fracture mode. At higher magnifications, cleavage step patterns are especially obvious, indicating that fractures are almost wholly restricted to certain crystallographic planes. The elongated nature of some of these cleavage steps strongly suggests that fractures occur parallel to the pyroxene chains, on the $\{110\}$ planes. It is this extensive transgranular cleavage-fracture mode that imparts toughness to *jadeite*.

Nephrite has a different fracture topography from *jadeite*, as is shown in the micrograph in Fig. 2. Fibers and bundles of fibers can be seen protruding from the *nephrite* fracture surface even at low magnification. The random orientation of individual fibers leads to interlocking, increasing the fracture strength. Fibrous microstructure such as this occurs also in ballas *diamond,* a particularly tough variety of carbonado used in drilling rock.

ROBERT E. NEWNHAM

References

Bradt, R. C.; Newnham, R. E.; and Biggers, J. V., 1973. The toughness of jade, *Am. Mineralogist*, **58**, 727-732.

Parsons, C. J., 1969. *Practical Gem Knowledge for the Amateur*. San Diego, Calif.: Lapidary Journal.

Zara, L., 1969. *Jade*. New York: Walker & Co.

Cross-references: *Electron Microscope (Scanning); Gemology; Mohs Scale of Hardness; Placer Deposits;* see also mineral glossary.

JOLLY BALANCE—*See* DENSITY MEASUREMENTS

K

KAOLINITE-SERPENTINE GROUP

The ***kaolinite-serpentine*** group has the general formula

$$M_{2-3}(Si,Al,Fe^{3+})_2O_5(OH)_4$$

where M=Mg,Fe^{2+},Ni,Mn^{2+},Al, and includes the following minerals:

amesite	*dickite*	*nacrite*
antigorite	*endellite*	*nepouite*
baumite	*fraiponite*	*orthochrysotile*
berthierine	*greenalite*	*parachrysotile*
brindleyite	*halloysite*	*pecoraite*
caryopilite	*kaolinite*	*zinalsite*
clinochrysotile	*kellyite*	
cronstedite	*lizardite*	

Layered silicates of this group are characterized by a ratio of 1:1 of tetrahedral:octahedral layers. ***Kaolinites*** are dioctahedral with trivalent cations occupying two-thirds of the available octahedral sites. The ***serpentines*** are trioctahedral with divalent cations occupying all the octahedral sites.

KEITH FRYE

Reference

Brindley, G. W., and Brown, G. (eds.), 1980. *Crystal Structures of Clay Minerals and their X-ray Identification*. London: The Mineralogical Society, 495p.

Cross-references: *Asbestos; Clays, Clay Minerals; Phyllosilicates; Soil Mineralogy;* see also mineral glossary.

L

LATTICE

A lattice, also called a Bravais lattice or a space lattice, is a three-dimensional array of points, each of which has an identical surrounding of neighboring points. As such, it forms a repetitious pattern of dimensionless points. Lattices extending infinitely in three dimensions are used in crystallography to describe the structural patterns found in all crystalline materials. The periodic arrangement of ions, atoms, molecules, or radicals in a crystal is described in terms of lattice and an asymmetric unit, the actual chemical units being assigned to each lattice point. Any lattice may be described in terms of translation vectors. A translation vector is any displacement that arrives repeatedly at an identical environment of points. As derived by Auguste Bravais in 1850 (see *Crystallography: History*), only 14 types of translational space lattices exist, based upon the symmetry (q.v.) that a periodic array of points may possess.

Nets

Some properties of lattices may be illustrated by nets, which are two dimensional analogs of lattices. Any two nonparallel vectors describe a net (Fig. 1). The two vectors may be of the same or of different length and they may be orthogonal or not. The same net may be represented by different pairs of translation vectors, the choice among these being dictated by convection or by convenience in crystallographic problems.

Repeating chosen translation vectors to form a parallelogram yields a unit cell for a particular net (Fig. 2). Repetition of a unit cell, a parallelogram, infinitely by translation vectors describes a net.

A unit cell is primitive (*P*) if it contains only one lattice point. This may be seen by displacing the unit-cell edges slightly in an arbitrary direction so that the unit cell encloses but one point (Fig. 2). (Alternatively, each point in a net may be "quartered," with one quarter assigned to each adjoining unit cell.)

An alternate choice of translation vectors such that a lattice point lies at the center as well as at each corner of a unit cell yields a nonprimitive, centered net with two points in each unit cell (Fig. 2). Other choices of translation vectors and unit cells can be envisioned but are not generally useful.

Note that a net is intrinsically neither primitive nor centered. Choice of translation vectors and unit cells determines how it is described. Obviously, a centered point has surroundings that are identical to those of all other points. All nonprimitive nets may be converted into

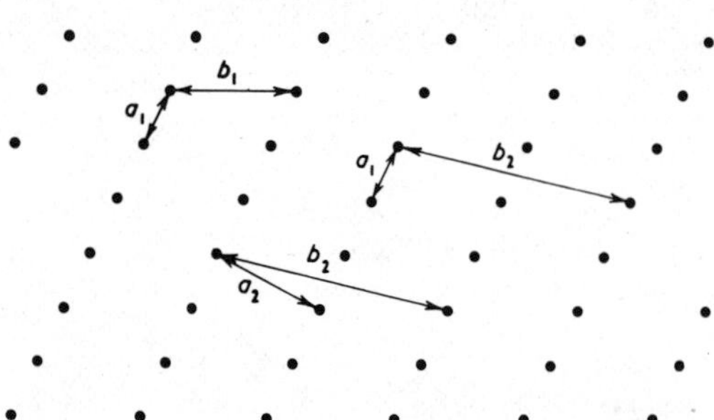

FIGURE 1. A net (lattice of two dimensions) with four different translation vectors—$\mathbf{a}_1$, $\mathbf{b}_1$, $\mathbf{a}_2$, and $\mathbf{b}_2$ (from H. D. Megaw, *Crystal Structures*, W. B. Saunders Company, Philadelphia, 1973; reprinted with permission). Note that the selection of $\mathbf{a}_1$ and $\mathbf{b}_2$ includes only half the points in the net.

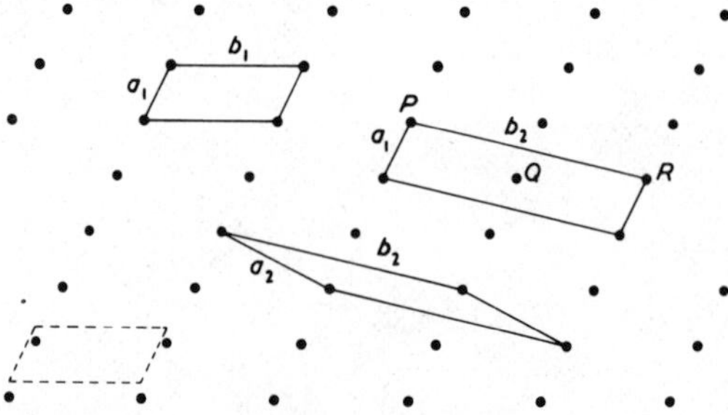

FIGURE 2. A net with unit cells determined by the translation vectors from Fig. 1 (from H. D. Megaw, *Crystal Structures*, W. B. Saunders Company, Philadelphia, 1973; reprinted with permission). Note that choosing $\mathbf{a}_1$ and $\mathbf{b}_2$ results in a face-centered unit cell. Dashed unit cell at lower left is displaced slightly to show that a primitive cell contains but one point.

primitive nets by selection of appropriate new translation vectors, although some of the symmetry of the net may be lost from the individual unit cell in the process. The same is true for lattices in three dimensions.

Bravais Lattice

A third translation vector, not in the plane of the net, describes a three-dimensional Bravais lattice. The directions of the three translation vectors describe the three reference axes of a lattice. By crystallographic convention, z is positive up or to the top of a page, y is positive to the right, and x is positive toward an observer or out of a page (toward lower left).

A unit cell is thus described by the lengths of its edges a, b, and c along the axes x, y, and z, respectively, and by the angles between the edges, α between b and c (y and z), β between a and c (x and z), and γ between a and b (x and y). These six values, three scalar and three angular, are the lattice parameters. Statement of the lattice parameters, of whether or not a lattice is primitive, and, if not primitive, of the location of points not at unit-cell corners completely and rigorously describes a lattice.

Nonprimitive lattice points may be located at the center of one face, at the centers of all faces, or at the interior or body center of a unit-cell parallelepiped. Unit cells may be selected in which other nonprimitive lattice points exist, but they are unconventional, not very useful, and reducible to one of the aforementioned unit cells by reselection of translation vectors (see Fig. 2). Since opposite faces of a unit cell are identical by definition, stipulation of a lattice point at one face center simultaneously locates a lattice point at the opposite face center.

If a nonprimitive lattice point at one face center is in the ab plane, it is termed C centered, in the ac plane B centered, and in the bc plane A centered (Fig. 3). (Centering of lattice points on two different faces simultaneously is demonstrably impossible since the resulting points would not have the same surroundings as the corner points.) Addition of a lattice point at one face center yields a unit cell with a total of two lattice points. (This may be visualized by displacing the unit cell slightly or by counting each corner point as being shared equally by eight adjoining unit cells and contributing one eighth to each. Similarly, each point at the center of a face is shared by two unit cells and contributes one half to each.)

A nonprimitive lattice having additional points at the centers of all six faces (designated F) has four lattice points assigned to each unit cell (Fig. 4). One having an additional point at

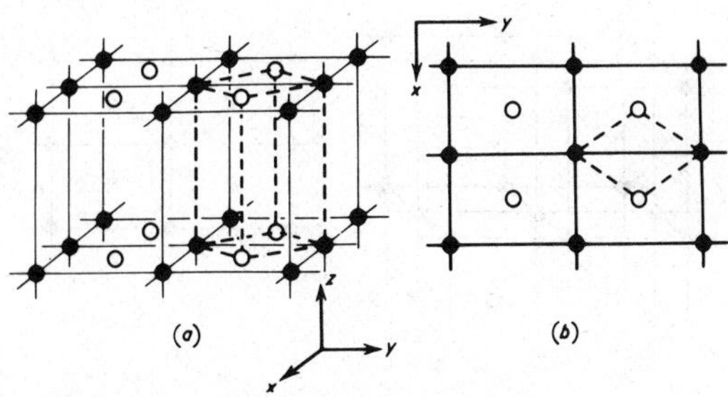

FIGURE 3. One-face-centered lattice (from H. D. Megaw, *Crystal Structures*, W. B. Saunders Company, Philadelphia, 1973, reprinted with permission). (a) Perspective diagram: solid lines outline unit cells of centered lattice; broken lines outline primitive unit cell of same lattice; black circles and open circles mark identical lattice points. (b) Same lattice in projection down z axis; all points are at height zero.

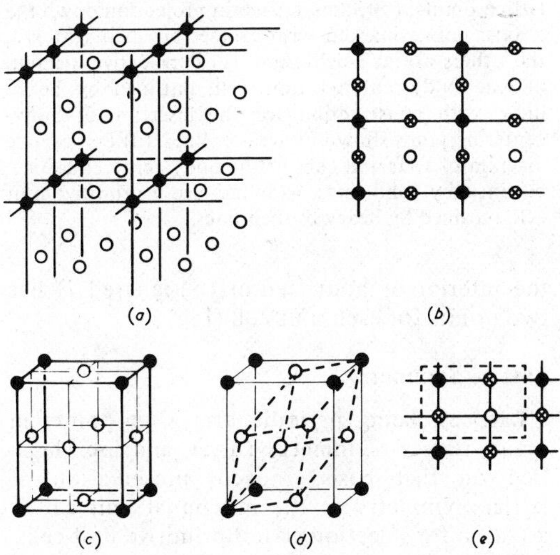

FIGURE 4. All-face-centered lattice. (from H. D. Megaw, *Crystal Structures*, W. B. Saunders Company, Philadelphia, 1973; reprinted with permission). (a) Perspective diagram: solid lines outline unit cells of the centered lattice; black circles and open circles mark identical lattice points (the open circles lying at face centers). (b) Same lattice in projection down the z axis; points marked with crosses are at height 1/2, the others are at height zero. (c) Perspective diagram of one-face-centered unit cell, outlined by heavy lines, with construction for the insertion of face centering points shown by weaker lines. (d) Perspective diagram of one unit cell of the face-centered lattice, outlined by solid lines, showing also a primitive unit cell, outlined by heavy broken lines. (e) Projection of one unit cell, showing displacement of origin (as indicated by broken lines) to demonstrate that there are four lattice points per cell.

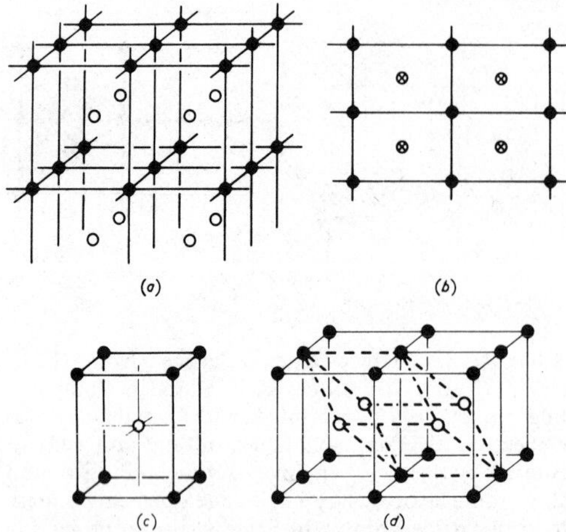

FIGURE 5. Body-centered lattice (from H. D. Megaw, *Crystal Structures*, W. B. Saunders Company, Philadelphia, 1973; reprinted with permission). (a) Perspective diagram: solid lines outline unit cells of centered lattice; black and open circles mark identical lattice points. (b) Same lattice in projection down the z axis; points marked with crosses are at height 1/2, the others are at height zero. (c) Perspective diagram of one body-centered unit cell, outlined by heavy lines, with construction for the insertion of body-centering point shown by weaker lines. (d) Perspective diagram of four unit cells of the body-centered lattice, outlined by solid lines, showing also a primitive unit cell, outlined by heavy broken lines.

the interior or body center (designated I) has two points for each unit cell (Fig. 5).

Lattice Symmetry

Lattices, being periodic arrays of points in space, possess symmetry (q.v.), and are classified on that basis. Indeed, preservation of lattice symmetry in the individual unit cell is a reason for selection of nonprimitive unit cells.

The most general lattice is a primitive one (P) in which $a \neq b \neq c$ and $\alpha \neq \beta \neq \gamma \neq 90^\circ$ ($\neq$, is not equal to by reason of symmetry). It is triclinic since its symmetry is $\bar{1}$ and, as a result, all three unit-cell edges and all three interior angles must be specified. If $a \neq b \neq c$ and $\beta \neq \alpha = \gamma = 90^\circ$, the symmetry is $2/m$, monoclinic. Reference axes for monoclinic lattices are selected such that x is inclined to z by as small an obtuse angle as is possible. Both primitive and C-centered unit cells exist (Fig. 6). (There are two accepted descriptions of monoclinic lattices; that given herein is called the "second setting" and is standard for crystallographic studies. See Henry and Lonsdale, 1965, ix and 7.)

If a lattice may be described by $a \neq b \neq c$, but $\alpha = \beta = \gamma = 90^\circ$, its symmetry is $2/m\ 2/m\ 2/m$, orthorhombic. An orthorhombic lattice may be primitive; one-face centered A, B, or C; all-face centered F; or body centered I (Fig. 6). Conventionally, reference axes are chosen such that $b > a > c$. (This convention is commonly waived in order to show structural relationships in real crystals.)

A lattice in which $a = b$ (then labeled a_1 and a_2) with c either longer or shorter than a and at right angles to the $a_1 a_2$ plane and in which $\gamma = 120^\circ$ has a z axis with threefold (trigonal) or sixfold (hexagonal) rotational symmetry. The unit cell is a rhombic prism and is primitive. A special trigonal lattice exists with interior points at $2/3\ a_1$, $1/3\ a_2$, $1/3\ c$ and at $1/3\ a_1$, $2/3\ a_2$, $2/3\ c$, termed rhombohedral (R). An alternative description of a rhombohedral lattice is $a = b = c$ and $\alpha = \beta = \gamma \neq 90^\circ$ (see Megaw, 1973, pp., 65, 208; Buerger, 1963, p. 106).

If $a = b \neq c$ and $\alpha = \beta = \gamma = 90^\circ$, the z axis has fourfold rotational symmetry and the lattice is tetragonal. It may be primitive or body centered. (Selection of alternate a-unit cell edges at 45° to the previous choice yields a C-centered lattice or an F-centered lattice, which may be preferable in the description of real crystals. The alternative description does not change the lattice, only its appearance and description.)

Three lattices exist in which $a = b = c$ ($a_1 = a_2 = a_3$) and $\alpha = \beta = \gamma = 90^\circ$. They are cubic (isometric) primitive, F centered, and I centered. There are, thus, a total of 14 types of space lattices as derived by Auguste Bravais in 1850 (Fig. 6). As indicated, selection of different reference axes produces different descriptions of them, but there are only 14 different lattices.

Asymmetric Unit

A lattice is not a structure, but a set of reference points that describe repetitive properties in three dimensions. To describe a real crystal structure, each lattice point is associated with a chemical unit (atom, ion, molecule, radical, or any collection of atoms or ions that are not symmetrically related to one another). This chemical unit, the asymmetric unit, is the smallest portion of a crystal structure that can be repeated by lattice operations to build or to describe an entire structure. Thus, a lattice point need not correspond to the position of an atom or an ion in a crystal structure; indeed, many complex crystal structures are best visualized when an asymmetric unit is selected in which the lattice points do not coincide with a particular atom or ion (e.g., see *Orthopyroxenes; Pyroxene Group*).

A lattice point has no dimensions; the symmetry of a lattice lies in its array of points. An

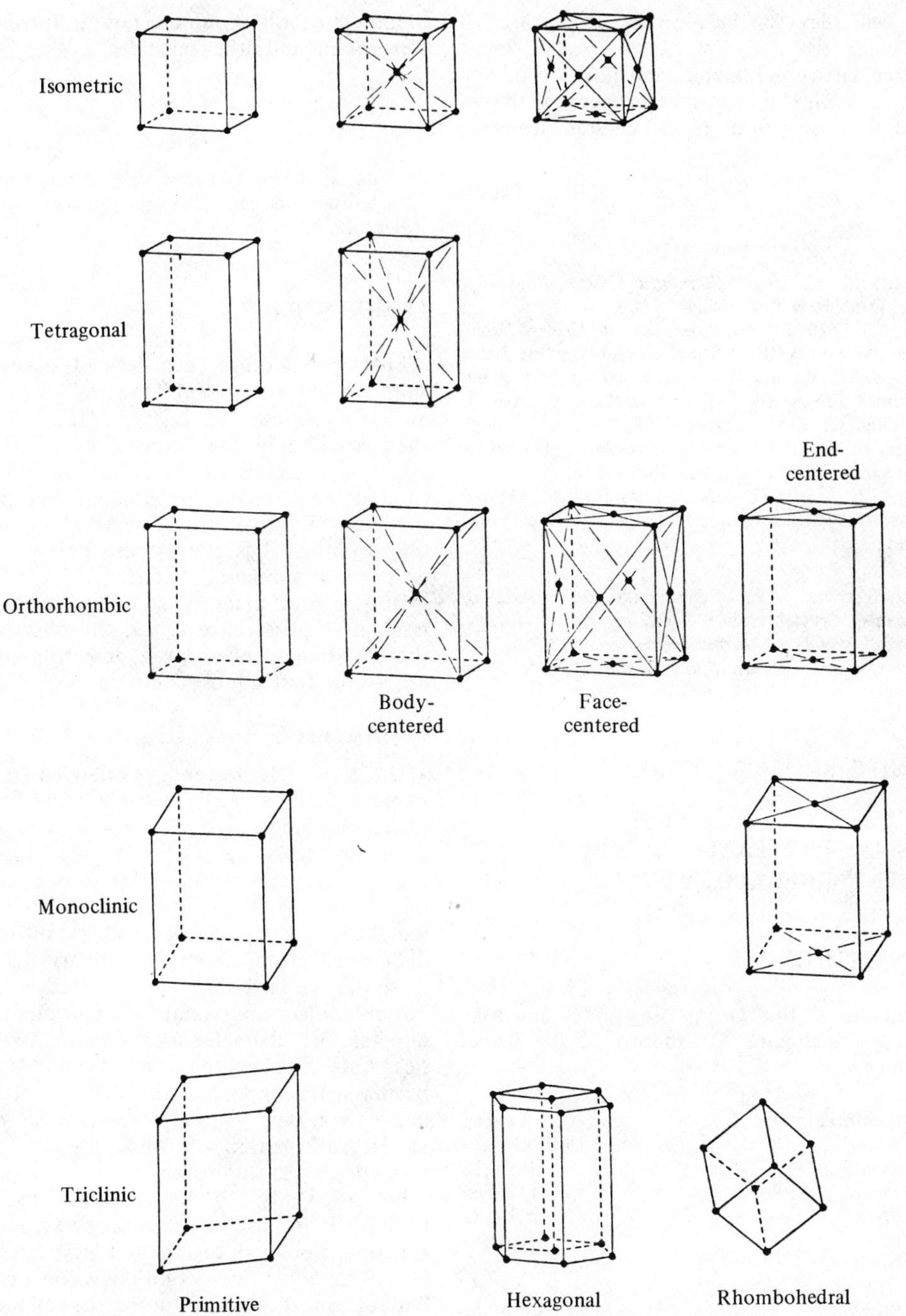

FIGURE 6. The fourteen Bravais space lattices arrayed according to lattice character and symmetry.

asymmetric unit, however, may have a symmetry of its own resulting in a crystal structure having a lower degree of symmetry than its corresponding lattice (compare the symmetry and structure of *pyrite* with that of *halite*).

In order to describe completely a crystal structure after identification of its associated lattice (including lattice type and unit-cell parameters), coordinates of chemical units within the asymmetric unit must be given. These are commonly given by atomic position parameters, or fractions of a, b, and c in the x, y, and z directions. When they are not simple fractions, e.g., ½, ½, ½ for the body center of a unit cell, they can be given as percentages of *a*, *b*, and *c*.

Given lattice parameters and position parameters for each chemical species, a complete geometrical description of a crystal structure results.

KEITH FRYE

References

Buerger, M. J., 1963. *Elementary Crystallography*, Rev. Print. New York: Wiley, 528p.

Bloss, F. D., 1971. *Crystallography and Crystal Chemistry*. New York: Holt, Rinehart and Winston, 545p.

Henry, N. F. M., and Lonsdale, K., eds., 1965. *International Tables for X-ray Crystallography*, vol. 1. Birmingham: Kynoch Press, 6-56.

Megaw, H. D., 1973. *Crystal Structure: A Working Approach*. Philadelphia: Saunders, 563p.

Phillips, F. C., 1971. *An Introduction to Crystallography*, 4th ed. London: Longman; New York: Wiley, 351p.

Cross-references: *Axis of Symmetry; Barker Index of Crystals; Crystal Habit; Crystallography: History; Crystallography: Morphological; Directions and Planes; Point Groups; Space-Group Symbol; Symmetry.*

LIGNITE–*See* Vol. VI: COAL

LIME–*See* PORTLAND CEMENT MINERALOGY

LINNAEITE GROUP

Minerals of this group of sulfides and selenides are analogous in structure to the ***Spinel*** group (q.v.):

- *bornhardtite:* Co_3Se_4
- *carrollite:* $Cu(Co,Ni)_2S_4$
- *daubreelite:* $FeCr_2S_4$
- *fletcherite:* $Cu(Ni,Co)_2S_4$
- *greigite:* $FeFe_2S_4$
- *indite:* $FeIn_2S_4$
- *linnaeite:* Co_3S_4
- *polydymite:* $NiNi_2S_4$
- *siegenite:* $(Ni,Co)_3S_4$
- *trüstedtite:* $NiNi_2Se_4$
- *tyrrellite:* $(Cu,Co,Ni)_3Se_4$
- *violarite:* Ni_2FeS_4

Linnaeites are isometric ($Fd3m$) and typically crystallize in octahedra with imperfect cubic cleavage. *Linnaeite, siegenite, carrollite, violarite*, and *polydymite* are relatively rare accessory minerals in sulfide ore deposits. *Daubreelite* is a meteoritic mineral (q.v.). The selenide members of the group are rare, occurring with *clausthalite* and other selenides.

KEITH FRYE

Reference

Ostwald, J., 1978. Linnaeite series minerals from the Kalgoorlie district, Western Australia, *Mineral. Mag.*, **42**, 93-98.

LUMINESCENCE

Luminescence has been defined as the emission of light from a substance by all processes except incandescence. Energy is first absorbed, then released in the form of light. Luminescence in minerals has been linked with structural defects or the presence of foreign ions that act as activators replacing major ions in the structure, but it may also be an intrinsic property of a mineral.

Of the four main types of luminescence in minerals–fluorescence (q.v.), phosphorescence, thermoluminescence (q.v.), and triboluminescence–the first is most common.

Fluorescence

The term "fluorescence" is derived from the mineral *fluorite* (CaF_2) which fluoresces blue under ultraviolet light if activated by rare-earth elements replacing Ca^{2+} in the structure. Fluorescence is a useful tool in prospecting and in ore dressing. The tungsten ore mineral *scheelite* ($CaWO_4$) is a colorless mineral difficult to distinguish from *quartz* in ordinary light, but it shows up strikingly in the dark under a portable ultraviolet lamp. Certain fluorescent minerals are characteristic of particular localities. An example is the *willemite* ($ZnSiO_4$) that occurs in the important zinc deposits of Franklin, New Jersey, which fluoresces green due to partial substitution of Mn^{2+} for Zn^{2+}; pure *willemite* is not fluorescent.

Recent study of luminescence has been prompted by use of this property in light-emitting diodes, in crystal and glass lasers, and in phosphors for color-television screens. Whereas most physical properties of minerals depend upon the crystal structure of the ground state, luminescent properties are determined by the structure in an excited state. As an example, elevating an electron to a normally unoccupied orbital increases the effective radius of the excited ion and the length of its chemical bonds to surrounding ions. Photon emission is then controlled by the crystal structure surrounding the excited ion.

Luminescence activators, known as "phosphors," may have an excited-state electronic configuration that is different from that of the ground state (e.g., $d \rightarrow f$); in which, case luminescent emission occurs within microseconds or less. On the other hand, fluorescence may take place between excited and ground states within the same orbital (e.g., $d \rightarrow d$ or $f \rightarrow f$), with decay times in the millisecond range.

As in mineral coloration (see *Color in Minerals*), bond length, coordination number, and crystal field strength (see Vol. IVA: *Crystal-Field Theory*) will influence both the wavelength and the band width of the luminescent spectral emissions. Thus, the phosphor Mn^{2+} fluoresces green in tetrahedral coordination, but red in octahedral coordination. However, emission wavelengths of rare-earth-element phosphors involving $f \rightarrow f$ transitions and some transition metal ions (e.g., Cr^{3+}, Mn^{4+}) are only weakly influenced by the host structure.

In addition to ion phosphors, complex ions, such as tungstates, niobates, vanadates, and titanates, may act as phosphors in crystals. For example, the luminescence of *scheelite* ($CaWO_4$) involves a molecular-orbital transition in the WO_4^{2-} tetrahedron. Efficiency of luminescence in minerals may be enhanced by sensitizers where the excitation energy is transferred through the structure from the sensitizer (energy absorber) to the activator (energy emitter). Alternatively, luminescence may be "poisoned" by the transfer of energy to ions, such as Fe^{2+} or Ni^{2+}, that have radiationless transitions to the ground state.

Phosphorescence

Phosphorescence differs from fluorescence in that the emission of light continues for a time after the irradiating source has been removed. *Sphalerite* (ZnS) phosphoresces when stimulated by alpha particles or by ultraviolet light.

Thermoluminescence

Thermoluminescence (q.v.) is the emission of visible light upon heating, due to release of energy stored as electron displacements in the crystal structure. It has been suggested that thermoluminescence may provide a relatively simple method of determining the relative age of carbonate sediments. This idea is based on the fact that radiation damage to the calcite crystal structure, which determines the thermoluminescence, seems to be a function of the age of the structure and the natural rate of alpha-particle radiation from radioactive elements.

Triboluminescence

Triboluminescence is the phenomenon in which the emission of light is induced by crushing, scratching, or rubbing. Both artificial and natural ZnS (*sphalerite*) exhibit triboluminescence.

CHARLES J. VITALIANO
DOROTHY B. VITALIANO

References

Berry, L. G., and Mason, B. H., 1959. *Mineralogy*. San Francisco: Freeman, 612p.

Przibram, K., 1956. *Irradiation Colours and Luminescence*. London: Pergamon, 332p.

White, W. B., 1975. Luminescent materials, *Trans. Am. Crystallogr. Assoc.*, **11**, 31–49 [48 references].

Zeller, E. J.; Wray, J. L.; and Daniels, F., 1957. Factors in age determination of carbonate sediments by thermoluminescence, *Bull., Am. Assoc. Pet. Geol.*, **41**(1), 121–129.

Cross-references: *Order-Disorder; Thermoluminescence;* see also mineral glossary.

LUNAR MINERALOGY

Extensive mineralogic and petrologic studies were performed on the lunar samples returned by the Apollo and Luna missions. Essentially all basic data are published in the proceedings of the annual lunar science conferences (Proceedings, 1970–1980). Taylor (1974) provides an integrated discussion of the geochemical aspects of the Moon. Presented here are the general features of lunar mineralogy, an enumeration of the recognized lunar minerals (until fall, 1980), and an extended description of the more common mineral groups. Although several new minerals have been identified, most are basically similar to terrestrial counterparts described in standard reference works (e.g., Deer et al., 1963).

General Mineralogic Features

With the examination of the first lunar samples in the fall of 1969, it was immediately realized that there are several important differences between lunar and terrestrial mineralogy. First, the moon is extremely depleted in volatiles, leading to the absence (with possibly several exceptions) of such mineral groups as hydrous silicates (e.g., ***micas, amphiboles, zeolites***), carbonates, borates and nitrates, to name only a few. Second, the lack of water almost eliminates alteration due to weathering and, thus, secondary minerals are not present except for some formed as a result of reduction, exsolution, or thermal metamorphism. Third, mineral-forming processes have occurred under highly reducing conditions leading to low oxidation states; hence Fe^{3+} is barely detectable by the most sensitive techniques and Ti^{3+} and Cr^{2+} are considered important. The above conditions greatly limit the number and diversity of lunar minerals. In fact, only about 100 mineral species have been recognized on the moon in contrast to approximately 2500 on Earth.

Two important points should be kept in mind. First, sampling on the moon is limited to eight locations; thus, large areas of the moon, where other minerals may occur, remain unsampled. Second, only surface samples (to several meters depth) have been collected directly. Although this region contains deeper material excavated by crater formation, the maximum sampled depth is probably limited to tens of kilometers.

Mineral Forming and Alteration Processes—Lunar Rock Types

Essentially all minerals on the moon originated by igneous processes; these primary crystallization products have since been modified by meteoritic impact (comminution) and metamorphism (including shock and thermal types plus sintering) producing a range of rock types, each characterized by mineralogy as well as composition and texture. Petrologic nomenclature is complicated and most workers do not agree on classifications, but three rock groups are generally recognized:

1. mare basalts—produced by near-surface crystallization of a variety of basaltic lavas, some very rich in Ti and Fe, from ≈3.9 to ≈3.1 b.y. ago. These are the youngest rocks returned from the moon and comprise the dark surface regions. The principal minerals are ***pyroxenes, plagioclases***, *ilmenite*, and ***olivines***; but rare minerals occur in the last-crystallizing material (mesostasis).
2. ANT (*A*northositic-*N*oritic-*T*roctolitic) rocks—dominated by ***plagioclases*** and ***pyroxenes*** with minor ***olivine***. These generally are the oldest rocks, formed between ≈4.6 and ≈3.9 b.y. ago and comprise the light-colored, highland regions. A general feature is severe textural modification by impact and metamorphism, which almost totally erased primary characteristics. Most ANT rocks are strictly breccias and may be mixtures of rocks formed quite independently. Thermal metamorphism resulting from meteorite-impact heating has often recrystallized these breccias with consequent modification of mineral chemistry.
3. KREEP rock suite—of enigmatic origin but rich in *K*, *R*are *E*arth, *E*lements, and *P*. Often occurs as metamorphosed breccias and glass; but rare fine-grained, igneous-textured samples have been recognized.

Several mineral-alteration processes have occurred on the moon. Frequently the oxidation conditions changed during or after crystallization, resulting in subsolidus reactions that produced free iron and oxygen to form a new mineral assemblage (e.g., $FeTiO_3 \rightarrow Fe^0 + TiO_2 + \frac{1}{2}O_2$). Metamorphism either at depth or as a result of impact heating produced a coarsening of texture and subsolidus reactions such as exsolution and cation ordering or disordering. Shock processes, due to impact, degraded crystal structures and, at the extreme, resulted in shock melting producing a glassy structure without morphological change. Often large volumes were melted to produce abundant glass or secondary "melt rocks" with igneous texture. Impact of comets or carbonaceous chondrites has been proposed to explain formation of some volatile-containing phases from a transient atmosphere; others consider possible alteration by terrestrial contaminants or outgassing of the Moon. No high pressure phases such as *coesite* have been recognized, although meteoritic impact is common.

Recognized Lunar Minerals

The minerals described to date are divided into three groups:

- Group I (Table 1), Rock-Forming Minerals (described below in detail)
- Group II (Table 2), Accessory Minerals—widespread but usually volumetrically insignificant
- Group III (Table 3), Trace Minerals—usually only one report of minute grains; most are not fully described; occurrence may be due to meteoritic or terrestrial contamination

References to original descriptions are given by Smith (1974) and Frondel (1975).

TABLE 1. Lunar Minerals: Group I, Rock-Forming Minerals

Mineral	General Description
armalcolite	Near $Mg_{0.5}Fe_{0.5}Ti_2O_5$ but may contain high Cr and Zr as well as rare-earth elements; several compositional types have been described.
ilmenite	$FeTiO_3$ but often with Mg substitution for Fe; common phase in mare basalts but accessory phase in ANT rocks.
olivine	$(Mg,Fe)_2SiO_4$ with minor Mn, Ca, and Cr; wide compositional range is common.
plagioclase	Solid solution of $NaAlSi_3O_8$ and $CaAl_2Si_2O_8$ but usually close to $Ca_{0.95}Na_{0.05}Al_{1.95}Si_{2.05}O_8$; minor Mg, K, and Fe.
pyroxene	$(Mg,Fe,Ca)SiO_3$ with minor Al, Cr, Ti, and Mn; wide compositional range with complex zoning and exsolution common.
spinel	Solid solutions of $MgAl_2O_4$, $FeCr_2O_4$, and Fe_2TiO_4 with minor Mn and V; complex zoning.

TABLE 2. Lunar Minerals: Group II, Accessory Minerals

Mineral	Ideal Composition	Comments
alkali ***feldspar***	$KAlSi_3O_8$	Occurs in breccia usually in association with silica-rich ("granitic") minerals; often contains Na, Ba, and Sr.
apatite	$Ca_5(PO_4)_3(Cl, F)$	Common phase in mesostasis of mare basalts; often contains rare-earth elements in KREEP materials; occurs in vugs of mare basalts and ANT breccia as vapor deposit.
kamacite *taenite*	α-Fe γ-Fe	Common phases resulting from crystallization under reducing conditions or vapor deposition; some definitely due to meteoritic contamination; *taenite* can contain high Ni.
pyroxferroite	$FeSiO_3$	New metastable lunar mineral related to terrestrial *pyroxmangite;* formed at last stages of crystallization of mare basalts, especially from Ti-rich liquid compositions.
quartz *tridymite* *cristobalite*	SiO_2	Found in mesostasis of mare basalts, in "granitic" clasts, and in ANT breccia; often polymorphic form is not known, but all three have been recognized.
rutile	TiO_2	Relatively rare, usually occurring in complex Ti-Zr mineral assemblages; probably due to subsolidus reduction of *armalcolite* and other phases; often contains high Nb.
tranquillityite	$Fe_8(Zr,Y)_2Ti_3Si_3O_{24}$	New mineral commonly found in mesostasis of mare basalts.
troilite	FeS	Common sulfide in all lunar rocks; some may be due to meteoritic contamination but most is indigenous; usually associated with Fe-metal.
whitlockite	$Ca_3(PO_4)_2$	Most common in KREEP breccia and can contain up to 10% rare earth elements; frequently intergrown with ***apatite.***
zircon	$ZrSiO_4$	Rare; occurs in mesostasis of mare basalts and as single grains in breccias.

Mineralogy of the Rock-forming Minerals

Armalcolite. *Armalcolite* is a new mineral close to $Mg_{0.5}Fe_{0.5}Ti_2O_5$, but with a variable Fe/Mg ratio (Anderson et al., 1970). The structure is similar to terrestrial *pseudobrookite* ($Fe_2^{3+}TiO_5$), but the absence of trivalent iron and the Mg content justify a new mineral name. Minor Cr,Al,Ca, and Zr are present. Since recognition in lunar samples, two terrestrial

TABLE 3. Lunar Minerals: Group III. Trace Minerals[a]

Mineral	Composition
Sulfides and Sulfates	
?*bornite*	Cu_5FeS_4
chalcopyrite	$CuFeS_2$
cubanite	$CuFe_2S_3$
mackinawite	$\approx FeS$
?*molybdenite*	MoS_2
?*niningerite*	(Mg,Fe)S
pentlandite	$(Fe,Ni)_9S_8$
sphalerite	ZnS
Zn,Fe,Cl-sulfate	
Silicates	
amphibole	$(Na,Ca)_2(Mg,Fe)_5Si_8O_{22}(OH,F)_2$
?*chondrodite*	$Mg(F,OH)_2 \cdot 2Mg_2SiO_4$
?*cordierite*	$Al_3(Fe,Mg)_2Si_5AlO_{18}$
garnet	$Fe_3Al_2Si_3O_{12}$
?*melilite*	$Ca_2(Mg,Fe,Al,Si)_3O_7$
?***mica***	$K_2Al_6Si_6O_{20}(OH,F)_4$
?sphene (*titanite*)	$CaTiSiO_4(O,F)$
Metals	
?brass	CuZn
copper	Cu
?*indium*	In
tin	Sn
Oxides and Hydroxides	
akaganéite	β-FeO(OH)
baddeleyite	ZrO_2
corundum	Al_2O_3
?*hematite*	Fe_2O_3
perovskite	$CaTiO_3$
zirconolite[b]	$Ca(ZrTi)_2O_7$
zirkelite[b]	$Ca(TiZr)_2O_5$
Miscellaneous	
?aluminum carbide	Al_4C_3
?*aragonite*	$CaCO_3$
?*cohenite*	Fe_3C
farringtonite	$Mg_3(PO_4)_2$
?*graphite*	C
?*lawrencite*	$FeCl_2$
?*moissanite*	SiC

[a] A question mark indicates that indigenous lunar origin or identification is particularly uncertain.

[b] There is considerable controversy over the exact formulas of zirkelite and zirconolite. Some evidence suggests that there is no difference.

occurrences in kimberlites have been described. *Armalcolite* was first observed as an early crystallizing phase in Apollo 11 high-Ti mare basalts and later in similar Apollo 17 mare basalts. Experimental studies have shown that it is a stable phase above about 1100°C in high-Ti compositions under reducing conditions; minor Cr and Al tend to stabilize it at lower temperatures. With progressive cooling, the primary *armalcolite* is unstable and reacts with the magma to produce an overgrowth of *ilmenite* (*armalcolite* + melt → *ilmenite*); with slow cooling no *armalcolite* will remain. If cooling is rapid enough, some *armalcolite* remains, but the subsolidus reaction, *armalcolite* → *rutile* (TiO_2) + Mg-rich *ilmenite*, often occurs.

In ANT breccia, several other *armalcolite*-type phases characterized by high Cr, Ca, and Zr have been observed (Haggerty, 1974). These are minor phases but appear important because they can contain high rare-earth (up to 10%) concentrations. Their chemical formulas, (Cr, Ca,Fe,Mg) $(Ti,Zr)_2O_5$, although complicated, are close to *armalcolite*; but no X-ray confirmation of structural similarity has been made. These compositions may be closer to the *zirconolite* mineral group except for the Ti:Zr ratio. These high-Zr *armalcolites* appear to have formed during recrystallization at relatively low temperatures, and frequently break down to assemblages of *chromite, baddeleyite* (ZrO_2), *rutile, iron*, and Mg-rich *ilmenite*, but details are uncertain.

Ilmenite. *Ilmenite* occurs in nearly all lunar rocks in amounts up to ≈10% in Ti-rich mare basalts, but usually at ≈1% levels in ANT rocks. X-ray studies have shown that lunar *ilmenite* is essentially identical with terrestrial samples (Raymond and Wenk, 1971). Variations from the ideal composition, $FeTiO_3$, are mainly Mg (up to 12% MgO) for Fe as well as small amounts of Mn, Cr, Zr, Ca, and Al; Ti^{3+} has been detected. In mare basalts and KREEP basalts, *ilmenite* usually crystallizes either early or very late in the crystallization sequence depending on a number of factors such as bulk composition and cooling rate. In ANT breccia the origin is not clear; some *ilmenites* may be relics from the parent rock, but some definitely have formed by thermal metamorphism and subsolidus reduction.

The Mg/(Mg+Fe) ratio in *ilmenite* has received considerable attention because it is thought to reflect the same ratio in the parent liquid (Nehru et al., 1974) now represented by the bulk rock. Data do show a positive correlation but with much scatter; quantitative use of this ratio, at least for mare basalts, is limited because of nonequilibrium crystallization. *Ilmenite* in breccias, however, probably in many cases, is in equilibrium with other phases (Steele, 1974) and quantitative use appears promising. Of particular interest is its use as a geothermometer (Anderson et al., 1972) in lunar breccia if suitable laboratory calibration can be obtained (see *Thermometry, Geologic*). ZrO_2 contents are higher than found in terrestrial *ilmenites* (Arrhenius et al., 1971), but no suitable explanation has been given.

Ilmenite often is a reaction product either of early crystals reacting with a liquid or of reduction of Ti-containing phases. Several observed reactions are:

1. *armalcolite* + melt → *ilmenite*
2. *ulvöspinel* → *ilmenite* + Fe^0 + ½O_2
3. *armalcolite* → Mg-rich *ilmenite* + TiO_2
4. Mg,Al,Ti *spinel* → Mg,Al-*spinel* + *ilmenite*

In each of these reactions the *ilmenite* apparently has inherited a telltale chemical characteristic or texture; for example, the *ilmenite* of reaction 3 contains high Mg and *rutile* exsolutions.

Plagioclase. ***Plagioclases*** are the most common minerals in the returned samples and probably dominate the near-surface (upper 50 km ?) rocks. Like terrestrial ***plagioclases***, they have been extensively studied and only the highlights of the results can be described here.

Modal amounts of ***plagioclase*** range from nearly 100% in several large anorthositic rocks and anorthositic breccias to about 30% in the mare basalts. ***Plagioclase*** is clearly a result of liquid crystallization in mare basalts but its ultimate source and subsequent history is seldom obvious in ANT breccia because of severe metamorphism. The anorthositic rocks are generally considered to form by ***plagioclase*** accumulation.

Major and minor element compositions in general conform to the above threefold rock classification (Fig. 1; Steele and Smith, 1973). Nearly all ***plagioclase*** is Ca-rich, mostly because of the low Na abundance in the moon; thus *anorthite* ($Ca_{0.95}Na_{0.05}Al_{1.95}Si_{2.05}O_8$) is probably close to the mean composition with $Ca_{0.8}Na_{0.2}Al_{1.8}Si_{2.2}O_8$ being the most common Na-rich extreme. Rare *albite* (Na-rich) has been described. ANT ***plagioclase*** is often close to pure *anorthite*, especially in the most ***plagioclase***-rich rocks. Similar compositions are seen in some early Archean terrestrial anorthosites and have led to suggestions concerning possible similarities between early terrestrial and lunar evolution (Windley, 1970). Minor elements in ANT ***plagioclase*** are low (Fig. 1), with frequently only traces of Mg, Fe, and K; probably this is due to long periods of metamorphism of the

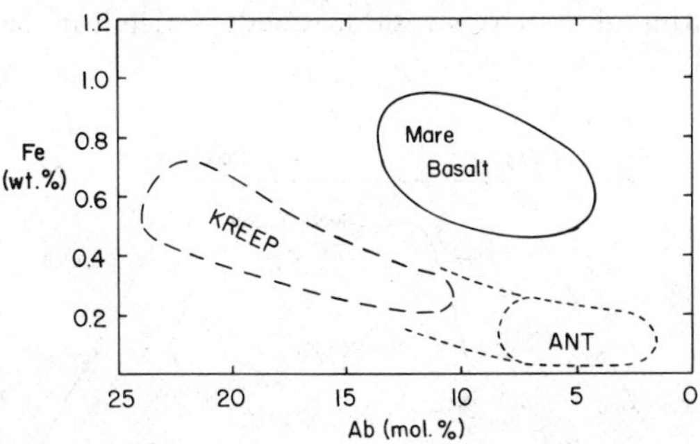

FIGURE 1. General range of lunar *plagioclase* compositions versus Fe content. ***Plagioclase*** composition is given as mol% *Ab* ($NaAlSi_3O_8$). Threefold rock classification described in text shows little overlap, and thus Fe and Na are useful for classifying ***plagioclase*** with respect to rock type.

ANT rocks, which tend to remove these elements. ***Plagioclase*** from mare basalts tends to be more Na-rich and commonly is zoned with Ca-rich cores and Na-rich rims. In contrast to ANT ***plagioclase***, minor elements are high with FeO values up to ≈1.0% and MgO up to 0.25%. These high contents are probably due to rapid, high-temperature crystallization in the low-viscosity mare magmas. KREEP ***plagioclase*** is the most Na-rich (as are the rocks) and average ***plagioclase*** compositions are close to $Ca_{0.85}Na_{0.15}Al_{1.85}Si_{2.15}O_8$; and, at least in the igneous-textured samples, zoning is common. Minor elements are low and often similar to ANT ***plagioclase***. Figure 1 summarizes these data for iron. These chemical trends enable individual ***plagioclase*** grains in soils or breccia to be classified based only on two or three elements.

Many other data exist for ***plagioclase***. Mössbauer and electron-spin-resonance studies have shown that Fe^{2+} is dominant with only traces of Fe^{3+} (Niebuhr et al., 1973). Solid-state recrystallization (metamorphism) has commonly caused apparent reduction and exsolution of iron to form minute, oriented *iron* rods (Bell and Mao, 1973). Steele and Smith (1973) suggested that metamorphism of nonstoichiometric ***plagioclase*** (common in mare basalts) may have resulted in exsolution of ***pyroxene***, silica and *ilmenite* in ANT ***plagioclase***. Crystal-structure studies have detected no basic structural differences in ***plagioclase*** although cation vacancies are indicated by one single-crystal X-ray study (Smyth, 1975). The rare-earth content, especially europium in ***plagioclase***, has been of particular interest. With extremely reducing conditions, Eu is divalent while the other rare-earths remain trivalent; this enables Eu^{2+} to be selectively incorporated into the ***plagioclase*** structure, depleting the parent liquid. The magnitude of the Eu anomaly in a rock becomes important in proposed differentiation processes because ***plagioclase*** separation, either floating or sinking, will affect this anomaly in the remaining liquid.

Olivine. ***Olivine*** is the third most abundant mineral in the returned samples and occurs in nearly all rocks. Some small clasts or fragments of breccia are nearly 100% ***olivine***, but most commonly it comprises less than 10% of any rock. ***Olivine*** is an early crystallizing phase in mare basalts and some samples show ***olivine*** accumulation; most however, crystallized relatively rapidly, preventing ***olivine*** from settling. KREEP rocks show little or no ***olivine***. ANT rocks show mainly single grains of ***olivine*** in a breccia, although some notable exceptions occur.

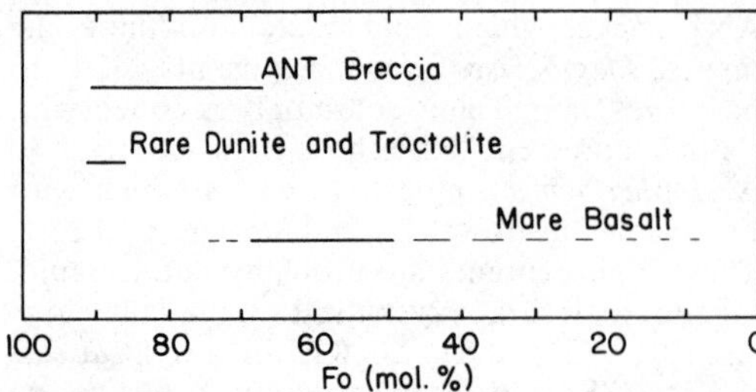

FIGURE 2. General range of *olivine* compositions as mol% Mg_2SiO_4 (*Fo*) in *olivine*. Note that ANT, dunite, and troctolite contain Mg-rich *olivine*, which is consistent with early crystallization from Mg-rich liquids. Solid lines indicate dominant compositions. KREEP rocks have little to no *olivine*.

As with *plagioclase*, minor- and major-element chemistry differs to some extent and can be used for classification; in addition, the well-determined crystal-liquid compositional relationships provide a good mineralogic test for petrogenetic models. Most important is the Mg/(Fe+Mg) ratio, which closely reflects that of the parent liquid (Roeder and Emslie, 1970). *Olivines* in mare basalts are relatively Fe rich (Fig. 2) with maximum Mg/(Mg+Fe) values of about 0.75; these are consistent with derivation from the Fe-rich mare basalt compositions. Zoning to more Fe-rich compositions is common and frequently nearly pure *fayalite* (Fe_2SiO_4) occurs at *olivine* rims. *Olivines* in ANT rocks (Fig. 2) are significantly more Mg rich [Mg/(Mg+Fe)=0.9–0.7], indicating an origin from a more primitive (Mg-rich) source, possibly during an early moon-wide differentiation. Rare dunite and troctolite rocks also have Mg-rich *olivine*.

Minor elements show important differences from those of terrestrial *olivines*. Cr_2O_3, barely detectable in terrestrial *olivines*, occurs at levels of <1 wt% in lunar *olivine*. The usual explanation is that, under reducing conditions, divalent chromium is common, and Cr^{2+} will enter the olivine structure while Cr^{3+} will not (as on earth). Nickel in lunar *olivine* is always low (<0.1 wt%), whereas on earth it is high but variable. Two explanations are, either that the outer region of the moon is depleted in Ni because of efficient extraction into a core or because of low bulk abundance, or that Ni is reduced to Ni^0 and preferentially enters the common iron metal. Ca in *olivine* proves to be a useful element to separate plutonic from extrusive rocks in terrestrial samples; a similar trend appears to hold on the moon. High CaO (>0.15 wt%) is common in *olivine* from mare basalts and other rocks that, from textural evidence, appear to be extrusive; low CaO (<0.15) occurs in *olivine* from coarse-grained rocks and in rocks that have Mg-rich *olivines* (interpreted as plutonic). Other elements such as Al_2O_3, MnO, and TiO_2 can be detected, but variations show little apparent significance.

Pyroxene. *Pyroxene* is probably the second most common mineral in the samples returned, and most studies of it have concentrated on major- and minor-element variations as well as on crystallography.

The important minor elements are Ti, Cr, Al, and Mn. Na and K, although significant in terrestrial *pyroxenes*, are in very low concentrations in lunar *pyroxenes* and are seldom determined.

In mare basalt, *pyroxene* zoning is the rule. There is only a small range in the maximum Mg/(Mg+Fe) value of the most Mg-rich *pyroxenes*, due to the relatively constant Mg/(Mg+Fe) ratio (≈.50) of all parental liquids (Dowty et al., 1974); with progressive crystallization, more Fe-rich *pyroxenes* crystallize (Fig. 3). Details in zoning trends differ for each rock and depend on such factors as cooling rate, bulk composition, and order of crystallization of other phases (Bence and Papike, 1972). For example, if high-Ca *pyroxene* (*augite*) is crystallizing, the major zoning trend will be from Mg- to Fe-rich compositions. If, however, Ca-rich *plagioclase* begins to crystallize, a sharp lowering in the Ca content of the *pyroxene* results because of depletion of Ca in the remaining liquid. This observed zoning trend can be correlated with textural details.

Minor element variations show similar features. Usually these are interpreted in terms of three pyroxene components: $R^{2+}TiAl_2O_6$, $R^{2+}Al_2SiO_6$, and $R^{2+}CrAlSiO_6$ (R^{2+}=Ca,Fe, Mg). Plots of Ti vs. Al, Al vs. Si, and ternary plots of the above molecules derived from a series of compositions taken from the core to rim of *pyroxenes* show trends which can be

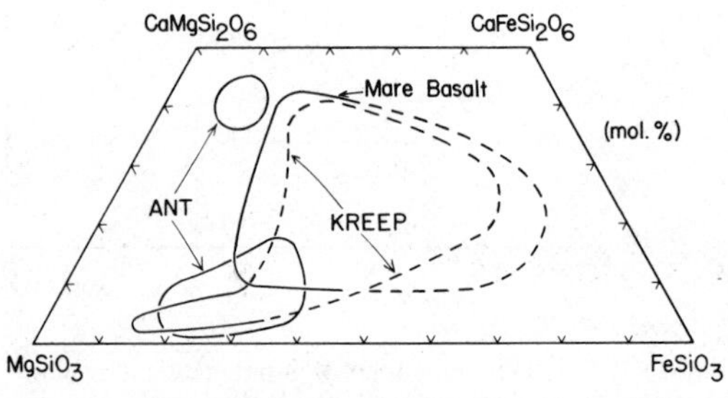

FIGURE 3. General range of *pyroxene* compositions as a function of rock type. ANT and KREEP have Mg-rich *pyroxenes* while mare basalts have Fe-rich compositions. Solid lines indicate dominant compositions.

correlated with changes in the liquid composition. Again, ***plagioclase*** crystallization tends to reduce $R^{2+}Al_2SiO_6$ substitution because Al tends to enter ***plagioclase***. Thus, on a Ti vs. Al plot the Ti/Al ratios would change from values less than ½ to values approaching ½ showing the increased effect of the $R^{2+}TiAl_2O_6$ molecule (Ti/Al=½) over $R^{2+}Al_2SiO_6$. These trends and their interpretations are similar to those for terrestrial basalts. In the lunar case, R^{2+} $TiSiAlO_6$ and $Cr^{2+}Si_2O_6$ end-member compositions must be considered because of the possible presence of Ti^{3+} and Cr^{2+}, respectively.

Pyroxenes in ANT rocks (Fig. 3) are dominated by low-Ca varieties—usually ***orthopyroxenes*** (q.v.)—with Mg/(Mg+Fe) ratios ≈0.9 to ≈0.7, which are higher than for mare basalts. Zoning is seldom present, probably because of a plutonic origin or because of metamorphic equilibration.

Exsolution in mare-basalt ***pyroxenes*** is limited to submicroscopic dimensions and observed only by X-rays or by the electron microscope. Rare examples, usually single grains with coarse exsolution, occur in ANT rocks and suggest that at least some rocks underwent slow enough cooling to develop these textures. At least two samples show a space group ($P2_1ca$) that has not been recognized in terrestrial ***pyroxenes*** (but is found in the Steinbach meteorite, see *Meteoritic Minerals*). Single-crystal X-ray techniques have shown that this results from extreme Mg-Fe ordering most easily interpreted as due to prolonged annealing. This suggests that these two samples are unique, and probably originated at depth in the moon.

Spinel Group. The chemistry (Mg, Al, Ti, Cr, Fe, and minor Mn and V) of the ***spinel*** group (q.v.) is similar to that of terrestrial samples except for the presence of divalent-only iron. Trivalent Ti and divalent Cr have been considered but there is no firm evidence for their existence. X-ray studies show no significant crystallographic differences, although nonisometric Ti-rich ***spinels*** have been reported based on reflection optics. Two ***spinel*** compositional series have been recognized (Haggerty, 1972): the normal ***spinel*** series is mostly confined to the ANT rocks and is characterized by low TiO_2 and mainly Fe-Mg and Cr-Al substitutions in the *A* and *B* sites, respectively; the normal-inverse ***spinel*** series is common in mare basalts and shows $Fe^{2+}+Ti^{4+}$ for Cr+Al substitution in the *B* site, while the *A* site is consistently occupied by Fe and minor Mg.

The normal ***spinels*** contain two compositional subgroups (Fig. 4), and detailed study suggests other possible subgroups. The first is

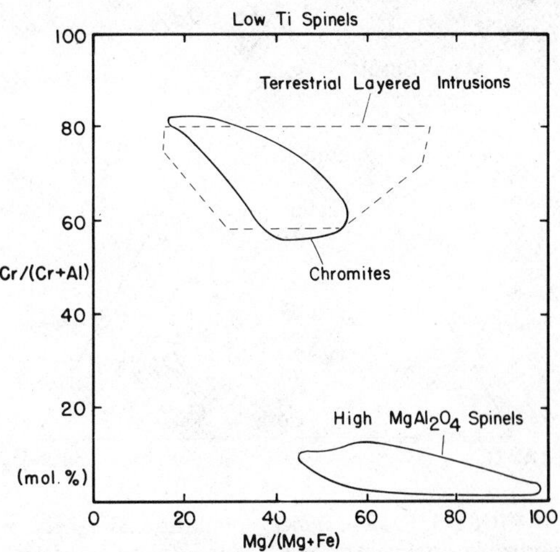

FIGURE 4. Range of ***spinel*** compositions in the low-Ti normal *spinel* series on Cr/Cr + Al) vs Mg/Mg + Fe) plot. Two subgroups are apparent, one with high $MgAl_2O_4$ and the other with high $FeCr_2O_4$ (*chromite*). The later subgroup is compositionally similar to *chromites* occurring in terrestrial layered intrusions.

dominated by $MgAl_2O_4$ and is commonly found as single grains in breccia or larger rocks of enigmatic but igneous origin (e.g., Prinz et al., 1973). The common characteristic of these rocks is that the associated silicates are Mg-rich ***olivine*** and ***plagioclase***. Crystallization of $MgAl_2O_4$-rich ***spinel*** is observed at high temperature in the experimental system ***anorthite-olivine***, but is not stable under equilibrium cooling. The presence in lunar rocks, thus, is due either to rapid cooling (i.e., near-surface conditions) or to accumulation by settling. The question is whether the parent Mg-Al-rich liquids were generated by internal thermal processes or by impact heating; evidence exists for both. The other subgroup is dominated by *chromite* ($FeCr_2O_4$) usually associated with ***olivine***-rich ("dunitic") breccia. The occurrence and compositions are similar to terrestrial *chromite* occurrences in layered intrusions, but, because of the brecciated nature of the lunar samples, little more can be determined.

The normal-inverse series is found in mare basalts; compositions usually lie between Fe_2TiO_4 and $FeAl_{0.5}Cr_{1.5}O_4$ (Fig. 5). This series is due to crystallization of an Fe-Ti-rich liquid producing Ti-poor ***spinel*** compositions early but zoning to Ti-rich compositions. Details of the zoning pattern can be used to infer crystallization conditions. For example, many rocks show a compositional gap between *chromite* and *ulvöspinel*, and this and other evidence suggest that crystallization was slow

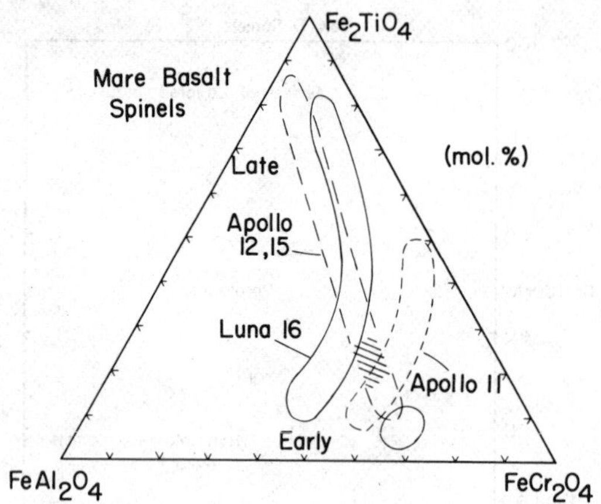

FIGURE 5. Fe-rich ***spinel*** compositional ranges in normal-inverse ***spinel*** series found in mare basalts. Samples from different sites show different trends, but all show early crystallizing $FeCr_2O_4$-rich compositions and late Fe_2TiO_4 compositions. Some rocks (especially of Apollo 12 and 15) show a compositional gap as indicated by cross-hatched pattern (see text).

enough for some early *chromite* to react with the liquid; the next ***spinel*** to crystallize was more Ti rich, thus producing a compositional gap. Other interpretations, including beginning of ***plagioclase*** crystallization, have been proposed to account for this gap.

The ***spinel*** group minerals commonly react to other assemblages. The Ti-rich varieties often undergo reduction, forming *ilmenite*, Fe metal, and a Ti-poor ***spinel*** at subsolidus temperatures. The Ti-poor ***spinels*** occurring in metamorphosed breccia show reaction rims adjacent to the breccia matrix because of the metastability of $MgAl_2O_4$-rich ***spinel*** at low temperature. ***Spinel*** phases are found as exsolutions in ***pyroxene*** and ***olivine***, probably due to low temperature equilibration.

IAN M. STEELE

References

Anderson, A. T., et al., 1970. Armalcolite: a new mineral from the Apollo 11 samples, *Proc. Apollo 11 Lunar Sci. Conf.*, vol. 1. New York: Pergamon, 55-64.

Anderson, A. T., et al., 1972. Thermal and mechanical history of breccias 14306, 14063, 14270, and 14321, *Proc. 3rd Lunar Sci. Conf.*, vol. 1. Cambridge: MIT Press, 819-835.

Arrhenius, G., et al., 1971. Zirconium fractionation in Apollo 11 and Apollo 12 rocks, *Proc. 2nd Lunar Sci. Conf.*, vol. 1. Cambridge: MIT Press, 169-175.

Bell, P. M., and Mao, H. K., 1973. Optical and chemical analysis of iron in Luna 20 plagioclase, *Geochim. Cosmochim. Acta*, **37**, 755-759.

Bence, A. E., and Papike, J. J., 1972. Pyroxenes as recorders of lunar basalt petrogenesis: chemical trends due to crystal-liquid interaction, *Proc. 3rd Lunar Sci. Conf.*, vol. 1, Cambridge: MIT Press, 431-469.

Deer, W. A.; Howie, R. A.; and Zussman, J., 1963. *Rock-Forming Minerals*, vols. 1-5. London: Longmans.

Dowty, E.; Keil, K.; and Prinz, M., 1974. Lunar pyroxene-phyric basalts: crystallization under supercooled conditions, *J. Petrol.*, **15**, 419-453.

Frondel, J. W., 1975. *Lunar Mineralogy*. New York: Wiley, 323p.

Haggerty, S. E., 1972. Apollo 14: subsolidus reduction and compositional variations of spinels, *Proc. 3rd Lunar Sci. Conf.*, vol. 1. Cambridge: MIT Press, 305-322.

Haggerty, S. E., 1974. Armalcolite and genetically associated opaque minerals in the lunar samples, *Proc. 4th Lunar Sci. Conf.*, vol. 1. New York: Pergamon, 777-797.

Nehru, C. E., et al., 1974. Spinel-group minerals and ilmenite in Apollo 15 rake samples, *Am. Mineralogist*, **59**, 1220-1235.

Niebuhr, H. H.; Zeira, S.; and Hafner, S. S., 1973. Ferric iron in plagioclase crystals from anorthosite 15414, *Proc. 4th Lunar Sci. Conf.*, vol. 1. New York: Pergamon, 971-982.

Prinz, M., et al., 1973. Spinel troctolite and anorthosite in Apollo 16 samples, *Science*, **179**, 74-76.

Proceedings (1970-1980). The Proceedings of the annual lunar science conferences as well as some special conferences are published as supplements to *Geochimica et Cosmochimica Acta*. Supplement 1, *Proceedings of the Apollo 11 Lunar Science Conference*, was published in 1970 by Pergamon. *Proceedings of the 2nd–and 3rd–Lunar Science Conference* were published in 1971 and 1972 by MIT Press. *Proceedings of the 4th* through *8th Lunar Science Conferences* were published from 1973 to 1977 and the *Proceedings of the 9th* through *11th Lunar and Planetary Science Conferences* from 1978-1980 by Pergamon Press.

Raymond, K. N., and Wenk, H. R., 1971. Lunar Ilmenite (Refinement of the Crystal Structure), *Contrib. Mineral. Petrol.*, **30**, 135-150.

Roeder, P. L., and Emslie, R. F., 1970. Olivine-liquid equilibrium, *Contrib. Mineral. Petrol.*, **29**, 275-289.

Smith, J. V., 1974. Lunar mineralogy: A heavenly detective story, *Am. Mineralogist*, **59**, 231-243.

Smith, J. V., and Steele, I. M., 1976. Lunar mineralogy: A heavenly detective story, pt. 2, *Am. Mineralogist*, **61**, 1059-1116.

Smyth, J. R., 1975. A crystal structure refinement of an anomalous lunar anorthite, in *Lunar Science*, vol. VI. Houston: The Lunar Science Institute, 756-758.

Steele, I. M., and Smith, J. V., 1973. Compositional and X-ray data for Luna 20 feldspar, *Geochim. Cosmochim. Acta*, **37**, 1075-1077.

Steele, I. M., and Smith, J. V., 1974. Intergrowths in

lunar and terrestrial Anorthosites with implications for lunar differentiation, *Am. Mineralogist*, **59**, 673–680.

Steele, I. M., 1974. Ilmenite and Armalcolite in Apollo 17 Breccias, *Am. Mineralogist*, **59**, 681–689.

Taylor, S. R., 1974. *Lunar Science: A post Apollo view*. New York: Pergamon.

Windley, B. F., 1970. Anorthosites in the early crust of the earth and on the moon, *Nature*, **226**, 333–335.

Cross-references: *Meteoritic Minerals; Rock-Forming Minerals;* see also mineral glossary.

M

MAGNATIC MINERALS—*See* ROCK-FORMING MINERALS

MAGNETIC MINERALS

Magnetic fields are generated by moving electric charge. A current I flowing in a coil of n turns per meter produces a magnetic field $H=nI$ amperes/meter. All materials respond to a magnetic field, producing a magnetization $M=\chi H$, expressed in weber/m^2. A weber is the unit of magnetic charge, weber-m of magnetic dipole moment, and weber/m^2 of magnetic dipole moment per unit volume (magnetization). Every dipole has its moment, just as every dog has his day. The magnetic susceptibility χ is in henry/m but is usually presented as a dimensionless quantity $\bar{\chi}=\chi/\mu_0$, where $\mu_0=4\pi\times10^{-7}$ henry/m is the permeability of vacuum. Observed values of the relative susceptibility $\bar{\chi}$ range from 10^{-5} in weakly magnetic materials to 10^6 in strong magnets (Chikozumi, 1964). The quantity $\bar{\chi}$ may be linear or nonlinear, positive or negative, and is often temperature-sensitive. It provides a convenient measure of the response of a material to an applied magnetic field, a response that is interpreted in terms of magnetic structure and electronic configurations.

Most materials are "diamagnetic"—i.e., have a weak magnetic response induced by an applied magnetic field. In a field gradient, diamagnetic materials experience a force driving them out of the field. In most pure compounds the electrons are paired in bonding, but magnetic fields cause small changes in orbital motion that result in a small negative susceptibility. For *quartz, calcite, fluorite*, and *halite*, $\bar{\chi}$ lies in the range -1.1×10^{-5} to 1.6×10^{-5} and is nearly independent of temperature.

Magnetization is also linearly proportional to field in "paramagnetic" materials, but $\bar{\chi}$ is positive and somewhat larger than diamagnetic susceptibilities. Paramagnetism is common in dilute transition-metal salts, in which the metal ions with unpaired electrons interact with one another only weakly. The spins are randomly oriented but align slightly when a field is applied. Alignment becomes more difficult at high temperatures, causing a decrease in susceptibility with temperature following the Curie law $\bar{\chi}=C/T$, where C is the Curie constant and T is the absolute temperature. Many ferromagnesian minerals contain small amounts of iron, maganese, and other ions with unpaired electrons. At room temperature, the susceptibility is about $+10\times10^{-5}$ for such materials. Another type of paramagnetism is found in metallic-like minerals where the conduction electrons create temperature-independent "Pauli paramagnetism." The effect is caused by small changes in the band structure for electrons of opposite spin when a magnetic field is applied. *Pyrite* is a Pauli paramagnet.

When spins interact appreciably, three types of ordered configurations occur: antiferromagnetism, ferromagnetism, and ferrimagnetism. All three show Curie-Weiss–Law behavior at high temperatures in the paramagnetic region. On cooling, the materials undergo a phase transition to a state in which the atomic dipoles are aligned, even in the absence of an applied field. Magnetic domains are observed in the low-temperature state.

Antiferromagnetism is the most common of the three phenomena. In a simple collinear antiferromagnet such as *wüstite*, adjacent spins are aligned in antiparallel directions, producing zero net moment at zero field. The magnetic susceptibility is small and field-independent, with a pronounced maximum near the transition temperature, called the Néel point. Canted, spiral, and other more complicated antiferromagnetic arrays have also been observed. A number of antiferromagnetic minerals are listed in Table 1.

The most important magnetic materials, however, are ferromagnets and ferrimagnets. Both possess a spontaneous magnetization that shows hysteresis under applied fields and disappears at the Curie temperature. The ordered spin array of a ferromagnet consists of parallel spins. Iron is ferromagnetic but the effect is rare among oxides, with only CrO_2, EuO, and a few other examples. In a simple ferrimagnet, neighboring spins are antiparallel but are either unequal in size or unequal in number. A number of useful ferrimagnetic oxides are found in the ***spinel*** and ***garnet*** groups. In *magnetite* and other ferri-

TABLE 1. Minerals with Long-Range Magnetic Order and Their Transition Temperatures

	Ferromagnetic		
cohenite (Fe_3C)	483°K	*nickel* (Ni)	627°K
iron (Fe)	1043°		
	Ferrimagnetic		
chromite ($FeCr_2O_4$)	80°K	*maghemite* (γFe_2O_3)	856°K
hausmannite (Mn_3O_4)	43°	*magnetite* (Fe_3O_4)	858°
pyrrhotite (Fe_7S_8)	593°	*ulvöspinel* (Fe_2TiO_4)	142°
	Antiferromagnetic		
akaganéite ($\beta FeOOH$)	295°K	*ilmenite* ($FeTiO_3$)	68°K
fayalite (Fe_2SiO_4)	65°	*lepidocrocite* ($\gamma FeOOH$)	73°
ferberite ($FeWO_4$)	76°	*siderite* ($FeCO_3$)	38°
franklinite ($ZnFe_2O_4$)	9°	*troilite* (FeS)	600°
goethite ($\alpha FeOOH$)	367°	*Wüstite* (FeO)	190°
hematite (αFe_2O_3)	953°	*vivianite* ($Fe_2(PO_4)_2 \cdot 8H_2O$)	9°
alabandine (MnS)	154°	*pyrolusite* (MnO_2)	90°
braunite (Mn_2O_3)	80°	*rhodochrosite* ($MnCO_3$)	32°
huebnerite ($MnWO_4$)	15°	*rhodonite* ($MnSiO_3$)	7°
manganite (MnOOH)	40°	*tephroite* (Mn_2SiO_4)	50°
chalcopyrite ($CuFeS_2$)	825°	*bunsenite* (NiO)	250°
dioptase ($CuSiO_2(OH)_2$)	21°	*millerite* (NiS)	264°
uraninite (UO_2)	30°	*tenorite* (CuO)	230°

magnetic minerals (Table 1), the exchange coupling is antiferromagnetic but the two sublattices are unbalanced in magnetic moment so that a net magnetization results. The minerals of greatest interest to rock magnetism are in the system $FeO-Fe_2O_3-TiO_2$ (Nagata, 1961). In addition to *magnetite*, these include *ulvöspinel, maghemite, hematite, ilmenite*, and *pseudobrookite.*

Superexchange Interaction

Magnetite has the spinel structure in which there are twice as many cations in octahedral positions as in tetrahedral (see *Spinel Group*). When transition metal ions occupy both sites, there is a strong antiferromagnetic superexchange coupling between the tetrahedral and octahedral ions. This is the dominant exchange interaction because it makes best use of the oxygen *p* orbitals, although there are other interactions as well.

Magnetic ordering occurs when the transition-metal atoms are nearest neighbors or next-nearest neighbors. Among oxides and fluorides, antiferromagnetism is much more common than ferromagnetism or ferrimagnetism. The reason is the superexchange interaction. Direct exchange seldom occurs in minerals because the transition-metal ions are not in direct contact, but interact via an intermediate anion. Superexchange is a strong interaction, leading to magnetic transition temperatures comparable to metals. Ferromagnetic ordering in Fe occurs at 1043°K, antiferromagnetism in αFe_2O_3 at 953°K, and ferrimagnetism in *magnetite* at 858°K.

In the superexchange interaction, two metal atoms M_1 and M_2 on opposite sides of an oxygen ion interact through a *p* orbital of oxygen (Fig. 1). Transition-metal ions with less than half-full *d* shells will be considered first. Since the oxygen ion is not fully ionized, the outer electrons spend time on the neighboring transition-metal ions. When an electron enters the *d* shell of a transition ion whose *d* orbitals are less than half full, the oxygen electron spin is parallel to that of the metal ion in accordance

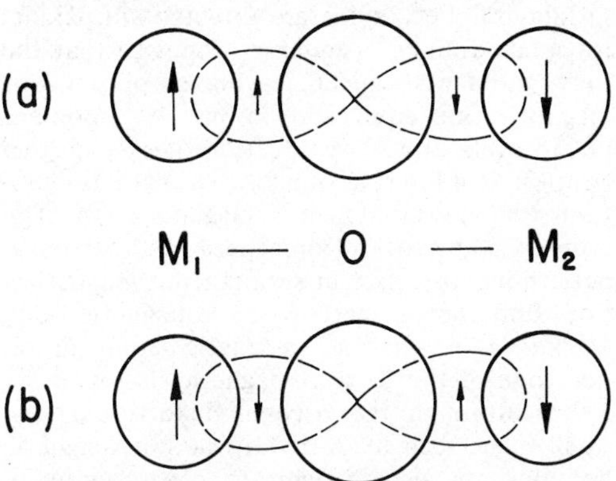

FIGURE 1. The 180° superexchange interaction responsible for ferrimagnetism and antiferromagnetism in many minerals. Diagram illustrates the interaction when the 3*d* shell of the transion-metal atoms is (a) less than half full and (b) half or more than half full.

with Hund's rule. Meanwhile, the other electron in the same oxygen p orbital is on the opposite side of the oxygen ion because of the coulomb repulsion between two electrons in the same p orbital. While there, the second electron (whose spin is antiparallel to the first electron because of the Pauli exclusion principle) also interacts with transition-metal ions—and its spin will again be parallel to that of the metal ion if its d shell is less than half full. The antiferromagnetic superexchange thus arises from three alignments as shown in Fig. 1: The first metal atom accepts an electron with parallel spin from an oxygen neighbor; the spins of the two electrons in the same oxygen p orbital are antiparallel; and the second electron spends part of its time in parallel alignment with the d electrons of the second metal ion.

A similar situation is obtained when the d-electron shell of the transition metal ion is more than half full, again resulting in antiferromagnetic superexchange (Fig. 1). The oxygen electrons enter the metal atom d shell antiparallel to the net spin, but since the same thing happens to the other electron, the interaction remains antiferromagnetic. Superexchange is strongest when the angle M_1-O-M_2 is 180°, allowing maximum overlap of the p orbital with the two metal ions. The interaction weakens as the angle approaches 90°, even though the metal–metal distance may be shorter.

Magnetic Domains

Magnetite is the most important magnetic mineral. Below 858°K, Fe_3O_4 is ferrimagnetic with a magnetic moment of 4 Bohr magnetons per molecule corresponding to the four unpaired electron spins associated with Fe^{2+} (Smit and Wijn, 1959). As shown in Fig. 2, tetrahedral Fe^{3+} spins are directed antiparallel to octahedral Fe^{3+} and Fe^{2+} spins so that the Fe^{3+} moments cancel, leaving a spontaneous magnetization equivalent to one Fe^{2+} moment per formula unit. The direction of easy magnetization is [111], giving rise to eight orientations of the spontaneous magnetization. The domain states of ferromagnetic and ferrimagnetic minerals differ in spontaneous magnetization, and can be switched by a magnetic field. *Magnetite* belongs to magnetic point group $\bar{3}m'$, one of the 21 pyromagnetic classes (Birss, 1964) although the trigonal distortion is too small to be seen by X-ray diffraction. Magnetic domains are also difficult to observe visually because *magnetite* is opaque to visible light even in thin section. The domains are probably similar to those in $MgFe_2O_4$, which has the same magnetic structure, but is transparent to red wavelengths. Sherwood, Remeika, and

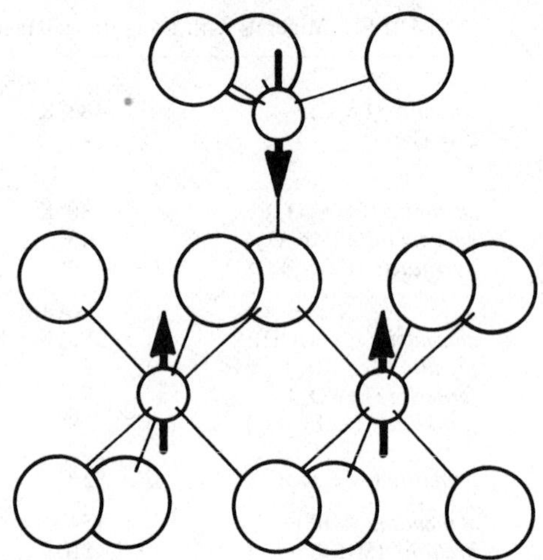

FIGURE 2. In *magnite*, the magnetic moments of tetrahedrally coordinated Fe^{3+} ions are antiparallel to those of the Fe^{3+} and Fe^{2+} ions in octahedral coordination. Spins are aligned along [111] or on the equivalent body-diagonal directions.

Williams (1959) observed snake-like domain patterns in a (111) thin section of magnesium ferrite.

Domains in transparent ferromagnetic and ferrimagnetic crystals are visible in polarized light because of the Faraday effect, a nonreciprocal rotation of the plane of polarization. The angle of rotation ϕ is given by $\phi=\rho t\cos\theta$, where t is the specimen thickness, ρ the rotation per unit thickness, and θ the angle between the magnetization vector and the direction of propagation. Faraday rotation coefficient for ferrites are typically about 1000°/cm (Sherwood et al., 1959).

Hematite, αFe_2O_3, exhibits both antiferromagnetism and weak ferromagnetism. From 250°K to 953°K, the Fe^{3+} spins lie in the rhomobohedral (111) plane and are nearly antiparallel, but with a small ferromagnetic component, also in (111). Mineralogists generally assign *hematite* to trigonal class $\bar{3}m$, but the magnetic point symmetry is $2/m$ at room temperature. Antiferromagnetic crystals often exhibit weak (parasitic) ferromagnetism when the ferromagnetic component does not violate the symmetry elements of the antiferromagnetic spin array (Birss, 1964). In *hematite*, weak spontaneous magnetization appears along the monoclinic twofold axis, corresponding to one of the three diad axes in $\bar{3}m$. At 250°K, the spin direction changes to the rhombohedral axis [111], and the weak ferromagnetic effect disappears. Below the spin-flop transition, the magnetic point group is $\bar{3}m$.

Magnetic domains in *hematite* have been ob-

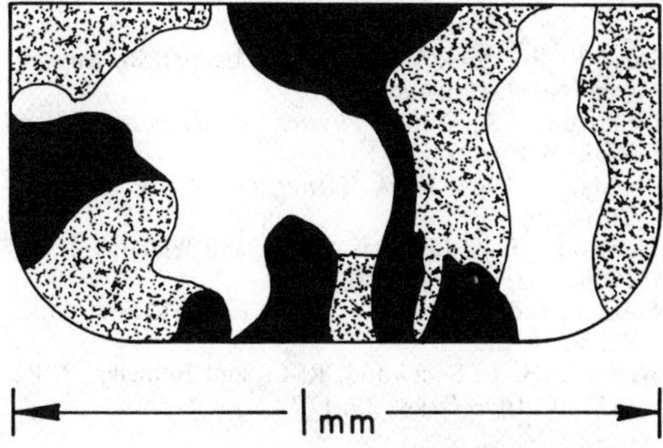

FIGURE 3. Magnetic domain pattern in a *hematite* crystal observed by means of the Faraday effect (after Birss, 1964). Three orientations of the parasitic ferromagnetism give rise to different light intensities. The specimen is a thin platelet (≈0.03 mm thick) with major faces parallel to the rhombohedral (111) plane. To observe the domains it was necessary to tilt the platelet with respect to the light beam; otherwise the Faraday effect is absent because the magnetization vectors are perpendicular to the beam.

served using the Faraday effect (Williams et al., 1958). The white, gray, and black regions in Fig. 3 correspond to domains with three different magnetic axes: small magnetic fields produce significant differences in the domain pattern. When cooled through the spin-flop transition at -120°C, the domains disappear, and then reappear in a different pattern on heating.

Magnetic Hysteresis

The saturation magnetization is a strong function of temperature. As Fig. 4 shows, the decrease in M_S as T approaches the Curie temperature becomes quite precipitous so that at temperatures well below T_C, the saturation magnetization is substantially equal to its value at absolute zero. If T_C is about 850°K, as for *magnetite*, most of the atomic spins are lined up at room temperature.

In spite of this, many specimens of *magnetite* do not exhibit the large spontaneous magnetization found in lodestones. The presence of magnetic domains results in cancellation of the spontaneous magnetization. The technical magnetization curve in Fig. 5 describes the behavior of ferromagnetic and ferrimagnetic minerals in a magnetic field.

There are a number of interesting features of the technical magnetization curve. Assume that the specimen is initially unmagnetized (Point O, Fig. 5). When a magnetic field is applied, M gradually increases, eventually reaching its saturation value M_S as H increases. The slow initial increase in M due to reversible domain-wall motion is followed by a rapid increase as domain walls move past the crystal imperfections that impede their movement at this stage in a polycrystalline specimen of *magnetite*—each grain is a single domain with the spontaneous magnetization in the [111] direction closest to the direction of the applied magnetic field. The final stage of saturation is caused by the rotation of magnetic moment away from the crystallographic direction toward the field direction. At saturation, all moments are aligned with the field.

If the applied magnetic field is turned off, some remanent magnetization remains as shown in Fig. 5. Magnetic spins rotate back to the preferred crystallographic direction, but domain walls do not return to their original positions.

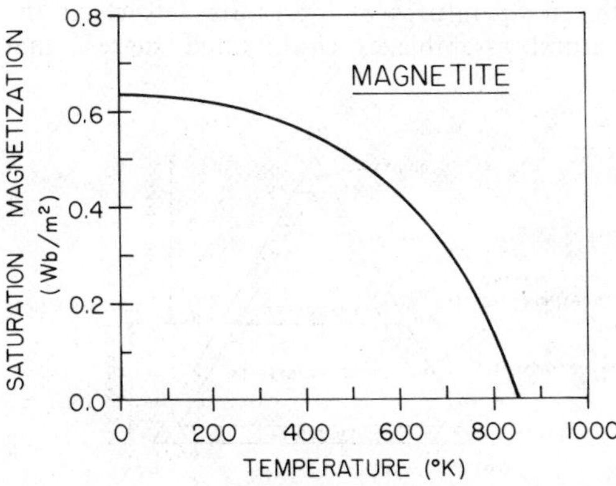

FIGURE 4. Saturation magnetization of *magnetite* as a function of absolute temperature (after Smit and Wijn, 1959). Spontaneous magnetization disappears above the Curie temperature where the mineral is paramagnetic.

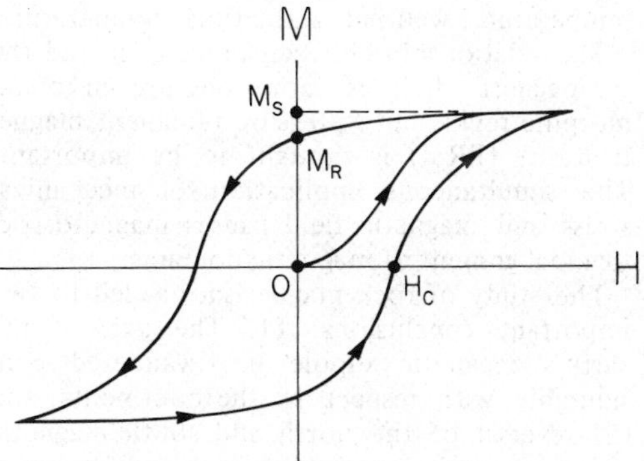

FIGURE 5. Magnetic hysteresis curve for a ferromagnetic or ferrimagnetic mineral. M_S is the saturation magnetization, M_R the remanent magnetization, and H_C the coercive magnetic field.

Often the domain-wall motion is small leaving a large remanent magnetization, M_R. The value of M_R is not large in most mineral specimens, because the earth's field is considerably smaller than that of a laboratory magnet and because there has been considerable time for the domain configuration to revert to a low-energy state.

Rock Magnetism

Many rocks contain a few percent *magnetite* and other magnetic minerals. Such rocks often exhibit a "natural remanent magnetization" (NRM), which is of considerable geologic interest and has several possible origins (Nagata, 1961). "Thermoremanent magnetization" (TRM) is the remanent magnetization after the specimen has been cooled from high temperatures in the earth's magnetic field. In igneous rocks, the temperature where the cooling starts is usually above the Curie point (T_C) of the magnetic minerals. When a magnetic grain cools through T_C, a small magnetization develops in the direction of the earth's field. The magnetization increases in size on further cooling but remains fixed in direction. At normal temperatures, the thermoremanent magnetization may be quite large; to produce a comparable magnetization isothermally would require a magnetic field much larger than that of the earth.

Remanent magnetization has a different origin in sedimentary rocks, where the grains are already magnetized and are then oriented by the geomagnetic field at the time of settling. This effect is sometimes referred to as "depositional remanent magnetization" (DRM). "Chemical remanent magnetization" (CRM) occurs during crystallization of a ferromagnetic mineral in a magnetic field. In many chemical deposits, the crystallization occurs near room temperature, well below the Curie temperature.

The relationship between remanence and the geomagnetic field is more obscure in metamorphic rocks, but "pressure remanent magnetization" (PRM) is thought to be important. The simultaneous application of mechanical stress and magnetic field causes magnetostrictive realignment of magnetic domains.

The study of rock magnetism has led to two important conclusions: (1) The axis of the earth's magnetic dipole has wandered considerably with respect to the continents; and (2) reversal of the north and south magnetic poles occurred several times in the ancient past. Attempts to explain this behavior in terms of models of the earth's interior have not been entirely successful.

ROBERT E. NEWNHAM

References

Birss, R. R., 1964. *Symmetry and Magnetism.* Amsterdam: North-Holland, 252p.

Chikozumi, S., 1964. *Physics of Magnetism.* New York: Wiley, 554p.

Nagata, T., 1961. *Rock Magnetism.* Tokyo: Maruzen, 217p.

Sherwood, R. C.; Remeika, J. P.; and Williams, H. J., 1959. *J. Appl. Phys.*, **30**, 217.

Smit, J., and Wijn, H. P. J., 1959. *Ferrites.* New York: Wiley, 369p.

Williams, H. J.; Sherwood, R. C.; and Remeika, J. P., 1958. *J. Appl. Phys.*, **29**, 1772.

MANTLE MINERALOGY

The earth's mantle may be subdivided into three distinct zones: upper mantle, transition zone, and lower mantle. The mineralogy of each zone is described separately.

Upper Mantle

Geophysical and petrologic evidence points to peridotite as the predominant rock type in the earth's upper mantle, extending from the Mohorovicic discontinuity at the base of the crust to a depth of about 350 km. The nomenclature of peridotites and other ultramafic rocks is shown in Fig. 1. Garnet lherzolites, spinel lherzolites, and eclogites (*pyrope* ***garnet*** + omphacitic *diopside*) are the most abundant mantle-derived xenoliths found in kimberlite pipes (Dawson, 1972).

Experimental investigation of the reaction, *enstatite* + *diopside* + *spinel* = forsteritic ***olivine*** + calcium-bearing *pyrope*, indicates that spinel lherzolite becomes unstable relative to garnet lherzolite at depths exceeding about 50 km (O'Hara et al., 1971). Determinations of the temperatures and pressures at which the natural assemblages equilibrated suggest that

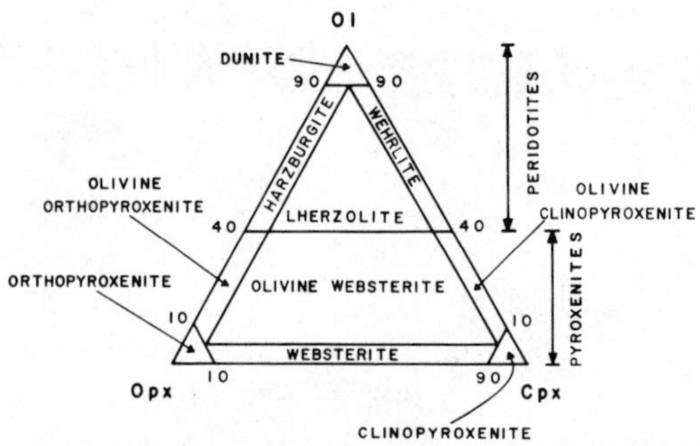

FIGURE 1. Nomenclature of ultramafic rocks composed of ***olivine***, ***orthopyroxene***, and ***clinopyroxene***, as recommended by the International Union of Geological Sciences, 1973.

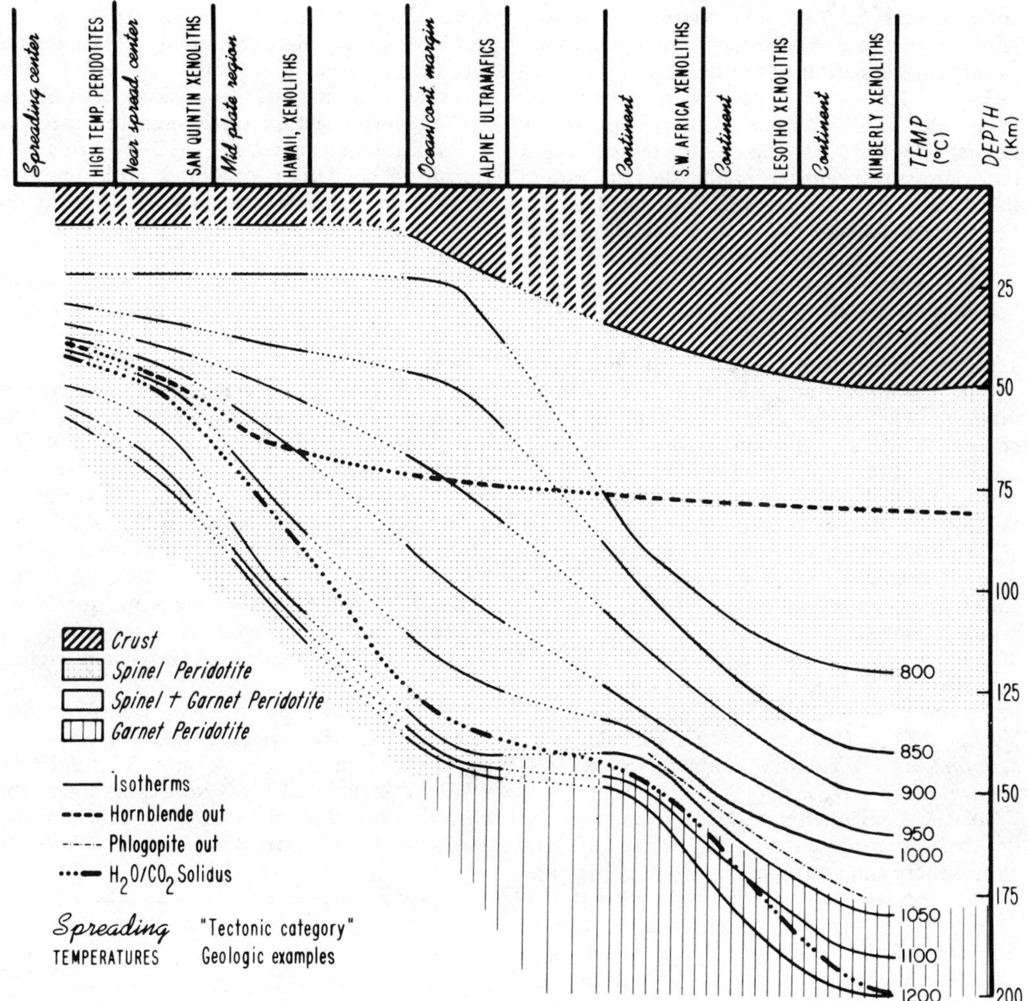

FIGURE 2. Petrologic and thermal structure of the uppermost mantle (from I. D. MacGregor and A. R. Basu, Thermal Structure of the lithosphere: A petrologic model, *Science* 185, 1975, p. 1010; copyright by the American Association for the Advancement of Science).

garnet lherzolites are derived from depths exceeding 130 km as shown in Fig. 2 (MacGregor and Basu, 1974).

The geothermometry used in this reconstruction is based upon the solubility of *enstatite* in *diopside* (Boyd, 1973). Mori and Green (1975) have subsequently shown that the *enstatite-diopside* solvus widens slightly with increasing pressure (see *Thermometry, Geologic*). The geobarometry is based upon the Al_2O_3 content of *enstatite* coexisting with *spinel* ($MgAl_2O_4$) or *pyrope* ($Mg_3Al_2Si_3O_{12}$). The Al_2O_3 content increases with temperature and decreases with pressure so that the equilibrium pressure may be estimated if the equilibrium temperature is known (see *Barometry, Geologic*; Lane and Ganguly, 1980; MacGregor, 1974a). Of course, there are problems inherent in applying experimental results from chemically simple systems to complex natural systems. The natural ***pyroxenes*** contain Fe^{2+}, Al^{3+}, and Na^+. Furthermore, *spinel* is a sink for chromium, and *chromite* ***spinel*** persists to high pressures with unknown effects on the composition of coexisting *enstatite*. Exsolution of *spinel* from *enstatite* is observed in xenoliths from kimberlites (Basu and MacGregor, 1975). The spinel compositions are intermediate between those of Cr-rich euhedral ***spinels*** restricted to kimberlite xenoliths and Al-rich interstitial ***spinels*** restricted to xenoliths in alkali olivine basalt, and are similar to the compositions of wormy symplectite intergrowths of *spinel* with ***clinopyroxene*** and other silicates in kimberlite xenoliths.

Garnet lherzolite xenoliths in kimberlite pipes exhibit a curious dichotomy. Those that are derived from the highest temperatures (and pressures) are generally sheared rather than granular; are relatively rich in Na, Ti, Fe, Ca, Al, Cs, and Ba; and are usually devoid of *graphite* and *phlogopite* (see Boettcher, 1974; and MacGregor, 1975).

Diamonds are stable only at depths exceeding about 130 km (Bundy et al., 1961; Kennedy and Nordlie, 1968). The inclusions preserved within *diamonds* are, therefore, of special interest. The most common inclusions are ***olivine*** (Fo_{92-95}), chrome *pyrope, chromite, enstatite*, and *diopside* (Meyer and Boyd, 1972). Some inclusions are eclogitic in character, *pyrope–almandine* and *omphacite* (Tsai et al., 1979). *Rutile, ilmenite, magnetite, corundum, pyrrhotite, zircon, hyanite, sanidine, phlogopite*, and *biotite* have also been noted as primary inclusions (MacGregor, 1974b; Giardini et al., 1974). Some of the inclusions are anomalously enriched in chromium: *pyrope* inclusions contain up to 63% *knorringite*, $Mg_3Cr_2Si_3O_{12}$ (Meyer, 1975); *diopsides* in *diamond* from Yakutia contain up to 34% *ureyite*, $NaCrSi_2O_6$ (Sobolev et al., 1971).

Volatile constituents within the upper mantle, besides representing the source of our hydrosphere and atmosphere, play a critical role in the formation of partial melts (Boettcher et al., 1975). The volatiles reside in pargasitic ***amphiboles***, *phlogopite,* titanian *clinohumite* (McGetchin and Silver, 1972), carbonates, and ubiquitous fluid CO_2 inclusions (Roedder, 1965). Most of the fluid inclusions are less than a micron in diameter and are attached to crystal defects and grain boundaries (Green and Radcliffe, 1975). The depths at which ***amphibole*** and *phlogopite* dehydrate are indicated in Fig. 2. Also shown in Fig. 2 is the depth to the solidus for a mantle having a $H_2O–CO_2$ gas phase in which the mole fraction of H_2O is 0.25. Kushiro et al. (1968) showed that the presence of water extends the incongruent melting of ***orthopyroxene*** to pressures above 35 kilobars. Partial fusion of the upper mantle may thus yield quartz-normative melts under wet conditions and olivine-normative melts under drier conditions (Mysen and Boettcher, 1975).

Transition Zone

Birch (1952) proposed that the rapid increases in seismic velocities occurring downward through the transition zone, extending from about 350 to 700 km in depth, are the result of phase transformations. This interpretation has been confirmed by static high-pressure experiments, shock-wave techniques, and the investigation of germanate analogues of silicates (Ringwood, 1972). Germanates transform to denser structures at lower pressures than their silicate isotypes because the germanium ion is slightly larger than the silicon ion.

Naturally shocked meteorites have yielded several examples of pressure-induced transformations, including ***olivine*** to a spinel structure (Binns et al., 1969), and ***pyroxene*** to ***garnet***, $Mg_3(MgSi)Si_3O_{12}$, in which one quarter of the silicon atoms are in sixfold coordination (Smith and Mason, 1970). Striking *diopside-ilmenite* intergrowths found in kimberlite xenoliths have been ascribed to the decomposition of ***garnet*** solid solutions shown to be stable above 100 kilobars (Ringwood and Lovering, 1970). However, the occurrence of *ilmenite-**orthopyroxene*** intergrowths in the same specimens supports the hypothesis of intergrowth by simultaneous crystallization (Boyd and Nixon, 1973).

Experimentally determined transformation pressures in the system $MgO-FeO-SiO_2$ are summarized in Figs. 3, 4, and 5 (Ringwood, 1972; Yagi et al. 1979). Note that forsteritic ***olivine*** passes through an intermediate β phase, a modified spinel structure, before transform-

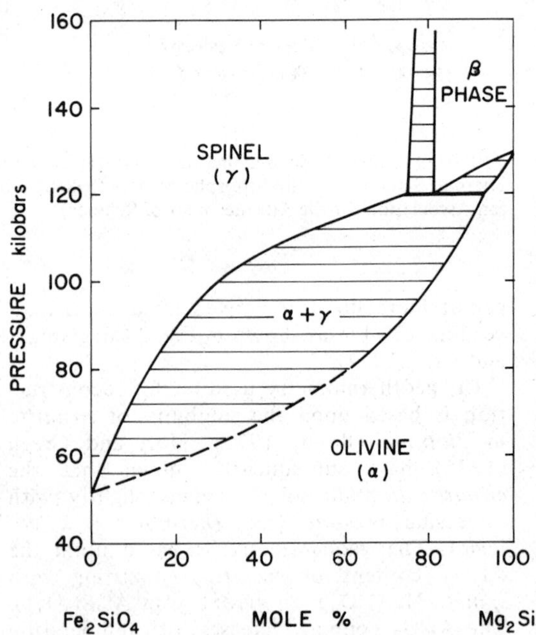

FIGURE 3. Phase diagram for the system $Mg_2SiO_4–Fe_2SiO_4$ at 1000°C (after Ringwood, 1972).

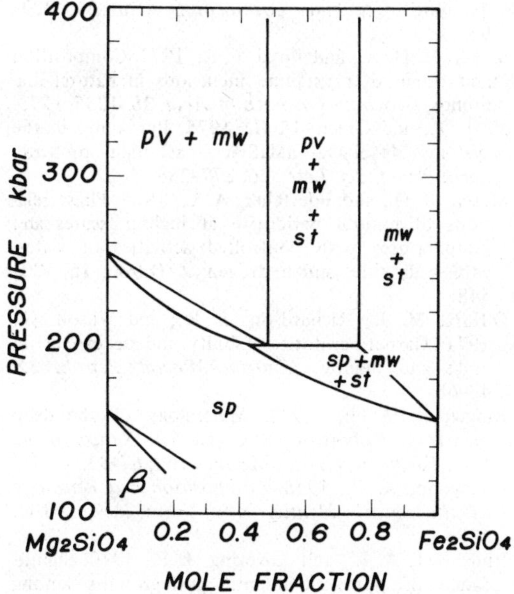

FIGURE 4. Pseudobinary phase diagram for olivine compositions at 1000°C and pressures exceeding 150 kbar. sp = *spinel*; pv = *perovskite*; mw = *magnesiowüstite*; and st = *stishovite* (after Yagi et al., 1979).

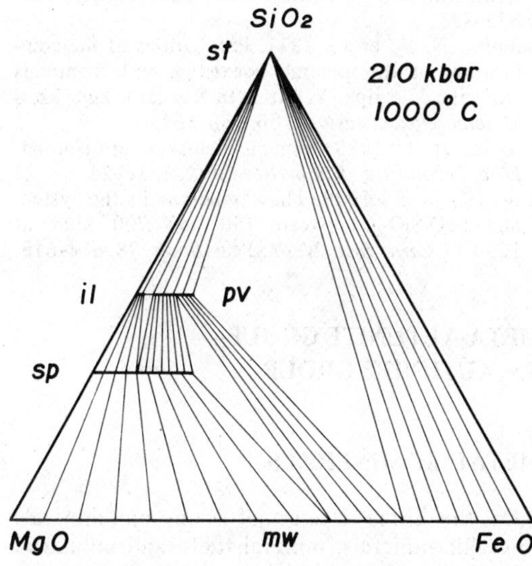

FIGURE 5. Phase diagram for the system SiO_2–MgO–FeO at 210 kilobars and 1000°C. il = *ilmenite*; other abbreviations as in Fig. 4 (after Yagi et al., 1979).

ing to the γ-spinel phase. Seismic velocity "discontinuities" near 400 km and 500 km are attributed to the collapse of ***olivine*** to the β- and γ-spinel structures, respectively (Burdick and Anderson, 1975).

Since both the olivine and spinel structures involve oxygen in closest packing, magnesium in six-coordination and silicon in four-coordination, it is necessary to resort to considerations of edge-sharing by coordination polyhedra in order to understand their 10% difference in densities (Kamb, 1968). The β-spinel phase differs from the γ-spinel phase by having each silica tetrahedron share one oxygen, and its formula may thus be written $Mg_4O \cdot Si_2O_7$.

Atomic substitution between magnesium and iron allows a phase transformation to progress gradually over a definite range of pressure or depth as is clear from Fig. 3 (Ringwood, 1972).

Lower Mantle

The lower mantle extends downward from a seismic velocity discontinuity, at a depth of about 650 km, which is attributed to the decomposition of the silicate spinel phase to a mixture of *magnesiwüstite*, (Mg,Fe)O, having a halite structure, and (Mg,Fe)SiO_3, having a perovskite structure (Yagi et al. 1979). *Perovskite* contains silicon and magnesium in 6-fold and 12-fold coordination, respectively.

Although silicate *perovskite* and *magnesiowüstite* may be the dominant minerals in the lower mantle, Bell (1979) has summarized the evidence that calcium is accommodated in a separate perovskite phase, and Ringwood (1975) has suggested that aluminum occurs in $MgSiO_3 \cdot Al_2O_3$, having an ilmenite structure, plus $NaAlSiO_4$, having a calcium ferrite structure.

Possible mantle differentiation resulting from Fe and Mg partitioning across phase transitions was discussed by Garlick (1969) and Kumazawa et al. (1974). The possible effects of Fe enrichment in *magnesiowüstite* relative to silicate *perovskite* were discussed by Mao et al. (1979). However, mantle models based upon seismic and shock-wave data have not revealed significant variations in Fe/Mg ratio within the mantle (Burdick and Anderson, 1975).

The recent advances in our understanding of mantle mineralogy owe a great deal to the development of the laser-heated, *diamond*-window, high-pressure cell in which megabar pressures can be produced in a tiny volume between *diamond* anvils (Yagi, et al., 1979).

G. DONALD GARLICK

References

Basu, A. R., and MacGregor, I. D., 1975. Chromite spinels from ultramafic xenoliths, *Geochim. Cosmochim. Acta*, **39**, 937–945.

Bell, P. M., 1979. Ultra-high-pressure experimental mantle mineralogy, *Rev. Geophys. Space Phys.*, **17**, 788–791.

Binns, R. A.; Davis, R. N.; and Reed, S. B. J., 1969. Ringwoodite, natural $(Mg,Fe)_2SiO_4$ in the Tenham meteorite, *Nature*, **221**, 943–944.
Birch, F., 1952. Elasticity and constitution of the earth's interior, *J. Geophys. Res.*, **57**, 227–286.
Boettcher, A. L., 1974. Review of symposium on deep-seated rocks and geothermometry, *Trans., Am. Geophys. Union*, **56**, 1068–1072.
Boettcher, A. L.; Mysen, B. O.; and Modreski, P. J., 1975. Phase relationships in natural and synthetic peridotite-H_2O-CO_2 systems at high pressures, *Phys. Chem. Earth*, **9**, 855–867.
Boyd, F. R., 1973. A pyroxene geotherm, *Geochim. Cosmochim. Acta*, **37**, 2533–2546.
Boyd, F. R., and Nixon, P. H., 1973. Origin of the ilmenite-silicate nodules in kimberlites, in P. H. Nixon, ed., *Lesotho Kimberlites*. Maseru: Lesotho National Development Corp., 254–268.
Bundy, F. P., et al., 1961. Diamond-graphite equilibrium line from growth and graphitization of diamond, *J. Chem. Phys.*, **35**, 383.
Dawson, J. B., 1972. Kimberlites and their relation to the mantle, *Phil. Trans. R. Soc. London.*, **271**, 297–311.
Garlick, G. D., 1969. Consequences of chemical equilibrium across phase changes in the mantle, *Lithos*, **2**, 325–331.
Giardini, A. A., et al., 1974. Biotite as a primary inclusion in diamond: Its nature and significance, *Am. Mineralogist*, **59**, 783–789.
Green, H. W., and Radcliffe, S. V., 1975. Fluid precipitates in rocks from the earth's mantle, *Geol. Soc. Am. Bull.*, **86**, 846–852.
Kamb, B., 1968. Structural basis of the olivine-spinel stability relation, *Am. Mineralogist*, **53**, 1439–1455.
Kennedy, S. C., and Nordlie, B. E., 1968. The genesis of diamond deposits, *Econ. Geol.*, **63**, 495–503.
Kumazawa, M., et al., 1974. Post-spinel phase of forsterite and evolution of the earth's mantle, *Nature*, **247**, 356–358.
Kushiro, I.; Yoder, H. S.; and Nishikawa, M., 1968. Effect of water on the melting of enstatite, *Geol. Soc. Am. Bull.*, **79**, 1685–1692.
Lane, D. L. and Ganguly, J., 1980. Al_2O_3 solubility in orthopyroxene. *Jour. Geoph. Res.*, **85**, 6963–6972.
MacGregor, I. D., 1974a. The system MgO-Al_2O_3-SiO_2: Solubility of Al_2O_3 in enstatite for spinel and garnet peridotite compositions, *Am. Mineralogist*, **59**, 110–119.
MacGregor, I. D., 1974b. First kimberlite conference, Republic of South Africa, *Geology*, **2**, 151–152.
MacGregor, I. D., 1975. Petrologic and geophysical significance of mantle xenoliths in basalts, *Rev. Geophys. Space Phys.*, **13**, 90–93.
MacGregor, I. D., and Basu, A. R., 1974. Thermal structure of the lithosphere: A petrologic model, *Science*, **185**, 1007–1011.
Mao, H. K., et al., 1979. Iron-magnesium fractionation model for the earth. *Carn. Inst. Wash. Year Book*, **78**, 621–625.
McGetchin, T. R., and Silver, L. T., 1972. A crustal-upper-mantle model for the Colorado Plateau based on observations of crystalline rock fragments in the Moses Rock dike, *J. Geophys. Res.*, **77**, 7022–7037.
Meyer, H. O. A., 1975. Chromium and the genesis of diamond, *Geochim. Cosmochim. Acta*, **39**, 929–936.
Meyer, H. O. A., and Boyd, F. R., 1972. Composition and origin of crystalline inclusions in natural diamonds, *Geochim. Cosmochim. Acta*, **36**, 1255–1273.
Mori, T., and Green, D. H., 1975. Pyroxenes in the system $Mg_2Si_2O_6$-$CaMgSi_2O_6$ at high pressure, *Earth Planet. Sci. Lett.*, **26**, 277–286.
Mysen, B. O., and Boettcher, A. L., 1975. Phase relations of natural peridotite at high pressures and temperatures with controlled activities of water, carbon dioxide, and hydrogen, *J. Petrol.*, **16**, 520–548.
O'Hara, M. J.; Richardson, S. W.; and Wilson, G., 1971. Garnet-peridotite stability and occurrence in crust and mantle, *Contrib. Mineral. Petrol.*, **32**, 48–68.
Ringwood, A. E., 1972. Mineralogy of the deep mantle, in Robertson, E. C., ed. *The Nature of the Solid Earth*. New York: McGraw-Hill, 67–92.
Ringwood, A. E., 1975. *Composition and Petrology of the Earth's Mantle*. New York: McGraw-Hill, 618p.
Ringwood, A. E., and Lovering, J. F., 1970. Significance of pyroxene-ilmenite intergrowths among kimberlite xenoliths, *Earth Planet. Sci. Lett.*, **7**, 371–375.
Roedder, E., 1965. Liquid CO_2 inclusions in olivine-bearing nodules and phenocrysts from basalts, *Am. Mineralogist*, **50**, 1746–1782.
Smith, J. V., and Mason, B., 1970. Pyroxene-garnet transformation in Coorara meteorite, *Science*, **168**, 832–833.
Sobolev, N. V., et al., 1971. Peculiarities of the composition of the minerals coexisting with diamonds from the Mir pipe, Yakutia (In Russian), *Zap. Vses. Mineral. Obschchestva*, **100**, 558–564.
Tsai H., et al., 1979. Mineral inclusions in diamond, *Proc. Second Int. Kimberlite Conf.*, **1**, 16–26.
Yagi, T., et al., 1979. Phase relations in the system MgO-FeO-SiO_2 between 150 and 700 kbar at 1000°C, *Carn. Inst. Wash. Year Book*, **78**, 614–618.

META-AUTUNITE GROUP–

See **AUTUNITE GROUP**

METALLIC MINERALS

Of the three classes of economic minerals (metallic minerals, mineral fuels, and industrial minerals), metallic minerals are those that may be used as a source of one or more metals. There are over 300 metallic minerals, the majority being sulfides and sulfosalts (60%) followed by oxides (25%) and native elements (10%).

If we ignore the early use of native elements (q.v.) the first use of metallic minerals as ores was about 4000 BC when reduction of "oxidized zone" copper ores, probably mainly *malachite*, to metal was first discovered. Later, the joint reduction of copper ore with *cassiterite* was found to give bronze. In about 2500 BC, *galena* could be treated to give silver and lead; and production of very small quantities of iron was possible. By 1000 BC, gold, silver, tin, brass, copper, and iron could probably be produced in quantity from suitable ores.

Some of the more important or typical metallic minerals occurring in each of Lindgren's classes of ore deposits (modified) are given below:

- Magmatic segregation deposits: *chromite, magnetite-ilmenite, magnetite*
- Pegmatites: *beryl, columbite-tantalite*
- Igneous metamorphic deposits: *sphalerite, galena, chalcopyrite, bornite, magnetite*
- Hypothermal deposits: native *gold, wolframite, scheelite, sphalerite, galena, pentlandite,* native *platinum*
- Mesothermal deposits: *chalcopyrite, galena, sphalerite, tetrahedrite, bornite, enargite, molybdenite*
- Epithermal deposits: *stibnite, cinnabar, argentite, calaverite*
- Telethermal deposits: *uraninite, coffinite, roscoelite, sphalerite*
- Chemical sedimentary deposits: *hematite,* "limonite," *pyrolusite*
- Mechanical sedimentary deposits: native *gold,* native *platinum, leucoxene, monazite, cassiterite*
- Weathering: "bauxite," *garnierite, pyrolusite, psilomelane, cerussite, malachite, azurite, smithsonite*
- Supergene sulfide enrichment: *chalcocite, covellite, acanthite*

Since the majority of metallic minerals are opaque, they may be studied by preparing polished mounts or polished thin sections and examining these with an ore microscope. Ore microscopy (q.v.) uses a wide variety of techniques including systematic etching tests, microchemical tests, and examination of hardness and optical properties. Physical and optical tests are being increasingly favored especially when combined with the quantitative measurement of properties such as reflectivity and Vicker's hardness. Final confirmation of the identity and composition of a mineral occurring in small grains may be made using X-ray powder photography either alone or in combination with the electron-probe microanalyzer. The former may be used on samples as small as 0.1 mg scratched from a polished surface. The latter will give the composition of grains only 0.001 mm diameter.

MICHAEL J. FROST

References

Cameron, C., 1961. *Ore Microscopy*. New York: Wiley, 293p.

Park, C. F., and MacDiarmid, R. A., 1975. *Ore Deposits* 3rd ed. New York: McGraw-Hill, 522p.

Schouten, C., 1962. *Determination Tables for Ore Microscopy*. Amsterdam: Elsevier, 242p.

Uytenbogaardt, W., and Burke, E. A. J., 1971. *Tables for Microscopic Identification of Ore Minerals*. Amsterdam: Elsevier, 430p.

Cross-references: *Electron Probe Microanalysis; Mineral and Ore Deposits; Native Elements and Alloys; Optical Mineralogy; Ore Microscopy;* see also mineral glossary.

METALLURGY

Metallurgy as a field of engineering is concerned with the art and science of producing and adapting metals to satisfy human wants. The science of metallurgy is involved with the study of metals, their physical and chemical properties. It is usually divided into two related areas, chemical and physical metallurgy.

Chemical metallurgy concerns itself with the extraction of metals from their ores, the refining of metals, and the thermodynamics that control these processes. Metals are mainly obtained from ores, naturally occurring aggregates of metallic oxide or sulfide minerals, which are present in the earth's crust. The chemical metallurgist develops processes for the making of a concentrate containing the ore minerals (ore dressing) of interest and the reduction of these ores to the metallic state. Three types of reduction processes are available:

1. *Pyrometallurgy* is the chemical reduction of ores by smelting at high temperatures (1000–2000°C). This process is by far the most important commercially. The reactions are carried out in furnaces with refractory linings of alumina-siliceous or siliceous material. Many of the most important commercial metals and alloys, iron and steel, nickel, copper, and zinc, are produced by this process.
2. *Hydrometallurgy* is the leaching of ores and concentrates with aqueous solutions to dissolve and recover primarily the precious metals.
3. *Electrometallurgy* is the use of electric current for chemical reduction. This process is employed to refine such metals as copper, zinc, and lead from impure metals. Aluminum is obtained electrolytically from molten salts of the metals.

The most important pyrometallurgical processes are those used for producing pig iron and steel. Iron ore is first reduced with coke in a blast furnace. Limestone is added to react with the siliceous impurities in the ore and coke ash to form a fusible slag. Such a slag is mainly lime, alumina, and silica with smaller amounts of magnesia and iron oxide. After the pig iron is produced, it is refined to make steel in an open hearth or basic oxygen furnace. The

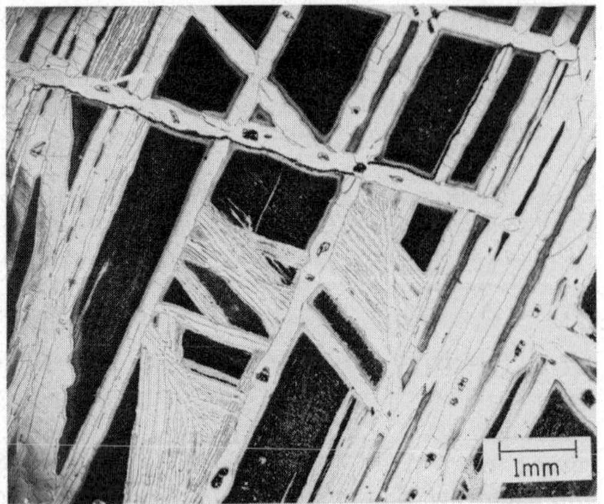

FIGURE 1. Polished surface of the Carlton iron meteorite showing Widmanstätten pattern (reprinted with permission from *Geochim. Cosmochim. Acta*, 27, J. I. Goldstein and R. E. Ogilvie, 1963, Electron microanalysis of metallic meteorites Part 1–Phosphides and sulfides. Copyright 1963, Pergamon Press, Ltd.).

basic oxygen process, which employs an oxygen lance positioned directly in the molten metal, is a recent development. This process is now used to produce most of the steel in the United States.

Metallurgists are just beginning to understand the thermodynamics that underlie these various chemical reduction processes. Particular attention is being paid to the thermodynamics and kinetics of slag-metal interactions. Computers are now used to model the processes and to calculate the thermodynamics and heat balances that control the processes.

Physical metallurgy concerns itself with the structure and properties of metals. This area of metallurgy probably had its beginning in 1808 when Alois de Widmanstätten polished the surface of an iron meteorite and discovered the pattern that bears his name. The Widmanstätten pattern of the Carlton meteorite is shown in Fig. 1. The pattern is made up of *kamacite* (α iron) in a parent matrix of *taenite* (γ iron). In and around 1890, mainly as a result of the work of Sorby in England, metallography, the observation of structural details of metals, began to be used to relate metallic structure to properties. Physical metallurgy has expanded greatly since the late 1800s and now consists of four main areas: (1) atomic theory and physical properties, (2) phase relationships, (3) mechanical properties, and (4) engineering alloys–materials.

Atomic Theory and Physical Properties

Several of the major topics of concern in physical metallurgy are bonding, atomic structure, and imperfections. Metallic bonding involves the attractive forces between the free outer-shell electrons and the positive-charged ions. The free-electron cloud is largely responsible for the high electrical and thermal conductivity of metals; and the free electrons absorb light energy, making them opaque to transmitted light. Also, because of the metallic bonding, the crystal structure of metals are mainly cubic.

Among the physical properties of interest to the metallurgist are the electrical, thermal, thermoelectric, magnetic, and optical properties. With respect to magnetic properties, iron, cobalt, nickel, and some specialty alloys exhibit ferromagnetism, an electrostatic interaction between adjacent atoms that aligns the resultant spins of the atoms. Only *magnetite* (Fe_3O_4) and *pyrrhotite* ($Fe_{(1-x)}S$) are common magnetic minerals. Most metals are paramagnetic; most nonmetals are diamagnetic, i.e., very weakly repelled by a magnetic field (see *Magnetic Minerals*).

Phase Relationships

The area of phase relationships in metallurgy concerns itself with phase diagrams, solidification, and solid state reactions. In most cases, the metallurgist is able to use binary (two-component) phase diagrams effectively, while the geologist is concerned with more complicated systems (see Vol. IVA: *Phase Equilibria*). Various phase relationships of interest to metallurgists are the binary solid-solution series, intermediate phases, and, in solidification, the eutectic and peritectic reactions.

In solidification processes, complete equilibrium can only be approached between solid and liquid under slow rates of cooling. Therefore, most alloys experience nonequilibrium solidification, where concentration gradients are developed in the solid phase. These gradients greatly influence the properties of the final alloy. The process of zone melting, however, has been used to purify metals (see *Crystal Growth*).

In solid-state transformations, the eutectoid reaction and precipitation from solid solution are very important. Solid-state reactions occur much more slowly than in solidification processes due mainly to low diffusion rates. They are subject to greater supercooling and rarely have the compositions given by the phase diagram. The crystal structure of the new phase has a definite orientation relationship to the crystal structure of the phase from which it formed, as in formation of the Widmanstätten pattern (Fig. 1).

The most serious problem in the commercial application of equilibrium diagrams is the fact that equilibrium is not often attained. An example of this can be found in the commercially important Fe base alloys (steels) which contain 0.1 to 1.5 wt % carbon. At a carbon content of 0.8%, a eutectoid reaction γ (austenite-*taenite*) $\rightarrow$ α (ferrite-*kamacite*) + Fe_3C (carbide-*cohenite*) occurs. If the cooling rate of the steel is slow, a platelike mixture of α and Fe_3C called pearlite is formed. If the cooling rate is increased, however, bainite or martensite or mixtures of these constituents may also be produced.

Mechanical Properties

The high strength and ductility of metals and alloys makes them ideally suited for structural applications. Mechanical properties of materials are often subdivided into the elastic and plastic ranges. In the elastic range, no permanent effect is produced by stresses; and in the plastic range, large deformations of metal parts occur. Since many finished products are to be used in the elastic range, a high yield strength and high Young's modulus is desired. Alloying and temperature can, however, affect these properties significantly.

Metal plasticity is particularly important in the forming operations that are carried out on all wrought metals. Rolling or forging of steel at high temperatures must be done at a rate determined by the deformation characteristics of the individual alloy. Cold forming operations, so important in the mass production of many metal parts, are even more restricted by plastic properties. Of particular concern to the metallurgist is the process of fracture, final failure. This physical phenomenon is often studied by various types of tests (tensile, impact, torsion, fatigue) to understand the mechanism and mode of failure.

Engineering Alloys and Materials

The engineering metallurgist focuses attention upon the relation between microstructures and properties, using the phase relationships (phase diagrams) as a guide in designing better alloys. However, much of the knowledge in this field still is on an empirical basis. Table 1 lists properties of many alloys in copper, aluminum, and steel alloys (Ruoff, 1973). The variety of mechanical properties obtained and the possible applications of these alloys is almost endless. In recent years, however, the metallurgist who must make selections of materials has become much more materials oriented and has extended his knowledge of materials selection to all types of materials; e.g., plastics, semiconductors, and ceramics.

Two areas of research in metallurgy have important bearing on the geochemical problems associated with the structure and composition of the earth and other planetary bodies. One of these is the effect of pressure on the phase relations of iron; the other is the effect of time, temperature, and pressure on the development of the structure of metal phases in meteorites and lunar samples. Pressure has a large and significant effect on the transition temperature for phase equilibria and phase transformations in metals (Kaufman, 1963). According to the Clapeyron-Clausius equation;

$$\left(\frac{dT}{dP}\right)_{T_m^\alpha}^{\alpha\rightarrow\mathrm{Liq}} = T_m{}^\alpha \left(\frac{\Delta V}{L}\right)_{T_m^\alpha}^{\alpha\rightarrow\mathrm{Liq}}$$

TABLE 1. Mechanical Properties of Various Metal Alloys

Material	Tensile Yield Strength, σ_0 psi	Tensile Yield Strength, σ_0 Kilobars (10^9 dynes/cm^2)
OFHC 99.95% annealed copper[a]	10,000	0.7
OFHC 99.95% cold-drawn copper	40,000	2.8
99.45% annealed aluminum	4,000	0.28
99.45% cold-drawn aluminum	24,000	1.7
2024 (3.8–4.9% Cu, 0.3–0.9% Mn, 1.2–1.8% Mg) heat treated aluminum alloy	50,000	3.4
1020 steel (0.2% C)	35,000–40,000	2.4–2.8
1095 steel (hardened) (0.95% C)	100,000–188,000	7–13
4340 annealed alloy steel[b]	65,000–70,000	4.5–4.8
4340 fully hardened alloy steel	130,000–228,000	9–16
Maraging (300) steel	290,000	20

Adapted from Ruoff, 1973.

[a]OFHC means oxygen-free high-conductivity.

[b]4340 steels are 1.65–2.0% Ni, 0.4–0.9% Cr, 0.2–0.3% Mo, 0.4% C, remainder Fe.

for the transition $\alpha \to$ Liq, solid to liquid, where T_m^{α} is the melting temperature and P is the pressure, L is the latent heat of phase change, and ΔV is the difference in volume of the phases.

$$\left(\frac{dT}{dP}\right)^{\alpha \to \gamma} = T_A^{\alpha} \left(\frac{\Delta V}{L}\right)^{\alpha \to \gamma}_{T_A^{\alpha}}$$

for the transition $\alpha \to \gamma$ where T_A^{α} is the transition temperature for the polymorphic reaction $\alpha \to \gamma$. The effect of increased pressure in pure iron, for example, is to raise the transition temperature for $\alpha(bcc) \to$ Liq and to decrease the transition temperature for the reaction $\alpha(bcc) \to \gamma(fcc)$.

The classical interpretation of the composition of the earth's core is that it is similar to the composition of iron meteorites, i.e., iron-nickel alloys with about 10 wt% nickel. Several significant effects have been reported from studies of Fe-Ni alloys at high pressures. At 25°C and 130 kilobars, the transition $\alpha \to \epsilon$ iron occurs, forming a new compact hexagonal phase (Mao et al., 1967). However, the addition of Ni to iron in the ϵ phase makes the material more compressible than pure iron at high pressures (McQueen and Marsh, 1966). Extrapolations to pressures and temperatures at the earth's core, assuming that the structure of the solid core is ϵ phase, indicate that the iron-nickel alloy would be about 10% heavier than pure iron.

However, density data determined by geophysical measurements indicates that the density of the earth's core is 10–15% less than the density of pure iron at comparable pressures, 1.4 to 3.4 megabars, and temperatures, greater than 4000°K. It has been suggested (Birch, 1952) that, if iron is the major component of the core, elements of lower density would be alloyed with iron. A number of light elements that would lower the density of iron or iron-nickel alloys in the core has been proposed, but the two most seriously considered at the present time are silicon and sulfur. A core of Fe-Ni-S has also been proposed for the moon (Brett, 1973). Since all the conclusions about the composition of the earth's core are derived from extrapolations from laboratory experiments, the nature of the earth's core is still very much open to question.

The Widmanstätten pattern in iron meteorites (Fig. 1) was developed as the meteorites cooled through the temperature range 700–300°C while the meteorites were yet a part of a planetary or asteroidal body. The α (*kamacite*) phase nucleated from the parent γ (*taenite*) phase, forming a Widmanstätten pattern, and grew by solid-state diffusion. Electron microprobe analysis of these phases have shown that they are not at equilibrium. However the Ni gradients measured by the microprobe in *kamacite* and *taenite* can be used to determine the rate at which equilibrium is approached.

It has been possible to simulate the growth process for the Widmanstätten pattern by a metallurgical analysis (Wood, 1964; Goldstein and Ogilvie, 1965), since the factors that control the growth process–the iron-nickel phase diagram and interdiffusion coefficients–are well known. These simulation studies have allowed determination of the cooling rates of the parent bodies of the iron meteorites. The cooling-rate values determined to date (Goldstein and Short, 1967) range between 0.5 and 500°C/m yr. They indicate that iron meteorites have formed in more than one parent body and have parent-body sizes in the asteroidal size range (20–350 km radius) and not in the planetary range. The internal pressures of these parent bodies are less than a few kilobars. Recent evidence has also suggested that not all of the iron meteorites form in the core of their respective parent bodies. It is apparent, however, that the effects of elements such as P, C, and S must be considered when discussing the development of the Widmanstätten structure in iron meteorites (Goldstein and Axon, 1973).

Until recently, the only source of extraterrestrial metal has been in the meteorites that have landed upon earth. It is now possible to study metal that has been returned from the moon. Two populations of lunar metal have been recognized (Frondel, 1975): (1) metal of meteoritic origin that has been subject to greater or lesser degrees of metamorphic change in the lunar crust or on the moon's surface, and (2) metal that is the result of igneous fractional crystallization of lunar rock, vapor condensation, or shock reduction of iron-bearing lunar minerals. The history and origin of lunar metal is presently a subject of intense research.

JOSEPH I. GOLDSTEIN

References

Birch, F., 1952. Elasticity and constitution of the earth's interior, *J. Geophys. Res.*, **57**, 227–286.

Brett, R., 1973. A lunar core of Fe-Ni-S, *Geochim. Cosmochim. Acta*, **37**, 165–170.

Frondel, J. W., 1975. *Lunar Mineralogy*. New York: Wiley, 323p.

Goldstein, J. I., and Axon, H. J., 1973. The Widmanstätten figure in iron meteorites, *Naturwiss.* **60**, 313–321.

Goldstein, J. I., and Ogilvie, R. E., 1963. Electron microanalysis of metallic meteorites Part 1–Phosphides and sulfides, *Geochim. Cosmochim. Acta*, **27**, 623–627.

Goldstein, J. I., and Ogilvie, R. E., 1965. The growth of Widmanstätten pattern in metallic meteorites, *Geochim. Cosmochim. Acta*, **29**, 893-920.

Goldstein, J. I., and Short, J. M., 1967. The iron meteorites, their thermal history and parent bodies, *Geochim. Cosmochim. Acta*, **31**, 1733-1770.

Kaufman, L., 1963, in Paul, W., and Warschauer, D. M., eds., *Solids Under Pressure*. New York: McGraw-Hill, 303-356.

Mao, H. K.; Bassett, W. A.; and Takahashi, T., 1967. Effect of pressure on crystal structure and lattice parameters of iron up to 300 kbar, *J. Appl. Phys.*, **38**, 272-276.

McQueen, R. G., and Marsh, S. P., 1966. Shock-wave compression of iron-nickel alloys and the earth's core, *J. Geophys. Res.*, **71**, 1751-1756.

Ruoff, A. L., 1973. *Materials Science*. Englewood Cliffs, N.J.: Prentice-Hall, 697p.

Wood, J. A., 1964. The cooling rates and parent planets of several iron meteorites, *Icarus*, **3**, 429.

METAMICT STATE

Broegger (1893) recognized that some minerals that show crystal form are, nevertheless, structurally amorphous but have attained this state in a different manner than glasses or rigid gels. He wrote: "For this third class of amorphous substances there is proposed the name *metamict* (from μεταμιγνυμ, mix otherwise, that is by a molecular rearrangement to another structure than the original crystalline) amorphous substance. The reason for the amorphous rearrangement of the molecules might perhaps be sought in the lesser stability which so complicated a crystal molecule as that of these minerals must have in the presence of outside influences."

Properties

Many of the features of metamict minerals had been recognized earlier; some were not recognized till a later date. The more significant features are:

1. Optically isotropic–Many metamict substances are heterogeneous, being partly isotropic and partly anisotropic.
2. Pyrognomic, i.e., readily becoming incandescent on heating–In this regard there is great variation, no glowing being observable in some cases. Even strongly pyrognomic minerals such as *gadolinite* can be annealed below the temperature of glowing with complete loss of the pyrognomic quality.
3. Lacking in cleavage, fracture conchoidal–Some are also particularly brittle.
4. Density increased by heating–The change may be slight.
5. Crystalline structure reconstituted by heating–Reconstitution may be a complex process. Even if a single phase results it is usually polycrystalline.
6. Resistance to attack by acids is increased by heating–This is just the opposite of the effect produced by heating in most substances.
7. Contain uranium (U) or thorium (Th)–The content may be low, e.g., 0.41% ThO_2 in *gadolinite* from Ytterby, Sweden. The presence of rare earths and related elements has been emphasized by some observers.
8. Some known in both the crystalline and metamict states–In these cases little chemical difference can be found. There is evidence of hydration attending isotropization but no direct or general correlation has been established.
9. X-ray amorphous–Vegard (1916) first reported the absence of X-ray diffraction by *thorite*. This has since been noted for many other metamict minerals and is now regarded as the crucial test. Some traces of X-ray diffraction may be observed though the minerals have become optically quite isotropic.

Causes

It is now generally agreed that α-particle bombardment is the cause of isotropization. In fact, some physicists have used the expression metamict in writing of radiation damage.

Many minerals are known in the metamict state, some only in this state; but many natural compounds of radioactive elements are never found in this state. Goldschmidt (1924) enumerated three conditions required for metamictization:

A. The original structure must be only weakly ionic and possibly susceptible to hydrolysis.
B. The structure must contain one or more kinds of ions that are readily susceptible to changes in the state of ionization.
C. In many cases, it may additionally be necessary that the crystal is subjected to relatively strong radiation, either from radioactive material within the crystal or from outside sources.

Mechanism

Metamictization is doubtless similar to the more rapid processes of radiation damage in technological materials, which have recently received much attention. Only a small part of the energy of the α particles is effective in producing structural damage. Directly or indirectly, through recoil atoms, α particles displace structural units, giving rise to vacancies and interstitial atoms. Some of the energy is converted to heat, which may locally bring about melting. Whether minerals become metamict may depend on the extent to which they recover from the damage by natural annealing.

Occurrence of Metamict Minerals

Scores of different mineral species are known in the metamict state. Many are characteristic of pegmatites but are also found in hydrothermal deposits. Some are known to have been concentrated in placers, possibly even in ancient placers. A few metamict minerals, notably *brannerite*, which is essentially UTi_2O_6, and *davidite*, are important constituents of uranium ores in Ontario and Australia. These and a number of other complex oxides such as *euxenite* and *fergusonite* are known only or mostly in a metamict condition. Others, such as *thorite, allanite, gadolinite*, and *pyrochlore* may or may not be metamict. *Zircon*, $ZrSiO_4$, one of the most widespread of minerals, is generally unaltered but has also been found metamict or in intermediate states in many places.

Degree of Metamictness

X-ray methods do not permit a close estimate of the degree of metamictness since the remnants of structure may be confined to small volumes that may be disoriented. Pellas (1954) estimates that a material will appear "X-ray amorphous" when 20% or more of its atoms have been displaced. In a few minerals, it has been possible to detect a lattice expansion in specimens that are partly metamict. This might be taken as a measure of the degree of metamictness in its early stages. The effect is pronounced in *zircons*, which vary between about 4.7 and 3.9 in density; and it has even been suggested that, in conjunction with measurements of radioactivity, cell dimensions of *zircon* might be indicators of age (Holland and Gottfried, 1955).

Reconstitution of Metamict Minerals

The pyrognomic effect observed in some metamict minerals is evidence of the exothermic character of the reconstitution, which, in these cases, proceeds rapidly. Faessler (1942) was able to measure in metamict *gadolinite* the stored energy that is released during reconstitution (see also Kurath, 1957).

Most metamict minerals can be recrystallized, usually reconstituted, at easily detectable rates at temperatures between 500 and 900°C. Some can be slowly reconstituted at lower temperatures. Though characteristic differential thermal analysis curves can be obtained from some metamict minerals, the peaks are controlled by heating conditions. Even if the original structure is reestablished, usually as a crystalline aggregate, the reconstitution may involve intermediate steps. For instance, when metamict *thorite* is heated, cryptocrystalline ThO is formed before or with the reconstituted $ThSiO_4$.

Due to the recognition of such side reactions and the fact that the results of annealing may be modified by the surrounding atmosphere, whether air or inert gas, some have doubted that the original structure or its cryptocrystalline equivalent is reconstituted. There should be no doubt in those cases in which it has been possible to reconstitute single crystals with lattices conforming strictly to the morphology. When material that is fully X-ray amorphous can be annealed to single crystals it must be presumed that sufficient remnants of the original structure persist to determine the orientation of the newly crystallized material. For a comprehensive discussion of metamict minerals, see Mitchell (1973).

A. PABST

References

Broegger, W. C., 1893. *Amorf. Salmonsens Store Illustrerede Konversationslexicon*, **1**, Copenhagen, 742-743.

Faessler, A., 1942. Untersuchungen zum Problem des metamikten Zustandes, *Zeitsch. Krist.*, **104**, 81-113.

Goldschmidt, V. M., 1924. Über die Umwandlung krystallisierter Minerale in den metamikten Zustand (Isotropisierung), *Norsk. Ak. Skr.*, ser. I, **1**(5), 51-58.

Holland, H. D., and Gottfried, D., 1955. The effect of nuclear radiation on the structure of zircon, *Acta Crystallogr.*, **8**, 291-300.

Kurath, Sheldon F., 1957. Storage of energy in metamict minerals, *Am. Mineralogist*, **42**, 91-99.

Mitchell, Richard S., 1973. Metamict minerals: A review. Pt. I, Chemical and physical characteristics, occurrence. Pt. II, Origin of metamictization, methods of analysis, miscellaneous topics, *Mineral. Rec.*, **4**, 177-182, 214-223.

Pellas, P., 1954. Sur la formation de l'état métamicte dans le zircon, *Bull. Soc. Fr. Mineral. Cristallogr.*, **77**, 447-460.

Vegard, L., 1916. Results of crystal analysis, *Philos. Mag.*, **32**, 65-96.

Cross-references: *Anisotropism; Crystal Growth; Isotropism; Optical Mineralogy; Order-Disorder; Pegmatite Minerals; Placer Deposits.*

METAMORPHIC MINERALS—*See* ROCK-FORMING MINERALS

METASILICATES—*See* mineral glossary. Vol. IVA: CRYSTAL CHEMISTRY

METEORITIC MINERALS

About 100 different minerals (see Table 1) have been found in meteorites. Of these, over

TABLE 1. Meteoritic Minerals (*indicates minerals not found in terrestrial rocks)

Elements and Alloys

Mineral	Formula	Mineral	Formula
awaruite	$FeNi_3$	kamacite	(Fe,Ni)
*barringerite	$(Fe,Ni)_2P$	*lonsdaleite	C
*carlsbergite	CrN	moissanite[b]	SiC
carbynes[a]	C	*osbornite	TiN
chaoite	C	*perryite	$(Ni,Fe)_2Si$
cohenite	$(Fe,Ni)_3C$	*schreibersite	$(Fe,Ni)_3P$
copper	Cu	sulfur	S
diamond	C	*taenite	(Fe,Ni)
graphite	C	*tetrataenite[c]	FeNi
*haxonite	$Fe_{23}C_6$		

Sulfides

Mineral	Formula	Mineral	Formula
alabandite	(Mn,Fe)S	mackinawite	FeS_{1-x}
*brezinaite	Cr_3S_4	*niningerite	(Mg,Fe)S
chalcopyrite	$CuFeS_2$	*oldhamite	CaS
cubanite	$CuFe_2S_3$	pentlandite	$(Fe,Ni)_9S_8$
*daubreelite	$FeCr_2S_4$	pyrite	FeS_2
djerfisherite	$K_3CuFe_{12}S_{14}$	pyrrhotite	$Fe_{1-x}S$
*gentnerite	$Cu_8Fe_3Cr_{11}S_{18}$	sphalerite	(Zn,Fe)S
heazlewoodite	Ni_3S_2	troilite	FeS
*heideite	$(Fe,Cr)1\text{-}x(Ti,Fe)_2S_4$	valleriite	$CuFeS_2$

Oxides

Mineral	Formula	Mineral	Formula
baddeleyite	ZrO_2	magnetite[e]	Fe_3O_4
chromite[d]	$FeCr_2O_4$	perovskite	$CaTiO_3$
corundum	Al_2O_3	quartz	SiO_2
cristobalite	SiO_2	rutile	TiO_2
hercynite	$(Fe,Mg)Al_2O_4$	*sinoite	Si_2N_2O
hibonite	$CaAl_{12}O_{19}$	spinel	$MgAl_2O_4$
illmenite	$FeTiO_3$	tridymite[f]	SiO_2

Oxalates / **Halides**

Mineral	Formula	Mineral	Formula
whewellite	$CaC_2O_4 \cdot H_2O$	*lawrencite	$(Fe,Ni)Cl_2$

Carbonates / **Sulfates**

Mineral	Formula	Mineral	Formula
calcite	$CaCO_3$	bloedite	$Na_2Mg(SO_4)_2 \cdot 4H_2O$
dolomite	$CaMg(CO_3)_2$	epsomite	$MgSO_4 \cdot 7H_2O$
magnesite	$(Mg,Fe)CO_3$	gypsum	$CaSO_4 \cdot 2H_2O$

Phosphates

Mineral	Formula	Mineral	Formula
*brianite	$Na_2CaMg(PO_4)_2$	*panethite	$(Ca,Na)_2(Mg,Fe)_2(PO_4)_2$
*buchwaldite	$NaCaPO_4$	sarcopside	$(Fe,Mn)_3(PO_4)_2$
chlorapatite	$Ca_5(PO_4)_3Cl$	*stanfieldite	$Ca_4(Mg,Fe)_5(PO_4)_6$
*farringtonite	$Mg_3(PO_4)_2$	whitlockite[g]	$Ca_3(PO_4)_2$
graftonite	$(Fe,Mn)_3(PO_4)_2$		

Silicates

Mineral	Formula
andradite	$Ca_3Fe_2Si_3O_{12}$
augite	$Ca(Mg,Fe,Al)(SiO_3)_2$
cordierite	$(Mg,Fe)_2Al_3Si_5AlO_{18}$
cronstedtite	$(Fe_2^{2+}Fe^{3+})(SiFe^{3+}O_5)(OH)_4$
diopside[h]	$Ca(Mg,Fe)(SiO_3)_2$
enstatite-hypersthene[i]	$(Mg,Fe)SiO_3$
fassaite	$(Ca(Mg,Fe,Al)(Si,Al)_2O_6$
ferric chamosite	$Fe_6Mg_3[(Si_4O_{10})(OH)_8]_2$
gehlenite	$Ca_2Al_2SiO_7$
grossular	$Ca_3Al_2Si_3O_{12}$
kaersutite	$Ca_2(Na,K)(Mg,Fe)_4TiSi_6Al_2O_{22}F_2$
*krinovite	$NaMg_2CrSi_3O_{10}$

Continued

TABLE 1. (Continued)

**majorite*	$Mg_3(MgSi)Si_3O_{12}$
melilite	$Ca_2(Mg,Al)(Si,Al)_2O_7$
**merrihueite*	$(K,Na)_2Fe_5Si_{12}O_{30}$
montmorillonite	$(Na,Ca)(Al,Mg)_6Si_{12}O_{30}(OH)_6 \cdot nH_2O$
nepheline	$NaAlSiO_4$
olivine	$(Mg,Fe)_2SiO_4$
pigeonite	$(Ca,Mg,Fe)SiO_3$
plagioclase[j]	$NaAlSi_3O_8$–$CaAl_2Si_2O_8$
potassium ***feldspar***	$(K,Na)AlSi_3O_8$
Rhönite	$CaMg_2TiAl_2SiO_{10}$
richterite	$Na_2CaMg_5Si_8O_{22}F_2$
**ringwoodite*	$(Mg,Fe)_2SiO_4$
**roedderite*[k]	$(K,Na)_2Mg_5Si_{12}O_{30}$
sanidine[l]	$KAlSi_3O_8$
serpentine	$(Mg,Fe)_3Si_2O_5(OH)_4$
sodalite	$Na_4ClSi_3Al_3O_{12}$
**ureyite*	$NaCr(SiO_3)_2$
wollastonite	$CaSiO_3$
**yagiite*	$(Na,K)_3Mg_4(Mg,Fe,Ti,Al)_8(Si,Al)_{24}O_{60}$
zircon	$ZrSiO_4$

[a]Group of carbon polymorphs containing triple bonds.
[b]Requires confirmation. May be a contaminant from carbide saws used to cut meteorites.
[c]Ordered 1:1 alloy.
[d]Includes magnesian variety.
[e]Possibly in part *trevorite,* $NiFe_2O_4$.
[f]Includes so-called "stuffed" varieties with minor amounts of Na and Al.
[g]The mineral merrillite has proved to be *whitlockite* in all cases.
[h]Includes chrome *diopside.*
[i]Orthorhombic and monoclinic forms both occur. In carbonaceous condrites some *enstatite* may be highly aluminous.
[j]Na-rich ***plagioclase*** occurs in both high and low structural states.
[k]Includes aluminous variety.
[l]Found in the structural state of high-*sanidine.*

60 have been described since 1960. This includes 15 that were described in the two years 1966 and 1967 alone. The remainder were discovered over a period of time ranging back to the mid-nineteenth century with, of course, a few of them having been known from antiquity.

This recent burst of activity is due to the development and availability of X-ray diffraction (q.v., Vol. IVA) methods and of the electron probe microanalyzer (see *Electron Probe Microanalysis*), which make it possible to obtain quite accurate structural and chemical data from mineral grains only a few microns in size. Consequently, all of the discoveries since the early part of this century have been of accessory minerals that occur usually in minor or trace quantities. Although some of these have been found in only a single meteorite, such limited reportings are usually only temporary. As minor phases in more different meteorites are examined, additional occurrences of new minerals are extended and some of them will eventually be reported as common accessories in certain types of meteorite.

Major Minerals

Three meteorite types—(1) ordinary chondrites, (2) iron, and (3) stony-iron meteorites—together comprise more than 95% of the approximately 3000 known meteorites (see Vol. II, *Meteorites*). Thus, the most abundant and common meteorite minerals are those found in these three types: they are ***olivines, pyroxenes, plagioclase***, metal, *troilite, schreibersite*, and *graphite*.

Olivine. The principal component of all the ordinary chondrites except enstatite chondrites (Fig. 1) is ***olivine*** (see *Olivine Group*); it is the only silicate in the stony-iron type of meteorite called pallasites (Fig. 2). ***Olivine*** is found as a major or minor component in some classes of achondrites and carbonaceous chondrites, and is often present in trace amounts in octahedrite iron meteorites.

Olivine compositions range from pure *forsterite* to *Fo* 31; however, the highest frequencies of compositions lie in the range from *Fo* 86 to *Fo* 68. This is the range covered by the ordinary chondrites, the largest single group

FIGURE 1. Tieschitz chondrite. Typical barred chondrule consisting of lamellar ***olivine*** crystals parallel to each other. This chondrule is about 1 mm across in the meteorite. (Photo courtesy of Field Museum of Natural History, Chicago.)

of the meteorite types. The most magnesium-rich and the most iron-rich ***olivine*** compositions are present in only a few achondrites.

Pyroxenes. ***Pyroxene*** (see *Pyroxene Group*) is a principal component in the chondrites (Fig. 3), the story-iron meteorites called mesosiderites, and most types of achondrites. It is found as a trace phase in some octahedrite irons.

Pyroxenes in meteorites can be grouped into seven types:

1. low calcium (orthorhombic)	$(Mg,Fe)SiO_3$
2. low calcium (monoclinic)	$(Mg,Fe)SiO_3$
3. *pigeonite*	$(Ca,Mg,Fe)SiO_3$
4. *diopside*	$Ca(Mg,Fe)(SiO_3)_2$
5. *augite*	$Ca(Mg,Fe,Al)(SiO_3)_2$
6. *ureyite*	$NaCr(SiO_3)_2$
7. *fassaite*	$Ca(Mg,Fe,Al)(Si,Al)_2O_6$

Types 1 and 2 are the main ***pyroxenes*** of the chondrites, among which compositions range from *En* 100 through *En* 75. In these meteorites, CaO is generally low in the ***pyroxenes*** (2 mol% or less). Although some chondrites contain only the orthorhombic form, most contain orthorhombic along with a minor amount of the monoclinic form. A few contain only the monoclinic type.

Among the mesosiderites and some achondrites, Type 1 ***pyroxenes*** are generally more iron rich, ranging from *En* 79 to *En* 50, with approximately 5 mol% CaO. Some mesosiderites and achondrites contain, in addition, *pigeonite* (Type 3) with approximately 10 mol% CaO, and magnesium contents ranging from *En* 50 to *En* 28.

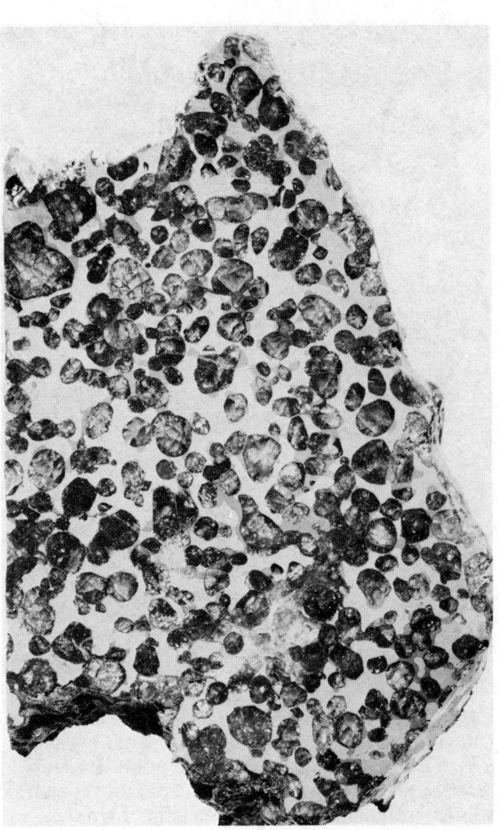

FIGURE 2. Springwater stony-iron meteorite of the type called pallasite. This photo is approximately one quarter natural size. White areas consist of *kamacite* and *taenite* enclosing nodules of ***olivine*** (gray to black in photo). Minor amounts of *farringtonite* occur along the rims of the ***olivine*** nodules, but these cannot be distinguished in a photograph. Some small (gray) specks of *troilite* occur in the metal. (Photo courtesy of Field Museum of Natural History, Chicago.)

Diopside is found as a major component only in two achondrites, and is otherwise quite rare. In some irons, it occurs in trace amounts and is usually a deep green chrome *diopside* with up to 1 wt% Cr_2O_3 content, and is quite magnesium rich (*Di* 95 and higher). *Augite* is found in only one single achondrite; while *ureyite* (a chromium analog of *acmite*) has been found as a trace mineral in two iron meteorites only. *Fassaite* is found in minor amounts in Type III carbonaceous chondrites.

Plagioclase. ***Plagioclase*** (see *Plagioclase Feldspars*) is present in small quantities in most

FIGURE 3. Dharmsala stone meteorite. Typical chondrule of radiating laths and blades of *hypersthene* ***pyroxene.*** Between the blades are minor amounts of ***plagioclase***. The matrix surrounding the chondrule consists of granular ***olivine*** with some ***pyroxene*** and ***plagioclase***. The chondrule is about 1.2 mm in the meteorite. (Photo courtesy of Field Museum of Natural History, Chicago.)

stony meteorites, and is found as a minor component of silicate inclusions in some iron and mesosiderite meteorites. Two fairly restricted compositional ranges occur: In the chondrites, enstatite achondrites, and irons composition ranges from *albite* to *oligoclase*; and in the remaining types of achondrites and mesosiderites, it ranges from *bytownite* to *anorthite*. Detailed studies of meteoritic ***plagioclases*** have been limited; however, it has been determined that the ***plagioclase*** in a number of chondrites is the high-temperature disordered form. Furthermore, many chondrites show small volumes of turbid glassy material that is partially devitrified ***plagioclase***.

Maskelynite is a ***plagioclase*** glass and is found in minor quantities in some chondrites and irons. Its presence is considered to be due to preterrestrial impact shock between meteorite bodies in space. Low temperatures and completely anhydrous conditions together retard any devitrification.

Iron-Nickel Alloys. Iron-nickel metal alloys occur in all meteorites with the exception of the Type I carbonaceous chondrites and a few achondrites. In the iron and stony-iron meteorites, metal is, of course, the major constituent (Figs. 2 and 4); in the stone meteorites, it may occur only in trace quantities (<1%) or as a major component (25%).

The metal occurs usually as two coexisting phases: *kamacite* (alpha iron) with a body-centered cubic structure, and *taenite* (gamma iron) with a face-centered cubic structure. The composition of the former is around 94% iron, 5% nickel, with <1% total of cobalt, carbon, phosphorus, etc. *Taenite* ranges from 73% iron, 27% nickel, to 35% iron, 65% nickel, depending upon the type of meteorite.

In the chondrites and the octahedrite iron meteorites (Fig. 4), both *kamacite* and *taenite* occur together. In iron meteorites with less than 5 wt% total nickel content (hexahedrites and nickel-poor ataxites) the metal is entirely *kamacite*. In the nickel-rich ataxites (with greater than 25 wt% total nickel) the metal is *taenite*. The reasons for these coexisting relations are well understood in terms of the Fe-Ni phase diagram, and are discussed in many of the references given at the end of this article.

An intimate, extremely fine-grained mixture of *kamacite* and *taenite* in meteorites is called "plessite" (Fig. 4).

Troilite. *Troilite* is found in minor quantities in all but a few meteorites (Fig. 4). The iron to sulfur ratio is exactly 1:1 in *troilite*, as opposed to the closely related terrestrial mineral, *pyrrhotite*, which is always slightly iron deficient due to its higher oxidation state. Meteoritic *troilite* contains a very low amount of impurities with generally less than 0.5 wt% total of elements like nickel and cobalt.

Schreibersite. *Schreibersite* is a 3:1 alloy of iron and nickel with phosphorus, and is commonly termed a "phosphide." It occurs in all iron meteorites (Fig. 4), many stony-irons, and many chondrites. It is always present in accessory amounts only. In irons it occurs in two forms: a massive form as distinct nodules, or rimming *troilite* modules; or as flat plates within the metal. In the latter case, it is referred to as "rhabdite." The nickel content of *schreibersite* ranges from about 12 mol% to 38 mol% relative to iron.

Graphite. *Graphite* is a common accessory in many iron meteorites, where it occurs as plates or massive nodules often associated intimately with *troilite*. Among the stone meteorites, it occurs only in a few rare instances, in one type of achondrite and one type of chondrite (enstatite chondrites).

In some iron meteorites, within massive graphite nodules, small cubic "crystals" of

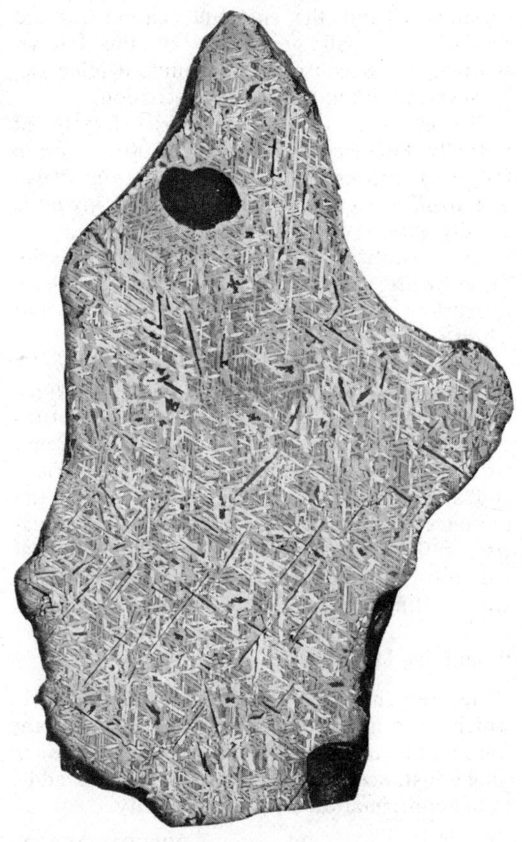

FIGURE 4. Breece octahedrite iron meteorite (1/6 natural size in this photo). Typical iron meteorite showing bright laths of *kamacite* (white and gray in photo). Between these laths are thin laths and areas of *plessite*, a fine grained mixture of *kamacite* and *taenite*. The large black rounded areas are *troilite*, and the long black streaks are laths of *schreibersite* (= ↑rhabdite). (Photo courtesy of Field Museum of Natural History, Chicago.)

graphite are found; these are considered by some workers to be possible *graphite* pseudomorphs after *diamond*.

In the carbonaceous chondrites *graphite*, as such, apparently does not occur; rather one or more members of the group of carbon polymorphs called "carbynes" are present.

Diamonds have been found in a few iron meteorites and in one type of achondrite. They appear to have been formed by mechanical impact shock, either preterrestrial or on impact with the earth. They apparently did not crystallize under equilibrium conditions at high hydrostatic pressure.

Other Accessory Minerals. *Cohenite* is found in small amounts in a relatively small number of iron meteorites. *Lawrencite* has been only rarely seen in some iron meteorites. However, it is completely unstable under normal atmospheric conditions; upon exposure to air it rapidly hydrates and oxidizes to ferric oxide and hydrochloric acid, which in turn attacks the metal of the meteorite. Thus, *lawrencite* may be more common than thought. The rapid rusting of some iron, stone, and stony-iron meteorites is often attributed to the initially unseen presence of minute grains of accessory *lawrencite*.

Common Trace Minerals

Spinel Group Minerals. *Spinel*, as such, has been found in trace amounts in only four chondrites. It has been found in several Type II and III carbonaceous chondrites, and may be present in all meteorites of these types. *Magnetite* (possibly in part *trevorite*) is an accessory in Type I, carbonaceous chondrites and a minor mineral in Type II, carbonaceous chondrites. *Chromite* is a very common trace mineral in all ordinary and some carbonaceous chondrites and is found in some iron meteorites that have silicate inclusions. In contrast to terrestrial *chromite*, meteoritic *chromites* have compositions very close to the ideal $FeCr_2O_4$. In one case, however, a fairly magnesian *chromite* was found within a *graphite* nodule in an iron meteorite.

Phosphate Minerals. Phosphates have been found as common accessories in stone, stony-iron, and iron meteorites. In all cases, only orthophosphates have been found. No hydrated or hydroxylated phosphates have been observed except for the presence of *vivianite*, $Fe_3(PO_4)_2 \cdot 8H_2O$, as film-like coatings on *sarcopside* and *graftonite* in several iron meteorites; these cases are products of terrestrial weathering of the original meteoritic phosphates.

Of the phosphates listed in Table 1 only *whitlockite* occurs with any significant frequency.

Significant Minerals in Special Meteorite Types

Carbonaceous Chondrites. The carbonaceous chondrites do not constitute a major class of meteorites. Thus, the unusual meteoritic minerals that are constituents of this group are not abundant when all meteorites are considered together. Nevertheless, because of their extraordinary mineralogy, the carbonaceous chondrites are very significant among meteorites in general, and have been considered the key to several theories of meteorite and planetary origin and development. The carbon found in carbonaceous chondrites has already been discussed.

The Type I and Type II carbonaceous chondrites contain major amounts of layered silicate phases that have often been referred to collectively as ***serpentine*** in the literature. Because of the very fine grain size (less than 100 nm), it has not been possible to separate the different phases for X-ray diffraction studies, however, because of advances in high resolution electron microscopy and electron diffraction, some identifications have been tentatively made of several ***clay*** and ***serpentine*** minerals. These are *chamosite* (probably nonaluminous ferric-*chamosite*), *cronstedtite, montmorillonite*, and one or more polytypes of ***serpentine***-group minerals.

As a group, carbonaceous chondrites contain from <1 wt% to 20 wt% H_2O in their bulk analyses, whereas all other meteorites are anhydrous except for possible trace amounts of water. The most water-rich carbonaceous chondrites contain accessory or trace amounts of hydrated phases such as *gypsum, epsomite*, and *bloedite*. The layered silicates, discussed above, are, of course, also hydrated. All these hydrated minerals are found in no other meteorite type. In addition, the Types I and II carbonaceous chondrites contain the only carbonate minerals known in meteorites: *dolomite, magnesite*, and *calcite*. These occur in trace amounts only.

The organic compounds found in carbonaceous chondrites are not minerals in the strict sense although it has become customary to treat them along with minerals in generalized discussions. As in the case of naturally occurring terrestrial organic compounds of low to medium mass, the problem of identification of molecular species and their abundances is difficult. Extraction and distillation procedures can easily destroy original molecules and create new ones. Furthermore, the porous nature of most carbonaceous chondrites permits easy contamination with the numerous organic molecular species found in terrestrial soil, dust, air, and(or) the hands of persons handling them.

In the analyses of organic compounds in carbonaceous chondrites, a variety of alkanes has been detected (in the C_{15} to C_{30} range) as well as aromatic hydrocarbons. Nitrogenated compounds have been found, including over 30 racemic amino acids. Thus far, no optical activity has been determined unequivocally in any organic material from any meteorite. A nonbiogenic origin is indicated for all the organic compounds observed.

Enstatite Chondrites and Enstatite Achondrites. The enstatite chondrites comprise only a small number of the large group of ordinary chondrites, and the enstatite achondrites are similarly a small percentage of the known achondrites. Because of their unusual minerals, however, they require special discussion.

The enstatite chondrites consist largely of virtually FeO-free *enstatite* (*En* 100), *diopside* (*Di* 100), and *kamacite.* ***Plagioclase*** (*oligoclase*) and *troilite* occur in minor amounts. ***Olivine*** is usually absent, or present in traces only.

The enstatite achondrites consist primarily of *enstatite* (*En* 100) with minor *oligoclase, forsterite*, or *diopside*, and a little metal and *troilite*.

Unusual minerals such as *oldhamite, daubreelite, albandite, osbornite, perryite*, and *sinoite* are found in trace amounts in these meteorites only. The presence of sulfides of calcium and chromium, as well as the only known nitrides and silicides in the mineral kingdom and the presence of up to 1.6% Si dissolved in *kamacite*, reflect the extreme degree of chemical reduction under which members of these groups of meteorites formed.

Remaining Minerals

The remaining minerals, listed in Table 1, which have not been specifically discussed, are found only in rare instances, generally in only single instances. A few of them require additional confirmation as to their validity.

EDWARD OLSEN

References

Cohen, E., 1894. *Meteoritenkunde*. Stuttgart: Schweizerbart'sche, 340p.

Farrington, O. C., 1915. *Meteorites*. Chicago: published by the author, 233p.

Heide, F., 1963. *Meteorites*. Chicago: Phoenix Science Series, University of Chicago Press, 144p.

Keil, K., 1969. Meteorite composition. Wedepohl, K. H., ed. *Handbook of Geochemistry*, **1**, 78–115.

Krinov, E. L., 1960. *Principles of Meteoritics*. New York: Pergamon, 535p.

Mason, B., 1962. *Meteorites*. New York: Wiley, 274p.

Mason, B., 1967. Extraterrestrial mineralogy, *Am. Mineralogist*, **52**, 307–325.

McCall, G. J., 1973. *Meteorites and Their Origins*. New York: Wiley, 352p.

Ramdohr, P., 1963. The opaque minerals in stony meteorites, *J. Geophys. Res.*, **68**, 2011–2036.

Wasson, J. T., 1974. *Meteorites*. New York: Springer-Verlag, 316p.

Wood, J., 1968. *Meteorites and the Origin of Planets*. New York: McGraw-Hill, 117p.

Cross-references: *Electron Probe Microanalysis; Plagioclase Feldspars; Pyroxene Group;* see also mineral glossary. Vol. II: *Carbonaceous Meteorites; Meteorites.* Vol. IVA: *X-ray Diffraction Analysis.*

MICA GROUP

The *micas* are the classic examples of silicates with layer or sheet structure (*Phyllosilicates*, q.v.). They are hydroxyl aluminosilicates of potassium with varied arrangements of lithium, magnesium, and iron. All are pseudohexagonal monoclinic, which permits stacking in "books." The general composition of the group is $(Si,Al)_4O_{10}(OH)_2$ with alkalis and magnesium. Disorder in the stacking permits a rather wide range of composition:

- *anandite*–$(Ba,K)(Fe,Mg)_3(Si,Al,Fe)_4O_{10}(O,OH)_2$
- *annite*–$KFe_3^{2+}(AlSi_3)O_{10}(OH,F)_2$
- *biotite*–$K(Mg,Fe)_3(Al,Fe)Si_3O_{10}(OH,F)_2$
- *bityite*–$CaLiAl_2(AlBeSi_2)O_{10}(OH)_2$
- *brammallite*–sodium *illite*
- *celadonite*–$K(Mg,Fe)(Fe^{3+},Al)Si_4O_{10}(OH)_2$
- *chernykhite*–$(Ba,Na)(V,Al)_2(Si,Al)_4O_{10}(OH)_2$
- *clintonite*–$Ca(Mg,Al)_3(Al_3Si)O_{10}(OH)_2$
- *ephesite*–$NaLiAl_2(Al_2Si_2)O_{10}(OH)_2$
- *glauconite*–$(K,Na)(Al,Fe^{3+},Mg)_2(Al,Si)_4O_{10}(OH)_2$
- *hendricksite*–$K(Zn,Mn)_3(Si_3Al)O_{10}(OH)_2$
- *hydromica*–$(K,Na,H_3O)Al_2(AlSi_3)O_{10}[(OH)_2,H_2O]$
- *illite*–$(K,H_3O)(Al,Mg,Fe)_2(Al,Si)_4O_{10}[(OH)_2,H_2O]$
- *kinoshitalite*–$(Ba,K)(Mg,Mn,Al)_3(Al_2Si_2)O_{10}(OH)_2$
- *lepidolite*–$K(Li,Al)_3(Si,Al)_4O_{10}(F,OH)_2$
- *margarite*–$CaAl_2(Al_2Si_2)O_{10}(OH)_2$
- *masutomilite*–$K(Li,Al,Mn^{2+})_3(Si,Al)_4O_{10}(F,OH)_2$
- *montdorite*–$(K,Na)_2(Fe^{2+},Mn,Mg)_5Si_8O_{20}(F,OH)_4$
- *muscovite*–$KAl_2(AlSi_3)O_{10}(OH)_2$
- *paragonite*–$NaAl_2(AlSi_3)O_{10}(OH)_2$
- *phlogopite*–$KMg_3(AlSi_3)O_{10}(F,OH)_2$
- *polylithionite*–$KLi_2AlSi_4O_{10}(F,OH)_2$
- *roscoelite*–$K(V,Al,Mg)_3(AlSi_3)O_{10}(OH)_2$
- *siderophyllite*–$K(Fe_2^{3+}Al)(Al_2Si_2)O_{10}(F,OH)_2$
- *taeniolite*–$KLiMg_2Si_4O_{10}F_2$
- *tarasovite*–$(Ca,Na)_{0.42}KNa(H_3O)Al_8(Si,Al)_{16}O_{40}(OH)_8 \cdot 2H_2O$
- *wonesite*–$(Na,K)(Mg,Fe,Al)_6(Si,Al)_8O_{20}(OH,F)_4$
- *zinnwaldite*–$KLiFe^{2+}Al(AlSi_3)O_{10}(F,OH)_2$

Usually the most striking attribute of the *micas* is their tendency to flake or cleave in thin sheets due to a perfect cleavage in the (001) plane. They also have very strong birefringence, and are optically negative. Depending upon the orientation of the optic axial plane, they can be subdivided into two classes; the plane being: (a) normal to (010), *muscovite* and *lepedolite*; (b) parallel to (010), *phlogopite* and *biotite*.

The principal minerals are *muscovite* (sericite is a variety with minute shred-like crystals; tuchsite or mariposite is a chromian variety), *illite* (hydromuscovite or "clay mica" is usually classified rather with the clays), *lepidolite* (*zinnwaldite* is a related lithium-iron mica), *phlogopite*, and *biotite*. *Glauconite* is usually a marine diagenetic replacement product (syndiagenetic; rarely anadiagenetic), and *celadonite* is a hydrothermal species.

RHODES W. FAIRBRIDGE

References

Berry, L. G., and Mason, B., 1959. *Mineralogy*. San Francisco: Freeman, 612p.

Deer, W. A.; Howie, R. A.; and Zussman, J., 1966. *An Introduction to the Rock-Forming Minerals*. London: Longmans, 528p.

Donnay, G., et al., 1964. Prediction of mica structures from composition and cell dimensions, *Carnegie Inst. Yearbook*, **63**, 228–232.

Heinrich, E. W., and Levinson, A. A., 1955. Studies in the mica group; polymorphism among the high-silica sericites, *Am. Mineralogist*, **40**, 983–995.

Kerr, P. F., 1959. *Optical Mineralogy*. New York: McGraw-Hill, 442p.

Levinson, A. A., 1955. Studies in the mica group; polymorphism among illites and hydrous micas, *Am. Mineralogist*, **40**, 41–49.

Smith, J. V., and Yoder, H. S., 1956. Experimental and theoretical studies of the mica polymorphs, *Mineral. Mag.*, **31**, 209–235.

Wyckoff, R. W. G., 1968. *Crystal Structures*, 2nd ed. New York: Interscience, 4.

Yoder, H. S., 1955. Synthetic and natural muscovites, *Geochim. Cosmochim. Acta*, **8**, 225–280.

Cross-references: *Chlorite Group; Clays, Clay Minerals; Crystallography: Morphological; Diagenetic Minerals; Mineral Industries; Optical Mineralogy; Pegmatite Minerals; Phyllosilicates; Pigments and Fillers; Plastic Flow in Minerals; Synthetic Minerals;* see also mineral glossary.

MICROPROBE—*See* ELECTRON PROBE MICROANALYSIS

MICROSCOPE—*See* ELECTRON MICROSCOPY (SCANNING); ELECTRON MICROSCOPY (TRANSMISSION); ORE MICROSCOPY (MINERAGRAPHY); POLARIZATION AND POLARIZING MICROSCOPE

MILLER INDICES—*See* DIRECTIONS AND PLANES

MINERAGRAPHY—*See* ORE MICROSCOPY (MINERAGRAPHY)

MINERAL—*See* MINERAL CLASSIFICATION: PRINCIPLES

MINERAL CLASSIFICATION: HISTORY

In order to understand the present method of grouping or classifying minerals, it is helpful to review briefly the history and development of methods of classification of minerals. Among the Greeks were several writers who showed a somewhat philosophical interest in minerals as well as other branches of early science. Aristotle's theory of the four primitive elements—earth, fire, air, and water—dominated scientific thinking for many centuries and this provided an elementary classification of substances.

Theophrastus, a pupil of Aristotle, although still believing that stones gave birth to young, presented many new concepts in his work *On Stones* written at the end of the fourth century, B.C. He classified minerals as metals, stones, or earths; and his work stood as the outstanding mineralogical contribution for about 2000 years. Pliny in about 77 A.D. in his *Natural History*, books 33 to 37, described minerals; but his writings were only an ill-digested compilation of the scientific knowledge of his time. From the time of Pliny, there was very little progress in mineralogy until Georgius Agricola published *De Natura Fossilium* in 1546. This might be considered the first textbook of mineralogy in the modern sense. Although it retained much of the metaphysics of the early writers, this work also contained a wealth of new material, much of it based on the author's own observations in the mining districts of Joachimsthal (Jachymov, C.S.S.R.) and Chemnitz (Karlmarxstadt, D.D.R.). Agricola is often conceded to be the father of modern mineralogy. His grouping of minerals listed them as simple, i.e., earths, congealed juices, stones, and metals; compounds, i.e., "mista" or intimately mixed; and "composita" or separable.

Abraham Gottlieb Werner, Professor at Freiberg, made a careful study of the external characters of minerals and in 1774 classified minerals into four groups—earthy, saline, combustible, and metallic. A very detailed classification of minerals was the work, in 1822–1824, of Friedrich Mohs, known better for his scale of hardness of minerals. Mohs developed a Natural History System of classification of minerals that included classes, orders, genera, and species and had little use for the "trivial name," that is, the name by species alone. The Mohs classification system, translated into English by W. Haidinger in 1825, had a profound effect on American mineralogy because it was adopted, in 1837, by James Dwight Dana in his *System of Mineralogy* who, in addition, used Latin names for genus and species, as suggested by Linnaeus. This was a period of conflict between proponents of "natural history" systems of classification, and others like the Swedish chemist, Jöns Jakob Berzelius, who believed that chemical composition would form a better basis for classification. In the third edition of his *System*, Dana gave up the natural history classification in favor of a chemical classification, which was found to be increasingly advantageous as better chemical techniques for analysis were developed.

Later, X-ray determinations of internal structure demonstrated the wisdom of using a chemical classification. The Dana classification has changed somewhat with each edition of the *System*, but in general its order begins with native elements, follows with anion groups such as oxides and halides, and ends with organic compounds. In early editions, the silicates, titanates, sulfates, phosphates, and carbonates were included under oxygen compounds; later, these were treated separately as special anisodesmic compounds. In these minerals, the anion groups act as units in the structure, with stronger internal bonding than to the external cations.

With the universal use of chemical classification in mineralogy, the emphasis of the "natural history" system on using classes, orders, genera, and species has been dropped. Taxonomy is no longer emphasized in textbooks on mineralogy, but an attempt is being made to define more clearly the grouping of minerals according to chemical, structural, or other physical similarities. Minerals are first divided into chemical *classes*, such as elements, carbonates, silicates. In some cases these are subdivided into *subclasses*. The next division is into chemical *types* usually based on the molar ratio of cations and anions. The types are divided into *groups* that comprise minerals with chemical or crystal structure similarities or occasionally paragenetic similarity. *Series* contain minerals with continuous variation in their properties related to change in composition. *Species* are the important units in mineralogy, synonymous with the definition of a mineral. For practical reasons, they are only roughly equivalent to "phase" as used by most other scientists. Further, there are *subspecies* and *varieties*. Other terms are occasionally used, such as *family* (usually synonymous with group or series). In this volume, names of mineral species are shown in *italic face* and names of mineral series and groups in ***italic boldface***.

A list of important minerals presented as mineral classes is given in Vol. IVA. Any plan for grouping minerals is, to some extent, dependent on arbitrary selection. In general, the grouping presented in this work is similar to that of Dana's *System* (7th ed.) where minerals of closely similar structure are grouped together; but this restriction is not rigidly enforced and the group sometimes contains minerals with similarities other than structural ones.

LLOYD W. STAPLES

References

Adams, F. D., 1938. *The Birth and Development of the Geological Sciences.* New York: Dover.

Ford, W. E., 1918. The growth of mineralogy from 1818 to 1918, *Am. J. Sci.*, **196**, 240-254.

Groth, Paul, 1926. *Entwicklungsgeschichte der Mineralogischen Wissenschaften.* Berlin: Springer, 147-156.

Mason, B., and Berry, L. G., 1968. *Elements of Mineralogy*, ch. 7. San Francisco: Freeman.

Palache, C.; Berman, H.; and Frondel, C., 1944. *The System of Mineralogy.* 7th ed., **1**. New York: Wiley, 1-3.

Staples, L. W., 1964. Friedrich Mohs and the Scale of Hardness, *J. Geol. Educ.*, **12**, 98-101.

Cross-references: *Crystallography: History; Mineral Classification: Principles.* Vol. IVA: *Mineral Classes: Nonsilicates; Mineral Classes: Silicates.*

MINERAL CLASSIFICATION: PRINCIPLES

Mineral classifications attempt to associate minerals that are similar, and to separate those that are dissimilar. Since minerals may be alike in some ways yet different in others, the basis selected for classification will determine which minerals are brought together and which are set apart.

Bases of Mineral Classifications

Minerals may be arranged in order of increasing specific gravities or hardness, according to luster, optical properties, or crystal structure. All minerals possessing the same chemical element as an important constituent may be grouped together. Minerals may be arranged according to chemical class. Each of the systems has its own uses.

A classification of the nonsilicate minerals founded upon both chemistry and crystal structure has become accepted as being of greatest use to the earth scientist. Conceived by Berzelius in 1814, this scheme is developed in its most elaborate form in the 7th edition of Dana's *System of Mineralogy* (Palache, et al.). The Berzelius-Dana system brings together minerals having similar chemical and structural properties with increasing degree of intimacy, is comprehensive, and provides a flexible numerical coding scheme that permits the placement of any mineral species in its logical position in relation to all other species.

Mineral Species

The fundamental unit of mineral classification is the mineral species. The Mineralogical Society of America in 1923 defined this unit as follows:

> A mineral species is a naturally occurring homogeneous substance of inorganic origin, in chemical composition either definite or ranging between certain limits, possessing characteristic physical properties and usually a crystalline structure.

In spite of certain shortcomings, this definition or a variation thereof is still useful.

It may be challenged on several grounds. The phrase "of inorganic origin" is questionable, for the minerals *graphite* and certain occurrences of *sulfur* and *calcite* are of organic origin, and Dana's *System* specifically provides for salts of organic acids and some hydrocarbon compounds. The word "usually" also poses difficulties. If omitted, such metamict substances as pitchblende, *thorite*, and some *zircons* as well as amorphous materials such as *opal* and *lechatelierite* would appear to be excluded. If included, even sea water and air might qualify as minerals. It seems most satisfactory to omit both these parts of the definition, but to be prepared to make exceptions for the relatively few solid substances of noncrystalline structure or organic origin that are generally recognized as minerals.

Most mineralogists would today reject the requirement that to qualify as a mineral, a substance must be "naturally occurring." Experimental techniques now permit the growth of crystalline material that is indistinguishable from its naturally occurring mineral counterpart, and a large number of the known mineral species have been synthesized in laboratories. Other substances have been grown in the laboratory that have structures analogous to those of known minerals, but for which there are no known natural equivalents; these too would be accepted as minerals by many mineralogists.

Scrutiny of the MSA definition must also be directed toward an interpretation of the phrase "certain limits." *Gold* invariably carries amounts of *silver*. Can the species name *gold* be applied equally to specimens one of which has less than 1% silver while the other has 21% silver? The ferrous, manganous, zinc, and magnesium ions may interchange in all proportions in combining with aluminum and oxygen. Can one species name be applied to this entire range of possible compositions? Almost without exception, mineral substances have a capacity for variation in chemical composition. Modern practice is to regard a mineral species as analogous to a naturally occurring solid phase in the physical-chemical sense. (A phase is a physically homogeneous and mechanically separable portion of a physicochemical system.)

Within the system $MgSiO_3$-$FeSiO_3$, a continuous series of natural phases may form, from pure $MgSiO_3$ up to about 90% $FeSiO_3$, 10% $MgSiO_3$. Any composition within these limits may crystallize as a homogeneous phase. Strict application of the concept that a mineral species is a phase would demand a name for each possible composition within the system. The proliferation of names that would result from such an interpretation would be unwieldy and unnecessary. The usage widely employed is to name the end components of a natural system, and perhaps arbitrarily subdivide the intervening compositional range into as many units as convenience demands. Whether the term species or subspecies is applied to the arbitrarily subdivided units is a matter for argument. When the $MgSiO_3$-$FeSiO_3$ system has crystallized at moderate temperatures, the names *enstatite* and *orthoferrosilite* are applied to the pure phases $MgSiO_3$ and $FeSiO_3$, respectively. Further arbitrary subdivision has produced the names bronzite (10–30% $FeSiO_3$), *hypersthene* (30–50% $FeSiO_3$), ferrohypersthene (50–70% $FeSiO_3$), and eulite (70–90% $FeSiO_3$). Mason and Berry (1968, p. 198) and Winchell and Winchell (1961, p. vi) consider that any arbitrary compositional unit should be termed a subspecies; hence they would call units of this system subspecies, including the end members. In the $MgAl_2O_4$-$FeAl_2O_4$-$ZnAl_2O_4$-$MnAl_2O_4$ system, however, Mason and Berry regard the end members as species. Dana's *System* (Palache et al., vol. 1, p. 43) states that a variety (syn. subspecies) is a "chemical deviation from the principal composition." The 7th edition of Dana's *System* has not been extended to include the silicate minerals, so no definite statement can be made about the placement in this classification of the units of the $MgSiO_3$-$FeSiO_3$ system. The end members of the $MgAl_2O_4$-$FeAl_2O_4$-$ZnAl_2O_4$-$MnAl_2O_4$ system are, however, designated as species in this encyclopedia.

It is clear that although the analogy to a physical-chemical phase is a useful one, a mineral species must be allowed more latitude than a definition as a phase would permit. It must be recognized that even in the selection of what constitutes a natural system, an arbitrary decision has been made, for in the definition of $MgSiO_3$-$FeSiO_3$ as a system, it is well understood that aluminum, calcium, manganous, ferric, titanium, chromium, and nickel ions are invariably present in the natural mineral substances formed. In the final analysis, the decision regarding the permissible range of the "certain limits" is man made.

Criteria have been established that must be met in order that a substance be accepted as a mineral species. The Commission on New Minerals and Mineral Names of the International Mineralogical Association is generally recognized as the final authority in decisions concerning new mineral species.

Classification

In spite of the difficulties encountered when a rigid definition of the term is attempted, "mineral classifications" are built up from the natural unit the mineral species. Once a substance has been accepted as a mineral species, it may be placed within the framework of a classification. Most mineralogy texts contain their own organizations of minerals, but all are modifications of the Berzelius-Dana system. The Commission on Mineral Data and Classification of the International Mineralogical Association is presently engaged in a review of the organization of minerals into a logical framework of classification. The classification outlined here is essentially that of Dana's *System*, but mention is made of corresponding divisions employed by some other standard works in mineralogy. Dana's *System* has not been extended to the silicate minerals, but the categories used in arranging the nonsilicates are equally applicable to the silicates. In the excellent five-volume work by Deer, Howie, and Zussman (1962–1965), the chief silicate minerals fall substantially into the subdivisions of the Dana *System*, and it has been necessary to introduce only one additional category, the subgroup. For details of the placement of minerals according to the Berzelius-Dana scheme, see Vol. IVA: *Mineral Classes: Nonsilicates; Mineral Classes: Silicates*; and articles on individual mineral groups. Table 1 illustrates the arrangement according to the Dana *System*, and gives the classification numbers for species of one type.

Class. The first two volumes of Dana's *System* place all nonsilicate minerals in 50 major categories, called classes. The divisions are chemical, and are made on the basis of the anion or anionic groups present. It is assumed that all silicate minerals will be placed in the fifty-first class. The 42 classes described in volume two are further aggregated into 13

TABLE 1. Abbreviated Listing of the Carbonates According to Dana's System

	Carbonates	
CLASS 13.	Acid carbonates (not detailed)	
CLASS 14.	Anhydrous normal carbonates	
Type 1.	$A(XO_3)$ (not detailed)	
Type 2.	$AB(XO_3)_2$	
14.2.1	Dolomite group	
14.2.1.1	*Dolomite*	$CaMg(CO_3)_2$
14.2.1.2	*Ankerite*	$Ca(Fe,Mg)(CO_3)_2$
14.2.1.3	*Kutnahorite*	$Ca(Mn,Mg)(CO_3)_2$
14.2.2	*Alstonite*	$CaBa(CO_3)_2$
14.2.3	*Barytocalcite*	$CaBa(CO_3)_2$
Type 3.	Miscellaneous (not detailed)	
CLASS 15.	Hydrated normal carbonates (not detailed)	
CLASS 16.	Carbonates containing hydroxyl or halogen (not detailed)	
CLASS 17.	Compound carbonates (not detailed)	

After C. Palache, H. Berman, and C. Frondel, *The System of Mineralogy of J. D. Dana and E. S. Dana,* 7th ed., vol. 2, John Wiley, New York, 1951, pp. 132–133. Reprinted with permission.

broader chemical generic categories, but this is done for indexing reasons, and these broader categories do not figure in the placement of a mineral species in the classification. Less detailed arrangements (e.g., Mason and Berry, 1968) recognize fewer classes (generally 8), and employ the chemical generic category names of Dana as class names.

Type. Classes are subdivided into types, which are made up of minerals having similar ratios of positive ions (or electropositive atoms) to negative ions (or electronegative atoms). Within a class, types are placed in order of decreasing ratios of numbers of cations to numbers of anions, e.g., Class 4 (Simple Oxides) may be subdivided into Type 1, A_2X; Type 2, AX; Type 3, A_3X_4; Type 4, A_2X_3, etc., where A represents positive ions, and X represents the negative oxygen ion. Although the basis for this subdivision is chemical, Dana's *System* recognizes the pronounced relationship that exists between type and crystal structure. It points out that minerals of different classes with like ionic ratios often show marked similarities in structure, e.g., *galena* (class 2), *periclase* (class 4), and *halite* (class 9) have similar ionic ratios of 1:1, and all exhibit a face-centered cubic structure. The native elements (class 1) subidivde into two types on the basis of bonding, and type 2 (semimetals and nonmetals) having essentially a homopolar bonding.

The silicate minerals (not covered by Dana's *System*) are most broadly subdivided on the basis of structure, not cation to anion ratios, in all modern classifications. (Such subdivision does, however, yield structural types that are made up of minerals of similar Si:O ratios.) It, therefore, seems prudent to broaden the basis for the initial division of the class to include structure as well as chemistry. Some classifications replace the term type with terms such as subclass or family, and so avoid the possible charge of misuse of the Dana term.

Group. Many mineral types contain a number of groups; some include a number of individual species or a single group together with one or more individual species. There are varied criteria for setting up a group. Most groups consist of minerals closely related structurally and chemically, e.g., the ***pyroxene, feldspar, galena***, and ***hematite*** groups. Minerals that have similar occurrences, but are not necessarily closely related either structurally or chemically, also may be considered to form a group, e.g., ***feldspathoids*** and ***zeolites***. The silica minerals are considered by some to constitute a group in spite of the lack of similarity in structure and paragenesis of these minerals. "When the characteristics of a number of minerals within a type can profitably be discussed together they form a group" (Palache et al., vol. 1, p. 3).

Subgroup. This category does not appear in Dana's *System*. Deer, Howie, and Zussman (1962–1965) have used it to recognize very close compositional (and structural) similarities among mineral series within groups showing great compositional variation. The ***amphibole*** group, for example, may be subdivided into subgroups on the basis of the character of the predominant ion in the X position of the structure given by the formula $X_{2-3}Y_5Z_8O_{22}(OH)_2$: The X position is occupied by (Mg,Fe) in subgroup 1, by Ca in subgroup 2, and by Na in subgroup 3. The very complex chemical substitution possible in some silicate minerals makes this category a useful one for silicate classification. It is unnecessary, however, in subdividing the nonsilicates.

Series. "Minerals showing a continuous variation in their properties with change in composition are called series. . . .In such instances, the natural mineralogical unit is the series, and an arbitrary segmentation does not give an adequate picture of any part of the series" (Palache et al., vol. 1, p. 3). Nearly all minerals show some variation in both composition and properties, so nearly all minerals might be considered to constitute series rather than species. Whether a natural unit is considered a series or a species hinges upon an interpretation of the compositional variation permissible, hence differences in usage are to be expected. Dana's *System* gives the ***plagioclases*** and the ***spinels*** as typical examples of a series.

Species. The fundamental unit of the classification. See discussion above.

Varieties and Subspecies. Species may be subdivided into varieties or subspecies, the former term being the one used in Dana's *System*. Varieties exhibit distinctive physical properties (amethyst is a transparent, purple type of *quartz*), or relatively minor chemical variations (electrum is a varietal name that has been applied to *gold* with 20% or more silver). One of the major contributions made by Dana's *System* is its detailed application of a proposal made by Schaller (see bibliography in Palache et al.) in 1930, to replace varietal names of species with ionic adjectives prefixed to the species' name, e.g., chromian *muscovite* replaces the varietal name fuchite, magnesian *fayalite* replaces hortonolite, and ferrian *biotite* replaces lepidomelane.

Classification Numbers

The Dana system provides for exact placement of any mineral species within the scheme by giving each class, type, and species a number consisting of one or two digits, and if a type contains groups, the groups are given numbers. The first number of the notation indicates the class, the next the type. The last number always refers to the individual species or series. If the species is a member of a group, the group number is placed between the type number and the species number. The classification number identifying a species contains a minimum of three, and a maximum of six digits, e.g., *atacamite* is represented by the notation 10.1.11.1, indicating that it is a member of class (10) (oxyhalides and hydroxyhalides), type 1 ($A_m[O,OH]_pX_q$), group 11 (***atacamite*** group), series 1 (*atacamite*). The practice of separating the numbers representing each category of the classification by a dot is consistently followed in the second volume of Dana's *System*, but not in the first; consequently, in a few cases some confusion results in the notation for the first eight classes of minerals.

JOHN T. JENKINS

References

Deer, W. A.; Howie, R. A.; and Zussman, J., 1962–1965. *Rock-Forming Minerals*. London: Longmans, 5 vols.

Hintze, C., and Linck, L., 1889–1939. *Handbuch der Mineralogie*. Berlin: De Gruyter, 6 vols.

Hurlbut, C. S., Jr., and Klein, C., 1977. *Dana's Manual of Mineralogy*, 19th ed. New York: Wiley, 532p.

Mason, B., and Berry, L. G., 1968. *Elements of Mineralogy*. San Francisco: Freeman, 550p.

*Palache, C.; Berman, H.; and Frondel, C. *The System of Mineralogy of J. D. Dana and E. S. Dana*, 7th ed. New York: Wiley, vol. 1, 1944; vol. 2, 1951; vol. 3, 1962.

Winchell, A. N., and Winchell, H., 1951. *Elements of Optical Mineralogy, Part II, Description of Minerals*, 4th ed. New York: Wiley, 551p.

Strunz, H., 1966. *Mineralogische Tabellen*, 4th ed. Leipzig: Akad. Verlag., 560p.

*Additional bibliographic references may be found in this work.

Cross-references: *Alkali Feldspars; Amphibole Group; Crystal Growth; Crystallography: Morphological; International Mineralogical Association; Mineral Classification: History; Mohs Scale of Hardness; Native Elements and Alloys; Optical Mineralogy; Plagioclase Feldspars; Pyroxene Group; Zeolites;* see also mineral glossary.

MINERAL DEFORMATION—*See* PLASTIC FLOW IN MINERALS

MINERAL DEPOSITS: CLASSIFICATION

Ore deposits are of such infrequent and random occurrence, of such modest size, and so different in composition in comparison with the generality of rocks, that they probably require an uncommon sequence of geologic happenings to bring about their formation. The study of such deposits in all quarters of the world over the last 150 years has, however, produced such a wide variety of information concerning them that certain ore geologists have attempted to classify them on a genetic basis; others have settled for classifications based on descriptive characteristics.

The first genetic classification of ore deposits to win wide acceptance was that initially proposed by Waldemar Lindgren in 1906 and last revised by him in 1933. As ideas on the genesis of ores have undergone change in the thinking of geologists, the Lindgren classification has come under attack mainly by two groups. The first of these groups is composed of geologists who doubt the fundamental genetic basis of Lindgren's work—namely, that a large fraction of the ore deposits known in the world were formed by processes directly related to the crystallization of magmas. Lindgren believed this group of magma-related deposits were developed in two principal ways: (1) as disseminations or segregations developed within the magma itself or from melts, implicitly poor in water, generated during the magma's crystallization cycle and either crystallized within the magma chamber or extruded from it into the surrounding rocks; or (2) as vein fillings and replacement masses deposited from hydrothermal (hot, water-rich) solutions also developed during the crystallization process. He emphasized that the processes by which these types of magma-related deposits were produced were directly interrelated and that many factors—such as the composition of the crystallizing magma, its position in relation to the surface of the earth (which in turn affected the rate of cooling and confining pressure), the character of its rock environment, the degree to which it reacted with that environment, and the extent to which that environment was affected by earth movements—all acted to determine what ores and ore-transporting media, if any, might be produced from a given magma.

Those who doubt Lindgren's fundamental thesis of the ultimate magmatic origin of ore deposits not formed by surface or near-surface processes have offered a variety of explanations for the methods by which ore concentration occurs. These explanations range from the reaction of volcanic gases with sea water to the diffusion, through solid rock, of the constituents to be concentrated under the influence of pressure or concentration gradients. No one of these, or other postulated mechanisms, has achieved sufficient acceptance to inspire the development of a scheme of classification of more than passing interest.

The second group of Lindgren's critics are those who accept his basic premises but believe that the advances in ore geology in the 30 years since the last version of the Lindgren classification was published require certain modifications in the classification itself. Probably the most commonly expressed disagreement with the Lindgren scheme is the argument that Lindgren placed too much emphasis on depth beneath the surface at the time of ore deposition as the most important factor in determining the chemical intensity of hydrothermal solutions. All high-temperature deposits, under the Lindgren system, are indicated as having formed at great depths beneath the surface (a defect he himself recognized). Yet, as Buddington pointed out (Noble, 1955, p. 168), many deposits whose mineral content strongly suggests that they were formed at high temperatures must, on the geologic evidence, have been emplaced at quite shallow depths beneath the earth's surface. For such deposits, he proposed the term "xenothermal," which has been extensively used in the geologic literature.

H. A. Schmitt (see Noble, 1955, p. 169), Ridge (1958, 1968, 1972), and Park and MacDiarmid (1964) have attempted to modify the Lindgren classification to include the term xenothermal; of these modifications, that of Ridge (Table 1) makes the greatest changes in the Lindgren concept in that he divides "hydrothermal" deposits into two general categories: (1) those formed with slow decrease of heat and pressure, and (2) those developed with rapid loss of heat and pressure. Into the first group, he places Lindgren's terms "hypothermal" and "mesothermal" but, for "epithermal," he substitutes two terms devised by L. C. Graton in 1933 (see Noble, 1955, p. 168). These terms are "leptothermal," which is, essentially, the less intense portion of Lindgren's mesothermal category, and "telethermal," which applies to deposits formed under low conditions of chemical intensity (such as the Mississippi Valley lead-zinc deposits). Where gradations from deposits formed under conditions of moderate (mesothermal) to low intensity are known, the gradation is from the mesothermal through (often minor) leptothermal mineralizations to telethermal and not from mesothermal to epithermal. Ridge, therefore, considered it sounder practice to remove epithermal from the hydrothermal-mesothermal sequence.

In Lindgren's time, none of the recognized epithermal deposits studied were known to grade downward into mesothermal-type mineralization; in fact, none was known to grade downward into any other type of mineralization. In even those epithermal deposits with the greatest vertical extent (>3000 ft beneath the surface in both Cripple Creek and the Comstock Lode), epithermal characteristics were exhibited through their entire range. Such deposits, however, carried minerals formed, almost certainly, under lower temperature and pressure (intensity) conditions than those of the mesothermal zone; therefore, Lindgren felt justified in placing the epithermal category above the mesothermal.

Since the Lindgren classification was last modified by its author, however, it has become clear that several epithermal deposits, such as Potosí and Oruro in Bolivia and Tombstone in Arizona, contain, not only epithermal minerals, but also mineral suites, typical of the mesothermal and hypothermal ranges that bear, however, every evidence of having been formed near the earth's surface. Of these near-surface, high-temperature mineral suites, those formed under the more intense conditions correspond to Buddington's xenothermal class; while they contain the minerals of the hypothermal range, they were certainly not formed at the great depths that Lindgren assigned to the hypothermal zone.

In several deposits, such as those just mentioned, Parral and Santa Eulalia in Mexico, Llallagua in Bolivia, and Akenobe in Japan, the xenothermal mineralization is associated with

TABLE 1. Modified Lindgren Classification of Ore Deposits

Type	Temperature (°C)	Pressure (atm)	Depth (ft)
	Conditions of Formation		
I. Deposits mechanically concentrated (placers)		Surface conditions	
II. Deposits chemically concentrated		Differ within wide limits	
A. In quiet waters	0–70	Low	Shallow (0–600)
1. By interaction of solutions (sedimentation)			
a. Inorganic reactions			
b. Organic reactions			
2. By evaporation of solvents (evaporation)	0–70	Low	Shallow (0–600)
3. By introduction of fluid igneous emanations (water-rich fluids)	100–300	Low–moderate (1–> 200)	Shallow–medium (low–> 6,000)
B. In rocks (with or without introduction of material foreign to rock affected)			
1. By rock decay and weathering (residual deposits)	0–100	Low	Shallow
2. By ground water circulation (supergene processes)	0–100	Low–moderate	Shallow–medium
C. In rocks by dynamic and regional metamorphism (with or without introduction of material foreign to rock affected)	⩽500	High–very high	Great
D. In rocks by hydrothermal solutions			
1. With slow decrease in heat and pressure			
a. Telethermal	50–150	Low–moderate (40–240)	Shallow (500–3,000)
b. Leptothermal	125–250	Moderate (240–800)	Medium (3,000–10,000)
c. Mesothermal	200–400	Moderate–high (400–1,600)	Medium (5,000–20,000)
d. Hypothermal			
1. In noncalcareous rocks (Lindgren's hypothermal)	300–600	High–very high (800–4,000)	Medium–great (4,000–50,000)
2. In calcareous rocks (contact metamorphic)	300–600	Very high (800–4,000)	Medium-great (4,000–50,000)
2. With rapid loss of heat and pressure			
a. Epithermal	50–200	Low–moderate (40–240)	Shallow–medium (500–3,000)
b. Kryptothermal	150–350	Low-moderate (40–280)	Shallow–medium (500–3,500)
c. Xenothermal (pressures initially appreciably higher than lithostatic pressure would produce	300–500	Low–moderate (80–700)	Shallow–medium (1,000–4,000)
E. By gaseous igneous emanations in rocks	100–600	Low	Shallow (100–600)
F. In magmas by differentiation or in adjacent country rocks by injection			
1. Early separation–early solidification	500–1,500	Very high (1,000+)	Great (15,000+)
a. Disseminations			
b. Crystal segregations			
c. Crystal segregations, plus injection as crystal mush			
2. Early separation–late solidification	500–1,500	Very high (1,000+)	Great (15,000+)
a. Early immiscible sulfide melt accumulation			

TABLE 1. (Continued)

Type	Conditions of Formation		
	Temperature (°C)	Pressure (atm)	Depth (ft)
b. Early immiscible sulfide melt accumulation, plus later fluid injection			
3. Late separation–late solidification, with or without fluid injection			
a. Silicate pegmatites			
i. Simple	575±	High–very high (800–4,000)	Great (1,000–50,000+)
ii. Complex	200–550	High–very high (800–4,000)	Great (10,000–50,000+)
iii. Barren quartz	100–300	High–very high (800–4,000)	Great (10,000–50,000+)
b. Immiscible (metal-oxygen-rich) melts	500–1,000	Very high (1,200+)	Great (15,000+)
c. Immiscible (carbonate-rich) melts	500–1,500	Low–very high (1–4,000)	Shallow–great (0–50,000+)
4. Late formation–deuteric alteration	<575	Moderate–very high (400–4,000)	Medium–great (5,000–50,000+)

one containing minerals characteristic of the mesothermal range. Their position relative to the earth's surface at their time of formation also, however, debars them from being a part of the much deeper mesothermal range in the Lindgren sense. Several deposits containing near-surface mesothermal mineralizations not only were formed under less intense conditions than the xenothermal deposits with which they are associated, but also are grouped with less intense epithermal minerals (e.g., Oruro, Potosí, and Tombstone), while others containing such intermediate mineral suites as the most intense development also contain epithermal mineralizations (e.g., Bor in Yugoslavia and Cerro de Pasco in Peru).

It would seem to follow from this evidence that the sequence of deep-seated ore categories (hypothermal, mesothermal, leptothermal, and telethermal) has a near-surface counterpart of which the two end members are xenothermal and epithermal; Ridge (1958, 1968, 1972) designated the intermediate group by the term "kryptothermal." Schmitt, although he developed no new term for the intermediate, near-surface category, had much the same idea of the positions of xenothermal and epithermal to the deep-seated sequence (see his diagram, ref. in Noble, 1955, p. 169). Park and MacDiarmid (1964, p. 210), on the other hand, have added the terms telethermal and xenothermal to the three basic Lindgren terms in such a way to suggest that telethermal is the upper terminus of the hydrothermal range, lying immediately above the epithermal. Park places xenothermal after telethermal in his categorization, but makes it clear that such deposits were formed under more intense conditions than epithermal and out of the main sequence of intensity zones.

Another modification of the Lindgren classification made by both Ridge and Park has been to remove the term "pyrometasomatic," which Lindgren applied to deposits formed "by direct igneous emanations from intrusive bodies." Lindgren (1933, p. 699) pointed out that these deposits are found mainly in limestone, dolomite, and calcareous shale; and pyrometasomatic, as used by Lindgren, was essentially synonymous with high-temperature deposition in calcareous rocks at great depth. However, work since introduction of the Lindgren classification has shown that most deposits found in such calcareous rocks, at or near igneous contacts, were not formed by "emanations" derived from the igneous body, but were deposited by hydrothermal solutions that used the igneous-sedimentary contact as a channelway, and carried out most of their deposition in carbonate rocks. In addition, several "pyrometasomatic" deposits were found at appreciable distances from any contact, showing that such formation of high-temperature deposits does not require the immediate influence of an igneous contact.

Studies of high-temperature hydrothermal deposits, in general, have also shown that the principal differences between Lindgren's hypothermal category and his "pyrometasomatic" group lies in the type of rock in which the deposit was formed. Except for the impressive development of calcium-rich minerals in "pyrometasomatic" deposits, the mineral content in the two types is much the same; there may be a variation in specific minerals from one type to the next, but there is little difference in the mineral species developed. From this, it follows that the ore-forming fluids that formed hypothermal and "pyrometasomatic" deposits were much the same, the differences between them having been caused largely by the rock types in which they emplaced their mineral load. For these reasons, Ridge dropped the term "pyrometasomatic" altogether and subdivided the hypothermal category into hypothermal in the noncalcareous and in calcareous rocks, respectively. The validity of this change is further demonstrated by work published by G. C. Kennedy on the pressure-volume-temperature relations of water, which shows that, at the temperature and pressures at which hypothermal deposits are postulated to have formed, the ore fluids, though technically in the gaseous state (being at temperatures above the critical temperature of water), were so highly compressed as to have sufficient density (about one-half of that of water at 25°C) to carry ore and gangue mineral ions in true solution. Thus, deposition from ore fluids in the temperature ranges immediately above their critical temperatures should not be expected to produce appreciably different mineral assemblages or mineral textures from those developed in the ranges immediately below those critical temperatures.

In his classification, Lindgren did not break down the category "in magmas by processes of differentiation" except to divide it into "magmatic deposits proper" and "pegmatites." Bateman (1950, p. 8), provided a subdivision of "magmatic deposits proper" into two major categories, "early-magmatic" and "late-magmatic" which recognized that some magmatic deposits develop early in the crystallization cycle and others late. Ridge (1958, p. 193; 1968; 1972) further broke down these two categories, using the terms "early separation–early solidification," "early separation–late solidification," "late separation–late solidification," and "late solidification–deuteric alteration" and then subdivided these (Table 1). He included pegmatites in this portion of the classification, which Lindgren and Bateman did not, because of the direct genetic relationship of all pegmatites of magmatic origin to the magmatic process, whether or not their ultimate place of solidification was within or without the magma chamber in which they were generated. In the same "late separation–late solidification" subcategory he included immiscible metal-oxygen-rich melts, the presumed parents of such deposits as Kiruna in Sweden, Allard Lake in Quebec, and Iron Mountain, Missouri, which appear to have been generated in the late stages of the crystallization of certain iron- or iron-titanium-rich magmas. The designation of the stages at which both separation and solidification occurred, instead of using only the general time categories employed by Bateman, made it possible to indicate more accurately the relationship of a given magmatic deposit to the genetic processes involved in magmatic ore production.

The recent prominence given by European geologists to the process of ore formation in shallow bodies of water from gaseous emanations of volcanic origin is missing from the Lindgren classification but was included by Ridge in the subcategory IIE of Table 1. There is still doubt that this process has the universality ascribed to it in Europe; but, for any deposits proved to have been formed in this manner, a space is available to incorporate them in the modified version of the Lindgren classification.

In Europe, in the last 25 years, there have been two major ore-deposit classifications in use, those of Schneiderhöhn (1941) and Niggli (1941). Schneiderhöhn's scheme has four main subdivisions: (1) intrusive and liquid-magmatic deposits that correspond roughly with the magmatic portion of the Lindgren classification, less pegmatites; (2) pneumatolytic deposits that contain the pegmatites and such deposits as Schneiderhöhn believes can be categorized as having formed above the critical temperature of the ore fluid; (3) hydrothermal deposits that he divides first on the basis of their mineral content and secondly by depth of formation, using the subheadings "hypabyssal" and "subvolcanic" with terms such as mesothermal and epithermal used on occasion as modifiers of the mineral association (content) types; and (4) exhalation deposits that correspond generally with the modified Lindgren gaseous emanation category (IIE of Table 1). His classification makes essentially no attempt, unlike Lindgren's, to include deposits formed by surface processes. Perhaps the greatest weakness in the Schneiderhöhn classification is the assumption that the mineral content of a given ore deposit furnishes a direct clue to the conditions under which it was formed. Antimony-mercury associations, so far as data are now available, are always formed under low-pressure, low-temperature conditions, whereas gold-silver associations may range from hypothermal to leptothermal and from xenothermal to epithermal; thus, the use of gold-silver association to categorize a deposit does little to locate it, genetically, in the intensity scale. The use of such terms as, e.g., mesothermal, as a modifier of an association designation helps in clarifying this problem; but even this device is not consistently used. The Lindgren classification, on the other hand, allows any mineral association to be included in any one of its categories, provided only that the minerals in question were formed under the proper range of temperature and pressure.

The Niggli classification (1941) is the only one of those in common use that makes a proper allowance for most of the variables that determine the conditions and results of ore deposition. Niggli includes the following variables in his classification: (1) place of origin of the ore-bearing solutions (deep plutonic, plutonic, subvolcanic, volcanic); (2) place of deposition of the ore minerals in relation to (a) depth in the earth's crust (abyssal, hypabyssal, epicrustal, subaquatic, aerial [subaerial]), (b) distance from the point of origin in the magma chamber (intramagmatic, perimagmatic, apomagmatic, kryptomagmatic, telemagmatic), and (c) character of wall rocks and their alteration products (not broken down because of the huge number of categories that would be needed); (3) physicochemical state of the ore fluids (orthomagmatic, pegmatitic, pneumatolytic, hydrothermal, exhalative); and (4) temperature during main period of mineralization (high, medium, low or kata, meso, and epi).

Normally, four variables are defined in each designation of a deposit in the Niggli classification, with the further use of the terms kata, meso, and epi in defining mineral associations that are used in much the same manner as is done in the Schneiderhöhn classification. Thus, a high-temperature *gold-pyrite* vein might be defined as plutonic, hypabyssal, apomagmatic, hydrothermal, katathermal *gold-pyrite*. As the terms plutonic and apomagmatic deal with concepts that cannot be determined from the deposit itself, but instead must be derived by geologic reasoning, they are less firmly established than hypabyssal and hydrothermal, the validity of which, many geologists think, can be established from the study of the deposit and its immediate surroundings. It is then readily apparent that more must be known, or conjectured, about a given deposit for it to be classified by the five variables of the Niggli system than if it were to be categorized under the depth (confining pressure) and temperature of the Lindgren scheme.

There is no geologist who has claimed, in print at least, that the ideal classification of mineral deposits of economic value has yet been achieved. The wide use, though perhaps less wide acceptance, of the Lindgren classification makes it, at present, the point of departure from which most thought on ore-deposit classification in North America appears to be based. That it can, however, be much refined should be readily evident. The prevalence of ores of more than one intensity range in a given rock volume (Butte, Magma, Noranda, Aberfoyle, and Oruro are outstanding examples) strongly suggests that factors other than depth beneath the surface bear a greater responsibility for the intensity range of the mineralization found in a given deposit than does depth. Temperature of, and confining pressure on, the ore fluid are of even greater importance than actual depth, but perhaps even more important than these are the pH and rate of change of pH of hydrothermal solutions, the pressure of oxygen (pO_2) and concentration of O^{2-} (pO) and of OH^- (pOH) of metallic and siliceous ore-parent melts, and the Eh (oxidation-reduction potential). None of these latter four factors has even been considered in the currently used classification nor, more importantly, have methods been suggested (much less developed) to permit the determination of these properties of the ore fluid from the ore deposits they have left behind them. Classification can, therefore, be improved only as fast as field and laboratory studies provide new bases for advances in the theories of ore formation, theories that define variables that can be incorporated in the classification scheme.

In 1973 and 1974, Ridge discussed the place of volcanic-exhalative deposits in the modified Lindgren classification. He pointed out that the depth of sea water overlying that portion of the sea floor where deposition occurred is the final determinant as to whether an ore fluid reaches that datum in a dense enough condition to carry material in true solution without boiling or begins to boil before it reaches that surface. As the volcanic-exhalative solution approaches nearer and nearer the sea floor, the confining effect of sea-water pressure becomes more and more important than that of the rock through which the solutions are passing. If, at any given depth, the combined confining pressure of rock plus sea water is less than the gas pressure of the solution, boiling will result. The first effect of such boiling would be to begin the precipitation of the least soluble constituents, mainly the sulfides. If continued long enough, boiling would result in the deposition of essentially all of the material carried in solution, of which the most abundant would be NaCl. Only the constituents about as readily converted to the gaseous phase as water would leave with it. If boiling did not continue long enough for NaCl to deposit, the remaining solution would be much higher in salt than it was initially, the increase being proportional to the time that boiling continued. Thus, the higher a fluid inclusion in a sulfide is in salt content, the more likely it is to have been derived from an ore fluid that boiled during the latter part of its journey toward the sea floor or the land surface.

For many stratabound deposits, it should be possible (1) to determine the temperature of deposition of one or more ore minerals and (2) to estimate the depth of sea water in which the sediment or volcanic material containing the ore deposit was laid down. The less the depth of sea water and the higher the temperature of formation, the less likely it is that the ore minerals in question were produced by an encounter between a volcanic exhalation and sea water. Any volcanic exhalation that boiled almost certainly would have left most of its sulfide content in solid rock and would not have had any left to deposit on the sea floor. On the other hand, the greater the depth of sea water and the lower the temperature of sulfide deposition, the more likely it is that the ore minerals in question were formed on the sea floor. Ridge, in his 1973 paper, gives curves that can be used to determine whether or not the ore minerals in question could have been formed syngenetically on the sea floor or epigenetically in solid rock.

JOHN D. RIDGE

References

Bateman, A. M., 1950. *Economic Mineral Deposits*, 2nd ed. New York: Wiley, 916p.

Lindgren, W., 1933. *Mineral Deposits*, 4th ed. New York: McGraw-Hill, 930p.

Niggli, P., 1941. Die Systematik der magmatischen Erzlagerstätten, *Schweiz. Mineral. Petrogr. Mitt.*, **21**, 161–172.

*Noble, J. A., 1955. The classification of ore deposits, *Econ. Geol. 50th Anniv. Vol.*, 155–169.

*Park, C. F., Jr., and MacDiarmid, R. A., 1964. *Ore Deposits*. San Francisco: Freeman, 475p.

*Ridge, J. D., 1958. Selected bibliographies of hydrothermal and magmatic ore deposits, *Geol. Soc. Am. Mem. 75*, 199p.

Ridge, J. D., 1968. Changes in concepts of ore genesis, 1933–1967, *Soc. Min. Engrs., A.I.M.E.*, Graton-Sales Vol. 2, 1713–1834.

Ridge, J. D., 1972. Annotated bibliographies of mineral deposits in the Western Hemisphere, *Geol. Soc. Am. Mem. 131*, appx. 1, 673–678.

Ridge, J. D., 1973. Volcanic exhalations and ore deposition in the vicinity of the sea floor, *Mineral. Dep.*, 8, 332–348.

Ridge, J. D., 1974. A note on boiling of ascending ore fluids and the position of volcanic-exhalative deposits in the modified Lindgren classification, *Geology*, **2**, 287–288.

Schneiderhöhn, H., 1941. *Lehrbuch der Erzlagerstättenkunde*, vol. 1. Jena: Gustav Fischer, 858p.

*Additional references will be found in these works.

Cross-references: *Blacksand Minerals and Metals; Metallic Minerals; Mineral and Ore Deposits; Pegmatite Minerals; Placer Deposits; Vein Minerals;* see also mineral glossary.

MINERAL FUELS–*See* Vol. VI: COAL; CRUDE OIL COMPOSITON AND MIGRATION

MINERAL INDUSTRIES

Minerals in History

Minerals have always played a prominent role in human activity and progress. Therefore, historians and anthropologists have frequently associated stages of civilization with various mineral materials: Stone Age, Copper Age, Bronze Age, Iron Age, or Age of Metals. Even the currently popular terms, Atomic Age and Nuclear Age, remind us of the existence of uranium and thorium. Each stage represents advances in man's ability to transform, process, and otherwise utilize minerals and materials.

Man's initial mineral needs centered around the use of rocks such as flint and obsidian and native metals such as *gold* and *copper*. This modest beginning led to one of the most significant technical and cultural events in human history–the successful recovery of metals from mineral ores. Copper was the first such metal reduced from ore. The impact and spread of iron metallurgy is well documented in historical accounts. The metals known by early metallurgists prior to the Christian era include gold, silver, copper, iron, lead, and mercury.

The use of fuels remained relatively unchanged prior to the industrial revolution. Wood was man's principal source of fuel from dim prehistory to the 19th century, when expanding economies required the superior thermal qualities of coal. Coal, in turn, dominated world fuel consumption from 1880 to 1950 and then was replaced by petroleum and natural gas.

Concomitant with the growing sophistication of mineral technology has been the increase in mineral requirements. The small quantities of rock and native metals consumed by prehistoric man for tools, arms, and shelter were dwarfed by the mineral and metal requirements of the ancient Mediterranean cultures. More recently, the industrial revolution has resulted in a dramatic increase of mineral demand. During the last half century, more mineral materials were consumed than in the whole of previous history. This consumption has been concentrated in a relatively few industrial nations, which represent a small fraction of the world's population.

Development of the Mineral Industry

The growth of the mineral industries has paralleled basic trends in the structure of the world economy. Prior to the industrial revolution, mining operations were small and quite labor intensive. Mineral trade was generally local or regional in nature. As the industrial revolution spread through Europe, the need for a wide variety of mineral raw materials increased significantly. However, these requirements were still met primarily from domestic and regional sources.

Most of the operations of the major mining and petroleum companies were originally located in their home country. However, by the early decades of this century, some of these companies were becoming international in scope as they discovered and developed mineral deposits abroad. Most large mining and petroleum companies of the United States currently have some foreign operations or interests. These foreign operations have been concentrated in Australia, Canada, the Republic of South Africa, and the developing nations of Africa, Asia, and South America. A number of factors have contributed to the rapid internationalization of the minerals industry: The further industrialization and expansion of many national economies; the depletion of high-grade, easily accessible mineral deposits in many major consuming nations; the generally favorable investment climate previously found abroad; the advances in bulk-cargo transportation methods, which facilitated the growth of international mineral trade.

The development of mineral resources abroad by mining companies and the subsequent increase in international mineral trade have served two important functions. The large mineral-importing nations have received the needed materials to support their diversified industrial base at a lower cost than would otherwise have been possible. In addition, mineral exporters have received the revenue, foreign exchange, capital, and technology that are necessary for continued economic expansion.

The structure of the world mineral industry has undergone considerable change in the last decade, as many mineral-exporting nations reexamined their role in the global mineral-supply system. These reevaluations have been prompted by such factors as the often cyclical nature of mineral demand and price, the dependency of many nations on one or two minerals for a large percentage of their foreign exchange and national output, and the increasing economic strength of these exporting nations. The nature and form of the actions taken by mineral exporters varies greatly. In general, they seek greater benefits from, and more control over, their domestic mining and mineral activities.

Mineral Industries in Relation to the National Economy

The proportion of employment and national output originating in the mining and quarrying industries varies among nations, but a few generalities are possible. Countries with a substantial portion of employment and domestic output originating in the extractive industries tend to be categorized as industrially developing nations. The converse is not necessarily true, that is, countries with low per capita incomes do not always have relatively large mineral industries. For some countries, such as the industrially developed nations where the employment and output attributable to mining are obscured by the employment and output of other industries, the contribution of the mining sector—along with that of agriculture—to the national economy is fundamental to the nation's welfare in peace and in war.

Figure 1 illustrates the underlying importance of the mineral industries to the economic well-being of the United States. In 1976, the supply of mineral raw materials consisted of domestically produced minerals valued at $68 billion and imported minerals and reclaimed scrap valued at $31 billion and $4 billion, respectively. These mineral raw materials were processed into energy and products valued at over $300 billion.

Mineral Supplies

The possibility of physically depleting most of our inorganic mineral resources is remote. There are vast quantities of raw materials potentially available from the earth's surface. It has been noted, for example, that a single cubic mile of average crustal rock contains 1 billion tons of aluminum, over 500 million tons of iron, 1 million tons of zinc, and 600,000 tons of copper. However, mining common crustal rock currently seems improbable because locally concentrated mineral deposits still exist. The degree of mineral concentration is a function of elemental crustal abundance and various physical and chemical processes. Economic and technological factors then combine to deter-

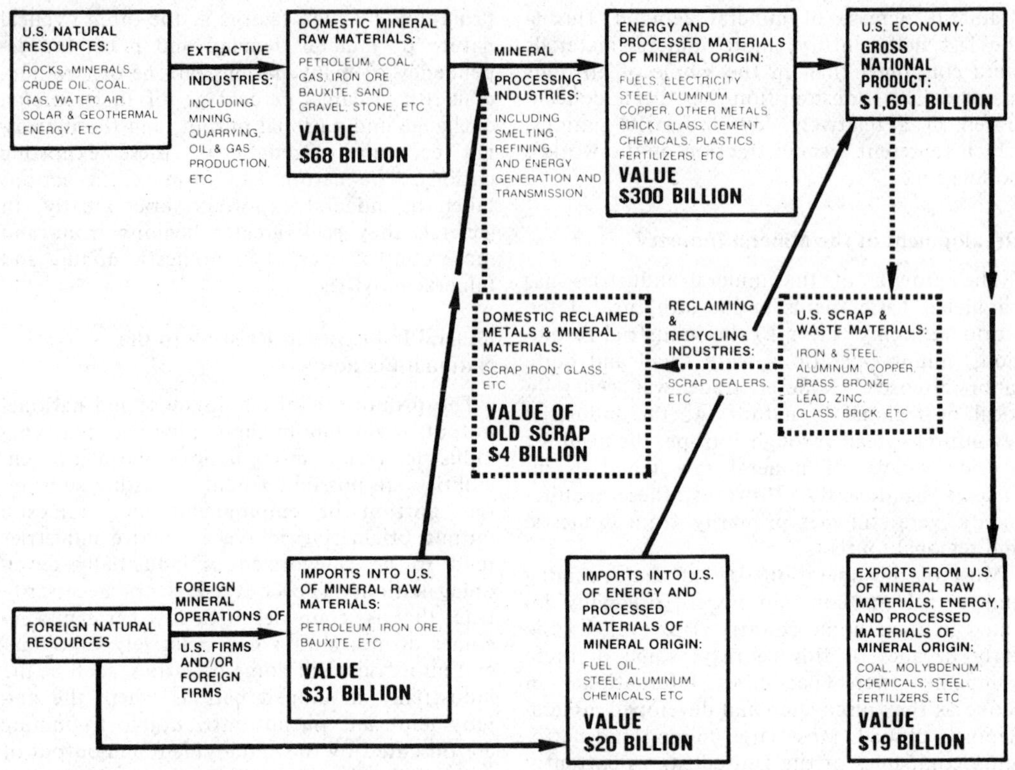

ECONOMIC ACTIVITY[1]	1971	1972	1973	1974	1975	1976
DOMESTIC MINERALS:						
RAW MATERIALS	31	32	37	55	63	68
RECLAIMED MATERIALS	2	2	4	4	3	4
IMPORTED MINERALS:						
RAW MATERIALS	4	4	7	20	21	31
ENERGY AND PROCESSED MATERIALS	6	10	12	22	19	20
DOMESTIC ENERGY & PROCESSED MATERIAL OF MINERAL ORIGIN	145	170	200	255	265	300
U.S. GROSS NATIONAL PRODUCT	1,063	1,171	1,306	1,415	1,516	1,691
EXPORTS OF MINERAL RAW MATERIAL, ENERGY AND PROCESSED MATERIALS	7	8	11	18	18	19

[1] ESTIMATED IN BILLIONS OF DOLLARS

FIGURE 1. Role of minerals in the US economy (1976 est.) (Bureau of Mines, U.S. Department of the Interior. Based in part on U.S. Department of Commerce data).

mine whether a known deposit can be profitably developed. Technological innovations have resulted in the mining of successively leaner ores. For example, the average grade of copper ore mined in the United States declined from approximately 3% in 1880 to 0.6% in 1970. The mining of lower-grade mineral deposits also illustrates that reserves are not fixed or static. Technological variables that lower costs or economic factors that raise prices allow the reclassification of previously uneconomic resources to reserves.

The resource availability of some fossil fuels, especially oil and gas, is not as optimistic as that of the inorganic mineral resources. There are large quantities of coal and oil shale and some deposits of tar sands in the United States. In addition, the technology exists for the conversion of these raw materials into synthetic fuels. However, the earth's deposits of fossil fuels are limited in amount and essentially nonrenewable.

The distinction between mineral resources and reserves is important when dealing with questions of supply estimation. Mineral resources, as defined by the Bureau of Mines and Geological Survey, U.S. Department of the Interior, are naturally occurring solids, liquids, or gases, discovered or only surmised, that might become economic sources of mineral raw materials. Mineral reserves are that portion of "mineral resources" that has actually been identified and that can be legally and economically extracted. The interrelationship between these two terms is illustrated in Fig. 2. The United States has a broad resource base and significant reserves of many commodities. Current US reserves of major mineral commodities are shown in Table 1.

Production

A large variety of minerals is produced in each of the six major land areas of the world, but with regional differences as to the quantity produced. The major industrial regions of the world, North America and Europe (including the USSR), produce a considerable portion of nearly every item listed in Table 2, thus fulfilling an important part of their own raw material requirements. The contribution of the other areas to world mineral production is significant and has increased as a result of increased domestic raw-material requirements and the enhanced ability to compete in international markets. The former is due to industrial development, and the latter to transportation advances and the depletion of commercial deposits in other areas.

The secondary production of metals makes a major contribution to US mineral supplies, as shown in Fig. 3. The increase in secondary recovery of mineral products due to economic, health, and aesthetic considerations has alleviated somewhat the dependence upon primary mineral sources.

U.S. Foreign Trade in Minerals

Mineral deposits are distributed without regard to national boundaries and markets. The

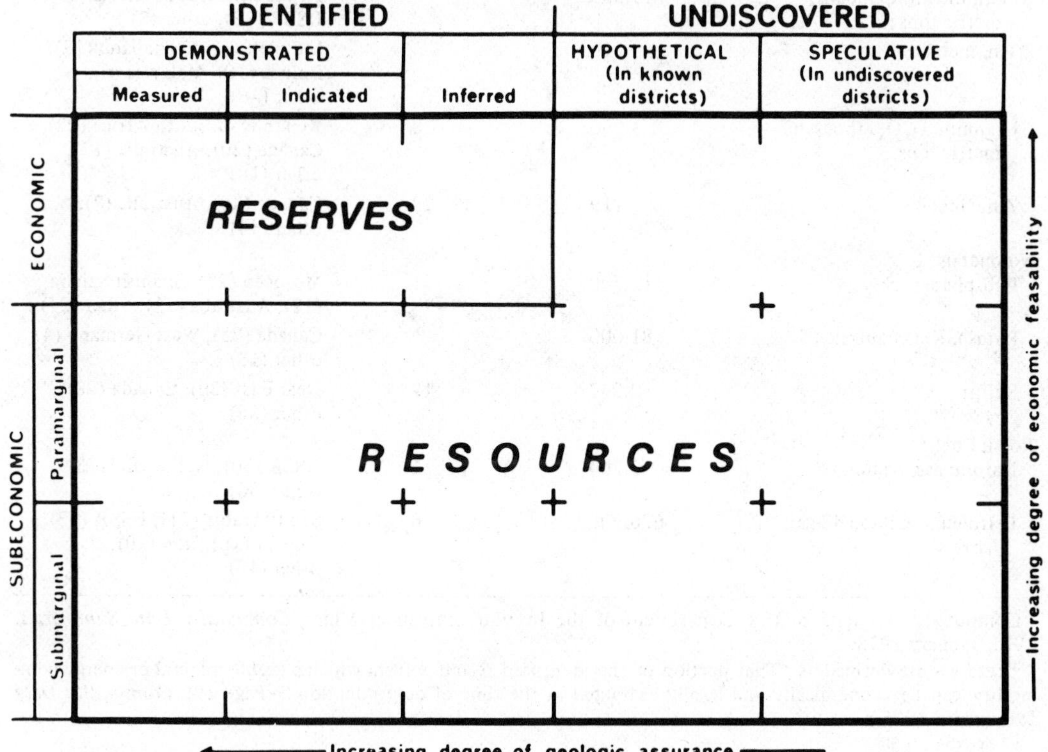

FIGURE 2. Definition of *reserves* and *resources* (Bureau of Mines and Geological Survey, U.S. Department of the Interior).

TABLE 1. Reserves of Selected Mineral Commodities, United States and Other World

Commodity	World Reserves[a] (million metric tons unless otherwise stated)	United States (percentage)	Other Countries (percentage)
Metals			
Bauxite	15,750	b	Australia (32), Guinea (23), Jamaica (6), other (39)
Chromite	1,730	–	Republic of South Africa (65), Southern Rohdesia (32), other (3)
Cobalt, metal	2.5	–	Zaire (28), New Caledonia and Australia (27), Zambia (14), other (31)
Copper, content	390	21	Chile (16), Canada (9), Peru (7), Zambia (7), other (40)
Iron, recoverable	88,000	2	Brazil (15), Canada (12), Australia (10), other (61)
Lead, metal	145	37	Australia (12), Canada (10), Mexico (3), other (38)
Manganese, gross weight	5,440	–	Republic of South Africa (37), Australia (6), Gabon (4) other (53)
Mercury, thousand 76-lb flasks	5,300	8	Spain (38), Yugoslavia (9), Italy (8), other (37)
Molybdenum, metal[c]	5	73	Chile (16), Canada (9), other (2)
Nickel, metal	44	b	New Caledonia (31), Canada (16), Cuba (9), other (44)
Platinum-group metals, metric tons	19,300	b	Republic of South Africa (65), USSR (32), other (3)
Tin, metal	10	b	Indonesia (24), Thailand (12), Bolivia (10), Malaysia (8), other (46)
Uranium, U_3O_8, thousand metric tons	1,090	28	Republic of South Africa (22), Canada (20), Australia (12), other (18)
Zinc, metal	119	23	Canada (26), Australia (7), other (44)
Nonmetals:			
Phosphate rock	12,700	21	Morocco (29), Spanish Sahara (12), Australia (11), other (27)
Potash, K_2O equivalent	81,000	b	Canada (83), West Germany (4) other (13)
Sulfur	2,030	12	Near East (30), Canada (20), other (38)
Fossil Fuels:			
Natural gas, trillion ft^3	2,200	11	USSR (30), Netherlands (3), other (56)
Petroleum, million 42-gal barrels	626,900	6	Saudi Arabia (21), USSR (13), Kuwait (10), Iran (10), other (40)

Computed from data in U.S. Department of the Interior, Bureau of Mines, *Commodity Data Summaries, 1975*, January 1975.

[a]Reserves are defined as "That portion of the identified resource from which a usable mineral or energy commodity can be economically and legally extracted at the time of determination." (Page 192 *Commodity Data Summaries, 1975*.)

[b]Less than 0.5%.

[c]Centrally planned economies not included.

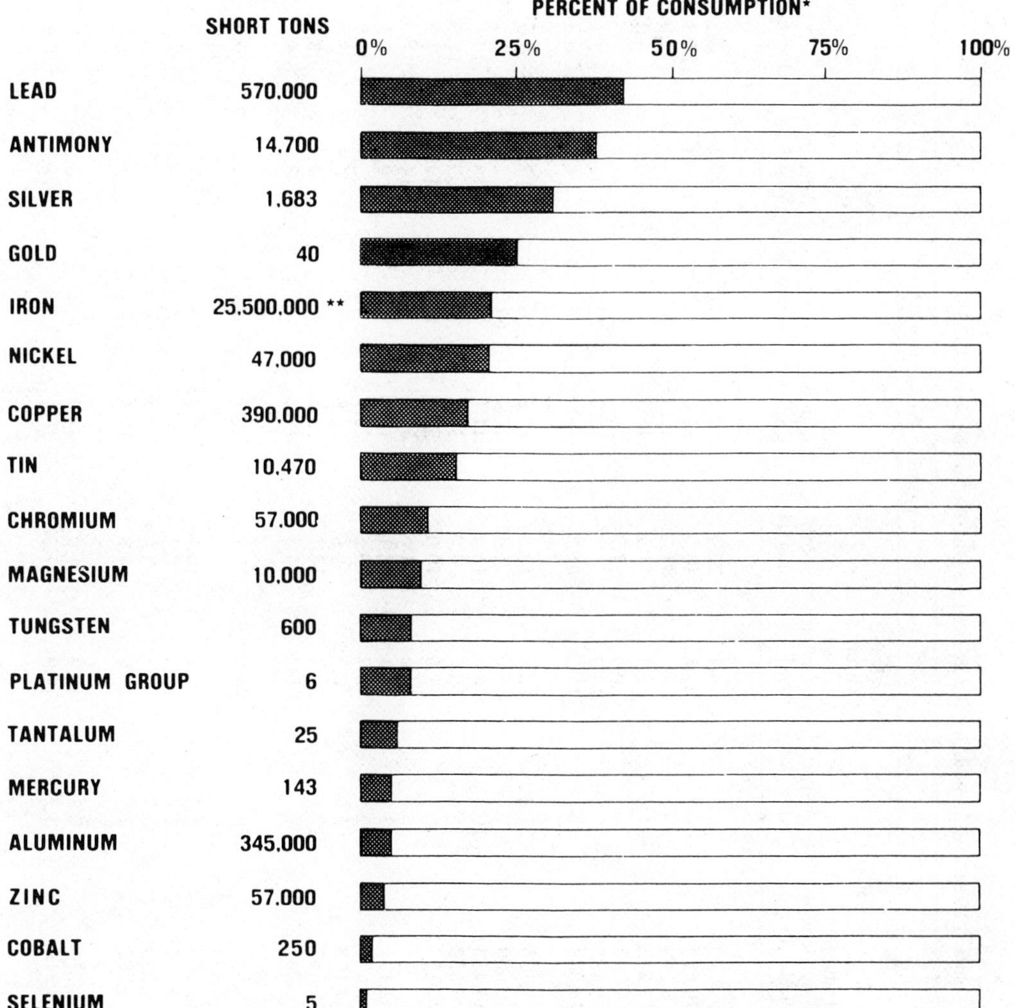

FIGURE 3. Old scrap reclaimed in the United States (1976) (U.S. Department of the Interior, Bureau of Mines).

quality of the deposit and the transportation costs to the market generally determine which deposits will be developed and to which destinations the output will be shipped. Although transportation costs can be a major impediment to mineral trade, tariffs, nontariff barriers, and export controls may also affect the volume and direction of trade flows.

Figure 4 shows that the United States, in its capacity as the world's leading industrial nation, imports minerals from all areas of the world. However, most US imports in 1975 were from other countries of the Western Hemisphere. This situation was due to the existence of mineral deposits in these countries, transportation advantages due to physical proximity, and the traditional economic ties maintained with many nations of this region.

TABLE 2. World Production of Selected Mineral Commodities and Percentage Distribution by Area, 1973[p]

Commodity	World Production (thousand metric tons unless otherwise stated)	Percentage Produced by Area						
		United States	North and Central America (excluding US)	South America	Europe (including USSR)	Africa	Asia and Near East	Oceania
Metals:								
Antimony, content of ore	70	0.7	6.1	22.2	17.1	24.3	27.4	2.2
Bauxite	70,694	2.7	21.8	16.9	21.6	5.7	6.1	25.2
Beryllium, *beryl* concentrate (metric tons)[b]	3,589	W	–	41.2	40.5	16.3	–	2.0
Cadmium, metal (metric tons)	17,013	21.6	6.4	1.2	44.5	2.3	20.0	4.0
Chromite (gross weight)[b]	6,701	–	0.3	1.9	50.1	27.4	20.3	–
Cobalt, content of ore (metric tons)	25,638	–	13.2	–	11.7	72.0	NA	3.1
Columbium-tantalum concentrates, gross weight (metric tons)[a]	24,039	–	11.2	81.3	–	6.2	0.5	0.8
Copper, content of ore	7,136	21.8	12.6	13.6	17.8	20.9	7.7	5.6
Gold, content of ore (kilograms)	1,340,000	2.7	5.0	1.4	17.4	67.1	2.7	3.7
Iron ore, gross weight	864,463	10.3	6.4	11.4	41.1	7.1	13.6	10.1
Lead, content of ore	3,532	15.5	16.6	8.0	32.4	6.1	9.9	11.5
Manganese ore, gross weight	22,153	c	1.6	10.0	37.3	31.2	12.9	7.0
Mercury (thousand 76-lb flasks)	276	0.8	14.6	1.5	63.3	5.1	14.7	c
Molybdenum, content of ore (metric tons)	82,191	63.9	15.1	8.1	10.7	–	2.1	0.1
Nickel, content of ore[b]	676	2.5	45.7	0.5	23.1	4.6	3.1	20.5
Platinum-group metals (kilograms)[b]	134,178	0.5	6.7	0.6	56.8	34.9	0.5	c
Silver, content of ore (kilograms)	9,558,387	12.3	29.4	18.1	23.3	4.2	4.7	8.0
Tin, content of ore	237	W	0.2	14.5	15.1	7.2	58.5	4.5
Titanium concentrates:								
Ilmenite, gross weight[a]	3,567	20.4	24.0	0.1	25.7	–	9.3	20.5
Rutile, gross weight[a]	334	–	–	c	–	–	1.7	98.3
Tungsten, content of ore (metric tons)	38,365	8.2	6.7	10.1	25.7	3.3	42.7	3.3
Uranium oxide, U_3O_8 (metric tons)[a]	23,154	51.8	18.3	0.2	9.5	20.2	NA	NA
Vanadium, content of ore (metric tons)[a]	19,223	20.7	–	5.0	28.2	46.1	–	–
Zinc, content of ore	5,703	7.6	27.0	9.3	31.4	4.6	11.7	8.4
Nonmetals:								
Asbestos	4,171	3.3	42.9	1.0	34.9	10.6	6.5	0.8
Barite	4,316	23.2	7.9	7.2	38.7	3.8	18.8	0.4

TABLE 2. (Continued)

Commodity	World Production (thousand metric tons unless otherwise stated)	Percentage Produced by Area: United States	North and Central America (excluding US)	South America	Europe (including USSR)	Africa	Asia and Near East	Oceania
Diamond:[b]								
Gem (thousand carats)	12,560	–	–	3.4	15.1	81.3	0.2	–
Industrial (thousand carats)	31,167	–	–	2.3	24.4	73.3	c	–
Feldspar[b]	2,594	27.7	3.7	3.8	57.5	1.6	5.6	0.1
Fluorspar (*fluorite*)	4,495	5.0	27.3	2.8	41.6	6.3	17.0	c
Graphite	370	W	17.7	1.1	34.6	4.1	42.5	–
Gypsum	60,575	20.3	16.0	1.8	48.2	2.3	9.7	1.7
Mica[a]	222	72.4	0.3	1.8	5.1	3.2	17.2	–
Peat, agricultural and fuel[b]	95,475	0.6	0.4	c	98.9	–	0.1	–
Phosphate rock	99,995	38.2	0.2	0.3	23.1	28.8	4.9	4.5
Potash, K_2O equivalent (marketable)	21,564	10.9	18.6	0.1	65.3	1.2	3.9	–
Pyrite, including cupreous, gross weight	22,110	2.6	0.6	–	71.1	4.8	20.0	0.9
Salt	150,749	26.5	7.3	3.3	38.0	1.4	20.7	2.8
Sulfur:								
Native, including Frasch	15,897	48.6	9.7	1.0	37.3	–	3.4	–
By-product elemental	16,670	14.7	45.2	c	27.0	0.3	10.8	2.0
Talc, soapstone, and *pyrophyllite*	5,232	21.6	2.0	2.1	24.8	0.3	48.1	1.1
Fossil fuels:								
Coal:								
Anthracite and bituminous (million metric tons)	2,340	22.7	0.9	0.3	45.1	2.8	25.5	2.7
Lignite (million metric tons)	820	1.6	0.4	–	93.7	–	1.3	3.0
Gas, natural, marketed (billion cubic feet)	44,862	50.5	8.3	2.2	33.9	0.8	4.0	0.3
Petroleum, crude (million barrels)	20,361	16.5	4.3	8.1	16.5	10.8	43.1	0.5

Derived from data in U.S. Department of the Interior, Bureau of Mines, "Minerals in the World Economy," *Minerals Yearbook,* Volume III, 1973.
[p]Preliminary.
NA Not available.
W Withheld to avoid disclosing individual company confidential data. Data are not included in the total from which percentages have been calculated.
[a]Excludes production in centrally controlled economies.
[b]Excludes production in centrally controlled economies of Asia.
[c]Less than 0.05%.

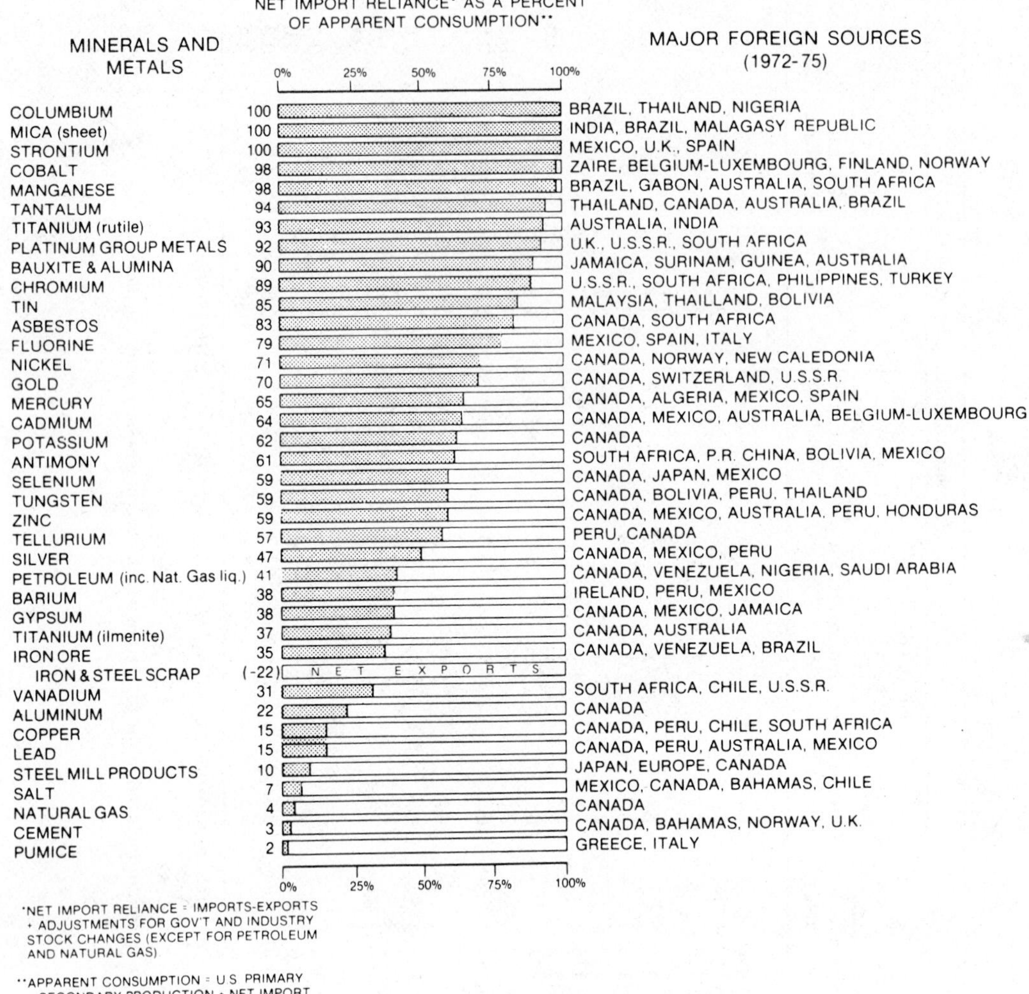

FIGURE 4. Imports that supplied a significant percentage of US minerals and metals consumption in 1976 (Bureau of Mines, U.S. Department of the Interior; import-export data from Bureau of the Census).

Most US mineral exports are in the form of processed commodities. However, as Fig. 5 illustrates, the value added by processing does not compensate for the large quantity of imported minerals. Consequently, the United States has a net deficit in the trade balance of raw and processed minerals.

Consumption

Table 3, which shows US consumption of selected mineral commodities and their end uses, gives some indication of the role played by minerals in meeting the consumptive needs of society. These data for the United States reflect the complexity and interdependence of a modern industrial society. Modern man has been fortunate to have had available at reasonable costs such a large number of materials to draw upon in support of his efforts toward attaining a higher standard of living. The unique characteristics (hardness, malleability, etc.) of the different materials and combinations of materials allow the production of many items that contribute to society's well-being. The ubiquitous nature of mineral materials in our society often goes unrecognized because the processed or finished product may bear little

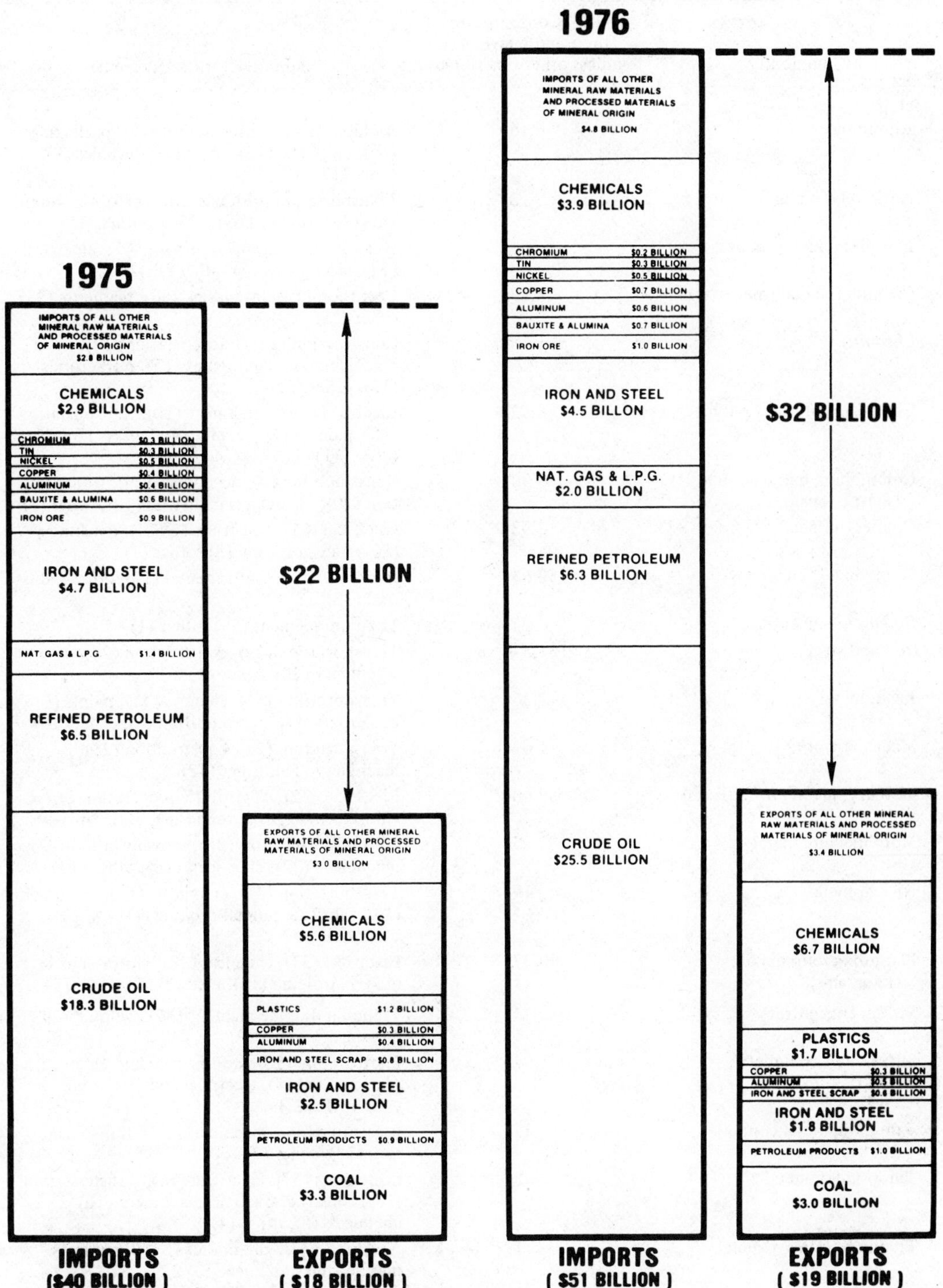

FIGURE 5. US imports and exports of raw and processed minerals for 1975 and 1976 (Bureau of Mines, U.S. Department of the Interior; data from Bureau of the Census).

TABLE 3. U.S. Consumption of Selected Mineral Commodities and End Use, 1974[p]

Commodity	U.S. Consumption (thousand metric tons unless otherwise stated)	End or Sector Use (percent)
Metals:		
Aluminum	6,169	Building (27), transportation (19), packaging (15), electrical (13), consumer durables (9), other (17)
Antimony, metal	38	Transportation (48), chemicals (16), fire retardants (12), rubber products (9), other (15)
Beryllium, metal (metric tons)	318	Nuclear reactors and aerospace (47), electrical (29), electronic (13), other (11)
Cadmium, metal (metric tons)	6,299	Corrosion-resistant plating (50), pigments (25), other (25)
Chromite	1,288	Construction (23), transportation (18), machinery and equipment (15), refractories (13), other (31)
Cobalt, metal (metric tons)	8,528	Electrical (29), machinery (20), transportation (18), paints (12), ceramics and glass (10), other (11)
Columbium, metal content (metric tons)	907	Construction (40), machinery (20), transportation (20), oil and gas industries (18), other (2)
Copper, refined	2,087	Electrical (61), construction (14), machinery (10), transportation (8), other (7)
Gold, metal (kilograms)	130,633	Jewelry and arts (49), dental (12), other industrial (39)
Ilmenite, concentrate[a]	953	Titanium pigment (99), other (1)
Iron and steel[b]	131,544	Transportation (29), construction (28), machinery (20), other (23)
Lead, metal	1,361	Transportation (64), electrical (8), paints (7) construction (6), ammunition (5), other (10)
Manganese ore	1,634	Transportation (22), construction (21), machinery (15), other (42)
Mercury, metal (76-lb flasks)	59,600	Electrical (32), electrolytic preparation of chlorine and caustic soda (28), other (40)
Molybdenum, metal	33	Oil and gas industry (28), transportation (27), machinery (22), chemicals (10), other (13)
Nickel, metal	191	Transportation (21), chemicals (15), electrical (13), fabricated metal products (10), other (41)
Platinum-group metals (kilograms)	60,278	Electrical (33), chemical (28), motor vehicle (13), petroleum (10), medical (7), other (9)
Rutile, concentrate	254	Titanium dioxide pigment (86), welding-rod coating (5), other (9)
Silver, industrial metal (metric tons)	5,526	Photography (26), sterlingware and electroplated ware (25), electrical and electronic (22) other (27)
Tantalum, metal content (metric tons)	424	Electronic (64), machinery (24), transportation (10), other (2)
Tin, primary metal	57	Containers (37), electrical (16), transportation (13), construction (12), machinery (10), chemicals (8), other (4)
Titanium, primary metal	19	Aircraft, space, and missile (87); chemical, marine, and ordnance (13)
Tungsten, metal (metric tons)	7,311	Machinery (74), transportation (11), lighting (7), electrical (4), chemicals (3), other (1)
Uranium, U_3O_8 (metric tons)	10,524	Nuclear fuel (100)
Vanadium, content (metric tons)	5,897	Machinery (27), transportation (26), construction (16), chemicals (6), other (25)
Zinc, slab	1,225	Construction (34), transportation (30), machinery and chemicals (13), electrical (11), other (12)

TABLE 3. (Continued)

Commodity	U.S. Consumption (thousand metric tons unless otherwise stated)	End or Sector Use (percent)
Nonmetals:		
Asbestos[c]	793	Flooring products (25), cement pipe (19), roofing products (10), friction products (9), other (37)
Barite, ground and crushed	1,542	Oil and gas well-drilling mud (83), other including paints, glass, rubber, and barium chemicals (17)
Clays[c]	53,307	Construction (79), refractories (8), paper products (4), other (9)
Diamond, industrial (million carats)	20	Transportation (20); shaping of stone, clay, and glass products and of abrasives (19); electrical (16); other (45)
Feldspar	662	Glass (50), ceramics (40), other (10)
Fluorspar (*fluorite*)	1,202	Steel fluxing (45), chemicals (33), aluminum (19), other (3)
Graphite, natural	62	Refractories (34), iron and steel (19), lubricants (6), clutch and brake linings (4), pencils (4), other (33)
Gypsum, crude[c]	18,039	Construction (91), agriculture (7), other (2)
Lime	20,229	Chemical and industrial (81), construction (10), refractories (8), agriculture (1)
Mica, natural:		
Scrap and flake	127	Joint cement (40), paint (30), roofing (20), rubber products (4), other (6)
Sheet[c]	3	Electrical and electronic (93), other (7)
Phosphate rock	29,938	Fertilizer (79), detergents (8), animal feed supplements (5), food products (4), other (4)
Potash, K_2O equivalent	5,552	Fertilizer (94), other (6)
Salt[c]	44,930	Chemicals (23), deicing (14), paper products (9), food products (7), other (47)
Sulfur	11,421	Fertilizer (55), synthetics, paper, paint, nonferrous metals, explosives (21), other (24)
Talc and related minerals	943	Ceramics (35), paints (18), paper (8), refractories (5), other (34)
Vermiculite, exfoliated	250	Concrete aggregate (30), insulation (28), plaster and cement aggregate (17), agriculture (17), other (8)
Fossil fuels:		
Coal:		
Anthracite[c]	4,717	Household and commercial (51), electric utilities (25), iron and steel (12), other (12)
Bituminous and lignite	489,888	Electric utilities (72), coke plants (17), other (11)
Natural gas (billion cubic feet)	22,500	Industrial (45), residential (22), commercial (10), electric utilities (15), pipeline fuel (3), other (5)
Petroleum products (miliion 42-gal barrels)	6,182	Transportation (50), household and commercial (20), industrial (18), electric utilities (9), other (3)

U.S. Department of the Interior, Bureau of Mines, *Commodity Data Summaries, 1975,* January 1975; U.S. Department of the Interior, Bureau of Mines, *Minerals in the U.S. Economy: Ten-Year Supply-Demand Profiles for Mineral and Fuel Commodities, 1975.*

[p] Preliminary.

[a] Includes titanium slag from Canada.

[b] Figure for consumption is production of raw steel. The end use distribution is domestic iron demand.

[c] End use breakdown is for 1973 and is computed from second data source.

resemblance to the material that is recovered from the earth.

THOMAS G. LANGTON
PHILLIP N. YASNOWSKY

Economists, U.S. Department of the Interior, Bureau of Mines. The statements and views contained in this paper are those of the authors, and should not be interpreted as representing the official position of the Bureau of Mines or of any other institution or government agency.

References

Brooks, David B., 1973. *Minerals: an Expanding or a Dwindling Resource*, Mineral Bulletin MR 134, Department of Energy, Mines and Resources, Ottawa, Canada.

Flawn, Peter T., 1966. *Mineral Resources*. New York: Rand McNally, 406p.

International Labour Office, 1974. *Yearbook of Labour Statistics, 1974*. Geneva.

McDivitt, James F., and Manners, Gerald, 1974. *Minerals and Men*. Baltimore: Johns Hopkins, 175p.

Mikesell, R., 1971. *Foreign Investment in the Petroleum and Mineral Industries*. Baltimore: Johns Hopkins, 459p.

National Academy of Sciences–National Research Council, Committee on Resources and Man, 1969. *Resources and Man*. 259p.

Sutulov, Alexander, 1972. *Minerals in World Affairs*. Salt Lake City: University of Utah.

United Nations, 1975. *Yearbook of National Accounts Statistics, 1973*, 3 vol. New York.

U.S. Department of the Interior, *Annual Report of the Secretary of the Interior Under the Mining and Minerals Policy Act of 1970*, annual.

U.S. Department of the Interior, Bureau of Mines, *Commodity Data Summaries, 1975*, January 1975, 193p.

U.S. Department of the Interior, *Minerals Yearbook*, Annual, 3 vols.

U.S. Department of the Interior, 1975. *Minerals in the U.S. Economy: Ten-Year Supply-Demand Profiles for Mineral and Fuel Commodities.*

Vernon, R., 1971. *Sovereignty at Bay*. New York: Basic Books, 326p.

Warren, Kenneth, 1973. *Mineral Resources*. New York: Halsted, 335p.

MINERAL AND ORE DEPOSITS

A "mineral" is a naturally occurring, homogeneous solid element or compound, of definite chemical composition, having an ordered atomic structure and, thus, in a crystalline condition. "Rocks," the solid materials of the earth, are aggregates of one or more minerals. "Ore" is a naturally occurring aggregate of minerals from which one or more metals may be extracted with profit or with hope of profit. This latter definition, being linked with economic factors, implies that what was not ore yesterday may become ore today—for example, as a result of the exhaustion of richer sources of a metal, or the development of large-scale, low-cost production methods; and what is of ore grade today may cease to be ore tomorrow—for example, as a result of a depression of market prices resulting from overproduction.

An "ore mineral" is a mineral that contains potentially valuable metal and is found in ore deposits. By original definition, that of the Arab alchemist Geber (Jabir ibn Hayyan), a "metal" is a miscible or fusible body that is extensible in all directions under the hammer (Russell, 1678). The six metals of antiquity were gold, silver, copper, iron, lead, and tin. Besides their ductility, all are of high density and all exhibit metallic luster. In modern chemical usage, however, the term metal is far more widely applied, being used for the electropositive elements generally. Many of these are of relatively low atomic weight (e.g., aluminum, magnesium); many are not ductile, but most exhibit metallic luster. Ore deposits, then, are worked for the recovery of these elements.

Another class of mineral deposits is worked for nonmetallic (i.e., electronegative) elements such as sulfur, phosphorous, fluorine, chlorine. In this class of nonmetallic deposits are also included deposits worked for compounds such as silica (SiO_2), *anhydrite* ($CaSO_4$), salt ($NaCl$), etc.

Metal Ores

Of the 96 elements, 27 are metallic constituents of ore deposits in major amount somewhere; at least 11 more are of minor importance. Table 1 summarizes the abundance of some of the most important metals in Henry S. Washington's "average igneous rock" approximately representative of the average composition of that part of the earth's crust accessible to man. The abundances, or clarkes, are compared with the minimum percentage of metal required to render a mineral deposit workable as ore under favorable conditions (Vernadsky, 1924). In all cases, mineral deposits are enriched relative to average crustal rock in the element sought. In the case of ores, the ore grade divided by clarke varies from 4 to 30,000, emphasizing that the deposits are special features of the earth's crust and limited in their distribution. This is illustrated by Fig. 1, which shows the distribution of lead mineralization in Australia which, although one of the world's leading producers of this metal, has only a minute fraction of its area underlain by lead deposits. Any ore deposit is the result of a combination of unusual geological processes, requiring a high degree of coincidence for its formation.

TABLE 1. Metals Concentrated in Ores

	Clarke (percent)	Minimum Ore (percent)	Concentration	Typical Minerals
Aluminum	8.13	30	4	*diaspore, boehmite* AlO(OH)
Titanium	0.44	1	3	*rutile* TiO_2, *ilmenite* $FeTiO_3$
Chromium	0.02	30	1500	*chromite* (Mg,Fe) Cr_2O_4
Manganese	0.1	35	350	*pyrolusite* MnO_2
Iron	5.0	30	6	*hematite* Fe_2O_3, *goethite* FeO(OH) *siderite* $FeCO_3$, *pyrite* FeS_2 *magnetite* Fe_3O_4
Cobalt	0.0023	b.p.		*skutterudite* $(Co,Ni)As_2$
Nickel	0.008	1.5	188	*pentlandite* $(Fe,Ni)_9S$
Copper	0.007	0.7	100	*chalcopyrite* $CuFeS_2$, *bornite* Cu_5FeS_4, *chalcocite* Cu_2S
Zinc	0.0132	4	300	*sphalerite* ZnS
Molybdenum	0.0015	1.5	1000	*molybdenite* MoS_2
Silver	0.00001	0.05	5000	*silver, Galena* (Pb,Ag)S, *argentite* Ag_2S
Cadmium	0.000015	b.p.		*sphalerite* (Zn,Cd)S
Tin	0.004	1	250	*cassiterite* SnO_2
Antimony	0.0001	3	30000	*stibnite* Sb_2S_3
Tungsten	0.0069	1	145	*wolframite* $(Fe,Mn)WO_4$
Platinum	0.0000005	0.0005	1000	*platinum* Pt
Gold	0.00000059	0.001	2000	*gold* Au, *calaverite* $AuTe_2$
Mercury	0.00005	0.5	10000	*cinnabar* HgS
Lead	0.0016	4	2500	*galena* PbS
Bismuth	0.00002	b.p.		*bismuthinite* B_2S_2
Uranium	0.0004	0.4	1000	*uraninite* UO_2

b.p. = by-product of mining for another metal.

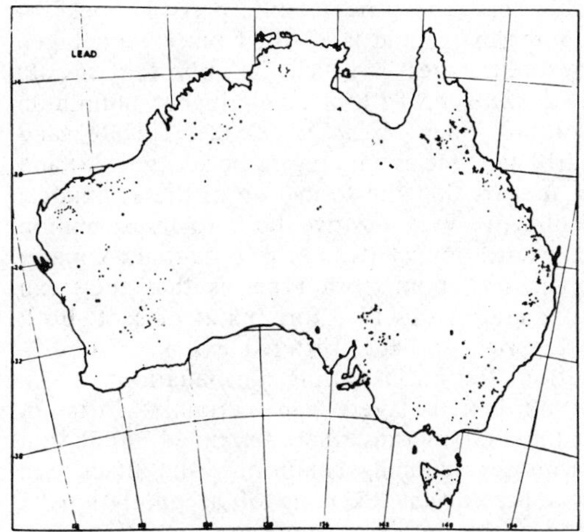

FIGURE 1. Distribution of lead mineralization in the Australian continent (from E. S. Hills, *Geology of Australian Ore Deposits*, 2nd ed.,The Australasian Institute of Mining and Metallurgy). Black dots indicate sites of lead deposits.

In Table 1, the commonest ore minerals are listed against the appropriate metals. A few elements, notably gold and platinum, are normally found in the native condition. Most metals, however, occur naturally as oxides (Ti, Cr, Fe, Mn, Sn, U), hydroxides (Al, Fe, Mn), or sulfides (Fe, Co, Ni, Cu, Zn, Ge, Mo, Ag, Cd, Sb, Hg, Pb, Bi). Carbonate and silicate ores are of main importance only in the case of iron. Some of the special processes leading to ore formation may be briefly considered here, but see also *Mineral Deposits: Classification*.

Magmatic Crystallization. Certain metallic oxides, notably *chromite, magnetite,* and *ilmenite,* are normal products of crystallization of mafic magmas at temperatures over 1000°C. *Chromite*, especially, separates early, and may accumulate under the influence of gravity into concentrated stratiform bands. Sulfides of iron, nickel, and copper are also known to separate from mafic magmas, probably as droplets immiscible with the magma; these may accumulate as heavy liquid differentiates, crystallizing after the associated silicate minerals. *Platinum* and the minor metals of the Pt groups are also differentiates of mafic or ultramafic magmas. Concentrations formed in this way accumulate in layers and pockets of ore minerals in mafic or ultramafic intrusions and may be classed as "syngenetic" since their origins are the same as those of the enclosing rocks (Beck, 1908). Some accumulative layers, though only a few inches or feet thick, are very extensive; the individual *chromite* seams in the Bushveld, north of Pretoria, South Africa, and in the "Great Dyke" of Zimbabwe can be followed for tens of miles along strike. Such concentrations may, however, migrate as a result of differential pressure, if the pressure is applied while interstitial liquid is still present. Similar considerations may explain some *magnetite* deposits, such as Kiruna in Sweden. Migrant differentiates may become "epigenetic" in their new positions, having an origin different from the rocks that enclose them.

Sedimentary Ores. The igneous syngenetic deposits mentioned above have their counterparts among the sedimentary rocks, where dense ore minerals—resistant to weathering processes—have accumulated under the control of gravity. Placer deposits (q.v.) formed in this way occur in river beds and along beaches of the present geomorphic cycle, and also formed during some geologically recent periods in the past. If the most popular hypothesis of the origin of the greatest gold deposits in the world, those of the Witwatersrand (South Africa, Fig. 2) is accepted, great placers also formed in Precambrian times; but some believe that the gold and uranium were introduced by solutions. The principal ore minerals concentrated in heavy mineral placers and seams in sediments, in addition to *gold*, are *platinum, cassiterite, rutile*, and *ilmenite*. *Magnetite* is concentrated in certain beach deposits in New Zealand; *chromite* placers are very rare. The bulk of the world's tin comes from alluvial workings in Malaysia and Thailand; most of its *rutile* from blacksand seams on the beaches (old and new) of Eastern Australia (see *Blacksand Minerals and Metals*). Up to the year 1860, when deep mining for *gold* began in Australia, it is likely that almost all the *gold* produced came from shallow alluvial deposits.

A second class of sedimentary syngenetic deposits must, however, be recognized—those where chemical precipitation, rather than density separation, was the concentrating process. To the chemical deposits belong the great iron ore deposits found in Precambrian rocks the world over, with the association *hematite*-chert or *magnetite*-chert as the commonest paragenesis, but including carbonate (*siderite*) and silicate (*greenalite, grunerite*) facies in addition to oxides. They are most valuable where residual enrichment by removal of silica has occurred. World reserves of these

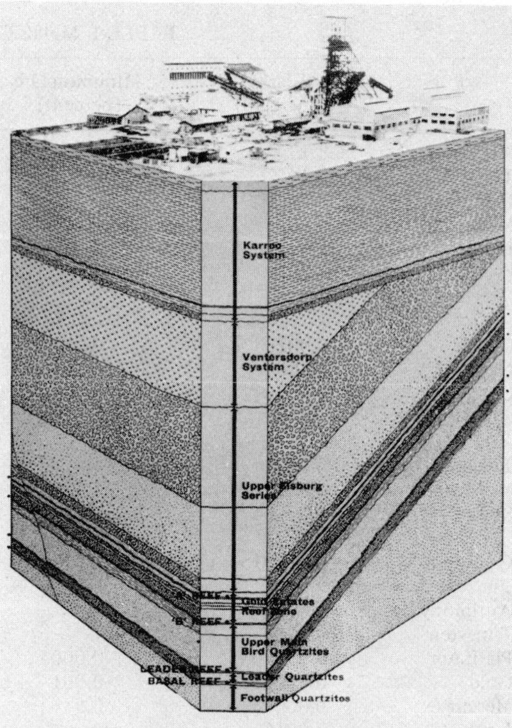

FIGURE 2. Mining inclined thin reefs of Witwatersrand type in the Orange Free State goldfield, South Africa (The Orange Free State Goldfield, 1957, privately published by Anglo-American Corporation Ltd., p. 14). Some geologists believe these to be fossil placers, others hydrothermally mineralized conglomerates.

ores amount to thousands of millions of tons. They have no strict counterparts in rocks of Cambrian age and later, their place being taken by oolitic ores in which the important silicate is *chamosite*, with a much higher aluminum content than *greenalite*. Oxide, silicate, and carbonate facies may again be recognized; and it appears that the formation of these chemical sediments was sensitive both to hydrogen-ion concentration and to redox potential. A major difficulty about both types is that they can nowhere be observed forming at present. Both categories probably suffered extensive modification after burial, during consolidation.

Sulfide-rich layers can certainly form in sedimentary basins where restricted circulation promotes reducing conditions; the Black Sea is a present-day example of accumulation of iron sulfide resulting from the action of anaerobic bacteria. There is equally no doubt that the reducing conditions accompanying black shale formation in some cases make it possible for abnormal amounts of the metals to accumulate in the shale. Some black shales, notably the

FIGURE 3. Outcrop of a mineral vein as pictured by Agricola (G. Bauer), *De Re Metallica,* 1551.

Kupferschiefer in the German Permian, carry enough metal to repay extraction. Whether sedimentary syngenesis is the whole explanation for these, and for other apparently concordant sulfide deposits such as the great copper deposits in shale and sandstone in Zambia and the limestones and dolomites carrying *galena* and blende in the southern Alps, in Poland, in the American Middle West, and in many other parts of the world, is a very controversial issue (Dunham, 1964; Nicolini, 1970).

Discordant Sulfide Ores. The normal occurrence of sulfide ores is, however, as discordant, nonstratified masses occupying fissure veins (Fig. 3) or other spaces produced during the postconsolidation tectonic deformation of the wall rocks as indicated in section of Butte, Montana, deposits (Fig. 4). Such deposits are unquestionably epigenetic, that is to say, of different origin from the enclosing rocks. Their mineralogy is usually complex, including in addition to sulfides, low *quartz* and silicates known from experimental studies to form at moderate or low temperatures, carbonates, *fluorite*, and *barite*. Various lines of evidence including fluid inclusions, the iron content of *sphalerite*, the presence or absence of *pyrrhotite*, and others show that temperatures of formation ranged from about 500°C to less than 100°C (see *Thermometry: Geologic*). These temperatures are far below the fusion temperatures of most of the minerals; and it has been clear, at least since Élie de Beaumont's time, that the metals were introduced by fluids, mainly aqueous solutions (de Beaumont, 1847). Evidence is slowly accumulating that in many cases these were brines more saline than sea water, and probably richer in potassium. The most significant evidence for this comes

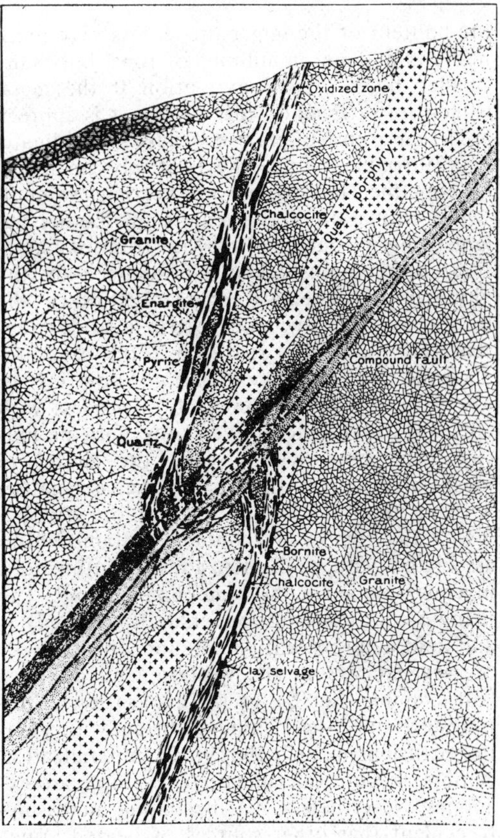

FIGURE 4. Cross section of a mineral vein that has invaded granite and a porphyry dyke, and that has subsequently been faulted (from Weed, 1912). The Anaconda Vein at Butte, Montana.

from deposits forming now at Salton Sea (Calif.), Cheleken (USSR), and in the Red Sea deeps, and also from the chemistry of fluid inclusions (Dunham, 1970).

In size, the epigenetic sulfide deposits form ore shoots that rarely exceed 2–3 km in length, and fail to exceed 1 km in vertical dimension except in a small minority of cases. Widths range from a meter or less to perhaps 30 m., except where extensive metasomatism of wall rocks has enabled broad replacements to form. Among the largest epigenetic deposits are the "porphyry coppers" where minute fracturing of felsic igneous rocks and adjacent wall rocks over substantial areas has permitted the ingress of sulfide-depositing solutions; ore reserves in these cases may reach hundreds of millions of tons, but the average grade is low, often in the range of 0.4–1% copper. In some of these, recovery of fractional percentages of molybdenum associated with the copper makes the operation economic; some also carry a little gold. In more concentrated sulfide deposits, the ore content of the larger ore shoots may reach millions or tens of millions of tons, but many smaller bodies repay extraction if the metal values are sufficiently high. When it is appreciated that 1 km^3 of granite weighs approximately 2700 million tons, the relatively small size of even the largest ore body becomes evident.

The formation of epigenetic ore bodies has largely been controlled by the structural situation, sometimes markedly influenced by the physical and chemical characteristics of the host rocks. The evidence often indicates that repeated tectonic reopening of fractures occurred during mineralization, pointing to association of the mineralization process with episodes of tectonic or even orogenetic periods (see *Vein Minerals*). In some, though by no means all instances, there is a spatial association with intrusive igneous rocks or (more strikingly) with zones of contact metamorphism associated with such intrusions, pointing to a juvenile source for the mineralizing fluids. The many associations of copper, lead, zinc, silver, and gold deposits with small monzonitic stocks of Nevadan age in the western United States (for example, Fig. 5), or the association of tin-tungsten-copper-lead mineralization with granite intrusions in such Hercynian mineralization centers as Cornwall and the Erzgebirge are examples. Nevertheless, it is becoming increasingly evident that other sources of heated brines need to be considered also, including ocean water circulated through submarine lava, connate water buried with sediments, and metamorphic water driven off during high grade metamorphism. Juvenile, oceanic, and meteoritic sources can be distinguished by the hydrogen and oxygen isotopes present in fluid inclusions (White, 1974). All three categories might derive their metal content from leaching of the rocks through which they passed, as Sandberger (1882) long ago suggested. Shallow meteoric groundwater is probably seldom an effective agent of primary concentration of sulfides owing to its oxygen content, but it is known to play an important part in some secondary enrichment processes both in the oxidation and sulfide-enrichment zones.

The world's resources of many of the less common metals are being used up at an alarming rate. More metal has been used in the past 30 years than in the whole of previous history (see *Mineral Industries*). Geological evidence, in many districts, favors the expectation that further concealed ore bodies will be discovered by modern methods of prospecting, including diamond drilling, geophysical methods—such as the electromagnetic, self-potential, and induced potential methods—and applied geochemical methods. The cost of prospecting is nevertheless rising steeply and may be expected to continue to do so.

Nonmetallic "Industrial-Mineral" Deposits

Considering, finally, the mineral deposits other than ores, a representative (but not comprehensive) list is assembled in Table 2. This omits rocks such as limestone, marble, sandstone, clay, which though quarried for various important purposes, especially for building and agriculture, are not normally thought of as mineral deposits. Those shown in the table again represent concentrations formed under various special geological conditions. Pegmatites, especially those associated with granites, carry concentrations of ***feldspar, mica, beryl, apatite*** and many rare minerals; very rarely they also carry *cassiterite* and sulfides (see *Pegmatite Minerals*). Three useful minerals, *fluorite, barite*, and *strontianite*, are derived from veins or replacements associated

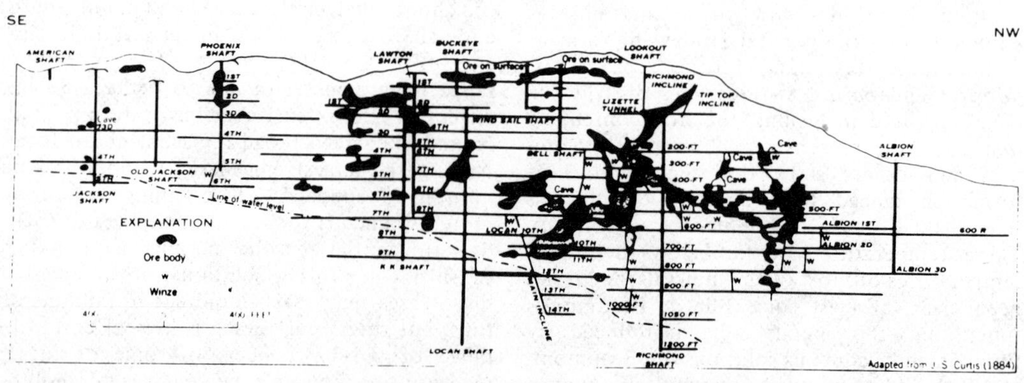

FIGURE 5. Longitudinal section showing irregular lead-silver-gold deposits in the form of epigenetic replacements of dolomite, Eureka District, Nevada (from Nolan, 1962).

TABLE 2. Mineral Deposits Other than Ores

Mineral	Use	Origin
Dolomite $CaMg(CO_3)_2$	refractories, chemicals	sedimentary, diagenetic
Gypsum $CaSO_4 \cdot 2H_2O$	plaster board	evaporite
Anhydrite $CaSO_4$	chemicals, fertilizers	evaporite
Halite NaCl	salt, chemicals	evaporite
Sylvite KCl	fertilizers	evaporite
Carnallite $KCl \cdot MgCl_2 \cdot 6H_2O$	fertilizers	evaporite
Barite $BaSO_4$	filler, paint, screen	gangue in ores
Fluorite CaF_2	chemicals, flux	gangue in ores
Celestite $SrSO_4$	chemicals	evaporite
Strontianite $SrCo_3$	chemicals	gangue in ores
Feldspars	ceramics	pegmatite
Spodumene $LiAlSi_2O_6$	chemicals	pegmatite
Muscovite (***mica***) $KAl_2(AlSi_3)O_{10}(OH)_2$	insulation	pegmatite
Diamond C	gems and industrial	kimberlite alluvial
Ruby, sapphire Al_2O_3	gems and industrial	metamorphic
Beryl, emerald $Be_3Al_2(SiO_3)_6$	gems and indus-	pegmatite
Zircon $ZrSiO_4$	refractories	beach alluvials
Monazite $(Ce,La)PO_4$	abrasives	beach alluvials
Apatite $Ca_3(PO_4)_2(OH,F)_2$	fertilizers	sedimentary, diagenetic, and pegmatite
Sulfur S	sulfuric acid	salt domes, "dirty" petroleum
Sillimanite, kyanite, andalusite Al_2SiO_5	refractories	metamorphic

with metalliferous ores. Evaporation in excess of inflow in large marine basins gives rise to the evaporite suite of minerals, including especially the calcium sulfate minerals, common salt, and the potash salts (Borchert and Muir, 1964); these are fundamental to the fertilizer and other branches of the chemical industry (see *Saline Minerals; Salt Economy*). Very great reserves of them, in thousands of millions of tons, remain, their magnitude being the result of their formation as beds in very large basins such as Texas-New Mexico, Manitoba, and the Zechstein (Stassfurt-Durham). The more important phosphate concentrations are specialized sedimentary deposits probably associated with rising cold currents. *Diamonds*, originally primarily from explosion of volcanic pipes of very deep-seated origin, are largely won from alluvials. *Sulfur* is obtained from the cap rocks of a limited number of salt domes, from replacements of limestone, and from deposits in modern volcanoes; at present, most *sulfur* is derived from petroleum production. Metamorphic rocks yield aluminum silicate minerals.

A great variety of processes have, thus, contributed to the formation of ores and mineral deposits. Few are adequately understood and there is scope for much future research, not only on the field relationship of known deposits, but on such fundamental issues as the chemistry and thermodynamics of ore formation, the nature and source of hydrothermal fluids, and the physical factors that have controlled their flow. The prediction of new deposits is a worthy aim for this work.

KINGSLEY DUNHAM

References

Bateman, A. M., 1950. *Economic Mineral Deposits*, 2nd ed. New York: Wiley, 916p.

Bauer, G. (Agricola), 1551. *De Re Metallica.* New York: Dover (transl., Hoover and Hoover), p. 63.

Beck, R., 1908. *Lehre von den Erzlagerstätten.* Berlin: Borntraeger, 724p.

Borchert, H., and Muir, R. O., 1964. *Salt Deposits.* London: Van Nostrand, 338p.

Boyle, R. W., 1965. *Geology, Geochemistry, and Origin of the Lead-Zinc-Silver Deposits of the Keno Hill-Galena Hill Area, Yukon Territory*, Geol. Surv. Canada, Dept. Mines and Tech. Surv., 302p.

de Beaumont, É., 1847. Note sur les émanations volcaniques et métallifères, *Bull. Soc. Géol. France*, **4**, 1249.

Dunham, K. C., 1964. Neputinist concepts in ore genesis, *Econ. Geol.*, **59**, 1-21.

Dunham, K. C., 1970. Mineralization by deep formation waters: a review, *Trans. Inst. Min. Metall.*, **79**, B 127-136.

Hills, E. S., 1953. Tectonic Setting of Australian Ore Deposits, in A. B. Edwards, ed., *Geology of Australian Ore Deposits*, Melbourne: Australasian Inst. Mining and Metallurgy, p. 51. (2nd ed., 1965.)

Hurlbut, C. S. Jr., 1969. *Minerals and Man.* London: Thames & Hudson, 303p.

Jancorić, S., 1967. *Wirtschaftsgeologie der Erze.* Vienna: Springer-Verlag, 347p.

Jenks, W. F., 1966. Some relations between Cenozoic Volcanism and ore deposition in northern Japan, *Trans. N.Y. Acad. Sci.*, **28**(4), 463-474.

Jones, W. R., 1963. *Minerals in Industry*, 4th ed. New York: Penguin, 149p.

Lindgren, W., 1933. *Mineral Deposits*, 4th ed. New York: McGraw-Hill, 930p.

Nicolini, P., 1970. *Gitologie des Concentrations Minérales Stratiformes.* Paris: Gauthier-Villars, 792p.

Nolan, T. B., 1962. The Eureka Mining District, Nevada, *U.S. Geol. Surv. Prof. Pap. 406*, 34p.

Park, C. F., Jr., and MacDiarmid, R. A., 1964. *Ore Deposits.* San Francisco: Freeman, 475p.

Petrascheck, W. E., 1961. *Lagerstättenlehre*, 2nd ed. Vienna: Springer-Verlag, 374p.

Raguin, E., 1961. *Géologie des Gites Mineraux*, 3rd ed. Paris: Masson et Cie, 686p.

Routhier, P., 1980. Où sont les métaux pour l'avenir? *Mem. du Bureau de Recherches Géologiques et Minières*, 105, 410p.

Russell, R., 1678. The Works of Gebar, the Most Famous Arabian Prince and Philosopher, in E. J. Holmyard, ed., *Alchemy.* London, 1957, 66-80.

Sandberger, F., 1882. *Untersuchungen über Erzgänge.* Weisbaden: Kreidel, 101p.

Schneiderhohn, H., 1955. Erzlagerstätten: Kurvorlesungen zur Einführung und zur Wiederholung, 3rd ed. Stuttgart: Fischer, 375p.
Smirnov, V. V., 1976. *Geology of Mineral Deposits.* Moscow: MIR Publishers, 620p.
Stanton, R. L., 1972. *Ore Petrology.* New York: McGraw-Hill, 713p.
Tatsumi, T. (ed.), 1970. *Volcanism and Ore Genesis.* Tokyo: University Press, 466p.
Vernadsky, V. I., 1924. *La Géochimie,* Vol. 1. Paris.
Washington, H. S., 1925. The chemical composition of the earth, *Amer. J. Sci.*, 9, 351-378.
Weed, W. H., 1912. Geology and ore deposits of the Butte District, Montana, *U.S. Geol. Surv. Prof. Pap. 74*, plate xiii.
White, D. E., 1974. Diverse origins of hydrothermal fluids, *Econ. Geol.*, **69**, 954-959.

Cross-references: *Blacksand Minerals: Fire Clays; Metallic Minerals; Mineral Deposit Classification; Mineral Industries; Native Elements and Alloys; Ore Microscopy; Pegmatite Minerals; Placer Deposits; Refractory Minerals; Saline Minerals; Salt Economy; Vein Minerals; Zeolites;* see also mineral glossary. Vol. IVA: *Earth's Crust Geochemistry; Hydrothermal Solutions–Sulfide Transport*; see also individual elements. Vols. XIX and XX.

MINERALOGIC PHASE RULE

Originally proposed by Goldschmidt (1911), the Mineralogic Phase Rule is a special boundary condition of the Gibbs Phase Rule, $F=C+2-P$ (see Vol. IVA, *Phase Equilibria*), which states that temperature and pressure are externally controlled and probably variable across the stability range of a particular assemblage. For $F=2$, $C=P$, i.e., the maximum number of phases present is equal to the number of chemical components. A common example is as follows: The common rock-forming oxides are SiO_2, Al_2O_3, MgO, FeO, CaO, Na_2O, and K_2O, which, if they are independent components, would produce a rock with a maximum of seven phases. However, (Si,Al) are interchangeable in ***plagioclase*** as are (Na,Ca). (Na,K) and (Mg,Fe) are interchangeable in ***alkali feldspar*** and the mafic minerals, respectively. This leads to a generalization that a maximum of four minerals using these components is found for a given rock type. Additional components (Ti,P) produce additional minerals (*ilmenite*, ***apatite***). The use of the Mineralogic Phase Rule is often misleading unless great care is taken. A full discussion of this limiting case of the Gibbs Phase Rule can be found in Korzhinskii (1959), Thompson (1970), and Weill and Fyfe (1967).

D. H. SPEIDEL

References

Goldschmidt, V. M., 1911. Die Kontaktmetamorphose im Kristianiagebeit, *Kristiana Vidensk Skr., I, Math-Nat. Kl., No. 11*, 483p.
Korzhinskii, D. S., 1959. *Physiochemical Basis of the Analysis of the Paragenesis of Minerals.* New York: Consultants Bureau, 142p.
Thompson, J. B., 1970. Geochemical reaction and open systems, *Geochim. Cosmochim. Acta*, **34**, 529-551.
Weill, D. F., and Fyfe, W. S., 1967. On equilibrium thermodynamics of open systems and the phase rule, *Geochim. Cosmochim. Acta*, **31**, 1167-1176.

MINERALOIDS—*See* OPAL; RESIN AND AMBER

MINERAL PROPERTIES

The properties of minerals are extremely important in understanding geological phenomena and in synthesizing new materials. Traditionally, the geologist has been concerned primarily with the properties of minerals for identification purposes. More recently, a greater emphasis has been placed on understanding the relationship of properties to the atomic structure of minerals and to the microstructure of polycrystalline materials.

Mineral properties change with temperature, pressure, and composition of the minerals (crystal solution).

The most important properties are chemical, mechanical, thermal, electrical, magnetic, and optical. These properties are related to the following geological problems: weathering, metamorphism, assimilation of materials by magmas, diastrophism, solidification of igneous rocks, and ore deposition.

In ceramics, a knowledge of mineral properties is essential to the development of new or improved materials having high mechanical strength and better resistance to high temperature, thermal shock, corrosion, or abrasion, and to gain insight into the manufacture of such materials.

Properties Generally Used for Identification

Color. Color depends upon the particular wavelengths of light that are reflected or transmitted by a mineral. All of the other wavelengths of light are absorbed by the mineral. Two extremes of color are white minerals, which reflect or transmit all colors of the spectrum, and black minerals which absorb all wavelengths. A green mineral reflects or transmits the green spectrum and absorbs the others (see *Color in Minerals*).

Luster. Luster is dependent upon the way light is reflected from the surface of a mineral. Many different types of luster are recognized: (1) metallic; (2) nonmetallic, (a) vitreous, (b) adamantine, (c) resinous, (d) greasy, (e) pearly, and (f) silky.

Optical. Optical properties are extremely useful for the identification of minerals. Much has been written about this subject and it is too broad to describe fully here (see *Minerals, Uniaxial and Biaxial; Optical Mineralogy*). Two major methods are used to observe substances through a light (generally polarized light) microscope: transmitted and(or) reflected light (see *Ore Microscopy*). Variations of light paths such as dark-field or phase-contrast illumination can also be employed for observing minerals. Substances to be studied can be small isolated grains, thin sections (transmitted light), or polished sections (reflected light).

Substances are classified under three groups, which depend upon the internal arrangement of the atoms: (1) an isotropic substance has only one refractive index; (2) a uniaxial substance has two significant refractive indices and (3) a biaxial mineral has three significant refractive indices.

Some of the other optical properties used to identify minerals are dispersion (q.v.), pleochroism, extinction angles, sign of elongation, interference figures, optic sign, and optic angle (2V).

Hardness. Hardness is the degree of resistance of a substance to abrasion (see *Mohs Scale of Hardness*). It is usually determined by observing the comparative ease or difficulty of indentation when one mineral is scratched by another mineral or substance of known hardness. In 1820, a standard scale of hardness was introduced by Mohs.

Microhardness. Microhardness data have been collected by metallurgists for years and, although not commonly used by most geologists, the information has been useful for industrial geologists. In either the Knoop or Vickers test, a *diamond* is loaded with a known weight and allowed to rest on a polished specimen for a short period of time. The length of the impression left by the *diamond* is measured and the microhardness is calculated in kilograms per square millimeter.

Specific Gravity. Specific gravity is the ratio of the weight of a certain volume of a substance to the weight of an equal volume of water. Specific gravities can be measured with any of the following devices: Jolly balance, Westphal balance, pycnometer, or liquids of high density (see *Density Measurements*).

Chemical Properties

A knowledge of chemical properties such as corrosion resistance, oxidation resistance, diffusion, reactivity, adsorption, absorption, catalysis, surface energies, and solubilities are important in understanding the origin of rocks and ore deposits. These properties are also useful in developing materials to be used in reactive environments.

Phase diagrams give some insight into the stability of minerals in a changing chemical environment (see Vol. IVA: *Phase Equilibria*). Generally, basic minerals [those containing alkali metal oxides and(or) alkaline earth oxides] are resistant to corrosion by basic slags and alkali-containing compounds. However, minerals rich in basic oxides are susceptible to attack by acids and siliceous liquids. Minerals containing ZrO_2, SiO_2, Al_2O_3, Cr_2O_3, ThO_2, HfO_2, TiO_2, and SnO_2 are more resistant to attack by acids or siliceous liquids, and less resistant to attack by bases and basic slags. In general, minerals have less resistance to corrosion in environments that produce low-melting eutectic phases. In environments where solid solutions form, they are more slowly corroded. In corrosion reactions, the rate of corrosion is affected by the viscosity of the liquids involved.

An increase in temperature greatly increases the rate of solution and corrosion. The rate of diffusion of the ions is also very important. In general, diffusion of ions into crystals will be greater if the crystal has large holes and channels (e.g., ***zeolites***, q.v.).

The resistance of materials to oxidation is important in many commercial applications. Two examples are the use of metals at high temperatures and the use of transition metal oxides during fluctuating temperatures and oxygen pressures. The mechanisms of oxidation, corrosion, and solution are interrelated. An important property of a material is whether the products of reaction adhere and protect or slough off. Aluminum forms a protecting, adhering oxide film; iron forms an oxide that sloughs off, exposing a fresh surface.

Oxidation, corrosion, solution, and hydration are important in forming ore deposits where secondary enrichment has taken place; these reactions are also important in the process of metamorphism, ore deposition, and the formation of sedimentary rocks.

Mechanical Properties

The following mechanical properties are important: yield point, elastic limit, tensile strength, elastic moduli, anelastic effects, creep, impact resistance, hardness, and fatigue. In understanding diastrophism, it is essential to know how these properties vary in minerals, and how the microstructures of the rocks containing these minerals affect these properties. It is also important to know how they vary with time, temperature, pressure, and composition.

Oxides are brittle materials having very little plastic deformation before fracturing (see *Rupture in Mineral Matter*). When loads are applied rapidly, they deform elastically and then fracture with a minimum of plastic deformation. The lack of ductility in ceramics has greatly limited their use commercially. The theoretical strength of minerals is much greater than measured values. This is due to imperfections in the crystal such as: dislocations, vacancies, and impurity concentrations (see *Defects in Crystals*). When loads are applied extremely slowly (e.g., during many geological processes), brittle materials can deform plastically by creep.

Flaws occurring at the surface (Griffith flaws) have the most pronounced effect in decreasing the strength of minerals. When the minerals are combined to form polycrystalline rocks or ceramics, their strength is even more greatly decreased. This is primarily due to the effect of grain boundaries (see *Jade*). The cohesive strength between crystals, across a grain boundary, is much less than the inherent strength of the crystal. In addition, dislocations pile up at grain boundaries causing crack formation (cleavage or conchoidal fractures). The tips of these microcracks are high-energy regions and points of stress concentration. Since the stress under these conditions is concentrated at the tips of the cracks, the cracks will propagate rapidly when the rocks are further stressed. Boundaries between anisotropic crystals are subjected to severe thermal stresses.

The strength of rocks and ceramics is greatly decreased by the occurrence of defects such as voids, cracks, foreign inclusions, and grain boundaries. The strength of rocks and ceramics can be increased by blunting the tips of incipient cracks, decreasing the grain size, decreasing the microporosity, and by decreasing the porosity and rounding the pores. Some recent investigations of ceramics indicate that their strength can be increased by solution and precipitation hardening. This retards the movement of dislocations decreasing their rate of buildup at grain boundaries and subsequent crack formation. Deformation can also occur in ceramics and in rocks by diffusional processes and by localized grain-boundary deformation.

There is a direct relationship between cohesive strength of a mineral and Young's Modulus. The cohesive strength is approximately equal to 10% of the elastic modulus (the ratio of stress to strain). Both of these properties are related to chemical bonding. The cohesive strength of crystals in a similar group of minerals (e.g., MgO, CaO, BaO) increases with decreasing interatomic distance. This is due to the fact that in covalent and(or) ionic crystals the primary cause of binding is an electrostatic attraction between oppositely charged ions. The electrostatic stress is approximately inversely proportional to the fourth power of the interatomic distance. The charge of the ions is also very important. Minerals that have small interatomic spacings and high valences tend to have high strength (e.g., *diamond* and *corundum*). The minerals with high cohesive strengths tend to have higher melting points and lower coefficients of expansion than similar minerals with low cohesive strengths.

In recent years, work done on the deformation of *periclase* and LiF crystals has shown that ionic crystals can be deformed plastically by movement along dislocations. Single crystals of these minerals have been permanently bent by this mechanism without rupture. However, ductility is greatly reduced in polycrystalline materials since dislocation buildup occurs at the grain boundaries. At high temperatures, movement can occur along more dislocation planes. Consequently, at high temperatures more plastic deformation through creep can be accommodated.

Thermal Properties

Thermal properties are extremely important in understanding and using minerals. In the natural geologic formation of inorganic materials, the thermal properties will affect the resulting phase assemblages of rocks and the resulting strength of these materials. Thermal properties also affect the degree of cracking that will occur upon solidification of igneous rocks and the amount of cracking that will take place in rocks when temperature fluctuations occur. Thermal expansion, thermal conductivity, heat capacity, thermal shock resistance, phase stability, and thermodynamic characteristics are some of the most important thermal properties.

Thermal Expansion. The coefficients of thermal expansion are related to the interionic distances. In minerals that are similar but have different interionic distances, the ones that have greater interionic separations will have a higher coefficient of expansion (e.g., NaI has a greater coefficient of expansion than NaF). Minerals that have similar crystal structure and a similar interionic spacing, but have ions with a greater charge, have a smaller coefficient of thermal expansion (e.g., NaF has a much greater coefficient of thermal expansion than CaF_2). The coefficient of thermal expansion is temperature dependent for many inorganic materials. Anisometric crystals have different coefficients of thermal expansion along their various axes. In some cases where this difference is great, the mineral may have a negative coefficient of expansion in one direction and the resulting volume expansion may be very low. Lithium aluminum silicates and *cordierite* are examples of low-volume expansion due to highly anisometric structures. These materials, because they have a very low coefficient of thermal expansion, are very resistant to thermal shock and are used where materials have to undergo rapid temperature fluctuations. Some silicates with open structures have the ability to absorb vibrational energy by transverse modes of vibration and by adjustment of bond angles that give them low coefficient of expansion values.

Thermal Conductivity. The thermal conductivity of a mineral is proportional to the rate at which heat is transferred through it under a particular temperature gradient. Solids can transfer thermal energy by three mechanisms: by electrons as in the metals; by thermal vibrations as in ionic or covalent solids; or by molecules as in organic solids. When thermal vibrations take place, the vibrations are anharmonic, which decreases the thermal conductivity. The conduction process in ionic-covalent minerals can be pictured as the movement of phonons, which is analogous with the movement of photons in radiation. When the phonons interact, due to the anharmonic nature of thermal vibrations, random scattering takes place. In general, thermal conductivity will be greater when the phonons have a greater mean-free path between collisions.

Crystalline materials have greater thermal conductivity at low temperatures than at high temperatures. This is due to an increase of the mean-free path resulting from the lower amplitude and greater harmonicity of the thermal vibrations at low temperatures. Less phonon scattering tends to take place when the cations approach the atomic weight of the anions. Consequently, BeO and MgO have greater thermal conductivity than NiO or CaO. Also, crystals with simple structure tend to have higher thermal conductivity than more complex structures (e.g., *spinel, mullite*, and ***olivines*** have lower thermal conductivity than Al_2O_3 or MgO).

The thermal conductivity of minerals is important in their use as either thermal insulators or thermal conductors. Also, materials with higher thermal conductivity will tend to have greater resistance to thermal spalling than similar minerals with lower thermal conductivity.

Phase Stability. Every mineral has stability limits and many of these can be found on phase diagrams (see Vol. IVA: *Phase Equilibria*).

Phase stabilities of individual minerals are also very important for the ceramist. For example, silica is employed in industry as an important ingredient in refractories. It undergoes phase transformations with temperature that affect its properties and wear rate. Spalling occurs with temperature fluctuation when SiO_2 undergoes transformations of *quartz*-to-*tridymite*-to-*cristobalite*.

M. J. Buerger classified transformations into two main groups: (1) "classical," and (2) "disordering." Classical transformations are further subdivided into: (a) "reconstructive transformation," where the polymorphic structures are so different that one structure must disintegrate into small units before these units join together to form a new structure; (b) "displacive transformation," which involves changes in structures by slight shifts and not a major rearrangement, the polymorphic structures of the two materials being quite similar; and (c) "semi-reconstructive transformation," which is intermediate between the displacive and reconstructive transformations. In the second major group, disordering transformation, the most important to the mineralogist is "substituted disorder" (see *Order-Disorder*). Here, transformations are due to solid solution of different atoms and to ordering or disordering of the atoms, which would cause a change in structure. Complete ordering would be unmixing of the atoms into two phases.

A. M. ALPER
R. C. DOMAN
R. N. McNALLY

References

Alper, A. M., 1970–1971. *High Temperature Oxides (Refractory Materials Series)*. New York: Academic Press.

Alper, A. M., 1970–1978. *Phase Diagrams: Materials Science and Technology (Refractory Materials Series)*. New York: Academic Press.

Azaroff, L. V., 1960. *Introduction to Solids*. New York: McGraw-Hill, 460p.

Bloss, F. D., 1960. *Introduction to the Methods of Optical Crystallography*. New York: Holt, Rinehart & Winston, 294p.

Buerger, M. J., 1948. The role of temperature in mineralogy, *Am. Mineralogist*, **33**, 101–121.

Campbell, I. E., 1956. *High Temperature Technology*. New York: Wiley, 526p.

Cortell, A. H., 1964. *The Mechanical Properties of Matter*. New York: Wiley, 421p.

Deer, W. A.; Howie, R. A.; and Zussman, J., 1962. *Rock-Forming Minerals*. London: Longmans, 5 vols.
Ford, W. E., 1951. *Danc's Textbook of Mineralogy*, 4th ed. New York: Wiley, 851p.
Kerr, P. F., 1959. *Optical Mineralogy*, 3rd ed. New York: McGraw-Hill, 442p.
Kingery, W. D., 1959. *Property Measurements at High Temperatures*. New York: Wiley, 415p.
Kingery, W. D., 1960. *Introduction to Ceramics*. New York: Wiley, 361p.
Klingsbury, C., 1963. *The Physics & Chemistry of Ceramics*. New York: Gordon and Breach, 361p.
Kraus, E. H.; Hunt, W. F.; and Ramsdell, L. S., 1959. *Mineralogy*. New York: McGraw-Hill, 686p.
Lynch, Charles T., 1974. *Handbook of Materials Science*. Cleveland, Ohio: CRC Press.
Newnham, R. E., 1975. *Structure-Property Relations*. New York and Heidelberg: Springer-Verlag, 234p.
Sinnott, M. J., 1961. *The Solid State for Engineers*. New York: Wiley, 522p.
Smoke, E. J., and Koenig, J. H., 1958. *Thermal Properties of Ceramics*. New Brunswick, N.J.: Rutgers, The State University, 53p.
Van Vlack, L. H., 1960. *Elements of Materials Science*. Reading, Mass.: Addison-Wesley, 528p.
Wahlstrom, E. E., 1951. *Optical Crystallography*. New York: Wiley, 247p.
Winchell, A. N., 1947. *Elements of Optical Mineralogy*, pt. I: *Principles and Methods*, 5th ed. New York: Wiley, 263p.

Cross-references: *Color in Minerals; Crystal Growth; Defects in Crystals; Metallurgy; Minerals, Uniaxial and Biaxial; Mohs Scale of Hardness; Optical Mineralogy; Order-Disorder; Polarization and Polarizing Microscope; Refractive Index; Refractory Minerals;* see also mineral glossary.

MINERAL RESOURCES—*See* BLACK-SAND MINERALS AND METALS; MINERAL DEPOSITS: CLASSIFICATION; MINERAL INDUSTRIES; MINERAL AND ORE DEPOSITS; PLACER DEPOSITS; SALT ECONOMY; SANDS, GLASS AND BUILDING; VEIN MINERALS. Vols. XIV, XIX, and XX

MINERALS, UNIAXIAL AND BIAXIAL

Light interacts with matter in several ways. A ray of light directed at an opaque substance, that is, having metallic luster, is reflected totally (see *Ore Microscopy*). Light incident upon nonopaque substances–gas, liquid, or solid–is reflected, refracted, and absorbed (see *Color in Minerals*).

Refractive Index

The velocity of light (v) while traveling through a substance is less than its velocity in vacuo ($c = 3 \times 10^{10}$ cm/sec). The refractive index (n) of a nonopaque material is the ratio of the velocity of light in vacuo to its velocity in a material ($n = c/v$). Most minerals have refractive indices (RI) between 1.4 and 2.5 (see *Refractive Index*). Isotropic materials (see *Isotropism*), which include gases, most liquids, most glasses, and minerals with isometric symmetry (see *Symmetry*) are characterized by a single RI (Table 1). Because of variations in chemical composition, some minerals exhibit a range of RI.

TABLE 1. Isotropic Minerals

Mineral	Refractive Index
Fluorite	1.433–1.435
Opal	1.435–1.460
Sodalite	1.483–1.487
Analcime	1.479–1.493
Leucite	1.508–1.511
Halite	1.544
Pyrope[a]	1.713
Spinel	1.719
Grossular[a]	1.734
Periclase	1.735–1.745
Spessartine[a]	1.799
Almandine[a]	1.830
Uvarovite[a]	1.863
Andradite[a]	1.887
Limonite	2.0–2.1
Diamond	2.4195
Sphalerite	2.37–2.50

Data from Phillips and Griffen, 1981.
[a]End-member refractive index for ***garnet*** group.

Anisotropism

When light is refracted doubly on passing through a mineral, the mineral is classified as anisotropic (see *Anisotropism*). Double refraction is best seen through a clear cleavage rhomb of the mineral *calcite*. When such a fragment is placed on a printed page, two images of whatever print is underneath can be seen.

The two images appear to be within the rhomb; hence the light carrying those images through the *calcite* has velocity less than that in air. The fact that the two images appear to be at different depths indicates that the light rays carrying the two images have different velocities. The two light rays are also polarized; a piece of polarizing plastic, such as is used in some sunglasses, cuts out (absorbs) one image or the other as it is rotated above the *calcite* rhomb.

One of the images appears to be directly over whatever print it is resting on and the other appears to be displaced to the side. As the rhomb is rotated about the first image, the second image rotates around the first. The ray carrying the first image is the ordinary ray, so called because it follows the laws of ordinary refraction–Snell's law ($n = \sin i/\sin r$, where i is the angle of incidence and r is the angle of refraction). The ray of light carrying the second image is extraordinary in that it does not follow this law.

All nonisometric, nonopaque minerals are optically anisotropic, but few have stronger double refraction than *calcite* and in most minerals the effect is much weaker. Anisotropic minerals are divided into two classes, based on their optical properties (see *Optical Mineralogy*).

Uniaxial Minerals

Minerals crystallizing with hexagonal, trigonal, or tetragonal symmetry are optically uniaxial and their optic axis coincides with the *c* crystallographic axis (see *Crystallography, Morphological*). Light traveling parallel to the optic axis is ordinary and behaves as though the mineral were isotropic in that direction. This RI is designated n_ω. Light traveling at right angles to the optic axis is resolved into two plane polarized rays, both of which are ordinary, but that are traveling at different velocities. One ray has RI n_ω and the other n_ϵ. If n_ϵ is greater than n_ω, the mineral is optically positive; if n_ϵ is less than n_ω, it is optically negative (Table 2). The numerical difference between n_ω and n_ϵ defines the birefringence of a uniaxial mineral.

Light traveling in any direction neither parallel to nor normal to the optic axis is resolved into two rays, one of which is ordinary with an RI of n_ω and the other extraordinary with an RI of $n_{\epsilon'}$. The value of $n_{\epsilon'}$ falls between those of n_ϵ and n_ω.

TABLE 2. Uniaxial Minerals

Mineral	n_ω	n_ϵ
Optically Positive Minerals		
Ice	1.30907	1.31052
Quartz	1.544	1.553
Brucite	1.559-1.590	1.580-1.600
Scheelite	1.9208	1.9375
Zircon	1.920-1.960	1.967-2.015
Cassiterite	1.990-2.010	2.091-2.100
Rutile	2.605-2.616	2.890-2.903
Optically Negative Minerals		
Hydrotalcite	1.511-1.531	1.495-1.529
Apophyllite	1.537-1.545	1.537-1.544
Vermiculite	1.545-1.583	1.525-1.564
Hydromuscovite	1.56 -1.60	1.53 -1.57
Beryl	1.568-1.602	1.563-1.594
Phlogopite	1.557-1.637	1.530-1.590
Paragonite	1.594-1.609	1.564-1.580
Dahllite (***apatite***)	1.603-1.628	1.598-1.619
Glauconite	1.61 -1.65	1.56 -1.61
Fluorapatite	1.633-1.650	1.629-1.646
Elbaite (***tourmaline***)	1.635-1.658	1.615-1.633
Biotite	1.605-1.696	1.565-1.625
Hydroxyapatite	1.643-1.658	1.637-1.654
Dravite (***tourmaline***)	1.631-1.658	1.610-1.633
Calcite	1.658	1.486
Stilpnomelane	1.576-1.745	1.543-1.634
Schorl (***tourmaline***)	1.658-1.698	1.633-1.675
Dolomite	1.679	1.500
Magnesite	1.700	1.509
Ankerite	1.690-1.750	1.510-1.548
Vesuvianite	1.702-1.752	1.698-1.746
Corundum	1.767-1.772	1.759-1.762
Rhodochrosite	1.816	1.597
Siderite	1.875	1.633
Anatase	2.561	2.488
Ilmenite	~2.7	
Hematite	3.15 -3.22	2.87 -2.94

Data from Phillips and Griffen, 1981.

Biaxial Minerals

Minerals crystallizing with orthorhombic, monoclinic, or triclinic symmetry are optically biaxial. Refractive indices of biaxial minerals are referred to three mutually orthogonal principal optic directions, *X*, *Y*, and *Z* (see *Optical Orientation*). Light traveling parallel to any one of these directions is resolved into two plane polarized rays, both of which are ordinary. For light traveling parallel to *Y*, one ray has the maximum RI (n_γ) for the mineral and the other has the minimum RI (n_α). (Alternatively, these are denoted n_Z and n_X.) The numerical difference between n_γ and n_α defines the birefringence for biaxial minerals.

Light traveling at right angles to *Y*, that is, parallel to the *XZ* plane, is resolved into two rays, one of which has an intermediate RI of n_β corresponding to the *Y* direction and the other an intermediate RI corresponding to the *XZ* plane. For two directions normal to *Y*, these two intermediate RIs are equal in value. Light traveling in either of these two directions behaves as though it were isotropic and, thereby, defines two optic axes. Light traveling in any direction other than that of the three principal axes and the optic axes is resolved into two plane polarized rays, both of which are extraordinary, having RIs designated $n_{\gamma'}$ and $n_{\alpha'}$.

If the value of n_β is closer to n_α than to n_γ, the mineral is characterized as biaxial positive (Table 3). If n_β is closer to n_γ than to n_α, it is biaxial negative (Table 4). Since optic properties of a mineral may vary substantially with the wavelength of light being refracted, the optic sign may be different for different wavelengths (see *Dispersion, Optical*).

Refractive indices of a mineral may be determined by matching the mineral powder with a liquid of known RI (see Hartshorne and Stuart, 1969; Kerr, 1977; Phillips, 1971; or Shelley, 1975, for methods). For gemstones, this deter-

TABLE 3. Biaxial Positive Minerals

Mineral	n_β	n_α	n_γ
Clinoptilolite	1.480	1.478	1.481
Mordenite	1.475–1.485	1.472–1.483	1.477–1.487
Natrolite	1.476–1.486	1.473–1.483	1.485–1.496
Heulandite	1.487–1.505	1.487–1.505	1.488–1.512
Phillipsite	1.484–1.509	1.483–1.504	1.486–1.514
Wairakite	~1.500	1.498	1.502
Gypsum	1.522–1.526	1.519–1.521	1.529–1.531
Albite	1.531–1.537	1.527–1.533	1.538–1.543
Oligoclase	1.537–1.541	1.533–1.537	1.543–1.547
Cordierite	1.524–1.574	1.522–1.560	1.527–1.578
Whewellite	1.553	1.489	1.649
Andesine	1.548–1.558	1.544–1.555	1.551–1.563
Labradorite	1.558–1.569	1.555–1.565	1.563–1.574
Dickite	1.560–1.566	1.558–1.564	1.563–1.571
Bytownite	1.569–1.572	1.565–1.567	1.574–1.577
Gibbsite	1.568–1.580	1.568–1.580	1.587–1.600
Anhydrite	1.576	1.570	1.614
Clinochlore	1.56 –1.60	1.56 –1.59	1.57 –1.60
Celsian	1.593–1.583	1.587–1.579	1.600–1.588
Amesite	1.59 –1.62	1.59 –1.60	1.61 –1.62
Pectolite	1.604–1.615	1.595–1.610	1.632–1.645
Topaz	1.609–1.631	1.606–1.630	1.616–1.638
Turquois	1.62	1.61	1.65
Celestite	1.623–1.624	1.621–1.622	1.630–1.633
Chondrodite	1.602–1.655	1.592–1.643	1.619–1.675
Prehnite	1.615–1.647	1.610–1.637	1.632–1.673
Anthophyllite	1.602–1.672	1.588–1.663	1.613–1.683
Humite	1.619–1.653	1.607–1.643	1.639–1.675
Barite	1.636–1.639	1.634–1.637	1.646–1.649
Pargasite	1.62 –1.66	1.61 –1.66	1.63 –1.67
Gedrite	1.622–1.676	1.610–1.667	1.632–1.684
Boehmite	1.65 –1.66	1.64 –1.65	1.65 –1.67
Sillimanite	1.654–1.670	1.653–1.661	1.669–1.684
Cummingtonite	1.640–1.676	1.628–1.658	1.652–1.692
Forsterite	1.651–1.673	1.635–1.653	1.670–1.690
Enstatite	1.655–1.669	1.654–1.664	1.665–1.675
Spodumene	1.655–1.670	1.648–1.663	1.662–1.679
Jadeite	1.659–1.674	1.654–1.665	1.667–1.688
Hornblende	1.618–1.714	1.610–1.700	1.630–1.730
Clinohumite	1.636–1.709	1.623–1.702	1.651–1.728
Lawsonite	1.672–1.676	1.665	1.684–1.686
Diopside	1.672–1.681	1.664–1.672	1.694–1.702
Clinoenstatite	1.654–1.706	1.651–1.705	1.660–1.726
Omphacite	1.670–1.700	1.662–1.691	1.688–1.718
Pigeonite	1.684–1.722	1.682–1.722	1.705–1.751
Zoisite	1.688–1.711	1.685–1.707	1.697–1.725
Riebeckite	1.690–1.712	1.690–1.702	1.702–1.719
Augite	1.672–1.741	1.671–1.735	1.703–1.761
Diaspore	1.705–1.725	1.682–1.706	1.730–1.752
Pumpellyite	1.675–1.754	1.674–1.748	1.688–1.764
Clinozoisite	1.707–1.725	1.703–1.715	1.709–1.734
Astrophyllite	1.703–1.746	1.678–1.740	1.733–1.765
Aegirine-augite	1.742–1.710	1.722–1.700	1.758–1.730
Hedenbergite	1.730–1.736	1.722–1.728	1.750–1.757
Chrysoberyl	1.734–1.749	1.732–1.747	1.741–1.758
Staurolite	1.740–1.754	1.736–1.747	1.745–1.762
Allanite	1.700–1.815	1.690–1.791	1.706–1.828
Piemontite	1.730–1.807	1.725–1.794	1.750–1.832
Monazite	1.777–1.801	1.774–1.800	1.828–1.851
Titanite (sphene)	1.870–2.034	1.840–1.950	1.943–2.110
Sulfur	2.038	1.958	2.245
Wolframite	2.22 –2.40	2.17 –2.31	2.30 –2.46
Brookite	2.584	2.583	2.700

Data from Phillips and Griffin, 1981.

mination is made with a refractometer (see *Gemology*).

KEITH FRYE

TABLE 4. Biaxial Negative Minerals

Mineral	n_β	n_α	n_γ
Natron	1.425	1.405	1.440
Borax	1.469	1.447	1.472
Trona	1.492	1.412–1.418	1.540–1.543
Nitratine	1.504–1.506	1.332–1.335	1.504–1.506
Laumontite	1.512–1.522	1.502–1.514	1.514–1.525
Microcline	1.518–1.519	1.514–1.516	1.521–1.522
Orthoclase	1.522–1.524	1.518–1.520	1.522–1.525
Sanidine	1.522–1.529	1.518–1.524	1.522–1.530
Anorthoclase	1.529–1.532	1.524–1.526	1.530–1.534
Albite	1.532–1.533	1.528–1.534	1.538–1.542
Oligoclase	1.541–1.548	1.537–1.544	1.547–1.551
Montmorillonite	1.50 –1.59	1.48 –1.57	1.50 –1.60
Chrysotile	1.530–1.564	1.529–1.559	1.537–1.567
Cordierite	1.524–1.574	1.522–1.560	1.527–1.578
Lizardite		1.538–1.554	1.546–1.560
Vermiculite	1.545–1.583	1.525–1.564	1.545–1.583
Kaolinite	1.559–1.569	1.553–1.565	1.560–1.570
Lepidolite	1.551–1.585	1.525–1.548	1.554–1.587
Antigorite	1.55 –1.59	1.54 –1.58	1.55 –1.59
Bytownite	1.572–1.580	1.567–1.573	1.577–1.585
Anorthite	1.580–1.585	1.573–1.577	1.585–1.590
Talc	1.575–1.594	1.538–1.550	1.565–1.600
Pyrophyllite	1.586–1.589	1.534–1.556	1.596–1.601
Illite	1.57 –1.61	1.54 –1.57	1.57 –1.61
Muscovite	1.582–1.610	1.552–1.574	1.587–1.610
Phlogopite	1.557–1.637	1.530–1.590	1.558–1.637
Paragonite	1.594–1.609	1.564–1.580	1.600–1.609
Tremolite	1.612–1.629	1.600–1.616	1.626–1.640
Margarite	1.625–1.648	1.595–1.638	1.627–1.650
Anthophyllite	1.602–1.672	1.588–1.663	1.613–1.683
Glaucophane	1.612–1.663	1.594–1.647	1.618–1.663
Andalusite	1.633–1.644	1.629–1.640	1.638–1.650
Wollastonite	1.627–1.659	1.615–1.646	1.629–1.662
Nimite	1.647	1.637	1.648
Gedrite	1.622–1.676	1.610–1.667	1.632–1.684
Biotite	1.605–1.696	1.565–1.625	1.605–1.696
Holmquistite	1.642–1.660	1.622–1.642	1.646–1.666
Actinolite	1.629–1.680	1.616–1.669	1.640–1.688
Chamosite	1.64 –1.67	1.64 –1.66	1.65 –1.67
Stilpnomelane	1.576–1.745	1.543–1.634	1.576–1.745
Monticellite	1.646–1.674	1.639–1.663	1.653–1.680
Strontianite	1.664	1.516	1.666
Magnesioarfved-sonite	1.652–1.677	1.638–1.672	1.654–1.684
Hornblende	1.618–1.714	1.610–1.700	1.630–1.730
Edenite-Ferroedenite	1.618–1.714	1.615–1.705	1.632–1.730
Witherite	1.676	1.529	1.677
Aragonite	1.680	1.530	1.685
Cummingtonite	1.676–1.684	1.658–1.664	1.692–1.700
Axinite	1.677–1.701	1.672–1.693	1.681–1.704
Arfvedsonite	1.677–1.710	1.672–1.700	1.684–1.715
Riebeckite	1.690–1.712	1.690–1.702	1.702–1.719
Hypersthene	1.695–1.722	1.686–1.710	1.699–1.725
Kaersutite	1.69 –1.74	1.67 –1.69	1.70 –1.77
Kyanite	1.720–1.725	1.712–1.718	1.727–1.734
Chloritoid	1.719–1.734	1.713–1.730	1.723–1.740
Epidote	1.725–1.784	1.715–1.751	1.734–1.797
Azurite	1.754–1.758	1.730	1.835–1.838
Allanite	1.700–1.815	1.690–1.791	1.706–1.828
Acmite (aegirine)	1.820–1.742	1.776–1.722	1.836–1.758
Tephroite	1.81	1.77	1.82
Fayalite	1.848–1.869	1.806–1.827	1.858–1.879
Malachite	1.875	1.655	1.909
Cerussite	2.074	1.803	2.076
Lepidocrocite	2.20	1.94	2.51
Goethite	2.393–2.409	2.260–2.275	2.398–2.515

Data from Phillips and Griffen, 1981.

References

Hartshorne, N. H., and Stuart, A., 1969. *Practical Optical Crystallography*, 2nd ed. New York: American Elsevier, 326p.

Kerr, P. F., 1977. *Optical Mineralogy*, 4th ed. New York: McGraw-Hill, 492p.

Phillips, W. R., 1971. *Mineral Optics: Principles and Techniques*. San Francisco: Freeman, 249p.

Phillips, W. R., and Griffen, D. T., 1981. *Optical*

Mineralogy: The Nonopaque Minerals. San Francisco: Freeman, 677p.
Shelley, D., 1975. *Manual of Optical Mineralogy*. Amsterdam: Elsevier, 239p.

Cross-references: *Anisotropism; Dispersion, Optical; Crystallography, Morphological; Isotropism; Optical Mineralogy; Symmetry.*

MOHS SCALE OF HARDNESS

The Mohs scale was devised by the Austrian mineralogist Frederick Mohs in 1820 for measuring hardness in minerals as a diagnostic property. It is based on the definition of hardness as resistance to scratching and defined by the use of ten common minerals as standards, each of which can scratch the mineral below it in hardness and can be scratched by the mineral above it. The Mohs hardness is an approximation to a simple function of indentation hardness (Fig. 1), which is itself related to the "plastic yield stress" and the "tensile strength" of the material. The standard abbreviation for hardness is H.

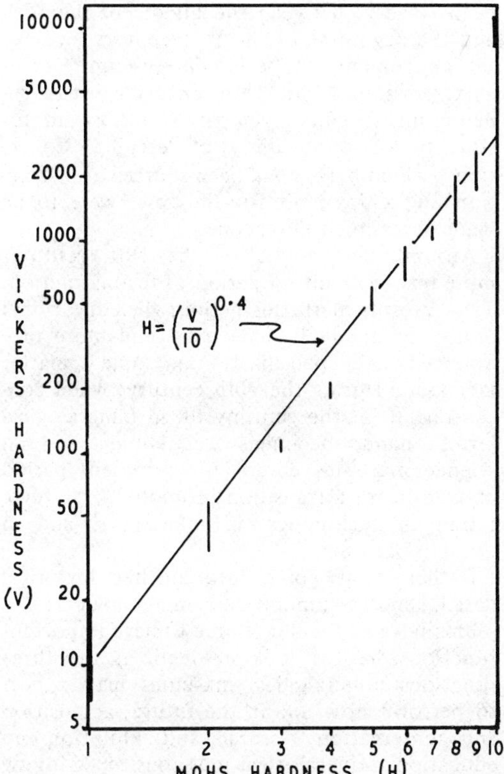

FIGURE 1. The Mohs hardness (log scale) plotted against the Vickers indentation hardness (log scale) in Kg/mm² (after Young and Millman, 1964).

TABLE 1. The Original Mohs Scale of Hardness

1. *Talc*	6. *Orthoclase*
2. *Gypsum*	7. *Quartz*
3. *Calcite*	8. *Topaz*
4. *Fluorite*	9. *Corundum*
5. *Apatite*	10. *Diamond*

TABLE 2. Mohs Scale of Hardness as Modified by Povarennykh (1963)

1. *Talc* (001)	6. *Magnetite* (111)	11. Crystallized boron
2. *Halite* (100)	7. *Quartz* (10$\bar{1}$1	12. B_4C
3. *Galena* (100)	8. *Topaz* (001)	13. $B_{12}C_2$
4. *Fluorite* (111)	9. *Corundum* (11$\bar{2}$0)	14. Carbonado *diamond*
5. *Scheelite* (111)	10. TiC	15. Bort *diamond* (111)

The Mohs scale is set forth in Table 1, *talc* being the softest mineral. The standard for hardness 2 is specified to be "of imperfect cleavage, and not perfectly transparent. Varieties perfectly transparent and crystallized, are commonly too soft." *Calcite* is not an ideal choice for 3, as the hardness varies by about half a unit between the hardest direction of the {10$\bar{1}$1} rhombohedron cleavage planes and the softer {0001} basal pinacoid planes.

Although other scales and modified versions of the Mohs scale have been suggested in the past, none has gained general acceptance. A recent suggestion, based on new studies (Povarennykh, 1963), gives 15 standards of which the first 9 define intervals similar to those of the Mohs scale but 5 new standards, 4 of which are artificial products, are introduced to divide the anomalously large interval between the hardness of *corundum* and *diamond*. This feature will add considerably to its value in the study of abrasive products. The revised Mohs scale as given by Povarennykh is set forth in Table 2.

Determination of the Mohs Hardness

Since the Mohs hardness is based on standards, rather than using those standards to define an independently measured variable, it can in theory only be used with reference to those standards. When testing hardness, care should be taken that a fresh surface is used and that neither separation of grains in granular material nor the streak left by a soft material on a harder is mistaken for a scratch. It should also be kept in mind that in a few minerals, e.g., *kyanite,* hardness varies significantly with direction.

Secondary standards may be used if they are first standardized against the primary standards. Mohs recommended the use of "a fine and very hard file" that has a hardness of about 6½. Other useful substances are: window glass 5½, pocket knife 5+, copper coin 3, and finger nail 2+. Minerals of hardness 1 have a greasy feel.

Factors Controlling Hardness

Hardness is basically controlled by the strength of the atomic bonding in a structure, and calculation of hardness from physicochemical data is possible (Povarennykh, 1963). In practice, it is found that hardness increases with decreasing size of the metallic ions, increasing valence or charge, and closer atomic packing.

MICHAEL J. FROST

References

Dana, E. S., 1932. *A Textbook of Mineralogy*, 4th ed. New York: Wiley, 851p.

Mohs, F., 1820. *The Characters of the Classes, Orders, Genera, and Species; or the characteristic of the natural history system of mineralogy*. Edinburgh: W. and C. Tait.

Povarennykh, A. S., 1963. *The Hardness of Minerals*. Kiev: Acad. Sci. Ukranian SSR, 304p. (In Russian)

Tabor, D., 1954. Hardness of solids, *Endeavour*, **13**, 27–32.

Young, B. B., and Millman, A. P., 1964. Microhardness and deformation characteristics of ore minerals, *Inst. Min. Metall., Trans.*, B **689**, 437–466; **691**, 690–694.

Cross-references: *Jade; Mineral Properties.*

MONOTAXY

A term "monotaxy" was proposed in 1972 by J. D. H. Donnay to designate the partial mutual orientation of two chemically and crystallographically distinct phases that are intergrown with only one lattice row in common. This case of one-dimensional structural control is rarely encountered: it was first reported by Bliznakov (1965), then by Morimoto, Koto, and Shinohara (1966) and by Gale, Donnay, and Winkler (1974).

J. D. H. DONNAY

Reference

Gale, R. J.; Donnay, Gabrielle; and Winkler, C. A., 1974. Crystal data for cupric penicillamine disulphide hydrates, *J. Appl. Crystallogr.*, 7, 309–310, where references are given.

Cross-references: *Epitaxy; Syntaxy; Topotaxy.*

MONTMORILLONITE GROUP—*See* SMECTITE GROUP

MUSEUMS, MINERALOGICAL

Mineralogical museums (Table 1) belong to the large group of museums called natural science museums, i.e., museums dealing with structures existing in nature independently of any intervention by man.

The origins of the museums of natural history were among the "cabinets of natural curiosities." When natural science began to take its own course, a result of the work of such scientists as E. Linnaeus, collecting of curiosities was gradually replaced by systematic collections. Arrangement and display of the objects came to be in rational, and thus instructive, groupings—arranged and displayed with scientific method.

Parallel with this evolution, western European society, between the middle of the 18th and the beginning of the 19th century, evolved the phenomenon "The Public Museum." It was an expression of the 18th-century spirit of enlightenment, which generated enthusiasm for equality of opportunity of learning. Collections, which before had been sources of instruction and enjoyment for the few, were to be made accessible to everyone.

Around the middle of the 19th century, museums went into a period of reconstruction. Two reforms mark this period: the educational motive gradually became more and more pronounced, and specialization in subject matter increased. During the 20th century, when concentration on the scrutiny of, in mineralogical terms, hand specimens was ebbing, natural science museums entered their present period of transition, a transition promoted by a revolution in techniques, in manpower, and in thought patterns.

Rather than going into further historical detail, and assuming that mineralogical museums have and will continue to have important functions, let us look in detail at the three functions mineralogical museums may expect to perform now and in the future—acquisition and conservation, research, and exhibition and education. Mineralogical museums must follow the general development of the society of which they are a part, but they also have important functions dictated by their own history, and these they cannot disregard.

TABLE 1. Mineralogical Museums

Name	Address	Total Specimens Specialties
Department of Geology and Mineralogy, University of Adelaide	Adelaide, South Australia, Australia	Total specimens: 25,790 Specialties: Antarctica display and stratigraphy of Australia (particularly South Australia)
The South Australian Museum	North Terrace, Adelaide, South Australia, Australia	Total specimens: ca. 46,000
Department of Minerals and Energy Bureau of Mineral Resources Australian Government	Canberra, A.C.T., Australia	Total specimens: 1,022,102
Geology Department Australian National University	Box 4, G.P.O., Canberra, A.C.T. 2600, Australia	Total specimens: ca. 22,000
Research School of Earth Sciences Australian National University	Box 4, G.P.O., Canberra, A.C.T. 2600, Australia	Total specimens: 317
Department of Earth Sciences Monash University	Wellington Road, Clayton, Victoria 3168, Australia	Total specimens: 3,500
Tasmanian Museum and Art Gallery	5 Argyle Street, Hobart, Tasmania, Australia	Total specimens: 7,329 Specialties: Tasmanian material, mainly from the mines in the west-coast area of the state
Queen Victoria Museum and Art Gallery	Wellington Street, Launceston 7250, Tasmania, Australia	Total specimens: 5,070 Specialties: Tasmanian minerals, rocks, meteorites, and tektites
National Museum of Victoria	Russell Street, Melbourne, Victoria, Australia	Total specimens: 43,820
Department of Geology University of Western Australia	Nedlands, Western Australia 6009, Australia	Total specimens: 73,254
Department of Geology University of Newcastle	Newcastle, New South Wales, Australia	Total specimens: ca. 10,300 Specialties: Rocks of the Hunter-Manning-Myall district
School of Earth Sciences Macquarie University	North Ryde, New South Wales 2113, Australia	Total specimens: ca. 7,500 Specialties: Broken Hill suite
Department of Geology School of Earth Sciences, University of Melbourne	Parkville, Victoria 3052, Australia	Total specimens: ca. 40,000 Specialties: Meteorites (Murchison and Crambourne) and Broken Hill minerals
West Australian Mines Department Government Chemicals Laboratories	30 Plain Street, Perth, West Australia, Australia	Total specimens: 10,200 Specialties: Tantalum-niobium minerals; phosphate minerals; Western Australian minerals
Western Australian Museum	Francis Street, Perth, West Australia 6000, Australia	Total specimens: 226 Specialties: West Australian meteorites including Mundrabilla, Bencubbin, and Millbillillie
Department of Geology and Mineralogy University of Queensland	St. Lucia, Queensland 4067, Australia	Total specimens: 71,250 Specialties: Granulites, charnockites, and Tertiary volcanic rocks
Australian Museum	College Street, Sydney, New South Wales, Australia	Total specimens: 51,000 Specialties: Broken Hill minerals; Australian gemstones, meteorites, and tektites; and Antarctic rocks.

TABLE 1. (Continued)

Name	Address	Total Specimens Specialties
Department of Geology and Geophysics University of Sydney	Edgeworth David Building, Sydney, New South Wales 2006, Australia	Total specimens: ca. 100,000 Specialties: Broken Hill minerals (H. Dixon Collection); rocks from various world-wide localities (Liversidge Collection); Antarctic rocks (Mawson's Collection); Tennyson-Woods Collection
Geological and Mining Museum	36 George Street, Sydney, New South Wales 2000, Australia	Total specimens: 49,850 Specialties: Ore minerals from New South Wales; gold ore from Australia; Broken Hill rocks and minerals; and tin, molybdenum and antimony ore from New England
West Coast Pioneers Memorial Museum	Main Street, Zeehan, Tasmania, Australia	Total specimens: 1,180 Specialties: *Crocoite* from Dundas, Tasmania
Amt der Steiermärkischen Landesregierung Landesmuseum Joanneum	Raubergasse 10 A–8010 Graz, Austria	Total specimens: 40,000
Institut für Mineralogie und Petrographie Universität Graz	Universitätsplatz 2 A–8010 Graz, Austria	Total specimens: >10,000
Tiroler Landesmuseum Ferdinandeum	Museumstrasse 15 A–6020 Innsbruck, Austria	Total specimens: 9,000 Specialties: Minerals and rocks of the North, South, and East Tyrol; historical collection of the Geognostisch-Montanistischer Verein.
Institut für Mineralogie und Petrographie der Universität Innsbruck	Universitätsstrasse 4 A–6020 Innsbruck, Austria	Total specimens: 3,800
Landemuseum für Karnten	Museumgasse 2 A–9010 Klagenfurt, Austria	Total specimens: ca. 20,000 Specialties: Specimens from the numerous old mining operations of Carinthia; large collections from the Hüttenberg iron mines and the Bleiberg lead-zinc-*wulfenite* deposits.
Institut für Mineralogie und Gesteinskunde Montanistische Hochschule	A–8700 Leoben, Austria	Total specimens: >23,000 Specialties: Eastern alpine ores
Oberösterreichisches Landesmuseum	Museumstrasse 14, A–4020 Linz, Austria	Total specimens: 12,000 Specialties: Regional displays Prambachkirchen meteorite of November 5, 1932
Haus der Natur	Museumsplatz 5, A–5020 Salzburg, Austria	Total specimens: 24,000 Specialties: Regional displays
Mineralogisch-Petrographisches Institut Universität Wien	Dr. Karl Luegerring 1 A–1010 Wien 1, Austria	Total specimens: 10,917
Institut für Mineralogie und Kristallographie der Universität	Dr. Karl Luegerring 1 A–1010 Wien, Austria	Total specimens: 12,200 Specialties: Austrian mineral deposits

TABLE 1. (Continued)

Name	Address	Total Specimens Specialties
Geologisches Institut Universität Wien	Universitätsstrasse 7 A-1010 Wien, Austria	Total specimens: 20,000–30,000
Naturhistorisches Museum Wien Mineralogisch-Petrographisches Abteilung	Postfach 417, Burgring 7 A-1014 Wien, Austria	Total specimens: 127,700
Private collection of Prof. Dr. Heinz Meixner Institute of Mineralogy and Petrography University of Salzburg	Akademiestrasse 26 A-5020 Salzburg, Austria	Total specimens: ca. 30,000 Specialties: Material described in approximately three hundred publications; European and Turkish specimens
Institut Royal des Sciences Naturelles de Belgique Koninklijk Belgisch Instituut voor Natuurwetenschappen	Rue Vautier 31/ Vautierstraat 31 1040 Brussels, Belgium	Total specimens: ca. 43,500 Specialties: *Quartz* and ***feldspar*** twins (Drugman's Collection)
Laboratoire de Minéralogie et de Pétrologie Université Libre de Bruxelles	Avenue F. D. Roosevelt 50 1050 Brussels, Belgium	Total specimens: 11,852
Mineralogical Collection of the Maredsous Abbey	B-5642 Denée, Belgium	Total specimens: 7,628 Specialties: *Calcite,* minerals from Villers-en-Fagne and Denée, Belgium; volcanic rocks from Eifel, Federal Republic of Germany
Chaire des Sciences de la Terre Faculté des Sciences Agronomiques de l'Etat	B-5800 Gembloux, Belgium	Total specimens: 6,000
Department of geography-geology Katholieke Universiteit Leuven	St. Michielstraat 6, B-3000 Leuven, Belgium	Total specimens: ca. 11,000 Specialties: Collection of polished sections of ore minerals
Laboratoire de Minéralogie et de Géologie Appliquée Université de Louvain	6 Sint Michielstraat Leuven, Belgium	Total specimens: >5,000
Institut de Minéralogie, Université de Liège	Place du 20 août, 9 B-4000 Liège, Belgium	Total specimens: 17,060 Specialties: Belgian minerals
Mineralogisch Museum	Churchillaan 59 2120 Schoten, Belgium	Total specimens: 596
Musée royal de l'Afrique centrale Départment de Géologie et Minéralogie	Steenweg op Leuven B-1980 Tervuren, Belgium	Total specimens: ca. 146,500 Specialties: Ore minerals, primary and secondary from Zaïre, Rwanda, Burundi, and Congo; copper silicates and uraniferous secondary minerals from Shaba (Zaïre); Pb, Zn, Cu minerals from the Niari (Congo); phosphates from Buranga (Rwanda)
Departamento de Quimica Facultade de Filosofía, Ciencias e Letras Araraquara	Araraquara, São Paulo, Brasil	Total specimens: 1,600
Departemento de Engenharia Metalúrgica Escola de Engenharia Universidade Federal de Minas Gerais	Rua Espírito Santo 35 Belo Horizonte Minas Gerais, Brasil	Total specimens: 1,550 Specialties: Local minerals

TABLE 1. (Continued)

Name	Address	Total Specimens Specialties
Museu de Minerais e Rochas Departemento de Geociências Universidade de Brasilia	Universidade de Brasilia Brasilia, Brasil	Total specimens: 1,601
Institute of Geosciences Universidade Federal do Paraná	Caixa Postal 5078 80.000 Curitiba, Paraná, Brasil	Total specimens: 950
Departemento Nacional de Produção Mineral	Rue Alberto Torres 7, Matatu, Salvador, Bahia, Brasil	Total specimens: 128
Museu de Mineralogía de Escola de Minas e Metalurgia	Praca Tiradentes 20, Ouro Preto, Minas Gerais Brasil	Total specimens: 25,000
Departemento Nacional da Produção Mineral	1° Distrito Porto Alegre, Rio Grande do Sul, Brasil	Total specimens: 173
Museu Luiz Englert Instituto de Geociências da UFRGS	Sala 511 do Novo Predio da Escola da Engenharia Av, Osvaldo Aranha, s/n, Porto Alegre Rio Grande do Sul, Brasil	Total specimens: 8,001
Departemento de Mineralogía e Geología Facultad de Filosofía, Ciencias, e Letras de Rio Claro	Rio Claro, São Paulo, Brasil	Total specimens: 3,900
Museu Rioclarense "Albertina P. Dias"	Rua Oito, 1693 Rio Claro, São Paulo, Brasil	Total specimens: 2,200
Labóratorio de Geologia Econômica Instituto de Geociências	Ilha do Fundão, Cidade Universitária Rio de Janeiro, Brasil	Total specimens: 950
Institute of Geoscience, University of São Paulo	Caixa Postal 20.899 01000 São Paulo, Brasil	Total specimens: 10,000
Museum of Mineralogy and Petrography Faculty of Geology and Geography University of Sofia "Kliment Okhridsky"	15 Russky Boulevard 1000 Sofia, Eulgaria	Total specimens: 19,816 Specialties: Bulgarian meteorites–Gumoshnik, Silistra, Konevo, Pavel; regional mineral, ore, and rock collections of Bulgaria
National Museum of Natural History, Bulgarian Academy of Science	1 Russky Boulevard 1000 Sofia, Bulgaria	Total specimens: 5,000 Specialties: Minerals from Vitosha Mt., Rila Mt., and Rhodopex Mt.
Glenbow-Alberta Institute	902-11th Avenue, S.W. Calgary, Alberta, Canada T2R OE7	Total specimens: 6,800 Specialties: Minerals of western Canada; *quartz* collection; radioactive mineral collection; rock-cycle chart
Riveredge Foundation	901-10th Avenue S.W. Calgary, Alberta, Canada T2R OB5	Total specimens: 1,000 Specialties: Has a mobile unit featuring crystals, gems, fluorescent minerals, native elements, and objects carved from stone. This unit is operated by Alberta Caravan Exhibits and annually covers a large portion of the province.
Department of Geology University of Alberta	University of Alberta Edmonton, Alberta, Canada	Total specimens: >230,000 Specialties: An almost complete suite of drill cores from the Precambrian of the western

TABLE 1. (Continued)

Name	Address	Total Specimens Specialties
		Canadian Basin; a suite of Precambrian rocks from northeastern Alberta collected by J. D. Godfrey on a grid system for the Research Council of Alberta.
Provincial Museum and Archives of Alberta	12845–102 Avenue Edmonton, Alberta, Canada T5N OM6	Total specimens: 7,103
Department of Geology Dalhousie University	Halifax, Nova Scotia, Canada	Total specimens: 16,115 Specialties: The Honeyman Collection (1860–1885)
Nova Scotia Museum	1747 Summer Street Halifax, Nova Scotia, Canada	Total specimens: 2,500
Kamloops Museum	207 Seymour Street Kamloops, British Columbia, Canada	Total specimens: Unknown Specialties: Local gem specimens
Department of Geological Sciences Queen's University	Queen's University Kingston, Ontario, Canada	Total specimens: >9,000 Specialties: Canadian ore mineral suites
Department of Geology University of Western Ontario	Department of Geology University of Western Ontario London, Ontario, Canada N6A 5B7	Total specimens: ca. 27,000 Specialties: Mineral deposits, world-wide
École Polytechnique	C.P. 501, Snowdon Montréal 248, P.Q. Canada	Total specimens: 5,010 Specialties: An excellent collection of minerals from Mont St. Hilaire, near Montréal, noted for its rare species
Redpath Museum McGill University	Montréal 110, P.Q., Canada	Total specimens: 25,623 Specialties: Reference specimens from worldwide localities collected during the 19th century and long since exhausted; minerals of Montréal area and Monteregian Hills; extensive collection of asbestos minerals; collection of ***micas*** and ***apatites*** from Québec localities
National Mineral Collection of Canada	Display Series: National Museum of Natural Sciences, Ottawa, Ontario, Canada K1A 0M8 Systematic Reference Series: Geological Survey of Canada, Ottawa, Ontario, Canada K1A 0E8	Total specimens: >30,000 Specialties: Specimens from Canadian mineral deposits that are now exhausted; mineral suites from Mont St. Hilaire, near Québec, and from the Bancroft area, Ontario; radioactive minerals; titanium minerals
Parrsboro Mineral, Gem, and Geological Museum	Parrsboro, Nova Scotia, Canada	Total specimens: 290 Specialties: Exhibits showing beach material before and after tumbling; stages of preparing a cabachon from beach material to the finished stone in a ring

TABLE 1. (Continued)

Name	Address	Total Specimens Specialties
Musée de Minéralogie et de Géologie. Université Laval	Québec, 10, P.Q., Canada	Total specimens: >17,100 Specialties: Haüy Collection. This is a collection of 429 specimens prepared under the guidance of Abbé René-Just Haüy and presented to the old Seminary of Québec in 1816
New Brunswick Museum	277 Douglas Avenue Saint John, New Brunswick, Canada	Total specimens: Not determinable at present but probably in the thousands Specialties: Primarily New Brunswick material
Department of Geological Sciences University of Saskatchewan	Saskatoon, Saskatchewan, Canada	Total specimens: >18,000
Department of Geology Memorial University of Newfoundland	St. John's, Newfoundland, Canada	Total specimens: >4,000
Royal Ontario Museum	100 Queen's Park Toronto, Ontario, Canada M5S 2C6	Total specimens: >162,000 Specialties: The gallery has a unique and effective teaching section
British Columbia Department of Mines and Petroleum Resources Mineralogical Branch	Victoria, British Columbia, Canada	Total specimens: 3,000 Specialties: Representative suites of minerals and rock from British Columbia mineral deposits and mines
Mineral Museum Department of Earth Sciences, University of Manitoba	Winnipeg 19, Manitoba, Canada	Total specimens: 5,000 Specialties: Pegmatite minerals, particularly from Manitoba
Slovenské banské múzeum	Banská Štiavnica, okres Žiar nad Hronom Czechoslovakia	Total specimens: 30,000 Specialties: Ore minerals of Slovakian mining districts
Katedra mineralógie a kryštálografie Prírodovedeckej fakulty Univerzity Komenského	Gottwaldovo námestí 2 Bratislava, Czechoslovakia	Total specimens: >10,000
Slovenské národné múzeum	Vajanského nábrežie 2 Bratislava, Czechoslovakia	Total specimens: 10,500
Department of Mineralogy and Petrography Purkyně University	Kotlářská 2 Brno, Czechoslovakia	Total specimens: 13,800 Specialties: Regional Czechoslovakian minerals
Moravian Museum Department of Mineralogy and Petrology	Náměstí 25, února 8 Brno, Czechoslovakia	Total specimens: 125,300 Specialties: Minerals of Moravia, Silesia, Bohemia, Slovakia
Laboratórium pre výskum nerastných surovín BF VŠT	Švermova 5c Košice, Czechoslovakia	Total specimens: 28,050 Specialties: Regional and local displays
Východoslovenské múzeum	Náměstí Maratónu mieru 2 Košice, Czechoslovakia	Total specimens: 9,500 Specialties: Minerals and ores from the mining districts of eastern Czechoslovakia; many are specimens no longer available, e.g., *evansite* from Nižná Slaná, *volynskite* from Rožňava, etc.

TABLE 1. (Continued)

Name	Address	Total Specimens Specialties
Department of Mineralogy and Geology, Palacký University	Leninova 26 Olomouc, Czechoslovakia	Total specimens: 4,022
Múzeum Vlastivědný ústav v Olomouci	Náměstí republiky 5,6 Olomouc, Czechoslovakia	Total specimens: 13,000 Specialties: Regional collection from Hrubý Jeseník Mts.; ore minerals; contact minerals; pegmatites; alpine minerals
Slezské múzeum	Vítězného února 35 Opava, Czechoslovakia	Total specimens: 5,009 Specialties: Opava iron meteorite (9 specimens)
Mining University Ostrava-Poruba	Ostrava-Poruba Czechoslovakia	Total specimens: 22,010 Specialties: Pošepuý collection from Příbram; ore minerals from Jeseníky Mts.; regional collection of coals of Czechoslovakia
Department of Mineralogy, Geochemistry and Crystallography. Charles University	Albertov 6 Praha 2, Czechoslovakia	Total specimens: ca. 20,000 Specialties: Czechoslovakian and European minerals
Institute of Chemical Technology Department of Mineralogy	Suchbatárova 5 Praha 6, Czechoslovakia	Total specimens: 30,250 Specialties: Ore minerals from various regions–Katanga, Tsumeb, Cornwall, Příbram, Binnental, Mexico; pegmatite minerals from Madagascar; alkaline rocks from Kola Peninsula; *zeolites* from Iceland and Faeroes
Národní múzeum	Václavské Náměstí 68 115 79 Praha 1, Czechoslovakia	Total specimens: 112,362 Specialties: Exhibit of minerals from Czech ore deposits; largest existing collection of tektites–ca. 20,000 specimens representing 10,132 tektites; collection of dynamic geology; collection of rock slices
Geologisk Museum Københavns Universitet	Øster Voldgade 5-7 1350 København K, Denmark	Total specimens: ca. 36,000 Specialties: Zeolites from Iceland and the Faeroes; native *silver* from Kongsberg, Norway, amber from Denmark; regiona 1 displays from Ivigtut, Ilímaussaq, and Narssârssuk in Greenland
Institut für Mineralogie und Lagerstättenlehre der RWTH Aachen	Wüllnerstrasse 2 D-5100 Aachen, Germany	Total specimens: 43,000 Specialties: Ore mineral collection
Institut für Lagerstättenforschung der Technischen Universität	Hardenbergstrasse 42 D-1000 Berlin 12, Germany	Total specimens: 11,000
Institut für Mineralogie der Freien Universität Berlin	Takustrasse 6 D-1000 Berlin 33, Germany	Total specimens: 5,755
Institut für Mineralogie und Kristallographie Technische Universität Berlin	Hardenbergstrasse 42 D-1000 Berlin 12, Germany	Total specimens: 19,150 Specialties: Minerals of eastern Bavaria, the Odenwald region, and of South Africa; pegmatite minerals; secondary minerals

TABLE 1. (Continued)

Name	Address	Total Specimens Specialties
Institut für Mineralogie Ruhr Universität Bochum	Postfach 2148 Universitätstrasse 150 D–463 Bochum, Federal Republic of Germany	Total specimens: 50,000 Specialties: Rock-forming minerals
Mineralogisch-Petrologisches Institut und Museum der Universität	Poppelsdorfer Schloss D–53 Bonn, Germany	Total specimens: 400,000 Specialties: Collections from the Siebengebirge, Eifel and Siegerland region
Mineralogisch-Petrographisches Institut der Technischen Universität	Konstantin-Uhde-Strasse 1 D–33 Braunschweig, Germany	Total specimens: New collection; no figures at this time Specialties: Igneous and metamorphic rocks
Mineralogisch-Petrographisches Institut Technische Universität Clausthal	D–3392 Clausthal-Zellerfeld, Germany	Total specimens: 53,900 Specialties: Regional collection from Harz Mts; systematic and genetic collections of minerals, rocks, and ores
Mineralogisches Institut der Technischen Hochschule	Alexanderstrasse 3 D–61 Darmstadt, Germany	Specialties: Display of minerals of the Odenwald region
Mineralogisches Institut der Universität Erlangen-Nürnberg	Schlossgarten 5 D–8520 Erlangen, Germany	Total specimens: 11,702
Institut für Kristallographie und Mineralogie Johann Wolfgang Goethe Universität	Senckenberganlage 30 D–6 Frankfurt am Main, Federal Republic of Germany	Total specimens: 3,010
Institut für Petrologie, Geochemie und Lagerstättenkunde Johann Wolfgang Goethe Universität	Senckenberganlage 28 D–6 Frankfurt am Main, Federal Republic of Germany	Total specimens: 6,100
Mineralogisches Institut der Universität Freiburg i. Br.	Hebelstrasse 40 D–78 Freiburg i. Br., Germany	Total specimens: 26,500 Specialties: Rocks of Germany, particularly the Kaiserstuhl region, and of Italy, altogether about 12,000 specimens
Mineralogisch-Petrologisches Institut der Justus Liebig Universität	Landgraf-Philipp-Platz 4–6 D–63 Giessen, Germany	Total specimens: 5,480
Mineralogische Anstalten der Georg-August-Universität	V. M. Goldschmidt-Strasse 1 D–34 Göttingen, Germany	Total specimens: 123,000 Specialties: Special collections of 1500 pseudomorphs, 570 meteorites, 500 slags, and other synthetics; minerals and rocks from Etna, Russia, Harz Mts., Norway, Japan, etc.; first synthetic *quartz* crystal from Spezia, donated in 1906
Mineralogisch-Petrographisches Institut der Universität Hamburg	Grindelallee 48 D–2000 Hamburg 13, Germany	Total specimens: 68,000 Specialties: Gibeon meteorite (424 kg); *antimonite* from Ichinokawa; *struvite* and *boracite* from Striegau; collections from the Harz Mts., Siebenbürgen, Cumberland, England
Mineralogisches Institut der Technischen Universität Hannover	Welfengarten 1 D–3 Hannover, Germany	Total specimens: 14,000

TABLE 1. (Continued)

Name	Address	Total Specimens Specialties
Mineralogisch-Petrographisches Institut der Universität	Berlinerstrasse 19 D–69 Heidelberg, Germany	Total specimens: 47,000
Mineralogisches Institut	Kaiserstrasse 12 D–75 Karlsruhe 1, Germany	Total specimens: 2,057
Mineralogisch-Petrographisches Institut der Universität zu Köln	Zülpicherstrasse 49 D–5000 Köln, Germany	Total specimens: 32,000
Mineralogisches Museum des Fachbereiches Geowissenschaften der Philipps-Universität	Deutschhausstrasse 10 D–3550 Marburg/Lahn, Germany	Total specimens: ca. 95,330
Institut für Mineralogie Technische Universität München	Arcisstrasse 21 D–8 München, 2, Germany	Total specimens: 40,000 Specialties: Minerals of Bavaria; granites from Germany and Scandinavia; systematic mineral and rock exhibit; genetic mineral exhibit
Mineralogische Staatssammlung	Theresienstrasse 41, 8000 München, Germany	Total specimens: ca. 16,000 Specialties: Groth's famous morphological collection of minerals; valuable specimens of the former Leuchtenberg collection; scientific exhibits on crystals
Institut für Mineralogie der Universität Münster Museum für Kristalle und Gesteine	Gievenbecker Weg 61 D–4400 Münster, Germany	Total specimens: 44,500 Specialties: Ore suites from Oberharzer veins, Rammelsberg, Meggen, Siegerland, Lahn-Dill region, Freiberg (Saxony), Erzgebirge, Alnö (Sweden), Kiruna Varuträsk, Trepča (Yugoslavia), Tsumeb (South West Africa); regional rock collections, e.g., Eifel, Schwarzwald, Egypt, Switzerland
Staatliches Forschungsinstitut für Angewandte Mineralogie	Kumpfmühlerstrasse 2 Dörnberg-Palais D–84 Regensburg, Germany	Total specimens: 13,020 Specialties: *Fluorite* and *pyrite* of the Ober Pfalz region
Mineralogisches Institut der Universität der Saarlandes	D–66 Saarbrücken Germany	Total specimens: 14,380 Specialties: Regional collection of Saarland minerals and rocks; collection of ultramafic rocks; collection of building stones; collection of ore minerals
Institut für Mineralogie und Kristallchemie der Universität Stuttgart	Paffenwaldring 55 D–7000 Stuttgart 80 Germany	Total specimens: 13,000 Specialties: Minerals of the Black Forest region
Museum und Sammlung des Mineralogisch-Petrographischen Institut der Universität Tübingen	D–74 Tübingen, Germany	Total specimens: 150,800 Specialties: Meteorites
Mineralogisches Institut der Universität Würzburg Museum	D–8700 Würzburg, Germany	Total specimens: 19,100
Kivimuseo, Eräjärvi-Seura ry	SF 35220, Eräjärvi, Finland	Total specimens: 200 Specialties: Pegmatite minerals of the Eräjärvi area; historical material related to local stone and mineral industry

TABLE 1. (Continued)

Name	Address	Total Specimens Specialties
Kivimuseo, Geologinen tutkimuslaitos	SF 02150, Espoo 15, Finland	Total specimens: 2,500 Specialties: Orbicular rocks of Finland
Taloudellisen geologian laboratorio Vuoriteollisuusosasto Helsingin teknillinen korkeakoulu	SF 02150, Espoo 15, Finland	Total specimens: 13,000 Specialties: Finnish ores; gems
Kivimuseo Helsingin Yliopisto	PL 115 SF 00171, Helsinki 17, Finland	Total specimens: 81,950
Geologian laitos Oulun Yliopisto	SF 90100 Oulu 10, Finland	Total specimens: 4,231 Specialties: Complete set of ores of Finland; regional display of bedrock of Finland, divided into 17 geological provinces; collection of sedimentary structures prepared by a lacquer film technique
Kivimuseo Rovaniemen Kaupunki	Hallituskatu 9, SF 96100, Rovaniemi, Finland	Total specimens: 530 Specialties: Minerals, rocks, and ores from Lapland
Geologian ja mineralogian laitos Turun Yliopisto	SF 20500, Turku 50, Finland	Total specimens: 10,035
Geologisk-Mineralogiska Institutionen. Åbo Akademi	SF 20500, Turku 50, Finland	Total specimens: ca. 11,000 Specialties: The Arrhennius collection of Pargas minerals, with a wealth of *pargasite*
Musée des Beaux-Arts	47015 Agen, France	Specialties: Gareau collection of granite, porphyry, limestone, etc.; Roton collection of jasper, onyx, agate from Sauterne (Gironde)
Muséum d'Histoire Naturelle d'Autun	14 Rue St. Antoine 71400 Autun, France	Total specimens: 100,000 (including prehistoric items) Specialties: Chaignon collection of rocks and minerals of Mowray
Janet Collection Musée Departemental de l'Oise	60006 Beauvais, France	Total specimens: 149
Viguier Collection Musée Lapidaire et d'Archeologie	Rue des Saintes-Maries 84200 Carpentras, France	Total specimens: ca. 1,000
Institut des Sciences de la Terre de l'Université de Dijon	6, Boulevard Gabriel 21000 Dijon, France	Total specimens: >19,000 Specialties: Regional minerals
Musée Départemental des Hautes-Alpes	Avenue de Maréchal Foch 05000 Gap, France	Total specimens: Several thousand Specialties: Rocks and fossils of the Province of Hautes-Alpes
Institut Dolomieu Département des Sciences de la Terre Université Scientifique et Médicale de Grenoble	Rue Maurice Gignoux 38031 Grenoble (Cedex), France	Total specimens: 4,000
Muséum d'Histoire Naturelle de la Ville de Grenoble	1, Rue Dolomieu 38000 Grenoble (Isère) France	Total specimens: large number Specialties: Collection of minerals of the Province of Dauphiné, the most important collection known for the Oisans region in Isère,

TABLE 1. (Continued)

Name	Address	Total Speciments Specialties
Muséum d'Histoire Naturelle de la Ville de Grenoble (cont.)		with very beautiful *quartz* specimens; excellent series of *malachite* and *azurite* specimens from the Chessy mine, near Lyon
Musée de Laval	Place de la Trémoille 53000 Laval, France	Total specimens: Several thousand
Musée Crozatier Jardin Henri Vinay	43000 Le Puy, France	Total specimens: 24,500 Specialties: Regional mineral and rock collection, with the minerals arranged according to Lacroix classification
Musée de Géologie et Minéralogie Musée Gosselet de la ville de Lille	59000 Lille, France	Total specimens: Several thousand
Muséum de Longchamp	13000 Marseille, France	Total specimens: 7,500 Specialties: Rare minerals of the Garonne area; many silicates and carbonates
Musée d'Histoire Naturelle	82000 Montauban, France	Total specimens: 600 Specialties: Orgueil meteorite; local rock display; common classic minerals
Laboratoire de Minéralogie et Cristallographie Université des Sciences et Techniques de Langudoc	Place Eugène Bataillon 34000 Montpellier, France	Total specimens: 3,000 (chiefly minerals)
Unité d'Enseignement et de Recherche Sciences Pharmaceutiques et Biologiques	7, Rue Albert Lebrun 54000 Nancy, France	Total specimens: 700 Specialties: Petrography of the Vosges region
Laboratoire de Pétrologie-Minéralogie Facultédes Sciences, Université de Nice	Parc Valrose 06000 Nice, France	Total specimens: 1,013 Specialties: Regional and local displays
Institut Catholique	21, Rue d'Assas 75006 Paris, France	
Minéralogie-Cristallographie Université Pierre et Marie Curie (ex Sorbonne)	Tour 16, 4, Place Jussieu, 75230 Paris France	Total specimens: 35,000
Musée de Minéralogie de l'École Nationale Supérieure des Mines	60, Boulevard Saint-Michel 75006 Paris, France	Total specimens: 85,000 Specialties: Regional and general displays; large collection of European and African minerals particularly for exchange purposes
Muséum d'Histoire Naturelle Laboratoire de Minéralogie	61, Rue de Buffon 75005 Paris, France	Total specimens: 142,400 Specialties: Haüy collection of gems; Lacroix collection of lithologic types, including thin sections and chemical analyses; meteorite collection; collection of polished surfaces of opaque minerals
Muséum d'Histoire Naturelle Ethnographie-Préhistoire	198, Rue Beauvoisine 76000 Rouen, France	Total specimens: Not inventoried Specialties: None

TABLE 1. (Continued)

Name	Address	Total Specimens Specialties
École Nationale Supérieure des Mines	158 bis, Cours Fauriel 42023 Saint-Etienne (Cedex) France	Total specimens: ca. 2,000 Specialties: None particularly–for teaching purposes, the broadest possible number of different minerals and rocks has been assembled
Laboratoire de Minéralogie et Cristallographie Université Paul Sabatier	39, Allées Jules-Guesde 31400 Toulouse, France	Total specimens: 8,000 Specialties: Regional displays
Museum d'Histoire Naturelle de Toulouse	35, Allée Jules Guesde 31400 Toulouse, France	Total specimens: 12,107 Specialties: Meteorites (Saint Sauveur and d'Orgueil, France); *opals* from Australia, *gold* from California
Museum für Naturkunde an der Humboldt Universität zu Berlin	DDR 104 Berlin, Invalidenstrasse 43	Total specimens: ca. 360,000 Specialties: Meteorites
Staatliches Museum für Mineralogie und Geologie zu Dresden	DDR–801 Dresden, Augustusstr. 2	Total specimens: ca. 60,200 Specialties: Meteorites; minerals: special collection of the historic region Saxony, Baldauf-collection; exhibition "gems"; rocks–special collection of the historic region Saxony
Bergakademie Freiberg, Sektion Geowissenschaften Bereich Sammlungen	DDR 92 Freiberg/sa., Brennhausgasse 14	Total specimens: ca. 250,000 Specialties: Collection of minerals, ores, and rocks of the territory of the GDR; private collection of A. G. Werner with its historical significance
E.-M.-Arndt-Universität Greifswald Sektion Geologische Wissenschaften	DDR–22 Greifswald, Fr.-L-Jahn-Str. 17a	Total specimens: ca. 27,000
Staatliche Kunstsammlungen Görlitz	DDR–89 Görlitz, Demianiplatz 1	Total specimens: 8,621
Mineralogical Museum of the division of Mineralogy Martin-Luther-Universität, Halle-Wittenberg, section of chemistry	DDR–402 Halle/Saale, Domstr. 5	Total specimens: ca. 40,000 Specialties: Minerals and rocks from "Harz"; minerals and rocks from "Thüringen"; porphyric and porphyritic rocks from "Halle/Saale"; copper schist from "Mansfeld"
Goethe-Nationalmuseum, Nationale Forschungs- und Gedenkstätten der klassischen deutschen Literatur	DDR–53 Weimar, Am Frauenplan 1	
Department of Petrography and Geochemistry Loránd Eötvös University	Müzeum krt 4/a Budapest VIII, Hungary	Total specimens: 15,000 Specialties: Regional collections of the Hungarian Basin and the Carpathians
Hungarian Geological Institute	Nepstadion ut 14 Budapest XIV	Total specimens: 26,000 Specialties: Ore deposit collection
Institute for Mineralogy and Geology Technical University of Budapest	Sztoczek-utca 2 Budapest 112, Hungary	Total specimens: 6,000
Mineralogical Institute Lórand Eötvös University	Muzeum krt 4/A Budapest VIII, Hungary	Total specimens: >65,000 Specialties: Gold telluride ores of Transylvania; meteorites: Magura, Mócs, Kňahyňa, Orgueil

TABLE 1. (Continued)

Name	Address	Total Specimens Specialties
M. N. M. Természettudományi Múzeum Asvanytar	Múzeum krt 14–16 Budapest VIII, Hungary	Total specimens: 62,908
Cathedra Mineralogica et Geologica Universitatis Scientiarum de L. Kossuth nominatae	Debrecen 10, Hungary Postal box 4	Total specimens: 6,005
Institute of Mineralogy, Geochemistry, and Petrography Attila József University	Táncsics M.u.2 Szeged, Hungary	Total specimens: 5,600 Specialties: Koch collection of minerals of the mines of the Carpathian Basin; industrial minerals and rocks of Hungary
Museum of Natural History Department of Geology	Laugavegur 105 Reykjavik, Iceland	Total specimens: ca. 6,800 Specialties: ***Zeolites*** from Iceland
Department of Geology Central College, Bangalore University	Bangalore 560001, India	Total specimens: 8,600 Specialties: Local and regional ores
Istituto di Mineralogia e Petrografia del'Università	Palazzo Ateneo, Piazza Umberto I 70121 Bari, Italy	Total specimens: 15,619 Specialties: Rare specimens from Tuscany, Sardinia, and mines in Sicily; rocks of southern Italy, especially Monte Vulture and Calabria; volcanic rocks from Dancalia, Assab area (Africa)
Museo Civico de Scienze Naturali	Piazza Cittadella 24100 Bergamo, Italy	Total specimens: 1,430 Specialties: Local collections
Museo dell'Istituto di Mineralogia e Petrografia dell'Università	Piazza di Porta San Donato 1 40127 Bologna, Italy	Total specimens: 15,130 Specialties: Regional collections from Emilia, Romagna, Predazzo
Istituto Giacimenti Minerari dell'Università Facoltà di Ingegneria	Piazza d'Armi 09100 Cagliari (Sardinia), Italy	Total specimens: 238 Specialties: Textures of ore minerals
Istituto di Mineralogia e Petrografia dell'Università	Corso Italia 55 95129 Catania, Italy	Total specimens: 3,520 Specialties: Evaporite minerals–*gypsum, aragonite, celestite, sulfur*; minerals and rocks of Mt. Etna
Collegio Mellerio Rosmini	Largo Madonna della Neve 21 28037 Domodossola (Novara), Italy	Total specimens: 1,244 Specialties: Collection of 322 specimens from the Simplon Tunnel
Fondazione Gian Giacomo Galletti	Piazza Convenzione 10 28037 Domodossola (Novara), Italy	Total specimens: 3,151 Specialties: Collection of 384 specimens from the Simplon Tunnel; minerals and rocks of Valle Vigezzo
Istituto di Mineralogia dell'Università	Corso Ercole 1° d'Este 32 44100 Ferrara, Italy	Total specimens: 320 Specialties: Complete collection of the main types of ***zeolites*** from Maharashtra State, India
Museo di Mineralogia, dell'Università	Via Lamarmora 4 50121 Firenze, Italy	Total specimens: 18,835 Specialties: Medici family collection of cut stones; collection of 5,000 specimens from Elba Island

TABLE 1. (Continued)

Name	Address	Total Specimens Specialties
Istituto di Mineralogia e Petrografia dell'Università	Palazzo delle Scienze, Corso Europa 16132 Genova, Italy	Total specimens: 6,100 Specialties: Local collections
Museo Civico di Storia Naturale "G. Doria"	Via Brigata Liguria 9 16121 Genova, Italy	Total specimens: 4,500 Specialties: Collections from the Alps, the Apennines, Sardinia, Burma, and Egypt
Istituto di Mineralogia e Geologia Applicata Politecnico, Facultà di Ingegneria	Piazza Leonardo da Vinci 20133 Milano, Italy	Total specimens: 1,520 Specialties: Industrial minerals and building stones of Italy
Istituto di Mineralogia, Petrografia, Geochimica e Giacimenti minerari dell'Università	Via Botticelli 23 20133 Milano, Italy	Total specimens: 19,000 Specialties: Specimens of Alpine localities, e.g., Baveno, Val d'Ala, Valtellina, Cuasso al Monte, etc.
Museo Civico di Storia Naturale	Corso Venezia 55 20121 Milano, Italy	Total specimens: 25,290 Specialties: Pegmatite minerals of Baveno and Central Alps; collection of Borromeo (1913), Mauro (1950–52), and Scaini (1972)
Istituto di Mineralogia e Petrografia dell'Università	Via S. Eufemia 19 41100 Modena, Italy	Total specimens: 10,500
Istituto di Mineralogia dell' Università	Via Mezzocannone 8 80134 Napoli, Italy	Total specimens: 20,000 Specialties: Collection of 6,000 mineral specimens from Vesuvius, made by Angelo Scacchi
Istituto di Mineralogia e Petrografia dell'Università	Corso Garibaldi 9 35100 Padova, Italy	Total specimens: 8,629 Specialties: Mineral displays from Veneto, Trentino-Alto Adige, and north Tyrol; collections of rocks from the Adamello massif and from Ethiopia; collection of crystalline rocks from the Alps
Istituto di Mineralogia dell'Università	Via Archirafi 36 90123 Palermo, Italy	Total specimens: 3,707
Istituto di Petrografia, Giacimenti Minerari e Minerologia dell'Università	Via Gramsci 9 43100 Parma, Italy	Total specimens: 7,020
Istituto di Mineralogia dell'Università	Via A. Bassi 4 27100 Pavia, Italy	Total specimens: 5,716 Specialties: Agate and opal collections dating from 1778 and 1802; mineral collection from Val Devero (Ossola, Piemonte)
Centro Italiano di Studio dei Meteoriti Istituto di Mineralogia dell'Università	Piazza Università 06100 Perugia, Italy	Total specimens: 86 Specialties: All types of meteorites
Istituto di Mineralogia dell'Università	Via Santa Maria 53 56100 Pisa, Italy	Total specimens: 10,000 Specialties: Regional collection from Tuscany
Istituto di Geochimica dell'Università	Città Universitaria 00100 Rome, Italy	Total specimens: 3,720 Specialties: Minerals of the Alban Hills and other volcanic regions in Italy

TABLE 1. (Continued)

Name	Address	Total Specimens Specialties
Museo di Mineralogia e Petrografia dell'Università	Città Universitaria 00100 Rome, Italy	Total specimens: 29,000 Specialties: Crystals, artificial crystals, meteorites; minerals and rocks of Lazio; a collection of rings with cut stones of Pope Leo XII
Accademia dei Fisiocritici	Piazza S. Agostino 4 53100 Siena, Italy	Total specimens: 3,930
Istituto di Mineralogia dell'Università	Via Mattioli 4 53100 Siena, Italy	Total specimens: 1,640
Istituto di Mineralogia dell'Università	Via San Massimo 24 10123 Torino, Italy	Total specimens: 15,350
Istituto di Mineralogia, Geologia e Giacimenti Minerari Politecnico di Torino	Corso Duca degli Abruzzi 24 101 Torino, Italy	Total specimens: 5,521 Specialties: Local and regional ore minerals (display in preparation)
Istituto di Mineralogia e Petrografia dell'Università	Piazzale Europa 1 34127 Trieste, Italy	Total specimens: 1,750
Museo Civico di Storia Naturale	Lungadige Porta Vittoria 9 37100 Verona, Italy	Total specimens: 2,782 Specialties: The Nicolis collection a complete petrographic set of the sedimentary series from the Lias to Miocene in the Verona area
Mineral Industry Museum Mining College, Akita University	28 Osawa, Tegata, Akita City, Japan	Total specimens: 7,400 Specialties: Japanese ore minerals
Geological Survey of Japan Kyusyu Branch	497-2 Shiobaru-Aizo Minamiku Fukuoka, Japan	Total specimens: 3,200
Department of Geology Kyushu University	Hakozaki, Fukuoka, 812 Japan	Total specimens: 9,000 Specialties: Ko Collection: large crystal ore specimens; Yoshimura collection of manganese minerals and ores
Department of Earth Sciences Faculty of Science, Kanazawa University	1-1 Marunouchi, Kanazawa, Japan	Total specimens: 4,900
Geological Survey of Japan	135 Hisamoto-cho, Takatsu-ku Kawasaki-shi, Japan	Total specimens: 30,315 Specialties: Japanese ore minerals; *diamond* cuts
Department of Geology and Mineralogy Faculty of Science, University of Kyoto	Kitashirakawa, Sakyo-ku, Kyoto, Japan	Total specimens: 1,610
Department of Earth Sciences Nagoya University	Chikusa-ku, Nagoya, Japan	Total specimens: 7,300 Specialties: Rock collections from Africa, Scotland, and Antarctica
Mitsubishi Mineral Collection Mitsubishi Metal Corporation	Kitabukuro-machi Omiya City, Saitama, Japan	Total specimens: 4,700 Specialties: Wada Collection of crystals of Japanese localities; *quartz* crystals, *stibnite, topaz, axinite*

TABLE 1. (Continued)

Name	Address	Total Specimens Specialties
National Science Museum	3–23–1 Hyakunin-cho, Shinjuku-ku Tokyo 160, Japan	Total specimens: 27,893 Specialties: Japanese manganese minerals
Mineralogical Laboratory Geological and Mineralogical Institute. Faculty of Science, Tokyo University of Education	3-chome, Otsuka, Bunkyo-ku Tokyo, Japan	Total specimens: 5,200 Specialties: Clay minerals (1000 specimens)
University Museum, University of Tokyo	7–3–1, Hongo, Bunkyo-ku Tokyo, Japan	Total specimens: 24,100 Specialties: Japanese ore minerals, e.g., manganese; Chinese and Korean specimens
Geologisch Instituut der Universiteit van Amsterdam	Nieuwe Prinsengracht 130, The Netherlands	Total specimens: 133,000
Instituut voor Aardwetenschappen. Vrije Universiteit	De Boelelaan 1085 Amsterdam 11, The Netherlands	Total specimens: 6,000 Specialties: Systematic and regional collections of about 1000 polished sections of opaque and ore specimens; complete core drilling of Panda Hill carbonatite in Tanzania; collections from Västervik region, southeast Sweden, and from southwest Tanzania
Mineralogisch-Geologisch Museum der Mijnbouwkunde. Technische Hogeschool	20 Mijnbouwstraat Delft, The Netherlands	Total specimens: 86,700 Specialties: Collection of rocks and fossils made by the Timor Expedition, 1911–1917, to Timor Island, Indonesia; collection of billitonites (tektites)
Geologisch Instituut, Rijksuniversiteit	Melkweg 1, Groningen, The Netherlands	Total specimens: 5,000
Teyler Museum	Spaarne 16 Haarlem, The Netherlands	Total specimens: >6,500
Rijksmuseum van Geologie en Mineralogie	Hooglandse Kerkgracht 17 Leiden, The Netherlands	Total specimens: 60,110 Specialties: Historical gem collection of King William the First of the Netherlands; collection of Indonesian *tektites*; private collection of late Dr. J. Erb of 1,500 high quality specimens (on loan)
Billiton N. V.	Louis Couperusplein 19 The Hague, The Netherlands	Total specimens: Not available at the moment because collection is being reorganized Specialties: Ores of tin, aluminum, and base metals; carbonatites
Auckland Institute and Museum	Private Bag Auckland, New Zealand	Total specimens: 9,000
Department of Geology, University of Auckland	Private Bag Auckland, New Zealand	Total specimens: 23,500 Specialties: Local regional collections
Canterbury Museum	Rolleston Avenue Christchurch, New Zealand	Total specimens: 5,035 Specialties: Display of local rocks and minerals; meteorite display

TABLE 1. (Continued)

Name	Address	Total Specimens Specialties
Geology Department, University of Canterbury	Christchurch, New Zealand	Total specimens: 7,800
Geology Department, University of Otago	Dunedin, New Zealand	Total specimens: >31,000
Otago Museum	Great King Street Dunedin, New Zealand	Total specimens: 2,600
Southland Museum	Victoria Avenue Invercargill, New Zealand	Total specimens: 1,000 Specialties: Southern New Zealand rocks and minerals
New Zealand Geological Survey	Andrews Avenue Lower Hutt, New Zealand	Total specimens: 40,000 Specialties: Large representative collection, with thin sections, of New Zealand and Antarctic rocks and minerals; reference collection of overseas minerals and rocks
Normandale Mineralogical Museum	168 Miro Miro Road Lower Hutt, New Zealand	Total specimens: >3,000 Specialties: New Zealand minerals rocks, and ores
Thames Borough Council Borough Council Chambers	Mary Street Thames, New Zealand	Total specimens: ca. 5,000, mainly minerals Specialties: Minerals and rocks of Coromandel Peninsula mining region; models of mining equipment
Geology Department Victoria University of Wellington	Box 196 Wellington, New Zealand	Total specimens: 11,200 Specialties: New Zealand volcanic and plutonic rocks, Antarctic rock collection
Department A, Geological Institute University of Bergen	Joach. Frielesgt. 1, N-5000, Bergen, Norway	Total specimens: 59,415 Specialties: *Gadolinite* from Hitra, *quartz* with *anatase* from Hardangarvidda and amethyst from Brazil; exhibition on Geology of Norway, and Norwegian building stones and ores
Mineralogisk-Geologisk Museum University of Oslo	Saragate 1 Oslo 5, Norway	Total specimens: ca. 210,000 Specialties: Norwegian minerals; pegmatite minerals
Tromsø Museum Department of Geology	9000 Tromsø, Norway	Total specimens: 13,655 (plus duplicates)
Geologisk Institutt Norges Tekniske Høјskole	7034 Trondheim-N.T.H., Norway	Total specimens: 47,300 Specialties: Slag mineral collection of J. H. L. Vogt
Norges Geologiske Undersøkelse	Leiv Eirikssons vei 39 7000 Trondheim, Norway	Total specimens: 190,000
Museu Municipal Da Beira	Beira, Box 1702, People's Republic of Mozambique	Total specimens: 219
Academy of Mining and Metallurgy Institute of Mineralogy and Mineral Deposits	al. Mickiewicza 30 30-059 Kraków, Poland	Total specimens: 5,200 Specialties: Minerals and rocks of Poland
Geological Institute	Rakowiecka 4 02-517 Warsaw, Poland	Total specimens: 11,325

TABLE 1. (Continued)

Name	Address	Total Specimens Specialties
Muzeum Ziemi Polish Academy of Science	Warsaw, Poland	Total specimens: 31,100 Specialties: The amber collection, founded in 1958, contains 12,750 specimens showing natural forms, organic and inorganic inclusions, and varieties; meteorites found between 1935 and 1951, near Lowicz and Pultusk in Poland; Tertiary lignite from the Turonow coal mine
Muzeum Mineralogiczne Uniwersytetu Wroclawskiego Zaklad Mineralogii i Petrografii Instytutu Nauk Geologicznych	ul. Cybulskiego 30 50–205 Wroclaw, Poland	Total specimens: >21,000 Specialties: The mineral collection, the largest in Poland, contains 6,500 silicates, 3,300 oxides, 3,100 sulfides and sulfosalts, 2,500 carbonates, 1,600 sulfates. The meteorite collection is also the largest in Poland.
National Museum of Ireland	Kildare Street, Dublin 2, Ireland	Total specimens: 15,000 Specialties: Leskean Collection, C. L. Giesecke, and other early Irish collections formerly in the museum of the Royal Dublin Society, transferred in 1877
Trinity College	Dublin 2, Ireland	Total specimens: 5,143
National Museum	P.O. Box 266, Bloemfontein Republic of South Africa	Total specimens: 2,330
South African Museum	P.O. Box 61, Cape Town 8000 Republic of South Africa	Specialties: Meteorites
Durban Museum	City Hall, Smith Street P.O. Box 4085, Durban 4000 Republic of South Africa	Total specimens: 2,723
East London Museum	319 Oxford Street, East London 5201 Republic of South Africa	Total specimens: 1,126
McGregor Museum	P.O. Box 316, Kimberley 8300 Republic of South Africa	Total specimens: 1,702
Port Elizabeth Museum	P.O. Box 13147, Humewood, Port Elizabeth Republic of South Africa	Total specimens: 961 Specialties: Hofmeyr meteorite Feb. 1914
Museum of the Geological Survey	Private Bag XI 12, Pretoria (0001) Republic of South Africa	Total specimens: 25,802 Specialties: Fluorescent minerals, which were first found in Southern Africa and described as such
Department of Geology University of Cape Town	7700 Rondebosch, Republic of South Africa	Total specimens: 7,725
National Museum and Monument of Rhodesia	P.O. Box 240, Bulawayo, Rhodesia	Total specimens: 10,500 (incl. palaeontological material) Specialties: Minerals from Broken Hill, Zambia
Department of Geological Survey Ministry of Mines	Box 8039, Causeway, Salisbury, Rhodesia	Total specimens: 28,550 Specialties: Exhibits: Geological History of Rhodesia, minerals

TABLE 1. (Continued)

Name	Address	Total Specimens Specialties
		and rocks from the Great Dyke and minerals from the Bikita Tin Fields
State Museum	P.O. Box 1203, Windhoek 9100 South West Africa	Total specimens: 7,833 Specialties: Southwest African minerals and semiprecious stones
Museo de Geología	Parque de la Ciudadela Barcelona 3, Spain	Total specimens: >24,200 Specialties: Collection of Catalonian minerals, especially sulfosalts of the Catalan potassic basin, general collection of Spanish rocks and minerals; collection of Scandinavian rocks and minerals
Departamento de Geología Facultad de Ciencias, Universidad Autónoma	Apartado 644 Bilbao, Spain	Total specimens: 230
Museo Nacional de Ciencias Naturales Sección de Mineralogía	Paseo de la Castellana 84 Madrid 6, Spain	Total specimens: 15,689 Specialties: *Pyromorphite* collection from Horcajo (Ciudad Real) Meteorite
Department of Geology Faculty of Sciences, University	Murcia, Spain	Total specimens: 770 Specialties: Regional collections
Bergskolan i Filipstad	S-68200 Filipstad, Sweden	Total specimens: 10,000–12,000 Specialties: Minerals from Långban
Department of Geology Chalmers University of Technology University of Göteborg	Fack, S-402 20 Göteborg, Sweden	Total specimens: 15,000 Specialties: Ore minerals, silicate minerals and regional rock and ore collections
Department of Mineralogy and Petrology University of Lund	Sölvegatan 13, S-223 62 Lund, Sweden	Total specimens: >18,883
Department of Geology Royal Institute of Technology	100 44 Stockholm 70, Sweden	Total specimens: 2,500
Department of Geology University of Stockholm	11 386 Stockholm, Sweden	Total specimens: ca. 20,000
Naturhistoriska Riksmuseet	S-10405 Stockholm 50, Sweden	Total specimens: ca. 150,000
Geologiska Institutionen Uppsala Universitet	Box 555 S-751 22 Uppsala, Sweden	Total specimens: >32,000
Naturhistorisches Museum Mineralogische Abteilung	CH-4051 Basel, Switzerland	Total specimens: 27,150 Specialties: Systematic collection of Swiss ore minerals; regional collections of Swiss Alpine minerals, Binnental, etc.
Mineralogisch-Petrographisches Institut der Universität Bern	Sahlistrasse 6 CH-3012 Bern, Switzerland	Total specimens: 19,000 Specialties: Swiss Alpine minerals
Naturhistorisches Museum, Bern	Bernastrasse 15 CH-3005 Bern, Switzerland	Total specimens: 28,860 Specialties: *Quartz* collection; Alpine fissure minerals; sulfosalts and other minerals from the Lengenbach quarry in the Binntal

TABLE 1. (Continued)

Name	Address	Total Specimens Specialties
Bündner Naturhistorisches und Nationalpark Museum	CH-7000 Chur, Switzerland	Total specimens: >1,000
Museum d'Histoire Naturelle Faculté des Sciences	CH-1700 Fribourg, Switzerland	Total specimens: >20,000 Specialties: A collection of ores of the Binntal, a systematic collection; diorama of an Alpine fissure with talking film show
Museum d'Histoire Naturelle	Route de Malagnou N° 1 CH-1211 Geneva 6, Switzerland	Total specimens: 19,000 Specialties: An exhibit of 60 luminescent mineral specimens, switch operated to demonstrate fluorescence and phosphorescence; a collection of the largest smoky *quartz* crystals found in the Alps, from a cleft at Tiefen Glacier, Uri canyon; 1,300 rocks collected by H. B. Saussure and described in his "Voyages dans les Alpes"
Musée Géologique	Palais de Rumine CH-1005 Lausanne, Switzerland	Total specimens: 6,200
Institute de Géologie, Université	Rue Emile-Argand 11 CH-2000 Neuchâtel 7, Switzerland	Total specimens: 2,825
Bally Museum-Stiftung	CH-5012 Schönenwerd, Switzerland	Total specimens: 10,600 Specialties: Alpine fissure minerals, ores, and meteorites
Solothurn Museum Mineralogisch-Geologische Abteilung	Reinertweg 10 CH-4500 Solothurn, Switzerland	Total specimens: 6,900
Naturwissenschaftliche Sammlungen der Stadt Winterthur	Museumstrasse 52 CH-8400 Winterthur, Switzerland	Total specimens: 6,260
Mineralogisches-Petrographisches Sammlung, Eidgenössisches Technische Hochschule	Sonneggstrasse 5 CH-8006 Zürich, Switzerland	Total specimens: 73,580 Specialties: Alpine regional collections of minerals
Geological Museum, Geological Sciences Institute Kazakh SSR Academy of Sciences	Kalinin Str., 69a, Alma-Ata 480100, USSR	Total specimens: ca. 27,300 Specialties: Minerals, rocks and ores from Kazakhstan
Museum of the Geological Institute Kola Filial of the USSR Academy of Sciences	Fersman Str., 14, Apatity, 184200, USSR	Total specimens: 3,500 Specialties: Regional collection of Kola Peninsula
Geological Museum	Promyshlennaya Str., 4, Asbest, 624060, USSR	Total specimens: 1,170 Specialties: Minerals and rocks of asbestos deposits
Geological Museum Geological Board, Azerbaijan SSR Council of Ministers	Vakhram Agaev Str., 100a, Baku 370073, USSR	Total specimens: 5,300 Specialties: Ore minerals and regional mineralogy
Geological-Mineralogical Museum Chita Polytechnic Institute	Gorky Str., 28, Chita 672000, USSR	Total specimens: 5,100 Specialties: Systematics of minerals; types of mineral deposits; regional mineralogy

TABLE 1. (Continued)

Name	Address	Total Specimens Specialties
Geological-Mineralogical Museum Grozny Oil Institute	Ordzhonikidze Sq. 100, Grozny 364902, USSR	Total specimens: 2,500 Specialties: Systematics of minerals; gemstones; regional mineralogy
Mineralogical Museum Irkutsk Polytechnic Institute	Lermontov Str., 83, Irkutsk 664028, USSR	Total specimens: 27,250 Specialties: Systematics of minerals; genesis of minerals; properties of minerals, crystals, gemstones, and synthetic minerals; regional mineralogy
Geological Museum Geological Sciences Institute Ukranian SSR Academy of Science	Lenin Str., 15, Kiev 252030, USSR	Total specimens: ca. 23,600 Specialties: Systematics of minerals; meteorites; gemstones; ores of Ukraine; systematics of rocks; properties of minerals
Y. S. Fedorov Geological Museum	Oktyabrskaya Str., 45, Krasnotur'insk 624460, USSR	Total specimens: 116,100 Specialties: Regional collection of the Sverdlovsk Region; Y. S. Fedrov memorial collection
F. N. Chernyshow Central Research Geological Museum	Srdny Prospect, 74, Leningrad 199026, USSR	Total specimens: ca. 435,000 Specialties: Regional mineralogy and geology; geology of mineral deposits; systematics of minerals; memorial collection
Mining Museum Leningrad Mining Institute	21 Line, 2, Leningrad 199026, USSR	Total specimens: 63,360 Specialties: Systematics of minerals; meteorites; gemstones; crystals; synthetic minerals; pseudomorphs; ontogeny of minerals; regional mineralogy
Mineralogical Museum Lvov State University	Stcherbakov Str., 4, Lvov 290005, USSR	Total specimens: ca. 13,500 Specialties: Systematics of minerals; genesis of minerals; meteorites; synthetic minerals; regional mineralogy; gemstones; ontogeny of minerals
Museum of V. I. Lenin Ilmen State Reservation	Miass 456301, USSR	Total specimens: ca. 1,800 Specialties: Regional collection of Ilmen Mountains
A. Y. Fersman Mineralogical Museum USSR Academy of Science	Leninsky Prospect 18-2, Moscow V–71, USSR	Total specimens: ca. 155,000 Specialties: Systematics of minerals; mineral forming processes; morphology of minerals; gemstones; synthetic minerals; meteorites; minerals discovered in the USSR; minerals of the Moscow region; colors of minerals; pseudomorphs; natural crystals; new specimens
Mineralogical Collection Institute of Mineralogy and Geochemistry of Rare Elements	Sadovnicheskaya Naberezhnaya, 71, Moscow 113127, USSR	Total specimens: 15,000 Specialties: Type specimens; mineral associations; systematics; morphology of growth of crystals and mineral aggregates in *gold* and thermal caves

TABLE 1. (Continued)

Name	Address	Total Specimens Specialties
Mineralogical Museum Moscow Geological Prospecting Institute	Marx Prospect, 18, Moscow 103012, USSR	Total specimens: ca. 50,000 Specialties: Systematics of minerals; genesis of minerals; regional mineralogy; pseudomorphs; crystals; gemstones; meteorites; properties of minerals
Museum of Earth Sciences Moscow State University	MGU, Leninskie Gory, Moscow 117234, USSR	Total specimens: 58,359 Specialties: Systematics of minerals; genesis of minerals; types of ore deposits; crystals; synthetic minerals; physical properties of minerals; gems; meteorites; regional collection
Petrographic Museum Institute of Ore Deposits Geology Petrography, Mineralogy and Geochemistry, USSR Academy of Sciences	Staromonetny, 35, Moscow 109017, USSR	Total specimens: 10,000 Specialties: Systematic of magmatic rocks; memorial (author's) collections
Central Siberian Geological Museum Siberian Branch of USSR Academy of Sciences	Universkity Prospect, 3, Novosibirsk 630090, USSR	Total specimens: ca. 32,500 Specialties: Systematics of minerals; meteorites; minerals from Siberia and Soviet Far East; synthetic minerals; gemstones; minerals of high pressures; agricultural raw minerals
Geological Museum Institute of Mineral Resources Ministry of Geology of Ukranian SSR	Kirov Prospect, 47/2, Simferopol 333620, USSR	Total specimens: 1,600 Specialties: Minerals, ores, and rocks of Ukraine; karst mineralogy, and building stones from the Crimea
Ural Geological Museum Sverdlovsk Mining Institute	Kuibyshev Str., 30, Sverdlovsk 620219, USSR	Total specimens: 25,190 Specialties: Systematics of minerals; gemstones; minerals discovered in the Urals; minerals and ores of Urals; meteorites
Mineralogical Museum Tomsk Polytechnic Institute	Lenin Prospect, 30, Tomst 634004, USSR	Total specimens: 12,650 Specialties: Ore and nonmetallic minerals; regional mineralogy
Geological Museum Buryat Geological Board	Lenin Str., 57, Ulan-Ude 670000, USSR	Total specimens: 6,000 Specialties: Systematics of minerals and crystals; industrial minerals of Buryaty
Geological Museum Geological Institute of Buryat Filial of the USSR Academy of Sciences	Pavlov Str., 2, Ulan-Ude 670015, USSR	Total specimens: 2,300 Specialties: Minerals, rocks, and ores from Buryaty
Geological Museum Institute of Geology of Yakut Filial of the USSR Academy of Sciences	Lenin Prospect, 39, Yatutsk 677891, USSR	Total specimens: ca. 1,400 Specialties: Systematics of minerals; minerals discovered in Yakuty; crystals; regional mineralogy
Museum of Board on Geology Council of Ministers of Armenian SSR	Charents Str., 46, Yerevan 375200, USSR	Total specimens: 1,000 Specialties: Building raw materials; gemstones; chemical raw materials; mineral dyes; ores

TABLE 1. (Continued)

Name	Address	Total Specimens Specialties
Ulster Museum Department of Geology	Botanic Gardens, Belfast BT9 5AB, Northern Ireland	Total specimens: ca. 10,000 Specialties: Principal representation–***garnets*** (> 1,000 specimens) and silica minerals (1,500 specimens); gemstones including 3 ***tourmalines*** and a peridot of exceptional quality
The Museums, Department of Geological Sciences University of Birmingham	P.O. Box 363, Birmingham B15 2TT, England	Total specimens: >5,000
National Museum of Wales Department of Geology	Cathays Park, Cardiff CF1 3NP, Wales	Total specimens: ca. 38,000 Specialties: Regional (Welsh) display, scientific exhibit
The Royal Scottish Museum Department of Geology	Chambers Street Edinburgh EH1 1JF, Scotland	Total specimens: 56,000 Specialties: Scottish mineral collection, about 20,000 specimens, is the most comprehensive such collection in the world; several carbonaceous meteorites
Glasgow Art Gallery and Museum	Kelvingrove, Glasgow G3 8AG, Scotland	Total specimens: ca. 12,250
Hunterian Museum University of Glasgow	Glasgow G12 8QQ, Scotland	Total specimens: ca. 38,500 Specialties: Scottish ***zeolites*** (Eck and Brown of Linfine collections); South American economic minerals (Eck collection); meteorites
Department of Geology University of Leicester	University Road, Leicester LE1 7RH, England	Total specimens: ca. 50,000
Leicestershire Museums, Art Galleries, and Records Service	96 New Walk, Leicester LE1 6TD, England	Total specimens: 12,780
Merseyside County Museum	Wililam Brown Street, Liverpool L3 8EN, England	Total specimens: ca. 5,700 Specialties: 19th Century collection of polished stones, especially agates, *nephrites,* and *labradorite*
British Museum (Natural History) Department of Mineralogy	Cromwell Road, London SW7 5BD, England	Total specimens: 282,500 Specialties: Russell Collection; Kingsbury Collection; Ashcroft Collections of Irish ***zeolites*** and of Swiss minerals; Pain collection of Burmese gemstones (on display); one of the most representative meteorite collections known. The mineral collection is widely known for species coverage and a wealth of type material.
Institute of Geological Sciences	Exhibition Road, London SW7 2DE, England	Total specimens: 225,000 Specialties: Regional displays from various mining areas; building-stone display; extensive gemstone collection–crystals, gems in matrix, and cut stones

TABLE 1. (Continued)

Name	Address	Total Specimens Specialties
Manchester Museum Department of Geology University of Manchester	Manchester M13 9PL, England	Total specimens: 24,000 (incl. dupl.) Specialties: ***Zeolites***, mainly from Caroline Birley Collection; unique suite of rocks and minerals from Alderley Edge, Cheshire
Hancock Museum	Newcastle upon Tyne NE2 4PT, England	Total specimens: >8,000 Specialties: William Hutton Collection (general, good British and local representation); Tsar Nicholas I Collection (1938, Russian rocks and minerals); N. Cookson Collection (general); African ore collection
University Museum Mineral Collections	Parks Road, Oxford OX1 3PR, England	Total specimens: 23,350 Specialties: Many specimens from classic localities, especially British; Corsi collection, 1,000 cut and polished slabs of decorative stones and marbles, particularly those used by the ancient Egyptians, Greeks, and Romans; exhibits of systematics; British regional mineralogy; rock-forming minerals; meteorites; gemstones; mineral fakes
Sheffield City Museum	Weston Park, Sheffield S10 2TP, South Yorkshire, England	Total specimens: ca. 3,600 Specialties: Minerals of the Derbyshire, England, orefield, including "Blue John" *fluorite* and industrial slags from metal refining. Half the exhibited specimens relate to South Yorkshire and central Derbyshire.
County Museum, Royal Institute of Cornwall	River Street, Truro, Cornwall, England	Total specimens: ca. 10,000 Specialties: Cornish minerals, particularly secondary copper minerals and especially the arsenates from the Gwennap mines; fine specimens from Derbyshire, Cumberland, and Durham
Yorkshire Museum	Museum Gardens, York Y01 2DR, England	Total specimens: ca. 6,000 Specialties: Historical interest, as much of the collection has connection with classical localities and 19th century collectors; coverage mainly Britain, central Europe, and North America
New York State Museum and Science Service	State Education Building Albany, New York 12234 USA	Total specimens: 38,934 Specialties: New York State rocks, minerals and ores; cores of New York State carbonate rocks; cutting and cores from oil and gas well drilling in New York State; gem display explains concepts of gemology

TABLE 1. (Continued)

Name	Address	Total Specimens Specialties
Department of Geology University of New Mexico	Albuquerque, New Mexico 87131 USA	Total specimens: ca. 5,000 Specialties: Meteorites
Institute of Meteorites University of New Mexico	Albuquerque, New Mexico 87131, USA	Total specimens: 17,775 Specialties: Meteorite collection
Colburn Memorial Mineral Museum Southern Appalachian Mineral Society	Box 1617, Asheville, North Carolina 28802 USA	Total specimens: ca. 5,000 Specialties: *Hiddenite, fluorite, torbernite, silver, copper, itacolumnite,* and local gemstones
Department of Geology Emory University	Atlanta, Georgia 30322 USA	Total specimens: 2,805 Specialties: Gems; local and regional display; scientific exhibit
Geoscience Department Belmont Abbey College	Belmont, North Carolina 28012, USA	Total specimens: 8,205 Specialties: Minerals of the Foote Quarry; minerals of North Carolina
Department of Geology and Geophysics. University of California	Berkeley, California 94720, USA	Total specimens: ca. 90,000 Specialties: California localities
Department of Geological Sciences Lehigh University	Bethlehem, Pennsylvania 18015, USA	Total specimens: 11,040 Specialties: Ore minerals
Virginia Polytechnic Institute and State University	Blacksburg, Virginia 24061, USA	Total specimens: 3,450
Cranbrook Institute of Science	500 Lone Pine Road Bloomfield Hills, Michigan 48013, USA	Total specimens: 10,800 Specialties: Displays of opals and other gem minerals, fluorescent minerals, transparencies; displays illustrating occurrences, chemistry, crystallography, physical properties of minerals; regional displays (Michigan, Ohio, Ontario)'
Department of Geology Indiana University	1005 East Tenth Street Bloomington, Indiana 47401, USA	Total specimens: 15,015 Specialties: Excellent representation of igneous (volcanic and plutonic) rock suites from all over the world
Geology Department Appalachian State University	Boone, North Carolina 28608, USA	Total specimens: 7,027
Boston University	725 Commonwealth Ave. Boston, Massachusetts 02215, USA	Total specimens: ca. 7,000 Specialties: General systematic collection
Museum of Sciences	Science Park, Boston, Massachusetts 02114, USA	Total specimens: 7,053
University of Colorado Museum	Boulder, Colorado 80302 USA	Total specimens: 11,619 Specialties: Uranium minerals
Department of Geology Bowdoin College	Brunswick, Maine 04011 USA	Total specimens: 8,009 Specialties: Maine minerals exhibit
Rand Collection Bryn Mawr College, Department of Geology	Bryn Mawr, Pennsylvania 19010, USA	Total specimens: 9,500

TABLE 1. (Continued)

Name	Address	Total Specimens Specialties
Vaux Collection Bryn Mawr College, Department of Geology	Bryn Mawr, Pennsylvania 19010, USA	Total specimens: ca. 10,000 Specialties: Extensive collection from French Creek and some other Pennsylvanian localities, New England, and Western South America
Buffalo Museum of Science	Humboldt Park, Buffalo, New York 14211, USA	Total specimens: ca. 15,000
Montana College of Mineral Science and Technology	Butte, Montana 59701 USA	Total specimens: ca. 14,000 Specialties: Minerals from Butte, Montana, minerals from Montana; display of fluorescent minerals
Harvard University Mineralogical Museum	24 Oxford Street Cambridge, Massachusetts 02138, USA	Total specimens: 126,500 Specialties: Meteorites; gems; gold; Franklin, N.J., minerals; New England minerals; all classic localities
Nevada State Museum	Capital Complex, Carson City, Nevada 89710 USA	Total specimens: ca. 15,000 Specialties: Gold and silver ores from the Comstock Lode
University of Northern Iowa Museum University of Northern Iowa	Cedar Falls, Iowa 50613 USA	Total specimens: ca. 10,000 Specialties: Scientific exhibits
Geology Department University of North Carolina	Chapel Hill, North Carolina 27514, USA	Total specimens: 8,500
Department of Environmental Sciences, Clark Hall University of Virginia	Charlottesville, Virginia 22903, USA	Total specimens: >17,000 Specialties: Virginia minerals and rocks
Field Museum of Natural History	Roosevelt Road at Lake Shore Drive Chicago, Illinois 60606 USA	Total specimens: 64,800 Specialties: The meteorite collection is one of the largest in the world
Cincinnati Museum of Natural History	1720 Gilbert Avenue Cincinnati, Ohio 45202, USA	Total specimens: ca. 3,000
Department of Geology Hamilton College	Clinton, New York 13323, USA	Total specimens: ca. 4,000 Specialties: Oren Root Mineral Collection: Adirondack and Canadian minerals with a few from other parts of the world
Department of Geology University of Missouri	Columbia, Missouri 65201, USA	Total specimens: 25,000 Specialties: Minerals from Crestmore, California; ores from Tri-State District, Old Lead Belt, New Lead Belt, Missouri
Laurence L. Smith Geology Museum LeConte Building Department of Geology, University of South Carolina	Columbia, South Carolina 29208, USA	Total specimens: ca. 8,500 Specialties: Hiddenite, corundum, and quartz with clay inclusions from North Carolina; radioactive minerals from Spruce Pine, North Carolina
Gillespie Museum of Minerals Stetson University	Deland, Florida 32720, USA	Total specimens: 26,112

TABLE 1. (Continued)

Name	Address	Total Specimens Specialties
Geology Department Denver Museum of Natural History	City Park, Denver, Colorado 80205, USA	Total specimens: 15,700 Specialties: Breckenridge gold; Colorado minerals and meteorites; Western United States classic localities; reference micromount collection
Geology Department Duke University	6665 College Station Durham, North Carolina 27708, USA	Total specimens: ca. 5,000 Specialties: ***Zeolites***
Lafayette College	Easton, Pennsylvania 18042, USA	Total specimens: ca. 11,000 Specialties: Minerals from Pennsylvania
Lizzadro Museum of Lapidary Art	220 Cottage Hill Avenue Elmhurst, Illinois 60126, USA	Total specimens: 4,750 Specialties: Raw lapidary material
El Paso Centennial Museum University of Texas at El Paso	El Paso. Texas 79968, USA	Total specimens: ca. 3,400 Specialties: Meteorites; ore and mineral specimens from New Mexico, Colorado, Texas, Arizona, and Chihuahua, Mexico
The University of Texas at El Paso	El Paso, Texas 79968, USA	Total specimens: 12,000 Specialties: Ore minerals
Department of Geological Sciences Northwestern University	Evanston, Illinois 60201, USA	Total specimens: ca. 9,000
University of Alaska	Fairbanks, Alaska 99701, USA	Total specimens: 3,400 Specialties: Ore specimens from Alaskan mining districts
Fall River Public Library	104 North Main Street Fall River, Massachusetts 02720, USA	Total specimens: 3,000
University Museum, University of Arkansas	Fayetteville, Arkansas 72701, USA	Total specimens: 2,000 Specialties: Hugh D. Miser *Quartz* Crystal Collection; collections made during thesis research on most mineral districts in Arkansas
Findlay College	1000 North Main Street Findlay, Ohio 45840, USA	Total specimens: ca. 10,000
Museum of Northern Arizona	P.O. Box 1389, Flagstaff, Arizona, 86001, USA	Total specimens: 1,776 Specialties: Minerals of Arizona
Franklin Mineral Museum, Inc.	Evans Street, P.O. Box 76 Franklin, New Jersey 07416, USA	Total specimens 2,565 Specialties: Franklin, New Jersey specimens; fluorescent display
The John L. Beal Memorial Collection The Schiele Museum of Natural History and Planetarium, Inc.	1500 East Garrison Blvd. P.O. Box 953 Gastonia, North Carolina 28252, USA	Total specimens: 10,600 Specialties: History of gold in North Carolina (educational exhibit); regional collection of minerals from North Carolina and the southeast U.S.
The Geology Museum, Colorado School of Mines	16th and Maple, Golden, Colorado 80401, USA	Total specimens: 40,700 Specialties: Colorado and Western U.S. minerals; Colorado ores

TABLE 1. (Continued)

Name	Address	Total Specimens Specialties
Department of Geology University of North Dakota	Grand Forks, North Dakota 58202, USA	Total specimens: 8,000
Neville Public Museum	129 South Jefferson Street Green Bay, Wisconsin 54301, USA	Total specimens: ca. 3,400 Specialties: Iron and copper minerals from Michigan; rocks, minerals and fossils from Wisconsin
William Penn Memorial Museum	Box 1026, Harrisburg, Pennsylvania 17120, USA	Total specimens: 13,231 Specialties: Minerals from Pennsylvania; minerals from French Creek Mine, Chester County
The A. E. Seaman Mineralogical Museum Michigan Technological University	Houghton, Michigan 49931, USA	Total specimens: >30,000 Specialties: Minerals of the "Copper Country," northern Michigan
Tri State Mineral Museum	Schifferdecker Park Joplin, Missouri 64801, USA	Total specimens: 9,000 Specialties: All specimens from local district
Department of Geosciences University of Missouri	Kansas City, Missouri 64110, USA	Total specimens: ca. 8,500 Specialties: Crystals containing megascopic fluid inclusions; geodes; Magnet Cove and Tristate minerals; Gypsum
Kansas City Museum of History and Sciences	3218 Gladstone Blvd. Kansas City, Missouri 64123	Total specimens: ca. 12,000 Specialties: The "Kansas City Meteorite"
Department of Geological Sciences The University of Tennessee	Knoxville, Tennessee 37916, USA	Total specimens: ca. 10,000
North Museum, Franklin and Marshall College	Lancaster, Pennsylvania 17604, USA	Total specimens: ca. 7,000 Specialties: Local and regional displays; scientific exhibits; ore minerals
University of Wyoming Geological Museum	Laramie, Wyoming 82071, USA	Total specimens: 8,012
St. Vincent College Museum	Latrobe, Pennsylvania 15650, USA	Total specimens: 1,877
Department of Geology University of Kansas	Lawrence, Kansas 66044, USA	Total specimens: 6,600
Washington and Lee University	Lexington, Virginia 24450, USA	Total specimens: ca. 5,500
University of Nebraska State Museum, Morrild Hall	14th and "U" Streets Lincoln, Nebraska 68588, USA	Total specimens: 36,750 Specialties: Nebraska minerals, rocks, and meteorites; Nebraska oil and gas; E. F. Schramm Collection; rocks from Antarctic
Geology Department University of California at Los Angeles	405 Hilgard Avenue Los Angeles, California 90024, USA	Total specimens: ca. 19,000
Los Angeles County Museum of Natural History	900 Exposition Blvd. Los Angeles, California 90007, USA	Total specimens: 20,354 Specialties: ***Tourmalines*** from southern California
Tufts University	Medford, Massachusetts 02155, USA	Total specimens: ca. 13,000

TABLE 1. (Continued)

Name	Address	Total Specimens Specialties
Wesleyan University	Middletown, Connecticut 06457, USA	Total specimens: ca. 8,000 Specialties: Minerals of Connecticut
Milwaukee Public Museum	800 West Wells Street Milwaukee, Wisconsin 53233, USA	Total specimens: 50,475
Department of Geology and Geophysics University of Minnesota, Twin Cities	108 Pillsbury Hall Minneapolis, Minnesota 55455, USA	Total specimens: 56,500 Specialties: Collection of radioactive minerals; analyzed rocks; samples from mineral districts
Zeitner Geological Museum	Mission, South Dakota 57555, USA	Total specimens: 8,012 Specialties: Meteorites; minerals and gems from South Dakota; minerals from Mexico; geodes; petrified wood; *barite.*
Department of Geology University of Montana	Missoula, Montana 59801, USA	Total specimens: 7,456
Morris Museum of Arts and Sciences	Normandy Heights and Columbia Rds. Morristown, New Jersey P.O. Box 125, Convent, N.J. 07961, USA	Total specimens: 4,336 Specialties: European minerals
Department of Geology University of Idaho, College of Mines	Moscow, Idaho 83843, USA	Total specimens: 3,018 Specialties: Excellent collector's "rock hound" specimens from Idaho localities; especially fine historical collection of early German mining artifacts and pictures; excellent Korean collection
Department of Geology Cornell College	Mount Vernon, Iowa 52314, USA	Total specimens: 9,860
Department of Geology Vanderbilt University	Nashville, Tennessee 37235, USA	Total specimens: ca. 14,000
New Almaden Museum	21570 Almaden Road, P.O. Box 1, New Almaden, California 95042, USA	Total specimens: 2,350 Specialties: Minerals from the first workable mercury mine in North America
Department of Geology University of Delaware	Newark, Delaware 19711, USA	Total specimens: ca 4,000 Specialties: The duPont Mineral Room contains a display of minerals from all localities, with some emphasis on the minerals of Delaware and southeastern Pennsylvania
Department of Geology Rutgers University	Newark, New Jersey 07102, USA	Total specimens: ca. 13,000
Geological Museum Rutgers University	New Brunswick, New Jersey 08903, USA	Total specimens: ca. 11,000 Specialties: Mines and their minerals of New Jersey; displays
Peabody Museum Yale University	New Haven, Connecticut 06520, USA	Total specimens: 31,000

TABLE 1. (Continued)

Name	Address	Total Specimens Specialties
American Museum of Natural History	Central Park West at 79th Street, New York, New York 10024, USA	Total specimens: 104,500 Specialties: Gem collection; meteorite collection; North American and classic mineral specimens
Department of Geology Columbia University	New York, New York 10027, USA	Total specimens: ca. 50,000
Carleton College	North Field, Minnesota 55057, USA	Total specimens: >8,000 Specialties: ca. 1,500 polished agates
Orton Geological Museum Department of Geology and Mineralogy Ohio State University	Ohio State University Columbus, Ohio 43210, USA	Total specimens: 4,785
Division of Geology and Planetary Sciences California Institute of Technology	Pasadena, California 91125, USA	Total specimens: 9,389
Paterson Museum	268 Summer Street Paterson, New Jersey 07505, USA	Total specimens: 6,275 Specialties: Minerals from Paterson and Franklin, New Jersey; fluorescent minerals
The Academy of Natural Sciences	19th and The Parkway Philadelphia, Pennsylvania 19103, USA	Total specimens: 309 Specialties: Probably the finest collection of Pennsylvania minerals. Notable specimens collected by Samuel G. Gordon on his expeditions to Greenland (Narssârssuk), Bolivia (Llallagua), Chile, and South Africa
Wagner Free Institute of Science	Seventeenth Street and Montgomery Avenue Philadelphia, Pennsylvania 19121, USA	Total specimens: 5,153
State of Arizona, Department of Mineral Resources	State Fairgrounds Phoenix, Arizona 85007, USA	Total specimens: 2,651 Specialties: Native *copper* (Arizona); manganese minerals; peridot; *quartz*; mercury ores; Ray Mine ores; Inspiration Mining Company minerals and ores; gold, silver, and uranium ores; meteorites
Carnegie Museum of Natural History	4400 Forbes Avenue, Pittsburgh, Pennsylvania 15213, USA	Total specimens: 26,648 Specialties: Pennsylvanian minerals
Diablo Valley College Museum	Golf Club Road, Pleasant Hill, California 94523, USA	Total specimens: 3,706
Vassar College	Poughkeepsie, New York 12601, USA	Total specimens: 5,565
Department of Geological Sciences Brown University	Providence, Rhode Island 02912, USA	Total specimens: 3,500
Brigham Young University	Provo, Utah 84601, USA	Total specimens: ca. 15,000
Museum of Geology South Dakota School of Mines and Technology	Rapid City, South Dakota 57701, USA	Total specimens: 15,730 Specialties: Pegmatite minerals; systematic collection; Black Hills minerals; *gold*; meteorites

TABLE 1. (Continued)

Name	Address	Total Specimens Specialties
Geology Museum, Department of Earth Sciences University of California	Riverside, California 92502, USA	Total specimens: ca. 20,000 Specialties: Minerals from Crestmore and Boron, California
Riverside Municipal Museum	3720 Orange Street Riverside, California 92501, USA	Total specimens: 7,500 Specialties: Minerals from nearby localities
Pinch Mineralogical Museum	82 Kensington Court Rochester, New York 14612, USA	Total specimens: 11,035 Specialties: Minerals from St. Hilaire, Tsumeb, Franklin, and Langban; uranium minerals from Congo; mercury, tellurium, and selenium minerals
Rochester Museum and Science Center	657 East Avenue Rochester, New York 14603, USA	Total specimens: 5,359 Specialties: Native *copper, calcite*
Fryxell Geology Museum and John Deere Planetarium Augustana College	Rock Island, Illinois 61201, USA	Total specimens: 5,756 Specialties: Illinois minerals and fossils; meteorites; geodes
University of Missouri-Rolla	Rolla, Missouri 65401, USA	Total specimens: 3,200
Department of Geology and Geophysics University of Utah	Salt Lake City, Utah 84112, USA	Total specimens: 29,001 Specialties: Ore suites from local mining districts
San Diego Natural History Museum	P.O. Box 1390 Balboa Park, San Diego, California 92112, USA	Total specimens: ca. 7,000 Specialties: Borates from California; minerals from Mexico; minerals from Arizona
California Academy of Sciences	Golden Gate Park San Francisco, California 94118, USA	Total specimens: 10,000 Specialties: Minerals from famous localities of California–San Diego County pegmatites, Crestmore minerals, California borates, glaucophane schist minerals; suites of minerals from Sterling Hill and Franklin, New Jersey
State of California, Division of Mines and Geology	Room 2022 Ferry Building San Francisco, California 94111, USA	Total specimens ca. 30,000
Geology Department San José State University	San José, California 95172, USA	Total specimens: ca. 5,500
Santa Barbara Museum of Natural History	2559 Puesta del Sol Road Santa Barbara, California 93105, USA	Total specimens: 3,220
Union College	Schenectady, New York 12308, USA	Total specimens: ca. 8,000
Burke Memorial Washington State Museum Division of Geology and Paleontology University of Washington	Seattle, Washington 98195, USA	Total specimens: 5,602 Specialties: *Quartz* minerals, especially amethyst, *wulfenite,* copper minerals, and iron minerals; regional display
New Mexico Bureau of Mines and Mineral Resources	Socorro, New Mexico 87801, USA	Total specimens: 6,000 Specialties: New Mexico minerals; C. T. Brown Collection

TABLE 1. (Continued)

Name	Address	Total Specimens Specialties
Department of Geology and Geography Mount Holyoke College	South Hadley, Massachusetts 01075, USA	Total specimens: ca. 18,000
Illinois State Museum	Spring and Edwards Street Springfield, Illinois 62706, USA	Total specimens: 6,600
Springfield Science Museum	236 State Street, Springfield, Massachusetts 01103, USA	Total specimens: 8,100 Specialties: Local (western Massachusetts and Connecticut) minerals
The Science Museum of Minnesota	30 East 10th Street St. Paul, Minnesota 55101, USA	Total specimens: ca. 15,000 Specialties: Iron ore minerals; opals
Arizona State University	Tempe, Arizona 85281, USA	Total specimens: 4,700 Specialties: Meteorites
Heidelberg College	Tiffin, Ohio 44883, USA	Total specimens: ca. 6,000
New Jersey State Museum	205 West State Street Trenton, New Jersey 08625, USA	Total specimens: ca. 4,000 Specialties: Economic geology specimens from New Jersey
Arizona-Sonora Desert Museum	P.O. Box 5607 Tucson, Arizona 85703, USA	Total specimens: 11,800 Specialties: Regional collection covering the states of Arizona, Sonora, Mexico, both states of Baja, Mexico, and the Islands of the Gulf of California
Mineralogical Museum, Geology Building University of Arizona	Tucson, Arizona 85721, USA	Total specimens: ca. 10,700 Specialties: Arizona minerals; Bisbee and Tiger minerals; *wulfenite* specimens; meteorites; gems; fossils
National Museum of Natural History Smithsonian Institution	Washington, D.C. 20560, USA	Total specimens: 411,500 Specialties: Minerals of the United States and Mexico; gems; meteorites
West Chester State College	West Chester, Pennsylvania 19380, USA	Total specimens: 5,000 Specialties: Minerals from Chester County, Pennsylvania
Purdue University	West Lafayette, Indiana 47907, USA	Total specimens: ca. 8,900

Acquisition and Conservation

One of the most important tasks of the curator of any mineralogical museum is to collect systematically, without regard to current academic fashion. Only by doing so will it be possible for the museum to satisfy future needs, the form of which, in many cases, may not be anticipated. Although such factors as the overall ideas of the museum administrator and the personal taste and education of the curator influence the policy of acquisition in the short term, there remain some basic, long-term purposes of procurement that no curator can afford to overlook.

The specimens sought by curators fall into several categories. One such is that of so-called "museum-specimens," which includes not only the show pieces but also specimens—whether they be beautiful or not, large or small, well crystallized or not—that in one way or another are unique.

Except for museums with only regional collections, any mineralogical museum will endeavour to complete its systematic collection, i.e., to acquire the best available specimens of each of the roughly 3000 known mineral species and as many of the varieties as possible.

In order to comply with the oft-repeated request for many specimens of a particular mineral or group of minerals from a wide variety of localities, curators of mineralogical museums often attempt to establish a collection that is as geographically representative as possible. A completely systematic collection would have, for instance, *galena* from every significant occurrence that has produced it.

The last group of mineral specimens to be found on a mineralogical museum's acquisitions list is specimens that form part of regional reference collections, i.e., collections designed to illustrate all mineralogical aspects of certain localities and(or) certain types of mineral deposits.

Mineralogical museums expand their collections in many different ways, four of which are collecting, exchange, buying, and donations. To which of these activities priority is given varies from museum to museum, and is determined by internal factors and by the total welfare of the society of which the museum is part.

The museum curator must collect as opportunity arises. In order that he may do so, he must develop friendly contacts with as wide a range of people as he can—professional geologists, amateur geologists, and discerning amateur collectors of any kind (most mineralogical museums owe large and important parts of their collections to enlightened amateurs).

Most mineralogical museums exchange specimens—primarily with cooperating institutions but also with amateur collectors and dealers. An effective, well-managed exchange program is an excellent way of converting accumulated surplus material into the valuable addition of specimens not represented or poorly represented in the collection.

Once important, but now universally reduced in significance, is acquisition by buying. Rapidly increasing prices on mineral specimens due to an explosively growing interest and a relative decrease in supply, and a continuous, though fluctuating, decrease in economic resources for this purpose have put many museums into a precarious financial situation, where they are unable even to maintain their present standards.

Donations have always been and still are an important way for museums to improve and enlarge their collections. Donations of mineral collections, single superb specimens or just ordinary material suitable for study purposes, are important to any museum. Their importance increases when restrictions regarding their use are minimal.

Acquisition, and the consequent accumulation of mineral specimens, is of little use if the minerals are not properly curated. Adequate storage in fire-proof and burglar-proof buildings with appropriate rooms—dust free, with constant temperature and humidity, and especially built cabinets—is essential. Complete, accurate, and entirely updated documentation of all specimens is just as essential. The aim should be that all information is easily available on demand; the solution probably lies in computerization of information so that sections of data can be printed out on demand. Although appropriate conservation should be applied to all specimens, type materials require especially careful treatment. For historical reasons, mineralogical museums formerly served as safe, permanent repositories for type material; however, the diminished research role of museums relative to the whole body of earth science activities has changed this. A recent decision by the International Mineralogical Association's Commission on New Minerals and Mineral Names has the aim that museum collections of minerals will recover their position as natural repositories of the actual homogeneous material selected for study of a new species, the holotype material. Conservation of mineral specimens also offers a number of technical problems—some are already solved; others require much more research.

Occasionally, museum mineral collections need to justify their existence. This can probably best be done by describing in detail the functions of research, exhibition, and education. It must, however, be emphasized that there are other justifications. Quite a number of museums contain several, a few are filled with, valuable mineralogical materials collected through tens of generations. The mere existence of these collections is one of the best reasons for curating and caring for them. Some collections, whether great collections or not, may be justified by their archival value. A small example may illustrate the special responsibilities and archival contribution of a museum with a long history, where unique mineral specimens are preserved.

In the Mineralogical Museum of Copenhagen, the M.M. No. 1 specimen of *cryolite* carries a label stating on one side "schwerspath," on the other side "kriolit/allanunerde" and "flusspathsäure/Groenland." Translation of the first side is "barite," of the other side "cryolite/a compound reacting like a solution of allunite in water = Al_2O_3 and the acid in fluorite = HF/Greenland." In 1795 and again in 1798, Regiment-Surgeon C. F. Schumacher examined the physical properties of some of the "cryolite" specimens brought to Copenhagen as curiosa, but reached the erroneous conclusion that he was dealing with barite. The veterinary-scientist P. C. Abbildgård in 1799 showed, as the label says, that the specimen contains, in modern terms, Al_2O_3 and HF, and suggested the name kriolit (*cryolite*) for this new species. In 1868, Abbildgård's collection found its final home in the Mineralogical Museum of Copenhagen. It was fortunate that the numerous caretakers there had the good sense and opportunity to keep this ugly specimen, which probably has been examined by Schumacher, handed over to Abbildgård, analyzed by him, and provided by him with the label stating his results and the name kriolit (*cryolite*) for the first time in print.

It should not be forgotten either that exhibits of mineral specimens, which are as good and as abundant as opportunities and aesthetics permit, have a high value for entertainment.

Research

Since the number of scientists working within the fields of earth science has increased dramatically during the last twenty years, and since museums must continue to supply the needs of geologists whose interests lead them to the examination of specimens, it is obvious that the importance of mineral collections as specimen banks, with possibilities for loan to qualified researchers on request, has also increased markedly. It is essential, however, that museum staff members themselves are active in research, whether prompted by the special characteristics of their respective collections or by their own special abilities. Only by being active are they able to provide a disciplined foundation upon which their judgments may be based. Descriptive mineralogy is appropriate to museum research, based as it is on mineral collections; but it is certainly not the only mineralogical discipline that should attract geologists privileged to work in a museum.

Exhibition and Education

The third major function of a mineralogical museum is exhibition and education.

What are the relevant objectives of a mineralogical museum and what sort of exhibits will lead to their attainment? Obviously, such complicated problems have no simple answers. However, there are perhaps three types of exhibits that no mineralogical museum of reasonable size (unless limited in scope) can freely ignore. Which of these shall be given priority must be decided by each museum in the light of public need and the history, nature, and size of the collections and resources of the museum.

The first group of exhibits could be called systematic displays, i.e., presentations of as many as possible of the treasures of the mineral kingdom arranged in some orderly and attractive way. Although strong feelings have developed today among museologists that this type of exhibit is not enough, one should not lose sight of the fact that such displays for many generations were the objective of all mineralogical museums. For quite a few museums, history has made it mandatory that they maintain exhibits of comprehensive systematic collections; they are the only ones with this capacity. Also, it should be remembered that really beautiful mineral specimens, by their mere existence, have a high entertainment value. Furthermore, such displays provide a fine source of helpful visual aids to education at almost all levels. The visitors' previous learning provides the key to the information assembled row upon row.

Educational centers today very often have excellent slides, models of crystals, and even casts of mineral specimens. Although television, for several reasons, is able to present things in a very powerful way, the unique object, the real thing seen in the museum, will be studied with much greater and more concentrated interest by the learner.

Actual as well as potential visitors to mineral collections should have easy access to ordinary specimens, e.g., *quartz*, but the beauty of the many fine specimens of *quartz* seen in the museum would greatly expand the meaning of *quartz*.

With the strongly growing feeling that museums must be educational institutions, exhibits are and will be designed to lead visitors into a learning situation. Mineral specimens easily lend themselves to a popularized explanation of mineralogical principles. Mineralogical specimens can be used to explain what they are, how they differ from each other, how they form, their physical and chemical nature, their crystallography, and so on. Their reaction to light, heat, and other forms of energy; their stability, their peculiarities; and a multitude of other topics are suitable educational fare. In order to spell out carefully these lessons, the intensive use of dioramas, photographs, graphics, models, open exhibits, participating exhibits, etc., must be intensively used. Care must, however, be taken not to degenerate this type of display so that the minerals are reduced to a few instructive specimens stuck in between the lessons; the specimens are the reason for the exhibits being there at all. These sorts of exhibits could probably be called topical exhibits. As it will appear from the following, there is no sharp, well-defined boundary between this second group of exhibits and the third.

The third group of exhibits are interdisciplinary displays. Mineralogy is intimately connected with geology which itself is part of the earth sciences. Science includes the natural sciences, of which the earth sciences are only one part. Mineral collections may contribute, in various ways, to displays within the entire field of science. The curator of a mineralogical museum must keep abreast of the field to insure that such interdisciplinary, integrated, educational exhibitions inform the public on the latest development in mineralogy. The increasing specialization in earth science, as well as in other scientific disciplines, requires increasingly the service of the popularizer to interpret mineralogy, not only to the general public, but also to colleagues in other sciences.

Whether mineral specimens should be used in the exposition of more remote themes, e.g., environment of man (air pollution) or activities of man (race discrimination), is a matter of taste or, perhaps more correctly, a political problem.

Quite a number of mineralogical museums curators, often in cooperation with colleagues in education, have gone a long way in developing integrated programs to assist earth science education on all levels from kindergarten to high school and sometimes including courses for adults. The results of the new thinking on exhibitions and education will become increasingly apparent in the coming years. It is to be expected that the expository functions of museums will increase in scope and effectiveness. It is to be regretted that so many mineralogical museums have not carried out, or have not had the opportunity to carry out, the many new ideas along this trend.

OLE V. PETERSEN

References

Auer, H., 1974. Museums of the natural and exact sciences, *Museum*, **XXVI**(2), 68–75.

Desautels, P., 1970–1975. The museum record, *Mineral. Rec.*, **1**(1) through **6**(2).

Encyclopedia Britannica, 1965, vol. 15, 1047.

Murray, D., 1904. *Museums, Their History and Their Use*. Glasgow, 3 vols.: James McLehore and Sons.

Stevenson, L. S., 1972. Geological museums as a part of earth science education. *Intern. Geol. Cong., 24th Sess. Canada*, sect. 17, 65–70.

Waterston, C. D., 1972. Geology and the museum, *Scott. J. Geol.*, 8, pt. 2, 129–144.

Wittlin, A. S., 1949. *The Museum: Its History and Its Task in Education*. London: Routledge and Kegan Paul.

Cross-references: *Collecting Minerals; Crystal Habit; Gemology; Mineral Classification: History; Naming of Minerals.*

N

NAMING OF MINERALS

The rules for the selection of names for newly discovered minerals are of interest, not only to the professional mineralogist, but also to the much larger group of hobbyists who are amateur mineral collectors and "rockhounds." It has come as a matter of surprise to some members of the latter group that there exists a formal code for the naming of minerals. It is necessary to understand and follow carefully the rules of this code in order to prevent confusion in mineralogical nomenclature. In the past, there has not always been strict adherence to rules of mineral nomenclature, but progress is being made in enforcing better cooperation. It is probably true that nearly as much time is now spent in correcting earlier errors and confusion and in discrediting improperly named or invalid species as in describing and naming new valid species. Each issue of *The American Mineralogist* has a list, prepared by Michael Fleisher and others, of new mineral names, new data, and discredited minerals; and frequently the contributions in the latter two categories outnumber those in the first.

The gratitude and respect of mineralogists go to James Dwight Dana (1813–1895) for the part he played in the development of rules for mineral nomenclature. In the first editions of his *System of Mineralogy*, Dana followed the suggestions of Friedrich Mohs (Staples, 1964) and Linnaeus by adopting the binomial latin nomenclature using genus and species. Later, in the third edition of *System of Mineralogy*, in 1850, Dana broke away from Mohs' natural history classification and adopted a classification based on chemical composition as recommended by Berzelius, with a single-word name for minerals. For minerals, unlike organic materials, the binomial nomenclature was without scientific basis and conservation of space and effort highly recommended the change. As an example, such a difficult name as "barulus ponderosus" was changed to "*barite*."

Although there was a simplification in the types of names used for minerals, neither Dana nor anyone else attempted to limit the choice of names to any particular category. This has led to a wide variety of name types, which some people of orderly minds have decried as producing an unacceptable potpourri, with names difficult to memorize and, in some cases, even more difficult to pronounce. Because mineral names are necessarily international in origin, there will always be difficulty in pronunciation for people of different nationalities. As an example, most English-speaking people have difficulty in pronouncing the phosphate mineral, *przhevalskite*, or the vanadate, *tyuyamunite*.

Derivation of Mineral Names

The mineralogist who first describes a new mineral has almost complete freedom in the selection of a name. The practice of naming minerals after distinguished people of many different callings has been very common. This practice was initiated in 1783 by A. G. Werner, who named *prehnite* after the Dutch Colonel Prehn, who is reported to have obtained the mineral at the Cape of Good Hope. Other examples are *goethite* after Goethe, the German poet; *wernerite*, after the famous German mineralogist, A. G. Werner; *scheelite* after K. W. Scheele, a Swedish chemist; *uvarovite* after Count Uvarov of Russia, who was an amateur mineral collector; and *alexandrite* after Alexander II, Emperor of Russia.

A problem has arisen when the discoverer of a new mineral has found that the surname of a person he wanted to honor is already in use for a mineral. In that case, the given name may have to be used. As an example, the writer wished to name a new calcium zinc arsenate after Professor Austin Flint Rogers of Stanford University (Staples, 1935). The name "rogersite" had been preempted in 1877, and again in 1928 for a second mineral. Both uses were later discredited (Fleischer, 1966, 1306). The use of the name a third time would have led to confusion; consequently the mineral was named *austinite*. Recently (Gaines, 1969) a mineral was named *cliffordite*, after Prof. Clifford Frondel of Harvard University, using his given name because previously (1949) the mineral *frondelite* had been named after him. There are also examples of full names being used, as in the case of the mineral *tombarthite*–the name "barthite" was already in the literature, although discredited (Fleischer, 1945). Using the given name, when it is as short as "Tom," would not have been satisfactory, and so mineralogists are happy to have the great Norwegian geologist, Tom F. W. Barth, honored by using his full name (Neumann and Nilssen, 1968).

It is also common practice to name minerals after the locality in which they are found. For example, countries have been honored, as in the name *brazilianite*, states as in the case of *oregonite*, counties as in *benitoite*, and towns or localities as *franklinite* for Franklin Furnace, N.J. Physical properties were used frequently, as illustrated by *azurite* for the color; *amblygonite* for the Greek word indicating a blunt angle between the cleavages; and *scorodite* from the Greek word for garlic, which is the odor given off when the mineral is heated. The mineral *tetrahedrite* was named after its crystal form (the converse is also true of *pyrite* where a crystal form, the pyritohedron, was named after the mineral). Several minerals have been named after the principal metal present, as in *zincite*; and, as a special warning, the name *sphalerite* meaning "treacherous" was given because of the difficulty in identification.

Werner started the practice of naming minerals after people, but he was criticized as being guilty "of creating a paternity, and providing the childless with children to hand down their names to posterity" (Dana, 1892, p. xli). To get away from the objection, and at the same time to provide a crutch for the student to use in remembering the chemical composition of a mineral, one can use a mnemonic name. An example of this is the new calcium vanadium silicate named by the writer *cavansite* from the first letters of the chemical constituents (Staples, Evans, and Lindsay, 1967).

Rules of Mineral Nomenclature

The rules of nomenclature, which were used by J. D. Dana and later updated, are stated in the introductory section of the 6th edition of his *System of Mineralogy*, and a further discussion of them is given in volume one of the 7th edition by Palache, Berman, and Frondel. A Committee on Nomenclature and Classification of Minerals of the Mineralogical Society of America made several important recommendations (*Am. Mineralogist*, **8**, 1923: 50; **9**, 1924: 60; **21**, 1936: 188), most of which have been generally adopted by mineralogists. In 1933 considerable progress was made in obtaining agreement on usage by the American and British Mineralogical Societies. International agreements on nomenclature have since been delegated to the Commission on New Minerals and Mineral Names of the International Mineralogical Association. An index of new mineral names was compiled by Michael Fleischer in 1966, based on papers published in *The American Mineralogist*.

It has been generally agreed that new mineral names should end with the suffix "ite." In 1923, a minority of the Committee on Nomenclature made the suggestion that all mineral names be required to end in "ite," while the majority of the committee recommended changing only 43 mineral names. This resulted in names such as "cinnabarite" and "galenite." Some textbooks followed this suggestion, although not always consistently. For example, Moses and Parsons, in *Mineralogy, Crystallography, and Blowpipe Analysis* (5th ed.), use "galenite" and "metacinnabarite" but retain *cinnabar*. Most texts now have logically dropped the "ite" from *metacinnabar*, as well as from all those names of long historical standing that did not originally end in "ite."

Other less common endings for mineral names are "ine," as in ***olivine***, "ase" as in *dioptase*, "ime" as in *analcime*, "ole" as in ***amphibole***. There are advantages in the use of such a variety of suffixes in making the nomenclature less monotonous. The obvious advantage in using "ite" to indicate that the reference is to a mineral is lost, in part, because of the use of the ending on rock names, for example, "andesite." It has been recommended that all rock names end in "yte," as in "trachyte," but this suggestion has not been generally adopted.

Priority in Nomenclature

In the naming of minerals, those names that have priority are generally accepted over names subsequently proposed for a mineral. Dana's *System of Mineralogy* (6th ed., p. xliii) gives 11 rules for setting aside or revoking a mineral name, even though it has priority. Most important of these are an inadequate or incorrect original description of the mineral, a description that gives a false impression of the physical properties, or the loss of the name of a mineral for more than 50 years.

Examples of the problems raised by the law of priority are numerous and two of them with which the writer has been involved will be briefly reviewed here. As mentioned above, the writer named the mineral *austinite* in 1935 (Staples, 1935). F. Ahlfeld had called attention in 1932 in the *Neues Jahrbuch für Mineralogie* to a mineral from Bolivia that he called "brickerite," which was a *nomen nudum* because he only indicated its general composition. In 1936, brickerite was analyzed chemically, but incorrectly, and it was not until 1938 that W.

Brendler of Hamburg, Germany, carefully analyzed the material and determined that *austinite* and brickerite were identical. He stated that priority should be given the name *austinite*. To go back even further, the name "barthite" was proposed in 1914 for a mineral that later proved to be *austinite*; but, as stated by Fleischer (1945), "the description of barthite, especially the chemical analysis, was so faulty that priority may be set aside and the name barthite (=cuprian austinite) should be dropped." The name *austinite* withstood these two challenges and it is now internationally accepted.

A priority problem that had a happy ending is illustrated by the case of *erionite*. This ***zeolite***, first discovered by Eakle (1898) from Durkee, Oregon, had its occurrence, unit cell, and structure described in detail by Staples and Gard (1959), when the writer rediscovered the locality, which had been "lost." As a result of the potential commercial use of *erionite* for "molecular sieves," and its occurrence in large quantities as a diagenetic mineral, the name became well established not only in mineralogy, but also in the literature of chemical and industrial minerals. In 1962, the British mineralogists, Hey and Fejer, in studying the little-known mineral *offretite*, found it to be identical with *erionite*. Because *offretite* was first described in 1890 by Gonnard, the name had priority and Hey and Fejer believed it should replace *erionite*. On the other hand, *offretite* had been inadequately described, it had been lost sight of for more than 50 years, and the name *erionite* was so thoroughly entrenched in the literature that replacing it would cause great confusion. Many letters were written expressing strong viewpoints on the matter of replacing *erionite*, and when the Commission on New Minerals and Mineral Names voted on the matter there was a split that seemed irreconcilable. At this time, good fortune entered the picture when it was determined by Bennett and Gard (1967) that the *c*-cell dimension of *offretite* is half that of *erionite*. This indicated that *offretite* and *erionite* are distinct species, and both names should be retained, thus solving a problem in nomenclature that otherwise would have defied a happy solution.

The New Mineral Dilemma

The above material has been written to underline the difficulties involved in naming a new mineral. The problem is really threefold: (1) determining if it is a new mineral; (2) adequately describing it; (3) naming it. As has frequently been pointed out by the writer (Staples, 1948, 1962), to name a mineral without properly completing the first two steps can only lead to confusion. Fleischer (1966) has stated that during the period 1941–1960 about half the new mineral names proposed were considered unnecessary. This leads to listing more minerals as discredited minerals. According to Permingeat (1961), the ideal description of a new mineral requires a listing of macroscopic properties, crystallographic properties, physical properties, optical properties, chemical properties, physical-chemical properties, methods of synthesis, description of the deposit, nomenclature and classification, location of the depository of the material, and a bibliography. To provide these data it is necessary to have an adequate library and laboratory facilities that may include X-ray diffraction equipment, differential thermal apparatus, polarizing microscope, universal stage, X-ray fluorescence, analytical chemical apparatus, absorption spectrometer, and other equipment.

It is evident that only a professional mineralogist is capable today of determining if a specimen is a new mineral, and then describing it properly. Consequently, any suspected new mineral material should be sent to a properly equipped laboratory or university for examination. If the material turns out to be a new mineral, the description of it may take several years, depending on the problems involved. Only after carefully determining the properties of the new mineral will a name be recommended for it and the mineralogist will then submit the name to the Commission on New Minerals and Mineral Names for approval before publication takes place. The amateur collector's role in this is the searching for and finding of new material, and his ability to advance the science of mineralogy in this way should be a great source of satisfaction. The amateur's chances of attaining success are actually much greater in mineralogy than in making contributions to other fields of science, such as physics and chemistry. Part of the thrill of mineral collecting for the amateur, as well as the professional mineralogist, is the chance that the next outcrop may yield a mineral that has never been found before.

LLOYD W. STAPLES

This article was originally published in the *Ore Bin*, **32**, 73–77, Oregon Dept. Geology and Mineral Industries, 1970.

References

Bennett, J. M., and Gard, J. A., 1967. Nonidentity of the zeolites erionite and offretite, *Nature*, **214**, 1005–1006.

Brendler, W., 1938. On the identity of austinite and brickerite, *Am. Mineralogist*, **23**, 347–349.

Dana, E. S., 1892. *System of Mineralogy*, 6th ed., New York: Wiley.

Eakle, A. S., 1898. Erionite, a new zeolite, *Am. J. Sci.*, ser. 4, **6**, 66.

Embrey, P. G., and Fuller, J. P., 1980. *A Manual of New Mineral Names, 1892-1978.* Oxford and the British Museum (N.H.), 467p.

Fleischer, M., 1945. Discredited minerals, *Am. Mineral.*, **30**, 550.

Fleischer, M., 1966. Index of new mineral names, *Am. Mineralogist*, **51**, 1247-1357.

Gaines, R. V., 1969. Cliffordite–a new tellurite mineral from Moctezuma, Sonora, Mexico, *Am. Mineralogist*, **54**, 697-701.

Gonnard, M. F., 1890. Sur l'offretite, espèce minérale nouvelle, *Comptes Rendus Acad. Sci.* (Paris), **111**, 1002-1003.

Hey, M. H., and Fejer, E. E., 1962. The identity of erionite and offretite, *Mineral. Mag.*, **33**, 66-67.

Mitchell, R. S., 1979. *Mineral Names: What Do They Mean?* New York: Van Nostrand Reinhold, 229p.

Neumann, H., and Nilssen, B., 1968. Tombarthite, a new mineral from Hogetveit, Evje, south Norway, *Lithos*, **1**, 113-123. (*Am. Mineralogist*, **54**, 327).

Permingeat, F., 1961. Description idéale d'un minéral, *Bull. Soc. Fr. Minéral. Cristallogr.*, **84**, 98-100.

Staples, L. W., 1935. Austinite, a new arsenate mineral, from Gold Hill, Utah, *Am. Mineralogist*, **20**, 112-119.

Staples, L. W., 1948. Oregon's new minerals and discredited species, *Mineralogist*, **16**, 470-476.

Staples, L. W., and Gard, J. A., 1959. The fibrous zeolite erionite; its occurrence, unit cell, and structure, *Mineral. Mag.*, **32**, 261-281.

Staples, L. W., 1962. The discoveries of new minerals in Oregon, *The Ore Bin*, **24**, (6), 81-87.

Staples, L. W., 1964. Friedrich Mohs and the scale of hardness: *J. Geol. Educ.*, **12**, 98-101.

Staples, L. W.; Evans, H. T.; and Lindsay, J. R., 1967. Cavansite, a new calcium vanadium silicate mineral [abst.], *Geol. Soc. Am. Spec. Paper 115*, abstracts for 1967, 211-212.

Cross-references: *Collecting Minerals; Mineral Classification: History; Mineral Classification: Principles; Museums, Mineralogical.*

NATIVE ELEMENTS AND ALLOYS

The few chemical elements that occur naturally in their elemental state or, in the case of metallic elements, alloyed with other metals, are referred to as "native elements" (Table 1). In some cases, the element is quite pure, e.g., *diamond*, whereas complete solid solutions exist among others, e.g., between *gold* and *silver*, which are isostructural, or between *osmium* and *iridium*, which are respectively hexagonal and isometric. In addition, distinct intermetallic compounds exist, e.g., *maldonite* and *moschellandsbergite*.

Several elements and alloys occur in two or more crystal structures which, according to

TABLE 1. Native Elements and Alloys

Mineral Name	Composition	Crystal System
Amalgam	(Ag,Hg)	isometric
Antimony	Sb	trigonal
Arsenic	As	trigonal
Arsenolamprite	As	orthorhombic
Bismuth	Bi	trigonal
Chaoite	C	hexagonal
Cohenite	$(Fe,Ni,Co)_3C$	orthorhombic
Copper	Cu	isometric
Diamond	C	isometric
Gold	Au	isometric
Graphite	C	hexagonal
Indium	In	tetragonal
Iridium	Ir	isometric
Iridosmine	(Os,Ir)	hexagonal
Iron	αFe	isometric
Kamacite	α(Fe,Ni)	isometric
Kolymite	Cu_7Hg_6	isometric
Lead	Pb	isometric
Lonsdaleite	C	hexagonal
Maldonite	Au_2Bi	isometric
Mercury	Hg	trigonal (−39°C)
Moissanite	SiC	hexagonal
Moschellandsbergite	Ag_2Hg_3	isometric
Nickel	Ni	isometric
Osbornite	TiN	isometric
Osmiridium	(Ir,Os)	isometric
Osmium	(Os,Ir)	hexagonal
Palladium	Pd	isometric
Platiniridium	(Ir,Pt)	isometric
Platinum	Pt	isometric
Polarite	Pd(Pb,Bi)	orthorhombic
Rhodium	Rh	isometric
Rosickyite	S	monoclinic
Rutheniridosmine	(Os,Ir,Ru)	hexagonal
Ruthenium	Ru	hexagonal
Ruthenosmiridium	(Ir,Os,Ru)	isometric
Schreibersite	$(Fe,Ni)_3P$	tetragonal
Selenium	Se	trigonal
Siderazot	Fe_5N_2	trigonal
Silver	Ag	isometric
Sobolevskite	PdBi	hexagonal
Sulfur	S	orthorhombic
Taenite	γ(Fe,Ni)	isometric
Tellurium	Te	trigonal
Tetraferroplatinum	PtFe	tetragonal
Tin	Sn	tetragonal
Zinc	Zn	hexagonal

mineralogical procedure, are accorded separate mineral names (see *Naming of Minerals*). Thus, orthorhombic S is *sulfur* but monoclinic S is *rosickyite*. The element carbon is *chaoite*, *diamond*, *graphite*, or *lonsdaleite*, depending upon its structure.

KEITH FRYE

Reference

Fleischer, M., 1980. *Glossary of Mineral Species 1980*. Tuscon, Arizona: Mineralogical Record, 192p.

Cross-references: *Naming of Minerals;* see also mineral glossary. Vol. IVA: *Native Elements;* also see individual elements.

NESOSILICATES—*See* mineral glossary. Vol. IVA: CRYSTAL CHEMISTRY; MINERAL CLASSES: SILICATES

NITRATE MINERALS

The nitrates are similar to the carbonates in structure; they are anisodesmic with $(NO_3)^-$ radicals, and isostructural with *calcite*. Nitrates can be grouped as minerals of the $A(XO_3)$ type, e.g.,

- *nitratine:* $Na(NO_3)$
- *niter:* $K(NO_3)$
- ammonia niter: $NH_4(NO_3)$

of the $A(XO_3)_2$ type, e.g.,

- *nitrobarite:* $Ba(NO_3)_2$
- *nitrocalcite:* $Ca(NO_3)_2 \cdot 4H_2O$
- *nitromagnesite:* $Mg(NO_3)_2 \cdot 6H_2O$

with a hydroxyl or halogen, e.g.,

- *gerhardtite:* $Cu_2(NO_3)(OH)_3$

and as compound nitrates

- *darapskite:* $Na_3(NO_3)(SO_4) \cdot H_2O$
- *buttgenbachite:* $Cu_{19}(NO_3)_2Cl_4(OH)_{32} \cdot 2H_2O$

The most important commercial nitrate deposits, chiefly *nitratine* (soda niter), are confined to desert regions, such as northern Chile. *Nitratine* is found with other water-soluble minerals such as iodates, sulfates, and chlorides.

LLOYD W. STAPLES

References

Dana, J. W., and Dana, E. S., 1944–1962. *System of Mineralogy*, 7th ed., C. Palache, H. Berman, and C. Frondel, eds. New York: Wiley, 3 vols.

Whitehead, W. L., 1920. The Chilean nitrate deposits, *Econ. Geol.*, **15**, 187.

Wyckoff, R. W. G., 1920. The crystal structure of sodium nitrate, *Phys. Rev.*, **16**, 149.

O

OLIVINE GROUP

Minerals of the ***olivine*** group are abundant in the earth, the moon, and the stony meteorites. They are usually the first silicate mineral to crystallize from ultramafic and mafic magmas. Much research effort is applied to determination of the physical, chemical, and mechanical properties of these minerals in order to understand the composition and physical state of the interior of the earth; the origin, composition, and crystallization of magmas; and the nature and history of the planetary bodies from which meteorites are derived.

For a treatise on the ***olivines*** the reader is referred to Deer, Howie, and Zussman (1962). Brown (1968) has reviewed knowledge of the ***olivines*** in ultramafic and basaltic rocks.

Crystal Structure

Minerals of the ***olivine*** group belong to the "nesosilicate" (or "orthosilicate") structural group (see Vol. IVA); individual tetrahedron-shaped SiO_4^{4-} ions are linked together by divalent cations in octahedral coordination with the oxygens of the silicate groups. The loci of the oxygen atoms conform to a series of sheets parallel to the (100) plane, resulting in a high degree of crystal symmetry (orthorhombic, space group *Pbnm*).

Members of the Olivine Group

All ***olivines*** conform to the composition $(M,N)_2SiO_4$, where M and N are divalent cations. The most abundant natural members of the ***olivine*** group (Table 1) belong to the *forsterite-fayalite* isomorphous series. These are the names given respectively to the pure magnesium (Mg_2SiO_4; column 1, Table 2) and pure iron (Fe_2SiO_4; column 2, Table 2) end members of the series. Names have been assigned to specific ranges of Mg:Fe ratio (Fig. 1), but it is current usage to indicate the composition by giving the mole proportion of either the *forsterite* (*Fo*) or the *fayalite* (*Fa*) component; for example, Fo_{82} (or Fa_{18}) contains 82 mol% of *forsterite* and 18 mol% of *fayalite*.

The remaining natural members of the ***olivine*** group (Table 1) have such rare occurrence that the term ***olivine*** has come to signify compositions of the *forsterite-fayalite* sequence. Other than the information in Table 1, the rare ***olivine*** group minerals will not be discussed further, and all subsequent comments refer to the members of the *forsterite-fayalite* series. The ***olivine***-structured compounds Ni_2SiO_4 and Mg_2GeO_4 have no natural occurrence (Table 1).

Chemistry of the *Fo–Fa* Olivines

Analytical Methods. Major elements are usually determined by X-ray fluorescence (q.v., Vol. IVA) or electron microprobe techniques (see *Electron Probe Microanalysis*). Because of the difficulty of removing ***plagioclase, pyroxene, spinel***, and *ilmenite* inclusions from ***olivines***, the minor and trace element analyses obtained by bulk analysis must be viewed with caution. Analyses of the minor-element contents of ***olivines*** are more reliably obtained by the electron microprobe. This instrument also permits

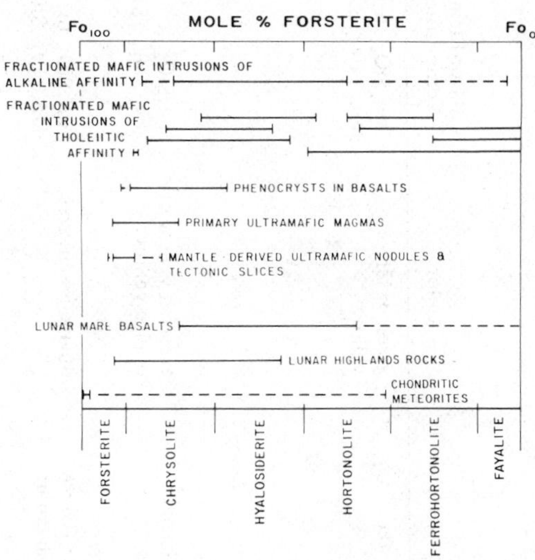

FIGURE 1. Nomenclature of ***olivines*** in the *forsterite-fayalite* series and the relation of natural ***olivine*** compositions to occurrence in igneous rocks. Dashed line indicates comparative scarcity of compositions. The four mafic intrusions of tholeiitic affinity are (from top to bottom): Okonjeje, SW Africa; Skaergaard, E. Greenland; New Amalfi Sheet, South Africa; Bushveld, South Africa.

TABLE 1. The Olivine Group Minerals

Mineral Name	Origin of Name	Color	Major Element Composition	Comments	Natural Occurrence: Igneous Rocks	Natural Occurrence: Metamorphic Rocks
Forsterite (gem quality is known as peridot)	J. Forster	green or lemon-yellow	Mg_2SiO_4 to $Mg_{1.8}Fe_{0.2}SiO_4$	Complete solid solution between Mg_2SiO_4 and Fe_2SiO_4.	Pure Mg_2SiO_4 never found. Compositions between Fo_{94} and Fo_0 are common in ultramafic and mafic rocks. Fo_{80-50} is the commonest range of ***olivine.***	In marbles formed by thermal metamorphism of impure dolomitic limestone or by regional metamorphism of siliceous dolomite. Intermediate compositions rare.
Fayalite	Fayal Island, Azores	greenish yellow or orange-yellow	Fe_2SiO_4 to $Fe_{1.8}Mg_{0.2}SiO_4$		Rare felsic and alkaline volcanic and plutonic rocks.	In thermally and regionally metamorphosed Fe-rich sediments.
Tephroite	*tephros* (Greek for "ash-colored")	olivine-green; grey	Mn_2SiO_4 to $Mn_{1.8}Fe_{0.2}SiO_4$		Hydrothermal veins; iron-manganese ore deposits and their skarns. (Tephroite and Knebelite)	In thermally and regionally metamorphosed Mn-rich sediments. (Tephroite and Knebelite)
Knebelite	Major von Knebel	brown-black	$Mn_{1.4}Fe_{0.6}SiO_4$ to $Mn_{0.6}Fe_{1.4}SiO_4$			
Glaucochroite	–	bluish green; violet or pale pink	$MnCaSiO_4$		–	High-grade thermal metamorphism of limestone in presence of Mn ions
Monticellite	T. Monticelli	colorless or grey	$CaMgSiO_4$		In silica-undersaturated rocks, e.g., alnöite.	In thermally and regionally metamorphosed siliceous limestones.
Kirchsteinite	–	colorless or yellow	$CaFeSiO_4$		No natural occurrence–known only in slags.	
–	–		Ni_2SiO_4	Complete solid solution with Mg_2SiO_4.	Known only from laboratory synthesis.	
–	–	–	Mg_2GeO_4	Complete solid solution with Mg_2SiO_4.		

TABLE 2. Composition of "Pure" Members of the *Forsterite-Fayalite* Series Compared with Two Natural Olivines

Oxide (wt %)	1	2	3	4	5	6
SiO_2	42.71	29.49	41.30	41.22	36.20	35.10
FeO	–	70.51	7.79	7.89	39.10	40.73
MgO	57.29	–	50.60	50.89	23.60	24.17
MnO	–	–	0.09	–	0.36	–
NiO	–	–	0.39	–	0.03	–
CaO	–	–	0.02	–	0.46	–
Total	100.00	100.00	100.20	100.00	99.75	100.00
Fo %	100	0	92	92	51.4	51.4
					Also contains 0.12 TiO_2 0.12 Cr_2O_3	

1 Pure *forsterite*.
2 Pure *fayalite*.
3 ***Olivine*** from peridotite nodule, Lashaine volcano, Tanzania.
4 Theoretical composition of pure Fo_{92}.
5 ***Olivine*** from lunar mare basalt, sample 12036.
6 Theoretical composition of pure $Fo_{51.4}$.

analysis of large numbers of crystals in a single sample, to check for homogeneity, and analysis of changing composition ("zoning") in single crystals. Neutron activation analysis has provided data on elements present in trace quantities in ***olivine***, but suffers from the same problems as bulk analysis. The ion microprobe has recently been used to obtain trace-element data (e.g., Hervig et al., 1980), but no such data are included here.

Composition. The compositions of two natural ***olivines*** are compared with the theoretical compositions of ***olivines*** of similar *Fo* contents in Table 2, columns 3–6; and it is seen that as much as 2.5% of the Mg and Fe is replaced by other ions. The "impurity" ions (Table 3) enter the crystal structure because natural ***olivines*** crystallize from melts, vapors, and solid media that invariably contain these ions. Fe^{2+} ions are believed to camouflage entry of many of the minor and trace elements into the ***olivine***, although some Fe^{3+} ions may form by oxidation of Fe^{2+} ions in the ***olivine*** structure.

Olivines grown in large, deep-seated bodies of magma have Ca contents of less than 1000 ppm, whereas those grown in dykes, sills, and lavas have larger contents (up to 5000 ppm). The Ca content is also sensitive to silica activity of the parent melt and to whether ***plagioclase*** accompanies ***olivine*** crystallization.

The Mn content increases with *Fa* content, whereas the Ni decreases (Simkin and Smith, 1970); this behavior is attributable to crystal-chemical control, i.e., Ni and Mg are favored in early-crystallizing (refractory) ***olivine*** and Mn and Fe in later-crystallizing (less refractory) ***olivine***.

Co is present only in trace amounts (Table 3).

Substitution of Fe^{3+}, Cr^{3+}, Al^{3+}, Ti^{4+}, and trace elements for Mg and Fe (Ti^{4+} is too large to replace Si^{4+}) results in charge imbalance and is, therefore, unfavorable.

Ferric iron (up to 10,000 ppm) is reported in wet chemical analyses of terrestrial ***olivines***, and the presence of exsolved *magnetite* and *chromite* dendrites in some ***olivines*** indicates that both Fe^{3+} and Cr^{3+} ions were originally in the ***olivine*** structure. Most terrestrial ***olivines*** examined by electron microprobe have Cr, Ti, and Al contents close to, or less than, the error in the determination ($\approx$200 ppm).

The Cr species in lunar ***olivines*** (Table 3) is assumed to be Cr^{2+}, present because of the highly reduced condition of lunar magmas (see *Lunar Minerals*). Rare Cr^{3+}-rich terrestrial ***olivines*** (see, e.g., Green et al., 1975) probably result from rapid crystal growth, high crystallization temperature, and large Cr content of the parent magma.

The larger Ti contents of ***olivines*** in lunar mare basalts relative to those in terrestrial basalts (Table 3) reflect the higher TiO_2 content of the lunar magmas and possibly also the presence of Ti^{3+} ions.

Other trace elements in ***olivine*** are present in exceedingly small amounts (Table 3).

Miscellaneous Properties of Olivine

Shape, Density, and Hardness. ***Olivines*** crystallize rapidly and adopt a wide variety of crystal shapes (Fig. 2), apparently related to the environment of crystallization (Drever and Johnston, 1957). With progressive increase in the cooling rate of a melt, the shapes progress from subequant, faceted crystals, to subequant,

TABLE 3. Range of Minor and Trace Element Contents (ppm) of Olivine

Element	Terrestrial	Lunar
Mn	1000–15,000	1800–7300
Ni	100–4400	<500
Ca	200–12,000	200–7800
Co	80–320	n.m.
Fe^{3+}	0–11,000	n.m.
Cr	10–500	<200–3000
Al	<200	<300
Ti	<200	400–2000
Cu	6–100	n.m.
V	20–300	n.m.
Ba	<7	n.m.
Sr	<10	n.m.
Ga	<3	n.m.
Li	<4	n.m.
Mo	<7	n.m.
U	<0.0025	n.m.

n.m. not measured

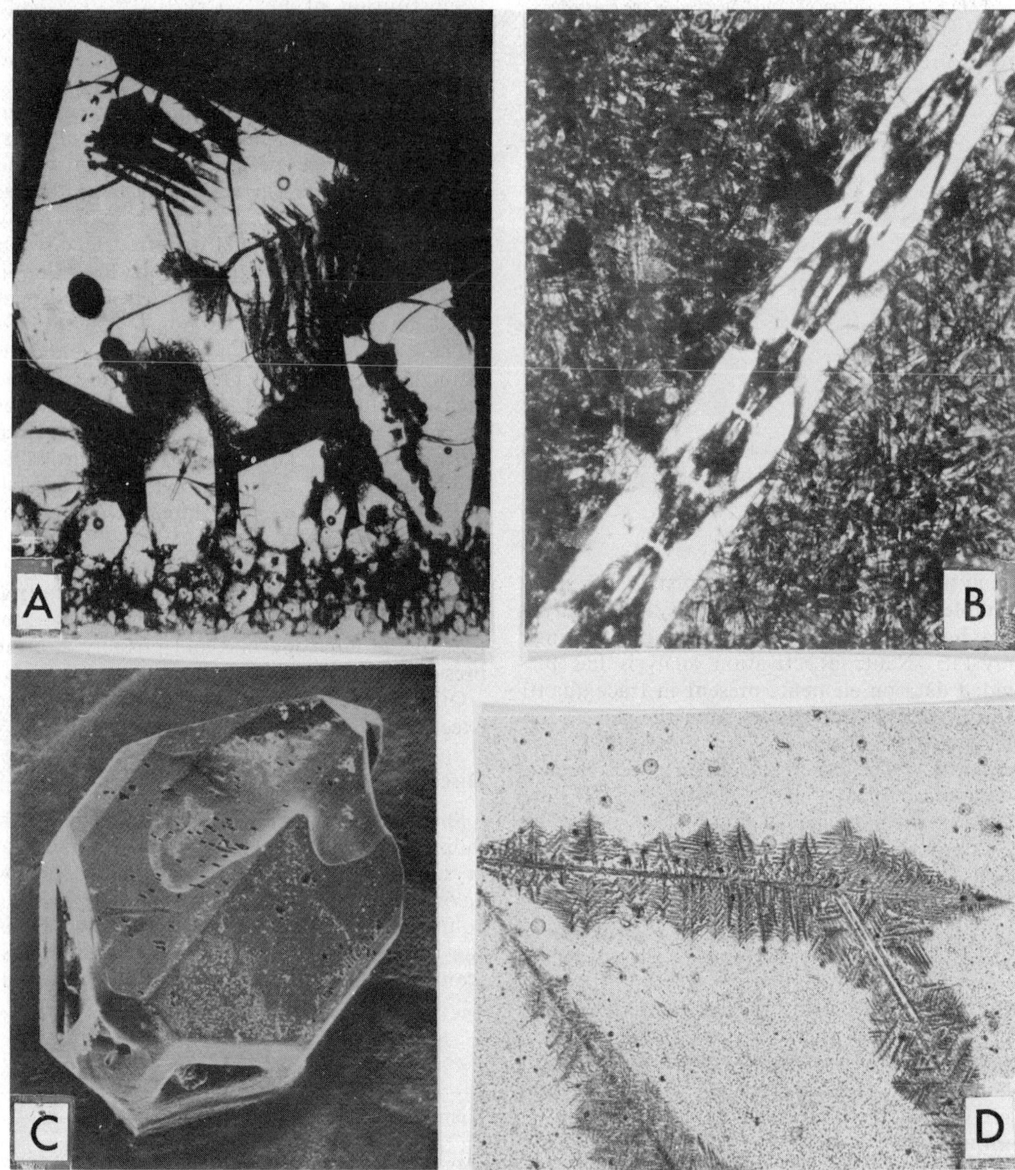

FIGURE 2. Variety in the shapes of ***olivine*** crystals. A, B, and D are thin sections of synthetic ***olivines*** grown from rock melts; C is a scanning electron microscope picture of a natural crystal. Each field of view is approximately 2 mm (A from Donaldson, 1977; B from Donaldson, 1976; C courtesy of G. Switzer).

indented crystals, to subequant skeletons, to elongate skeletons, to dendrites (see *Skeletal Crystals*).

Density increases linearly with composition from 3.2 g/cm^3 (Fo_{100}) to 4.4 g/cm^3 (Fo_0), and hardness is 7–6.5 on the Mohs scale (q.v.).

Stability. At 1 atm pressure, *forsterite* and *fayalite* melt respectively at 1890°C and 1205°C, and these values increase with pressure at the rates of 4.7 and 3.5°/kilobar. In natural silicate melts (magmas), the melting points are greatly depressed, e.g., at 1 atm, Fo_{88} crystals melt in a basalt magma at approximately 1250°C.

At high pressure and temperature, the ***olivine*** structure is unstable and transforms to a spinel-type, body-centered cubic structure, the change being known as the "olivine-spinel transition." The transition is believed to be the cause of the abrupt change in seismic velocity in the earth's mantle at 400–500 km depth (see *Mantle Mineralogy*).

Alteration. Over a wide range of temperatures, *olivine* readily undergoes alteration reactions to produce a variety of possible secondary minerals: *serpentine*, ***chlorite***, *magnetite*, *hematite*, NiS, *talc*, ***amphiboles***, carbonates, and ***smectites*** (montmorillonites). Formation of *serpentine* from *forsterite*, by removal of MgO and addition of H_2O, occurs at approximately 450°C and, in nature, only below 3 kilobars pressure (the temperature decreases with increasing Fe content of the ***olivine***).

Natural Occurrence

Table 1 summarizes the occurrence of ***olivine*** group minerals. The Mn-rich ***olivines*** are most commonly of metamorphic origin; *monticellite* is present in small amounts in rare ultramafic igneous rocks. *Forsterite* exists in metamorphosed siliceous dolomitic limestones and in limestones into which Si and Mg have been introduced by metasomatism. *Fayalite* is found in metamorphosed siliceous Fe-rich sediments, called "eulysites."

Minerals of the *forsterite-fayalite* series are common in ultramafic and mafic rocks and are a useful index of fractionation in basaltic rocks. Crystals in rocks derived from the upper mantle are characteristically magnesian (Fig. 1); those in mafic layered intrusions, minor intrusions, and lavas are never as magnesian, and may have a wide range in composition (e.g., Fo_{90}–Fo_0), either within an intrusion or in individual zoned crystals. In a sequence of differentiated rocks, the parent magma of which is of tholeiitic affinity, there is a compositional gap, reflecting temporary cessation of ***olivine*** crystallization (Fig. 1). This behavior is due to the fact that, at low pressures as ***olivine*** crystallizes from a melt and the temperature falls, the silica activity may increase sufficiently to cause reaction ("discontinuous reaction") between ***olivine*** and the melt to form ***pyroxene***, e.g.,

$$\underset{\text{crystal}}{Mg_2SiO_4} + \underset{\text{melt}}{SiO_2} \rightleftarrows \underset{\text{crystal}}{2MgSiO_3}$$

Olivines more Fe-rich than Fo_{30} do not undergo this reaction, because Fe-rich ***orthopyroxene*** is unstable at low pressures, hence the reappearance of Fe-rich ***olivine*** in the layered intrusions (Fig. 1).

Recent and Future Study

Some of the more unusual, recent investigations of ***olivine*** have been of the electrical properties (Duba, 1972); the deformation properties (Green and Radcliffe, 1972); the growth of crystals in cooling melts (Lofgren et al., 1974); the rate of diffusion of Fe^{2+} and Mg^{2+} ions in the ***olivine*** structure (Buening and Buseck, 1973); the ordering of Mg and Fe ions into two structurally nonequivalent octahedral sites in the structure (e.g., Shinno et al., 1974); and the partition of major and minor elements between ***olivine*** and a melt (Roeder and Emslie, 1970; Roeder, 1974). Use of the ion microprobe to analyze for trace elements may provide the petrologist with new indices of the pressure and temperature at which ***olivine*** crystals grow; the analysis of small inclusions of melt and vapor in ***olivines*** is likely to provide further understanding of the genesis and fractionation of magmas. All these studies will permit new insight into the thermal, mechanical, and chemical processing of the earth's crust and upper mantle.

COLIN H. DONALDSON

References

Brown, G. M., 1968. Mineralogy of basaltic rocks, in H. H. Hess and A. Poldervaart, eds., *Basalts*, vol. 1. New York: Interscience (Wiley), 103–162.

Buening, D. K., and Buseck, P. R., 1973. Fe-Mg lattice diffusion in olivine, *J. Geophys. Res.*, 78, 6852–6862.

Deer, W. A.; Howie, R. A.; and Zussman, J., 1962. *Rock-Forming Minerals*, vol. 1, *Ortho- and Ring-Silicates*. London: Longmans, 333p.

Donaldson, C. H., 1976. An experimental investigation of olivine morphology, *Contrib. Min. Pet.*, **57**, 187–213.

Donaldson, C. H., 1977. Laboratory duplication of comb layering in the Rhum pluton, *Mineral. Mag.*, **41**, 323–336.

Drever, H. I., and Johnston, R., 1957. Crystal growth of forsteritic olivine in magmas and melts, *Trans. Roy. Soc. Edinburgh*, **63**, 289–315.

Duba, A., 1972. Electrical conductivity of olivine, *J. Geophys. Res.*, **77**, 2483–2495.

Green, H. W., and Radcliffe, S. V., 1972. The nature of deformation lamellae in silicates, *Geol. Soc. Am. Bull.*, **83**, 847–852.

Green, D. H., et al., 1975. Experimental demonstration of the existence of peridotitic liquids in earliest Archean magmatism, *Geology*, **3**, 11–14.

Hervig, R. L., et al., 1980. Fertile and barren Al-Cr spinel hartzburgites from the upper mantle, *Earth Planet. Sci. Letts.*, **50**, 41–48.

Lofgren, G. E., et al., 1974. Experimentally reproduced textures and mineral chemistry of Apollo 15 quartz-normative basalts, *Proc. 5th Lunar Sci. Conf.*, **1**, 549–568.

Roeder, P. L., 1974. Activity of iron and olivine solubility in basaltic liquids, *Earth Planet. Sci. Lett.*, **23**, 397–410.

Roeder, P. L., and Emslie, R. F., 1970. Olivine-liquid equilibrium, *Contrib. Mineral. Petrol.*, **29**, 275–289.

Shinno, I.; Hayashi, M.; and Kuroda, Y., 1974. Möss-

bauer studies of natural olivines, *Mineral. J.* (Japan), 7, 344-358.
Simkin, T., and Smith, J. V., 1970. Minor-element distribution in olivine, *J. Geol.*, 78, 304-325.
Switzer, G.; Melson, W. G.; and Thompson, 1972. Olivine crystals from the floor of the Mid-Atlantic Ridge near 22°N latitude, *Smithson. Contrib. Earth Sci., No. 9*, 43-46.

Cross-references: *Crystal Growth; Crystallography: Morphological; Density Measurements; Electron Probe Microanalysis; Gemology; Glass, Devitrification of Volcanic; Lunar Mineralogy; Mantle Mineralogy; Meteoritic Minerals; Mineral Properties; Naming of Minerals; Plastic Flow in Minerals; Portland Cement Mineralogy; Refractory Minerals; Rock-Forming Minerals; Spinel Group; Topotaxy;* see also mineral glossary. Vol. IVA: *Crystal Chemistry; Mineral Classes: Silicates; Trace Elements in Silicate Minerals–Substitution.*

OPAL

The widespread occurrence and variable appearance of those natural forms of silica termed *opal* has resulted in a confusing nomenclature based in different instances on morphology, physical properties, or optical characteristics. In order to rationalize the classification of natural hydrous silicas, Jones and Segnit (1971) proposed the scheme set forth in Table 1.

Opal-C is a rather rare form. It has the typical appearance of vitreous *opals* and is commonly associated with volcanic rocks. *Opal-CT* embraces most occurrences of opal not showing a play of color. *Opal-A* includes most samples of precious *opal* (i.e., those showing a vivid play of color) and associated potch (see glossary below). The diagnostic differences of the three groups are illustrated by the X-ray diffraction patterns of Fig. 1 (Jones and Segnit, 1971).

Composition and Chemical Properties

Opal is essentially SiO_2 but contains nonstructural water, commonly in the range 4-9%.

TABLE 1. Classification of Natural Hydrous Silicas

	1. *Opal-C* (well-ordered low-*cristobalite*)	
Opal (compact and vitreous)	2. *Opal-CT* (disordered low-*cristobalite*, low-*tridymite*)	Opaline Silica (friable or dispersed)
	3. *Opal-A* (highly disordered, near amorphous)	

Some samples are remarkably pure silica, SiO_2+H_2O totaling 99.9%. Alumina may be present up to a few percent and iron oxide higher, especially in brown varieties. In some of these *opals*, discrete gel-like patches or iron oxides and hydroxides are visible with a microscope.

The greater part of the water is physically held but a small portion, particularly in *opal-A*, is present as hydroxyls. The water may be regarded as nonessential since its removal leaves the structural characteristics unaltered, although the appearance and some physical properties are much altered.

Opal is readily soluble in hot, concentrated alkaline solution and hydrofluoric acid.

Physical Properties

Opal consists of an aggregate of very fine, randomly oriented crystallites with a structure that varies from well-ordered low-*cristobalite* to near amorphous. Aggregates of lamellar crystals of *opal-CT* are sometimes found in small cavities (Fig. 2). Specific gravity is variable; commonly it is in the range 1.99-2.25 for compact varieties, but is dependent on water content in porous samples. Hardness is in the range 5.5-6 for compact varieties. Color is widely variable (see glossary below).

Optical Properties

Opal is commonly isotropic, but may be distinctly birefringent due to microscopic crystallites of *cristobalite-tridymite*.

Gem varieties show a brilliant play of color produced by diffraction from a regular three-dimensional array of close-packed spheres 1500-3000 Å in diameter (Sanders, 1968; Fig. 3). Under crossed polars, pseudo-birefringence and lamellar "twinning" (stacking faults in the close packing) produce striking color effects in thin sections. The geometry of this phenomenon has been described in detail by Baier (1932).

The refractive index varies from 1.111 to 1.459 (Frondel, 1962), but is commonly in the range 1.44-1.46. In general, for a particular specimen, increase in water content leads to an increase in refractive index.

Occurrence

Opal occurs as hydrothermal deposits, particularly associated with volcanic activity, and as sedimentary deposits, some of inorganic, some of organic, origin. Opaline silica is probably a common constituent of many clay deposits but is difficult to detect due to poor crystallinity and fine grain size. *Opal* common-

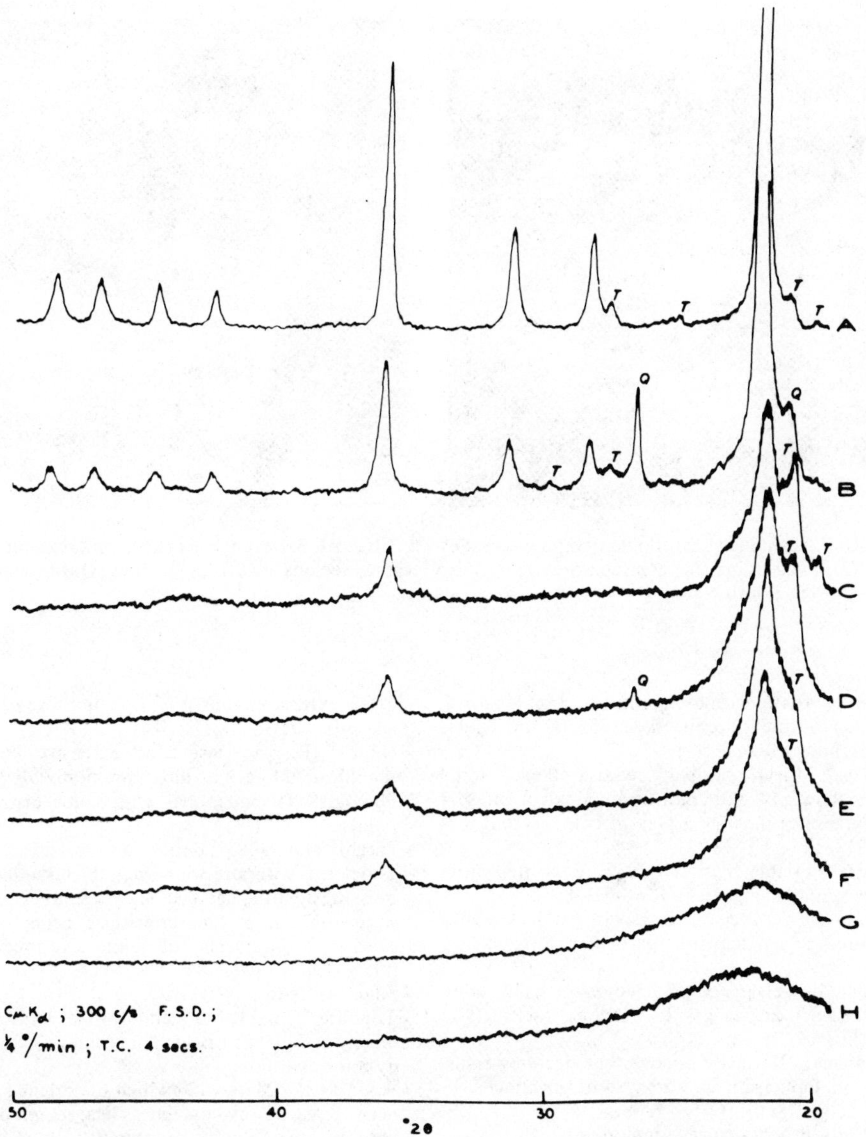

FIGURE 1. Diffractometer traces of forms of silica. A. Synthetic low-*cristobalite*. B. *Opal-C*. C–F. *Opal-CT*. G–H. *Opal-A*. Peaks marked *Q* are due to *quartz*, those marked *T* to *tridymitic* stacking.

ly replaces wood, often accurately pseudomorphing the cellular structure. It is also common in growing plants, particularly grasses, and as urinary calculi in grazing animals.

Uses

Varieties of *opal* showing a play of color are prized as gemstones (see *Gemology*). Diatomite and similar varieties are used as polishing agents, fillers, absorbents, and as insulating refractories (see *Abrasive Materials; Pigments and Fillers*).

Glossary

Some of the more commonly used names in connection with *opal* and *opaline* silica are:

- Black *Opal*: A precious variety, showing brilliant play of color against a dark body color. Chiefly from Lightning Ridge, Australia, but occurs also in parts of the United States.
- Common *Opal*: Shows no play of color. Deposits occur all over the world. Distinguished from other siliceous materials by a characteristic vitreous luster. It is mostly *opal-CT*.

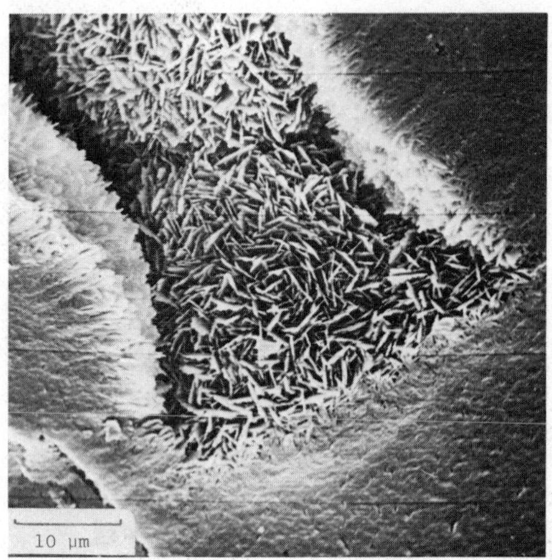

FIGURE 2. Scanning electron micrograph of crystals of *opal-CT* in vug in massive common *opal* from near Wentworth, New South Wales.

FIGURE 3. Scanning electron micrograph of a volcanic precious *opal* from Maleney, Queensland.

- Diatomite, Diatomaceous earth: A friable form, composed of the siliceous skeletons of unicellular diatoms; normally *opal-A*.
- Fire *Opal*: Clear, brilliant reddish-orange; some samples show a play of color. From Mexico. Unusual among varieties showing a play of color in that it is *opal-CT*.
- Geyserite: Friable hydrated silica, deposited from thermal springs and geysers. It is *opal-A*.
- Harlequin *Opal*: Precious *opal* with color regularly distributed as a patchwork; particularly prized as a gem.
- Hyalite: Glass-clear *opal*, botryoidal in habit, commonly occurs in vesicles in volcanic rocks. It is *opal-A*.
- Hydrophane: Normally opaque, but becomes transparent on immersion in water, some specimens revealing color.
- Jasp *Opal*: Brown-orange common *opal*.
- Jelly: Transparent, play of color inconspicuous as a consequence.
- Kieselguhr: German equivalent of diatomite.
- Liver *Opal*, Menilite: Gray or liver-colored material found originally as nodules at Menil-Montant near Paris. Some specimens available to the writers proved to be *quartz*; most samples also contained calcium and magnesium carbonates.
- Matrix *Opal*: Material in which precious *opal* has impregnated or replaced the host rock. In some specimens the porosity allows impregnation with suitable organic materials that can be decomposed to carbon, thus producing a dark background that enhances the play of color.
- Milk *Opal*: White, translucent to opaque.
- Noble *Opal*: Precious *opal*.
- *Opal* matrix: Ironstone concretions containing numerous small veins of *opal*.
- *Opal* claystone: Semilithified bedded deposits, often extensive, consisting of a mixture of *kaolinite* and *opal-CT*.
- Opoka: Russian term for extensive deposits of semilithified hydrous silica, commonly *opal-CT*.
- Pinpoint: Precious opal with color distributed as points.
- Potch: Australian term for worthless material associated with precious *opal*. Structurally similar, but due to irregular packing of spheres, irregularly sized spheres, or the interstices being filled with silica it shows no play of color. Like precious opal, it is normally *opal-A*.
- Siliceous Sinter: Geyserite.
- Tripolite: Friable to semilithified hydrous silica, mainly residual in origin. Sometimes used as a synonym for diatomite.
- Wood *Opal*: Material in which the original tissue has been replaced by opaline silica, commonly with preservation of woody structure.

J. B. JONES
E. R. SEGNIT

References

Baier, E., 1932. Optics of opal, *Zeit. Kristallogr.*, **81**, 183–218.

Frondel, C., 1962. *The System of Mineralogy*, vol. III. New York: Wiley, 287–306.

Jones, J. B., and Segnit, E. R., 1966. The occurrence and formation of opal at Coober Pedy and Andamooka, *Australian J. Sci.*, **29**, 129–133.

Jones, J. B., and Segnit, E. R., 1971. The nature of opal. I. Nomenclature and constituent phases, *J. Geol. Soc. Australia*, **18**, 57–68.

Leechman, F., 1961. *The Opal Book*. Sydney: Ure Smith.

Sanders, J. V., 1968. Diffraction of light by opals, *Acta Crystallogr.*, **A24**, 427–434.

Sanders, J. V., 1975. Microstructure and crystallinity of gem opals, *Am. Mineralogist*, **60**, 749–757.

Cross-references: *Authigenic Minerals; Diagenetic Minerals; Gemology; Human and Vertebrate Minerals; Invertebrate and Plant Mineralogy; Minerals, Uniaxial and Biaxial; Museums, Mineralogical; Quartz; Refractive Index; Soil Mineralogy.*

OPAQUE MINERALS—*See* METALLIC MINERALS; ORE MICROSCOPY (MINERAGRAPHY)

OPTICAL MINERALOGY

Optical mineralogy, the study of minerals by means of transmitted plane-polarized light, has been an important tool of mineralogists for more than 100 years. The groundwork had been laid as early as the 17th century by W. Snell, of Leyden, C. T. Bartolinus, of the University of Copenhagen, Christian Huygens, the Dutch mathematician, and others who studied the refraction, double refraction, and polarization of light. It was not, however, until the first half of the 19th century, when Sir David Brewster, a Scottish physicist, discovered the laws of polarization by reflection and refraction and William Nicol of Edinburgh invented the polarizing prism, that the science of optical mineralogy began to grow.

At first, mineralogists were slow to take advantage of this new discipline, but after publication of the Brooke and Miller edition of Phillips' *Mineralogy* in 1852, which gave the refractive indices and other optical data for the then-known common transparent minerals, optical mineralogy began to make rapid strides. Further developments followed regularly. In 1849, and again in 1857, H. C. Sorby demonstrated how the technique of preparing thin slices, first developed by Nicol for the study of petrified wood in 1828, could be applied to the study of rocks in order to identify the constituent minerals under the microscope. In 1893, F. J. Becké, then at the German University in Prague, described a method for determining the relative refractive indices of two contiguous transparent minerals in thin section. Further development of this technique led to the development of immersion oils for use in determination of refractive indices of minerals in fragments. The first version of the universal stage was introduced by the Russian E. Federow in 1894. In 1900, J. C. Schroeder van der Kolk published the first tables prepared for general use in determination of minerals by means of the immersion method. Later, in 1911, F. E. Wright of the Geophysical Laboratory in Washington, D.C., published *The Methods of Petrographic Microscopic Research*, in which he introduced to the petrographic microscopist a number of additional accessory devices for use with the polarizing microscope (see *Polarization and Polarizing Microscope*).

Improvements in technique and accessories for the polarizing microscope continue to be made. For example, Federow's two-axis universal stage has been developed so that now universal stages with as many as four or five axes are available. Also, a binocular research microscope is now available, and many excellent textbooks have been written on the subject. For a more exhaustive treatment of the theory of optical mineralogy than that given below, or for the data necessary for identifying particular minerals, the reader is referred to the works in the list of references.

Preparation of Samples

Minerals may be studied under the petrographic microscope either in the form of "fragments" or in thin slices known as "thin sections."

Fragments of convenient and uniform size are prepared with the aid of 100- and 120-mesh wire screens. The fragments may be permanently mounted on microscope slides or immersed in oils of known refractive index for temporary study.

Thin sections are prepared by mounting the smoothly polished side of a 3-mm slice of a rock or mineral on a microscope slide (by means of Canada balsam or synthetic cement of similar nature) and then cutting and grinding the slice to a thickness of 0.02 to 0.04 mm. When the proper thickness is reached (best determined by the interference color of a known mineral in the section), a cover glass is cemented over the slice by means of Canada balsam dissolved in xylene. After cleaning with xylene, washing, and drying, the slide is ready for use.

The Polarizing Microscope

The polarizing microscope, sometimes called the "petrographic microscope" (see *Polarization and Polarizing Microscope*), is the basic tool of the optical mineralogist. The polarizing microscope is equipped with a substage condenser, rotatable stage, two nicol prisms (the polarizer, below the stage, and the analyzer, above it), an auxiliary lens known as the Amici-Bertrand lens, and several accessory devices that can be inserted into a special slot in the microscope tube, of which the first-order red, $\lambda/4$, and quartz wedge are the commonest.

The nicol prism is a crystal of optical *calcite*, cut and prepared in such a way that light emerging from it is polarized in a single plane. The lower nicol prism is fixed, with its preferred vibration direction parallel to the north-south cross hair of the microscope; the upper nicol slides in and out of the microscope tube, and has its preferred vibration direction fixed parallel to the east-west cross hair. A mineral is said to be examined "in plane-polarized light" when only the lower nicol is in the optical train, and "under crossed nicols" when the upper nicol is also in the optical train. With the substage condenser and the auxiliary lens in place, the optical train is made "conoscopic" (convergent light); otherwise it is "orthoscopic."

The "universal stage" is an auxiliary device that can be mounted on the microscope stage. The first universal stage permitted rotation of the slide about only two axes; modern versions permit rotation about as many as six, so that the slide can be brought to almost any desired position in space. The mineral mount to be studied by means of the universal stage is placed between two small glass hemispheres, with the interfaces lubricated and sealed with cedar oil or crown oil. The glass hemispheres are made in sets with different refractive indices, the pair selected to have a refractive index most closely approaching that of the mineral to be studied. The effect of refraction is thus minimized.

By bringing the mineral into an orientation known with respect to the crystal symmetry, the optical properties and certain physical properties can be measured more accurately than in randomly oriented sections. Among these properties are cleavage and cleavage angle, refractive index, optic axial angle, and extinction angle (see below), all of which are diagnostic characteristics and particularly important for distinguishing between the individual members of mineral groups such as the ***plagioclase feldspars, amphiboles***, and ***pyroxenes***.

Optical Properties

The optical properties of a mineral–those phenomena produced when a fragment or a thin section is examined under the polarizing microscope–depend qualitatively upon the class of crystal symmetry of the mineral, and quantitatively on the lattice parameters of the crystal. Cleavege, relief, and color (including absorption and pleochroism) are determined by means of plane-polarized light; refractive index is also determined in plane-polarized light, but the conoscopic train may be used to locate preferred vibration directions (see below). Isotropism or anisotropism, birefringence (including retardation), and extinction angle and elongation are determined under crossed nicols with the orthoscopic train. Optical character (uniaxial or biaxial) and sign and dispersion are determined under crossed nicols with the conoscopic train; dispersion can also be determined from measurements of refractive indices.

Normally, only transparent or translucent minerals can be studied under the polarizing microscope; however, by shielding the source of transmitted light and using only the small amount of naturally reflected light, certain optical properties of opaque minerals–color, relief, and isotropism or anisotropism–can also be determined (see *Ore Microscopy*).

Cleavage. The tendency of a mineral to part readily along certain planes (cleavage) is a structural rather than an optical property. It is recognizable under the microscope as straight lines varying in distinctness and arrangement according to the strength and directions of the cleavage. Fragments of minerals that have good cleavage in one direction, ***mica***, for example, tend to form plates that lie flat on the slide. Viewed edgewise in a thin section, such cleavage appears as a series of thin, closely spaced, parallel lines. In minerals having good cleavage in two directions, the traces of the cleavage planes intersect; and the angle of intersection can be used to help identify the mineral. In a basal section of *hornblende*, for example, the cleavage planes intersect at angles of 56° and 124° (Fig. 1A), and in ***pyroxene***, at 89° and 91° (Fig. 1B).

Cleavage in three directions may form a rectangular (in cubic, tetragonal, or orthorhombic minerals) or rhombic (in hexagonal minerals with rhombohedral cleavage) pattern; rhombic cleavage is especially well displayed in *calcite* (Fig. 2A), and rectangular cleavage is developed in *anhydrite* (Fig. 2B). Cleavage in four directions, such as the octahedral cleavage of *fluorite* (in the isometric system) forms a characteristic triangular pattern (Fig. 2B). Patterns of cleav-

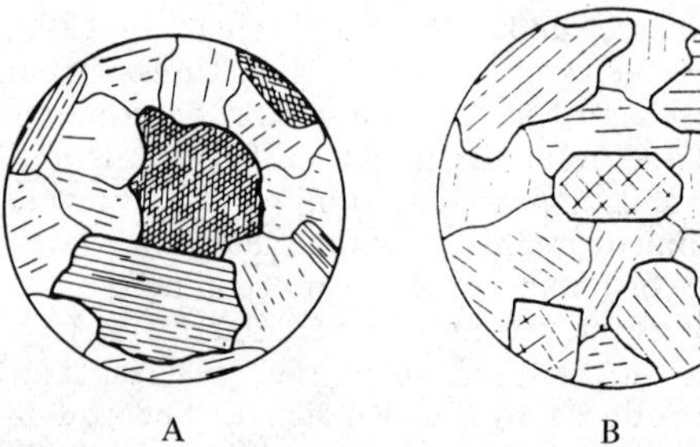

FIGURE 1. Cleavage intersecting at angles of 56° and 124° in *hornblende* (A) and at approximately 90° in ***pyroxene*** (B) (from Kerr, 1977).

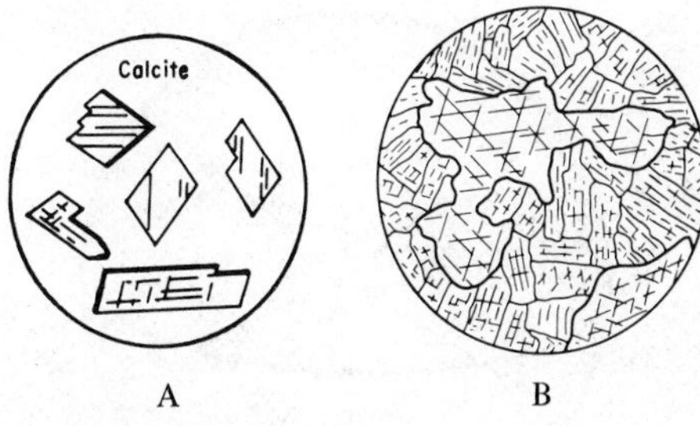

FIGURE 2. Cleavage in three directions; (A) rhombohedral cleavage in *calcite,* (B) octahedral cleavage in *fluorite* (from Kerr, 1977). Also shown in B is rectangular cleavage as developed in *anhydrite.*

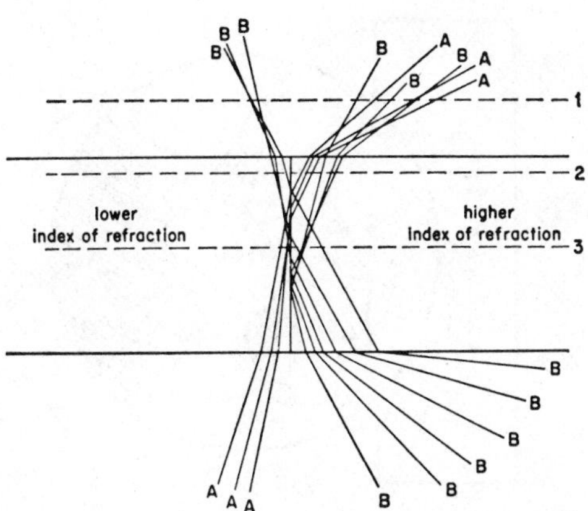

FIGURE 3. Origin of Becké line (from E. A. Wood, *Crystal and Light*, copyright © 1964 by D. Van Nostrand, New York; reprinted with permission). Dotted lines 1, 2, and 3 in figure represent different positions of the focal plane as the microscope tube is raised or lowered from focus (solid line).

age in six directions, such as the dodecahedral cleavage of *sphalerite*, may not be readily recognizable under the microscope.

Refractive Index. Light passing obliquely from one medium to another changes direction (is refracted) as it changes velocity. Snell discovered in 1621 that the ratio of the sine of the angle of incidence to the sine of the angle of refraction is a constant. The "refractive index" of a medium, n, is defined as the ratio of the velocity of light in a vacuum, V, to its velocity in that medium, v: $n = V/v$. For practical purposes, the velocity of light in air can be used as V. Since v is always less than V, refractive indices are always greater than one. Because n is a ratio of velocities, it is dimensionless. The refractive indices of two media, n_1 and n_2, are inversely proportional to the velocity of light in each, v_1 and v_2 ($n_1/n_2 = v_2/v_1$).

Given the refractive index of a substance, that of another substance may be determined relative to it. Thus, by means of a set of immersion oils of known refractive index (which can easily be calibrated to achieve an accuracy of ±0.001 or better, if necessary), the refractive index of a mineral fragment can be determined with any desired degree of precision.

The refractive index of mineral fragments relative to an immersion oil may be measured by the central illumination or oblique illumination methods (see *Refractive Index*).

Central Illumination. In the central illumination method, at normal magnifications, the intensity of transmitted light is reduced by closing down the substage diaphragm of the microscope. As the field darkens, a narrow zone of brightness develops at the periphery of each grain, exactly at the contact of the oil and the mineral when the microscope is in focus. If the microscope tube is raised slightly, this bright band moves toward the substance having the higher refractive index; if it is lowered, the band moves toward the substance having the lower index. This phenomenon is known as the "Becké-line" effect (Fig. 3). By using oils of progressively higher or lower refractive index, as the case may be, the oil with a refractive index matching that of the mineral can be found—under such conditions the white Becké line is replaced by a reddish-orange and a bluish band at the edge of the grain.

At high magnifications, it may be necessary to leave the substage diaphragm open and to insert the substage condenser, in order to provide sufficient illumination for the Becké-line test.

Oblique Illumination. In the method of oblique illumination, half of the microscope field is darkened by inserting a card between the reflecting mirror and the stage. Low magnification is preferable, in order to increase the number of grains visible at one time; and, for the best results, the condensing lens should be left out of the system. The mineral fragments in the field will now be illuminated on one side and shaded on the other. The field of the microscope will also be darkened on one side. If the refractive index of the mineral is higher than that of the oil, the shaded side of the grain is away from the darkened half of the field, and if the index of the mineral is lower than that of the oil, the shaded side will be toward the darkened half. When the refractive indices of mineral and oil are the same, the fragments appear to be bordered by bluish bands (Fig. 4).

Thin Section. The refractive index of a mineral can be determined relative to a con-

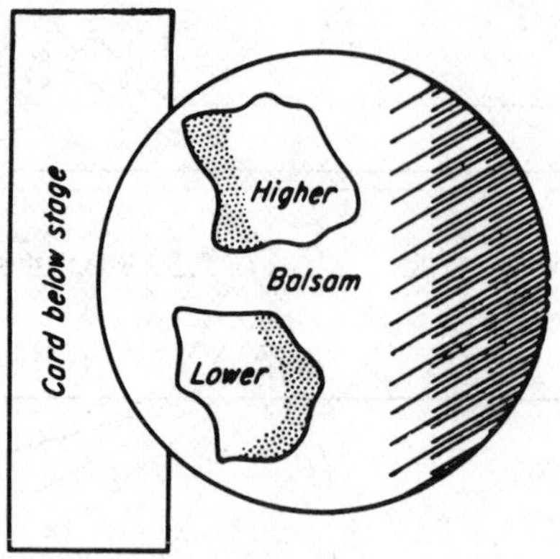

FIGURE 4. Illustrating oblique illumination with the condenser removed (from Kerr, 1977). Higher = mineral grain, $n >$ balsam. Lower = mineral grain, $n <$ balsam.

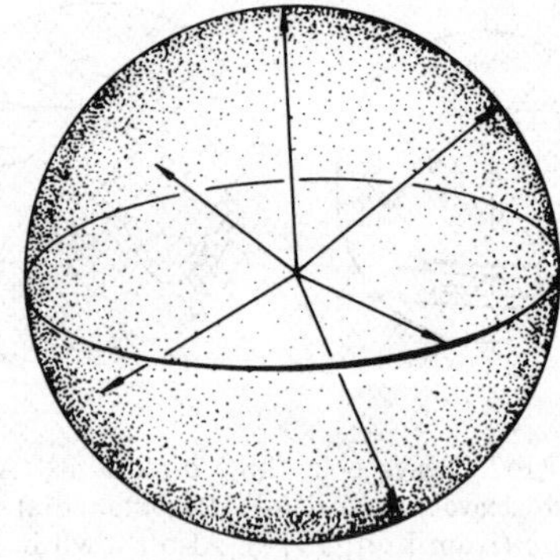

FIGURE 5. Indicatrix for isotropic mineral (from *Crystallography and Crystal Chemistry: An Introduction* by F. Donald Bloss. Copyright © 1971 by Holt, Rinehart and Winston. Reprinted by permission of Holt, Rinehart and Winston). The refractive index is constant in every preferred vibration direction for light. The indicatrix, therefore, is a sphere.

tiguous mineral grain or to Canada balsam, using the Becké-line effect in thin section.

Relief. The degree to which a mineral stands out against its background is a function of the difference between the respective refractive indices of mineral and background. In a colorless immersion oil of the same refractive index, a colorless mineral grain is hardly visible in plane-polarized light because it has no relief. Minerals having high birefringence (see below) may show a noticeable variation in relief, as the stage is rotated, due to the marked difference in the refractive indices in different preferred vibration directions. Relief is estimated quantitatively, as low, moderate, high, or extremely high, usually with respect to Canada balsam.

Isotropism and Anisotropism

Optically, all minerals are classified as isotropic or anisotropic on the basis of the manner in which light is transmitted through them, which in turn depends upon their crystal structure.

Isotropism. In isotropic minerals, light travels with equal velocity in all directions and, therefore, there is but a single refractive index. A three-dimensional surface generated by vectors proportional to refractive indices and parallel to vibration directions is called the "optical indicatrix"; in an isotropic mineral the optical indicatrix is a sphere (Fig. 5). Under crossed nicols, isotropic minerals remain dark as the microscope stage is rotated through 360° Minerals that crystallize in the isometric system and substances that lack internal structure, such as glass, are optically isotropic.

Anisotropism. In traveling through anisotropic minerals, light rays are doubly refracted, that is, they are split into two rays traveling with different velocities and vibrating in different planes (polarized) at right angles to each other. In uniaxial anisotropic minerals, there are two preferred vibration planes; in biaxial, there are three. Uniaxial minerals, therefore, have two significant refractive indices, a maximum and a minimum, corresponding to these preferred vibration planes. Biaxial minerals have three significant refractive indices, a maximum, a minimum, and that in the direction at right angles to the plane containing these two. The quantitative difference between the maximum and the minimum refractive indices of a mineral is called its "birefringence."

Minerals crystallizing in the tetragonal, trigonal, and hexagonal systems are uniaxial; those crystallizing in the orthorhombic, monoclinic, and triclinic systems are biaxial.

Birefringence

As the microscope stage is rotated with the nicols crossed, anisotropic minerals in most positions show a change from bright to dark and back every 90° When the mineral lies with its preferred vibration planes aligned with the preferred vibration planes of the polarizer and the analyzer, no light passes through the latter and the grain is said to be at the "extinction position." With the stage rotated to any other position, the grain exhibits "interference colors."

Interference colors result from the fact that rays doubly refracted in the mineral, vibrating at right angles to each other, are resolved in the analyzer to a single polarization plane; having traveled through the plane at different velocities, they are out of phase to an extent depending upon the birefringence, orientation, and thickness of the mineral, and also on the wavelength of the light used. The extent to which they are out of phase is called "retardation," designated Δ.

Optical Character; Sign and Dispersion

Uniaxial Minerals. In uniaxial minerals, the light rays corresponding to the two preferred vibration directions are known as the ordinary, or ω ray, and extraordinary, or ϵ ray. The ω ray travels with equal velocity in all directions, while the ϵ ray travels with different velocity in different directions. Along the optic axis of the mineral, which in all instances coincides with the vertical crystallographic axis (c axis), both rays travel with the same speed; under crossed nicols, therefore, sections cut normal to the c axis (basal sections) appear isotropic. In any other direction, the ϵ ray is either faster or slower than the ω ray, the difference being greatest in a direction perpendicular to the optic axis; therefore, interference colors are exhibited. If the ϵ ray is the slower, the mineral is said to be "optically positive"; if it is faster, the mineral is "optically negative." Correspondingly, in positive minerals the refractive index of the ϵ ray (n_ϵ) is the higher, and in negative minerals the lower.

The optical indicatrix of a uniaxial mineral (Fig. 6) is an ellipsoid of rotation, with the optic axis as the axis of rotation. A plane normal to the optic axis intersects the indicatrix to form a circle. All other sections are elliptical, and the semiaxes of a section containing the optic axis are n_ϵ and n_ω.

Biaxial Minerals. The three mutually perpendicular preferred vibration directions in biaxial minerals are designated X, Y, and Z; X and Z correspond respectively to the directions of minimum (n_α) and maximum (n_γ) refractive index; Y corresponds to some intermediate value n_β. The optical indicatrix of a biaxial mineral is a triaxial ellipsoid with minor intermediate, and major semiaxes of n_α, n_β, and n_γ, respectively. All plane sections passing through the center point of this figure are ellipses except two, which are circles with a radius of n_β. The optic axes are the two normals to these circular sections. They also lie in the XZ plane, which, therefore, is called the "optic plane"; the Y direction is the "optic normal." The acute angle between the optic axes is called the "optic axial angle," and designated $2V$. The principal vibration axis that bisects the optic axial angle is called the "acute bisectrix" (*Bxa*), and the axis that bisects the supplementary obtuse angle is called the "obtuse bisectrix" (*Bxo*) (Fig. 7). If Z is the acute bisectrix, the mineral is optically positive; if X is the acute bisectrix, it is optically negative.

In orthorhombic minerals, the three principal vibration directions coincide with the three crystallographic axes, but the relationships between X, Y, and Z and the crystallographic a, b, and c axes are not the same for all minerals. In monoclinic minerals, only one of the principal vibration directions coincides with a crystallographic axis—always the b (twofold) axis. In triclinic minerals there is no necessary coincidence (see *Optical Orientation*).

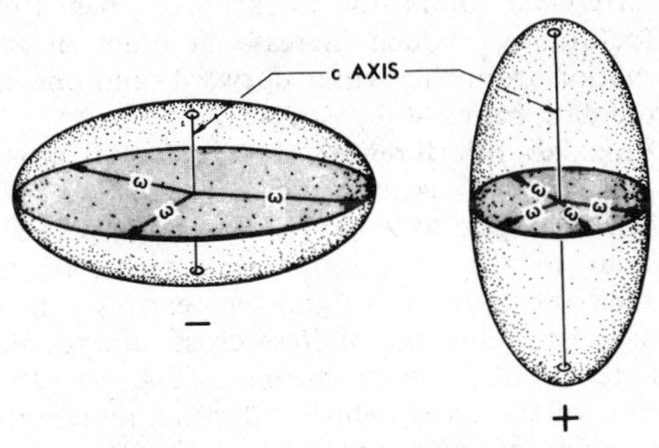

FIGURE 6. Indicatrices for negative and positive uniaxial minerals (from *Crystallography and Crystal Chemistry: An Introduction* by F. Donald Bloss. Copyright © 1971 by Holt, Rinehart and Winston. Reprinted by permission of Holt, Rinehart and Winston). The position of the crystallographic axis is shown in each case.

Extinction Angle and Elongation

Extinction is said to be parallel if the trace of a cleavage plane or crystal face coincides with the trace of a preferred vibration plane—that is, that cleavage or crystal face is parallel to one of the cross hairs when the mineral is in the extinction position. In some cases, when the angle of intersection of two prominent cleavage planes or crystal faces is bisected by either cross hair, extinction is said to be symmetrical. Otherwise it is said to be oblique or inclined (Fig. 8).

Extinction Angle. The angle between the extinction position and the cleavage trace or crystal face is a useful optical property for identifying minerals, particularly for distinguishing individual members within groups such as the ***plagioclase feldspars***, the ***amphiboles***, and the ***pyroxenes***. It is usually measured with respect to the trace of the preferred vibration plane of the slow ray. To determine the trace of the slow-ray plane, the stage of the microscope is rotated to bring the mineral grain 45° from its extinction position, which brings the trace of one preferred vibration plane parallel to the accessory slot and the other perpendicular to it.

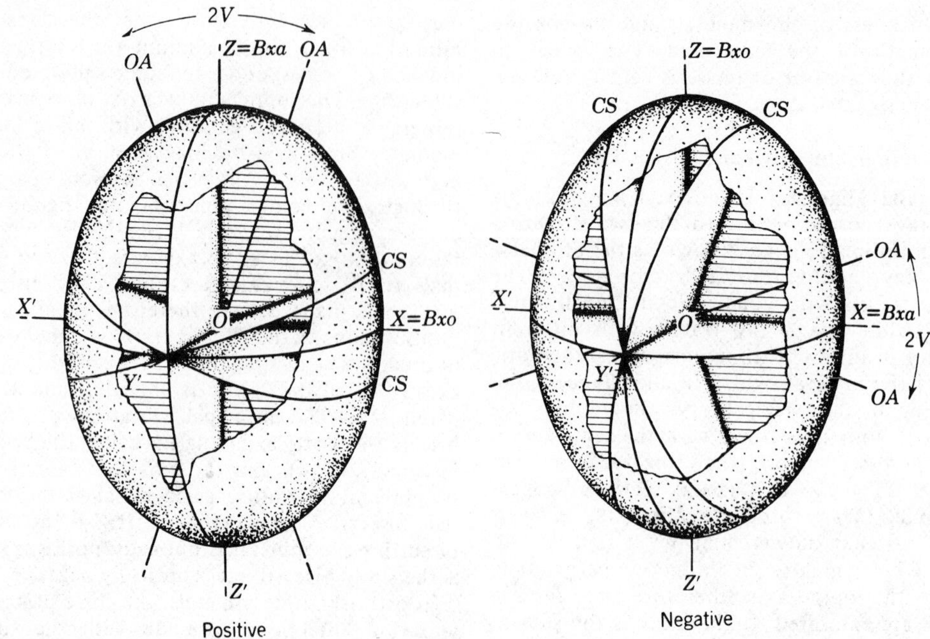

FIGURE 7. Indicatrices for positive and negative biaxial minerals: *OA*, optic axis; *CS*, circular section; *Bxa*, acute bisectrix; *Bxo*, obtuse bisectrix (from E. E. Wahlstrom, *Optical Crystallography*, copyright © 1960, John Wiley & Sons, Inc. Reprinted by permission).

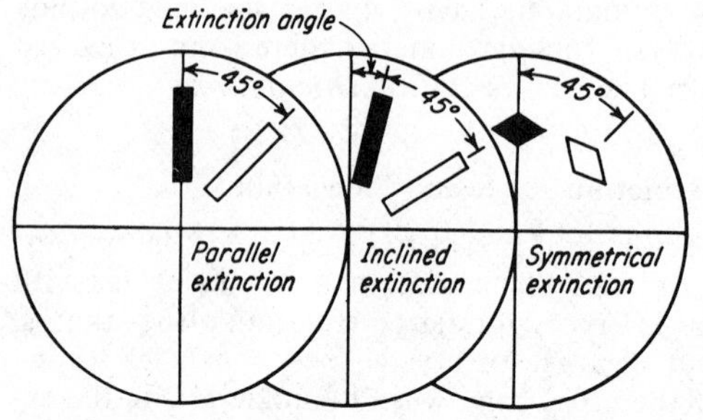

FIGURE 8. Relative positions of illumination and darkness in parallel, oblique, and symmetrical extinction (from Kerr, 1977).

If one of the accessory plates is inserted in the slot, the order of interference color will increase if the trace of the slow ray (higher refractive index) is perpendicular to the axis of the accessory plate (and the trace of the fast ray, therefore, parallel to it), and will decrease if the trace of the fast ray (lower refractive index) is perpendicular and the trace of the slow ray is parallel.

Elongation. If the mineral is developed in elongated crystals or cleavage fragments, the trace of the fast- and slow-ray vibration directions can be referred to the crystal habit or cleavage. If the trace of the slow-ray direction is parallel to the length of the crystal or fragment, the mineral is said to have "positive elongation" or to be "length-slow." If the trace of the fast-ray vibration direction is parallel to the length, the mineral has "negative elongation" or is "length-fast."

Interference Figures

When anisotropic minerals are viewed in convergent light, interference figures are produced. These consist of a series of concentric bands of interference colors—the isochromatic bands or "isochromes," which increase in order of retardation from the center outward—and one or more dark bands—the "isogyres."

Uniaxial Interference. The uniaxial interference figure is best developed (centered) when the optic axis of the mineral is perpendicular to the microscope stage. This figure (called the optic axis figure) appears as a dark cross resembling the Maltese cross, superposed on the concentric isochromes (Fig. 9). The center of the cross, which is darkest, represents the point of emergence (pierce-point) of the optic axis, the "melatope." The isogyres, forming the arms of the cross, are due to the removal of light from the image along the trace of the preferred vibration directions of the polarizer and the analyzer. The isochromes are lines joining points of equal retardation of the two

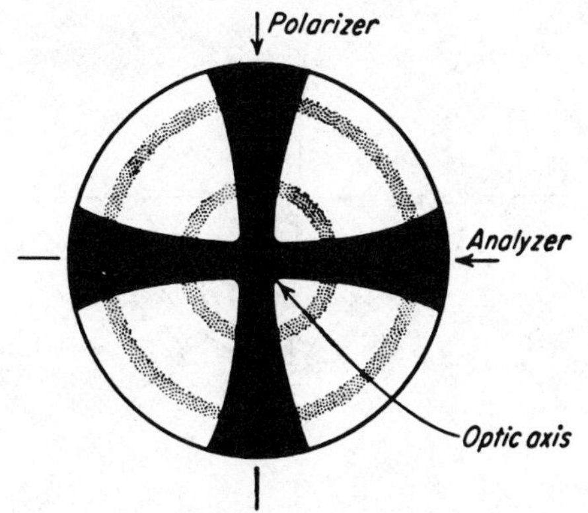

FIGURE 9. Centered optic axis uniaxial figure (from Kerr, 1977). The optic axis is perpendicular to the plane of the diagram.

interfering light rays traveling through the mineral. As the microscope stage is rotated, a centered optic axis figure remains stationary. If the optic axis of the mineral is inclined from the optimum position, the cross will be off-center; upon rotation of the stage, the center of the figure will rotate in the same direction, the whole figure moving with it while maintaining its orientation with respect to the cross hairs. If the optic axis is strongly tilted, the melatope may lie outside the field; in this case, the isogyres sweep alternately across the field (Fig. 10).

Flash Figures. If the optic axis is parallel or nearly parallel to the stage, a "flash figure" is obtained. In this case, when the optic axis is in the 45° position with respect to the cross hairs, the isochromes form two conjugate hyperbolae asymptotic to the cross hair directions; and when the optic axis is parallel to one of the cross hairs, a very broad, diffuse cross is produced which "flashes" out of the field upon slight rotation of the stage (Fig. 11).

Optic Sign. The optic sign is determined from the interference figure with the aid of one of the accessory plates. When one of these is inserted into the accessory slot, retardation is increased and decreased in alternate quadrants of the figure; if it is increased parallel to the trace of the slow ray of the interference figure, the mineral is positive ($n_\epsilon > n_\omega$), and if decreased, negative ($n_\epsilon < n_\omega$). The mica plate ($\lambda/4$) and the gypsum plate (first-order red) are commonly used for minerals with low to moderate birefringence. With the $\lambda/4$ plate, a dark dot will appear in the first and third quadrants in the case of a negative mineral, and in the second and fourth in the case of a positive mineral. With the first-order-red plate, the first and third quadrants appear orange and the second and fourth bluish if the mineral is negative, and vice versa if it is positive. The quartz wedge is used with minerals of high birefringence, in which case there are numerous isochromes; as the wedge is inserted from the thin edge, the isochromes in the first and third quadrant shift away from the center, and those in the second and fourth toward the center, if the mineral is negative; if it is positive, the shifts take place in the opposite sense.

Biaxial Interference. The appearance of the biaxial interference figure depends upon the orientation of the crystal, its optic angle, and birefringence, as well as the magnification used (amount of figure visible). Four main types are distinguishable: *acute bisectrix, obtuse bisectrix, optic normal*, and *optic axis figures.*

The acute bisectrix figure is seen when the acute bisectrix (*Bxa*) is normal to the stage. A centered *Bxa* figure for a mineral with moderate $2V$ and moderate birefringence is shown in Fig. 12. If $2V$ is 35° or less, the two melatopes lie within the field in a centered *Bxa* figure, and remain so throughout a complete rotation of the stage. If $2V$ is larger than 35°, the melatopes lie outside the field; but the optic axial angle can still be recognized from the pattern of isochromes and relative thickness of the arms of the cross. If $2V$ is very small, it

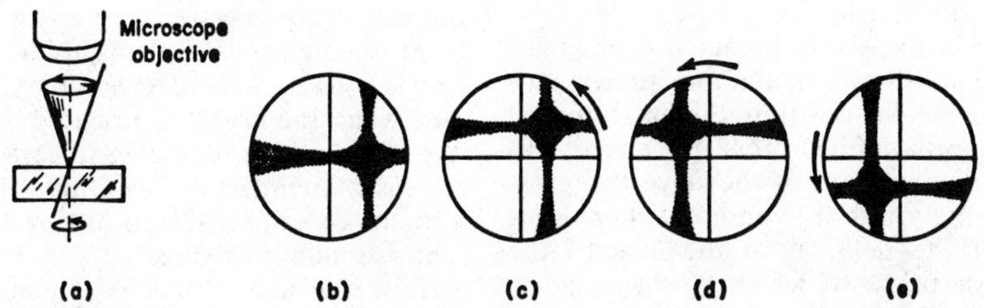

FIGURE 10. The position of the optic axis is shown is (a). Stages b, c, d, and e demonstrate the appearance of the off-centered uniaxial figure during rotation of the microscope stage through approximately 290° (from E. A. Wood, *Crystal and Light*, copyright © 1964 by D. Van Nostrand, New York; reprinted with permission).

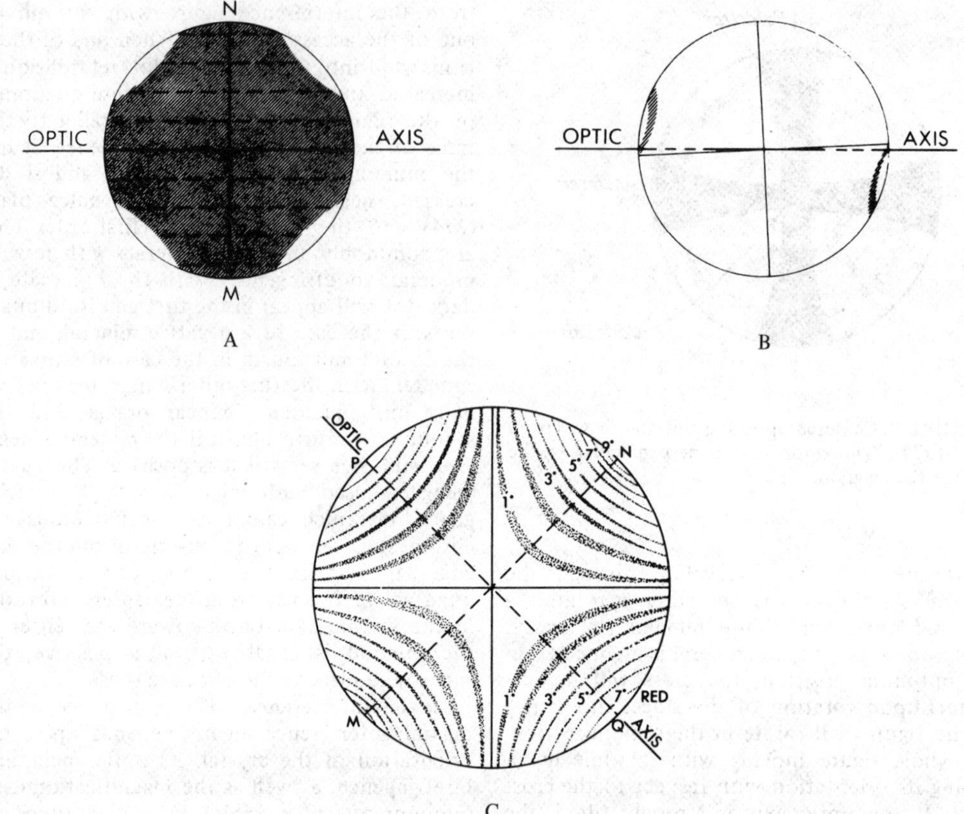

FIGURE 11. Uniaxial flash figure (from *Crystallography and Crystal Chemistry: An Introduction* by F. Donald Bloss. Copyright © 1971 by Holt, Rinehart and Winston. Reprinted by permission of Holt, Rinehart and Winston). (A) Figure produced when the mineral is at extinction. (B) Appearance of the figure upon slight rotation of the microscope stage. (C) showing the arrangement of the isochromatic bands when the optic axis is in the 45° position.

may be difficult to distinguish an acute bisectrix figure from a uniaxial figure.

The obtuse bisectrix figure, obtained when the obtuse bisectrix (*Bxo*) is normal to the stage, resembles the *Bxa* figure for crystals with large $2V$, and may be difficult to distinguish from the latter; methods have been developed for positively identifying *Bxa* and *Bxo* figures in such cases, the details of which are beyond the scope of this article.

Optic normal figures are obtained from grains or slices that show maximum interference colors. In the parallel position, the optic normal figure is a poorly defined cross that almost fills the field. Upon rotation of the stage this cross breaks up rapidly into a hyperbola whose arms flash out of the field, as in the case of the uniaxial flash figure to which it is analogous. The form and distribution of the isochromes is also similar to that in the uniaxial flash figure.

Optic axis figures are obtained when one of the optic axes is perpendicular to the stage. Such grains show minimum interference colors. In a perfectly centered optic axis figure, one melatope coincides with the point of intersection of the cross hairs. If $2V$ is small, the other melatope may lie within the field; otherwise, the interference figure in the 45° position consists of a single isogyre with its isochromes (Fig. 13). The magnitude of $2V$ may be estimated from the degree of curvature of the isogyre. The isogyre is convex toward the point of emergence of the acute bisectrix.

Off-centered biaxial interference figures are seen when the crystal is oriented in any but the directions discussed above. In some cases, they are easily identifiable; in others they are very difficult or impossible to distinguish from off-centered uniaxial figures.

The optic sign of a biaxial mineral can best be determined from an acute bisectrix figure. With the trace of the optic plane in the 45° position, the fast- and slow-ray directions can be determined from the color shifts that take

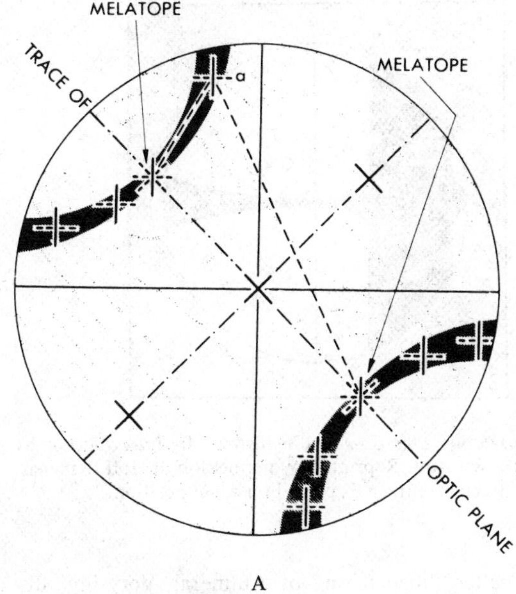

A

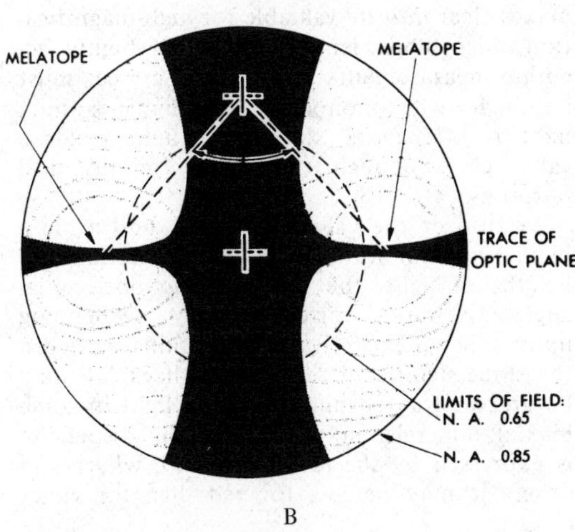

B

FIGURE 12. Acute bisectrix figure (from *Crystallography and Crystal Chemistry: An Introduction* by F. Donald Bloss. Copyright © 1971 by Holt, Rinehart and Winston. Reprinted by permission of Holt, Rinehart and Winston). (A) traces of the optic plane in the 45° position; (B) trace of the optic plane parallel to the crosshairs.

place when one of the accessory plates is inserted, in a manner similar to that described for uniaxial minerals. If the mineral is positive, Z is the acute bisectrix and X and Y lie in the plane of the stage; light vibrating parallel to the optic plane is then the fast ray (velocity $1/n_\alpha$) and that vibrating parallel to the optic normal, the slow (velocity $1/n_\beta$). If the mineral is negative, X is the acute bisectrix and the Y and Z directions lie in the plane of the stage. Light vibrating parallel to the optic plane is then the slow ray (velocity $1/n_\alpha$) and that vibrating parallel to the optic normal, is the fast ray (velocity $1/n_\beta$).

The optic sign can also be determined from an optic axix figure or off-centered figure, so long as the optic plane and the direction toward the acute bisectrix can be recognized.

Interference Figures and Refractive Index. For precise determinations of refractive indices in anisotropic minerals, it is necessary to measure the index in each of the preferred vibration directions. The ϵ and ω directions in uniaxial minerals, or the X, Y, and Z directions in biaxial minerals, can be located by means of interference figures. For instance, a grain showing a centered biaxial optic normal interference figure must have its X and Z directions in the plane of the stage; the maximum and minimum indices measured on such a grain therefore are the true values of n_γ and n_α, respectively. Similarly, n_α and n_β can be measured on a grain showing an optically positive acute bisectrix figure or optically negative obtuse bisectrix figure (X and Y in the plane of the stage), and n_β and n_γ from an optically negative acute bisectrix figure or optically positive obtuse bisectrix figure (Y and Z in the plane of the stage).

Optic Axial Angle (2V)

The value of the optic angle is one of the diagnostic characteristics of a biaxial mineral. Qualitative estimates can be made from the appearance of the acute bisectrix figure or optic axis figure (Fig. 14).

Methods have been developed for measuring $2V$ with an accuracy of the order of 1° or better, based on micrometer measurements of the intermelatope distance, that can be used even on somewhat off-centered acute bisectrix figures. Another, less precise, method permits measurement of $2V$ when both melatopes lie outside the field of view.

The value of $2V$ can also be calculated to within 2° from measured values of n_α, n_β, and n_γ, using the relationship

$$\cos 2V = 1 - 2\,[(n_\beta - n_\alpha)/(n_\gamma - n_\alpha)]$$

Optically positive minerals yield positive values of $\cos 2V$ and negative minerals yield negative values. A nomogram has been constructed for rapid determination of $2V$ on the basis of this relationship.

The angle $2V$ is the true optic axial angle within a mineral; due to refraction of light as it

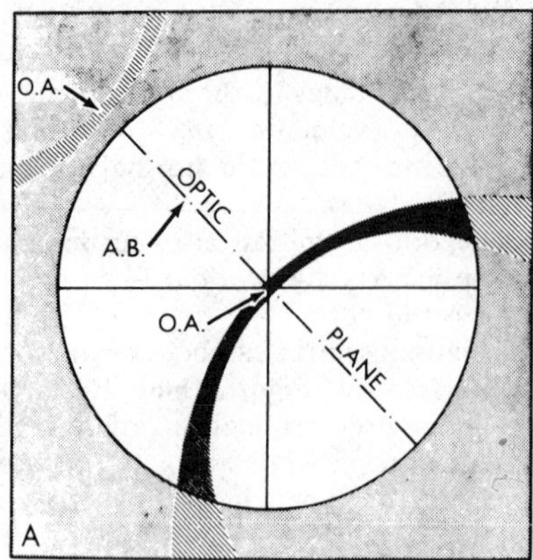

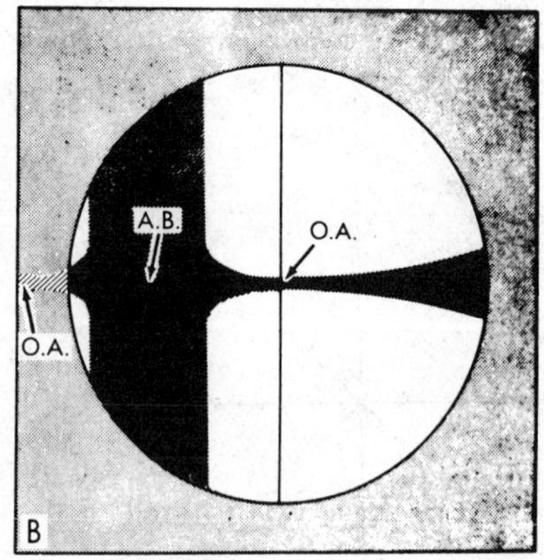

FIGURE 13. Centered optic axis figure (from *Crystallography and Crystal Chemistry: An Introduction* by F. Donald Bloss. Copyright © 1971 by Holt, Rinehart and Winston. Reprinted by permission of Holt, Rinehart and Winston). (A) trace of optic plane in 45° position; (B) trace of optic axis plane in parallel position.

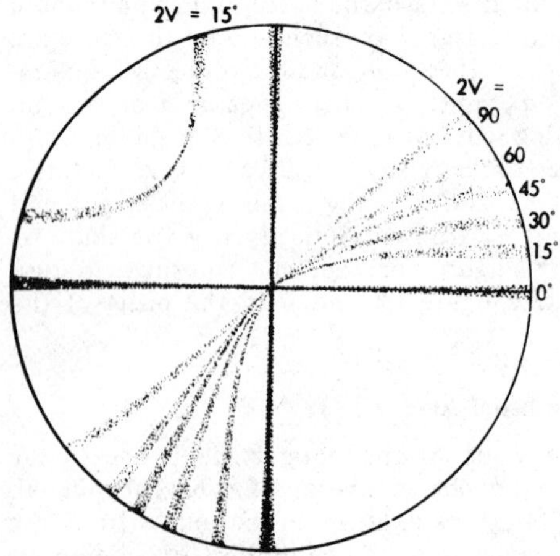

FIGURE 14. Guide for estimation of $2V$ from degree of curvature of isogyres in an optic axis figure (from *Crystallography and Crystal Chemistry: An Introduction* by F. Donald Bloss. Copyright © 1971 by Holt, Rinehart and Winston. Reprinted by permission of Holt, Rinehart and Winston).

passes from the mineral into air, the angle actually observed and measured (known as $2E$) is slightly larger. Measurements are corrected for this effect by calibrating each individual microscope against minerals of known $2V$.

Dispersion

The refractive index of a substance depends not only on its intrinsic properties but also on the wavelength of the light used; red light is less deflected than violet and hence has a lower refractive index. The difference between the refractive indices for red and for violet light is called "dispersion" of a mineral. Very high dispersion is responsible for the "fire" of a *diamond*; very low dispersion, on the other hand, makes clear *fluorite* valuable for high-magnification microscope lenses. The most highly accurate measurements of refractive indices must be made with monochromatic light; sodium light is often used, but white light yields a value close enough to yellow for practical purposes.

Because of the values of n_α, n_β, and n_γ in a biaxial mineral are different for different wavelengths of light, the value of the optic axial angle $2V$ varies correspondingly. Depending upon the quantitative relationships between the three significant refractive indices, $2V$ may be greater for red and for violet light in some biaxial minerals, in which case the dispersion is expressed by the formula ($r > v$), whereas in others it may be less for red than for violet ($r < v$).

The dispersion formula of a mineral is commonly determined from the colored fringes that appear at the edges of the isogyres in the vicinity of the melatopes, when an acute bisectrix figure is viewed in white light in the 45° position. If $r > v$, a reddish fringe is visible on the convex side and a bluish fringe on the concave side of the isogyres (see *Dispersion, Optical*). If $r < v$, the opposite effect is seen. As the isogyres are extinction phenomena, the observed effect is the reverse of what would be expected from the dispersion formula; in other words, when $r > v$, the red wavelengths are extinguished on the concave side of the isogyre (wider $2V$) and the blue comes through, and vice versa.

The dispersion formula can also be determined directly, by observation of the acute bisectrix in red and in blue monochromatic light.

The symmetry of the dispersion in a biaxial interference figure is governed by the class of crystal symmetry. In orthorhombic minerals, dispersion is usually *normal* (*symmetrical*), although the mineral *brookite*, in certain orientations, shows a type called *crossed axial plane dispersion*. Monoclinic crystals show *inclined, horizontal*, or *crossed* dispersion, depending on the relationship of the optic axes to the crystallographic axes. In triclinic crystals, dispersion is *asymmetrical*.

Color. Most minerals appear more or less colorless in thin section or fragments, but a dark mineral may show some of its natural color in plane-polarized light. If the mineral is isotropic, it maintains uniform coloration during a complete rotation of the microscope stage. If it is anisotropic, on the other hand, it may show a variation in the intensity of the color (absorption) or an actual change in color (pleochroism) as the stage is rotated.

Absorption is the result of conversion of some of the light energy to heat as the light rays pass through the mineral. The degree of absorption depends upon the orientation of the grain (i.e., of its preferred vibration directions) with respect to the preferred vibration directions of the polarizer. To describe the absorption pattern of a mineral, the vibration directions must be located and identified and the degree of absorption in each direction noted. A colored biaxial mineral, for example, may have an absorption formula $X<Y<Z$, indicating that light vibrating parallel to Z is most absorbed, parallel to Y less so, and parallel to X least.

Pleochroism is the result of selective absorption of different portions of the spectrum in the different vibration directions of the crystal. Uniaxial minerals show two colors ("dichroism"), one corresponding to the ϵ-ray vibration direction and the other to the ω-ray direction. Biaxial minerals are "trichroic," showing three colors corresponding to the X, Y, and Z directions. Pleochroism is determined in a manner similar to that used to determine absorption, and similarly is noted by a formula. For example, a uniaxial mineral with the pleochroic formula ϵ=yellow, ω=green shows a yellow color when the ϵ direction is parallel to the north-south cross hair of the microscope and a green color when the ω direction is parallel.

CHARLES J. VITALIANO
DOROTHY B. VITALIANO

References

Bloss, F. P., 1961. *An Introduction to the Methods of Optical Crystallography*. New York: Holt, Rinehart, & Winston, 294p.

Burri, C., 1950. *Das Polarisationmikroskop, Lehrb. Mon. Exakte Wissensch., Chem. Reihe., vol.* 5. Basel: Verlag Birkhausser.

Hallimond, A. F., 1953. *Manual of the Polarizing Microscope*, 2nd ed. York, England: Cooke, Troughton & Simms, 107p.

Hartshorne, N. H., and Stuart, A., 1964. *Practical Optical Crystallography*. London: E. Arnold, 326p.

Heinrich, E. W., 1965. *Microscopic Identification of Minerals*. New York: McGraw-Hill, 414p.

Kerr, P. F., 1977. *Optical Mineralogy*, 4th ed. New York: McGraw-Hill, 492p.

Larsen, E. S., and Berman, Harry, 1934. The microscopic determination of the non-opaque minerals, *U.S. Geol. Surv. Bull.*, 848 (2nd ed. revised 1960), 266p.

Marshall, C. E., 1953. *Introduction to Crystal Optics*, 2nd ed. York, England: Cooke, Troughton & Simms, 124p.

Oelsner, H. O., 1961. *Atlas der wichtigsten Mineralparagenesen im mikroskopishen Bild.* Bergakad: Freiberg-Fernstudium, 309p.

Roubault, M. E., 1963. *Determination des Minèraux des Roches.* Lamarre-Poinat. Editeur, 365p.

Wahlstrom, E. E., 1969. *Optical Crystallography*, 4th ed. New York: Wiley, 489p.

Winchell, A. N., 1956. *Elements of Optical Mineralogy*, pt. I–*Principles and Methods*, 5th ed., 263p. pt. II–*Description of Minerals*, 4th ed., 2nd printing, 551p. pt. III–*Determinative Tables*, 2nd ed., 231p. New York: Wiley.

Wood, E. A., 1964. *Crystals and Light.* New York: Van Nostrand, 160p.

Cross-references: *Anisotropism; Amphibole Group; Color in Minerals; Crystallography: Morphological; Isotropism; Minerals, Uniaxial and Biaxial; Polarization and Polarizing Microscope; Refractive Index.*

OPTICAL ORIENTATION

The orientation of the principal vibration axes of a crystal's optical indicatrix to its crystallographic axes constitutes the crystal's optical orientation. For isotropic media, the optical indicatrix is a sphere and thus possesses no principal axes to relate to the crystallographic axes. For hexagonal, trigonal, or tetragonal crystals, the unique principal vibration axis (optic axis) of the uniaxial indicatrix is inevitably parallel to the symmetry-defined c axis.

In orthorhombic crystals, the X, Y, and Z principal vibration axes of the biaxial indicatrix coincide for all wavelengths of light with the symmetry-defined crystallographic axes according to one of the six following permutations:

	(1)	(2)	(3)	(4)	(5)	(6)
a =	X	X	Y	Y	Z	Z
b =	Y	Z	X	Z	X	Y
c =	Z	Y	Z	X	Y	X

The orthorhombic mineral *brookite*, TiO_2, is unusual because the dispersion curve (see *Dispersion, Optical*) for light vibrating parallel to its a axis intersects (at λ=5500 Å) the curve for light vibrating parallel to the c axis. Consequently, *brookite* is uniaxial (+) for λ = 5500 Å, biaxial (+) corresponding to permutation (2) for longer wavelengths, and biaxial (+) corresponding to permutation (4) for shorter wavelengths. These features and the resultant crossed axial plane dispersion observed in acute-bisectrix-centered interference figures of *brookite* have been discussed in detail elsewhere (Bloss, 1961, 187–188).

For monoclinic crystals, one principal axis of the biaxial indicatrix will coincide, for all wavelengths of light, with the single 2- or $\bar{2}$-fold axis. This latter is conventionally denoted as the crystallographic c axis (first setting) or as the b axis (second setting). Adopting the second setting, the b axis will be a principal vibration axis—either X, Y, or Z depending on the mineral—for all wavelengths of light. The remaining two principal vibration axes will be mutually perpendicular and lie in the plane normal to the b axis, their positions changing slightly as different wavelengths are considered. The result is as if the biaxial indicatrix rotates slightly about the b axis as different wavelengths of light are used. To specify the optical orientation of a monoclinic crystal—that is, the attitude of its indicatrix to its crystallographic axes—it is necessary to state (1) the optical identity of the symmetry-defined crystallographic axis, for example, whether b equals X, Y, or Z, plus (2) the angle between one of the remaining principal vibration axes and one of the remaining crystallographic axes, this angle being located (for the second setting) in the (010) plane. To illustrate, the optical orientation of *gypsum*, $CaSO_4 \cdot 2H_2O$, is defined as $b=Y$ and $Z \measuredangle c = 52°32'$ (Fig. 1), where $Z \measuredangle c$ should be read as "the angle between Z and c." Some authorities use a convention whereby a positive sign indicates Z to be located in the obtuse angle β; thus the angle between Z and c $=+52°32'$ indicates the orientation in Fig. 1. An opposite convention is followed by Winchell and Winchell (1951), who extend Schuster's rule for the sign of ***feldspar*** extinction angles to all monoclinic and triclinic crystals. Using this convention, the orientation in Fig. 1 would be denoted as $Z \measuredangle c = -52°32'$.

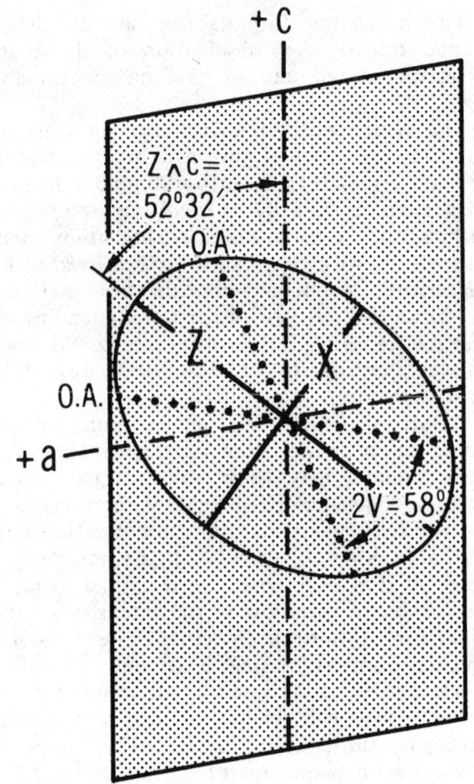

FIGURE 1. Section parallel to (010) of the monoclinic crystal, *gypsum*, to show the angles between its principal vibration directions X, Y (normal to page), and Z and its morphologically determined crystallographic axes a, b (normal to page), and c. Directions X, Y, and Z as well as the optic axes (dotted) are those for sodium light (λ = 5893 Å).

In triclinic crystals, the crystallographic a, b, and c axes do not generally coincide, except by chance, with the X, Y, and Z directions.

F. DONALD BLOSS

References

Bloss, F. D., 1961. *An Introduction to the Methods of Optical Crystallography*. New York: Holt, Rinehart & Winston, 294p.

Winchell, A. N., and Winchell, H., 1951. *Elements of Optical Mineralogy*. New York: Wiley, pts. I–III.

Cross-references: *Crystallography: Morphological; Directions and Planes; Symmetry.*

ORDER-DISORDER

Order–disorder phenomena occur in crystals in which two or more structurally and energetically nonequivalent sites are occupied by two or more atoms, ions, vacancies, or other particles. In the lower temperature range of stability of such crystals, each type of particle tends to concentrate in one or more of the available sites and this is known as the "ordered state." However, with increasing temperature, order generally decreases continuously as the particles distribute themselves more randomly among the sites, so that for any given crystal there exists a definite state of order for each temperature. In some cases order–disorder is associated with a first-order phase change or "inversion" that involves a change in crystal symmetry.

General Features

Long- and Short-Range Order. In the most familiar type of order, a particle is statistically concentrated at a site that is characterized by a definite repeat distance within the crystal structure. This type, which is clearly revealed by such experimental techniques as X-ray diffraction and infrared, Mössbauer, and electron-spin resonance spectroscopy, is referred to as "long-range" order. However, ordering may also occur in such a way that a certain particle B is statistically coordinated with a particle A as a nearest or next-nearest neighbor irrespective of the lattice position of A. This is referred to as "short-range" or "local" order and is much more difficult to study experimentally than is long-range order.

The Mechanism. If the ordered state corresponds to a concentration of particle A in site 1 and of particle B in site 2, the disordering process may be represented by the exchange reaction

$$A(1) + B(2) \rightleftarrows A(2) + B(1) \quad (1)$$

In terms of kinetic theory, the time rate of change of A on site 1 is given by the difference between two mass-action terms:

$$-\frac{dC_1^A}{dt} = K_{12}\phi_{12}C_1^A C_2^B - K_{21}\phi_{21}C_1^B C_2^A \quad (2)$$

In this differential equation the C's refer to the concentrations (in atoms or moles cm^{-3}) of the species indicated by the superscripts in the sites indicated by the subscripts. The K's, which are functions only of the temperature and volume (or pressure), are specific rate constants. The ϕ's are analogous to activity coefficient products in macroscopic chemical systems and are functions of the temperature, volume (or pressure), and the total composition of the crystal. However, the effect of volume and pressure changes are generally expected to be small as compared to the effect of temperature changes.

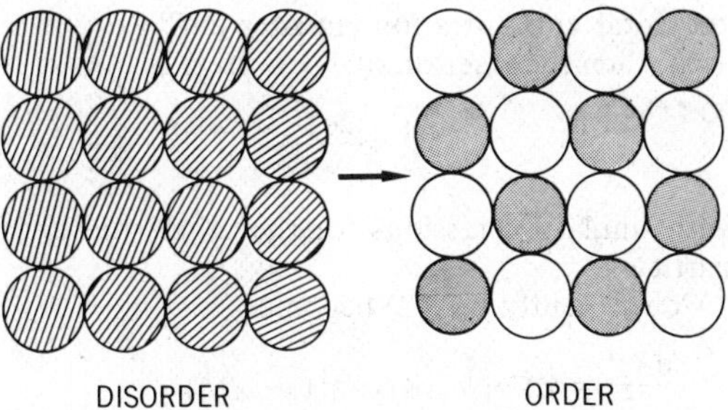

FIGURE 1. Order–disorder in a simple binary metallic alloy. In the high-temperature, disordered state (left) the two kinds of atoms are randomly distributed over both sites, which are indistinguishable. In the low-temperature, ordered form nearly complete segregation into two nonequivalent sites has occurred.

Simple Metallic Alloys. The theoretical basis for order–disorder equilibria in simple metallic alloys was provided by W. L. Bragg and E. J. Williams and their treatment was extended to rate phenomena by Dienes (1955). The ordering process in a simple binary alloy (Fig. 1) is characterized by the direct participation of all the atoms of the lattice so that the ϕ factors of Eq. (1) vary greatly with the degree of order as well as the temperature. Furthermore, at a certain critical temperature the different sites become equivalent, so that at all higher temperatures the crystal is completely disordered. Because of these characteristics, it is convenient to cast Eq. (2) and its limiting equilibrium expression into forms involving an ordering parameter O which is equal to unity at the hypothetical state of complete order and equal to zero when complete disorder is attained.

Complex Crystals. In complex crystals, the picture is distinctly different from that of simple alloys. In these crystals there are usually more than two kinds of sites and particles and some of these may not participate directly in the order–disorder phenomena. Rather they may form a relatively inert framework which stabilizes the structure and preserves the site nonequivalence at the highest temperatures of stability for the structure. Thus, it is reasonable to suppose that, if the exchangeable particles are similar in size, charge, and other characteristics, the site-preference energies (corresponding to the difference in binding energy of a given particle between the nonequivalent sites) will be relatively independent of the degree of order. For such crystals, it is convenient to cast Eq. (2) into a form involving the atomic or ionic fractions on each site (Mueller, 1967). Thus, if sites 1 and 2 are present in equal numbers, and if Co represents the total concentration (in sites cm^{-3}) of both sites, then the fraction of A on site 1 is

$$X_1^A = \frac{2C_1^A}{Co}$$

with similar expressions for the other sites and particles.

Consequently Eq. (2) becomes

$$-\frac{dX_1^A}{dt} = \tfrac{1}{2}Co[K_{12}\phi_{12}X_1^A(1 - X_2^A) - K_{21}\phi_{21}X_2^A(1 - X_1^A)] \quad (3)$$

since $X_1^A + X_1^B = 1$, etc. Also the equation of equilibrium obtained by setting

$$\frac{dX_1^A}{dt} = 0$$

is then

$$K_{21}^{\circ} = \frac{K_{21}}{K_{12}} = \frac{X_1^A(1 - X_2^A)\phi_{12}}{X_2^A(1 - X_1^A)\phi_{21}} \quad (4)$$

Here K_{21}°, the equilibrium constant of exchange for ordering, decreases toward unity as the temperature increases. However K_{21}° attains unity only when the two sites become equivalent as in the simple alloys.

The most interesting forms of Eqs. (3) and (4) are found by setting $\phi_{21} = \phi_{12} = 1$, which corresponds to ideal mixing on each structure site and to constant site preference and exchange energies at constant temperature.

Equation (3) can be integrated directly at constant temperature for any fixed total composition of the crystal. Thus if X^A is the fixed total fraction of A for both sites 1 and 2 we have

$$2X^A = X_1^A + X_2^A \quad (5)$$

The relation (5) may be used to eliminate either X_1^A or X_2^A from (3) so that it may be integrated.

The graphical form of the integrated rate equation with $\phi_{21} = \phi_{12} = 1$ is shown as curve (a) in Fig. 2. A hypothetical form for which the ϕ factors are not unity is also shown as curve (b) in the same figure.

Rate and equilibrium equations analogous to Eqs. (3) and (4) may also be derived for crystals with more than two structure sites and exchangeable particles. However the solutions are then more difficult and complicated (Mueller, 1967). Furthermore, it is not to be expected that the ϕ factors of complex crystals will always be small. Indeed, it is likely that, where exchange between sites involves particles of unlike charge and size, the deviation from ideal solutions might be as great as for simple alloys.

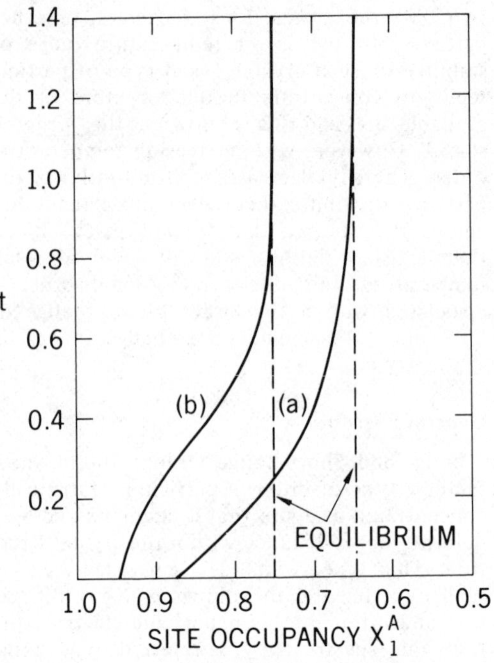

FIGURE 2. Isothermal disordering rate curves showing the change of the atomic fraction X_1^A of atom A on site 1 as a function of the time. Curve (a) corresponds to ideal solutions on the individual sites while (b) corresponds to one type of nonideal solution. The time scale is arbitrary.

Examples of Long-Range Order–Disorder

Pyroxenes and Amphiboles. In these complex crystals the exchangeable ions are sandwiched between Si-O chains. In ***orthopyroxene*** (q.v.) exchange occurs between Mg^{2+} and Fe^{2+} ions which are confined to M_1 and M_2 (metal positions) as shown in Fig. 3. The M_1 site occupies the center of a fairly regular octahedron in coordination with O^{2-} (oxygen) anions, but the M_2 octahedron with these same anions is much more distorted. As a result of these different degrees of distortion, the site preference energies are sufficiently different to produce a marked concentration of Fe^{2+} in the M_2 site (Ghose, 1965). Equations (3) and (4) are directly applicable here if we set $A = Mg^{2+}$ and $B = Fe^{2+}$. Then, if it is assumed that $\phi_{21} = \phi_{12} = 1$ in Eq. (4), ***orthopyroxenes*** from metamorphic rocks yield $K_{21}^{\circ} \approx 48$. The isothermal distribution curve corresponding to this value is shown as curve (1) in Fig. 4. For metamorphic ***orthopyroxenes*** heated to 1000° C and rapidly quenched, it is found that $K_{21}^{\circ} \approx 7$ (Ghose and Hafner, 1967). The graphical expression of the value is shown as curve (2) in

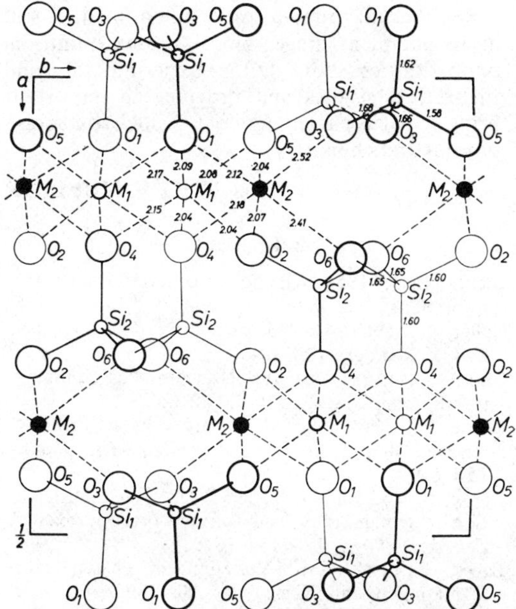

FIGURE 3. Structure of ***orthopyroxene*** viewed along [001] (after Ghose, 1965). Filled circles indicate concentration of Fe^{2+} in M_2 sites. Open circles indicate concentration of Mg^{2+} in M_1 sites. Numbers refer to interionic distances in angstrom units. O = oxygen, Si = silicon. Nonequivalent O and Si sites are indicated by subscripts.

Fig. 4. It is interesting that, as might be expected, the points corresponding to more slowly quenched volcanic ***orthopyroxenes*** fall between the two curves.

Order–disorder phenomena have also been observed by Ghose (1961) in the ***amphibole*** $(Fe,Mg)_7Si_8O_{22}(OH)_2$ (*cummingtonite*) in which Mg^{2+} and Fe^{2+} are distributed among the four nonequivalent ***amphibole*** sites M_1, M_2, M_3, and M_4. In this case, Fe^{2+} is again concentrated in the highly distorted M_4 polyhedron while Mg^{2+} is concentrated in the slightly distorted M_1, M_2, and M_3 octahedra. Other ***amphiboles*** (q.v.) also show a great tendency for ordering of a variety of ions such as Al^{3+}, Fe^{3+}, Na^+, and Ca^{2+} among the M sites. As was first noted by E. J. W. Whittaker and confirmed by Papike and Clark (1968), there is a strong tendency for coupling to occur such that Na^+ and Fe^{3+} tend to go into adjacent sites. From such behavior one should expect deviations in the ϕ values from unity. In fact, some deviation is even indicated for Mg^{2+} and Fe^{2+} in recent highly precise data for both ***orthopyroxene*** and *cummingtonite*. An extensive treatment of the thermodynamics of order–disorder in ***pyroxenes*** has been provided by Saxena (1973).

Feldspars. The ***feldspars*** (q.v.) are important examples of a variety of complex mineral phases that exhibit ordering. Laves (1960) proposed an equilibrium model for $KAlSi_3O_8$ ***feldspar*** in which Al^{3+} and Si^{4+} are highly ordered among four nonequivalent tetrahedral sites A_1, A_2, B_1, and B_2 in the low-temperature stability field of triclinic *microcline*, but in which these sites unite to form only two nonequivalent sites A and B at the temperature of inversion to monoclinic *sanidine* (Fig. 5). Disordering is thought to occur continuously with increasing temperature both above and below the inversion (Phillips and Ribbe, 1973). An analogous but more complex scheme of ordering may also be inferred for ***plagioclase feldspars*** (Taylor, 1965). In the case of $NaAlSi_3O_8$ (*albite*), the order–disorder relationships appear similar to those of $KAlSi_3O_8$, but $CaAl_2Si_2O_8$ (*anorthite*) is more highly ordered. In intermediate members of the series, the order–disorder relationships are functions of the variable Si/Al ratio and are coupled to the distribution of Na^+ and Ca^{2+} as well. At low temperatures, these factors result in the division of the series into several compositional discontinuities and in unmixing on a minute scale.

Garnets. In the case of ***garnets*** (see *Garnet Group*), order–disorder studies have centered on synthetic varieties of physical or economic

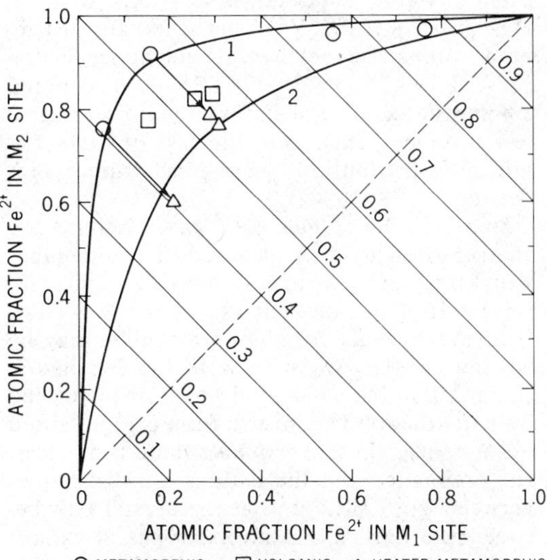

FIGURE 4. Experimental data and ideal isothermal distribution curves for Fe^{2+} and Mg^{2+} among the M_1 and M_2 sites of ***orthopyroxene*** (modified after Ghose and Hafner, 1967). The dashed line shows the completely random distribution which occurs in simple metallic alloys but only rarely in complex crystals of this type.

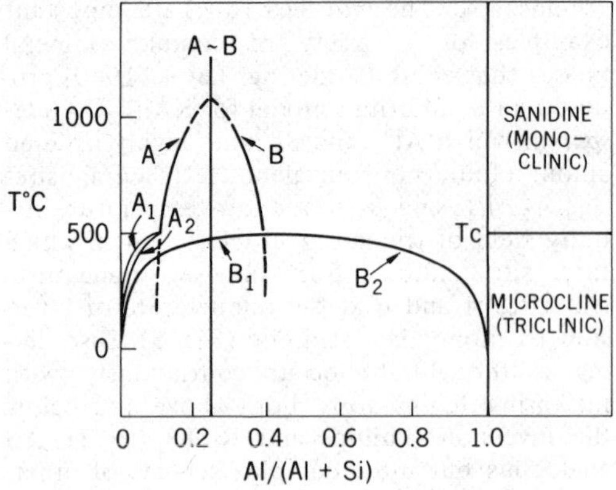

FIGURE 5. Diagrammatic sketch of the Al-Si equilibrium distribution in $KAlSi_3O_8$. The letters with subscripts refer to nonequivalent tetrahedral sites. Tc refers to the temperature of inversion of *microcline* to *sanidine*. (Modified after Laves, 1960).

interest. Geller (1967) and his associates have made extensive studies of the magnetic properties of varieties with yttrium, scandium, and other rare elements as major constituents. In these and in natural ***garnets*** as well there are nonequivalent sites of tetrahedral, octahedral, and dodecahedral coordination of cations with oxygen. A general structural formula may be written as $A_3^{VIII}B_2^{VI}C_3^{IV}O_{12}$ where *A, B,* and *C* are cation positions of the coordination indicated by the superscripts. Cations such as Mg^{2+}, Fe^{2+}, Mn^{2+}, Y^{3+}, and Zn^{2+} are distributed among the octahedral and dodecahedral sites while Al^{3+}, Fe^{3+}, and Ti^{4+} are distributed among the octahedral and tetrahedral sites. In a few instances, such as in the case of Co^{2+}, significant distribution among all three sites occurs.

Spinels. The *spinels* (see *Spinel Group*) are characterized by two idealized or end-member structures referred to as "normal" and "inverse." If *A* is a divalent cation and *B* is trivalent, the formula for the normal state may be written as $A(B_2)O_4$ with *A* in the tetrahedral site and *B* in the octahedral site as distinguished by parentheses. The inverse form is represented by $B(AB)O_4$. In many ***spinels*** there is no clear-cut preference and the cations are distributed between both sites so that the crystal falls between the normal and inverse types. A variant of Eq. (4) was suggested by L. Néel and his associates and was experimentally verified by Navrotsky and Kleppa (1967). These authors found the following sequence of increasing preference for the octahedral site: Zn^{2+}, Mn^{2+}, Fe^{2+}, Ga^{3+}, Co^{2+}, Mg^{2+}, Fe^{2+}, Cu^{2+}, Al^{3+}, Ni^{2+}, Mn^{3+}, Cr^{3+}. Mg^{2+} and Fe^{2+}, in particular, show little site preference.

Research in order-disorder phenomena is still in its incipient stage, but it presents unusual opportunities for both experimental and theoretical studies and provides an important bridge between mineralogy and solid-stage physics and chemistry.

ROBERT F. MUELLER

References

Dienes, G. J., 1955. Kinetics of order-disorder transformations, *Acta Metall.*, **3**, 549–557.

Geller, S., 1967. Crystal chemistry of the garnets, *Zeit. Kristallogr.*, **125**, 1–47.

Ghose, S., 1961. The crystal structure of cummingtonite, *Acta Crystallogr.*, **14**, 622–627.

Ghose, S., 1965. Mg^{2+}-Fe^{2+} order in an orthopyroxene, $Mg_{0.93}$, $Fe_{1.07}$, Si_2O_6, *Zeit. Kristallogr.*, **122**, 6–99.

Ghose, S., and Hafner, S., 1967. Mg^{2+}–Fe^{2+} distribution in metamorphic and volcanic orthopyroxenes, *Zeit. Kristallogr.*, **125**, 157–162.

Laves, F., 1960. Al/Si-Verteilungen, Phasen-Transformationen und Namen der Alkalifeldspate, *Zeit. Kristallogr.*, **113**, 265–296.

Mueller, R. F., 1967. Model for order–disorder kinetics in certain quasi-binary crystals of continuously variable composition, *J. Phys. Chem. Solids*, **28**, 2239–2243.

Navrotsky, A., and Kleppa, O. J., 1967. The thermodynamics of cation distributions in simple spinels, *J. Inorg. Nucl. Chem.*, **29**, 2701–2714.

Papike, J. J., and Clark, J. R., 1968. The crystal structure and cation distribution of glaucophane, *Am. Mineralogist*, **53**, 1156–1173.

Phillips, M. W., and Ribbe, P. H., 1973. The structure of monoclinic potassium-rich feldspars, *Am. Mineralogist*, **58**, 263–270.

Saxena, S. K., 1973. *Thermodynamics of rock-forming crystalline solutions.* New York: Springer-Verlag, 188p.

Taylor, W. H., 1965. Framework silicates: The feldspars, in W. L. Bragg, ed., *Crystal Structures of Minerals.* London: G. Bell and Sons, 293–339.

Cross-references: *Alkali Feldspars; Amphibole Group; Crystals, Defects in; Crystal Growth; Etch Pits; Feldspar Group; Garnet Group; Magnetic Minerals; Orthopyroxenes; Plagioclase Feldspars; Pyroxene Group; Spinel Group;* see also mineral glossary. Vol. IVA: *Crystal Chemistry; Infrared Analysis; Ion Exchange; Mössbauer Effect; Phase Equilibria; X-ray Diffraction Analysis.*

ORE DEPOSITS—*See* MINERAL AND ORE DEPOSITS

ORE GENESIS—*See* MINERAL DEPOSITS: CLASSIFICATION; VEIN MINERALS

ORE MICROSCOPY (MINERAGRAPHY)

Ore microscopy (mineragraphy or mineralography) is the study of polished surfaces of ores or of ore minerals by means of a polarizing, reflected-light microscope and the interpretation of the mineral associations and microtextures so observed.

The earliest reference to polished-ore techniques is that of Berzelius of Sweden who, about 1806, studied a suite of *pyrrhotite*-bearing ores with a microscope adapted for reflected light. H. C. Sorby of Sheffield in the 1820s and W. Campbell of the U.S. Geological Survey in the early 1900s developed the technique, which has since found numerous applications in both theoretical mineralogy and problems of an economic nature.

Crystal optics in reflected light has been the subject of important contributions by Voigt (1906), Cissarz (1932), Berek (1937, 1943), Capdecombe and Orcel (1941), Cameron and Green (1950), and Hallimond (1953). Some of these studies were aimed at devising means of mineral determination in polished section; others were concerned exclusively with optical theory.

The most comprehensive text on descriptive ore mineralogy in reflected light is that of Ramdohr (1960). Other texts on determinative methods are those of Schneiderhöhn (1952), Short (1940), Uytenbohaardt (1951), and an important paper by Bowie and Taylor (1958). Particular mention is warranted of the Peacock Atlas of X-ray diffraction data for the ore minerals (Berry and Thompson, 1962).

No major text specifically devoted to applied ore microscopy, especially in relation to ore dressing, is extant; but works by Fairbanks (1928), Gaudin (1939), Edwards (1954), and Cameron (1961) contain chapters on this important application. A review of the various economic applications of ore microscopy is given by Lawrence (1965).

Ore microscopy is consistently used to assess such matters as the sequence of ore-mineral precipitation, incidence of periods of ore deformation, degree of oxidation, relative temperatures of ore formation, and shape and size analysis of constituent minerals.

Basic Equipment

The basic equipment for mineragraphic work includes a polarizing ore microscope (or metallographic microscope), a high-intensity light source with transformer to vary light intensity, and a means of mounting and polishing small fragments of ore.

The essential features of the reflecting microscope and accessories are shown in Fig. 1, which also illustrates the passage of light from the light source to the observer. Numerous types of ore microscopes are available, ranging from simple bench microscopes, to which cameras may be fitted, to very elaborate projection microscopes with built-in photographic equipment adaptable for phase contrast work.

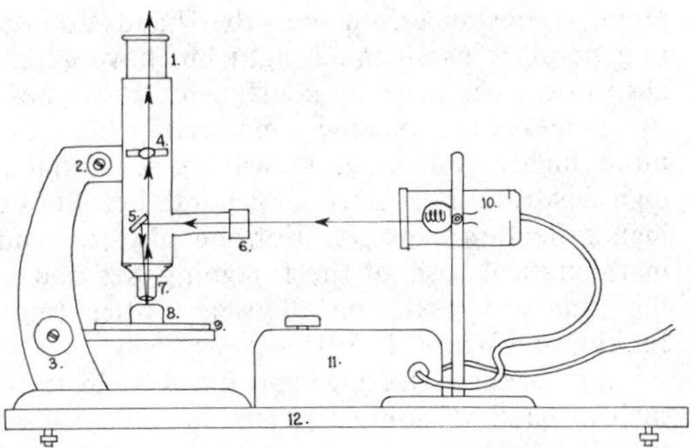

FIGURE 1. Ore microscope showing passage of light from light source to specimen, thence to observer. 1, ocular; 2, fine focus; 3, coarse focus; 4, analyzer; 5, reflector; 6, polarizer in light entrance tube; 7, objective; 8, polished specimen briquette; 9, rotating stage; 10, light source; 11, transformer, 12. alignment board.

Ore fragments may be broken off, or sawed off by a diamond wheel, from larger pieces. The pieces are then ground on one side to a smooth surface and mounted in bakelite or in cold-setting plastic so that the flat surface is uppermost. The resulting briquettes are then hand ground on glass plates with carborundum and water, in progressive stages, down to ultrafine powder. Several mechanical devices are available for imparting the final polish. The most modern technique, however, is that of a Syntron polishing machine. This machine facilitates mass production of polished sections with minimum labor. It consists of an electrically operated drum that vibrates in such a way that the briquettes, weighted onto the polishing cloth at the base of the drum by means of steel mounting cups, move around the drum in a circular fashion. Perfect polish is imparted to up to twenty specimens simultaneously in a few hours by the grinding fluid covering the polishing cloth.

Optical Principles

Of the light that is normally incident on a polished surface, some is transmitted, some is absorbed, and some is reflected. Substances that absorb only a small amount of light energy generally will transmit much of the incident light; these are the transparent or translucent substances and they behave as dielectrics.

Metallic substances, on the other hand, do not, in general, transmit much light but have a high absorption capacity or coefficient. It is these substances—the opaque materials—that are more highly reflecting. It will be seen that a high absorption capacity, in general, betokens a high reflection capacity. For the physical and mathematical basis of the foregoing, see one of the standard texts on physical optics (e.g., Jenkins and White, 1950).

Study of the reflecting capacity of dielectrics, such as basal sections of *quartz*, has shown that "reflectivity," i.e., the proportion of light reflected to that normally incident, is expressed by the equation of Fresnel:

$$R = \frac{(n - N)^2}{(n + N)^2} \tag{1}$$

where R = that fraction of light that is reflected
n = the refractive index of the transmitting substance
N = the refractive index of the immersion medium

For air, equation (1) becomes

$$R = \frac{(n - 1)^2}{(n + 1)^2} \tag{2}$$

In the case of *quartz* (taking n=1.54),

$$R = \frac{(1.54 - 1)^2}{(1.54 + 1)^2} = .045$$

Thus, only 4.5% of white light normally incident on a polished surface of *quartz* is reflected.

For the highly absorbing, poorly transmitting, nondielectric substances, the equation contains another term κ (kappa) to take account of absorption. The equation becomes

$$R = \frac{(n - N)^2 + (n^2\kappa^2)}{(n + N)^2 + (n^2\kappa^2)} \tag{3}$$

or in air

$$R = \frac{(n - 1)^2 + (n^2\kappa^2)}{(n + 1)^2 + (n^2\kappa^2)} \tag{4}$$

From equation (4) it may be noted that it is possible to calculate a refractive index of a metal or other opaque substance by determining, by actual measurement, the reflectivity of the substance in two different media (e.g., air and cedar oil) and then rearranging the equation so that n becomes the unknown.

Both absorption and reflectance are selective. A highly transmitting substance may allow certain wavelengths of white light to pass through while other wavelengths are absorbed and reflected. For example, the ruby silver ores *proustite* and *pyrargyrite*, or the copper oxide *cuprite* are all blood red in transmitted light but are pale blue in reflected light. The color of a mineral in reflected light, then, is related to the wavelengths of the light that are preferentially absorbed. Many ore minerals reflect essentially whitish color, indicating a very high absorption and very little transmission, e.g., a polished surface of metallic *silver* reflects 94% of incident white light. It follows that the observed color is also whitish since very few of the wavelengths are lost by transmission. After transmission, though, the intensity is reduced.

Light from the microscope lamp is polarized in passing through the light-entrance tube of the microscope (Fig. 1). If an anisotropic ore mineral on the microscope stage is turned 45° from an extinction position, polarization colors are observed. Unlike interference colors in thin-section study, the color observed in an ore mineral between crossed polars in reflected light is due to a "rotation of the plane of polarization." This may be illustrated for monochromatic light as follows (see Fig. 2). *A-A* and *P-P* are the directions of the analyzer and polarizer respectively. OH_1 and OH_2 are the principal reflecting directions for that particular crystal orientation. Upon reflection, light vibrating along principal reflection direction OH_1 is reduced in amplitude to Oh_1 and along OH_2 is reduced to Oh_2. The resultant ON no longer coincides with the polarizer direction, having been rotated through an angle α toward that principal direction that possesses the higher reflectivity. Light of greatly reduced intensity (I), compared to the incident-light

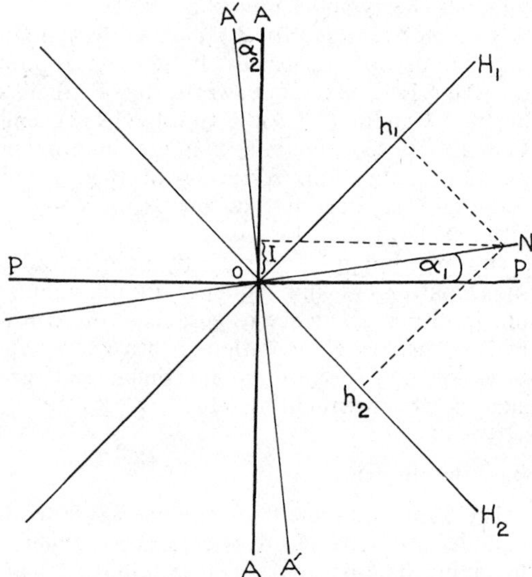

FIGURE 2. Vector diagram of rotation of plane of polarization of an anisotropic ore mineral in monochromatic light.

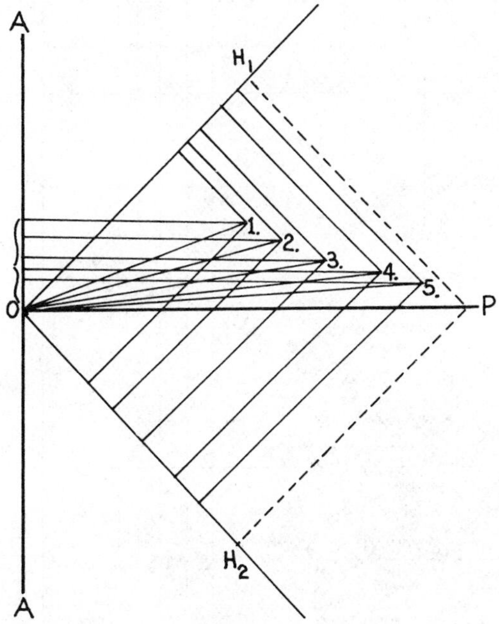

FIGURE 3. Vector diagram showing rotations from various wavelengths of white light. 1, rotation for red; 2, rotation for yellow; 3, rotation for orange; 4, rotation for green; 5, rotation for blue.

intensity, passes through the analyzer; it will be noted that

$$I \propto (\text{amplitude})^2$$

In white light, each wavelength undergoes a different rotation, the integration of the resolved parts of which, in the direction *OA* (Fig. 3), produces the familiar (and often diagnostic) polarization colors observed between crossed polars.

In monochromatic light (Fig. 2), the single rotation of the plane of polarization, A_r, can be compensated by an equal rotation of the graduated analyzer or Wright ocular. This recrosses analyzer and polarizer and restores extinction. The exact value of A_r is obtained by reducing the measured angle α by a small amount due to an additional rotation induced by the reflection of light from the reflector.

Determinative Methods

In its early stages, mineral identification in polished section was largely a matter of knowing the appearance of as many minerals as the observer's experience permitted—a technique, paradoxically enough, that is still quite effective.

The first attempts at systematic identification were those of Murdoch (1916) and Davy and Farnham (1920) who utilized etch tests using a series of standard etch reagents: HNO_3, HCl, KOH, $FeCl_3$, KCN, and $SnCl_2$ and observing, under the microscope, surface tarnish and other effects. The method was further developed by Short (1940) whose determinative tables utilize hardness tests and polarization phenomena as well as microchemical reactions. Short's etch tests are widely employed, apparently with some success; but many workers find them unreliable. The main difficulty seems to be that a freshly polished surface does not wet readily and reaction between droplet and mineral is inhibited.

Microhardness tests, since the early scratch tests of Talmage (1925), have been refined by utilizing "diamond indentation" testers. Improved measurements of surface reflectivity (Follinsbee, 1949) have provided useful determinative data. These two parameters—microhardness and reflectivity—plotted against one another (Fig. 4), have been used effectively by Bowie and Taylor (1958) and by Gray and Millman (1962) using photoelectric cells with galvanometers. Simple diamond indenters attached to the barrel of the microscope facilitate ready measurement of microhardness. The values so obtained are vectors so that, in practice, hardness tests should be averaged for several grains of the unknown mineral.

Reflectivity measurements require two assumptions: (1) that the unknown mineral possesses its maximum polish and (2) that the reference standard employed (e.g., *pyrite* with a theoretical reflectance of 54.5% for white light) does, in fact, possess the reflectance value assigned to it, since

$$R = \frac{A \times C}{B}$$

where R = percentage reflectance of unknown mineral
A = galvanometer deflection for the unknown mineral
B = galvanometer deflection for the standard
C = reflectance of the standard

Microhardness has been found to vary with distance from the center of the grain, within minerals with marked cleavage, and for other reasons related to the physical condition of the ore. Moreover, being vector quantities, both reflectivity and microhardness will vary with crystal orientation. However, the tables of Bowie and Taylor are valuable in placing an unknown mineral within a small range of possibilities, which other properties—polarization effects, surface color, paragenesis, etc.—may further limit. Unfortunately, certain

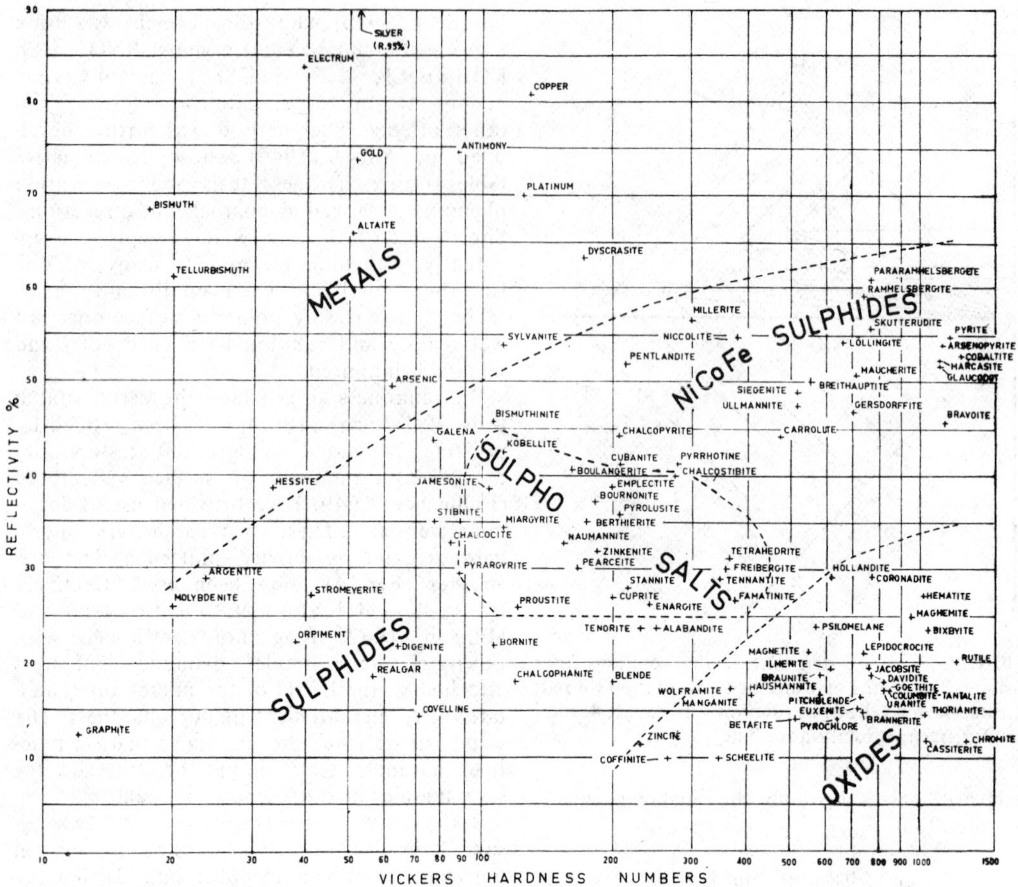

FIGURE 4. Graph of reflectivity and microhardness values (after Bowie and Taylor, 1958).

groups of minerals possess similar properties—e.g., the feather ores or the ruby silver ores—and are difficult to distinguish by this method.

Grains of less than ca. 0.2 mm must be isolated for reflectance measurements by closing down the microscope diaphragm to an extent that the reflected light is too small for sensitive registration on the galvanometer. Photomultipliers in circuit with the selenium cell, or the use of cadmium sulfide cells, will greatly increase the sensitivity. With very minute grains (e.g., 0.05 mm), reflectance measurements would always be suspect, if not insignificant.

Rotation properties, originally studied by Campbell (1906) and by Neuerberg (1947), were considered further by Cameron and Green (1950) and by Lawrence (1960). Cameron and Green sought to use certain dispersion phenomena, observed conoscopically, as diagnostic properties. These studies provided important theoretical data but do not appear to have fulfilled expectations. More recently Cameron, et al. (1961) have provided a set of tables of the "apparent angle of rotation of the plane of polarization" (see Fig. 2) measured for various wavelengths. But again, rotation properties vary slightly with crystal orientation; and the precise setting of the mineral to extinction prior to measuring rotation angles, even with the aid of sensitive tint plates, is problematical. The method does facilitate the grouping of minerals according to various amounts of rotation.

Other determinative methods have failed to stand the test of time. Among these are Davy and Farnham's electrical-conductivity method (Davy and Farnham, 1920); McKinstry's electrochemical and photochemical methods (McKinstry, 1927) and Gaudin's iridescent filming technique (Gaudin, 1938). There is no universally accepted method of mineral identification in reflected light.

Undoubtedly the most reliable method of identification is that of X-ray diffraction measurement. This necessitates, ideally, the drilling out, under the microscope, of a minute quantity of a mineral and applying to the drilled surface a droplet of collodium from which a filament suitable for powder photographs can be made. The drilled powder is of the exact fineness for sharply defined diffraction. Dental burrs fine enough to drill grains of 0.25 mm, while observing the operation under low-power magnification, are readily available for use with a foot-controlled electric dental unit mounted on the wall of the laboratory. Smaller burrs can be made if required.

Diffraction patterns of intimate intergrowths of two mineral phases can be distinguished by X-raying a composite powder.

Notwithstanding these aids, there is no substitute for a detailed knowledge of the ore minerals and their parageneses. Bulk assays of an ore are useful in indicating possible mineral constitution. For example, an ore containing copper, lead, and bismuth may be expected to contain *chalcopyrite, galena*, and native *bismuth*, but it may contain also rarer minerals such as *galenobismutite* ($PbBi_2S_4$), *wittichenite* (Cu_3BiS_3), or *aikinite* ($PbCuBiS_3$). Knowledge of paragenesis should extend as far as to include, for example, such minerals as *naumannite, klockmannite, clausthalite*, and *umangite*. These rare minerals have one thing in common—they are all selenides. They often occur together, and recognition of one of them under the microscope may point to the presence of one or more of the others. Paragenetic groupings of the ore minerals are given in most textbooks; the texts by Ramdohr (1960) and by Uytenbogaardt (1951) are invaluable in this regard.

Application

Brief mention has already been made of various optical properties exhibited by ore minerals in reflected light. A detailed account of the theoretical background is given by Hallimond (1953). Determining these optical properties by mineragraphic techniques constitutes an important (though little used) application of mineragraphy.

On a more practical level, the assessment of ore microtextures provides data on such matters as the depositional sequence and the subsequent history of an ore. Other applications include regional correlation of ore deposits, amenability to beneficiation of newly discovered ore, and as an aid in solving problems associated with ore dressing in an operating mine.

Microtextures of Ore. Much valuable information may be gleaned from a study of the microtextural relationships of the various components of an ore.

Ore textures and microtextures are many and varied, but may be considered under the following headings:

Simultaneous Deposition. The constituent minerals of many ores show evidence of having been precipitated from solution simultaneously. It has been the custom to search for criteria that would indicate a certain sequence of ore mineral precipitation—a procedure that became so subjective that many wrong assessments have been made. In modern practice, unless unambiguous textures to the contrary are observed, simultaneous precipitation of the constituent ore minerals may be presumed.

Such textures as interlaminations and myrmekitic intergrowths offer conclusive evidence that the components have crystallized together (Figs. 5a,b).

Notwithstanding the established depositional sequence for many base-metal sulfide deposits—i.e., *sphalerite-chalcopyrite-galena-tetrahedrite*—there are many instances where these minerals precipitated simultaneously. The main textural evidence is that any one or more of them may occur as more or less rounded areas within any other (Fig. 5c).

Replacement Textures. In contrast to textures indicative of simultaneous deposition are textures that indicate a certain sequence of ore-mineral precipitation. The most diagnostic is that of veinlets of one mineral transecting other minerals in such a way that earlier textures are truncated by the veinlet. Two main types of veinlets may be identified: replacement veinlets (Fig. 6a) and fracture fillings (Fig. 6b). In both instances, preexisting grain relationships are clearly interrupted by the later veinlet. Replacement veinlets are characterized by uneven walls whereas fracture fillings have parallel matching walls.

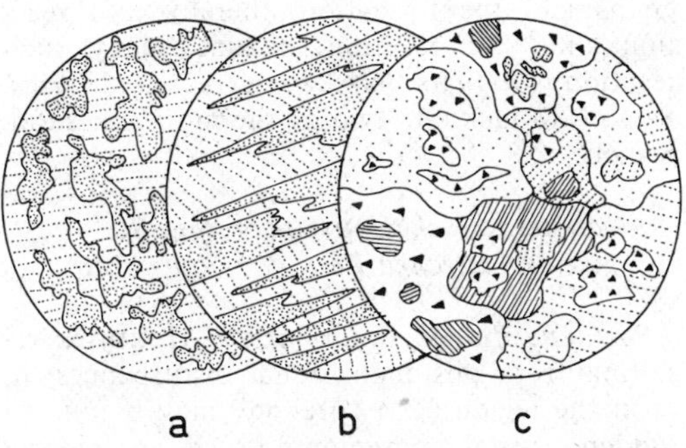

FIGURE 5. (a) Myrmekitic intergrowth of *chalcopyrite* (rows of dots) and *bornite* (close dots). (b) Interlamination of *chalcopyrite* (rows of dots and *bornite* (close dots). (c) Intergrowth of *chalcopyrite* (rows of dots), *sphalerite* (sparse dots), *galena* (triangular pits), and *tetrahedrite* (lines).

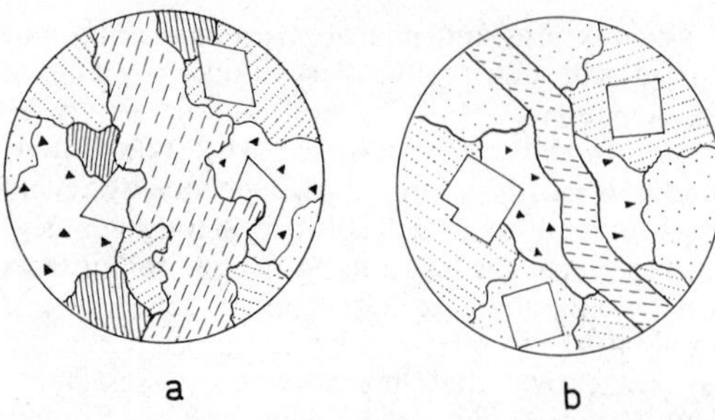

FIGURE 6. (a) Intergrowth of *arsenopyrite* (unshaded), *chalcopyrite* (rows of dots), *bornite* (sparse dots), *sphalerite* (close rows of dots), *galena* (triangular pits), and *tetrahedrite* (close lines) with a replacement veinlet of *chalcocite* (broken lines). (b) Intergrowth of *pyrite* (unshaded), *chalcopyrite* (rows of dots), *sphalerite* (sparse dots), and *galena* (triangular pits) with a fracture vein filling of *chalcocite* (broken lines).

Exsolution. Many ore minerals are capable of entering into solid solution in other ore minerals at the elevated temperatures at which ore deposition often commences. This requires that between "solute" and "solvent" there is similarity in atomic structure, in valency, and in cell dimension. Upon cooling, the solute diffuses ideally along crystal structure planes of the solvent where it is precipitated as an exsolution product. Very many examples of "exsolution pairs" are known in both the sulfides and the oxides. Some typical exsolution textures are indicated in Figs. 7a,b,c,d.

Solid-state Reactions. Some ore minerals, notably certain copper-lead sulfides, copper-lead-antimony sulfides, copper-bismuth sulfides, copper-lead-bismuth sulfides, and, particularly, the *stannite* series, are formed by solid-state reactions. This involves atomic diffusion between two juxtaposed ore-mineral grains and the formation, thereby, of "reaction rims" or reaction margins. Many such reaction products have been recorded; Figs. 8a and 8b illustrate two of the more common; namely,

cassiterite + *chalcopyrite* → *stannite*
galena + *tetrahedrite* → *bournonite*

Postdepositional Deformation. Textures resulting from postdepositional disturbances in, or in the region of, an orebody may be due to sudden crustal movements, such as faulting along the lode channel, or may be due to regional metamorphism. The former type of movement usually induces fragmentation of any brittle minerals, e.g., *pyrite* or *magnetite*; and, if ore fluids are still circulating, the fragments of early-formed minerals may be recemented by later-formed ore minerals. Thus, brecciated *pyrite* is often "healed" by later-formed *chalcopyrite, magnetite* by *hematite*, or *arsenopyrite* by *tetrahedrite* or by *pyrargyrite* (Fig. 9a).

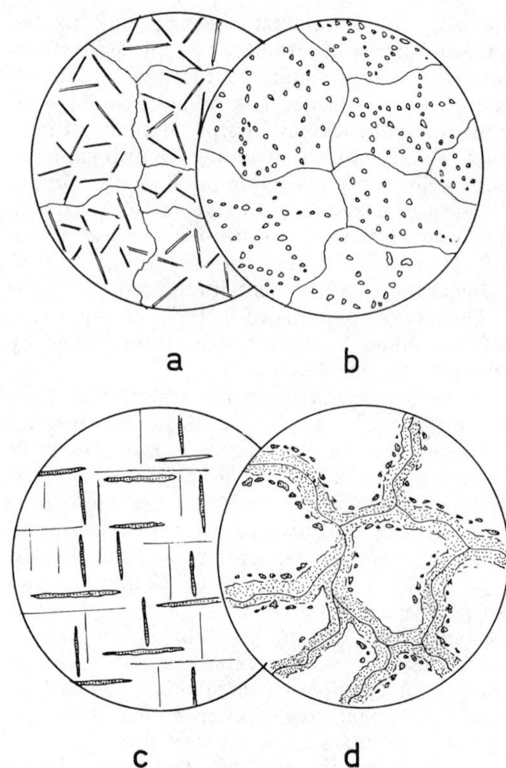

FIGURE 7. (a) *Magnetite* with exsolved *ilmenite* lamellae occupying the {111} structure planes of the *magnetite*. (b) *Sphalerite* with exsolution *chalcopyrite* "beads"—an emulsion texture. (c) *Galena* with exsolution rods of *galenobismutite* occupying the {100} structure planes of the *galena*. (d) *Pyrrhotite* (fine dots) unmixed from solid solution in *sphalerite* (unshaded) and having migrated to the *sphalerite* grain boundaries—"rim" or "grain-boundary" texture.

Textures due to metamorphism of an orebody are much more difficult to evaluate—indeed, not a great deal is known about metamorphism of ore. However, many orebodies contain *galena*, a mineral that does reveal the effects of metamorphism even where associated minerals do not. The initial indication that *galena* has undergone deformation can be seen both under the microscope and, even better, in hand specimen, where a gneissic texture is imparted to the *galena* as a result of plastic flow. In a more advanced stage of metamorphism, *galena* recrystallizes, losing its cuboidal aspect and taking on a fine-grained, polygonal grain arrangement. (Fig. 9b). This may require

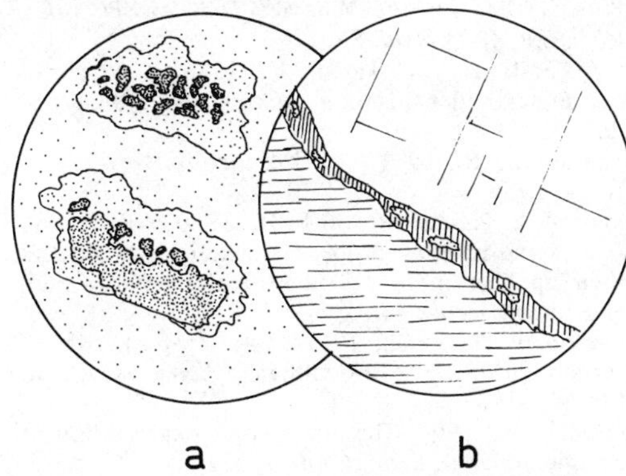

FIGURE 8. (a) "Reaction *stannite*" (medium sparse dots) formed by solid-state reaction between *cassiterite* (close dots) and *chalcopyrite* (sparse dots). (b) Reaction "rim" of *bournonite* (solid lines) formed by solid state reaction between *galena* (cubic cleavage) and *tetrahedrite* (broken lines). In this reaction small amounts of *chalcopyrite* (sparse dots) are formed from excess copper.

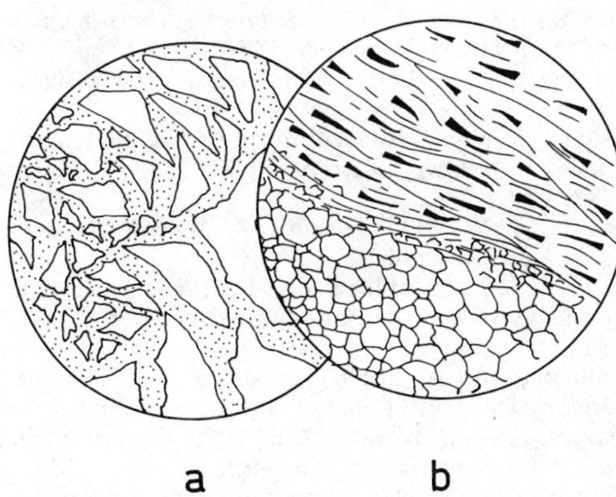

FIGURE 9. (a) Brecciated *pyrite* (unshaded) "healed" by later *chalcopyrite* (sparse dots). (b) *Galena* showing "plastic flowage" (with curved cleavage pits) ultimately recrystallizing into a granular aggregate.

etching for microscopic examination, but the sugary texture is usually quite apparent macroscopically.

Regional Correlation of Ore Deposits. Orebodies belonging to a particular metallogenetic epoch within a given region often possess certain mineralogical characteristics that distinguish them from other deposits within the region or from deposits of another period of metallogenesis in an adjacent region. These characteristics may include certain rare minerals as accessories, particular textures, or both, by which an ore type can be recognized quite clearly under the microscope. By this process, orebodies may be identified and correlated, and consanguinity established.

Mineragraphic Study of Newly Discovered Ore. With the advancement in knowledge of metallurgical processes, it may be assumed that very few orebodies of large tonnage and better-than-minimum-grade ore remain idle merely because of complex mineralogy. However, mineral constitution and textural relationships of an ore are of utmost importance in the economics of ore dressing as well as in the design of a treatment plant.

Modern prospecting for large disseminated orebodies usually involves, in the proving stages, systematic drilling on a grid system. Mineragraphic study of such drill cores is likely to yield valuable data to supplement developmental procedures.

EX and *AX* cores are ideally suitable for polished sections; special attention should be paid to parts of the drill core consisting of massive ore or to parts that are polymineralic, since these contain the greater diversity of mineral constituents.

In addition to mineral composition and textural intergrowths, information on the incidence of supergene enrichment, correlation with drill core assays, and the depth of incipient oxidation as it affects flotation can be obtained from polished sections of drill core well in advance of mining.

Mineragraphy in Ore Dressing. The greatest application of mineragraphy lies in the provision of data for ore-treatment plants. In operating mines, the particular application will depend on the problem in hand, but two aspects warrant particular mention.

The first of these is "size-shape distribution analysis." It is a feature of many large orebodies that the distribution of mineral types varies along strike and down dip, across width of the orebody, or in other ways. This variation may be a systematic one in a given direction or it may be a random or haphazard variation. Again, the variation may be with respect to one or more of the economically important mineral components or with respect to the occurrence of metallurgically deleterious minerals. Chemical assay, of course, will provide information of this type, but it will not throw light on how the various components are intergrown or what minerals are causing variation in the bulk composition of the ore.

Yet again, the texture of the ore, rather than its composition, may vary vertically, laterally, or in other directions. Such variation may involve the shape relationships of one mineral to another (replacements, inclusions, exsolutions) and(or) the grain size of economic and gangue minerals and of deleterious constituents.

Information of this type can be useful to both assay laboratory and mill superintendent if supplied early enough. Underground development of new ore offers good opportunity, along with sampling for assay, for the collection of mineragraphic specimens of ore that will eventually become mill feed.

It should be emphasized, however, that mineragraphic study of newly penetrated ore should not be expected to yield any more than broad trends in given directions. This information can be usefully recorded on assay sheets and underground plans.

Certainly the most important application of mineragraphy is in solving problems associated with ore dressing. This may be done on the site or in a central laboratory. In general, this involves the examination of particles of ore correctly sampled from the various stages of grinding or comminution–rod mill, ball mill, regrind ball mill–and from tailings discharge.

The degree of liberation of valuable ore from valueless gangue is assessed by grain counting. A series of polished sections of crushed ore is examined and, either by means of a graduated ocular or by an electric point counter attached to the microscope, a statistical analysis is made of the ore particles. Ideally, something approaching a thousand particles should be counted and the number of released ore particles, the number of particles of ore plus gangue, and the number of particles of gangue alone can be computed. An analysis of this type may determine whether it is worthwhile to recirculate the ground ore at any stage or whether to regrind, and may also reveal the cause of excessive middlings (particles consisting partly of gangue and partly of ore).

A study of tailings (material that is ultimately discharged to the tailings dam) is of considerable importance, especially with gold-bearing ore. Tailings should consist wholly of gangue; but, with a valuable metal such as gold, even minute particles locked within gangue can, over a period of time, constitute a serious loss if the situation is not corrected. Mineragraphic study of tailings will show to what extent, if at all, such losses are occurring and whether or not finer grinding is appropriate and economical.

L. J. LAWRENCE

References

Berek, M., 1937. Optische Mesamethoden in polarisiertem Auflicht, *Fortschr. Mineral. Kristallogr. Petrogr.*, **22**.

Berek, M., 1943. Zur Messung der Stärke des Anistropieeffektes, *Neues Jahrb. Mineral. Abh.*, **183**.

Berry, L. C., and Thompson, R. M., 1962. X-ray Powder Data for Ore Minerals, *The Peacock Atlas, Geol. Soc. Am., Mem.* **85**, 279p.

Bowie, S. H. U., and Taylor, K., 1958. A system of ore mineral identification, *Mining Mag.*, **96**, 265–267, 337–345.

Cameron, E. N., 1961. *Ore Microscopy*. New York: Wiley, 295p.

Cameron, E. N., and Green, L. H., 1950. Polarization figures and rotation properties in reflected light and their application to the identification of ore minerals, *Econ. Geol.*, **45**(8), 719–754.

Cameron, E. N., et al., 1961. Rotation properties of certain anisotropic ore minerals, *Econ. Geol.*, **56** (3), 596–583.

Campbell, W., 1906. The microscopic examination of opaque minerals, *Econ. Geol.*, **1**, 751.

Capdecombe, L., and Orcel, J., 1941. Determination des propriétés optiques des cristaux opaques, *Révue d'optique Théorique Instrument*, **20**, 47.

Cissarz, A., 1932. Reflexionsmessungen, *Zeit. Kristallogr.*, **82**, 438–450.

Davy, W. M., and Farnham, C. M., 1920. *Microscopic Examination of Ore Minerals*. New York: McGraw-Hill, 154p.

Edwards, A. B., 1954. *Textures of the Ore Minerals*. Melbourne: The Australian Inst. Min. Met.

Fairbanks, E. E., ed., 1928. *Laboratory Investigation of Ores*. New York: McGraw-Hill, 202p.

Follinsbee, R. E., 1949. Determination of reflectivity of ore minerals, *Econ. Geol.*, **44**, 425–436.

Gaudin, A. M., 1938. Identification of sulphide minerals by selective iridescent filming, *A.I.M.E. Tech. Pub. 912*.

Gaudin, A. M., 1939. *Principles of Mineral Dressing*. New York: McGraw-Hill, 554p.

Gray, I. M., and Millman, A. P., 1962. Reflection characteristics of ore minerals, *Econ. Geol.*, **57**(3), 325–349.

Hallimond, A. F., 1953. *Manual of the Polarizing Microscope*. York: Cooke, Troughton and Simms.

Jenkins, F. A., and White, H. E., 1950. *Fundamentals of Optics*. New York: McGraw-Hill.

Lawrence, L. J., 1960. Dispersion phenomena in the conoscopic study of the ore minerals, *Univ. N.S.W. Geol. Ser. No. 1*, 29p.

Lawrence, L. J., 1965. The application of ore microscopy to mining geology, in L. J. Lawrence, ed., *Exploration and Mining Geology, Proc. 8th Commonwealth Mining and Metallurgy Congress*.

McKinstry, H. E., 1927. Magnetic, electrochemical and photochemical tests in the identification of opaque minerals, *Econ. Geol.*, **22**, 669.

Murdoch, J., 1916. *Microscopical Determination of the Opaque Minerals*. New York: Wiley, 105p.

Neuerberg, C. J., 1947. Optical figures obtained with the reflecting microscope, *Am. Mineralogist*, **32**, 527–543.

Ramdohr, P., 1969. *The Ore Minerals and their Intergrowths.* New York: Pergamon, 1174p.

Schneiderhöhn, H., 1952. *Erzmikroskopisches Praktikum*. Stuttgart: Schweizerbart, 274p.

Short, M. N., 1940. Microscopic determination of the ore minerals, *U.S. Geol. Surv. Bull.*, **914**.

Talmage, S. B., 1925. Quantitative standards for hardness of the ore minerals, *Econ. Geol.*, **20**, 531.

Uytenbogaardt, W., 1971. *Tables for Microscopic*

Identification of Ore Minerals. 2nd ed. New York: Elsevier, 430p.

Voigt, W., 1906. *Lehrbuch die Kristaloptik.* Berlin.

Cross-references: *Metallurgy; Mineral and Ore Deposits; Optical Mineralogy; Pleochroic Halo; Polarization and Polarizing Microscope; Refractive Index; Vein Minerals.* Vol. IVA: *X-Ray Diffraction Analysis.* Vol. V: *Metamorphism; Reaction Rim.*

ORTHOPYROXENES

The ***orthopyroxenes*** are an important series of rock-forming minerals. They are the orthorhombic members of the ***pyroxene*** group of silicates and consist essentially of a simple chemical series of $(Mg,Fe)SiO_3$ minerals, in contrast with the larger subgroup of monoclinic ***pyroxenes*** (***clinopyroxenes,*** q.v.) which have a very wide range of chemical composition. ***Orthopyroxenes*** form a solid-solution series of general formula $(Mg,Fe^{2+})SiO_3$ in which Mg and Fe^{2+} are mutually replaceable between $Mg_{100}Fe^{2+}$ and approximately $Mg_{10}Fe^{2+}{}_{90}$. They occur commonly in mafic and ultramafic plutonic rocks, in mafic and intermediate volcanic rocks, and in high-grade thermally and regionally metamorphosed rocks of both igneous and sedimentary origin. The ***orthopyroxenes*** are particularly characteristic constituents of norites and charnockites.

Structure

$MgSiO_3$ and the magnesium-rich members of the series occur in at least three polymorphic forms, the low-temperature orthorhombic *enstatite* (the form almost universally found in rocks); the high-temperature protoenstatite, which also has orthorhombic symmetry; and *clinoenstatite*.

The essential features of the structure of *enstatite* and the other ***orthopyroxenes*** is the linkage of SiO_4 tetrahedra by sharing two of the four corners to form continuous chains of composition $(SiO_3)_n$, as in all ***pyroxenes***. These chains are linked laterally by the octahedrally coordinated cations Mg or (Mg,Fe) to produce an orthorhombic cell with approximately double the *a* dimension of the end-member ***clinopyroxene, diopside***: an idealized projection on (001) of the structure of *enstatite* is given in Fig. 1. There are two nonequivalent cation positions M_1 and M_2 in the structure, and research is currently being carried out to determine the distribution of Mg and Fe^{2+} between these sites. Present indications (from Mössbauer spectroscopy) are that ***orthopyroxenes*** from high-grade metamorphic rocks have a high degree of ordering, with Fe^{2+} preferring the M_2 position (see *Order–Disorder*). A detailed report was given by Virgo and Hafner (1970); more recent work has been summarized by Wood and Banno (1973).

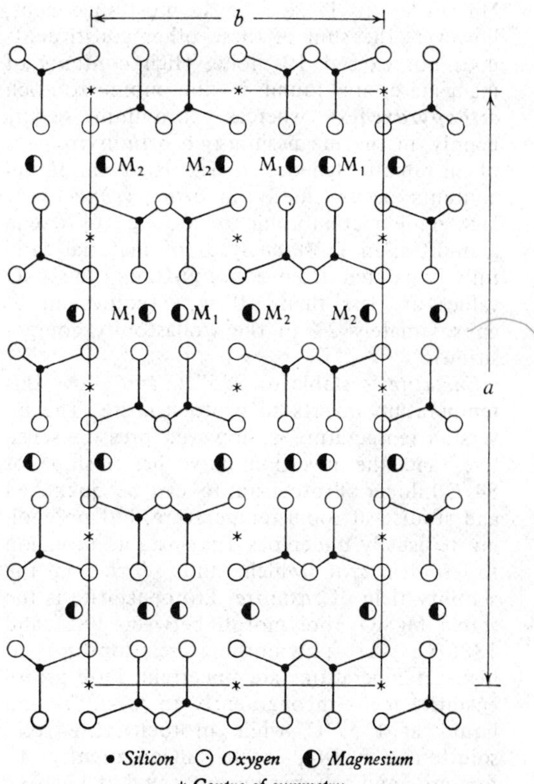

FIGURE 1. Idealized projection on (001) of the structure of *enstatite* (compare Fig. 3 of *Order-Disorder*) (from W. A. Deer, R. A. Howie, and J. Zussman, *Rock Forming Minerals, vol. IIA, Single Chain Silicates*, Longman Group Limited, London, 1978, p. 6).

The relationship between cell parameters and composition is somewhat complex in that, in addition to the major increase in all three dimensions due to the substitution of the larger Fe^{2+} ion for Mg across the series, minor amounts of Ca, Mn, and Al also have an appreciable effect. In particular, the entry of Al causes a considerable contraction in the unit cell. The end-member values are $a=18.228$, $b = 8.805$, $c = 5.185$Å for $MgSiO_3$ and $a= 18.433$, $b = 9.060$, $c = 5.258$Å for $(Fe^{2+}, Fe^{3+}, Mn)SiO_3$.

Chemistry

Although, theoretically, an ideal series from $MgSiO_3$ to $FeSiO_3$, cations other than Mg and Fe^{2+} are almost invariably present in the ***orthopyroxenes***. These commonly include Ca,

Mn, Fe^{3+}, Al, Ti, and Cr; in most specimens, however, the sum of these other constituents does not exceed 10 mol%. High contents of manganese are found in the more iron-rich ***orthopyroxenes***, whereas chromium occurs mainly in the magnesium-rich ***orthopyroxenes*** of ultramafic igneous rocks. High aluminum contents occur chiefly in ***orthopyroxenes*** of high-grade metamorphic rocks, e.g., pyroxene granulites. In most analyses of material carefully separated from ***clinopyroxene***, the CaO values are less than 1.0 wt% (equivalent to approximately 2% of the wollastonite composition).

Enstatite is stable to 985° C and above this temperature inverts to protoenstatite. The inversion temperature is, however, pressure sensitive, and the inversion curve has a slope of 84°C/kilobar. Protoenstatite can be quenched and studied at room temperature, but on cooling it usually undergoes a metastable inversion to *clinoenstatite* which, thus, forms in the stability field of *enstatite*. Protoenstatite is the stable $MgSiO_3$ polymorph between 985° and 1385°C; the polymorphic relationships at higher temperatures are uncertain. Pure protoenstatite melts incongruently to *forsterite* and liquid at 1557°C, while protoenstatite solid solution $En_{76}Di_{24}$ melts incongruently to *forsterite* and a more siliceous liquid at 1386°C. At 985°C, the maximum solubility of *diopside* in *enstatite* is 5 wt%, but this decreases at lower temperatures and is only 2% at 800°C. In many magnesium-rich ***orthopyroxenes*** of plutonic rocks, the bulk of the $Ca(Mg,Fe)Si_2O_6$ initially in solid solution at the time of crystallization is exsolved on cooling and forms lamellae of a Ca-rich monoclinic ***pyroxene*** as thin, parallel-sided sheets or as rows of flattened blebs in the (100) plane of the orthorhombic host (***pyroxenes*** of Bushveld type). Similar lamellae in the more iron-rich ***orthopyroxenes*** of plutonic rocks are inverted *pigeonites*, and the exsolved plates of the Ca-rich monoclinic ***pyroxene*** are oriented parallel to a plane that represents (001) of the original *pigeonite* (***pyroxenes*** of Stillwater type). The Ca content of the latter is about three times greater than that of the Bushveld type. ***Orthopyroxenes*** are sometimes altered to *serpentine*; where alteration is complete, the pseudomorphs show a characteristic bronze-like metallic luster or schiller and are known as "bastite."

In the synthetic system $MgO\text{-}SiO_2\text{-}H_2O$, *enstatite* crystallizes at temperatures above 650°C at water vapor pressures greater than 350 bars and is stable in the presence of water vapor below about 850°C. Various approaches have been made to the development of a pyroxene geothermometer (see *Thermometry, Geologic*). An empirical *enstatite* geothermometer was produced using the ratio Ca/(Ca+Mg) based on temperatures of equilibration estimated for coexisting *diopsides* (Boyd and Nixon, 1973), whereas a geothermometer using the ratio Ca/(Ca+Mg+Fe) of ***orthopyroxene*** coexisting with ***clinopyroxene*** was developed by Mysen (Fig. 2). No clear pressure dependence was observed over the range 10–30 kilobars. Similarly, an empirical approach to take account of the effect of Fe^{2+} on the ***orthopyroxene-clinopyroxene*** miscibility gap in natural systems has made possible the calculation of equilibration temperatures for two-***pyroxene*** assemblages (Wood and Banno, 1973). Experimental work on the use of the amount of $MgSiO_3$ dissolved in *diopside* to estimate temperature of equilibration of coexisting diopside ***clinopyroxene*** and enstatitic ***orthopyroxene*** was reported (Warner and Luth, 1974; Nehru and Wyllie, 1974).

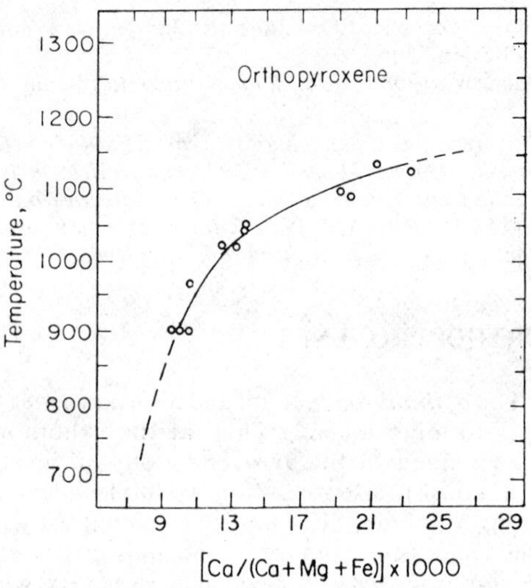

FIGURE 2. ***Orthopyroxene*** geothermometry (from Mysen, 1973). Ca/(Ca + Mg + Fe) in ***orthopyroxene*** coexisting with ***clinopyroxene*** versus temperature.

Work on the $MgSiO_3\text{-}Al_2O_3$ system led to the calculation of equilibration pressures of natural ***orthopyroxene-garnet*** assemblages provided temperatures are known (Wood and Banno, 1973). A similar approach to pressure determination has been reported by Macgregor (1974) using the Al_2O_3 content of ***orthopyroxene*** coexisting with an aluminous phase such as ***spinel*** or ***garnet***. These methods mostly involve the magnesian range of compositions; and, thus, extrapolations of values to natural phases—often containing Ti, Fe^{3+}, and Mn as well as Fe^{2+} and Al—may lead to ambiguous

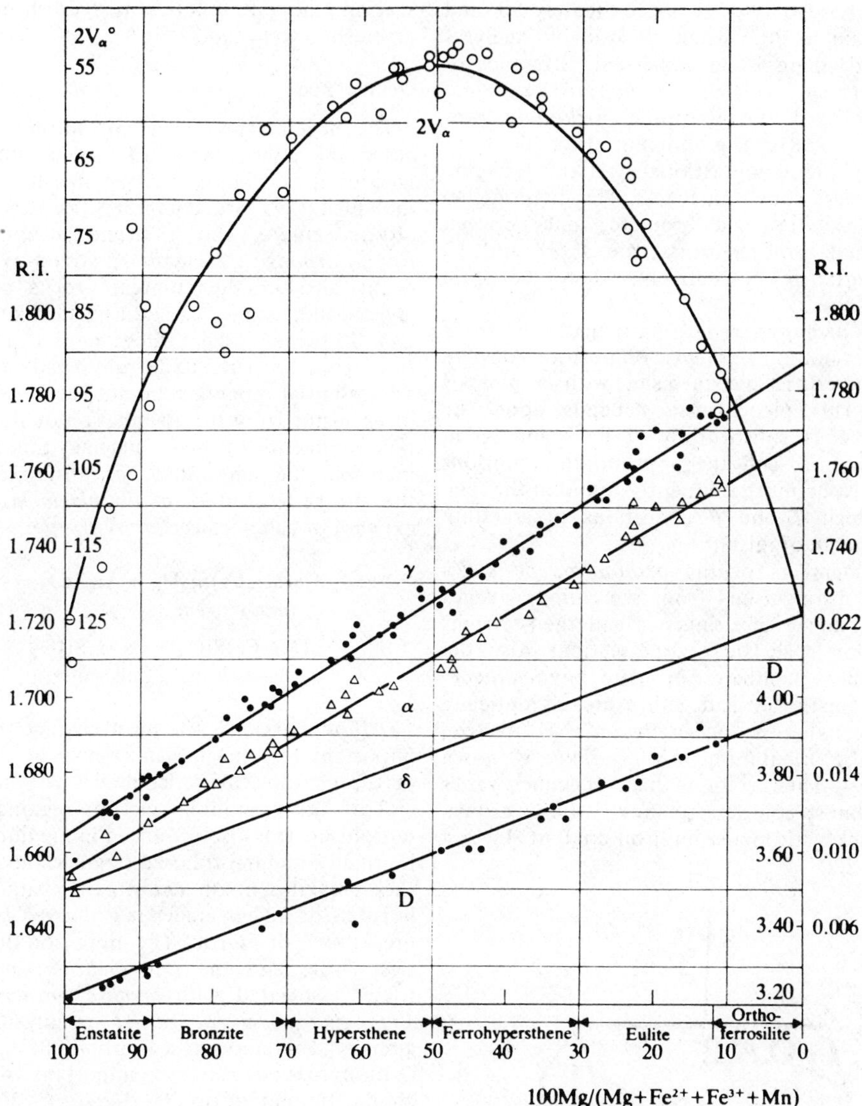

FIGURE 3. The relationship of the optical and physical properties to the chemical composition of ***orthopyroxenes*** (from W. A. Deer, R. A. Howie, and J. Zussman, *Rock Forming Minerals, vol. IIA, Single Chain Silicates*, Longman Group Limited, London, 1978, p. 109).

results (see *Barometry, Geologic*). Nevertheless one may quote Boyd (1973) who, after plotting a fossil geotherm using estimated temperatures and pressures, concluded, "Although crude, these estimates appear sufficiently accurate to be useful."

Optical and Physical Properties

There is very good correlation between the optical properties and chemical composition of the ***orthopyroxene*** series (Fig. 3). Deviations from the linear variation for the refractive indices are mainly due to variable contents of Al. Because many crushed fragments lie on a {210} cleavage plane, the γ index can be measured more readily than the α and β indices; and this is the most precise optical method of determining the Mg/Fe ratio of an ***orthopyroxene***.

The optic axial angles of the series vary continuously and symmetrically with increasing replacement of Mg by Fe^{2+}. *Enstatite* is optically positive as is the iron end member *orthoferrosilite*, whereas in the center of the series the optic sign is negative and the optic axial angle varies from 90° to approximately 50° and rises again to 90°. When all available analyses are used, there is no consistent difference of $2V$ between volcanic ***orthopyroxenes*** and plutonic and metamorphic ***orthopyroxenes*** (Leake, 1968). The nomenclature used to describe the compositional variations in the series is also shown in Fig. 3. The divisions between *enstatite* and bronzite, and between eulite and *orthoferrosilite*, are at 88 and 12 mol% *enstatite*, respectively, where the optic sign changes.

Many ***orthopyroxenes***, particularly those of bronzite and *hypersthene* composition, display a characteristic pleochroism, with α pink, γ green. This pleochroism depends upon the simultaneous substitution of Fe^{3+} and Al in the structure, and the appropriate conditions appear to be most frequently found in the relatively high T and P conditions of granulite facies metamorphism.

The members of the ***orthopyroxene*** series can be distinguished from the ***clinopyroxenes*** by their lower birefringence and their straight extinction in all [001] zone sections. Also, the commoner members, bronzite, *hypersthene*, ferrohypersthene, and eulite are all optically negative, whereas the *augite* series is positive. The optic orientation of *hypersthene* is shown in Fig. 4. Their color in hand specimen varies from almost colorless to grey, yellow, brown, and black with increasing iron content. In thin section, they are colorless to reddish brown or greenish.

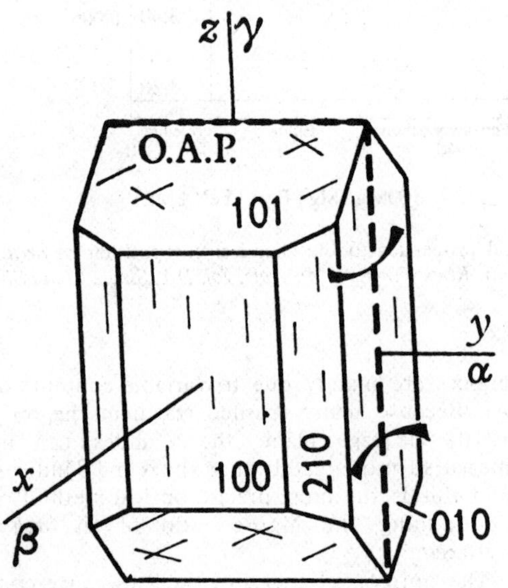

FIGURE 4. The optic orientation of *hypersthene* (from W. A. Deer, R. A. Howie, and J. Zussman, *Rock Forming Minerals, vol. IIA, Single Chain Silicates*, Longman Group Limited, London, 1978, p. 20).

Occurrence

The more magnesium-rich ***orthopyroxenes*** are common constituents of some ultramafic igneous rocks such as pyroxenites and picrites, in which they are commonly associated with ***olivine***, *augite*, and a magnesian ***spinel*** (see *Rock-Forming Minerals*). ***Orthopyroxenes*** occur also in the cumulate rocks of many layered intrusions, e.g., Bushveld, Skaergaard, and Stillwater, where they range in composition from bronzite to ferrohypersthene. They are essential minerals in norites, where they often result from the assimilation of aluminum-rich sediments by mafic magma; this reaction increases the amount of ***orthopyroxene*** and the anorthite content of the ***plagioclase*** at the expense of calciferous ***clinopyroxene***:

$$\underset{\textit{clinopyroxene}}{Ca(Mg,Fe)Si_2O_6} + Al_2SiO_5 \rightarrow$$

$$\underset{\textit{hypersthene}}{(Mg,Fe)SiO_3} + \underset{\textit{anorthite}}{CaAl_2Si_2O_8}$$

Orthopyroxene is the most characteristic and important ferromagnesian mineral of the rocks of the charnockite series and is a typical mineral of the granulite facies of regional metamorphism. It is also produced in medium-grade, thermally metamorphosed, argillaceous rock by the breakdown of ***chlorite***. In argillaceous hornfels of higher grades it is derived from the breakdown of *biotite*. The more iron-rich varieties, eulite and the rare *orthoferrosilite*, are found associated with *fayalite*, *grunerite*, and *almandine-spessartine* ***garnet*** in eulysite, a regionally metamorphosed iron-rich sediment. ***Orthopyroxenes*** are an important phase in chondritic meteorites (see *Meteoritic Minerals*).

ROBERT A. HOWIE

References

Boyd, F. R., 1973. A pyroxene geotherm, *Geochim. Cosmochim. Acta*, **37**, 2533–2546.

Boyd, F. R., and Nixon, P. H., 1973. Structure of the upper mantle beneath Lesotho, *Carnegie Inst. Washington, Geophys. Lab., Yearbook*, **72**, 431–445.

Burns, R. G., 1966. Origin of optical pleochroism in orthopyroxenes, *Mineralogical Mag.*, **35**, 715-719.

Cameron, M., and Papike, J. J., 1981. Structural and chemical variations in pyroxenes, *Am. Mineral.*, **66**, 1–50.

Deer, W. A.; Howie, R. A.; and Zussman, J., 1978. *Rock-Forming Minerals*, vol. IIA. New York: Wiley, 667p.

Fleet, M. E., 1974. Mg-Fe^{2+} site occupancies in coexisting pyroxenes, *Contr. Min. Petr.*, **47**, 207–214.

Howie, R. A., 1963. Cell parameters of orthopyroxenes, *Mineral. Soc. Am., Spec. Paper 1*, 213–222.
Howie, R. A., 1965. The geochemistry of the charnockite series of Madras, India, *Trans. Roy. Soc. Edinburgh*, **62**, 725–768.
Howie, R. A., and Smith, J. V., 1966. X-ray emission microanalysis of rock-forming minerals. V. Orthopyroxenes, *J. Geol.*, **74**, 443–462.
Kuno, H., 1954. Study of orthopyroxenes from volcanic rocks, *Am. Mineralogist*, **39**, 30–46.
Leake, B. E., 1968. Optical properties and composition in the orthopyroxene series, *Mineralog. Mag.*, **36**, 745–747.
Macgregor, I. D., 1974. The system MgO-Al_2O_3-SiO_2: solubility of Al_2O_3 in enstatite for spinel and garnet peridotite compositions, *Am. Mineralogist*, **59**, 110–119.
Mysen, B. O., 1973. Melting in a hydrous mantle: phase relations of mantle peridotite with controlled water and oxygen fugacities. *Carnegie Inst. Washington, Geophys. Lab., Yearbook, 72*, 467–478.
Mysen, B. O., 1976. Experimental determination of some geochemical parameters related to conditions of equilibration of peridotites in the upper mantle, *Am. Mineralogist*, **61**, 677–683.
Nehru, C. E., and Wyllie, P. J., 1974. Electron microprobe measurement of pyroxenes coexisting with H_2O-undersaturated liquid in the join $CaMgSi_2O_6$-$Mg_2Si_2O_6$-H_2O at 30 kilobars, with applications to geothermometry, *Contr. Mineralogy Petrology*, **48**, 221–228.
Smith, J. V., 1969. Crystal structure and stability of the $MgSiO_3$ polymorphs; physical properties and phase relations of Mg,Fe pyroxenes, *Mineral. Soc. Am., Spec. Paper 2*, 3–29.
Virgo, D., and Hafner, H. S., 1970. Fe^{2+}, Mg disorder in natural orthopyroxenes, *Am. Mineralogist*, **55**, 201–223.
Warner, R. D., and Luth, W. C., 1974. The diopside-orthoenstatite two-phase region in the system $CaMgSi_2O_6$-$Mg_2Si_2O_6$. *Am. Mineralogist*, **59**, 98–109.
Wells, P. R. A., 1977. Pyroxene thermometry in simple and complex systems, *Contr. Min. Petr.*, **62**, 129–139.
Wood, B. J., and Banno, S., 1973. Garnet-orthopyroxene and orthopyroxene-clinopyroxene relationships in simple and complex systems, *Contr. Mineralogy Petrology*, **42**, 109–124.

Cross-references: *Clinopyroxenes; Electron Probe Microanalysis; Mantle Mineralogy; Order-Disorder; Plastic Flow in Minerals; Pyroxene Group; Rock-Forming Minerals; Thermometry, Geologic;* see also mineral glossary.

ORTHOSILICATES—*See* individual mineral groups. Vol. IVA: **CRYSTAL CHEMISTRY; MINERAL CLASSIFICATION: SILICATES**

OSUMILITE GROUP

In these hexagonal cyclosilicate minerals, the silica tetrahedra form double six-membered rings with three of each four apical oxygen ions shared. The tetrahedral ring is doubled by inverting a second ring on top of the first. The resulting $(Si_{12}O_{30})^{12-}$ units are linked by other cations in the structure:

- *Armentite*(?): $BaCa_2Al_6Si_9O_{30}\cdot 2H_2O$
- *Brannockite*: $KSn_2Li_3Si_{12}O_{30}$
- *Darapiosite*: $KNa_2Li(Mn,Zn)_2ZrSi_{12}O_{30}$
- *Emeleusite*: $Na_4Li_2Fe_2^{3+}Si_{12}O_{30}$
- *Merrihueite*: $(K,Na)_2(Fe^{2+},Mg)_5Si_{12}O_{30}$
- *Milarite*: $K_2Ca_4Be_4Al_2Si_{24}O_{60}\cdot H_2O$
- *Osumilite*: $(K,Na)(Fe^{2+},Mg)_2(Al,Fe)_3(Si,Al)_{12}O_{30}\cdot H_2O$
- *Osumilite-(Mg)*: $(K,Na)(Mg,Fe^{2+})(Al,Fe)_3(Si,Al)_{12}O_{30}\cdot H_2O$
- *Roedderite*: $(Na,K)_2(Mg,Fe)_5Si_{12}O_{30}$
- *Sogdianite*: $(K,Na)_2Li_2(Li,Fe,Al)_2ZrSi_{12}O_{30}$
- *Sugilite*: $(K,Na)(Na,Fe^{3+})_2(Li_2Fe^{3+})Si_{12}O_{30}$
- *Yagiite*: $(Na,K)_3Mg_4(Al,Mg)_6(Si,Al)_{24}O_{60}$

Osumilite occurs in vugs and fissures in andesites and other volcanic rocks of intermediate composition. *Merrihueite, roedderite*, and *yagiite* are rare *meteoritic minerals* (q.v.). *Brannockite, milarite*, and *sogdianite* are rare *pegmatite minerals* (q.v.). *Darapiosite, emeleusite*, and *sugilite* occur rarely in alkalic igneous rocks.

KEITH FRYE

Reference

Fleischer, M., 1980. *Glossary of Mineral Species, 1980*. Tuscon, Arizona: Mineralogical Record, 192p.

P

PARAMORPH–*See* PSEUDOMORPHISM

PARTING–*See* RUPTURE IN MINERAL MATTER

PEGMATITE MINERALS

Granitic pegmatites are usually associated with large granite stocks, masses, and batholiths, as in New England, the Black Hills of South Dakota, and Governador Valadares in the state of Minas Gerais, Brazil. Compositionally, the pegmatites consist of the ***feldspars***, *quartz*, and *muscovite* with local concentrations of accessory minerals such as *beryl*, *spodumene*, ***tourmaline***, *columbite-tantalite*, *triphylite-lithiophilite*, *amblygonite*, and minor sulfides and arsenides. Although the bulk composition reflects that of the parent granite, important differences prevail. Pegmatites usually afford a deficit of Ti^{4+} with respect to the granite, and the Mn/Fe ratios in accessory phases are much greater, with values in the latter ranging from 0.10 to 0.95. The range Mn/Fe=0.20–0.40 is most frequently encountered. In addition, local concentration of minor to trace-quantity elements is encountered, such as Li^{1+}, Fe^{2+}, B^{3+}, P^{5+}, Nb^{5+} and Ta^{5+}. As with granites, the pegmatites are depleted in Cr^{3+}, Co^{2+}, Ni^{2+}, and Zn^{2+} with respect to ultramafic rocks.

The most important chromophores in pegmatite minerals are Mn^{2+}, Mn^{3+}, Fe^{2+}, and Fe^{3+} and the Fe^{2+}–Fe^{3+} valence transfer pair. Magnesium is usually a minor element in most granitic pegmatites, and Mg-rich mineral species are infrequently encountered. Although *muscovite* is the typical ***mica***, some pegmatites may provide abundant *biotite*, especially if the pegmatite is poor in B^{3+} and P^{5+}. Distinctly different are the highly alkalic agpaitic pegmatites that are associated with nepheline syenite massifs. Zirconium, titanium, niobium, and rare earths are important in these pegmatites. The accessory minerals are *analcime, sodalite, arfvedsonite, aegirine, biotite*, and a very long list of complex niobo-, titano-, and zirconosilicates. Giant crystals of *beryl, triphylite, amblygonite*, and ***tourmaline*** are practically unknown from alkalic pegmatites. A detailed account of the famous Lovozero alkali massif can be found in Vlasov, Kuz'menko, and Es'kova (1966) and of the Langesundfjord district in southern Norway in the classic treatise by Brögger (1890).

Structure and Origin

Structural features of pegmatites can be grotesque to the uninitiated. Generally, they occur concordantly emplaced in outlying country rocks such as schists and amphibolites although portions of outlying granite may merge into pegmatite. Pegmatite bodies range from inches to hundreds of feet in dimension. Typical shapes are lenses, ovoids, ellipsoids, dikes, and veins. More fantastic shapes appear as hoods, funnels, splashes, and turnips.

The most spectacular feature is the enormous grain size of the crystals, which range from centimeters to meters in dimension. Extreme local segregations of unusual ions are typical, such as are found in the minerals *beryl*, $Al_2[Be_3Si_6O_{18}]$; *triphylite-lithiophilite*, $Li(Fe,Mn)^{2+}[PO_4]$; *columbite-tantalite*, $(Fe,Mn)^{2+}[(Nb,Ta)_2O_6]$; and *hafnon*, $(Hf,Zr)[SiO_4]$. Simple granitic pegmatites are chemically and mineralogically monotonous with a rather uniform distribution of minerals throughout. Typical examples of commercial interest include the associations *quartz*-***plagioclase***-*spodumene*, or *quartz*-***plagioclase***-*lepidolite*.

More interesting are the complex granitic pegmatites, which are often sharply zoned both in composition and in texture. Typical zones include a narrow chill or border zone, wall zone, one or several intermediate zones, and a core or multiple cores. Although zonation is often symmetrical about a vertical axis, particularly if the pegmatite is spheroidal in shape, the core is usually displaced toward the top, representing the final stage of crystallization from an aqueous-rich fluid. For most pegmatites, the intermediate zones and the core occupy the greatest volume. Early crystallized are ***plagioclase*** and *quartz*; then ***plagioclase***-*quartz*-***mica***; coarse perthite-***plagioclase***-*quartz*; and finally *quartz*, which constitutes the bulk of the core. Accessory minerals such as *spodumene, beryl, triphylite, amblygonite*, and *columbite* crystallize near the junction between the intermediate zone and the core. The accessory minerals known from granitic pegmatites are listed in Table 1, where potentially important species are denoted by asterisks. The fluoroaluminates, known from but few pegmatites, did occur in enormous quantity at Ivigtut, Greenland. The list would easily double if retrograde products of these minerals were included.

The genesis of pegmatite minerals is still controversial. Jahns and Burnham (1969) have demonstrated that the melt of the Harding pegmatite at 5000 bars and 650°C becomes saturated with 11.2 wt% water. Beginning with a granite containing 1.0 wt% water, if only anhydrous phases (*quartz* and ***feldspars***) crystallize out, the separation of an aqueous-rich fluid appears when about 90% of the melt is crystallized. If 0.5 wt% water crystallizes out (=11% *muscovite* in the bulk solids), saturation occurs after 95% solid phases have formed.

The appearance of an aqueous-rich fraction during consolidation of a pegmatite is important in understanding replacement of earlier-formed anhydrous phases, concentration of unusual but relatively soluble complexes (such as tantalates, fluoroaluminates, and beryllosilicates), and the paragenesis of the gem-pocket minerals. It must be stressed, however, that pegmatites can vary widely in their mineralogy and chemistry and that evidence for resorption of wall rocks, metamorphic processes, and solid-state exchange reactions have been documented for some pegmatites. Most, however, evolved from igneous fluid under rather closed conditions.

Compositional Zoning

The vast bulk of a granitic pegmatite is constituted of *quartz*, ***feldspars***, and *muscovite*

TABLE 1. Accessory Minerals from Granitic Pegmatites

Carbonates	
**Calcite*	$CaCO_3$
**Siderite*	$FeCO_3$
Bastnäsite	$CeFCO_3$
Synchisite	$CaCeF(CO_3)_2$
Borates	
Jeremejevite	$Al_6B_5O_{15}(OH)_3$
Rhodizite	$CsAl_4[Be_4B_{11}O_{26}(OH)_2]$
Phosphates	
Berlinite	$[AlPO_4]$
Lithiophosphatite	$Li_3[PO_4]$
**Beryllonite*	$Na[BePO_4]$
**Hurlbutite*	$Ca[BePO_4]_2$
**Triphylite-lithiophilite*	$Li(Fe,Mn)[PO_4]$
**Heterosite-purpurite*	$(Fe,Mn)[PO_4]$
**Graftonite-beusite*	$Ca(Fe,Mn)_2[PO_4]_2$
Sarcopside	$(Fe,Mn)_3[PO_4]_2$
**Wyllieite*	$Na_2Fe_2Al[PO_4]_3$
**Alluaudites*	$(Na,Ca)_{<2}(Fe^{2+},Fe^{3+},Mn^{2+})_3[PO_4]_3$
**Arrojadite-dickinsonite*	$KNa_4Ca(Fe,Mn)_{14}Al(OH,F)_2[PO_4]_{12}$
Fillowite	$Na_2Ca(Mn,Fe)_7(PO_4)_6$
**Griphite*	$(Mn,Na,Li)_6CaFeAl_2F_2[PO_4]_6$
Whitlockite	$Ca_9MgH[PO_4]_7$
**Monazite*	$Ce[PO_4]$
Xenotime	$Y[PO_4]$
**Herderite-hydroxylherderite*	$Ca[Be(F,OH)(PO_4)]$
**Amblygonite-montebrasite*	$LiAl(F,OH)[PO_4]$
**Zwieselite-triplite*	$(Fe,Mn)_2F[PO_4]$
Wolfeite-triploidite	$(Fe,Mn)_2(OH)[PO_4]$
**Scorzalite-lazulite*	$(Fe,Mg)Al_2(OH)_2[PO_4]_2$
**Rockbridgeite-frondelite*	$(Fe,Mn)Fe_4(OH)_5[PO_4]_3$
Brazilianite	$NaAl_3(OH)_4[PO_4]_2$
**Fluorapatite-hydroxyapatite*	$(Ca,Mn)_5(F,OH)[PO_4]_3$
**Childrenite-eosphorite*	$(Fe,Mn)Al(OH)_2(H_2O)[PO_4]$
Morinite	$Ca_2Na[Al_2F_4(OH)(H_2O)_2(PO_4)_2]$
Elements	
**Bismuth*	Bi

TABLE 1. (Continued)

Sulfides, Arsenides	
**Pyrite*	FeS_2
**Löllingite*	$FeAs_2$
**Sphalerite*	ZnS
**Pyrrhotite*	$Fe_{1-x}S$
Bismuthinite	Bi_2S_3
Halogenides	
Fluorite	CaF_2
**Cryolite*	$Na_3[AlF_6]$
Thomsenolite	$NaCa[AlF_6]\cdot H_2O$
Chiolite	$Na_5[Al_3F_{14}]$
Weberite	$Na[MgAlF_7]$
Oxides, simple and complex	
Gahnite	Al_2ZnO_4
**Chrysoberyl*	Al_2BeO_4
Hematite	Fe_2O_3
**Pyrochlore-microlite*	$Ca_2(Nb,Ta)_2O_6(O,F)$
Westgrenite	$Bi(Ta,Nb)_2O_6(OH)$
Sukulaite	$Sn_2Ta_2O_7$
**Cassiterite*	SnO_2
Tapiolite	$(Fe,Mn)(Ta,Nb)_2O_6$
**Columbite-tantalite*	$(Fe,Mn)(Nb,Ta)_2O_6$
Samarskite	$(Y,U)(Nb,Fe)_2(O,OH)_6$
Aeschynite	$(Ce,Th,Ca)(Ti,Nb,Ta)_2O_6$
Fergusonite-formanite	$Y(Nb,Ta)O_4$
Bismutotantalite	$Bi(Ta,Nb)O_4$
Simpsonite	$Al_4Ta_3O_{13}(OH)$
Thorianite	ThO_2
**Uraninite*	UO_2
Silicates	
**Phenakite*	$[Be_2SiO_4]$
**Eucryptite*	$Li[AlSiO_4]$
**Spessartine-almandine*	$(Mn,Fe)_3Al_2[SiO_4]_3$
**Zircon*	$Zr[SiO_4]$
**Thorite*	$Th[SiO_4]$
Hafnon	$Hf[SiO_4]$
Euclase	$Al[Be(OH)(SiO_4)]$
**Topaz*	$Al_2(F,OH)_2[SiO_4]$
**Gadolinite*	$Y_2Fe[BeOSiO_4]_2$
Thortveitite	$Sc_2[Si_2O_7]$
**Thalenite*	$Y_2[Si_2O_7]$
**Bertrandite*	$Be_4(OH)_2[Si_2O_7]$
Tschevkinite	$(Ce,La)_2Ti_2O_4[Si_2O_7]$
**Allanite*	$(Ca,Ce)_2(Fe^{2+},Fe^{3+})Al_2O(OH)[SiO_4][Si_2O_7]$
**Beryl*	$Al_2[Be_3Si_6O_{18}]$
Bazzite	$Sc_2[Be_3Si_6O_{18}]$
Indialite	$Mg_2[Al_4Si_5O_{18}]$
****Tourmaline*** group	$Na(Fe,Al,Li)_3Al_6(OH)_4[BO_3]_3[Si_6O_{18}]$
Milarite	$KCa_2Al[Be_2Si_{12}O_{30}]\cdot H_2O$
**Spodumene*	$LiAl[Si_2O_6]$
Holmquistite	$Li_2(Mg,Fe)_3Al_2(OH)_2[Si_8O_{22}]$
Bavenite	$Ca_4(OH)_2[Al_2Be_2Si_9O_{26}]$
**Zinnwaldite*	$KLiFeAlF_2[AlSi_3O_{10}]$
**Lepidolite*	$KLi_2AlF_2[Si_4O_{10}]$
Bityite	$CaLiAl_2(OH)_2[AlBeSi_2O_{10}]$
**Bikitaite*	$Li[AlSi_2O_6]\cdot H_2O$
**Petalite*	$Li[AlSi_4O_{10}]$
**Pollucite*	$Cs[AlSi_2O_6]\cdot H_2O$

*Potentially important species.

mica. Lithium-rich ***micas***, along with many giant-crystal accessory minerals, crystallize toward the junction between intermediate zone and core. Coarse, platy *albite* (the variety "cleavelandite") is characteristic of a late-stage pocket or replacement paragenesis. Here, cleavelandite forms the matrix for clear crystals of ***tourmaline***, *beryl*, and *spodumene*, which, as pink-gem varieties, are called "rubellite," "morganite," and "kunzite," respectively (see *Gemology*). The pink color of these crystals is probably due to the presence of Mn^{3+} substituting for Al^{3+} in octahedral positions. The gem rubellites are compositionally distinct from the far more abundant black *schorl*. They fall within the compositional range for *elbaite*, $Na(Li,Al)_3Al_6(OH)_4[BO_3]_3[Si_6O_{18}]$, whereas *schorl* is $Na(Fe,Mn,Mg)_3^{2+}Al_6(OH)_4[BO_3]_3[Si_6O_{18}]$. *Schorl* often occurs in large quantities throughout several zones of the pegmatite and is the most ubiquitous boron-bearing mineral. It seems that iron is selectively separated off in the formation of *schorl* and accounts for the high Mn/Fe ratios for accessory minerals formed later in the sequence of certain pegmatites.

Replacement Units, Cotectic Crystallization Textures, and Hydrothermal Veins

Aqueous-rich fluids, through the accumulation of alkalies, alkaline earths, and minor elements, may react with earlier-formed crystals and either replace entire sections of pegmatite or selectively replace certain species. The Na- and Ca-metasomatism of earlier-formed minerals is common and manifests itself as albitized perthite with checkerboard texture; Na-rich phosphates replacing Li-rich species; apatitized Li-Fe phosphate; etc. Total corrosion of earlier-formed crystals is also well documented—porous Fe-Mn oxides after *triphylite; kaolinite* after *spodumene* and ***feldspars***; *cookeite* and *lepidolite* after gem-pocket minerals such as ***tourmaline*** and *spodumene; bertrandite*, $Be_4(OH)_2[Si_2O_7]$, after *beryl*. Noteworthy replacements are *albite* + *muscovite* after *topaz* (Stripåsen, Sweden) and *spodumene* (Branchville, Connecticut). Phosphate minerals are often partly replaced by members of the *siderite-rhodochrosite* series, indicating increasing activity of CO_2 during the formation and concentration of the aqueous-rich fluid. Less abundant alkalies and alkaline earths such as Cs^+, Rb^+, Ba^{2+}, and Sr^{2+} increase in concentration in these fluids and often form druses of exotic minerals upon earlier-formed Na^+- and Ca^{2+}-rich minerals. Local conditions were probably rather stagnant during the concentration of such rarer elements.

Quite distinct is the appearance of lamellar, checkerboard, and exsolution-like textures of at least two compositionally discrete species. Examples include ***feldspar****-quartz, triphylite-pyrrhotite, triphylite-graftonite, graftonite-sarcopside, arrojadite-albite,* ***apatite****-albite, triphylite-albite*, and *triphylite-spessartine*. Samples showing sharp morphological outlines of crystals of the predominant phase are common although easily overlooked. These appear to be examples of cotectic cocrystallization. Unmixing through exsolution from a host crystal is far less likely owing to the disparate compositions and gross structural differences of the phases involved. Such silicate-phosphate, phosphate-silicate, and phosphate-sulfide cotectic pairs were earlier believed to afford evidence of replacement processes, but crystal chemical arguments rule out obvious mechanisms for formation of this kind.

Typical hydrothermal vein minerals include *quartz; topaz; beryl; amblygonite-montebrasite;* ***apatite****; herderite-hydroxylherderite*, $CaBe(F,OH)[PO_4]$; *childrenite-eosphorite*, $(Fe,Mn)Al(OH)_2(H_2O)[PO_4]$; *fairfieldite*, $Ca_2Mn(H_2O)_2[PO_4]_2$; *brazilianite*, $NaAl_3(OH)_4[PO_4]_2$; *goyazite*, $SrAl_3(OH)_6[PO_4][PO_3OH]$; and many others. Such minerals have formed by the introduction of aqueous-rich fluid into transverse fractures that cut earlier-formed units of the pegmatite.

Certain accessory minerals are particularly susceptible to a long and complex series of retrograde reactions involving repeated exchange, reequilibration, and recrystallization with a progressively and often discontinuously changing aqueous rest-fluid composition. For example, primary *uraninite* may yield up to ten discrete species through the course of such reactions. Studies on these late-stage minerals have been given considerable attention by mineralogists, and even the briefest account of the myriad phases is impossible in this outline. However, the phosphates, owing to their inherent complexity and diversity, will be singled out as an example.

Regional Control in Pegmatite Compositions and Mineralogy

Although compositions of accessory phases may vary widely from pegmatite to pegmatite in a given province, significant trends prevail when pegmatites from various provinces are compared with each other. The pegmatites of Colorado afford fluorine-rich assemblages as found in fluoroaluminate minerals, *fluorite*, and *triplite*. Lithium-bearing phosphates are conspicuous in their absence. In the pegmatites of the Black Hills, Li-Fe phosphates are ubiquitous. Gem-pocket parageneses are absent. Na- and Ca-metasomatism of earlier-formed primary phases is typical. In the New England pegmatites, accessory phosphate phases, notably *triphylite-lithiophilite*, occur often in great abundance, but Na- and Ca-metasomatic products are missing. Mn-rich pegmatites in the New England region often afford gem-pocket minerals, notably "rubellite" ***tourmaline*** and purple ***apatite***. Phosphate minerals are relatively sparse in the pegmatites of North Carolina. Instead, rare-earth- and uranium-thorium-bearing complex oxides occur. Rare-earth complex oxides are also typical accessories of pegmatites of southern Norway, including one example of a curious scandium silicate, *thortveitite,* $Sc_2[Si_2O_7]$. In pegmatites of central Sweden, *allanite* and *thalenite*, $Y_2[Si_2O_7]$, are locally abundant. Pegmatites of Malagasy Republic are noted for rare-earth minerals and Be- and B-rich gem-pocket minerals.

These differences add credence to the argument that most pegmatite compositions are little influenced by their intruded country rocks and that their crystal-chemical evolution occurs under closed conditions. Compositions and abundances of the accessory phases more probably reflect the geochemistry of the parent granite sources.

Primary and Secondary Phosphate Minerals in Pegmatites

The paragenetic relationships among primary and secondary phosphate minerals are extraordinarily complex since no fewer than 140 distinct species are documented from pegmatites. Most occur in but minute quantities and require detailed crystal structure analysis to unravel their chemistry. The highly aquated, basic phosphates of the first transition-series metals are of considerable interest since their sequence in crystallization is duplicated in numerous novaculite deposits, in gossans and zones of oxidation surrounding phosphate-bearing iron ores, and in deposits of sedimentary origin, particularly those that are iron and aluminum rich. By analogy, study of the sequences of crystallized phases in pegmatites aids in understanding the crystal chemistry of sequences encountered in sediments and other low-temperature deposits. Owing to the tendency for the formation of compact polynuclear clusters, their structures have added much to our understanding of cluster models for polyions in solution, the relationships between structure and color, and other physical properties. The $Fe^{3+}-OH-Fe^{3+}$, $Fe^{3+}-O-P^{5+}$, and $Fe^{3+}-O-Fe^{3+}$ types of bonds occur in many of the structures; consequently, they are highly insoluble, and superior crystalline material of most phases has not been synthesized in the laboratory.

A compact review of the pegmatite phosphates can be found in Moore (1973). The general crystal-chemical formula for most of the phases can be written

$$M^{2+}_{r_1} M^{3+}_{r_2} (OH)^{-}_{2r_1+3(r_2-z)}(H_2O)_n (PO_4)^{3-}_z \cdot qH_2O,\ 2r_1 + 3r_2 = 3z$$

where M = octahedrally coordinated cations
n = water molecules directly bonded to these cations
q = water molecules beyond the sphere of octahedral coordination.

With but few exceptions, the octahedral cluster, $M_r\phi_s$ where r=octahedral centers and ϕ=octahedral vertices, can be related to the general formula via $r=r_1+r_2$, $s=2r_1+3r_2+n+z$. Important cations include $M=Mn^{2+}$, Fe^{2+}, Mn^{3+}, Fe^{3+}, Al^{3+}, and Mg^{2+}. Other phosphates include cations such as the alkalies and alkaline earths, but the same compact clustering of the octahedral fraction is observed in these as well.

The geochemistry of the phosphate minerals follows a fairly well-defined pattern, consistent within any pegmatite province. For example, the pegmatites of the Black Hills afford abundant accessory *triphylite*, which follows a sequence of metasomatic exchange:

$$\underset{\textit{triphylite}}{Li(Fe,Mn)^{2+}[PO_4]} \rightarrow \underset{\textit{ferrisicklerite}}{Li_{1-x}Mn^{2+}_{1-x}Fe^{3+}_x[PO_4]} \rightarrow$$

$$\underset{\textit{heterosite}}{(Fe,Mn)^{3+}[PO_4]} \rightarrow$$

$$\underset{\textit{alluaudite}}{Na_{1+x}Ca_{1-x}Mn^{2+}(Fe,Mn)^{2+}_{2-x}Fe^{3+}_x[PO_4]_3} \rightarrow$$

$$\underset{\textit{hydroxyapatite}}{Ca_5(OH)[PO_4]_3}$$

This can be interpreted as a progressive leaching of the alkali cations with concomitant oxidation of the transition metals, resulting in a well-ordered *heterosite* pseudomorph, followed by Na^+ metasomatic reaction and partial reduction of the metals with the aqueous-rich rest liquid, and finally Ca^{2+} metasomatic exchange, the transition metals being completely removed from the crystal. Pegmatite provinces that are noted for abundant accessory phosphate minerals include the Black Hills, South Dakota; Pala, southern California; New England; Mangualde, Portugal; Varuträsk, Sweden; the Bavarian Forest, West Germany; state of Minas Gerais and the Paraiba district, Brazil.

The phosphates can be classified paragenetically as follows:

Primary Phosphates

Subdivision I: *Subhedral to euhedral primary phases*
Enormous crystals, up to 6 m in dimension, are crystallized near the junction of the intermediate zone and the core. Anhydrous species include *triphylite-lithiophilite, graftonite-beusite, wyllieite*, with hydroxyl-fluoride-bearing species *amblygonite-montebrasite, arrojadite*, **apatite**, and *zwieselite-triplite*. Fifteen species are known.

Subdivision II: *Anhedral "primary" phases*
These species are metasomatic exchange products. They replace earlier-formed giant crystals of subdivision I and include *alluaudite, natrophilite, scorzalite-lazulite, wolfeite-triploidite*, and *heterosite-purpurite*. Ten species are known.

Secondary Phosphates

Subdivision I: *Ligand addition, alkali-leached products, oxidation of some metals*
Over thirty known species occur as small crystals in open cavities of primary phosphates and their metasomatic products.

Subdivision II: *Ligand addition products: no oxidation of metals*

a. *Derivatives of primary transition metal phosphates* Six species are known and directly replace earlier-formed *triphylite*.
b. *Hydrothermal vein and fracture products, derived from* amblygonite, beryl, triphylite, *etc.* Over thirty species are known. They occur sporadically as late-stage minerals. Large hydrothermal veins and cavities may afford gem-quality crystals of **apatite**, *brazilianite*, and *herderite*.

Further study on these curious minerals shall surely lead to a better understanding of the properties of the aqueous-rich fluid separate and its variation with time of temperature, metal abundances, hydrogen-ion concentration, and oxygen fugacity.

A paragenetic outline of the pegmatite phosphates appears in Fig. 1.

PAUL BRIAN MOORE

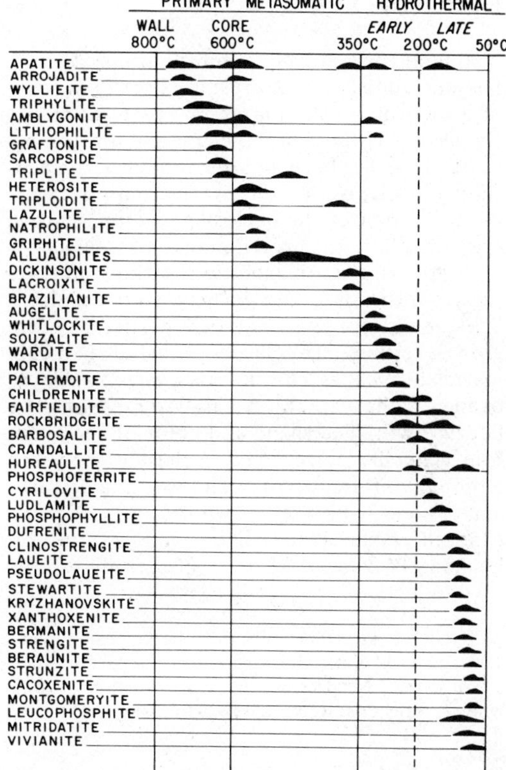

FIGURE 1. Paragenetic outline of phosphate minerals in pegmatites. The dashed line is the upper limit of stability of water ligands.

References

Brögger, W. C., 1890. Die Mineralien der Syenitpegmatitgänge der Süd norwegischen Augit-und Nephelinsyenite, *Zeit. Kristallogr. Mineral.*, **16**, 1–663.

Cameron, E. N., et al., 1954. Pegmatite investigations, 1942–1945, New England, *U.S. Geol. Surv. Prof. Paper 255*, 352p.

Fersman, A. E., 1931. Zur Geochemie der Granitpegmatite, *Mineral. Petrogr. Mitt.*, **41**, 200–213.

Heinrich, E. W., 1948. Pegmatites of Eight Mile Park, Fremont County, Colorado, *Am. Mineralogist*, **33**, 420–448, 550–588.

Jahns, R. H., 1953. The genesis of pegmatites. Occurrence and origin of giant crystals, *Am. Mineralogist*, **38**, 563–598.

Jahns, R. H., 1955. The study of pegmatites, *Econ. Geol.*, **50**, 1025–1130.

Jahns, R. H., and Burnham, C. W., 1969. Experimental studies of pegmatite genesis: I. A model for the derivation and crystallization of granitic pegmatites, *Econ. Geol.*, **64**, 843–864.

Landes, K. K., 1933. Origin and classification of pegmatites, *Am. Mineralogist*, **18**, 33–56, 95–103.

Moore, P. B., 1973. Pegmatite phosphates: Descriptive mineralogy and crystal chemistry, *Mineral. Rec.*, **4**, 103–130.

Page, L. R., et al., 1953. Pegmatite investigations, 1942–1945, Black Hills, South Dakota, *U.S. Geol. Surv. Prof. Paper 247*, 228p.

Vlasov, K. A., Kuz'menko, M. Z., and Es'kova, E. M., 1966. The Lovozero Alkali Massif (Trans. by D. G. Fry and K. Syers). New York: Hafner, 627p.

Cross-references: *Collecting Minerals; Museums, Mineralogical; Rock-Forming Minerals; Vein Minerals;* see also mineral glossary and individual mineral groups. Vol. IVA: *Phosphates, Arsenates, and Vanadates.*

PERIODICALS, MINERALOGICAL–

See Preface

PHANTOM CRYSTALS

Minerals occasionally exhibit outlines of crystals embedded within crystals (Fig. 1) because of growth in a solution whose composition has changed slightly from the initial fluid, or because there has been an interruption in the continuity of deposition due to fluctuations in the concentration of the enclosing medium. During crystallization, the crystal may often remain for a length of time in solution without further growth occurring, and it may even acquire a light coating of dust or other particles before growth resumes. Thin layers of included foreign material, such as ***chlorite*** in *quartz*, serve to delineate the earlier crystal outline from the later growth. The shape and color of the phantom generally correspond to the outer crystal, although differences in color (e.g., amethyst-*quartz*) and habit are known. As an example of changed crystal habit, *calcite* crystals from Chihuahua Province, Mexico, began growth as

FIGURE 1. Phantom crystal of quartz (from I. Vanders and P. F. Kerr, *Mineral Recognition*, John Wiley & Sons, Inc., New York. Copyright © 1967 by John Wiley & Sons, Inc. Reprinted by permission).

FIGURE 2. Phantom crystal of *calcite*. Dark scalenohedrons are overgrown by clear rhombohedrons.

trigonal scalenohedrons, coated with *hematite*; further growth continued as rhombohedrons (Fig. 2; Sinkankis, 1966, Fig. 23). The development of phantoms is often observed in artificially grown crystals, where the seed is clearly visible (e.g., ammonium dihydrogen phosphate, from the Bell Telephone Laboratories). In nature, phantom crystals appear in *quartz, calcite, fluorite,* ***tourmaline***, and *halite* (overgrowths of clear *halite* on pyramidal hoppers). The phenomenon of phantom growth is closely related to zoning, overgrowth, and parallel growth (epitaxy).

VIVIEN GORNITZ

References

Buckley, H. E., 1951. *Crystal Growth*. New York: Wiley, 571p.

Sinkankis, J., 1966. *Mineralogy: a first course*. Princeton, N. J.: Van Nostrand, p. 73.

Vanders, I., and Kerr, P. F., 1967. *Mineral Recognition*. New York: Wiley, 316p.

Cross-references: *Crystal Growth; Epitaxy; Skeletal Crystals.*

PHYLLOSILICATES

This is the name given to silicate minerals having a layer type of atom arrangement. The term derives from the Greek φυλλον (= sheet). Many of these minerals crystallize in sheet-like forms and have an excellent cleavage parallel to the structural layers. White ***mica***, *muscovite*, and black ***mica***, *biotite*, are the best known examples. Figure 1 shows a model of the crystal structure of *muscovite*.

The principal phyllosilicates can be classified on the basis of their layer structures and chemical compositions into the following groups: ***kaolinite-serpentine, pyrophyllite-talc, smectite, vermiculite, mica***, brittle ***mica***, and ***chlorite***. Within each group, the chemical compositions can vary within certain limits, and the structural layers can be stacked in a variety of ways. Extensive lists of phyllosilicate compositions have been given by Deer, Howie, and Zussman (1962).

Minerals with essentially the same chemical composition and the same layer structure but with different stacking arrangements are called "polytypes." Polytypic varieties are very common among the phyllosilicates and normally are distinguished by X-ray diffraction analysis. In well-crystallized phyllosilicates, the layers are stacked in a regular spatial sequence and develop structures with well-defined symmetry properties. In poorly crystallized phyllosilicates,

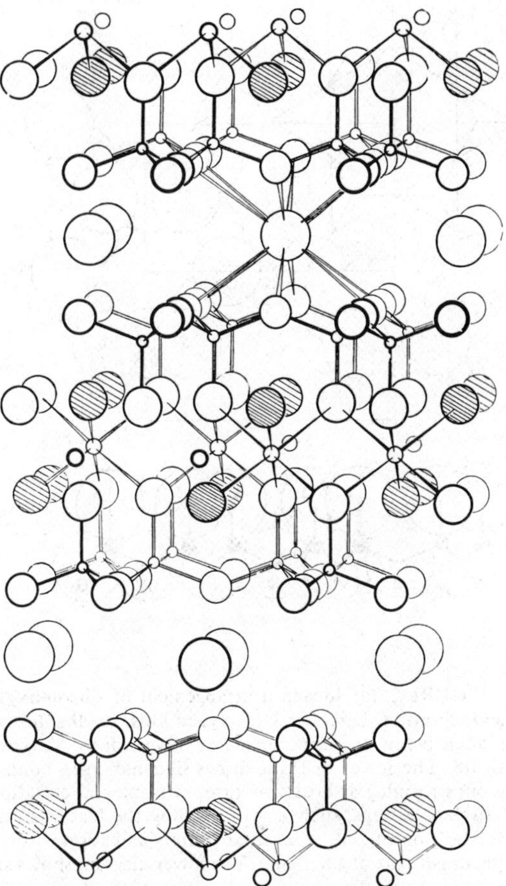

FIGURE 1. Drawing of structure of *muscovite* ***mica*** (from Brindley and MacEwan, 1953).

various kinds of disorder exist, including displacements and rotations of successive layers in a more or less random rather than regular manner, and interstratifications of layers of more than one kind in various proportions and with various degrees of order. Additionally, in many phyllosilicates more than one kind of atom occupies a particular structural position, e.g., Si and some Al atoms in tetrahedral positions. The question then arises whether these atoms are distributed randomly or in some ordered arrangement.

Atomic (or Crystal) Structures

The structural layers of phyllosilicates are composed of sheets of atoms arranged mainly in two rather simple geometrical patterns. The large, negatively charged atoms (anions) of oxygen, hydroxyl, and occasionally fluorine, are arranged in fourfold and sixfold groups, in the form of tetrahedra and octahedra, around the small positively charged atoms (cations).

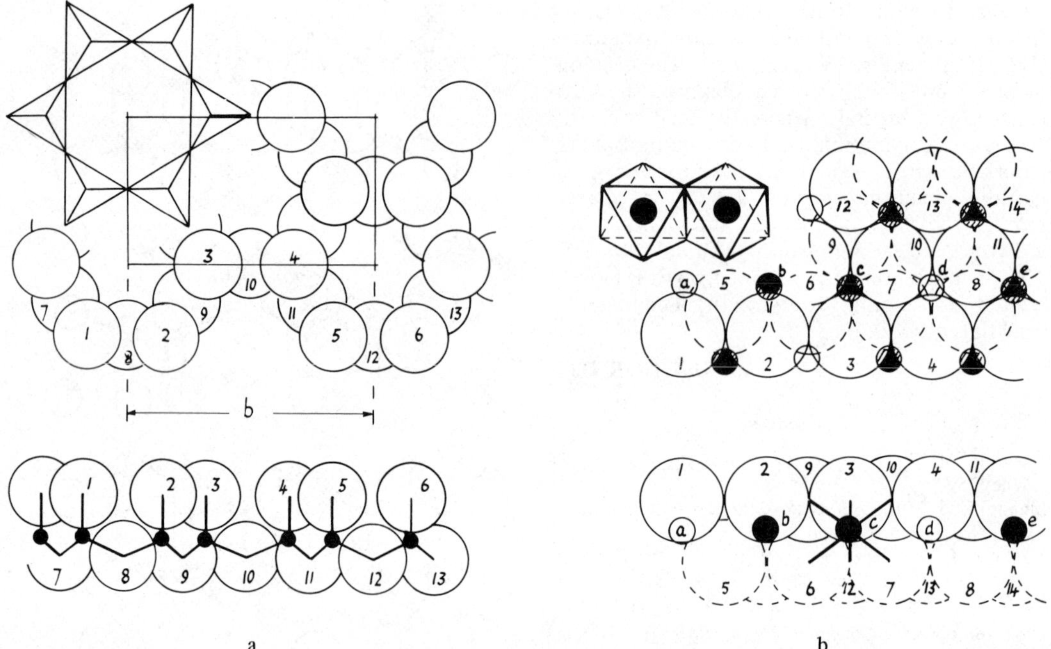

FIGURE 2. (a) Idealized arrangement of silicon-oxygen tetrahedral groups forming a sheet structure, in plan and elevation. Large circles, oxygen ions; small solid circles, silicon ions. In the upper diagram, the silicons are hidden below the upper plane of oxygen ions. Six corner-linked tetrahedra are shown in the upper left of the figure. The lower diagram shows silicon-oxygen bonds. (b) Idealized arrangement of cation-oxygen octahedral groups forming a sheet structure, in plane and elevation. Large circles, oxygen ions; small circles metal cations. Trivalent cations, such as Al^{3+}, occupy the black positions and form a dioctahedral sheet. Divalent cations, such as Mg^{2+}, occupy all octahedral centers and form a trioctahedral sheet. Two edge-linked octahedra are shown in the upper part of the figure. The lower diagram shows six metal-oxygen bonds.

Figure 2a shows the organization of tetrahedral groups XO_4 (X principally Si, but in some minerals also Al, and occasionally Fe^{3+}) in extended sheets of linked hexagonal rings.

Figure 2b shows the organization of octahedral groups $Y(O,OH,F)_6$ (Y principally Al, Fe^{3+}, Mg, Fe^{2+}) with the octahedra linked edge to edge in continuous sheets. The oxygen ions lie in two planes in a close-packed type of arrangement. Octahedral sheets are of two kinds, one with two-thirds of the possible cation positions occupied mainly by trivalent ions, and the other with three-thirds (i.e., all) of these positions occupied mainly (but not invariably) by divalent cations. They are called, respectively, "dioctahedral" and "trioctahedral" sheets and both kinds are illustrated in Fig. 2b.

The arrangements shown in Figs. 2a and 2b are idealized representations of the actual structures and for many purposes are useful approximations, but not all questions relating to phyllosilicates can be answered from these simplified representations. The combination of the tetrahedral and octahedral sheets into the various kinds of layers is illustrated schematically in Fig. 3, where di- and trioctahedral forms are distinguished in the same way as in Fig. 2. More detailed descriptions are given by Bailey (1980).

Chemical Compositions

The chemical compositions of the principal mineral groups can be considered on the basis of the structural diagrams given in Figs. 2 and 3.

Minerals of the ***kaolinite-serpentine*** group (Fig. 3a) have layers compounded of one tetrahedral and one octahedral sheet, called 1:1 type layers, with the following compositions:

	Kaolinite	*Serpentine*
octahedral	$Al_2O_2(OH)_4$	$Mg_3O_2(OH)_4$
tetrahedral	Si_2O_3	Si_2O_3
total	$Al_2Si_2O_5(OH)_4$	$Mg_3Si_2O_5(OH)_4$

Kaolinites are dioctahedral minerals; ***serpentines*** are trioctahedral minerals. Here and in other phyllosilicates, the composition of the tetrahedral sheet alone is Si_2O_5 per structural unit

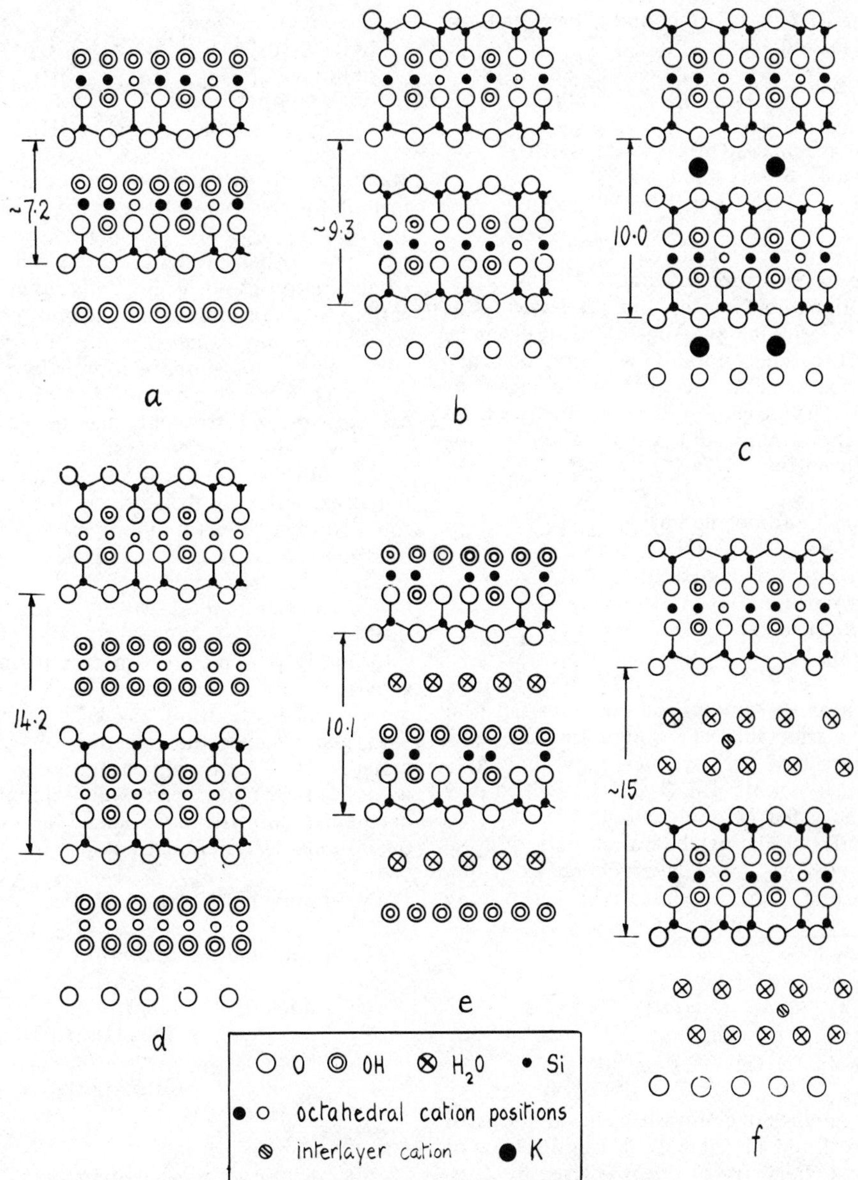

FIGURE 3. Schematic diagrams of some phyllosilicate (layer silicate) structures. (a) 1:1 ***kaolinite-serpentine.*** (b) 2:1 ***pyrophllite-talc.*** (c) *Mica* (di- and trioctahedral) forms. (d) ***Chlorite*** (trioctahedral). (e) *Halloysite* (hydrated form with a sheet of water molecules between the aluminosilicate layers). (f) *Montmorillonite.* Dioctahedral cation positions are indicated by small shaded circles; trioctahedral cation positions are indicated by small solid and additionally the small open circles. The interlayer H_2O positions in (e) and (f) and the interlayer cation positions in (f) are not known exactly and are shown in a purely schematic way. Approximate layer spacings in Ångstrom units (1 Å = 10^{-10} m).

of the sheet; but, when compounded with an octahedral sheet, two of the oxygens are held in common with the octahedral sheet. This is how the sheets are linked together. To avoid counting jointly held atoms twice over, they will be counted in the octahedral sheet and accordingly the tetrahedral sheet composition is shown as Si_2O_3.

The ***pyrophyllite-talc*** group of minerals, illustrated by Fig. 3b, has one octahedral sheet sandwiched between two tetrahedral sheets, called 2:1 type layers. *Pyrophyllite* is dioctahedral and *talc* is trioctahedral. Their compositions are as follows:

	Pyrophyllite	*Talc*
tetrahedral	Si_2O_3	Si_2O_3
octahedral	$Al_2O_4(OH)_2$	$Mg_3O_4(OH)_2$
tetrahedral	Si_2O_3	Si_2O_3
total	$Al_2Si_4O_{10}(OH)_2$	$Mg_3Si_4O_{10}(OH)_2$

Mica-group minerals (Fig. 3c) have similar 2:1-type layer structures but with important compositional differences; the 2:1 layer has an overall negative charge that is neutralized by interlayer cations, notably K^+ and less commonly Na^+, and in the brittle-***mica*** group by Ca^{2+} ions. Typical di- and trioctahedral ***micas*** have compositions approximating to the following formulae:

muscovite (dioctahedral)
$KAl_2(Si_3Al)O_{10}(OH)_2$
paragonite (dioctahedral)
$NaAl_2(Si_3Al)O_{10}(OH)_2$
biotite (trioctahedral)
$K(Mg,Fe^{2+})_3(Si_3Al)O_{10}(OH)_2$

For most natural ***micas***, compositions fall near a di- or a trioctahedral formulation, but tetrasilicic ***micas*** have been synthesized with compositions such as $KMg_{2.5}Si_4O_{10}(OH)_2$, which may have to be called "2½-octahedral."

In brittle ***micas***, tetrahedral cations approximate to (Si_2Al_2), the layer charge is -2 per formula unit, and the interlayer cations are Ca^{2+}. Typical formulae are:

margarite (dioctahedral)
$CaAl_2(Si_2Al_2)O_{10}(OH)_2$
clintonite (trioctahedral)
$Ca(Mg,Al)_3(Si_2Al_2)O_{10}(OH)_2$

Some important compositional variations of ***micas*** are listed in Table 1. A useful review of the crystal chemistry of ***micas*** is given by Zussman (1979).

Minerals of the ***chlorite*** group are mainly trioctahedral (see Fig. 3d) both in the 2:1 silicate layer and in the hydroxide interlayer. Dioctahedral ***chlorites*** do occur (e.g., *donbassite*), however, and ***chlorites*** that are dioctahedral in the 2:1 layer but trioctahedral in the hydroxide interlayer also occur. The latter are described as di,trioctahedral ***chlorites*** (e.g., *cookeite* and *sudoite*). No tri,dioctahedral ***chlorites*** are yet known.

Cationic substitutions occur at all levels of the chlorite structure. The simplest formulae for the two structural components can be written:

2:1 layer:
$[(Mg_{3-0.5x}Al_{0.5x})(Si_{4-x}Al_x)O_{10}(OH)_2]^{-x/2}$
interlayer: $[(Mg_{3-0.5x}Al_{0.5x})(OH)_6]^{+x/2}$
total composition:
$(Mg_{6-x}Al_x)(Si_{4-x}Al_x)O_{10}(OH)_8$

Many chlorite compositions approximate to a formula of this type with x in the range of 0.6–1.6. Commonly, Mg is replaced partially by Fe^{2+} and other divalent cations and Al^{3+} in octahedral positions by Fe^{3+} and other trivalent cations. These simple formulae presuppose that Al replaces Mg equally in the two structural components; this simplification is by no means always correct. The distribution of octahedral cations, di- and trivalent, can be determined only by highly precise crystal-structure analysis. A further complication is that ***chlorites*** initially containing iron principally in the ferrous (divalent) form may subsequently be oxidized so that ferric (trivalent) ions are formed; other compositional changes must occur to maintain electrical neutrality, e.g., $(OH)^- \rightarrow O^{2-}$.

Evidently the chlorite structure permits wide variations in chemical composition and, consequently, the literature has become burdened with many names that are now recognized as unnecessary. A simplification suggested by Bayless (1975) has been widely adopted. Trioctahedral ***chlorites*** are named according to the dominant divalent cation:

Mg-dominant:	*clinochlore* $(Mg_5Al)(Si_3Al)O_{10}(OH)_8$
Fe^{2+}-dominant:	*chamosite* $(Fe_5Al)(Si_3Al)O_{10}(OH)_8$
Ni-dominant:	*nimite* $(Ni_5Al)(Si_3Al)O_{10}(OH)_8$
Mn^{2+}-dominant:	*pennantite* $(Mn_5Al)(Si_3Al)O_{10}(OH)_8$

TABLE 1. Compositional Variations of Micas with General Formulae $RY_2X_4O_{10}(OH)_2$ and $RZ_3X_4O_{10}(OH)_2$

Dioctahedral	X_4	Y_2	R
celadonite	Si_4	AlMg	mainly K
phengite	$Si_{3.5}Al_{0.5}$	$Al_{1.5}Mg_{0.5}$	
muscovite	Si_3Al	Al_2	
margarite	Si_2Sl_2	Al_2	Ca
Trioctahedral	X_4	Z_3	R
polylithionite	Si_4	$AlLi_2$	mainly K
lepidolite	$Si_{3.5}Al_{0.5}$	$Al_{1.25}Li_{1.75}$	
trilithionite	Si_3Al	$Al_{1.5}Li_{1.5}$	
zinnwaldite	Si_3Al	$AlFe^{2+}Li$	
phlogopite	Si_3Al	Mg_3	
biotite	Si_3Al	$(Mg,Fe^{2+})_3$	
annite	Si_3Al	Fe_3^{2+}	
eastonite	$Si_{2.5}Al_{1.5}$	$Mg_{2.5}Al_{0.5}$	
clintonite	$SiAl_3$	Mg_2Al	Ca

These compositions correspond to $x = 1$ in the general formula given above. Important octahedral cations other than the dominant cations or unusual tetrahedral compositions are indicated by suitable adjectival modifiers to the four principal names.

The Clay Minerals

Although the clay minerals are principally phyllosilicate minerals, since they have special properties and are dealt with elsewhere (see *Clays, Clay Minerals*) they will receive no more than a passing reference here. *Kaolinite* itself is an important clay mineral and *serpentine* minerals also qualify occasionally to be called clays. Many clay minerals are intermediate in character and composition between the ***pyrophyllite-talc*** group and the ***mica*** group, namely the ***smectites*** and ***vermiculites***. Besides its clay-like forms, ***vermiculite*** also develops as large mica-like sheets and could properly be included in the present article, but to avoid repetition, ***vermiculites*** are treated in the article *Clays, Clay Minerals*. Two of the clay mineral structures are illustrated in Figs. 3e and f.

Relationships Between Unit-Cell Dimensions and Chemical Compositions

Figure 3 shows the perpendicular distances between layers of the principal phyllosilicate structures; approximate values only are given because the precise values depend on the ions, especially the cations, that are present. These distances are measured easily and accurately by X-ray diffraction methods and provide an easy identification procedure in circumstances where direct optical observation is not easy, for example if fine-grained shales and clays.

Much attention has been given to relationships between chemical composition and dimensions within the layers, particularly the parameter b shown in Fig. 2a. This parameter provides a useful indication of whether a phyllosilicate is di- or trioctahedral. For many dioctahedral minerals, $b \approx 8.90–9.05$ Å, and for most trioctahedral minerals, $b \approx 9.2–9.4$ Å, but there are some exceptions to these criteria and they must be used cautiously. An empirical, statistical approach (Radoslovich, 1962) led to the following relationships (the precise numerical values may need some modification in light of recent data):

Kaolinite-serpentine:

$$b = (8.923 + .125\text{Mg} + .229\text{Fe}^{2+} + .079\text{Fe}^{3+} + .28\text{Mn}^{2+} + .17\text{Ti}) \pm .014\text{Å}$$

Micas:

$$b = (8.925 + .099\text{K} - .069\text{Ca} + .062\text{Mg} + .116\text{Fe}^{2+} + .098\text{Fe}^{3+} + .166\text{Ti}) \pm .03\text{Å}$$

Chlorites:

$$b = (9.23 + .03\text{Fe}^{2+}) \pm .03\text{Å}$$

Smectites:

$$b = (8.944 + .096\text{Mg} + .096\text{Fe}^{3+} + .037\text{Al}^{\text{IV}}) \pm .012\text{Å}$$

The chemical symbols stand for the numbers of atoms per formula unit. Only for ***smectites*** is there a small term involving Al in tetrahedral positions, indicated by Al^{IV}. As an application of these formulae, consider a brittle ***mica*** with the composition

$$\text{Ca}(\text{Mg}_2\text{Al})(\text{SiAl}_3)\text{O}_{10}(\text{OH})_2$$

The calculated value of b is

$$8.925 - 0.069 + 0.124 = 8.980\text{Å}$$

which places this *tri*octahedral mineral within the b-parameter range for *di*octahedral minerals. Observed values agree with the calculated value so that the formula is indeed obeyed. Minerals of this kind are an exception to the general rule relating to b values for di- and trioctahedral phyllosilicates.

Detailed Structure Analyses

Many questions concerning phyllosilicates cannot be answered on the basis of idealized geometrical models and require highly accurate and detailed crystal structure analyses. Parenthetically, it can be remarked that this is not peculiar to the phyllosilicates, but applies to all groups of minerals.

If models are constructed of the tetrahedral and octahedral sheets illustrated in Figs. 2 and 3 using known interatomic distances, or if corresponding calculations of the sheet dimensions are made, then it becomes obvious that some modifications are necessary in order to fit the sheets together to form composite layers. Often the tetrahedral sheets are on a larger scale than dioctahedral sheets; they are contracted principally by a twisting of the tetrahedra through an angle α, alternatively + and − as one proceeds around the sixfold ring shown in Fig. 2a. Precise structure analyses have given values of α from near zero to about 23°. The sheet symmetry is lowered from hexagonal to trigonal by these rotations. In trioctahedral layer structures, the tetrahedral sheets are commonly on a smaller scale than the trioctahedral sheets and the two sets are matched mainly by compression and expansion of the octahedral sheets (they are "squashed" flatter, so to speak); the tetrahedral sheets retain more nearly their ideal hexagonal symmetry. Bailey (1980) has summarized the details of phyllosilicate layer symmetry. The actual deviations of the layer structures from the simple geometric models depend on the chemical compositions of the minerals and their overall structure (polytypic form).

In the mica structures (see Figs. 1, 3c), the K ions are between layers of 2:1 type, largely in "holes" of the silicon-oxygen networks (see Fig. 2a). Ideally the K ions are between six upper and six lower oxygens, but, because of the twisting of the tetrahedra, three upper and three lower oxygens are brought into close contact with each K ion; the remaining oxygens are moved further away. Consequently we see here another constraint on the overall structure.

As noted above, the formulae relating *b* parameters to composition seldom show a dependence on the extent of Si replacement by Al, yet Al–O distance in an AlO_4 tetrahedron is about 1.77Å as compared with 1.62Å for the Si–O distance, i.e., a 10% difference. Evidently the incorporation of the larger tetrahedra is made possible by the rotation mechanism. A further question is whether the replacements take place in an ordered, a partially ordered, or a random manner. This question is difficult to answer for two reasons. In the first place, Al and Si (adjacent atoms in terms of electron configuration and both with ten electrons if fully ionized) scatter X-rays almost identically and, therefore, cannot easily be distinguished in X-ray diffraction analysis. The difference in size of the two tetrahedral groups would distinguish them provided each occupies the same crystallographic positions throughout a crystal. Conventional X-ray diffraction analysis is concerned with long-range order (LRO). Conceivably there can be short-range order (SRO) within small domains (see *Order-Disorder*). Study of SRO requires evaluation of the whole X-ray diffraction scattering; the usual restriction to the sharp reflections is not adequate. One study of SRO in *muscovite* (Gatineau, 1964) has indicated that Al-for-Si replacements are segregated in bands that are in directions at 120° to each other in SRO domains.

Polytypism

Polytypism is concerned with the different stacking arrangements of layers having essentially the same structure and composition. The way in which the layers are arranged relative to each other determines the overall symmetry of the crystal structure. Therefore, the determination of the symmetry space group (see *Space Group Symbol; Symmetry*) is an essential step in the determination of the total structure and, normally, single-crystal diffraction methods must be used. X-ray powder diffraction, which is extensively used for mineral identification (see vol. IVA, X-ray Diffraction), will usually provide sufficient evidence for classifying a mineral as belonging to a particular polytypic variety. Beside a wide variety of ordered polytypes, many partially ordered and disordered forms also are recognized (see Bailey, 1980; Brindley, 1980).

In general terms, the layer stacking arrangements depend on the character of the interlayer forces. Relatively weak, nonspecific van der Waals forces are always present. When planes of oxygen and hydroxyl ions face each other as in 1:1-type structures (Fig. 3a) and in ***chlorites*** (Fig. 3d), OH•••O bonds, i.e., hydrogen bonds, are possible. These bonds are weak but specific between particular OH and O ions. In ***micas***, the interlayer K ions key the layers together in a specific way but various options are possible because of the hexagonal, quasi-hexagonal, or trigonal organization of the Si(Al)–O sheets. The bonds are relatively strong. Long-range ionic forces also play an important role. For example, Si^{4+} ions of one layer may attract appreciably O^{2-} ions of an adjacent layer and repel Al^{3+} or Si^{4+} ions of an adjacent layer. Thus, many interactions occur between layers. The stacking of the layers is a complex function of the kinds of forces that operate and their specific or nonspecific character. Against this general background, the polytypic forms of layer structures can be discussed.

The principal 1:1-type structures are the dioctahedral ***kaolinite***-group minerals and trioctahedral ***serpentine*** minerals. *Kaolinite*, the most commonly occurring member of the first group, has one layer per unit cell and the overall symmetry is triclinic; it is a 1Tc structure. The unit cell parameters, *a* 5.14, *b* 8.93, *c* 7.37 Å, α 91.6°, β 104.8°, γ 89.9°, are such that in passing from one layer to the next, the principal displacement is approximately $a/3$ parallel to the *x*-axis making $\beta = 104.8°$ and a much smaller displacement parallel to the *y*-axis so that α departs from 90°. These displacements are conditioned by bringing OH and O into favorable positions for OH•••O bonds, but long-range ionic forces play a role in determining the particular mutual arrangements and the tetrahedral rotations of about 11° also condition the OH•••O associations. Here, as in all the layer structures, there is a complex interplay of forces and simple descriptions tend to minimize inherent complexity. The mineral *dickite*, with essentially the same chemical composition and layer structure, has two layers per unit cell arranged to give overall monoclinic symmetry (therefore, a 2M structure) and the cell parameters, *a* 5.15, *b* 8.94, *c* 14.74 Å, β 103.6°, reflect similar layer dimensions. There is a similar displacement of layers in the *x*-direction (similar β angle), but no net displacement parallel to the *y*-axis; the small displacements found in *kaolinite* parallel to *y* are + and −, and cancel out in the two-layer cell. The tetrahedral rotations in *dickite* are about 7.5°. *Nacrite*, another ***kaolinite***-group mineral also with monoclinic symmetry and two layers per unit cell, has a different arrangement of the layers (see Bailey, 1980). Beside these regular polytypes, many disordered forms occur. Because the OH ions are distributed parallel to the *y*-axis at intervals near $b/3$, layers can be displaced relative to each other by $nb/3$ (*n* an integer) and retain a similar OH···O association. Many ***kaolinites*** show variable amounts of so-called "$b/3$ displacement" disorder. However, more general layer displacements are also found (Brindley, 1980).

In the ***serpentine***-group minerals, *chrysotile* is of special interest in that it is a fibrous mineral (see *Asbestos*). The inherent misfit of the trioctahedral Mg sheets and the Si tetrahedral sheets causes the layers to curl in a spiral manner that has been seen directly in high-magnification electron micrographs. When Al ions partly replace Mg and Si, the octahedral sheet is contracted, the tetrahedral sheet expanded, and the layers then develop in the usual planar manner.

Pyrophyllite and *talc* are the simplest 2:1-type layer silicates and appear commonly to have 1Tc type structures, but 2M type structures are not excluded. Bonding between successive layers is provided partly by van der Waals forces and partly by long-range ionic attractions. These weakly bonded layers are easily displaced by mechanical treatment and the structures commonly show appreciable or considerable layer stacking disorder.

Micas have probably received more detailed structural study than other groups of phyllosilicates. Bonding between the negatively charged 2:1 layers (due to Al-for-Si substitutions) is strong and specific in that the K ions occupy well-defined positions between the adjacent planes of oxygen ions. It is inherent in the 2:1 layer structure that the upper tetrahedral sheet is displaced with respect to the lower tetrahedral sheet by a built-in displacement of about $a/3$. In so far as the tetrahedral sheets have hexagonal symmetry, the displacement vector of magnitude $a/3$ from one 2:1 layer to the next 2:1 layer may be rotated by $n \times 60°$ (n = 0, 1, 2, 3, 4, 5). Consequently, by continued repetition from layer to layer of the same rotation, structures can be developed with various overall symmetries denoted by 1M, $2M_1$, 3T, 2Or, $2M_2$ and 6H (prefix numbers give layers per unit cell; letters give symmetry monoclinic (M), trigonal (T), orthorhombic (Or), hexagonal (H); and subscripts indicate different monoclinic arrangements). The Si(Al) tetrahedral sheets have tetrahedral rotation angles which range from 1° to 11° in trioctahedral ***micas*** and usually are near 10–12° in dioctahedral ***micas***. The relative abundance of different ***mica*** polytypes is shown in Table 2. The $2M_1$ structure evidently is more frequent among dioctahedral ***micas*** than trioctahedral ***micas***. This feature of the mica structures may be related, at least in part, to the generally smaller tetrahedral rotation angles found in trioctahedral than in dioctahedral ***micas***.

TABLE 2. Relative Abundances of Mica Polytypes

Dioctahedral	High	Medium	Low
muscovite	$2M_1$	1M, 1Md	3T
phengite	1M, 1Md, 3T	$2M_1$	$2M_2$
celadonite	1M, 1Md	–	–
Trioctahedral	**High**	**Medium**	**Low**
phlogopite	1M, 1Md	3T	$2M_1$
biotite	1M, 1Md	3T	$2M_1$
lepidolite	1M, $2M_2$	3T	$2M_1$
zinnwaldite	1M, 1Md	$2M_1$, 3T	–

After Bailey, 1980

Chlorites, with positively charged hydroxyl sheets between negatively charged 2:1 layers, have the potential for developing many more polytypic varieties than ***micas***. It appears, however, that most ***chlorites*** have either 1-layer structures or irregular layer sequences. As shown by S. W. Bailey and G. Brown (see Bailey, 1980), there can be twelve different 1-layer ***chlorite*** polytypes. The prevalence of layer stacking disorder in ***chlorites*** can be related to the weak OH···O bonding between the hydroxyl interlayer sheets and the planes of oxygen ions of the tetrahedral sheets. The overall bonding of the layers and interlayers involves the nonspecific ionic attractions between the negatively and positively charged structural units. The absence or weakening of particular groups of reflections in X-ray powder diffraction patterns, those with the *R* index $\neq 3n$, points to random layer displacements of $nb/3$; i.e., similar to those present in ***kaolinite***-group minerals.

G. W. BRINDLEY

References

Bailey, S. W., 1980. Structures of layer silicates, in G. W. Brindley and G. Brown, eds., *Crystal Structures of Clay Minerals and their X-ray Identification*. London: Mineralogical Society, 1-123.

Bayliss, P., 1975. Nomenclature of the trioctahedral chlorites, *Can. Mineral.*, **13**, 178-180.

Brindley, G. W., 1980. Order-disorder in clay mineral structures, in G. W. Brindley and G. Brown, eds., *Crystal Structures of Clay Minerals and their X-ray Identification*. London: Mineralogical Society, 126-195.

Brindley, G. W., and MacEwan, D. M. C., 1953. Structural aspects of the mineralogy of clays, in A. T. Green and G. H. Stewart, eds., *Ceramics–A Symposium*. Stoke-on-Trent: British Ceramic Society, 15-59.

Deer, W. A., Howie, R. A., and Zussman, J., 1962. *Rock-forming Minerals*, vol. 3, *Sheet Silicates*. London: Longmans, 270p.

Gatineau, L., 1964. Structure reélle de la muscovite. Répartitions des substitutions isomorphes, *Bull. Soc. fr. Minéral. Cristallogr.*, **87**, 321-355.

Radoslovich, E. W., 1962. Cell dimensions and symmetry of layer-lattice silicates. II. Regression relations, *Am. Mineral.*, **47**, 617-636.

Zussman, J., 1979. The crystal chemistry of the micas, *Bull. Minéral.*, **102**, 5-13.

Cross-references: *Chlorite Group; Clays, Clay Minerals; Mica Group; Mineral Properties; Polysomatism;* see also mineral glossary.

PIEZOELECTRICITY

Piezoelectricity was discovered by P. J. Curie in 1880. Crudely stated, piezoelectricity is electricity generated by "squeezing a solid," the solid being specifically a crystal of low symmetry. For example, if a block is properly cut from a rochelle-salt crystal ($KNaC_4H_4O_6 \cdot 4H_2O$), two specific opposite faces being made electrically conducting and electrically connected to a neon lamp, the lamp will be observed to emit a flash of light when the block is suddently compressed parallel to the conducting faces. The sudden compression is best supplied by a cushioned hammer blow. Rochelle salt is used because it has a large piezoelectric constant: conversely when these conducting faces are connected to a battery, the dimensions of the block are changed slightly. The first experiment illustrates the piezoelectric effect, the second, the "converse effect."

Symmetry Requirements for Activity

Piezoelectric activity is the exception, not the rule, among crystals. Roughly 2% of the known minerals exhibit the effect, but only *quartz* and ***tourmalines*** have found extensive use. However, several synthetic materials have found use. Rochelle salt, ammonium dihydrogen phosphate (ADP), potassium dihydrogen phosphate (KDP), and ethylene diamine tartrate (EDT) were the principal nonmineral piezo crystals, but synthetic quartz is also now being made on a large scale. Other synthetic piezoelectric crystals of current interest are lithium niobate ($LiNbO_3$), BGO ($Bi_{12}GeO_{20}$), $Sr_x Ba_{1-x}Nb_2O_6$, $LiGaO_2$, TeO_2, GaAs, and InSb.

A crystal cannot be piezoactive if its structure has a center of symmetry (see *Symmetry*). If all rows of atoms look the same from both ends, the structure is centro-symmetric. The sequence *ABCABC*,..., where *A*, *B*, and *C* represent three kinds of atoms, does not look the same from both ends. A homogeneous stress is centrosymmetric and cannot produce an unsymmetrical result, i.e., an electric polarization, unless the medium lacks a center of symmetry. One cubic class of crystals cannot be piezoactive even though it lacks a center of symmetry. A center of symmetry excludes the possibility of piezoelectric activity. The low symmetry of a crystal is often shown by the disposition of its faces: the two ends of a prism may be differently terminated (hemimorphism). The edges of a prism may be of two kinds as shown by neighboring faces. If natural faces do not exist, etch pits (q.v.) may show that opposite sides of a plate are crystallographically different because the pits on the two sides are of different shapes.

Besides centers of symmetry, other kinds of symmetry are important in the study of piezoelectric crystals. These are symmetry around an axis, reflection symmetry in a mirror plane, and some combinations of axial symmetry plus inversion through a center. In crystals, atom groups are repeated at regular intervals along all lines. This restricts the nature of axial symmetry to twofold (binary), threefold (trigonal), fourfold, and sixfold symmetries. Paralleling the structural symmetry is the symmetry of physical properties. If we made a three-dimensional plot of some vector property of a crystal (such a property as the growth rate of the crystal in a standard solution), we observe that the three-dimensional plot has symmetry. For example, if we make a mold around a model of the three-dimensional plot, we may find that the model will fit into the mold in several positions. If the model can be lifted out of the mold, turned through one-third of a turn about some axis and then fitted back into the mold, the crystal is said to have trigonal symmetry. Proceeding in this way we find that every possible crystal must fall into one of 32 symmetry classes. Of these, 20 classes may be piezoactive. To show this, we assume that each of the three components of electric polarization is linearly related to each of the six components of stress (three expansion stresses, three shearing stresses), i.e., there are 18 piezoelectric constants for a crystal of no symmetry. If we now introduce the idea that these constants will be unchanged by a rotation of 180° about a coordinate axis (because the crystal has an axis of twofold symmetry), we find that 10 of these constants must be zero.

Proceeding in this way we find that *quartz* (which has a trigonal axis and three binary axes perpendicular to the trigonal axis) has but two independent constants named d_{11} and d_{14}. However, these constants are used in more than one place in the linear relationships mentioned above. For example, in *quartz*, compressing along the X axis (taken as along a binary axis) with a stress of one Newton per square meter will release d_{11} ($=-2.25 \times 10^{-12}$) coulombs of electricity per square meter on the $+X$ and $-X$ faces (positive charge on one end, negative on the other). A compression along Z ($=c$, the prism axis) produces no charge; a similar compression along Y will deliver the same size charges to the X faces as did compression along X, but with opposite signs. A shear about X (i.e., in the Y-Z plane) produces d_{14} ($=0.853 \times 10^{-12}$) coulombs per Newton on the X faces, a shear about Y produces the same charge d_{14} on the Y faces per Newton, and a shear about Z produces $-2d_{11}$ coulombs per Newton on Y faces. Piezo charges never appear on Z faces of *quartz* (see Table 1).

Returning for a moment to the converse piezoelectric effect, we note that, if d_{11} is the electric charge released in coulombs per square meter for a stress of one Newton per square meter in the direct piezoelectric effect, this same number, d_{11}, gives the strain–voltage gradient relation in the converse effect. For example, a stress of one Newton per square meter on the Y face of a quartz block releases 2.25×10^{-12} coulombs per square meter on the X faces. Conversely, one volt per meter between the X faces causes a strain of 2.25×10^{-12} meters per meter length of the Y dimension.

In reducing the number of constants from 18 to 2 for *quartz*, we chose the axes along symmetry elements of the crystal. If the coordinate axes are chosen at random, there will again be 18 piezo constants but these constants will be complicated functions of the direction angles of the symmetry elements and of the minimal set of piezo constants—two for *quartz*.

In a way similar to the above, we can treat the elastic properties of crystals. (They are important in understanding many piezoelectric devices.) We assume that the six strain components are proportional to each of the six stress components. This gives 36 elastic constants for a crystal of no symmetry, but these always reduce to 21 because, if we write the strain energy as a function of stress or strain, this expression should be a perfect differential and its second derivative with respect to two components must be independent of the order of differentiation. Again, as before, symmetry operations reduce the 21 to a smaller number, 6 for *quartz*, **tourmaline**, etc. Again, these reductions hold only if the axes are chosen along symmetry elements; 21 result if the coordinate axes are chosen at random. We now find a very convenient fact: while most materials become more yielding as the temperature rises, a very few materials have at least one elastic modulus that becomes less yielding as the temperature rises. *Quartz* is one of these materials and EDT (ethylene diamine tartrate) is another. Consequently, it is possible to find orientations for which some natural vibrational modes have a zero temperature coefficient of frequency at some temperature or within some temperature range.

TABLE 1. Piezoelectric Constants for Quartz

	X_X	Y_Y	Z_Z	Y_Z	Z_X	X_Y
E_X	d_{11}	$-d_{11}$	0	d_{14}	0	0
E_Y	0	0	0	0	$-d_{14}$	$-2d_{11}$
E_Z	0	0	0	0	0	0

(E_X,E_Y,E_Z represent electric polarizations along X, Y, and Z respectively. X_X, Y_Y, and Z_Z represent expansive stresses in the X, Y, and Z directions, respectively. Y_Z, Z_X, and X_Y are shearing stresses in the Y_Z, Z_X, and X_Y planes, respectively.)

Quartz

Because of the possibility of making plates with a zero temperature coefficient of frequency, *quartz* has found great use as a standard of frequency. The piezoelectric effect allows a crystal plate to be used as an electric circuit element and an "equivalent circuit" is considered to represent the crystal in the circuit. In this circuit, resistance represents dissipative loss. If the dissipative loss is small (high Q), the natural response frequency is very sharp and the crystal can maintain a very precise frequency. This frequency is often used to control a clock of great precision. Carefully prepared *AT*-cut plates (see Fig. 1) are stable to about one part in 10^9 per month, 2 to 3 by 10^{-11} parts per day. Even wristwatches are now made that keep time very precisely because of a *quartz* crystal which replaces the balance wheel.

During World War II, every U.S. Army tank was expected to have about 50 such *quartz*-

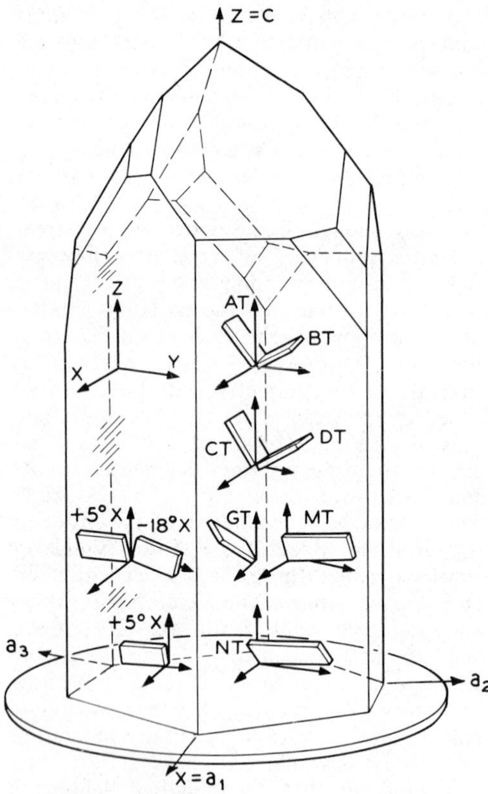

FIGURE 1. Idealized *quartz* crystal (right-handed) showing several orientations used for oscillator plates, filter plates, etc.

crystal plates on hand. These were plugged in at predetermined intervals so that the enemy would have difficulty following the frequency changes in the radio communications.

Quartz for piezoelectric work must be untwinned (see *Twinning*). Natural *quartz* is often twinned, and plates must be cut so as not to cross a twin boundary, otherwise the two parts of the crystal buck one another. The two most common kinds of twinning in *quartz* are 180° and Right-Left. In Right-Left twinning, portions of right-hand *quartz* are joined to left-hand portions. This becomes obvious if the piece is observed by polarized light. In 180° twinning, both portions are of the same hand but the plus-electric axis of one portion is in the direction of the minus-electric axis of the other. Any kind of twinning of *quartz* can be revealed by etching in hydrofluoric acid. *Quartz* is now grown on a large scale. Low-*Q* *quartz* for filter plates can be grown quickly and from this material quartz filters can be made much more cheaply than from natural *quartz*. This is because the raw material is very uniform, untwinned, and of convenient size—

TABLE 2. *Tourmaline d* Matrix

0	0	0	0	d_{15}	$-2d_{11}$
$-d_{22}$	d_{22}	0	d_{15}	0	0
d_{31}	d_{31}	d_{33}	0	0	0

($d_{15} = -3.6 \times 10^{-12}$, $d_{22} = 0.33 \times 10^{-12}$, $d_{31} = -0.34 \times 10^{-12}$, $d_{33} = 1.8 \times 10^{-12}$ coulombs per Newton.)

also, the orientation is controlled—something that is not true of all natural *quartz*. High-*Q* *quartz* must be grown much more slowly; and, at present, natural *quartz* is cheaper for high-*Q* uses (see *Crystal Growth; Synthetic Minerals*).

Tourmaline

Minerals of the ***tourmaline*** group have a trigonal axis and three planes of symmetry parallel to the trigonal axis. Analysis shows that four piezo constants exist and that they appear eight times in the stress-polarization equations. The *d* matrix is given in Table 2.

Tourmaline has no zero temperature coefficient of frequency possibilities, but has one feature that makes it valuable—it gives an electric response to hydrostatic pressure, which *quartz* does not do. Hence, ***tourmaline*** has been used as a microphone, particularly for standardization work. Since ***tourmaline*** is much more expensive than *quartz* and since it has not been found practical to synthesize it, ***tourmaline*** is justified only for very special tasks. Synthetic lithium niobate ($LiNbO_3$) is of the same class and is strongly piezoelectric. It has replaced ***tourmaline*** almost completely.

Tourmaline is the material in which pyroelectricity was first discovered. The two ends of a ***tourmaline*** crystal become oppositely charged if the temperature is changed. The charged ends attract dust particles much as a magnet attracts iron fillings. This property cannot exist for a crystal having a center of symmetry or having more than one axis of rotational symmetry. *Quartz*, having three binary axes perpendicular to a trigonal axis cannot be pyroelectric. Lithium niobate is pyroelectric.

Rochelle Salt

Rochelle salt has been used extensively for phonograph pickups and for higher-register loudspeakers (so-called "tweeters)." The material suffers from temperature sensitivity and is very water soluble. It melts at about 80°C but its largest piezo constant, d_{14}, changes greatly near room temperature. The crystals are orthorhombic above 20°C, become slightly monoclinic below 20°C, but again revert to orthorhombic below -18°C. There are three piezo constants: $d_{14}=2.3\times10^{-9}$, $d_{25}=56.3\times$

10^{-12}, and $d_{36} = 11.8 \times 10^{-12}$ coulombs per Newton. These are all such as to give electric polarization under shearing stress; but, by a rotation of axes, shears and expansions are partially interchanged. The crystals have to be well protected from moisture and overheating but many thousands of them have been used for devices. A rotation of 45° about the *X* axis (which is the crystallographic *Y* axis) produces a block that can be used to demonstrate the piezoelectric effect by the neon lamp flashing when the crystal end is struck a blow.

Ammonium Dihydrogen Phosphate

ADP, a tetragonal crystal, was extensively used for submarine detection in World War II. In France, Langevin had earlier (in 1918) developed such devices using *quartz* crystals, but the electronic art was too primitive in his day to make the greatest use of such things. There are two piezo constants here: d_{14} and d_{36}; $d_{14} = 1.7 \times 10^{-12}$, $d_{36} = 49 \times 10^{-12}$ coulombs per Newton. To get a simple compression along the length in order to have an electric polarization, the crystal length must be cut at 45° to *X* and *Y*. Banks of these crystals immersed in castor oil were placed in a blister on a ship's hull. The castor oil was retained and the sea kept out by a diaphragm. A sudden electric pulse sent to the crystals caused a sudden increase in length due to the converse piezoelectric effect. The sound wave so generated passed through the castor oil and the diaphragm into the sea and was transmitted as a fairly sharp beam. On striking some object such as a submarine, part of the wave bounced back where it struck the crystals and generated an electric pulse. This pulse was observed electronically and the time delay between the strong outgoing pulse and the feeble reflected pulse was interpreted in terms of the distance to the reflecting object.

Although it is water soluble, ADP is easily protected and is not very temperature sensitive in contrast to Rochelle salt. However, piezoelectric ceramics have replaced ADP in this sonar use.

Potassium Dihydrogen Phosphate

KDP is very similar to ADP but is a bit more expensive to produce. Here, $d_{14} = 1.4 \times 10^{-12}$ and $d_{36} = 23.2 \times 10^{-12}$ coulombs per Newton. It has an advantage over ADP in a related property. Its refractive index can be changed noticeably by an electric field. This property permits the use of KDP as a light valve and it is somewhat superior to ADP in this use at high frequencies.

Ethylene Diamine Tartrate

EDT, a monoclinic crystal, has eight piezoelectric constants and 13 elastic constants. There is a range of directions in EDT for which the temperature coefficient of expansion is negative. There are also cuts for which the temperature coefficient of frequency is zero over a convenient temperature range. This fact was used to advantage for the manufacture of telephone channel filters where large pieces were required for the rather low frequency filter networks that sort out each of many conversations going over a single wire. In designing filter networks, a larger *Q* (lower dissipation loss) allows sharper passbands with steeper sides. Networks made from coils and condensers alone will not give very steep sides, hence these channels cannot be packed closely together and frequency space is wasted. The lower loss in *quartz* allows for very steep sides—in fact, steeper than is needed, but this is easily compensated for by adding resistances in the circuit. The *Q* of *quartz* is higher than that of EDT, but the latter suffices for filter work—although it probably would not make a precision oscillator such as a primary time standard might require. As the supply of *quartz* crystals in such large sizes became more difficult, synthetic *quartz* supplied this need, so that EDT is not now used for filter networks. The *d* matrix for EDT is given in Table 3.

Barium Titanate

$BaTiO_3$ is a tetragonal crystal, but is cubic above 120°C. Although large untwinned crystals never occur, it has been found possible to make this high-melting-point crystal into a ceramic. As the ceramic is cooled through the transition temperature, an applied electric field forces the *C* axis to pop out of the one of the three available cubic directions that is closest to being parallel to the field and thus make a piezoelectric ceramic of considerable power. These units are superior to ADP for submarine detection. Much work has been done to improve performance by varying the composition through the introduction of small amounts of other metallic ions. One such substance is a ceramic made from a mixture of 47% to 52% lead zirconate, the remainder, lead titanate.

TABLE 3. The *d* Matrix for EDT

0	0	0	d_{14}	0	d_{16}
d_{21}	d_{22}	d_{23}	0	d_{25}	0
0	0	0	d_{34}	0	d_{36}

($d_{14} = -8.7 \times 10^{-12}$, $d_{16} = -8.3 \times 10^{-12}$, $d_{21} = 11.3 \times 10^{-12}$, $d_{22} = 7.0 \times 10^{-12}$, $d_{23} = -10.3 \times 10^{-12}$, $d_{25} = -13 \times 10^{-12}$, $d_{34} = -14 \times 10^{-12}$, and $d_{36} = 16.7 \times 10^{-12}$ coulombs per Newton, at 20°C.)

This is known by the trade name **PZT**. This ceramic handles greater energy than does a barium titanate ceramic of the same size. For $BaTiO_3$ (in coulombs per Newton) $d_{15} = 390 \times 10^{-12}$, $d_{31} = 35 \times 10^{-12}$, $d_{33} = 86 \times 10^{-12}$, while for the ceramic, effective values are $d_{15} = 90 \times 10^{-12}$, $d_{31} = -80 \times 10^{-12}$, $d_{33} = 190 \times 10^{-8}$. PSN, $(K,Na)_2Nb_2O_6$, is also a useful such ceramic.

Complicated shapes are cut from hard, brittle materials by forcing a die against the material surface with abrasive between. The die is moved at hypersonic frequency and crunches the abrasive into the material, thus eroding it away. Even sapphire is readily cut in this way, and piezoceramic units are often used to replace the original magneto-strictive units in some commercial tools. The intense sound is too high in frequency to be heard and, thus, is not objectionable.

To reach very high frequencies requires piezo units of very small dimensions. Vapor deposition of some piezo materials is possible and results in layers much thinner than could be achieved by grinding plates. Zinc oxide, cadmium selenide, cadmium sulfide, and aluminum nitride are used in this way to achieve frequencies exceeding 10^9 Hertz. By generating harmonics, frequencies approaching 10^{10} Hertz are obtained.

There are many more technical uses of the piezo effect. Driving crystals, mostly *quartz*, are used to send groups of sound waves through materials to measure the velocity of sound in the material, and hence the elastic constants. Quite small samples can be used for this, as was done in measuring the elastic constants of *diamond*. This has also been done on high-frequency viscosities of liquids. Of late, piezoelectric crystals have been used with intense light sources to produce light harmonics. The laser has provided the intense light and the piezoelectric effect assures the required nonlinearity of the polarization versus electric field relationship. A *quartz* crystal is often used to drive a delay line. Here, a signal from a crystal is sent into a body as an acoustic wave. The wave consumes time in the body before striking another crystal (or the same one again). The signal, now reconverted to electrical energy, can be amplified. This time delay has allowed time for some other action to take place. It can be used for information storage and fed back into the delay-line time after time. A half-and-half sodium niobate–potassium niobate ceramic (PSN) is used extensively in delay lines. Surface waves in bismuth germanium oxide (BGO) show promise for longer time delays. Because BGO is very dense ($\rho = 9.2$) and of low elasticity, waves travel slowly in it.

In 1940, domestic crystal production expanded from \$1.5 million to \$150 million. During World War II, over 30 million crystal plates were made in one year from about 1000 tons of *quartz*. Many tons of *quartz* were brought up from Brazil by ship and plane. Later, it was stockpiled for possible emergency use.

ADP crystals were made by the thousands. Large temperature-controlled rooms were used for holding the hundreds of trays in which the crystals grew, growth being caused by a dropping of the temperature a few tenths of a degree per day. Since these crystals (in fact most crystals) are less soluble in cold water than in warm water, this temperature drop forces crystal growth.

W. L. BOND

References

Cady, W., 1946. *Piezoelectricity*. New York: Dover, 822p.

Mason, W. P., 1950. *Piezoelectric Crystals and Their Application to Ultrasonics*. Princeton, N. J.: Van Nostrand, 508p.

Nye, John, 1957. *Physical Properties of Crystals*. Oxford: Clarendon, 322p.

Tiersten, H. F., 1969. *Linear Piezoelectric Plate Vibrations*. New York: Plenum, 212p.

Voight, W., 1910. *Lehrbuch der Kristal Physik*. Tuebner.

Cross-references: *Crystal Growth; Etch Pits; Quartz; Synthetic Minerals; Twinning; Tourmaline Group.*

PIGMENTS AND FILLERS

Fillers are fine particulate materials added to a substance to modify or enhance its physical properties or to extend more costly or scarce materials. Fillers do not dissolve or react with the host compound. Where color in a product is important, as it usually is, fillers with high whiteness are desired. Thus, fillers also function as coloring matter and in industry are commonly called pigments or functional pigments (see references in Patton, 1973; Brown, 1973). As used here, however, pigments are fine particulate materials added only for the purpose of imparting color, tint, or opacity to a product. Thus defined, a pigment is a one-purpose filler serving only to color.

Some industries have specialized terminology for fillers. For instance, the paper industry, a major consumer of fillers, calls them "fillers" or "coaters" depending upon whether the filler is "loaded" among the fibers of the paper sheet or is used as a surface coating. The paint industry uses fillers to help form the paint film and to extend or spread out more expensive pigments and hence calls them "extenders." The pesticide and fertilizer industries use the inert but absorbent qualities of some fillers to absorb the chemicals and hence call them "carriers." This carrier-chemical combination in turn is diluted for application by fillers known as "diluents."

Although many materials are used for fillers and pigments, only those made directly from rocks, minerals, ores, or gas and oil are discussed here. Most natural mineral pigments and fillers are processed simply by grinding to very fine sizes. In addition, some are calcined, and colored pigments are finished by mixing with white pigments to obtain desired tints.

Other pigments are manufactured from metallic ores by chemical processes. Carbon black, which is used as both pigment and filler, is produced from gas and oil by incomplete combustion or petrochemical methods.

Mining of pigments was one of the earliest mineral industries of primitive man and in America was done extensively by the Indians. Most notable among the mineral pigments are the earth colors produced by iron oxides. Other more exotic mineral pigments used in early times are ultramarine from lapis lazuli (*lazurite*), green earth from weathered basaltic tuffs, and a very unstable blue pigment from *azurite*. According to Ladoo and Myers (1951), alteration of the latter might be responsible for the now-green skies in some old Italian paintings.

Most pigments are now produced synthetically from ores or metals or as by-products of other industrial processes. The only natural mineral pigments produced in quantity today are iron oxides. However, synthetic iron oxides are becoming more important because color and particle size can be more easily controlled in the synthetic process.

Iron-Oxide Pigments

The principal iron-oxide pigments are ochers, siennas, and umbers. Compositions vary widely, but limonite and *hematite* are the major coloring agents. Mixed with these are various quantities of manganese oxide; clay; and, for Vandyke brown, organic material in the form of humates. Table 1 lists the sources and trade names of the natural iron-oxide pigments.

Micaceous iron oxide is a special-purpose pigment used to make anticorrosive paint and is not used for its color. Austria is the principal source; although this pigment is not commonly used in the United States, some of it is imported from Austria.

Most mining of iron-oxide pigments in the United States is on a small scale. The largest production is from mines in Marquette County, Michigan, where it is a coproduct of iron mining. More than 25,000 tons of *hematite* is produced there annually for the pigment, paint, and chemical industries. Other production is from small mining operations mainly in Georgia, Virginia, and Pennsylvania. The deposits in these states are limonite and ocherous material that have been leached from *pyrite*-bearing sedimentary rocks and redeposited in fractures or in residual soil by groundwater. At one time, these deposits were the sites of small-scale iron-mining operations, but now they are worked only for pigment.

TABLE 1. Natural Iron-Oxide Pigment Colors and Sources

Colors	Trade Names	Principal World Sources	United States Sources	Comment
Yellow and yellow-brown	Yellow ocher, Roman ocher, Chinese yellow	South Africa; France	Cartersville, Ga.; Henry, Va.; eastern Pa.	Permanent, opaque.
Red and red-brown	Persian red, Spanish red	Ormuz Island, Persian Gulf; Jaen, Spain	Mather iron mine, Marquette Co., Mich.	Permanent, opaque.
Brown	Raw umber (named for Umbria, Italy)	Cyprus	Imported	Permanent colors; raw umber is calcined to remove organic material and to oxidize the ferrous oxide to Fe_2O_3 and produce burnt umber
Dark brown	Raw umber	Cyprus	Imported	do
Dark yellow-brown	Raw sienna	Sienna, Italy	Imported (ochers from Georgia are sometimes called sienna)	Transparent, therefore can be used for staining; raw sienna calcined to produce burnt sienna
Bright reddish brown	Burnt sienna	do	do	do
Brown-black	Vandyke brown			Contains humates; fades to gray when exposed to bright light

From C. Ervin Brown, "Pigments and fillers," in *United States Mineral Resources,* U.S. Geol. Survey Prof. Paper 820, pp. 527–535.

Market for Natural Iron-Oxide Pigments. Most iron-oxide pigments are now produced synthetically or as a by-product of industrial chemical processes. The synthetic pigments have the advantage of greater consistency of color and composition and thus are strong competitors for mineral pigments.

In 1979, 156,000 short tons of finished

iron-oxide pigments valued at \$94.2 million were sold in the United States. Of these, 73,240 short tons were natural mineral pigments. The demand for synthetic iron oxides is increasing, partly because of their nonpigment use in the production of ferrites, which are used by the computer and electrical industries for their electromagnetic properties.

Prospecting and Field Tests. Almost any ocherous soil, vein, or deeply colored red, yellow, or brown sedimentary rock is potentially an iron-oxide pigment. Most such deposits are diluted with clay and other impurities; consequently their tinting strength is low. A simple empirical test for tinting strength is to grind the material in a mortar and mix it with 10 to 20 times its weight of powdered ZnO. Comparing this mixture with a commercial pigment treated similarly discloses its tinting strength. Dilution with white ZnO is called "letting down" the pigment and is done to obtain desired shades. Another test is to determine the effect of particle size on the color. Each pigment has an optimum particle size, close to 1 μm below which light scattering causes some undertones to be lost. On the other hand, particle size above 2 μm causes a rough paint film. Tests for other properties, such as oil absorption, hiding power, and mass color, are also important, as described by Siegel (1960).

Common Pigments Manufactured from Other Earth Materials

Sphalerite, ilmenite and *rutile*, and fossil fuels are used as raw material in the production of ZnO, TiO_2, and carbon black, respectively. All are major pigments.

Zinc Pigments. The principal zinc compounds used as pigments are zinc oxide, zinc chloride, zinc sulfate, leaded zinc oxide, and lithopone, a coprecipitate of zinc sulfide and barium sulfate. Shipments of zinc oxide and zinc sulfate, produced mainly from *sphalerite* have increased sharply as zinc oxide consumption has increased since 1960. The other zinc compounds are produced mainly from slab zinc and scrap and as by-products.

American consumption of zinc oxide alone was 206,000 short tons in 1979. This represents about 45% of the total American smelter production of zinc. Zinc oxide is used mainly in the manufacture of white rubber products such as white-sidewall tires. Other important uses are the manufacture of paint and photosensitive copying paper.

Titanium Dioxide. Titanium dioxide, a very important pigment, is made from *ilmenite*, from *rutile*, or from titanium slag that is the residue after the smelting of *ilmenite*. All titanium dioxide for use as pigment is recrystallized after chemical treatment of the mineral concentrates. The resulting TiO_2 is synthetic *rutile* or *anatase*, or a mixture of the two. Both have extremely high indices of refraction; hence their very high reflectance and opacity, or hiding power. They are, therefore, especially valuable for use in white paint or as paper coating.

Titanium dioxide for pigment is the principal use of titanium ores in the United States. In 1979, 724,887 short tons of synthetic *rutile* and anatase pigments were produced from *ilmenite* concentrates. This tonnage is about 75% of total titanium production for the United States. The principal consumers are the paint, paper, and plastics industries, which used 47, 22, and 12%, respectively. The remainder was divided among rubber, floor covering, textiles, ink, and ceramics industries, and exports.

Carbon Black. Carbon Black is a nearly pure, micron-sized carbon produced from natural gas or liquid petroleum by incomplete combustion or by thermal decomposition under carefully controlled conditions that determine the particle size, structure, and physical properties. Although carbon black is the main black pigment used in ink and paint, its chief use is as a filler that acts as a reinforcing agent in tire rubber. An average auto tire contains 6–7 lb of carbon black. The plastics industry used only a small amount of carbon black to color plastics until it was discovered that a larger percentage greatly improved the resistance of polyethylene to becoming brittle from sunlight aging. Now as much as 50% carbon black is added to polyethylene cable coverings for this purpose, another example of the overlapping functions of pigments and fillers.

A direct relationship exists between the rate of production of rubber and the consumption of carbon black. American production of carbon black is 1975 was 1.4 million short tons, valued at \$306 million. About 26,200 million cubic feet of natural gas and 517.4 million gallons of liquid hydrocarbons were consumed in its production. A process having an efficiency of only 1–5% and using natural gas to produce "channel blacks" is giving way to much more efficient processes as the price of natural gas climbs.

Miscellaneous Pigments. Small amounts of black pigments are made from *graphite, magnetite*, and carbonaceous slate. *Graphite* and *magnetite* are used in special-purpose paints. *Graphite*, because of its inertness, is used as a protective ingredient in paints used to protect metals in corrosive atmospheres. *Magnetite*, one of the few minerals that grind to a black powder, is an important black pigment for special uses.

Filler Minerals

Almost any pulverized rock can serve some of the many functions of a filler; however, certain rocks and minerals satisfy all of the rigid physical specifications of filler products particularly well. The principal filler materials are *kaolin*, limestone, diatomite, *talc*, asbestos, ***mica***, *barite*, fuller's earth (predominantly *montmorillonite*), *pyrophyllite*, and *wollastonite*. More than 9 million tons of these are consumed yearly in the United States. Many have specific properties, such as the fibrous particle shape of asbestos (q.v.), that make them uniquely suited for certain uses. Others, however, such as *kaolinite, talc*, and *pyrophyllite*, have similar properties and can be substituted for each other for some purposes; the utilization of these depends heavily on availability, transportation, and cost of mining and beneficiation (see *Clays, Clay Minerals*).

Table 2, mainly from Cummins (1960), gives some of the properties of the major filler minerals and rocks.

Industrial Filler Uses and Requirements

Each industry using mineral fillers has its own specific requirements for the fillers it consumes. A general discussion of industrial applications is given here. Much more specific information on the utilization of fillers by industry is given by Severinghaus (1975) and Patton (1973).

Table 3 lists the industries that are the principal users of the major mineral fillers.

Asphalt or Bituminous Compositions. Fillers are used in asphalt composition for road paving, roofing, reservoir linings, tile, joint sealers, battery boxes, and many other products, to increase the viscosity, melting point, resistance to mechanical stress, hardness, or resistance to weathering. In many products the filler is loaded to nearly 50%, and in items such as asphalt tile, filler is the major component. In asphalt products the filler particles form an interlocking skeletal lattice that provides high viscosity and strength. Therefore, fibrous or platy particle shape is a more important property of the filler for asphalt products than are color and smoothness.

Paint. Fillers used in paint are known as extenders because they hold more expensive pigments in suspension and upon drying form the paint film that supports and spreads out the pigment. Here the desirable mineral properties are platy or fibrous particle shape, smoothness and extreme fineness, high reflectance, and low oil absorption. The latter is important because most paints use oil and resins as vehicles; therefore, more extender can be used per volume of oil and drying time is reduced if the oil is not absorbed by the mineral particles. Desirable qualities sought through the proper choice of extender are consistency of film formation, film toughness, resistance to weathering, flatness, opacity, and a reduction of costs through decreased use of the prime pigment. Generally, a mixture of extenders is used in order to get the desired paint qualities. With the increased use of water-based latex paints, some consideration is given to the alkalinity of the filler in order to increase corrosion resistance. Most fillers are inert, but some, such as *wollastonite*, produce alkaline suspensions, which are desirable to decrease "rusting through" of nail heads and rusting of paint cans and which are compatible with pigments that would react if used in low-pH suspension. Uniform fineness of particles is extremely important for paint extenders, because coarse particles show on thin paint films. For most uses, fillers must pass a 325-mesh screen; however, as paint extenders they are ground to micron sizes, and techniques to grind particles even finer are actively sought.

Plastics. The plastic floor tile industry has long been a major consumer of asbestos and other fillers. Inasmuch as the plastics industry is rapidly expanding and developing new products, many more types of mineral fillers are being used. The consumption of *talc, wollastonite*, and other fillers is quickly expanding for this use because of the strength, electrical and heat resistance, and whiteness of these fillers. Carbon black is used to increase resistance of polyethylene to sunlight aging although it was used initially as a pigment. Properties desired from fillers in plastics are color, strength, electrical and heat resistance, increased melting point, stiffness, decreased brittleness, and low shrinkage. Many plastic products contain nearly 50% filler, which is responsible for most of the physical properties of the products.

Rubber. The rubber industry is a major consumer of fillers, which function as reinforcing agents to increase tear resistance, abrasion resistance, stiffness, heat dissipation, and electrical resistance and to intensify color. Much research has been done by the rubber industry to understand the actual function of filler particles in rubber because the amount, grain size, and type of fillers have a profound effect on the physical properties of rubber. They obviously do much more than simply occupy space. Particle size requirements for rubber fillers demand nearly 100% passing

TABLE 2. Some Properties of the Major Mineral Fillers

	Theoretical Chemical Composition	Sp. gr.	Bulk Density (lb/ft³)	Hardness (Mohs Scale)	Refractive Index	Reaction (pH)	Oil Absorption (cc/100 g)	Particle Characteristics
Asbestos (*Chrysotile*)	$Mg_3Si_2O_5(OH)_4$	2.5–2.6	10–40	2.5–4.0	1.51–1.55	8.5–10.3	40–90	Fibers fine, easily separable
Barite	$BaSO_4$	4.3–4.6	80–100	2.5–3.5	1.64	7	6–10	Generally equidimensional; heaviest of the major fillers
Bentonite (*montmorillonite*)	$(Na,Ca)_{0.33}(Al,Mg)_2Si_4O_{10}(OH)_2 \cdot nH_2O$	2.3–2.8	50–60	1.5+	1.55–1.56	6.2–9.0	20–30	Porous microaggregates, irregular shapes; ultimate plate structure
Carbon black	Fixed carbon, 85–96%	1.7–1.8	10–15	–	–	6–6.8	30–90	Quasigraphitic C particles, 50–5000 Å size
Diatomite	$SiO_2 \cdot nH_2O$	2.0–2.35	6–20	4.5–6.0	1.42–1.49	6–8.5	100–300	Unique diatom structure; micro and ultramicro porosity
Fuller's earth (*montmorillonite*)	$(Na,Ca)_{0.33}(Al,Mg)_2Si_4O_{10}(OH)_2 \cdot nH_2O$	2.2–2.6	27–38	4	1.50	7.5–8.2	30	Apparently equidimensional; electron-microscopically fibrous, lath-like
Gypsum	$CaSO_4 \cdot 2H_2O$	2.3	25–40	1.5–2.0	1.52	6.5–7	17–25	Irregular, roughly equidimensional
Kaolinite	$Al_2Si_2O_5(OH)_4$	2.6	20–40	2.0–2.5	1.56–1.58	4.5–7	25–50	Thin, flat hexagonal plates, 0.05–2 μm size and stacks of same
Limestone (*calcite*)	$CaCO_3$	2.7	40–60	3	1.63–1.66	7.8–8.5	12–30	Variable-size particles; ultimate rhombs
Mica (*Muscovite*)	$KAl_2Si_3O_{10}(OH)_2$	2.7–3.0	12–20	2.0–3.0	1.59±	7.4–9.4	25–50	Plate-like particles
Perlite	Like rhyolite	2.5–2.6	4–20	5.0	1.48–1.49	9	50–275	Expanded "glass" bubbles and fragments
Portland cement	Essentially Ca silicates	2.9–3.15	90–100	5.6	1.72±	11.0–12.6	20	Variable, smooth, rounded, angular and flake particles
Pumicite	A silicate, like rhyolite	2.2–2.63	40–50	5–6	1.49–1.50	7–9	30–40	Vesicular
Pyrophyllite	$Al_2Si_4O_{10}(OH)_2$	2.8–2.9	25–30	1–2	1.57–1.59	6–8	40–70	Minute foliated plates or scales and crystalline particles
Rock dusts	Variable	2.6–3.3	50–100	4–6.5	Variable	Usually >7	20–40	Variable
Silicas, crystalline	SiO_2	2.60–2.65	50–80	6.5–7.0	1.53–1.54	6–7	20–50	Variable sized, angular and equidimensional particles
Slate	Mixture of mineral silicates	2.7–2.8	40–80	4–6	–	6.8	20–25	Flat or wedge shaped, or spherical grains
Talc	$Mg_3Si_4O_{10}(OH)_2$	2.6–3.0	26–60	1–1.5	1.57–1.59	8.1–9.0	20–50	Lamellar, foliated, or microfibrous
Vermiculite	$(Mg,Fe^{2+},Al)_3(Al,Si)_4O_{10}(OH)_2 \cdot 4H_2O$	2.2–2.7	6–10; fines–20	1.5	1.56	Pract. neutral	–	Platelets or lamellar structure
Wollastonite	$CaSiO_3$	2.9–3.1	30–95	4.5–5	1.6–1.65	10	20–26	Acicular

Adapted from A. B. Cummins, "Mineral fillers," in *Industrial Minerals and Rocks,* 3rd ed., Am. Inst. Mining, Metall., and Petroleum Engineers, New York, 1960, pp. 567–587.

TABLE 3. Distribution of Industrial Fillers Use by (M, major use; m, minor use; p, used as pigment)

Filler	Use: Paint	Rubber	Paper	Fertilizer	Pesticides	Plastics	Roofing Compounds	Flooring Compounds	Other Filler Uses	Filler Consumption as Percent of Total Commodity Consumption (Approx.)	Remarks
Asbestos						m	m	M	m	25	Filler uses take mainly "fines" and "shorts"
Barite	M	M	m					m	m	5	About 80% of the world's annual production is used for heavy drilling mud
Bentonite and other clays except *kaolinite* and Fuller's earth					M				m	1	Not an important filler; mainly an insecticide diluent
Carbon black	p	M	p			m,p				95	Excellent rubber reinforcing filler; also a pigment
Diatomite	m		m	m	m	m			m	20	Used in raw state, milled or calcined for fillers; Large reserves. important filter aid
Fuller's earth					M				m	20	Important thixotropic and absorptive properties, mainly used for sweeping compound and pet litter, rotary drilling muds, and decolorizing agent
Gypsum	m		m			m				<5	Minor filler mineral; used raw or calcined
Kaolinite	m	M	M	m	m	m			M	65	Most important filler mineral; principal paper filler; important rubber filler
Limestone	M	M	m			M		M	M	5	Cheapest filler; has low oil absorption and good color; main filler for putty, caulking, sealants, and joint compound for wallboard; important filler
Mica, ground (includes sericite schist)	M	M	m		m	m	M		m	60	Scrap ***mica,*** including some *phlogopite* and *biotite* (Mica and sericite schists are large resources.)
Perlite	m					m			m	1	Not an important filler; mainly used for plaster and concrete aggregate and filtering aid
Pyrophyllite	M	m			m		m		m	65	Uses and properties similar to *talc*
Silica, ground	M	m			m	m			m	<1	For paint extender in deck paints, for high abrasive use
Talc	M	m	M	m	m	m	M	m	m	40	Some grades contain much *tremolite, anthopyllite,* or ***serpentine;*** important filler
Wollastonite	M					M		m		25	Mainly used in ceramics; newest important filler mineral

through a 325-mesh screen. Grit particles are undesirable as they tend to cause tearing and excess wear of cutting and extruding equipment. The percentage of fillers depends upon the desired quality of the finished rubber product, and a wide range up to more than 50% is common. Color is important for white rubber products, and *barite*, zinc oxide, and titanium dioxide are used with *kaolinite*. *Barite* is also used to increase specific gravity for uses such as the manufacture of rubber mallets. Carbon black is the major reinforcing pigment used by the rubber industry. *Kaolinite* also is an important rubber filler and is used to increase hardness and abrasion resistance. Very fine-grained *kaolinite* produces hard rubber and hence is known as "hard clay." "Soft clay," a coarser *kaolinite*, allows rubber to remain soft.

Paper. Another major consumer of fillers is the paper industry. This industry uses mineral fillers in two ways, one as a filler or loader, in which filler mixed with bonding agents fills the voids between the felted cellulose and textile fibers that are rolled out as the paper sheet. The other filler use is as a surfacing material or coating, where whiteness, opacity, and gloss are desired. Pigments and mineral fillers—for example, flux-calcined diatomite and TiO_2—are often used in combination for coating. In this combination, the diatomite is a pigment extender. *Kaolinite* is by far the principal mineral used for coating and loading paper.

Pesticides. Mineral fillers are used as diluents and carriers for insecticides and fungicides. Because the toxic elements cannot be distributed efficiently or effectively in the concentrated form, they must be diluted in liquids or dusts to ensure proper distribution and a concentration that will not have deleterious effects on the host, yet will destroy the harmful organisms. The toxic chemicals are first mixed with a mineral powder known as a carrier to make a concentrate that is simple to store, ship, and handle prior to use in the field. Mineral fillers suitable for carrier use, such as diatomite, fuller's earth, and some clays, are highly absorptive. These must be dispersable in liquids for sprays and must be inert, and compatible with the toxicant involved. Fineness and softness are important to avoid clogging and excess wear of nozzles.

Carrier-pesticide mixtures are usually further diluted for application from aircraft or ground equipment. The filler used for this application is known as a diluent. It must be of fine particle size and have low abrasiveness but need not be highly absorptive like the carrier. A good pesticide carrier also acts as a grinding aid in preparation of some pesticides and as a conditioning agent for the diluent. The resulting dust must be free-flowing, but capable of sticking to the surface of plants to provide maximum exposure to the harmful organisms.

One interesting aspect of diluents and carriers is that some mineral dusts kill some insects and larvae even without the toxic chemicals. Apparently these mineral dusts clog the digestive and respiratory systems.

Other Uses. Much mineral filler is used as a carrier and diluent for fertilizer, for most of the same reasons that it is used in pesticides. Fillers also have miscellaneous uses in adhesives, joint cement, and many other products, each of which demands specific physical properties of the filler.

C. ERVIN BROWN

References

Andrews, R. W., 1970. *Wollastonite*. London: Institute of Geological Sciences, 114p.

Brown, C. Ervin, 1973. United States mineral resources: Pigments and fillers, *U.S. Geol. Surv. Prof. Paper 820*, 527-535.

Cummins, A. B., 1960. Mineral fillers, *Industrial Minerals and Rocks*, 3rd ed. New York: American Institute of Mining, Metallurgical, and Petroleum Engineers, 567-584.

Hancock, K. R., 1975. Mineral Pigments, *Industrial Minerals and Rocks*, 4th ed. New York: American Institute of Mining, Metallurgical, and Petroleum Engineers, 335-357.

Industrial Minerals, 1968. Talc—mineral with a multitude of uses. (London) *Indus. Minerals*, no. 5, February, 9-16.

Kollonitsch, Valerie; Kliff, E. F.; and Kline, C. H., 1970. Functional pigments, pt. 1. (London) *Indus. Minerals*, no. 38, November, 35-40.

Kollonitsch, Valerie; Kliff, E. F.; and Kline, C. H., 1970. Functional pigments, pt. 2. (London) *Indus. Minerals*, no. 39, December, 46-51.

Ladoo, R. B., and Myers, W. M., 1951. *Nonmetallic Minerals*, 2nd ed. New York: McGraw-Hill, 605p.

Patton, Temple C., ed., 1973. *Pigment Handbook*, vol. 1, *Properties and Economics*. New York: Wiley, 987p.

Schreck, A. E., ed., 1977. Metals, minerals, and fuels, *1975 Minerals Yearbook*. Washington: U.S. Bureau of Mines Minerals Yearbook, vol. 1, 1550p.

Severinghaus, Nelson, Jr., 1975. Fillers, filters, and absorbents, *Industrial Minerals and Rocks*, 4th ed. New York: American Institute of Mining, Metallurgical, and Petroleum Engineers, 235-249.

Siegel, Alfred, 1960. Mineral pigments, *Industrial Minerals and Rocks*, 3rd ed. New York: American Institute of Mining, Metallurgical, and Petroleum Engineers, 585-593.

PLACER DEPOSITS

The word "placer" was used by early Americans, and others, to designate gold-bearing deposits of stream gravel. This term is thought to have been derived from the Spanish *plaza*, meaning "place" or, locally, "sand bank." The Germans used the word *Seife*, or *Erzseife*, for a placer type of mineral deposit, while the British, Australians, and South Africans, among others, referred to placers simply as "alluvials" or "alluvial deposits." In Malaysia, tin placers are called *kakas* by the Chinese. In 1895, Posepny proposed that the name "hysteromorphous deposits" be applied to placer concentrations. He called the minerals "hysterogenites" (secondary or later deposits). Because of widespread general use and new breadth of coverage, and in the interests of simplicity, the term "placer" should be used *to designate a natural residual or transported concentration of potentially valuable and recoverable detrital minerals or gemstones, which include sand and gravel, and often contain mixtures of clay, silt, boulders, and vegetation.* An "economic placer deposit" is one that contains marketable mineral materials recoverable in sufficient quantities, at a cost that will yield a profit. A blacksand placer is one in which dark minerals predominate (see *Blacksand Minerals and Metals*).

Distribution of Placers

Nearly every country in the world has some kind of workable placer deposits that produce gemstones or one or more minerals containing elements such as platinum, gold, gold with mercury (*amalgam*), mercury, tin, uranium, thorium, chromium, iron, niobium (columbium), tantalum, titanium, tungsten, and zirconium. This is why placer minerals and gemstones annually account for a substantial return to the world's mining industry, particularly if one includes the large returns received for commercial sand, gravel, and other nonmetallic mineral materials, commonly classified by some governments as placer commodities.

Excluding sand and gravel, placer deposits yield, have produced, or have reserves of, small to large quantities of *gold* in countries too numerous to mention (see *Blacksand Minerals and Metals*); tin in the states and territories of Malaysia and Indonesia, Thailand, Burma, China, Nigeria, Zaire, and the United States (Alaska); *platinum* and its associated elements at localities in the USSR, Colombia, Ethiopia, and the United States (Alaska and Oregon); industrial *diamonds* in South-West Africa (Namibia), Republic of South Africa, Sierra Leone, Ghana, Zaire, Angola, Guyana, Australia, and the United States; and zirconium and ***garnet*** in widespread world occurrences. Moderate to small placer deposits of niobium (columbium) and tantalum occur in Guyana, Nigeria, and the United States (Idaho); titanium and thorium occur in Australia and the southeastern and northwestern United States; tungsten placers occur in China; uranium in small quantities occurs in many placers on a worldwide basis; and commercial placers are known in many parts of the world that include gemstones such as rubies, sapphires, ***tourmalines***, agates, jade, *zircons, diamonds,* ***garnets***, and emeralds.

New placer deposits continue to be discovered. For example, in 1963 the *Mining Journal Review* of London cited reports of several new discoveries. Relatively new tin deposits are present in Queensland, Australia. They contain an estimated 100 million cubic yards of tin-bearing gravel of as yet undetermined grade. Over 36 million tons of iron mineral placer, averaging 55.6% iron and 11.64% titanium dioxide, have reportedly been discovered lying in Ariake Bay, off the coast of Kyushu, Japan. South Africa developed one of the world's first seaborne *diamond* mining units, designed to recover widely occurring deposits of diamondiferous gravel lying off the coasts of South Africa and Namibia. A rich *diamond* placer is reportedly present along the Caroni River near Cuidad Bolívar in Venezuela; while a second *diamond* deposit was reported in the Mato Grosso of Brazil, near Chapado dos Guimarces.

Ages of Placers

Most commercial placers are relatively young because they were developed on the earth's present topographic surface, that is, during the last 70 million years (Tertiary to present). However, there are good reasons for assuming that the heavy-mineral elements in these placers were originally uniformly dispersed through the earth's crust (Savage, 1961). Therefore, the first natural concentrations of detrital minerals must have been the result of weathering, erosion, and transportation more than 3 billion years ago (in the Precambrian). Furthermore, some of the earth's older placer deposits must have been subjected to physical and chemical changes brought about by diastrophism and metamorphism, which have altered the contained minerals. Some of these minerals probably were redispersed as accessory minerals in bodies of igneous and metamorphic rocks while others may have been reconverted or remobilized and emplaced in the bedrock as veins and lodes. The emplacement of workable ore as veins, lodes, and placer deposits could have been the result of repeated episodes of mineral concentration, dispersal, and reconcentration by means of degradational and aggradational

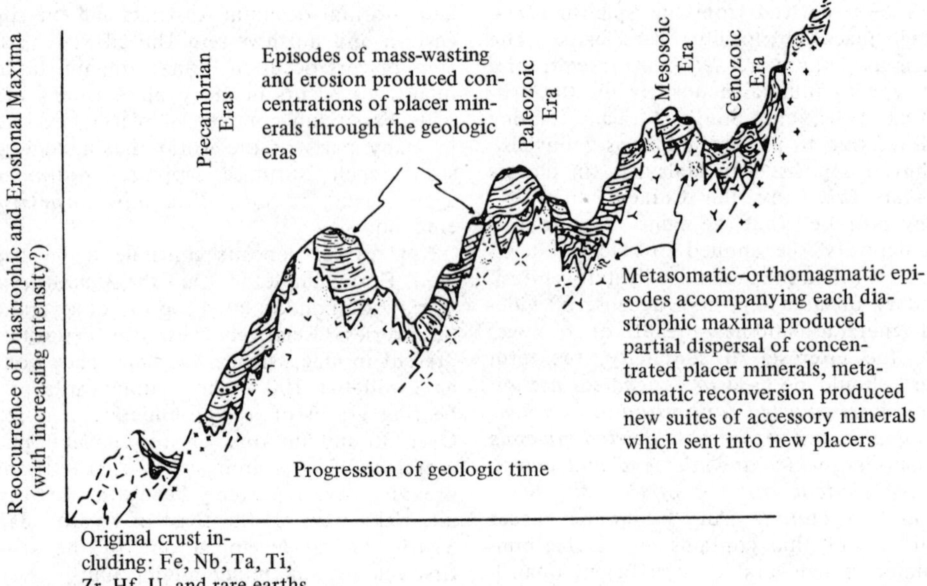

FIGURE 1. Diagrammatic representation of metasomatic reconversion or regeneration and reconcentration of detrital minerals (after Savage, 1960).

processes, and by metasomatic reconversion (Fig. 1). Such an explanation makes it easier to understand the tendency both for bedrock and for placer mineral deposits to be located near belts of persistent crustal unrest. It reveals why our economic mineral deposits are so often found where the many geological processes have worked long in their accumulation. This knowledge in turn can be used to advantage in mineral exploration. It must be reemphasized, however, that contemporary placer deposits of commercial value are chiefly those that have accumulated during the most recent, long erosion of the earth's crust. Furthermore, the only reason some areas have existing economically valuable *gold* accumulations is that widely scattered *gold*-bearing *quartz* veins have been weathered and eroded to contribute the *gold* to placer accumulations. Commonly, the widely dispersed *gold* in such *quartz* veins cannot be recovered economically by lode mining methods.

Regardless of early mineral origin, because placer accumulations are recent products of the geomorphic agencies, the immediate source of the minerals or gemstones therein contained must be either bedrock or previously existing, reworked placers. Some valuable, young gemstone placers apparently have been derived from the erosion of preexisting mineral pipes, lodes, veins, vugs, and so forth, contained in the local bedrock. In other occurrences, a broad range of mineral commodities are derived from swarms and widely dispersed, small, accessory mineral grains within superficially common-looking bedrock.

A study of major published works and field experience suggest that Table 1 offers a good methodology for placer classification. Sometimes detrital mineral products have not traveled far from their source before they are concentrated; such accumulations may be called "fixed-residual placers" or "in-transit placers" (Table 1). Some detrital minerals have traveled great distances before being deposited; very fine minerals called "flour" or "dust," for example, are known to have been transported over 1000 miles from their probable source before coming to rest in a placer trap (Snake River from Wyoming to the Pacific Ocean). All mineral accumulations that have been transported some distance from their mineral source may be called "trapped-in-transit placers." Of all the minerals contained in average sand and gravel placers, normally, less than 5% by weight consist of concentrated heavy minerals (Savage, 1961).

Conditions Favoring Concentration

At the earliest stages, when minerals are being concentrated, original mineral characteristics are very important. These include such things as hardness, texture, chemical composition,

TABLE 1. Classification of Detrital Mineral Placers and Related Deposits

I. FIXED-RESIDUAL PLACERS: Developed in any environment, directly above mineralized bedrock, veins, and lodes by mechanical and chemical weathering, eluviation, and wind action.
  1. *Eluvial placer* ("root deposit" or "seam digging")–Heavy, resistant minerals concentrated in place by eluviation and gravity (Fig. 2).
  2. *Eluvium placer* ("dry placer")–Minerals concentrated in place by wind and gravity (Rickard, 1899) (Fig. 2).

II. IN-TRANSIT PLACERS: Located on slopes. Transitional between fixed-residual and trapped-in-transit placers. Heavy minerals are colluvium, in-transit in a semi-mobile mixture; minerals are being reconcentrated and sorted by mass-wasting and slopewash (see Vol. III, *Mass-Wasting*).
  1. *Colluvial slope placer* ("hillside placer")–Mixture of alluvium and angular fragments in an environment of mass-wasting and slopewash. May occur in humid to semiarid climate (Fig. 2).
  2. *Colluvial gulch placer*–Colluvium accumulated at the heads of small, steep-walled gullies or creeks. May occur in any climate (Fig. 2).
  3. *Colluvial bajada placer* ("dry bajada" placer)–Largely angular fragments accumulated along a break in slope. Principally in arid regions (Weber, 1935).

III. TRAPPED-IN-TRANSIT PLACERS: Developed in a variety of environmental sites, but most important are fluviatile and marine placers. Resorted and concentrated heavy minerals deposited by numerous agencies; variously interrupted in-transit.
  1. *Fluviatile deposit placer*–Heavy minerals concentrated because of changes in the velocity of running water. Alluvial deposits (Fig. 3).
    a. *Stream bed, bar,* or *floodplain placer* ("creek placer"): Deposits result from changes in load capacity.
    b. *Stream terrace placer* ("high-" or "low-bench" placer): Remnants of former bed, bar, or floodplain placers; the result of a change in stream regimen.
    c. *Fan, cone,* or *delta placer:* Interruptions in stream capacity because of an abrupt change in gradient or velocity (Fig. 2).
    d. *Channel-fill placer* (type of "buried placer"): Former placer-bearing stream channels filled by sediments, lava flows, ash falls, etc.
    e. *Paludal placer:* Swampy traps filled by sediments.
    f. *Piedmont plain placer:* Placer deposits concentrated by irregular anastomosing and braided streams.
    g. *Glaciofluvial placer:* Any placer concentration that may be identified as being in materials deposited by glacial meltwater. Similar to other alluvial placers, except that the deposits are usually lean in minerals, and show evidence of heavy loads and strong stream action.
  2. *Marine* or *littoral deposit placer* –Heavy minerals concentrated by beach processes (Fig. 4) including wave, current, swash, and rip currents ("undertow"). Materials include alluvium and sometimes colluvium (Pardee, 1934).
    a. *Marine beach, submergent beach, emergent* ("raised") *beach* and *marine terrace* ("bench") *placer:* Marine beach environment and processes involved in mineral concentration.
    b. *Marine longshore* or *bar placer:* Offshore tidal and current action.
  3. *Lacustrine deposit placer*–Heavy minerals concentrated somewhat similarly to marine placers by beach processes and also by stream inwash. Largely alluvium.
    a. *Lake bed,* or *lake plain placer:* Formed by variable bottom currents and stream inwash, usually lean in minerals.
    b. Lake beach, and *lake terrace* ("raised" beach or "bench") *placer.*
  4. *Estuarine deposit placer*–Characteristics of both lacustrine and alluvial placers. Deposited by inwash in salt or brackish water.
  5. *Eolian deposit placer*–Result of wind action in any climate.
    a. *Desert plain-* or *plateau-lag deposit placer:* Concentrated not far from bedrock sources, chiefly by strong wind action, very little stream work (Rickard, 1899). Very old surfaces (Fig. 2).
    b. *Beach* and *floodplain eolian lag placer:* Humid to semiarid, or glacial climates. Wind agencies responsible for concentration.
    c. *Dune placer:* Concentration of "pay streaks" because of interference with the wind velocity and transporting power (Fig. 2). Relatively small concentrations of heavy minerals.

IV. ARTIFICIAL TRAPPED-IN-TRANSIT PLACERS: Alluvial-type placers occurring in mine and dredge tailings. They present special drill testing and mineral recovery problems (Savage, 1961, 116–119).

Note: Various types of placer are referred to as "recent" or "ancient" to differentiate between more recent and older, often "buried," placers; that is, ancient deposits may represent earlier geomorphic "cycles." Burial may be accomplished by lava flows, volcanic ash, or stream deposits, etc. However, few "ancient" deposits are older than Tertiary.

and density—all important to the separation of the minerals from their rock matrix. Of almost equal significance are size, shape, and degree of roundness of the mineral grains. Usually soft, brittle, or fractured dense minerals and large-sized fragments will be found in concentrations close to their bedrock sources while hard, more resistant, smaller mineral particles will be farther from their source. Relatively dense minerals are the ones most commonly found in placer concentrations. Minerals must be fairly hard in order to resist the attritional environment of transportation. Brittle minerals break up faster than tough minerals (see *Jade*). Minerals must also be resistant to solution in waters naturally charged with chemicals from the environment, particularly mild acids.

Placer concentration of minerals is especially favored by the right climatic and topographic environments and is closely related to changes in the energy systems of the transporting media and physical environments. Placers are more likely to develop in temperate climates where the topography is moderate to low in relief, where geologic history has included crustal unrest and igneous activity, and where climatic extremes have included wet, dry, and glacial phases. Deep weathering and disintegration of bedrock is most helpful to placer development. Glacial ice is not a good "concentrating" agent, but glaciation and meltwater runoff are effective agents in the preparation of large quantities of rock debris from which the heavy-mineral content later may be effectively "winnowed" out and concentrated by other geologic agents.

Lithologic and structural characteristics that weaken local bedrock are also important to rock disintegration and its subsequent removal by erosion. Later, the winnowing of the detrital minerals from other rock debris is effected by the erosion and transportation.

Concentration of detrital minerals begins just as soon as they are freed from the bedrock by weathering. Gravity and local relief are notable factors in starting the movement of mineral-bearing rock materials on the way to later concentration. The principal agency that separates and sorts minerals is their movement in a water medium. Running water is a major winnowing agent, while waves, currents, and wind action are effective to a lesser degree. Free mineral particles shift and move vertically to be concentrated in place, or they will move horizontally downslope to be concentrated in some sedimentary trap. Placers may almost always be considered temporary deposits in terms of the vastness of geological time scales. Even though placers may remain entrapped or buried for several thousand years, drainage changes will ultimately result in renewed degradation and subsequent removal of previous placer deposits. In some localities spring floods have been known to transport out or to bring in placer minerals. In one reach of the Snake River in Idaho, just below its confluence with the Salmon River, a suction dredge is recovering *gold* from eroded traps in the bedrock. Several sections have been gone over more than once because new placer is being deposited annually. On the basis of the above concept of mineral dispersal and reconcentration through geologic time, it is very likely that some deeply buried placers, in the past, have been caught up in diastrophic phases of earth history and subjected to metamorphism or metasomatism (Fig. 1).

The best developed placer ground is found where relatively sudden changes in environmental setting have been followed by less rapid change during which the erosional agents develop more "normal" energy regimens (Figs. 2, 3, and 4). Some of these sudden changes may be the result of geomorphic "catastrophes," that is, relatively sudden crustal uplift or subsidence, or eustatic or isostatic changes of water level. These major events may precede

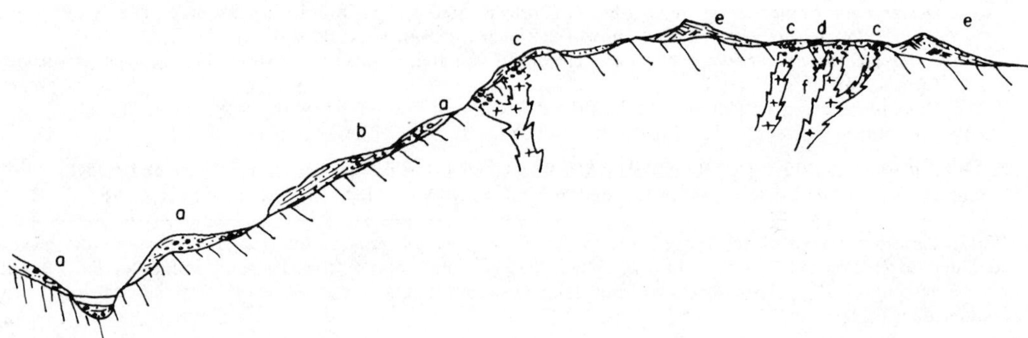

FIGURE 2. Eluvial, eluvium, and colluvial placers. (a) Colluvial slope and gulch; (b) fan; (c) eluvial; (d) eluvium; (e) dune; (f) heavy mineral-bearing dikes.

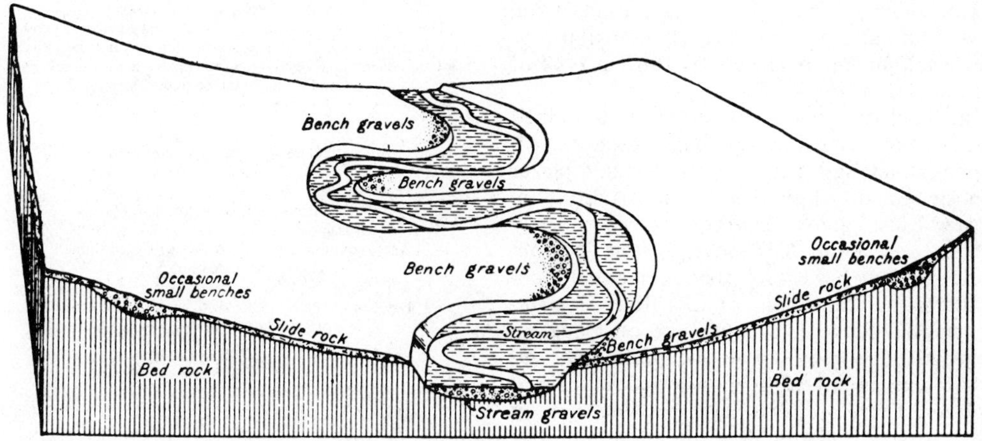

FIGURE 3. Fluviatile placers (from Collier et al., 1908).

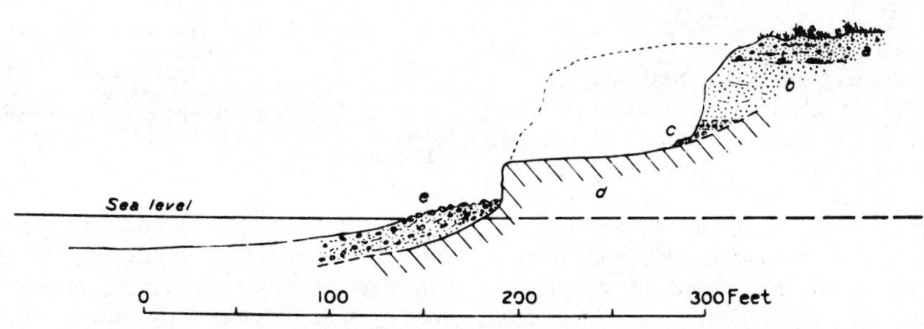

FIGURE 4. Marine or littoral placers (from Pardee, 1934). (a) Sandy soil; (b) gray sand, windblown?; (c) marine sand and cobbles, terrace placer; (d) bedrock; (e) beach placer.

actual placer development. Abrupt changes in the relative resistance of bedrock that is undergoing erosion may also lead to changes in the rate of denudation and, thereby, favor detrital mineral concentration.

Regional and local drainage derangements are probably most effective in producing placer deposits—for example, derangements that produce centripetal drainage around poorly drained piedmont plains, fan and delta deposits, terraces ("benches") and strandlines, and beaches and bars. Also suggesting situations favorable to heavy-mineral concentration is the presence of glaciofluvial and morainal deposits, sand dunes and lag gravels, prograded and retrograded shorelines, stream piracy and barbed tributaries, long narrow basins above narrow stream gorges, and swampy flats.

Decreases in the velocity of motion, or the cessation of motion of streams, currents, and winds may result in the accumulation of heavy minerals. Gradients where these fluid agents are at work, for example, may change rapidly from an upland to a lowland surface, or stream flow may be interrupted because of natural dams. Meandering decreases stream gradient by lengthening stream profiles; therefore, meandering commonly is accompanied by deposition of heavy-mineral grains in slack-water traps. Obstacles along a stream bed such as boulders, joint faces, resistant beds and dikes, bars, sharp bends in the channel, and vegetation may also cause slack-water deposition (Fig. 3; see also *Blacksand Minerals and Metals*, Fig. 2). The occurrence of floods, droughts, and stream diversions in a drainage system may cause deposition of sediments because of variation in stream velocity and flow. Almost any excessive load of stream detritus or a decrease in runoff will normally produce deposition.

Finally, when detrital minerals begin to accumulate in a placer trap, gravity further aids concentration by causing dense particles to work or settle downward. This settling is hastened by any agency that churns or "puddles" the accumulating debris, such as water or wind current action including "dust devil" turbulence and wave swash. Removal of less dense and finer particles by any winnowing action also increases the concentration of the heavier minerals that tend to lag behind. Wind, wave, and current action may all contribute to the development of placers by this lag type of concentration (Fig. 4).

The combined work of the above-described agencies of concentration may produce enriched zones or "pay streaks" in any placer deposit. Usually these richer concentrations are located just above bedrock; however, pay streaks frequently lie above a false bottom made up of hard silt, clay, cemented conglomerate, volcanic ash, or lava flow. Depending upon the geological history of a region, there may be more than one productive enriched level in any vertical or horizontal part of a placer. Therefore, one may conclude that, by the nature of placer deposition, commodity mineral values are erratic in their distribution.

Commercial Recovery

Commercial minerals have been recovered from placer deposits by many methods, most of which involve an attempt by man further to separate and to sort mineral concentrations that were begun by nature. It is possible effectively to separate about 65 to 75% of the detrital minerals from placer beds (W. W. Staley, oral communication). Such artificially concentrated minerals may be further sorted or "classified" with mechanical equipment into individual mineral types, in accordance with the different physical and chemical properties of the individual mineral grains. In general, mechanical concentration of minerals by man is accomplished by gold pan, batea, dulang, and plaque (Fig. 5); rockers, sluice boxes, long "toms," surf washers, dry washers, dip boxes, and puddling boxes; amalgamation; dragline and washing plant; suction dredging; jig plant; dredging boat; and electrostatic, magnetic, and hydraulic separation (Gardner and Johnson, 1934–35; Daily, 1973; Popov, 1971; Thomas, 1973; United Nations, 1972; and Wells, 1969) (Tables 2 and 3).

The accompanying list of selected references aspect of placer deposits and mineral recovery. They include information on classification, definitions, occurrence, prospecting methods, placer mineral suites, development or recovery methods, placer mineral economics, and environmental constraints. Popov (1971) describes and discusses large dredging and hydraulic recovery of placer minerals and recovery of buried, ancient placers by underground mining techniques. The United Nations (1972) publication discusses small-scale placer mining by artisanal and primitive recovery methods still found in some of the developing nations.

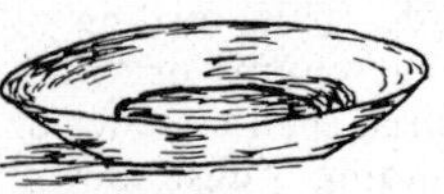
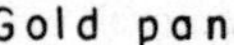

FIGURE 5. Pans.

TABLE 2. Prospecting Methods
The method most suitable for prospecting a placer depends on the type and characteristics of the deposit, and the valuable mineral of interest. The table summarizes the proven methods applicable to inland deposits.

Metallic Minerals–Au, Sn, Pt

I. Deposits other than beach sand, thawed ground
  A. Depth to about 15 ft
    1. Dry: Hand-dug pits
       Power digger, square or cylindrical bucket
       Occasionally placer churn drill, eased hole
    2. Wet: Hand-dug pits with caisson or lagged lining
       Placer churn drill, eased hole
  B. Depth 15–200 ft
    1. Dry or wet: Placer churn drill, eased hole
       (Rarely open hole)
       Becker percussion drill

II. Deposits other than beach sand, frozen ground
  A. Depths to 200 ft or more
     Placer churn drill, open hole
     Hand-dug shafts, usually little or no lagging
     Becker percussion drill–not yet proven (1969)

III. Beach sand deposits (at or above present sea level)
  A. Depth to about 15 ft–greater depth rarely encountered
    1. Above water level: Hand auger
       Occasionally placer churn drill
    2. To below water level: Hand auger with pipe easing
       Placer churn drill

A. F. Daily, 1973. Placer mining, in *SME Mining Engineering Handbook*, 2. New York: Society of Mining Engineers, A.I.M.E., p. 17-154. Reprinted with permission.

The classification of placers used herein (Table 1) has been developed and derived from many sources with modifications. It is intended to be inclusive, comprising common terms used all over the world. The broad designation "placer" is intended to encompass all the different terms used for detrital minerals accumulated by means of natural processes and agents.

C. N. SAVAGE

References

Collier, A. H., et al., 1908. The gold placers of parts of Seward Peninsula, Alaska, *U.S. Geol. Survey Bull.*, **328**, 343p.

Daily, A. F., 1973. Placer mining, *SME Mining Engineering Handbook*, vol. 2. New York: Society of Mining Engineers, A.I.M.E., 17-151 to 17-180.

Gardner, E. E., and Johnson, C. H., 1934–35. Placer mining in the Western United States, pts. I and II (1934), and pt. III (1935), *U.S. Bur. Mines Info. Circs., 6786, 6787*, and *6788*, 75p., 91p., and 83p.

Jenkins, O. P., 1935. New techniques applicable to the study of placers, *Calif. J. Mines Geol.*, **31**(3), 143–210.

Pardee, J. T., 1934. Beach placers of the Oregon Coast, *U.S. Geol. Survey Circ.*, 8, 41p.

TABLE 3. Placer Mining Methods by Method of Excavation

1. Land-based plant
   a. With hand tools and minor auxiliary equipment
      Drift mining
      Shallow open-cut surface mining
   b. With bucket scrapers and wire line
   c. With mobile powered equipment:
      Dozers
      Dragline
      Shovel plus trucks
      Small bucket-wheel excavator plus trucks
      Large powered scrapers plus dozers
   d. With water:
      Ground sluicing
      Hydraulicking
2. Floating plant
   a. Dragline and floating washing plant, "doodlebug"
   b. Bucket-line dredge
   c. Hydraulic dredge, barge- or ship-mounted

The principal methods currently being employed, the minerals being mined, and localities are listed below (except USSR).

1. Hand mining with auxiliary equipment:
   Large-scale for *diamonds* in Africa
   Small-scale for *diamonds* in South America and a few other places
   Small-scale for *gold* and a few for tin and other minerals in most continents
2. Mobile powered equipment:
   Very large-scale for *diamonds* in Africa
   Small-scale for *gold* and a few for tin in many countries; *rutile* in Sierra Leone
3. Bucket-line dredging:
   Very large-scale for tin in Southeast Asian countries
   Large-scale for *gold/platinum* in Colombia
   Small-scale for tin in Australia and Bolivia
   One (or more) dredges for *gold* in Ghana, Brazil, Bolivia, Peru, Greece, South Korea, Alaska; one for *platinum* in Alaska; two for *diamonds* in Brazil
4. Suction dredging and powered equipment mining of beach sand placers for industrial minerals:
   Large-scale in Australia, Florida
   Small-scale in Brazil, India, South Africa, Philippines
5. Off-shore suction dredging:
   Two operations for *diamonds* in South-West Africa (both terminated early 1971)

Note: Only a brief outline of this subject can be given because of space limitations. Comprehensive detailed information on most methods can be found in publications listed in the bibliography and further references in them.

A. F. Daily, 1973. Placer mining, in *SME Mining Engineering Handbook*, 2. New York: Society of Mining Engineers, A.I.M.E., p. 17-160. Reprinted with permission.

Popov, G., 1971. Placer deposits, *The Working of Mineral Deposits* (English translation). Moscow, USSR: MIR Publishers, 69–71.

Raeburn, C., and Milner, H. B., 1927. *Alluvial Prospecting, the Technical Investigation of Economic Alluvial Minerals.* London: Thomas Murby, 478p.

Rickard, T. A., 1899. The alluvial deposits of Western Australia, *AIME Trans.*, **28**, 490–537.

Savage, C. N., 1960. Nature and origin of central Idaho blacksands, *Econ. Geol.*, **55**, 789–796.

Savage, C. N., 1961. Economic geology of Central Idaho blacksand placers, *Idaho, Bur. Mines Geol., Bull.*, **17**, 160p.

Thomas, L. J., 1973. Placer mining, *An Introduction to Mining–Exploration, Feasibility Extraction and Rock Mechanics.* New York: Halsted, Sidney, Australia: Hicks, Smith and Sons, 83–96.

United Nations, 1972. *Small-scale Mining in the Developing Countries.* New York: Dept. of Economic and Social Affairs, United Nations, ST/ECA/155, 171p.

Weber, B. W., 1935. Bajada placers of the arid Southwest, *AIME Tech. Publ. 588*, 3–16.

Wells, J. H., 1969. *Placer Examination . . . Principles and Practices.* Washington: *U.S. Bur. Land Management, Tech. Bull.*, no. 4, 155p.

Cross-references: *Blacksand Minerals and Metals; Gemology; Mineral Deposits: Classification; Mineral and Ore Deposits;* see also mineral glossary. Vol. I: *Mean Sea Level Changes.* Vol. III: *Drainage Patterns; Erosion; Lava-Displaced Drainage; Regolith and Saprolite.*

PLAGIOCLASE FELDSPAR

Plagioclase is an aluminosilicate series with sodium and(or) calcium. The word ***plagioclase*** is from the Greek, *plagios* (oblique) and *clasis* (fracture), in reference to the two nonorthogonal cleavages; ***feldspar*** is from the Swedish, *feldt* (field) and *spat* (spar), supposedly a reference to the occurrence of the spar in fields (see *Feldspar Group*).

Composition

Ideally, chemical composition of the ***plagioclase*** group varies from $NaAlSi_3O_8$ (*albite*) to $CaAl_2Si_2O_8$ (*anorthite*); however, some potassium is usually present, especially in soda-rich ***plagioclase***. Specific names have been applied to different parts of the compositional range. These names and the range of composition over which they are usually applied are given in Table 1. It is usual to express feldspar compositions in terms of mole percent of the end members *albite* (*Ab*), *anorthite* (*An*), and *orthoclase* (*Or*, $KAlSi_3O_8$). The stoichiometric compositions of several ***plagioclase feldspars*** in terms of weight percent oxides are given in Table 2.

Since soda-rich ***plagioclases*** are richer in SiO_2 (hypothetical anhydrous silicic acid), they have been called "acid" ***plagioclases***. "Basic" ***plagioclase*** has been used to refer to calcium-rich compositions.

TABLE 1. Nomenclature of the *Plagioclase Feldspars* in Terms of Albite (*Ab*) and Anorthite (*An*) Content (%)

Albite $NaAlSi_3O_8$	*An* 0–10,	*Ab* 100–90
Oligoclase	*An* 10–30,	*Ab* 90–70
Andesine	*An* 30–50,	*Ab* 70–50
Labradorite	*An* 50–70,	*Ab* 50–30
Bytownite	*An* 70–90,	*Ab* 30–10
Anorthite $CaAl_2Si_2O_8$	*An* 90–100,	*Ab* 10–0

TABLE 2. Oxides in Plagioclase Feldspars (wt %)

Oxide	Albite	*An*10	*An*30	*An*50	*An*70	*An*90	Anorthite
SiO	68.74	66.05	60.76	55.59	50.54	45.61	43.19
Al_2O_3	19.44	21.26	24.82	28.30	31.70	35.02	36.65
Na_2O	11.82	10.57	8.12	5.73	3.40	1.12	0.00
CaO	0.00	2.12	6.30	10.38	14.36	18.25	20.16

Structure and Symmetry

Naturally occurring ***plagioclases*** are triclinic. Structurally the ***plagioclases*** are similar to other ***feldspars***. It is convenient to assign to ***plagioclase*** a unit cell that has parameters $a \approx 8.6$ Å, $b \approx 13.0$ Å, $c \approx 14$ Å, $\beta \approx 116°$, $\alpha \approx \gamma \approx 90°$. This results, for some ***plagioclases***, in the choice of nonprimitive triclinic unit cells larger than the smallest possible triclinic unit cell.

Plagioclases are framework silicates (see Vol. IVA: *Mineral Classes: Silicates*). The silicon and aluminum ions are linked to four oxygen ions to form tetrahedral coordination polyhedra. These tetrahedra share each vertex oxygen with adjacent tetrahedra to form a three-dimensional framework. The sodium and(or) calcium ions occupy large cavities in this framework. The distribution of silicon and aluminum among the tetrahedra is variable; it can be ordered, in which case the aluminum is concentrated in specific tetrahedra throughout the structure, or disordered when, statistically, all tetrahedral sites have the same aluminum/silicon occupancy; intermediate states also occur. The degree of order–disorder has been termed the structural state (see *Feldspar Group; Order–Disorder*).

In ordered *albite* (sometimes called low *albite*, or low-temperature *albite*), the aluminum occupies one particular tetrahedral site; the remaining three symmetrically independent tetrahedral sites are occupied by silicon. In disordered *albite* (often called high *albite*, high-temperature *albite*, or analbite) each of the four tetrahedral sites contains, statistically, 25% Al, 75% Si. A unit cell can be chosen for *albite* that has similar dimensions to *sanidine* and has $C\bar{1}$ symmetry (i.e., it has a *c*-face centered, centrosymmetric, triclinic cell).

Anorthite exists only in an ordered state; Al and Si occupy alternate tetrahedra throughout the structure, consistent with the 1:1 Al:Si ratio in *anorthite*. A unit cell can be chosen for *anorthite* that, in effect, consists of two sanidine-sized cells stacked one above the other in the *c* direction so that the *c* parameter is approximately twice the *c* parameter of *sanidine* (i.e., $c \approx 14.2$); this has symmetry $P\bar{1}$.

An anorthite-size cell with $I\bar{1}$ symmetry (i.e., with a body-centered, centrosymmetric, triclinic cell) has been identified from the systematic absences of $h + k + l$ odd Bragg reflections from some *bytownite* specimens. However, the presence of non-Bragg reflections in the X-ray diffraction patterns of most ***plagioclases*** with compositions intermediate between *albite* and *anorthite* suggests the presence of complex domain structure (sometimes observable by electron microscopy), which could account for at least some of the systematic absences.

The degree of order tends to increase as a crystal cools during geological time. In the case of ***feldspars*** with compositions between An_0 and An_{20}, the ordering processes may lead to the unmixing of two phases, *albite* (An_0) and *oligoclase* (approximately An_{20}), which segregate into domains observable in X-ray and electron-diffraction studies (Fig. 1). The term peristerite is applied to such ***plagioclase*** intergrowths, a reference to the pigeon-like (Greek, *peristeri*), irridescent colors that sometimes result from diffraction of light by lamellar intergrowths. Similar subsolidus miscibility gaps occur from $An \approx 48$ to $An \approx 58$ and from $An \approx 70$ to $An \approx 90$ (Fig. 1). The iridescence (labradorescence) of some *labradorite* (An_{50}–An_{60}) specimens results from optical diffraction by lamellae of different *An* contents.

Melting Relationships

Pure *albite* melts at 1118°C and *anorthite* at 1550°C at one bar pressure. In the system

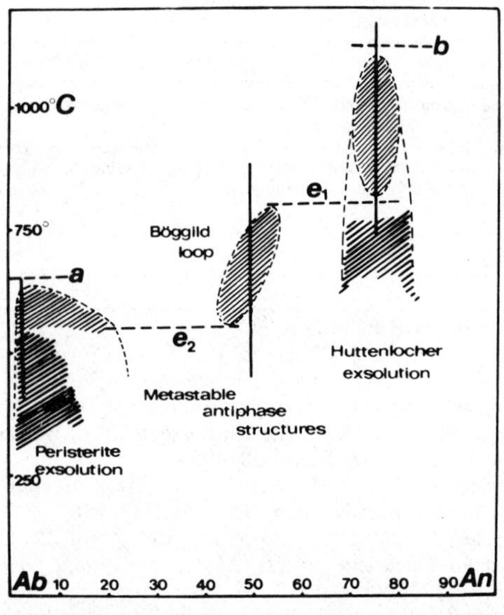

FIGURE 1. Phase diagram showing possible miscibility gaps (hatched areas) in ***plagioclases*** (from J. V. Smith, 1974, *Feldspar Minerals: Crystal Structure and Physical Properties*, vol. 1. New York: Springer-Verlag; reprinted with permission).

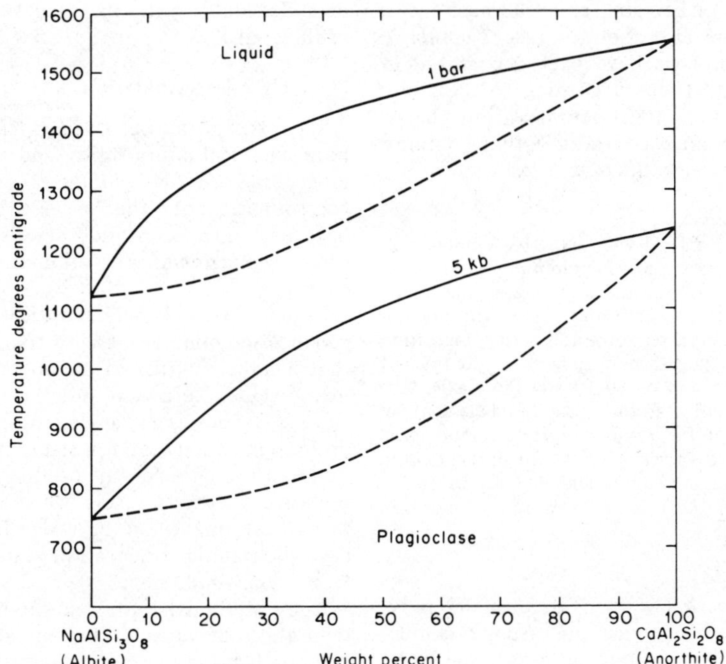

FIGURE 2. Melting relationships of ***plagioclases*** projected onto the anhydrous plane in the system $NaAlSi_3O_8$–$CaAl_2Si_2O_8$–H_2O at 1 bar and 5 kilobar pressure (from *The Interpretation of Geological Phase Diagrams* by Ernest G. Ehlers. W. H. Freeman and Company; copyright © 1972).

$NaAlSi_3O_8$–H_2O at 5 kilobar pressure, the melt point of *albite* is depressed to 750°C. *Anorthite* plus water melts at 1225°C at 5 kilobar (Fig. 2). Because of complete miscibility between *albite* and *anorthite* at temperatures near the solidus, the liquid field is separated from the plagioclase field by a liquidus-solidus loop bounding the two-phase field of liquid plus ***plagioclase***. The compositions of crystalline ***plagioclase*** on the solidus (dashed lines in Fig. 2) are connected to liquid compositions on the liquidus (solid lines) by isothermal tie lines at equilibrium. Thus, ***plagioclase*** crystals are always more calcic than the liquid with which they are in equilibrium, and crystallization of ***plagioclase*** from a melt always drives the liquid toward sodic compositions (see Vol. IVA: *Phase Equilibria*).

Recrystallization of ***plagioclases*** below the solidus leads to a decrease in disorder of Al and Si in the tetrahedral sites which results in the proposed miscibility gaps (hatched areas) in Fig. 1.

Physical Properties

Form. The ***plagioclase feldspars*** present triclinic prismatic crystals, often tabular parallel to (010) and with forms {010}, {1$\bar{1}$0}, {001}, {10$\bar{1}$}, and {101} developed; they may also be massive or granular. Cleavelandite is a variety of *albite* that occurs as coarse plates parallel to (010). The plates are often wedgeshaped and may form radiating groups. Pericline is a sodic variety that occurs as crystals, which are prismatic parallel to the *b* axis and in which simple twins after the Pericline law are commonly developed (see *Feldspar Group*).

Color. Usually white, ***plagioclase feldspars*** can also be grey or brown. *Labradorite* is commonly dark blue or black and shows iridescent blue-red colors. Peristerite (see above) may show iridescence in pink or blue. Aventurine or sunstone is a variety of soda-rich ***plagioclase*** with a reddish sheen caused by the presence of thin *hematite* flakes oriented parallel to structural planes within the crystal.

Luster. Minerals of the ***plagioclase feldspar*** group have a vitreous luster and are pearly on cleavage planes. They are often translucent or transparent, especially on thin edges.

Hardness. 6–6½.

Specific Gravity. 2.60 (*albite*) to 2.76 (*anorthite*).

Cleavage. (001) perfect, (010) good to perfect. The angle between the cleavages varies from 86.1° (*albite*) to 85.9° (*anorthite*).

Twinning. Polysynthetic twinning is commonly visible on crystal and cleavage faces; simple twins are also common (see *Twinning*). Numerous twin laws have been recognized in the ***plagioclases***. Twins according to the albite law–twin axis=b^* [010], composition plane=(010)–are extremely common. Other common twin laws observed in the ***plagioclases*** are:

1. *The Pericline Law*–Twin axis=b, composition plane=rhombic section. The rhombic section is an irrational plane that includes the b axis and makes equal intercepts on (110) and $(1\bar{1}0)$. The orientation of the rhombic section varies with composition and structural state. The angle between the trace of the rhombic section on (010) and the a axis, σ, is considered positive if the trace lies between the positive ends of the a and c crystallographic axes. The values of σ vary from +33° for ordered *albite* to -5° for disordered *albite* and -16° for *anorthite*.
2. *The Carlsbad Law*–Twin axis=c, composition plane = $\{hk0\}$.

Albite and pericline twins are usually polysynthetic; Carlsbad twins are usually simple. Twins according to more than one law may be present within an individual crystal.

Optical Properties

Plagioclase feldspars occur as euhedral, subhedral, or anhedral crystals; *albite* often displays lath-shaped cross-sections. They are colorless but may appear cloudy due to alteration; sodic varieties alter to sericite (a fine-grained white ***mica***); more calcic varieties alter to sausserite (a mixture of *albite*, an ***epidote***, a ***scapolite***, etc.). The relief is usually low except for calcic members which show moderate relief; sodium-rich varieties have negative relief, $n_\alpha = 1.527(Ab) - 1.577(An)$, $n_\beta = 1.531(Ab) - 1.585(An)$, $n_\gamma = 1.538(Ab) - 1.590(An)$. The birefringence colors are first order grey or pale yellow.

The optic sign is variable; it is positive for compositions near *albite* and *labradorite* and negative for composition near *oligoclase* and *anorthite*; $2V_\alpha = 50°–105°$; the orientation of the optic axial plane changes from nearly parallel to (001) in *albite* to close to $(1\bar{1}0)$ in *anorthite*. Dispersion is noticeable and varies from $r<v$ for *albite* to $r>v$ for *anorthite*.

Polysynthetic twinning is almost always present, sometimes in two nearly perpendicular orientations. Zoning is common and appears as a concentric variation in optical properties, noticeably extinction, from the center of a grain toward the periphery (in normal zoning the calcium content decreases outward; the opposite distribution is termed reverse zoning; in oscillatory zoning the calcium/sodium content fluctuates with distance from the center of the grain).

Plagioclase Determination

Since the properties of ***plagioclases*** vary with both chemical composition and structural state in general, no one parameter yields a unique determination of either variable. Many methods have been proposed and there is a large literature on ***plagioclase*** determination (see references).

Optical Methods. The optical properties of ***plagioclase*** minerals are sufficiently variable that a large number of parameters are of diagnostic value. Refractive index (best for composition), optic axial angle, optic orientation, optic sign, $2V$ (best for structural state), and extinction angles are all useful. Further determinative methods result when such measurements are made on crystals, for which the crystallographic orientation can be deduced from twins or cleavages.

X-ray Methods. X-ray methods based on the separation of various lines on powder diffraction patterns are extensively used for ***plagioclase*** structural-state determination when the composition is known. Such methods depend on variations of the lattice parameters of ***plagioclase***, particularly the direct cell angle γ or the reciprocal cell angle γ^*.

Occurrence

Plagioclase feldspars occur in igneous rocks of both plutonic and volcanic origin; sodic ***plagioclase*** is characteristic of felsic igneous rocks while mafic igneous rocks are typified by the presence of calcic ***plagioclase***. The initial ***plagioclase*** to crystallize in mafic plutonic rock is *bytownite*, which may settle out as a ***feldspar***-rich accumulate of essentially unzoned crystals or may react partially with the more sodic liquid to develop zoned crystals with compositions ranging from *bytownite* cores to, in some cases, *oligoclase* rims. ***Plagioclase***, An_{40}-An_{80}, is the only essential constituent of the rock anorthosite. Fractional crystallization leads to progressively more albitic ***plagioclase*** and the albite component of ***plagioclase*** is the first to melt during partial fusion. *Albite* and *oligoclase* occur in pegmatites (see *Pegmatite Minerals*).

The ***plagioclase*** phenocrysts of basalts are commonly zoned with a uniform *bytownite* core and a thin rim of more albitic composition. In other instances, particularly with rocks of intermediate composition, the plagioclase zoning may be oscillatory, extending over a wide range of compositions. Magmatic ***plagioclases*** commonly show both simple and polysynthetic twins; the latter may indicate change of phase (and symmetry) during cooling.

Plagioclases in regional metamorphic rocks show increasing calcium content with increasing grade of metamorphism from *albite* to *andesine* (rarely *labradorite*). *Albite* occurs with ***chlorite*** and *biotite* in schists and with *epidote* in metamorphosed basalts in the greenschist facies. ***Plagioclase*** in the composition range An_3–An_{22} (peristerite) is generally absent from metamorphic assemblages. Rocks in the amphibolite facies typically have *oligoclase* with a composition on the calcium-rich side of peristerite. *Andesine* occurs in the granulite facies. ***Plagioclase*** is not stable in the eclogite facies (see *Mantle Mineralogy*).

In sediments, the detrital ***plagioclase*** is usually *albite* or *oligoclase* except in some very immature sediments where more calcic varieties may be found. *Albite* is a common authigenic mineral (see *Authigenic Minerals*), where it is relatively pure ($Or < 3$ and $An < 1$) and typically free of polysynthetic twinning. It is also a common vein mineral (see *Vein Minerals*), sometimes in the varietal habit of cleavelandite or pericline.

Uses

Sodic ***plagioclase*** is used in the porcelain and pottery industries for the production of glazes and in the manufacture of opalescent glass. Peristerite is used in jewelry (see *Gemology*). Rocks rich in *labradorite* are used as facing stones.

JOHN STARKEY

References

Bambauer, H. V., ed., 1969, in W. E. Tröger, *Optische Besimmung der gesteinsbildenden Minerale,* 2nd ed. Stuttgart: E. Schweizerbart'sche Verlagsbuchhandlung, 645–762.

Burri, C.; Parker, R. L.; and Wenk, E., 1967. Die optische Orientierung der Plagioklase, *Mineralog. Geotechn. Reihe*, vol. 8. Basel and Stuttgart: Birkhauser Verlag, 334p.

Deer, W. A.; Howie, R. A.; and Zussman, J., 1963. *Rock-Forming Minerals*, vol. 4. London: Longmans, 94–165.

Marfunin, A. S., 1962. *The Feldspars. Proc. Academy of Sciences of the U.S.S.R.*, issue 78, Geol. Ore Deposits, Petrog., Mineral. and Geochem., (translated 1966, Israel Program for Scientific Translations), 317p.

Ribbe, P. H., ed., 1975. *Feldspar Mineralogy*, vol. 2. Washington: Mineralogical Soc. Am. Short Course Notes, 243p.

Smith, J. V., 1974. *Feldspar Minerals*, vol. 1, *Crystal Structure and Physical Properties*, 627p., and vol. 2 *Chemical and Structural Properties*, 690p. New York: Springer-Verlag.

Van Der Plas, L., 1966. *The Identification of the Detrital Feldspars*. Amsterdam: Elsevier, 305p.

Cross-references: *Alkali Feldspars; Authigenic Minerals; Electron Probe Microanalysis; Feldspar Group; Fire Clays; Glass, Devitrification of Volcanic; Lunar Mineralogy; Meteoritic Minerals; Mineral Classification: Principles; Mineralogical Phase Rule; Optical Mineralogy; Order-Disorder; Orthopyroxenes; Pegmatite Minerals; Refractive Index; Refractory Minerals; Rock-Forming Minerals; Soil Mineralogy; Staining Techniques; Thermoluminescence; Twinning.* Vol. IVA: *Crystal Chemistry.*

PLASTIC FLOW IN MINERALS

For many years structural geologists have been recording extensive evidence of plastic flow in rocks. The major macroscopic features take the form of folds with considerable bending and warping of strata, while on a microscopic scale, flow is apparent through undulatory extinction, kinking, and optical deformation lamellae. The former are expressions of a significant amount of bending of the crystal structure of the component minerals, whereas deformation lamellae are interpreted to be indicators of plastic deformation, which apparently follow stress and(or) strain axes or active slip planes. In the optical microscope they appear as thin bands 1 to 2 μm thick with slightly different refractive indices and birefringence to the host material (Fig. 1); under phase-contrast illumination they appear asymmetrical—dark on one side, bright on the other.

To the casual observer, however, most rocks appear to be brittle and incapable of sustaining any significant plastic deformation. This paradox was resolved when it became clear that rocks were normally deformed only very slowly (typically over many thousands of years) at high temperatures (several hundred degrees

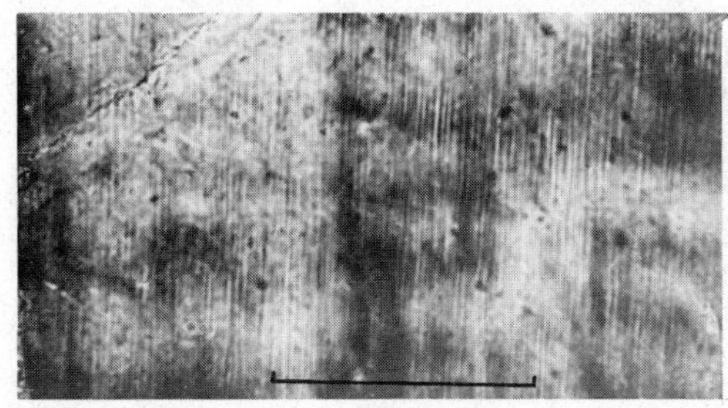

FIGURE 1. Deformation lamellae in experimentally deformed, synthetic *quartz*, as observed in the polarizing microscope. The lamellae are the vertical linear features running across the field of view. Scale marker represents 0.1 mm. (D. J. Morrison-Smith, Ph.D. Thesis, 1974.)

centigrade) and high hydrostatic pressures (several kilobars), and only when conditions approaching these could be attained in the laboratory would it be possible to deform rocks experimentally.

Experimental Techniques

Modern deformation equipment can duplicate the conditions of both temperature and pressure found in the crust and upper mantle, but the time factor is a major problem. Although some experiments have run for several months, the strain rates used have to be several orders of magnitude higher than those occurring naturally.

The earliest experiments were performed toward the beginning of this century, but not until the mid-60s were major advances in understanding the mechanisms of deformation made. These were made possible by the development by D. T. Griggs and co-workers at the University of California, Los Angeles, of an apparatus that could operate at temperatures up to 1000°C and confining pressures of 25 kilobars using a soft solid such as *talc* as the pressure-transmitting medium. Although this apparatus did not obtain useful quantitative data (because of severe temperature gradients across the specimen, inaccuracies in strength measurements due to the finite strength of the confining medium, frictional losses in the pressure seals, etc.), it led to the development of an apparatus, utilizing argon or other inert gas as the pressure medium, with which accurate stress-strain curves at well-defined temperatures, pressures, and strain rates were obtained for the first time.

The most recent major development has been the application of transmission electron microscopy to the study of defect microstructures in minerals (see *Electron Microscopy*; Wenk, 1975).

Experimentally and Naturally Deformed Minerals

Many of the more important rock-forming minerals (q.v.) have been investigated, and the structures observed in experimentally deformed examples are generally similar to those found in naturally deformed material.

Quartz. This mineral has received most attention, mainly because it is a major constituent of the crustal rocks, which show significant amounts of deformation. It has been studied in the field since the late nineteenth century.

Most of the work on *quartz* has been directed toward the study of deformation lamellae. It is now established that they do not necessarily reflect the orientation of active slip planes, although they appear to be reliable indicators of paleo-stress and(or) strain axes. Electron microscopy has shown that although a wide variety of features may produce the optical effects attributed to lamellae, most consist of some form of dislocation array containing a majority with the same Burgers vector. Despite the differences in structure, most lamellae tend to follow similar distributions of orientation over pressure, temperature, and strain rate.

A large number of slip systems have been identified by electron microscopy of experimentally deformed synthetic *quartz*. Generally these observations confirm earlier optical microscope findings, but they also include other slip systems not previously recorded. Table 1 lists all observed slip systems. In a given specimen only a few systems will be observed to operate—generally less than four; the choice of active slip systems depends on the orientation of the principal stress axes with respect to the crystallographic axes, the strain rate, temperature, and confining pressure.

Synthetic *quartz* crystals are grown from hydrothermal solution for use in the electronics industry but were used in deformation experiments in the hope that they would be more homogeneous than the natural material (see *Synthetic Minerals*). This was, unfortunately, not the case, and it has also been found that the impurities in the synthetic crystals have a significant effect on their deformation properties. Care must be taken, therefore, in attempting to apply the results obtained to the natural situation.

TABLE 1. Slip Systems Identified in Quartz by Optical and Transmission Electron Microscopy

Slip Plane	Slip Direction
(0001)	$\langle 11\bar{2}0\rangle$
$\{10\bar{1}0\}$	$[0001]$
$\{10\bar{1}0\}$	$\langle\bar{1}2\bar{1}0\rangle$
$\{10\bar{1}0\}$	$\langle\bar{1}2\bar{1}3\rangle$
$\{10\bar{1}1\}$	$\langle\bar{1}2\bar{1}0\rangle$
$\{10\bar{1}1\}$	$\langle 11\bar{2}\bar{3}\rangle$
$\{10\bar{1}2\}$	$\langle\bar{1}2\bar{1}0\rangle$
$\{10\bar{1}3\}$	$\langle\bar{1}2\bar{1}0\rangle$
$\{11\bar{2}0\}$	$[0001]$
$\{11\bar{2}1\}$	$\langle\bar{1}2\bar{1}\bar{3}\rangle$
$\{11\bar{2}2\}$	$\langle 11\bar{2}\bar{3}\rangle$

Note: The standard notation for slip systems is to list the slip planes first followed by the slip directions thus: $\{hkil\}$ $\langle uvtw\rangle$ or $(hkil)$ $[uvtw]$, where $\{hkil\}$ indicates the set of planes possible from any permutation of the indices h,k,i,l (including negative values) and $\langle uvtw\rangle$ means the directions possible from permutation of the indices u,v,t,w,; $(hkil)$ and $[uvtw]$ indicate those particular planes and directions and are necessarily components of the set $\{hkil\}$ and $\langle uvtw\rangle$. For example $\{10\bar{1}0\}\langle\bar{1}2\bar{1}3\rangle$ contains $(01\bar{1}0)$ $[2\bar{1}\bar{1}3]$, $(\bar{1}100)$ $[11\bar{2}\bar{3}]$, etc. in addition to $(10\bar{1}0)$ $[\bar{1}2\bar{1}3]$. For the case of hexagonal axes, (0001) and $[0001]$ are unique so that the notation $\langle 0001\rangle$ and $\{0001\}$ is in this instance redundant (see *Directions and Planes*).

One of the most important impurity effects is that due to water (hydrolytic weakening). This effect is also observed in other silicates but is best illustrated in synthetic *quartz*, which generally contains a higher concentration of water because of the growth technique. Hydrolytic weakening significantly reduces the strength of a mineral above a certain temperature (Fig. 2), apparently on account of the weakening effect of substituting a hydrogen bonded Si–OH···HO–Si for the normal Si–O–Si bonding arrangement. The temperature dependence of the weakening is related to the diffusion rates for water in the structure, as dislocation propagation is controlled by the rate at which bonds surrounding the dislocation become hydrolyzed (see *Crystals, Defects in*).

Olivine. During the past few years ***olivine*** has been the subject of a considerable amount of study because of its importance as a major constituent of the upper mantle and the relevance of any investigations to models of plate tectonics, which necessarily involve flow in the mantle (see *Mantle Mineralogy*).

It appears that slip is simpler than in *quartz*, occurring predominantly on the systems (100) [001] and {110} [001] at low temperatures and high strain rates (conditions similar to those that may be associated with shock deformation observed in many meteorites) changing to {0*kl*} [100] and (010) [100] with increasing temperatures and(or) decreasing strain rate (Fig. 3). The system {0*kl*} [100] is known as pencil glide, where slip is parallel to one particular direction but several slip planes containing this direction are operative (analogous to a pile of pencils slipping parallel to their long axes with their faceted faces acting as slip planes). In addition, there are several other minor slip components that have been observed by electron microscopy.

The experimental results suggest that slip parallel to [100] should predominate in the upper mantle, and observations of material that is believed to have undergone deformation in the mantle, such as alpine-type peridotites and xenoliths in alkali basalts, generally confirm these findings. It is clear, however, that under mantle conditions processes of recovery and recrystallization are very important, and it is conceivable that other processes too slow to be measured in laboratory experiments may also be operative. For instance, Herring-Nabarro creep may occur where grains deform by diffusion of material from parts of grains under compression to regions under tension.

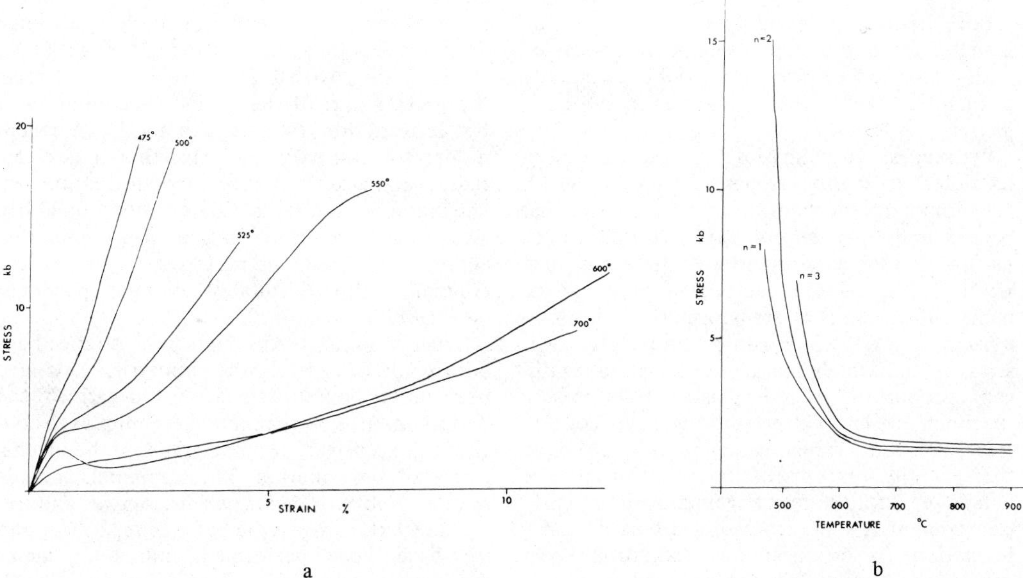

FIGURE 2. (a) Stress-strain curves for synthetic crystals deformed in a gas apparatus. Note the steepness of the curves at low temperatures and the shallow curvature for temperatures about about 550°C. The stress drop obtained just after the yield point in the 600°C experiment is a feature typical of experiments on synthetic *quartz*, its appearance being dependent on the crystal's water content. (b) Graph compiled from a series of experiments such as in (a). The stress to produce a given strain is plotted against temperature; the curves n = 1,2,3, are for 1%, 2%, 3% strain, respectively. Note how the stress drops significantly with increasing temperature up to about 600°C; this effect is an expression of the phenomenon of hydrolytic weakening. (D. J. Morrison-Smith, Ph.D. Thesis, 1974.)

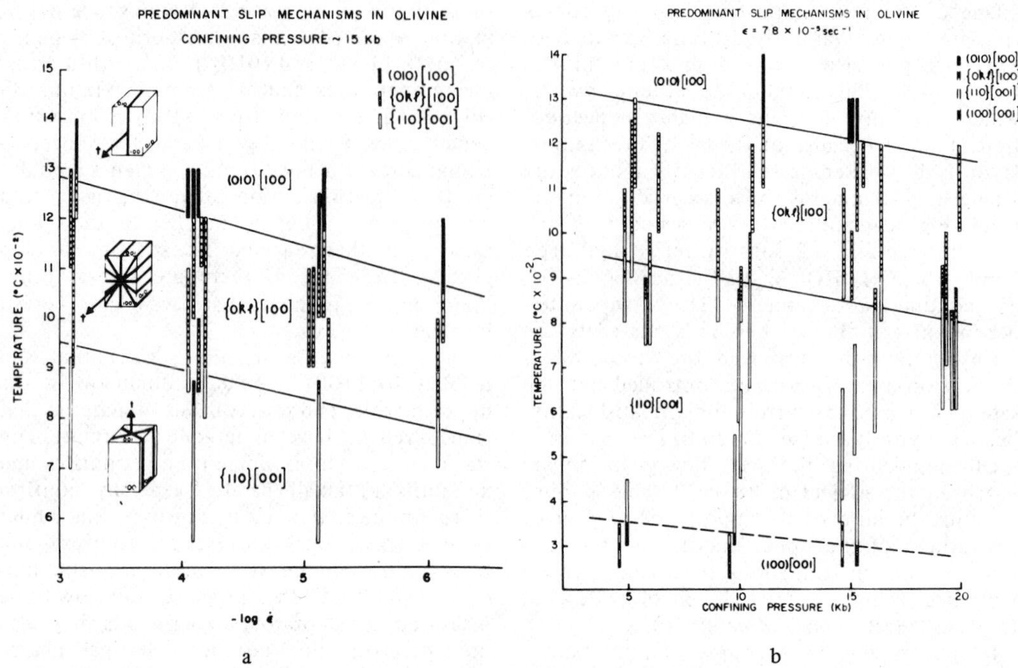

FIGURE 3. Predominant slip systems in experimentally deformed ***olivine*** as functions of temperature and strain rate (a) and pressure (b) (from Carter and Avé Lallemant, 1970). Vertical bars indicate the range of actual experimental observations. Schematic representation of each slip system is given in the left of (a).

Both hydrolytic weakening and deformation lamellae are commonly found to be associated with deformed ***olivine*** and exhibit essentially similar features to those observed in synthetic *quartz*.

Pyroxenes. Investigations so far have been restricted to ***orthopyroxenes*** (q.v.), in which the dominant mechanism of plastic flow has been found to be slip on (100) [001] in both naturally and experimentally deformed material. An important feature in many specimens, however, is a transformation to ***clinopyroxene*** (q.v.), which can be induced by shear stresses, and which can accommodate a significant proportion of the plastic strain in any specimen. Several investigations have concentrated on this transformation, and, although there is still some discussion as to the precise details of atomic rearrangements, it is fairly clear that it is some form of martensitic transformation (a homogeneous structural shear with only minor diffusion-controlled atomic shuffles required to complete the process). Little bond breaking is required by this process and it therefore has a low activation energy. In many cases of suitably oriented specimens this is the major deformation mechanism.

Amphiboles. Only limited attention has been given the ***amphiboles***, being restricted mainly to the establishment of the major deformation modes of experimentally deformed *hornblende*. It appears that, although (100) [001] is the favored slip system, in many cases the greater part of the strain is accommodated by deformation twinning on ($\bar{1}01$). A recent model for this twin indicates that it is essentially a martensitic-type transformation involving only a limited amount of bond breaking. This transformation has a low activation energy and should be preferred to dislocation-controlled slip in suitably oriented specimens (see *Amphibole Group*).

Other Minerals. Although some studies have been made of several other minerals, they have been either too superficial or too early in the development of experimental techniques to contribute extensively to understanding the mechanisms of deformation. These minerals include *calcite, dolomite, halite,* ***micas****, galena*, and *ice*.

Quite extensive studies of *calcite, halite*, and *ice* have been performed, and their major deformation properties have been established. However, these materials have been subjected to little study using electron microscopy in conjunction with well-defined deformation experiments; therefore, their mechanisms of slip and dependences on temperature, pressure, strain rate, etc., are not yet well understood. One of the major contributions made by the study of *ice* has been to show the importance of recovery and recrystallization to flow properties in minerals deforming at temperatures close to their melting points (as expected in parts of the upper mantle).

Most other minerals have been subjected to only preliminary or exploratory investigations.

Theoretical Work

A major objective of these studies is to be able to explain deformation processes occurring naturally. Some form of model is therefore required to bridge the gap between the conditions attainable in the laboratory and those operating in the earth. One method simply involves extrapolating the experimental results to natural strain rates as has been done for ***olivine*** (see above). In addition, several attempts have been made to fit the data to empirical flow laws relating stress and strain, and including various parameters depending on temperature, pressure, etc. So far the majority of the data fit a flow law of the type proposed by Weertman originally for metals:

$$\frac{d\epsilon}{dt} = A\sigma n \exp(-Q/RT)$$

ϵ being strain, σ stress, A a constant independent of temperature, Q the activation energy, R Boltzmann's constant, and T the absolute temperature. Results fitted to the equation give n varying from 2 to 10, A between 10^2 and 10^{10}, and Q between 20 and 130 kcal/mole depending on the material.

A different approach was taken by D. T. Griggs for the case of synthetic *quartz* (Griggs, 1974). Using the available data, he constructed a model based on the detailed dislocation dynamics and the relationship to diffusion of water in order to take account of hydrolytic weakening. This model becomes very complicated, but it appears to explain many of the experimental observations although several difficulties still exist that require modification to the model. In addition, the model may not be applicable at present to natural *quartz* because of the important differences in the deformation mechanisms in natural and synthetic *quartz* mentioned above.

Future Progress

Despite the large amount of effort that has been devoted to the study of flow in minerals, there remains much work to do, both experimental and theoretical, before the mechanisms of plastic flow in minerals are completely understood. For instance, it is not yet clearly known why the operative slip systems change with increasing temperatures in any particular case. Much of this work is continuing, and considerable progress should be made over the next few years.

D. J. MORRISON-SMITH

References

Blacic, J. D., and Christie, J. M., 1973. Dislocation substructure of experimentally deformed olivine, *Contr. Mineralogy Petrology*, **42**, 141–146.

Carter, N. L., and Avé Lallemant, H. G., 1970. High temperature flow of dunite and peridotite, *Geol. Soc. Am. Bull.*, **81**, 2181–2202.

Christie, J. M., and Ardell, A. J., 1974. Substructures of deformation lamellae in quartz, *Geology*, **2**, 405–408.

Coe, R. S., and Muller, W. F., 1973. Crystallographic orientation of clinoenstatite produced by deformation of orthoenstatite, *Science*, **180**, 64–66.

Griggs, D. T., 1967. Hydrolytic weakening of quartz and other silicates, *Geophys. J. Roy. Astron. Soc.*, **14**, 19–32.

Griggs, D. T., 1974. A model of hydrolytic weakening in quartz, *J. Geophys. Research*, **79**, 1653–1661.

Heard, H. C., et al., eds., 1972. Flow and fracture in rocks, the Griggs Volume, *Geophysical Monograph 16*, American Geophysical Union.

Kohlstedt, D. L., and Goetz, C., 1974. Low-stress, high-temperature creep in olivine single crystals, *J. Geophys. Research*, **79**, 2045–2051.

Kohlstedt, D. L., and Vander Sande, J. B., 1973. Transmission electron microscopy investigation of the defect microstructure of four natural orthopyroxenes, *Contr. Mineralogy Petrology*, **42**, 169–180.

Rooney, T. P.; Riecker, R. E.; and Ross, M., 1970. Deformation twins in hornblende, *Science*, **169**, 173–175.

Wenk, H.-R., ed., 1975. *Application of Electron Microscopy in Mineralogy*. New York: Springer-Verlag.

Cross-references: *Crystals, Defects in; Etch Pits; Mantle Mineralogy; Order-Disorder; Synthetic Minerals;* see also mineral glossary.

PLEOCHROIC HALOS

As early as 1880, petrologists such as Michel Levy, Lacroix, and Rosenbusch noted in certain minerals the presence of minute colored aureoles around microscopic inclusions of accessory minerals. These aureoles were referred to as "pleochroic halos" although they generally display a pleochroism weaker than that of their host mineral. They are known to occur in colored minerals such as *biotite*, ***tourmalines, amphiboles***, and ***chlorites***, as well as in colorless ones (*muscovite, cordierite, fluorite*). The inclusions most commonly identified are *zircon*, ***apatite***, *titanite, xenotime*, and *monazite*. The halos that develop around very small inclusions, with size of less than 1 μm generally show a concentric ring structure. The

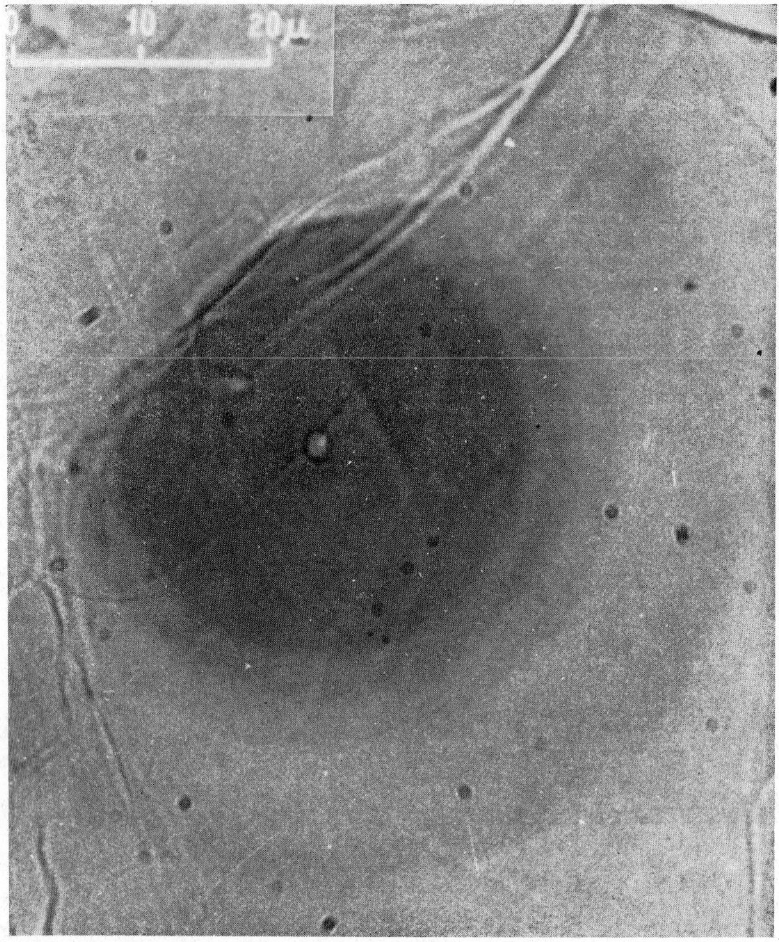

FIGURE 1. Flake of *biotite* with a ring-structured halo around a minute inclusion containing uranium (courtesy of Mrs. Sarah Deutsch). Four rings are visible, their radii and the corresponding alpha-emitters being 15.5μm (Ra, U_{II}, I_0); 19.5 μm (Rn, RaF); 23.5 μm (RaA); and 35 μm (RaC′). A first ring at 13 μm (U_I) is no longer visible as the zone between 0 and 15.5 μm is saturated.

radii of the rings range from about 5 to 45 μm (Fig. 1).

Mugge and Jolly independently in 1907 showed that the pleochroic halos were due to the effect of α particles emitted by the central inclusion. They invoked the following evidences:

1. All minerals around which halos are found are known to be radioactive and to contain uranium or thorium.
2. The radii of the rings correspond to the range of α particles emitted by natural radioactive nuclides. The ring structure occurs when the inclusion is small enough to make negligible the self-absorption of the α particles.
3. The coloration effect can be experimentally reproduced by irradiating the mineral with α doses comparable with the estimated natural dose.

Experiments and observations of natural halos both show that the intensity of halo coloration begins to be noticeable above a threshold in α-radiation dose, then increases with the radiation does until it reaches a saturation stage in which it remains at its maximum value independently of the radiation dose. A further increase in irradiation leads to a diminution in the coloration.

Much attention has been paid to the identification of the α emitters responsible for the various rings found in well-structured halos. The studies were carried out mainly by Henderson and his coworkers and have permitted the attribution of all the observed rings, but one, to known natural α emitters, including the members of the three radioactive series and

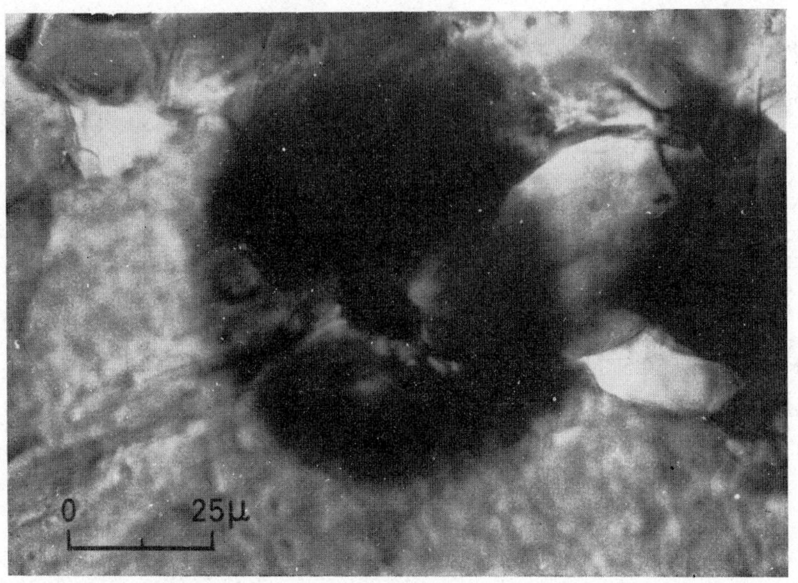

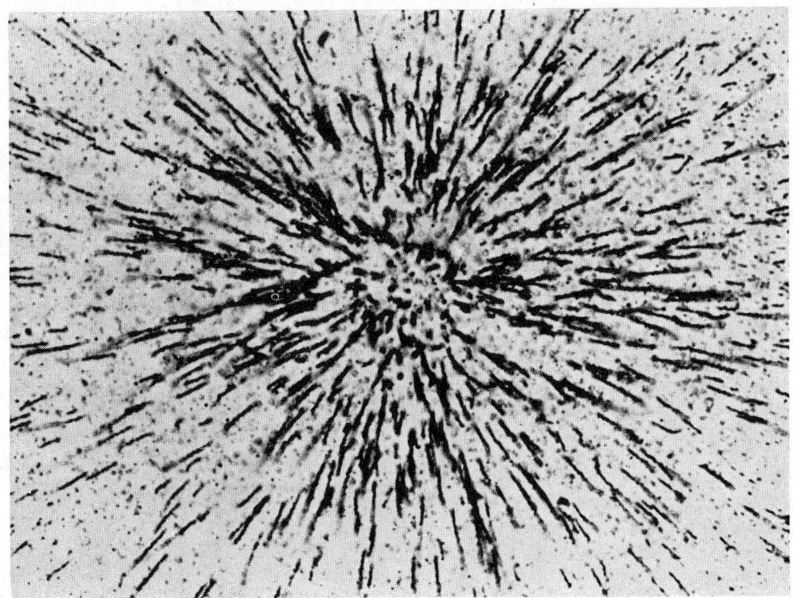

FIGURE 2. Above, opaque inclusion surrounded by a halo. Below, an Ilford C 2 nuclear emulsion applied on the thin section for three months. Innumerable alpha-particle traces emitted by the inclusion are to be seen.

samarium-147. Jolly and Henderson have reported the very rare occurrence of a ring of 8.6 μm radius in *biotite*. This range would correspond to an α particle with an energy of 3 MeV. In spite of various speculations, reviewed in Picciotto and Deutsch (1961), this ring, taking for granted its existence, cannot be attributed to any of the known natural α emitters.

Pleochroic halos have also attracted the attention of geologists because of their potential use in geochronology. The interest in this field was revived by the introduction of the nuclear photographic emulsion technique that permitted solution of the difficult problem of determining the radioactivity of individual microscopic inclusions (Fig. 2). Most of this work was carried out by Deutsch and various associates. They have shown the existence of a general correlation between the integrated α dosage and the intensity of the halo measured by microdensitometry. By measuring the specific α activity of the inclusion, one can determine the length of time during which the mineral has been irradiated in order to produce the observed halo intensity. However, the chronological information to be drawn from the halos is very inaccurate, being subject to a number of experimental errors.

On the other hand, the halo coloration is very sensitive to metamorphic actions, even a slight metamorphism being sufficient to erase the coloration partially or totally. In this respect, halos may provide interesting information on the thermal history of the rock.

E. PICCIOTTO

References

Henderson, G. H., 1939. A quantitative study of pleochroic haloes. V. The genesis of haloes, *Proc. Roy. Soc.*, (London), **A173**, 250.

Picciotto, E. E., and Deutsch, S., 1961. Pleochroic haloes, *Summer Course on Nuclear Geology*, Varenna, 1960. Pisa: Laboratorio di Geologia Nucleare.

POINT GROUPS

Elements or operations of symmetry, either singly or in combination with one another, may intersect at a point and constitute a point group. Thus, a point group is a group of symmetry operations that leaves a point invariant. The ten elements of crystallographic symmetry comprise five *n*-fold axes of rotation 1, 2, 3, 4, and 6, plus five *n*-fold axes of inversion (rotation accompanied by inversion through a center of symmetry, see *Symmetry*) $\bar{1}=i$, $\bar{2}=m$,$\bar{3}$,$\bar{4}$, and $\bar{6}=3/m$. Each element of symmetry constitutes a point group. In addition, there are 22 possible combinations of symmetry elements, each of which also constitutes a point group. The 32 point groups were deduced by Johann F. C. Hessal in 1830 and rediscovered independently by Axel Gadolin in 1867 (see Boisen and Gibbs, 1976, for derivation). Of two notations for point-group symmetry in current use, that of Artur M. Schoenflies is preferred by spectroscopists but the international notation based on the work of C. Hermann and C. Mauguin is preferred by crystallographers. The former notation suffers the disadvantage of arbitrariness when extended to space groups.

Combination of Axes

Since each axis of rotational symmetry alone may constitute a point group, these are designated simply by the fold of the axis in the Hermann-Mauguin notation, i.e., 1, 2, 3, 4, or 6. In the Schoenflies notation, they are designated *C* (cyclic), with the fold of the axis given as a subscript, i.e., C_1, C_2, C_3, C_4, or C_6. These operations are designated proper since they do not change the chirality (sense of handedness) of asymmetric units (see *Symmetry* for illustrations of axes).

The constraint that elements of symmetry themselves must be symmetrically arrayed according to the symmetry of a point group limits the combinations that are possible. Six point groups of combined proper rotation axes are: $222(D_2)$, $32(D_3)$, $422(D_4)$, $622(D_6)$, $23(T)$, and $432(O)$ (Fig. 1). In the Hermann-Mauguin notation, the initial numeral designates the fold of the main rotational axis, which is conventionally set vertically, to the top of the page in perspective drawings, or out of the page in projection. The subsequent numerals indicate the fold of axes at angles to the first. The Schoenflies notation (in parentheses) uses *D* (dihedral) to indicate that the principal or vertical rotation axis has diads (twofold rotational axes) at right angles to it. The subscript gives the fold of the principal rotation axis and, consequently, the number of perpendicular diads.

Thus, $222(D_2)$ is a point group with three mutually perpendicular diads (oriented parallel to the *x*, *y*, and *z* axes respectively). The point group $32(D_3)$ has one vertical triad and three diads at 120° ($2\pi/3$) to each other in the plane perpendicular to the triad. A second and a third numeral 2 are omitted from the Hermann-Mauguin notation since the diads are symmetrically equivalent by operation of the triad. In point groups $422(D_4)$ and $622(D_6)$, the initial number indicates the fold of the vertical axis. The first numeral 2 indicates two and three, respectively, symmetrically related horizontal diads and the second 2 a set of bisector diads in the same plane (that is, at 45° and 30°, respectively, to the first set of diads).

The remaining two point groups are isometric as indicated by the second-place numeral 3 in the Hermann-Mauguin notation. Point group $23(T)$ has three orthogonal diads parallel to the crystallographic axes and four triads oriented parallel to the body diagonals of a cube. Point group $432(O)$ has three orthogonal tetrads (fourfold axes), four diagonal triads, and six diads oriented parallel to the face diagonals of a cube. In the Schoenflies notation, *T* and *O* stand for tetrahedral and octahedral, respectively.

Each of the five inversion axes may constitute a point group, i.e., $\bar{1}(C_i)$, $\bar{2}=m(C_s)$, $\bar{3}(C_{3i})$, $\bar{4}(S_4)$, and $\bar{6}(C_{3h})$ (see *Symmetry*). The bar

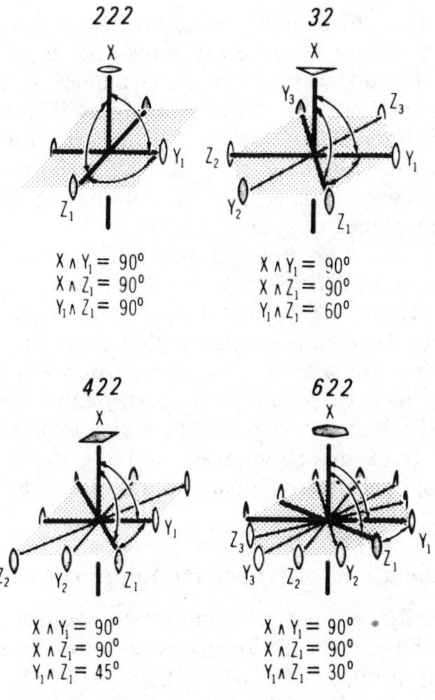

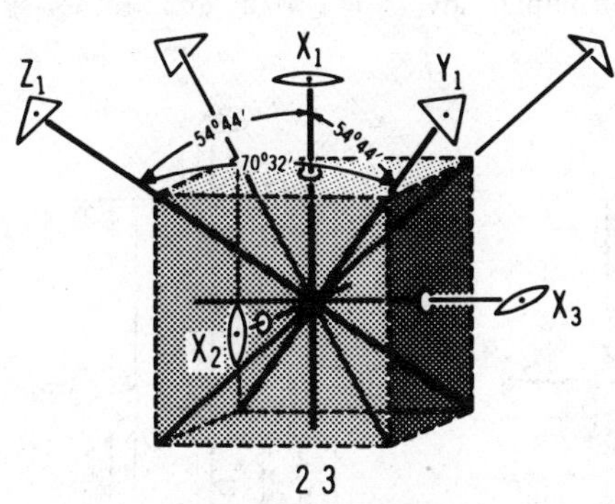

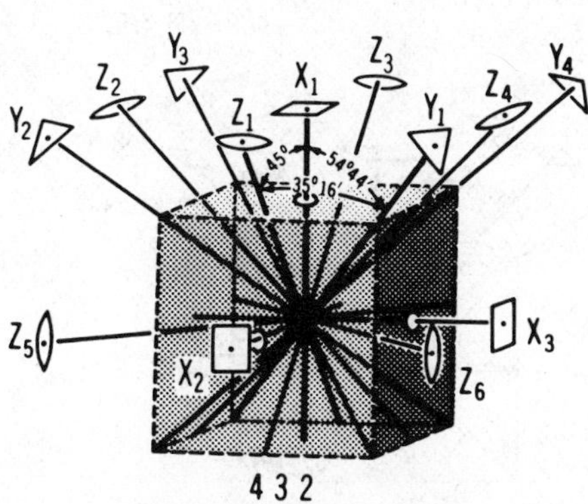

FIGURE 1. Point groups of combined proper rotation axes (from *Crystallography and Crystal Chemistry: An Introduction* by F. Donald Bloss. Copyright © 1971 by Holt, Rinehart and Winston. Reprinted by permission of Holt, Rinehart and Winston).

over the numeral indicating the fold of the axis in the Hermann-Mauguin notation signifies that rotation is followed by inversion through a center of symmetry. The axes are improper in that their operation causes reversal of chirality. In the Schoenflies notation, *i* stands for inversion and *S*, both capital and lower case, for *Spiegelung* (reflection). Each operation of an inversion axis is equivalent to an operation of rotation followed by reflection across a mirror plane at right angles to the axis and is indicated $\tilde{n}$. Thus, $\bar{1} = \tilde{2}$, $\bar{2} = \tilde{1}$; $\bar{3} = \tilde{6}$; $\bar{4} = \tilde{4}$; and $\bar{6} = \tilde{3}$. The subscript *h* (horizontal) indicates that there is a mirror plane of symmetry at right angles to the principal (vertical) rotation axis.

Axes Combined with Mirrors

Three additional point groups combine mirror planes with proper rotation axes at right angles; they are $2/m(C_{2h})$, $4/m(C_{4h})$, and $6/m(C_{6h})$ (Fig. 2). The point groups $1/m = m(C_s)$ and $3/m = \bar{6}(C_h)$ are already considered. Note that these three point groups each have a center of symmetry not designated in the group symbol but generated by the two operations given.

Four additional point groups combine vertical mirror planes of symmetry with a principal rotation axis at their intersection; they are $2mm$ (also $mm2$) (C_{2v}), $3m(C_{3v})$, $4mm(C_{4v})$, and $6mm(C_{6v})$ (Fig. 3). The subscript in the Schoenflies notation indicates both the fold of the vertical axis and the number of vertical mirror planes.

When both horizontal and vertical mirror planes are combined with an axis of rotation, the right-angle intersections of the mirrors generate additional horizontal diads. The five point groups generated are: $2/m\,2/m\,2/m(D_{2h})$, $4/m\,2/m\,2/m(D_{4h})$, $6/m\,2/m\,2/m(D_{6h})$, $2/m\,\bar{3}$ (T_h), and $4/m\,\bar{3}\,2/m(O_h)$ (Fig. 4). The first, consisting of three orthogonal mirror planes with diads at their three intersections, may be abbreviated mmm, since the three mirror planes generate the diads. Likewise, the second

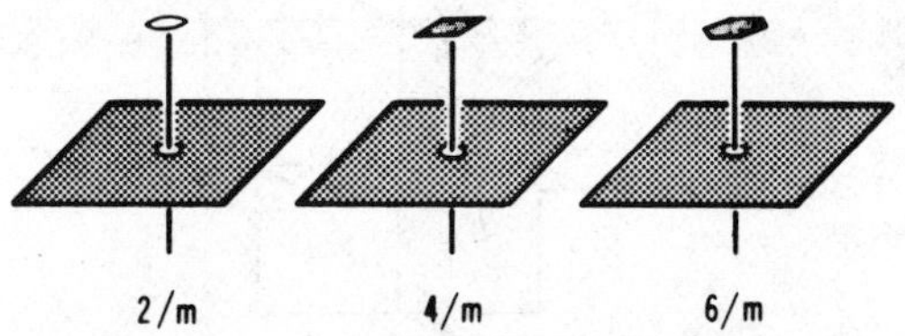

FIGURE 2. Point groups with mirror plane at right angles to symmetry axis (from *Crystallography and Crystal Chemistry: An Introduction* by F. Donald Bloss. Copyright © 1971 by Holt, Rinehart and Winston. Reprinted by permission of Holt, Rinehart and Winston).

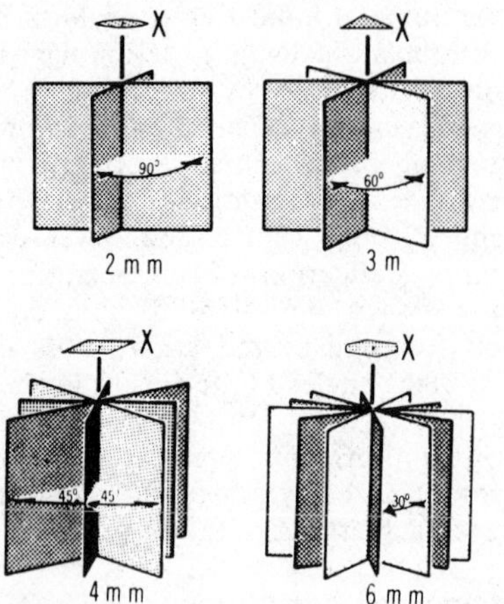

FIGURE 3. Point groups with mirror planes parallel to symmetry axis (from *Crystallography and Crystal Chemistry: An Introduction* by F. Donald Bloss. Copyright © 1971 by Holt, Rinehart and Winston. Reprinted by permission of Holt, Rinehart and Winston).

and the third may be abbreviated $4/m\ mm$ and $6/m\ mm$ because the intersections of horizontal and vertical mirror planes generate horizontal diads. As previously indicated for the Schoenflies notation, the subscript numeral gives the fold of the vertical axis, the number of horizontal diads, and the number of vertical mirror planes of symmetry.

The last two point groups are isometric. In point group $2/m\bar{3}(T_h)$, the three axes are diads with three orthogonal mirror planes of symmetry. Inversion triads parallel the four body diagonals of a cube. The point group $4/m\bar{3}2/m$ (O_h) has three mutually orthogonal tetrads (fourfold axes) with orthogonal mirror planes, four triad inversion axes, and six diads with orthogonal mirror planes parallel to the face diagonals of a cube.

Combinations of Proper and Improper Axes

Finally, proper and improper rotation axes may be combined to give the four remaining point groups: $\bar{3}2/m(D_{3d})$, $\bar{4}2m(D_{2d})$, $\bar{6}m2$ (D_{3h}), and $\bar{4}3m(T_d)$ (Fig. 5). The first one has a principal inversion triad and three sym-

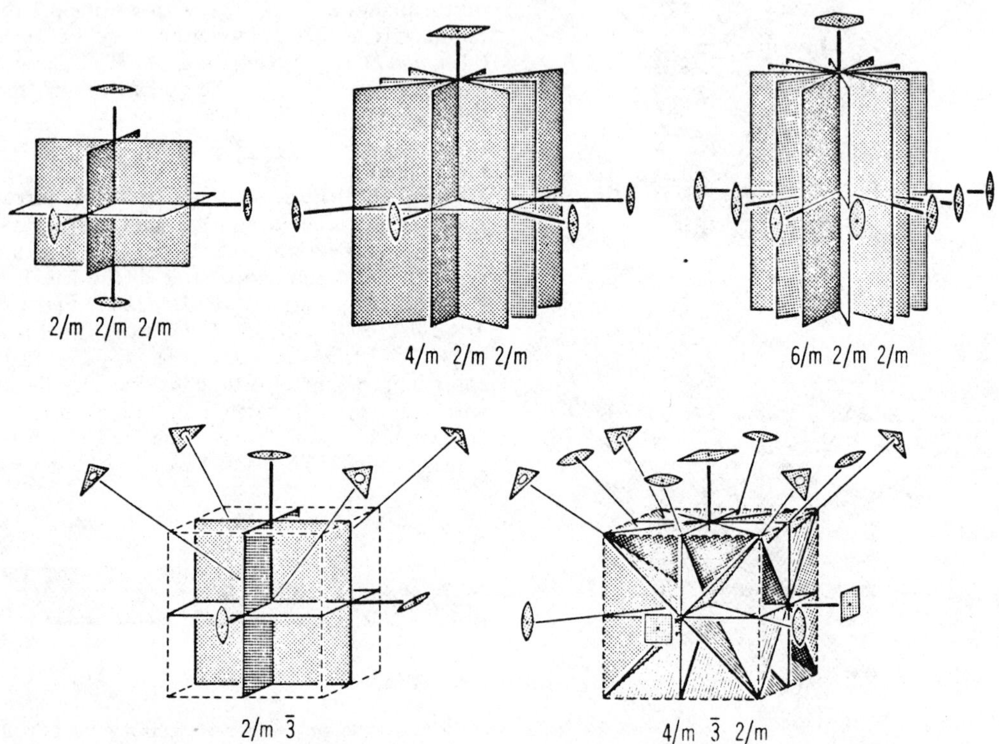

FIGURE 4. Point groups with mirror planes at right angles to and parallel to symmetry axes (from *Crystallography and Crystal Chemistry: An Introduction* by F. Donald Bloss. Copyright © 1971 by Holt, Rinehart and Winston. Reprinted by permission of Holt, Rinehart and Winston). Note that all symmetry elements intersect at a single point.

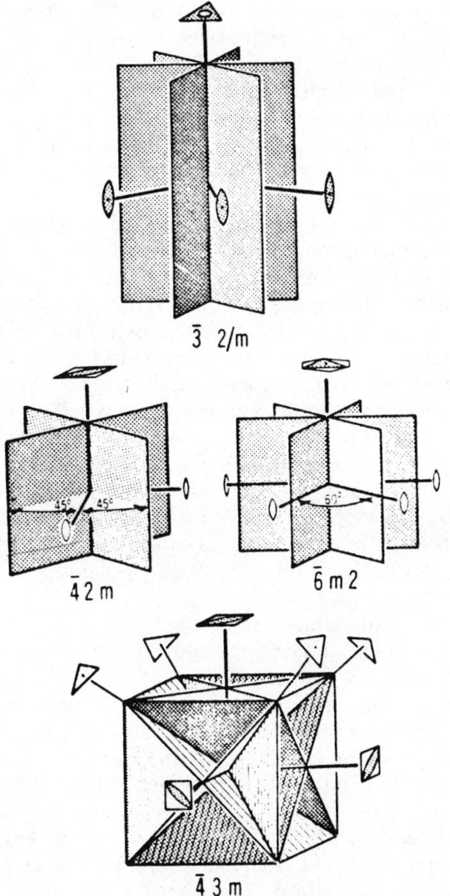

FIGURE 5. Point groups that combine proper and improper rotation axes with mirror planes (from *Crystallography and Crystal Chemistry: An Introduction* by F. Donald Bloss. Copyright © 1971 by Holt, Rinehart and Winston. Reprinted by permission of Holt, Rinehart and Winston).

metrically related horizontal diads with orthogonal mirror planes. The subscript d in the Schoenflies notation indicates that the vertical mirror planes are arrayed diagonally to the horizontal diads. The point groups $\bar{4}2m(D_{2d})$ and $\bar{6}m2(D_{3h})$ have principal inversion axes as indicated and two and three horizontal diads with orthogonal mirror planes of symmetry. Since $\bar{6}=3/m$, there is, in addition, a horizontal mirror plane of symmetry in the latter point group. The $\bar{6}$ notation is preferred for purposes of classification of the point group. The last point group is isometric with three mutually orthogonal inversion tetrads, four diagonal triads, and six diagonal mirror planes parallel to the face diagonals of the cube.

Point groups are classified according to the lattices in which they can reproduce spatial identity. All point groups with at least four diagonal triads of symmetry, as indicated by the numeral 3 in the second position in the Hermann-Mauguin notation and by the letters O and T in the Schoenflies notation, apply to lattices assigned to the isometric (cubic) crystal system (Table 1).

TABLE 1. Crystal-Class Symbols

Crystal System	Schoenflies Notation	Hermann-Mauguin Notation
Triclinic	C_1	1
	$C_i(S_2)$	$\bar{1}$
Monoclinic	C_2	2
	$C_s(C_{1h})$	m
	C_{2h}	$\frac{2}{m}$
Orthorhombic	$D_2(V)$	222
	C_{2v}	$mm2$
	$D_{2h}(V_h)$	$\frac{2}{m}\frac{2}{m}\frac{2}{m}$ (mmm)
Tetragonal	C_4	4
	D_4	422
	S_4	$\bar{4}$
	C_{4h}	$\frac{4}{m}$
	C_{4v}	$4mm$
	$D_{2d}(V_d)$	$\bar{4}2m$ or $\bar{4}m2$
	D_{4h}	$\frac{4}{m}\frac{2}{m}\frac{2}{m}$ $(4/mmm)$
Trigonal	C_3	3
	D_3	32
	$C_{3i}(S_6)$	$\bar{3}$
	C_{3v}	$3m$
	D_{3d}	$\bar{3}\frac{2}{m}$ $(\bar{3}m)$
Hexagonal	C_{3h}	$\bar{6}$
	D_{3h}	$\bar{6}m2$ or $\bar{6}2m$
	C_6	6
	D_6	622
	C_{6v}	$6mm$
	C_{6h}	$\frac{6}{m}$
	D_{6h}	$\frac{6}{m}\frac{2}{m}\frac{2}{m}$ $(6/mmm)$
Isometric	T	23
	O	432
	T_h	$\frac{2}{m}\bar{3}$ $(m3)$
	T_d	$\bar{4}3m$
	O_h	$\frac{4}{m}\bar{3}\frac{2}{m}$ $(m3m)$

The hexagonal system may be divided into hexagonal and trigonal subsystems according to whether the minimal lattice symmetry is a hexad or a triad of rotation or inversion as indicated by the initial numeral in the Hermann-Mauguin notation and by the subscript numeral in the Schoenflies notation (except C_{3h} and D_{3h} which are hexagonal since $3/m = \bar{6}$).

Point groups belong to lattices assigned to the tetragonal system if the principal axis is a tetrad and there are no diagonal triads. Orthorhombic point groups have at least three mutually orthogonal diads and monoclinic point groups have at least a single diad (note, $m = \bar{2}$). If a point group operates on a lattice having no more than a center of symmetry ($i = \bar{1}$), it belongs to the triclinic crystal system.

KEITH FRYE

References

Bernal, I.; Hamilton, W. C.; and Ricci, J. S., 1972. *Symmetry: A Stereoscopic Guide for Chemists.* San Francisco: Freeman, 182p.

Bloss, F. D., 1971. *Crystallography and Crystal Chemistry.* New York: Holt, Rinehart & Winston, 545p.

Boisen, M. B., and Gibbs, G. V., 1976. A derivation of the 32 crystallographic point groups using elementary theory, *Am. Mineralogist*, **61**, 145–165.

Buerger, M. J., 1963. *Elementary Crystallography.* New York & London: Wiley, 528p.

Fackler, J. P., 1971. *Symmetry in Coordination Chemistry.* New York & London: Academic, 139p.

Ferraro, J. R., and Ziomek, J. S., 1975. *Introductory Group Theory*, 2nd ed. New York & London: Plenum, 292p.

Henry, N. F. M., and Lonsdale, K., eds., 1965. *International Tables for X-ray Crystallography*, vol. 1. Birmingham, England: Kynoch Press.

Megaw, H. D., 1973. *Crystal Structures: A Working Approach.* Philadelphia: Saunders, 563p.

Nussbaum, A., 1971. *Applied Group Theory for Chemists, Physicists and Engineers.* Englewood Cliffs, N. J.: Prentice-Hall, 384p.

Cross-references: *Crystal Habit; Crystallography: History; Crystallography: Morphological; Directions and Planes; Lattice; Space-Group Symbol; Symmetry.*

POLARIZATION AND POLARIZING MICROSCOPE

Electromagnetic theory assumes that light may be considered as consisting of pulsating energy waves characterized by simultaneous interdependent variations of intensities of electrical and magnetic fields. Visible effects of light waves are related to the magnitudes of variations in the intensity of the electrical field and to the frequencies of such variations, and it is convenient to represent the behavior of the electrical field in time and space by means of electric vectors, also called "light vectors."

The simplest representation of a light wave combines propagation with uniform velocity in a given direction with an oscillating light vector, the magnitude of which periodically increases from zero to some positive maximum, decreases to zero, decreases still further to a negative maximum, and then increases to zero. If the light vector is thought of as oscillating in a plane, the light is described as "plane-polarized" or "linearly polarized," and the tip of the light vector describes a sine wave as the wave advances in the direction of energy flow. Other systematic motions of the tip of the light vector define either "circularly polarized light" or "elliptically polarized light."

The polarizing microscope is equipped with two linear polarizers, also called "polars," constructed of *calcite* or of Polaroid. The "polarizer" intercepts nonpolarized light from a source of illumination before it is incident on an object on the stage of the microscope and converts it into linearly polarized light vibrating in a fixed azimuth. Light from the polarizer incident on a birefringent object on the microscope stage, in general, is resolved into two linearly polarized components vibrating in mutually perpendicular planes and a phase difference is produced between the two components during either reflection or transmission.

The light reflected from or transmitted by the birefringent object now passes through an "analyzer" which normalizes the mutually perpendicular light vibrations from the birefringent object, renders the vibrations quasi-coherent, and enables the vibrations to interfere in the analyzer so as to produce visible interference effects. The particular interference effects depend on the orientation of the object on the microscope stage and the phase differences produced by it.

When the polarizing microscope is used as an "orthoscope" using both polars, a magnified image of an object as modified by interference effects is seen on the microscope stage. Insertion of an auxiliary Amici-Bertrand lens in the tube of the microscope converts the polarizing microscope into a "conoscope," a low-power telescope, and permits viewing of "interference figures," which form by interference in the analyzer of light that has been brought to a focus in the back (upper) plane of the objective lens system.

The polarizing microscope has an important use in the identification of crystalline or noncrystalline substances in single grains or aggregates by measurement or estimation of optical properties. By judicious observation and interpretation of interference effects and correlation with crystal morphology, it is possible to ascertain that the orientation of a birefringent substance on the microscope stage is or is not appropriate for the measurement of critical optical properties.

Fundamental optical properties of colorless, nonopaque substances are the principal refractive indices and the crystallographic orientations of the vibration directions of transmitted light waves corresponding to the refractive indices, all stated for one or more standard wavelengths (colors) of light. An additional property in colored substances is the amount and degree of absorption of light of various frequencies as a function of direction of vibration. Absorption more or less modifies interference effects observed under the orthoscope or conoscope.

A typical polarizing microscope is equipped with a substage condensing-lens system containing a rotatable linear polarizer (the lower polar or polarizer), iris diaphragms for controlling intensity of illumination and effective angular aperture of the objective lens system, and auxiliary removable or vertically adjustable lenses providing a capability for varying the numerical aperture of the lens system. Above a rotatable stage, provisions are made for interchanging a variety of objective lenses of different magnifications. A removable, rotatable linear polarizer (the upper polar or analyzer) intercepts light from an objective lens before it reaches the ocular lenses. For viewing interference figures, an Amici-Bertrand lens may be inserted between the ocular and the upper polar.

For viewing objects in reflected light, a vertical illuminator containing a linear polarizer is attached to the microscope tube (see *Ore Microscopy*).

Many so-called "optical accessories" are available to assist in measurements. The most widely used accessories are plates or wedges of birefringent substances that permit increase or decrease, under controlled conditions, of the phase difference produced by an object on the microscope stage; these are called "compensators." Stage accessories include mechanical stages and universal stages, which permit rotation of stage objects into desired positions.

Microscope illumination is provided by a "white-light" source, or for special purposes, by spectral or filtered light sources or a monochromater. Precision measurements require the use of illumination that has been rendered as nearly monochromatic as feasible or possible.

ERNEST E. WAHLSTROM

References

Cameron, E., 1961. *Ore Microscopy*. New York: Wiley, 293p.

Kerr, P. F., 1977. *Optical Mineralogy*, 4th ed. New York: McGraw-Hill, 492p. (Previous editions by Rogers and Kerr.)

Wahlstrom, E. E., 1969. *Optical Crystallography*, 4th ed. New York: Wiley, 489p.

Wood, E. A., 1964. *Crystals and Light*. Princeton: Van Nostrand, 160p.

Cross-references: *Anisotropism; Color in Minerals; Crystallography: Morphological; Dispersion, Optical; Isotropism; Minerals, Uniaxial and Biaxial; Optical Mineralogy; Ore Microscopy; Refractive Index.*

POLYCRYSTAL

A polycrystal is a heterogeneous crystalline edifice that simulates a single crystal but is composed of two (or more) structurally or chemically distinct species, syntactically intergrown (Donnay, 1953). Example: In the system Na_2SO_4–NaF–NaCl, *schairerite* and *galeite*, two forms of $Na_3SO_4(F,Cl)$, form polycrystals (Pabst, Sawyer, and Switzer, 1963). Two chemically different species may form polycrystals under the following conditions: (1) absence of complete solid solution, (2) similarity of crystal structures, and (3) a common pseudocell whose dimensions may vary by a few percent from one compound to the other (the true cells may vary widely both in dimensions and in symmetry).

J. D. H. DONNAY

References

Donnay, G., 1953. The "polycrystal," a product of syntaxic intergrowth, *Am. Cryst. Assoc. Abstracts*, p. 16.

Pabst, A.; Sawyer, D. L.; and Switzer, G., 1963. Galeite and related phases in the system Na_2SO_4–NaF–NaCl. *Am. Mineralogist*, **48**, 485–510.

Cross-reference: *Syntaxy.*

POLYMORPHISM

Polymorphism is a phenomenon of the solid crystalline state in which a single chemical species (element or compound) can exist with more than one arrangement of atoms or ions—with more than one crystalline structure. Many well-known substances occur in the earth's crust as polymorphs identified by different mineral names—carbon as *graphite* or *diamond*, titanium dioxide as *rutile, anatase*, or *brookite*, for example.

The occurrence of a particular polymorph of a substance is determined by the conditions of temperature, pressure, or composition present at the time of crystallization. When silica crystallizes from a solution within the temperature range 573° to 870°C at normal pressure the crystalline phase is known as high *quartz*; if at 870° to 1470°C *tridymite* forms, and at 1470° to 1720°C *cristobalite*. If formed below 573°C, it is low *quartz*. If high *quartz* is cooled below 573°C, it transforms rapidly to low *quartz*, and the reverse occurs when low *quartz* is heated above 573°C (Fig. 1).

The low-high (alpha-beta) transformation in *quartz* is termed "enantiotropic" since it is reversible when temperature is raised or lowered through the inversion point at constant pressure. This transformation is often known as "displacive" because the change involves changes in bond angle. In such cases, a whole crystal can change from one form to the other, remaining as a single crystal. The high-temperature forms usually have higher symmetry, lower density, and smaller coordination numbers. *Tridymite* and *cristobalite* also show temperature transformations of the alpha-beta type. The alpha forms lie in the phase field of *quartz* but can be observed because the transformations *cristobalite–tridymite–quartz* are of the "reconstructive" type and proceed very slowly in the absence of fluxes (see Vol. IVA: *Phase Equilibria*).

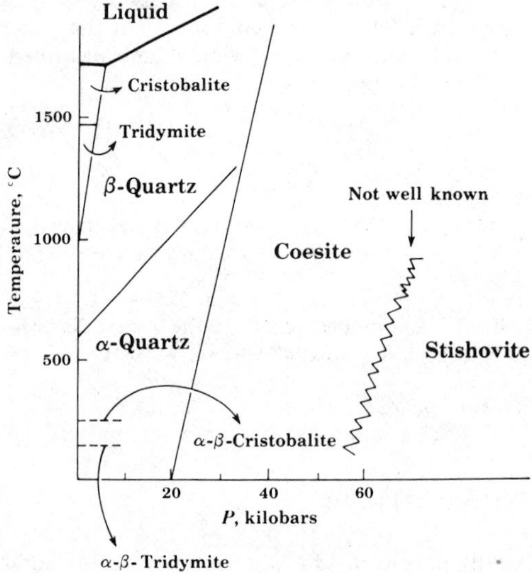

FIGURE 1. Regions of stability of the various polymorphs of SiO_2 (from Fyfe, 1964). The $\alpha = \beta$ transitions are rapid, but all other types tend to be sluggish.

Reconstructive transformations involve the breaking of existing bonds and, at times, a change in the coordination number. In *diamond*, C occurs in tetrahedral coordination and in *graphite* with trigonal coordination.

The low-high transformation temperature for *quartz* increases with pressure. At pressures of 35 kilobars and temperatures of 500° to 800°C, SiO_2 crystallizes as *coesite*, monoclinic but pseudohexagonal, with density 3.01 g/cm^3. *Coesite* has been identified in meteor craters, where it was probably formed under an impact shock wave in excess of 20 kilobars. At higher pressures (about 100 kilobars), SiO_2 forms tetragonal crystals of *stishovite* in which Si is in sixfold coordination with oxygen (the *rutile* structure), with a density of 4.28 g/cm^3. High pressures favor compact structures with high density and large coordination numbers.

Some substances can occur in a structure that lies in a region where another form is more stable. The less stable form is a "monotropic" polymorph. *Marcasite* is a good example. The transformation *marcasite*→*pyrite* is monotropic; it can proceed only toward *pyrite* in the solid state. With monotropic polymorphs, the unstable form always tends to change to the stable form but the stable structure cannot be changed into the unstable unless completely destroyed by melting, vaporization, or solution.

In some substances with significant solid solution, a high-temperature form occurs in which the elements of solid solution are present in disordered relation while at lower temperature these elements may take on an ordered array (see *Order–Disorder*). The ordered phase tends to be more compact and lower in symmetry. The disordered and ordered structures may be considered polymorphs. In other systems, the disordered phase breaks up (exsolves) on cooling to an intergrowth of two phases.

Sanidine, a monoclinic alkali **feldspar** (see *Alkali Feldspars*), crystallizes at about 1000°C in volcanic rocks. In this phase, $(K,Na)AlSi_3O_8$, sodium and potassium are disordered with respect to each other in the K structural site and Al and Si are disordered in the tetrahedral silicon sites, with 25% Al randomly distributed in each of the two eightfold Si sites. *Orthoclase*, monoclinic $KAlSi_3O_8$ with reduced substitution of sodium for potassium, shows partial ordering of Al with 32% randomly distributed in one of the eightfold Si sites and 18% in the other. *Microcline*, which crystallizes in plutonic, metamorphic, or sedimentary rocks, is triclinic. Structural studies show that the Al is highly ordered and concentrated in one of the

TABLE 1. Polymorphs of Alkali Feldspars

	Symmetry	γ^*	Tetrahedral	Al	Si
Sanidine	Monoclinic	90°	$8T_1$	2	6
$Or_{92}Ab_7An_1$ [a]			$8T_2$	2	6
Orthoclase	Monoclinic	90°	$8T_1$	2.4	5.6
$Or_{92}Ab_7An_1$			$8T_2$	1.6	6.4
Microcline	Triclinic	90° 50′	$4T_1(O)$	2.6	1.4
$Or_{85}Ab_{13}An_1$	(intermediate)		$4T_1(m)$	1.0	3.0
			$4T_2(O)$	0.12	3.88
			$4T_2(m)$	0.04	3.96
Microcline	Triclinic	92° 13′	$4T_1(O)$	3.68	0.32
(authigenic)	(maximum)		$4T_1(m)$	0.08	3.92
$Or_{97}Ab_3$			$4T_2(O)$	0.24	3.76
			$4T_2(m)$	0.00	4.00
Microcline	Triclinic	92° 13′	$4T_1(O)$	3.76	0.24
(igneous)	(maximum)		$4T_1(m)$	0.12	3.88
$Or_{95}Ab_5$			$4T_2(O)$	0.04	3.96
			$4T_2(m)$	0.08	3.92
Albite (high)	Triclinic	88° 01′	$4T_1(O)$	1.20	2.80
"nearly pure"[b]			$4T_1(m)$	0.84	3.16
			$4T_2(O)$	0.92	3.08
			$4T_2(m)$	1.04	2.96
Albite (low)	Triclinic	90° 28′	$4T_1(O)$	3.76	0.24
"nearly pure"			$4T_1(m)$	0.00	4.00
			$4T_2(O)$	0.16	3.84
			$4T_2(m)$	0.08	3.92

[a]Heated at 1075°C.
[b]Heated at 1050°C.

four fourfold Si sites. *Microcline* with "intermediate" triclinic character shows about 60% ordering, but with "maximum" triclinic character shows over 90% of the aluminum in one fourfold site (Table 1). The maximum *microclines* contain very little Na. Thus *sanidine, orthoclase*, and *microcline* are polymorphs of $KAlSi_3O_8$ (see *Feldspar Group*).

A high-sodium *sanidine* or *orthoclase*, on recrystallizing at lower temperature, separates into a perthitic or microperthitic intergrowth of *microcline* and *albite* (or high-soda ***plagioclase***). *Albite,* $NaAlSi_3O_8$, also shows polymorphism with a disordered form at high temperature and an ordered form at low.

LEONARD G. BERRY

References

Azaroff, L. V., 1960. *Introduction to Solids.* New York: McGraw-Hill.

Dana, E. S., 1962. *System of Mineralogy*, vol. 3. New York: Wiley, 334p.

Fyfe, W. S., 1964. *Geochemistry of Solids.* New York: McGraw-Hill, 199p.

McKinstry, H. A., 1965. Precision crystal structure analysis, key to understanding phase transitions, *Miner. Ind. J.*, **34**(7), 1–8.

Smith, J. V., 1974. *Feldspar Minerals*, vol. 1. New York: Springer-Verlag, 627p.

Verma, A. R., and Krishna, P., 1966. *Polymorphism and Polytypism in Crystals.* New York: Wiley, 341p.

Cross-references: *Coesite and Stishovite; Feldspars; Order–Disorder; Quartz;* see also mineral glossary.

POLYSOMATISM

Some crystal structures may be regarded as being made up of modules, or structural chemical units, taken from other crystals or crystal structures. If all the modules are identical layers and the different structures differ only in the way in which modules are stacked, the phenomenon is called "polytypism" (see *Phyllosilicates*). If more than one kind of module is required conceptually to build a three-dimensional crystal structure, it is called "polysomatism."

The modules used in polysomatic series may be rows, "beam modules," or they may be layers, "layer modules." A beam or a layer module need not, however, consist of rows or layers of complete unit cells, although it may. A single chain of silica tetrahedra, $(SiO_3)_n$, in a pyroxene structure (see Vol. IVA: *Mineral Classes: Silicates*) or a row of unit cells parallel to the *c* axis of the pyroxene structure (see *Pyroxene Group*) may be taken as an example of a beam module.

A modular description of *kaolinite* (see *Clays, Clay Minerals*) could consist of half a gibbsite layer $[Al_2(OH)_6]$ adjoined to half a pyrophyllite layer $[Al_2Si_4O_{10}(OH)_2]$ to yield the *kaolinite* structure and composition $[Al_4Si_4O_{10}(OH)_6]$. Other ***phyllosilicates*** may be described in a similar fashion.

The structures of the minerals of the ***amphibole*** group (q.v.) may be described as consisting of mica-group modules and pyroxene-group modules. For example, the ***clinoamphibole*** *tremolite* $[Ca_2Mg_5Si_8O_{22}(OH)_2]$ consists structurally and chemically of a talc module $[Mg_3Si_4O_{10}(OH)_2]$ and a diopside module $[Ca_2Mg_2Si_4O_{12}]$. Other members of this polysomatic series are tabulated in *Biopyriboles* (q.v.).

Other polysomatic series include the ***humite*** group, consisting of norbergite and forsterite modules, and ***pyroxenoids***, consisting of pyroxene and wollastonite modules. The concept of polysomatism, in addition to being a useful pedagogical device, serves to point up gaps where new mineral structures and compositions may be sought by a mineralogist.

KEITH FRYE

Reference

Thompson, J. B., Jr., 1978. Biopyriboles and polysomatic series, *Am. Mineralogist*, **63**, 239–249.

Cross-references: *Amphibole Group; Clays, Clay Minerals; Phyllosilicates; Polymorphism; Pyroxene Group.* Vol. IVA: *Mineral Classes: Silicates.*

POLYTYPY–*See* PHYLLOSILICATES

PORTLAND CEMENT MINERALOGY

Portland cement is a nearly universal building material possessing relatively few physical limitations; it can be mixed with aggregate and water and used successfully under extremely adverse conditions. However, its relative insensitivity to its environment masks a complex set of chemical reactions that lead to its ultimate strength and durability. Portland cement is the most important of the hydraulic cementing materials (i.e., will set up under water); this article, therefore, will limit itself to discussing the manufacture and hydration reactions of Portland cement and its derivatives as it finds application in the construction industry and in oil-well technology.

History

The use of a Portland-like cement can be traced to the ancient Egyptians. They seem to have been the first to burn native stone to a high enough temperature to make it more reactive to the later addition of water. ("Burn" or "burnt" is used in cement technology to name the process of heating raw materials to high enough temperatures so that definite chemical changes occur whereby a new reactive product is formed.) Their mortar seems to have been dehydrated *gypsum* ($CaSO_4 \cdot 2H_2O \rightarrow CaSO_4 + H_2O$) and was a direct forerunner of present-day plaster. The first people to use burnt lime ($CaCO_3 \rightarrow CaO + CO_2$) with or without sand aggregate were the Greeks. This process was then adopted by the Romans, whose buildings even today reflect the sophistication of their knowledge about the properties of lime-sand cements. During the Middle and Dark Ages, this technologic sophistication was lost, only to reappear in about the 14th to 17th century. Portland cement is a direct descendent of these early lime-sand mixtures and has its origin in the pioneering experiments of John Smeaton (1756), James Parker (1796), L. J. Vicat (1818), Joseph Aspdin (1824), and Isaac C. Johnson (1851) (see Lea, 1970).

Initially, Portland cement was manufactured from naturally occurring clay-rich limestone deposits having the proper $CaO:SiO_2$ ratio and minor element concentrations. However, as the demand for Portland cement increased and the number of cement plants spread, these natural deposits were soon exhausted. It was then realized that Portland cement could also be manufactured from individual clays and limestones located in other parts of the country. Transportation of one or both raw materials to existing plants became an increasing cost consideration. Needless to say, locally available raw materials were used whenever possible.

Nature of Portland Cement

Composition and Properties. Portland cement consists of four main compounds listed in order of their importance: β-dicalcium silicate (β-Ca_2SiO_4), tricalcium silicate (Ca_3SiO_5), tricalcium aluminate ($Ca_3Al_2O_6$), and calcium aluminoferrite (Ca_2AlFeO_5).

These compounds are formed by carefully controlling the composition of the raw materials (limestone and clay) such that, when these materials are burnt (heated to 1300° to 1500°C), they form a solid material (clinker) consisting of crystals and interstitial liquid. After burning, the solid clinker is cooled and ground to a fine powder, with the addition of about 5% *gypsum* ($CaSO_4 \cdot 2H_2O$) to aid in the hydration process of the cement. As more sophisticated needs arose, different kinds of Portland cements were developed, and at present there are a total of five Portland cements described in ASTM (*American Society for Testing and Materials*, q.v.) standards as types I through V. Types I and II are general-purpose cements. Type III is a rapid-hardening cement having more C_3A and C_3S (see Table 1), which have properties of rapid hydration. Type IV is a low-heat Portland cement with very little C_3A and a relatively high $C_2S:C_3S$ ratio to reduce the exothermic heat buildup that would occur in large poured structures, such as dams, if a type I or II cement were

TABLE 1. Commonly Used Abbreviations and Compositions of Portland-Cement Clinker Materials

C	CaO
S	SiO_2
A	Al_2O_3
F	Fe_2O_3
$\bar{S}$	SO_3
H	H_2O
C_2S	Ca_2SiO_4
C_3S	Ca_3SiO_5
C_3A	$Ca_3Al_2O_6$
C_4AF	$2(Ca_2AlFeO_5)$
C_2A	$Ca_2Al_2O_5$
C_2F	$Ca_2Fe_2O_5$
$C_{12}A_7$	$Ca_{12}Al_{14}O_{33}$
CA_2	$CaAl_4O_7$
CA	$CaAl_2O_4$

used. Type V is a sulfate-resisting Portland cement, low in C_3A content, which makes it resistant to the deterioration effects caused by salt water on Type I and III cements.

In addition to cements used in construction, there exists a separate API (American Petroleum Institute) classification of oil-well Portland cements that conform to the specialized requirements of conditions encountered within the earth's crust. Instead of numerals (Type I, II, etc.), these cements are classified by letters (A, B, C, etc.). The main difference between types of oil-well cement is the rate at which they will harden under pressure and temperature conditions encountered at depth. The classification includes cements having a normal Portland cement–like composition that are used under near surface conditions, as well as cements of a hydrothermal nature, which are resistant to the effects of elevated pressures and temperatures so that they can be successfully pumped to great depths (in the neighborhood of 5000 feet or greater) without setting up and hardening on the way down.

Whereas in the Type I to V cements temperature was the only variable, in cementing an oil well, varying pressure as well as temperature becomes an important factor to be considered. Class A, B, and C cements are essentially similar to Types I, II, and III, respectively, and are used in holes up to 1830m (6000 ft) deep. The main difference between the Class A to C and Type I to III cements are the former's resistance to possible attack by sulfate-containing groundwater. Classes D and E are used from 1830 to 3050 m and from 3050 to 4720 m (6000 to 12,000 ft and 6000 to 14,000 ft) in depth, respectively, and are available in moderate or high sulfate-resistant types. Class F is intended for use in 3050 to 4880 m (10,000 to 16,000 ft) holes and is also available in moderate or high sulfate-resistant types. In recent years, due to deeper oil well drilling techniques, new classes of oil well cements are being developed. There now exist Classes G, H, and J cements, which are used for ultradeep hole cementing. Class J is the newest cement, additionally classified as a hydrothermal cement, which implies that ultimate strength is developed only under high pressure and temperature conditions.

Raw Materials. Commonly occurring rocks are the source of raw materials used in Portland cement manufacture. At first Portland cement was manufactured from cement rock, an argillaceous limestone containing the source of both lime and silica, which has been found in various parts of the world. However, as more and more cement plants were constructed, less and less suitable sites having the necessary cement rock were available. As a result, it was soon discovered that a mixture of limestone and clay or shale would serve just as well. Cement plants were then constructed near the source of a single raw material. Limestones having only a small amount of magnesium became the major source of lime for Portland cement, while clays and blast furnace slag became the main source of aluminosilicates. Essentially any source of silica and lime may be used, along with small amounts of alumina and iron oxide, as long as the defined compositional limits are met, but, as always, processing and economic considerations favor certain materials over others.

Manufacture. The manufacturing process is roughly divided into four steps. The raw materials (limestone and clay) are first ground, either wet or dry (after beneficiation, if necessary), and then intimately mixed, after which they pass through a rotary kiln. The kiln itself consists of a steel cylindrical shell, lined with refractories, up to 200 m (600 ft) long and 7 m (23 ft) in diameter. The temperature of the mixture is slowly raised in the kiln to 1300° to 1500°C. During this slow heating, any water that adheres to the sample is driven off—between 100° and 110°C. With increasing temperatures up to about 600°C, the water of hydration and hydroxyl water begin to be slowly driven off. At about 800° to 850°C, calcium carbonate begins to decompose to yield lime (CaO) and CO_2. This is the beginning of the clinkering process. The reactivity of the components starts to increase. The CaO begins to combine with aluminate and silicate to produce the final required compounds. As the temperature approaches 1300° to 1500°C, a small amount of melt is produced. This melt wets the crystals and produces an easy path for the interdiffusion of various cations, enabling more rapid reaction. Finally, the semisolid mass is cooled in a blast of incoming air, producing the clinker. The nodules of cement clinker, now having the predetermined combination of C_3A, C_3S, β-C_2S, and C_4AF, are passed through a grinding mill where *gypsum* is added, and they are reduced to the final cement particle size (1–10 μm). This process is represented schematically in Fig. 1. To meet modern air pollution standards, it is often necessary to fit the kiln with cyclone arrestors, filter bag systems, or electrostatic precipitators to minimize dust emission.

Chemistry of Kiln Reactions. The basic phase diagram of cement chemistry is that of the system CaO-Al_2O_3-SiO_2 (Figs. 2, 3). It illustrates approximately the course of melting of cement compositions. The first melting of liquids in the composition triangle C_3S-C_2S-

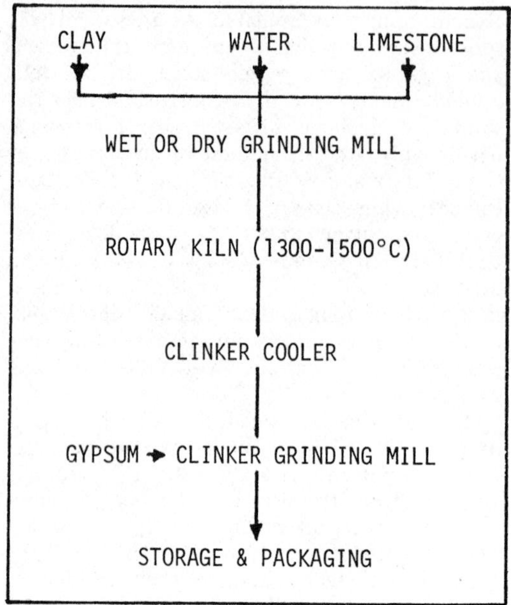

FIGURE 1. Manufacturing process for Portland cement.

C_3A (see Fig. 3), which would be characteristic of a "white," low-iron, cement composition, takes place at the ternary peritectic *Y* at a temperature of 1455°C.

The addition of small percentages of Fe_2O_3 allows a lower kiln temperature to be used, as the ternary peritectic is replaced by a quaternary invariant point in equilibrium also with C_4AF. First liquid formation from compositions in the tetrahedron C_3S-C_2S-C_3A-C_4AF takes place at 1338°C. This lower melt-formation temperature permits use of a lower kiln temperature, affording considerable fuel economies and also allowing greater leeway in raw material composition. The total kiln chemistry is, of course, much more complex since all three of the cement compounds exist in polymorphic modifications, all form solid solutions within the C-A-F-S system, and all incorporate additional minor components into solid solution. The composition C_4AF, for example, is a simplified expression for a solid-solution series between C_2A and C_2F.

A new family of "regulated set" cements can be derived from compositions in the triangle C_3S-C_2S-C_3A, with the addition of CaF_2 to produce quaternary compositions to forming a new, rapid-hydrating compound, $11CaO \cdot 7Al_2O_3 \cdot CaF_2$. Another relatively recent development is that of expansive cements that contain significant amounts of sulfate and aluminate. One way in which they are made involves formation of the compound $3CaO \cdot 3Al_2O_3 \cdot CaSO_4$, although the aluminate and sulfate component can be incorporated by using other compounds as well.

A class of cements having refractory properties, called high-alumina cements, is based on calcium aluminate compositions but also contains lesser quantities of silica and iron oxide. They hydrate rapidly at room temperature to form relatively high-strength cements, yet when they are heated, they do not disintegrate, but sinter to form a refractory body. $CaO \cdot Al_2O_3$ is the main component, usually mixed with C_4AF, another aluminate [either $C_{12}A_7$ (similar crystal structure to $11CaO \cdot 7Al_2O_3 \cdot CaF_2$, above) or CA_2], and a silicate phase (a complex compound, pleochroite). The high-alumina cements are commonly known by the French name *ciment fondu*, referring to the fact that appreciable melting takes place during clinkering. The temperature of the $C_{12}A_7$–C_2S-CA invariant point is quite low, 1335°C (point *Z* in Fig. 3) (lower with addition of Fe_2O_3), and compositions are such that a large portion is melted during manufacture.

Mineralogy of Clinker. The composition Ca_3SiO_5 (C_3S) exists in six allotropic forms, three triclinic, two monoclinic, and one trigonal, related by displacive transitions. Solid solution with other components present in the clinker tends to stabilize one of the higher-temperature monoclinic or trigonal modifications, preventing inversion to one of the lower-temperature forms. The resultant clinker phase is called alite. Pure C_3S is unstable below about 1250°C; therefore, it can disproportionate on prolonged heat treatment at lower temperatures.

Six crystalline forms of C_2S (Ca_2SiO_4) exist between ambient and 1500°C. The monoclinic β form, which is the most common modification in clinker, also occurs as the natural mineral *larnite*. All except the lowest-temperature γ (olivine structure) form have been identified as solid solutions in clinker and are collectively called belite.

Pure C_3A ($Ca_3Al_2O_6$) is cubic and hence undergoes no polymorphism. Solid solutions with alkali, however, result in formation of one orthorhombic, one tetragonal, and one monoclinic form.

Brownmillerite (Ca_2AlFeO_5, C_4AF) is the mineralogical name given to the phase found in cement clinker. It is a member of the solid solution series represented by $C_2A_pF_{1-p}$, where p = 0–0.7. The structure of C_2F is orthorhombic, pseudotetragonal. Additional solid solution of Mg, Si, Ti, Mn, and Cr exists in the phase in clinker.

Other components, SO_3, P_2O_5, MgO, TiO_2, and alkalis, are the common minor components

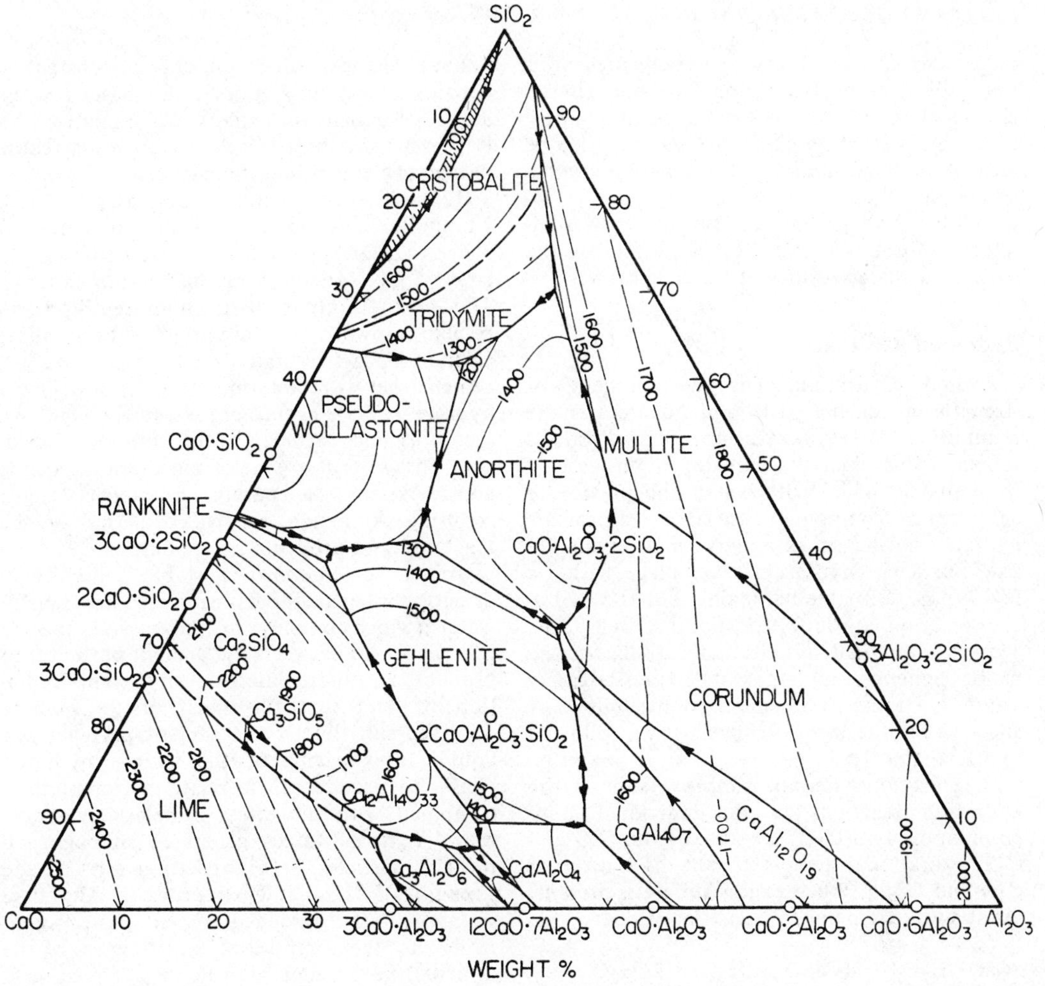

FIGURE 2. Diagram of the system $CaO–Al_2O_3–SiO_2$ (from A. Muan and E. F. Osborn, *Phase Equilibria Among Oxides in Steelmaking*, Reading, Mass.: Addison-Wesley, 1965, p. 95).

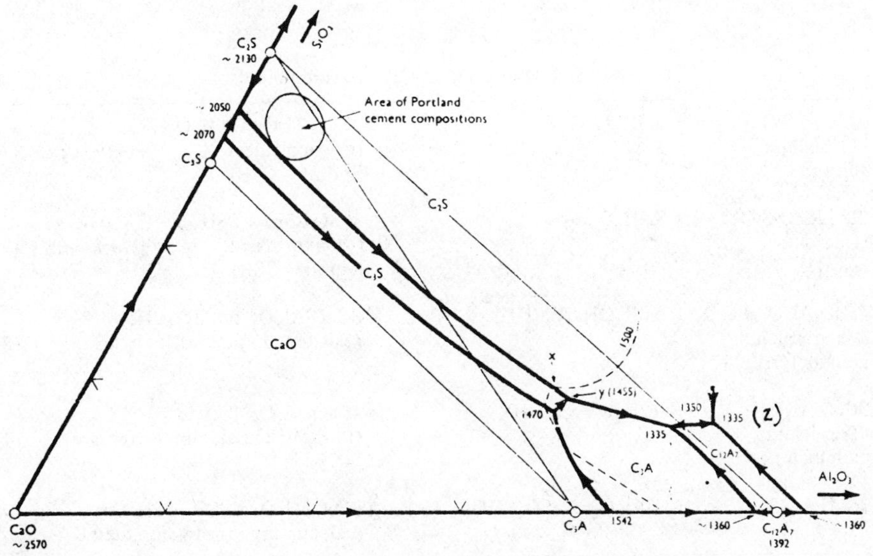

FIGURE 3. Diagram of the lime-rich corner of the system in Fig. 2 (from R. B. Williamson, Solidification of Portland Cement, *Prog. Materials Sci.*, 15, p. 194).

of cement clinker. They are usually present as solid solution in the major Portland cement clinker minerals but may exist in small quantity as double salts. Free MgO or CaO may also be present in small quantity as a separate crystalline phase.

Each crystallographic modification or solid-solution formation may have an effect on the hydration characteristics of the final cement.

Hydration Reactions

Normal Hydration. The development of strength in cement paste and concrete is the result of a hydration reaction. When Portland cement (plus "inert" aggregate, in the case of concrete) is mixed with water, the previously anhydrous compounds react to produce hydrates having an extremely small crystallite size. Sufficient hydration takes place within a few hours to enable hardening, but the hydration reaction continues for months, even years. Typical simplified hydration reactions for the major cement compounds are expressed as in Table 2. Of the four cement compounds, C_3A and C_3S are the most highly reactive, followed by C_4AF and C_2S.

The reactions are not simply a sum of the hydration reactions of the individual compounds; e.g., $Ca(OH)_2$ liberated from C_3S and C_2S hydration interacts in the hydration of C_3A and C_4AF. When excess sulfate is present, the trisulfate mineral *ettringite* is formed:

$$3CaO \cdot Al_2O_3 + 3(CaSO_4 \cdot 2H_2O) + 26H_2O \rightarrow 3CaO \cdot Al_2O_3 \cdot 3CaSO_4 \cdot 32H_2O$$

This latter reaction produces a net positive volume change; when present in substantial, controlled amounts, this is the basis for the action of expansive cements. When inadvertently present in considerable amount, it may cause destructive expansion of concrete.

Hydrates of calcium aluminate, calcium aluminoferrite, and calcium aluminosulfate as well as $Ca(OH)_2$ may form crystals sufficiently large to give ordered X-ray diffraction patterns. The major component of hydrated Portland cement, however, is an impure subcrystalline calcium silicate hydrate called *tobermorite* gel, cement gel, or amorphous calcium silicate hydrate (CSH), having an extremely high surface area and particle size of colloidal dimension. In addition to kinetic and thermodynamic considerations, a purely geometric factor resulting from the relative densities of the reactants and products favors hardening of the hydrated cement into a stable solid: High-density clinker materials react with low-density water (which initially in the plastic "paste" fills the space between the cement particles) to produce an intermediate-density solid hydration product that fills most of the volume occupied initially by the cement particles and liquid. Interweaving of the various hydrated phases to produce an intricate microstructure, combined with the effect of relatively strong van der Waals forces generated among such fine-size particles, is believed to give rise to the strength of the hardened product. The final strength of the hardened cement paste is controlled by the composition and fineness of the cement, the amount of water used in mixing, the use of minor amounts of additives, the efficiency of the mixing process, and the curing conditions.

TABLE 2. Hydration of Cement Compounds

Compound	Reactants	Products	
$2(3CaO \cdot SiO_2)$ (Tricalcium silicate)	$+ 5.5H_2O$	$= 3CaO \cdot 2SiO_2 \cdot 2.5H_2O + 3Ca(OH)_2$ (calcium silicate hydrate–CSH) (*Portlandite*)	(1)
2β=$(2CaO \cdot SiO_2)$ (Dicalcium silicate)	$+ 3.5H_2O$	$= 3CaO \cdot 2SiO_2 \cdot 2.5H_2O + Ca(OH)_2$ (calcium silicate hydrate–CSH) (*Portlandite*)	(2)
$4CaO \cdot Al_2O_3 \cdot Fe_2O_3$ ("Tetracalcium aluminoferrite")	$+ 10H_2O + 2Ca(OH)_2$	$= 6CaO \cdot Al_2O_3 \cdot Fe_2O_3 \cdot 12H_2O$ (calcium aluminoferrite hydrate)	(3)
$3CaO \cdot Al_2O_3$ (Tricalcium aluminate)	$+ 12H_2O + Ca(OH)_2$	$= 4CaO \cdot Al_2O_3 \cdot 13H_2O$ (tetracalcium aluminate hydrate)	(4)
$3CaO \cdot Al_2O_3$	$+ 10H_2O + CaSO_4 \cdot 2H_2O$ (*Gypsum*)	$= 3CaO \cdot Al_2O_3 \cdot CaSO_4 \cdot 12H_2O$ (calcium monosulfo-aluminate)	(5)

Hydrothermal Hydration. Hydrothermal cements differ from normally hydrated cements with respect to the process of hydration, total composition, and structure of product. The elevated temperatures and pressures of hydrothermal reactions enable rapid hardening and result in a thermodynamically stable (or nearly so) material, in contrast to the typically slow-reaction kinetics of normal cement hydration in which amorphous, thermodynamically metastable materials are produced. The hydration product of normal cement has been called "tobermorite gel," alluding to its similarity to the natural mineral *tobermorite*; but, as indicated in the hydration formulas (Table 2), the composition of the amorphous reactant, CSH gel, has a formula $\approx 3CaO \cdot 2SiO_2 \cdot 2.5H_2O$. The usual end product of hydrothermal cement reaction is a crystalline mineral essentially identical to the natural mineral *tobermorite*, which has the formula $5CaO \cdot 6SiO_2 \cdot 5H_2O$. Either cements or lime may be used as reactant in hydrothermal cements, but in order to produce *tobermorite*, additional silica is always required to achieve the desired stoichiometry [the cement clinker minerals have $CaO:SiO_2$ ratios of 2:1 (C_2S) and 3:1 (C_3S) and, therefore, substantial amounts of fine-grained silica are added, usually in the form of the mineral *quartz*, to adjust to the proper $5CaO:6SiO_2$ stoichiometry]. The hydrothermal reaction products obtained by treating cement alone do not have satisfactory properties.

Autoclaving or so-called high-pressure steam curing of concrete block, employed by some manufacturers, has the advantage that curing is complete in a few hours, while normal concrete curing takes weeks. Sand-lime bricks are produced through compaction of sand-lime mixtures in a press to a dense form, followed by a few hours of precuring at room temperature, and then autoclave curing for a few hours at temperatures up to 200°C at water pressure of the saturated steam curve. Usually the cementing phase is either crystalline, *tobermorite*, or a less well crystallized precursor, CSH (I).

Oil-well cements, used at great depth where they are subject to both elevated temperatures and pressures, make use of the principles of hydrothermal reaction too. Due to geothermal gradient, temperatures in the range 110° to 200°C or higher are encountered, and hydrostatic load generates pressures exceeding that of the saturated steam curve. Fine-particle-size *quartz* is added to cement compositions used for temperatures above 110°C in order to enable early crystallization of *tobermorite*; otherwise the CSH gel formed from cement alone undergoes undesirable volume shrinkage on recrystallization. Additives are sometimes used in oil-well cements in order to control the setting time and to provide sufficient handling time, since one of the greatest problems encountered in using cement in an oil well is the tendency of the cement to harden or set before it is in its final position within the oil well.

Summary

The mineralogy of the individual crystalline cement components is reasonably well known. Each of the before-mentioned compounds has been studied extensively by X-ray diffraction, optical microscopy, electron microprobe, and numerous other methods used to study crystalline materials. The clinker-mineral phase equilibria are also quite well understood. The hydration process, on the other hand, is very complex, and parts remain a mystery. The addition of water to a crystalline cement results in a myriad of simultaneous reactions that produce very finely crystalline hydrates of nearly colloidal size; these cannot be well characterized by techniques used for crystalline materials. We suspect that the mineralogy of the hydration process will remain somewhat uncertain for some time to come and will continue to demand sophisticated studies in an attempt to observe the hydration reaction directly.

MICHAEL W. GRUTZECK
DELLA M. ROY

References

Heller, L., and Taylor, D. F. W., 1956. *Crystallographic Data for the Calcium Silicates.* London: Her Majesty's Stationery Office.

Lea, F. M., 1970. *The Chemistry of Cement and Concrete*, 3rd ed. London: Edward Arnold.

Muan, A., and Osborn, E. F., 1965. *Phase Equilibria Among Oxides in Steelmaking.* Reading, Mass.: Addison-Wesley.

Nurse, R. W., 1969. Phase equilibria and formation of Portland cement minerals, *5th Intl. Symp. Chem. Cement, Proc.*, vol. 1. Tokyo: Cement Assn. Japan, 77-89.

Robson, T. D., 1962. *High Alumina Cements and Concretes.* London: Contractors Record.

Roy, D. M., and Harker, R. I., 1962. Phase equilibria in the system $CaO\text{-}SiO_2\text{-}H_2O$, *4th Intl. Symp. Chem. Cement, Proc.*, vol. 1. Washington, D.C.: National Bureau of Standards, 196-203.

Taylor, H. F. W., ed., 1964. *The Chemistry of Cements.* New York: Academic, 2 vols.

Taylor, H. F. W., 1969. The calcium silicate hydrates, *5th Intl. Symp. Chem. Cement, Proc.*, vol. 2. Tokyo: Cement Assn. Japan, 1-26.

Williamson, R. B., 1972. Solidification of Portland cement, *Prog. Materials Sci.*, **15**(3), 189-286.

PROJECTIONS–*See* CRYSTALLOGRAPHY: MORPHOLOGICAL

PSEUDOMORPHISM

Pseudomorphism (*pseudos* = false, *morphe* = form) is the crystallization of a mineral with a crystal shape foreign to that mineral.

Mineral crystals, when grown in an unrestricting environment, display crystal shapes or forms (a group of crystal faces related by symmetry), which are a reflection of the internal arrangement of atomic particles, and may be influenced by the environment of growth. Crystal shapes are often sufficiently unique to be significant factors in the identification of the mineral (see *Barker Index of Crystals*) and the assumption of a false form is, therefore, important.

Pseudomorphism occurs when a new mineral takes the place of an original crystal. It presupposes that the volumes of the original and the new mineral are the same, since even the minute details of the original crystal are reproduced in the pseudomorph.

Pseudomorphism is of three kinds: (1) substitution, (2) replacement, and (3) alteration.

Substitution is the simultaneous removal of the original mineral and deposition of the pseudomorph. Solution and crystallization may have (a) direct chemical dependence or (b) no apparent chemical dependence. Native *silver* occurring associated with native *copper* in the deposits of the Keweenaw district of Michigan may be of type (a). It can be demonstrated in the laboratory that when a solution containing silver comes into contact with copper, the copper is dissolved and silver is crystallized. Natural solutions containing dissolved silver, coming into contact with already crystallized *copper*, will dissolve the copper and crystallize *silver*. That any particular crystal of *silver* is a pseudomorph of *copper*, may be difficult to determine, since both *silver* and *copper* crystallize in the hexoctahedral class of the isometric system, but the crystal forms with Miller indices {140}, {120}, and {350} are common in *copper* but not in *silver* (see *Crystallography: Morphological*). If several *silver* crystals exhibit these forms, they can be reasonably assumed to be pseudomorphic. Chemical reaction does not take place when *pyrite* (FeS_2) is pseudomorphic after *fluorite* (CaF_2) or *calcite* ($CaCO_3$) since the same elements are not present before and after substitution, and since the compounds involved have different properties. This substitution involves only solution and crystallization.

Replacement is the exact filling of a cavity formed by the removal of the original crystal, so that the new mineral is pseudomorphic. Criteria that may be used to distinguish replacement from substitution are (a) the presence of a partial cavity showing incomplete filling and (b) the requirement that the material surrounding the pseudomorph be different from the original crystal, ensuring that it withstood solution and maintained the original crystal shape in the cavity formed. It is clear that replacement may be recognized only some of the time.

Alteration resulting in pseudomorphism can take place (a) without a change in composition, (b) with the addition of one or more elements, (c) with the loss of one or more elements, and (d) with both the loss and addition of elements. Alteration without change in chemical composition may take place only when a chemical compound has paramorphic forms, i.e., when different minerals have the same chemical composition. Examples of this are low *quartz*, the species of silicon dioxide stable at room temperature and pressure with one threefold and 3 twofold symmetry axes, displaying the hexagonal dipyramids of high *quartz* (one sixfold and 6 twofold symmetry axes), which is stable above 573°C. *Rutile, anatase*, and *brookite*, all dioxides of titanium, may occur as pseudomorphs of each other. The mineral often identified as *marcasite* (FeS_2) because of its characteristic crystal shape, is often really *pyrite* pseudomorphic after *marcasite*, since the latter disintegrates readily on exposure to a moist atmosphere and would soon break down to a whitish powder, destroying the crystal shape. *Pyrite*, being stable under conditions of room exposure, maintains the brassy color, the metallic luster, and the shape of the crystal. Generally, volume changes between paramorphs are small, so that minor cracks, defects, or strain may occur to accommodate the pseudomorphic change. An exception is the case of *rutile* pseudomorphic after *anatase* or *brookite*, when volume changes are about 40%.

Chemical changes of the type indicated in (b), (c), and (d) are usually considered from the point of view of the number of molecules or ions involved, and not the volume of materials exchanged in the reaction. It is not likely that the volume occupied by the number of reacting molecules of original and pseudomorphic mineral will be the same, particularly in (b) and (c) where material is added or removed. For example, in the change from one unit cell of *cuprite* ($2Cu_2O$) to one unit cell of *copper* ($4Cu$) there is a volume loss of approximately 40%. Therefore, additional pseudomorphic mineral must be introduced, perhaps from an identical reaction taking place in an adjacent location, or, for other reactions, must be disposed of, perhaps by crystallization elsewhere.

In nature, the addition of elements may occur as oxidation, carbonation, or hydration; all of which can be pseudomorphic. "Martite" is the name given to octahedra of *hematite* (Fe_2O_3) resulting from the pseudomorphic oxidation of *magnetite* (Fe_3O_4). At the famous locality of Chessy, France, octahedra of *malachite* [$Cu_2(OH)_2CO_3$] occur pseudomorphically after *cuprite*, and *gypsum* ($CaSO_4 \cdot 2H_2O$) is frequently pseudomorphic after *anhydrite* ($CaSO_4$). These changes generally result in a volume increase per molecule or atom of original mineral. The change from *cuprite* to *copper*, which is often pseudomorphic in the oxidized zone of copper deposits, results in a volume loss and requires additional copper to be introduced from other sources. Both addition and loss of elements may take place in reactions such as the oxidation of sulfides—*goethite* [$FeO(OH)$] from *pyrite*, for example—or the weathering of silicates—*pyrophyllite* [$Al_2Si_4O_{10}(OH)_2$] from *orthoclase* ($KAlSi_3O_8$), for example—all of which may be pseudomorphic.

Pseudomorphism is much more frequent than is commonly acknowledged, and almost every mineral is listed in some reference as pseudomorphic after a variety of other minerals. Because they occur commonly, the development of pseudomorphs appears to be more than just chemical change in which the shape of the original mineral just happens to be preserved. Whatever special conditions enable pseudomorphism, they appear to be confined to nature, since pseudomorphism has not been recorded for industrial or laboratory reactions.

HAROLD B. STONEHOUSE

References

Dana, E. S., 1932. *Textbook of Mineralogy*, 4th ed. New York: Wiley, 851p.

Mason, B. and Berry, L. G., 1968. *Elements of Mineralogy*. San Francisco: Freeman, 550p.

Cross-references: *Crystal Growth; Crystallography: Morphological; Enhydro, Enhydrite, or Water-Stone; Polymorphism.*

PUMICE—*See* Vol. V

PYRIBOLE—*See* BIOPYRIBOLE

PYRITE GROUP

Compositions of members of this group can be characterized by the general formula AX_2, where A = Fe,Co,Ni,Cu,Mn,Pt,Pd,Os,Ru,Au, and X = S,Se,As,Sb,Bi.

aurostibite: $AuSb_2$
bravoite: $(Ni,Fe)S_2$
cattierite: CoS_2
erlichmanite: OsS_2
fukuchilite: Cu_3FeS_8
geversite: $PtSb_2$
hauerite: MnS_2
insizwaite: $Pt(Bi,Sb)_2$
krutaite: $CuSe_2$
laurite: RuS_2
malanite: $Cu(Pt,Ir)_2S_4$
maslovite: PtBiTe
michenerite: (Pd,Pt)BiTe
penroseite: $(Ni,Co,Cu)Se_2$
pyrite: FeS_2
sperrylite: $PtAs_2$
testibiopalladite: Pd(Sb,Bi)Te
trogtalite: $CoSe_2$
vaesite: NiS_2
villamininite: $(Cu,Ni,Co,Fe)S_2$

The structure of the *pyrite* group is based upon a cubic unit cell with A ions at each corner and at the centers of each face (face-centered cubic). The X ions occur in pairs, the pair center being located along the cube edges and at the cube center. This ion pair reduces the crystallographic axis from fourfold to twofold symmetry resulting in the unique crystal form, the pyritohedron {210}.

Sulfides of this group are notably hard (*pyrite*, 6½; *laurite*, 7½). Cleavage is cubic, ranging from indistinct to perfect. Luster of fresh samples is metallic.

Except for *pyrite*, minerals of this group are quite rare, occurring mainly in a few *placer deposits* (q.v.) and as *vein minerals* (q.v.). *Pyrite* is the most abundant sulfide mineral; it is found in veins, in magmatic and volcanic igneous rocks, in pegmatites (see *Pegmatite Minerals*), in regional and contact metamorphic rocks, and in sedimentary rocks of all ages.

KEITH FRYE

Reference

Fleischer, M., 1980. *Glossary of Mineral Species 1980*. Tuscon, Arizona: Mineralogical Record, 192p.

PYROCHLORE GROUP, STIBICONITE GROUP

Members of the ***pyrochlore*** and ***stibiconite*** groups are isometric ($Fd3m$) niobates, tantalates, titanates, and antimonates having the general formula

$$A_{1-2}B_2O_6(O,OH,F)$$

where A = Ca,Na,K,Ba,Sr,Ag,Pb,Sb,Fe,Mn,Ce, Y,U,Th,Sn,Bi,Zr
B = Nb,Ta,Ti,Sb

Solid solution is extensive among many end members of these groups.

Members of the ***pyrochlore*** group are assigned to one of three subgroups according to B-site occupancy; and species within the subgroups are defined by A-site occupancy (Table 1). These minerals are commonly found in the metamict state (q.v.), but can be recrystallized by heating. Crystals are typically octahedra of various shades of yellow, brown, or black with vitreous to resinous luster. Streak is yellow to brown. They occur mostly in pegmatites or in blacksands.

Members of the ***stibiconite*** group, namely,

Bindheimite: $Pb_2Sb_2O_6(O,OH)$
Lewisite: $(Ca,Fe,Na)_2(Sb,Ti)_2O_6$
Partzite: $Cu_2Sb_2(O,OH)_7$
Romeite: $(Ca,Fe,Mn,Na)_2(Sb,Ti)_2O_6(O,OH,F)$
Stetefeldtite: $Ag_2Sb_2(O,OH)_7$(?)
Stibiconite: $SbSb_2O_6(OH)$

are characterized by the predominance of Sb in the B site. These minerals are formed by oxidation of antimonial ores and occur as incrustations or small crystals. They are commonly ocherous in various shades of red, yellow, and brown.

KEITH FRYE

Reference

Hogarth, D. D., 1977. Classification and nomenclature of the pyrochlore group, *Am. Mineralogist*, **62**, 403–410.

Cross-references: *Blacksand Minerals and Metals; Metamict State; Pegmatite Minerals.*

PYROELECTRICITY–*See* PIEZOELECTRICITY

PYROXENE GROUP

The name ***pyroxene*** was first used by Haüy to describe the greenish crystals found in lavas. The term is derived from two Greek words, *pyro*, a fire and *xenos*, a stranger.

The ***pyroxene*** structure is similar to that of ***amphiboles***. Considerable isomorphous substitution takes place. There is an abundance of nomenclature in the group, but the system used here is, with few exceptions, similar to

TABLE 1. The *Pyrochlore* group

Subgroups Defined by B atoms			*Pyrochlore* Nb+Ta>2Ti Nb>Ta	*Microlite* Nb+Ta>2Ti Ta⩾Nb	*Betafite* 2Ti⩾Nb+Ta
Species defined by A atoms	Na+Ca, but no other A atoms > 20% total A atoms		*pyrochlore*	*microlite*	
	One or more A atoms (other than Na or Ca) > 20% total A atoms. Species named by most abundant A atom, other than Na or Ca	K	*kalipyrochlore*		
		Sn		*stannomicrolite*	
		Ba	*bariopyrochlore*	*bariomicrolite*	
		REE[a]	*yttropyrochlore* (ΣY>ΣCe)[b] *ceriopyrochlore* (ΣCe>ΣY)		*yttrobetafite* (ΣY > ΣCe)[b]
		Pb	*plumbopyrochlore*	*plumbomicrolite*	*plumbobetafite*
		Bi		*bismutomicrolite*	
		U	*uranpyrochlore*	*uranmicrolite*	*betafite*

After Hogarth, 1977.
[a]*REE* = Y + (La → Lu), and for purposes of species definition, *REE* counts as one A atom.
[b] ΣY = Y + (Gd → Lu); ΣCe = La → Eu.

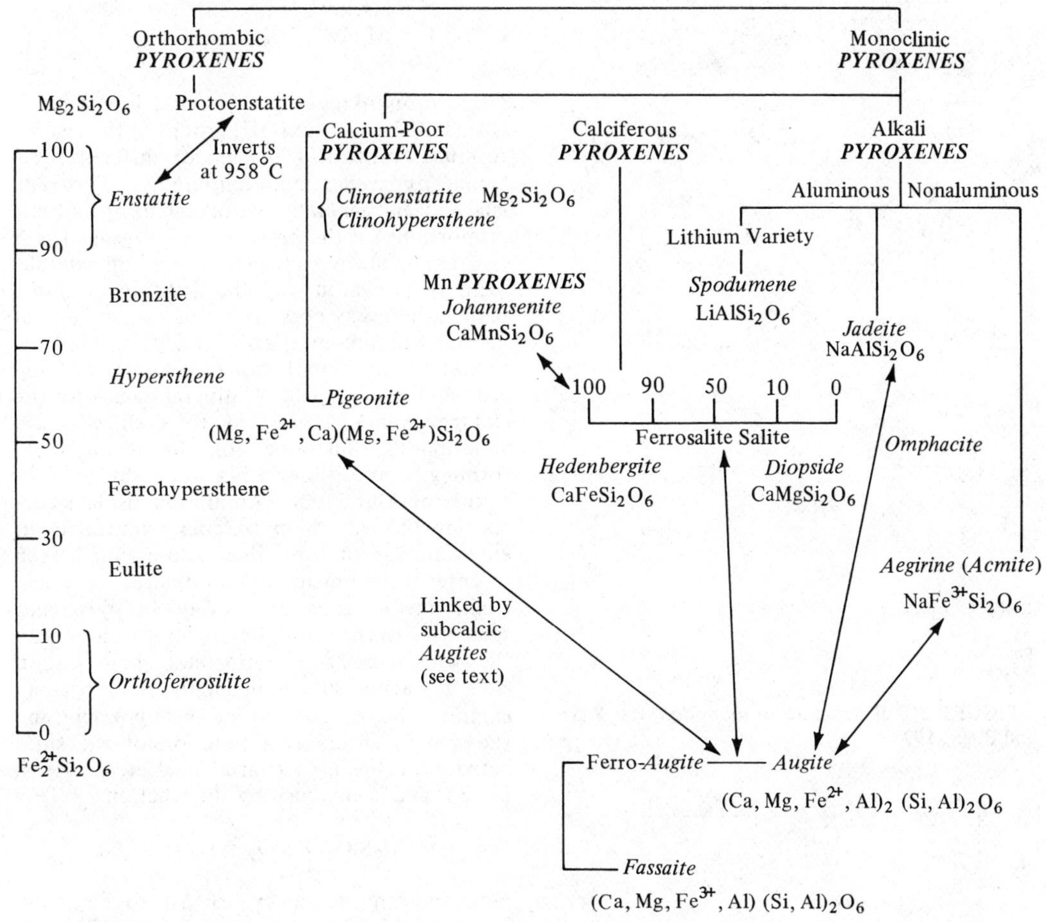

FIGURE 1. The ***Pyroxene*** group.

that of Deer, Howie, and Zussman (1963). It is summarized in Fig. 1.

Structure

Pyroxenes have either orthorhombic or monoclinic symmetry. They are classified as inosilicates and have a structural arrangement of continuous single chains of silicon-oxygen tetrahedra linked by common oxygen atoms. The ***pyroxene*** unit cell is shown in Fig. 2 projected down the *c* axis with the elevation of each ion given in percent of the *c* parameter.

The chains repeat at approximately 5.3 Å and this defines the *c* parameter of the unit cell. Cations of various sizes act as spacers between the chains and occupy two nonequivalent sites M_1 and M_2 in Fig. 2). The presence of calcium, sodium, lithium, or manganese on M_2 sites establishes the structure as monoclinic.

The cell dimensions of ***pyroxenes*** depend upon both bulk composition and site preferences. In ***orthopyroxenes*** (q.v.), the *a* parameter is increased by Fe, markedly increased by Ca, and little affected by Al substitution. The *b* dimension is increased by Fe and little affected by Al and Ca. The *c* dimension is increased by Fe and hardly affected by Al or Ca (Smith et al., 1969). Table 1 shows that *pigeonite* and *clinoenstatite* at room temperatures have space groups $P2_1/c$ compared with $C2/c$ for other low-temperature forms. Various substitutions for silicon and the metal ions produce distortions in the chains, kinking of the chains and causing changes in cell sizes. These unit-cell dimensions are also shown in Table 1. The differences between monoclinic and orthorhombic ***pyroxene*** structures is shown in Fig. 3, where the *a* parameter is double that in Fig. 2.

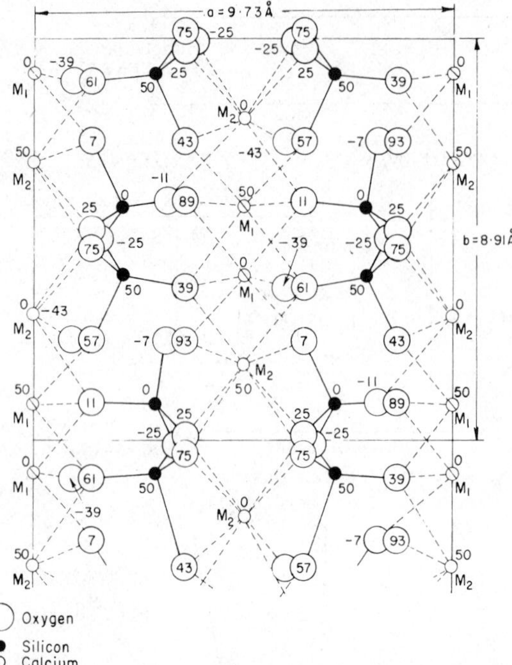

FIGURE 2. Cell structure of *diopside* (after Warren and Bragg, 1928).

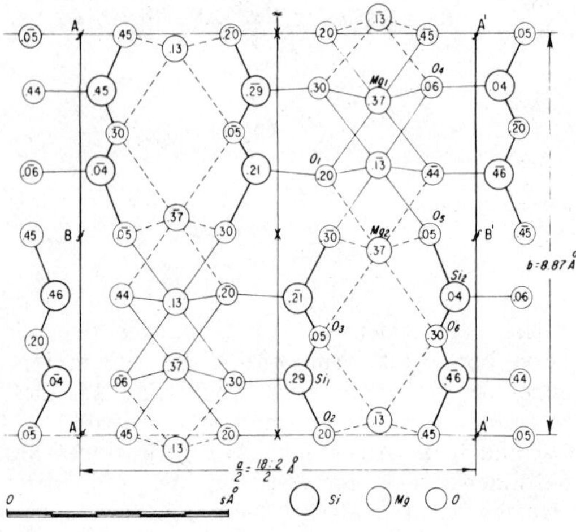

FIGURE 3. Cell structure of *enstatite* (after Warren and Modell, 1930).

Chemistry

The general formula of the group as developed by Berman, in 1937, and Hess, in 1949, (see Deer et al., 1978, p. 14) is illustrated thus:

$$(W)_{1-p}(X,Y)_{1+p}Z_2O_6$$

where W = Na,Ca

X = Mg,Fe^{2+},Mn,Ni,Li
Y = Al,Fe^{3+},Cr,Ti
Z = Si,Al

In ***orthopyroxenes,*** $p \leqslant 1$, hence $W \approx 0$ and Y is small. The lack of (OH) groups in the general formula constitutes the major difference between ***pyroxenes*** and ***amphiboles. Pyroxene*** analyses are generally recalculated into atomic proportions on the basis of six oxygens. The Z position is always adjusted to its full complement of two atoms by the addition of aluminum wherever possible. For a survey of methods of presentation of analyses, reference should be made to Bown (1964) and to Cameron and Papike (1981). Mineral names for the ***clinopyroxenes*** as related to their chemical variation is illustrated for the major rock-forming compositions in Fig. 4.

Role of Aluminum. Aluminum displays contrasting behavior in ***pyroxenes*** from different environments of formation. Aluminum is seen to enter the octahedrally coordinated, between-chain cation sites exclusively in ***pyroxenes*** such as *enstatite* and *jadeite* or to enter into the tetrahedrally coordinated, chain cation sites replacing silicon in *augites* and titanian *augites*. The Al content of the ***pyroxene*** and the site it enters is a function of the silica activity of the melt (Carmichael et al., 1974, p. 274) and is governed by the reaction

$$CaAl_2SiO_6 + SiO_2 = CaAl_2Si_2O_6$$

indicating the tendency for Al to be incorporated in Z sites under conditions of low Si activity. Titanium was formerly thought to replace Si in the chain structure, but is now known to enter the Y sites and its presence in Y sites also favors Al entry into Z sites.

Pressure also affects the behavior of aluminum in ***pyroxenes***. Studies of ***pyroxene*** solid-

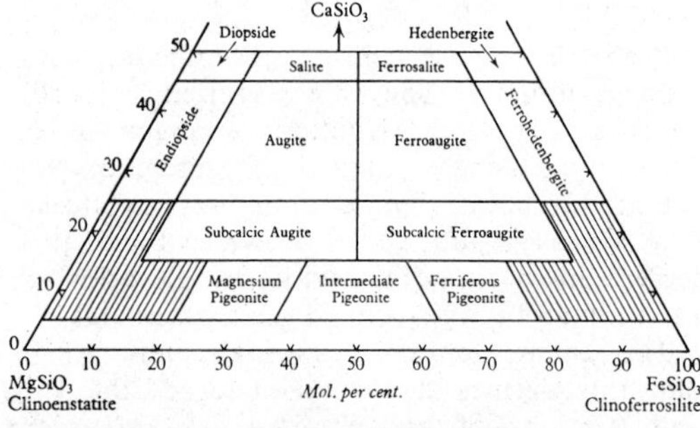

FIGURE 4. Nomenclature of ***clinopyroxenes*** (from A. Poldervaart and H. H. Hess, Pyroxenes in the crystallization of basaltic magma, *Journal of Geology* **59**, 1951, p. 427, copyright © 1951 by the University of Chicago).

solution relationships over a range of pressures (Kushiro, 1969) show that at low pressures ***clinopyroxenes*** (q.v.) are relatively $CaAl_2SiO_6$-rich and *jadeite*-poor and at higher pressures $CaAl_2SiO_6$-poor and *jadeite*-rich, i.e., entry of Al into octahedral sites is favored by high pressures.

Aluminum is most abundant in the relatively rare ***pyroxene*** *jadeite* in which Al_2O_3 may reach as much as 24% by weight of the analysis. The position of the aluminum in the M_1 sites is unusual in this mineral because there is no replacement of silicon in the chain. In addition, aluminum is not replaced by ferric iron. In some respects, the mineral is similar to the ***amphibole*** *glaucophane*, but the ferric iron analog, the equivalent of *riebeckite*, is unknown.

In *augite* the Al_2O_3 content is usually limited to a maximum of 4%, but in titanian *augites* this figure may be raised to 14%. The aluminum is largely sited in the Z position. The Al_2O_3 is present in amounts of up to 6% in some ***orthopyroxenes*** (Deer et al., 1978). It is equally common in some *diopside-hedenbergite* samples but rarely exceeds 2% in *pigeonites*.

Ferric Iron. Ferric iron compares with aluminum in some respects insofar as entry into the Y or Z sites occurs under specific conditions. *Acmite* (*aegirine*) $NaFe^{3+}Si_2O_6$, is the iron analog of *jadeite* but there is little solid solution between them and their stabilities are controlled by contrasted physical conditions of origin. Ferri-*diopside,* $CaFe_2^{3+}SiO_6$, is the analog of aluminous *augite*.

Both *acmite* (*aegirine*) and ferri-*diopside* are stabilized under conditions of high f_{O_2} in natural rocks and generally form at low pressures compared with *jadeite* (Huckenholz et al., 1969; Carmichael et al., 1974).

Other Elements. Generally, orthorhombic inosilicate minerals do not contain significant amounts of eightfold coordinated elements. In ***orthopyroxenes***, the weight of CaO rarely exceeds 2%. According to L. Atlas the quantity is generally related to the temperature at which the mineral crystallized (see Deer et al., 1978, p. 34).

Spodumene is formed in a specialized environment of lithium-bearing granite pegmatites (see *Pegmatite Minerals*). The calcium content of these minerals is very low and analyses show that amounts of Na_2O are rarely much in excess of 1.5% by weight. The Li_2O content varies between 6% and 7% by weight. In *spodumene*, there is little variation from the ideal formula.

Diopsides from mafic and ultramafic rocks often contain appreciable amounts of chromium, which is essential in the rare meteoritic ***pyroxene*** *ureyite*. Nickel-bearing ***pyroxenes*** do not occur in nature, but have been synthesized, although traces of nickel occur in many natural ***pyroxenes***. Titanium is generally not important in the ***pyroxene*** group except in titanian *augites* where the TiO_2 content may reach 6% by weight.

Manganese is present in many ***pyroxenes***, usually in small amounts. The mineral *johannsenite*, however, may contain up to 27.7% MnO and forms the manganese end member of the *diopside-hedenbergite-johannsenite* series. Hutton suggested in 1956 the further subdivision *hedenbergite*-manganoan, *hedenbergite*-ferroan, *johannsenite*-johannsenite (Deer et al., 1978, p. 416).

Zirconium, cerium, and the rare-earth elements are not uncommon in certain of the alkali ***pyroxenes***. *Aegirine augites* frequently also show an appreciable vanadium content.

Solid Solution. The ***pyroxene*** group represent perhaps one of the best studied groups of minerals experimentally. A wealth of information is available on solid-solution relationships and phase equilibria for most ***pyroxene*** subgroups. The relationships between the ***orthopyroxene*** *pigeonite* and *augite* have been especially well investigated. These studies are critical to an understanding of the genesis of igneous rocks (see Cameron and Papike, 1981).

As with the ***amphiboles***, there is a multiplicity of solid-solution relationships within the various ***pyroxene*** subgroups. There is also some intergroup solid solution. There are, however, definite breaks in the solid solution and *spodumene* and *jadeite* are commonly relatively pure.

When substitutions take place within the subgroups, they are often relatively simple, for example, a Mg/Fe interchange. Complications to this simple exchange occur by exchanges between Na/Ca or $(Mg,Fe)/(Fe^{3+},Al)$. These substitutions are often coupled, i.e., Ca(Mg,Fe) for $Na(Fe^{3+},Al)$. However, exchanges of the type $CaFe^{3+}$ for CaMg occur to give the ferri-*diopside* series, which is accompanied by Al for Si substitution to maintain change balances. There are, thus, extensive solid solutions between the *diopside-hedenbergite* and *augite*-ferri-*diopside* series.

Although ***clinopyroxenes*** of intermediate calcium compositions are named on the ***pyroxene*** quadrangle (Fig. 4), the coexistence of *augite* with *pigeonite* in the same rock indicates an extensive miscibility gap between the two. However, sodium can replace calcium giving *augite–aegirine-augite–acmite*.

The Pigeonite Problem

Pigeonite shows variations in compositions due to the replacement of magnesium by iron.

TABLE 1. Physical Properties of the Pyroxene Group

Mineral Series	Symmetry	D	Hardness	Cleavage	Twinning	Color	Sign	Optic Orient	Disp.	Pleochroism
Enstatite	orthorhombic	3.209	5–6	$(210):(2\bar{1}0)$ $\approx 88°$	{100} simple or lamellar	colorless, gray, green, yellow, brown	+	$\alpha = y$	$r < V$	none
Orthoferrosilite	orthorhombic	3.96	5–6	$(210):(2\bar{1}0)$ $\approx 88°$	{100}	green, dark brown, reddish	+	$\alpha = y$	$r < V$	α-pale reddish, β-pale green, yellow, γ-greens, blue
Pigeonite	monoclinic	3.17–3.46	6	$(110):(1\bar{1}0)$ $\approx 87°$	{100} or {001} simple or lamellar	brown, green, black	+	α, $\beta = y$	$r \gtrsim V$	may be absent, α-colorless, yellow, green, β-browns, γ-colorless, greens, yellow
Clinoenstatite	monoclinic			$(110):(1\bar{1}0)$ $\approx 86° 30'$	{100} lamellar		+	$\beta = y$		
Diopside	monoclinic	3.22–3.38	5½–6½	$(110):(1\bar{1}0)$ $\approx 87°$	{100} or {010} simple or multiple	white, pale green	+	$\beta = y$	$r > V$	not pleochroic
Hedenbergite	monoclinic	3.50–3.56	6	$(110):(1\bar{1}0)$ $\approx 87°$	{100} or {010} simple or multiple	brownish green, green, black	+	$\beta = y$	$r > V$	α-pale green, blue, β-green, blue, γ-green yellow-green
Johannsenite	monoclinic	3.27–3.54	6	$(110):(1\bar{1}0)$ $\approx 87°$	{100} simple or lamellar	clove-brown, green, blue, colorless	+	$\beta = y$	$r \gtrsim V$	colorless
Fassaite	monoclinic	2.96–3.60	6	$(110):(1\bar{1}0)$ $\approx 87°$	{100} simple or lamellar	pale to dark green, brown, crimson	+	$\beta = y$	$r > V$	α-pale green, β-pale yellow-green, γ-pale green, yellow gray
Augite	monoclinic	3.19–3.56	5½–6	$(110):(1\bar{1}0)$ $\approx 87°$	{100} simple or multiple {001} multiple	pale brown to black, green, purplish	+	$\beta = y$	$r > V$	α-pale browns, green, yellow, β-as for α or violet γ-pale green, grayish, violet
Aegirine augite	monoclinic	3.40–3.55	6	$(110):(1\bar{1}0)$ $\approx 87°$	{100} simple or lamellar	dark green, black, green, yellow-green, brown	±	$\beta = y$	$r > V$	α-strong greens, β-yellow greens, γ-yellow browns, brownish green
Acmite Aegirine	monoclinic	3.50–3.60	6	$(110):(1\bar{1}0)$ $\approx 87°$	{100} simple or lamellar	dark green to greenish black, reddish brown	–	$\beta = y$	$r > V$	α-emerald green, β-strong green, yellow, γ-browns, green, yellow
Jadeite	monoclinic	3.24–3.43	6	$(110):(1\bar{1}0)$ $\approx 87°$	{100} simple or lamellar	colorless, green, blue white	+	$\beta = y$	$r > V$	colorless
Omphacite	monoclinic	3.16–3.43	5–6	$(110):(1\bar{1}0)$ $\approx 87°$	{100} simple or lamellar	greens	+	$\beta = y$	$r > V$	α-colorless, β-very pale green, γ-very pale green, blue green
Spodumene	monoclinic	3.03–3.23	6½–7	$(110):(1\bar{1}0)$ $\approx 87°$	{100} common	green, blue, lilac, white, gray, colorless	+	$\beta = y$	$r < V$	kunzite α-purple, γ-colorless hiddenite: α-green, γ-colorless
Ureyite	monoclinic	3.60	≈ 6	$(110):(1\bar{1}0)$ $\approx 87°$	–	dark emerald green	–	$\beta = y$	–	α-yellowish green, β-grass green, γ-emerald green

TABLE 1. (Continued)

Mineral Series	α	β	γ	2Vγ[a]	γ:Z[b]	δ	Å a_0	Å b_0	Å c_0	β	Z	Space Group
Enstatite	1.650–1.662	1.653–1.671	1.658–1.680	55°–90°	–	0.007–0.011	18.223	8.815	5.169	–	16	*Pbca*
Orthoferrosilite	1.755–1.768	1.763–1.770	1.772–1.788	55°–90°	–	0.018–0.020	18.431	9.080	5.238	–	16	*Pbca*
Pigeonite	1.682–1.732	1.684–1.732	1.705–1.757	0°–30°	32°–44°	0.023–0.029	9.7	8.9	5.2	108°	4	$P2_1/c$
Clinoenstatite	1.650–1.662	1.652–1.654	1.658–1.660	25°–56°	22°	0.008–0.010	9.6065	8.8146	5.1688	108.335°		$P2_1/c$
Diopside	1.664[c]	1.6715[c]	1.694[c]	59.3°[c]	38.5°[c]	0.030[c]	9.752[c]	8.926[c]	5.248[c]	105.83°[c]	4	*C*2/*c*
Hedenbergite	1.732[c]	1.723–1.730	1.755[c]	52°–62°	47°–48°	0.024[c]	9.844[c]	9.028[c]	5.246[c]	104.80°	4	*C*2/*c*
Johannsenite	1.699–1.710	1.710–1.738	1.725–1.738	58°–72°	46°–55°	0.022–0.029	9.83–9.98	9.04–9.16	5.26–5.29	105–105.5°	4	*C*2/*c*
Fassaite	1.672–1.730	1.682–1.735	1.702–1.750	51°–64°	33°–47°	0.018–0.028	9.71–9.80	8.85–8.91	5.26–5.36	105.5–106°	4	*C*2/*c*
Augite	1.671–1.735	1.672–1.741	1.703–1.774	25°–61°	35°–48°	0.018–0.033	≈9.8	≈9.0	≈5.25	≈105°	4	*C*2/*c*
Aegirine augite	1.700–1.760	1.710–1.800	1.730–1.813	70°–110°	0°–20°	0.028–0.050		between *acmite* and *augite*			4	*C*2/*c*
Acmite (Aegirine)	1.750–1.776	1.780–1.820	1.795–1.836	60°–70°	0°–10°	0.04–0.06	9.658[c]	8.795[c]	5.294[c]	107.42°	4	*C*2/*c*
Jadeite	1.640–1.681	1.645–1.684	1.652–1.692	60°–96°	32°–55°	0.006–0.021	9.418	8.562	5.219	107°97′	4	*C*2/*c*
Omphacite	1.662–1.701	1.670–1.712	1.685–1.723	56°–84°	34°–48°	0.012–0.028	9.45–9.68	8.57–8.90	5.23–5.28	105–108°	4	*C*2/*c* or *P*2/*n*
Spodumene	1.648–1.663	1.655–1.669	1.662–1.679	58°–68°	20°–26°	0.014–0.027	9.45	8.39	5.215	110°	4	*C*2/*c*
Ureyite (Cosmochlore)	1.740–1.766	1.756–1.778	1.762–1.781	53°	8°–22°	0.015–0.022	9.54–9.58	8.69–8.73	5.26–5.28	107.3–107.5°	4	*C*2/*c*

All data from Deer et al., 1978.
[a]Except *aegirine augite, acmite* (*aegirine*) where the value is 2Vα
[b]Except *aegirine augite, acmite* (*aegirine*), *ureyite,* where the value is α:Z.
[c]Synthetic end member

Published data show that this does not take place to the fullest extent ideally possible. *Pigeonites* can contain up to 2.0% Al_2O_3 by weight in their structure but, like the ***amphibole*** equivalent *cummingtonite*, their Al_2O_3 content is usually low, and nearly always in the tetrahedral sites. As in *cummingtonite*, the amount of manganese present may be appreciable, especially in iron-rich samples.

A comparison of the composition of *pigeonite* and calcium-poor *augites*, crystallized under plutonic conditions, shows that there is a field of low-temperature immiscibility between the two ***pyroxenes***. Even in the volcanic environment, complete solid solution may not exist between the two series (Deer et al., 1978, p. 174).

In *pigeonite*, some 10% of the total divalent ions may be calcium. This is about three times as much as the ***orthopyroxene*** structure can accept. In volcanic and other quickly chilled rocks, optically homogeneous *pigeonite* may be preserved, although X-ray diffraction and electron-probe studies indicate submicroscopic inhomogeneity. On slow cooling, however, *augite* is exsolved parallel to (001); further exsolution of calcic ***pyroxene*** would take place at the inversion temperature of monoclinic to orthorhombic ***pyroxene***.

Many ***pyroxenes*** show exsolution relationships between the calcic and subcalcic members. Exsolution textures are often well developed in slowly cooled, large mafic intrusions such as the Bushveld. Exsolution of *clinohypersthene* from *augite*, *augite* from *pigeonite*, and *augite* from ***orthopyroxene*** are common. Inversion of earlier crystallized phases may complicate the interpretation of the exsolution relationships. *Pigeonite* may exsolve *augite* parallel to (001) and then become subcalcic *pigeonite*. A second exsolution of *augite* parallel to (001) may then occur before subcalcic *pigeonite* inverts to an ***orthopyroxene*** (Deer et al., 1978, p. 178).

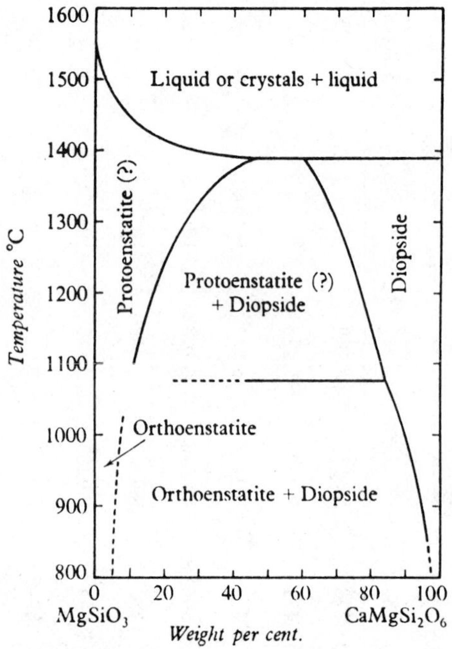

FIGURE 5. Subsolidus equilibrium in the system $MgSiO_3$-$CaMgSi_2O_6$ (from J. F. Schairer and F. R. Boyd, Pyroxenes, the join $MgSiO_3$-$CaMgSi_2O_6$, in *Annual Report to the Director of the Geophysical Laboratory, 1956-57*, Washington, D.C.: Carnegie Institute, 1957, p. 223).

Experimental Work

Experimental work on the ***pyroxene*** group has been extensive and varied. On ***orthopyroxenes***, much has been done on inversion points in the protoenstatite–*enstatite*–*clinoenstatite* series. At about 1560°C protoenstatite melts incongruently to *forsterite* plus liquid; it is the stable polymorph down to around 1000°C, where *enstatite* becomes the stable phase. The stability field of *enstatite* increases with increasing pressure, however, and this ***orthopyroxene*** becomes the stable phase at the liquidus above 1600°C and 7 kilobars. Conditions of the inversion of *enstatite* to *clinoenstatite* remain somewhat enigmatic, however, due to uncertainty about the role of shear stresses in the transformation.

Another line of experimental work involves the phase relationships at 1 atmosphere pressure between *enstatite* and *diopside* (Fig. 5). With increasing temperature, the amount of Ca increases in *enstatite* and decreases in *diopside*. More recent work (see Deer et al., 1978, p. 84–90) reveals that *pigeonite* with 10–20 wt. % $CaMgSi_2O_6$ becomes a stable phase above 1250°C and the assemblage *diopside*-protoenstatite is no longer stable above that temperature.

The solubility of aluminum in *enstatite* provides information on pressure-temperature conditions for the formation of ultramafic rocks (see *Barometry, Geologic; Mantle Mineralogy*). The amount of Al in *enstatite* in equilibrium with *pyrope* ($Mg_3Al_2Si_3O_{12}$) is sensitive both to pressure and to temperature, whereas in equilibrium with *spinel* ($MgAl_2O_4$) the Al content is relatively pressure insensitive.

The monoclinic ***pyroxene***, ***diopside***, is an important phase in the synthetic basalt system used as a model of the Bowen reaction series.

Its relationships with other minerals in this series have been studied experimentally. Bowen's initial work in 1915 on the system *diopside–anorthite–albite* recorded the *diopside–anorthite* eutectic at 1274°C and the *diopside–albite* eutectic at 1085°C. Subsequently, Yoder (in 1954) illustrated the effects of water on the *diopside–anorthite* system when, at 5000 bars, the eutectic is lowered to 1095°C.

Phase relationships in the system *diopside-forsterite*-silica were first reported by N. L. Bowen in 1914. Experimental work has shown that *diopside* may contain up to 5 wt% *forsterite,* 2 wt% *enstatite,* and 4 wt% excess silica. *Diopside-forsterite* mixtures, with the exception of those very rich in *diopside*, melt incongruently to *forsterite* and liquid. The system *diopside*-silica is not truly binary, since a *diopside* solid solution may coexist with *tridymite* and a liquid over a range of temperature and pressure. In the system *forsterite*–silica, *enstatite* melts incongruently below 1300 bars to *forsterite* plus liquid. The incongruent melting of compositions along the join *enstatite–diopside* illustrate the production of silica saturated melts from initially undersaturated melts by fractional crystallization.

Lamellar Structures

Orthorhombic and monoclinic ***pyroxenes*** commonly show characteristic lamellar structures. In some orthorhombic hosts, exsolution of a calcium-rich ***clinopyroxene*** (*augite*) lamellae occurs parallel to (100) of the host *hypersthene*. These associations are called ***orthopyroxenes*** of the Bushveld type. From the [010] direction, these lamellae may appear under a microscope as fine, straight-ruled lines or rows of blebs (Deer et al., 1978, Frontispiece). Rapid cooling generally prevents the formation of visible exsolution lamellae, but such structures ranging from 40 to 3200 Å have been found by means of electron microscopy in ***pyroxenes*** of volcanic and dike rocks.

Among the ***clinopyroxenes***, many *augites*, particularly those from slowly cooled, mafic plutonic rocks, contain lamellae of a calcium-poor phase, *hypersthene* or *pigeonite*. The former is exsolved parallel to (100) and the latter parallel to the (001) plane of host *augite*; they may occur separately or together in the same host. The exsolved *pigeonite* itself may also invert to ***orthopyroxene***. Inverted *pigeonite* is *hypersthene* with two sets of *augite* lamellae–one parallel to (100) and the other parallel to an (*h0l*) plane. Grains with this lamellar structure are also known as ***orthopyroxenes*** of the Stillwater type.

Crystallization and Paragenesis

There are important differences in composition between igneous and metamorphic ***orthopyroxenes***. In the igneous rocks, the early-formed ***pyroxenes*** are the magnesian varieties in ultramafics and early-formed rocks of layered intrusions. However, in norite gabbro, the ***orthopyroxene*** is either bronzite (ferroan *enstatite*) or *hypersthene*. In special circumstances, for example from the iron-rich diabase at Beaver Bay, iron-rich ***orthopyroxene***, eulite, has been described.

In the metamorphic rocks, the felsic or mafic nature of the rock sometimes controls the level of iron content in the ***pyroxene***. Howie, in 1955, described ***pyroxenes*** from Madras, India, where the mafic rocks contain bronzites whereas the felsic and intermediate rocks contain ferro-*hypersthene*.

Much has been written on the composition of ***pyroxenes*** crystallizing from basaltic melts. It is now recognized that the silica activity of the magma (Carmichael et al., 1974) has a fundamental control on the ***pyroxenes*** that crystallize from the melt.

In tholeiitic magmas, silica activity is such that the crystallization of ***orthopyroxenes*** rather than ***olivine*** is stabilized by the reaction

$$\underset{\textit{forsterite}}{Mg_2SiO_4} + \underset{\text{in melt}}{SiO_2} = \underset{\textit{clinoenstatite}}{2MgSiO_3}$$

The silica activity defined by this reaction subdivides the tholeiitic and alkali basalt fields.

Tholeiites are then characterized by the crystallization of two ***pyroxenes***, a calcic and a subcalcic variety from the melts, whereas magmas with silica activities lower than that defined by the above equation crystallize one ***pyroxene***.

The sequential crystallization of two ***pyroxenes*** from igneous melts has also been the subject of detailed studies. These results are perhaps best illustrated by Fig. 6, (Deer, et al., 1963). This figure shows that two ***pyroxene*** phases often form during the early stages of crystallization from such magmas. In the early stages, one of the pair is orthorhombic. These ***pyroxene*** pairs are typical in the crystallization of tholeiite magma types.

Figure 6 also shows that, as crystallization proceeds, the residual magmas become enriched in iron, and *pigeonite* crystallizes in place of ***orthopyroxene***. The two phases may continue to recrystallize until, with change in composition, only a simple phase of calcium-rich ferro-*augite* crystallizes.

W. LAYTON
M. M. WILSON

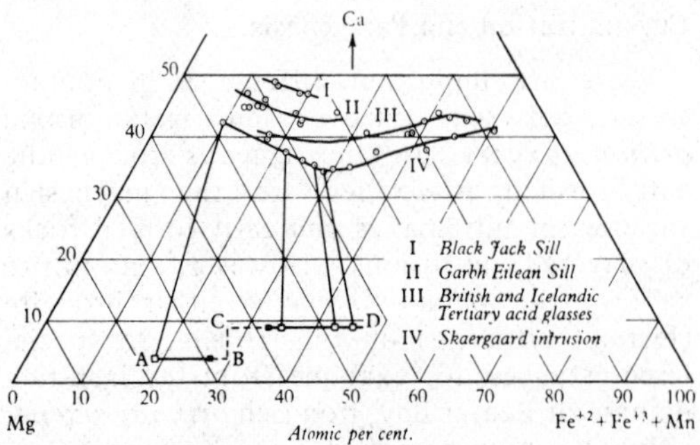

FIGURE 6. Trends of pyroxene crystallization (from W. A. Deer, R. A. Howie, and J. Zussman, *Rock-Forming Minerals,* vol. 2. London: Longman Group Limited, 1963, p. 127).

References

Bown, M. C., 1964. Recalculation of pyroxene analyses, *Am. Mineralogist*, **49**, 190–194.

Boyd, F. R., and Brown, G. M., 1969. Electron-probe study of pyroxene exsolution, *Mineral. Soc. Am. Spec. Paper*, **2**, 211–216.

Cameron, M., and Papike, J. J., 1981. Structural and chemical variations in pyroxenes, *Am. Mineralogist*, **66**, 1–50.

Carmichael, I. S. E.; Turner, F. J.; and Verboogen, J., 1974. *Igneous Petrology*. New York: McGraw-Hill, 739p.

Deer, W. A.; Howie, R. A.; and Zussman, J., 1963. *Rock-Forming Minerals*, vol. 2. London: Longman, 379p.

Deer, W. A.; Howie, R. A.; and Zussman, J., 1978. *Rock-Forming Minerals: Single-Chain Silicates*, vol. 2A. London: Longman, 668p.

Huckenholz, H. G.; Schairer, J. F.; and Yoder, H. S., Jr., 1969. Synthesis and stability of ferri-diopside. *Mineral. Soc. Am. Spec. Paper*, **2**, 163–177.

Kushiro, I., 1969. Clinopyroxene solid solutions formed by reactions between diopside and plagioclase at high pressures, *Mineral. Soc. Am. Spec. Paper*, **2**, 179–191.

Prewitt, C. T., ed., 1980. *Pyroxenes*. Reviews in Mineralogy, vol. 7. Washington, D.C.: Mineralogical Society of America.

Smith, J. V.; Stephenson, D. A.; Howie, R. A.; and Hey, M. H., 1969. Relations between cell dimensions, chemical composition and site preference of orthopyroxene, *Mineralog. Mag.*, **37**, 90–114.

Warren, B. E., and Bragg, W. L., 1928. The structure of diopside $CaMg(SiO_3)_2$, *Zeit. Kristallogr.*, **69**, 168.

Warren, B. E., and Modell, D. I., 1930. Structure of enstatite $MgSiO_3$, *Zeit. Kristallogr.*, **75**, 1.

Cross-references: *Amphibole Group; Clinopyroxenes; Crystal Growth; Glass, Devitrification of Volcanic; Isomorphism; Lunar Mineralogy; Mineral Classification: Principles; Optical Mineralogy; Order-Disorder; Orthopyroxenes; Plastic Flow in Minerals; Polysomatism; Refractory Minerals; Rock-Forming Minerals; Skeletal Crystals; Soil Mineralogy; Twinning;* see also mineral glossary.

PYROXENOIDS

The ***pyroxenoid*** minerals are single-chain silicates that do not have the pyroxene structure. Included are the minerals *bustamite, pectolite, pyroxmangite, rhodonite, serandite*, and *wollastonite*.

KEITH FRYE

References

Deer, W. A.; Howie, R. A.; and Zussman, J., 1978. *Rock-Forming Minerals: Single-Chain Silicates*, vol. 2A. London: Longman, 547–613.

Chashi, Y., and Finger, L. W., 1978. The role of octahedral cations in pyroxenoid crystal chemistry. I. Bustamite, wollastonite, and the pectolite-schizo lite-serandite series, *Am. Mineralogist*, **63**, 274–288.

Q

QUARTZ

The name *quartz* is derived from "quertz," itself a contraction of "Querklufterz" (cross-vein ore), a mining term that was in use in Saxony. The ancient term *crystallos*, from the Greek word for frozen water, is still used for the variety rock crystal. Otherwise, the term "crystal" has now (since M. A. Cappeler, 1723, and Romé de l'Isle, 1772) a more general significance.

Historical Notes

Quartz has been known, and made use of, since the palaeolithicum. Great technical skills were developed in early times, as is evidenced by the carved gems of Crete, Greece, Rome, and Alexandria. There are, however, few meaningful indications on *quartz* before the late Renaissance. From that time on, fundamental observations have been made on *quartz*.

The law of constancy of angles is based on a work by Nicolaus Steno, 1669, which is preponderantly on *quartz*. The *Prodromus Crystallographiae* by Moritz Anton Cappeler, 1723, first to use the term crystallography and first to attempt a correlation of shape, physical properties, and structure, is also largely a study of *quartz*. Optical activity and piezoelectricity were first observed on *quartz*.

Occurrence

Quartz is one of the most widespread minerals. It occurs in felsic volcanic rocks (e.g., rhyolite), plutonic rocks (e.g., granite), metamorphic rocks (e.g., mica schist), sedimentary material (marine, limnic, and aeolian), and also crystallized at low temperature (authigenic) and formed by organogenic processes. *Quartz* has also been identified in small amounts in some stone meteorites and in lunar material. Other polymorphs are given in Table 1. *Quartz* varieties are as follows:

Phanerocrystalline varieties
- rock crystal
- smoky *quartz*
- morion (dark brown)
- amethyst (violet, purple, colorless, yellow)
- citrine (yellow)
- rose *quartz* (pink)
- blue *quartz*

Cryptocrystalline varieties
- chalcedony (microcrystalline fibrous *quartz*)
- lussatite (microcrystalline, fibrous *cristobalite*)
- agate (chalcedony in layers with different colors)
- chrysoprase (light-green)
- plasma (greyish-green)
- prase (with actinolite)
- heliotrope (green with red spots)
- flint
- chert (var. of chalcedony)
- novaculite
- sard (light-dark brown; var. of chalcedony)
- carnelian (yellow-red)
- onyx (black)
- *opal*, $SiO_2 \cdot nH_2O$ (minute crystals of *cristobalite* in an amorphous matrix)

Composition

Main Components. *Quartz* is chemically silicon dioxide, SiO_2, with the theoretical content of Si=46.757 and O=53.249 percent.

Trace Elements. The elements Li, Na, Al, H, K, Fe, Mn, Ti, and a number of others including Rb, Ca, Cs, Ba, Pb, Ag, Sn, Cu, Zn, V, Cr, and U may be present in small but significant amounts.

The presence of Fe_2O_3, TiO_2, MnO, or carbon may account for the rose, blue, purple, or black color of *quartz*. Two types of *quartz* with markedly different content of trace elements have been distinguished in *quartz* from Alpine vugs. Similar observations have been made on *quartz* from other parts of the world (Bambauer, 1961; Poty, 1969; Bukanov, 1974).

Oxygen Isotopic Composition. The O^{18}/O^{16} ratios of authigenic minerals are generally 10 to 30 per mill greater than those of detrital minerals of high-temperature origin. In detrital *quartz*, $\sigma O^{18}/O^{16}$, referred to Standard Mean Ocean Water, was found to be 10.9 to 16 ‰; in silicified fossils (Devonian brachiopods from Roberts Mtn., Nevada) up to 27.1 ‰ (Savin and Epstein, 1970).

Solubility. *Quartz* resists acids, except hydrofluoric acid; on the other hand, it is strongly attacked by alkaline solutions.. Solubility in water is dependent upon pressure, temperature, and the presence of other substances. At atmospheric pressure and room temperature, it

TABLE 1. Polymorphs of Silica

Name	Thermal Stability Range (°C at 1 atm)	Symmetry
Quartz (*α-quartz*)	below 573°	trigonal trapezohedral $D_3^4\text{–}P3_1 21$ and $D_3^6\text{–}P3_2 21$
High *quartz* (*β-quartz*)	573°–870°	hexagonal and trapezohedral $D_6^4\text{–}P6_2 22$ and $D_6^5\text{–}P6_4 22$
Tridymite (low *tridymite*)	below 117°, metastable below 870°	orthorhombic, pseudohexagonal
Middle *tridymite*	117°–163°	hexagonal
High *tridymite* (*β-tridymite*)	163°–1470°	hexagonal $D_6^4 h\text{–}C6/mmc$
Cristobalite	below 200°	tetragonal trapezohedral $D_4^4\text{–}P4_1 2_1 2$ and $D_4^8\text{–}P4_3 2_1 2$
High *cristobalite* (*β-cristobalite*)	200°–1720°, stable at 1470°–1720°	isometric $O_h^7\text{–}Fd3m$
Keatite (artificial only)	metastable at ordinary conditions	tetragonal trapezohedral $D_4^4\text{–}4_1 2_1 2$ and $D_4^8\text{–}P4_3 2_1 2$
Coesite	metastable at ordinary conditions	monoclinic, pseudohexagonal $C_{2h}^6\text{–}C2/c$
Stishovite	metastable at ordinary conditions	tetragonal $D_{4}^5\text{h–}P4/mm$
Melanophlogite	below 800°	isometric $O^2\text{–}P4_2 32$
Lechatelierite (silica glass amorphous)		
Opal (crystalites of *cristobalite* in amorphous matrix		

lies in the range of 6 to 30 ppm for *quartz* and between 100 and 140 ppm for cryptocrystalline or amorphous silica. At a hydrostatic pressure of 1000 atm, the solubility of *quartz* is 68 ppm at 100°C; 330 ppm at 200°C; and 924 ppm at 300°C, (Morey et al., 1962). Directional pressure increases the solubility.

Crystallographic Properties

Structure. Low *quartz* is the polymorph of SiO_2 that is stable at ordinary temperature and pressure conditions. It belongs to the space groups D_3^6; $P3_22$ (right-handed) or D_3^4; $P3_12$ (left-handed).

The dimensions of the unit cell may be given as

$$a_0 = 4.9130 \text{ Å} \qquad \text{and} \qquad c_0 = 5.4045 \text{ Å}$$

The unit cell contains three Si atoms. The atoms of oxygen are grouped around the Si in a tetrahedral arrangement. The SiO_4 tetrahedra are linked by sharing of each of the corner oxygen atoms and form in this way a three-dimensional network (Figs. 1,2,3).

The atoms of Si lie in the planes 0*c*, 1/3*c*, and 2/3*c*; the atoms of O in the planes 1/6*c*, 3/6*c*, and 5/6*c*.

As the shortest distance, Si–Si links particles lying in planes at + or - 1/3*c*, these directions are polar. In reverse direction, an Si atom is met at only five times that distance.

In the direction of the zones [1$\bar{2}$10], [0001], and [1$\bar{2}$13], on the other hand, the Si–Si bond is not polar. Consequently, the average Si–Si distances will be shortest in these directions.

The ratio of the mean Si–Si distances are listed for a few zones:

[1$\bar{2}$10] :	[0001] :	[1$\bar{2}$13] :	[1$\bar{1}$00] :	[$\bar{2}$112] :	[10$\bar{1}$1]
1	1.1	1.45	1.73	1.85	2

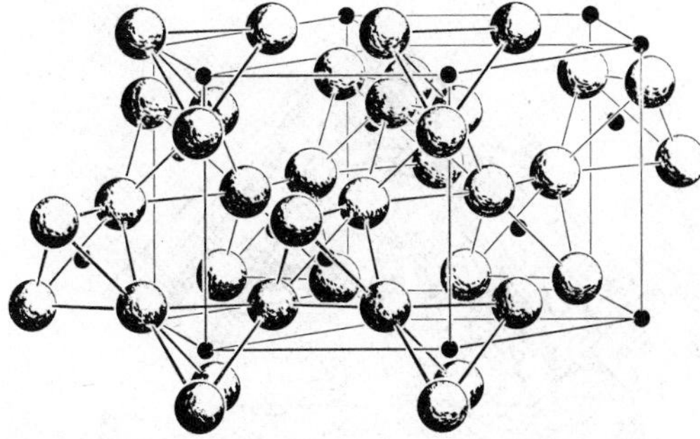

FIGURE 1. Structure of *β-quartz* (T > 573°C) (from Ernst, 1955, p. 35). Si represented by small black spheres.

The three directions [$1\bar{2}10$], [0001], and [$1\bar{2}13$] are, in fact, the main zones of *quartz* and the intersection of these zones determines the principal forms of the mineral, namely {$10\bar{1}1$} and {$10\bar{1}0$}.

Symmetry. *Quartz* crystallizes in the trigonal trapezohedral class (Schoenflies symbol, D_3; Mauguin, 32; elements of symmetry, one triad axis and, normal to it, three polar diad axes.)

The triad axis is a screw axis, with either right or left gyration. We have, therefore, enantiomorphic right and left forms (Fig. 4).

Forms. About 500 forms of *quartz* have been described, but many of them are rare or even doubtful. A rather small number of forms are of common occurrence and only three are really persistent, namely, the hexagonal prism, *m* {$10\bar{1}0$}; the positive rhombohedron, *r* {$10\bar{1}1$} and the negative rhombohedron, *z* {$01\bar{1}1$}.

Still very frequent are the trigonal pyramid, *s* {$11\bar{2}1$} and trapezohedron, *x* {$51\bar{6}1$}. The base, *c* {0001} is conspicuously absent.

Conventionally, a *quartz* crystal with such forms as *s* and *x* at the right below the positive

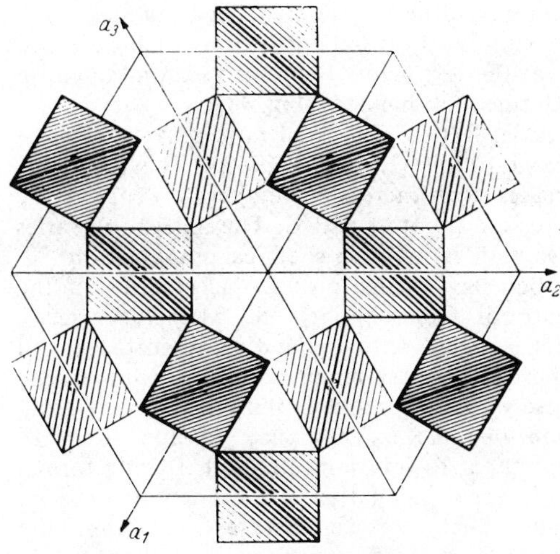

FIGURE 3. *β-quartz* (from Ernst, 1955, p. 35). Projection on (0001).

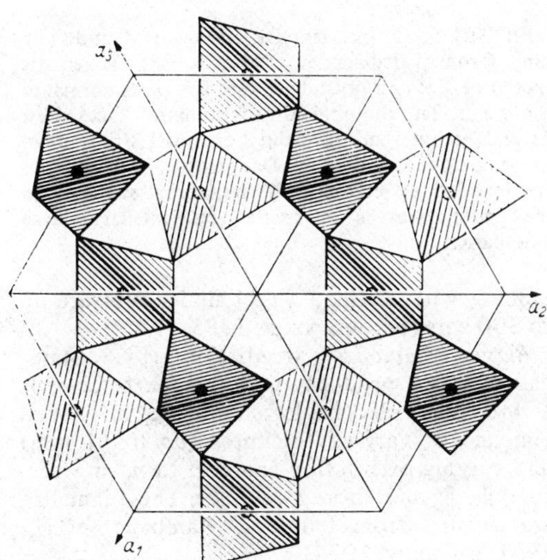

FIGURE 2. *α-quartz* (from Ernst, 1955, p. 35). Projection on (0001). Si small spheres in the center of the tetrahedra.

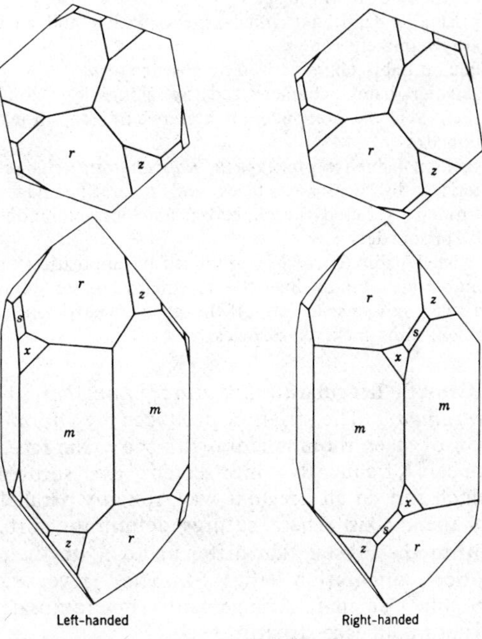

FIGURE 4. (a) Left-handed *quartz*; (b) right-handed *quartz* (from C. Frondel, 1962, ***Dana's System of Mineralogy***, 7th ed., ***Silica Minerals***, John Wiley & Sons, New York, p. 74, Figs. 7, 8).

rhombohedron is termed right-handed and, conversely, a crystal with forms such as *'s* and *'x* at the left below the positive rhombohedron is termed left-handed. (Fig. 4).

Axial Ratio. The axial ratio may be given as $a:c=1.10009$ (Frondel, 1962). The value decreases with increasing temperature up to the inversion point at 573°C. The axial ratio varies also with variation in chemical composition.

Vicinals. Vicinal planes ("accessories" in the sense of Goldschmidt) and vicinal surfaces—that is, planes with high indices or surfaces with irrational crystallographic indices approaching closely planes of simple indices—are very common on *quartz*. They may be due either to growth or to solution (etching). During formation, their inclination does not remain constant. Such surface features will limit the precision of goniometrical measurements.

Habit. The habit, that is the shape of the crystal as determined by the forms present and their relative development, depends upon physicochemical conditions of formation. Its characterization may therefore be significative.

Habit and aspect vary widely but only a relatively few forms are common in *quartz* and only a small number of combinations of forms are persistent.

One may distinguish

- prismatic habit, rhombohedral aspect (with *r* and *z* distinctly different in size)
- prismatic habit, hexagonal aspect (with *r* and *z* of approximately same size)
- equant habit, characterized by absence of *m*
- acicular habit, characterized by elongation along the *c* axis and preponderant presence of steep rhombohedra
- skeletal habit (scepter *quartz, Kappenquarz*) characterized by depressed faces and separating ridges formed by raised rhombohedral and, less commonly, prism edges
- twisted habit (*Gwendel*) produced by apposition of numerous subindividuals that are slightly offset along a twist axis so as to form, if the steps are sufficiently small, smooth curved surfaces.

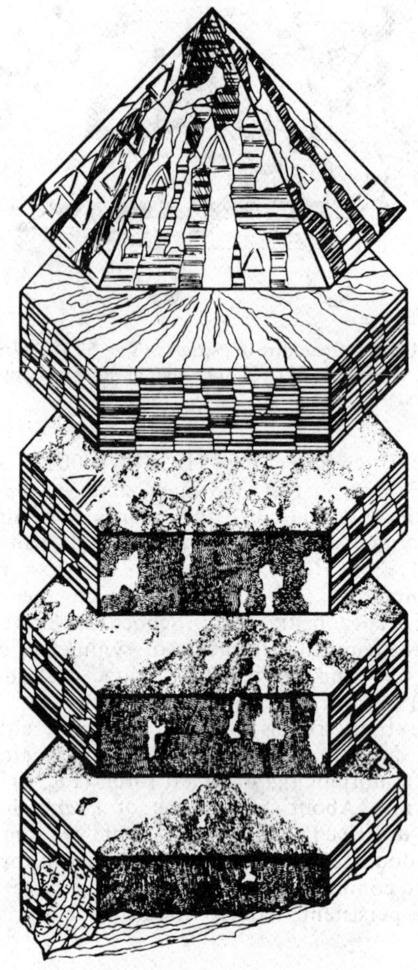

FIGURE 5. Sutures, lamellar arrangement, and twinning (from Friedlaender, 1951, p. 88). Schematic sketch of a *quartz* crystal cut into five parts normal to the *c* axis. The rhombohedral termination faces show natural etching, sutures, and horizontal striation. In the successive sections, the Dauphiné twin boundaries are revealed by etching. The horizontal striation on the prism faces is intersected by lamellar lineage boundaries.

Growth Irregularities. *Sutures, Lamellar Arrangement.* The striation produced by alternation of steep rhombohedra with the prism faces appears frequently intersected by sutures which run, in an irregular way, roughly parallel to the *c* axis. These sutures delimitate parts within the crystal that differ up to 3° in their optical orientation (Fig. 5). They reveal a roughly lamellar arrangement (macromosaic structure, lineage structure).

Mosaic Structure. Most natural *quartz* crystals are made up of blocks, of about 10^{-4} to 10^{-7} cm, in slightly oscillating orientation. A "shingle" appearance may become visible by etching with distilled water under pressure up to 300 bars (Frederickson, 1955).

Biaxial Lamellae. Irradiation (by X-rays, gamma rays) produces in some *quartz* crystals a brownish pleochroic coloration in biaxial lamellae of varying thickness (up to ±1 mm) and varying orientation (parallel to *m* in 3 sectors, ||*r*, ||*z* and, more rarely, ||*s*). These lamellae are distinct from twinning (Bambauer et al., 1961; Bukanov, 1974).

Zonal and Sectoral Arrangement. Concentric zones (particularly ||*m*, *r*, or *z*) with a width of up to 1 mm and sectoral arrangement due to variations in trace-element content are fairly common in *quartz*. Irradiation may make this apparent through selective coloration.

Turbidity, Inclusions. In *quartz* crystals, there are all transitions from colloidal suspension to inclusions of minute crystallites and distinct individual inclusions. Sizable inclusions—solid, liquid, or gaseous—are very common. The distribution of cloud-like turbid parts within a crystal frequently reveals growth stages (phantom *quartz*). Blue and rose color of *quartz* has been recognized to be due to inclusion of swarms of needles of *rutile*.

Deformation Features. *Undulatory Extinction.* Slight, more or less gradational, variation in the extinction position may be due to bend gliding.

Recrystallization. Normal to the greatest stress, the solubility will be strongest. In the stress shadow, the solution will be oversaturated and there may be a directional recrystallization, in fracture-free and apparently plastic behavior. In *quartz*, this crystalloblastic mobility is very marked.

Cleavage, Fracture, Faulting. In addition to conchoidal fracture, *quartz* may show cleavage ||*r* and *z* and, in order of decreasing frequency, also ||*m*, *c*, *a*(11$\bar{2}$0), *s*, and *x*. Clastic to kataclastic processes may result in the development of subgrains or mortar structure.

Deformation Lamellae. Narrow, closely spaced subplanar features are common in *quartz* from most tectonic environments. Slightly brownish lamellae with alignment of minute inclusions are known as Böhm lamellae.

Planar Features. Ascribed to Impact Metamorphism. The hypervelocity impact of a body such as a meteorite may result in the formation of multiple sets of closely spaced planar features. Lamellae parallel (0001) are widely considered to be indicative of shock metamorphism, but the effects of shock metamorphism are not always uniquely distinguishable as such.

Twinning. Twinning (q.v.) is widespread in *quartz*. We distinguish twins with parallel axes and twins with inclined axes.

a. Twins with parallel axes
   i. Simple twins, made up of two crystals (Fig. 6)
      (1) Dauphiné Law (Swiss Law, "electrical twin-

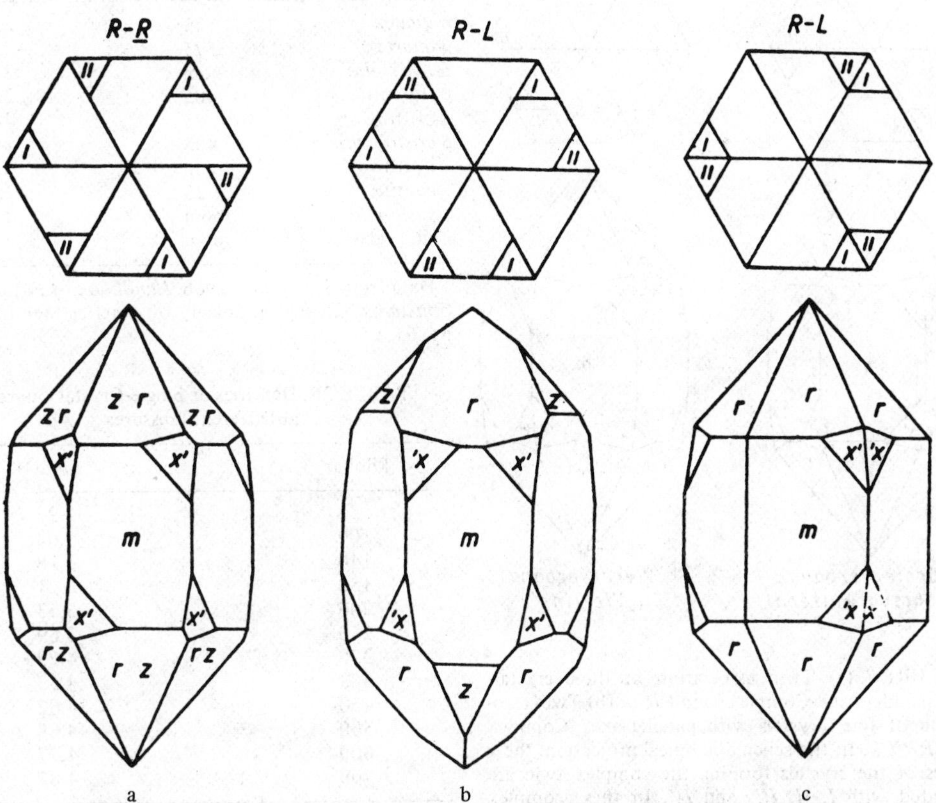

FIGURE 6. (a) Dauphiné twin; (b) Brazil twin; (c) Liebisch twin (from Gross, 1969, p. 30). The sketches show the position of the positive trigonal trapezohedron *x*. In the schematic apical projection, the signs *I* and *II* indicate the position of the *x* planes in the two crystals forming the twin.

ning"): part *I* rotated 180° around the *c* axis [0001] relative to part *II*. Both parts are of the same hand (*RR'* or *LL'*). Twinning boundaries curved. Symmetry: D_6–622.

(2) Brazil Law (chiral twin, "optical twinning"): part *I* mirror image of part *II*, twinning plane {11$\bar{2}$0}; part *I* enantiomorphous to part *II* (*R-L*). Twinning boundaries straight. Symmetry: D_{3d}–$\bar{3}m$.

(3) Liebisch Law (combined Dauphiné-Brazil, Leydolt Law, Law "a"): part *I* mirror image of part *II*, twinning plane (0001) and {10$\bar{1}$0}; part *I* enantiomorphous to part *II*(*R–L'*). Symmetry: D_{3h}–$\bar{6}2m$.

ii. Twin associations, made up of three or four crystals (Fig. 7)

(4) Complex twin made up of three crystals, *RR'L* or *LL'R*. Symmetry: D_3–32.

(5) Complex twin made up of four crystals, *RR'LL'*. Symmetry: D_{6h}–6mmm.

b. Twins with inclined axes

Japan Law: the *c* axes of part *I* and part *II* form an angle of 84°33'; the plane of hemitropy {11$\bar{2}$2}; twin axis normal to (11$\bar{2}$2). Numerous other twins with inclined axes have been described but they are less common. Furthermore, some intergrowths of this type may be considered accidental appositions (Engel, Gross, Nowacki, 1964; Gross, 1969; for older references see Frondel, 1962).

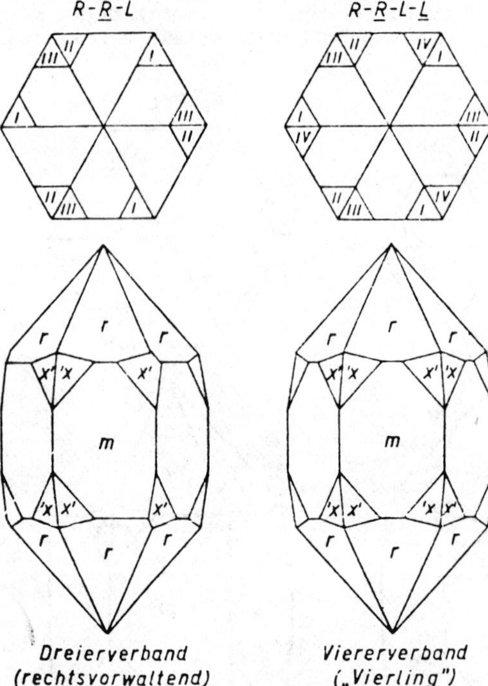

FIGURE 7. (a) Twin association of three crystals with parallel axes. Complex twin *RR'L*. (b) Twin association of four crystals with parallel axes. Complex twin *RR'LL*. In the schematic apical projection, the *x* planes of the crystals forming the complex twin are indicated with *I*, *II*, *III*, and *IV*. In these complex twins, all three simple twinning laws, Dauphiné (*D*), Brazil (*B*), and Liebisch (*L*), are present: I–II = D; I–III = B; I–IV = L; II–III = L; II–IV = B; and III–IV = D (from Gross, 1969, p. 41).

Physical Properties

Inversion Low Quartz to High Quartz. Low *quartz* changes at atmospheric pressure to high *quartz* at a temperature of 573°C. The temperature of inversion of low quartz to high *quartz* as a function of pressure has been determined up to 10 kilobar (Clark 1966, p. 348):

pressure (kilobar)	0	2	4	6	8	10
temperature (°C)	573	626	679	728	773	815

Density. See Table 2.

TABLE 2A. Densities of Silica Minerals at Atmospheric Pressure

	temperature °C	g/cm³
α-quartz	25	2.648
β-quartz	575	2.533
α-tridymite	room	2.265
β-tridymite	405	2.192
α-cristobalite	25	2.334
β-cristobalite	405	2.194
keatite	room	2.503
coesite	25	2.911
stishovite	room	4.287
silica glass	room	2.203

Data from S. P. Clark, 1966. *Handbook of Physical Constants,* Geological Society of America Memoir 97, pp. 65, 95.

TABLE 2B. Densities of Single-Crystal Quartz at Different Pressures

kilobar[a]	g/cm³
50	2.92
100	3.08
144	3.18
150	3.21
200	3.43
250	3.69
350	4.29
383	4.53
450	4.62
500	4.68
600	4.77
700	4.85

Data from S. P. Clark, 1966. *Handbook of Physcial Constants,* Geological Society of America Memoir 97, p. 158.

[a]Values above 400 kb probably refer to *stishovite.*

Hardness. The hardness of a crystal is dependent upon orientation and direction. In the Mohs scale of hardness, *quartz* scratch hardness is given the arbitrary value of 7. Some values of Rosiwal hardness (which is inversely proportional to the loss in volume after a grinding time of 8 minutes on a grinding surface of 4 cm^2 referred to mean hardness of *corundum*= 1000) are given in Table 3. Abrasion hardness is strongly dependent upon the contact liquid. In a general way, *quartz* is harder in nonpolar molecules, such as hexane and benzole, than in water; it is softer in asymmetric polar molecules such as alcohols and fatty acids than in water.

Plasticity. There is no uncontroversial evidence of plastic behavior of *quartz* under laboratory conditions. *Quartz* remains brittle and elastic up to its fracture point. The pressure-induced twinning in *quartz* (Schubnikow and Zinserling, 1932), devoid of visible gliding and with rotation only, may be interpreted as a type of gliding. This "rotation plasticity" is distinct from the translation (rock-salt type) and the simple twin gliding (*calcite* type; see *Plastic Flow in Minerals*); it may be termed "*quartz* plasticity." *Quartz* is so far the only mineral in which such a rotatory gliding has been ascertained.

Elasticity. Within the limits of elastic behavior, stress is proportional to strain: $p=E\lambda$ (Hooke's law). The elasticity modulus E, Young's modulus, for *quartz* at normal pressure is (10^{11}dynes/cm^2)

	0°C	570°C	600°C
$\parallel c$	7.9	3.0	10.7
$\perp c$	10.3	5.9	9.5

(see Gray 1957, 2-103).

Compressibility, compressive breaking strength, thermal expansion, and thermal conductivity data are summarized in Tables 4, 5, 6, and 7.

Electrical Resistivity. At room temperature, the electrical resistivity of *quartz* is on the

TABLE 3. Rosiwal Hardness of Quartz

	Grinding Loss with 100 mg *Corundum* (V in mm^3)	Relative Hardness (*Corundum* = 1000)	*Quartz* (%)	Absolute Hardness in m.kg/cm^3
Corundum (mean)	1.16	1000	949	–
(0001)	10.98	105.5	100	5.250
(10$\bar{1}$0)	12.75	91.0	86.3	4.535
(10$\bar{1}$1)	15.0	77.4	73.4	3.890

Data from H. Tertsch, 1949. *Die Festigkeitserscheinungen der Kristalle.* Vienna: Springer-Verlag, p. 207.

TABLE 4. Compressibility of Quartz

Pressure (kg/cm^2)	5,000	10,000	15,000	20,000	25,000	30,000
Linear $\parallel c$	.00334	.00642	.00920	.01170	.01406	.01622
$\perp c$	.00480	.00909	.01308	.01688	.02056	.02411
Volume	.01289	.02440	.03495	.04478	.05418	.06308

Data from S. P. Clark, 1966. *Handbook of Physical Constants,* Geological Society of America Memoir 97, p. 138.

TABLE 5. Compressive Breaking Strength of Quartz Under Extreme Hydrostatic Pressures (kilobars)

Confining pressure	32	38	39
Breaking stress	16	23	25
Remarks	no flow	cut $\parallel$ *c* axis	reduced to powder

Data from S. P. Clark, 1966. *Handbook of Physical Constants,* Geological Society of America Memoir 97, p. 275.

TABLE 6. Thermal Expansion of Quartz Referred to Dimensions at 20°C

Temperature (°C)	Change of Length $\parallel c$ axis (%)	Change of Length $\perp c$ axis (%)	Change of Volume (%)
100	.08	.14	.36
200	.18	.30	.78
300	.29	.49	1.27
400	.43	.72	1.87
500	.62	1.04	2.70
570	.84	1.46	3.76
and after transition to *β-quartz* at 573°			
580	1.03	1.76	4.55
600	1.02	1.76	4.54
700	1.01	1.75	4.51
800	.97	1.73	4.43
900	.92	1.71	4.34
1000	.88	1.69	4.26

Data from S. P. Clark, 1966. *Handbook of Physical Constants,* Geological Society of America Memoir 97, p. 91.

TABLE 7. Thermal Conductivity of Quartz (10^{-4} cal cm/sec cm^2 °C)

Temperature (°C)	$\parallel c$ axis	$\perp c$ axis
-252	–	6800
-250	–	5100
-190	1170	586
-78	467	240.9
0	325	173.1
100	215	133.3
145	174	–
236	153	–
260	–	89
308	137	–
383	–	104
468	–	106
475	123	–

Data from D. E. Gray, ed., 1957. *Handbook, American Institute of Physics.* New York; McGraw-Hill, p. 4-73.

order of 10^{14} ohm cm $\|c$ and 10^{16} ohm cm $\perp c$. It decreases with rising temperature.

Optical Properties. *Quartz* is transparent in the range of visible and ultraviolet light. Normally, it is colorless, but there are colored varieties (yellow, blue, pink, brown). Irradiation (by gamma radiation, electrons, or X-rays) induces brown to black color in a nonuniform distribution. The colored varieties show pleochroism. *Quartz* has vitreous luster; it is uniaxial positive.

Refraction. The refractive indices vary with wavelength. At 18°C,

		λ	ω	ϵ
D	Na	5892.9 Å	1.544246	1.553355
F	H	4861.33Å	1.549683	1.558979
H	Ca	3968.48Å	1.55813	1.56772

The birefringence decreases with rising wavelength. The refractive indices decrease with rising temperature. Unilateral pressure causes *quartz* to become biaxial.

The extinction is frequently undulous or wavy. This may be due either to strain (metamorphism) or to primary divergence in orientation within the crystal, (growth irregularities).

Infrared Absorption. In infrared light, between 2900 and 4000 cm^{-1}, 11 distinct bands of absorption are known in *quartz* in the temperature range of -190° and +550°C. Other bands of marked absorption are at approximately 2500, 1900, 1660–1430, 1175–1110, 835–770, and 475–385 cm^{-1}. Infrared spectroscopy provides a means of studying different structural defects in *quartz* (Brunner et al., 1961).

Optical Activity. The property of rotating the plane of polarization is very marked in *quartz*. The amount of rotation (1) is proportional to thickness of slab, (2) depends upon wavelength, and (3) depends upon inclination β to the optic axis. Rotation of *R quartz*→*r*; for *L quartz*→*l*.

Specific rotation ρ (for 1 mm thickness) in section normal to the optical axis, at 20°C, is (per mm)

	λ	ρ
Na	5892.9 Å	21.724°
H	4861.33Å	32.761°
Ca	3968.48Å	51.115°

At the inclination β=56°10′, *quartz* is optically inactive for all wavelengths (Szivessy, 1937).

Verdet's Constant

The magnetic field affects the rotatory power. This phenomenon is termed the Faraday effect. $\alpha = VlH$, where α is the angle of rotation of the linear polarized light of wavelength λ; l is the length passed by the polarized light in direction of the external magnetic field; H is the strength of the magnetic field; V the Verdet constant. A proportionality coefficient characteristic for the substance (Gray, 1957; d'Ans-Lax, 1970), in angular minutes per cm. Gauss is

	°C	λ Å	V min/cm^{-1}Gauss^{-1}	n_ω
quartz	20	5893	0.01644	1.5443
		5780	0.01714	–
		5461	0.01952	1.5462
silica glass	20	5780	0.01479	–
		5460	0.0165	–
		4360	0.0261	–

Pyroelectricity; Piezoelectricity

Change of temperature or strain produce electrical polarization on *quartz*. Conversely, heat and strain are produced by electrical polarization. Pyroelectricity and piezoelectricity may be of diagnostic value in crystal-structure analysis as they occur only in crystalline substances with polar axes and lacking center of symmetry. Piezoelectricity has technological importance, particularly in the telecommunication industry (Cady, 1946).

Use

Different types of *quartz* material are being used for a variety of industrial applications. *Quartz* sand is used in the manufacturing of glass and silica brick and in the chemical industry. Fused *quartz* is used for manufacturing of industrial and laboratory equipment. Because of its hardness, *quartz* is used as an abrasive, for bearing stones, and its colored varieties are used as ornamental material. The optical and piezoelectric properties find a wide range of use in optical devices (lenses and prisms for UV, monochromators, spectrographic instruments, *quartz* reflector mirrors) and resonators and oscillators, to generate ultrasound and to control frequency of vibration (radio, multichannel telephony, *quartz* clock).

C. G. I. FRIEDLAENDER

References

Bambauer, H. U., 1961. Spurenelementgehalte und Farbzentren in Quarzen aus Zerrklüften der Schweizer Alpen, *Schweiz Mineral. Petrogr. Mitt.*, **41**, 335–369.

Bambauer, H. U., et al., 1961. Beobachtungen über Lamellenbau an Bergkristallen, *Zeit. Kristallog.*, **116**, 173–181.

Boyle, R. W., 1953. On the colour of black and grey quartz from Yellowknife, Northwest Territories, Canada, *Am. Mineralogist*, **38**, 528–535.

Brunner, G. O.; Wondratschek, H.; and Laves, F., 1961. Ultrarotuntersuchungen über den Einbau von H in natürlichem Quarz, *Zeit. Elektrochemie*, **56**, 735–750.

Bukanov, V. V., 1974. Rock crystal from the subpolar Urals, (in Russian), *Akad. Nauk SSSR, Komi Filial, Inst. Geol.*, 212p.

Cady, W. G., 1946. *Piezoelectricity*. New York; McGraw-Hill, 000p.

Clark, S. P., 1966. *Handbook of Physical Constants. Geol. Soc. Am. Mem.*, **97**, 587p.

d'Ans-Lax, J., 1970. In K. Schafer and C. Synowietz, eds., *Taschenbuch für Chemiker u. Physiker*, vol. 3. Berlin, Heidelberg, New York: Springer-Verlag, p. 365.

Deer, W. A.; Howie, R. A.; and Zussman, J., 1962. *Rock-Forming Minerals*, vol. 4, *Framework Silicates*. London: Longmans, 179–230.

Donnay, J. D. H., and Le Page, Y., 1975. Twin Laws versus electrical and optical characters in low quartz, *Canadian Mineral.*, **13**, 83–85.

Engel, P.; Gross, G.; and Nowacki, W., 1964. Alpine Quarzkristalle mit Einschlüssen als (R-L) Zwillinge, *Schweiz. Mineral. Petrogr. Mitt.*, **44**, 485–488.

Ernst, T., 1955. *Kristallphysik*. In Landolt-Börnstein, vol. **I**, pt. 4. Berlin: Springer-Verlag.

Flörke, O. W., 1967. Die Modifikationen von SiO_2, *Fortschr. Mineral.*, **44**, 181–230.

Frederickson, A. F., 1955. Mosaic structure in quartz, *Am. Mineralogist*, **40**, 1–9.

French, B. M., and Short, N. M., eds., 1968. *Shock Metamorphism of Natural Materials*. Baltimore: Mono Book Corp., 644p.

Friedlaender, C., 1951. Untersuchung über die Eignung alpiner Quarze für piezoelektrische Zwecke, *Beitr. Geol. Schweiz, Geotech. Ser.*, **29**, 98p.

Frondel, C., 1962. *Dana's System of Mineralogy*, vol. 3, *Silica Minerals*, 7th ed. New York: Wiley, 333p.

Gray, D. E., ed., 1957. *Handbook, American Institute of Physics*. New York: McGraw-Hill.

Gross, G., 1969. Ueber Flächenverteilung und Symmetrie bei Quarzzwillingen, *Aufschluss*, **20**, 29–53.

Michel-Lévy, A., and Meunier-Chalmas, S., 1892. Mémoire sur diverses formes affectées par le réseau élémentaire du quartz, *Bull. Soc. Fr. Mineral.*, **15**, 159–190.

Morey, G. W.; Fournier, R. O.; and Rowe, J. J., 1962. The solubility of quartz in water in water in the temperature interval from 25° to 300°C, *Geochim. Cosmochim. Acta*, **26**, 1029–1043.

Niggli, P., 1926. Beziehungen Zwischen Struktur und äusserer Morphologie am Quarz, *Zeit. Kristallogr.*, **63**, 295–311.

Poty, B., 1969. La croissance des cristaux de quartz dans les filons sur l'exemple du filon de La Gardette (Bourg d'Oisans) et des filons du Mont Blanc, *Thèse Univ. Nancy, Sciences Terre, Mém. 17*, 162p.

Rosenbusch, H., and Mügge, O., 1927. *Mikroskopische Physiographie der petrographisch wichtigen Mineralien*, vol. 1, (2). Stuttgart: E. Schweizerbart, 174–213.

Savin, S. M., and Epstein, S., 1970. The oxygen isotopic composition of coarse grained sedimentary rocks and minerals, *Geochim. Cosmochim. Acta*, **34**, 323–329.

Schubnikow, A., and Zinserling, K., 1932. Ueber die Schlagund Druckfiguren und über die mechanischen Quarzzwillinge, *Zeit. Kristallogr.*, **83**, 243–264.

Siever, R., and Scott, R. A., 1963. Organic Geochemistry of Silica, in I. A. Breger, ed., *Organic Geochemistry*. Oxford: Pergamon, 579–595.

Sosman, R. B., 1965. *The Phases of Silica*, 2nd ed. New Brunswick, N.J.: Rutgers University Press, 389p.

Szivessy, G., 1937. Neuere Untersuchungen über die optischen Erscheinungen bei aktiven Kristallen, *Fortschr. Mineral., Kristallogr. Petrogr.*, **21**, 111–168.

Tertsch, H., 1949. *Die Festigkeitserscheinungen der Kristalle*. Vienna: Springer-Verlag, 310p.

Tröger, W. E., 1967. *Optische Bestimmung der gesteinsbildenden Minerale*, vol. 2. Stuttgart: E. Schweizerbart, 154–166.

Turner, F. J., and Weiss, L. E., 1963. *Structural Analysis of Metamorphic Tectonites*. New York: McGraw-Hill, 545p.

White, S., 1973. The dislocation structures responsible for the optical effects in some naturally deformed quartzes, *J. Mater. Sci.*, **8**, 490–499.

Cross-references: *Abrasive Materials; Alkali Feldspars; Authigenic Minerals; Beudantite Group; Blacksand Minerals; Clinopyroxenes; Collecting Minerals; Crystal Growth; Crystal Habit; Crystallography: History; Crystallography: Morphological; Diagenetic Minerals; Electron Microscopy (Scanning); Electron Microscopy (Transmission); Etch Pits; Fire Clays; Gemology; Gossan; Green River Mineralogy; Human and Vertebrate Minerals; Inclusions, Fluid; Isotopic Variations in Mineral Deposits; Jade; Lunar Mineralogy; Magnetic Minerals; Mantle Mineralogy; Meteoritic Minerals; Mineral Classification: Principles; Mineral Properties; Minerals, Uniaxial and Biaxial; Mohs Scale of Hardness; Museums, Mineralogical; Optical Mineralogy; Ore Microscopy; Pegmatite Minerals; Phantom Crystals; Piezoelectricity; Plastic Flow in Minerals; Pleochroic Halos; Polymorphism; Portland Cement Mineralogy; Refractory Minerals; Rock-Forming Minerals; Sands, Glass and Building; Skeletal Crystals; Soil Mineralogy; Staining Techniques; Synthetic Minerals; Thermoluminescence; Thermometry, Geologic; Twinning; Vein Minerals; Wolframite Group.*

R

RATIONAL INTERCEPTS–*See* DIRECTIONS AND PLANES

RECIPROCAL LATTICE–*See* Vol. IVA: X-RAY DIFFRACTION ANALYSIS

REFRACTIVE INDEX

The refractive index (n) of an isotropic substance is the ratio of the speed of light in air to that in the substance. It may also be expressed in terms of the angle of incidence (i) and refraction (r) of a ray of light passing from air obliquely into the substance (Fig. 1); thus, $n = \sin i / \sin r$.

Theoretical Considerations

Substances that are anisotropic have refractive indices which vary with the direction of vibration and transmission of the light (see *Optical Mineralogy*). Those belonging to the tetragonal and hexagonal (including rhombohedral) systems have two principal refractive indices associated with the "ordinary" and "extraordinary" rays, designated ω and ϵ

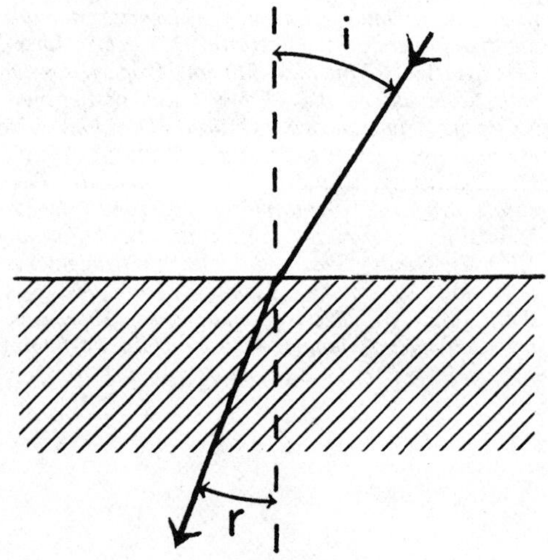

FIGURE 1. Angle of incidence (i) of a ray of light is shown in relation to the angle of refraction (r).

respectively. Those belonging to the orthorhombic, monoclinic, and triclinic systems have three principal refractive indices, the least (α), the intermediate (β), and the greatest (γ). When only one is stated, it is commonly either ω or β, but it may be any principal refractive index or an index related to an easily recognized crystallographic direction (see *Minerals, Uniaxial and Biaxial*).

General. The refractive index of natural minerals transparent to yellow light varies from about 1.309 (*ice*) to about 3.201 (*cinnabar*). Roughly 54% of known minerals have refractive indices between 1.475 and 1.700.

Transparent minerals with subvitreous and vitreous luster commonly have refractive indices between 1.3 and 1.9; those with adamintine luster, between 1.9 and 2.6; and those with submetallic luster, between 2.6 and 3.0. Minerals with metallic luster have calculated refractive indices generally in excess of 3.0. For metals and opaque minerals other factors are involved.

Variation with Wavelength. The refractive indices of most substances vary considerably with light of different wavelengths. When no wavelength is given, the refractive index is assumed to be for the wavelength of the yellow sodium *D* "line" (589.3 nm). The difference between the refractive indices for two different wavelengths is known as the "dispersion" (q.v.). Dispersion is commonly stated as the difference of refractive index for the blue hydrogen line H_β or F (486.13 nm) and the red hydrogen line H_α or C (656.28 nm), expressed as $n_F - n_C$.

The dispersion ($n_F - n_C$) of transparent minerals varies from about 0.004 (*fluorite*) to about 0.23 (iron-rich *sphalerite*). Most minerals have dispersions between 0.006 and 0.020. Dispersion tends to increase with increase in refractive index (see *Dispersion, Optical*).

Use of Refractive Index. Refractive index is one of the more valuable diagnostic properties of minerals. Tables of the refractive indices of minerals are available (e.g., Larsen and Berman, 1934), and most standard textbooks of mineralogy include optical properties of minerals (see *Minerals, Uniaxial and Biaxial*).

Several functions of refractive index and density have been shown to be approximately additive properties of compounds, dependent only on their chemical composition and structure. A discussion of these is given by Allan (1956).

When minerals form solid-solution series, the refractive index is usually found to vary simply with the composition, the relationship often appearing to be linear. Refractive index may thus be used to estimate the composition of those minerals the composition of which may be approximated by two end members and may be used as one variable in the estimation of the composition of minerals showing more complex solid solution. In the ***plagioclase*** series, the refractive index of the glass formed by fusing and quenching the mineral may be a more useful indication of the composition than the refractive indices of the mineral itself. The refractive indices of some minerals in metamorphic rocks, e.g., ***garnets***, have been correlated with the grade of metamorphism. The refractive indices of natural glasses or of glasses formed by fusing and quenching igneous rocks have been used as a rough indication of their silica content.

Determination of Refractive Index

In Thin Section. The refractive indices of grains in thin section may be estimated by comparison with those of surrounding grains or with the refractive index of the mounting medium using the Becké line and relief (see *Optical Mineralogy*). If the mounting medium is Canada balsam, its refractive index is usually between 1.533 and 1.541; if Lakeside 70 cement, about 1.539.

When higher accuracy is required, it is often possible to adapt the methods below to the determination of the refractive indices of grains in thin sections.

In Crystals. When perfect crystals or crystals on which polished surfaces may be prepared are available, the refractive indices may be determined by the prism methods described in most physics textbooks or by measuring directly on a refractometer of the Abbé, Pulfrich, or Rayner design. The last is particularly useful (see *Gemology*).

In Grains. The refractive indices of crushed minerals or separated grains are best determined by comparison with those of a series of refractive index liquids in which they are immersed. Comparison may be made by the use of either of two methods.

Central Illumination. The method of central illumination is based on the fact that a lenticular grain immersed in a medium of different refractive index behaves like a lens. When the tube of the microscope used to observe such a grain is raised, the center of the grain becomes brighter if the grain has a *higher* refractive index than the medium. Best results are obtained using a narrow cone of light.

Oblique Illumination. The method of oblique illuminations consists in observing which side of a grain appears bright when the light from a condenser is blocked on one side. This method, with certain refinements, is capable of giving high accuracy, but it has disadvantages when used with anisotropic minerals in which the extreme refractive indices differ considerably. Using monochromatic light, either method can give results accurate to at least ±0.001.

Immersion Liquids. The basis of the above methods is adjusting the immersion medium so that its refractive index matches that of the mineral grain. This may be done in one of the following ways:

Changing the Immersion Liquid. Sets of media of known refractive indices may be purchased or made up. These may be in steps from 0.002 to 0.010 depending on the accuracy required. Tables 1 and 2 list liquids suitable for making up such sets. The refractive indices of individual liquids, especially those of higher refractive index, should be checked regularly

TABLE 1. Liquids for Immersion-Media Mixtures

	Liquid	*n*
1.	Perfluorotributylamine	1.292
2.	Water	1.333
3.	Chlorotrifluoroethylene	1.406
4.	Kel-F(R) polymer oil No. 10	1.411
5.	Decane	1.411
6.	Tributyl phosphate	1.422
7.	Glyceryl triacetate	1.429
8.	2-(β-Butoxyethoxy) ethanol (=butyl "carbitol")	1.430
9.	Kerosene distillate	1.45±
10.	1,2,3,4-Tetrahydronaphthalene (=tetralin)	1.543
11.	Ethyl cinnamate (trans)	1.558
12.	1-Chloronaphthalene	1.633
13.	1-Bromonaphthalene	1.659
14.	Diiodomethane (-methylene iodide)	1.738
15.	Diiodomethane + sulfur	1.785
16.	Diiodomethane + sulfur + tetraiodoethylene	1.81
17.	Arsenic tribromide + 10% sulfur	1.814
18.	Arsenic tribromide + 20% sulfur + 20% arsenic disulfide	2.003
19.	Diiodomethane + 80% phosphorus + 10% sulfur	2.06

TABLE 2. Sets of Immersion Media Made with the Liquids in Table 1

Series	Range	Liquids	Alternative	Remarks
Standard series	1.292–1.406	1 & 3 or 1 & 4	or 1.333–1.430 2 & 8	The liquids of this series are neither highly poisonous nor corrosive to refractometer glass. Those above 1.738 are rather dark and change rapidly with evaporation of the volatile component. Liquids 5 and 9 may probably be replaced by any of a number of organic plasticizers.
	1.406–1.411	1 & 4		
	1.411–1.45±	5 & 9		
	1.45±–1.659	9 & 13	1.45±–1.633 9 & 12 or	
	1.659–1.738	13 & 14	1.633–1.738 12 & 14	
	1.738–1.785	14 & 15 or		
	1.785–1.810	14 & 16		
High-index series (Meyrowitz, 1951)	1.738–1.814	13 & 18	(can be made from 1.66)	These liquids are poisonous, affect refractometer glass and hydrolyze in the presence of water. The higher series is dark in color.
	1.814–2.003	17 & 18		
(West, 1936)	1.738–2.060	14 & 19		These liquids are highly poisonous, flammable, and liable to spontaneous combustion. They should be stored under water and used with *great care.* They are clear yellow in color.
High-dispersion series (Wilcox, 1964)	1.50–1.55	6 & 11		Suggested for refractive index determination by dispersion methods.
Clay series (after Vendl, 1945)	1.45±–1.543	9 & 10		Suggested as minimizing absorption errors in determining refractive indices of clay minerals.

on a refractometer at the temperature at which they are used.

Mixing Liquids on the Slide. The liquids are mixed until matching occurs when the refractive index is measured on a Jelley microrefractometer.

Changing the Physical Conditions. The refractive indices of most liquids vary with temperature and wavelength of light. Either or both may be varied until matching occurs. This maneuver forms the basis of the highly accurate single and double variation methods of Emmons (1943). Media suitable for these methods are given in Table 3.

Orientation of Grains. Anisotropic grains must be oriented before measurement so that the required index is measured. Orientation may be done by using optical properties to find grains having the required orientation or by rotating the grains to the required position. Many methods, of which the universal stage techniques of Emmons (1943) are the most sophisticated, are available for the latter purpose. The indices given by sections showing centered interference figures are given in Table 4.

MICHAEL J. FROST

References

Allan, R. D., 1956. A new equation relating index of refraction and specific gravity, *Am. Mineralogist*, **41**, 245–257.

Emmons, R. C., 1943. The universal stage, *Geol. Soc. Am., Mem.*, 8, 205p.

Hartshorne, N. H., and Stuart, A., 1970. *Crystals and the Polarising Microscope*, 3rd ed. London: Edward Arnold, 614p.

*Kerr, P. F., 1977. *Optical Mineralogy*, 4th ed. New York: McGraw-Hill, 492p.

Larsen, E. S., and Berman, H., 1934. The microscopic determination of the nonopaque minerals, *U.S. Geol. Surv. Bull.*, 848, 266p.

*Phillips, W. R., and Griffen, D. T., 1980. *Optical Mineralogy: The Nonopaque Minerals.* San Francisco: Freeman, 677p.

Vendl, M., 1945. Zur Bestimmung der Licht silikatischer Tonminerale, *Chem. Erde*, **15**, 325–370.

TABLE 3. Single and Double Variation Liquids

Liquid	n_D^{25}	Disp $n_F - n_C$ ($\times 10^5$)	$\frac{dn}{dT}$ ($\times 10^5$)	Temperature or Double Variation
Ethyldiiodoarsine[a]	1.807	49±	81	*D*
Diiodomethane (methylene iodide) and sulfur	1.775	38±	70±	*D*
Diiodomethane (methylene iodide)	1.737	37	68	*T & D*
Diiodomethane (methylene iodide) and 1-iodonaphthalene	1.716	37	56	*T*
I-Iodonaphthalene	1.699	37	47	*T & D*
I-Iodonaphthalene and 1-bromonaphthalene	1.675	36	55	*T & D*
1-Bromo-2-iodobenzene (*o*-bromoiodobenzene)	1.661	26	52	*T & D*
Isothiocyanatobenzene (phenylisothiocyanate)	1.647	35	58	*T & D*
sym-Tetrabromoethane (acetylene tetrabromide)	1.634	18	53	*T*
1-Chloronaphthalene	1.633	30	47	*D*
Iodobenzene	1.616	25	55	*T & D*
1-Ethoxynaphthalene (1-naphthylethylether)	1.599	29	50	*D*
Bromoform	1.594	18	60	*T*
1,2,3-Tribromopropane (glyceryl tribromohydrin)	1.582	25	55	*D*
1-Bromo-3-chlorobenzene (*m*-chlorobromobenzene)	1.568	19	55	*D*
1,2,4-Trichlorobenzene	1.565			*T*
o-Bromotoluene	1.553	18	50	*D*
Nitrobenzene[b]	1.549	24	48	*T*
o-Nitrotoluene	1.544	23	49	*D*
Benzenecarbonitrile[c] (benzonitrile)	1.527			*T*
Ethyl salicylate	1.523	33	46	*D*
3-Ethoxytoluene (*m*-cresyl ethyl ether)	1.511			*T*
Ethyl benzenecarboxylate (ethyl benzoate)	1.503	29	47	*D*
Cymene	1.490	13	49	*D*
s-Butylbenzene	1.488			*T*
Methyl thiocyanate	1.466	11	54	*T & D*

After Emmons, 1943.
[a]Poison and vesicant.
[b]Toxic.
[c]Highly toxic.

TABLE 4. Indices Given by Grains Showing Centered Interference Figures

Figure	Sign	Indices Obtainable
Acute bisectrix	+	α, β
Acute bisectrix	−	γ, β
Obtuse bisectrix	+	γ, β
Obtuse bisectrix	−	α, β
Flash (biaxial)	+ or −	α, γ
Flash (uniaxial)	+ or −	ω, ϵ
Optic axis (biaxial)	+ or −	β
Optic axis (uniaxial)	+ or −	ω

Wilcox, R. E., 1964. Immersion Liquids of relatively strong dispersion in the low refractive index range (1.46 to 1.52), *Am. Mineralogist*, 49, 683–688.

*Additional bibliographic references may be found in this work.

Cross-references: *Anisotropism; Gemology; Minerals, Uniaxial and Biaxial; Optical Mineralogy; Polarization and Polarizing Microscope.*

REFRACTOMETER—*See* GEMOLOGY

REFRACTORY MINERALS

Refractory minerals can be defined as minerals or synthetic inorganic crystal phases that have high melting points. They should also be resistant to deformation and to softening at high temperatures. Even though their primary function is resistance to elevated temperatures, in many applications they must resist other destructive forces, such as chemical corrosion and abrasion (see *Abrasive Materials*). In this article, only the major refractory minerals and their uses will be discussed.

Refractories are usually classified as acid, basic, or other. "Acid refractories" are mainly composed of silica or alumina-silica minerals. "Basic refractories" are produced from alkaline-earth oxides, such as *dolomite*, magnesia, lime, and magnesia-chrome ore mixtures. Other refractory minerals are listed with their important uses.

The major consumers of refractories are the metals industry, especially the steel industry, and the glass industry (see *Sands, Glass and Building*). Smaller industrial furnaces, such as cement kilns, ceramic kilns, and induction furnaces, also use refractory linings (see *Portland Cement Mineralogy*).

Refractory minerals are usually made into bricks by one of three methods: chemical bonding, sintering, or fusion casting. A short description of the minerals or inorganic substances, sources of the raw materials, and applications are listed below. For a more detailed description of the minerals or deposits, the references at the end of this article should be consulted. Many of the same refractory minerals can be used in different applications.

Acid Refractory Minerals

Quartz. *Quartz* (SiO_2, M.P.—melting point—1710°C), is used in the manufacture of silica bricks, semi-silica bricks, chrome-silica bricks, and fireclay bricks. These refractories are used in open-hearth furnaces, gas retorts, coke ovens, copper furnaces, and glass furnaces.

Sand molds (as thermal insulators), porcelains, and earthenwares also contain *quartz*. Fused silica is used in a variety of applications, such as lamps, tubing, and jackets.

The major sources of *quartz* are sedimentary deposits (e.g., Berea sandstone of Ohio) or metamorphic quartzites (e.g., Tuscarora of Pennsylvania or Baraboo of Wisconsin) (see Vol. VI: *Sands and Sandstones*).

Corundum. Other refractory, high-alumina minerals are βAl_2O_3, which has the composition $Na_2O \cdot 11Al_2O_3$ and a melting point of 1800°C, and $CaO \cdot 6Al_2O_3$, which has a melting point of 1850°C.

Refractory bricks containing *corundum* (αAl_2O_3, M.P. 2043°C) and(or) other refractory alumina phases are particularly useful in glass-contact refractories, many metal furnaces, nose cones, laboratory ware, spark plugs, and high-temperature lamps.

Bauxite (essentially $Al_2O_3 \cdot 2H_2O$), the major source of alumina, occurs in semitropical climates as residual products of weathering. Economic deposits are located in Arkansas, South Carolina, Cuba, and West Central Africa.

Mullite. *Mullite* ($3Al_2O_3 \cdot 2SiO_2$, M.P. 1810°C) is utilized in spark-plug cores, electrical porcelains, laboratory porcelains, glass ceramics, saggers, and high-grade refractories.

Sillimanite ($Al_2O_3 \cdot SiO_2$), *andalusite* ($Al_2O_3 \cdot SiO_2$), *kyanite* ($Al_2O_3 \cdot SiO_2$), and *dumortierite* ($8Al_2O_3 \cdot B_2O_3 \cdot 6SiO_2 \cdot H_2O$?) all convert to *mullite* and glass upon heating. These minerals occur in metamorphic rocks and in pegmatites. Commercial occurrences can be found in California, Nevada, Virginia, North Carolina, and Georgia.

Basic Minerals

Periclase. *Periclase* (MgO, M.P. 2825°C) is the most important refractory mineral in the metals industry. It can be found in the following refractories: magnesia, chrome-magnesia, *forsterite*-magnesia, *spinel*-magnesia, and dead-burned *dolomite*. Because of its high melting point, low cost (≈$270/ton), and resistance to corrosion by basic slags, it is used widely in the basic steel industry. In addition, *periclase* is a primary raw material for crucibles, muffles, refractory tubes, and kiln furniture.

At the present time, most of the *periclase* is obtained from seawater or salt brines. *Magnesite* ($MgCO_3$), which is dead-burned (>1450°C) to form *periclase*, is also a raw material. *Magnesite* deposits can be formed by (1) hydrothermal replacement—Washington, Austria; (2) veins—California, Greece, India, Russia; and (3) sedimentary rocks—Nevada.

Lime. Lime (CaO, M.P. 2625°C) is a very important ingredient in tar-bonded or tar-impregnated, sintered (calcined) *dolomite* or tar-bonded lime bricks. These bricks are employed mainly in the basic oxygen furnaces in the steel industry. Other lime ceramics have limited application because of their tendency to hydrate.

Limestone ($CaCO_3$) and dolomite [$CaMg(CO_3)_2$] are the major sources of CaO. These chemically deposited sedimentary rocks are very common, and commercial occurrences are located in Ohio, Pennsylvania, West Virginia, Michigan, Missouri, and other areas.

Strontium Oxide. The use of strontium oxide (SrO, M.P. 2420°C) as a refractory has been limited because (1) it is less refractory, (2) it hydrates more rapidly, and (3) it costs more than MgO or CaO. Some industrial materials containing SrO are glasses, glazes, and ceramics.

Strontium oxide is obtained from *celestite* ($SrSO_4$) and from *strontianite* ($SrCO_3$), which occur in sedimentary and hydrothermal deposits. Important localities are at Gloucester and Somerset, England; Westphalia, Germany; and Sicily.

Barium Oxide. Barium oxide (BaO, M.P.

1920°C) is not an important refractory material since it is not as refractory as MgO or CaO and it hydrates more rapidly. It is used as a batch ingredient to form other phases, such as barium titanate for electronic applications.

The major source of BaO is *barite* ($BaSO_4$), and some is also recovered from *witherite* ($BaCO_3$). *Barite* occurs in residual deposits, hydrothermal deposits (see *Vein Minerals*), and sedimentary deposits. It is found in large quantities in the Appalachians, California, Virginia, Arkansas, Missouri, Germany, England, and other localities.

Chromium Oxide. Chromium oxide (Cr_2O_3, M.P. 2330°C) refractories are used in glass furnaces that produce fibers. Single crystals of αAl_2O_3 with small amounts of Cr_2O_3 in solid solution are grown for synthetic gems and for ruby lasers. The high cost, high rate of vaporization at high temperatures, and strong coloring effect on other materials have limited its use.

Ores containing chrome ***spinel*** [(Fe,Mg)O·$(Cr,Fe,Al)_2O_3$] are the chief source of chromium oxide (see *Spinel Group*).

Spinels. ***Spinels*** ($RO{\cdot}R_2O_3$, Table 1) are particularly important in magnesia-chrome ore, magnesia-spinel, and spinel refractories. A complex ***spinel*** phase, e.g., (Mg,Fe)O·$(Cr,Al,Fe)_2O_3$, usually occurs in refractories containing chrome ore. They are primarily used in bricks for steel furnaces (open-hearth, electric, or Kaldo), copper furnaces, cement kilns, and glass furnace checkers. Refractories containing MgO-$MgAl_2O_4$ are also used but to a lesser extent. Crucibles and other specialty ceramic items are often constructed of ***spinel***.

Chrome ore is found associated with peridotites and serpentinite bodies. Rich deposits are located in the Transvaal, Rhodesia, Iran, Turkey, Cuba, and the Philippines. *Spinel* (MgO·Al_2O_3) is found in contact metamorphic deposits and as an accessory mineral of many mafic rocks, particularly the peridotite group. It occasionally occurs in placer deposits (q.v.) associated with *magnetite*. Many of the ***spinels*** are made by reacting, at high temperatures, their oxide components, e.g., MgO and Al_2O_3.

Forsterite. *Forsterite* (2MgO·SiO_2, M.P. 1890°C) refractories have been used for furnace linings in many industries. *Forsterite* can take iron and calcium in solid solution, and the solid solutions often occur in the matrix of magnesia refractories, magnesia-chrome refractories, and chrome-magnesia refractories.

Forsterite and other ***olivines*** (q.v.) occur in mafic igneous rocks, and a rock composed mainly of ***olivine*** is called dunite. Relatively low-iron ***olivine*** is obtained from North Carolina. Smaller quantities have been mined in the state of Washington. It is often produced by reacting either magnesia and silica or magnesia and ***serpentine***.

TABLE 1. Refractory Spinels

Mineral	Composition	Melting Point
Spinel	(MgO·Al_2O_3)	2105°C
Picrochromite	(MgO·Cr_2O_3)	2400°C
Chromite	(FeO·Cr_2O_3)	2200°C
Magnesioferrite	(MgO·Fe_2O_3)	1750°C
Hercynite	(FeO·Al_2O_3)	1800°C

Other Refractory Minerals

Beryllium Oxide. The use of beryllia (BeO, M.P. 2500°C) has been limited by its high cost, poisonous effects, and self-destructive phase transformations near its melting point. Because of its high thermal conductivity and refractoriness, it is being investigated for aerospace applications. Beryllia has resistance to radiation damage and is used in ceramic moderators for nuclear reactors.

Beryllia is extracted from the mineral *beryl* [$Be_3Al_2(SiO_3)_6$], which occurs in pegmatites (see *Pegmatite Minerals*). Deposits of *beryl* are located in South Dakota, New England, North Carolina, California, Brazil, and Namibia.

Cobalt Oxide. Although cobalt oxide (CoO, M.P. 1800°C) is refractory, it is not often used because of its cost and its tendency to oxidize to Co_3O_4 when heated in air. CoO forms a solid solution with MgO, and some experimental work has been done using either CoO or NiO solid solution to harden MgO. Its most important use is a dark blue coloring agent in glazes (see *Pigments and Fillers*).

Smaltite (*Skutterudite*, $CoAs_{2-3}$) and *cobaltite* (CoAsS) are the major sources of cobalt oxide. Both of these minerals occur in veins in igneous rocks or in metasomatic contact deposits. Economic deposits are located in Coleman Township, Ontario, Canada.

Nickel Oxide. Nickel oxide (NiO, M.P. 1990°C) is rarely used as a refractory because of its high cost. It is mainly utilized as a green or purple coloring agent.

Pentlandite (Fe,Ni)S, *millerite* (NiS), and *nickeline* (NiAs), the major sources of nickel, are associated with mafic igneous rocks. The major occurrence in North America is the Sudbury district of Canada.

Rutile. Since *rutile* (TiO_2, M.P. 1825°C) has a low dielectric loss at radio frequencies, it is sometimes used as a raw material in electrical-porcelain bodies. Small amounts of *rutile* are added to basic refractories to improve the spalling properties. Some glasses, glass ceramics, and ceramic specialty items also contain TiO_2.

Rutile is obtained from beach and stream deposits resulting from the weathering of igneous and metamorphic rocks (see *Placer Deposits*). Commercial deposits occur in Florida, India, Australia, and Brazil.

Baddeleyite. Zirconium dioxide (*baddeleyite*, ZrO_2, M.P. 2720°C) is an extremely important phase in fused-cast or sintered refractories (with or without Al_2O_3) used in glass and metal furnaces. Zirconia-containing bricks are quite refractory and are resistant to corrosion by siliceous slags, molten glass, and metals. Zirconia is often stabilized to the cubic structure by additions of CaO, MgO, or Y_2O_3. Crucibles, tubes, muffles, kiln furniture, refractory insulating brick, and aerospace materials are also prepared from ZrO_2.

The primary zirconia minerals are *zircon* ($ZrSiO_4$) and *baddeleyite* (ZrO_2), which occur as accessory minerals in siliceous igneous rocks. They are concentrated in beach placer deposits located in India, Brazil, Florida, and New South Wales.

Zircon. Mixtures of *zircon* ($ZrSiO_4$, M.P. 2550°C?) and alumina are fused-cast to form glass refractories and abrasion-resistant materials. Sintered blocks are used in furnaces that produce glass fibers.

The geological occurrence of *zircon* is the same as that of *baddeleyite*.

Thorium Oxide. Thorium oxide (ThO_2, M.P. 3220°C) has found limited use as a high-temperature specialty ceramic. Its high cost, radioactive properties, and poor thermal shock resistance have prevented widespread usage as a refractory.

The primary source of thorium is *monazite*, which is a rare-earth phosphate containing 3 to 8% thoria. *Monazite* is concentrated in sands derived from weathering of granites, gneisses, and pegmatites. Deposits occur in the Carolinas, Brazil, India, and Ceylon.

Graphite. *Graphite* (C, sublimes at 3600°C in neutral atmosphere) is an extremely important high-temperature material. Some of the major applications are in electrodes, mold materials, crucibles, refractory blocks, tubes, furnace parts, and aerospace materials.

Graphite occurs naturally in some granites, gneisses, schists, quartzites, limestone, and mafic extrusive igneous rocks. Deposits are located in Sri Lanka (Ceylon), Siberia, Cumberland (England), and Sonora (Mexico). Practically all of the commercial *graphite* used today is produced by heating coke, obtained from coal and petroleum, in electric resistance furnaces.

TABLE 2. Miscellaneous Ceramic Materials

Plagioclase feldspars: $NaAlSi_3O_8-CaAl_2Si_2O_8$
Potash ***feldspars:*** $KAlSi_3O_8$
Barium ***feldspars:*** $BaAl_2Si_2O_8$
Feldspathoids: (Na,K,Al silicates)
Calcined clays: Al silicates, see *Fire Clays*
Beryl: $Be_3Al_2Si_6O_{18}$
Barite: $BaSO_4$
Anhydrite: $CaSO_4$
Apatite: $Ca_5F(PO_4)_3$ (can substitute OH and(or) Cl for F)
Pyroxenes: (primarily silicates of Mg, Ca, and(or) Fe)
Cordierite: $(Mg,Fe)_2Al_4Si_5O_{18}$
Garnets: $3RO{\cdot}R_2O_3{\cdot}3SiO_2$
Talc: $Mg_3(OH)_2Si_4O_{10}$
Asbestos minerals: hydrous silicates of Mg, Ca, and(or) Fe (see *Asbestos*)
Micas: hydrous and(or) halogen-containing Al silicates K,Li,Na,Mg,Fe, etc.
Chlorite: $(Mg,Fe,Al)_6(OH)_8(Si,Al)_4O_{10}$
Fluorite: CaF_2
β-Spodumene: $LiAlSi_2O_6$
β-eucryptite: $LiAlSiO_4$
Perovskite: $CaTiO_3$

Refractory Hard Metals

Carbides, borides, nitrides, and silicides, particularly of Ti, Zr, Hf, V, Nb, Ta, Cr, Mo, W, V, Th, and Si, comprise a very important group of refractory materials. They are used primarily in special high-temperature applications (missile parts, crucibles, and refractory electrical parts) and in abrasion-resistant applications (abrasives, cutting tools, and tire studs). Refractory hard metals are usually prepared by heating the elements or chemical compounds in the proper proportions to obtain the desired product (see *Synthetic Minerals*).

Other Minerals Useful in Ceramics

The minerals and mineral groups listed in Table 2 are not discussed in detail because they are not very refractory. Their melting points are less than 1600°C. However, they are used in making important ceramic articles.

R. C. DOMAN
A. M. ALPER

References

Alper, A. M., 1970. High Temperature Oxides, *Refractory Materials Monograph*, no. 5. New York: Academic.

Alper, A. M., 1970. Phase Diagrams: Materials Science and Technology, *Refractory Materials Monograph*, no. 6. New York: Academic.

Amer. Ceram. Soc., Bull., 1975. Current Ceramic Research Issue, **54** (2).

Bateman, A. M., 1950. *Economic Mineral Deposits.* New York: Wiley, 916p.
Dana, E. S., and Ford, W. E., 1951. *A Textbook of Mineralogy.* New York: Wiley, 851p.
Levin, E. M.; Robbins, C. R.; and McMurdie, H. F., 1964. *Phase Diagrams for Ceramics.* Columbus, Ohio: American Ceramic Society, 601p.
Lindgren, W., 1933. *Mineral Deposits.* New York: McGraw-Hill, 930p.
Lynch, C. T., 1974. *Handbook of Materials Science.* Cleveland: CRC Press.
Rigby, G. R., 1953. *The Thin-Section Mineralogy of Ceramic Materials.* London: British Ceramics Research Assoc., 231p.
Shaffer, P. T. B., 1964. *High-Temperature Materials.* New York: Plenum, 740p.
Winchell, A. N., and Winchell, H., 1964. *The Microscopical Characteristics of Artificial Inorganic Solid Substances.* New York: Academic, 439p.

Cross-references: *Abrasive Materials; American Society for Testing and Materials; Fire Clays; Metallurgy; Mineral Industries; Mineral Properties; Portland Cement Mineralogy; Sands, Glass and Building; Synthetic Minerals*; see also mineral glossary.

RESIDUAL MINERALS–*See* BLACKSAND MINERALS; PLACER DEPOSITS: SOIL MINERALOGY

RESIN AND AMBER

Natural resins, secreted by trees or shrubs upon injury, appear either as balsam (a resin solution in esters) or as an oleoresin (a resin solution in essential oils). Resins are volatile and nonvolatile terpenoids and may occur with alcohols, aldehydes, esters, resenes, and other minor nonterpenoid substances. Amber and other fossil or subfossil resins of geologic interest are apparently derived chiefly from nonvolatile terpenoids through oxidation and polymerization. The details are poorly known, but the process seems to require a relatively long lapse of time.

Kauri Gum and Copal

Kauri gum and copal are solid resins from trees but may be found in soils and alluvial deposits. Both characteristically contain agathic acid, melt at from 150° to 210°C, range from almost black to clear white, and have a specific gravity of 1.07 to 1.08.

Copals that are geographically distinguished may be chemically or botanically distinct. Copals range from colorless and transparent to yellow-brown or black, are resinous, have conchoidal fracture, and have a specific gravity of about 1.07. They melt at from 180° to 245°C.

The plant source of Kauri gum and most copals is known, and most of the physical and chemical intermediates between fresh resin and subfossil or fossil resins have been found in nature. Subfossil and fossil Kauri gum and copal occur in lumps and irregular masses from a few inches to a few feet below the soil surface. Both may be found in areas where there is no other trace of the source tree and where the plants have not been known during recorded history. It is generally assumed that these resins are at least several thousand years old. It should be noted, however, that at one time commercially significant quantities of Sierra Leone copal were gathered from streambeds and beaches as "pebble gum." This fact suggests that deposits not related to forests may be alluvial rather than relict.

Copal and Kauri gum were formerly important raw materials for varnish manufacture, but depletion of reserves and competition from synthetics have greatly reduced the trade. Kauri, in addition to being specially prized for its ready solubility in linseed oil, was also used in linoleum manufacture. Both gums have found extensive use in plastics, paper sizing, adhesives, fireworks and other products.

Resinite

Coal petrographers refer to fossil resins and waxes as "resinite." Much of this material occurs as cylindrical, hollow, or solid rods called resin rodlets. These rodlets presumably were secreted in canals or ducts of the coal-forming plant. Less commonly, coal beds as much as 0.75 m thick may be more than half composed of resin bodies, and the material is referred to as "resinite" coal. Chemical composition of resinites ranges widely, depending upon the original plant source, the degree and mode of decay before coalification, the age of the coal, and the maximum temperature to which the coal has been subjected. In general, Early Cretaceous coals have a higher proportion of coniferous material and a higher ratio of resin to wax than do Late Cretaceous and Tertiary coals, which contain more angiosperm material. Older resins also tend to be less soluble in organic solvents, reflecting a higher degree of polymerization and oxidation.

Resinite and resinite coals from Utah, Germany, and elsewhere have been exploited as chemical raw material in the varnish, enamel, and plastic industries.

Amber

Resins are widespread in unmetamorphosed sedimentary rocks containing woody debris as well as occurring in almost all coals ranking below medium bituminous. Resins associated with coal or carbonaceous sediments were probably deposited at or near their point of origin. Much amber, however, occurs in sandstone, claystone, or the so-called blue earth and is associated with marine fossils. This material has most probably been transported either as fresh or as subfossil resin. The Baltic amber has yielded several specimens with coral inclusions, and one specimen with a partially embedded oystershell is known from Chiapas, suggesting transport as unsolidified resin or secretion by plants growing in or over salt water. Isolated amber gragments in marine sedimentary rocks, however, may be reworked, either from subfossil accumulations or from older rocks, and the amber in the Pleistocene deposits of northern Europe and Arctic Alaska is definitely redeposited.

Any lump resin in coal or sedimentary rock may be referred to as amber, but it should be noted that the name is restricted by many authorities to succinite—the well-known Baltic amber. The almost infinite range of chemically and botanically different ambers has been divided among a succinite group characterized by succinic acid, a retinite group lacking succinic acid, and a tasmanite series of sulfur-bearing resins. In addition, many resin minerals have been distinguished by chemical, physical, botanical, geographic, or temporal criteria. Numerous mineral names have been applied to fossil resins in the past, but this practice has ceased because most mineralogists no longer include resins among minerals.

The highly oxygenated and polymerized ambers contain organic acids, the most important of which are succinic, formic, butyric, and cinnamonic acid. The specific composition of individual specimens from the same or different localities may differ greatly, however, because of the wide range of possible botanic origin and subsequent geologic history for any specimen. Amber generally ranges from 1 to 3 on the Mohs scale of hardness, with most specimens falling approximately at 2. Specific gravity ranges from 1.00 to 1.25, but is usually between 1.05 and 1.08. The most prized amber is clear and yellow, with conchoidal fractures and resinous luster. Most specimens are pale to golden yellow, but red to orange-red amber is well known, and yellow-green, deep red, brown, and black ambers are also known. In addition, there are all gradations from transparent resinous material to opaque, dull, or earthy substances.

Fossil resins are brittle, but most varieties soften at around 150°C and melt at from 250° to 300°C. They burn readily, generally but slightly above the melting point, and give off a variety of smoke or fumes depending upon their composition. Amber is noted for the readiness with which it becomes negatively charged with static electricity; in fact, the word "electricity" has its root in the classic Greek word for amber—*elektron*. Fossil resins differ widely in chemical composition, and their solubility in alcohol, turpentine, acetone, ether, benzene, chloroform, or other organic solvents ranges from essentially insoluble to nearly completely soluble. Finally, most ambers fluoresce in ultraviolet light. Most amber occurs as irregularly rounded lumps with brown to black opaque crusts. Although lumps as large as 8 kg have been reported, most pieces are small, and a fist-sized lump is considered large. Many pieces have ropy, swirled surface patterns, and frozen runnels and droplets are widespread.

Until recently, ideas as to the plant source for fossil resins has been derived from associated fossil plant remains. Thus, Baltic amber was long thought to be derived from an associated fossil pine, *Pinus succinifera*, so named because of the presumed relationship to the amber.

More recently, however, infrared spectrophotometry has been extensively applied to fossil and recent resins by J. H. Langenheim, who finds that the infrared spectra of ambers in many cases closely match those of resins derived from living trees. This observation allows direct comparison of fossil and living resins and has served to relate numerous ambers, chiefly Cretaceous but including the Early Tertiary Baltic amber, to resin of modern *Araucaria* and(or) *Agathis*. Other, chiefly Tertiary, ambers have been related to a wide variety of tropical trees, including *Liquidambar, Shorea, Copaifera*, and *Hymenaea*. In addition, X-ray diffraction analysis has indicated an angiospermous origin for the fossil resins, Highgate copalite, glessite, and guayaquillite.

Amber has long been noted for fossil inclusions. Bachofen-Echt lists more than 100 plant species and several hundred animals from the Baltic amber. Insects are preponderant, but corals (*Hydraulis* and *Caryophyllia*), a number of worms, crustacea, myriopods, spiders, and gastropods also occur, and higher animals are represented by feathers and hair. In addition, most of the major plant groups are represented by bits of foliage, woody fragments, spores, and even a few flowers and seeds.

Baltic amber has been an article of trade in Europe almost since the dawn of history. Specimens have been recovered from Egyptian tombs, and the ancient Greeks and Romans wrote of Baltic amber. In the New World, Chiapanecan amber has been found at Mayan archaeological sites in Mexico and Central America. Burmese amber has long been mined and exported to China for use in jewelry and medicine. During the 19th and early 20th centuries, Baltic amber was widely used for jewelry, cigarette holders, pipe stems and other ornamental objects but has largely been displaced by plastics. In addition, Baltic amber formerly was extensively mined, especially at the great open pits at Palmnicken, Paniken, and elsewhere in Sammland (East Prussia). This material has been used mostly in the chemical and varnish industry. The trade is now reduced greatly, but production at more than one million pounds per year was maintained from 1895 to 1914.

RALPH L. LANGENHEIM

References

*Bachofen-Echt, A., *Der Bernstein und seine Einschlusse*. Vienna: Springer-Verlag, 204p.

Carpenter, F. M., et al., 1937. Insects and arachnids from Canadian amber, *Univ. Toronto Stud., Geol. Ser.*, **40**, 7-62.

Durham, J. W., 1957. Amber through the ages, *Pacific Discovery*, **10**(2), 3-5.

*Francis, W., 1961. *Coal, Its Formation and Composition*. London: Edward Arnold, 806p.

Hill, A. F., 1952. *Economic Botany, a Textbook of Useful Plants and Plant Products*. New York: McGraw-Hill, 560p.

Hollick, C. A., 1905. The occurrence and origin of amber in the eastern United States, *Am. Naturalist*, **39**(459), 137-145.

*Howes, F. N., 1949. *Vegetable Gums and Resins*. Waltham, Mass: Chronica Botanica, 188p.

Hurd, P. D.; Smith, R. F.; and Durham, J. W., 1962. The fossiliferous amber of Chiapas, Mexico, *Ciencia*, **21**(3), 107-118.

Kirchner, G., 1950. Amber inclusions, *Endeavour*, 9(34), 70-75.

*Langenheim, J. H., 1969. Amber: A botanical inquiry, *Science*, **163**, 1157-1169.

Langenheim, R. L., Jr.; Smiley, C. J.; and Gray, Jane, 1960. Cretaceous amber from the Arctic Coastal Plain of Alaska, *Geol. Soc. Am. Bull.*, **71**(9), 1345-1356.

Ley, W., 1951. *Dragons in Amber; Further Adventures of a Romantic Naturalist*. New York: Viking, 328p.

*Paclt, J., 1953. A system of Caustolites, *Tschermaks Mineral. Petrog. Mitt.*, 3(4), 332-347.

Rice, P.C., 1980. New York: Van Nostrand Reinhold, 289p.

Sanderson, M. W., and Farr, T. H., 1960. Amber with insect and plant inclusions from the Dominican Republic, *Science*, **131**(3409), 1313.

White, D., 1914. Resins in Paleozoic plants and in coals of high rank, *U.S. Geol. Surv. Prof. Paper 85*, 65-96.

*Williamson, G. C., 1932. *The Book of Amber*. London: Ernest Benn, 268p.

*Extensive bibliographies are found in the starred references.

Cross-references: *Collecting Minerals; Invertebrate and Plant Mineralogy*.

ROCK-FORMING MINERALS

The rocks constituting the crusts of all the inner planets so far explored are composed of a relative few of the nearly 3000 mineral species and varieties known to mineralogists. The most common rock-forming minerals are silicates (see Vol. IVA: *Mineral Classes: Silicates*), but they also include oxides, hydroxides, sulfides, sulfates, carbonates, phosphates, and halides (see Vol. IVA: *Mineral Classes: Nonsilicates*). The major rock-forming minerals and mineral groups listed in Table 1 are classified after Deer et al., 1966.

Three main factors contribute to the relatively small number of minerals that are actually common to planetary crusts: First, not all chemical elements are equally abundant in crustal rocks (see Vol. IVA: *Elements; Elements: Planetary Abundances and Distributions*). This inequality results first from variations in cosmic abundances and second from intrinsic characteristics of the elements involved in planetary formation. Crustal rocks may contain concentrations of those cosmically available elements that have not been depleted from the crust by concentration in the planetary core (e.g., the chalcophile elements; see Vol. IVA: *Core Geochemistry*). Concentration may also occur by selective removal because of solubility in water or because of volatility, escaping to the atmosphere or to outer space. The net result is that eight elements (oxygen, silicon, aluminum, iron, magnesium, calcium, sodium, and potassium) constitute more than 98% by weight of the terrestrial continental crust. In terms of atomic or of volume percentages, the remaining elements constitute less than 0.01% of the terrestrial continental crust. Oceanic, lunar, and Martian crustal rocks have elemental concentrations that differ significantly in detail from terrestrial continental abundances but follow the same general pattern—only a few of the elements are geochemically significant (see Vol. IVA: *Earth's Crust Geochemistry*).

The second major factor limiting the number of rock-forming minerals results from chemical restrictions on the ways in which elements may combine into naturally occurring chemical compounds, i.e., into minerals. Electropositive elements, those from the left and center of a standard periodic chart of the elements, tend to

TABLE 1. Rock-Forming Minerals

Ortho- and Ring Silicates

Olivine group
Humite group
Zircon
Titanite (sphene)
Garnet group
Kyanite, andalusite, sillimanite, mullite
Topaz
Staurolite, chloritoid
Datolite
Larnite, merwinite, spurrite, eudialyte, rosenbuschite
Epidote group
Lawsonite, pumpellyite
Melilite group
Rankinite, tilleyite, lavenite, catapleiite
Tourmaline group
Axinite group

Chain Silicates

Pyroxene group
Pyroxenoid group
Sapphirine
Amphibole group
Aenigmatite, astrophyllite

Sheet Silicates

Mica Group
Stilpnomelane
Pyrophyllite, talc
Chlorite group
Serpentine group
Clays
Apophyllite, prehnite

Framework Silicates

Feldspar group, ***Alkali Feldspars, Plagioclase Feldspars***
Quartz, tridymite, cristobalite
Feldspathoid group
Beryl, cordierite
Sodalite group
Cancrinite, ***Scapolite*** group
Analcime, ***Zeolite*** group

Nonsilicates

Oxides: *periclase, cassiterite, corundum, hematite, ilmenite, rutile, anatase, brookite, perovskite,* ***Spinel*** group
Hydroxides: *brucite, gibbsite, diaspore, boehmite, goethite, lepidocrocite,* limonite
Sulfides: *pyrite, pyrrhotite, chalcopyrite, sphalerite, galena*
Sulfates: *barite, celestite, gypsum, anhydrite*
Carbonates: *calcite, magnesite, rhodochrosite, siderite, dolomite, ankerite, huntite, aragonite, strontianite, witherite*
Phosphates: ***Apatite*** group, *monazite*
Halides: *fluorite, halite*

After Deer et al., 1966.

react with electronegative elements from the right-hand side of the chart. Since oxygen is the most abundant element on the right, it follows that most rock-forming minerals are compounds with oxygen. Exceptions to this rule are the sulfides (compounds with sulfur) and halides (compounds with fluorine or chlorine), which are locally abundant in terrestrial rocks.

The third factor limiting the diversity of rock-forming minerals is that, under conditions of chemical equilibrium, certain compounds tend not to occur together independently, but react to form intermediate compounds to the exclusion of one or more of the end members (see Vol. IVA: *Phase Equilibria*). This reactivity of compounds restricts not only the number of common rock-forming minerals but, more importantly, the number of mineral assemblages that exist under conditions of chemical equilibrium. For example, both *quartz* and *corundum* are rock-forming minerals, but they do not occur together in equilibrium assemblages of minerals because reaction between them forms minerals of intermediate chemical composition–*andalusite, sillimanite, kyanite*, or *mullite*, depending upon pressures and temperatures of assemblage equilibria.

A rock is an assemblage of one or more minerals constituting a portion of a planetary lithosphere. The term is also applied, by extension, to rock samples. Rocks are classified according to genesis: They are called "igneous" if they crystallized from a melt of essentially the same chemical composition; "sedimentary" if they are laid down as a surficial deposit by wind, water, ice, or meteoritic impact; or "metamorphic" if they have recrystallized in the solid state to new mineral assemblages in response to changes in pressure, temperature, and(or) chemical environment (see Vol. IVA: *Paragenesis*; also, Vols. V, *Petrology*; VI, *Sedimentology*; and XXII, *Volcanoes and Volcanology*).

The most common rock in the oceanic crust and in the lunar maria is basalt, an extrusive igneous rock, the predominant minerals of which belong to the ***plagioclase feldspars*** but include lesser amounts from the ***olivine*** group and from the ***pyroxene*** group, chiefly ***orthopyroxenes***. The terrestrial continental crust is predominantly granite or granodiorite, composed mostly of minerals of the ***feldspar*** group, both ***plagioclase feldspars*** and ***alkali feldspars***. These rocks also contain *quartz* and certain mafic minerals (containing magnesium and iron) from the ***amphibole*** group, the ***clinopyroxenes***, and the ***mica*** group of the ***phyllosilicates***. Lunar highland rocks are anorthosites composed essentially of ***plagioclase feldspars***.

Terrestrial sedimentary rocks include sandstones, composed principally of *quartz*, in some instances with lesser amounts of ***alkali feldspars*** (arkoses); shales, composed chiefly of ***clays***; limestones and dolomites (dolostones), composed of *calcite* and *dolomite*; and rocks formed by evaporation of saline water leaving behind water-soluble halides and sulfates, usually in combination with sodium, calcium, or magnesium (see *Saline Minerals*).

Lunar sedimentary rocks are limited to impact breccias (see *Lunar Minerals*) composed of fragments of anorthosite and basalt mixed with glass and meteoritic materials (see *Meteoritic Minerals*). Martian sediments appear to include impact breccias, ferroan clays of the ***smectite*** group, sulfates (*kieserite*?), carbonates (*calcite*?), and oxides (*hematite*?) (Baird et al., 1976).

Terrestrial metamorphic rocks include slates, composed chiefly of ***clays*** in which *pyrite* is locally abundant; schists, characterized by minerals from the ***mica*** and the ***chlorite*** groups, with lesser amounts of ***alkali feldspars*** and *quartz* and local abundances of ***garnet***-group minerals and *staurolite, sillimanite*, and *kyanite*; amphibolites, composed chiefly of minerals from the ***amphibole*** group with ***plagioclase feldspar***; marbles, made up mostly of *calcite* or *dolomite* with lesser amounts of minerals from the ***epidote*** and ***chlorite*** groups; and gneisses, composed chiefly of minerals of the ***feldspar*** group associated with *quartz* and mafic minerals from the ***mica*** and the ***amphibole*** groups (see Vol. IVA: *Metamorphic Environments–Chemical Mobility*).

Terrestrial soils are composed of various mixtures of primary minerals, those derived from parent rocks without chemical alteration, and secondary minerals, those produced by chemical reaction between primary minerals and the earth's air and water (see *Soil Mineralogy*). The most abundant primary minerals are *quartz* and ***feldspars; clays***, *calcite*, and a variety of hydroxides of iron and aluminum dominate among the secondary minerals.

Meteoritic minerals are dominated by ***plagioclase feldspars***, minerals from the ***olivine*** group, *hypersthene*, and the nickel-iron minerals, *kamacite* and *taenite* (see *Meteoritic Minerals*).

Minerals not found in Table 1 are generally insignificant in terms of crustal abundances although many are rather important to human activities (see *Metallic Minerals; Mineral Industries; Mineral and Ore Deposits; Salt Economy*).

KEITH FRYE

References

Baird, A. K., et al., 1976. Mineralogic and petrologic implications of Viking geochemical results from Mars: Interim report, *Science*, **194**, 1288–1293.

Blatt, H.; Middleton, G.; and Murray, R., 1972. *Origin of Sedimentary Rocks*. Englewood Cliffs, N.J.: Prentice-Hall, 634p.

Carmichael, I. S. E.; Turner, F. J.; and Verhoogen, J., 1974. *Igneous Petrology*. New York: McGraw-Hill, 739p.

Deer, W. A.; Howie, R. A.; and Zussman, J., 1962–63. *Rock-Forming Minerals*, vols. **1-5**. London: Longman.

Deer, W. A.; Howie, R. A.; and Zussman, J., 1966. *An Introduction to the Rock-Forming Minerals*. London: Longman, 528p.

Frondel, J., 1975. *Lunar Mineralogy*. New York: Wiley, 323p.

Mason, B., and Melson, W. G., 1970. *Lunar Rocks*. New York: Wiley, 179p.

Ringwood, A. E., 1975. *Composition and Petrology of the Earth's Mantle*. New York: McGraw-Hill, 618p.

Winkler, H. G. F., 1979. *Petrogenesis of Metamorphic Rocks*, 5th ed. New York: Springer-Verlag, 348p.

Cross-references: *Gossan; Lunar Mineralogy; Mantle Mineralogy; Meteoritic Minerals; Soil Mineralogy*; see also mineral glossary.

RUPTURE IN MINERAL MATTER

When stressed, solid matter deforms. Metals are ductile and flow plastically before rupturing. Most minerals are brittle and deform elastically, but when stressed beyond their elastic limits, such materials rupture, or break.

If a crack in a breaking mineral follows a particular crystallographic plane, breakage is by cleavage; if not, it is by fracture. Minerals that develop crystallographically oriented planes of weakness through twinning or exsolution in particular specimens at specific locations in the crystal may break by parting.

Cleavage is determined primarily by crystal structure. Thus, minerals with the halite structure (*halite*, NaCl; *periclase*, MgO; *galena*, PbS) have cubic {100} cleavage (three cleavage directions at mutual right angles). Minerals with the calcite structure (*calcite*, $CaCO_3$; *magnesite*, $MgCO_3$; *rhodochrosite*, $MnCO_3$; *siderite*, $FeCO_3$; *nitratine*, $NaNO_3$), which may be characterized as a distorted halite structure, have rhombohedral $\{10\bar{1}0\}$ cleavage (three intersecting cleavage planes at equal, but not at right, angles).

In discussing cleavage, direction, not position, is noted. Thus, successive parallel cleavages count as one cleavage direction or plane. Minerals may have one, two, three, four, or six cleavages. Direction of cleavage indicates weaker or fewer chemical bonds across the cleavage plane than in other crystallographic directions.

Phyllosilicates cleave into thin sheets because

the stronger Si–O bonds are arrayed in sheets held together by weaker chemical bonds. Thin cleavage sheets of ***mica***-group (q.v.) minerals are used for electric parallel-plate capacitors. The smoothness and covering properties of *talc* (talcum powder) result from the gliding of particles along cleavage planes. Likewise, the plasticity of ***clay*** minerals when mixed with water derives from the sliding of particles between cleavage layers. Other minerals with layered structures and one cleavage direction include *graphite* (C) and *molybdenite* (MoS_2), both of which are useful as dry lubricants. *Topaz* ($Al_2SiO_4F_2$) has a {001} cleavage because the silica tetrahedra are arrayed in layers, and, thus, cracks tend to propagate across the relatively weak Al–O and Al–F bonds.

Minerals of the ***amphibole***, the ***pyroxene***, and the ***feldspar*** groups cleave along two intersecting planes. The ***amphiboles*** and the ***pyroxenes*** are chain silicates with $(Si_4O_{11})_n$ double chains and $(SiO_3)_n$ single chains, respectively, oriented parallel to the *c* axis. Prismatic cleavage occurs along {110} planes between, but not across, chains of silica tetrahedra. The angles between {110} cleavages in ***pyroxenes*** are 93° and 87°, whereas they are 124° and 56° in ***amphiboles*** because of doubling of the chain width. The toughness of *jadeite* (see *Jade*) results from cracks that follow cleavages through minute, randomly oriented grains.

Although minerals of the ***feldspar*** group are framework silicates, fewer chemical bonds per unit area across their {001} and {010} planes permit prismatic cleavage at right angles or near right angles.

Fluorite (CaF_2) has octahedral cleavage (four intersecting directions) because propagating cracks encounter fewer Ca–F bonds across the {111} planes than across any other plane. *Diamond* (C) also has octahedral cleavage. *Sphalerite* (ZnS) has fewer Zn–S bonds across the {110} planes than across any other planes, resulting in dodecahedral cleavage.

Brittle minerals having no crystallographically oriented planes of weaker or of fewer bonds rupture by fracture. Fracture patterns are determined by defects (see *Crystals, Defects in*) and by other minute flaws and microfractures in the crystal structure. The distribution and orientation of these flaws differ from mineral to mineral resulting in fracture patterns that may be characteristic of a particular mineral species, e.g., the conchoidal fracture of *quartz*.

KEITH FRYE

Reference

Newnham, R. E., 1975. *Structure-Property Relations*. New York: Springer-Verlag, 234p.

Cross-references: *Abrasive Materials, American Society for Testing Materials; Crystals, Defects in; Jade; Mineral Properties; Mohs Scale of Hardness; Plastic Flow in Minerals.*

RUTILE GROUP

Minerals of the ***rutile*** group are related by structure and crystallography but have little in common in their paragenesis owing to the great dissimilarity of the cations. The rutile structure is composed of chains of MO_6 octahedra distorted by the sharing of octahedral edges. The *M* site may be occupied by Sn (*cassiterite*), Pb (*plattnerite*), Mn (*pyrolusite*), Ti (*rutile*), or Si (*stishovite*). *Rutile* is trimorphous with *anatase* and *brookite*. *Stishovite* is polymorphous with *quartz, tridymite, cristobalite*, and *coesite*.

Crystals are typically tetragonal prisms ranging from stubby to slender, capped by tetragonal bipyramids, and not uncommonly twinned. *Rutile* crystals are typically striated parallel to the prism axis. *Pyrolusite* is frequently dendritic (see *Skeletal Crystals*). *Stishovite* grains are in the submicron size range.

KEITH FRYE

References

Deer, W. A.; Howie, R. A.; and Zussman, J., 1966. *An Introduction to the Rock-Forming Minerals*. London: Longman, 415–417.

Phillips, W. R., and Griffen, D. T., 1981. *Optical Mineralogy: The Nonopaque Minerals*. San Francisco: Freeman, 19–20.

S

SALINE MINERALS (SOLUBLE EVAPORITES)

Evaporite minerals are sedimentary minerals that have crystallized during the solar evaporation of aqueous solutions, predominantly solutions of strong electrolytes. The term "saline minerals" refers to a large subgroup consisting of soluble salts, the formation of which includes not only evaporation but also cooling and "salting out."

Classification

A rough genetic classification of saline minerals may be based upon the environment of formation and broadly divided into (1) marine or marginal evaporites and (2) continental or nonmarine evaporites. However, since many of the same minerals can be and have been formed in either group, this designation is only broadly useful. Nevertheless, since a large share of the world's saline-mineral deposits were derived from sea water, such a classification is convenient and will be used (Table 1).

There is a wide variety of continental solutions, the compositions of which are influenced primarily by the weathering of surface rocks and additional factors such as volcanic contributions. Many of these are similar to sea water and deposit the same salts upon evaporation or cooling. Also, a large number of the soluble minerals are rare and of only scattered local occurrence, consisting of many chemical elements. A more refined classification might include different source materials (Lotze, 1957), but, since most salines derive their salt content from various sources, a predominant chemical classification is definitely preferable. This can be by phase relationship groupings or, more simply, by chemical species as will be used in the nonmarine classification.

TABLE 1. Genetic Classification of Saline Minerals

1. Marine Deposits
   a. Direct Seawater Deposits
   b. Modified Marine Deposits

2. Continental Deposits
   a. Chloride
   b. Sulfates
   c. Carbonates
   d. Borates
   e. Nitrates

3. Brines

Minerals of Marine Salt Deposits

Descriptive Mineralogy. The main mineral components of seawater are Na^+, K^+, Mg^{2+}, Ca^{2+}, Cl^-, SO_4^{2-}, Br^-, and $B_4O_7^{2-}$. A very large number of minor constituents, such as Fe^{2+}, CO_3^{2-}, and I^-, are also present. More than thirty minerals containing these elements are known from soluble marine evaporite deposits, and a much larger number of insoluble minerals. Only a few, however, are common: *halite, sylvite, carnallite, kainite, langbeinite, kieserite, polyhalite*, and *syngenite* (the last two are only sparingly soluble). *Glauberite, thenardite*, and *epsomite* are sometimes associated with marine salts. All of the other soluble marine minerals are relatively rare.

Some crystallographic and diagnostic criteria of most of the evaporite minerals are presented in Table 2. Very rare (*hydrohalite, erythrosiderite*), metastable (*bassanite*, leonardite, *pentahydrite*), or doubtful minerals (chlorocalcite, *douglasite*) are omitted. The data for Table 2 were collected from several sources (mainly Palache et al., 1951) as compiled by Braitsch (1962) with a few additions. Color is not indicated because pure evaporite minerals are nearly always colorless. The characteristic red color of *sylvite, carnallite, polyhalite*, and others is due to *hematite* admixtures, partly in oriented intergrowths. Other evaporite minerals show rather variable colors caused by diadochic replacement in trace amounts or by other factors.

A few additional remarks concerning origin, mode of occurrence, and properties not listed in Table 2 will follow for the important evaporite minerals. Detailed descriptions of evaporite minerals and occurrences are given by Gorgey, Riedel, Schaller, and Henderson and by Stewart (see bibliography in Stewart, 1963).

Halite (NaCl) is the most common of the halides. It occurs as a rule interbedded with *anhydrite*, but it may be associated with nearly

TABLE 2. Physical Constants of Saline Minerals

Mineral Formula	Symmetry Space Group[a]	Cell Const.[b] a_0 b_0 c_0	α β γ	Cleavage[c] Sp. Gr.[d] Hardness[e]	Indices of Refraction[f]	2V[g] Dispersion
HALIDES						
halite	cub.	5.640		{100}*v*	1.5443	
(rock salt)	*Fm3m*			2.168		
NaCl				2		
sylvite	cub.	6.293		{100}*v*	1.4903	
KCl	*Fm3m*			1.99		
				2		
bischofite	mono.	9.92		–	1.495	(+)79
$MgCl_2 \cdot 6H_2O$	*C2/m*	7.16	93.7	1.604	1.507	$r>v$
		6.11		1–2	1.528	
carnallite	ortho.	9.56		–	1.466	(+)66
$KMgCl_3 \cdot 6H_2O$	*Pban*	16.05		1.602	1.475	$r<v$
		22.56		2½	1.494	
tachyhydrite				$\{10\bar{1}1\}v$	1.520	
$CaMg_2Cl_6 \cdot 12H_2O$	hex.			1.667	–	
	rh.			2	1.512	(–)
rinneite	hex.	11.98		$\{11\bar{2}0\}g$	1.5886	(+)
$K_3NaFeCl_6$	rh.	($a_{rh}8.31_4$)	($\alpha_{rh}92.2$)	2.347	–	
	R3c	13.84		3	1.5894	
SULFATES						
thenardite	ortho.	9.821		{010}*v*	1.471	(+)82
Na_2SO_4	*Fddd*	12.304		{101}*g*	1.477	
		5.863		2.664	1.484	
				2½–3		
mirabilite	mono.	11.51		{100}*v*	1.394	(–)76
(Glauber's salt)	$P2_1/c$	10.38	107.75	1.490	1.396	$r<v$
$Na_2SO_4 \cdot 10H_2O$		12.83		1½–2	1.398	
aphthitalite	hex.	5.66		$\{10\bar{1}0\}d$	1.491	(+)
(glaserite)	$P\bar{3}m$	–		2.66	–	
$K_3Na(SO_4)_2$		7.30		3	1.499	
kieserite	mono.	6.89		{110} {111}*v*	1.520	(+)55
$MgSO_4 \cdot H_2O$	*C2/c*	7.61	116.3	2.571	1.533	$r>v$
		7.53		3½	1.584	
hexahydrite	mono.	10.06		{100}*v*	1.426	(–)38
(sakiite)	*C2/c*	7.16	98.6	1.75	1.453	
$MgSO_4 \cdot 6H_2O$		24.39		?	1.456	
epsomite	ortho.	11.86		{010}*v*	1.432	(–)51
(reichardtite)	$P2_12_12_1$	11.99		1.677	1.455	
$MgSO_4 \cdot 7H_2O$		6.86		2–2½	1.461	
vanthoffite	mono.	9.79		?	1.485	(–)84
$Na_6Mg(SO_4)_4$	$P2_1/c$	9.22	113.5	2.69	1.488	$r<v$
		8.20		3½	1.489	
bloedite	mono.	11.09		none	1.483	–71
(astrakhanite)	$P2_1/a$	8.20	100.65	2.25	1.486	$r<v$
$Na_2Mg(SO_4)_2 \cdot 4H_2O$		5.50		2½–3	1.487	strong
loeweite	hex.	18.96		–	1.490	
$Na_{12}Mg_7(SO_4)_{13} \cdot 15H_2O$	$R\bar{3}$(R3?)	–		2.38	–	
		13.47		2½–3	1.471	
langbeinite	cub.	9.920		none	1.5347	
$K_2Mg_2(SO_4)_3$	$P2_13$			2.83		
				3½–4		

TABLE 2. (Continued)

Mineral Formula	Symmetry Space Group[a]	Cell Const.[b] a_0 b_0 c_0	α β γ	Cleavage[c] Sp. Gr.[d] Hardness[e]	Indices of Refraction[f]	2V[g] Dispersion
leonite	mono.	11.78		none	1.479	(+)≤90
$K_2Mg(SO_4)_2\cdot 4H_2O$	*C2/m*	9.53	95.4	2.20	1.483	
		9.88		2½–3	1.487	
picromerite	mono.	9.06		$\{\bar{2}01\}v$	1.4607	(+)48
(schoenite)	$P2_1/a$	12.26	104.8	2.03	1.4629	$r>v$
$K_2Mg(SO_4)_2\cdot 6H_2O$		6.11		2½	1.4755	
glauberite	mono.	10.10		$\{001\}v$	1.515	(–)7
$Na_2Ca(SO_4)_2$	*C2/c*	8.28	112.2	2.85	1.535	$r>v$
		8.51		2½–3	1.536	very strong
syngenite	mono.	9.72		$\{100\}$ $\{110\}v$	1.501	(–)28
$K_2Ca(SO_4)_2\cdot H_2O$	$P2_1/m$	7.16	104.1	2.57	1.517	$r<v$
		6.21		2½	1.518	very strong
goergeyite	mono.	17.47_5		$\{100\}d$	1.560	(+)79
(mikheevite)	*C2/c*	6.83_2	113.2	2.77	1.569	
$K_2Ca_5(SO_4)_6\cdot H_2O$		18.22_6		3½	1.584	
polyhalite	tricl.	11.68	90.6	$\{10\bar{1}\}v$	1.547	(–)64
$K_2MgCa_2(SO_4)_4\cdot 2H_2O$	$1(\bar{1}?)$	16.33	90.1	2.78	1.560	
		7.60	91.9	3½	1.567	
MIXED SALTS						
kainite	mono.	19.76		$\{001\}v$	1.494	(–)≤90
$MgSO_4\cdot KCl\cdot 2.75H_2O$	*C2/m*	16.26	94.9	2.15	1.505	$r>v$
		9.57		2½–3	1.516	
dansite	pseudo-	15.90		none	1.503	
$MgSO_4\cdot 3NaCl\cdot 9Na_2SO_4$	cub.			2.59		
	$I\bar{4}3d$(?)			?		
NITRATES, Miscellaneous						
nitratine	hex.	5.0696(25°C)		$\{10\bar{1}1\}v$	1.5874	
$NaNO_3$	$R\bar{3}c$	(a_{rh}=6.327, α_{rh}=47.8)		2.266	1.3361	
		16.829		1½–2	–	
niter	ortho.	5.414		$\{011\}v$-g	1.332	(–)7
KNO_3	*Pmcn*	9.164		2.109	1.504	$r<v$
		6.431		2	1.504	
darapskite	mono.	not det.		$\{010\}v$	1.391	(–)27
$Na_3NO_3SO_4\cdot H_2O$	*2/m*			2.20	1.481	$r>v$
				2½	1.486	
lautarite	mono.	7.19		$\{011\}g$	1.792	(+)90
$Ca(IO_3)_2$	$P2_1/c$	11.40	106.37	4.59	1.840	$r>v$
		7.33		3½–4	1.888	
dietzeite	mono.	10.18		$\{100\}b$	1.825	(–)86
$Ca_2CrO_4(IO_3)_2$	$P2_1/c$(?)	7.31	106.53	3.62	1.842	$r>v$
		14.06		3½	1.857	very strong
CARBONATES						
trona	mono.	20.11		–	1.417	(–)74
$Na_3H(CO_3)_2\cdot 2H_2O$	*I2/c*	3.49	103.13	2.147	1.494	$r<v$
		10.31		2½–3	1.543	
thermonatrite	ortho.	6.45		$\{100\}b$	1.420	(–)48
$Na_2CO_3\cdot H_2O$	*Pmmm*	10.74		2.255	1.506	$r<v$
		5.25		1–1½	1.524	

TABLE 2. (Continued)

Mineral Formula	Symmetry Space Group[a]	Cell Const.[b] a_0 b_0 c_0	Cell Const.[b] α β γ	Cleavage[c] Sp. Gr.[d] Hardness[e]	Indices of Refraction[f]	2V[g] Dispersion
nahcolite	mono.	7.525		none	1.375	(−)74
$NaHCO_3$	$P2_1/n$	9.72	93.3	2.238	1.498	$r<v$
		3.53		2½	1.583	
pirssonite	ortho.	11.32		2.382	1.504	(+)27
$Na_2Ca(CO_3)_2 \cdot 2H_2O$	*Fdd*2	20.06		3–3½	1.509	$r<v$
		6.00			1.573	
gaylussite	mono.	11.57		none	1.445	(−)40
$Na_2Ca(CO_3)_2 \cdot 5H_2O$	*I*2/*a*	7.765	102.0	2.037	1.514	$r<v$
		11.20		2½–3	1.524	
shortite	ortho.	4.961		–	1.531	(−)75
$Na_2Ca_2(CO_3)_3$	*Amm*2	11.03		2.629	1.555	$r<v$
		7.12		3½	1.570	
northupite	cub.	14.08		none	1.513	
$Na_2Mg(CO_3)_2 \cdot H_2O$	*Fd*3*m*			2.407		
				3½–4		
bradleyite	mono.	8.85		none	1.487	(−)50
$Na_3MgCO_3PO_4$		6.63	90.4	2.720	1.546	$r<v$
		5.16		3½	1.560	
BORATES						
sassolite	tricl.	7.039	92.58	{001}*vv*	1.340	(−)
H_3BO_3	$P\bar{1}$	7.053	101.17	1.48	1.456	very small
		6.578	119.83	1	1.459	
borax (tincal)	mono.	11.84		{100}*v*	1.4466	(−)40
$Na_2B_4O_7 \cdot 10H_2O$	*C*2/*c*	10.63	106.6	1.715	1.4687	$r>v$
		12.32		2–2½	1.4717	strong
tincalconite	trig.	11.3		?	1.461	(+)
$Na_2B_4O_7 \cdot 5H_2O$	*R*32	(a_{rh}=9.58, α_{rh}=71.7)		1.880	–	
		20.9		?	1.474	
kernite	mono.	15.68		{100}*v*	1.454	(−)80
$Na_2B_4O_7 \cdot 4H_2O$	*P*2/*a*	9.09	108.87	{001}*v*	1.472	$r>v$
		7.02		1.91	1.488	
				2½		
ulexite	tricl.	8.73	90.27	{010}*v*	1.493	(+)73
$Na_2B_4O_7 \cdot Ca_2B_6O_{11} \cdot 16H_2O$	$P\bar{1}$	12.75	109.15	1.96	1.506	
		6.70	105.12	2½	1.519	
priceite	tricl. (?)	not det.		none	1.572	(−)43
(pandermite)				2.42	1.591	$r<v$
$Ca_4B_{10}O_{19} \cdot 7H_2O$(?)	(?)			3–3½	1.594	
colemanite	mono.	8.74		{010}*v*	1.5863	(+)55
$Ca_2B_6O_{11} \cdot 5H_2O$	$P2_1/a$	11.26	110.12	2.423	1.5920	$r>v$
		6.10		4½	1.6140	
inyoite	mono.	10.63		{001}*g*	1.495	(−)70
$Ca_2B_6O_{11} \cdot 13H_2O$	$P2_1/a$	12.06	114.03	1.875	1.51	$r<v$
		8.40_5		2	1.520	
hydroboracite	mono.	11.76		{010}*v*	1.521	(+)60
$MgCaB_6O_{11} \cdot 6H_2O$		6.68	102.8	2.167	1.534	$r<v$
		8.20		2(–3)	1.570	
boracite	ortho.	8.54		none	1.662	(+)82
$(Mg,Fe)_3ClB_7O_{13}$	$Pca2_1$	8.54		2.9–3.0	1.667	
		12.07		7–7½	1.673	

TABLE 2. (Continued)

Mineral Formula	Symmetry Space Group[a]	Cell Const.[b] a_0 b_0 c_0	α β γ	Cleavage[c] Sp. Gr.[d] Hardness[e]	Indices of Refraction[f]	2V[g] Dispersion
szaibelyite	ortho.	10.42		–	1.58	(-)24
(ascharite)	$A2_122$	25.05		2.65	1.646	
$MgHBO_3$		3.14		3	1.65	

[a]The space group is indicated with Hermann-Mauguin symbols only.
[b]Lattice parameters are given in angstrom units (10^{-8} cm).
[c]Only the most prominent cleavage forms are indicated: *vv* = very perfect, *v* = perfect, *g* = good, *d* = distinct.
[d]Specific gravities are given at 20°C in g/cm³.
[e]Hardnesses are given in terms of the Mohs scale.
[f]Indices of refraction are given for sodium light in the order n_ω, n_ϵ, or n_α, n_β, n_γ.
[g]Optic axial angle $2V$ is given for the acute bisectrix, the sign being indicated (+) or (-).

all other evaporite minerals. On slow evaporation *halite* sometimes forms hopper crystals on the surface of salt lakes or salinas (see *Skeletal Crystals*). Crystals growing on the surface of older salt often form coarse, vertically oriented or radiating aggregates, distinctly elongated parallel [111]. Both types may contain liquid inclusions on the cube faces, particularly near the corner. On parting planes in porous rocks such as shales, *halite* occurs in fibrous aggregates. The individual fiber grows from its base during the widening of the vein walls. In association with other minerals, one occasionally observes blue *halite*, the color of which is due to structural imperfections (color centers) caused by radioactivity (see *Crystals, Defects in*). The crystals will become colorless upon the application of moderate (100°C) heat.

Sylvite (KCl) is the most important potash ore mineral, but it is much less common than *halite*. It occurs in granular, bedded deposits, always mixed with *halite* (to form sylvinite). In some deposits it is deep red in color, as in Canada, the color being due to minute inclusions of *hematite*. *Sylvite* is also commonly a whitish-gray color because of inclusions of gray clay.

Carnallite ($KMgCl_3 \cdot 6H_2O$) may occur in massive crystalline sedimentary beds but also often occurs as a minor constituent with sylvinite. Sometimes the grains show a metallic schiller texture that is due to oriented intergrown flakes of *hematite*. On slow evaporation *carnallite* may form hopper crystals with pseudohexagonal forms, but these are not preserved in massive rocks. *Carnallite* squeaks under the twisting action of a pointed hammer or knife. The noise is due to easily produced secondary twinning lamellae.

Kieserite ($MgSO_4 \cdot H_2O$) is one of the most abundant magnesium sulfate minerals and is commonly found in the German potash deposits. It is not found in the present-day evaporation of sea water and is assumed to be an alteration product of the prevalent hydrate *epsomite* ($MgSO_4 \cdot 7H_2O$), or to have resulted from the decomposition of *kainite* in strong magnesium chloride brines. It is usually intergrown with other salts, forming rounded grains, and only rarely occurs as crystals of dipyramidal habit.

Polyhalite [$K_2Ca_2Mg(SO_4)_4 \cdot 2H_2O$] often occurs in small quantities with other potash deposits, but by itself in some extremely large deposits (as in Kansas). It is frequently associated with *syngenite*, and always with *anhydrite* and *halite*. It usually forms fibrous aggregates or foliated beds. Well-formed crystals are rare and are restricted to *halite*-rich associations. *Polyhalite* always shows polysynthetic twinning. It is often red in color because of minute inclusions of *hematite*.

Kainite ($KCl \cdot MgSO_4 \cdot 2.75H_2O$) and *halite* are probably the only major marine salts that were originally deposited from sea water and not later altered. Large deposits occur in Sicily, while the ones in Germany appear to be a secondary capping on potash salts as a product of solution metasomatism effected by invading meteoric waters. *Kainite* occurs primarily in sugary-grained masses or sometimes in fibrous aggregates. Good crystals are found in vugs and fissures. *Kainite* is always associated with *halite*, usually with *sylvite*, and sometimes with *langbeinite*, *leonite*, or *kieserite*.

Langbeinite [$K_2Mg_2(SO_4)_3$] may also have been directly deposited from hot solutions but only occurs in commercial zones in Carlsbad and E. Germany. It is always associated with *halite* or *sylvite*. It generally occurs in nodular masses but occasionally forms coarse-grained, bedded deposits containing as much as 90%

langbeinite. Individual crystals intergrown with *halite* and *sylvite* show tetrahedral habit. These are commonly replaced by *kieserite* plus *sylvite*. The conchoidal fracture and glassy appearance facilitate the identification of *langbeinite*.

Geologic and Economic Distribution of Marine Evaporites. Some of the most notable saline-mineral formations and deposits are discussed, on the basis of the data of Krumbein (1953), Ivanov and Levitskii (1960), and other sources (see references in Braitsch, 1962). Among the marine evaporates, the Canadian deposit undoubtedly represents the world's largest potash reserves and contains the highest grade of ore. The Russian Urals deposit has not been as extensively described in the scientific literature but is also estimated to have tremendous reserves. There are many large deposits, but none of a similar magnitude to the Canadian or the Russian.

Evaporites have been identified in strata as old as Early Cambrian and occasionally in the Precambrian, but none of the soluble salt deposits are that old. In many of the older deposits only the earlier stages of the evaporitic sequence are represented. Apparently the optimum conditions for soluble evaporite deposition occurred during Permian time. The formation of large deposits of marine evaporites was restricted to warm, arid belts. However, small deposits and incrustations may be formed in all latitudes.

Minerals of Nonmarine Evaporites

Soluble nonmarine evaporites are common in many intracontinental basins. The tonnage of these deposits is generally small in comparison with that of marine evaporites, and they are highly variable in their geochemical composition. However, some deposits, such as the Green River soda-ash formation (see Milton and Eugster, 1959), the Searles Lake mixed-salts basin, and the Laguna del Rey *bloedite* deposit, are very large by any comparison. Table 2 gives some crystallographic and diagnostic criteria for the more important minerals. The anions present will be used for further classification of the mineral systems.

Sulfates. This group consists of a very large number of salts: *thenardite, mirabilite, glauberite, bloedite, epsomite*, and other metallic sulfates.

Thenardite (Na_2SO_4) forms good dipyramidal crystals but often occurs as powdery crusts or efflorescences in soil of arid regions. It sometimes exists in purverulent pseudomorphs and casts with a hollow core after *mirabilite.*

Mirabilite ($Na_2SO_4 \cdot 10H_2O$) forms short prismatic crystals sometimes elongated parallel to the *b* axis, or is occasionally thin and tabular. *Mirabilite* is the typical mineral deposited by seawater, brines, or alkali-sulfate lakes during the winter season because of the strong negative-temperature coefficient of *mirabilite* solubility.

Glauberite [$Na_2Ca(SO_4)_2$] is variable in habit. It occurs in basal tablets or in prisms parallel to [$\bar{1}01$] or to *c* (in the latter case striated parallel to *a*). It dissolves incongruently in water, depositing *gypsum. Glauberite* is often associated with *thenardite* and *halite*, forming solid crusts or incrustations on playa muds.

Bloedite (astrakhanite) ($Na_2SO_4 \cdot MgSO_4 \cdot 4H_2O$) occurs as short prismatic crystals or as large, well-shaped granular masses.

Sulfate minerals are found in many salt pans and playas of arid and semiarid regions. Many intermittent and some perennial continental salt lakes belong to this group. Other deposits are formed in glacial basins or near alkali springs in any latitude (i.e., northwestern USA and south-central Canada). Deposits may also form by the cooling of brines or sea water and have been found in many areas of the world (Antarctica, Russia, etc.). Sulfates are found in all ground waters, mineral springs, and brines, originating from the leaching of sulfate minerals (*gypsum*, etc.), the oxidation of sulfides, or the absorption of sulfur gases from the atmosphere into rain. High-sulfate brine solutions, and then frequently solid deposits, result when the content of soluble ions (Na, K, Mg) is in excess of that of the precipitating ion, calcium. As would be expected from this method of origin, usually the brine or deposit is of mixed salts, but the sulfate ion is so common that sometimes other anions are absent, and sulfates (usually Na_2SO_4) predominate.

Because of its prolific occurrence, and the ease and selectivity that Glauber's salt (*mirabilite*) crystallizes upon cooling, many examples are known. Some of the large commercial deposits are in Laguna del Rey, Mexico; Searles Lake, California; Antofagasta and María Elena, Chile; several lakes in south-central Canada; and, in numerous small deposits, other countries such as Turkey, South Africa, Rumania, Russia, Spain, and Argentina. References to these deposits are given in Lotze (1957).

Nitrates. These salts are exceedingly rare, with only one commercial deposit and a few occurrences reported in the world. Nitrates are highly soluble, their salts are usually deliquescent, and they are subject to reduction by organic matter. The only significant occurrence is in the rocks called "caliche" in the Chilean

Atacama desert. This deposit consists of roughly equal parts of salt (NaCl), salt cake (Na_2SO_4), and *nitratine* ($NaNO_3$), as well as commercial quantities of iodine (iodates), magnesium, potassium, borates, and other components. Sodium nitrate has the same type of crystal structure as *calcite*. It usually occurs as the cementing material for a conglomerate structure or in granular masses, or it forms coarsely crystalline white crusts. The Chilean deposit occurs in the upper 5 meters of much of the Taramgul Plains, covering an area roughly 300 X 80 km (200 X 50 miles). The only other reported deposit is a small lake bed in Libya, but there are numerous very small bat guano or bird dropping nitrate occurrences. *Niter* (KNO_3) and *darapskite* ($Na_2SO_4 \cdot NaNO_3 \cdot H_2O$) are rare, but occur accessorily in tabular, colorless crystals in cavities or crevices in massive nitrate rocks.

Although subordinate in amount, iodates, chlorates, and chromates occur typically with the nitrates. *Lautarite* [$Ca(IO_3)_2$] is often radially arranged in stellate aggregates or in short prismatic crystals. *Dietzeite* [$Ca_2CrO_4(IO_3)_2$], usually in deep golden-yellow fibrous or columnar crusts, is restricted to the Chilean "caliche jaune" or "caliche azufrado." They afford iodine as a valuable by-product. The yellowish color of these rocks, excepting *dietzeite*, is due to the admixture of *tarapacaite* (K_2CrO_4), which occurs in thick, tabular, bright yellow crystals, and of *lopezite* ($K_2Cr_2O_7$), which forms tiny ball-like aggregates that are orange-red to brick-red in color. All these minerals, as well as the nitrates themselves, indicate a strong oxidation potential of about +1 to +1.3 volts, which in nature is characteristic of extreme deserts without reducing organic substances.

The origin of the nitrate is unknown, but it appears certain to have migrated from higher ground to its present location by sparse local water movements (even including heavy fogs). The original source could have been bird droppings, volcanic springs, decomposed ocean or animal residues, or all of these.

Carbonates. This group consists entirely of different sodium carbonate salts. *Trona* ($Na_2CO_3 \cdot NaHCO_3 \cdot 2H_2O$) occurs normally in white, yellow, or greenish-gray coarsely crystalline masses. It forms spherulitic and radiating aggregates, beds composed of fibrous, acicular, or bladed crystals, and fine-grained aggregates composed of microscopic acicular crystals. *Thermonatrite* ($Na_2CO_3 \cdot H_2O$) forms powdery aggregates originating from the disintegration of earlier soda. *Nahcolite* ($NaHCO_3$) occurs as friable crystal aggregates, often as "fishtail" twinned crystals and porous masses. The decahydrate *natron* ($Na_2CO_3 \cdot 10H_2O$) is somewhat less common because its stability range is restricted to comparatively low temperatures. In addition, Fahey (1962) has described in detail the complex sodium-calcium carbonates *pirssonite*, *gaylussite*, and *shortite*, together with other complex minerals, *northupite* and *bradleyite* (see Table 2). The following minerals of the carbonate subgroup are only locally abundant, as at Searles Lake, California: *hanksite* ($9Na_2SO_4 \cdot 2Na_2CO_3 \cdot KCl$), hexagonal dipyramidal ($C6_3/m$), uniaxial (-), n_ω=1.481, n_ϵ=1.461, occurs in large euhedral crystals, sometimes similar to *quartz* in habit, sometimes in short or stubby prisms with prominent basal pinacoids, sometimes without prisms and bounded entirely by pyramidal faces; and *burkeite* ($2Na_2SO_4 \cdot Na_2CO_3$), orthorhombic, biaxial (-), n_α=1.448, n_β = 1.489, n_γ=1.493, occurs in tabular, cushion-like, uneven crystals, often twinned to an X shape. Hard beds up to 0.3 m and thick pockets and nodules sometimes consist of pure *burkeite*, but most are intermixed with *trona, borax*, or other saline minerals. Rare associated minerals in the same deposits are *tychite* ($2Na_2CO_3 \cdot 2MgCO_3 \cdot Na_2SO_4$), cubic; *sulfohalite* ($2Na_2SO_4 \cdot NaCl \cdot NaF$), cubic; and *schairerite* [$Na_2SO_4 \cdot Na(F,Cl)$], hexagonal.

Occurrences and Origin. Many of the world's numerous soda-ash deposits were originally carbonate-containing lakes such as occur in Egypt, the Sahara, South West Africa, South and North America, and elsewhere. They can be formed by any of three different mechanisms: the leaching of carbonatites, a soluble carbonate rock crystallized directly from volcanic flow or magmatic intrusion; bacterial reduction of sulfates from stratified closed lakes with a high organic content; and ordinary drainage and evaporation. The world's largest sodium-carbonate deposit is of the closed-lake type. It is of Eocene age and occurs as intercalations in the bituminous marls (oil shales) of the Green River formation (Wyoming, Utah, Colorado, USA; see *Green River Mineralogy*). *Trona, shortite* (wedge-shaped or short prismatic idiomorphic crystals), and locally *northupite* are the most abundant sodium-containing carbonates in Wyoming. The Wyoming *trona* formation is exceedingly large, and very important from an economic point of view, forming nearly monomineralic beds up to 4 meters thick over a vast area. *Nahcolite* is abundant in Colorado (in the center of the Piceance basin) and occurs in Utah (Uinta Basin). The Green River deposits are expected in the near future to supply the major part of the free world's sodium-carbonate demand. An approximate *trona* reserve figure for the Green

River formation is over one billion tons of Na_2CO_3; perhaps 100 million tons of *nahcolite* exists in the oil shale (Piceance) basins.

The dependency of the phase relationships of the sodium carbonates on temperature, water activity (as controlled by salinity), and carbon-dioxide activity was summarized by Eugster (Milton and Eugster, 1959; Eugster and Smith, 1965). The main factors responsible for the formation of sodium carbonate-rich deposits are the excess of soluble over precipitating carbonates in a drainage area, the dissolving of reactive alkali carbonatites or volcanic glasses, and the precipitation of CO_2 from organic decay or volcanic sources as the sparingly soluble *trona*. After precipitation of *calcite* and *dolomite*, the residual brines, on evaporation, reach saturation in sodium-containing carbonates. The intimate connection of some carbonate deposits with volcanism is clearly established in the case of Lake Natron, which derives its sodium-carbonate content from leaching of the recent, alkali-rich carbonatite lavas of the Oldoinyo Lengai, Tanzania (Lotze, 1957, p. 113). The Egyptian deposit is a good example of bacterial reduction.

Borates. These minerals are almost exclusively associated with volcanic springs. The borate content of seawater is small; therefore, marine evaporites usually contain very small amounts of boron, commonly present in *anhydrite* and illitic ***clay*** minerals. Since borates are quite soluble, they crystallize only in the later stages of the evaporation process–*boracite*, ascharite (*szaibelyite*); see below. On the other hand, the boron content of inland evaporites is generally thought to be intimately connected with boron-containing thermal springs. Calcium borates are the most common compounds, although a few large deposits of sodium salts are known. Among the less common salts, magnesium borates are usually found with marine evaporites, and the very rare strontium borates are formed from nonmarine deposits, or occasionally in modified marine evaporites. Numerous sodium, calcium, and magnesium borates, and borates with additional anions, are known, but only about eleven (Table 2) are important, four of which merit additional description.

Borax ($Na_2B_4O_7 \cdot 10H_2O$) is the most important primary borate, but it is not stable in a dry atmosphere or at elevated temperatures. *Borax* commonly occurs at outcrops but is always associated with *tincalconite*. When protected from the weather, it forms short prismatic or elongated euhedral crystals similar to ***pyroxene***, sometimes with a distinct oscillatory zonal structure, or mostly subhedral, sometimes anhedral crystalline masses. Fresh crystals show a colorless or yellowish to greenish color with a resinous luster.

Kernite ($Na_2B_4O_7 \cdot 4H_2O$) is found in the massive Kramer deposit (Kern County, California) as a major component. It occurs in huge colorless crystals and shows two perfect cleavages. It forms from *borax* in the temperature range of about 53° to 63°C.

Ulexite ($Na_2B_4O_7 \cdot Ca_2B_6O_{11} \cdot 16H_2O$) generally forms transparent, silky fibers with a radiated or random arrangement or with a massive structure. From their appearance as dull white aggregates in playa muds they are usually called "cotton balls." In massive deposits, *ulexite* forms large well-shaped crystals, intermixed with other boron salts.

Colemanite ($Ca_2B_6O_{11} \cdot 5H_2O$) usually forms massive layers and crudely spherulitic aggregates. In vugs very perfect crystals may be observed. The crystals commonly show a short prismatic or pseudorhombohedral habit. They locally appear to have been formed as an alteration product of *ulexite* and *borax* under the action of circulating ground waters. In the Inder deposit, Kazakhstan, USSR, *colemanite* is replaced mainly by *hydroboracite*.

Most Prominent Occurrences. The world's largest and most important source of boron is found in the Kramer district in the Mojave desert at Boron, California, supplying about 90% of the free world's borate demand. It consists of nearly pure, large crystals of various borate salts. The deepest part of the ore body consists of coarse-grained *kernite* with traces of *probertite* ($NaCaB_5O_9 \cdot 5H_2O$). Laterally and vertically the *borax* beds pass into ***clays*** containing *ulexite* and, in marginal parts, *colemanite*. The *borax* ore body is unique in that it is nearly completely free of other saline minerals. Its association with numerous tuff beds and with the low-temperature, hydrothermal sulfides, *realgar* and *stibnite*, points to a volcanic source of the boron, in particular to hot springs (Bowser, 1965. Ph.D. Dissertation, Univ. of California, Los Angeles). The large deposits in Turkey are somewhat similar, but more complex and scattered.

A much more complex volcanic-spring-related evaporate deposit containing borax is found at Searles Lake, California, supplying about 5% of the free world's borate demand (see Eugster and Smith, 1965). *Borax* is locally abundant, averaging less than 1% of the total salts (Bowser, 1965). The boron appears to have been derived from thermal springs, with some of the other salts being derived from normal chemical weathering of rocks in a large drainage area. Many other occurrences of borax of minor economic importance are known–for instance, from salt pans in Chile, Argentina, Tibet, and

Kashmir, and from efflorescences on the soil in arid regions.

The other borates given in Table 2 are of local importance only. The last two (*boracite* and ascharite) are restricted to marine evaporites. However, in some deposits, boron is much in excess in comparison with the boron content of concentrated normal sea water. A well-known example with many other different borate minerals and with a complex genetic history is the Inder deposit in Kazakhstan, USSR (Godlevsky, 1937; see reference in Braitsch, 1962). Details and references on borate deposits are given by W. C. Smith (1960), and a complete list of borate minerals known up to 1962 is given by C. Tennyson (1963).

O. BRAITSCH
Revised by D. E. GARRETT

References

Braitsch, O., 1962. *Entstehung und Stoffbestand der Salzlagerstatten*. Berlin: Springer, 232p.

Eugster, H. P., and Smith, G. I., 1965. Mineral equilibria in the Searles Lake evaporites, California, *J. Petrology*, **6**, 473–522.

Fahey, J. J., 1962. Saline minerals of the Green River Formation, *U.S. Geol. Surv. Prof. Paper 405*, 1–50.

Lotze, F., 1957. *Steinsalz und Kalisalze* (I). Berlin: Gebr. Borntraeger, 465p.

Milton, C., and Eugster, H. P., 1959. Mineral assemblages of the Green River Formation, in P. H. Abelson, ed., *Researches in Geochemistry*. New York: Wiley, 118–150.

Palache, C.; Berman, H.; and Frondel, C., 1951. *Dana's System of Mineralogy*, 7th ed., vol. 2. New York: Wiley, 1224p.

Smith, W. C., 1960. Borax and borates, in Seeley W. Mudd Series, *Industrial Minerals and Rocks*, 3rd ed. New York: American Institute of Mining, Metallurgical and Petroleum Engineers, 103–118.

Stewart, F. H., 1963. Marine evaporites, *Data of Geochemistry*, 6th ed. *U.S. Geol. Surv. Prof. Paper 440-Y*, 53p.

Tennyson, C., 1963. Eine Systematik der Borate auf kristallchemischer Grundlage, *Fortschr. Mineral.*, **41**, 64–91.

Wetzel, W., 1961. Die Hypothese der kapillaren Konzentration und die geologische Realität der chilenischen Nitrat-Lagerstatten, *Chem. Erde*, **21**, 203–209.

Cross-references: *Anhydrite and Gypsum; Green River Mineralogy; Mineral Industries; Salt Economy; Skeletal Crystals*; see also mineral glossary. Vol. III: *Caspian Sea; Kara-Bogaz Gulf; Playa*. Vol. IVA: *Evaporite Processes; Seawater, Chemistry*. Vol. VI: *Diagenesis; Evaporites.*

SALT ECONOMY

Salt is indispensable for human and animal diet. Depending on the salt content of drinking water and the food available (bread contains 1 to 2% salt) between 2 and 8 g of additional salt per person per day is required. With a vegetarian diet, rainwater for drinking, and hard work in a hot climate, salt can become the limiting factor for animal and human population density. The physiological cause is that, with very rare exceptions, mammals tolerate practically no variation in the salt content of their body liquids (4.8 ± 0.3 g/liter). Water and salt input is regulated by thirst and salt hunger; salt output is regulated by hormones, which change the salt content of urine and sweat. Overconsumption of salt is habit forming.

Uses

Food Preservation. A saturated salt solution in water (350 g/liter) has a water pressure considerably lower than that of water in animal and vegetable tissues; in consequence, salt and salt brine attract water from such tissues when brought together with them. Salted meat, salted fish, salted olives, salads, pickles, etc., are so dehydrated by salt that their proteins and other nutrient contents cannot be decomposed by microbes. (Up to 2% salt is used to preserve cheese and butter, 6% for meat, and 20% for fish.) Thus food can be preserved from times of surplus to seasons of want and can be transported from places of easy growth to places of agricultural insufficiency. Civilizations with uneven distribution of population are consequently sometimes limited by salt supply. A typical combination of agricultural and town life before the Industrial Revolution (80% of population farming and 20% in towns) needed about 25 g of salt per person per day for survival: e.g., the Roman Empire or France before 1789.

Traditional Uses for Industry. The main use in traditional industry is the preservation of hides, for which the equivalent of 1/3 of the final leather weight of salt is used. Again it is the desiccating influence of the salt of which use is made. (Embalming of corpses is based on the same technique.) Other early industrial uses of salt were for soap, glass, and glazing ceramics. These industries are small when compared with today's, but they played almost as important a role as the modern steel industry does at present. Transport and defense depended as much on leather then as they do on metals today. It was necessary not only for shoes but also for armor, protective shields, reins, saddles, and light building.

Present Uses for Industry. The uses of salt for nutritional purposes have remained constant, though a great deal of food preservation is now being done by refrigeration. The industrial uses of salt in modern society have increased to such an extent that in the USA 568 g of salt per person per day are consumed (1979). Table 1 shows the distribution of salt for the different uses at different times and in different countries.

Today the tanning industry has lost its relative importance, but glass, the inorganic polymers and cellulose, paper pulps, vinylchloride, and the organic polymers (for which soda and chlorine from salt are required) are now of paramount significance. World consumption of salt is about 110 million tons per year per 3 billion people with an average consumption of about 75 g per day. The reduction of salt supply at any level of civilization would soon limit population density. On the highest consumption level, there would be a lack of glass windows for housing as well as paper for printing, rayon, covers for plants growing under plastic, etc. On the lowest level, people would lose weight and die of salt starvation.

Salt Sources and Technology

Quarries and Mines. Salt is found as such (almost pure NaCl, the mineral *halite*) in mountains, where it is quarried and mined. Quarries are possible only in very arid climates; anywhere else, exposed salt is dissolved by rainwater. For this reason, salt quarries are rather rare and quite far away from agricultural areas, which, by their nature, need rain. In Asia, the mountain of Sedom at the Dead Sea and a salt mountain near Hom in Iran have been quarried for more than 7000 years; in Europe, there are salt quarries in Spain at Cardona and in Sicily that have been in use for at least 2500 years.

Salt mines are not confined to arid climates but are generally difficult to work efficiently. Salt mines near Lungro and Volterra (Italy), in Hallstatt (Austria), and in Camp Verde (Arizona) have been worked since Neolithic times. Only a few hundred years ago a very shallow salt mine in Africa, Taodeni, was opened. Until modern technology increased the productivity of a salt miner from 15 tons/yr to 1000 tons/yr, only a small portion of the world's demand for salt could be satisfied by mining. Table 2 gives a survey of the efficiency of saltmaking with methods used before and after the Industrial Revolution.

Salt Springs and Wells. Salt formations yield their salt not only to the miner and quarrymen, they often dissolve naturally in aquifers underground and surface as salt springs. Such springs have very often had a surprisingly good effect on the health of salt-starved people and have retained their name for "healing" in times of easier salt supply. Halle, Hallstatt, Heilbronn, Droitwich, Vichy, and Salinas are only a few names of places connected with salt springs (*Hal*, *Vic*, *Sal*, and *Wich* are the important syllables in this context).

Ocean Water. The biggest single salt source is ocean water. Although it contains only about 1/10 of a saturated salt brine (35 g salt/liter ocean water against 360 g salt/liter in a saturated brine), it is still used for more than 1/3 of the world's salt supply. All ocean coasts in arid climates are suitable for saltmaking by solar evaporation of ocean water wherever they are sufficiently flat and impermeable. Such places are good for diking in as evaporation pans (Fig. 1). These are several acres each, with dikes 90 to 170 cm high and interconnected by canals. When the ocean water evaporates, 40 m^3 yield about one ton of salt. Like the farmer, the saltmaker depends on the weather. The sea water has to be pumped into the salterns if they cannot be built below sea level or at least below high tide, a rare and somewhat risky position. Since about 500 A.D., pumps for this purpose have been worked by wind

TABLE 1. Uses of Salt

	USA 1950	Germany 1900	France 1750
Uses:			
Food	700,000 tons	450,000 tons	80,000 tons
Meat, Fish } Preserve	1,000,000 tons		
Livestock	800,000 tons	320,000 tons	100,000 tons
Mixcel	3,250,000 tons		
Leather	250,000 tons	30,000 tons	
Soda (glass polymers) Chlorine	10,000,000 tons	400,000 tons	
Overall use	16,000,000 tons	1,200,000 tons	180,000 tons
Inhabitants	150,000,000	60,000,000	25,000,000

TABLE 2. Salt Production per Man Employed

Period	Locality	Tons/Year	Men Employed	Method of Production	Tons/Man/Year
1900	Taodeni (Sahara)	4,000	250	primitive mining	16
1900	Coserra (Italy)	6,000	250	primitive mining	24
1890	Sicily	17,000	400	primitive solar pans, ocean	43
1660	Tirol (Austria)	12,000	250+300[a]	brining	48 (22)
1700	Rhe (France)	4,000	250	solar	16
1960	Reichenhall (Germany)	100,000	400	brining	250

[a]For gathering wood.

FIGURE 1. Dikes in solar evaporation ponds at Macau, in Rio Grande do Norte province, Brazil.

power (with sails on a horizontal or the vertical axis), or, before, by man or animal. Much inventive genius was employed for pump construction; chain pumps, the Archimedes screw, the helix, and many other devices came into use. The strength of the brines had to be checked as they concentrated and it was probably with the control of these brines that the concept of "specific gravity" arose.

Peat-Ash Extraction. Until the last century, ocean water was evaporated for salt, not only with the help of solar energy, but also with the peat that formed in the marginal marshes along the coasts of the North Sea in Essex, Jutland, Friesland, and Holland (Fig. 2). There and in the Black Sea area, the making of peat salt can be traced back to Neolithic times by the characteristic remnants this method has left.

Crude ceramic caldrons were filled with the filtered extract from the ash of peat fires, and the remaining salt mud was formed into cakes, which were dried between the boiling pots. The salt cakes were sold (outside black, inside whitish gray) in widespread consumer areas and even used as "money" or better as standardized means of payment (the word "salary" is connected with this use). The fire that produced the ash to be extracted was the one that kept the kettles boiling. The broken shards of the crude half-burned pots can be found today along all these northern coasts as "red hills," or "bricketage"; they can also be found in the Moselle, Saale, and Doubs river valleys near those springs where the salty spring water was concentrated by being boiled over wood fires.

Fuel for Evaporation

Almost wherever salt springs are found, drilling down to the salt body and pumping water into the drill hole will lead to artificial wells of very high salt concentration (Fig. 3). This technique is called "brining" and seems to have been used for the first time by the Chinese in about 400 A.D. The brines were conveyed by bamboo pipes to evaporation equipment (Fig. 4). Today salt bodies for brining (salt domes, etc.) are located together with fossil fuel, and the search for both has made many hundreds of diapiric salt structures well known to the oil and evaporite geologists all over the world.

Salty spring water and brine from artificial wells have to be evaporated by some form of heat. In arid climates, solar energy can be used. When peat was at hand, such salt brine was concentrated in caldrons. If wood was near, this was used, as in places like Droitwich,

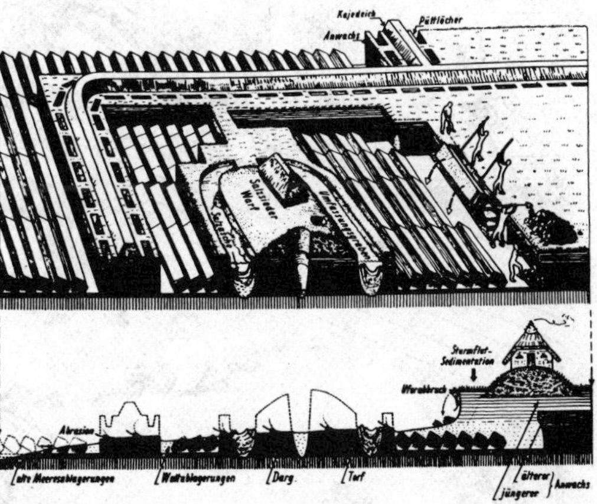

FIGURE 2. Characteristic scheme of peat saltmaking as reconstructed for the German North Sea coast from remnants of islands showing changing shorelines.

FIGURE 3. "Brining" through drill-hole techniques (China about 400 A.D.).

FIGURE 4. Making of bamboo pipes for piping "brine" (China about 400 A.D.).

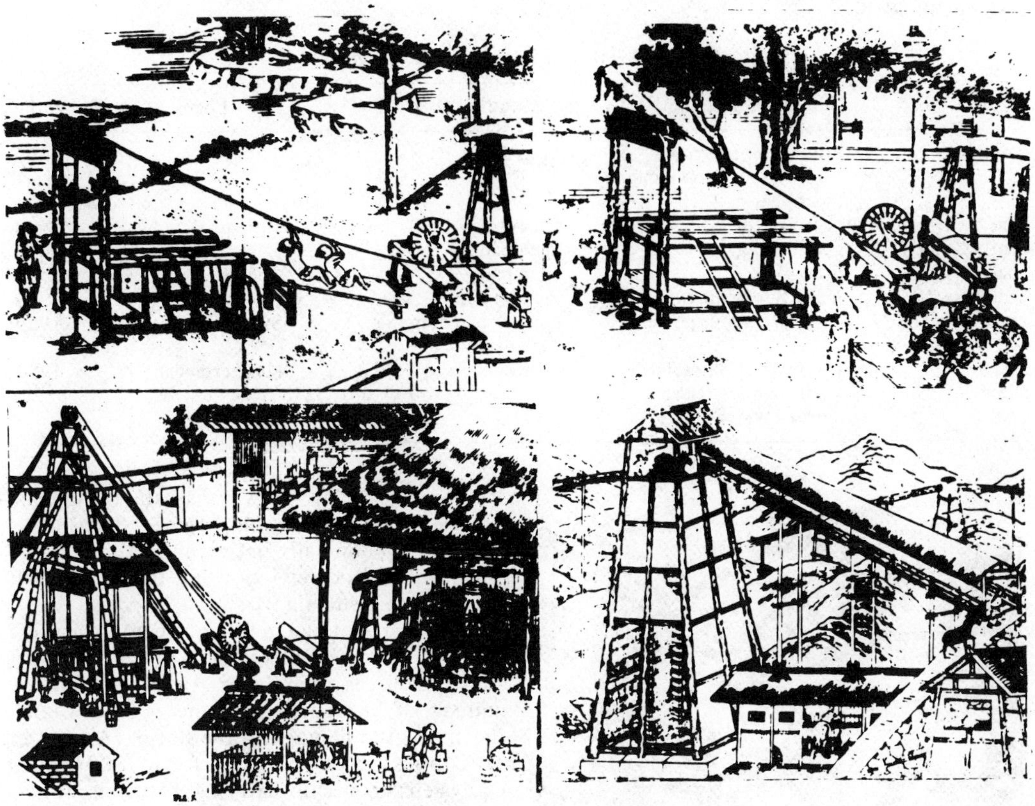

FIGURE 5. China salt industry (about 400 A.D.).

Halle, and Hallstatt. In Kimmeridge (Dorset), oil shale served as a fuel long before Roman times. The Chinese had fossil gas, which was not too far from their salt wells; this they piped for use under their salt boilers (Fig. 5). Coal replaced peat in Preston-Pans, Scotland, and wood in Cheshire in the 17th century, making the United Kingdom the biggest salt supplier in the world during the 19th century. In the present century, more and more salt brine is evaporated by the use of coal, oil and electric power with highly sophisticated vacuum equipment (Figs. 6 and 7). The greatest supplier in the world is now the USA. As soon as desalination of ocean water with atomic energy becomes economical, many places may use the resulting concentrate for saltmaking. How much fuel was used for making a ton of salt by different methods from different brines is shown in Table 3.

Transport

Ocean and River. Since salt is made in relatively few places only, often far from urban and agricultural areas where it is required and consumed, transport is an important part of its cost. Salt derived from ocean water can mostly be put on ships, and it pays to carry salt over great distances by sea. During the Middle Ages, for instance, salt was brought from Spain and France to England, Holland, Scandinavia, and Russia by the vessels (up to 800 tons) of the Hansa. Cheshire salt was shipped from England to the Indies, and from Portugal to Brazil dur-

FIGURE 6. Primitive salt boiling by Kelabit natives, Borneo, using metal pan, clay oven, and wood for evaporating salt spring water (photo by Barbara Harrisson, courtesy Sarawak Museum).

TABLE 3. Fuel for Production of 1 Ton Salt

Fuel	Ton	Source of Salt and Fuel	Method Used	Locus	Period
wood	4	brining, wood growing on 10,000 m² woodland per year	open fire	Tirol	1600
peat	14	ocean water	ash extraction	Jutland	1800
coal	6	ocean water	open pan	England east coast	1650
coal	1	brining	open pan	Cheshire	1700
coal	0.2	+ 20KWh brining	vacuum pan	Germany	1950
coal	0.02	200 KWh water power	vacuum pan compression still	Switzerland	1950
solar energy		100 m² flat, impermeable pan area near ocean water	solar pan	Mediterranean	1950

FIGURE 7. Modern vacuum salt boilers.

ing the 18th and 19th centuries; about 2 million tons of salt were sent from Spain, Sardinia, Turkey, and Egypt to Japan, and 3 million tons more went from Mexico to Australia to the Far East. Where river transport was possible, inland salt sources were always cheap enough; the salt of Charleston, W. Va., was carried on the Ohio River. In central Europe, the Danube (see Fig. 8) carried salt from the Alps to Austrian and Hungarian agricultural lands and the Elbe to northern Bohemia from the salt mines of Saxony.

Land Transport. In Europe, mule caravans were replaced by railways about 1830 (first horse drawn, later steam driven)—for instance, by the wooden rails between Linz and Česke Budejovice (Budweis) carrying the Alpine salt to southern Bohemia.

Those parts of the great continents that were not easily connected by river or rail transport to salt sources, like inner Africa, had to use animals of burden such as the ass, the camel, the mule, the llama, and the horse. (A caravan of a thousand camels cared for by 100 men is equivalent to a ship carrying 150 tons of salt under conditions before the Industrial Revolution.)

The salt swamps and quarries of Bilma and Taodeni are still supplying traders who make caravan return journeys of about 700 km through the African desert in order to supply salt to the salt-poor agricultural areas of the Sudan.

Table 4 shows what it meant to transport 10,000 tons of salt at different places and at different times of historical development.

Trade and Monopolies

Transport of salt over great distances entailed great cost; the transporters had to ensure that enough agricultural and industrial products for themselves, their animals, and the salt producers would be given in exchange for the salt to make further supplies profitable or at least possible. This required protection of stores in the areas where the salt was delivered, the transportation routes had to be safeguarded against theft, and provisions and rest places had to be provided. At the site of salt production, agricultural and industrial products, which had been brought from far away, also had to be protected. The greater the distance, the higher the reward for powerful forces that had to be organized to protect the traded goods and the means of transportation. Many wars had to be fought for the maintenance of vital salt supplies.

FIGURE 8. River transport of salt–horses on towpath (from an aquarelle of around 1770, on the Danube).

TABLE 4. Transport of Salt

Means of Transport	Quantity	Men Employed	Velocity	Period
1 Liberty ship	10,000 tons	50	20 km/h	present
60,000 camels	10,000 tons	5,000	5 km/h	ancient desert transport up to present
30 ships (Hansa)	10,000 tons	1,000	10 km/h	1400 A.D.
100 river trains (barges), 5000 horses	10,000 tons	5,000	5 km/h	Germany 18th century
200 motor trucks, rail or road	10,000 tons	250	50 km/h	present

It also seems that the salting of meat and fish was better done in well-protected enclosures where priests looked after the salt stores and the clean slaughter and salting of meat and hides. Powerful political organizations developed through salt monopolies that even today exist in many places; they were the basis for financing the great as well as the small empires until the 19th century.

The number and importance of urban settlements originally intimately connected with salt trade are considerable. Rome on the Via Salaria, Jerusalem on the main way from the Dead Sea (the salt Sea), Venice, Munich, Cologne, Axum in Ethiopia, the Aztec capital where Mexico City is now, and Bombay in India are only a few of the many that can be named as examples.

In Africa, salt was so expensive that its supply seriously limited the agricultural population. Not only was gold exchanged for salt, but also freedom–the slave trade was significantly an exchange of man for salt.

For the preservation of salt in the life cycle, everything possible was done in salt-poor areas. Even the urine of men and animals was returned to food by admixing it to the milk-and-blood diet. The ash of excrements and plants was also reused for its small salt content.

Timbuktu and Kano, old centers of civilization in Africa, are clearly seen in Fig. 9 with their network of salt trade routes. There, for thousands of years, slaves, gold, ebony, and flour were exchanged for salt and salt products.

Changing Availability of Salt Through Geophysical Changes

Ocean-borne salt is cheaper than most other salt because of its transport advantages. Nevertheless, inland salt sources have been developed at certain times with great ingenuity and against all odds. The explanation lies in the abrupt changes of sea level through which the flat seacoasts suitable for sun-salt and peat-salt production were intermittently flooded. After the steep rise of the world oceans due to the melting of the glaciers covering the Northern Hemisphere during the last deglaciation period (late Wisconsin to Holocene, 18,000 B.P. to 6000 B.P.), smaller eustatic oscillations of the sea level occurred, as originally shown by Fairbridge (1961). These oscillations have influenced the vital salt trade, and from historical facts, archaeological findings, and dating of beach lines, an approximate record of sea levels can be inferred (Fig. 10).

The most widely recognized facts are those concerning the low sea level at the height of Greek civilization (about 500 B.C.) and the high sea level at the end of the Western Roman Empire (about 500 A.D.). The maxima of sea level coincide with the decline of maritime

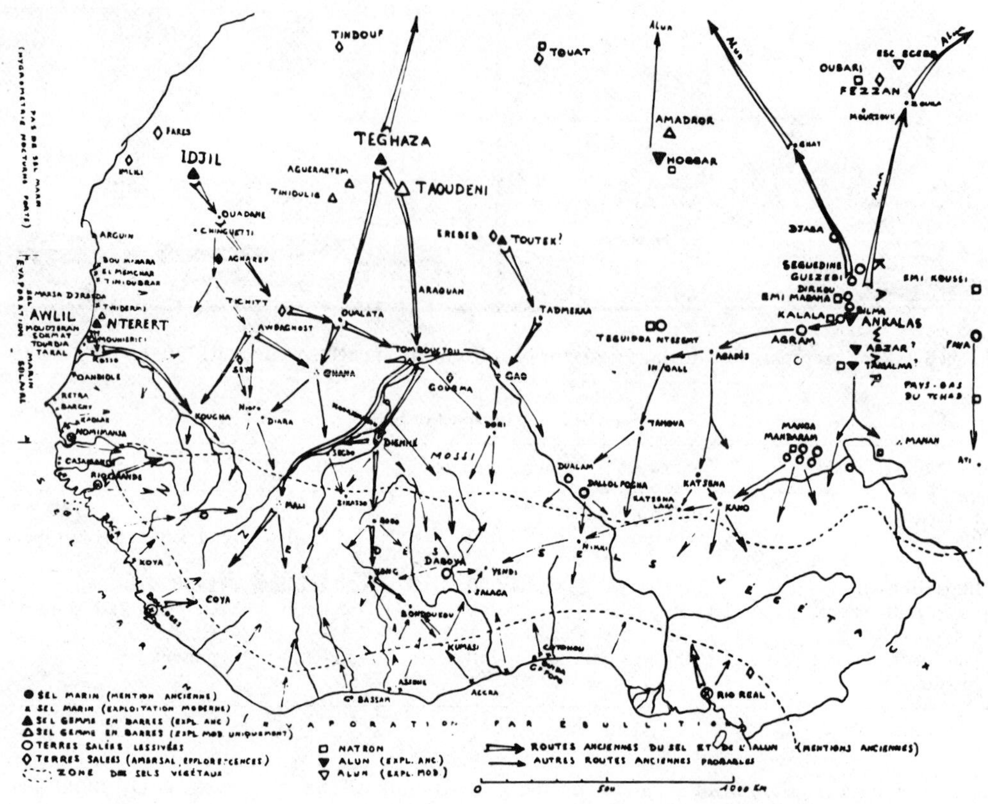

FIGURE 9. Network of African salt-trade ways (from a French survey).

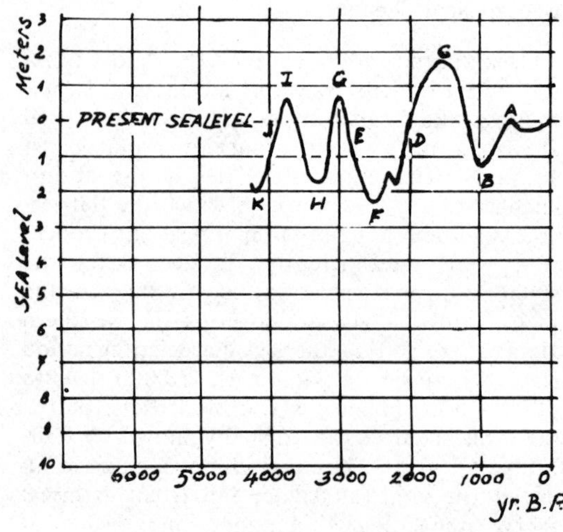

FIGURE 10. Eustatic sea-level changes.

civilizations, like the Mycenaean (1200 B.C.) and the Greco-Roman, and the rise of the continental civilizations, like those in Persia, in Syria, and in Asia Minor. The Roman-Celtic civilization (which used ocean-derived peat salt at the North Sea coast and solar ocean salt from its other coasts) declined completely during the Dark Ages (about 400–650 A.D.); a renaissance in this area occurred at about the same time as the Doomsday Book (1070 A.D.) appeared, recording hundreds of peat salt-makers along the English east coast. The ocean was then lower than it is today.

A slight rise of the oceans caused salt famine at the end of the Middle Ages (1350 A.D.) and a growth of continental powers like Austria, where salt mining in the Alps developed new technologies for salt-brine transportation in wooden pipes, for pumping, and for forestry, surpassed only in the 19th century by coal and oil technologies.

All these changes of sea level were probably caused by changing quantities of ice stored up in mountain glaciers of Greenland and the Antarctic continent; it is thought that volcanic outbreaks, which at times darkened the surface of the ice, were partly responsible for the sudden periods of melting. Fine ash in the upper atmosphere also reduces back radiation from the earth and raises mean air temperature.

Ash layers were found in an ice core dated at between 2000 and 3000 B.P. Since in the Antarctic more than 90 kcal are radiated from the sun on a square centimeter of snow surface per summer season, and since up to 90% of this amount is reflected by the normally white snow, a change in color due to ash might effect considerable warming and melting of polar caps. Thus a geological phenomenon, namely, a volcanic outbreak, might have a decisive influence on world history, shifting the vital patterns of salt supply through eustatic sea-level changes.

M. R. BLOCH

References

Bloch, M. R., 1953. History of salt (NaCl) technology, *Actes du Septième Congrès International d'Histoire des Sciences, Jérusalem*, 221–225.

Bloch, M. R., 1964. Dust-induced albedo changes of polar ice sheets and glacierization, *J. Glaciology*, 5(38), 241–244.

Bloch, M. R., 1971. *Mitteilungen der List Gesellschaft zur Entwicklung der vom Salz abhängigen Technologien*, 253–293.

Borchert, H., and Muir, R. O., 1964. *Salt Deposits.* Princeton, N.J.: Van Nostrand, 338p.

Buschman, Freiherr von J. O., 1906. *Das Salz, dessen Vorkommen und Verwertung in sämtlichen Staaten der Erde.* Leipzig: Verlag Wilhelm Engelmann.

Fairbridge, R. W., 1961. Eustatic changes in sea-level, *Physics and Chemistry of the Earth*, vol. 4. London: Pergamon, 99–185.

Fairbridge, R. W., 1962. World sea level and climatic changes, *Quaternaria*, **6**, 111–134.

Fürer, F. A., 1900. *Salzbergbau- und Salinenkunde.* Braunschweig: von Friedrich Vieweg und Sohn, 1124p.

Hauser, H., 1927. Le sel dans l'histoire, *Rev. Econ. Intern. Brussel.*, **3**, 270–287.

Hehn, V., 1873. *Das Salz*, 2nd ed. Berlin: O. Schrader.

Houghton, F. T. S., 1932. Saltways, *Birmingham Archeol. Soc. Trans.*, 1–17.

Kaufmann, D. W., 1960. *Sodium Chloride, The Production and Properties of Salt and Brine, Am. Chem. Soc. Mono. Ser.* New York: Reinhold, 743p.

Nenquin, J., 1961. *Salt, a Study in Economic Prehistory, Dissertationes Archaeologicae Gandenses*, vol. 6. Bruges: De Tempel, 159p.

Pierce, W. G., and Rich, E. I., 1962. Summary of rock salt deposits in the United States as possible storage sites for radioactive waste materials, *U.S. Geol. Surv. Bull. 1148*, 91p.

Samuel, A. M., 1918. *The Herring; Its Effect on the History of Britain.* London: John Murray, 199p.

Springer, A., 1918. *Die Salzversorgung der Eingeborenen Afrikas vor der neuzeitlichen europäischen Kolonisation.* Weida: von Thomas & Hubert, 223p.

Cross-references: *Mineral Industries; Saline Minerals.* Vol. I: *Salinity in the Ocean.* Vol. IVA: *Evaporite Processes; Natural Brines; Seawater, Chemistry.* Vol. VI: *Evaporites.*

SANDS: GLASS AND BUILDING

Sand and gravel together constitute the largest volume of mineral raw materials used in our industrial society. In the United States alone the quantity produced is over a billion tons annually and, although of low unit value, is exceeded only by cement, crushed stone, and mineral fuels in total dollar value.

Commercial sand and gravel may be defined as continuously graded, unconsolidated products of the natural disintegration of rocks. If the grains are cemented together, they are known as the sedimentary rocks, sandstone and conglomerate, clastic rocks of grain size in excess of 1/16 mm. Clastic rocks are crushed to sand and gravel size. Some *quartz*-rich products are produced as by-products from other industrial mineral operations, especially from pegmatitic deposits.

Commercial sand and gravel deposits are classed in four main genetic groups: (1) fluvial, (2) glacial, (3) marine and lake, and (4) residual. Since only rapidly moving water can carry the coarse grade of sand desired, most of the fluvial deposits are confined to upland areas. In the northern portions of North America and Europe, most of the deposits are glacial or fluvioglacial in origin.

Characteristic Properties

There are a number of characteristic properties that make sand and gravel useful commodities, the most important of which are their almost universal availability and accessibility in a naturally sorted state. Most sand and gravel is composed primarily of *quartz*, and the properties are governed by the strength, chemical inertness, and abrasion resistance of that mineral.

The largest use of sand and gravel is as an aggregate for roads, for airfields, and in concrete manufacture. The qualities that are most important for a deposit to be commercially valuable are grain size (coarse sands above 16 mesh and medium-sized gravels, 2.5–6.5 cm in diameter being preferred), impact and abrasion resistance of the grains, and freedom from deleterious or reactive impurities such as sulfides or amorphous silica (chert).

Finer sands have use in plaster or mortars. In Europe, a great deal of fine sand is used as a reactive aggregate in autoclave-cured lime-silica products. In this system (see *Portland Cement Mineralogy*), part of each grain acts as a source of silica for the production of calcium-silicate-hydrate cement between grains but, for the most part, it remains as an aggregate. In the USSR, because of an acute shortage of coarse aggregates in many areas, lime-silica products are gradually replacing concrete products as precast building components.

If a deposit is sufficiently pure (98 to 100% *quartz*), there are a number of possible applications other than aggregate. Pure sandstones, quartzites, and igneous *quartz*, when crushed and sorted, can be used in similar applications. The applications can be classified as: (1) glass and chemical use, (2) abrasives (see *Abrasive Materials*), (3) metallurgical pebble, (4) refractory sand, and (5) permeable media.

Glass- and Chemical-Use Sands

Sand for glassmaking must be pure. Glass is transparent and any impurities in the sand will show in the finished product. Inhomogeneity of product, abnormal grain-size material, and excessive contaminants, such as iron, chrome, and titania, to mention a few, can result in glass defects.

Natural deposits and sandstones that analyze over 99% silica are found. In the present-day manufacturing of glass, high-purity silica sands are used almost exclusively as the major constituent of most common varieties of glass. Ordinary soda-lime container glass and plate glass contain from 65 to 75% silica.

Most glass sands occur in nature as loose, unconsolidated deposits or as individual grains bound together by some cementing agent so as to form sandstone. A commercial deposit must be amenable to washing the cement from the crushed rock. *Quartz* and quartzite deposits, as well as by-product *quartz* obtained from ***feldspar*** operations, also form a source of glass sand. Some of these latter sources offer some of the highest purity glass sands available. While deposits of silica are abundantly distributed, deposits that are uniformly constant and relatively free from deleterious contaminants that cannot be economically removed are rare.

The quality of a glass sand may best be defined in terms of the quality of the resultant glass desired. Controlled grain size is very important. Coarse grain sizes are preferred in soda-lime container glasses and fine grain sizes are preferred in borosilicate glasses.

Glass sands are normally divided into three groups: optical, flint, and amber grades. Optical sands normally contain less than 0.02% Fe_2O_3. Flint sands range between 0.02 and 0.10% Fe_2O_3. Amber sands contain more than 0.10% Fe_2O_3 and may contain as much as 1.0% Fe_2O_3.

The silica content may range from 95.0+% (amber grade) to 99.8+% (optical grade). The remainder is commonly Al_2O_3, CaO, and MgO. The Al_2O_3 and CaO + MgO content may be present in varying amounts; however, the amount present should be uniform with a variation of ±0.30 for alumina, lime, and magnesia.

Generally speaking, coloring oxides such as Cr_2O_3, Co_3O_4, CuO, MnO, NiO, SnO_2, TiO_2, and V_2O_5 should not be present in amounts greater than 0.01% each. The amounts of coloring oxides that can be tolerated depend on the individual glass being melted. Varying amounts of a coloring oxide may be added to a glass, depending on the coloring effect desired.

Industrial silica for chemical and ceramic use must be of comparable purity to that of glass sand and the products are shipped interchangeably to manufacturers of sodium silicate or silicon carbide. For silicon carbide, as an example, there are other restrictions such as Al_2O_3 being less than 0.2% for black product and less than 0.05% for the high-purity green product.

Abrasive Sands

In today's industry, few natural abrasives can compete with synthetic products and sand is no exception (see *Abrasive Materials*). The most important use of sand is in sandblasting, a process whereby closely sized, preferably angular, sand is propelled in a high-velocity air or water stream against metal or masonry surfaces to clean them. Clean, angular sands were once used extensively in plate-glass grinding and in sawing and rubbing soft building stone. Since the development of float glass, which requires no grinding, there is little need for such sand in other than building stone applications.

Metallurgy

Metallurgical pebble is clean-graded *quartz* gravel used in the preparation of silicon alloys such as ferrosilicon and silicon copper (see *Metallurgy*). Purity is as critical as in the glass industry. Impure pebble is used as a flux in electric furnace reduction of phosphorus.

Refractories

Refractory uses of sand take advantage of its high permeability and resistance to sintering at elevated temperatures (see *Refractory Minerals*). Used as a mold or core around which molten metal is cast, the sand allows the escape of steam and gases generated by the action of the hot metal with binders and additives in the core.

Filters

Clean sand is used as filter medium for removal of turbidity and bacteria from municipal and industrial water supply systems. Well-rounded grains of high uniformity are used as a hydraulic fracturing medium in order to open, and to keep open, fractures from which valuable liquids, such as petroleum, can be drained from a rock formation.

There are numerous miscellaneous uses of specialty sands such as coal-washing, traction, and hydraulic fracturing. The latter has recently become a large-volume, relatively high-value utilization. Well-rounded, closely sized sand is immersed in a suitable fluid and pumped under high pressure into a hydrocarbon-containing formation to create fractures in the formation. The fluid then drains from the formation leaving the spherical sand grains to keep the fractures propped open. Quantities of sand as large as 2.8 million pounds have been used in a single well.

Although the industrial uses of sand and pebble are varied and diverse, they represent only a tiny fraction of the volume or dollar value of the sand and gravel used as aggregates.

DAVID ROSTOKER

References

Bessey, G. E., 1964. *Chemistry of Cements*, vol. 2. New York: Academic.

S. J. Lefond (Ed.) *Industrial Minerals and Rocks*, 4th ed. New York: American Institute of Mining, Metallurgical and Petroleum Engineers, Seeley W. Mudd Volume.

Cross-references: *Abrasive Materials; Glass: Theory of Crystallization; Metallurgy; Mineral Industries; Portland Cement Mineralogy; Refractory Minerals*, Vol. VI: *Sands and Sandstones.*

SEDIMENTARY MINERALS–*See* ROCK-FORMING MINERALS

SHEET STRUCTURES–*See* PHYLLOSILICATES

SILICATES–*See* Vol. IVA

SKELETAL CRYSTALS

A skeletal crystal is one that develops under conditions of rapid growth and high degree of supersaturation. Atoms are added more rapidly to the edges and corners of a growing crystal than to the centers of crystal faces, resulting in either branched, tree-like forms or hollow, stepped depressions. Branched crystals are considered to have a "dendritic" habit; the hollow stepped crystals are referred to as "hoppers."

While the concentration gradient of matter to a growing crystal tends to be highest at corners and edges of faces, at low levels of supersaturation the overall growth rate is uninfluenced by local fluctuations in the degree of supersaturation, since flat, polyhedral faces are the general rule among crystals. Only above a critical level of supersaturation will the high corner and edge concentration gradients promote most rapid growth at these positions with the consequent development of skeletal forms.

The term "skeletal growth" may be used in two slightly different ways. During rapid edge growth, internal cavities form on some or all crystal faces. Subsequent growth seals the cavities, but commonly the hollows are filled with fluid, presumably of the same composition as the solution from which the crystal grew. *Quartz* from Minas Gerais, Brazil, and *halite* from the Dead Sea, Israel, illustrate this phenomenon (Figs. 1, 2).

More generally, the term "skeletal growth" refers to the development of crystals with a branching, dendritic (tree-like) habit (see *Crystal Habit*). Examples of dendritic crystals include *ice* on a windowpane, *pyrolusite* on agate (moss agate), and native *gold, silver*, and *copper*.

Hoppers

A hopper crystal is produced when atoms are deposited more rapidly at the edges and corners than at the centers of crystal faces,

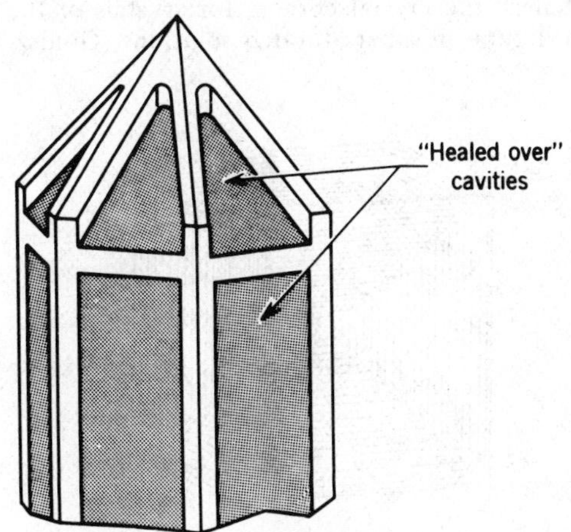

FIGURE 1. Skeletal *quartz* (from I. Vanders and P. F. Kerr, 1967. *Mineral Recognition*. New York: John Wiley. Copyright © by John Wiley & Sons, Inc. Reprinted by permission).

FIGURE 2. Hopper crystal of *halite* from the southwestern shore of the Dead Sea near Sodom (width of crystal 6 cm).

forming deep, stepped depressions in the centers of the affected faces (Fig. 3). Hopper crystals generally develop under conditions of a high degree of supersaturation (such as at elevated temperatures and high concentration gradients) and, consequently, a rapid rate of crystallization.

By use of an interferometric technique, the concentration gradient was shown to be greatest near the crystal corners for crystals of the NaCl type in supersaturated solutions (Goldsztaub and Kern, 1953). Since diffusion of matter to the growing crystal is proportional to the concentration gradient, matter can be transported to edges and corners faster than to the centers of faces.

Under usual conditions of crystallization, the cube face normally develops parallel to itself, even though the level of supersaturation is higher near the apexes. This occurs because the most rapidly growing faces (i.e., [111] or [110]) quickly disappear, whereas the slower-growing cube faces are those that ultimately survive. On the other hand, the presence of crystal imperfections near face centers increases the growth rate there relative to the corners (Chernov, 1974), thus counteracting the effect of diffusion. Above a critical level of supersaturation, however, the importance of diffusion predominates and those defects located near the position of greatest supersaturation (i.e., near the apexes) will then promote rapid deposition. Once hopper-type growth is initiated, the projecting edges may restrict diffusion of solute to the face centers, thus slowing deposition there to an even greater extent. The presence of impurities, such as included mud, may also inhibit development at face centers as seen in the formation of the Dead Sea *halite* hoppers (Gornitz and Schreiber, 1981; and Fig. 2).

Halite, *pyromorphite*, and *vanadinite* are minerals that commonly develop hopper crystals (Figs. 2, 3). Less frequently, hopper crystals may be observed on *chalcopyrite, diamond*, and *gold* (Desautels, 1968).

Metallic bismuth, although produced artificially, also exhibits hoppered pseudocubic rhombohedrons. Other synthetic crystals displaying hopper-like habits include ZnS, CdS, CdTe, HgSe, and HgTe (which crystallize with

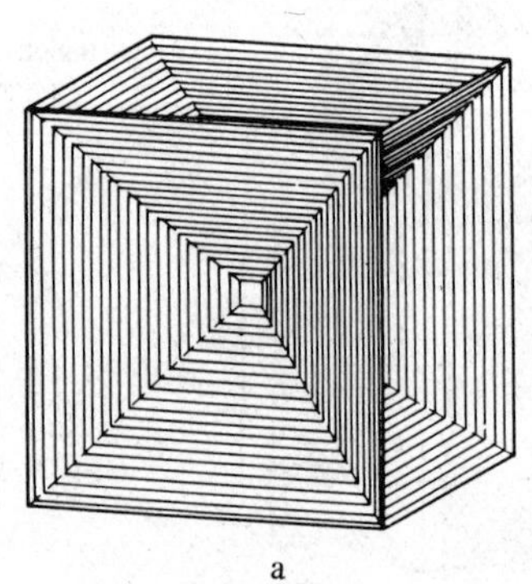

a

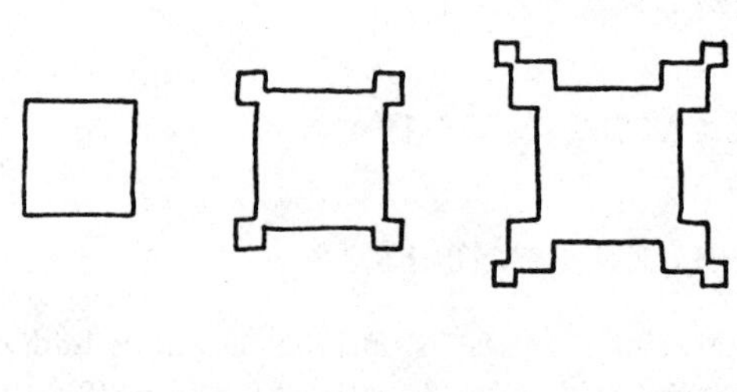

b

FIGURE 3. (a) Hopper Crystal (from I. Vanders and P. F. Kerr, 1967. *Mineral Recognition*. New York: John Wiley. Copyright © 1967 by John Wiley & Sons, Inc. Reprinted by permission). (b) Idealized hopper growth (from *Crystals and Crystal Growing* by Alan Holden and Phylis Singer. Copyright © 1960 by Educational Services Inc. Reprinted by permission of Doubleday & Co., Inc.).

the wurtzite structure). These substances form hollow, sometimes scroll-like hexagonal prisms, in some cases perched on whiskers or thin hair-like growths (Fig. 4). A possible growth mechanism is helicoidal growth along a screw dislocation with growth concentrated along the periphery, developing a pyramid (see *Crystal Growth*). Upon restriction of continued outward expansion (because of interference with neighboring crystals), the growth becomes predominantly vertical, leading to a hollow hexagonal prism (Simov et al., 1974). In ZnO, clusters of hollow, hexagonal prisms and hoppers grow upward in the *c*-axis direction, from a substrate crystal that lies parallel to the $(2\bar{1}\bar{1}0)$ plane. The hoppers are composites of several hollow prisms that have nucleated simultaneously and have grown radially from a common point on the substrate (Iwanaga et al., 1978).

Halite not only forms hoppered cubes; it also occurs in pyramidal hoppers. Pyramidal *halite* hoppers represent growth in shallow bodies of brine under conditions of high temperatures, high evaporation rates, and, hence, high degree of supersaturation. According to Dellwig (1955), growth is initiated at the surface of the evaporating water. Small, flat, platy crystals form and float on the surface because of the surface tension of the brine. Growth continues downward, as pyramidal hoppers. Eventually plates aggregate into crusts, which sink to the bottom but continue to grow in a cubic habit. The presence of "phantom" pyramidal hoppers is seen in the zoning and banding of natural salt (see *Phantom Crystals*). The same mechanism applies to grainer salt, produced artificially.

An unusual occurrence of hopper development comes from the mare basalts of Oceanus Procellarum on the moon (see *Lunar Minerals*). The ***pyroxenes*** (see *Pyroxene Group*) have an extremely complex crystallization history in which core *hypersthene* has been epitaxially overgrown by a sheath of *pigeonite*, and that in turn is rimmed by *augite* (see *Epitaxy*). Electron microprobe traverses suggest, however, that the (001) sector of *pigeonite* remained hollow and was filled in later by *augite* in epitaxial overgrowth (Hollister et al., 1971). Similarly, electron microprobe traverses across ***plagioclase*** (see *Plagioclase Feldspars*) crystals from another specimen indicates increases in Na and K toward both the outer rim and the core, implying that ***plagioclase*** grew outward, and inward as a hollow crystal, trapping residual liquid that ultimately formed a core of ***pyroxene*** and other minerals (Walter et al., 1971).

Dendrites

While most dendritic crystals exhibit irregular branching, a snowflake provides a beautiful

a

b

FIGURE 4. Hollow CdTe crystals. (a) Hollow hexagonal prism and pyramid perched on whisker. (b) Hollow hexagonal prism with scroll-like opening.

illustration of the control exerted by the hexagonal symmetry on the overall shape of the crystal. In a snowflake, the dendrite branches are all parallel to the crystallographic axes a_1, a_2, and a_3. The exact cause for the regularity in dendrite spacing and branching of a snowflake is still a matter of speculation. The even spacing of the branches could result from variations in the level of supersaturation. Growth on the side branches reduces the supersaturation in the immediate vicinity and prevents further growth except beyond a certain distance. According to one point of view (Fletcher, 1973), the overall symmetry of a snowflake derives from anisotropy in surface free energy and kinetics, which leads to the simultaneous development of dendritic growth at all six corners of the *ice* hexagon. This process is critically dependent on the excess water vapor density.

Experiments have shown that the habit of a snow crystal is a function both of temperature and of vapor density (Fig. 5). Extreme dendritic forms develop in regions of highest vapor density. Hollow hexagonal prisms are restricted to a narrow temperature range between -6° and -8°C. Hexagonal prisms capped by thin hexagonal plates (Tsuzumi crystals) result from transport of the growing *ice* crystal from the prism thermal regime to cooler conditions that favor growth of flat plates (Knight and Knight, 1973). An unusual type of snowflake is a three-dimensional dendrite, restricted to snow crystals grown from highly supercooled water droplets, below -20°C (Kobayashi and Furukawa, 1978). The angle between the *c*-axis of the planar snowflake and the secondary branches is close to 70°. This orientation allows the formation of a superlattice in which the misfit of lattice points between the platy crystal and its branches is minimized.

VIVIEN GORNITZ

References

Chernov, A. A., 1974. Stability of faceted shapes, *J. Cryst. Growth*, **24/25**, 11–31.

Dellwig, L. F., 1955. Origin of the Salina salt of Michigan, *J. Sed. Petrology*, **25**, 83–110.

Desautels, P. E., 1968. *The Mineral Kingdom*. New York: Grosset & Dunlap, 251p.

Fletcher, N. H., 1973. Dendritic growth of ice crystals, *J. Cryst. Growth*, **20**, 268–272.

Foster, R. J., 1960. Origin of embayed quartz crystals in acidic volcanic rocks, *Am. Mineralogist*, **45**, 892–894.

Goldsztaub, S., and Kern, R., 1953. Study of the concentration of solution around a growing crystal, *Acta Crystallogr.*, **6**, 842–845.

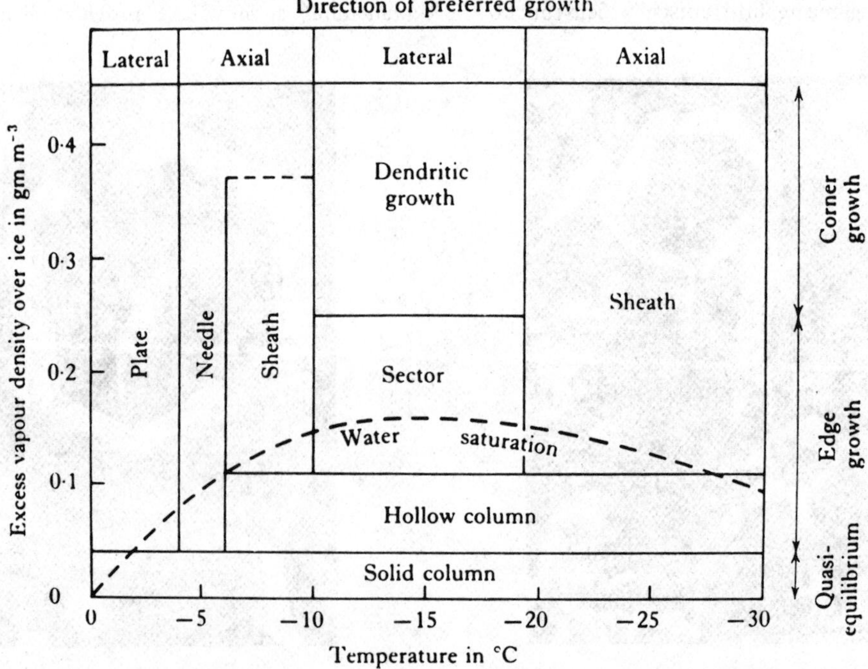

FIGURE 5. The Nakaya-Kobayashi diagram, illustrating the various crystal habits of ordinary *ice* as a function of both temperature and excess vapor density (from N. H. Fletcher, 1973. Dendritic growth of ice crystals, *J. Crystal Growth*, **20**, p. 268).

Gornitz, V., and Schreiber, C., 1981. Displacive halite hoppers from the Dead Sea: some implications for ancient evaporite deposits, *J. Sed. Petr.*, in press.
Holden, A., and Singer, P., 1960. *Crystals and Crystal Growing*. New York: Doubleday, 320p.
Hollister, L. S., et al., 1971. Petrogenetic significance of pyroxenes in two Apollo 12 samples, *Proc. 2nd Lunar Sci. Conf.*, **1**, 529–557.
Iwanaga, H.; Yamaguchi, T.; Shibata, N.; and Hirose, M., 1978, Growth mechanism of hollow ZnO crystals from ZnSe. II, *J. Cryst. Growth*, **43**, p. 71–76.
Knight, C., and Knight, N., 1973. Snow crystals, *Sci. American*, **228**, 100–107.
Kobayashi, T., and Furukawa, Y., 1978. Epitaxial relationships during the formation of three-dimensional snow dendrites, *J. Cryst. Growth*, **45**, 48–56.
Lendvay, E., and Kovàcs, P., 1970. Hollow single crystals of ZnS, *J. Cryst. Growth*, **7**, 61–64.
Nakaya, U., 1954. *Snow Crystals: Natural and Artificial*. Cambridge: Harvard University Press.
Simov, S. B., et al., 1974. Observations of hollow cadmium telluride crystals with the scanning electron microscope, *J. Cryst. Growth*, **26**, 294–300.
Strickland-Constable, R. F., 1968. *Kinetics and Mechanism of Crystallization*. New York: Academic, 335p.
Vanders, I., and Kerr, P. F., 1967. *Mineral Recognition*. New York: Wiley, 316p.
Van Hook, A., 1961. *Crystallization: Theory and Practice*, New York: Reinhold, 325p.
Walter, L. S., et al., 1971. Mineralogical studies of Apollo 12 samples, *Proc. 2nd Lunar Sci. Conf.*, **1**, 343–358.

Cross-references: *Crystal Habit; Crystal Growth; Crystallography: Morphological; Crystals, Defects in; Epitaxy; Phantom Crystals; Polycrystal; Symmetry*. Vol. II: *Snowflakes.*

SMECTITE GROUP

Smectites are a group of hydroxyl aluminosilicates containing alkalis and alkaline earths (see *Clays, Clay Minerals*). The earliest use of the term appears to have been by Cronstedt in 1758. Subsequently the term "*montmorillonite*" gained popularity as both an individual mineral name and the term for all the ***smectite*** minerals. More recently the name "*montmorillonite*" has been restricted to the dioctahedral mineral that contains both aluminum and magnesium. The group includes the following minerals:

- *Montmorillonite*
- *Beidellite*
- *Nontronite*
- *Hectorite*
- *Saponite*

which share the general formula

$$X_{0.33}Y_2Si_4O_{10}(OH)_2 \cdot 4H_2O$$

where X may be Ca or Na and Y may be Al, Fe^{3+}, Cr, Mg, Ni, Zn, or Li (Ross and Hendricks, 1945).

The ***smectite*** group is monoclinic with a hardness of 1 to 2 and density variable between 2 and 3. The color ranges from colorless to white, yellow, yellow-green, green or brown depending on iron content. Pinkish ***smectites*** result from the presence of small amounts of manganese (Ross and Kerr, 1931). ***Smectites*** have perfect basal {001} cleavage. They have extremely fine particle size (usually less than one micron). They take up considerable amounts of water or other polar liquids between individual layers (see *Phyllosilicates*).

Optical properties are difficult to obtain for this group because of the extremely fine particle size. Optically (−), they are colorless or faintly colored (yellow or green) in thin section; $\alpha = 1.48–1.61$, $\beta = 1.50–1.64$, $\gamma = 1.50–1.64$, $\delta = 0.01–0.04$; and 2V is usually small (Deer, et al., 1962).

Smectites and most easily studied by electron microscopic methods which show them to be composed of individual sheets as thin as one unit cell in thickness or aggregates of many unit cells. The boundaries of the flakes are irregular. Calcic ***smectites*** tend to occur in thicker aggregates than do sodic ***smectites***. When dispersed in water the sheets tend to weld themselves together by edge-to-edge bonding.

Smectites are attacked by acids and have very high cation exchange capacities. Calcium is the exchange cation most commonly, but sodium is the exchange cation in Wyoming bentonites and a few other occurrences.

They occur as alteration products of volcanic glass and ash (bentonite) and of other igneous rocks (Fuller's earth). They are also found as products of hydrothermal alteration of ***plagioclases***, in soils, and in metamorphic rocks (see *Rock-Forming Minerals; Soil Mineralogy*).

The dioctahedral minerals of the ***smectite*** group are *montmorillonite*, which contains both Al^{3+} and Mg^{2+}; *beidellite*, which is Al-rich; and *nontronite*, which is Fe-rich. The trioctahedral members of the group are *saponite*, which is Mg-rich; *hectorite*, which contains lithium; and *sauconite*, which contains zinc. Volchonsoite is a dioctahedral variety that contains chromium (chromian *nontronite*).

Smectites are used in drilling muds (see *Thixotropy*); as ceramic raw material; as a binder in taconite pelletization; as a filler in paper, rubber, and paints (see *Pigments and Fillers*); in cosmetics; as a catalyst; to adsorb impurities; and in molding sands.

EMMY BOOY

References

Cronstedt, A., 1758. *Mineralogie*. Stockholm. English translation by Magellan, J. H., *An Essay Toward a System of Mineralogy*, London, 1788, p. 98.

Deer, W. A.; Howie, R. A.; and Zussman, J., 1962. *Rock-Forming Minerals: Sheet Silicates*, vol. 3. New York: Wiley, 270p.

Ross, C. S., and Hendricks, S. B., 1945. Minerals of the montmorillonite group: *USGS Prof. Paper 205-B*, 23–79.

Ross, C. S., and Kerr, P. F., 1931. The clay minerals and their identity, *J. Sed. Pet.*, 1, 55–65.

Cross-references: *Clays; Clay Minerals; Phyllosilicates; Rock-Forming Minerals; Soil Mineralogy, Thixotropy*; see also mineral glossary.

SODALITE GROUP

The ***sodalites*** are a group of isometric framework aluminosilicates with a structure composed of SiO_4 and AlO_4 tetrahedra linked by corner sharing in an open cage-like array:

- *Hauyne*: $(Na,Ca)_{4-8}Al_6Si_6O_{24}(SO_4,S)_{1-2}$
- *Lazurite*: $(Na,Ca)_8(Al,Si)_{12}O_{24}(S,SO_4)$
- *Nosean*: $Na_8Al_6Si_6O_{24}(SO_4)$
- *Sodalite*: $Na_4Al_3Si_3O_{12}(Cl)$

Channels in the cages intersect to form large cavities at the corners and the centers of the unit cell. The SO_4, Cl, and S ions are located in these cavities, each coordinated to four Na or Ca ions situated along the body diagonals of the unit cube. Each Na or Ca is surrounded by one SO_4, Cl, or S ion and the O^{2-} ions of the framework.

Sodalites typically occur with ***feldspathoids*** and ***feldspars*** in silica undersaturated rocks such as syenites and phonolites, often in the pegmatite phase as large masses (see *Pegmatite Minerals*). *Lazurite* and *sodalite* may also be found in marbles as a result of contact metamorphism. *Nosean* is uncommon except in volcanic rocks. Lapis lazuli, a gem variety of *lazurite*, is almost always associated with *pyrite* and commonly with *calcite*. Ultramarine is synthetic *lazurite*.

KEITH FRYE

References

Deer, W. A.; Howie, R. A.; and Zussman, J., 1966. *An Introduction to the Rock-Forming Minerals*. London: Longmans, 375–379.

Phillips, W. R., and Griffen, D. T., 1981. *Optical Mineralogy: The Nonopaque Minerals*. San Francisco: Freeman, 369–370.

SOIL MINERALOGY

Soil mineralogy is concerned with the inorganic minerals found in the pedosphere and to the depth of weathering. Soils and weathering profiles result from weathering of parent materials at the surface of the earth, (1) by physical disintegration of minerals originally occurring at the site, (2) by chemical and biological degradation of less resistant minerals, and (3) by recombinations and synthesis of new minerals (see Vol. XII: *Weathering*). Even though any mineral can theoretically occur in soils, the actual number of minerals commonly found in soils is rather limited. They are important because they greatly influence physical properties, exert a strong influence on chemical processes, and affect soil fertility and productivity through release of and reactions with plant nutrients. The occurrence of trace elements in soils is relevant to mineral exploration, as some soils themselves are important mineral deposits (e.g., bauxites), and to the monitoring of the environment (pedosphere) for pollution control. The mineral composition of soils, thus, has an important bearing on ecology and on the exploitation of soil resources.

Soil Minerals

The minerals in soils can be classified (1) as primary minerals, formed at elevated temperatures and(or) pressures, that are inherited from igneous and metamorphic rocks, sometimes through a sedimentary cycle, and (2) as secondary minerals, formed by low-temperature reactions, that are inherited by soils from sedimentary rocks or formed in situ by weathering.

Primary Minerals. Primary minerals comprise the main part of the sand and silt fractions of most soils; the coarse fraction, therefore, essentially governs the percentages of primary minerals. The most abundant primary minerals in soils are *quartz* and ***feldspars***, which dominate in the rocks making up the lithosphere (see *Rock-Forming Minerals*). Free *quartz* frequently comprises 50 to 90% of the sand fraction, but silica is also present in some soils as opal (phytoliths), chalcedony, agate, flint, chert, or *cristobalite*. ***Feldspars*** normally range from 5 to 25% of the sand and silt fraction, though they may account for as much as 40% in clays derived from glacial rock flour. Potassium ***feldspars*** of soils include *microcline, orthoclase*, adularia, and *sanidine*, the former being most abundant. Though the more albitic varieties of ***plagioclase*** predominate in weathered tropical and subtropical soils, they may also be present in soils formed from slightly weathered regoliths.

Accessory (heavy) minerals include a wide variety of primary minerals occurring in small but significant amounts. ***Pyroxenes*** include *enstatite, hypersthene, diopside*, and *augite*, of which the latter is most important in amounts as well as compositional significance in soils. ***Amphiboles*** and ***olivenes*** normally account for less than 1% of the soil mass. Other important accessory minerals are ***apatites, tourmalines,*** *zircon, topaz,* ***garnets,*** *titanite (sphene), kyanite, sillimanite, andalusite, staurolite*, and *rutile*.

The occurrence of heavy detrital minerals in soils can be a reliable indicator of the provenance of transported soil parent materials and can facilitate the recognition of soil stratigraphic layers in soil profiles (see summaries in Finkl, 1980). These minerals can also be useful in estimating volume losses due to weathering as well as the degree of weathering; ***garnets,*** *zircon*, and ***tourmalines***, and the ***amphiboles*** and ***pyroxenes*** have been particularly useful in this regard (Haseman and Marshall, 1945; Bear, 1964). Soil fertility and productivity are in general related to mineralogy as young soils usually contain a high proportion of weatherable (unstable) minerals that can supply essential plant nutrients. The ***micas***, for example, are the only common heavy minerals that supply large amounts of potassium to plants. Significant amounts of calcium and magnesium are also released by weathering of *hornblende, hypersthene*, and *biotite* (Russell, 1966).

Secondary Minerals. Among the secondary minerals found in soils are carbonates, sulfur-bearing minerals, layer silicates, oxides, and phosphate minerals. Their occurrence has been reviewed by Van Der Plas and Van Reeuwijk (1974) and by Gieseking (1975). Extensive summaries in Dixon and Weed (1977) further consider structure and composition, modes of formation, chemical and physical properties, and reactions of secondary minerals in soils.

Calcite, the most abundant carbonate mineral in soils, accumulates in subsoils of subhumid and semiarid regions. It is commonly disseminated, but also occurs in indurated forms (Goudie, 1973) as nodules or as hard calcareous (travertinous) layers termed calcrete (caliche). *Dolomite* and, more rarely, *siderite, aragonite*, and *magnesite* also occur in soils where the environment is suitable for carbonate accumulation.

Although *gypsum* commonly occurs as disseminated crystals in soils, it may also form an indurated horizon beneath a calcrete layer in subsoils of arid regions. Drainage of some tidal marshes or the exposure of acid-forming underclays results in acid sulfate soils (cat clays) that contain *pyrite, jarosite, mackinawite*, and *alunite* (Dost, 1973; Ivarson and Hallberg, 1976).

Phyllosilicates (q.v.) are extremely important constituents of most soils because they can determine physical properties such as plasticity and structure, can influence soil-water relationships, especially the water-holding capacity, and can affect ion-exchange phenomena. The various phyllosilicates characteristic of the clay-size fraction of soils are listed in Table 1.

Although local parent materials, drainage, and climate largely determine the kinds of soil, ***clays***, and ***clay*** minerals (see *Clays, Clay Minerals*), a few broad generalizations relating to their geographic distribution are possible. *Kaolinite* dominates in strongly weathered soils of tropical and subtropical regions whereas ***smectites, vermiculites, illites***, and ***chlorites*** tend to be associated with middle-latitude soils. If the parent material or soil solution contains a relatively high proportion of potassium, ***illite***-group minerals are likely to be formed. Parent materials containing much magnesium or soil drainage that favors retention of bases will encourage the formation of ***smectites; illites*** and ***smectites*** are, thus, prevalent in soils of cool semiarid and arid regions (see vol. XII, *Clay Mineral Formation*).

Clay minerals also differ among soils of different ages. Soils derived from parent materials low in ***clay*** tend to form stable mineral assemblages. Soils formed from clayey parent materials containing a wide variety of ***clay*** minerals, however, frequently show alteration from the surface downward to a more stable assemblage (Birkeland, 1974).

Iron and aluminum oxides, in various polymorphs and states of hydration, are involved in processes that are unique to the pedosphere. The term *pedosesquioxides* (q.v. in Vol. XII) is used to distinguish the neogenetic products of weathering and soil formation from the sesquioxides of geochemical usage. Oxide minerals influence the physical behavior of soils by affecting properties such as consistency, microstructure, aggregate stability, porosity, and color. They also participate in physicochemical reactions such as ion exchange, fixation of elements, and buffering. Oxides occur in soils in the form of amorphous hydrous oxides of Al, Fe, Mn, and Si; as disseminated individual crystals; as coatings, as concretions (pisolites); or as massive indurated layers of ferruginous duricrust, ferricrete (Goudie (1973). Some of the more common soil oxides, hydroxides, and oxyhydroxides are given in Table 2.

Amorphous hydrous oxides of aluminum

TABLE 1. Phyllosilicates Expected in Soil Materials[a]

Type	Group	Subgroup	Species
2:1	***Pyrophyllite-talc***	dioctahedral	*pyrophyllite*
		trioctahedral	*talc, minnesotaite**
	Smectite[b]	dioctahedral	*montmorillonite, beidellite,* nontronite, volkonskoite**
		trioctahedral	*saponite, sauconite,* hectorite**
	Vermiculite	dioctahedral	dioctahedral *vermiculite*
		trioctahedral	*vermiculite*
	Mica[c]	dioctahedral	*muscovite, hydromica, paragonite, illite, margarite**
		trioctahedral	*phlogopite, biotite, lepidolite,* zinnwaldite,* clintonite* (seybertite)*
2:1:1	***Chlorite***[e]	dioctahedral	*sudoite**
		trioctahedral	leptochlorites and orthochlorites
1:1	***Kaolinite-Serpentine***	dioctahedral	*kaolinite, dickite,* nacrite,* halloysite,* metahalloysite
		trioctahedral*	*amesite, antigorite, berthierine, chrysotile, cronstedtite, lizardite*
	Hormite[f]	mixoditrioctahedral	*palygorskite*
		trioctahedral	*sepiolite,* xylotile*

*Less common occurrences in soils.

[a]Compiled from Warshaw and Roy (1961), Grim (1968), Van Der Plas and Van Reeuwijk (1974), and Kretz (1976).

[b]The term ***smectite*** encompasses the group of ***phyllosilicates*** including ***montmorillonites*** and related ***clay*** minerals. The term is not universally accepted and the group is also referred to as the ***montmorillonite*** or ***montmorillonite-saponite*** group.

[c]Various polymorphs exist, depending on the manner in which the layers are superposed. The so-called brittle ***micas,*** with which *margarite* and *clintonite* are sometimes grouped, have even greater Al for Si substitution than do the ***micas.***

[d]The ***illites*** give rise to much controversy, and it is not clear whether they should be separated from the ***micas*** (Mackenzie, 1965). Most reputed ***illites*** are dioctahedral, but trioctahedral analogues are not uncommon in soils, and for these the name "ledikite" has been suggested (Brown, 1955).

[e]Because the ***chlorite*** group contains layers of the 2:1 type interstratified with brucite or gibbsite layers which are banded together by isomorphous substitutions in the 2:1 layer balanced by opposite charges from substitutions in the brucite or gibbsite layer, the structure may be expressed using the notation 2:1:1 (Mackenzie, 1975). Even though ***chlorites*** and ***serpentines*** are readily interconvertible, they are listed here as separate groups.

[f]*Palygorskite* and *sepiolite* are so closely related that it may be desirable to consider them as one group. R. H. S. Robertson (quoted in Mackenzie, 1965) suggested the term ***hormite*** for this group.

occur as gels in intensely weathered wet soils that also contain iron. Amorphous aluminosilicate gels, such as sesquioxidic and halloysitic *allophanes*, are common in soils derived from volcanic ash. *Gibbsite*, the most abundant hydrous oxide of aluminum in soils, is characteristic of strongly weathered tropical and subtropical soils. Iron oxides in soils range from less than 0.5 to more than 80%. Although *hematite* and *goethite* are most abundant, other oxides such as *pyrolusite*, limonite, *lepidocrocite, magnetite*, and *maghemite* are also present in some soils, especially those that have been subjected to weathering over a long period of time. Titanium oxides, such as *rutile* and *anatase*, can occur in amounts as great as 1% in certain soils; they may be both pedogenic and residues from weathering.

Phosphates constitute less than 0.05% of the total soil mass. Although phosphates account for only a small percentage of the soil minerals, they greatly influence the phosphorus cycles of plants and other organisms in the soil. *Fluorapatite* is the original source of much of the soil's phosphorus supply. Combinations of the phosphate ion with iron and aluminum in the form of definite minerals occurring in alkaline calcareous soils include *hydroxyapatite, fluorapatite, chlorapatite, wagnerite*, and *wavellite*. Other phosphates that are present in soils, in-

TABLE 2. Common Oxides and Hydroxides Formed in Soils

Type	Mineral	Formula	Occurrence
Monohydrates (Oxyhydroxides)	*Boehmite*	AlO(OH)	Major constituent of some bauxites
	Goethite	α-FeO(OH)	Common weathering product in many soils, especially those containing ironstone
Trihydrates (Hydroxides)	*Lepidocrocite*	γ-FeO(OH)	Featured in waterlogged soils
	Gibbsite	$Al(OH)_3$	Major constituent of bauxites and some laterites, present in many lateritic soils
Oxides	*Hematite*	α-Fe_2O_3	Characteristic of oxidized soils
	Maghemite	γ-Fe_2O_3	Mainly associated with laterites

cluding those that can reasonably be expected to occur even in minute amounts, are given in Gieseking (1975).

Methodologies

The analytical instruments and procedures of soil-mineral analysis (q.v. in Vol. XII) and their applications are briefly indicated in what follows. Techniques of X-ray diffraction (q.v. in Vol. IVA) are used to characterize the fine fractions of the soil and, in particular, to identify ***clay*** minerals (Brown, 1961; Thorez, 1975). Application of electron optical techniques in studies of crystal morphology, using transmission electron microscopy as described by Gard (1971), has provided a better understanding of the mechanisms of mineral alteration and of chemical processes occurring on mineral surfaces within soils [see *Electron Microscopy (Transmission)*]. The scanning electron microprobe is particularly useful in the study of soil structure and mineral morphology because it permits the direct observation of the shape and arrangement of small crystals and pores in soils (see *Electron Probe Microanalysis*). Infrared absorption spectrometry (q.v. in Vol. IVA) is widely used to study isomorphous substitution, hydroxyl configuration, and clay-organic complexes. Although this method is not a routine analytical technique for the identification of minerals, it may provide great sensitivity for the detection of *gibbsite, goethite, quartz, kaolinite*, and *halloysite* in some mixtures (Farmer, 1974). Thermal techniques are generally more useful in detailed investigations of mineral properties than in the identification of the components of complex mixtures. Differential thermal analysis, thermogravimetric analysis, and differential thermogravimetric analysis (q.v. in Vol. IVA) have their widest application as an analytical tool for identifying ***clay*** minerals (Gard, 1971). Total chemical analysis and cation exchange capacity (q.v. in Vol. XI) are both used to complete identification of minerals. The former permits the calculation of a structural formula indicating which ions occupy the structural sites, whereas the latter technique is a guide to the types of ***clay*** minerals present.

It is perhaps worth noting that the detection of certain minerals in soils is related to the sensitivity of the analytical procedure. Logan et al. (1976), for example, indicated that solids with very fine particle sizes, such as most secondary iron minerals, appear amorphous by X-ray diffraction. Using Fe Mossbauer spectroscopy they found that *akaganeite*, previously considered to be rare in soils, was a major iron mineral in New Zealand soils.

Applications

The multidisciplinary aspect of soil mineralogy provides a broad base of operations for the soil mineralogist. The concise review of the applications of soil mineralogy presented by Stelly and Dinauer (1968) emphasizes, in particular, the relationships between soil mineralogy and engineering properties, between the physics and chemistry of soil and its fertility. Rich (1968), for example, draws attention to the important role of soil mineralogy in the development of ion-exchange models, to the determination of the structure of water near exchange surfaces, and to the importance of anion and trace-element reactions with soil minerals. Winterkorn (1968), in discussing the engineering use of soil minerals, emphasizes their actual and potential interactions with aqueous solutions. The importance of mineralogical data in soil stratigraphic studies has been underscored by Brewer (1964). Studies of mineralogical properties of soils, especially those dealing with particle-size variations with depth, nature of ***clay*** distribution with depth, and weathering differentials, help differentiate geogenetic and pedogenetic discontinuities within weathering profiles. Mineralogical data are also used as a taxonomic criterion at several levels in the new American *Soil Taxonomy* (1975), an important comprehensive system. The mineralogy families are based on the approximate mineralogical composition of selected size fractions in control sections of soil profiles. Mineralogy is further used, together with other features in this system, to differentiate diagnostic subsurface horizons as well as certain taxa at other categorical levels (Allen, 1977). In studies of soil genesis (q.v. in Vol. XII), the mineralogy of the clay fraction may be used to indicate the type and degree of weathering that has occurred during soil formation.

CHARLES W. FINKL, JNR.

References

Allen, B. L., 1977. Mineralogy and soil taxonomy, in Dixon and Weed, 771–796.

Bear, F. E., 1964. *Chemistry of the Soil*. New York: Reinhold, 515p.

Birkeland, P. W., 1974. *Pedology, Weathering, and Geomorphological Research*. New York: Oxford University Press, 285p.

Brewer, R., 1964. *Fabric and Mineral Analysis of Soils*. New York: Wiley, 470p.

Brown, G., 1955. Report of the Clay Minerals Group sub-committee on nomenclature of clay minerals. *Clay Minerals Bull.*, **2**, 294–302.

Brown, G., ed., 1961. *The X-ray Identification and Crystal Structures of Clay Minerals*. London: Mineralogical Society, 544p.

Dixon, J. B., and Weed, S. B. (eds.), 1977. *Minerals in Soil Environments*. Madison, Wisconsin: Soil Science Society of America, 948p.

Dost, H., ed., 1973. *Proceedings of the International Symposium on Acid Sulphate Soils*, vol. 1. Wageningen: International Institute of Land Reclamation and Improvement, 295p.

Farmer, V. C., ed., 1974. *The Infrared Spectra of Minerals*. London: Mineralogical Society, 539p.

Finkl, C. W., Jnr., 1980. Stratigraphic principles and practices as related to soil mantles, *Catena*, **7**, 169–194.

Gard, J. A., ed., 1971. *The Electron-Optical Investigation of Clay*. London: Mineralogical Society, 383p.

Gieseking, J. E., ed., 1975. *Soil Components*; vol. 2, *Inorganic Components*. New York: Springer-Verlag, 684p.

Gorbunov, N. I., and Gradusov, B. P., 1967. Achievements and future development of soil mineralogy, *Sov. Soil Sci.*, 1968, 1186p.

Goudie, A., 1973. *Duricrusts in Tropical and Subtropical Landscapes*. Oxford, England: Clarendon, 174p.

Grim, R. E., 1968. *Clay Mineralogy*. New York: McGraw-Hill, 596p.

Haseman, J. K., and Marshall, C. E., 1945. Use of heavy minerals in studies of the origin and development of soils, *Missouri Agricultural Experiment Station Research Bulletin 387*.

Hendrick, J., and Newlands, G., 1927. The mineralogical composition of the soil as a factor in soil classification, *1st Intl. Congr. Soil Sci., Trans.*, **5**, 104–107.

Ivarson, K. C., and Hallberg, R. O., 1976. Formation of mackinawite by microbial reduction of jarosite and its application to tidal sediments, *Geoderma*, **16**, 1–7.

Kretz, R., 1976. Physical constants of minerals, in C. Weast, ed., *Handbook of Chemistry and Physics*. Cleveland: Chemical Rubber Company, B214–219.

Logan, N. E.; Johnston, J. H.; and Childs, C. W., 1976. Mossbauer spectroscopic evidence for akaganeite (β = FeOOH) in New Zealand soils, *Australian J. Soil Res.*, **14**, 217–224.

Mackenzie, R. C., 1959. The classification and nomenclature of clay minerals, *Clay Minerals Bull.*, **4**, 52–66.

Mackenzie, R. C., 1965. Nomenclature sub-committee of CIPEA, *Clay Minerals*, **6**, 123.

Mackenzie, R. C., 1975. The classification of soil silicates and oxides, in Gieseking (1975), 1–25.

Mitchell, B. D., 1975. Oxides and hydrous oxides of silicon, in Gieseking (1975), 395–432.

Oades, J. M., 1963. The nature and distribution of iron compounds in soils, *Soils Fert.*, **26**, 69–80.

Rich, C. I., 1968. Applications of soil mineralogy in soil chemistry and fertility investigations, in Stelly and Dinauer (1968), 61–90.

Russell, J. E., 1966. *Soil Conditions and Plant Growth*. London: Longmans, 688p.

Soil Survey Staff, 1975. *Soil Taxonomy: A Basic System of Soil Classification for Making and Interpreting Soil Surveys*. U.S. Dept. Agric. Handbook no. 436, 754p.

Stelley, M., and Dinauer, R. C., eds., 1968. *Mineralogy in Soil Science and Engineering*. Madison, Wis.: Soil Science Society of America, 106p.

Thorez, J., 1975. *Phyllosilicates and Clay Minerals: A Laboratory Handbook for Their X-Ray Diffraction Analysis*. Dison, Belgium: G. Lelotte, 579p.

Van Der Plas, C., and Van Reeuwijk, L. P., 1974. From mutable compounds to soil minerals, *Geoderma*, **12**, 385–405.

Warshaw, C. M., and Roy, R., 1961. Classification and a scheme for the identification of layer silicates, *Geol. Soc. Am. Bull.*, **72**, 1455–1492.

Winterkorn, H. F., 1968. Engineering applications of soil mineralogy, in Stelly and Dinauer (1968), 35–51.

Cross-references: *Authigenic Minerals; Clays, Clay Minerals; Phyllosilicates; Rock-Forming Minerals*; see also mineral glossary. Vol. IVA: *Infrared Analysis; Thermal Analysis; X-Ray Diffraction Analysis*. Vol. VI: *Clay Minerals; Duricrust*. Vol. XII: (Pt. 1): *Hydrous Oxides; Pedosesquioxides; Mineralogy, Soil; Mineral Analysis; Clay Decomposition; Clay Formation; Clay Minerals*. Vol. XII (Pt. 2): *Ferricrete; Soil Pans; Weathering*. Vol. XIII: *Clays, Engineering Geology; Mineralogy, Applied; Soil Mechanics.*

SOLID SOLUTION—*See* Vol. IVA: SOLID SOLUTION

SPACE GROUP SYMBOL

A space group (also known as a Schoenflies group or a Federov group) is a possible combination in space of symmetry operations or elements (see *Symmetry*). In three-dimensional space, all possible combinations of symmetry are classified into 230 space groups. A space group differs from a point group (see *Point Groups*) in that symmetry operations involving translation—lattices, glide planes, and screw axes—are involved in addition to the point-group operators, axes of rotation and inversion. A space group is thus extended in space instead of leaving a point invariant.

Crystallographers use the Hermann-Mauguin notation to designate space groups as set forth in the *International Tables for X-ray Crystallography*. The first letter in the symbol designates the lattice (q.v.) as follows: P = primitive; A, B, C = end centered; I = body (interior) centered; F = all-face centered; and R = rhombohedral. Note that, with the exception of R, lattice symmetry is not designated by the first letter of the symbol.

The remainder of the space group symbol designates symmetry as follows: 1, 2, 3, 4, 6 = axes of rotation; $\bar{1}$, $\bar{2}$, $\bar{3}$, $\bar{4}$, $\bar{6}$ = axes of inversion; 2_1, 3_1, 3_2, 4_1, 4_2, 4_3, 6_1, 6_2, 6_3, 6_4, 6_5 = screw axes; m = mirror plane; and a, b, c, n, d = glide planes.

Derivation of point-group symmetry from a space-group symbol may be accomplished by dropping the lattice symbol, replacing the glide-plane notation by m, and dropping the subscripts from the screw axes. Thus, the space groups $C2/c$, $Pban$, $I\bar{4}2d$, $P6_3mc$, and $I4_132$ are contained in point groups $2/m$, mmm (= $2/m$ $2/m$ $2/m$), $\bar{4}2m$, $6mm$, and 432 and in the monoclinic, orthorhombic, tetragonal, hexagonal, and isometric crystal systems, respectively.

KEITH FRYE

References

Buerger, M. J., 1963. *Elementary Crystallography*, revised printing. New York: Wiley, 528p.

Henry, N. F. M., and Lonsdale, K., 1952. *International Tables for X-ray Crystallography*, vol. 1. Birmingham: Kynoch Press.

Megaw, H. D., 1973. *Crystal Structures: A Working Approach*. Philadelphia: Saunders, 563p.

Phillips, F. C., 1956. *An Introduction to Crystallography*, 2nd ed. London: Longmans.

Cross-references: *Crystallography: Morphological; Directions and Planes; Lattice; Point Groups; Symmetry.*

SPACE LATTICE—*See* LATTICE

SPINEL GROUP

Since the disclosure of the cubic structure of *spinel*, $MgAl_2O_4$, by Bragg in 1915, the ***spinel*** group of minerals that display the AB_2O_4 formula and an $Fd3m$ space group have become ever more important in mineralogy, as well as in many other diverse fields. To give but a few examples, the mineralogy and crystal chemistry of ***spinels*** have been utilized in extractive metallurgy (zinc, iron, and chromium ores; see *Mineral and Ore Deposits*); in electrical engineering (ferrite memory cores and switching devices; see *Metallurgy*); in refractories (slag-resistant furnace linings; see *Refractory Minerals*); in meteoritics (chemical and textural classification schemes; see *Meteoritic Minerals*); in lunar science (genesis of the early moon deduced from spinel crystallization rates and exsolutions; see *Lunar Minerals*); in geophysics (olivine to spinel phase transitions at pressures equivalent to 400 km terrestrial depth; see *Mantle Mineralogy*); in paleomagnetic studies (magnetochemical and magnetostrictive properties of the inverse spinels; see *Magnetic Minerals*); and in economic geology (petrogenesis of an ore body as deduced by equilibria studies involving spinel phases; see *Mineral Deposits: Classification*). While the simple mineralogical and crystal chemical details of the ***spinel*** group are given below, related data for ***spinels*** are also to be found in the suggested cross-references. In addition, several review articles on the properties of the ***spinel*** group are available: Deer, Howie, and Zussman (1962) reviewed the available mineralogical literature for the entire natural ***spinel*** group; Ulmer (1970) summarized the references from the preceding five to ten years for the iron, chromium, and aluminum ***spinels***; and several good bibliographic lists and review articles for ***spinels*** are to be found within the proceedings of a Carnegie Institution symposium on the geochemistry of chromium (Irving, 1975).

Descriptive Mineralogy

Table 1 presents the conventional data (after Deer, et al., 1962, and Ulmer, 1970) for the aluminum, iron, chromium, and titanium series of the ***spinel*** group. This table is limited to

TABLE 1. Properties of Natural Spinels

	n^a	D (g/cc)	a (Å)	AB_2O_4 A site (idealized)	B site
Aluminum Series					
Spinel	1.719	3.55	8.084	Mg^{2+}	Al^{3+}
Hercynite	1.835	4.40	8.153	Fe^{2+}	Al^{3+}
Gahnite	1.805	4.62	–	Zn^{2+}	Al^{3+}
Galaxite	1.92	4.04	–	Mn^{2+}	Al^{3+}
Iron (Fe^{3+}) Series					
Magnesioferrite	2.38	4.52	8.391	Fe^{3+}	Mg^{2+},Fe^{3+} [b]
Magnetite	2.42	5.20	8.397	Fe^{3+}	Fe^{2+},Fe^{3+} [b]
Maghemite	2.52–2.74	4.88	8.34	γ-Fe_2O_3 (cubic)	
Franklinite	2.36	5.34	8.420	Zn^{2+}	Fe^{3+}
Jacobsite	2.3	4.87	8.51	Fe^{3+}	Mn^{2+},Fe^{3+} [b]
Trevorite	2.3	5.2	8.43	Fe^{3+}	Ni^{2+},Fe^{3+} [b]
Chromium Series					
Picrochromite[c]	2.00	4.43	8.337	Mg^{2+}	Cr^{3+}
Chromite	2.16	5.09	8.381	Fe^{2+}	Cr^{3+}
Titanium Series					
Ulvöspinel	–	4.78	8.53	Fe^{2+}	Fe^{2+},Ti^{4+} [b]

[a]*n* = Index of refraction.
[b]Inverse ***spinels*** according to the scheme of Barth and Posnjak, 1932. See text for details.
[c](Magnesiochromite).

naturally occurring ***spinels***, but, in many cases, in order to utilize existing phase-pure materials, data were obtained on synthetic preparations of the naturally occurring ***spinels***. In addition to the minerals in Table 1, there exists an extremely large number of spinel compounds that are known only from synthetic work (for examples, see Smit and Wijn, 1959), but these synthetic ***spinels*** are not further discussed here (see *Synthetic Minerals*). Properties of the group not listed in Table 1 are as follows:

- No natural ***spinel*** displays cleavage, although a (111) parting is common and is probably related to the common (111) twinning (the "spinel law").
- Color, even within a given cation type, may be highly variable and is often the result of the redox condition of the major cations or trace elements within the spinel structure.
- Hardnesses ranging from 5 to 8 are reported, but unweathered single-crystal material always has a hardness that is nearly 8 on the Mohs scale.
- Pure end members (Table 1) are rare in nature, with considerable solid solution being observed within and(or) among the various series (Palache, 1935, for the zinc ***spinels***; Stevens, 1944, for the iron-chrome-aluminum ***spinels***).
- All the ***spinels*** in Table 1, except $MgAl_2O_4$ and $MgCr_2O_4$, can be dissolved in strong acid–some require the triprotic ionization of H_3PO_4 at elevated temperatures (Ingamells, 1960).
- Fusion-dissolution with $KHSO_4$ or $NaCO_3$ dissolves all ***spinels***, and is the only technique for ***spinels*** rich in either $MgAl_2O_4$ or $MgCr_2O_4$.
- Large single crystals (>2 to 3 mm) are extremely rare; large gem-quality natural crystals are common only from Sri Lanka (see *Gemology*).

Crystal Chemistry

Structure. As shown in Table 1, the "normal" ***spinel*** coordination may be represented by AB_2O_4 in which *A* represents the divalent-cation-tetrahedral sites and *B* represents the trivalent-cation-octahedral sites. The $Fd3m$ unit cell contains 8 such units; and within this unit cell ($A_8B_{16}O_{32}$) the 32 oxygens are cubic close packed, which provides 64 tetrahedral cation sites and 32 octahedral cation sites. Of these 96 cation sites per unit cell, only 24 are filled in this face-centered cubic cell.

In 1932, Barth and Posnjak disclosed that in addition to the "normal" arrangement of the *A* and *B* cations, an "inverse" arrangement is also possible in which the *A* cations (divalent in normal natural ***spinels***) actually occur in the octahedral sites, displacing–in the ideal limit–one half of the *B* cations (trivalent in normal natural ***spinels***) into the tetrahedral sites. In Table 1, γ-Fe_2O_3 and Fe_2TiO_4 are additionally exceptional in that the former has cation deficiency and the latter has only divalent and quadrivalent cations (the so-called 2-4 ***spinel*** common in synthetic compounds). Either normal or inverse structure is demonstrated by the perspective view of the unit cell shown in Fig. 1. Currently, data has accumulated for $MgAl_2O_4$ (Datta and Roy, 1967); for the ***spinel*** solid-solution series, $MgFe_2O_4$-$MgCr_2O_4$ (Ulmer and Smothers, 1968; Virgo and Ulmer, 1973); and for the ***spinel*** solid-solution series, Fe_3O_4-$FeCr_2O_4$ (Wasilewski et al, 1975) to support the idea that normal or inverse arrangements in natural ***spinels*** are ideal limits and that the coordination may best be represented by

$$(B_x A_{1-x})_{\text{tetrahedral}}(A_x B_{2-x})_{\text{octahedral}}O_4$$

where either in a single compound (as a function of temperature) or in a solid solution (as an isothermal function of composition), x may have any value between 0 and 1. With an x value of zero, an ideal normal ***spinel*** results; and with an x value of one, an ideal inverse ***spinel*** results.

For any ***spinel*** solid solution between a normal and inverse ***spinel***, the coordination inversion may not be Vegardian, i.e., the change in coordination may not be linear and may not even be a smooth function of changes in composition along the solid solution. This non-Vegardian behavior manifests itself for example as nonlinear lattice parameters in some binary ***spinel*** compositional joins or as high temperature (> 600°C) exsolutions in other binary ***spinel*** compositional joins.

In Fig. 1B the roles of inverse and normal coordinations on natural ***spinel*** solid solutions are summarized. The behavior of the more complex ***spinel*** solid solutions within the body of the triangular prism has received only limited study but enough is known to speculate that the exsolutions observed on joins c and d degenerate to solid solutions with non-Vegardian physical properties as one decreases the amount of aluminum cation and(or) increases the amount of chromium cation within the complex solid solutions in the body of the triangular prism (See exsolution section).

There are 72 unused cation sites per unit cell; order-disorder (q.v.) is, therefore, a definite theoretical possibility. Despite a search for such evidence as super-lattice lines

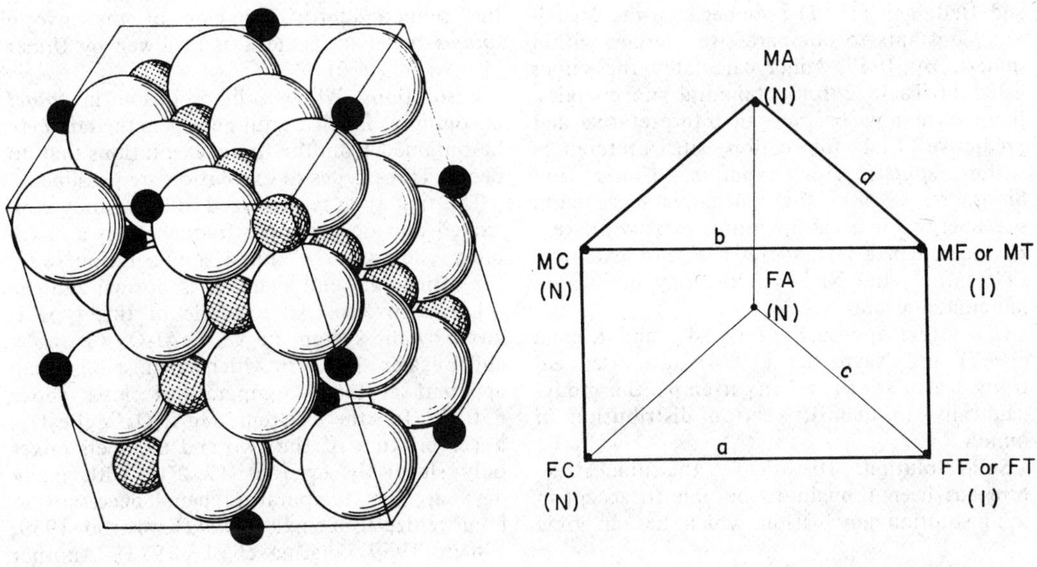

FIGURE 1. Crystal chemistry of ***spinels.*** A. Perspective of the *spinel* structure (from Verwey and Heilman, 1947). Large spheres represent oxygen, shaded spheres represent octahedral (*B*) cation sites, and dark spheres represent tetrahedral (*A*) cation sites. B. Summary of solid solution behavior as a function of composition as reviewed by Ulmer (unpublished). MF = *Magnesioferrite;* FF = *Magnetite;* MT = Magnesiotitanate (The magnesium analog of *ulvöspinel* has been used repeatedly to describe spinel compositions for example in terrestrial kimberlites and in lunar rocks; however, the IMA official position is that this end member composition is not a mineral.); FT = *Ulvöspinel;* MA = *Spinel;* FA = *Hercynite;* MC = *Picrochromite;* FC = *Chromite;* N = normal coordination (see text); I = inverse coordination (see text). For the ferrite or titanate joins c and d, the literature reports high temperature sub-solidus exsolutions, whereas for the ferrite or titanate joins a and b, the literature reports non-Vegardian behavior in the lattice parameters but no exsolutions. All other joins in the figure have complete solid solutions, for example at 1000° or higher.

in diffraction patterns (Allen, 1964), or absorption shifts in near and far infrared (Ulmer and Smothers, 1968; Preudhomme and Tarte, 1971a, 1971b, 1971c, 1972), or Mossbauer coordination studies (for example, Virgo and Ulmer, 1973), no ordering on these "extra" cation sites has been found to be thermodynamically reproducible (i.e., not reversible) in the natural ***spinels*** nor in annealed synthetic preparations of the ***spinels*** listed in Table 1. Studies of the Curie Point behavior of inverse $MgFe_2O_4$ have been used to argue for cation ordering in this ***spinel*** (Walters and Wirtz, 1971), but again this hysteresis was nonreproducible.

Site Preference Calculations. Because of the complexity of possible cation positioning within the ***spinel*** unit cell, Verwey and Heilmann (1947) attempted to quantify the spatial relationships by defining the oxygen parameter, μ, as being $\frac{1}{4}+\delta$, where δ is the distance from the center to the corner of a cation-filled tetrahedron divided by the body diagonal of the unit cell. When perfect cubic close-packing results, $\mu = 0.375$; but most ***spinels*** show values of $\mu > 0.375$, including tetrahedral distortions caused by imperfectly fitting cations. Verwey and Heilmann (1947) also began using Madelung Constants to compare site energies within ***spinels***. By 1959, Miller calculated the values listed in Table 2 for octahedral site energies; these have now become an interpretative and predictive tool for cation site preferences within ***spinels*** (for example, Ulmer and Smothers, 1968). High negative octahedral site energy for a cation indicates it will likely be located in a tetrahedral site. For example, Cr^{3+}, Mn^{3+}, and Ni^{2+} would likely be located in octahedral sites.

In another approach, Navrotsky and Kleppa (1967) and Navrotsky (1974) have used entropy calculations in an attempt thermodynamically to quantify cation distribution in ***spinels***.

Solid Solution. Historically, the spinel structure has been thought to be able to accept in solid solution any cation which has an ionic radius between 0.44 Å and 1.00 Å (Goldschmidt values). ***Olivine***, Mg_2SiO_4, presumably does not have spinel structure (except at mantle pressures) because Si^{4+} has a radius of 0.39 Å; and dicalcium silicate, Ca_2SiO_4, has neither olivine nor spinel structure, since Ca^{2+} has a radius of 1.06 Å. However, larger polarizable cations such as Cd^{2+}, 1.06 Å, have been incorporated in synthetic ***spinels*** at one atmosphere pressure (Muller et al., 1969). The pervasiveness of natural ***spinel*** solid solutions has even prohibited the use of certain *magnetite* as iron ore because, for example, the 5 to 15 wt% of TiO_2 in solid solution in the *magnetite* causes chemical problems in furnace slag. In some cases, the pervasiveness of solid solution in ***spinel*** is beneficial as exemplified by *magnetite* in the Bushveld Complex where the 0.5 to 7.0 wt% of V_2O_5 in solid solution is economically worth more than the iron (Willemse, 1969). In these natural ***spinel*** solid solutions, charge balance is usually maintained by balanced substitutions, but defect ***spinels*** are known with missing cations (in the extreme case, γ-Fe_2O_3; see *Crystals, Defects in*). Delocalization of electrons between di- and trivalent sites has also been documented extensively by the semiconductor behavior of the inverse ***spinels*** listed in Table 1 as reviewed by Ulmer (1969 and 1970).

TABLE 2. Octahedral Site Preference Energies for Various Cations

Zn^{2+}	-31.6	Ti^{3+}	-21.9
Mn^{2+}	-14.7	Fe^{3+}	-13.3
Fe^{2+}	- 9.9	Al^{3+}	- 2.5
Mg^{2+}	- 5.0	Mn^{3+}	+ 3.1
Ni^{2+}	+ 9.0	Cr^{3+}	+16.6

After Miller, 1959.

Values are in Kcal/gram-atomic-weight. High positive values indicate octahedral coordination stability.

Exsolution. While solid solution in ***spinel*** is common, much useful petrogenetic data may be obtained from the ***spinel*** exsolutions that do occur. Three types of exsolution are possible.

The first type is theorized to be entropy controlled and occurs most frequently as a miscibility gap in those ***spinel*** solid-solution series for which one end member is normal and the other is inverse. An example of this type is given by the system Fe_3O_4-$FeAl_2O_4$ (Turnock and Eugster, 1962) in which the miscibility gap apparent at lower temperatures closes above 850°C. In the system $MgAl_2O_4$-$MgFe_2O_4$, a 1:1 mixture of the two end members mixes only sluggishly even at 1250°C, with many days at that temperature being necessary to homogenize to a single ***spinel*** (Kwestroo, 1959; Ulmer, 1969; Sharma et al., 1973). Another example of immiscibility is given by Muan et al. (1972) for the ***spinel*** series $MgAl_2O_4$-Mg_2TiO_4, which shows exsolution even at 1300°C whereas their study of $MgCr_2O_4$-Mg_2TiO_4 does not show exsolution even at a temperature as low as 1000°C. While Cremer (1969) suggests that Fe_3O_4 and $FeCr_2O_4$ exsolve below 900°C and that $FeCr_2O_4$ and $FeAl_2O_4$ also exsolve below 950°C, attainment of equilibria in this study is highly suspect.

A second type of exsolution is related to nonstoichiometry; in general, high temperature favors a few weight percent of solid solution in excess of the exact ***spinel*** stoichiometry. For example, at 1300°C, *ulvöspinel* (Fe_2TiO_4) may have a small excess of iron and titanium in solid solution which exsolves as the ***spinel*** cools (Taylor, 1964):

Nonstoichiometric *ulvöspinel* solid solution	$\xrightarrow{\text{cooling}}$	Fe_2TiO_4 stoichio-metric	$+ FeTiO_3$ lamellae

And finally, a third type of exsolution can result from redox reactions; this is in reality a subsolidus oxidation or reduction. For example, stoichiometric *ulvöspinel* can easily oxidize upon cooling to produce an assemblage of *ilmenite-hematite* solid solution and *ulvöspinel* enriched in *magnetite*. The *ilmenite-hematite* appears to be exsolved as laths within the ***spinel***. Natural cooling of lunar basalts took place in a more reduced environment than the cooling of terrestrial basalts; therefore, associated ***spinels*** in lunar melts could be reduced. Interlocking textures of *iron* plus ***spinel*** plus either sesquioxide or *ilmenite*, all in a ***spinel*** matrix are, therefore, inferred to be reduction-exsolution assemblages for lunar ***spinels*** (Haggerty, 1971).

Geologic Occurrences

Many of the ***spinels*** listed in Table 1 are of very limited occurrence in nature; on the other hand, the natural ***spinels*** containing zinc, chrome, iron, and titanium are abundant enough to be considered important sources for these metals. For igneous ***spinels***, the electron microprobe has provided many data that indicate that, at least within any given igneous complex, the spinel chemistry changes from Cr- to Ti- to V- to Fe-rich compositions as bulk SiO_2 increases in a differentiating magma.

Economic deposits of the zinc-rich ***spinels*** are associated with *willemite*, Zn_2SiO_4, and occur in metamorphic belts, such as the world-famous Franklin Furnace and Sterling Hill localities in New Jersey (e.g., Palache, 1935).

Sizable deposits of chrome-rich ***spinels*** are to be found as cumulate stratiform layers in igneous complexes, such as the Bushveld in the Republic of South Africa, the Great Dike in Zambia and Zimbabwe, or the Stillwater in Montana. Large lenticular masses of podiform *chromite* are commonly associated with Alpine peridotites and serpentinized belts. Such deposits are known, for example, in Cuba, the Philippines, and the eastern Mediterranean. Thayer and Ulmer (1976) have compiled known reserve estimates and geologic-geographic data for the chrome-rich ***spinels***.

The iron-rich ***spinels*** are ubiquitous. They occur in igneous massifs as cumulate strata (e.g., the Bushveld of South Africa), in metamorphic terrain (e.g., Magnet Cove, Arkansas, or the Mesabi Range in the Great Lakes Precambrian of North America), and in sedimentary deposits either as heavy mineral resistates (e.g., South Carolina long-shore-drifted beach sands) or in association with lateritic-type deposits (e.g., Venezuelan coastal deposits).

Titaniferous ***spinels***, such as the ulvöspinel-containing stratiform *magnetite* of the South African Bushveld Complex or the podiform *ilmenite*-titanian *magnetite* deposit associated with metagabbro in the Adirondack Mountains of New York (Tarawas), are typical of the larger terrestrial occurrences. The overall higher percentages of titanium in lunar rocks is at least in part due to the disseminated lunar opaques rich in *ilmenite* and in the ***spinels*** Mg_2TiO_4 and Fe_2TiO_4 (see *Lunar Mineralogy*).

Other Properties

Beyond the scope of classical mineralogy, a huge body of data exists for the natural ***spinels*** owing to the extensive ***spinel*** researches available from many diverse technologies. Summary tables of infrared spectral bands, thermodynamic properties, electrical conductivities, specific heats, pertinent phase equilibria, and cation diffusion coefficients have been compiled for the iron-chrome-aluminum ***spinels*** (Ulmer, 1970). The magnetic properties of many ***spinels*** have also been tabulated by Schieber (1967) and by Wasilewski et al. (1975).

GENE C. ULMER

References

Allen, W. C., 1964. *Solid Solution in the Refractory Magnesium Spinels*. Ph.D. thesis, School of Ceramics, Rutgers, The State University, New Brunswick, N. J.

Barth, T. F. W., and Posnjak, E., 1932. Spinel structure: With and without variate atom equipoints, *Zeit. Kristallogr.*, **82**, 325-341.

Bragg, W. H., 1915. The structure of the spinel group of crystals. *Philos. Mag.*, **30**, 305-315.

Cremer, V., 1969. Mixed crystal formation in the system chromite-magnetite-hercynite between 1000° and 500°C, *Neues Jahrb. Mineral. Abh.*, **111**, 184-205.

Datta, R. K., and Roy, R., 1967. Equilibrium order-disorder in spinels, *J. Am. Ceram. Soc.*, **50**, 578-583.

Deer, W. A.; Howie, R. A.; and Zussman, J., 1962, *Rock-Forming Minerals*, vol. 5, *The Non-silicates*. Longmans: London, 56-88.

Haggerty, S. E., 1971. Subsolidus reduction of lunar spinels, *Nature (London), Phys. Sci.*, **234**, 113-117.
Ingamells, C. O., 1960. A new method for ferrous iron and excess oxygen in rocks, minerals, and oxides, *Talanta*, **4**, 268-273.
Irving, N., 1975. Chromium: Its physicochemical behavior and petrologic significance, *Geochim. Cosmochim. Acta*, **39**, 779-1078.
Kwestroo, W., 1959. Spinel phase in the system $MgO-Fe_2O_3-Al_2O_3$, *J. Inorg. Nucl. Chem.*, **9**, 65-70.
Miller, A., 1959. Distribution of cations in spinels, *J. Appl. Phys.*, **30**, 24S-25S.
Muan, A.; Hauck, J.; and Löfall, T., 1972. Equilibrium studies with a bearing on lunar rocks. Proc. 3rd Lunar Sci. Conf., *Geochim. Cosmochim. Acta Suppl. 3*, vol. 1, 185-196.
Muller, O.; White, W. B.; and Roy, R., 1969. Infrared spectra of the chromates of magnesium, nickel, and cadmium, *Spectrochim. Acta*, **25A**, 1491-1499.
Navrotsky, A., 1974. Thermodynamics of binary and ternary transition metal oxides in the solid state, in D. W. A. Sharp, ed., *MTP International Reviews of Science*, Inorganic Chemistry Ser. 2, 5. London: Butterworth, 29-70.
Navrotsky, A., and Kleppa, O. J., 1967. The thermodynamics of cation distributions in simple spinels, *J. Inorg. Nucl. Chem.*, **29**, 2701-2714.
Palache, C., 1935. The minerals of Franklin and Sterling Hill, Sussex County, New Jersey. *U.S. Geol. Surv. Prof. Paper 180*.
Preudhomme, J., and Tarte, P., 1971a. Infrared studies of spinels–I. A critical discussion of actual interpretations, *Spectrochim. Acta*, **27A**, 961-968.
Preudhomme, J., and Tarte, P., 1971b. Infrared studies of spinels–II. The experimental bases for solving the assignment problem, *Spectrochim. Acta*, **27A**, 845-851.
Preudhomme, J., and Tarte, P., 1971c. Infrared studies of spinels–III. The normal II-III spinels, *Spectrochim. Acta*, **27A**, 1817-1835.
Preudhomme, J., and Tarte, P., 1972. Infrared studies of spinels–IV. Normal spinels with a high-valency tetrahedral cation, *Spectrochim. Acta*, **28A**, 69-79.
Schieber, M., 1967. *Experimental Magnetochemistry*. New York: Wiley/Interscience, 572 p.
Sharma, K. K.; Langer, K.; and Seifert, F., 1973. Some properties of spinel phases in the binary system $MgAl_2O_4-MgFe_2O_4$, *Neues Jahrb. Mineral., Monatsh.*, **10**, 442-449.
Smit, J., and Wijn, H. P. J., 1959. *Ferrites*. New York: Wiley.
Stevens, R. E., 1944. Composition of some chromites of the Western Hemisphere, *Am. Mineralogist*, **29**, 1-34.
Taylor, R. W., 1964. Phase equilibria in the system $FeO-Fe_2O_3-TiO_2$ at 1300°C., *Am. Mineralogist*, **49**, 1016-1031.
Thayer, T. P. and Ulmer, G. C., 1976. Some problems of chromite Supply by 1985. *Bull. Am. Ceram. Soc.*, **55**(4), 443.
Turnock, A. C., and Eugster, H. P., 1962. Fe-Al oxides: Phase relationships below 1000°C., *J. Petrology*, **3**, 533-565.
Ulmer, G. C., 1969. Experimental investigations on chromite spinels, *Soc. Econ. Geol. Symp. Monograph 4*. Stanford University, 114-131.
Ulmer, G. C., 1970. Chromite spinels, in A. Alper, ed., *High Temperature Oxides*, pt 1: *Magnesia, Lime, and Chrome Refractories*, New York: Academic Press, 251-314.
Ulmer, G. C., and Smothers, W. J., 1968. The system $MgO-Cr_2O_3-Fe_2O_3$ at 1300°C in air, *J. Am. Ceram. Soc.*, **51**, 315-319.
Verwey, E. J. W., and Heilmann, E. L., 1947. Physical properties and cation arrangements of oxides with spinel structures. I. Cation arrangements in spinels, *J. Chem. Phys.*, **15**, 174-180.
Virgo, D., and Ulmer, G. C., 1973. The Fe^{3+} site preference in the solid solution series $MgCr_2O_4-MgFe_2O_4$, *Carnegie Institution Yearbook*, **72**, 567-569.
Walters, D. S., and Wirtz, G. P., 1971. Kinetics of cation ordering in magnesium ferrite, *J. Am. Ceram. Soc.*, **54**, 563-566.
Wasilewski, P., et al., 1975. Magnetochemical characterization of Fe (Fe_xCr_{2-x}) O_4 spinels. *Geochim. Cosmochim. Acta*, **39**, 889-902.
Willemse, J., 1969. The geology of the Bushveld Igneous Complex, the largest repository of magmatic ore deposits in the world, *Soc. Econ. Geol. Symp. Monograph* 4. Stanford University, 1-22.

Cross-references: *Crystal Growth; Lunar Minerals; Magnetic Minerals; Mantle Mineralogy; Meteoritic Minerals; Mineral Classification: Principles; Order-Disorder; Refractory Minerals*; see also mineral glossary.

STAINING TECHNIQUES

In the fields of prospecting, exploration, mining, commercial extraction and concentration plants, quality control, and laboratory research, investigators are frequently confronted with problems of distinguishing rapidly between mineral mixtures in grain or rock forms. Visual identifications if reliable are preferred because of their speed and low cost.

Mineral identifications based on the production of differently colored surface stains (coatings) have been gradually developed, and now represent a separate field of mineral identification. Although a great number of mineral-specific staining methods have been described, their use is not as widespread as their obvious applications would indicate (see below).

Differential staining provides a means of rapid identification for: (1) mineral species that appear essentially the same and (2) minerals that can be distinguished but by a process that is difficult and time consuming. With few exceptions, differential staining techniques have been developed to solve particular identification problems. Thus the published work is essentially composed of a series of individual stains specifically developed to differentiate between the constituents of two main types of mineral mixtures: (1) a mixture in which one mineral is present with a limited number of other minerals, i.e., *witherite* in *witherite-barite-fluorite-quartz* mixtures; and (2) a mixture containing mineral species of similar appearance—the minerals rarely number more than two or three, i.e., *calcite* and *aragonite* or *calcite, dolomite*, and *gypsum*.

The development of staining tests for specific identification problems has led to a piecemeal approach and coverage and the widespread distribution of the published results. This fragmentation would appear to be mainly responsible for the surprising under-application of staining techniques in general.

Although staining techniques have found their main applications in mineral identification, these colored effects are increasingly being used as aids in the textural and structural studies of rocks, minerals (including zoning), and fossils; in diagenesis, paragenesis, and percentage evaluations—particularly in conjunction with point counting; and to increase the visual contrast in sedimentary structures and photomicroscopy.

It should always be borne in mind that individual staining procedures were often developed and tested on relatively pure end members of mineralogical series. Away from these pure examples, the stains may give anomalous results which, although to be expected, cast strong but usually unfounded doubts on the validity of the original staining procedures. However, once recognized and understood, such apparently abnormal results may prove of considerable value in evaluating more correctly the real composition of these minerals. The lack of published work in this direction represents a serious gap in our knowledge. It is hoped therefore that the valuable contributions of Walger (1961, on isomorphous series), Warne (1962, on *ankerites, dolomites*, and *siderites*) and Evamy (1963, on the ferrous iron contents of *calcite* and *dolomite*) will stimulate further studies in this area.

The value of differential staining methods lies in their speed and ease of application, together with the bare minimum of cheap simple equipment, reagents and operative skill required for their effective application. Generally, they are easily applied in the field, office, or laboratory to freshly broken hand specimens, drilling chips, uncemented fragments, loose grains, ground and polished surfaces, in many cases uncovered thin sections, and even to cellulose and acetate peels (Davies and Till, 1968).

The number of published stains for specific minerals groups them in their probable order of importance, i.e., carbonates, ***clays, feldspars***, sulfates, detrital, and other minerals. These are the groupings used herein.

Carbonates: Evamy (1963), Friedman (1959), Hügi (1945), Walger (1961), Warne (1962), and Wolf et al. (1967)

Clay minerals: Grim (1953), Hambleton and Dodd (1953), and Dodd (1955)

Feldspars: Bailey and Stevens (1960), Broch (1961), Hayes and Klugman (1959), and Norman (1974)

Sulfates and detrital and other minerals: Hosking (1964), Warne (1962), Reid (1969), and Norman (1974)

Carbonates

The anhydrous carbonate minerals are notoriously difficult to differentiate and identify by the usual macro- and microscopic methods employed. This factor has greatly stimulated work on their identification by various staining techniques that, since the classic work of Lemberg in 1887, have been published in considerable numbers.

This difficulty has resulted in five major reviews of the subject (Rodgers, 1940; Hügi, 1945; Le Roy, 1950; Friedman, 1959; and Wolf et al., 1967). Furthermore, it has led to the formulation of at least four comprehensive staining schemes. These use several staining reagents in turn, and, by a systematic process of elimination, the identification of a specific carbonate from a group of other carbonates may be made.

The scheme of Hügi (1945) positively identifies *calcite, aragonite, dolomite, ankerite, siderite*, and *magnesite* by using only five reagents. Friedman (1959) formulated two schemes for the identification of *aragonite, calcite*, high-Mg *calcite, anhydrite, dolomite*, and *magnesite* plus *gypsum*. The first (Fig. 1) utilized only two reagents; alizarin red S in solutions of various degrees of acidity and alkalinity, together with the specific stain for *aragonite* of Feigl. Friedman's second scheme is shown in Fig. 2. By utilizing Friedman's alizarin red S scheme together with three new staining reagents, Warne (1962) produced a much enlarged version whereby all the minerals covered by Friedman plus *witherite, siderite, rhodochrosite, smithsonite, ankerite, strontianite*, and *cerussite* could be specifically identified (Fig. 3).

Friedman's work (1959) is of additional importance because of the considerable number of staining reagents discussed, their reliability and application to carbonate-rock analysis having been evaluated by an independent laboratory study. Over forty staining agents for detailed coverage, including data on reagent preparation are selected. Complementary to Friedman's publication is the detailed tabulation by Wolf et al. (1967), which lists chemi-

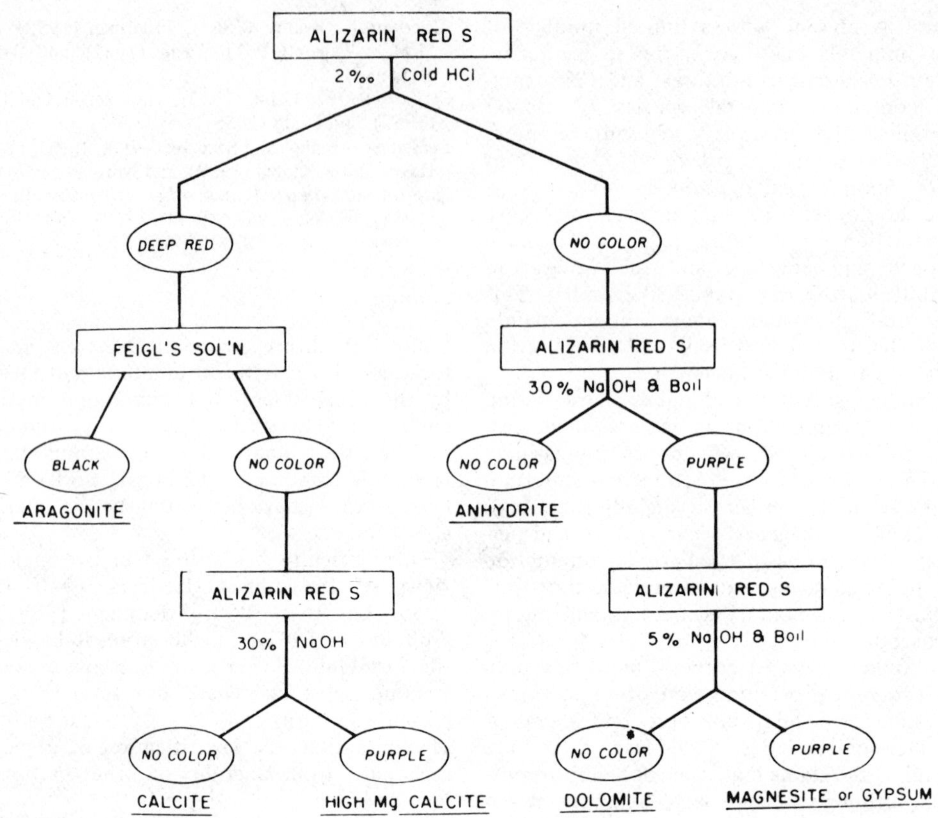

FIGURE 1. Staining procedure utilizing alizarin red S and Feigl's solutions (from G. M. Friedman, 1959. Identification of carbonate minerals by staining methods, *J. Sed. Petrology*, **29**, 89. Reprinted with permission of the Society of Economic Paleontologists and Mineralogists).

cals, preparation and methods, results, and remarks. Valuable contributions, which should serve as models for others, have been made by Choquette and Trusell (1978; stain permanency), Winland (1971; detection of magnesian *calcite*), and Lindholm (1974; detection of ferroan *calcite*).

Clays

To date, dyestuffs and other reagents, which, when adsorbed on ***clay*** particles, exhibit characteristic colors, have been widely employed for the identification of ***clays***. The characteristic colors produced have been attributed to pleochroic effects and oxidation-reduction or acid-base mechanisms.

The evaluation of aromatic diamines for the identification of ***smectites*** culminated with the development of a qualitative test using benzidine. Improved multistaining procedures use a combination of malachite green, safranine "Y," and benzidine. The effects of these stains on *kaolinite, halloysite, dickite, nacrite, montmorillonite, nontronite, hectorite,* ***illite***, *palygorskite*, and *pyrophyllite* have been very clearly set out in tabular form by Grim (1953). Subsequently, a method whereby alcoholic solutions of *p*-aminophenol in varying concentrations were followed by treatment with hydrochloric acid was put forward by Hambleton and Dodd (1953). These authors indicated that this test was most sensitive for ***smectites*** and a little less so for ***illites*** and hydrous ***micas***, while, although *kaolinites* were identified with ease, confirmatory aniline dye tests often proved useful.

The whole subject was further critically evaluated and the major methods and mechanisms of staining were discussed by Dodd (1955) and Skawinski (1965), while Yarin et al. (1976) clarified the "mechanism of staining *montmorillonite* by benzidine."

Feldspars

Initially the staining of ***feldspar*** minerals was stimulated by the necessity of differentiating

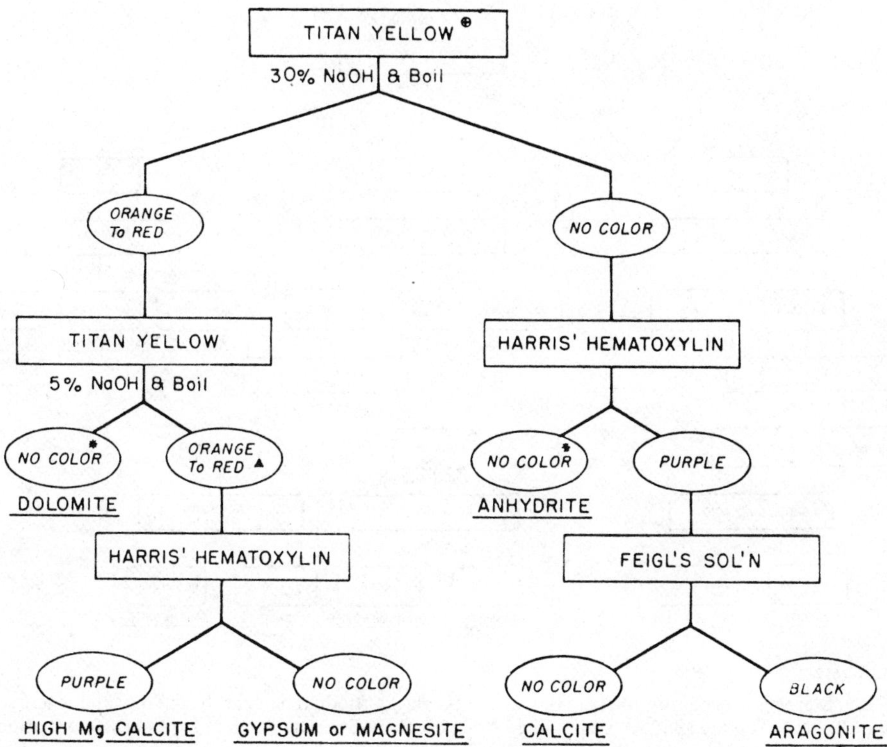

⊕*Dolomite* has a wide range of colors, [other] stains may be substituted for titan yellow to provide color contrast between dye and *dolomite* tested.
*Or faint stain (light orange).
▲High-magnesium *calcite* used in this study was very fine grained. The behavior of coarse-grained high-magnesium *calcite* was not studied.

FIGURE 2. Staining procedure utilizing titan yellow, Harris' hematoxylin, and Feigl's solutions (from G. M. Friedman, 1959. Identification of carbonate minerals by staining methods, *J. Sed. Petrology*, **29**, p. 90. Reprinted with permission of the Society of Economic Paleontologists and Mineralogists).

rapidly among potash ***feldspars, plagioclase feldspars***, and *quartz*. However, subsequent developments have been directed toward textural and accurate petrographic analyses.

Potash Feldspars. The most widely used and reliable stain for the identification of potash ***feldspars*** is based directly on a method originally described by Gabriel and Cox (1929). Subsequent valuable, but minor, modifications to the actual application techniques have followed (see Broch, 1961). The result is a very reliable, distinctive, and simply applied staining technique.

Essentially, the method involves lightly etching the surface to be stained (uncovered thin section, faced slab, grain mount, or loose grains) with hydrofluoric acid (HF), vapor or liquid. This is followed, after careful washing in distilled water and drying, by a short period of immersion in a concentrated solution of sodium cobaltinitrite.

The product of this treatment is an intense yellow surface stain on the potash ***feldspars***, while ***plagioclase***, *quartz*, and ***apatite*** remain completely unstained, although ***plagioclase*** assumes a whitish opacity. Under these conditions, ***clay*** minerals sometimes become stained while the ***micas***, if weathered, assume a greenish-yellow coloration.

Clearly, it is the etch residues remaining after the HF treatment that are stained and not the actual minerals. Thus, at the ***feldspar*** surface, silicon is removed as a volatile fluoride, while the other constituent elements are left exposed in a thin gelatinous surface film of aluminum fluorosilicate. In this way, the residual potassium from the potash ***feldspars*** reacts with the sodium cobaltinitrite to form the yellow potassium cobaltinitrite stain.

Plagioclase Feldspars. If required, ***plagioclase,*** after the HF etching and the sodium cobaltinitrite treatment for potash ***feldspars***, may also

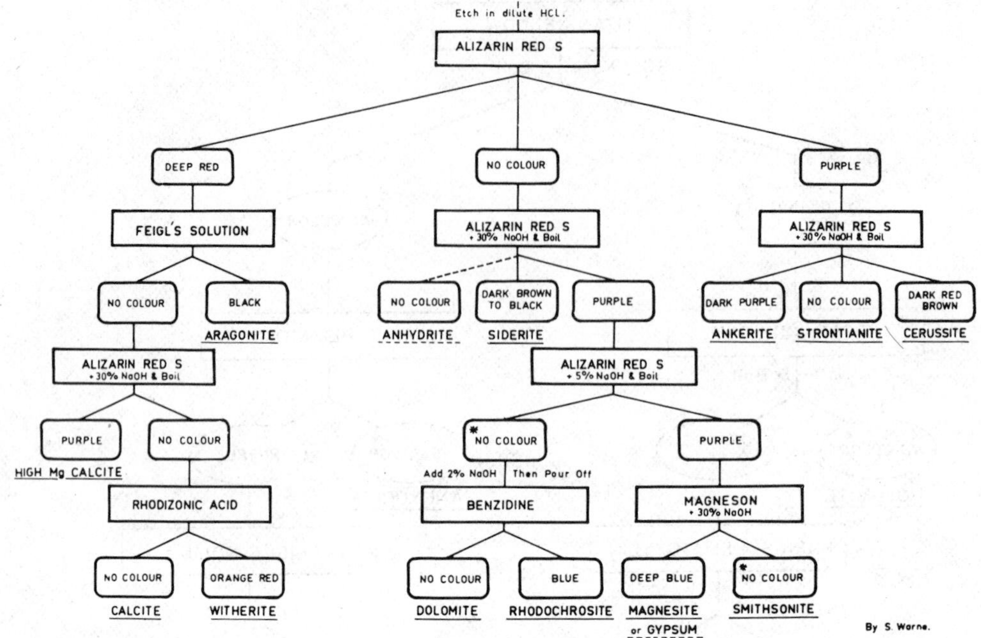

*Or faint stain.

FIGURE 3. Staining procedure utilizing alizarin red S, Feigl's rhodizonic acid, benzidine, and magneson solutions (from S. St. J. Warne, 1962. A Quick Field or Laboratory Staining Scheme for the Differentiation of the Major Carbonate Minerals, *J. Sed. Petrology*, 32, 30. Reprinted with permission of the Society of Economic Paleontologists and Mineralogists).

be stained by immersion in solutions of water-soluble dyes such as eosine B or malachite green.

Alternatively, the method of Bailey and Stevens (1960) yields excellent results. After preparative HF etching, the specimen is dipped into a 5% solution of barium chloride, rinsed, and covered with a saturated solution of potassium rhodizonate. When the ***plagioclase*** has assumed a pink coloration, the slide is carefully washed and allowed to dry.

After this treatment, ***plagioclase*** ($An \geqslant 3$, but not pure albite) exhibits a bright red surface stain of barium rhodizonate.

Although the potash ***feldspars*** may previously have been stained yellow, none of the staining procedures described for ***plagioclase***, if applied subsequently alters in any way the yellow potassium cobaltinitrite stain. The resultant color contrast between the yellow- and red-stained potash and ***plagioclase feldspars*** is particularly well marked.

The advantages of such a two color staining method in relation to petrographic analyses, textures, distribution and association and even zoning of ***feldspars*** are quite clear.

An alternative method for staining ***plagioclase*** including *albite* using amaranth, has been described by Norman (1974).

Anorthoclase Feldspar. *Anorthoclase feldspar* may be stained by amaranth using the method of Ford and Boudette (1968).

Feldspar Minerals Collectively. A general method has been recommended in many sedimentology text books for the differentiation of all ***feldspars*** particularly from *quartz*. The intensity of the stain retained reaches a maximum on *anorthite*, becomes progressively lighter on the less calcic and is weakest on the potassium ***feldspars***. Suitable dyes are fuchsine, methylene blue, safranine, malachite green, and eosine. However, this method has attracted strong criticism on the grounds of poor reproducibility from Hayes and Klugman (1959).

Sulfates

Gypsum and Anhydrite. By application of the staining schemes of Friedman (1959) and Warne (1962), *gypsum* may be identified together with, but not separated from, *magnesite*, while *anhydrite* is indicated only by its failure to stain at a particular stage in the proceedings (Figs. 1 to 3).

Immersion in cold methanol solutions of rhodamine B base, barium eosinate, titan yellow, or alizarin cyanine green provide surface stains of purplish red, pinkish red, orange, and bluish green, respectively, which are specific for *gypsum*. In addition, Megnien (1957, see in Friedman, 1959) described the use of an acidified mercuric nitrate solution that stains *gypsum* yellow.

Except by processes of elimination (see Figs. 1 to 3) no effective stains appear to have been developed specifically for *anhydrite*.

Barite. *Barite* may be successfully stained after the initial development of a suitable reactive coating, which is then stained. The sample is boiled for a few minutes with excess sodium carbonate, which causes the surface of the *barite* ($BaSO_4$) to change to barium carbonate. Washing with water is followed by immersion in a warm solution of potassium chromate; this changes the previously formed surface layer of barium carbonate to yellow barium chromate.

Detrital and Other Minerals

Bauxite. Gaudin (1935, see in Hosking, 1964) recommended the direct use of a solution of malachite green for staining hydrous oxides, particularly bauxite and limonite, and noted that "the action seems to be specific for minerals with hydroxyl radicals."

Other Minerals. The differential staining of a large number of other minerals may, on occasion, be of value. Space precludes giving details, but staining procedures for the identification of *amblygonite, anglesite, arsenopyrite, beryl, cassiterite, columbite, fluorite, galena, pyrite, sphalerite, spodumene, scheelite* (and other tungstates), and *thorite* are included, together with those for many other minerals, in extremely informative reviews by Hosking (1964) and Reid (1969). More recently, Norman (1974) has applied the reagent amaranth to the selective staining of a wide range of minerals.

S. ST. J. WARNE

References

Bailey, E. H., and Stevens, R. E., 1960. Selective staining of K-feldspar and plagioclase on rock slabs and thin sections, *Am. Mineralogist*, **45**, 1020–1025.

Broch, O. A., 1961. Quick identification of potash felspar, plagioclase and quartz for quantitative thin section analysis, *Am. Mineralogist*, **46**, 752–753.

Choquette, P. W., and Trusell, F. C., 1978. A procedure for making the titan-yellow stain for Mg-calcite permanent, *J. Sed. Petrology*, **48**, 639–641.

Davies, P. J., and Till, R., 1968. Stained dry cellulose peels of ancient and recent impregnated sediments. *J. Sed. Petrology*, **38**, 234–237.

Dodd, C. G., 1955. Dye adsorption as a method of identifying clays. Proc. 1st. Natl. Conf. Clays and Clay Min., *Calif. Div. Mines., Bull.*, **169**, 105–111.

Evamy, B. D., 1963. The application of a chemical staining technique to a study of dedolomitisation, *Sedimentology*, **2**, 164–170.

Ford, A. B., and Doudette, E. L., 1968. On the staining of anorthoclase, *Am. Mineralogist*, **53**, 331–334.

Friedman, G. M., 1959. Identification of carbonate minerals by staining methods, *J. Sed. Petrology*, **29**, 87–97.

Gabriel, A., and Cox, E. P., 1929. A staining method for the quantitative determination of certain rock forming minerals, *J. Mineral. Soc. Amer.*, **14**, 290–292.

Grim, R. E., 1953. *Clay Mineralogy*. London: McGraw-Hill, 384p.

Hambleton, W. W., and Dodd, C. G., 1953. A qualitative color test for rapid identification of the clay mineral groups. *Econ. Geol.*, **48**, 139–146.

Hayes, J. R., and Klugman, M. A., 1959. Feldspar staining methods. *J. Sed. Petrology*, **29**, 227–232.

Hosking, K. F. G., 1964. Rapid identification of mineral grains in composite samples. *Mineral. Mag.*, **110**(1), 30–37; (2), 111–121.

Hügi, T., 1945. Gesteinsbildend wichtige Karbonate und deren Nachweis mittels Färbmethoden: *Schweiz. Mineral. Petrogr. Mitt.*, **25**, 114–140.

Le Roy, L. W., 1950. Stain analysis. In *Subsurface Geologic Methods*. Golden, Colorado: Colorado School of Mines, 193–199.

Lindholm, R. C., 1974. Fabric and chemistry of pore filling calcite in septarian veins: Models for limestone cementation. *J. Sed. Petrology*, **44**, 428–440.

Norman, M. B., 1974. Improved techniques for selective staining of feldspar and other minerals using amaranth. *J. Res. U.S. Geol. Surv.*, **2**, 73–79.

Reid, W. P., 1969. Mineral staining tests. *Colorado School Mines, Mineral Indust. Bull.*, **12**, 1–20.

Skawinski, R., 1965. Investigation of factors bearing on staining of clay minerals. *Bull. Acad. Polou. Sci. Ser. Sci. Geol. Geogra.*, **13**, 39–42, 43–48.

Walger, E., 1961. Zur mikroskopischen Bestimmung der gesteinsbildenden Karbonate im Dünnschliff. *Neues Jahrb. Mineral. Monatsh.*, **8**, 182–187.

Warne, S. St. J., 1962. A quick field or laboratory staining scheme for the differentiation of the major carbonate minerals. *J. Sed. Petrology*, **32**, 29–38.

Wolf, K. H., Easton, A. J., and Warne, S. St. J., 1967. Techniques of examining and analyzing carbonate skeletons, minerals and rocks, in *Carbonate Rocks*, pt. B. Amsterdam: Elsevier, 253–341.

Winland, H. D., 1971. Nonskeletal deposition of high-Mg calcite in the marine environment and its role in the relation of textures, in *Carbonate Cements*. Baltimore: John Hopkins Univ., Studies in Geol. No. 19, 278–284.

Yarin, S., Lahav, N., and Lacher, M., 1976. On the mechanism of staining montmorillonite by benzidine. *Clays, Clay Minerals*, **24**, 51–52.

Cross-references: *Alkali Feldspars; Calcite Group; Clay, Clay Minerals; Plagioclase Feldspars*; see also mineral glossary. Vol. VI: *Diagenesis.*

STIBICONITE GROUP—*See* PYROCHLORE GROUP

STREAK–*See* COLOR IN MINERALS

SYMMETRY

Symmetry is a special kind of spatial identity. A symmetry operation, therefore, reproduces an identical pattern in space. Although symmetry is described in terms of successive operations, the elements of symmetry–the planes, lines, and points of symmetry across, around, and through which operations are executed–are intrinsic to a structure.

Elements of Symmetry

Two hands held forth in a prayerful position represent symmetry across a mirror plane (m) between them (Fig. 1a). A pair of hands turned in such a way that palms are parallel but fingers point in opposite directions has no mirror relationship, but a new element of symmetry is created. At opposite ends of lines that pass through a midpoint located between the palms are identical points on identical fingers–thumb opposite thumb, etc. (Fig. 1b). This relationship is variously called a center of symmetry or an inversion and is given the symbol i or $\bar{1}$ (crystallographers read the latter as "bar one").

Symmetry about a line involves an operation of rotation (Fig. 2). Since an operation produces identity upon any number of repetitions, only those rotations that return by repetition to the original position may be considered. The fold (n) of a symmetry axis represents the number of repetitions required to complete a rotation of exactly 360° about an axial element. (Note in Fig. 2 that only a left hand is involved in the identity operation.)

In Figs. 1 and 2, symmetrically identical points (equipoints) are labeled a_1, a_2, and b_1, b_2. Symmetry operations, thus, may be divided into two kinds: (1) rotations about an axis in which the chirality (sense of handedness) does

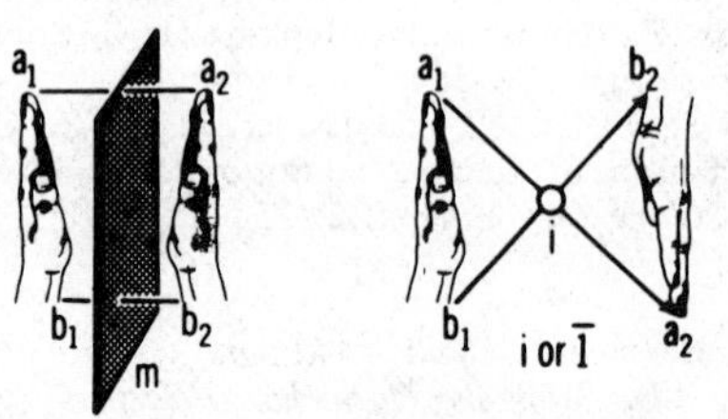

FIGURE 1. Elements of symmetry. (a) Mirror plane. (b) Center of symmetry or inversion (from F. Donald Bloss, *Crystallography and Crystal Chemistry*. Copyright © 1971 by Holt, Rinehart and Winston, Inc. Reprinted by permission).

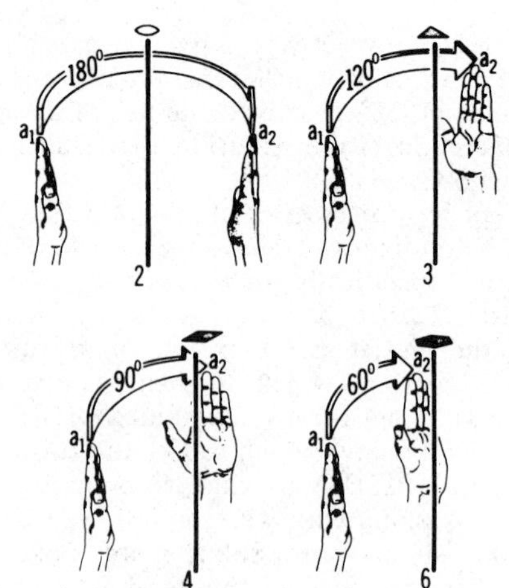

FIGURE 2. Rotational axes of symmetry showing angular rotation and the fold of the axis. The symbol for each axis is at its upper end. (from F. Donald Bloss, *Crystallography and Crystal Chemistry*, copyright © 1971 by Holt, Rinehart and Winston, Inc. Reprinted by permission). Equipoints are labeled a_1 and a_2.

not change with each successive repetition of an operation; and (2) mirror reflections and inversions through a center that reverse chirality. No amount of rotation or other contortion can change one's left hand into one's right.

The rotational symmetry of crystals may be represented by onefold axes (called monads), twofold axes (diads), threefold axes (triads), fourfold axes (tetrads), or sixfold axes (hexads) (Fig. 2). (The fivefold axis does not appear in simple crystals since pentagonal unit cells cannot be used to fill space.) These five axes represent proper rotations in that their operations do not change the chirality of objects.

A second set of five axes is called improper in that they involve chirality reversal, i.e., invert left hands to right and vice versa. These axes describe rotation followed by inversion through a center located on the axis and are indicated by $\bar{1}$, $\bar{2}$, $\bar{3}$, $\bar{4}$, and $\bar{6}$ (read "bar one, bar two," etc.).

A $\bar{1}$ axis describes a symmetrical relationship in which two objects or points are related to one another and are thus identical by rotating through 360° about an axis and then inverting through a center of symmetry on that axis. Since $\bar{1}$ is identical to a simple inversion i, no new symmetrical relationship is developed and $\bar{1}$ is used to represent a center of symmetry (compare Fig. 1b with Fig. 3).

A $\bar{2}$ axis describes rotation by 180° followed by inversion, an operation that is identical to a mirror reflection through a plane at right

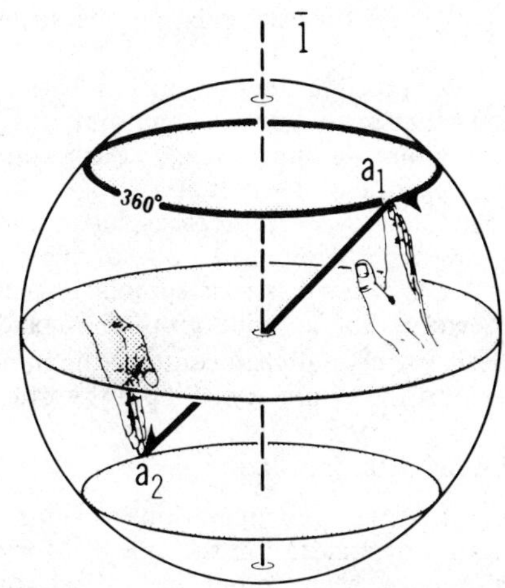

FIGURE 3. Center of symmetry or one-fold rotoinversion axis of symmetry (from F. Donald Bloss, *Crystallography and Crystal Chemistry*, copyright © 1971 by Holt, Rinehart and Winston, Inc. Reprinted by permission).

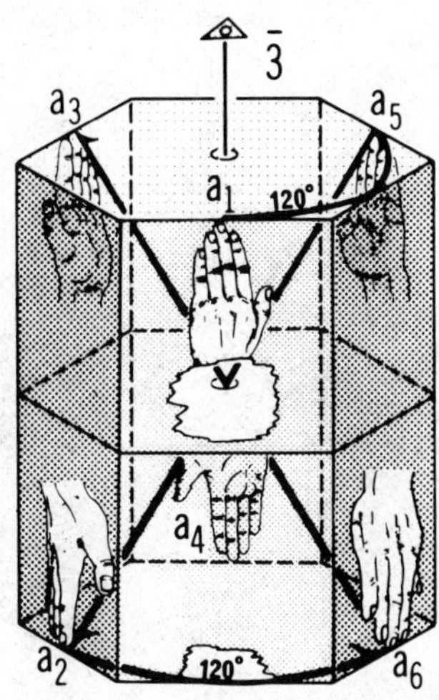

FIGURE 5. Threefold rotoinversion axis of symmetry (from F. Donald Bloss, *Crystallography and Crystal Chemistry*. Copyright © 1971 by Holt, Rinehart and Winston, Inc. Reprinted by permission).

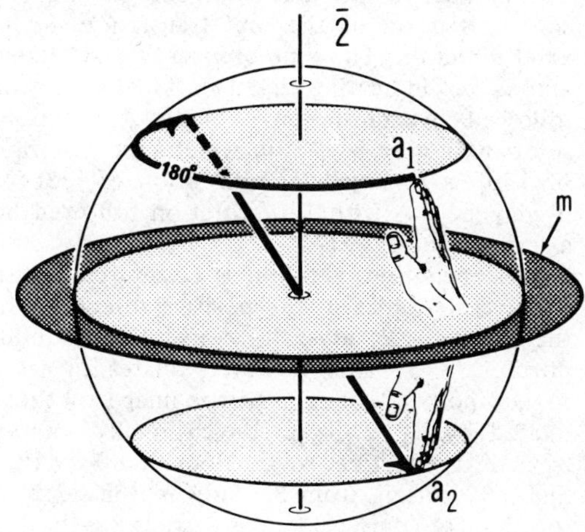

FIGURE 4. Mirror of symmetry or two-fold rotoinversion axis of symmetry (from F. Donald Bloss, *Crystallography and Crystal Chemistry*, Copyright © 1971 by Holt, Rinehart and Winston, Inc. Reprinted by permission).

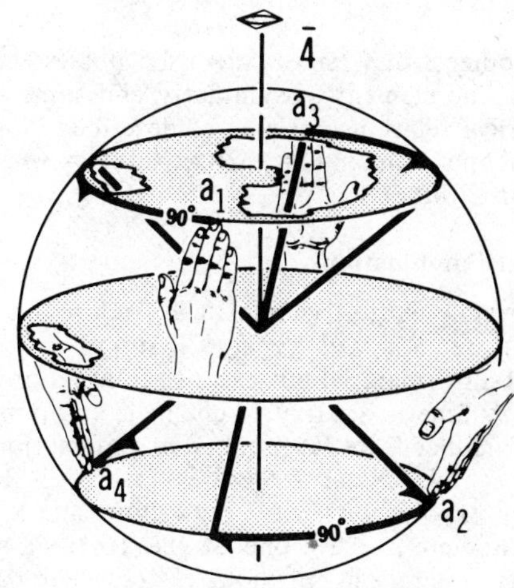

FIGURE 6. Fourfold rotoinversion axis of symmetry (from F. Donald Bloss, *Crystallography and Crystal Chemistry*. Copyright © 1971 by Holt, Rinehart and Winston, Inc. Reprinted by permission).

angles to the axis (compare Fig. 1a with Fig. 4). Again, no new symmetry is described and the symbol *m* is used to represent this relationship.

Sets of two equipoints each are developed by the operations $\bar{1}$ and *m* so long as the point lies neither at the center nor on the plane.

A $\bar{3}$ axis describes rotation through 120° followed by inversion through a center generating or describing equipoint sets of six (Fig. 5). (The number of equipoints related by symmetry operations is systematically reduced for points lying on an element of symmetry.)

A $\bar{4}$ axis describes rotation through 90° followed by inversion and relates four equipoints (Fig. 6).

Symmetry represented by a $\bar{6}$ axis (Fig. 7) is equivalent to symmetry represented by a threefold proper axis set at right angles to a mirror plane (noted as $3/m$). For purposes of classification, the $\bar{6}$ notation is preferred.

As is evident from the foregoing discussion,

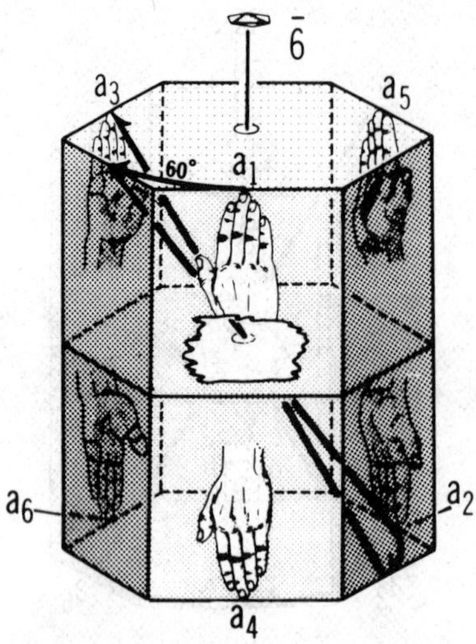

FIGURE 7. Sixfold rotoinversion axis of symmetry (from F. Donald Bloss, *Crystallography and Crystal Chemistry*. Copyright © 1971 by Holt, Rinehart and Winston, Inc. Reprinted by permission).

an object may simultaneously possess more than one element of symmetry and some symmetrical relationships may be described in more than one manner ($\bar{1}=i$, $\bar{2}=m$, and $\bar{6}=3/m$; see *Point Groups*).

Axial Combinations

Crystals exhibit geometrically related crystal faces (see *Crystallography: History*) and each crystalline material possesses an internal orderly and symmetrical array of chemical components as evidenced by the coherent diffraction of X-rays by crystal structures (see Vol. IVA: *X-ray Diffraction*). Any crystalline shape may be characterized by one of the 10 basic symmetry operations previously described or by any one of a possible 22 combinations. When two or more symmetry axes combine, however, their intersection at a single point defines the center of the shape that is being described.

Two restrictions limit to 32 the number of possible unique combinations of symmetry elements passing through a point. First, duplication results from the aforementioned equivalency of some combinations of operations of symmetry. Second, the elements of symmetry themselves must be symmetrically disposed so that each operation returns all other elements of symmetry to identity. For example, two threefold axes cannot intersect at an angle of 10°, since operation of the first would generate two more identical threefold axes and operation of each newly generated axis would produce two more for each existing one ad infinitum.

The 32 possible unique symmetries with elements that may intersect at a point represent the 32 point groups (q.v.). Each mineral species, regardless of whether or not a particular specimen shows crystal faces, has symmetry permitting its assignment to one of the crystal classes, each of which corresponds to a point group. Note that a point group is a specific symmetry. A crystal class contains the minerals that exhibit the symmetry of a point group.

Operations with Translation

Internal elements or operations of symmetry, which are not manifest in the external forms of crystals, have been found by diffraction of X-rays by crystal structures. These operations, the screw axis, the glide plane, and lattice translations, symmetrically relate equipoints in three-dimensional space (see *Lattice*).

A screw axis combines rotation about an n-fold axis (n=2,3,4, or 6) followed by translation parallel to that axis by an amount that is a simple fraction of the unit translation in the axial direction. The symbol for a twofold screw axis is 2_1, indicating that rotation by 180° is followed by translation of 1/2**t**. A threefold screw axis may be 3_1; in which case, rotation of 120° is followed by translation of 1/3**t**; or it may be 3_2, with 120° rotation followed by a translation of 2/3**t**.

By convention, rotation is counterclockwise and translation is in the positive direction. In the case of a 2_1 axis, this convention is unimportant. A 3_1 axis, however, relates an array of equipoints that is a mirror image of those related by 3_2. Thus, a conventionally rotated 3_1 axis is identical to a "clockwise 3_2 axis." Since 3_1 differs from 3_2 only in the sense of rotation of the screw axis, structures having these symmetries are enantiomorphous (see *Quartz*).

External symmetry represented by a fourfold rotation axis may result from internal symmetry represented by the screw axes 4_1, 4_2, or 4_3 with rotation of 90° followed by translation of 1/4**t**, 2/4**t**, or 3/4**t**, respectively. The axes 4_1 and 4_3 are enantiomorphous.

A sixfold screw axis may be 6_1, 6_2, 6_3, 6_4, or 6_5 with rotation of 60° followed by translation or 1/6**t**, 2/6**t**, 3/6**t**, 4/6**t**, or 5/6**t**, respectively. The screw axis 6_1 is enantiomorphous to 6_5 and 6_2 is so related to 6_4.

A glide plane is a mirror reflection accompanied by a translation **t** parallel to the plane of the mirror. Glide planes with translations of 1/2**t** in a crystallographic axial direction are labeled *a, b,* or *c*, depending upon whether the translation direction is parallel to x, to y, or to z, respectively. A diagonal glide n has translation components of $1/2\mathbf{t}$ parallel to two or three of the crystallographic axes, depending upon the orientation of the mirror plane with respect to the crystallographic axes. The diamond glide d is a diagonal glide with translation components of $1/4\mathbf{t}$.

The application of the elements of symmetry intersecting at a point (see *Point Group*) to three dimensionally infinite lattices (see *Lattice*) with the operation of translation being taken into account results in the 230 space groups (see *Space-Group Symbol*).

KEITH FRYE

References

Bernal, I.; Hamilton, W. C.; and Ricci, J. S., 1972. *Symmetry*. San Francisco: Freeman, 182p.

Bloss, F. D., 1971. *Crystallography and Crystal Chemistry*. New York: Holt, Rinehart & Winston, 545p.

Buerger, M. J., 1963. *Elementary Crystallography*, rev. printing. New York: Wiley, 528p.

Cotton, F. A., 1971. *Chemical Applications of Group Theory*, 2nd ed. New York: Wiley/Interscience.

Fackler, J. P., 1971. *Symmetry in Coordination Chemistry*. New York: Academic, 139p.

Cross-references: *Barker Index of Crystals; Crystal Habit; Crystallography: History; Crystallography: Morphological; Directions and Planes; Lattice; Point Groups; Space-Group Symbol.*

SYNTAXY

In its original meaning, the term "syntaxy," coined by Ungemach (1935b) by analogy with "epitaxy" (q.v.), referred to the oriented intergrowth of two polymorphic forms of one and the same chemical substance in which the ratios of corresponding cell edges are rational. As an example of syntactic intergrowth, two (or more) polymorphs of silicon carbide can crystallize together in a single edifice, in which a cell (simple or multiple) of one polymorph nearly coincides in shape and size with a cell of another polymorph. This is a case of three-dimensional structural control. Another example is provided by *coquimbite* and *paracoquimbite*, $Fe_2(SO_4)_3 \cdot 9H_2O$.

Syntactic intergrowth also occurs with constituent substances that are chemically different, and the meaning of syntaxy has been extended accordingly (Donnay and Donnay, 1953). In the series *bastnaesite*($CeFCO_3$)–*parisite* ($2CeFCO_3 \cdot CaCO_3$) – *roentgenite* ($3CeFCO_3 \cdot 2CaCO_3$) – *synchisite* ($CeFCO_3 \cdot CaCO_3$)–*vaterite* ($CaCO_3$), syntactic intergrowth has been observed for every pair of fluorocarbonates, even for *bastnaesite-synchisite* (Iitaka and Stalder, 1961). Such intergrowths are examples of polycrystals (see *Polycrystal*).

When the mutual orientation of two species is the result of a solid-state reaction or transformation, as in the dehydration of *goethite* (α = FeOOH) described by Goldsztaub (1935), the term "topotaxy" (q.v.) is used.

J. D. H. DONNAY

References

Donnay, G., and Donnay, J. D. H., 1953. The crystallography of bastnaesite, parisite, roentgenite, and synchisite, *Am. Mineralogist*, **38**, 932–963.

Goldsztaub, S., 1935. Déshydratation de la goethite, *Bull. Soc. fr. Minéral.*, 58, 42–43.

Iitaka, Y., and Stalder, H. A., 1961. Synchisit and Bastnäsit aus dem Druckschacht des Kraftwerkes Oberaar, *Schweiz. Mineral. Petrogr. Mitt.*, **41**, 485–488.

Ungemach, H., 1935a. Sur certains minéraux sulfatés du Chili, *Bull. Soc. fr. Minéral.*, **58**, 97–221.

Ungemach, H., 1935b. Sur la syntaxie et la polytypie, *Zeit. Kristallogr.*, **91**, 1–22.

Cross-references: *Epitaxy; Polycrystal; Polymorphism; Topotaxy.*

SYNTHETIC MINERALS

From the beginning of civilization, man has been interested in natural crystals such as gems. Later, many different natural minerals (*diamonds, mica, quartz, corundum*, etc.) were used in a great variety of applications (cutting, abrasion, electrical, etc.). During World War II, the supply of many of these minerals was endangered and the demand for them in technical applications was greatly increased. This situation, together with technological advances, led to the birth of a new and exciting industry, that of growing synthetic crystals. This is a field where geology, chemistry, physics, metallurgy, and ceramics merge. Knowledge of the methods in which natural minerals form and knowledge of the crystallography and crystal chemistry of natural minerals have helped man synthesize a large variety of synthetic minerals.

Solid crystals are grown from a donor phase, which can be a gas, liquid, or solid. To initiate the crystal-growing process, the donor phase is frequently seeded with a small piece of the crystal to be grown. Crystals are grown from a melt, hydrothermally, from low-temperature solutions, and at high pressures and(or) temperatures.

Synthetic crystals are used for a multitude of applications, as in oscillators, polarizers, infrared optics, lenses, diodes, lasers, and masers. These and other uses of synthetic minerals have extended the field far beyond the classical use of gems (see *Gemology*). The materials included are elements, oxides, metals, intermetallic compounds, organics, refractory hard metals, and other inorganics. Also, the synthesis of polycrystalline inorganic materials (ceramics) by sintering and fusion is important in a wide variety of applications (refractories, consumer ware, electronic ceramics, etc.).

Single-Crystal Techniques

Numerous techniques for growing single crystals are described in detail in books by Gilman, Buckley, Shubnikov, and many others. Crystals can be grown from the vapor phase either by condensation of a supersatured vapor or by chemical reaction (see, especially, Gilman, 1963). In addition, preparation of crystals can be accomplished by precipitation from liquids at relatively low temperatures and pressures because of the high concentration of many compounds in liquid solution and their high mobilities. Another technique often used is hydrothermal synthesis, which is the use of aqueous solvents under high temperature and high pressure to dissolve and recrystallize materials that are relatively insoluble under ordinary conditions.

Crystals can also be formed from a melt, with or without a container, as in the Czochralski and Bridgeman techniques, by flame fusion (Verneuil), and by means of dry systems at high pressures as described in Gilman and others.

Vapor-Phase Technique

As mentioned, crystals can be grown from the vapor phase either by condensation of a supersaturated vapor or by chemical reaction. Condensation takes place by a combination of adsorption and migration of the vapor phase across the crystal surface and is either incorporated into the structure or reevaporated. Often the batch ingredients are heated in a zone where volatile species form. Then a carrier gas transports them to a cooler zone, where they are precipitated. The batch, the gas phases, the temperature, and the substrate are important in the crystal formation. Crystals grown by the use of chemical reaction have a vapor phase chemically different from that of the crystal being grown.

Vapor-phase growth is used commercially in the production of luminescent crystals (ZnS and CdS) and semiconducting crystals (silicon, germanium II–VI compounds). Furthermore, high-strength whiskers are formed this way (Al_2O_3 and SiC). This method is also applicable for growing organic crystals but is limited by the molecular weight of the compound.

Precipitation of Crystals from Solution

Low Temperatures (<200°C) and at Atmospheric Pressure. Crystals are grown from solution by formation of a precipitate in a liquid. This is accomplished because many compounds can achieve high concentration in liquid solution and high mobilities, which permit the growth of large crystals. The advantages of this method are (1) control provided over temperature of growth, which makes possible the growth of crystals that are unstable at their melting points or exist in several crystal forms at different temperatures, and (2) control of viscosity, which allows crystals that tend to form glasses when cooled from their melts to be grown. In addition, crystals grown from solution usually have well-defined faces and fewer structural imperfections (dislocations, vacancies, etc.—see *Crystals, Defects in*) as compared to those grown from melts. This method is used for growth of crystals of elements, ionic salts, and silver halides.

Hydrothermal Synthesis. This technique consists in dissolving constituents in an aqueous solution under high temperature and high pressure. The constituents are then precipitated in a cooler region onto a seed crystal. The process is very similar to the formation of pegmatite minerals (q.v.). A key to determining how readily a synthetic mineral can be grown hydrothermally is the observation of natural pegmatites. If huge crystals are observed in many locations in nature, the mineral should be easy to synthesize; if a mineral occurs naturally in very small crystals, it will be difficult to grow large crystals synthetically. One mineral made commercially by this method is *quartz*. Other materials, such as alpha alumina (*corundum*), *magnetite*, nickel ferrite, aluminum phosphate, zinc oxide, and zinc sulfide, have been made this way, but the crystal size has been rather small and the process difficult to perfect. Emerald and *beryl* have also proved to be exceptions to the rule.

In crystallizing *quartz*, the usual dissolving temperature is 400°C. The crystallizing temperature is 360°C. The degree of fill of the dissolving region is 80% and a solution of 1.0 M NaOH is used, and a baffle is placed between the solution region and the precipitating region. Generally, the baffle has a 5% opening and the pressure is 21,000 psi (1500 kg/cm^2). Single crystals weighing up to 800 g (Western Electric Co.) have been grown under these conditions. One of the advantages of hydrothermal synthesis is that it permits crystal growth at relatively low temperatures and with fewer imperfections than are produced at higher temperatures from a melt. The rate of growth is rather rapid and *quartz* can be grown in the [0001] direction as fast as 0.250 inch per day (0.64 cm/day). Also, the atmosphere can be controlled. The low viscosities of the solutions and rapid convection permit large growth rates. An excellent article on this method is written by Ballman and Laudise in Gilman (1963).

Crystal Growth from the Melt

Controlled Slow Crystallization from a Melt. Tammann (1925) found that single crystals could be grown from a melt, providing the rate of cooling was slow. He grew metal crystals by cooling the melt extremely slowly, with only a slight degree of supercooling. He also found that, by crystallizing in a container that was very narrow, one crystal would emerge with an orientation having a more rapid growth than that of its neighboring crystals. This led to the growth of a single crystal. Often the tube was tapered to a fine capillary at one end. After a single crystal formed in the narrow tub, growth in an adjoining, wider tube was permitted at a more rapid rate. Tammann also observed that crystallization from a melt tended to cause impurities to be repelled at the solid-melt interface, permitting the formation of single crystals with a higher purity than the starting material had.

Tammann's method has been modified and refined by other scientists. In the Bridgeman technique, the sample is lowered gradually into a cool zone, often starting with an oriented seed. Kapitza's method is a modification of the Bridgeman method. These methods are discussed in detail by Buckley (1951) and by Gilman (1963).

Pulling Crystals from a Melt. This technique, initiated by Czochralski, is now used to grow many single crystals, particularly for electronic applications. Usually a seed crystal is dipped into the top surface of a melt and then slowly withdrawn at a rate approximating the rate of crystallization. Pull rates are dependent on the temperature gradient at the crystal-melt interface and can vary from 0.1 to 4.0 cm/h. The growth rate is controlled by the pull rate to a maximum. Factors important in obtaining crystals with a minimum of imperfections are temperature-gradient control, correct rate of pull, and a minimum of thermal and mechanical stresses.

In recent years, more refractory-type crystals have been grown by this method at temperatures significantly below their melting points. The refractory mineral is fluxed with a low-melting phase, generally a salt. A composition is chosen such that the primary phase to crystallize upon cooling will be the high-melting phase. Then a seed of the desired mineral is placed in the melt and withdrawn slowly to make crystals that might be unstable at higher or lower temperature (i.e., incongruently melting phases). Crystals such as barium titanate and yttrium-iron garnet are grown in this way. The method also permits the growth of a refractory mineral with fewer imperfections than would have been possible if it had been grown at its melting point.

Crystals are likewise grown by allowing a salt-oxide melt to cool extremely slowly, resulting in the formation of large oxide crystals surrounded by a salt-rich matrix. After solidification has occurred, the salt can be leached, leaving crystals of the desired phase. Many phases with limited stability regions cannot be grown this way. The method also allows greater chance of crystal contamination.

Crystallization Under Very High Pressure. The most valuable material to be made this way is artificial *diamond*. The *graphite-diamond* phase diagram illustrates the temperature and pressures required for the process. *Graphite* is mixed with catalyst metals, such as iron, manganese, nickel, and platinum, and subjected to great pressure (>45 kilobars at temperatures greater than 1750°C). Under these conditions, a *graphite*-metal eutectic occurs.

With time, a film of the molten catalyst sweeps through the *graphite* precipitating *diamond* in its wake. The greater the pressure above that required for equilibrium, the greater the rate of *diamond* nucleation and growth. Low pressures tend to produce cubic habit and high pressures favor the octahedral habit (see *Crystal Habit*). At low temperatures the *diamonds* tend to be darker than those produced at high temperatures.

The apparatus often used to produce these high temperatures and pressure conditions was designed by H. T. Hall in 1960 and named "Belt." It is also referred to as "Anvil." The sample is compressed by two tapered pistons and is confined horizontally by a tapered cylinder resembling a belt or girdle. The apparatus is capable of maintaining pressures up to 100 kilobars and temperatures up to 3000°C. The material is heated directly by electrical resistance heating. It is difficult to measure the pressure and temperature accurately.

A similar high-pressure technique uses a hydraulic press, and the sample is heated either by a graphite-tube heater or by induction. The growth of *diamond* and other elements is discussed by Wentorf and Bovenkerk (1962).

Flame Fusion Technique. In the early 20th

century, crystal growth, produced by feeding powders through a flame, was introduced by Verneuil. This technique consisted in seeding a single crystal by the flow of powder through a flame, traditionally using an oxygen-hydrogen burner. The powder is melted and deposited on a setter, which is placed in the flame. A single crystal is usually favored at the end of the polycrystalline material. A boule is built up by careful control of the growing crystal, feeding, and flow of gases (Shubnikov and Sheftal, 1959). The crystal is continually and slowly withdrawn into a cool zone as crystallization occurs in the hotter region.

This method has been extended to include the use of plasma and induction heating. Rubies, sapphires, *rutile*, and other gems have been made this way. Sapphires are the most important commercially, because of their wide application as gems, abrasives, optical windows, etc. (see *Abrasive Materials; Gemology*). This method has also been extended to make doped crystals such as chromium oxide–doped *corundum*, which is used in lasers. The method eliminates melt contamination by crucible materials, which is experienced in many other melting techniques. These crystals tend to have a greater imperfection concentration than crystals produced at a lower temperature.

Other Methods. Another method used for growth of crystals is recrystallization of a fine-grained polycrystalline body into a single crystal or, at most, a few large crystals. There is also double-compound thermal-imaging technique, in which a high-intensity heat source such as a xenon lamp is used with a series of mirrors.

Uses of Synthetic Single Crystals. As mentioned, synthetic minerals have many contemporary applications. Of particular interest is their use as lasers and masers. Some laser materials include U^{3+} in CaF_2, Cr^{3+} in Al_2O_3, and Nd^{3+} in $CaWO_4$. Some maser materials are Gd^{3+} in $CaWO_4$ doped with Er^{3+} and Cr^{3+} in Y_2O_3. Much work has been done in the field of growing *diamonds* to be used in cutting equipment, sapphires for gem uses, yttrium-iron garnets because of their ferrimagnetism, barium titanate for application of its dielectric properties, group III and V compounds because of their thermoelectric properties, and halides such as NaCl, CaF_2, and KBr for infrared optics. A great variety of single crystals is being used in the electronic industry for electrical conduction, semiconduction, and other purposes.

Polycrystalline Techniques

Sintering. Most ceramic articles are manufactured by a method involving sintering, a process through which heating at a suitable temperature causes agglomeration of individual fine particles into a cohesive compact by increasing the area of particle contact. The agglomeration is usually accompanied by a decrease in porosity and an increase in strength and bulk density. During the process, the surface area decreases so that the surface free energy, and hence the free energy of the system, decreases. In many respects this process is similar to the formation of sedimentary rocks and metamorphic rocks, except that the times involved are minutes or hours instead of the centuries and millennia of geologic time.

Inception of Sintering. Significant irreversible changes may take place in a crystalline material at temperatures well below its melting point. The lowest temperature at which short-term sintering can be detected was first investigated in detail by Tammann and the ratio of this temperature to the melting temperature is described as the "Tammann ratio." For oxides, the ratio is usually in the range of 0.5 to 0.6. It is important that these ratios should refer to pure stoichiometric compounds in the absence of chemical or physical reaction such as dehydration or inversion.

By use of proper techniques, many materials can be sintered to nearly theoretical density at a temperature only two thirds of the melting point without the presence of a liquid phase. In the presence of a liquid phase, significant increases in the speed of reaction can often be achieved.

Effect of Liquid Phase on Densification

If a monolayer of eutectic liquid between two phases is formed on one of the phases, a phenomenon termed "activated sintering" occurs. The thin film of eutectic liquid is a region in which diffusion coefficients for the host material are much higher than the diffusion coefficient for the host material in pure grain boundaries. As an example, pure tungsten must be sintered at approximately 2700°C to achieve nearly theoretical density, but if 0.01% nickel is added, theoretical density is achieved at 1400°C by activated sintering. The rate of sintering increases linearly up to the content at which a monolayer appears, but it is unaffected by larger additions.

When more than a monolayer of liquid phase is formed, but insufficient to separate the dominant-phase particles during heating, "recrystallization sintering" occurs. For small liquid contents (approximately 5 to 15% of liquid by volume), grain-shape change is required for complete densification. The two possible rate-controlling mechanisms are solution rate at the crystal-to-liquid interface and diffusion rate in the liquid.

For liquid-phase contents of 15 to 30% by volume, there is little need for solution and reprecipitation of the crystal phase, and complete densification can be achieved principally by particle rearrangement, pore removal being largely governed by viscosity of the liquid phase.

For still greater liquid contents, of the order of 30 to 50% by volume, particles of the crystal phase are separated by liquid during the firing process and no sintering is required for complete densification. This is the traditional firing process by which relatively coarse-grained ceramics such as porcelains, sanitary ware, electrical insulators, and artware are manufactured.

The temperature at which complete densification can be achieved on any of the foregoing processes may be reduced by application of sufficient pressure at the lower temperature. These processes, termed "hot pressing" and "hot isostatic pressing," are increasing in application.

Modern sintered products are nuclear fuel elements, metal-cutting tools, diamond-cutting wheels, electrical contacts, magnets, ferrite cores, semiconductors, porous self-lubricating bearings, machine cams and gears, refractory crucibles, filters, dielectrics, cermets, electrodes, and a host of others that are essential to modern industry.

Solid-State Reactions in Electric Resistance Furnaces

Massive electric resistance furnaces consuming many hundreds of kilowatts of power are industrially useful for the manufacture of *graphite* and silicon carbide in commercial quantities.

Silicon carbide is formed by the reaction $SiO_2 + 3C = SiC + 2CO$ upon heating a mixture of carbon and silica sand in a horizontal resistance furnace, such as is shown schematically in Fig. 1. The central core is a relatively conductive material that heats rapidly to a suitable temperature to promote reaction in the surrounding furnace charge.

Similar furnaces are used for the manufacture of *graphite* from amorphous carbon. Silicon carbide and *graphite* are used extensively as refractories and electrical conductors (see *Refractory Minerals*). Boron carbide can also be similarly made in a resistance furnace from boric oxide and carbon. Operating data for such a furnace is given in Table 1.

Calcination

Calcination, the operation of heating or roasting, has always been one of the most important operations in the chemical laboratory. Industrially, calcination is useful to prereact materials to provide stable products as improved raw materials (see *Portland Cement Mineralogy*). By heating at a suitable temperature, the water of hydration can be removed from *kaolinite* ($Al_2O_3 \cdot 2SiO_2 \cdot 2H_2O = Al_2O_3 \cdot 2SiO_2 + 2H_2O$), *magnesite* can be converted to *periclase* ($MgCO_3 = MgO + CO_2$), or the inversion expansion of *kyanite* can be accomplished to provide a dimensionally stable *mullite* and glass grog [$3(Al_2O_3 \cdot SiO_2) = 3Al_2O_3 \cdot 2SiO_2 + SiO_2$]. A rotary kiln is often used to provide a continuous flow of calcined raw materials.

Fusion Casting

Fusion casting is an important type of process for producing refractory materials that are used in glass furnaces and steel plants. They are also used as skid rails in metal-reheat furnaces, in cement kilns, and in abrasion-resisting applications. In general, these refractories have excellent corrosion and erosion resistance. Their chemical durability is chiefly attributed to such microstructural characteristics as large crystal size, tight crystal bond, usually interlocking crystals, high density, and low permeability and porosity. They also have very good spall

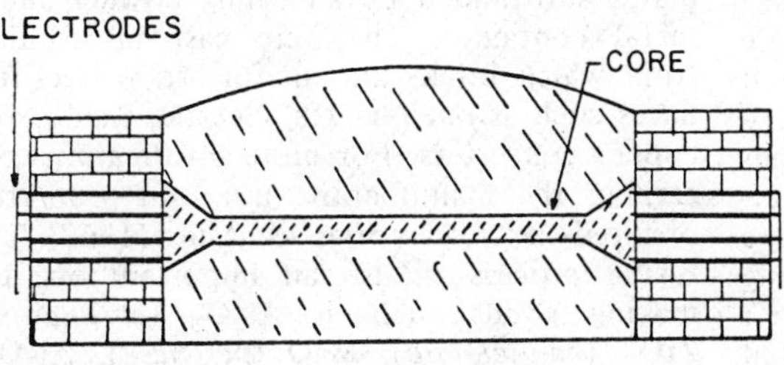

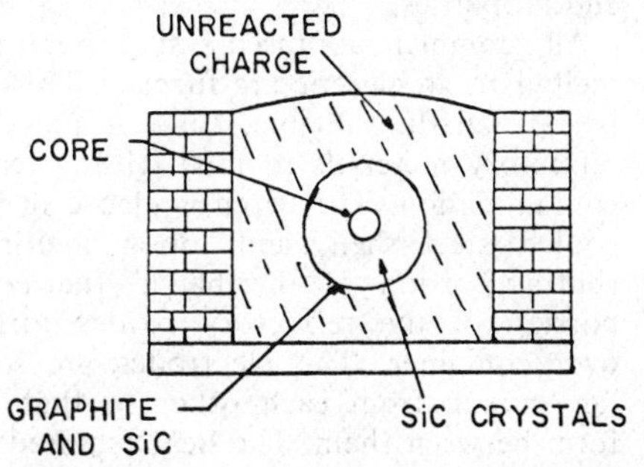

FIGURE 1. Horizontal resistance furnace

TABLE 1. Products of Resistance Furnaces

Factors	Graphite	Silicon Carbide
Raw materials	Low-ash anthracite or petroleum coke	Coke, 98% silica sand
Additions	–	Sawdust and salt
Furnace:		
Type	Resistance	Resistance
Size (hp)	1000	2000
Length (ft)	30	30
Cross section (ft)	–	10 × 10
Cross section of charge, (ft dia.)	2	3
Walls	Refractory brick or concrete blocks	Refractory brick, castiron, or steel supports
Initial voltage	200	230
Final voltage	80	75
Initial current (amp)	–	6000
Maximum current at 200 volts (amp)	3700	–
Final current (amp)	9000	20,000
Current density across furnace charge (amp/ft^2)	900–2250	650–2200
Core temperature (°C)	–	2350
Furnace temperature (°C)	2200	1820–2220
Length of run (hr)	24	36
Conversion of material (%)	90–100	70–80
Energy consumption (kw-hr/lb)	1.5	3.2–3.85
Energy efficiency (%)	25–30	55–70

From Perry, 1950.

resistance, which is related to their high-temperature strength and high degree of bonding by very refractory phases.

The properties of fused-cast refractories are related to their microstructure, which can be greatly varied by controlling such factors as composition, dissolved gases, degree of superheat, and rate of cooling of the molten material. A phase diagram of the particular refractory system is very important in determining how these variables should be modified in order to obtain the optinum microstructure and properties.

All commercial fused-cast refractories are melted in an electric arc furnace. This method is particularly valuable, since it permits very refractory materials to melt without container contamination. The batch is placed in a water-cooled steel shell, and, upon melting of a central portion of the batch, the remaining portion of the refractory oxides form their own container. The electrodes are separated far enough from each other so that no arcs form between them. The heat required to melt the charge is developed by the direct flow of current from one electrode through the molten charge to the next electrode. Heat is also generated from the electric arc developed between the tips of the electrodes and the bath.

Glass-furnace fused-cast refractories (ZrO_2-Al_2O_3-SiO_2, $Al_2O_3 \pm CaO \pm Na_2O$, Al_2O_3-SiO_2, Al_2O_3-Cr_2O_3) are generally poured into molds made from sand plates surrounded by annealing powder and a metal container. Since steel-plant refractories (MgO-chrome ore, MgO, etc.) melt at very high temperatures, they are usually poured into molds made of *graphite* plates surrounded by annealing powder and a metal container. They are cast into billets from which bricks are cut for use in steel furnaces such as open-hearth, electric, Kaldo, and copper converters. For more information concerning the manufacture, use, and properties of fused-cast refractories see Alper (1967). The fusion process is also an important one for making grains such as Al_2O_3 (*corundum*), ZrO_2 (*baddeleyite*), MgO (*periclase*), Al_2O_3-$2SiO_2$ (*mullite*), and MgO plus chrome ore. After fusion and crystallization, the materials are crushed and sized for use in sintered ceramics and refractories, for use as abrasive grain, and for use in sheathed electrical resistance heating elements for stoves.

R. N. McNALLY
A. M. ALPER
R. K. SMITH

References

Alper, A. M., 1967. *Science of Ceramics*, vol. 3. New York: Academic.

Alper, A. M., 1970. *High Temperature Oxides, Refractory Materials Monographs*, vol. 5. New York: Academic, 358p.

Alper, A. M., 1976. *Phase Diagrams: Materials Science and Technology, Refractory Materials Monographs*, vol. 6. New York: Academic.

Buckley, H. E., 1951. *Crystal Growth*. New York: Wiley, 571p.

Bundy, F. P., 1964. Diamond synthesis and the behavior of carbon at very high pressures and temperatures, *Ann. N.Y. Acad. Sci.*, **105**, 951-982.

Coble, R. L., and Burke, J. E., 1963. *Sintering in Ceramics, Progress in Ceramic Science*, vol. 3. New York: Macmillan.

Current Ceramic Research Issue, *Am. Ceramic Soc. Bull.*, **54**(2).

Gilman, J. J., 1963. *The Art and Science of Growing Crystals*. New York: Wiley, 493p.

Hausner, H. H., 1964. *Bibliography on Fundamental and Practical Aspects of Sintering in Powder Metallurgy*. Riverton, N.J.: Hoeganaes Sponge Iron Corp.

Lynch, Charles T., 1974. *Handbook of Materials Science*; vol. 1, *General Properties*; vol. 2, *Metals, Composites, and Refractory Materials*. Cleveland: CRC Press.

Perry, J. H., 1950. *Chemical Engineers Handbook*, 3rd ed. New York: McGraw-Hill, 1942p.

Shubnikov, A. V., and Sheftal, N. N., 1959. *Growth of Crystals*. New York: Consultants Bureau, 291p.

Suits, C. G., 1964. Man-made diamonds–A progress report, *Am. Scientist*, **52**, 395–408.

Tammann, G., 1925. *Metallography*. New York: Chemical Catalog Co., 26p.

Wentorf, R. H., Jr., and Bovenkerk, H. P., 1962. *J. Chem. Phys.*, **36**, 1987.

White, J., 1962. *Science of Ceramics*, vol. 1. New York: Academic.

Cross-references: *Crystal Growth; Crystallography; Metallurgy*. Vol. IVA: *Crystal Chemistry*.

T

TEKTOSILICATES—*See* **ALKALI FELDSPARS, FELDSPAR GROUP, FELDSPATHOID GROUP, PLAGIOCLASE FELDSPARS, QUARTZ, SODALITE GROUP, ZEOLITES.** Vol. IVA: **CRYSTAL CHEMISTRY; MINERAL CLASSES: SILICATES**

THERMOLUMINESCENCE

Thermoluminescence is an emission of light in the near-ultraviolet and visible range, which occurs when certain crystalline and glassy materials are heated to temperatures up to 400 or 450°C. Such materials can be considered as semi-insulators, in which the valence (outer shell) electrons of some atoms in the vicinity of defects are in a metastable state of higher than normal energy (excited state). When such materials are heated, this excess energy is released, partly as photons and partly as nonradiative transitions. In most cases, the photon emission is probably less than the actual stored (or trapped) energy.

Peak emission takes place at temperatures that are more or less specific for a particular set of trapping conditions and indirectly reflect the activation energy of the traps. The traps are usually considered to be structural defects (vacancies, impurity ions, and dislocations; see *Crystals, Defects in*), or more correctly, combinations of defects. The area under a glow curve can be an indication of the total number of traps. Figure 1 shows a typical natural glow curve with two peaks and an accentuated glow curve with four peaks from the same material after it has been artificially excited with ionizing radiation (alpha, beta, gamma, X-ray, or ultraviolet radiation). Unless the irradiated sample is refrigerated, normal ambient temperature will cause the artificially excited peaks occurring at 180°C and below to decay due to thermal drainage. Peaks above 180°C will be unaffected unless the sample is raised to a temperature above ambient temperature. Incandescence (red heat) starts to appear at about 250°C and seriously interferes with observations of thermoluminescence at temperatures above 400–450°C. Since the emission is dependent on a metastable energy state, it cannot be repeated after heating unless the electrons are reexcited with ionizing radiation. However, heating may change the defect concentration and exact reproduction of the natural glow curve may not be possible.

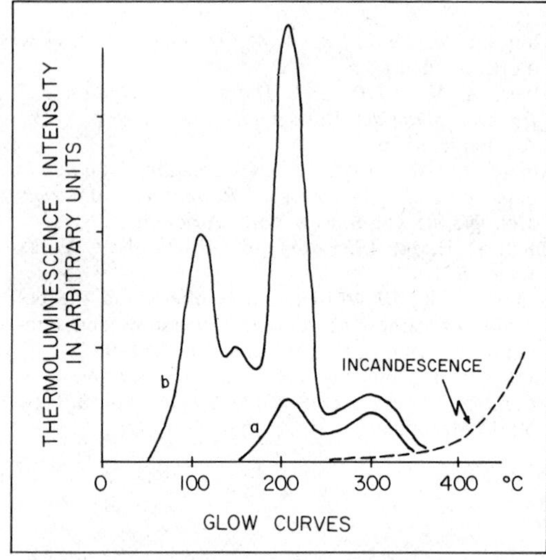

FIGURE 1, (a) Natural glow curve. (b) Glow curve enhanced by ionizing radiation.

Although natural thermoluminescence in minerals is often considered to have been produced by ionizing radiation from small amounts of radioactive elements (uranium, thorium, potassium 40) in or near the thermoluminescent minerals, many cases can be cited of thermoluminescence-like emissions that are not due to radioactivity. Such nonradiation-induced thermoluminescence includes that due to polymorphic transitions (*aragonite* to *calcite*); thermal decomposition (monoclinic *pyrrhotite* to hexagonal *pyrrhotite* with loss of sulfur); annealing of strain in crystals that have been physically deformed but not ruptured by high-velocity shock, or hydrostatic or directed pressure; and metastable order–disorder of the crystal structure due to elevated temperatures of formation or thermal metamorphism

(McDougall, 1968b). It should be noted, however, that all polymorphic transitions do not cause thermoluminescence and some transitions may cause a decrease in thermoluminescence (Bettinali et al., 1967). In general, nonradiation-induced thermoluminescence appears to result from elections that have been raised to an excited state due to requirements for maintaining electroneutrality in metastable crystals.

The principal rock-forming minerals that usually exhibit thermoluminescence are *quartz, calcite, dolomite,* and the *feldspars.* Where partial or complete isomorphous substitution occurs in these minerals, the level of emission may be a function of the chemical composition. However, in many thermoluminescent minerals the amount and type of trace elements (exclusive of the radioactive elements) may have an equal or greater bearing on the level of emission than does the gross chemical composition. For example, in the carbonate minerals, certain trace elements, notably manganese and lead, will activate or enhance thermoluminescence while others, including iron, nickel, and cobalt, will suppress or reduce thermoluminescence.

The simplest technique for observing thermoluminescence is by visual means in a dark room, although measurement of the intensity of the emitted light can, at best, be only semiquantitative. Several versions of thermoluminescence readers have been devised. These consist essentially of a heater, a photomultiplier tube, and recording instrumentation. Several versions of thermoluminescence readers are available from manufacturers of electronic equipment (see Appendix 1 in Cameron et al., 1968). Techniques of sample preparation and handling, selection of photo tubes and heating rates, and problems due to anomalous thermoluminescence are discussed in Cameron et al. (1968).

Although thermoluminescence of minerals was once considered to be an erratic and unpredictable phenomenon, advances in the understanding of the physics and chemistry of the solid state have shown that variations in thermoluminescence often reflect the physical and chemical conditions of formation and the subsequent history of many minerals and rocks. Comprehensive bibliographies have been compiled covering investigations of thermoluminescence in the fields of physics, chemistry, and the earth sciences (Angino, Grogler, and McCall, 1965, cited in McDougall, 1968); geology, archaeology, and the study of meteorites (McDougall 1968b); and medical dosimetry (Cameron et al., 1968). To a remarkable extent, the recent developments in these fields stem from work initiated in the late 1940s by Farrington Daniels at the University of Wisconsin.

Geochronology

A number of early investigators appear to have recognized a possible relationship between thermoluminescence and geological age; and a systematic study of the relationships of thermoluminescence, radioactivity, and age was initiated at the University of Wisconsin in 1950. Two basic methods for determining geologic age have been evolved, the earlier depending on a radiation-damage concept, and the more recent based on a saturation (or dosimetry) concept. In the radiation-damage method, the initial assumption is that alpha radiation from uranium or thorium in crystals create structural vacancies, thereby exciting electrons to a higher energy state. Some of this energy remains trapped until released by thermal energy as thermoluminescence. With increasing age of a material, radiation damage increases, so that the older the specimen, the greater the thermoluminescence. It has been recognized that complications are introduced into the radiation-damage method by impurity content, thermal history, crystallization, and pressure. On the other hand, the dosimetry method assumes that ionizing radiation from radioactive elements raises the trapped energy from zero to a condition of saturation, after which there will be no further increase in the level of thermoluminescence. Inherent in this concept is the idea of little or no radiation damage.

In its simplest form, the time necessary for natural thermoluminescence to accumulate can be calculated if the natural thermoluminescence, the saturation thermoluminescence, and the amount and kind of radioactivity to which the sample has been subjected are known. For limestones, the dosimetry method is probably only useful to attempt to date events younger than 10^5 years and may thus serve to bridge the gap between potassium-argon and carbon 14 dating (the Pleistocence-Pliocene period or about 10^4 to 10^6 yr B.P.). Thermoluminescent dating has been applied to a variety of materials including carbonate rocks, lava flows, dikes, meteorites, *quartz*, and *fluorite* (McDougall, 1968b). However, since defect concentrations in rock-forming minerals may be drastically changed by natural processes—such as deformation, thermal effects, recrystallization, and metasomatic processes—most of the rock ages cannot be considered to be very reliable. Archaeologists have had some success in using thermoluminescence for dating ancient pottery, although complications are introduced by the composition of the original materials, by firing temperature, and by radiation and(or) thermal history.

Geothermometry

The early approach to thermoluminescence as a geothermometer assumed that an increase in temperature around igneous intrusives and hydrothermal ore deposits would produce a fossil-heat aureole of low or nil thermoluminescence (see *Thermometry, Geologic*). This concept was applied with some apparent success to a zone of hydrothermal dolomitization and sulfide replacement in carbonate rocks. However, this approach was criticized on several grounds, one of the most important criticisms being the fact that, due to natural radioactivity, suppression of thermoluminescence by natural heating cannot persist in carbonate rocks for more than 10^4 to 10^5 years. A more elegant treatment, which combines dosimetry techniques and heat-condition theory, has been applied to carbonate rocks that have been affected by Pleistocene lava flows and dikes (McDougall, 1968b). A suggested third method involves differences between the level of thermoluminescence artificially induced by ionizing radiation in unheated and heated *calcites* and the level in *calcites* formed at high and low temperatures (for example, from carbonatites and stalactites). Similar effects have been observed in *quartz*, initially by both Fleming and Thompson (1970) and McDougall (1971) and most recently by Han (1975). Since both increases and decreases in thermoluminescence have been observed in close proximity to igneous contacts and hydrothermal ore deposits, this third approach may prove to be the most promising. As a special case of this third approach, it may be possible to distinguish between epizone and mesozone grades of metamorphism in quartzites.

Radiation-Temperature Environments

The concept of thermoluminescence saturation resulting from exposure to ionizing radiation has been extensively applied in medical dosimetry. Lithium fluoride and other compounds, which are used as personnel dosimeters, have been the subject of painstaking research (Cameron et al., 1968). This research has substantially expanded the understanding of the basic processes of thermoluminescence, and much of this understanding can be applied to geological investigations.

The dosimetry technique has not only been applied to geochronology and geothermometry, it has found direct applications in determining the radiation environment of marine invertebrate shells and the radiation history of meteorites. The same general technique has been applied to paleoclimate and to microclimate studies in attempts to gain information on rates of deglaciation and the surface temperatures of terrestrial rocks and of meteorites. This application is based on a concept of secular equilibrium between the radiation environment (constant dose rate) and the ambient temperature at or very close to the rock surface (McDougall, 1968b).

Deformation of Rocks

The probability of errors due to stratigraphic loading, tectonic events, and crushing during sample preparation has led a number of investigators to investigate the effects of strain on thermoluminescence. Experimental deformation of rocks and minerals by means of uniaxial, triaxial, and impact techniques has demonstrated that deformation can either increase or decrease thermoluminescence (Douglas and McDougall, 1973). Various lines of evidence indicate that increases in thermoluminescence are due to the formation of dislocations whereas decreases are due to their annihilation. These changes take place as the material is strained, so that, although strain-induced glow-curve peaks may be enhanced by ionizing radiation, a substantial part of the energy associated with the dislocations (or combinations of the dislocations and other defects) results from requirements for electroneutrality in the crystals.

Field investigations have been carried out in the vicinity of meteorite craters, underground nuclear explosions, faults, and fold structures. Figure 2 is a schematic synthesis of the results of laboratory and field investigations of the effect of deformation on the thermo-

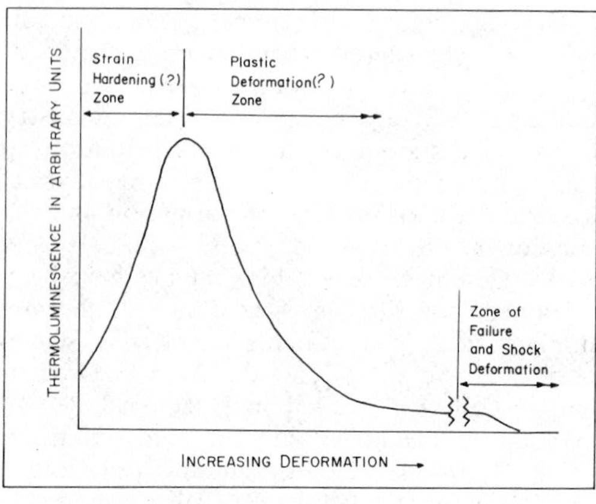

FIGURE 2. Effect of mineral deformation on thermoluminescence.

luminescence of rocks and minerals. This curve can be related approximately to the strain-hardening and the plastic-deformation portions of a stress-strain curve, with the effects of creep, anelasticity, and failure (particularly at very high stress) not clearly defined.

Ore Deposits

Variations in thermoluminescence of rocks and minerals from the vicinity of hydrothermal and contact metasomatic ore deposits can be grouped into the types illustrated in Fig. 3, which shows that thermoluminescence is (a) high immediately adjacent to the ore, but declining away from it to a more-or-less uniform "background" (or to nil values)–most typical of wall rocks that have suffered some type of recrystallization; (b) low immediately adjacent to the ore, but rising to high values some distance away and then declining to more-or-less uniform background (or nil) values at still greater distances–most typical of wall rocks where there has been introduction of supressing trace elements related to replacement-type ore deposits; (c) low immediately adjacent to the ore, but rising to a more-or-less uniform background–most typical in the vicinity of veins where there has been an introduction of suppressing elements into the wall rocks; (d) present, but with no distinctive change near the ore; and (e) not detectable.

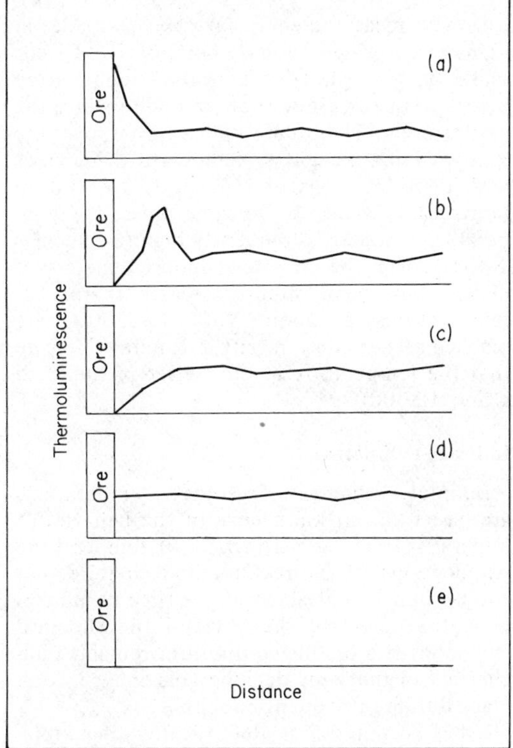

FIGURE 3. Various thermoluminescent responses of minerals in the vicinity of ore deposits (see text).

In general, thermoluminescence highs appear to reflect concentrations of defects and above-normal solid state and chemical reactivities. In many cases, high values appear due either to deformation (dislocations) or to recrystallization at elevated temperatures (grain growth or the formation of new minerals such as *albite* or *wollastonite*). In rare instances, highs appear to be related to above-normal radioactivity. On the other hand, many thermoluminescent lows in close proximity to ore deposits appear to be due to trace amounts of iron, nickel, or cobalt in the wall rocks (McDougall, 1968b). Studies of thermoluminescence near ore deposits have provided a very basic approach to the investigation of ore genesis (see *Mineral* and *Ore Deposits*).

Stratigraphic Correlation

An early attempt to use thermoluminescence for geological investigations was in the stratigraphic correlation of carbonate rocks. Attempts at correlation in a given basin of deposition rest on two basic assumptions: (1) if the physical and chemical conditions of a limestone bed are laterally persistent, artificially excited glow curves should be nearly identical throughout the bed; and (2) where there are lateral variations in the physical and chemical conditions, the thermoluminescence of samples taken from sections through a sedimentary sequence should serve to identify marker beds. In practice, thermoluminescence characteristics of individual limestone beds vary from virtually no persistence to persistence over reasonably long distances. However, local abnormalities are sufficiently common that most investigators have found it necessary to apply some form of statistical smoothing to their data in order to demonstrate correlation. In some cases, correlation or subdivision of limestone sequences has been found to be impractical. In addition to the work on limestones, at least one study has utilized thermoluminescence for correlation of volcanic tuffs.

DAVID J. McDOUGALL

References

Bäcktiger, K., 1967. Die Thermolumineszenz Einiger Skandinavischer und Nordamerikanisher Plagioklase, *Schweitz Mineral. Petrog. Mitt.*, 47(1), 366–370.

Bettinali, C.; Ferrasso, C.; and Manconi, J. W., 1967. Behavior of glow curves during phase transitions in inorganic compounds, *Rend. Cl. Sci. fis. mat. natur., Accad. Naz. Lincei, seri.* 8, **43**, 536–544.

Cameron, J. R.; Suntharalingam, N.; and Kenney, G. N., 1968. *Thermoluminescence Dosimetry.* Madison: University of Wisconsin Press, 231.

Douglas, G., and McDougall, D. J. 1973. Strain energy build-up in fatigue cycling, in *New Horizons in Rock Mechanics*, 14th symposium on Rock Mechanics. New York: American Soc. Civil Engineers, 121–126.

Fleming, S. J., and Thompson, J., 1970. Quartz as a dosimeter, *Health Physics*, **18**, 567.

Han, M. C., 1975. Effects of alpha and X-ray doses and annealing temperatures upon pottery dating by thermoluminescence, *MASCA Newsletter*, **II**(1), 1–2, University Museum, University of Pennsylvania.

Mazess, R. B., and Zimmerman, D. W., 1966. Pottery dating from thermoluminescence, *Science*, **152** (3720), 347–348.

McDougall, D. J., 1967. Thermoluminescence of geological materials, *Science*, **156**(3778), 1137.

McDougall, D. J., 1968a. A lattice defect-free energy, approach to replacement processes in ore deposition, *Econ. Geol.*, **63**, 671–681.

McDougall, D. J., ed., 1968b. *Thermoluminescence of Geological Materials.* New York: Academic, 678p.

McDougall, D. J., 1971. Comments on quartz as a dosimeter, *Health Physics,* **20**, 452–453.

Michael, H. N., and Ralph, E. K., eds., 1971. *Dating Techniques for the Archaeologist*. Cambridge: MIT Press, 227p.

Cross-references: *Crystals, Defects in; Luminescence; Order-Disorder; Rock-Forming Minerals; Thermometry, Geologic.*

THERMOMETRY, GEOLOGIC

Geologic thermometry concerns itself with the temperatures attained during geological processes. Such processes have been going on ever since the earth was born (about 4.75 billion years ago) at temperatures ranging from well below zero (glaciers) to several thousand degrees Centigrade (earth's core).

Numerous methods have been proposed and tried in efforts to measure or estimate geologic temperatures. Information is obtained by direct measurements when the geological processes can be observed and studied as they take place (e.g., hot springs, lava flows). Very accurate determinations of geologic temperatures can be made by these methods. More commonly, however, information is sought regarding conditions that existed during the geologic past. The textures (phenocrysts, mineral zoning, glass, replacements, etc.) produced in rocks and ores during cooling may, when observed in thin and polished sections, yield information on the sequence of mineral formation. Thus, relative temperatures may be established.

Absolute temperatures can only be obtained through deductions based on established data bearing on the effect of temperature on the physical and chemical properties of minerals and mineral systems (see *Mineral Properties*). The temperatures obtained through indirect methods are less accurate than those measured directly and depend, not only on experimental knowledge of mineral systems, but also on the validity of certain assumptions bearing on attainment of equilibrium at the time of rock formation or metamorphism and on preservation of mineral assemblages.

Direct Measurements

Direct measurements have been made of the temperature of hot springs, with results ranging from the mean regional annual temperature to the boiling point of water, which depends upon elevation of outlets and composition of the water. More sophisticated methods have been developed for direct measurements of fumarole temperatures, which vary from that of the boiling point of water to about 800°C. Temperatures ranging from 1050° to 1200°C have been measured directly in basaltic lava flows. In more felsic flows, temperatures ranging from 700° to 900°C have been recorded during extrusion.

Direct measurements have been made in many areas where mines, boreholes, or wells make it possible to determine temperature distribution for a few thousand meters into the earth's crust. These data indicate that, over the narrow known regions, temperature increases with depth at a rate of 1°C per 15 to 50 m in nonvolcanic areas. In volcanic areas, the temperature gradient is frequently high (as much as 2°C/m) over the first few hundred meters of depth, but then often decreases sharply as temperatures of about 250° to 300°C are reached. The most recent estimates indicate that the temperature at the center of the earth is about 3000°C.

Indirect Estimates

Indirect estimates of geologic temperatures are based on our knowledge of the behavior of mineral systems as a function of temperature. Applications of indirect methods require that equilibrium was attained at the time of mineral formation and that the physical and chemical characteristics of this equilibrium remained unchanged or that only decipherable changes took place during subsequent cooling.

It has been demonstrated, through phase relationships determined by experiments on synthetic systems, that minerals commonly are not preserved as originally deposited. Instead, when exposed to periods of cooling or metamorphism, minerals usually reequilibrate at widely differing rates to temperatures that may extend, particularly in sulfide systems, even below 200°C.

During the last decade, geothermometry has become one of the most active areas of research in petrology. Much attention has been given to pressure-temperature conditions in the upper mantle (see *Mantle Mineralogy*).

The partial pressures of volatiles such as oxygen and sulfur determine which oxides and sulfides are stable in a system at any specified temperature and total pressure. Since equilibrium conditions are assumed for deposition of the minerals studied for thermometric purposes, certain conclusions can be drawn concerning the partial pressure if the temperature of formation is known; or conclusions can be drawn about the temperature if the partial pressure is known. For instance, at very low partial pressures of oxygen, the phase Fe_2SiO_4 is stable; but at higher P_{O_2}, it breaks down to Fe_3O_4 (*magnetite*) and SiO_2 (*quartz*). In a pressure (P_{O_2}) versus temperature diagram, one field contains Fe_3O_4 + SiO_2 and another Fe_2SiO_4. The two fields are separated by a *P-T* curve along which all three phases are stable. This curve relates temperature of formation to partial pressure of oxygen for an assemblage containing all three minerals. Similar curves have been determined for numerous mineral assemblages.

In all systems applicable to geological temperature measurements, rock pressure significantly affects the stability fields and chemical compositions of the phases (for instance, the α–β inversion in *quartz* is at 573°C at 1 bar and at 815°C at 10 kilobars). Fractionation of isotopes among stable phases is practically unaffected by pressure. Therefore, deductions of temperatures, except those based on isotopic distributions, that existed during geological events, require knowledge of the influence of both pressure and temperature on the pertinent natural systems.

Because of the influence of both rock pressure and temperature on mineral equilibria, unique determinations of geologic temperatures cannot be obtained by applications of only one mineral pair unless the rock pressure can be estimated independently (see *Barometry, Geologic*). When two or more different geological thermometers can be used simultaneously, unique solutions for both temperature and pressure can be obtained provided the *P-T* curves of the two systems have significantly different slopes. Since the fractionation of isotopes between coexisting minerals apparently is not influenced measurably by pressure, methods of this kind may yield accurate geologic temperatures. When isotope data are combined with data obtained from pressure-influenced thermometers, accurate pressure estimates may result. The natural systems commonly contain several more components and, therefore, are more complicated than the systems that so far have been investigated in the laboratory. The number of components that can be systematically managed in laboratory studies is steadily increasing and the synthetic systems are gradually becoming more analogous to the natural ones. Thus, the possibilities of estimating geologic temperatures accurately are also increasing. Indirect methods for estimation of geologic temperatures are discussed below.

Melting Points. The *P-T* curve for the melting point of a mineral of fixed composition gives the maximum temperature of formation. Frequently, volatile components soluble in the liquid may lower the melting "point" drastically; and if this effect is not appreciated, wrong conclusions may result. (For instance, *diopside* melts at 1457°C and 5 kilobars in a dry system but at 1295°C under 5 kilobars of water pressure.)

Eutectics. Eutectic melting relationships exist between many common minerals, especially ores. Provided reactions with other minerals do not take place, the *P-T* curve of a eutectic indicates the maximum temperature of stability of a mineral association. Volatile components may drastically influence the temperature as well as the composition of the eutectic. Textures earlier considered to be due only to eutectic relationships have been demonstrated to form as a result of many other types of reactions, such as breakdown of one mineral to form two or more other minerals.

Breakdown Curves. Many minerals break down to form two or more other minerals when temperature exceeds certain limits. The *P-T* curves of such breakdowns give the maximum or minimum conditions for the stable existence of the minerals in question. As examples: *pentlandite*, on heating in the presence of vapor, breaks down at 610°C to *pyrrhotite* and a high-temperature modification of *heazlewoodite; polydymite*, in the presence of vapor, breaks down at 356°C to *millerite* and *vaesite*.

Inversion Points. Most minerals may exist in more than one polymorphic form (see *Polymorphism*). Pressure-temperature curves delineating polymorphic inversions have been of help in geologic thermometry. The initially deposited, high-temperature modifications often invert to low-temperature polymorphs on cooling. The low-temperature forms, which we are able to observe, frequently appear as pseudomorphs after the high-temperature modifications or retain other evidence, such as twinning, of the minerals having possessed the structure of the high-temperature forms. Thus, studies of such minerals frequently reveal whether mineral formation occurred above or below known *P-T* curves.

Mineral Assemblages. Coexisting minerals offer a variety of possibilities for estimation of geological temperatures. Laboratory studies have demonstrated that, in most systems representative of rocks and ores, common mineral assemblages are stable over limited temperature ranges. At these limits, reactions take place and new minerals form without change in the composition of the systems. As an example: *pyrite* (FeS_2) and *magnetite* (Fe_3O_4) in the pure Fe–S–O system form a stable mineral pair below 675°C. At this temperature, FeS_2 + Fe_3O_4 react to give $Fe_{1-x}S$ (*pyrrhotite*) and Fe_2O_3 (*hematite*). In pure systems under their own vapor pressure, such a reaction is invariant and, therefore, can only occur at one defined temperature. Under higher confining pressures, such as encountered in rocks and ores, the temperature of such reactions varies as a function of pressure. Numerous *P-T* curves of mineral reactions (tie-line switches) have been determined in laboratory experiments. They serve as very valuable indicators of the *P-T* conditions (facies) which existed when rocks and ores formed or were metamorphosed.

Solid solutions are common among coexisting minerals. Some solid solutions are extensive or even complete between mineral end members at elevated temperatures. The equilibrium solubilities in a given system depend on temperature and to a lesser extent on pressure. Promising possibilities exist for the purpose of geologic thermometry when the limit of solid solution of one phase in another is very sensitive to temperature and when the solid solutions remain unchanged (in a frozen condition) during the cooling of the rocks or ores in which they occur or when the original compositions can be inferred from those observed in geologic examples. The compositions of solid solutions between coexisting minerals vary with temperature and, when applicable, provide continuous temperature scales. Some examples of mineral pairs that form solid solutions useful in geologic thermometry are *pyrrhotite-sphalerite, pyrrhotite-pyrite, pyrite-vaesite, calcite-dolomite, magnetite-ilmenite*, and *albite-orthoclase*.

Geologic thermometers may also be based upon unsaturated solid solutions. The distribution ratio of a minor element (or elements) between two minerals in equilibrium is determined primarily by temperature and secondarily by pressure. Thus, for example, the distribution ratio of nickel between *pyrrhotite* and *pyrite* can be referred to a laboratory-determined ratio (Ni_{po}/Ni_{py})-versus-*T* curve to estimate the temperatures of formation of *pyrrhotite-pyrite* pairs in ores, provided equilibrium existed at the time of mineral formation and provided redistribution of nickel did not occur during the cooling of the minerals. The effects of rock pressure on such distribution ratios have yet to be determined.

The elements that form the compounds and minerals in synthetic and natural systems almost without exception are composed of two or more isotopes. For instance, sulfur in nature consists of a mixture of ^{32}S, ^{33}S, ^{34}S, and ^{36}S, of which ^{32}S accounts for about 95% and ^{34}S for more than 4%. Since isotopes are elements that occupy the same place in the periodic system and have the same nuclear charge but differ in atomic weight, the standard free energy of formation of minerals of identical "chemical" composition will vary slightly depending upon the isotopic conditions. (The standard free energy of formation of, for example, $^{208}Pb^{32}S$ is different from that of $^{208}Pb^{34}S$.) Such differences are, however, too small to influence measurably the thermal and compositional stabilities of minerals or mineral assemblages. However, the variations and distributions of isotopes among coexisting minerals can be measured by mass spectrometric methods and may provide valuable information on temperature of formation of mineral assemblages. It is possible, for example, that the fractionation of sulfur isotopes among coexisting sulfide phases under low-temperature equilibrium conditions may vary sufficiently with temperature to serve as indicators of geologic temperatures. It has been demonstrated that the fractionation of ^{18}O and ^{16}O between coexisting phases is sufficient to determine the temperatures of localities at which marine animals deposit their calcareous skeletons. Such studies have produced valuable information on paleotemperatures in the oceans and on Pleistocene climates. Some absolute thermometers utilizing $^{18}O/^{16}O$ ratios among coexisting minerals in rocks have now been developed: minerals are grown in equilibrium with one another at known temperatures and the isotope fractionation over a wide temperature range is determined in order to establish calibration curves. Much laboratory work remains to be done. So far, temperature trends in rocks and ores usually have been estimated from isotope studies of certain minerals that coexist in nature.

Other Indirect Methods. When minerals crystallize, samples of the solutions from which they precipitate are commonly trapped and retained as inclusions in crystal imperfections (see Vol. IVA: *Fluid Inclusions*). During cooling, the trapped solution contracts and a mixture of vapor and liquid exists. On heating at controlled rates under the microscope, it is possible to determine the temperature at which the solution on expansion again fills the cavity. Provided a number of assumptions (concerning filling of the cavities, volumes, primary and secondary inclusions, leakage, composition of the trapped solution, and *P-V-T* relationships) are valid, estimates can be made of the temperature existing at the time of mineral formation.

Several additional properties of minerals, such as conductivity, thermoluminescence, radiation colors, and metamictization, have been studied in efforts to establish methods for estimation of geologic temperatures. It has been shown that conductivity, as exemplified by numerous measurements on *pyrite*, does not convey information that can be related to the temperature of mineral formation. The thermoluminescence (q.v.) displayed by many minerals disappears when the minerals are heated. The temperature of its disappearance has sometimes been referred to as the temperature of formation of the minerals involved, but represents in reality an upper temperature limit beyond which the minerals could not have been heated in nature after becoming thermoluminescent.

Radiation colors also are dissipated on heating and are also acquired after formation of the minerals. The temperature of their disappearance indicates the maximum temperature to which a mineral could have been exposed in nature without losing its radiation color, but is not related to the temperature of mineral formation.

Certain minerals containing radioactive elements suffer destruction of crystal structure and are rendered metamict owing to emanations from these elements (see *Metamict State*). Since heating of such minerals restores the original crystal structures, it can be concluded that these minerals, after losing their structures, were not exposed long enough in nature to temperatures sufficiently high to promote recrystallization. As with thermoluminescence and radiation colors, however, this procedure gives no information bearing on the temperature of formation of minerals.

GUNNAR KULLERUD

References

Bowen, N. L., 1928. Geologic thermometry, in E. E. Fairbanks, ed., *The Laboratory Investigation of Ores*. New York: McGraw-Hill, chap. 10, 172–199.

Bowen, N. L., 1940. Geologic temperature recorders, *Sci. Monthly*, **51**, 5–14.

Buddington, A. F., and Lindsley, D. H., 1964. Iron-titanium oxide minerals and synthetic equivalents, *J. Petrology*, **5**, 310–357.

Carmichael, I. S. E.; Turner, F. J.; and Verhoogen, J., 1974. *Igneous Petrology*. New York: McGraw-Hill, 78–98.

Ingerson, E., 1955a. Geologic thermometry, *Geol. Soc. Am. Spec. Paper 62*, 465–488.

Ingerson, E., 1955b. Methods and problems of geologic thermometry, *Econ. Geol.*, 50th Anniv. vol., 341–410.

Kracek, F. C., 1942. Melting and transformation temperatures of mineral and allied substances, in *Handbook of Physical Constants, Geol. Soc. Am. Spec. Paper 36*, 139–174.

Kudo, A. M., and Weill, D. F., 1970. An igneous plagioclase thermometer, *Contr. Mineralogy Petrology*, **25**, 52–65.

Kullerud, G., 1959. Sulfide systems as geological thermometers, in *Researches in Geochemistry*. New York: Wiley, 301–335.

Kullerud, G., 1970. Sulfide phase relations, *Mineral. Soc. Am., Spec. Paper 3*, 199–210.

Seifert, H., 1930. Geologische Thermometer, *Fortschr. Mineral., Kristallogr. Petrogr.*, **14**, 167–291.

Smith, F. G., 1953. *Historical Development of Inclusion Thermometry*. Toronto: University of Toronto Press, 149p.

Spicer, H. C., 1942. Observed temperatures in the earth's crust, *Geol. Soc. Am. Spec. Paper 36*, 279–292.

Stormer, J. C., 1975. A practical two-feldspar geothermometer, *Am. Mineralogist*, **60**, 667–674.

Taylor, H. P., Jr., 1967. Oxygen isotope studies of hydrothermal mineral deposits, in H. L. Barnes, ed., *Geochemistry of Hydrothermal Ore Deposits*. New York: Holt, Rinehart & Winston, 109–142.

Williams, R. J., 1971. Reaction constants in the system Fe-MgO-SiO_2-O_2 at 1 atm between 900° and 1300°C. Experimental Results, *Am. J. Sci.*, **270**, 334–360.

See also *Am. Mineralogist*, **61**(7–8), for papers presented at the International Conference on Geothermometry and Geobarometry convened at The Pennsylvania State University, October 5–10, 1975.

Cross-references: *Barometry, geologic; Crystal Growth; Metamict State; Mineral Properties; Polymorphism; Thermoluminescence;* Vol. IVA: *Core Geochemistry; Crystal Chemistry; Fluid Inclusions; Mass Spectrometry; Solid Solution.*

THIXOTROPY

Thixotropy is an isothermal, reversible, time-dependent process occurring at constant composition and volume whereby a material stiffens while at rest and softens or liquefies upon remolding. This definition, is sufficiently general to account for the behavior of both

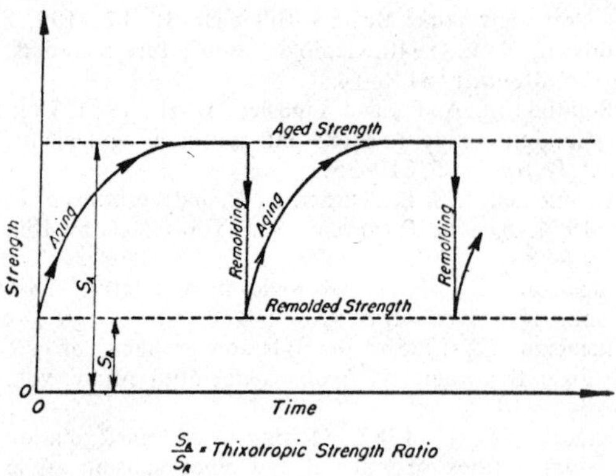

FIGURE 1. Properties of a purely thixotropic material (from Mitchell, 1961).

dilute and concentrated suspensions of clays or other colloidal materials (Mitchell, 1961).

The strength properties of a purely thixotropic material are indicated by Fig. 1. The "thixotropic strength ratio" provides a convenient measure of the intensity of thixotropic effects.

Thixotropy in Dilute Suspensions

Early studies of thixotropic behavior were concerned mainly with the reversible sol-gel transformations that can be observed in many colloidal suspensions. The stiffening of thixotropic materials during a period of rest is associated with a tendency for particles to flocculate, apparently into a linked, random array (see references in Mitchell, 1961). Suspensions of platy or elongated particles such as ***clays*** are particularly likely to exhibit thixotropic effects, and the extent to which thixotropy develops is greatly influenced by the electrolyte concentration of the suspending water. Thixotropic stiffening of suspensions can be likened to slow coagulation in a dilute sol, with the stiffening rate dependent on the size of the residual energy barrier in the particle interaction potential curve (van Olphen, 1963).

Thixotropy in Concentrated Suspensions

Several-fold increases in strength may result from aging remolded soils at constant water content, and some degree of thixotropy may be the rule rather than the exception in most sediments (Boswell, 1949). Even in cases where the ultimate strength is not increased appreciably, the stiffness of the material may be greatly increased as a result of thixotropic effects.

Thixotropic hardening has been suggested as a mechanism for the development of sensitivity to disturbance (strength loss by remolding) in natural ***clays***. Because of other post-depositional changes however, such as compression and leaching, it is virtually impossible to isolate the contribution of thixotropy in any given case (Mitchell and Houston, 1969). Nonetheless, almost all fine-grained soils are probably susceptible to thixotropic effects, and these effects may be of practical significance in both geological and engineering problems (see *Soil Mineralogy*).

Mechanism of Thixotropic Action

Thixotropic hardening is a natural response of a clay-water-electrolyte system to a change in ambient conditions. When a gel is stirred, or a ***clay*** at natural water content is remolded, a structure is induced that is compatible with the applied shearing stresses. During this process, the curves representing attractive and repulsive energies of interaction are as indicated in Fig. 2(a), with the net energy of interaction versus distance between particles relationship showing an energy barrier such that flocculation is prevented. When shearing stops upon removal of the externally applied energy, the system is left with an excess of internal energy and an interparticle force balance that favors flocculation, as shown schematically by Fig. 2(b).

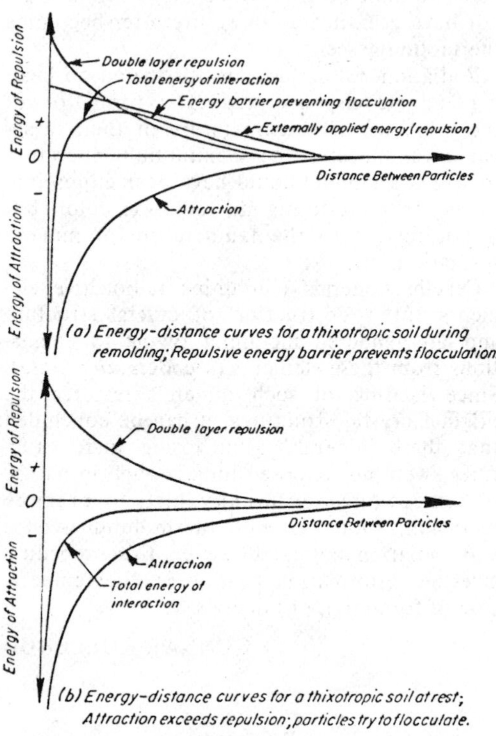

FIGURE 2. Energy-distance curves for a thixotropic soil (from Mitchell, 1961).

This excess internal energy is dissipated by means of small particle movements into a more flocculent array and changes in the adsorbed water structure, both of which lead to an increased strength. Because these changes require the movement of clay particles, water molecules, and solution ions, they are time and temperature dependent. From studies of the rate of thixotropic strength gain as a function of temperature (Mitchell, 1961), it has been determined that the activation energy for structural adjustment is of the same order as that for viscous flow of water. That small movements of particles occur during thixotropic hardening has been demonstrated by George (see Mitchell and Houston, 1969) using X-ray diffraction techniques. The pore-water tension (negative pore pressure) increases with time during the aging of thixotropic clay pastes (Ripple and Day, 1966). The increase in pore-water tension provides a direct measure of the dissipation of energy and of change in adsorbed water properties during thixotropic hardening.

It would appear that any fine-grained material will exhibit thixotropic behavior provided (1) the interparticle force balance is such that the material will flocculate if given a period of rest, and (2) this flocculation tendency can be overcome by mechanical working of the material.

JAMES K. MITCHELL

References

Boswell, P. G. H., 1949. A preliminary examination of the thixotropy of some sedimentary rocks, *Quart. J. Geol. Sci. London*, **104**, 499.

Fahn, R.; Weiss, A.; and Hofmann, H., 1953. Thixotropy in clays, *Ber. Dtsch. Keram. Ges.*, **30**, 21.

Fruendlich, H., 1935. *Thixotropy*. Paris: Hermann et Cie.

Mitchell, J. K., 1961. Fundamental aspects of thixotropy in soils, *Trans. Am. Soc. Civ. Eng.*, **126** (I), 1586–1626.

Mitchell, J. K., and Houston, W. N., 1969. The causes of clay sensitivity, *J. Soil Mech. Found. Div.*, **95** (SM3), 845–871.

Ripple, C. D., and Day, P. R., 1966. Suction responses due to homogeneous shear of dilute montmorillonite-water pastes, *Clays and Clay Minerals*, vol. 14. Pergamon, 307–316.

Roeder, H. L., 1939. *Rheology of Suspensions: A Study of Dilatency and Thixotropy*. Amsterdam: H. J. Paris.

van Olphen, H., 1963. *Introduction to Clay Colloid Chemistry*. New York: Wiley/Interscience, 137–139.

TOPOTAXY

The term "topotaxy" was applied by Lotgering (1959) to "all chemical solid state reactions that lead to a material with crystal orientations which are correlated with crystal orientations in the initial product." Various degrees of orientational control can be recognized. In displacive polymorphic transitions, a single crystal may be transformed into another single crystal, with preservation of the lattice modified only by a dilatation. In exsolution, the newly formed intergrown phases frequently bear a simple structural relationship to the parent phase, and may share a plane of atoms at the phase boundary determining their mutual relationships. Thermal dehydration or dehydroxylation may proceed with retention of certain features of the structure fixing the orientation of the newly formed phase. If the transformation involves great change in chemical composition, or if the polymorphic transition is reconstructive, there may be no preferred orientation, or only a low degree of orientation, in the product. In general, it may be expected that retention of symmetry elements, small volume change, and small activation energy will favor a high degree of orientation. The criteria used in interpreting topotactical reactions have been succinctly set forth by Shannon and Rossi (1964).

Though some of these phenomena have long been known, the coining of a name for them coincided with a resurgence of interest, and the word "topotaxy" has been redefined several times (Mackay, 1960; Bernal, 1960; Glasser, Glasser, and Taylor, 1962). The apparent need for redefinition arises in part from the varying degrees of topotaxy, and in part from diverging views as to the mechanism. In any case, a distinction is to be made from epitaxy (q.v.) and from syntaxy (q.v.).

Topotaxy is common among minerals, occurring either under natural conditions, or when minerals are treated in the laboratory. One of the most thoroughly studied cases is the conversion of single crystals of *goethite* into single crystals of *hematite* with the following correspondence of crystallographic features:

	Goethite		*Hematite*
a	4.65	*c*/3	4.59
b	10.02	2*a*	10.08
c	3.04	*a*T3/3	2.91
volume	141.64 Å^3		134.28 Å^3
content	4(FeOOH)		5-1/3(Fe)8(O)

This oriented conversion was first described by Böhm (1928). He suggested a mechanism involving oriented growth of the newly formed crystallites upon the retreating surface of the hydroxide that is being destroyed, and referred to this as a "topochemical" reaction. This transformation has since been studied by many others. According to current views, it may be pictured as involving the retention of oxygen atoms in that part of the original crystal that is transformed to *hematite*. Hydrogen atoms from this part combine with OH and O from a part that is being destroyed and are driven off as H_2O, while Fe atoms migrate into the part being transformed to *hematite*. During these changes, the oxygen arrangement in the preserved part suffers little change. Exactly similar relations exist in the transformation of *diaspore* to *corundum*.

More complicated relationships have been described in a number of cases. It may be that several orientations are produced, as when oriented ***olivine*** is formed as the product of thermal transformation of ***chlorite*** (Brindley and Ali, 1950), or that orientation is maintained to some degree during a sequence of changes as in the dehydration of *hemimorphite* (Taylor, 1962).

The retention of structural features during a transformation, the essence of topotaxy, does not always require that crystal axes of the initial material and the product coincide. When *rhodonite* is transformed to the *wollastonite* structure, a change that occurs rapidly at 1100°C., the paramorph formed has its *b* axis, the direction of the silicate chains, inclined ≈13° to the *c* axis of the original *rhodonite*, the direction of the initial silicate chains. This shift of direction is a key to the mechanism of the transformation, which involves breaking and reconstitution of the chains in a systematic manner (Glasser and Glasser, 1961).

The phenomena now referred to as topotaxy have generally not been described under this designation; but the term is finding wide acceptance, and the importance of the relationships to which it applies is generally recognized.

A. PABST

References

Bernel, J. D., 1960. On Topotaxy, *Schweiz. Arch.*, **26**, 69–75.

Böhm, J., 1928. Röntgenographische Untersuchung der mikrokristallinen Eisenhydroxidminerale, *Zeit. Kristallog.*, **68**, 567–585.

Brindley, G. W., and Ali, S. Z., 1950. X-ray study of thermal transformations in some magnesian chlorite specimens, *Acta Crystallogr.*, **3**, 25–30.

Dent Glasser, L. S., and Glasser, F. P., 1961. Silicate transformations: rhodonite-wollastonite, *Acta Crystallogr.*, **14**, 838–822.

Dent Glasser, L. S.; Glasser, F. P.; and Taylor, H. F. W., 1962. Topotactic reactions in inorganic oxy-compounds, *Quart. Rev., Chem. Soc., London*, **16**, 343–360.

Lotgering, K. F., 1959. Topotactical reactions with ferrimagnetic oxides having hexagonal crystal structures, I, *J. Inorg. Nucl. Chem.*, **9**, 113–123.

Mackay, A. L., 1960. Some aspects of the topochemistry of the iron oxides and hydroxides, *Proc. 4th Intl. Symp. on Reactivity of Solids*. Amsterdam: Elsevier, 571–583.

Royer, M. L., 1928. Recherches expérimentales sur l'epitaxie ou orientation mutuelle de cristaux d'espèces différentes, *Bull. Soc. fr. Minéral.*, **51**, 7–156.

Shannon, R. D., and Rossi, D. C., 1964. Definition of topotaxy, *Nature*, **202**(4936), 1000–1001.

Taylor, H. F. W., 1962. The dehydration of hemimorphite, *Am. Mineralogist*, **47**, 932–944.

Cross-references: *Crystal Growth; Epitaxy; Optical Mineralogy; Polymorphism; Syntaxy.*

TOURMALINE GROUP

A group of complex cyclosilicates of boron and aluminum, the structure of which is based on six-membered rings of SiO_4^{4-} tetrahedra. ***Tourmaline*** compositions are quite variable, but can be expressed by the general formula

$$XY_3Z_6(BO_3)_3Si_6O_{18}(OH,F)_4$$

where X is usually Na^+ often partially replaced by Ca^{2+}, K^+, Mg^{2+}, or a vacancy; Y is Mg^{2+}, Fe^{2+}, Fe^{3+}, Li^+, or Al^{3+}; and Z is predominantly Al^{3+} or Fe^{3+}. Rock-forming species include: brown, Mg-rich *dravites*; black, Fe^{2+}-rich *schorls*, and red, green, or blue, Li-bearing *elbaites* which are often used as gem stones (see *Gemology*). Crystals are usually prismatic, vertically striated, and either 3, 6, or 9-sided in cross section.

Tourmaline (Fig. 1), which is by far the most common borosilicate, occurs in granites, granite pegmatites, and in associated reaction zones (greisen) as a product of the crystallization of late-stage fluids and(or) their reaction with preexisting rocks (see *Pegmatite Minerals*). In metamorphic rocks (i.e., schists, gneisses), ***tourmaline*** is a product of boron metasomatism or recrystallization of boron-bearing sediments; in sedimentary rocks, it occurs as detritus and as a result of authigenic crystallization.

A review of ***tourmaline*** mineralogy with many references to the earlier literature is to be found in Deer, Howie, and Zussman (1962).

Crystallography and Habit

Tourmaline belongs to the trigonal subsystem of the hexagonal system. Its space group $R3m$ denotes a rhombohedral unit cell, the symmetry of which conforms to a threefold axis paralleling the intersection of three symmetrically equivalent mirror planes.

Tourmaline occurs as isolated crystals, colum-

FIGURE 1. ***Tourmaline***, *elbaite*, San Diego County, California (from P. E. Desautels, *Mineral Kingdom*; New York: Ridge Press, 1968, p. 102; reprinted by permission of the author).

nar masses, or radiating groups. The crystals are often terminated by the rhombohedrons $r\{10\bar{1}1\}$ or $o\{02\bar{2}1\}$ or by the pedion $c\{0001\}$ (Fig. 2). Doubly terminated crystals usually show different forms at opposite ends of the threefold or c axis (hemimorphism), reflecting the acentric symmetry of the internal structure. Cross sections of crystals frequently appear as spherical triangles due to the oscillatory development of prism faces belonging to the forms $a\{11\bar{2}0\}$ and $m\{10\bar{1}0\}$.

Physical and Optical Properties

The hardness of ***tourmalines*** is between 7 and 7½ while density varies from 3.0 to 3.25, increasing with iron and manganese content. The $\{11\bar{2}0\}$ and $\{10\bar{1}1\}$ cleavages are poorly developed; but ***tourmaline***, being brittle, often shows conchoidal fracture. Strong pyroelectric and piezoelectric properties are characteristic of ***tourmaline*** (see *Piezoelectricity*). The principal X-ray powder diffraction reflections for the Mg-rich end member, *dravite*, correspond to d-values of 6.38, 4.22, 3.99, 3.48, 2.96, 2.58, 2.04, and 1.92 Å.

Tourmaline is optically anisotropic (uniaxial negative) and displays the following ranges of refractive indices; $n_\epsilon = 1.610–1.675$, $n_\omega = 1.635–1.675$. Crystals may be of almost any color and color zoning commonly appears in cross sections or along the length of the crystal. Under the polarizing microscope, the colored varieties frequently exhibit two different colors (dichroism) corresponding to differences in spectral transmission for rays vibrating parallel to ϵ and ω. In general, the transmission of light is greater for rays vibrating parallel to ϵ(parallel to the threefold axis) than for rays vibrating parallel to ω(perpendicular to the threefold axis).

Structure

Early ***tourmaline*** structural investigations were carried out by Hamburger and Buerger, by Belov and Belova, and by Ito and Sadanaga (see references in Buerger et al., 1962). The structure (Fig. 3) basically consists of six-membered rings of SiO_4^{4-} groups interlayered with triangular BO_3^{3-} groups along the threefold symmetry axis (axis of crystal elongation). Octahedrally coordinated cations (Y) within the ring bind the rings into tubular columns while other octahedrally coordinated cations (Z), together with three hydroxyl groups which lie outside of each ring, link adjacent columns. The X cations alternate with a fourth hydroxyl group along the threefold axis in the center of the rings. All members of this group are thought to be isostructural. The distribution of Y and Z cations between the two octahedral sites and

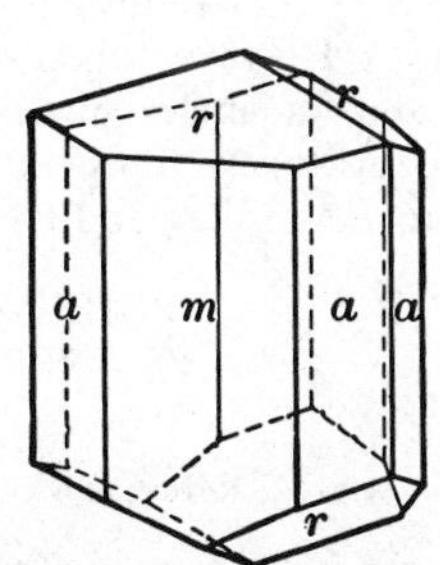

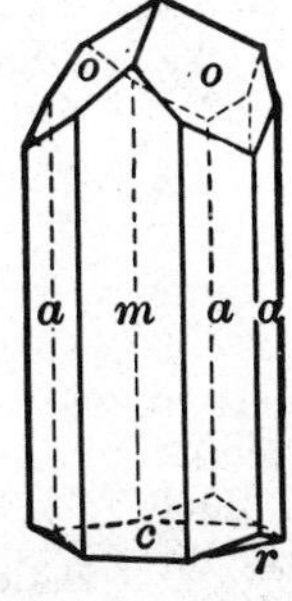

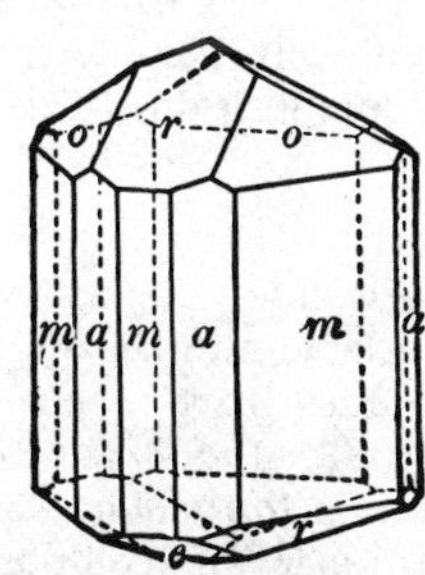

FIGURE 2. ***Tourmaline*** morphology (from E. S. Dana and W. E. Ford, 1954. *Textbook of Mineralogy*, 4th ed.; New York: John Wiley, p. 635; copyright © 1954 by John Wiley & Sons, Inc.; reprinted by permission).

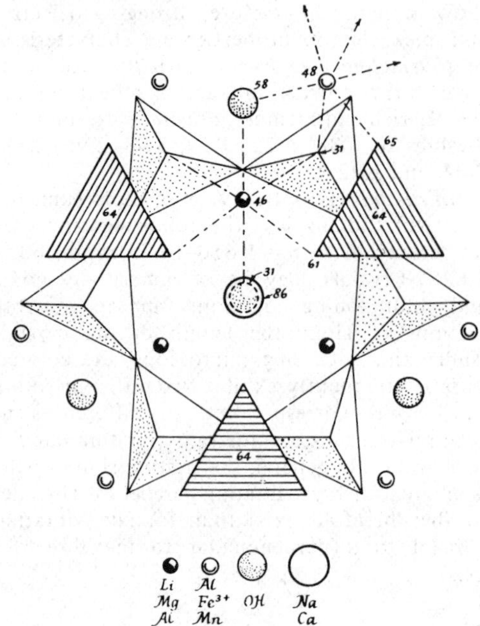

FIGURE 3. Structure of ***tourmaline*** projected on (0001) (from M. H. Battey, 1972, *Mineralogy for Students*; New York: Longman Group, p. 255; reprinted by permission).

the hydrogen-atom locations are known for *buergerite* (Barton, 1969; Tippe and Hamilton, 1971), *elbaite* (Barton and Donnay, 1971), *schorl* (Fortier and Donnay, 1975), and V-bearing *dravite* (Foit and Rosenberg, 1979). Some evidence for minor substitution of Al^{3+} for Si^{4+} has been obtained from infrared spectral studies.

Tourmaline crystals exhibit a marked pyroelectric and piezoelectric effect in compliance with the acentric symmetry resulting from a preferred orientation of the SiO_4^{4-} tetrahedra and the tetrahedral rings in the structure. The unit-cell dimensions (referred to hexagonal axes) of the ***tourmaline*** group minerals show the following ranges: a=15.84–16.03 Å, c= 7.10–7.25 Å; unit-cell parameters vary systematically with composition.

Chemistry

Although ***tourmaline*** chemistry is quite variable in that a large number of compositionally diverse varieties have been synthesized (Taylor and Terrell, 1967) and reported in nature, rock-forming ***tourmalines*** have a much more restricted range of compositions. Important end members of natural solid-solution series, both partial and complete, are as follows:

- *dravite:* Na Mg_3 Al_6 $(BO_3)_3Si_6O_{18}(OH,F)_4$
- *ferridravite:* Na Mg_3 Fe_6^{3+} $(BO_3)_3Si_6O_{18}(OH,F)_4$
- *schorl:* Na Fe_3^{2+} Al_6 $(BO_3)_3Si_6O_{18}(OH,F)_4$
- tsilaisite: Na Mn_3^{2+} Al_6 $(BO_3)_3Si_6O_{18}(OH,F)_4$
- *elbaite:* Na $(Li,Al)_3$ Al_6 $(BO_3)_3Si_6O_{18}(OH,F)_4$
- *uvite:* Ca Mg_3 $(MgAl_5)(BO_3)_3Si_6O_{18}(OH,F)_4$
- *buergerite:* Na Fe_3^{3+} Al_6 $(BO_3)_3Si_6O_{18}O_3(OH,F)$
- *liddicoatite:* Ca $(Li,Al)_3$ Al_6 $(BO_3)_3Si_6O_{18}O_3(OH,F)$

Complete ranges of composition exist between *schorl* and *dravite* and between *schorl* and *elbaite*, but intermediate compositions between *dravite* and *elbaite* have not been reported.

Foit and Rosenberg (1977) have shown that substitution of the type $(Mg^{2+},Fe^{2+})+H^+=Al^{3+}$ is very significant in natural ***tourmalines***. The extent of this substitution may be expressed in terms of a hypothetical Al-rich end-member analogous to the rare Fe^{3+}-rich ***tourmaline***, *buergerite*. Replacement of (Mg^{2+},Fe^{2+}) by Al^{3+} may also be partially compensated for by the creation of vacancies at the alkali-cation site. *Elbaite* participates in the buergerite-type substitution, but its substitutional relationships are more complex than those of *schorl* or *dravite* in that most specimens are Al-rich, Li-deficient, and many contain excess H_2O. Fe^{3+} may also substitute for Al^{3+} as in *ferridravite* (Walenta and Dunn, 1979). Small, but significant, amounts of Ca^{2+} and Mn^{2+} in some ***tourmalines*** may be expressed in terms of the end members *uvite* (Dunn et al., 1977b) and tsilaisite, respectively. The Ca analog of *elbaite*, *liddicoatite*, has recently been described (Dunn et al., 1977a).

Although ***tourmaline*** has been synthesized many times and crystal growth studies have been successfully attempted, little is known of the compositional variability or stability relationships of ***tourmaline*** in synthetic systems. *Schorl, dravite*, and *elbaite* have been synthesized under hydrothermal conditions in the temperature range 350–850°C at pressures up to 2.5 kilobars (e.g., Tomisaka, 1968). The upper thermal stability of *dravite* is approximately 865°C at 2 kilobars (Robbins and Yoder, 1962). However, alkali-free ***tourmalines***, with Mg and vacancies at alkali sites, which decompose at lower temperatures, have been synthesized (Rosenberg and Foit (1979).

PHILIP E. ROSENBERG
FRANKLIN F. FOIT, JR.

References

Barton, Jr., R., 1969. Refinement of the crystal structure of buergerite and the absolute configuration of tourmaline, *Acta Crystallogr.*, **B25**, 1524.

Barton, Jr., R., and Donnay, G., 1971. Refinement

of the crystal structure of elbaite, *Am. Mineralogist*, **56**, 356.

Buerger, M. J.; Burnham, C. W.; and Peacor, D. R., 1962. Assessment of the several structures proposed for tourmaline, *Acta Crystallogr.*, **15**, 583.

Deer, W. A.; Howie, R. A.; and Zussman, J., 1962. *Rock-Forming Minerals*, Vol. 1, Ortho-and ring silicates. London: Longmans, 300–319.

Dunn, P. J.; Appleman, D. E.; and Nelen, J. E., 1977a, Liddicoatite, a new calcium end-member of the tourmaline group, *Am. Mineralogist*, **62**, 1121–1124.

Dunn, P. J.; Appleman, D. E.; and Norberg, J., 1977b. Uvite, a new (old) common member of the tourmaline group and its implications for collectors, *Mineral. Record*, **8**, 100–108.

Foit, Jr., F. F., and Rosenberg, P. E., 1977. Coupled substitutions in the tourmaline group, *Contr. Mineral. Petrol.*, **62**, 109–127.

Foit, Jr., F. F., and Rosenberg, P. E., 1979. The structure of vanadium-bearing tourmaline and its implications regarding tourmaline solid solutions, *Am. Mineralogist*, **64**, 180–186.

Fortier, S., and Donnay, G., 1975. Schorl refinement showing compositional dependence of the tourmaline structure, *Can. Mineralogist*, **13**, 173–177.

Robbins, C. R., and Yoder, H. S., 1962. Stability relations of dravite, a tourmaline, *Carnegie Inst. Wash. Ann. Rep. 1961-1962*, 106.

Rosenberg, P. E., and Foit, Jr., F. F., 1979. Synthesis and characterization of alkali-free tourmaline, *Am. Mineralogist*, **64**, 180–186.

Taylor, A. M., and Terrell, B. C., 1967. Synthetic tourmalines containing elements of the first transition series, *J. Cryst. Growth*, **1**, 238.

Tippe, A., and Hamilton, W. C., 1971. A neutron diffraction study of the ferric tourmaline, buergerite, *Am. Mineralogist*, **56**, 101.

Tomisaka, T., 1968. Synthesis of some end-members of the tourmaline group, *Mineral J.* (Japan), **5**, 355.

Walenta, K., and Dunn, P. J., 1979. Ferridravite, a new mineral of the tourmaline group from Bolivia, *Am. Mineralogist*, **64**, 945–948.

Cross-references: *Authigenic Minerals; Crystal Growth; Fire Clays; Gemology; Pegmatite Minerals; Phantom Crystals; Piezoelectricity; Placer Deposits; Pleochroic Halos; Rock-Forming Minerals; Soil Mineralogy*; see also mineral glossary.

TRIBOLUMINESCENCE–*See* LUMINESCENCE

TRIPOLITE–*See* ABRASIVE MATERIALS. Vol. VI: TRIPOLI, TRIPOLITE

TWINNING

A "twin crystal" is a composite crystal of a single substance in which adjacent parts have different, but crystallographically related, orientations. The term "twinning," beside referring to the creation of twin crystals, also includes "translation gliding" (see below) which, although related to other processes of twinning, does not itself produce twins.

If two parts of a twin crystal meet in a plane, that plane is a "composition plane" and the twin crystal is a contact twin. If, however, the parts meet along a complex and irregular surface, there is no composition plane and the twin crystal is a "penetration twin" or "interpenetration twin."

Twin crystals consisting of only two parts are "simple twins." If more than two parts are present, either there are only two orientations, a "polysynthetic twin" (if the parts are thin plates, "lamellar twin"), or there are more than two orientations, a "multiple twin."

Geometry of Twin Crystals

The relationship between two parts of a twin crystal is expressed in terms of two concepts, the twin axis and the twin plane. When this relationship is that between an object and the same object rotated through 180° (or, rarely, 120° or 60°) about a line having a given orientation, that line is known as a "twin axis" and the twin crystal as a "rotation twin" or "hemitrope" (note that twins never actually form by such rotation; see *Symmetry*). When the relationship between the parts is that of an object and its reflection in a reflecting plane having a given orientation, such a plane is known as a "twin plane" and the twin crystal as a "reflection twin" or "symmetric twin." Twinning in crystals belonging to any of the centro-symmetrical classes can be described geometrically by reference either to a twin plane or to a 180° rotation about a twin axis normal to the twin plane. In other classes, the results of the two operations will be different. In about 90% of known twins, a twin axis is normal to a twin plane that is a possible crystal face and is identical with the composition plane. These are known as "normal twins." Other possibilities are that the twin axis is a possible crystal edge (in "parallel twins") or is perpendicular to a possible crystal edge (in "complex twins"). In both of these, the twin axis is parallel to the composition plane.

Fig. 1 illustrates a simple-parallel contact twin of *orthoclase* in which the composition plane is (010) and the twin axis is [001] which is the *c*-crystallographic axis. This type of twinning is known as twinning according to the Carlsbad law and the twin as a "Carlsbad twin" after the locality where such twin crystals were found (see *Alkali Feldspars; Feldspar Group*).

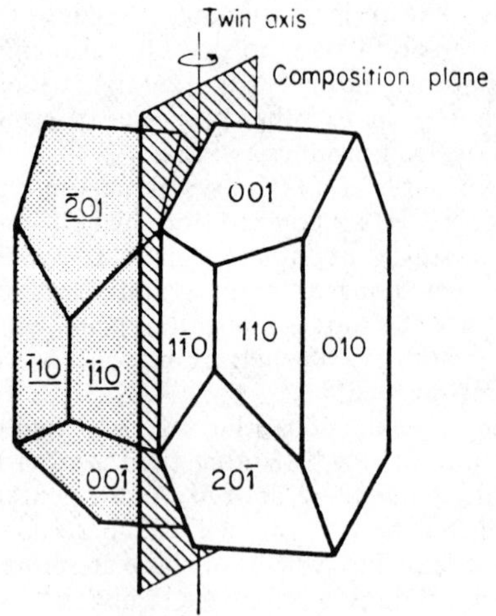

FIGURE 1. Carlsbad twin of *orthoclase* showing twin axis and composition plane (from A. C. Bishop, 1967, *An Outline of Crystal Morphology*; London: Hutchinson Publishing Group Limited, p. 272).

Genesis of Twin Crystals

Twin crystals may be formed by growth twinning, transformation twinning, or gliding.

Growth twins have grown as twins from an early stage. They occur because the energy stage of the twinned crystal is sufficiently close to that of the untwinned crystal for twinning to occur due to random fluctuations in growth. They probably persist because rapid growth in a supersaturated solution, or the large size of the crystal makes this energy difference small in comparison with the surface energy of the growing faces (see *Crystal Growth*).

Transformation twins may be formed when a mineral is transformed from a high temperature polymorph to one stable at a lower temperature. Such transformation usually begins at a number of points, and, if the orientation of the developing polymorph is not the same at all points, complete transformation may result in interpenetration twins. A well-known example is the development of Dauphiné twins, with the *c*-crystallographic axis as the twin axis, in *quartz* (q.v.).

Deformation twinning is the result of the application of stress after the crystallization of the mineral. Deformation twinning and transformation twinning are often known as "secondary twinning" as opposed to growth twinning which is "primary twinning." There are two types of deformation twinning. "Translation gliding" occurs when one part of the structure is displaced laterally (glides) by a whole but unlimited number of interatomic distances with respect to the other. It does not result in any structural break across the glide plane nor in the development of twin crystals. "Twin gliding" occurs when one part of the structure is displaced laterally by a fraction of an interatomic distance. Repeated on a number of adjacent glide planes, it produces a visibly twinned crystal. (For further discussion see Vol. V, *Petrofabrics* and *Rock Deformation*.)

Importance of the Study of Twinning

Growth twinning, besides its importance in crystallography, has long been important as a diagnostic property in optical mineralogy (q.v.). It is especially used in the ***feldspar*** group of minerals. Gorai (1951) has studied the relative frequency of twinning according to different laws in ***feldspars*** of igneous and metamorphic origin and this may form the basis of method in the study of rock genesis. Transformation twinning is of importance in mineral thermometry in that it may indicate whether a mineral grew as a low- or high-temperature form. Deformation twinning is the key to the study of rock deformation and petrofabric analysis (see *Plastic Flow in Minerals*).

MICHAEL J. FROST

References

Bishop, A. C., 1967. *An Outline of Crystal Morphology*, London: Hutchinson, 314p.

Buerger, M. J., 1945. The genesis of twin crystals, *Am. Mineralogist*, **30**, 467–482.

Gorai, M., 1951. Petrological studies on plagioclase twins, *Am. Mineralogist*, **36**, 884–901.

Cross-references: *Crystal Habit; Crystallography: Morphological; Directions and Planes; Feldspar Group; Plastic Flow in Minerals; Point Groups; Quartz; Symmetry.*

V

VEIN MINERALS

Vein minerals originated mostly (1) by deposition in free spaces within a fracture (fissure) that either was opened or was gradually opening during the period of mineral deposition, or (2) by replacement of the host rock along an unopened fracture. In the first case, we deal with true veins (fissure veins), in the second case with metasomatic veins. The term "deposition" is used here to cover both immediate growth of aggregates by crystallization from solutions (or, generally, from fluids) and colloidal deposition with later recrystallization.

There also exist intermediate types of veins or, generally, ore bodies with a combination of infilling of open spaces and replacement of the surrounding host rock. And there are still other, less common, types of vein fillings (see Bateman, 1951; Lindgren, 1933; and Park and MacDiarmid, 1975).

On a microscopic scale, the infilling of minor fractures gives rise to "veinlets with matching walls" (Bastin 1950), typically composed of one or two minerals without any apparent crustification. This type of veinlets is characterized by straight boundaries and similar changes in the course of their opposite walls. By contrast, the boundaries of the metasomatic veinlets are usually irregular. The minerals composing the microscopic veinlets are either hypogene or supergene.

On a megascopic scale, veins, both true and metasomatic, with a thickness of up to several meters, or tens of meters in case of a number of closed-spaced veins, sometimes extend horizontally for hundreds of meters or even for a few kilometers. In a vertical section, ore veins may extend for less than one hundred meters, but commonly for a few to several hundred meters. In the case of a system of vertical veins, especially when they are branching in an upward direction, the whole vein system may extend for more than one or two kilometers (e.g., the Ag-Pb-Zn veins at Příbram, Czechoslovakia; the Au-bearing veins in the Kolar mining district, Mysore state, India; and the Pb-Zn veins in the Coeur d'Alene mining district in Idaho, among others).

Zonal Distribution of Vein Minerals

The greatest concentration of ore-bearing veins is in areas manifesting deep-seated tectono-magmatic processes. The distribution of vein minerals changes with time and space, depending on physicochemical and tectonic conditions at the time of mineral deposition; and these changes cause mineral zoning.

Zoning appears both regionally and in individual veins. One of the most common features is the tendency of Sn and W minerals to concentrate in or very close to granitic massifs, the Pb and Zn minerals further away, and the Hg minerals most remotely. A general discussion of the zonal phenomena has been presented in the English literature by Park and MacDiarmid (1975) and extensively analyzed in two volumes of the symposium on *Postmagmatic Ore Deposition* (1963 and 1965).

The Source of Elements in Vein Minerals

The elements composing vein minerals come from different sources. Isotopic composition suggests a mantle origin of some metals, while several other elements evidently come from the crust. Some elements may be derived by magmatic differentiation; others might have been generated from rock suits through which the hot solutions penetrated.

The water of the hydrothermal solutions need not always be magmatic, but may be derived from rocks subjected to metamorphism at distinct depths. Some vein fillings, especially in easily soluble rocks in the near-surface portions of the Earth's crust, contain minerals deposited from descendent, low-temperature solutions, with water of atmospheric origin (see *Mineral Deposits: Classification*).

Descending solutions are also responsible for secondary changes in primary ore veins within the zone of oxidation and most of the cementation zones. A special kind of deposition of minerals from low-temperature brines may also be partly guided by fractures in some sedimentary rocks.

One of the best examples of the dependence of the vein components on the composition of the country rock is the Alpine paragenesis consisting of *quartz*, adularia (*orthoclase*), *epidote, zoisite,* ***chlorite,*** *sphene* (*titanite*), *anatase, brookite, rutile, prehnite*, some ***zeolites***, and others in open fractures in the metamorphic rocks of the Alps and other regions (see Niggli et al., 1940).

A historical dispute between Sandberger and Pošepný at the end of the last century (Pošepný, 1902) concerned the question of whether the metal contents, namely Pb and Ag, of the Middle-European deep-reaching veins (i.e., Příbram, Freiberg) were derived from the country rock by lateral secretion as advocated by Sanderberger, or carried from greater depth by ascending solutions. The quantitative analyses and calculations presented by Pošepný confirmed the ascendent nature of the respective metals.

Similarly, the high Mn contents of the *siderite* in the Příbram ore veins (up to 18% MnO) could not be derived from the surrounding diabases (with about 0.3% MnO) when the total amount of *siderite* in the veins is compared to the amount of leached diabase (Kutina et al., 1970). Generation of manganese from some Mn-rich rock suits beneath the ore deposit or derivation from a magmatic source have been proposed as two possible explanations.

The results of quantitative calculations such as the above do not mean, of course, that the respective elements must have been supplied by magmatic differentiation. A good example comes from the most recent discussion of the possible source of sulfur in endogenic sulfide ore deposits. A considerable amount of sulfur may be generated from the sulfate-*meionite* component ($3CaAl_2Si_2O_8 \cdot CaSO_4$) of ***scapolites*** that now appear to be a common constituent of the granulites in the lower crust (Goldsmith, 1976). The sulfur, released by retrograde metamorphism, may then be available for reaction with metals that may be generated from the rocks through which the solution penetrates and not necessarily from the wall-rock in the places of ore deposition.

The source of metals remains controversial in a number of ore deposits. One that is the subject of a long-standing discussion is the low-temperature, Mississippi-Valley-type deposit of lead, zinc, *fluorite*, and *barite* and its European counterparts (Barnes, 1967; Park and MacDiarmid, 1975; and Routhier, 1963).

Monoascendent and Polyascendent Vein Fillings

In the course of the formation of an ore deposit or an ore vein, tectonic movements often reopen old fractures, brecciate the old infillings, and permit the ascent of new solutions depositing a new assemblage of minerals. If such events proceed on a broad scale within an ore deposit and if the composition of the new mineral assemblage is considerably different from the preceding one, a new stage of mineralization is usually distinguished.

Systematic studies have shown that the process of mineral deposition during the individual stages of mineralization did not proceed continuously at all places (Fig. 1). Consequently, several infilling periods (*die Zufuhrperiode*) can often be distinguished, the number of which differs from place to place depending on the process of repeated reopening and infilling of fissures (Kutina, 1955; 1965).

Distinguishing between monoascendent and polyascendent character of mineral deposition

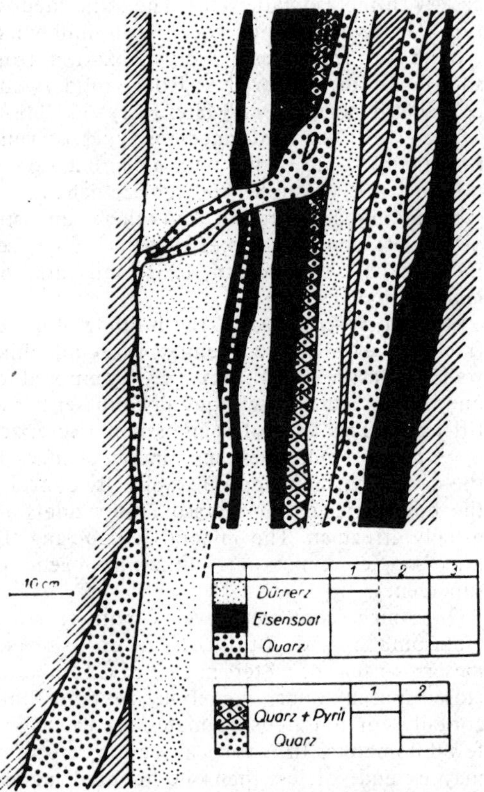

FIGURE 1. An example of a polyascendent vein filling in which the younger, barren *quartz* intersects the bands of ore-bearing *Dürrerz*, *siderite* (*Eisenspat*), and *quartz* + *pyrite*; Main Ethelbert Vein (Adalbert-Hauptgang), 38th level of the Anna mine, Příbram, Czechoslovakia (from Kutina, 1955). *Dürrerz* is a fine intergrowth of *siderite* and *quartz* and contains a number of ore minerals. The *siderite* in the *Dürrerz* is being replaced by *quartz* with the formation of numerous *quartz* metacrysts, and the *siderite* is being subsequently replaced by *sphalerite*, *galena*, and *boulangerite*. *Pyrite* and *arsenopyrite* are common.

is of great importance for predicting changes with depth of mineral composition in individual ore veins and ore deposits. For instance, the amount of younger mineral(s) may increase with increasing depth as a consequence of a downward widening of the reopened space within a distinct depth interval resulting in reverse polyascendent zoning (Kutina, 1957a; 1965; Fig. 2).

Sequence of Deposition of Vein Minerals

Vein minerals usually show a distinct sequence of deposition, known as "paragenesis," in the English literature. In non-English speaking European countries, paragenesis has usually been used, since the time of Breithaupt, for the association (assemblage) of minerals of a common origin (see Kutina et al., 1965).

Age relationships of minerals are especially complicated in the polyascendent vein fillings where solutions of younger infilling periods penetrated into older infillings, giving rise to complicated replacement textures. (For criteria of the age relationships of minerals, criteria for recognizing the metacrysts of minerals, and multiple selective replacements, see Bastin, 1950; Ramdohr, 1960; Betekhtin et al., 1958, 1964; and Kutina, 1956, 1963a,b).

Deposition of minerals symmetrically away from the walls of a fissure gives rise to symmetrical banded textures, composed of many bands in some cases as it illustrated in the classical works by Pošepný (1902) and Beck (1909) and in the newer treatises on ore deposits by Lindgren (1933) and by Jensen and Bateman (1979).

Repeated reopening and infilling of fissures is responsible for most asymmetric banded textures (Fig. 1).

Subsequent deposition of mineral bands around rock fragments accumulated in the open fissures, and their cementing by another component, gives rise to cockade textures. In some instances, however, the band(s) surrounding the fragments is(are) younger than the cement due to penetration of younger solutions along the boundaries between fragments and the cementing mineral due to partial replacement of the latter (Kutina and Sedlačková, 1961).

Generations of Vein Minerals

Commonly, deposition of some minerals is repeated two or even more times during the formation of an ore vein or an ore deposit. More than one generation have been observed in *quartz, siderite, calcite, barite, pyrite, sphalerite, galena*, and some other minerals.

The chemical composition of the individual generations is usually different. For instance, chemical changes have been observed in the different generations of *siderite* (Mn contents),

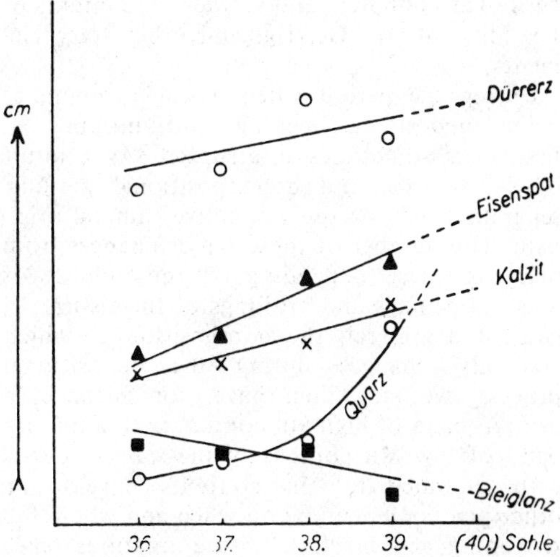

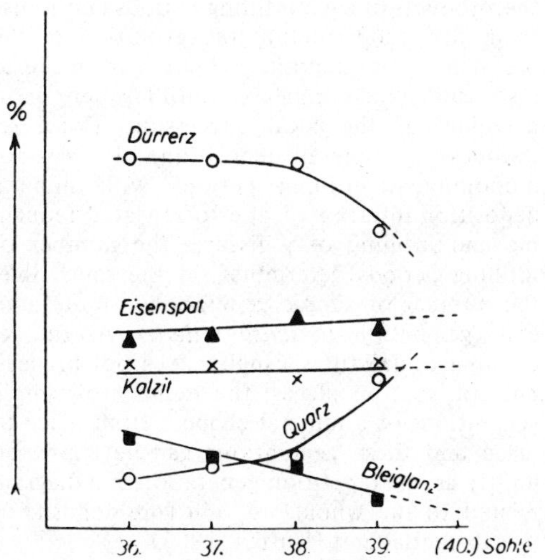

FIGURE 2. Results of a systematic measuring of the widths of the principal vein fillings, accomplished at the Main Ethelbert Vein (36th through 39th level) at Příbram, Czechoslovakia (from Kutina, 1957).

The amount of the younger, barren *quartz* (compare Fig. 1) increases rapidly with increasing depth. This depth change is conditioned by a downward widening of the newly reopened space of the fissure which was filled with the barren *quartz*. This feature represents, within the depth interval of the 36th through 39th levels, a case of reverse polyascendent zoning. An extrapolation predicts probable changes in the composition of the vein at the 40th level (at a depth of about 1 1/2 km). *Bleiglanz = galena, Dürrerz = siderite + quartz, Eisenspat = siderite, Kalzit = calcite, Sohle* = level.

barite (Sr contents), and *sphalerite* (contents of Fe, Mn, Cd, In, Ge, Hg, and other trace elements).

During the period when a single mineral is being deposited, opening and infilling of fissures is sometimes interrupted. As a result of depositional and nondepositional periods, separate bands of the respective mineral originate. The number of these bands changes from place to place depending on the number of local reopenings and infillings of the fissure. In case of a mineral, the composition of which gradually changes during the depositional process, we may then have, for instance, a *siderite* band of high Mn content with a *siderite* band of low Mn content, while *siderite* bands with intermediate Mn contents develop at other places, depending on when and where the reopening and infilling of the fractures takes place.

It is important to distinguish systematically the products of such infilling periods and to use them for reconstructing the evolution of the ore vein or ore deposit. Attempts are made to distinguish the number of infilling periods in diagrams of the local successions (local sequences of mineral deposition). In case of monomineral infilling periods with mineral deposition interrupted due to repeated reopening and infilling of a fissure, the number of infilling periods determines, at the same time, the number of simple generations of the mineral (generation in *sensu stricto*). As the reopening of fractures usually does not proceed uniformly at all places, the number of simple generations of a mineral changes from place to place and these generations can be expressed jointly as one repetition generation in a diagram related to the whole ore vein (ore deposit) or its substantial part (Kutina, 1955).

Similarly, a rhythmic generation may be distinguished as the case of another process of mineral deposition.

Nucleation, Initiation, Growth and Changes of the Individual Crystals and Aggregates of Vein Minerals

These processes have been studied systematically especially by Grigor'ev and his school (Grigor'ev, 1961). Figure 3 shows three nucleations and initiations within one and the same generation of *quartz* from the Polar Ural Mts. The large crystal of *quartz* belongs to the first nucleation.

Grigor'ev treats questions such as spontaneous nucleation, nucleation and initiation on previously formed crystallites, interaction with the substrate, gravitational settling of crystallites, fragments acting as seeds for initiation of crystals, zonal growth of mineral individuals, absolute and relative rate of growth, alteration of mineral individuals, and recrystallization, among others.

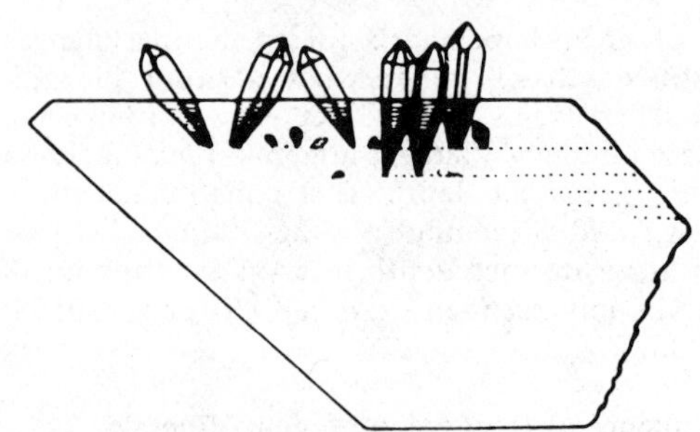

FIGURE 3. Nucleation and initiation of small crystals of *quartz* during the growth of a large *quartz* individual—that is, during a single *quartz* generation (from Grigor'ev, 1961). Polar Ural Mountains (x 2.25).

The discussion of mineral aggregates includes questions such as the geometric selection during the growth of differently oriented individuals (Lemmlein, 1941) among others.

Very important information about the process of the genesis of vein minerals and the fluids from which they have crystallized, is supplied by the fluid and gas inclusions (Ermakov, 1969; Roedder, 1972). The voluminous literature is systematically reviewed by the IAGOD Commission on Mineral Forming Fluids in Inclusions organized jointly by E. Roedder of the U.S. Geological Survey, N. P. Ermakov of the Moscow State University and G. Deicha of the Université de Paris.

Acknowledgements. Best thanks to Dr. Charles F. Park, Jr., for critical reading of the manuscript and valuable comments.

JAN KUTINA

References

Barnes, H. L., ed., 1967. *Geochemistry of Hydrothermal Ore Deposits.* New York: Holt, Reinhart & Winston.

Bastin, E. S., 1950. Interpretation of ore textures, *Geol. Soc. Am., Mem. 45.*

Bastin, E. S., et al., 1931. Criteria of age relations of minerals with special reference to polished sections of ores, *Econ. Geol.*, **26**, 561–610.

Beck, R., 1909. *Lehre von den Erzlagerstatten*, 3rd ed. Berlin.

Betekhtin, A. G., et al., 1958. [*Textures and Structures of Ores.*] Moscow: Gosgeoltekhizdat, 435p. (In Russian)

Betekhtin, A. G., et al., 1964. [*Structural-Textural Features of the Endogenic Ores.*] Moscow: Nedra, 598p. (In Russian)

Edwards, A. B., 1947. *Textures of the Ore Minerals and their Significance.* Melbourne: Institute of Mining and Metallurgy.

Emmons, W. H., 1924. Primary downward changes in ore deposits, *Trans. Am. Inst. Min., Metall. Eng.*, **70**, 964–992.

Emmons, W. H., 1927. Relations of metalliferous lode systems to igneous intrusives, *Trans. Am. Inst. Min. Metall. Eng.*, **74**, 29–70.

Emmons, W. H., 1933. On the mechanism of the deposition of certain metalliferous lode systems associated with granitic batholites, *Ore Deposits of the Western States* (Lindgren vol.). New York: Am. Inst. Min. Met. Eng., 327–349.

Emmons, W. H., 1936. Hypogene zoning in metalliferous lodes, *Int. Geol. Cong.* 16th Sess., **1**, 417–432.

Ermakov, N. P., 1960. [*Investigation of Mineral-Forming Solutions.*] Moscow: Kharkov, 460p. (In Russian)

Genkin, A. D., 1958. [On metacrysts,] in A. G. Betekhtin, et al. eds., [*Textures and Structures of Ores.*] Moscow: Gosgeoltekhizdat, 193–221. (In Russian)

Goldsmith, J. R., 1976. Scapolites, granulites, and volatiles in the lower crust, *Geol. Soc. Am. Bull.*, **87**, 161–168.

Grigor'ev, D. P., 1961. [*Ontogeny of Minerals.*] Lvov: Lvov State University. (in Russian). Engl. transl.: Y. Brenner. Jerusalem: Israel Program for Sci. Transl., 1965, 250p.

Jensen, M. L., and Bateman, A. M., 1979. *Economic Mineral Deposits*, 3rd ed. New York: Wiley, 105–150.

Kutina, J., 1955. Genetische Diskussion der Makrotexturen bei der geolchemischen Untersuchung des Adalbert-Hauptganges in Přibram, *Chem. Erde*, **17**, 241–323.

Kutina, J., 1957a. Studium der Steingunsrichtung erzführender Lösungen und der Zonalität am Adalbert-Hauptgang in Přibram, *Chem. Erde*, **19**, 1–37.

Kutina, J., 1957b. A contribution to the classification of zoning in ore veins, *Acta Univ. Carolinae, Geol.*, Prague, **3**, 1975.

Kutina, J., 1963a. On quartz metacrysts in siderite and on selective replacement in monomineral aggregates, *Problems of Postmagmatic Ore Deposition*, 1. Prague: Geol. Surv. Czech., 536–542.

Kutina, J., 1963b. A case of threefold selective replacement with taking over of metacrysts from the replaced mineral, in *Problems of Postmagmatic Ore Deposition*, **1**, Prague: Geol. Surv. Czech., 567–571.

Kutina, J., 1965. The concept of monoascendent and polyascendent zoning, *Problems of Postmagmatic Ore Deposition*, 2. Prague: Geol. Surv. Czech., 47–55.

Kutina, J., and Sedlackova, J., 1961. The role of replacement in the origin of some cockade textures, *Econ. Geol.*, **56**, 149–176.

Kutina, J.; Park, C. F., Jr.; and Smirnov, V. I., 1965. On the definition of zoning and on the relation between zoning and paragenesis, *Problems of Postmagmatic Ore Deposition*, 2. Prague: Geol. Surv. Czech., 589–595.

Kutina, J.; Adam, J.; and Pačesová, M., 1970. On the superabundant content of manganese in Přibram ore veins with respect to wall-rock leaching, in Z. Pouba, and M. Stemprok, eds., *Problems of Hydrothermal Ore Deposition, Int. Union Geol. Sci., Ser. A.*, vol. 2. Stuttgart: Schweizerbart, 294–298.

Lemmlein, G. G., 1941. [Process of geometric selection in the growing crystal aggregates,] *Doklady Acad. Sci. USSR*, **48**(3), 177–180. (In Russian)

Lindgren, W., 1933. *Mineral Deposits*, 4th ed. New York: McGraw-Hill.

Niggli, P.; Koenigsberger, J.; and Parker, R. L., 1940. *Die Mineralien der Schweizeralpen*. Basel.

Park, C. F., Jr., and MacDiarmid, R. A., 1975. *Ore Deposits*, 3rd ed. San Francisco: Freeman.

Pošepný, F., 1902. *The Genesis of Ore Deposits*, 2nd ed. New York: Am. Inst. Min. Eng.

Postmagmatic Ore Deposition. Symposium, 1, 1963; 2, 1975. Prague: Geol. Surv. Czechoslovakia.

Ramdohr, P., 1974. *Die Erzmineralien und ihre Verwachsungen*, 4th ed. Berlin.

Roedder, E., 1972. Composition of fluid inclusions, 6th ed., *U.S. Geol. Surv. Prof. Paper 440-JJ.*

Routhier, P., 1963. Les gisements métallifères, *Géologie et Principes de Recherche*, 1 and 2. Paris: Masson.

Smirnov, S. S., 1937. [On the question of zoning in ore deposits,] *Izvestiya Acad. Sci., USSR, Ser. Geol.*, No. **6**, Reprinted in: *Academician Smirnov, Collected Papers*, Moscow. (In Russian)

Cross-references: *Gangue Minerals; Gossan; Metallic Minerals; Mineral Deposits: Classification; Mineral and Ore Deposits; Pegmatite Minerals*; see also mineral glossary.

VOLCANIC MINERALS—*See* ROCK-FORMING MINERALS

WEISS SYMBOL—*See* CRYSTALLOGRAPHY: MORPHOLOGICAL

WIDMANSTÄTTEN FIGURES

When cut surfaces of most iron meteorites are etched with 6% nitric acid or other suitable reagent, a pattern of interesting bands is revealed. This pattern is known as the "Widmanstätten figure" (see *Metallurgy*, Fig. 1). It was discovered independently by F. Thomas and A. von Widmanstätten in 1808 and, although first published by the former, it bears the name of the latter who studied the phenomenon in detail.

The Widmanstätten figure or pattern is due to the differential etching of *kamacite* (α-iron with about 5.5% nickel) and *taenite* (γ-iron with between 13% and 48% nickel). The *kamacite* is in the form of plates that have exsolved, parallel to the faces of the octahedron, from *taenite*. Larger areas of *taenite* between the *kamacite* plates may show fine-grained exsolution of *kamacite* giving the intergrowth known as "plessite." Widmanstätten figures may also be shown by the iron in a few stony-iron meteorites and on a microscale in some artificial alloys (see *Metallurgy*).

Iron meteorites in which Widmanstätten figures can be reproduced, i.e., those with Widmanstätten texture, are known as octahedrites and are classified according to the thickness of the *kamacite* plates. A commonly used classification is: finest octahedrite, *Off*, < 0.2 mm; fine octahedrite, *Of*, is 0.2–0.5 mm; medium octahedrite, *Om*, 0.5–1.5 mm; coarse octahedrite, *Og*, 1.5–2.5 mm; coarsest octahedrite *Ogg*, 2.5 mm. The *kamacite* width is mainly controlled by the nickel content and the cooling rate of the meteorite.

It is believed that the *kamacite* has very slowly lost nickel by diffusion to the *taenite*, but without loss of the original crystal structure. Such slow diffusion could have taken place in the cores of bodies of asteroid size (less than 250 km in radius) or near the surface of larger bodies. The composition of any meteoritic *iron* is probably near that of the original melt from which it crystallized, usually in the form of giant grains many meters across.

Besides the Widmanstätten pattern, other bands were later discovered by F. E. Neumann. These Neumann bands are narrow, straight bands representing sections through lamellae parallel to the faces of a cube (hexahedron). They occur in *kamacite* and are prominent in meteorites with less than 6% nickel that are composed entirely of *kamacite*. Such meteorites are, for this reason, called "hexahedrites," in contrast to those meteorites—called "octahedrites"—in which *taenite* lamellae occur parallel to the faces of an octahedron. Neumann bands are due to shear within the crystal structure, such shear being one of a number of effects that are probably due to violent impact. Other evidence suggests that many meteorites suffered shock pressures up to 10^6 bars, which could have been due to collisions between asteroids. The asteroids move in planetary orbits in the asteroid belt, which represents the "missing planet" according to the Titius-Bode Rule (see vol. II, *Planetary Intervals*); and these collisions could also have had the effect of knocking fragments out of their planetary orbit and into earth-crossing orbits.

MICHAEL J. FROST

References

Axon, H. J., 1967. Metallurgy of Meteorites, *Prog. Mater. Sci.*, **13**, 183–228.

Goldstein, J. I., and Short, J. M., 1967. The iron meteorites, their thermal history and parent bodies, *Geochim. Cosmochim. Acta* **31**, 1733–1770.

McCall, G. J. H., 1973. *Meteorites and their origins.* Newton Abbott: David and Charles, 352p.

Tackett, S. L., et al., 1966. Electrolytic dissolution of iron meteorites, *Science*, **153** (3738), 877–880.

Cross-references: *Etch Pits; Metallurgy; Meteoritic Minerals*. Vol. II: *Meteorites; Planetary Intervals.*

WOLFRAMITE GROUP

There are only three ideal end-member compositions possible in the ***wolframite*** group.

These are $MnWO_4$ (*huebnerite*), $FeWO_4$ (*ferberite*), and $ZnWO_4$ (*sanmartinite*). *Wolframite*, the mineral for which the group is named, comprises the range of compositions from $(Mn_{0.2}Fe_{0.8})WO_4$ to $(Mn_{0.8}Fe_{0.2})WO_4$. In other words, the Mn:Fe ratio in *huebnerite* is at least 4:1; the Fe:Mn ratio in *ferberite* is at least 4:1; and all intermediate compositions are considered *wolframite*. In this respect the ***wolframite*** series differs from many others where only two names are given, one to each end of the series. While a complete isomorphous series exists between Mn^{2+} and Fe^{2+}, there is no evidence to indicate that *sanmartinite* forms an isomorphous series with either *huebnerite* or *ferberite*.

Minerals of the ***wolframite*** group have very similar physical properties. They are all found in prismatic to tabular crystals that tend to be striated. The colors vary from yellowish and reddish brown to dark gray to black, and the streak colors are much the same. The Mn- and Zn-rich members are lightest in color and most transparent, the color becoming darker and the crystals opaque as the iron content increases. Specific gravity, hardness, refractive indices, and the tendency toward prismatic rather than tabular crystal shape also increase as the iron content increases.

The ***wolframite***-group minerals occur mainly in veins, usually associated with *cassiterite*, and in *quartz* and pegmatite veins of hydrothermal origin genetically related to granitic intrusive rocks. A common occurrence is where such veins traverse metamorphosed sediments in the form of phyllites and schists. *Wolframite* is found in numerous occurrences around the world and is the most important ore of tungsten. *Ferberite* and *huebnerite* are far less significant as ore minerals.

Wolframite $(Fe,Mn)WO_4$. Examples of *wolframite* compositions in the literature show Mn:Fe ratios of 4:1 (Portugal), 1:1 (England), and 1:2 (Canada). The most important localities for specimens include Panasqueira, Portugal; Chicote Grande, Cochabamba, and Amutara, Bolivia; Schlaggenwald and Zinnwald, Czechoslovakia; several mines in Korea; and Trumbull, Connecticut.

Huebnerite ($MnWO_4$). The Uncompahgre district, Ouray County, Colorado, is known to produce *huebnerite* that contains only 0.24% FeO, very close to the pure Mn end member. Excellent specimens come from near Silverton, San Juan County, Colorado; the Hamme tungsten mine, near Henderson, North Carolina; near Ely, Nevada; and from Pasto Bueno and Quiruvilca, Peru.

Ferberite ($FeWO_4$). *Ferberite* from Riddarhyttan, Sweden, has been reported to contain only 0.19% MnO, and thus has a nearly pure end-member composition. The best specimens come from Boulder County, Colorado, and Oruro, Bolivia.

Sanmartinite ($ZnWO_4$). A one locality mineral, *sanmartinite* has been found only at Los Cerrillos, near San Martín, San Luis Province, Argentina, with *willemite* altering from *scheelite*. Several analyses of this mineral reveal a variable calcium content (1.48 to 6.32% CaO) and a more regular iron and manganese content (7.24 to 8.28% FeO, 0.74 to 1.73% MnO).

JOHN SAMPSON WHITE

Reference

Palache, C.; Berman, H.; and Frondel, C., 1951. *The System of Mineralogy*, vol. 2. New York: Wiley, 1064-1073.

Z

ZEOLITES

A *zeolite* is an aluminosilicate with a framework structure enclosing cavities occupied by large ions and water molecules, both of which have considerable freedom of movement, permitting ion exchange and reversible dehydration. This definition proposed by Smith (1963) is generally valid and still acceptable, when loosely applied, for any zeolitic material, both natural and synthetic.

The ease with which water is given off from *zeolite* minerals on gentle heating has given rise to the name, from the Greek ζεν = to boil and λίθος = stone. But this was only the first of many astonishing properties to be noted in these peculiar compounds. Several other phenomena that take place within a zeolite structure under different physical-chemical conditions have led to many industrial applications and many others will probably be found as studies of *zeolites* proceed.

Zeolite minerals are found in the three geologic environments of volcanic, sedimentary, and metamorphic rocks. They form under mild conditions of temperature and pressure and have a limited field of stability so that they are rarely found in rocks older than Mesozoic. Sedimentary *zeolites* are by far the most abundant, occurring in some cases in layers several kilometers wide and a few kilometers thick, as in volcanic ash deposits. *Zeolites*, however, were first noted in vescicles and cavities of volcanic rocks (mostly basalts) where they usually occur as well-developed and handsome specimens.

Chemistry, Structure, and Classification

Attempts to describe a *zeolite* by a well-defined chemical formula may result, as in the case of many other minerals, to be more misleading than useful due to wide variability of composition within the same family.

The framework structure of a *zeolite* is made up of tetrahedrally coordinated cations (typically Si and Al) which share all the oxygen atoms located at the vertices of the tetrahedra. These cations are denoted with the letter *T*, therefore, in a *zeolite* there will be twice as many oxygens as there are *T* atoms.

The cavities present in the tetrahedral framework are occupied, in natural *zeolites*, by large mono- or divalent cations such as Na, K, Mg, Ca, and Ba. Other species can be placed in these positions by ion exchange or direct synthesis in the laboratory. These cations balance, with their positive charges, the excess negative charges due to the presence of tetrahedrally coordinated Al^{3+} in the framework.

Ideally, a *zeolite* can be represented by the formula

$$M_xD_y[Al_{x+2y}Si_zO_{2x+4y+2z}]\cdot nH_2O$$

where *M* and *D* represent monovalent and divalent cations respectively.

Chemically related to ***feldspars*** (see *Feldspar Group*), ***zeolites*** have much more open structures. This relationship is evidenced by the ease with which ***zeolites*** transform into ***feldspars*** under thermal metamorphism.

The channels present in the structure of a *zeolite* are formed by different combinations of linked rings of tetrahedra and, according to their size, allow the passage of single cations or even organic molecules. The minimum free diameter of the largest channels ranges from 2.2 to 7.5 Å (Table 2) in the different species.

One of the classification criteria used for *zeolites* takes into account the structural features as they are displayed by the external morphology of the crystals. According to this classification, *zeolites* belong to three main groups:

1. Three-dimensional framework structures as in *phillipsite* and *chabazite*.
2. Fibrous structures formed by chains of tetrahedra weakly linked laterally as in *natrolite* and *scolecite*.
3. Lamellar structures formed by tetrahedra strongly linked in one plane with fewer bonds at right angle to this plane, as in *heulandite* and *clinoptilolite*.

This classification, however, is misleading because all *zeolites* belong, in effect, to the larger group of framework silicates even though their three-dimensional structures may display fewer bonds in certain crystallographic directions.

A more recent and complete *zeolite* classification (Breck, 1974) is based on the framework topology. Each framework can be thought of as consisting of structural units such as rings of tetrahedra or other motifs. Furthermore, these configurations may be grouped so as to give shape to easily identifiable polyhedra or "cages" within the structure. The identification of one or another type of motif in a ***zeolite*** structure also provides immediate information on some of the physical properties of a ***zeolite***. Thus, such a classification has importance from the viewpoint of industrial applications.

In order to consider the topological distribution of the atoms in a structure, certain simplifications are advisable. The fundamental and invariable building unit is the tetrahedron. All tetrahedra are linked throughout the structure via shared vertices; every position occupied by a tetrahedron can, thus, be considered a four-connected nodal point. The oxygen atoms do not need direct representation; their positions correspond approximately to the midpoints of the connecting branches. Such a skeletal representation allows the characteristic building features of each framework to be identified (Meier & Olson, 1978). A set of seven "Secondary Building Units" (*SBU*) has been chosen (Fig. 1) to represent every known structure (Meier, 1968).

Breck's classification divides ***zeolites*** into seven groups according to the type of *SBU* contained in the structure (Table 1).

Some examples of polyhedral cages found in ***zeolite*** structures are given in Fig. 2. Three examples of ***zeolite***-structure representation by the use of polyhedral building blocks are given in Fig. 3.

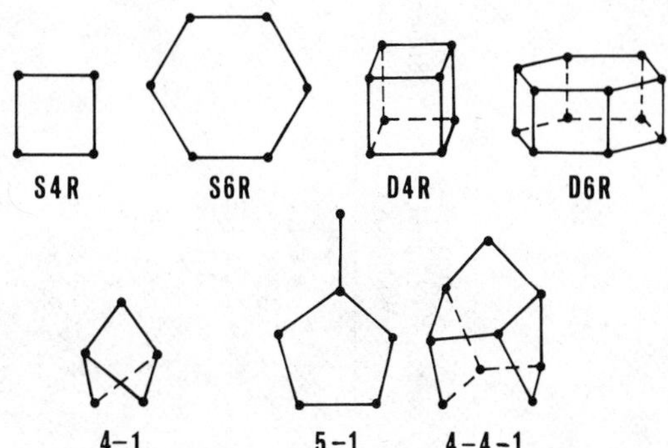

FIGURE 1. The seven Secondary Building Units (SBU) corresponding to the structural classification of ***zeolites***. Every corner or vertex (*T* position) is actually a four-connected nodal point in the three-dimensional framework. Oxygen atoms occupy approximately the midpoints of the branches. For explanation of symbols see text.

TABLE 1. Breck's Classification of Zeolites

Group	Secondary Building Unit (*SBU*)
1	Single 4-ring, S4R
2	Single 6-ring, S6R
3	Double 4-ring, D4R
4	Double 6-ring, D6R
5	Complex 4-1, T_5O_{10} unit
6	Complex 5-1, T_8O_{16} unit
7	Complex 4-4-1, $T_{10}O_{20}$ unit

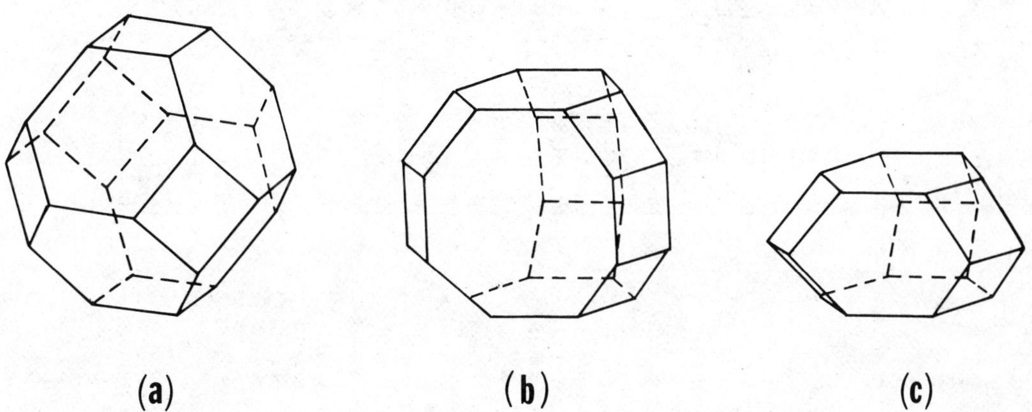

FIGURE 2. Some of the cages found in ***zeolite*** structures. (a) *Sodalite*-type cage (truncated octahedron), one of the space-filling Archimedean polyhedra, found in *faujasite*, Type X, Y, and A. See also Fig. 3(a) and (b). (b) *Gmelinite*-type cage found in *gmelinite*; *mazzite*, see also Fig. 3(c); *offretite*; and Type Ω. (c) *Cancrinite*-type cage (ϵ cage) found in *erionite*, *offretite*, and Type *L*.

TABLE 2. Zeolites

Group (*SBU*)	Species	Typical Unit Cell Content	Space-group	Structural Type[a]	Type of Dehydration Behavior	Approx. Max. Channel Width (Å)	Sedimentary Occurrence
1	*Phillipsite*	$(1/2Ca,Na,K)_6 [Al_6 Si_{10} O_{32}] \cdot 12H_2O$	$P2_1/m$	PHI	1	3.9 x 4.4	x
(S4R)	*Harmotome*	$Ba_2 [Al_4 Si_{12} O_{32}] \cdot 12H_2O$	$P2_1/m$	PHI	1	3.9 x 4.4	x
	Merlinoite	$(K_2,Ca,Ba,Na_2)_{4.7} [Al_9 Si_{23} O_{64}] \cdot 23H_2O$	$Immm$	MER	3a	3.3 x 5.0	
	Gismondine	$Ca_4 [Al_8 Si_8 O_{32}] \cdot 16H_2O$	$P2_1/c$	GIS	1	2.8 x 4.9	
	Amicite	$K_4 Na_4 [Al_8 Si_8 O_{32}] \cdot 10H_2O$	$I2$	GIS	3a	3.8 x 3.1	
	Garronite	$NaCa_{2.5} [Al_6 Si_{10} O_{32}] \cdot 14H_2O$	$I4_1/amd$	GIS	1	3.5	
	Paulingite	$(K_2,Na_2,Ca,Ba)_{76} [Al_{152} Si_{525} O_{1354}] \cdot 700H_2O$	$Im3m$	PAU		3.9	
	Laumontite	$Ca_4 [Al_8 Si_{16} O_{48}] \cdot 16H_2O$	Am	LAU	2	4.6 x 6.3	x
	Yugawaralite	$Ca_2 [Al_4 Si_{12} O_{32}] \cdot 8H_2O$	Pc	YUG	3a	2.8 x 3.6	
	Type P (synth.)	$Na_6 [Al_6 Si_{10} O_{32}] \cdot 15H_2O$	$I4_1/amd$	GIS		2.8 x 4.9	
	Type W (synth.)	$K_{42} [Al_{42} Si_{76} O_{326}] \cdot 107H_2O$	$Immm$?	MER			
2	*Erionite*	$(K_2,Ca,Mg,Na_2)_{4.5} [Al_9 Si_{27} O_{72}] \cdot 27H_2O$	$P6_3/mmc$	ERI	3a	3.6 x 5.2	x
(S6R)	*Offretite*	$(K_2,Mg,Ca,Na_2)_{2.5} [Al_5 Si_{13} O_{36}] \cdot 15H_2O$	$P\bar{6}m2$	OFF	3a	6.9	
	Levyne	$Ca_3 [Al_6 Si_{12} O_{36}] \cdot 18H_2O$	$R\bar{3}m$	LEV	2	3.2 x 5.1	
	Mazzite	$K_{2.5} Mg_2 Ca_{1.5} [Al_{10} Si_{26} O_{72}] \cdot 28H_2O$	$P6_3/mmc$	MAZ	3b	7.4	
	Type Ω (synth.)	$Na_{6.8} (TMA)_{1.6} [Al_8 Si_{28} O_{72}] \cdot 21H_2O$[b]	$P6mmm$?	MAZ	3b	7.4	
	Type T (synth.)	$Na_{1.2} K_{2.8} [Al_4 Si_{14} O_{36}] \cdot 14H_2O$	$P6m2$	ERI+OFF?	3b	5.2	
	LOSOD (synth.)	$Na_{12} [Al_{12} Si_{12} O_{48}] \cdot 19H_2O$	$P6_3/mmc$	LOS		2.2	
3	Type A (synth.)	$Na_{12} [Al_{12} Si_{12} O_{48}] \cdot 27H_2O$	$Pm3m$	A	3b	4.2	
(D4R)	Type N-A (synth.)	$Na_4 (TMA)_3 [Al_7 Si_{17} O_{48}] \cdot 21H_2O$[b]	$Pm3m$	A	3b	4.2	
	Type ZK-4 (synth.)	$Na_8 (TMA) [Al_9 Si_{15} O_{48}] \cdot 28H_2O$[b]	$Pm3m$	A	3b	4.2	
4							
(D6R)	*Chabazite*	$Ca_2 [Al_4 Si_8 O_{24}] \cdot 13H_2O$	$R\bar{3}m$	CHA	3a	3.7 x 4.2	x
	Gmelinite	$Na_8 [Al_8 Si_{16} O_{48}] \cdot 24H_2O$	$P6_3/mmc$	GME	1	7.0	
	Faujasite	$Na_{12} Ca_{12} Mg_{11} [Al_{59} Si_{133} O_{384}] \cdot 235H_2O$	$Fd3m$	FAU	3b	7.4	
	Type X (synth.)	$Na_{86} [Al_{86} Si_{106} O_{384}] \cdot 264H_2O$	$Fd3m$	FAU	3b	7.4	
	Type Y (synth.)	$Na_{56} [Al_{56} Si_{136} O_{384}] \cdot 250H_2O$	$Fd3m$	FAU	3b	7.4	

TABLE 2. (Continued)

Group (*SBU*)	Species	Typical Unit Cell Content	Space-group	Structural Type[a]	Type of Dehydration Behavior	Approx. Max. Channel Width (Å)	Sedimentary Occurrence
	Type ZK-5 (synth.)	$Na_{30}[Al_{30}Si_{66}O_{192}]\cdot 98H_2O$	*Im3m*	ZK5		3.9	
	Type L (synth.)	$K_9[Al_9Si_{27}O_{72}]\cdot 22H_2O$	*P6/mmm*	L	3b	7.1	
	Type P-L (synth.)	$K_{23}[Al_{33}Si_{26}P_{13}O_{144}]\cdot 42H_2O$	*P6/mmc*	L	3b	7.1	
5	*Natrolite*	$Na_{16}[Al_{16}Si_{24}O_{80}]\cdot 16H_2O$	*Fdd2*	NAT	1	2.6 x 3.9	x
(4–1)	*Scolecite*	$Ca_8[Al_{16}Si_{24}O_{80}]\cdot 24H_2O$	*Cc*	NAT	3a	2.6 x 3.9	
T_5O_{10}	*Mesolite*	$Na_{16}Ca_{16}[Al_{48}Si_{72}O_{240}]\cdot 64H_2O$	*Fdd2*	NAT	3a	2.6 x 3.9	x
	Thomsonite	$Na_4Ca_8[Al_{20}Si_{20}O_{80}]\cdot 24H_2O$	*Pnna*	THO	2	2.6 x 3.9	x
	Gonnardite	$Na_4Ca_2[Al_8Si_{12}O_{40}]\cdot 14H_2O$	*Pbmn*	THO	2	2.6 x 3.9	
	Edingtonite	$Ba_2[Al_4Si_6O_{20}]\cdot 8H_2O$	$P2_12_12$	EDI	2	3.1 x 4.1	
6	*Mordenite*	$Na_8[Al_8Si_{40}O_{96}]\cdot 24H_2O$	*Cmcm*	MOR	3b	6.7 x 7.0	x
(5–1)	*Dachiardite*	$Na_5[Al_5Si_{19}O_{48}]\cdot 12H_2O$	*C2/m*	DAC		3.7 x 6.7	
T_8O_{16}	*Ferrierite*	$Na_{1.5}Mg_2[Al_{5.5}Si_{30.5}O_{72}]\cdot 18H_2O$	*Immm*	FER		4.3 x 5.5	
	Epistilbite	$Ca_3[Al_6Si_{18}O_{48}]\cdot 16H_2O$	*C2/m*	EPI	2	3.2 x 5.3	
	Bikitaite	$Li_2[Al_2Si_4O_{12}]\cdot 2H_2O$	$P2_1$	BIK	3a	3.2 x 4.9	
	ZSM-5 (synth.)	$Na_3[Al_3Si_{93}O_{192}]\cdot 16H_2O$	*Pnma*	MFI	3b	5.4 x 5.6	
	Silicalite (synth.)	SiO_2	$Pn2_1a$	MFI		5.2 x 5.8	
7	*Heulandite*	$Ca_4[Al_8Si_{28}O_{72}]\cdot 24H_2O$	*C2/m*	HEU	1	4.4 x 7.2	x
(4–4–1)	*Clinoptilolite*	$Na_6[Al_6Si_{30}O_{72}]\cdot 24H_2O$	*C2/m*	HEU	3b	4.4 x 7.2	x
$T_{10}O_{20}$	*Stilbite*	$Na_2Ca_4[Al_{10}Si_{26}O_{72}]\cdot 34H_2O$	*C2/m*	STI	1	4.1 x 6.2	x
	Stellerite	$Ca_4[Al_8Si_{28}O_{72}]\cdot 28H_2O$	*Fmmm*	STI	2	4.1 x 6.2	
	Barrerite	$Na_6CaK[Al_8Si_{28}O_{72}]\cdot 26H_2O$	*Amma*	STI	1	4.1 x 6.2	
	Brewsterite	$(Sr,Ba,Ca)_2[Al_4Si_{12}O_{32}]\cdot 10H_2O$	$P2_1/m$	BRE	3a	2.3 x 5.0	
	Cowlesite	$Ca[Al_2Si_3O_{10}]\cdot 6H_2O$	–	–	1		

[a]Structural types are designated following the recommendations by IUPAC (International Union for Pure and Applied Chemistry), 1978.
[b]Tetramethylammonium ion = TMA

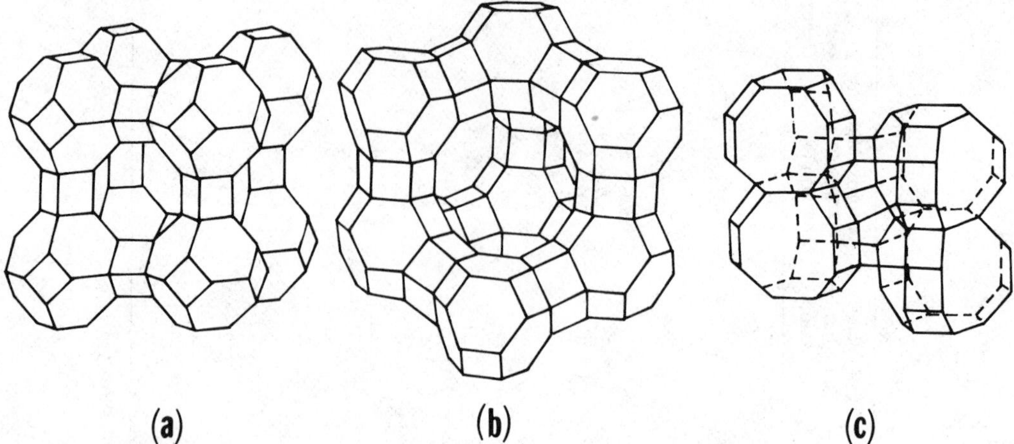

FIGURE 3. Arrays of building cages found in ***zeolites***: (a) type *A*; (b) *faujasite*; (c) *mazzite*. The *T* atoms are at the intersections of the lines, oxygen atoms are approximately at the midpoints of the lines.

Dehydration, Sorption, Ion Exchange, and Other Zeolitic Phenomena

Removal of water from a ***zeolite*** is usually a reversible process although the temperature at which a ***zeolite*** dehydrates, at least partially, and maintains the capability to rehydrate, varies considerably from one species to another.

The structural changes a ***zeolite*** undergoes upon dehydration are being investigated by means of such techniques as X-ray crystal structure analysis, differential thermal analysis (DTA), thermogravimetric analysis (TG), infrared spectroscopy (IR), nuclear magnetic resonance (NMR), and dielectric measurements. This work, much of which is currently under way, also tends to establish the role of water in a ***zeolite*** structure. The structural changes involved in the process vary from the simple thermal expansion found in *mordenite*, to the complex patterns of cation migration investigated in *chabazite, faujasite, mazzite*, etc., and finally to phase transformations into more stable anhydrous structures.

There are three main modalities of dehydration (Van Reeuwijk, 1974):

- *Type 1.* Sudden dehydration over a relatively short temperature range characterized by a single, sharp DTA endothermic peak, steep weight loss in the TG curve, marked change in unit-cell dimensions and(or) phase transformation. This pattern is seen in *gismondine, natrolite, phillipsite, gmelinite, heulandite*, and *stilbite*.
- *Type 2.* Gradual dehydration over a wide temperature range consisting of small consecutive dehydration steps revealed by broad DTA and TG reactions and accompanied by small stepwise structural adaptations. This behavior is typical of *edingtonite* and also occurs in *levyne, gonnardite*, and *laumontite*.
- *Type 3a.* Multiple-step dehydration with a first major reaction followed by smaller ones; therefore, a combination of Types 1 and 2. The crystal-structure adaptations follow the first major reaction. This behavior is characteristic of *chabazite, offretite, erionite, yugawaralite*, and *brewsterite*.
- *Type 3b.* Virtually the same thermal behavior as Type 3a, but with the absolute predominance of the first broad endothermic peak and the absence of structural changes (*mordenite*) or the occurrence of very modest ones (*faujasite* and *clinoptilolite*).

The different thermal dehydration behaviors correspond to the different roles played by the water molecules in the structure. Where there is Type 1 dehydration, water is evidently a necessary component of the structure, perhaps in the same way as in a hydrous chemical. For Type 2 dehydration, water is differently held according to the position in the structure, but its presence is not necessary for stability; it must be held mostly by absorptive forces within the channels and the cavities of the framework. The same naturally applies for Type 3.

The complexity of the actual situation is governed by the interactions between the large nonframework cations, the water molecules, and the aluminosilicate framework. These cations are solvated (surrounded by water molecules) in the hydrated state; as dehydration proceeds, they produce different effects according to their polarizing strength and electrostatic energy. In the case of a strongly polarizing cation, the water molecule will divide into a hydroxyl bonded to the cation plus a proton attached to a framework oxygen. Weakly polarizing cations will lose the surrounding water readily and will minimize their free energy by binding directly to framework oxygens. The two behaviors, thus, produce different framework distortions that tend to satisfy the resulting electrostatic fields.

A dehydrated, or partially dehydrated, *zeolite* is in an "activated" state and will rehydrate very rapidly, adsorbing water from the atmosphere if exposed to air.

The sorptive properties of an activated *zeolite* are not restricted to rehydration, but various other molecular species, both polar and nonpolar (hydrocarbons), may be admitted within the *zeolite*. The "free diameter" of the channels is the main discriminatory factor in admission of "guest species"; but cations already present in the pores may also act as "sentinels" by inhibiting the admission to insufficiently polar, active species. Furthermore, temperature affects the sorptive characteristics, not only in the quantitative sense, but also qualitatively by changing the effective pore size.

Because of their selective and sorptive properties, *zeolites* are also called "molecular sieves." This name is preferentially used for synthetic *zeolites* that are produced in large quantities for utilization in many industrial applications.

An increase of temperature in a hydrous environment favors the mobilization of the sorbed cations by providing the energy necessary to surmount the weak potential barriers that oppose cation migration. Substitution of a mobilized cationic species with a different one present in a solution is readily accomplished, thereby producing an "exchanged *zeolite*."

Due to the three-dimensional sieve nature of *zeolites*, the molecules adsorbed within their structure are in a state of maximum subdivision. This naturally favors the catalytic activity displayed by *zeolites* for a number of reactions. However, most *zeolites* used in industrial catalytic reactions have undergone activation processes usually consisting of the introduction of a catalytically active metal by ion exchange and subsequent calcination.

The chemistry of zeolite catalysis is under intensive investigation (Rabo, 1976; Naccace and Taarit, 1980), but the complexity of the phenomenon (heterogeneous catalysis) does not permit broad generalizations. From the structural viewpoint, catalysis by *zeolite* molecular sieves can be regarded as the combined effect of electrostatic fields, reversible proton and electron transfer, and the permanence of the sorbed molecules in a given location for a given time.

Genesis and Occurrence

The genesis and paragenesis of *zeolites* has been studied quite extensively in recent years, especially in view of the industrial utilization of large sedimentary deposits. The definition of a *zeolite* facies with characteristic mineral assemblages is still the object of investigation and discussion. Being hydrous minerals of low specific gravity (ranging from 1.9 to 2.3 g/cm^3), *zeolites* are relatively sensitive to temperature and pressure. A sequence of *zeolite* crystallization should then proceed from less hydrous to more hydrous minerals with decreasing temperature and pressure. Such a simple picture, however, is very seldom observed in nature, where other factors, such as water pressure, permeability, porosity, osmotic effects, and silica content of the host rock, may influence the process.

In Igneous Rocks. The best known *zeolite* mineralizations occur in the form of well developed crystalline aggregates in the vescicles and vugs of effusive rocks, mostly basalts (Fig. 4). They are most probably the result of late deposition from fluids that permeated the basalts after the extrusion of the rock. This hypothesis is confirmed by an extensive study of the Tertiary lavas of Northern Ireland where the *zeolites* zones cut discordantly across the basalt layers. The zones are probably related to variations of temperature and pore-water chemistry.

A correlation between the silica content of the rock and that of *zeolites* is generally possible, although the alkalinity of the solution is also a decisive factor in the process. *Zeolites* with decreasing Si:Al ratio have been found to form in the laboratory from solutions of increasing pH.

In some cases, several successive *zeolite* mineralizations may be found in the same cavity or within the same hand specimen, the compositions varying from one grain to another and even within the same crystal (Rinaldi, 1976; Černý et al., 1977).

The origin of the fluids that led to the formation of *zeolites* is, in most cases, problematic. Two mechanisms are possible. One involves rising hydrothermal solutions caused by late-stage volcanism either through a stationary, but still liquid, magma or through the pores of a consolidated rock. A second mechanism involves permeation of the porous rock from above by ground water. Combinations of the two may occur.

In Sediments. All *zeolite* species have been found in volcanic rocks; some have been recognized, mostly recently, as the main constituents of large sedimentary formations. The latter usually have a very small grain size and can be identified only by X-ray diffraction and electron microscopy.

The most common sedimentary *zeolites* are

FIGURE 4. Radial bundle of acicular crystals of *mazzite* (a very rare ***zeolite***) associated with *phillipsite* twins (upper right) and *chabazite* crystals (upper left). Magnification x15.

chabazite, clinoptilolite, erionite, mordenite, and *phillipsite*.

Sedimentary ***zeolite*** deposits are subdivided according to the depositional environment in which they formed. The mineral assemblages are not strictly characteristic of the environment, although the chemistry of the single species correlates with it. Recent comprehensive discussions of sedimentary ***zeolites*** are given by Hay (1966), Sheppard (1971 and 1973), Kossowskaya (1973), Mumpton (1973), Olson (1975) and Iijima (1980).

Saline, Alkaline Nonmarine Deposits—Here, ***zeolites*** occur as alteration products of volcanic glass that may constitute most or only part of the original sediments. They may form relatively pure beds throughout lacustrine sequences 600 or more meters thick and up to 5000 km^2 in extension (see *Green River Mineralogy*). ***Zeolites*** are also known to form at the surface of saline alkaline soils (see *Soil Mineralogy*).

These environments are characterized by high pH (commonly 9.4) and very high salinities. ***Zeolites*** are forming at present in the saline lakes of western United States and eastern Africa (Searles and China Lakes, California; Lake Natron, Tanzania; Lake Magadi, Kenya).

Marine and Freshwater Ash Deposits—Here, ***zeolites*** occur as alteration products of volcanic glass, and are most abundant in thick accumulations of tuffaceous sediments. *Phillipsite* was first recognized in the recent sediments of the ocean floor in 1891 by the Challenger Expedition. Several recent deep-sea drillings located vast ***zeolite*** beds at the bottom of all oceans. *Phillipsite* occurs over wide areas of the upper portion of the Pacific Ocean sediments; *clinoptiloite* occurs deeper; *clinoptilolite* dominates in the Atlantic Ocean sediments; whereas *clinoptilolite* and *phillipsite* occur in almost equal amounts in the Indian Ocean sediments.

Zeolites have been found in many lithic and feldspathic sandstones and associated detrital rocks of Paleozoic to Miocene age. They form cements and replace mineral grains (mostly ***plagioclase***) or ***clay*** matrix. Only rarely do they account for more than 25% of a sandstone.

A third occurrence of ***zeolites*** in this environment is represented by chemical and biogenic rocks of marine origin. *Clinoptilolite* and *heulandite* are most widespread in Mesozoic limestones (chalk and chalky marls) in Germany and the Ukraine. *Analcime* (a ***feldspathoid*** closely related to ***zeolites***) has been found in Russian limestones of Middle Cretaceous, Triassic, and Permian age. Miocene phosphorites of the eastern United States contain *clinoptilolite.*

Continental Ash Deposits—Here, ***zeolites*** are found in vitric tuff beds that may reach thicknesses of several kilometers. One such occurrence is *clinoptilolite* in tuff and tuffaceous claystone in the lower part of the John Day Formation in central Oregon. Subsurface pore water percolating through the formation altered the glass to *montmorillonite* with consequent increase in pH and alkali concentration, thereby providing a chemical environment favorable to the formation of ***zeolites***.

The concentration of ***zeolites*** in these deposits usually varies with depth of burial. In the Tertiary tuffs of Southern Nevada, a zone rich in *clinoptilolite* underlies the upper zone of unaltered glass with local *chabazite* concentrations and is succeeded downward by zones rich in *mordenite* and *analcime*.

Certain beds composed almost entirely of one ***zeolite*** are particularly useful for commercial exploitation. Figure 5 shows an example of *chabazite* mineralization in a tuff from central Italy.

In Metamorphic Rocks. Diagenetic processes

FIGURE 5. ***Zeolite*** mineralization (*chabazite* rhombohedra) in tuffaceous ash deposit from Latium (central Italy). Length of white bar 10 μm.

represent the first stage of burial metamorphism. As deposition proceeds, the increasing thickness of the sediments generates a vertical pressure-temperature gradient which favors reactions between aluminosilicates and pore water. Metamorphic ***zeolites*** are formed at the expenses of highly reactive volcanic glasses, ***plagioclases***, ferromagnesian minerals, and diagenetic ***zeolites*** that may be present in the original sediments.

Ideally, the mineral assemblages form by equilibrium reactions and although the real situation is complicated by many factors, generally, there is a transformation of glass into water-rich ***zeolites*** for a depth of about 1 km, followed by transformation into water-poor ***zeolites*** or anhydrous ***feldspars*** at greater depths. Water-rich ***zeolites*** also tend to be supplanted by water-poor ***zeolites*** with increasing age. Further complexities arise with the introduction of hydrothermal fluids in areas of thermal activity and with the metasomatic depletion or enrichment of certain elements within the sediments. Very commonly, there is the addition of sodium that occurs mainly in eugeosynclinal sequences. Calcium metasomatism is represented by the reaction of silicic tuffs to form *heulandite* or *laumontite*.

Some studies on low-grade metamorphism and the ***zeolite*** facies have been reported in *The Canadian Mineralogist* (1974). A first comprehensive description of the ***zeolite*** facies was given by Coombs et al. (1959). The ***zeolite*** facies bridges the gap between sedimentary processes and the *prehnite-pumpellyite* facies of regional metamorphism.

The primary control of mineral association with depth of burial appears to be temperature rather than pressure.

Low-grade metamorphic processes involve a vast portion of the Earth's crust and very large amounts of interstitial water that is forced to migrate upward through the sediments as these increase in thickness.

Synthetics

The first experiments on the separation of mixtures using dehydrated natural *chabazite*, performed by Barrer in 1945, were followed by an industrial research effort culminating with the successful synthesis, in 1948, of molecular sieves that had never been found in nature (see *Synthetic Minerals*). Their widespread industrial utilization as adsorbents began in 1954; since then molecular-sieve technology has come a long way as can be attested by more than 10,000 patents and several thousand scientific reports concerning these sieves (Flanigen, 1980).

The multidirectional scientific effort on ***zeolites*** has been systematized in important monographs recently published (Breck, 1974; Rabo, 1976). Other important references for ***zeolite*** studies are the proceedings of the International Conference on Molecular Sieves held at three-year intervals since 1968.

The synthesis of ***zeolites*** is achieved in industrial processes at low temperatures (usually below 200°C) that are consistent with the formation conditions of ***zeolites*** in sediments. Other synthesis conditions include: highly reactive starting materials such as freshly coprecipitated gels or amorphous solids, high pH (up to 14) achieved by the introduction of an alkali-metal hydroxide or other strong base, saturated water-vapor pressure, supersaturation of the components of the gel to achieve the nucleation of a large number of crystals.

More than 100 ***zeolites*** have been produced in the laboratory so far, but not all the natural species are represented by a synthetic analogue. Conversely, some of the most easily synthesized ***zeolites*** do not have natural counterparts. Although the reasons for such discrepancies are not fully known, the time involved in the formation of ***zeolites*** in the two cases may well be a determinant factor. Synthetic ***zeolites*** are likely to crystallize as highly metastable phases by kinetic reactions over a short period of time, whereas natural ***zeolites*** may approach equilibrium conditions of reaction over geologic time. Demonstration of this could be the rare attainment of Si–Al ordering in synthetic frameworks, a condition often displayed by natural ***zeolites*** (see *Order-Disorder*).

Hypothetically, the crystallization of a ***zeolite*** would start from the nucleation of basic building units, such as four- and six-membered rings of tetrahedra, which could then polymerize around solvated cations that would, thus, act as templates for the open aluminosilicate frameworks. This view is sustained by synthesis experiments in which the presence of certain large organic cations has produced structures with "cages" occupied by them. The tetramethylammonium (TMA) and tetrapropylammonium (TPA) ions have been used successfully to produce large-pore ***zeolites*** such as *N-A*, *ZK*-4, *N-X*, *N-Y*, ZSM-5 and the synthetic forms of *mazzite* and *offretite*.

Synthetic ***zeolites*** have also been produced with Ga^{3+}, P^{5+}, Ge^{4+}, and Fe^{3+} as tetrahedral ions in place of Si and Al.

Further studies on ***zeolite*** synthesis are aimed toward the production of ***zeolites*** tailored according to industrial needs. Such investigative effort has recently produced a new class of compounds characterized by a very high silica content—Si/Al ratios ranging from 10 to 100—and having silicalite as the pure silica end member. Silicalite is obtained by calcining a precursor crystalline phase with composition $4(TPA)OH \cdot 96SiO_2$. High silica ***zeolites*** have hydrophobic-organophilic selectivity, which makes them ideal products for use in separation of and catalytic processes on hydrocarbons, in contrast to the more common low- and intermediate-silica ***zeolites***, which readily adsorb water.

Due to the absence of tetrahedral Al in its structure, silicalite does not have ion exchange properties and has a remarkably hydrophobic character. It, therefore, cannot strictly be considered a ***zeolite***, although the term molecular sieve is certainly appropriate. On the other hand, synthetic zeolite ZSM-5, which is isostructural with silicalite, contains variable amounts of Al (from 0 to 2.5 atoms per unit cell) within the same crystal (Ballmoos and Meier, 1981), thereby indicating the existence of a complete compositional series for this family of ***zeolites***.

Commercial Applications

The combination of selective, sorptive, and catalytic properties has led to the use of ***zeolites***, both synthetic and natural, in a large number of practical applications and there is reason to believe that this number will increase as ***zeolite*** studies proceed.

The development of ***zeolite*** molecular-sieve technology has received an exceptional boost in recent years due to their utilization in fields as important as energy conservation and pollution control.

Some of the most important utilizations include:

Hydrocarbon separation processes: The synthetic Ca Type *A* molecular sieve has the ability to separate n-paraffins from a mixture of paraffins by adsorption. The n-paraffins are then removed from the **zeolite** by the use of displacing agents or other techniques. This process is used in octane-rating enhancement of gasoline (thus avoiding the introduction of tetrahethyl lead), and in the production of solvents and raw material for biodegradable detergents, fire retardants, plasticizers, alcohol, fatty acids, synthetic proteins, lubricating-oil additives, and α-olefins.

Synthetic *faujasite*, sodium *mordenite*, and Type Y ***zeolites*** are being or will be utilized in p-xylene separation, a necessary step in polyester production. Branched-chain olefins are separated from straight-chain olefins by means of a synthetic *faujasite*.

Drying of gases and liquids: Due to their highly hydrophilic and extremely large internal surface, ***zeolites*** are the most efficient dessicants. Their adsorption capacity is high at low concentration of water.

The many applications of this property range from the removal of water from alcohol bottles for laboratory use to the dessication of industrial streams of hydrocarbons to be cryogenically separated into different fractions. In the cryogenic fractionation, carbon dioxide, together with water vapor, is removed by the ***zeolite***.

Oxygen enrichment of air: Through a pressure-swing method, the use of a **zeolite** adsorbent has advantage over the cryogenic air-separation process at capacity below 25 tons per day.

Pollution control: Molecules such as SO_2, H_2S, and NO_x can be efficiently removed from industrial exhausts by means of acid-resistant ***zeolites***. Other processes include removal of mercury vapors, recovery of radioactive ions such as Cs and Sr from nuclear reactor wastes, and removal of toxic ammonia from agricultural and urban waste waters. *Clinoptilolite*, *phillipsite*, and *mordenite* are being used for the last two applications and many more such applications may be expected of the high-silica *zeolites*.

Dual function catalysis: ***Zeolites*** may act both as molecular sieves and as catalysts in oil-cracking processes. This property is widely utilized in oil-processing plants in the USA with considerable savings of energy and reduction in plant size.

All of these applications are, at least to some degree, regenerative processes. Recovery of both the ***zeolite*** and the adsorbate is usually achieved by heating the ***zeolite*** bed, by the use of another adsorbate as displacive agent, or by a combination of the two.

Nonregenerative applications: Drying of refrigerant fluids in refrigerators and air conditioners; manufacture of dual-pane windows; conditioning of agricultural soils by ion exchange; dietary supplements for pigs and poultry; production of special types of cement; carriers of one of the components of epoxy-type resins; and soft white fillers in paper industry are examples.

Large deposits of natural ***zeolites*** are presently being mined in Japan, the USA, and other countries. The obvious advantage of natural over synthetic ***zeolites*** is their lower cost, although lower purity is usually the limiting factor to their utilization.

ROMANO RINALDI

References

Breck, D. W., 1974. *Zeolites Molecular Sieves, Structure, Chemistry and Use*. New York: Wiley, 771p.

Ballmoos, R. von, and Meier, W. M., 1981. Zoned aluminium distribution in synthetic zeolite ZSM-5, *Nature*, **289**, 782–783.

Canadian Mineralogist, 1974. Low Grade Metamorphism, **12**(7), 437–543.

Černý, P.; Rinaldi, R.; and Surdam, R. C., 1977. Wellsite and its status in the phillipsite-harmotome group, *N. Jb. Miner. Abh.*, **128**, 312–330.

Coombs, D. S., et al., 1959. The zeolite facies, with

comments on the interpretation of hydrothermal syntheses, *Geochim. Cosmochim. Acta*, **17**, 53–107.

Flanigen, E. M.; Bennett, J. M.; Grose, R. W.; Cohen, J. P.; Patton, R. L.; Kirchner, R. M.; Smith, J. V., 1978. Silicalite, a new hydrophobic crystalline silica molecular sieve, *Nature*, **271**, 512–516.

Flanigen, E. M., 1980. Molecular sieve zeolite technology. The first twenty-five years, in L. V. C. Rees, ed., *Proceedings of the Fifth International Conference on Zeolites*. London: Heyden & Sons, 760–780.

Hay, R. L., 1966. Zeolites and zeolitic reactions in sedimentary rocks, *Geol. Soc. Am. Spec. Paper 85*, 130p.

Kossowskaya, A. G., 1973. Genetic associations of sedimentary zeolites in the Soviet Union, *Molecular Sieves, Adv. Chem. Ser.*, **121**, 200–208.

Iijima, A., 1980. Geology of natural zeolites and zeolitic rocks, in L. V. C. Rees, ed., *Proceedings of the Fifth International Conference on Zeolites*. London: Heyden & Sons, 103–118.

Lee, H., 1973. Applied aspects of zeolites adsorbents, *Molecular Sieves, Adv. Chem. Ser.*, **121**, 311–318.

Meier, W. M., 1968. *Zeolites Structures, Molecular Sieves*. London: Society of the Chemical Industry, 10–27.

Meier, W. M., and Olson, D. H., 1978. *Atlas of Zeolite Structure Types*. Pittsburgh, Pa: Polycrystal Book Service, 99p.

Mumpton, F. A., 1973. Worldwide deposits and utilization of natural zeolites, *Industrial Minerals No. 73*, Oct., 30–45.

Naccace, C., and Taarit, Y. B., 1980. Recent developments in catalysis by zeolites, in L. V. C. Rees, ed., *Proceedings of the Fifth International Conference on Zeolites*, London: Heyden & Sons Ltd., 529–606.

Olson, R. H., 1975. Zeolites, Industrial Minerals and Rocks, *Am. Inst. Mining, Metall. and Petroleum Engineers*, 1235–1274.

Rabo, J. A., ed., 1976. *Zeolite Chemistry and Catalysis*. Washington: The American Chemical Society, 796p.

Rinaldi, R., 1976. Crystal chemistry and structural epitaxy of offretite-erionite from Sasbach, Kaiserstuhl, *N. Jb. Miner. Mh., H.4*, 145–156.

Sheppard, R. A., 1971. Zeolites in sedimentary deposits of the United States—a review, *Molecular Sieves, Adv. Chem. Ser.*, **101**, 279–310.

Sheppard, R. A., 1973. Zeolites in sedimentary rocks, *U.S. Geological Survey Prof. Paper 820*, 689–695.

Smith, J. V., 1963. Structural classification of zeolites, *Mineral. Soc. Am., Spec. Paper*, **1**, 281–290.

Uytterhoeven, J. B., ed., 1973. *Molecular Sieves*. Leuven, Belgium: University Press, 484p.

Van Reeuwijk, L. P., 1974. *The Thermal Dehydration of Natural Zeolites*. Wageningen, Netherlands: Madedelingen Landbouwhogeschool, 88p.

Walker, G. P. L., 1960. The amygdale minerals in the Tertiary lavas of Ireland. III. Regional distribution, *Min. Mag.*, **32**, 503–527.

Cross-references: *Authigenic Minerals; Collecting Minerals; Diagenetic Minerals; Mineral Classification: Principles; Museums, Mineralogical; Naming of Minerals; Soil Mineralogy; Synthetic Minerals; Vein Minerals*; see also mineral glossary.

MINERAL GLOSSARY

The mineral glossary contains approximately 3000 entries. In addition to mineral species, names of mineral groups and crystal solution series, names of varieties, and names of mineraloids are given. Rock names, however, are excluded even though many of them end in "ite" and, thus, look like mineral names.

In each entry, the level of meaning of a mineral name is indicated. Mineral species are defined by chemical composition and by crystal system and have at least one reference where further information about the mineral may be found. An attempt was made to list recent English-language references rather than references for the most complete mineral description because interested users always can work their way back through the mineralogical literature.

In an effort to conserve space, and, thus, keep the cost of the volume down, very brief, nonstandard abbreviations of journal titles and reference books are used. These are listed in Table 1. More complete information about each journal or book may be found in the list of references at the front of this volume. Following each definition is a list of entries in which the particular mineral is mentioned; thus, the glossary, serves as the mineral index for this volume.

Over 500 of the more common mineral species are defined in greater detail in signed entries within the glossary. The affiliation of the authors of these entries may be found in the list of contributors at the beginning of the volume.

Varietal names are defined as "variety of . . ." and are also followed by a list of articles in which they are mentioned. A few very common mineral varieties, e.g., chalcedony, are given separate treatment.

Names of mineral series and groups are defined by a list of the mineral species that comprise the group or series. Mineral names that apply both to a species and to a group are entered twice. Major mineral groups are treated in articles in the main body of this volume and are indicated in the cross references that follow each group entry.

A mineral series occurs when two or more elements may substitute one for another to give a continuous range of mineral compositions. Modern mineral nomenclature gives

TABLE 1. Reference Abbreviations in the Glossary.

Abbreviation	Reference
Acta Geol. Sin.	*Acta Geologica Sinica*
AJS	*American Journal of Science*
AM. Min.	*American Mineralogist*
Ark. Min.	*Arkiv för Mineralogi*
Bull. Min.	*Bulletin de Minéralogie* (since 1978)
Bull. Soc. fr.	*Bulletin de la Société francais de Minéralogi et de Cristallographie* (before 1978)
Can. Min.	*Canadian Mineralogist*
Chem. Geol.	*Chemical Geology*
Cont. Min. Pet.	*Contribution to Mineralogy and Petrology*
CR	*Comptes rendus*, Académie des Sciences (Paris)
Dana I, II, III	*Dana's System of Mineralogy* 7th ed., volumes I, II, or III
DHZ 1, 2, 2A, 3, 4, 5	Deer, Howie, and Zussman, *Rock-Forming Minerals*, volumes 1, 2, 2A, 3, 4, or 5
Econ. Geol.	*Economic Geology*
GCA	*Geochimica et Cosmochimica Acta*
Geochim.	*Geochimica* (Chinese)
GSA Abst.	Geological Society of America, *Abstracts with Programs*
Hey	Hey, *An Index of Mineral Species & Varieties Arranged Chemically*, 2nd ed.
JGR	*Journal of Geophysical Research*
Min. Abst.	*Mineralogical Abstracts*
Min. Dep.	*Mineralium Deposita*
Min. Mag.	*Mineralogical Magazine*
Naturw.	*Naturwissenschaften*
N.J. Min. Mh.	*Neues Jahrbuch Mineralogie, Monatsheft* [Abhandlungen]
RRW	Roberts, W. L., Rapp, G.R., and Weber, J., 1974. *Encyclopedia of Minerals.* New York: Van Nostrand Reinhold, 693p.
Tsch. Min. Pet.	*Tschermak's Mineralogische und Petrographische Mitteilungen*
USGS Bull.	*U.S. Geological Survey Bulletin*
Winchell II	Winchell and Winchell, *Elements of Optical Mineralogy*, volume II
Zap. Vses. Min. Ob.	*Zapiski Vsesoyuznyi Mineral Obschechestva*
Z. Krist.	*Zeitschrift für Kristallographie, Kristaligeometrie, Kristallphysik, Kristallchemie*

mineral names to the pure end members of each series and often even a name to the series, but tends to discard names given to particular intermediate composition ranges. However, in some, such as the ***plagioclase*** series, the traditional names for intermediate compositions are still so widely used by mineralogists and nonmineralogists alike, that they are retained in this volume.

In addition to purely chemical information, mineral formulae often are cast so as to indicate additional mineralogic information. The mineral *kaolinite* has two formula units per unit cell, which may be indicated as either $2[Al_2Si_2O_5(OH)_4]$ or $Al_4Si_4O_{10}(OH)_8$.

Structural information may be incorporated into the chemical formula, as well. Mineral structures consisting of infinite chains, sheets, or three-dimensional frameworks with a net negative (anionic) electrostatic charge have charge-balancing cations occupying void spaces between or within the structural units. In these cases, elements of the structural unit may be grouped together as the last part of the formula. The formula for *muscovite,* thus, may be expressed $KAl_2(OH)_2[Si_3AlO_{10}]$, indicating that one-third of the aluminum ions are located in the tetrahedral sheet and two-thirds between the sheets. In the case of *vauxite,* the structural formula

$$Fe_2^{2+}(H_2O)_4[Al_4(OH)_4(H_2O)_4(PO_4)_4]\cdot 4H_2O$$

indicates that water molecules (H_2O) occupy three different kinds of sites in the structure, although the formula could also be rendered in the simpler form. Thus, there is no "one way" to cast a mineral formula.

The main sources for mineral names used in this glossary were the 1971, 1975, and 1980 *Glossary of Mineral Species* by Michael Fleischer. Mineral names introduced or discarded since publication of the 1980 *Glossary* have been gleaned from "Additions and Corrections to Glossary of Mineral Species," *Mineralogical Record,* **12**, 61–63; and from subsequent issues of *The American Mineralogist, Mineralogical Abstracts, Canadian Mineralogist,* and *Mineralogical Magazine.*

The glossary does not include, however, all existing mineral names. Michael Fleischer, in answering the question, "How many minerals?," (*Am. Mineralogist,* **54**, 960), estimated that there are "perhaps 10,000–20,000" mineral names in existence. These names are in addition to all rock and fossil names ending in "ite." For these latter names, see *The Encyclopedia of Paleontology* and *The Encyclopedia of Petrology.*

When a mineral name is found not to refer to a valid mineral species or to refer to an already named mineral, that name is discredited (see *Naming of Minerals*), although many such names live on as synonyms, as varietal names, or as group names. For mineral names not found in the glossary, the reader is referred to *The American Mineralogist,* **51**, 1247–1357; Palache, Berman, and Froudel's *Dana's System of Mineralogy,* 7th edition; Fleischer's *Glossary of Mineral Species* (updated annually in *The Mineralogical Record*); Hey's *Chemical Index of Minerals* (see *References* at the front of this volume), or *A Manual of Mineral Names* by P. G. Embrey and J. P. Fuller (Oxford Press, 1980).

As in the main body of the *Encyclopedia,* the level of a mineral name is indicated by the type face it is set in. Thus, the green variety of the mineral species *microcline,* widely referred to as amazonite, is set in roman type, the species name in italic type, and the group name, ***feldspar,*** in bold italic. No one level of name is superior to another, but the level of the name should be recognized. Discredited species names are also set in roman type.

A

ABELSONITE — *Am. Min.* **63**, 930
$C_{31}H_{32}N_4Ni$ — tric. — *GSA Abst.* **7**, 1221

Cross-reference: *Green River Mineralogy*

ABERNATHYITE (***meta-autunite*** gr.)
K(UO$_2$)(AsO_4)·$4H_2O$ — *Am. Min.* **41**, 82
tetr. *Min. Mag.* **31**, 952

ACANTHITE A silver sulfide of the ***argentite*** group, first named by Kenngott (1855) from the Greek meaning "thorn," in allusion to the shape of the crystals.

Composition: $4[Ag_2S]$; Ag, 87.06%; S, 12.94% by wt.

Properties: Monoclinic, $P2_1n$; in slender prisms, often twinned and pseudohexagonal; G=7.2–7.3; sectile; iron-black; often confused with *argentite;* most, if not all, natural silver sulfide specimens are pseudomorphs of *acanthite* after *argentite* (Hey, 1962).

Occurrence: Jáchymov, Czechoslovakia; Annaberg, E. Germany (Saxony); Georgetown and Rice, Colorado, and Owyhee County, Idaho, USA.

Uses: As an ore of silver with *argentite.*

R. W. FAIRBRIDGE

References

Dana, I, 191.
Hey, 10.
Min. Abst., **28**, 266, no. 77-2768.

Cross-reference: *Metallic Minerals.*

ACETAMIDE — hexa. — *Am. Min.* **61**, 338
CH_3CONH_2 — *Zap. Vses. Min. Ob.* **104**, 326

ACHAVALITE — Hey, 17
FeSe — RRW, 2

ACHREMATITE–Mixture of *mimetite* and *wulfenite* — *Am. Min.* **62**, 170

ACHROITE–Colorless *elbaite*

ACMITE (AEGERINE) A silicicate of sodium and iron of the pyroxene group.

Composition: $NaFe^{3+}(Si_2O_6)$; Na_2O, 13.4%; Fe_2O_3, 34.6%; SiO_2, 52.0%. When the composition varies due to ionic substitution of calcium for sodium and of magnesium and aluminum for ferric iron, the mineral is termed *aegirine.*

Properties: Monoclinic, $C2/c$; often in fibrous aggregates. H=6–6.5; G=3.40–3.44. Vitreous luster; translucent; color: brown or green.

Occurrence: A relatively rare rock-forming mineral found in sodium-rich, silica-poor rocks such as nephelene syenite and phonolite. Associated with *orthoclase,* ***feldspathoids, augite,*** and sodium-rich ***amphiboles.***

J. A. GILBERT

Reference

DHZ, 2A, 482–519.

Cross-references: *Clinopyroxenes; Green River Mineralogy; Minerals, Uniaxial and Biaxial; Pyroxene Group.*

ACTINOLITE A hydrated calcium, magnesium, iron silicate of the ***amphibole*** group. The name is from the Greek *aktis,* ray, and *lithos,* stone, in allusion to its common occurrence as bundles of radiating needles.

Composition: $2[Ca_2(Mg,Fe^{2+})_5Si_8O_{22}(OH,F)_2]$. The Fe^{2+} substitution for Mg commonly ranges from 50 mol% down to the iron-free end member of the series, *tremolite* (q.v.). When $Fe^{2+}/(Mg+Fe^{2+})>.50$, the name is *ferro-actinolite.*

Properties: Monoclinic ($C2/m$); short- to long-bladed crystals, columnar, fibrous (byssolite), or massive (nephrite, jade), granular; H=5–6; G=3.02–3.44; translucent to rarely transparent; pale to dark green in color.

Occurrence: Common rock-forming mineral in low-grade (greenschist facies) regionally metamorphosed carbonate, mafic, and ultramafic rocks where it is associated with *calcite, talc, antigorite, quartz, albite, epidote, hornblende, chlorite,* and *lawsonite. Actinolite* pseudomorphous after ***pyroxene*** is called uralite.

Use.: Nephrite is a jade (q.v.).

K. FRYE

References

Ernst, W. G., 1968. *Amphiboles.* New York: Springer-Verlag, 125p.
DHZ, 2, 249–262.
Am. Min. **63,** 1023-1052.

Cross-references: *Amphibole Group; Asbestos; Gemology; Human and Vertebrate Mineralogy; Jade, Rock-Forming Minerals.*

ADAMITE A hydrous zinc arsenate of the *tarbuttite-adamite* group, named by Friedel (1866) after Gilbert-Joseph Adam (1795–1881), French mineralogist.

Composition: $4[Zn_2AsO_4OH]$; copper, cobalt, and ferrous iron partially substitute for zinc.

Properties: Orthorhombic, often in groups and crusts; G=4.34–4.35; H=3.5; yellow, red, green, violet, to colorless.

Occurrence: In France, Italy, Greece, Turkey, Mexico, Chile. As a secondary mineral, in the oxidized zones of zinc- and arsenic-rich ore deposits. Associated with *smithsonite, calcite, malachite, azurite, limonite, hemimorphite,* and *quartz.*

Varieties: Cobaltoadamite and cuproadamite, respectively, cobalt-rich and copper-rich varieties.

R. W. FAIRBRIDGE

References

Dana, II, 864.
Am. Min. **61,** 979–986

ADELITE — *Dana,* II, 804
$CaMg(AsO_4)(OH)$ orth. *Can. Min.* **18,** 191

ADMONTITE — *Am. Min.* **65,** 205
$Mg_2B_2O_5 \cdot 15H_2O$ mono. *Min. Mag.* **43,** 1057

ADULARIA A colorless, low-temperature, translucent to transparent, and relatively pure variety of potassium ***feldspar*** with a distinctive {110} habit, named by Pini (1783) from the Adular Mountains of the Alps in the St. Gothard region, Switzerland, where it was first found.

J. A. GILBERT

References

Am. Min. **35,** 285
DHZ 4, 1-93.

Cross-references: *Alkali Feldspars; Gemology; Soil Mineralogy; Vein Minerals.*

AEGIRINE = Acmite

Cross-references: *Clinopyroxenes; Pegmatite Minerals; Pyroxene Group.*

AENIGMATITE Essentially a titanosilicate of ferrous iron and sodium, but may contain aluminum and ferric iron, related to *sapphirine,* named by Breithaupt, 1865, from the Greek word meaning "enigma."

Composition: $Na_2Fe_5^{2+}TiO_2[Si_6O_{18}]$. Triclinic.

Occurrence: In the sodalite–syenite of the Julianehaab district in Southern Greenland. As minute crystals in the liparite larvas of the island Pantellaria; also widespread in the rocks of E. Africa (cossyrite). In the basaltic rocks in the Rhön district and elsewhere in Germany.

K. FRYE

References

Am. Min. **56,** 427–446.
Min. Abst. **27,** 249, no. 76–2761.

Cross-references: *Rock-Forming Minerals.*

AERUGITE — *Am. Min.* **50,** 2108
$Ni_9As_3O_{16}$ mono. *Min. Mag.* **35,** 72

AESCHYNITE (ESCHYNITE) A niobate titanate of rare-earth elements and iron named by Berzelius (1828) from a Greek word for shame because some of the chemical elements could not at that time be separated.

Composition: $(Ce,Th,Fe,Ca)(Nb,Ti)_2O_6$; with Ce replaced by Y toward *aeschynite-(Y) (priorite).* If Ta replaces (Nb,Ti) it is called *tantalaeschynite.*

Properties: Orthorhombic, *Pmnb;* brownish black color, reddish brown in section.

Occurrence: Associated with minerals of the ***euxenite*** group; found in pegmatites in Miask, USSR.

D. H. SPEIDEL

References

Dana, I, 793–797
Am. Min. **60,** 309–315.

Cross-references: *Blacksand Minerals; Pegmatite Minerals.*

AESCHYNITE-(Y) orth. *Dana,* I, 793
$(Y,Ca,Fe,Th)(Ti,Nb)_2(O,OH)_6$

AFGHANITE *Am. Min.* **53**, 2105
(***cancrinite*** gr.) hexa. *Can. Min.* **17**, 47
$(Na,Ca,K)_{12}(Si,Al)_{16}O_{34}(Cl,SO_4,CO_3)_4 \cdot H_2O$

AFWILLITE A hydroxyl silicate of calcium.

Composition: $4[Ca_3Si_2O_4(OH)_6]$. Monoclinic.

Occurrence: From the Dutoitspan *diamond* mine at Kimberley, South Africa; as an alteration product of *spurrite* at Scawt Hill, Larne, Co. Antrim, Ireland; and from Crestmore, California.

K. FRYE

References

Am. Min. **10,** 447; **38,** 629–633.
Min. Abst. **27,** 299, no. 76-3274.
Min. Mag. **20,** 444; **29,** 838.

AGARDITE hexa. *Am. Min.* **55**, 1447
$(Y,Ca)Cu_6(AsO_4)_3(OH)_6 \cdot 3H_2O$

AGATE—Variety of chalcedony

Cross-references: *Gemology; Museums, Mineralogical; Placer Deposits, Skeletal Crystals; Soil Mineralogy.*

AGRELLITE *Am. Min.* **62**, 173
$NaCa_2Si_4O_{10}F$ tric. *Can. Min.* **14**, 120

AGRINIERITE *Am. Min.* **58**, 805
$(K_2,Ca,Sr)U_3O_{10} \cdot 4H_2O$ orth. *Min. Mag.* **38**, 781

AGUILARITE *Am. Min.* **35**, 377
Ag_4SeS orth. *Dana,* I, 178

AHLFELDITE *Am. Min.* **54**, 448
$NiSeO_3 \cdot 2H_2O$ mono.

AIKINITE *Dana,* I, 412
$PbCuBiS_3$ orth. *Am. Min.* **61**, 15

Cross reference: *Ore Microscopy.*

AJOITE *Am. Min.* **66**, 201
$(K,Na)Cu_7AlSi_9O_{27} \cdot 6H_2O$
tric. *Min. Mag.* **32**, 942

AKAGANÉITE *Am. Min.* **48**, 711
$FeO(OH,Cl)$ tetr. *Min. Mag.* **33**, 270

Cross-references: *Lunar Minerals; Magnetic Minerals; Soil Mineralogy.*

AKATOREITE *Am. Min.* **56**, 416
$Mn_9(Si,Al)_{10}O_{23}(OH)_9$ tric.

AKDALAITE *Am. Min.* **56**, 635
$4Al_2O_3 \cdot H_2O$ hexa.

AKERMANITE A silicate of calcium with magnesium (isomorphous with *gehlenite* and *melilite)* belonging to the ***melilite*** group; named by Vogt, 1884.

Composition: $2[MgCa_2Si_2O_7]$. Tetragonal.

Occurrence: Found only in certain slags that have cooled rapidly.

K. FRYE

References

Carnegie Inst. Wash. Publ. Yr. Bk. **72,** 449.
DHZ 1, 236–255.

AKROCHORDITE mono. *Dana* II, 927
$Mn_4Mg(AsO_4)_2(OH)_4 \cdot 4H_2O$

AKSAITE *Am. Min.* **48**, 930
$MgB_6O_{10} \cdot 5H_2O$ orth.

AKTASHITE *Am. Min.* **58**, 562
$Cu_6Hg_3As_5S_{12}$ trig.

ALABANDINE (ALABANDITE) A manganese sulfide named by Beudant (1832) from the locality, Alabanda, Aïdin, Turkey.

Composition: $4[MnS]$.

Properties: Isometric, $Fm3m$; commonly granular and massive; G=3.95–4.04; H=3.5–4; color: iron-black; streak green.

Occurrence: Turkey, Hungary, Rumania, Peru, Mexico, Colorado, and Arizona. As epithermal vein deposits, associated with *sphalerite, galena, pyrite, quartz, calcite, rhodocrosite,* and *rhodonite.*

Use: An ore of manganese.

R. W. FAIRBRIDGE

References

Dana, I, 207.
Min. Abst. **11,** 402; **12,** 229.
U.S. Bur. Mines Rept. Inv. 6495, 10.

Cross-references: *Magnetic Minerals; Meteoritic Minerals; Ore Microscopy.*

ALABASTER = Variety of *gypsum*

ALAMOSITE *Z. Krist.* **126**, 98
$PbSiO_3$ mono.

ALASKAITE—Mixture of sulfosalts *Am. Min.* **58**, 349

ALBERTITE–Hydrocarbon

Cross-reference: *Green River Mineralogy.*

ALBITE A sodium aluminosilicate of the ***feldspar*** group, named by Gahn and Berzelius (1814) from the Latin *albus,* white, the common color of the mineral.

Composition: $NaAlSi_3O_8$; with Na partially replaced by Ca and K and Si by Al. The composition represents the Na end member, *Ab,* of the series *albite-anorthite,* the ***plagioclase feldspars,*** and the mineral may contain up to 10 mol% of the anorthite composition, *An.*

Properties: Triclinic, $C\bar{1}$; platy crystals parallel to (010) (cleavelandite), commonly in cleavable masses or as irregular grains; prismatic {001} {010} cleavage; commonly twinned according to the albite, Carlsbad, or pericline laws; H=6; G=2.62; vitreous to pearly luster; translucent to transparent; colorless, white, or gray, rarely tinted green, yellowish, or reddish; commonly intergrown with potash ***feldspar*** (crypto- to macroperthite), with *oligoclase* (peristerite), or with bleby *quartz* (myrmekite).

Occurrence: A common rock-forming mineral, particularly in felsic volcanic and plutonic rocks associated with potash ***feldspar,*** *quartz,* and ***feldspathoids,*** in granite and syenite pegmatites; in spillites, in veins; in low-grade metamorphic (greenschist facies) with *epidote* and ***chlorites;*** as a result of sodium metasomatism of argillaceous sediments; as a detrital mineral in arkoses; and as an authigenic mineral (then untwinned and rather pure).

K. FRYE

References*

DHZ, 4, 1–165.

*See *Feldspar Group* and *Plagioclase Feldspars* for additional references.

Cross-references: *Alkali Feldspar; Clinopyroxenes; Feldspar Group; Gemology; Glass, Devitrification of Volcanic; Green River Mineralogy; Minerals, Uniaxial and Biaxial; Order-Disorder; Pegmatite Minerals; Plagioclase Feldspars; Polymorphism; Rock-forming Minerals; Staining Techniques; Thermometry, Geologic.*

ALBRITTONITE — *Am. Min.* **63**, 410
$CoCl_2 \cdot 6H_2O$ — mono.

ALDERMANITE — orth. — *Min. Mag.* **44**, 59
$Mg_5Al_{12}(PO_4)_8(OH)_{22} \cdot 32H_2O$

ALDZHANITE — *Am. Min.* **56**, 1122
$CaMgB_2O_4Cl \cdot 7H_2O$(?) — orth.

ALEKSITE — *Am. Min.* **64**, 652
$PbBi_2Te_2S_2$ — ps. trig. — *Min. Mag.* **43**, 1057

ALEXANDRITE–See *chrysoberyl*

Cross-references: *Gemology; Naming of Minerals.*

ALGODONITE — *Dana,* I, 171
Cu_6As — orth. — *Min. Abst.* **27**, 236

ALIETTITE–Regularly interstratified *talc-saponite* — *Am. Min.* **57**, 598

ALKALI FELDSPARS–*Albite, anorthoclase, microcline, orthoclase,* and *sanidine*

ALLACTITE — *Am. Min.* **53**, 733
$Mn_7(AsO_4)_2(OH)_8$ — mono.

ALLANITE (ORTHITE) A hydroxyl aluminosilicate of the rare-earth elements belonging to the ***epidote*** group and named by Thomson (1810) after T. Allan who discovered (in 1808) the mineral in East Greenland.

Composition:
$(Ca,Fe^{2+},Ce,La,Na)_2(Al,Fe^{3+},Mn,Be,Mg)_3(SiO_4)(Si_2O_7)(OH)$
Variable composition with the rare-earth elements partially substituting for Mn and Mg. If Y>Ce, the mineral is *allanite-(Y).*

Properties: Monoclinic, $P2_1/m$; commonly massive or granular; H=5.5–6; G=3.5–4.2; pitchy, resinous to submetallic luster, subtranslucent, slightly radioactive; color: brown to black; streak: greenish or brownish-gray.

Occurrence: In many igneous rocks as a minor accessory mineral. May also be found as a contact mineral in limestone. Associated with *epidote.* Found in Greenland; at Criffel, Scotland; at Jotun Feld, Norway; near Dresden, E. Germany; in Sweden; and at Miask in the Ural Mountains.

J. A. GILBERT

References

Am. Min. **47**, 1327; **49**, 1159.

DHZ, 1, 211–220.

Cross-references: *Blacksand Minerals; Metamict State; Pegmatite Minerals.*

Cross-reference: *Minerals, Uniaxial and Biaxial.*

ALLARGENTUM — *Am. Min.* **56**, 638
$Ag_{1-x}Sb_x$ — hexa. — *Can. Min.* **10**, 163

ALLCHARITE = *Goethite* *Am. Min.* **54**, 1498

ALLEGHANYITE (***Hunite*** gr.)
$Mn_5(SiO_4)_2(OH)_2$
mono. *Am. Min.* **54**, 1392

ALLEVARDITE = *Rectorite*
Am. Min. **51**, 1252

ALLOCLASITE orth. or mono. *Am. Min.* **57**, 1561
(Co,Fe)AsS *Can. Min.* **14**, 561

ALLOPALLADIUM = *Stibiopalladinite*
Am. Min. **63**, 796

ALLOPHANE A hydrous aluminum silicate, first named by Stromeyer and Hausmann in 1816, from the Greek for "other" and "appear," in allusion to its change in appearance under the blowpipe.

Composition: $SiO_2,Al_2O_3 \cdot nH_2O$. SiO_2 ranges from 13.89 to 39.00%, Al_2O_3 from 31.94 to 46.36%, and H_2O+ from 13.40 to 23.26%. SiO_2/Al_2O_3 ranges from 0.7 to 2.0; phosphates up to and in excess of 10% are common.

Properties: A gel, amorphous to light and to X-rays; G=1.90±0.25, H=1–3; color; white, blue, colorless, green, pink, and yellow; refractive index = 1.398–1.494; optically isotropic; structure: massive or granular; occurs as submicroscopic pellets or bullets; streak: white.

Chemical Properties: Soluble in strong acids and bases. Cation exchange increases with increasing *p*H and anion exchange increases with decreasing *p*H. The *p*H at which uptake of cations equals that of anions varies with the nature of the salt solution. For the salts of weak acids, the isoelectric point is higher than for salts of strong acids; the reverse holds for salts of bases. Allophane is a reactive mineral and forms identifiable compounds with many leaching solutions. Allophane will fix bases and it will form compounds with phosphate, molybdate, and fluoride.

Occurrence: Soils derived from weathering of volcanic ash and volcanic glass, in areas of hydrothermal alteration of other minerals, and in areas where sulfate-bearing meteoric waters react with other clay minerals and aluminum silicate minerals, e.g., allophane from Lawrence County, Indiana. First described from Grafental, S of Saalfeld, Thuringia, E. Germany (as riemannite).

Associated Minerals: *Halloysite, gibbsite,* phosphate-allophane.

W. A. WHITE

References

Am. Min. **38**, 634–642; **61**, 379–390.
Izv. Akad. Nauk SSSR, Ser. Geol. **30**(3), 51–57.
New Zealand J. Sci. **4**, 393–414; 9(3), 599–607.
U.S. Geol. Surv. Prof. Paper 185-G, 135–148.

Cross-references: *Clay Minerals; Soil Mineralogy.*

ALLUAUDITE mono. *Am. Min.* **56**, 1955
$(Na,Ca)Fe^{2+}(Mn,Fe,Mg)_2(PO_4)_3$
Min. Mag. **43**, 227

Cross-reference: *Pegmatite Minerals.*

ALMANDINE (ALMANDITE) An iron aluminum silicate of the ***garnet*** group, the name possibly derived from the ancient locality Alabanda in Asia Minor.

Composition: $8[Fe_3^{2+}Al_2Si_3O_{12}]$ with Fe^{2+} replaced by Mg toward *pyrope,* by Mn^{2+} toward *spessartine,* and by Ca toward *grossular.*

Properties: Isometric (*Ia*3*d*); crystals dodecahedra and trapezohedra, single and mutually modified, also massive or granular; H=7.5; G=4.0–4.2; red to brownish black.

Occurrence: In schists from low to medium grade regionally metamorphosed argillaceious sediments, also in granulites and in aureoles from contact metamorphism. Not uncommon in granites and in felsic volcanic rocks. A common and widespread detrital mineral.

Use: A gemstone or an abrasive material (q.v.).

Reference: DHZ, 1, 77–112.

K. FRYE

Cross-references: *Clinopyroxenes; Gemology; Minerals, Uniaxial and Biaxial; Orthopyroxenes; Pegmatite Minerals.*

ALSTONITE A carbonate of barium and calcium, named by Breithaupt (1841), from Alston Moor, Cumberland, England.

Composition: $BaCa(CO_3)_2$.

Properties: Triclinic, pseudo-orthorhombic. *Barytocalcite,* of the same composition, is monoclinic and *paralstonite* is trigonal.

Occurrences: Found as a low-temperature hydrothermal deposit. Associated with *calcite, barite,* and *witherite.*

J. A. GILBERT

Reference: *Dana,* II, 218.

ALTAITE A lead telluride isostructural with *galena,* named for the Altai Mountains of Siberia, USSR.

Composition: 4[PbTe]; Pb, 61.9%; Te, 38.1%.

Occurrence: With *gold,* sulfides, and other tellurides in veins. In addition to Siberia, it is found at Salzburg, E. Germany; Kalgoorlie, W. Australia; Coquimbo, Chile; and several localities in British Columbia, Canada. In the USA: Gaston Co., North Carolina; Goldhill, Colorado; Organ Mtns, New Mexico; and Carson Hill, Tuolamne Co., and Nevada City, California.

K. FRYE

References

Dana, II, 205-207.
Min. Abst. **27**, 306, no. 76-3346.

Cross-references: *Ore Microscopy.*

ALTHAUSITE *Am. Min.* **65**, 488
$Mg_2(PO_4)(OH,F,O)$ orth. *Lithos* **8**, 215

ALUM A group of hydrous alkali aluminum sulfates.

ALUMINITE mono., ps. orth. *Dana,* II, 600
$Al_2(SO_4)(OH)_4 \cdot 7H_2O$ *Min. Abst.* **27**, 77

ALUMINOCOPIAPITE *Am. Min.* **52**, 1220
$AlFe_4^{3+}(SO_4)_6O(OH) \cdot 2OH_2O$ tric.

ALUMINO-KATOPHORITE (***Amphibole*** gr.)
$Na_2Ca(Fe^{2+},Mg)_4AlSi_7AlO_{22}(OH)_2$
mono. *Am. Min.* **63**, 1023

ALUMOHYDROCALCITE A hydrated carbonate of calcium and aluminum named by Bilibin (1926).

Composition: $CaAl_2(CO_3)_2(OH)_4 \cdot 3H_2O$. Triclinic.

Occurrence: Rare, in the Khakassy District, Siberia, with allophane and other minerals, where it apparently formed by the action of carbonated waters on allophane. Also Nowej Rudy, Poland; Bergisch-Gladbach, Germany; and Chitral, Pakistan.

J. A. GILBERT

References

Am. Min. **13**, 569; **63**, 795.
Dana, II, 280.
Min. Mag., **21**, 557

ALUMOTUNGSTITE
$(W,Al)_{16}(O,OH)_{48} \cdot H_2O$ trig.

ALUNITE (ALUMSTONE) A hydroxyl sulfate of potassium and aluminum belonging to the ***alunite*** group. Named by Beudant in 1824 after aluminilite of Delamétherie (1797).

Composition: $KAl_3(OH)_6(SO_4)_2$; K_2O, 11.4%; Al_2O_3, 37.0%; SO_3, 38.6%; H_2O, 13.0%.

Properties: Trigonal crystal system but commonly massive or disseminated; H=4, G=2.6–2.8; transparent to translucent; color: white, gray, reddish.

Occurrence: Formed from the action of sulfuric acid in volcanic regions on rocks containing potash-***feldspar*** by the process known as "alunitization," which may result in the formation of large masses. Also found around fumaroles, usually in small amounts. Known from Nevada, Utah, Colorado, Italy, Hungary, Spain, Greece, and Australia.

Use: Widespread, but low-grade, potential aluminum ore.

J. A. GILBERT

References

Am. Min. **47**, 127; **65**, 953-956.
Dana, II, 556-560.
Min. Abst. **27**, 337, no. 76-3657.
U.S. Geol. Surv. Prof. Paper 1076A, 1-35.

ALUNITE GROUP–*Alunite, ammoniojarosite, argentojarosite, hydronium jarosite, jarosite, natroalunite, natrojarosite, osarizawaite,* and *plumbojarosite.*

ALUNOGEN A hydrous aluminum sulfate, the name meaning to make alum.

Composition: $Al_2(H_2O)_{12}(SO_4)_3 \cdot 5H_2O$.

Properties: Triclinic, $P\bar{1}$; prismatic crystals rare, commonly fibrous masses or crusts; H=1.5-2; G=1.65; silky to vitreous luster; white or tinged yellow to reddish by impurities; sharp acide taste.

Occurrence: In environments made acid from *pyrite* oxidation or volcanic activity. Filling crevices in coal, slates, or gossans. Also associated with fumarolic *sulfur* and *gypsum.*

K. FRYE

References

*Dana,*II, 537-540.
Am. Min. **61**, 311.

ALVANITE *Am. Min.* **44**, 1325
$Al_6(VO_4)_2(OH)_{12} \cdot 5H_2O$ mono.

AMAKINITE *Am. Min.* **47**, 1218
$(Fe,Mg)(OH)_2$ trig.

AMALGAM Alloy of Ag and Hg – see ***moschellandsbergite.***

Cross-references: *Native Elements and Alloys; Placer Deposits.*

AMARANTITE tric. *Min. Mag.* **42**, 144
$Fe_2(H_2O)_4(SO_4)_2O \cdot 3H_2O$ *Z. Krist.* **127**, 261

AMARILLITE *Dana* II, 468
$NaFe(SO_4)_2 \cdot 6H_2O$ mono.

AMAZONITE–A green variety of *microcline.*

Cross-reference: *Gemology.*

AMBER–Amorphous hydrocarbon.

Cross-references: *Gemology; Museums, Mineralogical; Resin and Amber*

AMBLYGONITE A lithium aluminum fluorine phosphate first named by Breithaup, 1817 from the Greek for "blunt angle." Earliest record of production is from Montebras (Creuse), France (1886).

Composition: $(Li,Na)Al(OH,F)PO_4$: Li_2O, 10.1%; Al_2O_3, 34.4%; F, 12.9%; P_2O_5, 47.9%.

Properties: Triclinic, $P\bar{1}$; crystals rare, usually massive; G=3.0–3.1; H=6; Luster is vitreous, but pearly on cleavage face, {001}; color: white, pale green, blue, brownish; habit: in slabs; cleavage: perfect on {100} (basal), good on {110} ; usually occurs in coarse cleavable masses; polysynthetic twinning common in two directions at 90°; fracture is uneven; streak: white; refractive index = 1.578–1.598; translucent to opaque.

Occurrence: Lithium-rich granite pegmatites, associated with Li minerals. Masses up to hundreds of tons occur in the Black Hills, South Dakota. Associated with *spodumene,* ***tourmaline,*** *lepidolite,* ***apatite,*** etc.

Uses: Source of lithium; third in importance to *spodumene* and *lepidolite.* Occasionally occurs as clear, transparent crystals of gem quality. In this form, it is fairly constant in composition and has one direction of cleavage, parallel to the basal plane. Its color is commonly pale to golden yellow.

R. W. FAIRBRIDGE

References

Am. Min. **47**, 387.
Gems and Gemology, **8**, 208.

Cross-references: *Naming of Minerals; Pegmatite Minerals; Staining Techniques.*

AMEGHINITE *Am. Min.* **60**, 879
$Na[B_3O_3(OH)_4]$ mono.

AMESITE A hydroxyl aluminosilicate of iron and magnesium. A ***septechlorite*** of the ***kaolinite-serpentine*** group.

Composition: $(Mg_4Al_2)(Si_2Al_2)O_{10}(OH)_8$.

Occurrence: Associated with emery at Chester, Massachusetts; Mn-ores at Gloucester, South Africa; Cr-ores, North Urals, USSR; and in the Pensacola Mtns., Antarctica.

K. FRYE

References

Am. Min. **61**, 497–499.
DHZ, 3, 166–167.

Cross-reference: *Minerals, Uniaxial and Biaxial; Soil Mineralogy.*

AMETHYST–Variety of ***quartz***

Cross-reference: *Mineral Classification: Principles.*

AMICITE (***zeolite*** gr.) *Am. Min.* **65**, 808
$K_2Na_2Al_4Si_4O_{16} \cdot 5H_2O$
mono. *Min. Abst.* **31**, 229

AMINOFFITE *Am. Min.* **53**, 1418
$Ca_3Be(OH)[Si_3O_{10}]$ tetr.

AMMONIA ALUM = *Tschermigite*

AMMONIOBORITE *Am. Min.* **44**, 1150
$(NH_4)Fe_3(SO_4)_2(OH)_6$ mono.

AMMONIOJAROSITE (***Alunite*** gr.)
$(NH_4)Fe_3(SO_4)_2(OH)_6$
trig. *Dana,* **II**, 652

AMOSITE = Asbestiform *gedrite, grunerite,* or *anthophyllite*

AMPANGABEITE = *Samarskite*
Am. Min. **46**, 770

AMPHIBOLE GROUP–*Anthophyllite, gedrite; cummingtonite, grunerite, tirodite, dannemorite; clinoholmquistite; tremolite, actinolite, edenite, pargasite, hastingsite, tschermakite, hornblende, kaersutite; richterite, winchite, barroisite, katophorite, taramite; glaucophane, riebeckite, eckermannite, arfvedsonite, kozulite;*

each mineral name may be preceded by one or more of the prefixes *alumino-*, *ferri-*, *ferro-*, or *magnesio-*, which then becomes part of the mineral name.

Reference: *Am. Min.* **63**, 1023–1052.

ANALCIME (ANALCITE) A hydrated sodium aluminosilicate named by Haüy, 1801 (from Greek "weak," in allusion to the weak electrostatic charge when heated or rubbed).

Composition: $Na[AlSi_2O_6] \cdot H_2O$; SiO_2, 54.5%; Al_2O_3, 23.2%; Na_2O, 14.1%; H_2O, 8.2% by wt.

Properties: Cubic, *Ia3d*; a=13.7 Å, G=2.24–2.29; H=5.5; luster: vitreous; color: colorless, white, gray, yellowish, greenish, or pink; transparent to nearly opaque; habit in icositetrahedra, in radiating aggregates (probably pseudomorphed after ***zeolites***), or in irregular granular masses; cleavage very poor on {100}; twinning on {001} and {110} (lamellar). Optically, it is isotropic; refractive index n = 1.479–1.493; many specimens show slight birefringence, and optical properties may vary even within a single grain. In its chemistry there may be a complete solid-solution series between *analcime* and *wairakite*, Ca analog of *analcime;* appreciable amounts of K sometimes occur in *analcime;* H_2O may increase with the increase in SiO_2 content; dehydration gradual, losing 8.4% H_2O at maximum; rehydration good; minimum diameter of the largest channel is 2.2 Å; molecular sieve properties selective for NH_3, H_2, N_2, Ar, and He; Na is freely exchanged with K, NH_4, Ag, Tl, and Rb, but only slightly exchanged with Li, Cs, Mg, Ca, and Ba; gelatinized by HCl.

Occurrences: Primary mineral of later formation in some intermediate and mafic igneous rocks such as teschenite, essexite, phonolite, nepheline-syenite pegmatite, dolerite, and basalt; hydrothermal crystallization in vesicles of basalt, monzonite, serpentinite, dacite, and rhyolite, often associated with *natrolite, chabazite, thomsonite, stilbite,* etc.; alteration product of precursor *clinoptilolite* and *mordenite,* which were derived from silicic glass shards, in tuffs and tuffaceous sandstone in association with *quartz, heulandite,* and *laumontite* as the *analcime* zone of low-grade metamorphism or burial diagenesis; direct precipitate from alkaline saline lake water in arid regions; alteration product of precursor potassium-bearing ***zeolites*** such as *phillipsite, chabazite, clinoptilolite,* and *erionite,* which are formed by reaction of silicic vitric ash with alkaline saline lake or interstitial water; in cement of palagonite tuffs on Oahu, Hawaii, associated with *phillipsite, chabazite, gonnardite,* and *natrolite,* which were formed by reaction of basaltic glass with percolatry water; in alkaline soil in arid regions; alteration product of *leucite* and leucitic glass in the recent marine sediment of Naples Bay.

A. IIJIMA
M. UTADA

References

Acta, Cryst. **7**, 357–359.
Am. Min. **44**, 300–313; **51**, 736-753; **53**, 184–200; **63**, 448–460.
Geol. Soc. Am. Bull. 78, 259-268.
Jour. Fac. Sci. Univ. Tokyo **19**, 133–147.

Cross-references: *Authigenic Minerals; Green River Mineralogy; Naming of Minerals; Pegmatite Minerals; Rock-Forming Minerals.*

ANANDITE (***Mica*** gr.) *Am. Min.* **52**, 1586
$(Ba,K)(Fe,Mg)_3(Si,Al,Fe)_4O_{10}(O,OH)_2$
mono. *Min. Mag.* **36**, 1

ANAPAITE *Dana*, II, 731
$Ca_2Fe(PO_4)_2 \cdot 2H_2O$ tric. *Min. Abst.* **27**, 256

ANARAKITE - Zincian *paratacamite*(?)
Am. Min. **58**, 560

ANATASE (OCTAHEDRITE) A titanium oxide usually credited to Haüy who named it in 1801 after the Greek meaning "erection" which he says alludes to the fact that the octahedron is longer than other tetragonal species.

Composition: $4[TiO_2]$. The chemical composition varies little from TiO_2.

Properties: Tetragonal; $I4_1/amd$; color: brown, greenish, gray, black; adamantine luster; transparent; H=5.5–6; G=3.90; the low-temperature polymorph of *rutile;* transforms to *rutile* at temperatures above about 600°C.

Occurrence: Found as a minor constitutent in hydrothermal veins of granitic pegmatites, scattered in igneous and metamorphic rocks, as a detrital mineral, as a rare authigenic mineral, or as an alteration product of *titanite* or *ilmenite* in Germany, Brazil, Colorado, and the Urals.

D. H. SPEIDEL

Cross-references: *Blacksand Minerals; Electron Microscopy (Transmission); Fire Clays; Green River Mineralogy; Minerals, Uniaxial and Biaxial; Pigments and Fillers; Polymorphism; Pseudomorphism; Rock-Forming Minerals; Vein Minerals.*

ANAUXITE–Mixture of *kaolinite* and amorphous silica *Clays Clay Miner.* **16,** 425

ANCYLITE (ANKYLITE) A hydrated carbonate of strontium and rare-earth (*RE*) elements, named by Flink (1900), from the Greek word meaning "curved," in allusion to the rounded and distorted character of the crystals.

Composition: $(RE)_x(Sr,Ca)_{2-x}(CO_3)_2(OH)_x \cdot (2-x)H_2O$.

Properties: Orthorhombic; H=4.5; G=3.9.

Occurrence: Rare, found in druses in pegmatite veinlets in nepheline syenite associated with aegirine (*acmite*), *albite*, and *microcline*, at Narssarssuaq in the Julianehaab district, Greenland, and Mont St. Hilaire, Quebec, Canada.

J. A. GILBERT

References

Am. Min. **60,** 280–284.
Dana, II, 291.
Min. Mag. **12,** 379.

ANDALUSITE An Aluminosilicate named for Andalusia, Spain, where it was first noted by Delamémetherie in 1789.

Composition: Al_2SiO_5; Al_2O_3, 63.2%; SiO_2, 36.8%.

Properties: Orthorhombic, *Pnnm;* H=7.5; G=3.16–3.20; vitreous luster, transparent to translucent; a variety called chiastolite has a dark cruciform design in the center of the prismatic habit formed by the ordered arrangement of dark-colored carbonaceous inclusions.

Occurrence: A product of the metamorphism of aluminum-bearing shales and slate under high temperature and low stress. Found in Andalusia, Cornwall, the Alps, the Urals, Brazil, California, Nevada, and Transvaal (the Republic of South Africa).

Use: In the manufacture of highly refractory porcelains for spark plugs and electric furnace linings. Transparent green varieties found in Brazil and Ceylon are used as gem stones.

J. A. GILBERT

References

Am. Min. **45,** 808; **46,** 1191.
DHZ, 1, 129–136.
Min. Abst. **3,** 250.

Cross-references: *Crystallography: Morphological; Mineral and Ore Deposits; Minerals, Uniaxial and Biaxial; Refractory Minerals; Soil Mineralogy.*

ANDERSONITE trig. *Am. Min.* **36,** 1
$Na_2Ca(UO_2)(CO_3)_3 \cdot 6H_2O$

ANDESINE A sodium calcium aluminosilicate of the ***feldspar*** group, named by Abich (1841) for occurrences in the Andes Mountains.

Composition: $(Na,Ca)[(Si,Al)AlSi_2O_8]$; with some K replacing (Na,Ca). In the ***plagioclase feldspars,*** the mineral is assigned the range *Ab* 50–70, *An* 50–30.

Properties: Triclinic, $C\bar{1}$; commonly as cleavable masses or irregular grains; twinning according to the albite, Carlsbad, or pericline laws; prismatic {001} {010} cleavage; vitreous to pearly luster; white or light to dark gray.

Occurrence: A rock-forming mineral as phenocrysts and ground mass in andesites; in the differentiated phases of some gabbroic intrusions; some anorthosites in their entirety; in metamorphic rocks of the granulite facies; and as detrital grains in some arkoses.

K. FRYE

References

DHZ, 4, 94–165.
*See *Feldspar Group* and *Plagioclase Feldspars* for additional references.

Cross-references: *Feldspar Group; Minerals; Uniaxial and Biaxial; Plagioclase Feldspars.*

ANDORITE *Am. Min.* **60,** 621
$PbAgSb_3S_6$ orth. *Min. Abst.* **27,** 306

ANDRADITE A calcium, iron silicate of the ***garnet*** group, named for the Portuguese mineralogist, J. B. d'Andrada.

Composition: $8[Ca_3Fe_2^{3+}Si_3O_{12}]$; SiO_2, 35.5%; Fe_2O_3, 31.5%; CaO, 33.0%; with variable amounts of Al and Ti replacing Fe^{3+}, Ti and Al replacing Si, and Mn^{2+} replacing Ca. Forms a series with *grossular* (***grandite***) and with *spessartine* (***spandite***).

Properties: Isometric (*Ia*3*d*); crystals dodecahedra or trapezohedra or in combination, massive or granular; H=6–7.5; G=3.86; color: yellow, green, reddish or yellowish brown, brown, or black; vitreous luster.

Occurrence: Calcareous metasediments and skarns with *calcite, hedenbergite,* and *mag-*

netite, in alkaline plutonic and volcanic rock, and in placers (see *Placer Deposits*).

K. FRYE

References

Am. Min. **63**, 378
DHZ, 1, 77–112.

Cross-references: *Garnet Group; Gemology; Meteoritic Minerals; Minerals, Uniaxial and Biaxial.*

ANDREMEYERITE *Am. Min.* **59**, 381
$BaFe_2Si_2O_7$ mono.

ANDREWSITE A hydroxyl phosphate of copper and iron, named by Maskelyne (1871), after Thomas Andrew (1813–1885), English chemist.

Composition: $(Cu,Fe^{2+})Fe^{3+}_3(PO_4)_3(OH)_2$.

Properties: Orthorhombic; radiating fibrous aggregates. Isostructural with *laubmannite.*

Occurrence: Rare, found in West Phoenix Mine, Cornwall, England, associated with "limonite," *dufrenite, chalcosiderite,* and *cuprite.*

K. FRYE

References

Am. Min. **34**, 534.
Dana, II, 802.

ANDUOITE *Am. Min.* **65**, 808
$(Ru,Os)As_2$ orth. *Min. Mag.* **43**, 1057

ANGELELLITE *Am. Min.* **44**, 1322
$Fe^{3+}_4O_3(AsO_4)_2$ tric. *Min. Abst.* **29**, 401

ANGLESITE A lead sulfate named by Beudant (1832), from the original locality on the Island of Anglesey.

Composition: $4[PbSO_4]$; PbO, 73.6%; SO_3, 26.4%.

Properties: Orthorhombic, frequently in concentric layers around an unaltered core of *galena* or in earthy form; H=3, G=6.2–6.4; conchoidal fracture, adamantine luster, transparent to translucent; color: white, gray (dark when impure), yellow, or colorless.

Occurrence: A supergene mineral found in the oxidized upper portions of lead veins and formed through the oxidation of *galena;* thus, it is a common secondary lead mineral. Associated with *galena, cerussite, smithsonite, sphalerite, hemimorphite,* and iron oxides.

Use: Minor ore of lead

J. A. GILBERT

References

Am. Min. **63**, 506–510.
Dana, II, 420–424.

Cross-references: *Gossan; Museums, Mineralogical; Staining Techniques.*

ANHYDRITE A calcium sulfate, and so named by Werner (1804) because it is the anhydrous member of the calcium sulfate compounds, the others being *bassanite,* $2CaSO_4 \cdot H_2O$, and *gypsum,* $CaSO_4 \cdot 2H_2O$.

Composition: $CaSO_4$; CaO, 41.2%; SO_3, 58.8%. Small amounts of strontium (up to about 0.7% Sr) can substitute for calcium (Ca), although in most natural samples the ranges are 0.1–0.3% strontium (Sr). The transition metal content [i.e., copper (Cu), cobalt (Co), chromium (Cr), lead (Pb), zinc (Zn)] of most natural *anhydrite* (and *gypsum*) samples is generally very low, less than 1 ppm, which is significant in view of the proposed origin of economic concentrations of these metals from redissolved evaporite beds.

Properties: Orthorhombic, rarely well crystallized; massive, fibrous, granular or compact; H=3–3.5; G=2.89–2.98; vitreous luster; transparent to translucent colorless to white to sometimes gray-black in massive form from impurities.

Occurrence: Although *anhydrite* occurs associated with volcanic rocks by precipitation from hot sulfate waters, and as a gangue mineral in metalliferous veins, its primary occurrence is as a sedimentary mineral. Extensive evaporite beds of *anhydrite* or *anhydrite* and *calcite* or *halite* occur throughout the post Precambrian geological column. In the Permian of West Texas and New Mexico and the Zechstein of Europe, beds of *anhydrite* more than 1000 feet thick occur within the evaporite column. These are considered to be formed from evaporation of saline waters, generally marine in origin. Extensive nonmarine *anhydrite* occurs in the Namib desert of Southwest Africa (see Vol. IVA: *Evaporite Processes*).

Associated Minerals: Other evaporites, particularly *halite, calcite, dolomite, gypsum,* and sometimes reducing-environment minerals such as *pyrite* or *sulfur* are often formed by bacterial reduction of sulfate in the bed.

E. F. CRUFT

References

Cruft, E. F., and Chao, P. C., 1969. Kinetic considerations of the gypsum-anhydrite transition, *3rd Int. Salt Symp.*, Cleveland, Ohio.
Econ. Geol. **60**, 942-954.
J. Sed. Petrology **34**, 512-523.

Cross-references: *Anhydrite and Gypsum; Authigenic Minerals; Diagenetic Minerals; Mineral and Ore Deposits; Minerals, Uniaxial and Biaxial; Optical Mineralogy; Pseudomorphism; Refractory Minerals; Rock-Forming Minerals; Saline Minerals; Staining Techniques.* Vol. IVA: *Evaporite Processes.*

ANILITE *Am. Min.* **62**, 107
Cu_7S_4 ortho. *Min. Abst.* **27**, 335

ANKERITE A calcium, iron and magnesium carbonate of the ***dolomite*** group, named (1825) for the Styrian (Austrian) mineralogist M. J. Anker.

Composition: $Ca(Fe^{2+},Mg)(CO_3)_2$; Fe>Mg; (Mn>Fe, *kutnahorite;* Mg>Fe, *dolomite*).

Properties: Trigonal, $R\bar{3}$; crystals rhombohedral, also crystalline massive, granular, or compact, commonly twinned polysynthetically; rhombohedral cleavage $\{10\bar{1}1\}$; H=3.5–4; G=3; reddish or brown; some varieties are fluorescent.

Occurrence: In veins associated with iron ores.

K. FRYE

References

Am. Min. **63**, 779-781.
Dana, II, 208-217.

Cross-references: *Minerals, Uniaxial and Biaxial; Rock-Forming Minerals; Vein Minerals.* Vol. VI: *Dolomite.*

ANNABERGITE (NICKEL BLOOM, NICKEL OCHER) A hydrous nickel-cobalt arsenate of the ***vivianite*** group. Named by Brooke and Miller (1852), from the locality in Saxony; it is called "nickel bloom" by prospectors who use it as a guide to ore.

Composition: $2[(Ni,Co)_3(AsO_4)_2 \cdot 8H_2O]$.

Properties: Monoclinic ($I2/m$); crystals striated; usually as fine-crystalline coatings or earthy; H=1.5–2.5; G=3.07; sectile, dull to earthy luster, transparent to translucent; color: light green; Ni and Co substitute mutually to form a complete series; the name *annabergite* applies when Ni>Co, the name *erythrite* applies when Co>Ni.

Occurrence: A secondary nickel mineral usually formed by the oxidation of arsenides of cobalt and nickel. Found as pale green crusts (nickel bloom) at Cobalt, Ontario; in Annaberg and Scheenberg, E. Germany.

Use: Its pale apple-green crusts are a distinctive guide to the presence of nickel ore.

J. A. GILBERT

References

Can. Min. **14**, 414–421.
Dana, II, 746.

ANNITE (***Mica*** gr.)
$KFe_3^{2+}AlSi_3O_{10}(OH,F)_2$ mono.

ANORTHITE A calcium aluminosilicate of the ***feldspar*** group, named by Rose (1832) from the Green *an,* not, and *orthos,* upright, in reference to its oblique crystal form.

Composition: $CaAl_2Si_2O_8$; with Ca partially replaced by Na and K, Al by Si. The composition represents the Ca end member, *An,* of the series *albite-anorthite,* the ***plagioclase feldspars,*** and the mineral may contain up to 10 mol% of the albite composition, *Ab.*

Properties: Triclinic, $P\bar{1}$; as cleavable masses and as irregular grains; twinning according to the albite, Carlsbad, or pericline laws; prismatic {001} {010} cleavage; virtreous to pearly luster; white to gray; transparent to translucent.

Occurrence: Rare, as an essential constituent of a few ultramafic intrusives and anorthosites, in some meteorites.

K. FRYE

References*

DHZ, 4, 94-165.
*See *Feldspar Group* and *Plagioclase Feldspars* for additional references.

Cross-references: *Feldspar Group; Glass, Devitrification of Volcanic Glass; Meteoritic Minerals; Minerals, Uniaxial and Biaxial; Order-Disorder; Orthopyroxene; Plagioclase Feldspars; Portland Cement Mineralogy; Staining Techniques.*

ANORTHOCLASE A sodium potassium aluminosilicate of the ***feldspar*** group, named by Rosenbusch (1885) from the Greek *an,* not, *orthos,* upright, and *klasis,* fracture in allusion to the slight deviation from 90° of the prismatic cleavage.

Composition: (Na,K)$AlSi_3O_8$; generally with small amounts of Ba or Ca replacing (Na,K).

Properties: Triclinic, $C\bar{1}$; twinned tabular crystals; H=6; G=2.6; colorless or white; prismatic cleavage.

Occurrence: In volcanic rocks as on Pantelleria and Sicily, Italy; in Rhineland, West Germany; on Mt. Kenya, Kenya; at Ropp, Nigeria; in Victoria, Australia; on Ross Island, Antarctica; and on Grande Caldeira, the Azores. Also in a dike in Argyllshire, Scotland, and in augite syenite at Larvik, Norway.

K. FRYE

References*

DHZ, 4, 1–93.

*See *Alkali Feldspars* and *Feldspar Group* for additional references.

Cross-references: *Alkali Feldspars; Minerals; Uniaxial and Biaxial; Plagioclase Feldspars.*

ANTARCTICITE *Am. Min.* **54**, 1018
$CaCl_2 \cdot 6H_2O$ trig. *Science* **149**, 975

ANTHOINITE *Am. Min.* **43**, 384
$Al_2W_2O_9 \cdot 3H_2O$ tric.

Cross-reference: *Museums, Mineralogical.*

ANTHONYITE *Am. Min.* **48**, 614
$Cu(OH,Cl)_2 \cdot 3H_2O$ mono.

ANTHOPHYLLITE A hydroxyl magnesium silicate of the ***amphibole*** group with iron frequently substituting for part of the magnesium. Named by Schumacher (1801), from the Latin *anthophyllum*, meaning clove, referring to the clove-brown color.

Composition:

$$Na_x(Mg,Mn,Fe^{2+})_{7-y}Al_y(Al_{x+y}Si_{8-x-y})O_{22}(OH,F,Cl)_2$$

where $x+y<1.0$ (if $x+y \geqslant 1.0$, the mineral is *gedrite*). If $Na \geqslant 0.5$, the mineral is *sodium anthophyllite.* The magnesian and iron end members are called *magnesio-* and *ferro-anthophyllite*, respectively.

Properties: Orthorhombic; rarely in crystals; usually lamellar or fibrous; H=5.5–6; G=2.85–3.2; vitreous luster, translucent; color: green and brown.

Occurrence: A relatively rare mineral thought to be derived from the metamorphism of ***olivine*** and found in crystalline schists. Occurs in mica schist associated with *hornblende* in Norway, in ***mica*** aggregates at Hermannschlag, Moravia, and with *corundum* in Franklin, North Carolina.

Use: Iron-rich fibrous variety is used as a nonspinning grade of asbestos.

J. A. GILBERT

References

Am. Min. **36**, 615; **63**, 1023.
Carnegie Inst. Wash. Yr. Bk., **61**, 85.
DHZ 2, 211–229.

Cross-references: *Amphibole Group; Minerals, Uniaxial and Biaxial.*

ANTIGORITE (***Kaolinite-Serpentine*** gr.)
$(Mg,Fe)_3Si_2O_5(OH)_4$ DHZ, 3, 170
mono. *Min. Abst.* **28**, 275

Cross-references: *Minerals, Uniaxial and Biaxial; Soil Mineralogy.*

ANTIMONITE = *Stibnite*

ANTIMONPEARCEITE *Am. Min.* **50**, 1507
$(Ag,Cu)_{16}(Sb,As)_2S_{11}$ mono. *Min. Abst.* **27**, 254

ANTIMONY The element Sb.

Properties: Trigonal, $R\bar{3}2/m$; massive, one good cleavage {0001}; H=3–3 1/2; G=6.6–6.7; metallic; tin white; gray streak.

Occurrence: A vein mineral (q.v.) commonly associated with silver, antimony, and arsenic ores; also with sulfides.

K. FRYE

References

Dana, I, 132–133.
Min. Mag. **43**, 1029.

Cross-references: *Native Elements and Alloys; Ore Microscopy.* IVA: *Antimony: Element and Geochemistry.*

ANTLERITE A hydroxyl copper sulfate named by Hillebrand (1889), from the Antler mine, Arizona.

Composition: $4[Cu_3SO_4(OH)_4]$; CuO, 67.3%; SO_3, 22.5%; H_2O, 10.2%.

Properties: Orthorhombic; crystals often vertically striated; also in parallel aggregates, reniform or massive; H=3.5–4; G=3.9; vitreous luster, transparent to translucent; color: emerald green to blackish green.

Occurrence: Found in the oxidized portions of copper veins, especially in arid regions such as the Atacama and Mohave deserts. It may form as a secondary mineral on *chalcocite* or may go into solution and be deposited, usually filling cracks. Associated with *atacamite, chalcocite, linarite, malachite,* and *gypsum.*

Use: An ore of copper. The chief copper mineral at Chuquicamata, Chile, the largest copper mine in the world.

J. A. GILBERT

References

Dana, II, 544, 592.
Min. Abst. **27**, 279.

APACHITE *Am. Min.* **65**, 1065
$Cu_9Si_{10}O_{29} \cdot 11H_2O$ mono. *Min. Mag.* **53**, 639

APATITE GROUP–*Belovite, chlorapatite, dahllite (carbonate-hydroxyapatite), dehrnite, fermorite, fluorapatite, francolite (carbonate-fluorapatite), hedyphane, hydroxyapatite, johnbaumite, lewistonite, mimetite, pyromorphite, strontium-apatite, svabite, vanadinite,* and *wilkeite. Britholite* and *ellestadite* are silicates with an apatite-type structure.

Cross-references: *Apatite Group; Human and Vertebrate Mineralogy; Lunar Minerals; Minerals, Uniaxial and Biaxial; Mohs Scale of Hardness; Refractory Minerals; Rock-Forming Minerals, Soil Mineralogy; Staining Techniques.*

APHTHITALITE *Dana,* II, 400
$(K,Na)_3Na(SO_4)_2$ hexa.

Cross-reference: *Saline Minerals.*

APJOHNITE (***Halotrichite*** gr.)
$MnAl_2(SO_4)_4 \cdot 22H_2O$ *Dana,* II, 527
mono. *Min. Mag.* **40**, 599

APLOWITE *Am. Min.* **50**, 809
$(Co,Mn,Ni)SO_4 \cdot 4H_2O$ mono. *Can. Min.* **8**, 166

APOPHYLLITE. A hydrated potassium calcium fluoro-silicate named by Haüy, 1805 (from the Greek, "away from leaf," in allusion to its exfoliation on heating). The name is for the series and for undifferentiated members of the series, *fluorapophyllite – hydroxyapophyllite.*

Composition: $2[KCa_4Si_8O_{20}F \cdot 8H_2O]$, *fluorapophyllite,* to $2[KCa_4Si_8O_{20}(OH) \cdot 8H_2O]$, *hydroxyapophyllite.* Small amounts of Na may substitute for K; the water is most probably hydrogen-bonded to oxygen atoms in the Si_8O_{20} network; dehydration stepwise, losing 17.8% H_2O maximum; most water is lost at about 250°C; rehydration poor; decomposed by HCl, leaving a skeleton of silica.

Properties: Tetragonal, $P4/mnc$; a=9.00 Å c=15.84 Å; G=2.33–2.37. H=4.5–5; luster: vitreous to pearly; colorless, white, pale yellow, pink, or green, transparent; habit: tabular, prismatic, or granular; cleavage perfect on {001}, poor on {110}. Optically, ω=1.534–1.542, ϵ=1.535–1.543; birefringence=0.002; uniaxial (+); dispersion high, sometimes anomalous.

Occurrences: Mainly as a secondary mineral in amygdules or druses in basalt, often associated with ***zeolites,*** *datolite, pectolite,* and *calcite;* in cavities in granite; in fissures in metamorphic rocks in contact with granite; in limestone or calc-silicate rock associated with *calcite;* in *natrolite* veins in syenite.

A. IIJIMA
M. UTADA

References

Am. Min. **26**, 565-567; **63**, 196.
DHZ, 3, 258–262.
Doklady Akad. Nauk SSSR **114**, 876.
Z. Krist. **77**, 146.

Cross-references: *Green River Mineralogy; Rock-Forming Minerals.*

APUANITE *Am. Min.* **64**, 1230
$Fe^{2+}Fe^{3+}_4Sb^{3+}_4O_{12}S$ tetr. *Min. Mag.* **43**, 1057

AQUAMARINE–Gem variety of blue-green *beryl*

Cross-reference: *Gemology; Minerals, Uniaxial and Biaxial.*

ARAGONITE Calcium carbonate in which strontium, lead, and zinc may substitute in part for the calcium. Named by Werner (1796), from Aragon, Spain.

Composition: $4[CaCO_3]$; CaO, 56.0%; CO_2, 44.0%.

Properties: Orthorhombic, acicular pyramidal, tabular, and pseudohexagonal twins are the three common habits; may also be found in reniform, columnar, and stalactitic masses; H=3.5–4; G=2.95; vitreous luster, transparent to translucent; color: white, colorless, tinted with a wide variety of colors; it is dimorphous with *calcite*, but less common and less stable.

Occurrence: *Aragonite* is formed as a precipitate from carbonated waters containing calcium when they are hot (*calcite* when cold). It is deposited from hot springs and is found as fibrous crusts on ***serpentine,*** in amygdaloidal cavities in basalt, in the pearly layer in many shells, "mother of pearl," which is the inorganic constituent of pearls.

J. A. GILBERT

References

Am. Min. **47**, 700.
Dana, II, 182.
GCA, **23**, 295.

Cross-references: *Authigenic Minerals; Calcite Group; Cave Minerals; Crystallography: Morphological; Diagenetic Minerals; Electron Microscopy (Transmission); Green River Mineralogy; Human and Vertebrate Mineralogy; Invertebrate and Plant Mineralogy; Lunar Minerals; Minerals, Uniaxial and Biaxial; Rock-Forming Minerals; Soil Mineralogy; Staining Techniques; Thermoluminescence.*

ARAGONITE GROUP–*Aragonite, cerussite, strontianite, witherite*

ARAMAYOITE — *Dana* I, 427
$Ag(Sb,Bi)S_2$ — tric.

ARCANITE — *Dana* II, 399
K_2SO_4 — orth.

ARCHERITE — *Min. Mag.* **41**, 33
$(K,NH_4)H_2PO_4$ — tetr. — *Am. Min.* **62**, 1057

ARCUBISITE — *Lithos* **9**, 253
Ag_6CuBiS_4 — *Am. Min.* **63**, 424

ARDEALITE — *Am. Min.* **17**, 251
$Ca_2(SO_4)(HPO_4)\cdot 4H_2O$ — mono.

ARDENNITE — orth. — Winchell II, 529
$Mn_4(Al,Mg)_6(As,V)O_4(SiO_4)_2(Si_3O_{10})(OH)_6$

Cross-reference: *Museums, Mineralogical.*

ARFVEDSONITE A hydroxyl iron, calcium, and alkali silicate of the ***clinoamphibole*** group. Named by Brooke (1823), after Professor J. A. Arfvedson, Swedish chemist.

Composition: $NaNa_2Fe_4^{2+}Fe^{3+}Si_8O_{22}(OH)_2$, with Fe^{2+} replaced by Mg toward *magnesioarfvedsonite* and by Mn^{2+} toward *kozulite.* Ca may replace Na.

Properties: Monoclinic, *C2/m*; found in crystal aggregates as well as in individual crystals; H=6; G=3.45; vitreous luster, translucent; color: dark green to black.

Occurrence: In igneous and metamorphic rocks poor in silica such as nepheline syenite and metabasalts. Specifically, in the Julianehaab district of southern Greenland; in Norway in the nepheline syenites of the Langesund fjord district, and near Oslo; in the United States at Red Hill, New Hampshire, and at St. Peter's Dome near Pike's Peak, Colorado.

J. A. GILBERT

References

Am. Min. **63**, 413; 1023.
DHZ, 2, 364–374.

Cross-references: *Clinopyroxenes; Green River Mineralogy.*

ARGENTITE (SILVER GLANCE) A silver sulfide, named by Haidinger (1845), from the Latin *argentum*, meaning silver.

Composition: Ag_2S; Ag, 87.1%; S, 12.9%.

Properties: Orthorhombic at ordinary temperatures, isometric above 180°C; commonly massive or as a coating; H=2–2.5; G=7.3; easily fusible; very sectile, metallic luster; color: lead-gray, opaque.

Occurrence: In veins, associated with native *silver*, ruby silvers, *polybasite, stephanite, galena,* and *sphalerite.* Although *argentite* is often listed as an important ore of silver, most natural silver sulfide is in fact *acanthite* (q.v.), pseudomorphous after *argentite.*

J. A. GILBERT

Reference: *Dana*, II, 176–178.

Cross-references: *Mineral and Ore Deposits; Ore Microscopy.*

ARGENTOJAROSITE (***Alunite*** gr.)
$AgFe_3(SO_4)_2(OH)_6$ — trig. — *Am. Min.* **8**, 230

ARGENTOPENTLANDITE — *Min. Abst.* **30**, 71
$Ag(Fe,Ni)_8S_8$ — iso. — *Min. Mag.* **43**, 1058

ARGENTOPYRITE — *Am. Min.* **39**, 475
$AgFe_2S_3$ — orth.

ARGYRODITE — *Dana*, I, 356
Ag_8GeS_6 — orth.

ARISTARAINITE — mono. — *Am. Min.* **59**, 647; **62**, 979
$Na_2Mg[B_6O_8(OH)_4]_2\cdot 4H_2O$

ARIZONITE–Mixture of *rutile* and *hematite*(?) — *Am. Min.* **35**, 117; *Min. Mag.* **15**, 416

ARMALCOLITE — *Am. Min.* **55**, 2136
$(Mg,Fe^{2+})Ti_2O_5$ — orth. — *Min. Abst.* **28**, 180

Cross-reference: *Lunar Minerals.*

ARMANGITE — trig. — *Am. Min.* **64**, 748
$Mn_{26}^{2+}[As_6^{3+}(OH)_4O_{14}][As_6^{3+}O_{18}]_2[CO_3]$

ARMENITE — ortho., ps. hexa.
$BaCa_2Al_6Si_9O_{30}\cdot 2H_2O$ — *Am. Min.* **26**, 235

ARMSTRONGITE Am. Min. **59**, 208
$CaZrSi_6O_{15} \cdot 2½H_2O$ mono.

ARNIMITE (Perhaps = *antlerite*)
$Cu_5(SO_4(OH)_6 \cdot 3H_2O$ orth. *Am. Min.* **39**, 851

ARROJADITE mono. *Am. Min.* **55**, 135
$(K,Ba)(Na,Ca)_5(Fe^{2+},Mn,Mg)_{14}Al(PO_4)_{12}(OH,F)$
Min. Mag. **43**, 227

Cross-reference: *Pegmatite Minerals.*

ARSENATE-BELOVITE (Talmessite)
$Ca_2Mg(AsO_4)_2 \cdot 2H_2O$ *Am. Min.* **50**, 583
tric. *Min. Abst.* **27**, 80

ARSENBRACKEBUSCHITE
$Pb_2(Fe^{3+},Zn)(OH,H_2O)(AsO_4)_2$
mono. *Am. Min.* **63**, 1282
Min. Abst. **29**, 481

ARSENDESTINEZITE = *Bukovskyite*
Am. Min. **54**, 994

ARSENIC Native semimetal, As. The name is derived from the Greek word meaning masculine, from the belief that metals were of different sexes.

Properties: Trigonal, $R\bar{3}m$; usually massive; reinform and stalactitic; H=3.5; G=5.7; nearly metallic luster on fresh surface; color: tin-white on fresh fracture, tarnishes to dark gray on exposure; brittle.

Occurrence: Often alloyed with antimony and traces of iron, silver, gold, and bismuth. A rare species found in veins in crystalline rocks associated with silver, cobalt, or nickel ores.

Use: Very minor ore of arsenic.

J. A. GILBERT

Reference: *Dana,* I, 128–130.

Cross-references: *Native Elements and Alloys; Ore Microscopy.* vol. IVA. *Arsenic: Element and Geochemistry.*

ARSENIOSIDERITE mono. *Am. Min.* **59**, 48
$Ca_6(H_2O)_6[Fe_9O_6(AsO_4)_9] \cdot 3H_2O$
Min. Abst. **27**, 253

ARSENOBISMITE *Am. Min.* **28**, 536
$Bi_2(AsO_4)(OH)_3$ iso.

ARSENOCLASITE *Am. Min.* **53**, 1779
$Mn_5(AsO_4)_2(OH)_4$ orth.

ARSENOHAUCHECORNITE *Am. Min.* **66**, 436
Ni_9BiAsS_8 tetr. *Min. Mag.* **43**, 877

ARSENOLAMPRITE *Am. Min.* **45**, 479
As orth.

Cross-reference: *Native Elements and Alloys*

ARSENOLITE An arsenic oxide named by Dana (1854).

Composition: As_2O_3.

Properties: Isometric, $Fd3m$; G=3.8; H=1.5; fusability = 1; white or tinted, semi-metallic molecular structure; polymorphic with *claudetite;* isostructural with *senarmontite* (Sb_2S_3) and *sillenite* (Bi_2O_3).

Occurrences: An oxidation product of arsenic ores and a smelter fume product.

D. H. SPEIDEL

Reference: *Dana,* I, 543–544.

ARSENOPALLADINITE *Am. Min.* **59**, 1332
$Pd_8(As,Sb)_3$ tric. *Can. Min.* **15**, 70

ARSENOPYRITE (Mispickel) An arsenic sulfide of iron, named by Glocker in 1847 as a contraction of "arsenical pyrites."

Composition: 8[FeAsS]; Fe, 34.3%; As, 46.0%; S, 19.7% by wt. Co and Ni may replace Fe; S and As may substitute for each other.

Properties: Monoclinic, $P2_1/c$; prismatic parallel to the *c* axis, occasionally pseudomorphous after *pyrrhotite;* H=5.5–6.0. G=6.07±0.15; luster: metallic; color: silver-white; streak: black. Invariably twinned according to several different laws.

Optically, it is opaque; strongly anisotropic, showing blue-green, or brownish-red anisotropy colors. Vicker's hardness: hard (1100); color: white (colorless) to pale shades of cream or pink; bireflectance: weak (for the colorless variety) to strong (for the weakly colored variety). Idiomorphic crystals with rhomb shape are common; some lamellar twinning. Reflectivity: 54%

Occurrence: Ubiquitous. Can form at very low temperature, but more commonly occurs with tin and tungsten minerals in hydrothermal veins, and as high-temperature deposits. Occasionally found in metasomatic aureoles, in pegmatites, and disseminated in crystalline limestones.

Arsenopyrite is an ore of arsenic. Chief areas of recovery are Roxbury, Connecticut; Franklin, New Jersey; Cornwall, England; and Freiberg and Munzig in Saxony. The major uses of arsenic are in the form of arsenious oxide for medicines, insecticides, pigments, etc. It is associated with *gold, pyrite, chalcopyrite, galena, sphalerite,* ores of Cu and Ag, tellurides, *marcasite,* etc.

R. W. CAMERON

References

Bragg, L., et al., 1965. *Crystal Structures of Minerals,* vol. 4. Ithaca: Cornell Univ. Press, 409p.

Hurlbut, C. S., Jr., and Klein, C., 1977. *Manual of Mineralogy,* 19th ed. New York: Wiley, 254–255.

Jensen, M. L., and Bateman, A. M., ed., 1979. *Economic Mineral Deposits,* 3rd ed. New York: Wiley, p. 445.

Schouten, C̄., 1962. *Determinative Tables for Ore Microscopy.* New York: Elsevier, 242p.

Cross-references: *Blowpipe Analysis: Ore Microscopy; Staining Techniques.*

ARSENOSULVANITE Am. Min. **40**, 368
$Cu_3(As,V)S_4$ iso. *Min. Abst.* **28**, 205

ARSENPOLYBASITE *Am. Min.* **50**, 1507
$(Ag,Cu)_{16}(As,Sb)_2S_{11}$ mono. *Can. Min.* **8**, 172

ARSENURANOSPATHITE *Min. Mag.* **42**, 117
$(HAl)_{0.5}(UO_2)_2(AsO_4)_2 \cdot 2OH_2O$
tetr. *Min. Abst.* **29**, 200

ARSENURANYLITE orth. *Am. Min.* **44**, 208
$Ca(UO_2)_4(AsO_4)_2(OH)_4 \cdot 6H_2O$

ARTHURITE mono. *Am. Min.* **55**, 1817
$Cu_2Fe_4(AsO_4,PO_4,SO_4)_4(O,OH)_4 \cdot 8H_2O$
Min. Mag. **37**, 519

ARTINITE A hydrated magnesium carbonate named (1902) in honor of Italian mineralogist Ettore Artini.

Composition: $2[Mg_2(CO_3)(OH)_2 \cdot 3H_2O]$.

Properties: Monoclinic, *C*2; acicular crystals in crusts, in botryoidal and spherical aggregates, and across veinlets; one perfect {100} and one good {001} cleavage; H=2½; G=2; silky or satiny; colorless or white.

Occurrence: In low-temperature veins and crusts in serpentinized ultramafic rocks with other carbonates.

K. FRYE

Reference: *Dana,* II, 263–264.

ASBECASITE trig. *Am. Min.* **55**, 1818
$Ca_3(Ti,Sn)As_6Si_2Be_2O_{20}$ *Min. Abst.* **28**, 208

ASBESTOS–Fiberous ***serpentine*** or ***amphibole***

Cross-references: *Asbestos; Electron Microscopy (Transmission); Human and Vertebrate Mineralogy.*

ASHANITE *Am. Min.* **66**, 217
$(Nb,Ta,U,Fe,Mn)_4O_8$ orth. *Min. Mag.* **43**, 1058

ASCHARITE = *Szaibelyite* *Am. Min.* **51**, 1257

ASHCROFTINE A hydrated potassium sodium calcium aluminosilicate named by Hey and Bannister, 1933, in honor of F. N. Ashcroft.

Composition: $KNaCaY_2Si_6O_{12}(OH)_{10} \cdot 4H_2O$.

Properties: Tetragonal; a=34.11 Å, c=17.52 Å; G=2.61; color: pink; habit: fibrous; cleavage: perfect on {100}, good on {001}; refractive indices, ω=1.536, ϵ=1.545; birefringence = 0.009; uniaxial (+).

Occurrence: Very rare; in pegmatitic pockets in augite syenite at Narssarssuaq, Greenland; in vesicular in fillings in alkali mafic and ultramafic lavas of southwest Uganda, associated with *phillipsite* and *natrolite.*

A. IIJIMA
M. UTADA

References

Min. Mag. **23**, 305–308; **24**, 408-420.

Proc. Acad. Natl. Sci. Philadelphia **76**, 294 (*Min. Abst.* **2**, 385).

ASHTONITE = Strontian *mordenite*
Min. Mag. **38**, 383

ASTROLITE = *Muscovite* *Am. Min.* **57**, 993

ASTROPHYLLITE (***Astrophyllite*** gr.)
$(K,Na)_3(Fe,Mn)_7Ti_2Si_8O_{24}(O,OH)_7$
tric. *Min. Abst.* **27**, 287
Winchell, II, 480

Cross-reference: *Minerals, Uniaxial and Biaxial.*

ASTROPHYLLITE GROUP–*Astrophyllite, cesium-kupletskite, hydro-astrophyllite, kupletskite, magnesium-astrophyllite, niobophyllite,* and *zircophyllite.*

ATACAMITE A hydroxyl copper chloride named by Gallitzen (1801), from the province of Atacama, Chile.

Composition: $4[Cu_2Cl(OH)_3]$; Cu, 14.88%; CuO, 55.87%; Cl, 16.60%; H_2O, 12.65%.

Properties: Orthorhombic; usually in fine crystal aggregates, fibrous or granular; H=3–3.5; G=3.75–3.77; adamantine to vitreous luster, transparent to translucent; grass green color and an apple-green streak.

Occurrence: In arid regions as a supergene

mineral in the oxidized zone of copper deposits.

Use: A minor ore of copper.

J. A. GILBERT

References:

Dana, II, 69–73.
Min. Abst. **28**, 148, no. 77-1562.

Cross-reference: *Mineral Classification: Principles.*

ATELESTITE *Dana* II, 792
$Bi_8(AsO_4)_3O_5(OH)_5$ mono.

ATHABASCAITE *Am. Min.* **56**, 632
Cu_5Se_4 ortho. *Can. Min.* **10**, 207

ATHENEITE *Am. Min.* **59**, 1330
$(Pd,Hg)_3As$ hexa. *Min. Mag.* **39**, 528

ATOKITE *Am. Min.* **61**, 340
$(Pd,Pt)_3Sn$ iso. *Can. Min* **13**, 146

ATTAKOLITE (ATTACOLITE)
$(Ca,Mn,Sr)_3Al_6(PO_4,SiO_4)_7 \cdot 3H_2O$
orth. *Am. Min.* **51**, 534

ATTAPULGITE = *Palygorskite*

AUBERTITE *Am. Min.* **65**, 205
$CuAl(SO_4)_2Cl \cdot 14H_2O$ tric. *Bull. Min.* **102**, 348

AUGELITE *Am. Min.* **53**, 1096
$Al_2(PO_4)(OH)_3$ mono. *Am. Min.* **61**, 409

Cross-reference: *Pegmatite Minerals.*

AUGITE A calcium, sodium, magnesium, iron aluminosilicate. Named by Werner (1792) from the Greek word meaning luster, it is the commonest member of the ***pyroxene*** group (q.v.).

Composition: $8[(Ca,Na)(Mg,Fe^{2+},Fe^{3+},Al)(Si,Al)_2O_6)]$, variable composition.

Properties: Monoclinic, $C2/c$; often lamellar and granular; in crystals, twinning common; H=5–6; G=3.2–3.4; vitreous luster; translucent; color: dark green to black.

Occurrence: *Augite* is a very common rock-forming mineral (q.v.) between *diopside* and *hedenbergite* in composition. It is found chiefly in dark igneous rocks, rich in iron, magnesium, and calcium, and is associated with *orthoclase*, ***plagioclase feldspar***, *nepheline*, ***olivine***, *hornblende*, *magnetite*, and *leucite*.

J. A. GILBERT

References

Am. Min. **49**, 599.
DHZ, 2A, 294–398.
Min. Mag. **29**, 897.

Cross-references: *Clinopyroxenes; Crystallography: Morphological; Meteoritic Minerals; Minerals, Uniaxial and Biaxial; Orthopyroxenes; Pyroxene Group; Rock-Forming Minerals; Skeletal Crystals; Soil Mineralogy.*

AURICHALCITE A hydroxyl carbonate of zinc and copper. It was named by Böttger (1845), probably from the Greek meaning mountain brass.

Composition: $4[(Zn,Cu)_5(CO_3)_2(OH)_6]$, with Cu : Zn ratio about 2 : 5.

Properties: Orthorhombic, usually in crusts of soft scales; H=1–2; G=3.6; pearly luster; pale greenish-blue color and streak.

Occurrence: Secondary mineral found in the oxidized zone of copper zinc ore deposits associated with *malachite, azurite, cuprite*, and *smithsonite*.

Use: Good guide to zinc ore.

J. A. GILBERT

Reference: *Dana*, II, 249–250.

AURICUPRIDE
Cu_3Au iso.

AURORITE tric.(?) *Am. Min.* **52**, 1581
$(Mn^{2+},Ag,Ca)Mn_3^{4+}O_7 \cdot 3H_2O$
Econ. Geol. **62**, 186

AUROSTIBITE (***Pyrite*** gr.)
$AuSb_2$ *Am. Min.* **37**, 461
iso. *Min. Abst.* **27**, 254

AUSTINITE (***Adelite*** gr.)
$CaZn(AsO_4)(OH)$ *Am. Min.* **56**, 1359
orth. *Can. Min.* **18**,191

Cross-reference: *Naming of Minerals.*

AUTUNITE A hydrous phosphate of calcium and uranium named by Brooke and Miller (1852), from Autun, France.

Composition: $2[Ca(UO_2)_2(PO_4)_2 \cdot 10\text{–}12H_2O]$.

Properties: Tetragonal; may be in foliated and scaly aggregates; H=2–2.5; G=3.1–3.2; vitreous luster, pearly on cleavage faces, fluorescent in ultraviolet light; color yellow to pale green.

Occurrence: A secondary mineral formed from the oxidation or hydrothermal alteration of *uraninite* or other uranium minerals. Found in Autun, France; Johanngeorgenstadt district, Germany; Katanga district, Zaire; Mount Painter, South Australia; Sabugal, Portugal; Spruce Pine, North Carolina, and near Spokane, Washington.

Use: An ore of uranium.

J. A. GILBERT

References

Dana, II, 980.
Geol. Soc. Am. Bull. **69**, 1694.
Min. Abst. **28**, 263.

Cross-references: *Autunite (Torbernite) Group; Meta-autunite (Metatorbernite) Group.*

AUTUNITE GROUP–*Autunite, fritzscheite, heinrichite, kahlerite, novacekite, sabugalite, saleeite, sodium-autunite, torbernite, troegerite, uranocircite, uranospinite,* and *zeunerite.*

AVENTURINE–Red soda-rich ***plagioclase feldspar***

AVICENNITE *Am. Min* **44**, 1324
Tl_2O_3 iso.

AVOGADRITE A fluoride of potassium, cesium, and boron. Named by Zambonini (1926), after Amadeo Avogadro (1776-1856), Professor of Physics, University of Turin, Italy.

Composition: $4[(K,Cs)BF_4]$, with up to about 20% $CsBF_4$.

Occurrence: Found as fumarolic incrustations on Vesuvius, admixed with *sassolite* and other salts.

J. A. GILBERT

References

Am. Min. **12**, 232.
Dana, II, 97.
Min. Mag. **21**, 558.

AWARUITE = *Nickel-iron* *Min. Mag.* **43**, 647

Cross-reference: *Meteoritic Minerals.*

AXINITE GROUP A hydroxyl calcium, iron, magnesium, aluminum borosilicate mineral group named by Haüy (1797), from the Greek meaning "axe," referring to the wedgelike shape of the crystals.

Composition:
Ferroaxinite: $Ca_2Fe^{2+}Al_2BSi_4O_{15}(OH)$
Magnesioaxinite: $Ca_2MgAl_2BSi_4O_{15}(OH)$
Manganaxinite: $Ca_2MnAl_2BSi_4O_{15}(OH)$
Tinzenite: $(Ca,Mn,Fe)_3Al_2BSi_4O_{15}(OH)$

Properties: Triclinic, $P\bar{1}$; crystals common but may be massive, lamellar to granular; H=6.5–7; G=3.27–3.35; vitreous luster, transparent to translucent; color: violet, green, gray, yellow, or brown; streak: white.

Occurrence: In cavities in granites or in contack zones around granitic intrusions, and in contact-metamorphosed impure limestones. Associated minerals include *calcite,* ***pyroxenes,*** *epidote,* ***tourmaline,*** *quartz, andradite, actinolite, zoisite,* and *prehnite.* Crystalline forms occur at Bourg d'Oisans, France; at Obira, Japan; at Cornwall, England; in the Vinagrillos hills, Durango, Mexico; at Riverside, California; at Lunig, Nevada; at Franklin, New Jersey, and Bethlehem, Pennsylvania, USA.

Use: As a minor gem stone.

J. A. GILBERT

References

Am. Min. **64**, 635; **66**, 428.
DHZ, 1, 320–327.
J. Gemmology **14**, 368–375.

Cross-references: *Crystallography: Morphological; Minerals, Uniaxial and Biaxial; Rock-Forming Minerals.*

AZOPROITE orth. *Am. Min.* **56**, 630
$(Mg,Fe^{2+})_2(Fe^{3+},Ti,Mg)BO_5$

AZURITE A copper hydroxyl carbonate named by Beudant (1824) in reference to its color.

Composition: $2[Cu_3(CO_3)_2(OH)_2]$.

Properties: Monoclinic, $P2_1/a$; complex habits and often malformed; also found in radiating spherical aggregates; H=3.5–4; G=3.77; vitreous luster, transparent to translucent; color: intense azure blue; streak: blue.

Occurrence: Found in the oxidized portions of copper veins as a supergene mineral, also rarely an anthigenic mineral (q.v.). Associated with *malachite, cuprite,* native *copper,* iron oxides, and sulfides of iron and copper. Spectacular crystalline forms occur in Tsumeb, SW Africa; in Chessy, France; and at Bisbee, Arizona. Other deposits are located in Cornwall, Devonshire, and Derbyshire, England; at Porto Cabello, S.A.; in the Cobar mines, New South Wales; in S. Australia; at Perkiomen lead mine, Pennsylvania; in Ossining, New York; near New Brunswick, New Jersey; and in Calaveras County, California.

Use: A minor ore of copper; sometimes used as a decorative material.

G. C. MOERSCHELL

References

Am. Min. 49, 111.
Dana, II, 264.

Cross-references: *Authigenic Minerals; Color in Minerals; Gossan; Metallic Minerals; Minerals, Uniaxial and Biaxial; Museums, Mineralogical; Naming of Minerals; Pigments and Fillers.*

B

BABEFPHITE *Am. Min.* **51**, 1547
$BaBe(PO_4)(OH,F)$ tetr.

BABINGTONITE tric. *Am. Min.* **17**, 295
$Ca_2(Fe^{2+},Mn)Fe^{3+}Si_5O_{14}(OH)$ *Min. Abst.* **27**, 212

BADDELEYITE Zirconium oxide, named by Fletcher (1892) after Joseph Baddeley who brought the original specimens from Ceylon.

Composition: ZrO_2

Properties: Monoclinic; $P2_1/c$; G=5.4–6.0; H=6.5; perfect basal cleavage; colorless to black; strongly pleochroic: *X*=yellow, brown green; *Y*=green, brown; *Z*=brown, light brown.

Occurrence: Found in gem sands; associated with *cassiterite* and *rutile*. In Sri Lanka at Rakwana found as rounded crystals in gem gravels associated with *zircon, spinel, corundum*, and ***tourmaline***. At Jacupiranga, Brazil it is present in the contact zone between jacupirangite (a *magnetite*-pyroxenite rock) and marble.

D. H. SPEIDEL

References

Acta Cryst. **12**, 507–511.
Dana, I, 608.

Cross-references: *Lunar Minerals; Meteoritic Minerals; Refractory Minerals; Synthetic Minerals.*

BADENITE (doubtful validity) *Dana*, I, 266
$(Co,Ni,Fe)_3(As,Bi)_4$(?)

BAFERTISITE mono. *Am. Min.* **57**, 1005
$Ba(Fe,Mn)_2TiSi_2O_7(O,OH)_2$

BAHIANITE *Am. Min.* **64**, 464
$Sb_3Al_5O_{14}(OH)_2$ mono. *Min. Abst.* **29**, 481

BAKERITE mono. *Am. Min.* **47**, 919
$Ca_4B_4(BO_4)(SiO_4)_3(OH)_3 \cdot H_2O$
Min. Abst. **27**, 333

BALIPHOLITE orth. *Am. Min.* **61**, 338
$BaMg_2LiAl_3(Si_2O_6)_2(OH)_8$ *Sci. Geol.* **1**, 100

BALKANITE *Am. Min.* **58**, 11
$Cu_9Ag_5HgS_8$ orth.

BALYAKINITE *Am. Min.* **66**, 436
$CuTeO_3$ orth.

BAMBOLLAITE *Am. Min.* **58**, 805
$Cu(Se,Te)_2$ tetr. *Can. Min.* **11**, 738

BANALSITE (***Feldspar*** gr.) *Am. Min.* **30**, 85
$BaNa_2(Al_4Si_4)O_{16}$ orth. *Min. Abst.* **27**, 213

BANDYLITE A hydrated chloroborate of copper, named by Palache and Foshag (1938) after Mark C. Bandy, mining engineer, who collected the mineral.

Composition: $CuCl[B(OH)_4]$.

Properties: Tetragonal, $P4/n$; crystals tabular or equant; very flexible; H=2.5; G=2.81; vitreous luster, with cleaved surfaces pearly; deep blue color with greenish lights; pale blue streak.

Occurrence: Rare, secondary mineral from Mina Quetena, near Calama, Chile, associated with *atacamite* and *eriochalcite*.

G. C. MOERSCHELL

References

Am. Min. **23**, 85.
Dana, II, 373.
Min. Mag. **25**, 263.

BANNISTERITE mono. *Am. Min.* **54**, 577
$(Na,K)(Mn,Fe,Al)_5(Si,Al)_6O_{15}(OH)_5 \cdot 2H_2O$
Min. Mag. **36**, 893

BOATITE tetr. *Am. Min.* **47**, 987
$Ba_4(Ti,Nb)_8Si_4O_{28}Cl$

BARARITE A fluoride of ammonium and silicon, named by Frondel (1951), from the Indian locality Barari, where it was first fully described.

Composition: $[(NH_4)_2SiF_6]$.

Properties: A hexagonal, low-temperature form of the cubic, high-temperature dimorph *cryptohalite*; H=2.5; G=2.15; color: white; vitreous luster; saline taste.

Occurrence: Found originally as a sublimation product at Vesuvius, Italy, associated with *salammoniac* and *cryptohalite*.

G. C. MOERSCHELL

References

Am. Min. **37**, 361.
Dana, II, 106

BARATOVITE mono. *Am. Min.* **61**, 1053;
$KLi_3Ca_7(Ti,Zr)_2[Si_6O_{18}]_2F_2$ **64**, 383

BARBERTONITE A hydrated magnesium chromium carbonate named by Frondel (1940) from its occurrence in the Barberton district, Transvaal.

Composition: $[Mg_6Cr_2(OH)_{16}CO_3 \cdot 4H_2O]$.

Properties: Hexagonal; lamellar masses or fibers with perfect basal cleavage; color: rose-pink to intense lilac; pale lilac to white streak; H=1.5–2; G=2.1; waxy to pearly luster; polymorph of *stichtite*.

Occurrence: Found at Dundas, Tasmania and at the Kaapsche Hoop, Transvaal, closely associated with *stichtite, chromite*, and *antigorite* as veinlets and small masses in serpentine rock.

D. H. SPEIDEL

References

Am. Min. **26**, 295–315.
Dana, I, 659.

BARBIERITE = *Microcline*

BARBOSALITE mono. *Am. Min.* **40**, 952
$Fe^{2+}Fe_2^{3+}(OH)_2(PO_4)_2$

Cross-reference: *Pegmatite Minerals.*

BARIANDITE mono. *Am. Min.* **57**, 1555
$V_2O_4 \cdot 4V_2O_5 \cdot 12H_2O$

BARICITE mono. *Am. Min.* **61**, 1053
$(Mg,Fe)_3(PO_4)_2 \cdot 8H_2O$ *Can. Min.* **14**, 403

BARIOMICROLITE (***Pyrochlore*** gr.)
$Ba(Ta,Nb)_2(O,OH)_7$ *Am. Min.* **48**, 1415;
iso. **62**, 403

BARIOPYROCHLORE (***Pyrochlore*** gr.)
$(Ba,Sr)_2(Nb,Ti)_2(O,OH)_7$
iso. *Am. Min.* **62**, 403

BARITE (BARYTE) Barium sulfate, named by Karsten (1800), from the Greek word meaning heavy, referring to its high specific gravity.

Composition: $4[BaSO_4]$; BaO, 65.7; SO_3, 34.3%.

Properties: Orthorhombic (*Pnma*); dipyramidal habit; may also be granular, laminated or earthy; H=3–3.5; G=4.5; vitreous luster, transparent to translucent; color: white and tints of blue, yellow, or red; streak: white.

Occurrence: As an accessory mineral in metallic veins, in veins in limestone with *calcite*, as a residual mass in clay overlying limestone, as an authigenic mineral, and in sandstone with copper ores; in places it acts as a matrix for sandstone. A few important locations are: in large crystals at Cumberland, Cornwall, Westmoreland, and Northumberland, England; with *stibnite* at Felsöbánya, Rumania; with marl as concretions at Mount Paterno near Bologna, Italy; massive deposits are mined in California, Georgia, Tennessee, Missouri, and Arkansas; and as beautiful "desert-roses" (containing much sand) in Norman, Oklahoma, and in Salina, Kansas.

Use: In oil and gas-well drilling (as a drilling mud, an aqueous suspension of the finely divided mineral), in the manufacture of lithopone (a white pigment), as a source of barium, and as a filler in glazed paper.

J. A. GILBERT

References

Carnegie Inst. Wash. Yr. Bk., **61**, 130.
Dana, II, 408.

Cross-references: *Authigenic Minerals; Blowpipe Analysis; Cave Minerals; Crystal Habits; Crystallography: Morphological; Green River Mineralogy; Human and Vertebrate Mineralogy; Mineral Industries; Mineral and Ore Deposits; Minerals, Uniaxial and Biaxial; Museums, Mineralogical; Naming of Minerals; Pigments and Fillers; Refractory Minerals; Rock-Forming Minerals; Staining Techniques; Vein Minerals.*

BARKEVIKITE = Ferro-pargasitic *hornblende*
Am. Min. **63**, 1049

BARNESITE *Am. Min.* **48**, 1187
$Na_2V_6O_{16} \cdot 3H_2O$ mono.

BARRERITE (***Zeolite*** gr.)
$Na_6(Ca,K)[Al_8Si_{28}O_{72}] \cdot 26H_2O$
orth. *Am. Min.* **61**, 1053
Min. Mag. **40**, 268

BARRINGERITE *Am. Min.* **55**, 317
$(Fe,Ni)_2P$ hexa. *Science* **165**, 169

Cross-reference: *Meteoritic Minerals.*

BARRINGTONITE *Am. Min.* **50**, 2103
$MgCO_3 \cdot 2H_2O(?)$ tric. *Min. Mag.* **34**, 370

BARROISITE (***Amphibole*** gr.)
$CaNa(Mg,Fe^{2+})_3(Al,Fe^{3+})_2(Si_7Al)O_{22}(OH)_2$
mono. *Am. Min.* **63**, 1034

BARSANOVITE = *Eudialyte* *Am. Min.* **54**, 1499

BARTONITE *Am. Min.* **66**, 369
$K_3Fe_{10}S_{14}$ tetr. *Min. Mag.* **43**, 1058

BARYLITE A rare barium and beryllium silicate, named by Blomstrand (1876) from the Greek meaning "heavy" and "stone."

Composition: $4[BaBe_2Si_2O_7]$.

Properties: Orthorhombic, *Pnma*; aggregates of tabular prismatic crystals; H=7; G=4.03; greasy luster; colorless; translucent.

Occurrence: Långban, Vermland, Sweden, and Franklin, New Jersey.

G. C. MOERSCHELL

References

Am. Min. 47, 924; **62**, 167.
Min. Abst., 28, 256.

BARYSILITE *Am. Min.* **54**, 510
$Pb_8Mn(Si_2O_7)_3$ trig.

BARYTE = BARITE *Min. Mag.* **43**, 1053

BARYTOCALCITE A carbonate of barium and calcium, named by Brooke (1824) in allusion to its composition.

Composition: $2[BaCa(CO_3)_2]$.

Properties: Monoclinic; trimorphous with *alstonite*, which is orthorhombic, and *paralstonite*, which is trigonal; short to long prismatic; H=4; G=3.64–3.66; brittle; vitreous to resinous luster; color: grayish, greenish, or yellowish white; streak: white; transparent to translucent.

Occurrence: First described from Alston Moor, Cumberland, England, in association with *barite* and *fluorite* in limestone. Found also with *quartz* and *barite* pseudomorphs after *barytocalcite* at Mies in Bohemia; in Badenweiler, Baden, Germany; and at Frieberg, Saxony.

G. C. MOERSCHELL

Reference: *Dana*, II, 220.

Cross-reference: *Green River Mineralogy.*

BARYTOLAMPROPHYLLITE
$(Na,K)_2(Ba,Ca,Sr)_2(Ti,Fe)_3(Si_2O_7)_2(O,OH)_4$
mono. *Am. Min.* **51**, 1549

BASALUMINITE hexa.(?) *Min. Abst.* **27**, 77
$Al_4(SO_4)(OH)_{10}\cdot 5H_2O$ *Min. Mag.* **43**, 931

BASILITE = *Hausmannite + feitnechite*
Am. Min. **58**, 562

BASSANITE *Dana*, II, 476
$2CaSO_4\cdot H_2O$ trig.

Cross-reference: *Saline Minerals.*

BASSETITE (***Meta-autunite*** gr.)
$Fe(UO_2)_2(PO_4)_2\cdot 8H_2O$
mono. *Am. Min.* **39**, 683
USGS Bull. **1064**, 200

BASTINITE = *Hureaulite* *Am. Min.* **49**, 398

BASTNAESITE A fluorocarbonate of lanthanum and cerium, named by Huot (1841), from Bastnäs Mine, Riddar-hyttan, Sweden.

Composition: $6[(Ce,La)FCO_3]$.

Properties: H=4.5; G=4.9–4.8; vitreous, greasy or pearly luster; wax-yellow to reddish brown color.

Occurrence: An alteration product of *tysonite*; also found in pegmatites and in contact zones. It was first described at Bastnäs, Västmanland, Sweden, associated with *allanite, cerite*, and *tysonite* in a contact metamorphic ***amphibole*** skarn. Found at Jamestown, Colorado, with *cerite, fluorite*, and *allanite* in a contact zone and with *tysonite* in the pegmatite of the Pike's Peak region, Colorado.

G. C. MOERSCHELL

References

Am. Min. 38, 932–963.
Dana, II, 289.
Econ. Geol., 59, 226.
U.S. Geol. Surv. Prof. Paper 254, 424C, C292.

Cross-references: *Blacksand Minerals; Pegmatite Minerals; Syntaxy.*

BASTNAESITE-(La)
$(La,Ce)(CO_3)F$ hexa.

BASTNAESITE-(Y) *Am. Min.* **57**, 594
$(Y,Ce)(CO_3)F$ hexa.

BATISITE *Am. Min.* **45**, 908, 1317
$Na_2BaTi_2Si_4O_{14}$ orth.

BAUMHAUERITE *Dana*, I, 460
$Pb_{11}As_{17}S_{36}$ tric.

BAUMITE (***Serpentine*** gr.) *Am. Min.* **61**, 174
$(Mg,Mn,Fe^{2+},Zn,Al,Fe^{3+})(Si,Al)O_{20}(OH)_{16}$
N.J. Min. Mh. **123**, 111

BAURANOITE *Am. Min.* **58**, 1111
$BaU_2O_7 \cdot 4\text{-}5H_2O$

BAUXITE A rock composed of aluminum hydroxides and oxyhyroxides, name usually credited to Deville (1861) from the locality at Baux, near Arles, France.

Composition: Essentially, $Al_2O_3 \cdot 2H_2O$.

Properties: Amorphous to microcrystalline; usually massive but also pisolitic (in round concretionary grains), earthy, and claylike; color: white, gray, yellow, and red; dull to earthy luster; H=1–3; G=2–2.5; bauxite is a group term for a mixture of primarily *gibbsite* $Al(OH)_3$; *bayerite*, $Al(OH)_3$; *boehmite* AlO (OH); and *diaspore* AlO(OH).

Occurrence: Found as weathered surface deposits formed from the prolonged leaching of silica from alumina-bearing rocks and from the weathering of clay-bearing limestones, commonly under conditions of tropical to subtropical weathering. Important producers are Surinam (Dutch Guiana), British Guiana, France, Hungary, Jamaica, Brazil, USSR, and the USA (Arkansas, Georgia, Alabama).

Use: Bauxite is the chief ore of aluminum, is used in the manufacture of artificial abrasives, and occurs in a number of chemicals and refractories.

D. H. SPEIDEL

References

Lamey, C. A., 1966. *Metallic and Industrial Mineral Deposits*, New York: McGraw-Hill, 402p.
Vanders, I., and Kerr, P., 1967. *Mineral Recognition*, New York: Wiley, 206p.

Cross-references: *Crystal Habits; Metallic Minerals; Mineral Industries; Refractory Minerals; Staining Techniques.*

BAVENITE A hydroxyl aluminosilicate of beryllium with calcium, named by Artini (1901) from the locality (Baveno, Italy) where it was first found.

Composition: $Ca_4Al_2Be_2(OH)_2(Si_9O_{26})$.

Properties: Orthorhombic; with radiating fibrous prismatic crystals; earthy luster; white color and streak.

Occurrence: In pegmatite druses in the granite of Baveno, Piedmont, Italy, and in Mesa Grande, California.

G. C. MOERSCHELL

References

Am. Min. 38, 988.
Min. Mag. 25, 475.

Cross-reference: *Pegmatite Minerals.*

BAYERITE *Am. Min.* **49**, 819
$Al(OH)_3$ mono. *Min. Mag.* **33**, 723

BAYLDONITE mono. *Am. Min.* **66**, 148
$PbCu_3(HOAsO_3)_2(OH)_2$ *Dana*, II, 929

BAYLEYITE mono. *Am. Min.* **36**, 1
$Mg_2(UO_2)(CO_3)_3 \cdot 18H_2O$

BAYLISSITE *Min. Abst.* **28**, 208
$K_2Mg(CO_3)_2 \cdot 4H_2O$ mono.
Schw. Min. Pet. Mitt. **56**, 187

BAZIRITE *Min. Abst.* **29**, 200
$BaZrSi_3O_9$ hexa. *Min. Mag.* **42**, 35

BAZZITE *Am. Min.* **40**, 370
$Be_3(Sc,Al)_2Si_6O_{18}$ hexa.

Cross-reference: *Pegmatite Minerals.*

BEARSITE mono. *Am. Min.* **48**, 210
$Be_2(AsO_4)(OH) \cdot 4H_2O$

BEAVERITE (***Alunite*** gr.)
$Pb(Cu,Fe,Al)_3(SO_4)_2(OH)_6$ *Dana*, II, 568
trig. *Min. Abst.* **28**, 79

BECKELITE (***Apatite*** gr.)
$(Ca,Ce)_5(O,OH)(SiO_4)_3$
hexa. *Min. Mag.* **14**, 395

BECQUERELITE A hydrated oxide of uranium named by Schoep (1922) after A. Henri Becquerel (1852–1908), French physicist, who discovered radioactivity.

Composition: $Ca(UO_2)_6O_4(OH)_6 \cdot 8H_2O$. Extensive Ba and K substitution occurs.

Properties: Orthorhombic, *Pnam;* perfect {010} cleavage or massive; dark amber yellow to colorless; yellow streak; H=2–3; G=5.2; adamantine to greasy luster; transparent.

Occurrence: Found at Kasolo, Katanga, Zaire, with *anglesite, soddyite, ianthinite, curite, schoepite,* and other secondary uranium minerals. It is a radioactive product of *uraninite* and *ianthinite*.

D. H. SPEIDEL

References

Am. Min. 38, 1019; **42**, 920; **45**, 334, 1026.
Dana, I, 625.

BEHIERITE *Am. Min.* **46**, 767; **47**, 414
(Ta,Nb)BO_4 tetr.

BEHOITE *Am. Min.* **55**, 1; **63**, 664
$Be(OH)_2$ orth.

BEIDELLITE (***Smectite*** gr.)
$(Na,Ca_{1/2})_{0.33}Al_2(Al,Si)_4O_{10}(OH)_2 \cdot nH_2O$
mono. *Am. Min.* **47**, 137

Cross-references: *Clay Minerals; Soil Mineralogy.*

BELLIDOITE *Am. Min.* **60**, 736
Cu_2Se tetr. *Econ. Geol.* **70**, 384

BELLINGERITE *Acta Cryst.* **B30**, 965
$Cu_3(IO_3)_6 \cdot 2H_2O$ tric. *Am. Min.* **25**, 505

BELLITE (***Apatite*** gr.) *Am. Min.* **43**, 798
$(Pb,Ag)_{10}Cl_2(Cr,As,Si)_6O_{24}$
hexa. *Min. Mag.* **14**, 395

BELOVITE (***Apatite*** gr.) *Am. Min.* **40**, 367
$(Sr,Ce,Na,Ca)_5(PO_4)_3(OH)$ hexa.

BELYANKINITE A hydrated tantaloniobate of titanium, calcium, and zirconium named by Gerasimovsky and Kazakova (1950) for D. S. Belyankin, Russian mineralogist and petrographer.

Composition:
$(Ti,Nb,Zr,Ca)(O,OH)_2 \cdot 1.5\text{–}2H_2O$.
Amorphous.

Occurrence: In a nepheline syenite pegmatite (of the Kola peninsula, USSR) with *microcline, nepheline,* and *aegirine.* Also in *aegirine* and *microcline,* associated with *eudialyte, lorenzenite* (ramsayite), and *lamprophyllite.*

G. C. MOERSCHELL

References

Am. Min., **37**, 882.
Min. Abst., **11**, 123.
Min. Mag., **29**, 976.

BEMENTITE *Am. Min.* **49**, 446; **65**, 335
$Mn_8Si_6O_{15}(OH)_{10}$ mono.

BENITOITE A barium titanosilicate named by Louderback (1907) after the San Benito Mine, San Benito County, California, where it was first found.

Composition: $2[BaTiSi_3O_9]$.

Properties: Hexagonal, small, flat triangular double pyramids; H=6–6.5; G=3.6; color; sapphire blue, colorless, or white; vitreous luster; brittle; transparent to translucent; luminesces blue under ultraviolet light.

Occurrence: Rare, found associated with black elongated *neptunite* crystals in compact *natrolite* veins in green *serpentine* and schist, near the headwaters of the San Benito River in California.

Use: As a minor gemstone (see *Gemology*).

G. C. MOERSCHELL

References

Min. Abst. **4**, 367.
Min. Mag. **15**, 417.
Vanders, I., and Kerr, P., 1967. *Mineral Recognition,* New York: Wiley, p. 284.

Cross-reference: *Naming of Minerals.*

BENJAMINITE *Can. Min.* **17**, 607
$Cu_{0.50}Pb_{0.40}Ag_{2.30}Bi_{6.80}S_{12}$ mono.

BENSTONITE A carbonate of calcium, magnesium, and barium named after O.J. Benston, metallurgist, National Lead Company (USA).

Composition:
$3[(Ca,Mg,Mn)_7(Ba,Sr)_6(CO_3)_{13}]$.

Properties: Trigonal; rhombohedral crystals, unmodified or interlocking, also massive; rhombohedral cleavage; H=3–4; G=3.60–3.65; vitreous luster; white, pale yellow, or brown; fluoresces yellow or red.

Occurrence: A vein mineral (q.v.) commonly associated with *calcite.*

K. FRYE

References

Am. Min. **47**, 585–598.
Min. Rec., **1**, 140–141.

BENTONITE – a rock consisting mainly of ***smectites.***

Cross-reference: *Pigments and Fillers:*

BERAUNITE A hydrated ferrous and ferric iron phosphate, named (1841) for the locality Beraun, Bohemia (now Beroun, Stredočesky, Czechoslovakia).

Composition:
$Fe^{2+}Fe^{3+}_5(OH)_5(PO_4)_4 \cdot 4H_2O$.

Properties: Monoclinic, *C*2/*c*; druses, foliated globules, concretions; one good cleavage {100}; H=3.5–4; G=2.8–3.0; vitreous luster; color: blue to red to reddish brown; streak: yellow to olive.

Occurrence: In secondary iron ore deposits

and as an alteration of primary phosphates in pegmatites (q.v.).

K. FRYE

References

Am. Min. **55**, 135–169.
Min. Rec. **5**, 241–244.

BERBORITE *Am. Min.* **53**, 348
$Be_2(BO_3)(OH,F){\cdot}H_2O$ trig.

BERGENITE orth. *Am. Min.* **45**, 909
$Ba(UO_2)_4(PO_4)_2(OH)_4{\cdot}8H_2O$

BERLINITE *Am. Min.* **61**, 409
$AlPO_4$ trig. *Dana,* II, 696

Cross-reference: *Pegmatite Minerals.*

BERMANITE mono. *Am. Min.* **61**, 1241
$Mn^{2+}Mn_2^{3+}(PO_4)_2(OH)_2{\cdot}4H_2O$ *Dana,* II, 967

Cross-reference: *Pegmatite Minerals.*

BERNDTITE, BERNDTITE-C6
SnS_2 *Am. Min.* **51**, 1551;
trig. and hexa. **63**, 289

BERRYITE *Am. Min.* **52**, 928
$Pb_2(Cu,Ag)_3Bi_5S_{11}$ mono. *Can. Min.* **8**, 407

BERTHIERINE A hydroxyl aluminosilicate of iron and magnesium of the ***kaolinite-serpentine*** group, named by Beudant (1832) after Berthier.

Composition: $(Fe^{2+},Fe^{3+},Mg)_{2-3}(Si,Al)_2O_5(OH)_4$

Properties: Monoclinic.

Occurrence: It constitutes a valuable bed of iron ore at Hayanes in Moselle, and is found in the ores of Champagne, Bourgogne, and Lorraine.

Use: An ore of iron.

G. C. MOERSCHELL

References

Bull. Soc. fr. 78, xlviii.
Min. Abst. **11**, 104, 210.
Min. Mag. **30**, 57, 279.

Cross-reference: *Soil Mineralology.*

BERTHIERITE *Am. Min.* **40**, 226
$FeSb_2S_4$ orth.

Cross-reference: *Ore Microscopy.*

BERTOSSAITE orth. *Am. Min.* **52**, 1583
$(Li,Na)_2CaAl_4(PO_4)_4(OH,F)_4$
Can. Min. **8**, 668

BERTRANDITE A hydroxyl silicate of berylium named by Damour (1883) after E. Bertrand, the French mineralogist.

Composition: $4[Be_4Si_2O_7(OH)_2]$; SiO_2, 50.3%; BeO, 42.1%; H_2O, 7.6%.

Properties: Orthorhombic, $Ccm2_1$; hemimorphic character as heart-shaped twins; H=6–7; G=2.59–2.6; vitreous to pearly luster; colorless to slightly yellow; transparent.

Occurrence: Most commonly in pegmatite in intimate association with *beryl.* Found at Irkutka and Altai Mountains, Siberia; in Czechoslovakia; originally discovered near Nantes, France. Also found in Norway, at various points in Cornwall, and in the USA in Maine, Virginia, and Colorado.

Use: In the Spor Mountain area of Utah, as an ore of beryllium.

G. C. MOERSCHELL

References

Am. Min. **47**, 67; **63**, 664.
Min. Abst. **5**, 324; **11**, 52; **27**, 67.

Cross-references: *Blacksand Minerals; Pegmatite Minerals.*

BERYL An aluminosilicate of beryllium named (about 315 BC), from the Greek word, which was applied to green gem stones.

Composition: $Be_3Al_2Si_6O_{18}$; BeO, 14.0%; Al_2O_3, 19.0%; SiO_2, 67.0%.

Properties: Hexagonal, *P6/mmc;* crystals frequently striated; H=7.5–8; G=2.75–2.8; vitreous luster, transparent to translucent; color: commonly blue-green or light yellow but may be deep green, pink, white, or colorless.

Occurrence: A widely distributed mineral found in granitic rocks, mica schists, and dark limestones. Associated with tin ores, *phenacite, chrysoberyl, rutile, spodumene,* and ***tourmaline,*** Important deposits of *beryl* are at: Minas Gerais, Brazil; Cándoba and San Luis, Argentina; Bikita tin mines, Rhodesia; Black Hills, South Dakota; Taos County, New Mexico.

J. A. GILBERT

References

Am. Min. **50**, 1783; **61**, 100; **63**, 664.
DHZ, 1, 256–267.

Cross-references: *Blacksand Minerals; Clinopyroxenes; Crystal Growth; Crystallography; Morphological; Gemology; Jade; Metallic Minerals; Mineral and Ore Deposits; Minerals, Uniaxial and Biaxial; Pegmatite Minerals; Refractory Minerals; Rock-Forming Minerals; Staining Techniques.*

BERYLLITE *Am. Min.* **40**, 787; **63**, 664
$Be_3SiO_4(OH)_2 \cdot H_2O$ orth.(?)

BERYLLONITE *Am. Min.* **39**, 397
$NaBePO_4$ mono.

Cross-reference: *Pegmatite Minerals.*

BERZELIANITE *Am. Min.* **35**, 337
Cu_2Se iso.

BERZELIITE iso. *Dana,* II, 681
$NaCa_2(Mg,Mn)_2(AsO_4)_3$ *Min. Abst.* **28**, 259

BESSMERTNOVITE *Min. Abst.* **31**, 495
$Au_4Cu(Te,Pb)$ orth.

BETA-FERGUSONITE *Am. Min.* **46**, 1516
$YNbO_4$ mono.

BETA-FERGUSONITE-(Ce) *Am. Min.* **60**, 485
$(Ce,La)NbO_4$ mono.

BETAFITE A tantaloniobate of uranium of the ***pyrochlore*** group, named by Lacroix (1912) after Betafo, Madagascar.

Composition: $(Ca,Na,U)_2(Ti,Nb,Ta)_2O_6(O,OH,F)$.

Properties: Isometric, *Fd3m*; greenish brown color; H=4–5.5; G=3.7–5; metamict; brittle.

Occurrence: Found in pegmatite in Madagascar, Siberia, and Norway; associated with ***pyrochlore.***

J. A. GILBERT

References

Am. Min. **46**, 1519; **62**, 403.
Can. Min. **6**, 610.
Dana, I, 803.

Cross-reference: *Blacksand Minerals.*

BETA-ROSELITE *Am. Min.* **40**, 828
$Ca_2Co(AsO_4)_2 \cdot 2H_2O$ tric.

BETA-URANOPHANE *USGS Bull.* **1064**, 307
$Ca(UO_2)_2Si_2O_7 \cdot 6H_2O$ mono.

BETEKHTINITE *Am. Min.* **41**, 371
$Cu_{10}(Fe,Pb)S_6$ orth. *Min. Abst.* **28**, 447

BETPAKDALITE mono. *Am. Min.* **47**, 172
$CaFe_2H_8(AsO_4)_2(MoO_4)_5 \cdot 10H_2O$

BEUDANTITE A basic sulfate-arsenate of lead and ferric iron. Named by A. Levy (1826) in honor of the French mineralogist Francois Sulpice Beudant.

Composition: $PbFe_3(AsO_4)(SO_4)(OH)_6$; minor amounts of Al may substitute for Fe^{3+}.

Properties: Trigonal, $R\bar{3}2/m$; H-3.5–4.5; G=4–4.3; luster: vitreous to resinous; color: black, dark green, or brown; cleavage {0001} easy; uniaxial negative; transparent to translucent; soluble in HCl.

Occurrence: Rare. A secondary mineral formed by the alteration of lead ores, especially lead sulfosalts.

Associated minerals: *Pharmacosiderite, scorodite, hematite,* and *quartz.*

Related minerals: *Corkite* and other members of the ***beudantite*** group (see *Alunite Group*).

A. PABST

Reference: *Min. Mag.* **35**, 1013-1016.

BEUDANTITE GROUP—*Beudantite, corkite, hidalgoite, hinsdalite, kemmlitzite, schlossmacherite, svanbergite, weilerite,* and *woodhouseite.*

BEUSITE *Am. Min.* **53**, 1799
$(Mn,Fe,Ca,Mg)_3(PO_4)_2$ mono.

Cross-reference: *Pegmatite Minerals.*

BEYERITE A carbonate of calcium and bismuth, named by Frondel (1943) after Adolf Beyer (1743–1805), mining engineer and mineralogist of Schneeberg, Saxony.

Composition: $2[Ca(BiO)_2(CO_3)_2]$; Pb may replace Ca. Tetragonal.

Occurrence: A secondary mineral found associated with *bismutite,* as tiny yellow platy crystals at Schneeberg, Saxony, East Germany, and as greenish gray compact masses at Stewart Mines, Pala, California.

J. A. GILBERT

References

Am. Min. **28**, 521; **32**, 660.
Min. Mag. **27**, 367.

BIANCHITE *Am. Min.* **15**, 538
$(Zn,Fe)SO_4 \cdot 6H_2O$ mono.

BICCHULITE *Am. Min.* **59**, 1330; **63**, 58
$Ca_2Al_2SiO_6(OH)_2$ iso.

BIDEAUXITE *Am. Min.* **57**, 1003
$Pb_2AgCl_3(F,OH)_2$ iso. *Min. Mag.* **37**, 637

BIEBERITE *Dana,* II, 505
$CoSO_4 \cdot 7H_2O$ mono.

BIKITAITE A hydrated lithium aluminosilicate, named for the Rhodesian locality (Bikita) where it was discovered.

Composition: $2[LiAlSi_2O_6 \cdot H_2O]$.

Properties: Monoclinic, $P2_1/m$; pseudo-orthorhombic crystals, also massive; vitreous luster, colorless.

Occurrence: A pegmatite mineral (q.v.).

K. FRYE

References: *Am. Min.* **42**, 792–797; **43**, 768.

BILIBINSKITE *Am. Min.* **64**, 652
$Au_3Cu_2PbTe_2$ ps. iso. *Min. Mag.* **43**, 1058

BILINITE (***Halotrichite*** gr.) *Dana,* II, 529
$Fe^{2+}Fe_2^{3+}(SO_4)_4 \cdot 22H_2O$ mono.

BILLIETITE *Am. Min.* **45**, 1026
$BaU_6O_{19} \cdot 11H_2O$ orth.

BILLINGSLEYITE *Am. Min.* **53**, 1791
$Ag_7(Sb,As)S_6$ orth.

BINDHEIMITE A lead antimonate of the ***stibiconite*** group, named (1868) after the German chemist Johann Jacob Bindheim.

Composition: $8[Pb_2Sb_2O_6(O,OH)]$ with Pb replaced by Ca toward *romeite.*

Properties: Isometric, $Fd3m$; dense to earthy masses and crusts; H=4–4.5; G=4.6–5.6; luster resinous to earthy; yellow, brown, gray, white, or greenish.

Occurrence: Common in the oxidized zone of antimonial lead ores.

K. FRYE

Reference: *Dana,* II, 1018–1020.

BINNITE = *Tennantite*

BIOTITE A hydroxyl aluminosilicate of potassium with magnesium and iron of the ***mica*** group, named by Hausmann (1847) after the French physicist, J. B. Biot.

Composition: $K(Mg,Fe)_3(AlSi_3O_{10})(OH)_2$.

Properties: Monoclinic; crystals rare; usually in foliated masses, scaly aggregates, or disseminated scaled; perfect basal cleavage; H=2.5–3; G=2.8–3.2; splendent luster; transparent to translucent; color: dark brown, green, or black, usually smoky when thin.

Occurrence: A very common rock-forming mineral (q.v.), occurring chiefly in igneous rocks in which ***feldspar*** is prominent. May also be found in gneisses, schists, pegmatite dikes, and basalts.

Use: Dehydrated expanded *vermiculite* a K-poor alteration product of *biotite*) is used for insulation, light-weight concrete aggregate, and in potting soil.

J. A. GILBERT

References

Am. Min. **50**, 1228
DHZ, 3, 55–84.
Min. Mag. **29**, 936.

Cross-references: *Crystal Growth; Diamond; Green River Mineralogy; Mantle Mineralogy; Mica Group; Mineral Classification: Principles; Minerals, Uniaxial and Biaxial; Pegmatite Minerals; Phyllosilicates; Pleochroic Halos; Soil Mineralogy.*

BIPHOSPHAMMITE *Min. Mag.* **38**, 965
$(NH_4,K)H_2PO_4$ tetr.

Cross-references: *Cave Minerals; Human and Vertebrate Mineralogy.*

BIRINGUCCITE mono. *Am. Min.* **48**, 709
$Na_4B_{10}O_{16}(OH)_2 \cdot 2H_2O$

BIRNESSITE *Am. Min.* **62**, 278
$(Na,K,Ca)MgMn_6O_{14} \cdot 3H_2O$
orth. *Science* **212**, 1024

BISBEEITE–Mixture of copper silicates(?) *Am. Min.* **57**, 1005

BISCHOFITE A hydrated magnesium chloride, named by Ochsenius (1877), after Gustav Bischof (1792–1870), German mineral chemist and geologist.

Composition: $2[MgCl_2 \cdot 6H_2O]$.

Properties: Monoclinic, $C2/m$; prismatic; H=1–2; G=1.6; colorless to white; vitreous to dull; deliquescent; bitter taste.

Occurrence: A rare constituent of saline deposits (see *Saline Minerals*), occurs with *carnallite* and *halite* in the *kieserite*-rich zones, and is partly of secondary origin, forming by the alteration of *carnallite* by water. Found in salt deposits of northern Germany at Leopoldshall, Anhalt, and at Stassfurt and Vienenburg, Saxony.

R. W. FAIRBRIDGE

Reference: *Dana,* II, 46.

BISMITE Bismuth trioxide, named by Dana (1868) in allusion to its composition. The name is restricted to the monoclinic polymorphy; *sillenite* is isometric.

Composition: Bi_2O_3.

Properties: Monoclinic; $P2_1/c$; G=8.6; massive; grayish green to yellow. Isostructural with *claudetite.*

Occurrence: An oxidation product of bismuth ores; as a crust on native *bismuth* at Colavi, Bolivia.

D. H. SPEIDEL

References

Am. Min. **28,** 521.
Dana, I, 599.
Min. Abst. 8, 216; 9, 5.

BISMOCLITE A chloride oxide of bismuth named by Mountain (1935) for its composition.

Composition: 2[BiOCl]. A closely related mineral is *daubréeite,* BiO(OH,Cl), named by Domeyko (1876).

Occurrence: Secondary mineral formed by the alteration of *bismuthinite* or native *bismuth,* associated with *jarosite, alunite, cerussite, bismutite,* and *iodargyrite.*

J. A. GILBERT

References

Am. Min. **30,** 813.
Dana, II, 60.
Min. Mag. **24,** 603.

BISMUTH Native semimetal, bismuth, named possibly from the Greek meaning "lead white."

Composition: 2[Bi].

Properties: Trigonal ($R\bar{3}m$) crystal system; usually laminated and granular; H=2–2.5; G=9.8; sectile, brittle, metallic luster; color: silver-white with a decided reddish tone.

Occurrence: A rare mineral occurring in hydrothermal veins or in pegmatites, usually in connection with ores of silver, cobalt, nickel, lead, and tin. Abundant deposits are located in Saxony, E. Germany, at Schneeberg, Annaberg, Johanngeorgenstads; Bolivia, at San Baldomero and LaPaz; Australia, at Chillagoo and Kingsgate; Canada, at Cobalt, Ontario, and Great Bear Lake, Mackenzie.

Use: Most *bismuth* is used in medicine and cosmetics, and as a major ore of the element that is used in low-melting-point alloys.

J. A. GILBERT

Reference: *Dana,* II, 134.

Cross-references: *Native Elements and Alloys; Ore Microscopy; Pegmatite Minerals; Skeletal Crystals.* Vol. IVA. *Bismuth: Element and Geochemistry.*

BISMUTHINITE Bismuth sulfide, named, by Beudant (1832) in allusion to its composition, bismuthine; spelling was modified by Dana (1868).

Composition: $4[Bi_2S_3]$; Bi, 81.2%; S, 18.8%.

Properties: Orthorhombic *Pbnm;* acicular striated crystals, but usually massive with foliated or fibrous texture; H=2, G=6.78± 0.03; metallic luster, opaque; color: lead-gray.

Occurrence: A rare mineral occurring commonly in high-temperature hydrothermal veins and in granite pegmatites. In the veins it is associated with native *bismuth, quartz, arsenopyrite,* and sulfides. The most important deposits are in Bolivia, at San Baldamero and at Llallagua.

Use: An ore of bismuth.

J. A. GILBERT

References

Am. Min. **61,** 15–20.
Dana, I, 275.

Cross-references: *Mineral and Ore Deposits; Ore Microscopy; Pegmatite Minerals.*

BISMUTITE *Dana,* II, 259
$Bi_2(CO_3)O_2$ tetr.

BISMUTOFERRITE *Am. Min.* **43,** 656
$BiFe_2(SiO_4)_2(OH)$ mono.

BISMUTOHAUCHE-CORNITE
$Ni_9Bi_2S_8$ tetr. *Am. Min.* **66,** 436
Min. Mag. **43,** 873

BISMUTOMICROLITE (**Pyrochlore** gr.)
$(Bi,Ca)(Ta,Nb)_2O_6(OH)$ iso. *Am. Min.* **48,** 215; **62,** 405

BISMUTOTANTALITE A tantalate of bismuth, named by Wayland and Spencer (1929) in allusion to its composition.

Composition: $BiTaO_4$, with replacement of Ta by Nb. Orthorhombic.

Occurrence: A rare mineral found in pegmatite with ***tourmaline,*** *cassiterite,* and *muscovite* at Gamba Hill, Uganda, Africa.

J. A. GILBERT

References

Am. Min. **15,** 201; **42,** 178.
Dana, I, 769.
Min. Mag. **22,** 616.

Cross-reference: *Pegmatite Minerals.*

*BITYITE (**Mica** gr.)* mono. RRW, 75
$CaLiAl_2(AlBeSi_2)O_{10}(OH)_2$

Cross-reference: *Pegmatite Minerals.*

BIXBYITE Maganese oxide named by Penfield and Foote (1897) for Maynard Bixby of Salt Lake City.

Composition: γ-Mn_2O_3.

Properties: Isometric, believed to have a defect spinel structure; H=6–6.5; G=4.9; black; polymorphic with *braunite.*

Occurrence: With *topaz* in cavities in rhyolites in the Thomas Range, Utah.

D. H. SPEIDEL

References

Am. Min. **61,** 1226–1240; **65,** 756–765.
Dana, I, 550.
Gazz. Chem. Ital. **70,** 145.

Cross-reference: *Ore Microscopy.*

BJAREBYITE mono. *Am. Min.* **59,** 873
$(Ba,Sr)(Mn,Fe,Mg)_2Al_2(PO_4)_3(OH)_3$
Min. Rec. 4, 282

BLACKJACK = *Sphalerite*

BLAKEITE—ferric tellurite *Dana,* II, 643

BLANCHARDITE = *Brochantite*
Min. Rec. 3, 229

BLENDE = *Sphalerite*

BLIXITE *Am. Min.* **45,** 908
$Pb_2Cl(O,OH)_2$ orth.

BLOEDITE A hydrated sodium magnesium sulfate named (1821) after the German chemist Carl August Bloede.

Composition: $4[Na_2Mg(SO_4)_2 \cdot 4H_2O]$, see *Nickelbloedite.*

Properties: Monoclinic, $P2_1/a$; highly modified short prismatic crystals, massive granular or compact; conchoidal fracture; H=2.5–3; G=2.25; colorless, bluish or reddish; faintly saline bitter taste.

Occurrence: An evaporite mineral (q.v.) and a cave mineral (q.v.).

K. FRYE

Reference: *Dana,* II, 447–450.

Cross-references: *Cave Minerals; Green River Mineralogy; Meteoritic Minerals; Saline Minerals.*

BLOMSTRANDINE = *Uranpyrochlore*(?)
Am. Min. **62,** 406

BLOODSTONE—Variety of *Quartz*

Cross-reference: *Gemology.*

BOBIERRITE *Am. Min.* **48,** 635
$Mg_3(PO_4)_2 \cdot 8H_2O$ mono.

Cross-reference: *Human and Vertebrate Mineralogy.*

BOEHMITE An aluminum oxyhydroxide of the ***lepidocrocite*** family, one of the principal constituents of bauxite (q.v.). Named after J. Böhm, a German chemist, by Lapparent (1927).

Composition: γ- AlO(OH).

Properties: Orthorhombic; *Aman;* H=3.5–4; G=3.0; good {010} cleavage; structure is based on cubic close packed layers of O with Al in octahedral coordination between them; H bonds only half of the oxygen atoms. Optical properties are: α=1.64–1.65; β=1.65–1.66; γ=1.65–1.67; white when pure. Chemically, it converts to γ=Al_2O_3 (spinel structure) at about 300°C. Structurally similar to *lepidocrocite,* γ–FeO(OH), and polymorphic with *diaspore.*

Occurrence: In bauxites, laterites, fire clays, associated with ***clays,*** *gibbsite, diaspore,* and *corundum.* Important localities are: Gánt, Hungary; Harz, Germany; Ayrshire, Scotland; Recoux, France.

Use: Main source of Al.

D. H. SPEIDEL

References

Ber. Deut. Keram. Ges., **42** (5), 167–184.
DHZ, 5, 111–117.
de Weisse, G., 1963. Lateritic and karstitic bauxites, *Symp. Bauxites, Oxydes, Hydroxydes Aluminum, Zagreb* **(1)** 220p.

Cross-references: *Mineral and Ore Deposits; Minerals, Uniaxial and Biaxial; Rock-Forming Minerals; Soil Mineralogy.*

BOGDANOVITE *Min. Abst.* **31**, 85
$Au_5(Cu,Fe)_3(Te,Pb)_2$ orth.(?) ps. iso.

BØGGILDITE *Am. Min.* **41**, 959
$Na_2Sr_2Al_2(PO_4)F_9$ mono.

BOHDANOWICZITE *Can. Min.* **18**, 353
$AgBiSe_2$ hexa. *Min. Mag.* **43**, 131

BOKITE *Am. Min.* **48**, 1180
$KAl_3Fe_6V_6^{4+}V_{20}^{5+}O_{76} \cdot 30H_2O$(?)

BOLEITE A hydroxyl chloride of lead, copper, and silver, named by Mallard and Cumenge (1891), from Boleo, Baja California, Mexico.

Composition: $Pb_{26}Cu_{24}Ag_9Cl_{62}(OH)_{47} \cdot H_2O$. Isometric.

Occurrence: A relatively rare secondary mineral originally described at Boleo near Santa Rosalia, Baja California, Mexico, associated with *cumengite, pseudoboleite, anglesite,* and *phosgenite.*

J. A. GILBERT

Reference: *Min. Mag.* **44**, 101–104.

BOLIVARITE amorph. *Min. Mag.* **38**, 418
$Al_2(PO_4)(OH)_3 \cdot 4\text{–}5H_2O$

BOLTWOODITE *Am. Min.* **46**, 12
$K_2(UO_2)_2(SiO_3)_2(OH)_2 \cdot 5H_2O$

BONACCORDITE *Am. Min.* **61**, 502
Ni_2FeBO_5 orth.

BONATTITE *Am. Min.* **51**, 276
$CuSO_4 \cdot 3H_2O$ mono.

BONCHEVITE (dubious species)
$PbBi_4S_7$(?) *Am. Min.* **43**, 1221
orth. *Min. Mag.* **31**, 821

BOOTHITE *Dana,* II, 504
$CuSO_4 \cdot 7H_2O$ mono.

BORACITE A magnesium chloroborate, named by Werner (1789) in allusion to its composition.

Composition: $8[Mg_6B_{14}O_{26}Cl_2]$; MgO, 25.71%; $MgCl_2$, 12.14%; B_2O_3, 62.15%.

Properties: Orthorhombic, isometric above 265°C; cubes with or without tetrahedral truncations, also massive; H=7; G=2.9–3.0; vitreous luster, transparent to translucent; color: white, green, colorless, gray. Soluble in HCl.

Occurrence: Forms in evaporites, associated with *halite, anhydrite,* and *gypsum.*

Use: Minor source of boron.

J. A. GILBERT

References:

Dana, II, 378.
Min. Abst., **11**, 424.

Cross-references: *Museums, Mineralogical; Saline Minerals.*

BORAX (TINCAL) A hydrated sodium borate, named by Wallerius (1748), from the Arabic name for the substance. Tincal is from the Malaysian name for the substance in crude, native state.

Composition: $Na_2[B_4O_7(OH)_4] \cdot 8H_2O$; Na_2O 16.2; B_2O_3, 36.6%; H_2O, 47.2%.

Properties: Monoclinic; $A2/a$; as encrustations or as cellular masses; H=2–2.5; G=1.7; vitreous luster, translucent; sweet, alkaline taste; color: white or colorless.

Occurrence: Forms as a precipitate from the evaporation of salt lakes and as an efflorescence on the surface of the ground in arid regions (see *Saline Minerals*). Associated with *anhydrite, kernite, ulexite, hanksite, gypsum, halite,* and *colemanite.* Some of the most famous deposits are at Furnace Creek and Resting Springs, Death Valley, California, where the "twenty-mule teams" hauled the ore to the railroad at Mojave.

Use: For washing, as an antispetic, a preservative, and a source of boron. Used in industry as a solvent for metallic oxides in soldering and welding and as a flux in various smelting and laboratory operations (see *Blowpipe Analysis*).

J. A. GILBERT

References

Dana, II, 339.
Smith, W. C., 1960. Borax and borates, *Industrial Rocks and Minerals,* 3rd ed. New York: Am. Inst. Min., Met. Petr. Engineers.

Cross-reference: *Saline Minerals.*

BORCARITE mono. *Am. Min.* **50**, 2097
$Ca_4Mg[B_4O_6(OH)_6](CO_3)_2$ *Min. Abst.* **28**, 142

BORICKYITE=*Delrauxite(?)* *Am. Min.* **65**, 813

BORISHANSKIITE *Am. Min.* **61**, 502
$Pd_{1+x}(As,Pb)_2$, $x<0.2$ orth.

BORNEMANITE orth. *Am. Min.* **61**, 338
$BaNa_4Ti_2NbSi_4O_{17}(F,OH)\cdot Na_3PO_4$(?)
Min. Abst. **27**, 81

BORNHARDTITE (*Linnaeite* gr.)
Co_3Se_4 iso. *Am. Min.* **41**, 164

BORNITE A sulfide of copper and iron, named by Haidinger (1845), after Ignatius von Born, Austrian mineralogist (1742–1791).

Composition: $8[Cu_5FeS_4]$; Cu, 63.3%; Fe, 11.2%; S, 25.5%.

Properties: Isometric, usually massive; H=3; G=5.06–5.08; metallic luster; color: brownish bronze on fresh surface but quickly tarnishes to variegated purple and blue and finally to almost black on exposure, hence the terms peacock and purple copper ores.

Occurrence: Alters readily to *chalcocite* and *covellite;* usually found in hypogene deposits; also disseminated in mafic rocks, in contact metamorphic deposits, in replacement deposits, in pegmatites, and as a rare authigenic mineral (q.v.). Commonly associated with *chalcopyrite* and *chalcocite.*

Use: An ore of copper but not quantitatively significant.

J. A. GILBERT

References

Am. Min. **46,** 1062; **46,** 1270; **49,** 1084.
Dana, I, 195.

Cross-references: *Color in Minerals; Lunar Minerals; Metallic Minerals; Mineral and Ore Deposits; Ore Microscopy.*

BOROVSKITE *Am. Min.* **59**, 873
Pd_3SbTe_4 iso.

BORT–granular to cryptocrystalline *diamond.*

Cross-reference: *Mohs Scale of Hardness.*

BOTALLACKITE *Am. Min.* **36**, 384
$Cu_2Cl(OH)_3$ mono.

BOTRYOGEN A hydrated magnesium iron sulfate named by Haidinger (1828), for its botryoidal habit.

Composition: $MgFe^{3+}(SO_4)_2(OH)\cdot 7H_2O$.

Properties: Monoclinic, $P2_1/n$; prismatic habit or reniform, botryoidal, or globular aggregates; H=2–2.5; G=2.14; vitreous; color: light to dark orange-red; streak: yellow-ochre.

Occurrence: Associated with secondary sulfates capping sulfide ore deposits in arid regions. Found at Chuquicamata, Queteña, and Alcaparrosa, Chile; San Juan, Argentina; Falun, Sweden; and in the USA near Knoxville and Calistoga, California, and Cornwall, Pennsylvania.

K. FRYE

Reference: *Dana,* II, 617–618.

BOULANGERITE A lead antimony sulfide named for the French mining engineer C. L. Boulanger.

Composition: $8[Pb_5Sb_4S_{11}]$.

Properties: Monoclinic, $P2_1/a$; habit prismatic, often in compact fibrous masses; one good cleavage {100}; H=2.5–3; G=6.2; color: bluish lead-gray; streak: brown to brownish gray; metallic luster.

Occurrence: Associated with sulfosalts, sulfides, carbonates, and *quartz* in low- to moderate-temperature veins.

K. FRYE

Reference: *Dana,* I, 420–423.

Cross-references: *Ore Microscopy; Vein Minerals.*

BOURNONITE A sulfide of lead, copper and antimony, named by Jameson, in 1805, after the French mineralogist, Count J. L. de Bournon.

Composition: $4[CuPbSbS_3]$; Pb, 42.4%; Sb, 24.9%; Cu, 13.0%, and S, 19.7% by weight. Arsenic may substitute for the antimony (Sb) toward *seligmannite.* The unit-cell dimensions are a=8.16 Å, b=8.71 Å, and c=7.81 Å.

Properties: Orthorhombic, $Pnm2_1$, pseudotetragonal; H=2.5–3.0; G=5.8–5.9; luster: metallic; color and streak: steel-gray to black; habit prismatic to tabular, also massive; twinning common. Optically opaque; weakly anisotropic; Vicker's hardness: medium (185–200); reflectivity: 32–36%; color strong next to other minerals, such as *galena,* which makes it appear blue, and colorless to pale gray by itself; bireflectance weak.

Occurrence: Typical of moderate temperature, hydrothermal veins. Commonly as microscopic inclusions in *galena.* Associated with *galena, pyrite, sphalerite, chalcopyrite, tetrahedrite, etc.*

Use: *Bournonite* is a common ore of lead, antimony, and copper. Among the more important areas of recovery are the Harz Mountains of Germany, Australia, Mexico, and USA.

R. W. CAMERON

References

Bragg, L., et al., 1965, *Crystal Structure of Minerals,* vol. 4. Ithaca: Cornell University Press, 409p.

Hurlbut, C. S., Jr., 1963, *Dana's Manual of Mineralogy,* 17th ed. New York: Wiley, 609p.

Schouten, C., 1962, *Determinative Tables for Ore Microscopy.* New York: Elsevier, 242p.

Cross-reference: *Ore Microscopy.*

BOUSSINGAULTITE *Dana,* I, 455
$(NH_4)_2Mg(SO_4)_2 \cdot 6H_2O$ mono.

BOYLEITE *Am. Min.* **64**, 464
$(Zn,Mg)SO_4 \cdot 4H_2O$ mono. *Min. Mag.* **43**, 1058

BRABANTITE *Min. Mag.* **43**, 1058
$CaTh(PO_4)_2$ mono.

BRACEWELLITE *Am. Min.* **62**, 593
$CrO(OH)$ orth. *USGS Prof. Pap. 887*

BRACKEBUSCHITE *Am. Min.* **40**, 597
$Pb_2(Mn,Fe)(VO_4)_2 \cdot H_2O$ mono.

BRADLEYITE *Am. Min.* **26**, 646
$Na_3Mg(PO_4)(CO_3)$ mono.

Cross-references: *Green River Mineralogy; Saline Minerals.*

BRAGGITE *Am. Min.* **63**, 832
$(Pt,Pd,Ni)S$ tetr. *Dana,* I, 259

BRAITSCHITE hexa. *Am. Min.* **53**, 1081
$(Ca,Na_2)_7(Ce,La)_2B_{22}O_{43} \cdot 7H_2O$

BRAMMALLITE (***Mica*** gr.) *Am. Min.* **29**, 73
Na-analog of *illite* mono.

BRANDISITE = *Clinotonite*
Am. Min. **52**, 1122

BRANDTITE mono. *Dana,* II, 725
$Ca_2(Mn,Mg)(AsO_4)_2 \cdot 2H_2O$

BRANNERITE An oxide of titanium, uranium, and calcium containing minor amounts of yttrium, thorium, and ferrous iron, named by Hess and Wells (1920) after J. C. Branner (1850–1922), an American geologist.

Composition: $(U,Ca,Th,Y)(Ti,Fe)_2O_6$.

Properties: Monoclinic, *I2/m*; black with greenish brown streak; radioactive; metamict; H=4.5; G=4.5–4.53.

Occurrence: Found in gold placers near pegmatites at the head of Kelly Gulch, Custer County, Idaho; associated with *euxenite.*

D. H. SPEIDEL

References

Am. Min. **39**, 109–117; **46**, 1086–1096.

Can. Min. **6**, 483–490.

Palache, C., et al., 1951. *Dana's System of Mineralogy,* vol. I, 7th ed. New York: Wiley, 774.

Cross-reference: *Ore Microscopy.*

BRANNOCKITE (***Osumilite*** gr.)
$KSn_2LI_3Si_{12}O_{30}$ *Am. Min.* **58**, 1111
hexa. *Min. Rec.* **4**, 73

BRASSITE *Am. Min.* **60**, 945
$MgHAsO_4 \cdot 4H_2O$ orth. *Min. Abst.* **28**, 261

BRAUNITE Manganese oxide, named by Haidinger (1831) after Kammerath Braun of Gotha.

Composition: $Mn^{2+}Mn^{4+}O_3$, Si replaces Mn^{4+} up to 40%, ferric iron replaces Mn up to 20%.

Properties: Tetragonal, *I4/acd*; H=6–6.5; G=4.75–4.82; color: dark brownish black; polymorphic with *bixbyite.*

Use: An ore of manganese.

D. H. SPEIDEL

References

Am. Min. **65**, 756–765.

AJS, **257**, 297-313.

(The question of the composition and structure of *braunite* is not entirely without controversy; the first reference presents an alternative interpretation of the mineral. Editor)

Cross-reference: *Magnetic Minerals.*

BRAVAISITE–Interlayer mixture of *illite* and *montmorillonite.*

BRAVOITE A disulfide of nickel and iron of the ***pyrite*** group. Named by Hillebrand (1907), after José J. Bravo (1874–1928) of Lima, Peru.

Composition: $4[(Ni,Fe)S_2]$; Ni>Fe; a complete series with increasing substitution of Fe extends from *bravoite* through nickeloan *pyrite* to *pyrite.*

Properties: Isometric, *Pa*3; H=5.5–6; G=4.62; color is steel-gray with a metallic luster.

Occurrence: As crusts or nodular masses with a radially fibrous or columnar structure. Associated minerals include *pyrite, galena, sphalerite,* and *chalcopyrite.*

J. A. GILBERT

References

Carnegie Inst. Wash. Yr. Bk., **61**, 144.
Dana, I, 290.
Min. Mag., **15**, 418.

Cross-reference: *Ore Microscopy.*

BRAZILIANITE *Dana,* II, 841
$NaAl_3(PO_4)_2(OH)_4$ mono.

Cross-references: *Naming of Minerals; Pegmatite Minerals.*

BREDIGITE A metastable, high-temperature phase of calcium silicate, named by Tilley and Vincent (1948) for M. A. Bredig, physical chemist, because of his studies on the polymorphism of Ca_2SiO_4 (see *Am. Mineralogist,* **28**, 594).

Composition: $(Ca,Ba)Ca_{13}Mg_2(SiO_4)_8$.

Occurrence: Found at Scawt Hill, Antrim County, Ireland, associated with *gehlenite, larnite,* and *spurrite* at one deposit and with *larnite, melilite, magnetite,* and *perovskite* at another. In the USA, at Marble Canyon, Culberson County, Texas, in a contact zone near a syenite-monzonite intrusion. Also occurs as an important phase in Portland cement (q.v.), clinkers, slags, and fertilizers.

G. C. MOERSCHELL

References

Am. Min. **51**, 1766; **61**, 74.
Min. Mag. **28**, 255; **29**, 180, 875.

Cross-reference: *Portland Cement Mineralogy.*

BREITHAUPTITE *Dana,* I, 238
NiSb hexa.

Cross-reference: *Ore Microscopy.*

BREUNNERITE = Ferroan *magnesite*

BREWSTERITE A hydrous aluminosilicate of strontium and barium of the ***zeolite*** group, named by Brooke (1822) after Sir David Brewster (1781–1868).

Composition: $(Sr,Ba)Al_2Si_6O_{16} \cdot 5H_2O$; usually contains some Ca. Monoclinic.

Occurrence: First observed at Strontian in Argyll, Scotland. Occurs also at the Giant's Causeway, Ireland. From the lead mines of St. Turpet, in the Black Forest, Baden. At the Col du Bonhomme, Savoie, and other localities in France.

R. W. CAMERON

References

Acta Cryst. **17**, 857.
DHZ, 4, 379, 383, 413.

BREZINAITE *Am. Min.* **54**, 1509
Cr_3S_4 mono.

Cross-reference: *Meteoritic Minerals.*

BRIANITE *Am. Min.* **53**, 508; **60**, 717
$Na_2CaMg(PO_4)_2$ mono.

Cross-reference: *Meteoritic Minerals.*

BRIARTITE *Am. Min.* **51**, 1816
$Cu_2(Fe,Zn)GeS_4$ tetr. *Min. Abst.* **27**, 41

BRINDLEYITE (***Kaolinite-serpentine*** gr.)
$(Ni,Al)_{2.7-2.8}(Si,Al)_2O_5(OH)_4$
mono., trig. *Am. Min.* **63**, 484

BRITHOLITE A hydroxyl calcium phosphosilicate, containing fluoride, sodium, cerium, and the rare earths. Named by Winther and Böggild (1899), it is a member of the ***apatite*** group.

Composition:
Near $2[(Ca,Ce)_5(SiO_4PO_4)_3(OH,F)]$.

Occurrence: Prior to 1956, *britholite* was known only from nepheline-syenite in a region of Julianehaab, S. Greenland. Now verified from Transbaikalia, Tuva, Vishnevye Gory, Siberia, and the Azov Sea, USSR. In nepheline-syenite, associated with *epistolite,* manganoan *pectolite,* and *steenstrupine.*

G. C. MOERSCHELL

References

Am. Min. **49**, 937.
Min. Mag. **12**, 380.
Medd. Grönland 1900, **24**, 190.

Cross-reference: *Apatite Group.*

BRITHOLITE-(Y) (***Apatite*** gr.) *RRW,* 89
$(Ca,Y)_5(SiO_4,PO_4)_3(OH,F)$ hexa.

BROCENITE = *Beta-fergusonite-(Ce)*
Am. Min. **60**, 485

BROCHANTITE *Dana,* II, 541
$Cu_4(SO_4)(OH)_6$ mono.

BROCKITE Am. Min. **47**, 1346
$(Ca,Th,Ce)(PO_4)\cdot H_2O$ hexa.

BROMARGYRITE A bromide of silver, related to *chlorargyrite,* named by Leymerie (1859) in allusion to its composition.

Composition: $4[AgBr]$. Isometric.

Occurrence: Secondary silver mineral which forms as a result of the surface oxidation of silver ores in regions of deep weathering where there is an abundance of chlorine and bromine, especially in arid regions. Associated with native *silver, iodargyrite, jarosite,* and wad.

J. A. GILBERT

Reference: *Dana,* II, 11.

Cross-reference: *Gossan.*

BROMELLITE A beryllium oxide related to *zincite,* named by Aminoff (1925) after Magnus von Bromell (1679–1731), a Swedish physician and mineralogist.

Composition: BeO.

Properties: Hexagonal, $P6_3mc$; würtzite structure; G=3.02; H=9; distinct $\{10\bar{1}0\}$ cleavage; color: white; transparent.

Occurrence: Found at Långban, Sweden, associated with *swedenborgite, richterite,* and manganoan *biotite* in veins of *calcite* that are in a skarn-forming *hematite* rock.

D. H. SPEIDEL

References

Am. Min. **11**, 135; **50**, 22; **63**, 664.
Dana, I, 506.

Cross-reference: *Refractory Minerals.*

BROMYRITE = *Bromargyrite*

BRONZITE = Ferroan *enstatite*

Cross-references: *Mineral Classification: Principles; Orthopyroxenes.*

BROOKITE Titanium dioxide, polymorphous with *anatase* (q.v.) and *rutile* (q.v.), and named by Lévy (1825) after H. J. Brooke (1771–1857), the English crystallographer and mineralogist.

Composition: $8[TiO_2]$; small amounts of Fe^{3+} may substitute for Ti.

Properties: Orthorhombic, *Pcab;* H=5.5–6; G=4.14; brittle; metallic adamantine luster; color: yellow-brown to iron-black; streak: white to gray or yellowish; transparent; an extremely high birefringence, 0.120.

Occurrence: As crystals, as an accessory mineral in some igneous and metamorphic rocks, as a detrital mineral associated with *anatase, rutile, hematite, titanite, quartz,* adularia, *albite,* and ***chlorite,*** and rarely as an authigenic mineral (q.v.). This association is found also in an Alpine-type deposit which is derived from hydrothermal leaching of gneisses and schists. Found commonly as placer deposits in the Urals, USSR and in Minas Gerais, Brazil.

D. H. SPEIDEL

References

Dana, I, 588.
DHZ, 5, 44–47.
Rocks Minerals **26**, 510.

Cross-references: *Blacksand Minerals; Dispersion, Optical; Minerals, Uniaxial and Biaxial; Optical Mineralogy; Optical Orientation; Polymorphism; Pseudomorphism; Rock-Forming Minerals; Vein Minerals.*

BROWNMILLERITE Am. Min. **50**, 2106
$Ca_2(Al,Fe)_2O_5$ orth.

Cross-reference: *Portland Cement Mineralogy.*

BRUCITE Magnesium hydroxide, named by Beudant (1824) after American mineralogist A. Bruce (1777–1818) who first described the mineral.

Composition: $Mg(OH)_2$; some Fe, ferrobrucite; Mn and Zn substitution. Readily alters to *hydromagnesite,* $4MgCO_3\cdot Mg(OH)_2\cdot 4H_2O$.

Properties: Trigonal, $P\bar{3}m1$; H=2.5; G=2.39; waxy luster; perfect basal cleavage, sometimes fibrous when ferroan (nemalite); structure composed of two sheets of OH in hexagonal close packing with layer of Mg ions between; Mg in sixfold coordination. Optically, ω=1.560–1.59, ϵ=1.58–1.60; color: white, greenish, or brownish; streak: white; transparent to translucent.

Occurrence: As an alteration product of *periclase* in contact metamorphosed limestone, and associated with *dolomite, magnesite, chromite, calcite, periclase* and ***serpentine.*** Large deposits are found east of Gabbs, Hye County, Nevada.

Uses: Magnesia refractory raw material; in welding rod coatings.

D. H. SPEIDEL

References

AJS, **263**, 668–677; **264**, 223–233.
Beitr. Mineral. Petrog. **11**(4), 393–397.
DHZ, 5, 89–92.

Cross-references: *Minerals, Uniaxial and Biaxial; Rock-Forming Minerals.*

BRUGGENITE *Am. Min.* **57**, 1911
$Ca(IO_3)_2 \cdot H_2O$ mono.

BRUGNATELLITE A hydrated carbonate-hydroxide of magnesium and iron, named by Artini (1909) after Luigi Brugnatelli (1859–1928), late Professor of Mineralogy, University of Pavia.

Composition: $[Mg_6FeCO_3(OH)_{13} \cdot 4H_2O]$; small amounts of Mn^{2+} substitute for Mg.

Properties: Trigonal, $P\bar{3}$ or $P3$; massive or in lamellar, flaky crystals; perfect basal cleavage; H=2; G-2.14; pearly luster; color: flesh-pink to yellow to brownish white; streak: white; transparent; pleochroic (ω=yellowish red, ϵ=colorless).

Occurrence: Found as crusts and coatings along cracks in serpentine rock, associated with *artinite, hydromagnesite, magnesite, chrysotile, aragonite, brucite,* and *pyroaurite.* Originally described at Ciappanico in the Val Malenco, Lombardy, Italy.

J. A. GILBERT

References

Dana, I, 660.
Min. Abst. **9**, 267.
Min. Mag. **15**, 418.

BRUNOGEIERITE (***Spinel*** gr.)
$(Ge,Fe)Fe_2O_4$ iso. *Am. Min.* **58**, 348

BRUNSVIGITE A hydroxyl aluminosilicate of iron and magnesium member of the ***chlorite*** group, named by Fromme (1902), presumably from the locality Brunswick in Germany.

Composition: $2[(Fe^{2+},Mg,Al)_6(Si,Al)_4O_{10}(OH)_8]$, $Fe^{2+}:(Mg + Fe^{2+}) = 0.5–0.8$. Monoclinic.

Occurrence: In cavities as fine scaly masses in the gabbro of the Radauthal in the Harz Mts. (Brunswick), West Germany. In the USA at Goose Creek, Loudon County, Virginia, and Westfield, Massachusetts.

R. W. CAMERON

References

DHZ, 3, 131–163.
Min. Mag. **13**, 365; **30**, 377.

BRUSHITE *Dana*, II, 704
$CaHPO_4 \cdot 2H_2O$ mono. *Min. Abst.* **27**, 256

Cross-references: *Cave Minerals; Human and Vertebrate Mineralogy.*

BUCHWALDITE *Am. Min.* **62**, 362
$NaCaPO_4$ orth.

BUDDINGTONITE (***Feldspar*** gr.)
$(NH_4)AlSi_3O_8 \cdot \frac{1}{2}H_2O$
mono. *Am. Min.* **49**, 831

Cross-reference: *Alkali Feldspars.*

BUERGERITE (***Tourmaline*** gr.)
$NaFe^{3+}Al_6Si_6B_3O_{30}F$
trig. *Am. Min.* **51**, 198

BUETSCHLIITE *Am. Min.* **59**, 353
$K_2Ca(CO_3)_2$ trig. *Min. Mag.* **28**, 725

BUKOVITE *Am. Min.* **57**, 1910
$Cu_{3+x}Tl_2FeSe_{4-x}$ tetr.

BUKOVSKYITE *Am. Min.* **54**, 991
$Fe_2(AsO_4)(SO_4)(OH) \cdot 7H_2O$ mono.

BULTFONTEINITE A hydroxyl calcium silicate containing fluorine, named by Parry, Williams, and Wright (1932), from the Bultfontein mine at Kimberley, South Africa, where it was originally found.

Composition: $Ca_2F(SiO_3OH) \cdot H_2O$.

Occurrence: Rare, found at the Bultfontein mine, associated with *calcite, apophyllite, natrolite,* dolerite, and shale fragments in kimberlite. Found also in the Dutoitspan mine and the Jagersfontein mine. In the USA, found at Crestmore, California, in a contact zone in association with *afwillite* and *scawtite.*

G. C. MOERSCHELL

References

Am. Min. **40**, 900.
Min. Mag. **30**, 569.

BUNSENITE *Dana*, I, 500
NiO iso.

Cross-reference: *Magnetic Minerals.*

BURANGAITE mono. *Am. Min.* **63**, 793
$(Na,Ca)_2(Fe,Mg)_2Al_{10}(OH,O)_{12}(PO_4)_8 \cdot 4H_2O$
Min. Abst. **29**, 83

BURBANKITE hexa. *Am. Min.* **62**, 158
$(Na_{5.0}Ca_{.62})(Ca_{1.48}Sr_{2.22}Ba_{.56}REE_{1.74})(CO_3)_{10}$
Min. Abst. **12**, 301

Cross-reference: *Green River Mineralogy.*

BURCKHARDTITE *Am. Min.* **64**, 355
$Pb_2(Fe,Mn)AlTeSi_3O_{10}(OH)_2O_2 \cdot H_2O$
mono., ps. hex. *Min. Mag.* **43**, 1058

BURKEITE A sulfate-carbonate of sodium, named by Teeple (1921) after W. E. Burke, chemical engineer who discovered the artificial salt.

Composition: $Na_6(SO_4)_2CO_3$; with a wide range of SO_4 possible.

Occurrence: Rare, found in drill cuttings from a clay stratum in Searles (Borax) Lake, San Bernardino County, California, associated with *calcite, halite, gaylussite, tychite, trona, sulfohalite,* and *borax.*

J. A. GILBERT

References

Am. Min. **20**, 50.
Dana, **II**, 633.
Min. Mag. **21**, 559.

Cross-references: *Green River Mineralogy; Saline Minerals.*

BURSAITE *Am. Min.* **41**, 671
$Pb_5Bi_4S_{11}$(?)

BUSTAMITE A silicate of manganese and calcium, named by Brongniart (1826) after M. Bustamente, the discoverer.

Composition: $3[(Mn,Ca,Fe)SiO_3]$, Ca:Mn> 1:2; passing into manganoan *wollastonite.*

Occurrence: *Bustamite* typically occurs in manganese ore bodies with other manganese-bearing minerals. It has been reported from Franklin, New Jersey; Camas Malag, Skye; Broken Hill, New South Wales: Novara, Italy; Altarnum, Cornwall; Långban, Sweden; Iwate Prefecture, Japan; and in the Harz Mountains, Germany.

R. W. CAMERON

References

Am. Min. **63**, 274.
DHZ, 2A, 575-585.

Cross-references: *Clinopyroxenes; Pyroxenoids; Rock-Forming Minerals.*

BUTLERITE *Dana,* II, 608
$Fe^{3+}(SO_4)(OH) \cdot 2H_2O$ mono.

BUTTGENBACHITE *Dana,* II, 572
$Cu_{19}Cl_4(NO_3)_2(OH)_{32} \cdot 2H_2O$ hexa.

BYSSOLITE—Asbestiform variety of *tremolite-actinolite*

BYSTRÖMITE *Am. Min.* **37**, 53
$MgSb_2O_6$ tetr.

BYTOWNITE A calcium sodium aluminosilicate of the ***feldspar*** group, named by Thomson (1835) after Bytown (now Ottawa), Canada.

Composition: $(Ca,Na)[(Al,Si)AlSi_2O_8]$; with some K replacing (Ca,Na). In the ***plagioclase feldspars,*** *bytownite* is assigned the composition range *Ab* 10–30 and *An* 90–70.

Properties: Triclinic, $P\bar{1}$; irregular grains or cleavable masses; twinning according to the albite, Carlsbad, or pericline laws; prismatic cleavage {001} {010}; H=6; G=2.7; vitreous to pearly luster; color: white to dark gray; streak: white.

Occurrence: As phenocrysts in some basalts, in some layered mafic and ultramafic intrusives, and in some stony meteorites.

K. FRYE

References

DHZ, 4, 94–165.
*See *Feldspar Group* and *Plagioclase Feldspars* for additional references.

Cross-references: *Meteoritic Minerals; Minerals; Uniaxial and Biaxial; Plagioclase Feldspars.*

C

CACOXENITE hexa. *Am. Min.* **51**, 1811; **55**, 135
$Fe_9^{3+}(OH)_{15}(PO_4)_4 \cdot 18H_2O$

Cross-reference: *Pegmatite Minerals.*

CADMIUM hexa. *Min. Abst.* **31**, 495
Cd

CADMOSELITE *Am. Min.* **43**, 623
CdSe hexa.

CADWALADERITE *Dana,* II, 77
$Al(OH)_2Cl \cdot 4H_2O$ amorph.

CAFARSITE iso. *Am. Min.* **63**, 795
$(CaO)_8(Ti,Fe)_{6.6}O_{10}(As_2O_3)_6 \cdot 2H_2O$(?)
Min. Abst. **29**, 139

CAFATITE orth. *Am. Min.* **45**, 476
$Ca(Fe,Al)_2Ti_4O_{12} \cdot 4H_2O$

CAHNITE *Am. Min.* **46**, 1077
$Ca_2B(OH)_4(AsO_4)$ tetr.

CAIRNGORM = Smokey *quartz*

CALAMINE = *Hemimorphite*
Min. Mag. **43**, 1053

CALAVERITE Gold ditelluride, named by Genth in 1861 from Calaveras County, California, where it was originally found in the Stanislaus Mine.

Composition: $2[AuTe_2]$; Au, 44.03%; Te, 55.97%; sometimes argentiferous, with Au:Ag=6:1.

Properties: Monoclinic; crystals commonly twinned, also granular; H=2.5; G=9.35; metallic luster; opaque; very brittle; color: brass yellow to silver white.

Occurrence: Found in veins and associated with *sylvanite.*

Use: An ore of gold.

J. A. GILBERT

References

Dana, I, 335.
Min. Abst. **27**, 306.

Cross-references: *Metallic Minerals; Mineral and Ore Deposits.*

CALCIBORITE *Am. Min.* **49**, 820
CaB_2O_4 mono.(?)

CALCIOCOPIAPITE tric. *Am. Min.* **47**, 807
$CaFe_4^{3+}(SO_4)_6(OH)_2 \cdot 19H_2O$

CALCIOFERRITE mono. *Dana,* II, 976
$Ca_4Mg(H_2O)_{12}[Fe_4^{3+}(OH)_4(PO_4)_6]$

CALCIOTANTALITE = *Microlite* + *tantalite* *Min. Mag.* **43**, 1054

CALCIOURANOITE *Am. Min.* **60**, 161
$(Ca,Ba,Pb)U_2O_7 \cdot 5H_2O$ amorph.

CALCIOVOLBORTHITE *Dana,* II, 817
$CaCu(VO_4)(OH)$ orth.

CALCITE Calcium carbonate, a member of the ***calcite*** group (q.v.), named by Haidinger in 1845 by derivation from the Greek root *calc,* burnt lime.

Composition: $CaCO_3$; commonly very pure, but capable of forming a complete solid-solution series with *rhodochrosite* ($MnCO_3$), and partial series toward *siderite* ($FeCO_3$), *smithsonite* ($ZnCO_3$), and *sphaerocobaltite* ($CoCO_3$).

Properties: Trigonal, $R\bar{3}c$; crystal habits (q.v.) extremely varied, also granular, stalactitic, nodular, or earthy; commonly twinned across $\{0001\}$ or $\{01\bar{1}2\}$; perfect rehmbohedral $\{10\bar{1}1\}$ cleavage; parting across twin planes; H=3 (Mohs scale of hardness, q.v.), but approximately 2.7 on base to 3.2 on prism faces; G=2.71; colorless to white except impurities cause pale to strongly colored specimens; transparent to translucent showing extreme double refraction of light; may be lumenescent when impure; optically ω=1.658, ϵ=1.486, birefringence= 0.172; dissolves with strong effervescence in cold, dilute HCl.

Occurrence: *Calcite* is a very common and widely distributed rock-forming mineral (q.v.), authigenic mineral (q.v.), and crystalline constituent of living things (see *Human and Vertebrate Mineralogy* and *Invertebrate and Plant Mineralogy*). Localities where *calcite* crystals may be collected are worldwide and too numerous to mention.

Use: Raw material in the manufacture of Portland cement (q.v.) and lime for agriculture (see *Soil Mineralogy*) and other purposes, a flux in the reduction of iron and other metallic ores, as limestone or marble in dimension stone and for statuary, and as aggregate for concrete.

K. FRYE

References*

Dana, II, 142–161.
DHZ, 5, 229–255.
*See *Calcite Group* for additional references.

Cross-references: *Authigenic Minerals; Blowpipe Analysis; Calcite Group; Cave Minerals; Clinopyroxenes; Collecting Minerals; Crystal Growth; Crystal Habits; Crystallography: History; Crystallography: Morphological; Diagenetic Minerals; Electron Microscopy (Transmission); Etch Pits; Fire Clays; Green River Mineralogy; Human and Vertebrate Mineralogy; Invertebrate and Plant Mineralogy; Magnetic Minerals; Meteoritic Minerals; Mineral Classification: Principles; Minerals, Uniaxial and Biaxial; Mohs Scale of Hardness; Nitrates; Optical Mineralogy; Pegmatite Minerals; Phantom Crystals; Plastic Flow in Minerals; Polarization and Polarizing Microscope; Pseudomorphism; Rock-Forming Minerals; Soil Mineralogy; Staining Techniques; Thermoluminescence; Thermometry, Geologic Vein Minerals.*

CALCITE GROUP–*Calcite, gaspeite, magnesite, otavite, rhodochrosite, siderite, smithsonite,* and *sphaerocobaltite.*

CALCIUM-CATAPLEIITE *Am. Min.* **49,** 1153
$CaZrSi_3O_9 \cdot 2H_2O$ hexa.

CALCJARLITE mono. *Am. Min.* **59,** 873
$Na(Ca,Sr)_3Al_3(F,OH)_{16}$

CALCLACITE *Dana,* II, 1107
$CaCl_2 \cdot Ca(C_2H_3O_2)_2 \cdot 10H_2O$
mono. or tric.

CALCSPAR = *Calcite*

CALCURMOLITE *Am. Min.* **49,** 1152
$Ca(UO_2)_3(MoO_4)_3(OH)_2 \cdot 11H_2O$

CALDERITE (***Garnet*** gr.) *Can. Min.* **17,** 569
$Mn_3Fe_2^{3+}Si_3O_{12}$ iso.

CALEDONITE orth. *Dana,* II, 630
$Pb_5Cu_2(CO_3)(SO_4)_3(OH)_6$

CALIFORNITE–Variety of *vesuvianite*

CALKINSITE orth. *Am. Min.* **38,** 1169
$(Ce,La)_2(CO_3)_3 \cdot 4H_2O$

CALLAGHANITE mono. *Am. Min.* **39,** 630
$Cu_2Mg_2(CO_3)(OH)_6 \cdot 2H_2O$

CALOMEL A mercury chloride. The name is from ancient alchemy.

Composition: $2[Hg_2Cl_2]$. Tetragonal.

Occurrence: A secondary mineral, forms by the alteration of *cinnabar,* mercurian *silver, amalgam,* mercurian *tetrahedrite,* selenian *metacinnabar,* and other mercury-containing minerals; associated with native *mercury, eglestonite, terlinguaite,* and *montroydite.*

J. A. GILBERT

References

Dana, II, 25.
Min. Abst. **20,** 510.

CALUMETITE *Am. Min.* **48,** 614
$Cu(OH,Cl)_2 \cdot 2H_2O$ orth.

CALZIRTITE *Am. Min.* **52,** 1880
$CaZr_3TiO_9$ tetr.

CANASITE mono. *Am. Min.* **45,** 253
$(Na,K)_6Ca_5Si_{12}O_{30}(OH,F)_4$

CANAVESITE mono. *Am. Min.* **64,** 652
$Mg_2(CO_3)(HBO_3) \cdot 5H_2O$ *Can. Min.* **16,** 69

CANCRINITE An aluminosilicate and carbonate-sulfate of sodium and calcium named by Rose in 1839 after Count Cancrin, Russian Minister of Finance. Also a group name.

Composition: $(Na,Ca)_{6\text{-}8}\,[(Al_6Si_6O_{24})(Co_3,SO_4,Cl)]_{1.5\text{-}2.0} \cdot 1\text{–}5\ H_2O$. When $SO_4 > CO_3$, the mineral is *vishnevite.* Hexagonal.

Occurrence: West of Irkutsk in Siberia and in the Ilmen Mts., USSR; in Transylvania of Rumania; in the Langesund fjord district of southern Norway. Also at Litchfield, Maine; in the syenites of Dungannon, Ontario. Sometimes as a primary mineral in silica-poor igneous rocks, but usually as a secondary mineral through the alteration of *nepheline* by CO_2-rich residual fluids. Associated with the ***feldspathoids,*** including *sodalite* and *nepheline,* and *calcite.*

G. C. MOERSCHELL

References

Can. Min. **8,** 134.
DHZ, 4, 310–320.

Cross-reference: *Rock-Forming Minerals.*

CANCRINITE GROUP—*Afghanite, cancrinite, dayvne, franzinite, liottite, microsommite, vishnevite,* and *wenkite.*

CANFIELDITE Dana, I, 356
Ag_8SnS_6 orth. *Min. Abst.* **28**, 204

CANNIZZARITE *Am. Min.* **38**, 536; **64**, 244
$Pb_4Bi_6S_{13}$ mono.

CAPPELENITE hexa. RRW, 106
$(Ba,Ca,Ce,Na)_3(Y,Ce,La)_6(BO_3)_6Si_3O_9$

CARACOLITE *Dana,* II, 546
$Na_3Pb_2(SO_4)_3Cl$ mono.

CARBOBORITE mono. *Am. Min.* **50**, 262
$Ca_2Mg(CO_3)B_2O_3(OH)_4 \cdot 8H_2O$

CARBOCERNAITE *Am. Min.* **46**, 1202
$(Ca,Ce,Na,Sr)CO_3$ orth.

CARBONADO = Black *diamond*

Cross-references: *Diamond, Jade; Mohs' Scale of Hardness.*

CARBONATE ***APATITE,*** see ***Apatite*** group.

CARBONATE-CYANOTRICHITE
$Cu_4Al_2(CO_3,SO_4)(OH)_{12} \cdot 2H_2O$
orth. *Am. Min.* **49**, 441

CARBONATE *FLUORAPATITE = Francolite*

CARBONATE *HYDROXYAPATITE = Dahllite*

CARBORUNDUM = Synthetic *moissanite*

CARLETONITE tetr. *Am. Min.* **56**, 1855
$KNa_4Ca_4Si_8O_{18}(CO_3)_4(OH,F) \cdot H_2O$

CARLFRIESITE *Am. Min.* **63**, 847
$CaTe_3O_8$ mono. *Min. Mag.* **40**,127

CARLHINTZEITE *Can. Min.* **17**, 103
$Ca_2AlF_7 \cdot H_2O$ tric. *Min. Rec.* **10**, 301

CARLINITE *Am. Min.* **60**, 559
Tl_2S trig. *Min. Abst.* **27**, 81

CARLSBERGITE *Am. Min.* **57**, 1311
CrN iso. *Nat. Phys. Sci.* **233**, 113

Cross-reference: *Meteoritic Minerals.*

CARMINITE *Am. Min.* **48**, 1; **55**, 135
$PbFe^{3+}_2(OH)_2(AsO_4)_2$ orth.

CARNALLITE A hydrated chloride of potassium and magnesium, named by Rose in 1856 after Rudolph von Carnall (1804–1874), Prussian mining engineer.

Composition: $12[KMgCl_3 \cdot 6H_2O]$; KCl, 28.81%; $MgCl_2$, 34.19%; H_2O, 39.0%.

Properties: Orthorhombic, *Pnna;* usually massive or granular; H=1; G=1.6; shining nonmetallic greasy luster; transparent to translucent; bitter taste; deliquescent; color: milk white but may have reddish tones due to included *hematite.*

Occurrence: Found associated with *halite, sylvite,* and other halides in the salt deposits at Stassfurt, East Germany. Also in potash deposits of western Texas and eastern New Mexico.

Uses: Source of potassium compounds and magnesium.

J. A. GILBERT

References

Dana, II, 920.
Min. Mag. **29**, 687.

Cross-references: *Mineral and Ore Deposits; Saline Minerals.*

CARNELIAN—Red or reddish brown variety of *quartz.*

Cross-reference: *Gemology.*

CARNOTITE A hydrated vanadate of potassium and uranium, named by Friedel and Cumenge in 1899 after Marie-Adolphe Carnot (1839–1920), French mining engineer and chemist.

Composition: $2[K_2(UO_2)_2(VO_4)_2 \cdot 3H_2O]$; the water is zeolitic.

Properties: Monoclinic, $P2_1/a$; usually disseminated or as a powder; luster: dull to earthy; color: yellow to greenish yellow.

Occurrence: A secondary mineral formed from the weathering of uranium and vanadium minerals. Characteristically found around petrified trees.

Uses: Ore of vanadium and uranium.

J. A. GILBERT

References

Am. Min. **39**, 323; **43**, 799; **50**, 825.
Dana, II, 1043.
Min. Abst. **11**, 92.
USGS Bull. **1064**, 243–247.

CAROBBIITE *Am. Min.* **42**, 117
KF iso.

CARPHOLITE *Min. Abst.* **27**, 20; **28**, 412
$MnAl_2Si_2O_6(OH)_4$ orth.

CARRBOYDITE hexa. *Am. Min.* **61**, 366
$Ni_7Al_{4.5}(SO_4,CO_3)_{2.8}(OH)_{22} \cdot 3.7H_2O$
Min. Abst. **27**, 179

CARROLLITE (***Linnaeite*** gr.)
$Cu(Co,Ni)_2S_4$ iso. *Min. Abst.* **27**, 253
Min. Mag. **43**, 733

Cross-reference: *Ore Microscopy.*

CARYINITE mono. *Dana,* II, 683
$(Na,Ca,Pb)_2(Mn,Mg,Fe)_3(AsO_4)_3$(?)

CARYOCERITE *Min. Abst.* **29**, 471
$(Ca,RE)_8(B,Si)_6(F,OH)_{24}$ hexa.
$\cdot nH_2O$

CARYOPILITE (***Friedelite*** gr.)
$(Mn,Mg)_8Si_6O_{15}(OH)_{10}$ *Am. Min.* **49**, 446;
mono. **65**, 335

CASSIDYITE *Am. Min.* **52**, 1190
$Ca_2(Ni,Mg)(PO_4)_2 \cdot 2H_2O$ tric.

CASSITERITE (Tin stone) Tin dioxide named by Beudant in 1832 from the Greek, *kassiteros,* tin. Member of the ***rutile*** group.

Composition: SnO_2, always contains impurities that can be manganese, ferrous and (or) ferric iron, probably tantalum and niobium; up to 25% *tapiolite* $Fe(Ta,Nb)_2O_6$ in solid solution.

Properties: Tetragonal, $P4_2/mnm$; H=6–7; G=6.98–7.02; adamantine luster; color is reddish brown to black, can be colorless in thin section; cleavage, {100} and {110} poor, {111} parting; rutile-type structure; optically, ω=1.990–2.010; ϵ=2.093–2.100; pleochroism very weak to strong, yellow, brown, or red.

Occurrence: Veins closely associated with granite and granite pegmatites. The Malayan peninsula deposits are alluvial in origin. Commonly associated with *quartz, topaz, fluorite,* ***apatite, tourmaline,*** and *baddeleyite.*

Use: Most important source of tin.

D. H. SPEIDEL

References

Dana, I, 574.

DHZ, 5, 5–10.

New Zealand J. Geol. Geophys. 8(3), 440–452.

Mulligan, J., and Gragy, W., 1965. Tin-lode investigations, Potato Mt., Seward Peninsula, Alaska, *U.S. Bur. Mines Rept. Inv. 6587,* 85p.

Cross-references: *Blacksand Minerals; Blowpipe Analysis; Crystallography: Morphological; Gossan; Metallic Minerals; Mineral and Ore Deposits; Minerals, Uniaxial and Biaxial; Ore Microscopy; Pegmatite Minerals; Rock-Forming Minerals; Staining Techniques; Wolframite Group.*

CATAPLEIITE A hydrous zirconium silicate with sodium and calcium named by Weibye and Sjögren in 1850 from the Greek meaning "rare minerals" because of its association with a number of rare minerals.

Composition: $2[Na_2ZrSi_3O_9 \cdot 2H_2O]$; forms a series with *calcium-catapleiite.* Hexagonal.

Occurrence: From various islands in the Langesundfjord district of southern Norway. *Catapleiite* is found only on the island Lille-Arö. Also reported from the Julianehaab district of southern Greenland and from Magnet Cove, Arkansas. In Norway, it is associated with *zircon, leucophanite, mosandrite,* and *tritomite;* and on Arö island, it is associated with ***feldspar,*** *sodalite, aegirine, eudiadyte,* and *astrophyllite.*

G. C. MOERSCHELL

References

Can. Min. 8, 120.

Min. Abst. 6, 180; 7, 94, 239, 398; 9, 45; 27, 49.

Cross-references: *Green River Mineralogy; Rock-Forming Minerals.*

CATOPHORITE = *Katophorite*
(***Amphibole*** gr.) *Am. Min.* **63**, 1050

CATTIERITE (***Pyrite*** gr.) *Am. Min.* **30**, 483
CoS_2 iso.

CAVANSITE orth. *Am. Min.* **58**, 405
$Ca(VO_4)Si_4O_{10} \cdot 4H_2O$ *Min. Abst.* **28**, 139

Cross-reference: *Naming of Minerals.*

CAYSICHITE *Am. Min.* **61**, 174
$(Y,Ca,RE)_4(Si,Al)_4O_{10}(CO_3)_3 \cdot 4H_2O$
orth. *Can. Min.* **12**, 293

CEBOLLITE *Min. Mag.* **17**, 346
$Ca_4Al_2Si_3O_{14}(OH)_2$(?) orth.

CELADONITE (***Mica*** gr.) *Am. Min.* **49**, 1031
$K(Mg,Fe)(Fe,Al)Si_4O_{10}(OH)_2$ mono.

Cross-reference: *Phyllosilicates.*

CELESTITE (CELESTINE) Strontium sulfate, named by Werner in 1799 from the Latin meaning "celestial," in reference to the faint blue color of the first specimens described. A complete solid solution exists between *barite* and *celestite.*

Composition: 4[$SrSO_4$]; SrO, 56.4%; SO_3 43.6%.

Properties: Orthorhombic, *Pnma;* may be fibrous in radiating groups or granular: H=3–3.5; G=3.95–3.97; luster is vitreous to pearly; transparent to translucent; color: white or colorless with tints of blue or red.

Occurrence: Found disseminated through limestone and sandstone or as cavity linings in such rocks, where it is associated with *calcite, dolomite, gypsum, halite, sulfur,* and *fluorite;* a common authigenic mineral.

Uses: For the preparation of strontium nitrate for fireworks and tracer bullets and in the preparations of other strontium salts used in the refining of beet sugar.

J. A. GILBERT

References

Am. Min. **45,** 1257.
Can. Mining J., **81,** 68.
Dana, II, 415.
Rocks Minerals, **25,** 348.

Cross-references: *Authigenic Minerals; Cave Minerals; Green River Mineralogy; Invertebrate and Plant Mineralogy; Mineral and Ore Deposits; Minerals, Uniaxial and Biaxial; Refractory Minerals; Rock-Forming Minerals.*

CELSIAN A barium aluminosilicate of the ***feldspar*** group, named by Sjögren in 1895 after A. Celsius, Swedish naturalist.

Composition: $BaAl_2Si_2O_8$; with Ba replaced by K, Na, and Ca. Forms a series with K-***feldspar*** and is dimorphous with *paracelsian.*

Properties: Monoclinic, $I2_1/c$; twinned stout or acicular crystals, commonly cleavable masses; H=6–6.5; G=3.1–3.4; colorless, white, or yellow.

Occurrence: Commonly assiciated with manganese minerals; localities include Jakobsberg, Sweden; Tatiki Prefecture, Japan; Broken Hill, New South Wales, Australia; Rhiw, North Wales, UK; Otjosondu, Namibia; and in the Alaska Range, USA.

K. FRYE

References*

DHZ, 4, 166–178.
*See *Alkali Feldspars* and *Feldspar Group* for additional references.

Cross-reference: *Minerals, Uniaxial and Biaxial.*

CERARGYRITE = *Chlorargyrite*
Am. Min. **49,** 224

CERIANITE *Am. Min.* **40,** 560
$(Ce,Th)O_2$ iso.

CERIOPYROCHLORE (***Pyrochlore*** gr.)
$(Ce,Ca,Na)_2(Nb,Ta,Ti)_2O_6(O,OH,F)$
iso. *Am. Min.* **62,** 405

CERITE A hydroxyl silicate of cerium with calcium, named by Hisinger and Berzelius in 1804 because it contains cerium, which was named by them after the asteroid, Ceres.

Composition:
$(Ce,Ca)_9(Mg,Fe)Si_7(O,OH,F)_{28}$(?). Trigonal.

Occurrence: Occurs at Bastnäs mine near Riddarhyttan, Västmanland, Sweden.

G. C. MOERSCHELL

References

Am. Min. **43,** 460.
Min. Abst. **8,** 339.

CERNYITE *Can. Min.* **16,** 139
$Cu_2(Cd,Zn)SnS_4$ tetr.

CEROLITE–Mixture of a ***serpentine*** and *stevensite* *Am. Min.* **50,** 2111

CEROTUNGSTITE *Am. Min.* **57,** 1558
$CeW_2O_6(OH)_3$ mono.

CERULEITE *Am. Min.* **62,** 598
$Cu_2Al_7(OH)_{13}(AsO_4)_4 \cdot 11½H_2O$
tric. *N.J. Min. Mon.* (1976), 418

CERUSSITE Lead carbonate of the ***aragonite*** group, named by Haidinger in 1845, from the Latin meaning white lead.

Composition: 4[$PbCO_3$]; PbO, 83.5%; CO_3 16.5%; but may contain small amounts of silver.

Properties: Orthorhombic, *Pmcn;* often twinned, also in fibrous, granular, compact, and earthy masses; H=3–3.5; G=6.55; adamantine luster; transparent to translucent; color: white, gray, or colorless.

Occurrence: Formed as a supergene ore by the action of carbonated waters on *galena* in the upper zone of lead veins, associated with *galena, sphalerite, anglesite, pyromor-*

phite, smithsonite, and limonite; rarely as an authigenic mineral (q.v.).

Use: An important ore of lead.

J. A. GILBERT

References

Am. Min. **44,** 1314; **47,** 1011; **49,** 1184.
Dana, II, 200.

Cross-references: *Authigenic Minerals; Blowpipe Analysis; Gossan; Metallic Minerals; Minerals, Uniaxial and Biaxial; Staining Techniques.*

CERVANTITE An anhydrous oxide of antimony, named by Dana in 1850 after Cervantes, Spain.

Composition: $Sb^{3+}Sb^{5+}O_4$.

Properties: Orthorhombic; *Pna;* G=6.6; color: yellow to white.

Occurrence: An oxidation product of antimony ores, *stibnite,* etc. Also as pseudomorphs after *stibnite* at Felsöbánya, Hungary, and Pocca, Bolivia.

D. H. SPEIDEL

References

Am. Min. **37,** 982; **39,** 406; **47,** 1221.
Dana, I, 595.
Min. Mag. **30,** 100.

CESAROLITE A hydrated lead manganese oxide.

Composition: $PbMn_3O_7 \cdot H_2O$ or $PbMn_6O_{12} \cdot 2H_2O$; *woodruffite* substitutes Zn for Pb and *todorokite* substitutes Ca for Pb.

Properties: H=5; steel-gray cellular masses.

Occurrence: Found in Tunisia.

D. H. SPEIDEL

Reference: *Dana,* 7, I, 744.

CESBRONITE orth. *Am. Min.* **64,** 653
$Cu_5(TeO_3)_2(OH)_6 \cdot 2H_2O$ *Min. Mag.* **39,** 744

CESIUM-KUPLETSKITE (***Astrophyllite*** gr.)
$(Cs,K,Na)_3(Mn,Fe)_7(Ti,Nb)_2Si_8O_{24}(O,OH,F)_7$
tric. *Am. Min.* **57,** 328

CEYLONITE—Variety of *spinel*

CHABAZITE A hydrated calcium aluminosilicate of the ***zeolite*** group. First named by Bosc d'Antic in 1788, from the Greek "chabazios," an ancient name of a stone.

Composition: $2[Ca(Al_2Si_4O_{12}) \cdot 6H_2O]$; SiO_2, 47.5%; Al_2O_3, 20.2%; CaO, 10.9%; H_2O, 21.4% by wt.

Properties: Trigonal $R\bar{3}m$; a=13.8 Å, c=15.0 Å; H=4.5; G=2.05–2.10; vitreous; color: colorless, white, yellowish, greenish, or reddish white; transparent to translucent; habit rhombohedral (pseudocubic); cleavage poor on $\{10\bar{1}1\}$; twinning on $\{0001\}$ (interpenetrant) and $\{10\bar{1}1\}$. Optically, refractive indices, ϵ=1.470 in minimum, ω=1.494 at maximum; birefringence=0.002–0.005; uniaxial (−). Chemically, considerable replacement of the types (Na,K)Si$\gtrless$CaAl and Ca$\gtrless$$Na_2$$\gtrless$$K_2$; strontium may replace calcium to a small extent; dehydration is gradual, losing 22% H_2O at maximum; rehydration is good; minimum diameter of the largest channel in the aluminosilicate framework is 3.9 Å; molecular sieve properties are very selective; the cations in *chabazite* can be exchanged for Li, Na, K, Rb, Cs, NH_4, Ag, Ca, Sr, Ba, and Pb; gelatinized by HCl.

Occurrence: In cavities in basalt and andesite associated with *heulandite, stilbite, phillipsite,* and other ***zeolites;*** in cavities in the Tertiary basalts of Iceland as the upper *chabazite-thomsonite* zone; in hydrothermal veins in gneiss; alteration product of silicic vitric ash in alkaline saline lake deposits; in the cement of palagonite tuffs associated with *phillipsite* and other ***zeolites*** which were formed by reaction of basaltic glass with percolating water; from the wall-work of Roman baths at Plombières emerging hot springs (maximum 70°C) together with *phillipsite;* likely as a primary mineral in the groundmass of alkali basalt in southeastern California.

A. IIJIMA
M. UTADA

References

Am. Min. **51,** 909–915.
Min. Mag. **29,** 773–791.
Nature **181,** 1794.
Smith, J. V., 1963. Structural classification zeolites, *Mineral Soc. Am. Spec. Paper 1,* 281–290.

Cross-references- *Authigenic Minerals; Zeolites.*

CHABOURNÉITE *Am. Min.* **64,** 242
$(Tl,Pb)_5(Sb,As)_{21}S_{34}$ tric. *Min. Mag.* **43,** 1059

CHALCANTHITE Copper sulfate pentahydrate (blue vitriol) named by Kobell in 1853 from two Greek words meaning brass and flower.

Composition: $2[CuSO_4 \cdot 5H_2O]$; CuO, 31.8%; SO_3, 32.1%; H_2O, 36.1%.

Properties: Triclinic, $P\bar{1}$; may be massive, reniform, stalactitic, or fibrous; H=2.5; G=2.12–2.30; vitreous luster; transparent to translucent; color: deep azure blue; soluble in water, metallic taste.

Occurrence: Found only in arid regions, it is a supergene mineral derived from the oxidation of copper sulfides. Found near the surface of copper veins or as a deposit from the waters in copper mines. The only commercial deposit is at Chuquicamata, Chile.

Use: A minor ore of copper.

J. A. GILBERT

Reference: *Dana,* II, 488.

CHALCEDONY A fine grained or cryptocrystalline variety of low-*quartz,* or *α-quartz.* Named by Pliny in 77 AD from the ancient town of Chalcedon on the Sea of Marmora.

Composition: SiO_2, 90–99%; the remainder primarily metallic oxides and water.

Properties: Microfibrous; H=6.5; G=2.57–2.64; waxy to subvitreous luster; color: pale blue, gray, yellow, green, or brown; secondary filling seams, cracks, and cavities in rock; fracture: subconchoidal, hackly, or splintery; no distinct cleavage; streak: white; fibrous, fibers elongated on *c* axis; refractive index: ω=1.526–1.535, ϵ=1.539–1.543; translucent to milky; soluble in HF and alkaline solutions; has been synthesized from obsidian and silica by heating in alkaline solutions at elevated pressures for short periods.

Occurrence: Widespread; secondary deposition often rounded, botryoidal, or irregular masses, crack and cavity fillings, sinter-like, in geodes, amygdule fillings in mafic igneous rocks, replacement of fossil shells. Well-known localities include Oregon and Pacific beaches, Montana, Wyoming, the Deccan traps of India. Often associated with felsic and mafic lavas, carbonates, ***zeolites*** and *quartz.*

Uses: In jewelry as semiprecious gemstone; for carvings; objects of art. Banded varieties are agate; varieties named by color are prase, chrysoprase, carnelian, and sard; by inclusions, moss agate and plume agate; also beekite.

W. E. HILL

References

AJS **254,** 32.
Am. Min. **46,** 112; **63,** 17.
Dana, III, 195.
Dake, H. C.; Fleener, F. L.; and Wilson, B. H., 1938. *The Quartz Family Minerals,* New York: Whittlesey House.
Lapidary Journal, **15,** 242, 250, 296.

Cross-references: *Authigenic Minerals; Diagenetic Minerals; Gemology; Green River Mineralogy; Soil Mineralogy.*

CHALCOALUMITE *Min. Rec.* **2,** 126
$CuAl_4(SO_4)(OH)_{12} \cdot 3H_2O$ mono.

CHALCOCITE A cuprous sulfide, named *chalcosine* by Beudant in 1832 and modified by Dana in 1868.

Composition: $96[Cu_2S]$; S, 20.2%; Cu, 79.8%; by wt.

Properties: Monoclinic; often twinned, pseudohexagonal, also massive; conchoidal fracture; H=2.5–3; G=5.5–5.8; sectile; color: gray-black; blue to green tarnish; metallic luster.

Occurrence: A secondary vein mineral, associated with copper and iron sulfides and their carbonates. Notable amounts at Butte, Montana; Kennecott, Alaska; also in Connecticut, New Jersey, Arizona, and Quebec.

Uses: An important ore of copper.

R. W. FAIRBRIDGE

References

Carnegie Inst. Wash. Yr. Bk., **61,** 157.
Dana, I, 187.
Min. Abst. 9, 97, 224; **11,** 85; **27,** 335, 28, 259.

Cross-references: *Mineral and Ore Deposits; Ore Microscopy.*

CHALCOCYANITE *Am. Min.* **46,** 758
$CuSO_4$ orth.

CHALCOMENITE *Am. Min.* **49,** 1481
$CuSeO_3 \cdot 2H_2O$ orth.

CHALCONATRONITE *Am. Min.* **40,** 943
$Na_2Cu(CO_3)_2 \cdot 3H_2O$ mono. *Science* **122,** 75

CHALCOPHANITE A hydrated zinc manganese oxide.

Composition: $(Zn,Mn,Fe)Mn_2O_5 \cdot 2H_2O$.

Properties: Triclinic, pseudotrigonal; perfect basal cleavage; bluish to black.

Occurrence: An alteration product at Franklin, New Jersey, associated with *hetaerolite* and *hydrohetaerolite;* on Groote Eylandt, NT, Australia, as an alteration of ***clay.***

D. H. SPEIDEL

References

Dana, I, 739–740.

Min. Mag. **44,** 109–111.

Cross-reference: *Ore Microscopy.*

CHALCOPHYLLITE trig. *Dana,* II, 1008
$Cu_{18}Al_2(AsO_4)_3(SO_4)_3(OH)_{27}\cdot 33H_2O$

CHALCOPYRITE Copper iron sulfide, first named by Henckel in 1725 from the Greek, meaning "brass" and "pyrites."

Composition: $4[CuFeS_2]$; 34.5% Cu; 30.5% Fe; and 35.0% S by weight. Cu/Fe varies from 0.5 to 1.5, being 1.0 for the low-temperature, tetragonal polymorph. Se replaces S toward *eskebornite.*

Properties: Tetragonal, $I\bar{4}2d$, but isometric above 700°C; usually massive; H=3.5–4.0; G=4.1–4.3; brass-yellow color; greenish black streak; poor cleavage on {011} and {111}; metallic luster; lamellar and interpenetrating twinning; brittle. *Chalcopyrite* is frequently intergrown with other sulfides of copper, and often contains trace amounts of many other elements. It oxidizes on heating or on exposure to air and water to sulfates of Cu and Fe. Opaque; weakly anisotropic; Victer's Hardness medium (190); strongly colored golden yellow, sometimes shows slight greenish tint; reflectivity = 41–43%; twinning lamellae difficult to see.

Occurrence: *Chalcopyrite* is the most widespread copper sulfide, occurring with other sulfides of late magmatic origin, either disseminated or in metalliferous veins, especially of the hydrothermal type. It is occasionally formed by metamorphic processes. It also may be found in sedimentary facies, and, in some of these cases, is believed to have formed syngenetically (see *Authigenic Minerals*), perhaps having been leached from the surrounding rock and then precipitated out. It is fairly common in secondary enrichment zones of ore deposits. Significant source areas include, among many, Freiberg, Saxony, E. Germany, and Sudbury, Ontario, Canada; it is associated with *pyrite, pyrrhotite, sphalerite, galena, quartz, calcite, siderite, dolomite,* etc.

Use: As a source of copper, it is one of the most important ore minerals.

R. W. CAMERON

References

Am. Min. **50,** 301–314.

Econ. Geol. **61,** 97–136, 897–903.

Habashi, F., 1978. *Chalcopyrite–Its Chemistry and Metallurgy.* New York: McGraw-Hill, 165p.

Hurlbut, C. S., and Klein, C., 1977. *Manual of Mineralogy (after James D. Dana).* New York: Wiley, 243–244.

Jensen, M. L., and Bateman, A. M., 1979. *Economic Mineral Deposits,* 3rd ed. New York: Wiley, 312–356.

Cross-references: *Blowpipe Analysis; Lunar Minerals; Magnetic Minerals; Metallic Minerals; Meteoritic Minerals; Mineral and Ore Deposits; Ore Microscopy; Rock-Forming Minerals; Skeletal Crystals.*

CHALCOSIDERITE A hydrated basic phosphate of copper and ferric iron, in the ***turquoise*** group, named by Ullmann in 1824.

Composition: $CuFe_6(PO_4)_4(OH)_8\cdot 4H_2O$. It is the iron-rich end-member of the iron-aluminum substitution in the ***turquoise*** series.

Occurrence: As a secondary mineral found in a gossan associated with *dufrenite, andrewsite,* and *goethite* at the West Phoenix mine, Cornwall, England.

J. A. GILBERT

References

Am. Min. **50,** 227–231.

Dana, II, 947.

Kostov, I., 1968. *Mineralogy.* London: Oliver and Boyd, 473.

CHALCOSINE = CHALCOCITE
Min. Mag. **43,** 1053

CHALCOSTIBITE *Can. Min.* **17,** 601
$CuSbS_2$ orth. *Min. Abst.* **28,** 45

Cross-reference: *Ore Microscopy.*

CHALCOTHALLITE *Am. Min.* **53,** 1775; **64,** 658
$Cu_6Tl_2SbS_4$ tetr.

CHALCOTRICHITE – Variety of *cuprite*

CHALLANTITE *Am. Min.* **60,** 736
$6Fe_2(SO_4)_3\cdot Fe_2O_3\cdot 63H_2O$
Min. Abst. **27,** 257

CHAMBERSITE *Am. Min.* **47,** 665
$Mn_3B_7O_{13}Cl$ orth. *Min. Abst.* **28,** 287

CHAMOSITE A hydroxyl aluminosilicate of iron and magnesium, named by Berthier in 1820 for Chamoson, near St. Maurice, Valais, Switzerland. Belongs to the ***chlorite*** group. There are two distinct polymorphs, monoclinic and orthorhombic, the latter called *orthochamosite.*

Composition: $2[(Fe^{2+},Mg,Fe^{3+})_5Al(Si_3Al)O_{10}(OH,O)_8]$

Occurrence: In Switzerland at various localities in limestone; rarely as a diagenetic mineral (q.v.).

Use: An important constituent of European oolitic iron ores.

G. C. MOERSCHELL

References

Am. Min. **49,** 993.
DHZ, 3, 164.
J. Sed. Petrol. **30,** 585.
Min. Mag. **30,** 277.

Cross-references: *Chlorite Group; Diagenetic Minerals; Meteoritic Minerals; Minerals, Uniaxial and Biaxial.*

CHANGBAIITE *Am. Min.* **64,** 242
$PbNb_2O_6$ trig. *Min. Abst.* **29,** 481

CHANTALITE *Am. Min.* **63,** 1282
$CaAl_2SiO_4(OH)_4$ tetr. *Min. Abst.* **29,** 341

CHAOITE *Am. Min.* **54,** 326
C hexa. *Science* **161,** 362

Cross-references: *Allotropy; Native Elements and Alloys.*

CHAPMANITE *Am. Min.* **49,** 1499
$SbFe_2(SiO_4)_2(OH)$ mono. *Min. Abst.* **28,** 197

CHAROITE mono.(?) *Am. Min.* **63,** 1282
$(K,Ba,Sr)(Ca,Na)_2Si_4O_{10}(OH,F)\cdot H_2O$
Min. Abst. **29,** 481

CHELKARITE orth. *Am. Min.* **56,** 1122
$CaMgB_2O_4Cl_2\cdot 7H_2O$(?)
Min. Abst. **27,** 80

CHENEVIXITE mono. *Am. Min.* **62,** 1058
$Cu_2Fe_2(AsO_4)_2(OH)_4$ *Min. Mag.* **41,** 27

CHERALITE (***Monazite*** gr.) *Am. Min.* **52,** 13
$(Ca,Ce,Th)(P,Si)O_4$ mono. *Min. Mag.* **43,** 885

CHERNOVITE *Am. Min.* **53,** 1777
$YAsO_4$ tetr. *Min. Abst.* **28,** 208

CHERNYKHITE (***Mica*** gr.)
$(Ba,Na)(V,Al)_2(Si,Al)_4O_{10}(OH)_2$
mono. *Am. Min.* **58,** 966

CHERT—Variety of chalcedony

Cross-references: *Electron Microscopy—Transmission; Soil Mineralogy.*

CHERVETITE *Am. Min.* **48,** 1416
$Pb_2V_2O_7$ mono.

CHESSYLITE = *Azurite* *Min. Mag.* **43,** 1053

CHESTERITE orth. *Am. Min.* **63,** 1000
$(Mg,Fe)_{17}Si_{20}O_{54}(OH)_6$ *Min. Abst.* **29,** 341

Cross-reference: *Biopyribole.*

CHEVKINITE mono. *Am. Min.* **53,** 1558; **63,** 499
$(Ce,La)_2Ti_2O_4(Si_2O_7)$

Cross-reference: *Pegmatite Minerals.*

CHIASTOLITE—Variety of *andalusite*

CHILDRENITE A hydrated iron aluminum phosphate named in 1823 after the English mineralogist J. G. Children.

Composition: $8[Fe^{2+}Al(PO_4)(OH)_2\cdot H_2O]$; with Fe^{2+} replaced by Mn^{2+} ($Mn>Fe$, *eosphorite).*

Properties: Orthorhombic, *Bba*2; crystals equant, pyramidal, prismatic, or tabular; H=5; G=3.2; vitreous to resinous luster; yellow brown to brown.

Occurrence: A pegmatite mineral (q.v.) or vein mineral (q.v.).

K. FRYE

References

Am. Min. **35,** 793–805.
Dana, II, 936–939.
Min. Mag. **43,** 1015.

CHILE-LOEWEITE = *HUMBERSTONITE*
Am. Min. **55,** 1518

CHILE SALTPETER = *Nitratine*
Min. Mag. **43,** 1053

CHIOLITE *Dana,* I, 123
$Na_5Al_3F_{14}$ tetr.

Cross-reference: *Pegmatite Minerals.*

CHKALOVITE *Am. Min.* **25,** 380
$Na_2BeSi_2O_6$ orth. *Min. Abst.* **28,** 140

CHLORALUMINITE *Dana,* II, 50
$AlCl_3 \cdot 6H_2O$ trig.

CHLORAPATITE (***Apatite*** gr.)
$Ca_5(PO_4)_3Cl$ mono. *Can. Min.* **10,** 252

Cross-references: *Meteoritic Minerals; Soil Mineralogy.*

CHLORARGYRITE A silver chloride named for its composition by Weisbach in 1875. The chloride may be replaced by bromide ion (Br>Cl, *bromargyrite*).

Composition: 4[AgCl]; Ag, 73.3%; Cl, 24.7%.

Properties: Isometric, *Fm3m*; usually massive, resembling wax or in plates and crusts; H=2–3; G=5.5; sectile; transparent to translucent; color: pearl-gray to colorless but darkens to violet on exposure to light.

Occurrence: A supergene mineral found only in the enriched zone of silver veins. Associated with other silver minerals, native *silver, cerussite, argentite, proustite, pyrargyrite,* limonite, and *pyromorphite.* Large deposits are found at Broken Hill, New South Wales; Atacama, Chile; and Potosi, Bolivia.

Uses: An ore of silver.

J. A. GILBERT

Reference: *Dana,* II, 11.

Cross-reference: *Gossan.*

CHLORITE GROUP–*Brunsvigite, chamosite, clinochlore, cookeite, diabandite, donbassite, grovesite, manandonite, nimite, orthochamosite, penninite, ripidolite, sheridanite, sudoite,* and *thuringite.*

Cross-references: *Diagenetic Minerals; Green River Mineralogy; Orthopyroxenes; Phyllosilicates; Pleochroic Halos; Refractory Minerals; Rock-Forming Minerals; Topotaxy; Vein Minerals.*

CHLORITOID A hydroxyl aluminosilicate of ferrous iron and magnesium, named by Rose in 1837 from its resemblance to ***chlorite.***

Composition: $4[(Fe^{2+},Mg,)_2Al_4Si_2O_{10}(OH)_4]$. Monoclinic and triclinic.

Occurrence: The original *chloritoid* is from Kosoibrod in the Ural Mountains. Also from Izmir, Turkey; Tyrol, Austria; St. Marcel, Piedmont, Italy (sismondine); in Ardennes and Morbihan, France; from Vielsalm, Luxembourg, Belgium (salmite); from the Shetland Islands, Scotland; from Natick, Rhode Island (masonite); from Michigamme Lake, Michigan; and from Chibougamau, Quebec.

R. W. CAMERON

References

Am. Min. **65,** 534–539.
DHZ, 1, 161–170.
J. Petrology, **2,** 49.

Cross-reference: *Rock-Forming Minerals.*

CHLORMAN-ASSEITE *Min. Abst.* **31,** 226
$(Mg,Fe^{2+})_5Al_3(OH)_{16}(Cl,OH,\frac{1}{2}CO_3)_3$
hexa. *Min. Mag.* **43,** 1059

CHLORMANGANOKALITE *Dana,* II, 41
K_4MnCl_6 trig. *Min. Mag.* **14,** 397

CHLOROARSENIAN = *Allactite*
Am. Min. **58,** 562

CHLOROCALCITE *Dana,* II, 91
$KCaCl_3$

Cross-reference: *Saline Minerals.*

CHLOROMAGNESITE *Dana,* II, 41
$MgCl_2$ trig.

CHLOROPHOENICITE *Am. Min.* **53,** 1110
$Mn_3Zn_2(AsO_4)(OH)_7$ mono.

CHLOROTHIONITE *Dana,* II, 547
$K_2Cu(SO_4)Cl_2$ orth. *Z. Krist.* **144,** 226

CHLOROTILE hexa. RRW, 128
$(Cu,Fe)Cu_6(OH)_6(AsO_4)_3 \cdot 3H_2O$

CHLOROXIPHITE *Dana,* II, 84
$Pb_3CuCl_2O_2(OH)_2$ mono. *Min. Mag.* **43,** 901

CHOLOALITE *Min. Mag.* **44,** 55
$CuPb(TeO_3)_2 \cdot H_2O$ iso.

CHONDRODITE A hydroxyl magnesium silicate containing fluorine; of the ***humite*** group. Named by D'Ohsson in 1817 from the Greek word for a grain in reference to its occurrence as isolated grains.

Composition: $2[Mg_5(SiO_4)_2(F,OH)_2]$; may contain Fe^{2+}; F : OH usually 1 : 1.

Properties: Monoclinic; usually in grains but may be massive; H=6; G=3.1–3.2; resinous to vitreous luster; translucent; color: yellow to red.

Occurrence: Most commonly found in Precambrian dolomitic marble associated with *phlogopite, spinel, pyrrhotite,* and *graphite.*

J. A. GILBERT

References: *Am. Min.* **55**, 1182; **64**, 1027.

Cross-references: *Lunar Minerals; Minerals, Uniaxial and Biaxial.*

CHRISTITE *Am. Min.* **62**, 421
TlHgAsS$_3$ mono. *Z. Krist.* **144**, 367

CHROMATITE *Am. Min.* **49**, 439
CaCrO$_4$ tetr.

CHROMINIUM = *Phoenicochroite*
Min. Mag. **43**, 1054

CHROMITE Any mineral of the *chromite* series in the ***spinel*** group (q.v.) that has the composition (Mg,Fe^{2+})(Cr,Al,Fe^{3+})$_2$O$_4$. Also commonly described as FeCr$_2$O$_4$, which, strictly speaking, is an end member (*ferrochromite*) that is found only in meteorites. *Chromite* is so-named because Cr usually is the dominant element. Several ambiguous varietal names such as picotite and beresovite have been used; Stevens (1944) proposed a systematic nomenclature based on composition.

Composition: Unaltered *chromite* varies in composition about as follows (weight %): 15–65% Cr$_2$O$_3$, 9–50% Al$_2$O$_3$; 5–18% MgO, and 10–40% Fe as FeO; TiO$_2$, MnO, NiO, V$_2$O$_3$, and ZnO are the chief minor constituents. Most altered *chromite* is enriched in Fe and, to a lesser extent, in Cr (Evans and Frost, 1975). Exsolution textures are extremely rare, if present at all, in primary *chromite* but characterize some highly metamorphosed *chromites.*

Properties: Cubic, *Fd*3*m;* octahedral crystals, rarely dodecahedral; G-3.9–4.8; H=5.5–7.5; black, with dark- to pale-brown streak; in thin section is opaque or deep cherry-red to pale-brown; no cleavage; n=2.1± weakly to moderately magnetic.

Occurrence and paragenesis: *Chromite* occurs as a primary accessory mineral in amounts <5% in most ultramafic rocks, in some closely related anorthositic rocks, and in some basalts. In all deposits, *chromite* is segregated as a cumulus mineral cotectic with ***olivine*** in peridotite and with ***olivine, pyroxene*** and ***plagioclase*** in feldspathic and pyroxenitic parts of peridotite-gabbro complexes. In stratiform complexes the *chromite* characteristically is euhedral except in massive ore, <1 mm in grain size, and forms layers that are uniform in grade and thickness for tens of km. In the Bushveld complex in South Africa and in the Fiskenaesset-Sittampundi type of stratiform anorthosite complexes the main *chromite* layers are interstratified with pyroxenite, norite, leucogabbro, and anorthosite. In ophiolitic complexes *chromite* forms lenticular to tabular–podiform–deposits in dunite units and interlayered with dunite in harzburgite and wehrlite, rarely in troctolite; crystals are mostly subhedral to anhedral, average 1–5 mm in size, and range up to 5–6 cm in massive ore; nodular textures are common and orbicules are rare. *Chromite* in stratiform and podiform deposits differs sharply in two respects. In stratiform deposits Cr and Fe vary inversely, whereas in podiform deposits Cr and Al vary inversely. The nodular textures that characterize many podiform deposits are unknown in the stratiform type.

Podiform ore bodies range in size from a few kg to >1,000,000 tons. Because of their lateral extent and continuity, stratiform deposits constitute about 99% of world chromite resources. South Africa, Zimbabwe, and the USSR posses the principal world *chromite* resources; resources of the other major producing countries–Albania, the Philippines, Turkey, India, Finland, Brazil, Malagasy and Iran–are small in comparison.

Uses: *Chromite* as the only economic source of chromium for metallurgical purposes, chemicals, and pigments; also it is used directly as a high-temperature refractory.

T. P. THAYER

References

Evans, B. W., and Frost, B. R., 1975. *GCA* 39, 959–972.

Jackson, E. D., 1961. Primary textures and mineral associations in the Ultramafic zone of the Stillwater complex, Montana. *USGS Prof. Paper 358,* 106p.

Jackson, E. D., 1969. Chemical variation in coexisting chromite and olivine in chromitite zones of the Stillwater Complex. *Econ. Geol. Mon. 4,* 41–71.

Stevens, R. E., 1944. *Am. Min.* **29**, 1–34.

Thayer, T. P., 1969. Gravity differentration and magmatic reemplacement of podiform chromite deposits. *Econ. Geol. Mon. 4,* 132–146.

Thayer, T. P., 1970. Chromite segregations as petrogenetic indicators. *Geol. Soc. South Africa Spec. Pub. 1,* 380–390.

Cross-references: *Blacksand Minerals; Blowpipe Analysis; Diamond; Magnetic Minerals; Mantle*

Mineralogy; Metallic Minnerals; Meteoritic Minerals; Mineral Industries; Mineral and Ore Deposits; Ore Microscopy; Refractory Minerals.

CHRYSOBERYL A beryllium aluminum oxide named by Werner (1790) meaning "golden beryl."

Composition: $BeAl_2O_4$.

Properties: Orthorhombic, *Pmnb;* straited faces and commonly twinned; H=8.5; G=3.65–3.8; vitreous luster; color is shades of green, brown, yellow, or red (alexandrite is an emerald green variety that appears red by incandescent light; cat's eye, cymophane, has an opalescent luster, chatoyancy).

Occurrence: Rare in granites, granite pegmatites, and mica schists; also of gem quality in alluvium in Sri Lanka and Brazil.

Use: A gemstone.

D. H. SPEIDEL

References

Am. Ceram. Soc. J. **74,** 274, 491.
AJS, **249,** 308.
Am. Min. **43,** 427; **63,** 664.
Dana, I, 718.

Cross-references: *Gemology; Jade; Minerals, Uniaxial and Biaxial; Pegmatite Minerals.*

CHRYSOCOLLA A hydrated silicate of copper, named by Theophrastus ca. 315 BC from two Greek words meaning gold and glue, which was the name of a similarly appearing material used for soldering gold.

Composition: $(Cu,Al)_2H_2Si_2O_5(OH)_4 \cdot nH_2O$, variable.

Properties: Monoclinic, cryptocrystalline, or amorphous; usually compact massive; H=2–4; G=2.0–2.4; vitreous to earthy luster; conchoidal fracture; color: bluish green to green, may be black or brown when impure.

Occurrence: A secondary mineral found in the oxidized portions of copper veins, associated with *azurite, malachite, cuprite,* native *copper,* and other copper minerals.

Use: As a minor ore of copper and as an ornamental stone.

J. A. GILBERT

References

Am. Min. **54,** 993.
Min. Abst. 9, 231; **11,** 453.

Cross-reference: *Gossan.*

CHRYSOLITE–*See* **Olivine** group

Cross-reference: *Gemology.*

CHRYSOPRASE – Green variety of chalcedony

Cross-reference: *Gemology.*

CHRYSOTILE A magnesium hydroxyl silicate of the ***serpentine*** group, named from the Greek *chrusos,* golden, and *tilos,* fiber, in allusion to its color and asbestiform habit.

Composition: $2[Mg_3Si_2O_5(OH)_4]$

Properties: Monoclinic (*clinochrysotile*) or orthorhombic (*orthochrysotile* and *parachrysotile*); flexible fibers; H=2.5; G=2.55; color: yellow, white, green, or gray.

Occurrence: Snarnum, Norway; Cornwall, England; Thetford mines, Megantic County, Quebec, Canada; in the USA in Morris County, New Jersey; Gila County, Arizona; Putnam, St. Lawrence, and Orange Counties, New York; and near Easton, Pennsylvania.

Use: The chief asbestos mineral.

K. FRYE

References

DHZ, 3, 164–190.
Can. Min. **13,** 227–258.
Min. Abst. **28,** 50.

Cross-references: *Asbestos; Pigments and Fillers; Soil Mineralogy.*

CHUDOBAITE *Am. Min.* **45,** 1130; **62,** 599
$(Na,K,Ca)(Mg,Zn,Mn)_2H(AsO_4)_2 \cdot 4H_2O$(?)
tric.

CHUKHROVITE iso. *Am. Min.* **66,** 932
$Ca_3(Y,Ce)Al_2(SO_4)F_{13} \cdot 10H_2O$

CHUKHROVITE-(Ce) *Am. Min.* **65,** 1065
$Ca_3(Ce,Y)Al_2(SO_4)F_{13} \cdot 10H_2O$ iso.

CHURCHITE (WEINSCHENKITE)
$YPO_4 \cdot 2H_2O$ *Am. Min.* **29,** 92; **39,** 851

CINNABAR Mercury sulfide, first named by Theophrastus ca. 315 BC. The name is supposed to have come from India, where it was applied to a red resin.

Composition: 3[HgS]; with Hg vacancies coupled with interstitial S.

Properties: Trigonal, *P*321; usually fine granular, massive, also earthy as incrustations and disseminations through a rock;

H=2.5; G=8.10; adamantine luster when pure to dull earthy when impure; transparent to transluscent; red to brownish red color. Polymorphic with *metacinnabar* and *hypercinnabar.*

Occurrence: As impregnations and vein fillings near recent volcanic rocks and hot springs. Associated with *pyrite, marcasite, stibnite,* and sulfides of copper in a gangue of opal, chalcedony, *quartz, barite, calcite,* and *fluorite.*

Use: The only important source of mercury.

J. A. GILBERT

References

Am. Min. **63,** 1143–1152.
Dana, I, 251.

Cross-references: *Blacksand Minerals; Blowpipe Analysis; Human and Vertebrate Mineralogy; Metallic Minerals; Mineral and Ore Deposits; Naming of Minerals; Refractive Index.*

CITRINE–Variety of *quartz.*

CLARINGBULLITE *Min. Abst.* **29,** 83
$Cu_4Cl(OH)_7 \cdot nH_2O$ hexa. *Min. Mag.* **41,** 433

CLARKEITE A uranium-hydroxide oxide (with sodium and calcium) of the ***curite*** group named by Ross, Henderson, and Posnjak in 1931 after F. W. Clarke (1847–1931), American mineral chemist of the U. S. Geological Survey.

Composition: $(Na_2,Ca,Pb)U_2(O,OH)_7$, usually with some potassium.

Properties: Massive; H=4–4.5; G=6.39; waxy luster; dark reddish brown color; yellowish brown streak; transparent orange in thin section.

Occurrence: As a hydrothrmal alteration product of *uraninite* at Spruce Pine, Mitchell County, North Carolina; where it is also associated with *becquerelite,* gummite, *ianthinite, schoepite,* and *uranophane.*

J. A. GILBERT

References

Am. Min. **16,** 212–220; **39,** 836.
Min. Mag. **22,** 617.
USGS Bull. **1064,** 95–98.

CLAUDETITE Arsenic trioxide, named by Dana (1868) after F. Claudet, French chemist who first described the species in nature.

Composition: As_2O_3.

Properties: Monoclinic, $P2_1/n$; chain type structure; polymorphic with *arsenolite; valentinite* (Sb_2O_3) has the same structure; *bismite* (Bi_2O_3) is closely related; H=2.5; G=4.2; colorless to white; vitreous to pearly luster.

Occurrence: An oxidation product of arsenic ores and smelter fumes. Observed as a sublimation product at Schmölnitz, Hungary; San Domingo mines, Portugal; and United Verde mine, Jerome, Arizona. Often associated with *arsenolite, realgar, orpiment,* and native *sulfur.*

D. H. SPEIDEL

References

Am. Min. **35,** 185.
Dana, I, 545.

CLAUSTHALITE *Dana,* I, 204
PbSe iso. *Min. Abst.* **27,** 355

Cross-references: *Linnaeite Group; Ore Microscopy.*

CLAY–See *Clays and Clay Minerals*

CLEAVELANDITE–Lamellar variety of *albite*

Cross-references: *Crystal Habits; Pegmatite Minerals; Plagioclase Feldspars.*

CLEVEITE–Helium-rich variety of *uraninite*

CLIFFORDITE *Am. Min.* **54,** 697; **57,** 597
UTe_3O_9 iso.

Cross-references: *Naming of Minerals.*

CLIFTONITE = Meteoritic *graphite* cubes *Min. Abst.* **27,** 63

CLINOBISVANITE *Min. Mag.* **39,** 847
$BiVO_4$ mono.

CLINOCHALCOMENITE *Am. Min.* **66,** 217
$CuSeO_3 \cdot 2H_2O$ mono. *Min. Abst.* **33,** 495

CLINOCHLORE A hydroxyl aluminosilicate of iron and magnesium belonging to the ***chlorite*** group. Named by W. P. Blake in 1851 in reference to its composition and structure.

Composition: $2[(Mg,Fe^{2+})_5Al(Si,Al)_4O_{10}(OH)_8]$. Sometimes substantially free from iron. Monoclinic.

Occurrence: Probably the most common of the ***chlorite*** minerals; typical mineral in greenschists. From the Ural Mountains; Tyrol,

Austria; Brewster, New York; West Chester, Pennsylvania; and at Wood's mine, Texas.

G. C. MOERSCHELL

Reference: DHZ, 3, 131–163.

Cross-references: *Chlorite Group; Minerals, Uniaxial and Biaxial.*

CLINOCHRYSOTILE (***Kaolinite-Serpentine*** gr.)
$Mg_3Si_2O_5(OH)_4$ mono. DHZ, 3, 170

CLINOCLASE A copper hydroxylarsenate, so named by Dana in 1868 for its inclined cleavage.

Composition: $4[Cu_3(AsO_4)(OH)_3]$

Properties: Monoclinic, $P2_1/a$; tabular or elongate crystals; one perfect cleavage {001}; H=2.5–3; G=4.4; vitreous luster; color: green; bluish green streak.

Occurrence: With other secondary copper minerals.

K. FRYE

Reference: *Dana*, II, 787–789.

CLINOENSTATITE (***Pyroxene*** gr.)
$MgSiO_3$ mono. DHZ, 2A, 3

Cross-references: *Clinopyroxenes; Mantle Mineralogy; Meteoritic Minerals; Minerals, Uniaxial and Biaxial; Orthopyroxenes; Pyroxene Group.*

CLINOFERROSILITE (***Pyroxene*** gr.)
$FeSiO_3$ mono. DHZ, 2A, 3

Cross-references: *Clinopyroxenes; Mantle Mineralogy; Pyroxene Group.*

CLINOHEDRITE *Min. Abst.* **29**, 19
$CaZnSiO_4 \cdot H_2O$ mono. *Z. Krist.* **144**, 377

CLINOHOLMQUISTITE (***Amphibole*** gr.)
$Li_2Mg_3Al_2(Si_4O_{11})_2(OH)_2$
Am. Min. **52**, 1585;
mono. **63**, 1030

CLINOHUMITE A magnesium silicate hydroxyl-fluoride of the ***humite*** group, named by Des Cloizeaux in 1876 from its monoclinic symmetry and compositional similarity to *humite.*

Composition: $4Mg_2SiO_4 \cdot Mg(OH,F)_2$, with F:OH usually ca. 1:1, generally contains some Ti and Fe replacing Mg.

Occurrence: In carbonate rocks metamorphosed or metasomatized by fluoride-bearing emanations from silicic or alkaline intrusive igneous rocks; commonly associated with *wollastonite, grossular, monticellite, forsterite,* and *diopside.*

G. C. MOERSCHELL

Reference: Phillips, W. R., 1981. *Optical Mineralogy: The Nonopaque Minerals.* San Francisco: Freeman, 142–144.

Cross-reference: *Mantle Mineralogy; Minerals, Uniaxial and Biaxial.*

CLINOHYPERSTHENE DHZ, 2A, 30
$(Mg,Fe)SiO_3$ mono.

Cross-references: *Clinopyroxenes; Meteoritic Minerals; Pyroxene Group.*

CLINOJIMTHOMPSONITE *Am. Min.* **63**, 1000
$(Mg,Fe)_{10}Si_{12}O_{32}(OH)_4$
mono. *Min. Abst.* **29**, 341

Cross-reference: *Biopyribole.*

CLINOPTILOLITE A hydrated alkali aluminosilicate of the ***zeolite*** group (q.v.) closely related to *heulandite* in structure but higher in alkalies and silica.

Composition: $(Na,K,Ca)_{2-3}Al_3(Al,Si)_2Si_{13}O_{36} \cdot 12H_2O$. Monoclinic.

Properties: See *heulandite.*

Occurrence: Alteration product of volcanic rocks.

K. FRYE

References

Am. Min. **45**, 341–369.
DHZ, 4, 351–428.

Cross-references: *Green River Mineralogy; Minerals, Uniaxial and Biaxial; Zeolite Group.*

CLINOSAFFLORITE *Am. Min.* **57**, 1552
$(Co,Fe,Ni)As_2$ mono. *Can. Min.* **10**, 877

CLINOSTRENGITE = *Phosphosiderite*

Cross-reference: *Pegmatite Minerals.*

CLINOTYROLITE *Min. Abst.* **31**, 495
$Ca_2Cu_9[(As,S)O_4]_4(O,OH)_{10} \cdot 10H_2O$
Min. Mag. **43**, 1059

CLINOUNGEMACHITE *Dana*, II, 597
$K_3Na_9Fe(SO_4)_6(OH)_3 \cdot 9H_2O(?)$
mono.

CLINOZOISITE A calcium aluminum hydroxyl silicate of the ***epidote*** group (q.v.) named by Weinschenk (1896) as the monoclinic modification of *zoisite* (which is named after the Austrian Baron Zois von Edelstein).

Composition: $2[Ca_2Al_3O(SiO_4)(Si_2O_7)OH]$; CaO, 14.3%;$Al_2O_3$, 25.1%; SiO_2, 59.1%; H_2O, 1.5%. A solid solution exists from *clinozoisite* through *epidote*, to *piemontite*.

Properties: Monoclinic, $P2_1/m$; crystals have striations; may be columnar or granular massive; H=6–6.5; G=3.25–3.37; vitreous luster; transparent to translucent; color: green, pink, or gray.

Occurrence: Found in metamorphic rocks which are the product of the metamorphism of mafic igneous rocks containing calcic ***plagioclase feldspar.*** It may be found in igneous rocks as an alteration of ***plagioclase.*** Associated with ***amphibole.***

J. A. GILBERT

References

DHZ, 1, 193–210.

Min. Mag. **11**, 325; **28**, 505.

Vanders, I., and Kerr, P. F., 1967. *Mineral Recognition.* New York: Wiley, 294.

Cross-reference: *Minerals, Uniaxial and Biaxial.*

CLINTONITE (***Mica*** gr.) DHZ, 3, 99
$Ca(Mg,Al)_3(Al_3Si)O_{10}(OH)_2$
mono.

Cross-references: *Phyllosilicates; Soil Mineralogy.*

COALINGITE hexa. *Am. Min.* **50**, 1893; **54**, 437
$Mg_{10}Fe_2^{3+}(CO_3)(OH)_{24}\cdot 2H_2O$(?)

COBALTITE Cobalt sulfarsenide named by Beudant in 1832 (cobaltine) in allusion to its composition.

Composition: 4[CoAsS]; Fe may substitute for Co up to 5%. A complete solid solution exists between *cobaltite* and *gersdorffite* (NiAsS).

Properties: Isometric; *Pa*3; may be granular; H=5.5; G=6.33; brittle; metallic luster; color: silver-white, inclined to red.

Occurrence: In high-temperature deposits, as disseminations in metamorphic rocks, or in vein deposits with other cobalt and nickel minerals. Skutterud, Norway is a good locality for high-temperature deposits and Cobalt, Ontario, Canada has a good arsenide vein deposit.

Use: An important ore of cobalt.

J. A. GILBERT

References

Am. Min. **50**, 1002.

Dana, I, 296.

Cross-references: *Crystallography: Morphological; Ore Microscopy; Refractory Minerals.*

COBALTOCALCITE = *Spherocobaltite*

COBALTOMENITE *Am. Min.* **48**, 1183
$(Co,Ni)SeO_3\cdot 2H_2O$ mono.

COBALT PENTLANDITE *Am. Min.* **44**, 897; **50**, 2107
Co_9S_8 iso.

COBALT-ZIPPEITE orth. *Can. Min.* **14**, 429
$Co_2(UO_2)_6(SO_4)_3(OH)_{10}\cdot 16H_2O$
Min. Mag. **43**, 1059

COCHROMITE iso. *Am. Min.* **65**, 811
$(Co,Ni,Fe^{2+})(Cr,Al)_2O_4$ *Min. Mag.* **43**, 1059

COCONINOITE mono.(?) *Am. Min.* **51**, 651
$Fe_2^{3+}Al_2(UO_2)_2(PO_4)_4(SO_4)(OH)_2\cdot 20H_2O$

COERULEOLACTITE tric. *Dana,* II, 961
$(Ca,Cu)Al_6(PO_4)_4(OH)_8\cdot 4\text{–}5H_2O$

COESITE *JGR* **67**, 419
SiO_2 mono. *Science* **140**, 991

Cross-reference: *Coesite and Stishovite.*

COFFINITE A uranous silicate of the ***uranophane*** group, often with extensive hydroxyl substitution. Named by Rosenweig, Gruner, and Gardiner in 1954 in honor of Reuben C. Coffin, an early investigator of the Colorado uranium and vanadium deposits.

Composition: $U(SiO_4)_{1-x}(OH)_{4x}$.

Properties: Tetragonal, rarely in crystals, most botryoidal, mixed with other uranium minerals; metamict; highest density observed is 5.1 but ideal density of pure $USiO_4$ is much higher; H=5–6; isostructural with *zircon* and *thorite.*

Occurrence: Widespread, often associated with *uraninite* as a constituent of uranium ores of the Colorado Plateau type. Commonly found as a fine-grained matrix in a mixture of organic matter and other minerals, such as *pyrite, barite, galena, cobaltite,* and *wurt-*

zite. Nearly all of the ore at Ambrosia Lake, New Mexico, is composed of this type of "organically dirty" sandstones impregnated with *coffinite.*

Use: An important ore of uranium in the United States.

K. FRYE

Reference: *Am. Min.* **41**, 675–688; **47**, 26.

Cross-references: *Metallic Minerals; Ore Microscopy.*

COHENITE *Dana,* I, 122
$(Fe,Ni,Co)_3C$ orth. *Min. Abst.* **27**, 64

Cross-references: *Lunar Minerals; Magnetic Minerals; Metallurgy; Meteoritic Minerals; Native Elements and Alloys.*

COLEMANITE A hydrated borate named by Evans (1884) after William T. Coleman, a San Francisco merchant who marketed the product of the *colemanite* mines and founded the California borax industry.

Composition: $2[Ca_2B_6O_{11}\cdot 5H_2O]$; CaO, 27.2%; B_2O_3; 50.9%; H_2O, 21.9%.

Properties: Monoclinic; may be found in granular to compact cleavable masses; H= 4–4.5; G=2.42; vitreous luster; transparent; white to colorless.

Occurrence: Found in beds interstratified with lake deposits of Tertiary age. Large deposits are along Furnace Creek, Death Valley, California. Associated with *ulexite, borax,* and other evaporites (see *Saline Minerals*).

Use: A source of *borax.*

J. A. GILBERT

References

Am. Min. **38**, 411.
Dana, II, 349.

Cross-reference: *Saline Minerals.*

COLLINSITE tric. *Dana,* II, 722
$Ca_2(Mg,Fe)(PO_4)_2\cdot 2H_2O$

COLLOPHANE–Massive, fine-grained or amorphous phosphate.

Cross-references: *Apatite Group; Authigenic Minerals; Green River Mineralogy; Minerals, Uniaxial and Biaxial.*

COLORADOITE (***Sphalerite*** gr.) *Dana,* I, 218
HgTe iso. *Min. Abst.* **27**, 306

COLUMBITE-TANTALITE GROUP–*Ferrocolumbite, ferrotantalite, manganocolumbite, manganotantalite,* and *magnocolumbite.*

Cross-references: *Blacksand Minerals; Metallic Minerals; Ore Microscopy; Pegmatite Minerals; Staining Techniques.*

COLUSITE *Am. Min.* **55**, 1787
$Cu_3(As,Sn,V,Fe)S_4$ iso.

COMBEITE trig. *Am. Min.* **43**, 791
$Na_4Ca_3Si_6O_{16}(OH,F)_2$ *Min. Mag.* **31**, 503

COMPREIGNACITE *Am. Min.* **50**, 807
$K_2U_6O_{19}\cdot 11H_2O$ orth.

CONGOLITE *Am. Min.* **57**, 1315
$(Fe,Mg,Mn)_3B_7O_{13}Cl$ trig.

CONICHALCITE *Am. Min.* **36**, 484
$CaCu(AsO_4)(OH)$ orth. *Can. Min.* **18**, 191

CONNELLITE hexa. *Dana,* II, 572
$Cu_{19}Cl_4(SO_4)(OH)_{32}\cdot 3H_2O$
Min. Abst. **30**, 320

COOKEITE A hydroxyl aluminosilicate of lithium belonging to the ***chlorite*** group; named by G. J. Brush in 1866 after J. P. Cooke, a former professor at Cambridge.

Composition: $4[LiAl_4(Si_3Al)O_{10}(OH)_8]$.

Occurrence: Often occurs as a pearly coating on crystals of pink *elbaite,* of which it appears to be an alteration product. Frequent alteration product of ***tourmaline*** and late-stage mineral in pegmatites throughout the world. Occurs with *tourmaline* and *lepidolite* at Hebron, Mt. Mica, and at Buckfield, Maine. Also at Haddam Neck, Connecticut.

J. A. GILBERT

References

Am. Min. **60**, 1041
Min. Mag. **24**, 515.

Cross-references: *Chlorite Group; Clays and Clay Minerals; Pegmatite Minerals.*

COOPERITE *Dana,* I, 258
(Pt,Pd)S tetr. *Am. Min.* **63**, 832

COPIAPITE A hydrated ferrous and ferric sulfate, named (1845) for the locality Copiapó, Chile.

Composition: $1[Fe^{2+}Fe_4^{3+}(SO_4)_6(OH)_2\cdot 20H_2O]$ with Fe^{2+} replaced by Mg (Mg> Fe, *magnesiocopiapite*) or by Cu (Cu> Fe, *cuprocopiapite*) and Fe^{3+} replaced by Al toward *aluminocopiapite.*

Properties: Triclinic, $P\bar{1}$; tabular {010} crystals, commonly in loose aggregates or crusts; one perfect cleavage {010}; H= 2.5–3; G=2.1–2.2; pearly luster; yellow.

Occurrence: Oxidation product of *pyrite;* precipitate from acid mine waters.

K. FRYE

Reference: *Am. Min.* **65**, 961–967.

COPIAPITE GROUP–*Aluminocopiapite, calciocopiapite, copiapite, cuprocopiapite, ferricopiapite, magnesiocopiapite, zinccopiapite.*

COPPER Native metal, named by Pliny in 77 AD from the Greek meaning "cyprus," where it was found.

Composition: 4[Cu]. Commonly alloyed with small amounts of Ag, As, Fe, Bi, Hg.

Properties: Isometric, *Fm*3*m*; commonly in irregular masses, plates, and scales; H= 2.5–3; G=8.9; highly ductile and malleable; fracture hackly; color: copper-red on fresh surface; dissolves readily in nitric acid.

Occurrence: Commonly found in the oxidized zone of copper deposits. Most notable deposit is on the Keweenaw Peninsula in upper Michigan. Often associated with *cuprite, malachite,* and *azurite.*

Use: A minor ore of copper.

J. A. GILBERT

Reference: *Dana,* I, 99.

Cross-references: *Crystal Habits; Lunar Minerals; Meteoritic Minerals; Mineral Industries; Native Elements and Alloys; Ore Microscopy; Pseudomorphism; Skeletal Crystals.* Vol. IVA: *Copper: Element and Geochemistry; Copper: Mineralogy of Compounds.*

COPPERAS = *Melanterite*

COPPER GLANCE = *Chalcocite*

COQUIMBITE *Am. Min.* **55**, 1534
$Fe_2(SO_4)_3 \cdot 9H_2O$ trig.

Cross-reference: *Syntaxy.*

CORDEROITE *Am. Min.* **59**, 652
$Hg_3S_2Cl_2$ iso.

CORDIERITE A magnesium aluminosilicate named by Lucas in 1813 after the French geologist P. L. A. Cordier (1777–1861).

Composition: $4[Mg_2Al_4Si_5O_{18}]$; MgO, 13.8%; Al_2O_3, 34.9%; SiO_2, 51.3%. Mg replaced by Fe^{2+} toward *sekaninaite.*

Properties: Orthorhombic, *Cccm;* crystals often twinned; may be granular or massive; H=7–7.5; G=2.66; vitreous luster; transparent to translucent; pleochroic blue; white streak; alters to ***mica, chlorite,*** and *talc.*

Occurrence: Found as an accessory mineral in peraluminous granite, schist, gneiss, and in contact metamorphic zones.

Use: Gem-quality material called saphir d'eau.

J. A. GILBERT

References

Am. Min. **35**, 173; **62**, 395.
Carnegie Inst. Wash. Yearbook 1959–60 **60**, 91.
DHZ, 1, 268–299.
Min. Mag. **33**, 226.

Cross-references: *Lunar Minerals; Meteoritic Minerals; Mineral Properties; Minerals, Uniaxial and Biaxial; Pleochroic Halos; Refractory Minerals; Rock-Forming Minerals.*

CORDYLITE *Dana,* II, 285
$Ba(Ce,La)_2(CO_3)_3F_2$ hexa.

CORKITE A hydroxyl sulfate-phosphate of lead and ferric iron of the ***beaudantite*** group. Named by Adam in 1869 from County Cork, Ireland.

Composition: $[PbFe_3(PO_4)(SO_4)(OH)_6]$. Analyses show some substitution of Cu for Pb and some preponderance of SO_4 over PO_4.

Properties: Trigonal, *R*3*m*; H=3.5–4.5; G= 4.295; vitreous to resinous luster; dark green, yellowish green to pale yellow; cleavage {0001} perfect, uniaxial negative, birefringence weak, mean index = 1.93–1.96; easily soluble in HCl.

Occurrence: A secondary mineral formed by the alteration of lead ores. Often found associated with limonite on chert in County Cork, Ireland, and with *pyromorphite* and limonite at Dernbach, near Montabaur, West Germany.

K. FRYE

Reference: *Dana,* II, 1002.

CORNETITE *Dana,* II, 789
$Cu_3(PO_4)(OH)_3$ ortho.

CORNUBITE *Am. Min.* **44**, 1321
$Cu_5(AsO_4)_2(OH)_4$ tric. *Min. Mag.* **32**, 1

CORNWALLITE *Am. Min.* **36**, 484
$Cu_5(AsO_4)_2(OH)_4 \cdot H_2O$ mono.

CORONADITE tetr.(?) *Dana*, I, 742
$Pb(Mn^{4+},Mn^{2+})_8O_{16}$ *Min. Abst.* **27**, 253

Cross-reference: *Ore Microscopy*

CORRENSITE A hydroxyl aluminosilicate of magnesium and iron believed to be related to the ***chlorites,*** but showing swelling behavior in glycerol. Named by Lippman in 1954 for Dr. Carl W. Correns, professor of sedimentary petrography at Göttingen.

Composition: The exact identity of *corrensite* is in doubt, but Vivaldi and MacEwan (*Clay Minerals Bull.*, 1960, **4**, 173) now consider true *corrensite* to be defined as a 1 : 1 regular interstratification of ***chlorite*** and a ***vermiculite***, as found in Germany, Italy, and France (see Occurrence, below).

Occurrence: Zaiserweiher and Hünstollen, Germany, in a ***chlorite*** association. Monte Chiavo, Italy, with ***vermiculite.*** Also from Middle Keuper, Périgny, Jura, France.

G. C. MOERSCHELL

References

Am. Min. **46**, 769.
Clay Minerals Bull. **4**, 173.
Heidel. Beitr. Mineral. Petrog., 1954, **4**, 130.

Cross-reference: *Clays and Clay Minerals.*

CORUNDOPHILITE (***Chlorite*** gr.)
$(Mg,Fe^{2+},Al)_6(Si,Al)_4O_{10}(OH)_8$ DHZ, 3, 137
Min. Mag. **30**, 277

CORUNDUM Aluminum oxide, named by Greville in 1798 from the Sanskrit word for it, *kauruntaka.*

Composition: Al_2O_3. Ruby, transparent, and red; contains moderate (0.7–2.6 wt%) amounts of Cr_2O_3; sapphire is transparent blue *corundum;* oriental emerald is green *corundum.*

Properties: Trigonal, $R\bar{3}c$; H=9 (by definition of the Mohs Scale); G=3.98–4.01; color: white, gray, blue, red, yellow, green; streak: uncolored; weakly colored in thin section; no cleavage, but basal or rhombohedral parting of some specimens; well-developed crystals common as hexagonal bipyramids and prisms; optically, ω=1.759–1.763; ϵ=1.767-1.772; γ=Al_2O_3 has the cubic spinel structure; minor amounts of ferric iron are common.

Occurrence: Nepheline-syenite pegmatites in Bancroft, Ontario; emery, a mixture of *magnetite* and(or) *spinel* with *corundum,* is considered an alteration product of mafic rocks, Cortlandt complex, New York. Most gem *corundum* is from placer deposits. *Corundum* is often associated with *magnetite,* ***mica, chlorite,*** *nepheline, serpentine, spinel, boehmite, diaspore,* and *gibbsite.*

Use: Abrasive materials (q.v.); gem purposes. Doped synthetic Al_2O_3 is used in lasers.

D. H. SPEIDEL

References

Am. Min. **50**, 1982–2022.
Dana, I, 520.
DHZ, 5, 11–20.

Cross-references: *Abrasive Materials; Blacksand Minerals; Color in Minerals; Crystal Growth; Crystal Habits; Crystallography: Morphological; Gemology; Jade; Lunar Minerals; Mantle Mineralogy; Meteoritic Minerals; Mineral Properties; Minerals, Uniaxial and Biaxial; Mohs Scale of Hardness; Portland Cement Mineralogy; Refractory Minerals; Rock-Forming Minerals; Synthetic Minerals; Topotaxy.*

CORVUSITE *Dana*, I, 602
$V_2O_4 \cdot 6V_2O_5 \cdot nH_2O$(?)

COSALITE *Dana*, I, 445
$Pb_2Bi_2S_5$ orth. *Min. Abst.* **27**, 306

COSMOCHLORE = *Ureyite*
Min. Mag. **43**, 1053

COSTIBITE *Am. Min.* **55**, 10
CoSbS orth.

COTUNNITE Lead chloride named by Monticelli and Covelli in 1825 after Domenico Cotugno (1736–1822), Professor of Anatomy at the University of Naples.

Composition: $4[PbCl_2]$. Orthorhombic.

Occurrence: Found originally at Vesuvius as a product of sublimation, it also occurs associated with *cerussite, anglesite, matlockite,* and other secondary minerals and as an alteration product of *galena* under arid, saline conditions; it has also been found as an alteration product of ancient leaden objects immersed in seawater.

J. A. GILBERT

References

Dana, II, 42.
Min. Abst. **28**, 140, no. 77–1497.

COULSONITE (***Spinel*** gr.)
$Fe^{2+}V_2^{3+}O_4$ iso. *Am. Min.* **47**, 1284

COUSINITE *Am. Min.* **44**, 910
$MgU_2Mo_2O_{13}\cdot 6H_2O$(?)

COVELLITE (COVELLINE) Copper sulfide named by Beudant in 1832 after N. Covelli (1790–1829), the discoverer of the Vesuvian *covellite.*

Composition: 6[CuS]; Cu, 66.4%; S, 33.6%, by wt.

Properties: Hexagonal $P6_3/mmc$; usually massive as coatings or disseminations through other copper minerals; H=1.5–2; G=4.6–4.76; metallic luster; opaque; often iridescent especially when wet; color: indigo-blue or darker.

Occurrence: Found in small amounts in most copper deposits as a supergene mineral, usually as a coating in the zone of sulfide enrichment. Associated with *chalcocite, chalcopyrite, bornite,* and *enargite* and derived from them by alteration.

Use: A minor copper ore.

J. A. GILBERT

References

Am. Min. **61,** 996–1000.
Dana, I, 248.

Cross-references: *Metallic Minerals; Ore Microscopy.*

COWLESITE (***Zeolite*** gr.) *Am. Min.* **60**, 951
$(Ca,Na)Al_2Si_3O_{10}\cdot 5\text{–}6H_2O$
orth. *Min. Abst.* **27**, 257

CRANDALLITE (***Crandallite*** gr.)
$CaAl_3[PO_3(O,OH)]_2(OH)_6$ *Am. Min.* **65**, 953
trig. *Min. Abst.* **27**, 229

CRANDALLITE GROUP—*Crandallite, dussertite, eylettersite, florencite, florencite-(Nd), morceixite, goyazite, lusungite, plumbogummite, waylandite,* and *zairite.*

CREASEYITE orth. *Am. Min.* **61**, 503
$Pb_2Cu_2Fe_2Si_5O_{17}\cdot 6H_2O$ *Min. Mag.* **49**, 227

CREDNERITE A copper manganese oxide named by Rammelsberg in 1847 after C. F. Credner (1809-76), American mining geologist and mineralogist.

Composition: $CuMn_2O_4$.

Properties: Monoclinic (pseudohexagonal) plates; H=4; G=4.9–5.1; metallic luster; color: iron-black to steel-gray; streak: black or brownish.

Occurrence: From Friedrichsrode, Germany, intergrown with psilomelane, *hausmannite, malachite, volborthite, barite, calcite,* and wad.

D. H. SPEIDEL

References

Am. Min. **51,** 1819.
Dana, I, 722.

CREEDITE A hydrated sulfate fluoride of calcium and aluminum named by Larsen and Wells in 1916 for the original locality in Creede quadrangle, Mineral County, Colorado.

Composition: $4[Ca_3Al_2(SO_4)(F,OH)_{10}\cdot 2H_2O]$.

Properties: Monoclinic, $C2_1/c$; as prismatic crystals, as radiating aggregates, as druses, or as embedded grains; one perfect cleavage {100}; vitreous; colorless to white, rarely purple; transparent.

Occurrence: Found in tin veins of Colquiri, Bolivia; in the USA in veins with *barite* at the original locality and with *fluorite* and *quartz* from Nye County, Colorado.

K. FRYE

Reference: *Dana,* II, 129.

CRESTMOREITE – Mixture of
torbermorite and *wilkeite* *Am. Min.* **39**, 405

CRICHTONITE (***Crichtonite*** gr.)
$(Sr,RE,Pb)(Ti,Fe,Mn)_{21}O_{38}$ *Am. Min.* **61**, 1203;
trig. **63**, 28

CRISTOBALITE A high-temperature polymorph of *quartz* forming when *tridymite* is heated above 1470°C, and existing as a metastable form when cooled. Named by vom Rath in 1886 from Cerro San Cristobal, Mexico.

Composition: $4[SiO_2]$.

Properties: *Cristobalite* may exist in either of two forms: high *cristobalite* (*β-cristobalite*), which is isometric, and forms from 1470°C to the melting point from *tridymite*; becomes metastable on cooling to 268°C, where it inverts to low *cristobalite* (*α-cristobalite*), which is tetragonal; H=6.5; G= 2.2–2.3; no cleavage; color: white; complex twinning; streak: white; mean refractive index of *α-cristobalite* = 1.482–1.487; soluble in HF and fused alkalies (opaline varieties soluble in alkali solutions).

Occurrence: Coarsely crystalline in high-temperature igneous rocks. In massive and encrusting cryptocrystalline hydrous opal deposits and siliceous sinters. A fibrous variety of *α-cristobalite,* lussatite, is common in rhyolites and felsic lavas. Often associated with *tridymite, quartz,* or *sanidine.*

W. E. HILL

References

Am. Min. **47**, 897.
Compass **42**, 30.
Dana, III, 273.

Cross-references: *Lunar Minerals; Meteoritic Minerals; Mineral Properties; Opal; Polymorphism; Portland Cement Mineralogy; Quartz; Rock-Forming Minerals; Soil Mineralogy.*

CROCIDOLITE–Asbestiform *riebeckite.*

CROCOITE Lead chromate, named by Beudant (1832) from the Greek word meaning "saffron" in reference to its color.

Composition: 4[$PbCrO_4$]; PbO, 68.9%; CrO_3, 31.1%.

Properties: Monoclinic, $P2_1/n$; commonly in slender vertically striated and columnar aggregates, may be granular; H=2.5-3; G= 5.9–6.1; adamantine luster; translucent; color: bright orange-red.

Occurrence: A rare mineral found in gossans (oxidized zones) of lead deposits where the veins have traversed country rock containing *chromite.*

Use: No commercial use, too rare, but it is collected for its very attractive specimens.

J. A. GILBERT

References

Dana, II, 646.
Lapidary J., **17**, 1092, 1096.

CRONSTEDTITE An iron hydroxyl silicate of the ***kaolinite-serpentine*** group named by Steimann in 1821 after A. Fr. Cronstedt, Swedish mineralogist and chemist.

Composition: $(Fe_2^{2+}Fe^{3+})(SiFe^{3+})O_5(OH)_4$.

Occurrence: At Pribram and Kuttenberg, Czechoslovakia; Cornwall, England; in Brazil in Conghonas do Campo, Minas Gerais; commonly associated with limonite and *calcite* in veins containing ores of silver.

K. FRYE

References

Am. Min. **47**, 781.
DHZ, 3, 167.
Min. Abst. **5**, 39; **6**, 333; **7**, 496.

Cross-reference: *Soil Mineralogy.*

CROOKESITE *Am. Min.* **35**, 337
$(Cu,Tl,Ag)_2Se$ tetr.

CROSSITE–An ***amphibole*** (q.v.) intermediate in the *glaucophane-riebeckite* series. DHZ, 2, 334

Cross-reference: *Clinopyroxenes.*

CRYOLITE Sodium aluminum fluoride named by Abildgaard in 1799 from two Greek words meaning frost and stone in reference to its icy appearance.

Composition: 2[Na_3AlF_6]; Na, 32.8%; Al, 12.8%; F, 54.4%, by wt. Related minerals are *cryolithionite* which is isostructural with the ***garnets,*** and *chiolite.*

Properties: Monoclinic, $P2_1/n$; usually massive; H=2.5; G=2.95–3.0; vitreous to greasy luster; transparent to translucent; index of refraction close to that of water so the powdered mineral appears almost invisible when immersed in water; colorless to white.

Occurrence: The only important deposit is at Ivigtut, Greenland, where it is found in a large vein-like mass in granite, associated with *siderite, galena, sphalerite,* and *chalcopyrite.*

Use: For the manufacture of some glass and porcelain products, as a flux in aluminum-metal reduction, and for cleaning metal surfaces.

J. A. GILBERT

References

Am. Min. **35**, 149.
Can. Min. **13**, 377.

Cross-references: *Green River Mineralogy; Museums, Mineralogical; Pegmatite Minerals.*

CRYOLITHIONITE *Dana,* II, 99
$Na_3Li_3Al_2F_{12}$ iso. *Min. Abst.* **28**, 46

CRYPTOHALITE *Dana,* II, 104
$(NH_4)_2SiF_6$ iso.

CRYPTOMELANE mono. *Am. Min.* **27**, 607
$K(Mn^{4+},Mn^{2+})_8O_{16}$ *Min. Abst.* **27**, 253

CSIKLOVAITE *Am. Min.* **35**, 333
$Bi_2Te(S,Se)_2$(?) trig.

CUBANITE *Dana*, I, 243
$CuFe_2S_3$ orth.

Cross-references: *Lunar Minerals; Meteoritic Minerals; Ore Microscopy.*

CUMENGÉITE tetr. *Dana*, II, 79
$Pb_{19}Cu_{24}Cl_{42}(OH)_{44}$ *Min. Mag.* **43**, 901

CUMMINGTONITE A magnesium iron hydroxyl silicate, a member of the ***amphibole*** group. Named by Dewey in 1824 from the locality Cummington, Massachusetts; the name "amosite" is derived from the initial letters of the Asbestos Mines of South Africa, and "montasite," the trade name for fine quality fiber, is named after the Montana mine, Pietersburg, Transvaal.

Composition: $(Mg,Fe^{2+})_7Si_8O_{22}(OH)_2$. The name *grunerite* is applied to members of the series with high iron contents. MgO, 25–0%; FeO, 6–47%; Si, 47–53%; H_2O, 1.5–2.5% by wt.

Properties: Monoclinic, $C2/m$; H=5.6; G=3.10–3.60; dull or pearly luster, rarely vitreous; color: gray, green, or brown; prismatic or fibrous habit; good cleavage on {110}, $(110):(1\bar{1}0) \approx 55°$; simple or lamellar twinning very common on {100}, white streak. Refractive indices, α=1.635–1.696, β=1.644–1.709, γ=1.655–1.729; $2V_\gamma$=65°–96°; $\gamma:c$= 10°–21°; colorless to pale green or pale brown in thin section; pleochroism, $\alpha=\beta$ colorless, γ pale green or pale brown. Insoluble in acids except HF, soluble in molten Na_2CO_3. Some Mn may substitute for Fe^{2+}; minor Al may substitute for Si. If the iron-rich compositions are referred to *grunerite*, the division is taken at 70% of the $Fe_7Si_8O_{22}(OH)_2$ molecule (approximately the position of the change of optic sign from positive for *cummingtonite* to negative for *grunerite*).

Occurrence: Common in amphibolites derived by regional metamorphism of mafic igneous rocks; not uncommon in hybrid rocks of intermediate composition; occurs as a primary mineral in some dacites. *Grunerite* is a mineral characteristic of metamorphosed iron-rich siliceous sediments. Often found in association with *hornblende, anthophyllite,* ***plagioclase;*** *grunerite* occurs with *fayalite, hedenbergite* and *almandine.*

Uses: The fibrous varieties are used as asbestos (q.v.); amosite and montasite are names given respectively to the harsher more iron-rich and softer more magnesium-rich fibers of economic importance.

R. A. HOWIE

References

Am. Min. **38**, 862; **49**, 963.
DHZ, 2, 234–248.
Geol. Mag. **75**, 76.
Trans. Geol. Soc. S. Africa **55**, 1.

Cross-references: *Amphibole Group; Minerals, Uniaxial and Biaxial; Order–Disorder.*

CUPRITE (Ruby copper) Copper oxide named by Haidinger in 1845 from the Latin *cuprum*, copper. Variety with elongated cubic capillar crystals is called chalcotrichite or plush copper.

Composition: Cu_2O.

Properties: Cubic, $Pn3m$; anti-flourite structure; H=3.5–4; G=6.1; color: various shades of red.

Occurrence: The upper oxidized portions of copper veins associated with limonite, native *copper, azurite, malachite,* and *chrysocolla.*

Use: An ore of copper.

D. H. SPEIDEL

Reference: *Dana*, I, 491.

Cross-references: *Blowpipe Analysis; Museums, Mineralogical; Ore Microscopy; Pseudomorphism.*

CUPROARTINITE *Am. Min.* **64**, 886
$(Cu,Mg)_2(CO_3)(OH)_2 \cdot 3H_2O$
mono. *Min. Abst.* **31**, 229

CUPROAURIDE = *Auricupride*

CUPROBISMUTITE *Min. Abst.* **27**, 23, 300
$Cu_{10}Bi_{12}S_{23}$ mono.

CUPROCOPIAPITE tric. *Dana*, II, 623
$CuFe_4(SO_4)_6(OH)_2 \cdot 20H_2O$

CUPROHYDRO-MAGNESITE
$(Cu,Mg)_2(CO_3)_4(OH)_2 \cdot 4H_2O$
Am. Min. **64**, 886
mono. *Min. Abst.* **31**, 229

CUPROPAVONITE *Bull. Min.* **102**, 351
$PbAgCu_2Bi_5S_{10}$ mono. *Can. Min.* **18**, 181

CUPRORIVAITE *Am. Min.* **47**, 409
$CaCuSi_4O_{10}$ tetr.

CUPROSKLODOWSKITE A hydrated silicate of uranium oxide with copper, named by Buttgenbach in 1933 for its supposed relationship to *sklodowskite* with CuO in place of MgO.

Composition: $Cu(UO_2)_2Si_2O_6(OH)_2 \cdot 5H_2O$. Triclinic. Not isomorphous with *sklodowskite*. Strongly radioactive.

Occurrence: At Great Bear Lake, Northwest Territory, Canada, as a secondary mineral produced by the alteration of pitchblende, and associated with *kasolite* and *schoepite*. Originally found in Shaba Province, Zaire. Also from Joachimstal, Czechoslovakia.

Use: A minor ore of uranium.

G. C. MOERSCHELL

References

Am. Min. **60**, 448.
Can. Min. 7, 331.
USGS Bull. **1064**, 304.

CUPROSPINEL Copper iron oxide of the *magnetite* series of the ***spinel*** group (q.v.).

Composition: $CuFe_2O_4$. Isometric.

Occurrence: Found intergrown with *hematite* in heavily oxidized material in an ore dump in Newfoundland, Canada.

D. H. SPEIDEL

References

Am. Min. **59**, 381.
Can. Min. **11**, 1003.

Cross-reference: *Spinel Group.*

CUPROSTIBITE *Am. Min.* **55**, 1810
$Cu_2(Sb,Tl)$ tetr.

CUPROTUNGSTITE *Dana,* II, 1091
$Cu_2(WO_4)(OH)_2$ *Min. Mag.* **43**, 448

CUPROURANITE = *Torbernite* *Min. Mag.* **43**, 1053

CURETONITE *Min. Abst.* **31**, 495
$Ba_4Al_3Ti(PO_4)_4(O,OH)_6$ mono. *Min. Rec.* **10**, 219

CURIENITE orth. *Am. Min.* **54**, 1220
$Pb(UO_2)_2(VO_4)_2 \cdot 5H_2O$

CURITE A hydrated oxide of lead and uranium, named by Schoep in 1921 after Pierre Curie (1859–1906), French physicist known for work on radioactivity.

Composition: $2PbO \cdot 5UO_3 \cdot 4H_2O$.

Properties: Orthorhombic; usually pseudomorphic after *uraninite;* color: orange red; streak: orange; H=4–5, G-7.2; adamantine luster; strongly radioactive.

Occurrence: Rare, found as an oxidation product of *uraninite,* associated with *torbernite, soddyite, fourmarierite, sklodowskite,* and additional secondary uranium minerals, at Kasolo, Shaba, Zaire.

D. H. SPEIDEL

References

Am. Min. 7, 128.
Dana, I, 269.
USGS Bull. **1064**, 92.

CURTISITE *Am. Min.* **61**, 1055
$C_{22}H_{14}$ orth. *Chem. Geol.* **16**, 245

CUSPIDINE RRW, 159
$Ca_4Si_2O_7(F,OH)_2$ mono. Winchell, II, 480

CUSȚERITE = *Cuspidine* *Am. Min.* **33**, 100

CYANITE = *Kyanite* *Min. Mag.* **43**, 1053

CYANOCHROITE *Dana,* II, 454
$K_2Cu(SO_4)_2 \cdot 6H_2O$ mono.

CYANOTRICHITE *Am. Min.* **44**, 839
$Cu_4Al_2(SO_4)(OH)_{12} \cdot 2H_2O$ orth. *Min. Abst.* **27**, 279

CYCLOWOLLASTONITE *Am. Min.* **58**, 560
$CaSiO_3$ tric.

CYLINDRITE *Dana,* I, 482
$Pb_3FeSn_4Sb_2S_{14}$ tric. *Min. Abst.* **27**, 301

CYMRITE orth. *Am. Min.* **52**, 1885
$Ba_2Al_5Si_5O_{19}(OH) \cdot 3H_2O$

CYRILOVITE *Am. Min.* **42**, 586
$NaFe_3^{3+}(PO_4)_2(OH)_4 \cdot 2H_2O$ tetr. *Min. Abst.* **28**, 230

Cross-reference: *Pegmatite Minerals.*

CYRTOLITE–Variety of *zircon*

Cross-reference: *Blacksand Minerals.*

D

DACHIARDITE A hydrated calcium, potassium, sodium, aluminosilicate of the ***zeolite*** group. First named by D'Achiardi in 1906 dedicated to his father.

Composition: Near
$(Ca,K_2,Na_2)[Al_4Si_{18}O_{45}]\cdot 14H_2O$
SiO_2, 62.6%; Al_2O_3, 11.8%; CaO, 6.5%; Na_2O, 1.8%; K_2O, 2.7%; H_2O, 14.6% by wt. Here, $Ca:K_2:Na_2=4:1:1$.

Properties: Monoclinic; $C2/m$; a=18.73 Å, b=10.30 Å, c=7.52 Å, β=107°54'; H=4–4.5; G=2.16; colorless; transparent; habit prismatic; cleavage perfect on {100} and {001}; twinning sector; optically, refractive indices, α=1.491, β=1.496, γ=1.499; birefringence= 0.008; $2V$=65°–73°(+); optic axial plane ⊥(010); $\alpha=y$, γ:Z=38°. X-ray patterns indicate a structural relationship with *epistilbite*, but *dachiardite* has higher alkali and slightly higher Si contents; a Na variety was reported from Alpe di Siusi; decomposed by HCl.

Occurrence: Rare; known from granitic pegmatite of San Piero, Campo, Elba, from porphyrite of Alpe di Siusi, Italy, and from basalts of southwestern Washington and Oregon.

A. IIJIMA
M. UTADA

References

Am. Min. **10**, 421–428; **46**, 769.
Cont. Mineral. Petrol. **49**, 63–69.
Soc. Toscana Sci. Nat. **50**, 14 (*Min. Abst.* **10**, 293).
Wise, W. S., and Tschernich, R. W., 1978. Dachiardite-bearing zeolite assemblages in the Pacific Northwest, in L. B. Sand and F. A. Mumpton (eds.), *Natural Zeolites.* Oxford: Pergamon, 105–112.
Z. Krist. **119**, 53–64.

Cross-reference: *Zeolites.*

DADSONITE *Can. Min.* **17**, 601
$Pb_{21}Sb_{23}S_{55}Cl$ mono. *Min. Mag.* **37**, 437

DAHLLITE (***Apatite*** gr.) *Am. Min.* **45**, 209
$Ca_5(PO_4,CO_3)_3(OH)$ hexa. *Dana*, II, 879

Cross-references: *Apatite Group; Authigenic Minerals; Human and Vertebrate Mineralogy; Invertebrate and Plant Mineralogy; Minerals, Uniaxial and Biaxial.*

DALYITE *Am. Min.* **37**, 1071
$K_2ZrSi_6O_{15}$ tric. *Min. Mag.* **29**, 850

DANALITE An end-member iron beryllosilicate sulfide, named by Cooke in 1866 after J. D. Dana and belonging to the ***helvite*** group.

Composition: $Fe_4Be_3(SiO_4)_3S$; with substantial Mn toward *helvite* and Zn toward *genthelvite*. Isometric.

Occurrence: In Essex County, Massachusetts; near Bartlett, New Hampshire; Iron Mountain, New Mexico, Hiroshima Prefecture, Japan; and at Redruth, Cornwall.

K. FRYE

References

Am. Min. **65**, 355.
Min. Mag. **40**, 627.

DANBURITE A calcium borosilicate named by Sheppard in 1839 after Danbury, Connecticut, where it was first found.

Composition: $CaB_2Si_2O_8$. Orthorhombic.

Occurrence: In marbles and low-temperature veins with *calcite* and *quartz* in Kyushu, Japan, and San Luis Potosi, Mexico; in gravels at Mogok, Burma.

Use: Minor gemstone, cut for collectors.

K. FRYE

Reference: Winchell, II, 258.

DANNEMORITE (***Amphibole*** gr.)
$Mn_2Fe_5Si_8O_{22}(OH)_2$ mono. *Am. Min.* **63**, 1030

D'ANSITE *Am. Min.* **43**, 1221
$Na_{21}Mg(SO_4)_{10}Cl_3$ iso.

Cross-reference: *Saline Minerals.*

DAOMANITE *Am. Min.* **61**, 184
$(Cu,Pt)_2AsS_2$ orth. *Min. Abst.* **30**, 170

DARAPIOSITE (***Milarite*** gr.)
$(K,Na)_3Li(Mn,Zn)_2ZrSi_{12}O_{30}$
Am. Min. **61**, 1053
hexa. *Min. Abst.* 27, 257

DARAPSKITE A hydrated sulfate and nitrate of sodium named by Dietze in 1891 after L. Darapsky of Santiago, Chile.

Composition: $Na_3NO_3SO_4 \cdot H_2O$. Monoclinic.

Occurrence: Found abundantly in certain of the Chilean nitrate deposits, especially those rich in sulfates; associated with *nitratine, bloedite, halite*, and *anhydrite*.

J. A. GILBERT

References

Am. Min. **55**, 1500.
Dana, II, 309.

Cross-references: *Nitrates; Saline Minerals.*

DASHKESANITE = Chlor potassian *hastingsite*
Am. Min. **63**, 1050

DATOLITE A calcium hydroxyl borosilicate named by Esmark in 1806 from the Greek meaning "to divide," referring to the granular character of a massive variety.

Composition: $CaB(SiO_4)(OH)$.

Properties: Monoclinic, $P2_1/c$; usually in crystals but may be granular or massive, resembling unglazed porcelain; H=5–5.5, G=2.8–3.0; vitreous luster; transparent to translucent; color: white to colorless or with a green tint.

Occurrence: A secondary mineral usually found in cavities in basalt or basaltic lavas, associated with ***zeolites***, *prehnite, apophyllite*, and *calcite*. Found at Andreasburg in the Harz Mountains; at Trentino, Italy; in Arendal, Norway, at Aust-Agder in a botryoidal *magnetite* bed.

Use: A minor gem material.

J. A. GILBERT

References

Am. Min. 58, 909.
DHZ, 1, 171.

Cross-reference: *Rock-Forming Minerals.*

DAUBREEITE *Dana*, II, 60
$BiO(OH,Cl)$ tetr.

DAUBREELITE (***Linnaeite*** gr.) *Dana*, I, 265
$FeCr_2S_4$ iso.

Cross-reference: *Meteoritic Minerals.*

DAVIDITE A complex oxide of rare earths (*RE*), titanium, and iron with many substituents and of the ***crichtonite*** group. First named by Sir Douglas Mawson in 1906 in honor of Sir T. W. Edgeworth David (1858–1934), an Australian geologist.

Composition: $AM_{21}O_{38}$, in which A is *RE*, Fe^{2+}, Mg, U, etc. and M is Ti, Fe^{3+}, etc. UO_2 contents from about 3.5% to about 20% have been reported.

Properties: Trigonal; H=5; G=4.3–4.9; submetallic to resinous luster; nearly opaque except on thin edges; color: black, in thin splinters brownish; found only in the *metamict state* (q.v.); can be reconstituted by heating, rarely to single crystals.

Occurrence: Mostly in pegmatites and hydrothermal deposits. Principal localities in Australia and Mozambique, also found in Arizona and Norway. Often found in close association with *rutile* and *ilmenite*; sometimes coated by *anatase*.

Uses: Originally mined for radium in South Australia. An important constituent of some uranium ores.

A. PABST

References: *Am. Min.* **46**, 700–718; **53**, 869–879; **64**, 1010–1017.

Cross-references: *Metamict State; Ore Microscopy.*

DAVISONITE *Am. Min.* **37**, 362
$Ca_3Al(PO_4)_2(OH)_3 \cdot H_2O$ hexa.

DAVREUXITE *Am. Min.* **63**, 795
$Mn_2Al_{12}[(SiO_4)_7O_3(OH)_6]$ mono.

DAVYNE An aluminosilicate of sodium, potassium, and calcium, containing sulfate, carbonate, and chloride and of the ***cancrinite*** group. Named by Monticelli and Covelli in 1825.

Composition: $[(Na,K,Ca)_8Al_6Si_6O_{24}(Cl_2, SO_4,CO_3)_{2-3}]$. Hexagonal.

Occurrence: From Monte Somma, Vesuvius, with *nepheline*.

G. C. MOERSCHELL

Reference: DHZ, 4, 310.

DAWSONITE A hydroxyl carbonate of sodium and aluminum named by Harrington in 1874 after John William Dawson (1820–1899), Canadian geologist and Principal of McGill University.

Composition: $4[NaAlCO_3(OH)_2]$.

Occurrence: A low-temperature hydrothermal mineral probably formed by the decomposition of aluminous silicates; associated with *calcite, dolomite, pyrite, fluorite, galena*, and *quartz*.

J. A. GILBERT

References

Dana, II, 276.
Min. Abst. **29**, 264, 78–2742.

Cross-reference: *Green River Mineralogy.*

DAYINGITE *Am. Min.* **61**, 184
$Cu(Co,Pt)_2S_4$ iso.

DEERITE mono. *Am. Min.* **50**, 278; **62**, 990
$Fe_6^{2+}Fe_3^{3+}O_3[Si_6O_{17}](OH)_5$

DEHRNITE = Carbonate *fluorapatite* *Min. Mag.* **42**, 282

DELAFOSSITE A copper iron oxide of the ***zincite-tenorite*** group named by Friedel (1873) after G. Delafosse (1796–1878), a French mineralogist.

Composition: $CuFeO_2$.

Properties: Trigonal, $R\bar{3}m$; H=5.5; color and streak are black.

Occurrence: As tabular crystals from the Ural Mountains and Bisbee, Arizona.

D. H. SPEIDEL

References

Am. Min. **53**, 1779.
Dana, I, 674.
Min. Mag. **36**, 643.

DELHAYELITE *Am. Min.* **44**, 1321
$(Na,K)_{10}Ca_5Al_6Si_{32}O_{80}(Cl_2,F_2,SO_4)_3\cdot18H_2O$
orth. *Min. Mag.* **32**, 6

DELLAITE *Am. Min.* **50**, 2104
$Ca_6Si_3O_{11}(OH)_2$ *Min. Mag.* **34**, 1

DELRIOITE *Am. Min.* **55**, 185
$CaSrV_2O_6(OH)_2\cdot3H_2O$ mono.

DELVAUXITE *Am. Min.* **65**, 813
$CaFe_3^{3+}(PO_4,SO_4)_2(OH)_8\cdot4\text{–}6H_2O$(?)

DEMESMAEKERITE tric. *Am. Min.* **51**, 1815
$Pb_2Cu_5(UO_2)_2(SeO_3)_6(OH)_6\cdot2H_2O$

DENNINGITE *Am. Min.* **48**, 1419
$(Mn,Zn)Te_2O_5$ tetr. *Can. Min.* **7**, 443

DERBYLITE mono. *Am. Min.* **62**, 396
$Fe_4^{3+}Ti_3^{4+}Sb^{3+}O_{13}(OH)$ *Dana*, II, 1025

DERRIKSITE orth. *Am. Min.* **57**, 1912
$Cu_4(UO_2)(SeO_3)_2(OH)_6\cdot H_2O$

DESAUTELSITE *Am. Min.* **64**, 127
$Mg_6Mn_2(CO_3)(OH)_{16}\cdot H_2O$
trig. *Min. Mag.* **43**, 1060

DESCLOIZITE A lead, zinc hydroxyl vanadate named in 1854 for the French mineralogist Alfred L. O. L. Des Cloizeau.

Composition: $4[PbZn(VO_4)(OH)]$ with Zn replaced by Cu (Cu>Zn, *mottramite*).

Properties: Orthorhombic, *Pnam*; in crystal druses and large groups, commonly massive or mamilliform; H=3–3.5; G=5.9–6.2; greasy luster; color: red, orange, light to dark brown, or black, green when cuprian; streak: yellowish, brownish red, or brown.

Occurrence: A secondary mineral in the oxidized zone of ore deposits, some crystals pseudomorphous after *vanadinite*.

Use: Mined for vanadium at Berg Aukas, Namibia.

K. FRYE

Reference: *Dana*, II, 811–815.

DESMINE = *Stilbite* *Min. Mag.* **43**, 1053

DESPUJOLSITE *Am. Min.* **54**, 326
$Ca_3Mn(SO_4)_2(OH)_6\cdot3H_2O$ hexa.

DEVILLINE (DEVILLITE)
$CaCu_4(SO_4)_2(OH)_6\cdot3H_2O$ mono. *Dana*, II, 590

DEWEYLITE—Disordered mixtures of ***serpentine*** and *talc* *Min. Mag.* **42**, 75

DEWINDTITE *Am. Min.* **39**, 444
$Pb(UO_2)_2(PO_4)_2\cdot3H_2O$
orth. *USGS Bull.* **1064**, 230

DIABANTITE A magnesium iron hydroxyl aluminosilicate of the ***chlorite*** group named by Liebe in 1870 from its association with diabase.

Composition:
$2[(Mg,Fe^{2+},Al)_6(Si,Al)_4O_{10}(OH)_8]$; Si>3.1; $Fe^{2+}:(Mg+Fe^{2+})$=0.2–0.5.

Occurrence: Filling cavities and seams in mafic igneous rocks.

G. C. MOERSCHELL

Reference: DHZ, 3, 137.

Cross-reference: *Chlorite Group.*

DIABOLEITE A hydroxyl chloride of lead and copper named by Spencer in 1923 from the Greek meaning "apart or distant from" and *boleite*.

Composition: $[Pb_2CuCl_2(OH)_4]$. Tetragonal.

Occurrence: Rare, found originally at Higher Pitts farm, Mendip Hills, Somerset, England, in oxidized iron and manganese ores associated with *mendipite, chloroxiphite, hydrocerussite*, and *cerussite*.

J. A. GILBERT

Reference: *Min. Mag.* **43**, 901.

DIADOCHITE tric.	*Am. Min.* **55**, 135
$Fe_2(OH)(PO_4)(SO_4)\cdot 5H_2O$	*Dana*, I, 1011

DIALOGITE = *Rhodochrosite*

DIAMOND A high-pressure polymorph of carbon named by Pliny in 77 AD from a corruption of the Greek word meaning invincible. (See main entry in text.)

Composition: 8[C]; may contain small amounts of Si, Fe, Al, N, Mg, or carbonaceous material.

Properties: Cubic, *Fd3m*; octahedral habit, less commonly dodecahedral or cubic, with curved striated crystal faces; commonly twinned; H=10; G=3.50–3.53; color: clear, bluish, reddish, yellowish, brown, gray, or black; adamantine to greasy luster; n=2.42; strong dispersion.

Occurrence: In kimberlite pipes and dikes; also placer deposits.

Polymorphs: *Chaoite, lonsdaleite, graphite.*

Use: Abrasive, gemstone, heat sink (electronics industry).

J. B. DAWSON

References

Bruton, E., 1978. *Diamonds*, 2nd ed. London: Nat. Assoc. of Gemology Press, 532p.

Field, J. E. (ed.), 1979. *The Properties of Diamond.* London: Academic; 674p.

Smith, G. F. H., 1958. *Gemstones*, 13th ed. London: Methuen, 560p.

Cross-references: *Abrasive Materials; Allotropy; Blacksand Minerals; Color in Minerals; Crystal Growth; Crystallography: Morphological; Diamond; Gemology; Isomorphism; Jade; Mantle Mineralogy; Meteoritic Minerals; Mineral Industries; Mineral and Ore Deposits; Mineral Properties; Minerals, Uniaxial and Biaxial; Mohs Scale of Hardness; Native Elements; Optical Mineralogy; Piezoelectricity; Placer Deposits; Polymorphism; Skeletal Crystals; Synthetic Minerals.*

DIAPHORITE		*Am. Min.* **60**, 621
$Pb_2Ag_3Sb_3S_8$	orth.	*Dana*, I, 414

DIASPORE An aluminum oxyhyroxide, one of the principal constituents of bauxite. Named by Haüy in 1801 from Greek *diaspora*, "scattering," alluding to usual decrepitation before the blowpipe.

Composition: α-AlO(OH).

Properties: Orthorhombic, *Pbnm*; H=6.5–7; G=3.2–3.5; perfect {010} cleavage; structure is based on hexagonal close-packed layers of O, with Al in octahedral coordination between them; H bonds occur between the oxygen atoms. Optically, α=1.682–1.706, β=1.705–1.725, γ=1.730–1.752. $2V$=84–86°. White or colorless, but Fe and Mn varieties are green, brown, pink. Converts to α-Al_2O_3, *corundum*, on heating at about 575°C. Structurally similar to *goethite*, FeO(OH); *groutite*, MnO(OH); *montroseite*, (Fe,V)O(OH); and *paramontroseite*, VO_2.

Occurrence: Bauxite and emery deposits, at Chester, Massachusetts; Campolongo, Switzerland; near Mramorskoi, Urals, USSR; Julianehaab district, Greenland. Commonly associated with *gibbsite* and *boehmite* in bauxite; *corundum, margarite, spinel,* ***chlorite***, and *magnetite* in emery.

Use: Bauxite is the main source of aluminum; diaspore is used as a refractory and an abrasive.

D. H. SPEIDEL

References

AJS **262**, 709–712.

Am. Min. **63**, 326–329.

Ber. Deut. Keram. Ges. **42**(5), 167–184.

DHZ, 5, 102–110.

Cross-references: *Mineral and Ore Deposits; Rock-Forming Minerals; Topotaxy.*

DICKINSONITE mono.	*Am. Min.* **50**, 1647
$(K,Ba)(Na,Ca)_5(Mn,Fe,Mg)_{14}$	
$Al(PO_4)_{12}(OH,F)\cdot H_2O$	*Min. Mag.* **43**, 227

Cross-reference: *Pegmatite Minerals.*

DICKITE A clay mineral of the ***kaolinite*** group differing from *kaolinite* by having a two-layer unit cell and monoclinic symmetry. Occurrence restricted to a hydrothermal origin with *quartz* and sulfide minerals.

K. FRYE

Reference: DHZ, 3, 194–212.

Cross-references: *Clays, Clay Minerals; Minerals, Uniaxial and Biaxial; Phyllosilicates; Soil Mineralogy; Staining Techniques.*

DIDYMOLITE = ***Plagioclase*** Am. Min. **50**, 2111

DIENERITE *Dana*, I, 175
Ni_3As iso.

DIETRICHITE (***Halotrichite*** gr.)
$(Zn,Fe,Mn)Al_2(SO_4)_4 \cdot 22H_2O$
mono. *Dana*, II, 528

DIETZEITE *Dana*, II, 318
$Ca_2(IO_3)_2(CrO_4)$ mono.

Cross-reference: *Saline Minerals.*

DIGENITE A copper sulfide of the ***argentite*** group named from the Greek *di*, two, and *genus*, kinds, for the presence of cupric and cuprous ions.

Composition: Cu_9S_5.

Properties: Isometric *Fm*3*m*; massive, conchoidal fracture; H=2.5–3; G=5.5–5.7; blue to black.

Occurrence: With *chalcocite, bornite*, etc. in copper ores.

K. FRYE

References

Am. Min. **62**, 107–114.
Dana, I, 180–182.

Cross-reference: *Ore Microscopy.*

DIMORPHITE *Z. Krist.* **138**, 161
As_4S_3 ortho.

DIOPSIDE A calcium magnesium silicate of the ***pyroxene*** group named by d'Andrada in 1800 from the Greek words meaning "double" and "appearance" because the vertical prism zone can be oriented in two ways.

Composition: $4[CaMg(Si_2O_6)]$; CaO, 25.9%; MgO, 18.5%; SiO_2, 55.6%. A complete solid solution exists between *diopside* and *hedenbergite*, $CaFe(Si_2O_6)$.

Properties: Monoclinic, *C*2/*c*; commonly polysynthetically twinned; may be granular massive, lamellar, or columnar; H=5–6; G=3.2–3.3; vitreous luster; transparent to translucent; color: white to light green.

Occurrence: Found commonly in contact metamorphic zones in crystalline limestone, and also in regionally metamorphic rocks; associated with ***scapolites***, *idocrase, titanite,* ***garnets***, and *tremolite*.

Use: Transparent variety may be used as a gem stone.

J. A. GILBERT

References

Am. Min. **66**, 1.
DHZ, 2A, 198.

Cross-references: *Clinopyroxenes; Diamond; Mantle Mineralogy; Meteoritic Minerals; Minerals, Uniaxial and Biaxial; Orthopyroxenes; Polysomatism; Pyroxene Group; Soil Mineralology; Thermometry, Geologic.*

DIOPTASE A copper hydroxyl silicate, named by Haüy in 1797 from the Greek "to see through" because internal cleavage planes can be seen.

Composition: $Cu_6(Si_6O_{18}) \cdot 6H_2O$

Physical Properties: Trigonal, $R\bar{3}$; H=5; G=3.28–3.35; vitreous luster; perfect cleavage $\{10\bar{1}1\}$; emerald-green color; commonly in stubby prismatic crystals and crystal aggregates, massive; conchoidal to uneven fracture; pale green streak; refractive index, ω=1.644–1.685, ϵ=1.697–1.709; transparent to translucent; becomes gelatinous in HCl.

Occurrence: Associated with copper ores such as *malachite, brochantite, chrysocolla* in the oxidized zone; found in Russia; Shaba Province, Zaire; Tsumeb, Namibia; Arizona; Chili. Gangue minerals frequently present are *calcite, quartz*, and limonite.

Use: Formerly quite rare, but recently found in Africa in quantities large enough to use as a copper ore.

W. E. HILL

References

Am. Min. **62**, 807–811.
Dana, E. S., 1945. *Textbook of Mineralogy.* New York: Wiley.
DHZ, 1, 257.

Cross-references: *Magnetic Minerals; Naming of Minerals.*

DIPYRE—Variety of ***scapolite*** DHZ 4, 322

DISTHENE = *Kyanite* *Min. Mag.* **43**, 1053

DITTMARITE orth. *Am. Min.* **57**, 1316
$(NH_4)Mg(PO_4) \cdot H_2O$
USGS Prof. Pap. 750-A, A115

DIXENITE trig. *Am. Min.* **63**, 155
$Mn_{11}^{2+}Mn_4^{3+}(OH)_8(AsO_3)_6(SiO_4)_2$

DJALMAITE = *Uranmicrolite*
Am. Min. **62**, 406

DJERFISHERITE *Am. Min.* **51**, 1815
$K_6(Cu,Fe,Ni)_{23}S_{26}Cl$ iso. *Science* **153**, 166

Cross-reference: *Meteoritic Minerals.*

DJURLEITE *Am. Min.* **48**, 215
$Cu_{1.94}S$ mono. *Science* **203**, 356

DOG-TOOTH SPAR = *Calcite*

DOLEROPHANITE *Dana,* II, 551
$Cu_2(SO_4)O$ mono.

DOLOMITE Calcium magnesium carbonate named by Saussure in 1792 after the French chemist, Dolomieu (1750–1801).

Composition: $[CaMg(CO_3)_2]$. In ordinary *dolomite* $CaCO_3:MgCO_3$ is 1:1. However, it may vary so that the ratio of calcium to magnesium ranges between 58:42 and 47.5:52.5. Forms a series with *ankerite,* Ca(Fe, Mg,Mn)$(CO_3)_2$, and with *kutnohorite,* Ca (Mn,Mg,Fe)$(CO_3)_2$.

Properties: Trigonal, $R\bar{3}$; saddle-shaped rhombohedral crystals, may also be found in coarse granular to compact masses; H= 3.5–4; G=2.85; vitreous luster, may be pearly (pearl spar); transparent to translucent; color: commonly some shade of pink but may be colorless, white, gray, green, brown, or black.

Occurrence: Found chiefly in large beds as dolomitic limestone or dolomitic marble, often intimately associated with *calcite* and has similar occurrences. Also found as a vein mineral (q.v.) associated with lead and zinc minerals, and an authigenic mineral (q.v.).

Uses: Building stone and ornamental stone; a source of magnesium.

J. A. GILBERT

References

Am. Min. **62**, 772.
DHZ, 5, 278.
J. Geol., **70**, 659.

Cross-references: *Authigenic Minerals; Cave Minerals; Diagenetic Minerals; Electron Microscopy (Transmission); Fire Clays, Green River Mineralogy; Meteoritic Minerals; Mineral and Ore Deposits; Minerals, Uniaxial and Biaxial; Plastic Flow in Minerals; Refractory Minerals; Rock-Forming Minerals; Soil Mineralogy; Staining Techniques; Thermoluminescence; Thermometry, Geologic; Vein Minerals.*

DOLORESITE *Am. Min.* **42**, 587; **45**, 1144
$H_8V_6O_{16}$ mono.

DOMEYKITE *Min. Abst.* **30**, 19, 294
Cu_3As iso. and hexa.

DONATHITE *Am. Min.* **54**, 1218
$(Fe,Mg)(Cr,Fe)_2O_4$ tetr.

DONBASSITE (**Chlorite** gr.)
$Al_4(Si,Al)_4O_{10}(OH)_8$(?) *Min. Mag.* **26**, 336

DONNAYITE tric. *Am. Min.* **64**, 653
$Sr_3NaCaY(CO_3)_6\cdot 3H_2O$ *Can. Min.* **16**, 335

DORFMANITE *Am. Min.* **66**, 217
$Na_2HPO_4\cdot 2H_2O$ orth. *Min. Mag.* **43**, 1060

DOUGLASITE *Dana,* II, 100
$K_2Fe^{2+}Cl_4\cdot 2H_2O$ mono.

Cross-reference: *Saline Minerals.*

DOWNEYITE *Am. Min.* **62**, 316
SeO_2 tetr. *Min. Abst.* **29**, 83

DRAVITE A magnesium-rich aluminum borosilicate of the ***tourmaline*** group named after the Drave district, Corinth, Greece.

Composition: $3[NaMg_3Al_6(BO_3)_3Si_6O_{18}(OH,F)_4]$; Na_2O, 3.2%; MgO, 12.6%; Al_2O_3, 31.9%; B_2O_3, 10.9%; SiO_2, 37.6%; H_2O, 3.8% by wt.

Properties: Trigonal, $R3m$; prismatic crystals with triangular or hexagonal cross section, also massive, columnar, and in parallel and radiating groups; poorly developed cleavage; H=7–7.5; G=3.03–3.15; vitreous to resinous luster; color: brown to black; transparent to opaque; strongly pyroelectric and piezoelectric; refractive indices, ω=1.635–1.661, ϵ=1.610–1.632; unit-cell dimensions a(Å)= 15.94–15.98, c(Å)=7.19–7.23.

Occurrence: Usually found in metamorphic or metasomatic lime-rich rocks including crystalline limestones and dolomites and mafic igneous rocks where *dravite* may be associated with *datolite* and ***axinite***. Fine crystals are found at Dobrava, Corinthia, Austria, and Gouverneur, St. Lawrence County, New York.

Uses: Because of its piezoelectric properties, ***tourmaline*** is used in the manufacture of pressure gauges to measure transient blast pressures. Plates are cut perpendicular to the three-fold axis for this purpose.

P. E. ROSENBERG
F. F. FOIT

References

Acta Cryst., **15**, 583–590.

Battey, M. H., 1972. *Mineralogy for Students.* Edinburgh: Oliver & Boyd, 255–256.

DHZ, 1, 300-319.

Sinkankas, J., 1966. *Mineralogy: A First Course,* 2nd ed. Princeton, N. J.: Van Nostrand, 510–514.

DRESSERITE orth. *Am. Min.* **55**, 1447
$Ba_2Al_4(CO_3)_4(OH)_8 \cdot 3H_2O$
Can. Min. **10**, 84

DROOGMANSITE = *Kasolite*
Am. Min. **64**, 1334

DRUGMANITE *Am. Min.* **65**, 809
$Pb_2(Fe,Al)(PO_4)_2(OH) \cdot H_2O$
mono. *Min. Mag.* **43**, 463

DRYSDALLITE *Am. Min.* **59**, 1139
$Mo(Se,S)_2$ hexa.

DUFRENITE mono. *Am. Min.* **55**, 135
$CaFe^{2+}_2Fe^{3+}_{10}(OH)_{12}(H_2O)_4(PO_4)_8$
Min. Rec. **9**, 388

Cross-reference: *Pegmatite Minerals.*

DUFRENOYSITE *Dana,* I, 442
$Pb_2As_2S_5$ mono.

DUFTITE *Am. Min.* **42**, 123
$PbCu(AsO_4)(OH)$ orth. *Can. Min.* **18**, 191

DUGGANITE hexa. *Am. Min.* **63**, 1016
$Pb_3Zn_3(TeO_6)_x(AsO_4)_{2-x}(OH)_{6-3x}$

DUMONTITE mono. *USGS Bull.* **1064**, 236
$Pb_2(UO_2)_3(PO_4)_2(OH)_4 \cdot 3H_2O$

DUMORTIERITE An aluminum borosilicate named by Gonnard in 1881 after the French paleontologist, Eugene Dumortier.

Composition: $Al_7(BO_3)(SiO_4)_3O_3$. Al may be replaced by Fe^{3+} in major amounts and by Ti^{4+} in minor amounts.

Properties: Orthorhombic, *Pcmn;* fibrous or radiating aggregates; H=7; G=3.26–3.36; vitreous luster; transparent to translucent; color: blue, violet, pink, or greenish-blue.

Occurrence: Embedded in ***feldspar*** in a gneiss near Lyons and Beaunan, France. Found in schist and gneiss, and occasionally in pegmatitic dikes, on Manhattan Island, New York. Common at Oreana, Nevada; Dehesa, California; Minas Geraes, Brazil; and Madagascar.

Use: In the manufacture of spark plug porcelain.

J. A. GILBERT

References

Am. Min. **61**, 1016–1019.

Geol. Soc. Am. Bull., **64**, 1504.

Cross-reference: *Refractory Minerals.*

DUNDASITE A hydroxyl carbonate of lead and aluminum named by Petterd in 1893 from the locality in Dundas, Tasmania.

Composition: $Pb_2Al_4(CO_3)_4(OH)_8 \cdot 3H_2O$. Orthorhombic.

Occurrence: Found originally associated with *crocoite, pyromorphite*(?), and limonite in gossan in the Adelaide Proprietary mine, Dundas, Tasmania.

J. A. GILBERT

References

Dana, II, 279.

Min. Mag., **14**, 167.

DURANGITE *Can. Min.* **12**, 262
$NaAl(AsO_4)F$ mono. *Dana,* II, 827

DURANUSITE *Am. Min.* **60**, 945
As_4S orth.

DUSSERTITE (***Crandallite*** gr.)
$BaFe_3[AsO_3(O,OH)]_2(OH)_6$
Am. Min. **55**, 135
trig. *Dana,* II, 839

DUTTONITE *Am. Min.* **42**, 455
$VO(OH)_2$ mono.

DYPINGITE mono.(?) *Am. Min.* **55**, 1457
$Mg_5(CO_3)_4(OH)_2 \cdot 5H_2O$

DYSCRASITE *Can. Min.* **14**, 139
Ag_3Sb orth. *Dana,* I, 173

Cross-reference: *Ore Microscopy.*

DZHALINDITE *Am. Min.* **49**, 439
$In(OH)_3$ iso.

E

EAKERITE mono. *Am. Min.* **61**, 956
$Ca_2SnAl_2Si_6O_{18}(OH)_2 \cdot 2H_2O$
Min. Rec. **1**, 92

EARLEYITE = *Takovite* *Am. Min.* **62**, 449

EARLANDITE (Calcium citrate)
$Ca_3(C_6H_5O_7)_2 \cdot 4H_2O$ *Dana,* II, 1105

ECDEMITE *Dana,* II, 1036
$Pb_6As_2O_7Cl_4$ tetr.

ECKERMANNITE A hydroxyl aluminoferrosilicate of sodium, magnesium, and lithium of the ***amphibole*** group. Named by Adamson in 1942 after Professor C. W. H. von Eckermann of Stockholm.

Composition: $Na_3(Mg,Li)_4(Al,Fe^{3+})Si_8O_{22}(OH,F)_2$. Monoclinic. *Eckermannite* is the magnesium-rich end member of the *eckermannite-arfvedsonite* series.

Occurrence: In the pectolite-nepheline syenite from Norra Kärr, Sweden.

Use: An asbestos mineral.

G. C. MOERSCHELL

References

Am. Min. **63**, 1036.
DHZ, 2, 364.
Min. Mag. **27**, 268.

Cross-reference: *Asbestos.*

EDENITE (***Amphibole*** gr.)
$NaCa_2Mg_5(AlSi_7)O_{22}(OH)_2$
Am. Min. **63**, 1032
mono. DHZ, 2, 263

EDINGTONITE A hydrated barium aluminosilicate of the ***zeolite*** group. First named by Haidinger in 1825 in honor of Mr. Edington, who found it in 1823 near West Kilpatrick, Dumbartonshire, Scotland.

Composition: $Ba_2[Al_4Si_6O_{20}] \cdot 8H_2O$; SiO_2, 35.5%; Al_2O_3, 20.1%; BaO, 30.2%; H_2O, 14.2% by wt.

Properties: Orthorhombic (pseudo-tetragonal), $P2_12_12$; a=9.60 Å, b=9.60 Å, c=6.54 Å; G=2.7–2.8; H=4–4.5; vitreous luster; colorless, white, brown, or pink; translucent to opaque; habit minute pseudotetragonal or in slabs; cleavage on {110} and {$1\bar{1}0$}; shows pyroelectricity. Optically, refractive indices, α=1.541, β=1.553, γ=1.557; birefringence = 0.015; $2V$=54°(−); optic axial plane (010); $\alpha=z$, $\beta=y$, $\gamma=x$; dispersion $r<v$, strong. Some specimens from Old Kilpatrick contain appreciable Ca; Ba is exchanged with Na, K, Ag, and Tl; gelatinized by HCl.

Occurrence: In mafic igneous rock from the Kilpatrick Hills, Scotland, associated with *harmotome, analcime, thomsonite, prehnite,* and *calcite;* at Böhlet mine. Sweden.

A. IIJIMA
M. UTADA

References

Acta Cryst. **B32**, 1623.
DHZ, 4, 354–376.
Min. Mag. **23**, 483–494.
Z. Krist. **88**, 53.

Cross-references: *Minerals, Uniaxial and Biaxial; Zeolites.*

EGLESTONITE *Am. Min.* **62**, 396
$Hg_6Cl_3O_2H$ iso. *Dana,* II, 51

EIFELITE (***Osumulite*** gr.) *Am. Min.* **66**, 218
$K_2Na_4Mg_9Si_{24}O_{60}$ hexa.

EITELITE *Am. Min.* **58**, 211
$Na_2Mg(CO_3)_2$ hexa.

Cross-reference: *Green River Mineralogy.*

EKANITE tetr. *Am. Min.* **46**, 1516
$(Th,U)(Ca,Fe,Pb)_2Si_8O_{20}$ *Can. Min.* **11**, 913

EKATERNITE hexa. *Am. Min.* **66**, 437
$Ca_2B_4O_7(Cl,OH)_2 \cdot 2H_2O$

ELBAITE A lithium-bearing, aluminum borosilicate of the ***tourmaline*** group named after the island of Elba in the Mediterranean Sea.

Composition: $3[Na(Li,Al)_3Al_6(BO_3)_3Si_6O_{18}(OH,F)_4]$. Na_2O, 3.2%; Li_2O, 4.2%; Al_2O_3, 40.1%; B_2O_3, 10.9%; SiO_2, 37.8%; H_2O, 3.8% by wt.

Properties: Trigonal, $R3m$; prismatic crystals with triangular or hexagonal cross sections, also massive, columnar, and in parallel and radiating groups; cleavage poorly developed; H=7–7.5; G=3.10–3.25; vitreous to resinous luster; color: red, green, yellow, blue, or colorless, often zoned within a crystal; strongly pyroelectric and piezoelectric; refractive indices ω=1.640–1.655, ϵ=1.615–1.620; unit-cell dimensions a(Å)=15.84–15.9, c(Å)=7.10–7.13.

Occurrence: Found in granites, granite pegmatites, and in late-stage granitic veins; near the core or in pockets in pegmatites associated with *quartz, lepidolite, spodumene,* and other minerals. Famous localities include Mount Mica, Oxford County, Maine; Haddam Neck, Connecticut; Mesa Grande and Pala, San Diego County, California; and Minas Gerais, Brazil.

Uses: Often used as a gem stone (rubellite, pink; indicolite, blue; verdelite, green; or achroite, colorless). Plates cut perpendicular to the threefold axis are used in the manufacture of pressure gauges to measure sudden transient changes in pressure.

P. E. ROSENBERG
F. F. FOIT

References

Battey, M. H., 1972. *Mineralogy for Students,* Edinburgh: Oliver & Boyd, 255–256.
DHZ, 1, 300–319.
Sinkankas, J., *Mineralogy: A First Course,* 2nd ed. Princeton, N.J.: Van Nostrand, 510–514.
Tsch. Min. Pet. **18**, 273–286.

Cross-references: *Gemology; Pegmatite Minerals.*

ELECTRUM
(Au,Ag) iso.

Cross-references: *Mineral Classification: Principles; Native Elements and Alloys; Ore Microscopy.*

ELLESTADITE A member of the ***apatite*** group and named by D. McConnell in 1937 after Dr. R. B. Ellestad of the University of Minnesota.

Composition:
$2[Ca_5(SiO_4,PO_4,SO_4)(OH,Cl,F)]$.
Hexagonal.

Occurrence: From Crestmore, California, where it is closely associated with *wilkeite, idocrase, calcite, diopside,* and *wollastonite.*

G. C. MOERSCHELL

References

Am. Min. **22**, 977.
Min. Abst. 7, 14, 88.
Min. Mag. **25**, 62.

Cross-reference: *Apatite Group.*

ELLISITE *Am. Min.* **64**, 701
Tl_3AsS_3 trig. *Min. Mag.* **43**, 1060

ELLSWORTHITE = *Uranpyrochlore*
Am. Min. **62**, 406

ELPASOLITE *Dana,* II, 114
K_2NaAlF_6 iso.

ELPIDITE A hydrated zirconium silicate with sodium named by Lindström in 1894.

Composition: $Na_2ZrSi_6O_{15}\cdot 3H_2O$; with some Ti, but Zr>Ti. Monoclinic.

Occurrence: From Narsarsuk, Greenland, and from Mount Chibina, Kola Peninsula, Russian Lapland (titanian *elpidite*) associated with pink *albite* in a nepheline-syenite pegmatite.

G. C. MOERSCHELL

References

Min. Mag. **11**, 326.
Min. Abst. **30**, 113.

Cross-reference: *Green River Mineralogy.*

ELYITE *Am. Min.* **57**, 364
$Pb_4Cu(SO_4)(OH)_8$ mono.

EMBOLITE A halide of silver named by Breithaupt (1849) from the Green meaning "an intermediate" because it is between the chloride (*chlorargyrite*) and bromide (*bromargyrite*) of silver in composition.

Composition: $4[Ag(Cl,Br)]$; ratio of chlorine to bromine varying extensively. Isometric.

Occurrence: A secondary mineral, found in the oxidized zone of silver deposits, especially in arid regions, such as Chañarcillo, Tres-Puntas, and Rosilla, Chile; Eulaia, in Chihuahua, Mexico; and Sunny Corner, New South Wales.

Use: An important silver ore in Chile.

J. A. GILBERT

References

Dana, II, 11–15.
Min. Record 7, 25–33.

EMBREYITE mono. *Am. Min.* **58**, 806
$Pb_5(CrO_4)_2(PO_4)_2 \cdot H_2O$ *Min. Mag.* **38**, 790

EMELEUSITE orth. *Min. Abst.* **29**, 200
$Li_2Na_4Fe_2^{3+}Si_{12}O_{30}$ *Min. Mag.* **42**, 31

EMERALD–Green gem variety of *beryl*

Cross-reference: *Color in Minerals; Gemology; Placer Deposits.*

EMERY A mixture of *corundum* and *magnetite*

Cross-reference: *Abrasive Materials.*

EMMONSITE *Dana,* II, 640
$Fe_2Te_3O_9 \cdot 2H_2O$ tric. *Min. Rec.* **3**, 82

EMPLECTITE *Dana,* I, 435
$CuBiS_2$ ortho. *Min. Abst.* **27**, 300

Cross-reference: *Ore Microscopy.*

EMPRESSITE *Am. Min.* **49**, 325
AgTe orth.

ENARGITE A copper arsenic sulfide named by Breithaupt in 1850 from the Greek word meaning "distinct" in reference to the cleavage.

Composition: Cu_3AsS_4; Cu, 48.3%; As, 19.1%; S, 32.6% by wt.

Properties: Orthorhombic, $Pnm2_1$; commonly columnar, bladed, or massive; H=3; G=4.43–4.45; metallic luster; color: grayish black to iron black.

Occurrence: A rare mineral found in vein and replacement deposits. Associated with *pyrite, sphalerite, bornite, galena, tetrahedrite, covellite,* and *chalcocite.* Principal localities are Butte, Montana; Bingham and Tintic, Utah; Chuquicamata, Chile; Morococha and Cerro de Pasco, Peru.

Use: Ore of copper and arsenic.

J. A. GILBERT

References

Am. Min. **49**, 1458.
Dana, I, 289.

Cross-references: *Metallic Minerals; Ore Microscopy.*

ENDELLITE (***Kaolinite-Serpentine*** gr.)
$Al_2Si_2O_5(OH)_4 \cdot 2H_2O$
mono. *Am. Min.* **48**, 214

ENDLICHITE (***Apatite*** gr.) *Dana,* II, 897
$Pb_5[(V,As)O_4]_3Cl$ hexa.

ENGLISHITE mono. *Dana,* II, 957
$K_4Na_2Ca_9Al_{18}(H_2O)_8$ *Min. Mag.* **40**, 863
$(OH)_{36}(PO_4)_6(PO_3OH)_{12}$

ENIGMATITE = *Aenigmatite*

ENSTATITE Magnesium silicate of the ***pyroxene*** group named by Kenngott in 1855 from the Greek word meaning "opponent" because of its refractory nature.

Composition: $Mg_2Si_2O_6$; MgO, 40.4%; SiO_2; 59.6%. Replacement of Mg by Fe. If the amount of Fe is > 13%, the mineral is called *hypersthene.* Pure iron silicate, $Fe_2Si_2O_6$, is *ferrosilite.*

Properties: Orthorhombic, *Pbca;* usually massive, fibrous, or lamellar; H=5.5; G=3.2–3.5; vitreous to pearly luster on fresh surface; translucent; color: white with tints of gray, yellow or green, to olive green and brown.

Occurrence: Found in igneous rocks such as pyroxenite, periodites, gabbro, norite, and basalt, and in meteorites. Associated with ***garnet*** and *chondrodite* in the Tilly Foster *magnetite* mine at Brewster, New York.

J. A. GILBERT

References

Am. Min. **66**, 1–50.
DHZ, 2A, 20–161.

Cross-references: *Crystallography: Morphological; Diamond; Mantle Mineralogy; Meteoritic Minerals; Mineral Classification: Principles; Minerals, Uniaxial and Biaxial; Orthopyroxenes; Pyroxene Group; Soil Mineralogy.*

EOSPHORITE A hydrated manganese aluminum hydroxyl phosphate of the ***childrenite*** series, named in 1878 from the Greek *eos,* dawn, and *phoros,* bearing, because of its pink color.

Composition: $8[MnAl(PO_4)(OH)_2 \cdot H_2O]$ with Mn^{2+} replaced by Fe^{2+} toward *childrenite.*

Properties: Orthorhombic; crystals prismatic, also in crusts; H=5; G=3.1–3.2; vitreous; color: pink, brown.

Occurrence: A pegmatite mineral (q.v.) or a vein mineral (q.v.).

K. FRYE

References

Am. Min. **35,** 793–805.
Dana, II, 936-939.
Min. Mag. **43,** 1015–1023.

EPHESITE A sodium lithium hydroxyl aluminosilicate belonging to the ***mica*** group (q.v.). Named after Ephesus, Turkey, where it occurs.

Composition: $Na(LiAl_2)[Al_2Si_2O_{10}](OH)_2$. Monoclinic.

Occurrence: From near Ephesus, Turkey, and in the Postmasburg district, South Africa. Commonly associated with emery deposits and Mn ores.

K. FRYE

References

Am. Min. **47,** 599; **52,** 1689.
Min. Mag. **22,** 482.

EPIDIDYMITE A sodium hydroxyl beryl-liosilicate named by Flink in 1893 in allusion to its dimorphic relationship to *eudidymite.*

Composition: $HNaBeSi_3O_8$.

Occurrence: In pegmatite on the island of Arö, Langesundfjord, Norway, and at Narsarsuk, Greenland.

K. FRYE

References

Am. Min. **55,** 1541.
Min. Mag. **11,** 326.

EPIDOTE A calcium iron hydroxyl aluminosilicate of the ***epidote*** group named by Haüy in 1801 from the Greek meaning "increase" because the base of the vertical prism has one side longer than the other.

Composition: $2[Ca_2(Al,Fe)_3O(SiO_4)(Si_2O_7)OH]$. If $Fe^{3+} < 15\%$ (Fe + Al), the mineral is *clinozoisite;* Mn replaces (Fe,Al) toward *piemontite.*

Properties: Monoclinic, $P2_1/m$; striated crystal faces; usually granular, may be fibrous; H=6–7; G=3.35–3.45; vitreous luster; translucent; color: commonly green or may be gray to black.

Occurrence: Common in metamorphic rocks, such as gneiss, schist, marble, and amphibolite, in which it was formed from the metamorphism of ***feldspars, pyroxenes, amphiboles,*** and *biotite.* Commonly associated with ***chlorite,*** *albite, actinolite, scheelite,* and ***garnet*** (in California). Particularly well-formed crystals are found at Untersulzbachtal, Austria, and Prince of Wales Island, Alaska.

Use: Minor gem stone.

J. A. GILBERT

References

Am. Min. **44,** 720; **56,** 467.
DHZ, 1, 193–210.
Kostov, I., 1968. *Mineralogy.* Edinburgh: Oliver & Boyd, 308–314.

Cross-references: *Diagenetic Minerals; Minerals; Uniaxial and Biaxial; Plagioclase Feldspars; Rock-Forming Minerals; Vein Minerals.*

EPIDOTE GROUP–*Allanite, clinozoisite, epidote, hancockite, mukhinite, piemontite,* and *zoisite.*

EPIGENITE *Dana,* I, 361
$(Cu,Fe)_5AsS_6$(?) orth.

EPISTILBITE A hydrated calcium aluminosilicate of the ***zeolite*** group. First named by Rose in 1826 from the Greek meaning "in addition to *stilbite.*"

Composition: $Ca[Al_2Si_6O_{16}]\cdot 5H_2O$; SiO_2, 59.2%; Al_2O_3, 16.8%; CaO, 9.2%; H_2O, 14.8% by wt.

Properties: Monoclinic $C2/m$; a=8.92 Å, b=17.73 Å, c=10.21 Å, β=124°20′; H=4; G=2.2; vitreous luster; colorless, white or yellowish; transparent to translucent; prismatic habit or in radiated spherical aggregates; cleavage very good on {010}; twinning cruciform; interpenetrant; shows piezoelectricity. Optically, refractive indices, α= 1.485–1.505, β=1.497–1.515, γ=1.497–1.519; birefringence = 0.010–0.014; 2V=44°(–); optic axial plane (010); β=y, γ:z=–10°; disperson $r<v$. Coombs et al. (1959) suggest that the normal compositional range is $(Ca,Na_2)_{2.85}Al_{5.7}Si_{18.3}O_{48}\cdot 16H_2O$ to $(Ca,Na_2)_{3.5}Al_7Si_{17}O_{48}\cdot 16H_2O$; decomposed by HCl.

Occurrence: In cavities in basalt and andesite, associated with *laumontite, heulandite,* and *mordenite;* on *bertrandite* on *beryl* in a New York pegmatite; in a Precambrian mudstone. Common localities are Isle of Skye, Scotland; Rhone Valley, Switzerland; Bombay, India; and Margaretsville, Nova Scotia.

A. IIJIMA
M. UTADA

References

Am. Min. **18**, 369–385; **21**, 264–265; **59**, 1055–1061.
GCA **17**, 53–107.

Cross-reference: *Zeolites.*

EPISTOLITE RRW, 193
$Na_2(Nb,Ti)_2Si_2O_9 \cdot nH_2O$ tric.

EPSOMITE A hydrated magnesium sulfate of the ***epsomite*** group named by Delamétherie in 1806 from the original locality of Epsom, England.

Composition: $MgSO_4 \cdot 7H_2O$.

Properties: Orthorhombic, $P2_12_12_1$; commonly in botryoidal masses or fibrous crusts; H=2–2.5; G=1.68; vitreous to earthy luster; transparent to translucent; bitter taste; white to colorless; soluble in water.

Occurrence: Usually found as an efflorescence on walls of caves or rocks in mines. Found in lacustrine deposits. Large salt deposits are found at Kruger Mountain, Oroville, Washington; Albany County, Wyoming; Carlsbad, New Mexico; and Stassfurt, E. Germany. Commonly associated with other soluble salts.

Uses: Very limited; medical purposes, dyeing, tanning, and as a cotton filler. First processed from natural springs, but now commercial epsom salt is a manufactured product.

J. A. GILBERT

References

Acta Cryst. **17**, 1361.
Dana, II, 509.
Min. Mag. **22**, 510

Cross-references: *Cave Minerals; Halotrichite Group; Meteoritic Minerals; Saline Minerals.*

EPSOM SALT = *Epsomite*

ERDITE *Am. Min.* **65**, 509
$NaFeS_2 \cdot 2H_2O$ mono.

EREMEYEVITE = *Jeremejevite*

ERICAITE *Am. Min.* **41**, 372
$(Fe,Mn)_3B_7O_{13}Cl$ orth.

ERICSSONITE mono. *Am. Min.* **56**, 2157
$BaMn_2(Fe^{3+}O)Si_2O_7(OH)$ *Lithos* **4**, 137

ERIOCHALCITE A hydrated copper chloride named by Scacchi in 1884 from the Greek words meaning "wool" and "copper" referring to the habit of the original material on Vesuvius.

Composition: $2[CuCl_2 \cdot 2H_2O]$. Orthorhombic.

Occurrence: Rare, a secondary mineral, associated with *atacamite* and *bandylite,* originally described on Vesuvius in fumaroles during the eruption of 1869.

J. A. GILBERT

Reference: *Dana,* II, 44.

ERIONITE A hydrated sodium potassium calcium magnesium aluminosilicate of the ***zeolite*** group. First named by Eakle in 1898 from Greek, "wool."

Compositon: $(Na_2,K_2,Ca,Mg)_{4.5}[Al_9Si_{27}O_{72}] \cdot 27H_2O$. SiO_2, 57.0%; Al_2O_3, 16.1%; MgO, 1.4%; CaO, 2.9%; Na_2O, 2.2%; K_2O, 3.3%; H_2O, 17.1% by wt. Here $Na_2:K_2:Ca:Mg$ is 2:2:3:2.

Properties: Hexagonal, $P6_3/mmc$; a=13.26 Å, c=15.12 Å; G=2.02; luster pearly; color: white; habit in radiating groups of crystals or in fine wooly fibers elongated parallel to z-axis. Optically, refractive indices, ω= 1.468–1.472, ϵ=1.473–1.476; birefringence = 0.003; uniaxial (+). The cations are exchangeable and the water molecules are easily removed and resorbed; K, Ag, and Tl ions are exchanged for the alakli ions; minimum diameter of the largest channel in the aluminosilicate framework is 3.6 Å; cation sieve properties very selective. Soluble in HCl.

Occurrence: Alteration product of silicic vitric tuffs accumulated in the Cenozoic saline, alkaline lakes in the western USA, associated with *clinoptilolite;* rarely as alteration product of Tertiary silicic tuffs from Sado Island, Japan; in cavities in basalt from the Faeroe Islands and Maze and Iki Island, Japan, associated with other ***zeolites;*** in alkaline soils of Olduvai Gorge, Tanzania.

A. IIJIMA
M. UTADA

References

Am. Min. **44**, 501–509.
Bull Soc. fr. **92**, 250–256.
GSA Bull. **74**, 1281–1286.
Min. Mag. **32**, 261–281; **33**, 66.
Nature **214**, 1005.

Cross-references: *Naming of Minerals; Zeolites.*

ERLICHMANITE (*Pyrite* gr.)
OsS_2 — *Am. Min.* **56**, 1501
iso. — *Min. Abst.* **28**, 323

ERNSTITE mono. *Am. Min.* **56**, 637
$(Mn^{2+}_{1-x}Fe^{3+}_{x}Al(PO_4)(OH)_{2-x}O_x(x=0-1)$

ERUBESCITE = *Bornite*

ERYTHRITE A hydrated cobalt arsenate named by Beudant in 1832 from the Greek word meaning "red."

Composition: $2[Co_3(AsO_4)_2 \cdot 8H_2O]$; CoO, 37.5%; As_2O_5, 38.4%; H_2O, 24.1% by wt. A complete series to *annabergite*, $Ni_3(AsO_4)_2 \cdot 8H_2O$, is formed as Ni substitutes for Co.

Properties: Monoclinic, $C2/m$; as crusts in spherical and reniform shapes, may be earthy; H=1.5–2.5; G=3.06; adamantine to vitreous luster, pearly on cleavage faces; color: pink to crimson; translucent.

Occurrence: Rare, a weathering product of cobalt arsenides. Important localities are Cobalt, Ontario; Great Bear Lake, Canada; Schneeburg, East Germany; and Joachimstaal, Czechoslovakia.

Use: As an indicator of an environment in which other cobalt deposits and associated native *silver* might be present.

J. A. GILBERT

Reference: *Dana*, II, 746.

ERYTHROSIDERITE *Dana*, II, 101
$K_2Fe^{3+}Cl_5 \cdot H_2O$ ortho.

Cross-reference: *Saline Minerals.*

ESCHYNITE = *Aeschynite*

ESKEBORNITE *Am. Min.* **57**, 1560
$CuFeSe_2$ iso. *Can. Min.* **10**, 786

ESKOLAITE *Am. Min.* **43**, 1098; **46**, 998
Cr_2O_3 trig.

ESPERITE *Am. Min.* **50**, 1170
$(Ca,Pb)ZnSiO_4$ mono.

ETTRINGITE hexa. *Am. Min.* **45**, 1137
$Ca_6Al_2(SO_4)_3(OH)_{12} \cdot 26H_2O$

EUCAIRITE *Am. Min.* **35**, 337
CuAgSe orth.

EUCHLORINE orth. *Dana*, II, 571
$(K,Na)_8Cu_9(SO_4)_{10}(OH)_6$(?)

EUCHROITE *Dana*, II, 934
$Cu_2(AsO_4)(OH) \cdot 3H_2O$ orth.

EUCLASE A hydroyxl beryllium aluminosilicate named by Haüy in 1792 from the Greek words meaning "easily" and "fracture" in allusion to its easy cleavage.

Composition: $4[BeAlSiO_4(OH)]$. Monoclinic.

Occurrence: Found in the auriferous sands in the Orenburg district of the southern Ural Mts., USSR; also found in Austria; in Bavaria; and in Minas Gerais, Brazil.

Use: A minor gem stone.

R. W. CAMERON

References

Am. Min. **46**, 1505; **63**, 664; **65**, 183.
Z. Krist. **112**, 275; **117**, 16.

Cross-reference: *Pegmatite Minerals.*

EUCRYPTITE A lithium aluminosilicate of the ***nepheline*** group. Derived from the alteration of *spodumene*. Named by Brush and Dana in 1880 from the Greek words meaning "well" and "concealed."

Composition: $3[LiAlSiO_4]$. Trigonal.

Occurrence: At Branchville, Connecticut, intimately associated with *albite* and presumably derived from the alteration of *spodumene*.

Use: Synthetic as a constituent of special ceramics having zero net thermal expansion.

G. C. MOERSCHELL

References

Am. Min. **38**, 353; **47**, 557.
Thrush, Paul, 1968. *A Dictionary of Mining, Mineral, and Related Terms.* Washington: U.S. Bureau of Mines, 397p.

Cross-references: *Pegmatite Minerals; Refractory Minerals.*

EUDIALYTE A hydroxyl silicate of sodium, calcium, iron, and zirconium named by Stromeyer in 1819 from the Greek meaning "easily" and "to dissolve" in allusion to its easy solubility in acids.

Composition: $Na_4(Ca,Ce,Fe^{2+})_2ZrSi_6O_{17}(OH,Cl)_2$. Trigonal.

Occurrence: In syenites and associated pegmatites, in the Julianehaab district of Greenland; on the island Sedlovatoi and on the Kola Peninsula, USSR; from Langesundfjord, southern Norway (eucolite); from Madagascar; on Mount Gooneringerringgi, Queensland, Australia; and at Magnet Cove, Arkansas, USA.

K. FRYE

References

Can. Min. **6**, 297.
Min. Abst. **28**, 72.

Cross-reference: *Rock-Forming Minerals.*

EUDIDYMITE A sodium hydroxyl beryllosilicate named by Brögger in 1887 from the Greek meaning "well" and "twin," alluding to its common occurrence as twin crystals.

Composition: $NaBeSi_3O_7(OH)$. Monoclinic.

Occurrence: Very sparingly in nepheline syenite on the island Övre-Arö, Norway; also at Narsarsuk, Greenland.

G. C. MOERSCHELL

Reference: Phillips, W. R., and Griffin, D. T., 1981. *Optical Mineralogy.* San Francisco: Freeman, p.486.

EULYTITE A bismuth silicate named by Breithaupt in 1827 from the Greek meaning "easily dissolved" or "fusible."

Composition: $4[Bi_4Si_3O_{12}]$. Isometric.

Occurrence: Found in Saxony with native bismuth near Schneeberg, E. Germany; also at Johanngeorgenstadt in crystals on *quartz.*

G. C. MOERSCHELL

References

Am. Min. **28**, 526.
Winchell II, 494.

EUXENITE A columbate-tantalate-titanate containing uranium and the rare earths, named by Scheerer in 1840 from the Greek meaning "friendly to strangers, hospitable," in allusion to its rare-earth constituents.

Composition: $(Y,Ca,Ce,U,La,Th)(Nb,Ta,Ti)_2O_6$. If Ta is dominant, named *tanteuxentie;* if Ti is dominant, named *polycrase.*

Properties: Orthorhombic *Pcan*; color: black to brown in mass; streak: yellow, gray, or reddish brown.

Occurrence: Found in pegmatites and placers in Norway and Brazil.

D. H. SPEIDEL

References

Am. Min. **47**, 812.
Can. Min. **14**, 111–119.
Dana, I, 787.

Cross-references: *Blacksand Minerals; Metamict State.*

EVANSITE amorph. *Dana,* II, 923
$Al_3(PO_4)(OH)_6 \cdot 6H_2O$(?) *Min. Mag.* **40**, 609

EVEITE *Am. Min.* **55**, 319
$Mn_2(AsO_4)(OH)$ orth.

EVENKITE *Am. Min.* **50**, 2109
$C_{24}H_{50}$ mono.

EWALDITE *Am. Min.* **56**, 2156
$Ba(Ca,Ce,Y,Na)(CO_3)_2$ hexa.

Cross-reference: *Green River Mineralogy.*

EYLETTERSITE (***Crandallite*** gr.)
$(Th,Pb)_{1-x}Al_3(PO_4,SO_4)_2(OH)_6$(?)
trig. *Am. Min.* **56**, 1366; **59**, 208

EZCURRITE *Am. Min.* **52**, 1048
$Na_4B_{10}O_{17} \cdot 7H_2O$ tric.

F

FABIANITE — *Am. Min.* **48**, 212
$CaB_3O_5(OH)$ mono. *Can. Min.* **10**, 108

FAHEYITE hexa. *Am. Min.* **49**, 395
$(Mn,Mg)Fe_2^{3+}Be_2(PO_4)_4 \cdot 6H_2O$

FAIRBANKITE — *Am. Min.* **65**, 809
$PbTeO_3$ tric. *Min. Mag.* **43**, 453

FAIRCHILDITE A carbonate of potassium and calcium named by Milton and Axelrod in 1947 after John G. Fairchild, analytical chemist of the U.S. Geological Survey.

Composition: $K_2Ca(CO_3)_2$; dimorphous with *buetschliite.* Hexagonal.

Occurrence: Found with *buetschliite* and *calcite* as clinkers formed by the fusion of wood ash in partly burned trees.

J. A. GILBERT

References

Am. Min. **32**, 607.
Min. Mag. **28**, 728.

FAIRFIELDITE — *Dana,* II, 720
$Ca_2(Mn,Fe)(PO_4)_2 \cdot 2H_2O$ tric.

Cross-reference: *Pegmatite Minerals.*

FALCONDOITE orth. *Can. Min.* **14**, 407
$(Ni,Mg)_8Si_{12}O_{30}(OH)_4(H_2O)_4 \cdot 8H_2O$
Min. Abst. **29**, 83

FAMATINITE — *Am. Min.* **42**, 766
Cu_3SbS_4 tetr. *Min. Abst.* **27**, 254

Cross-reference: *Ore Microscopy.*

FARRINGTONITE — *Am. Min.* **46**, 1513
$Mg_3(PO_4)_2$ mono.

Cross-reference: *Lunar Mineralogy; Meteoritic Minerals.*

FASSAITE (***Pyroxene*** gr.) DHZ, 2A, 399
$Ca(Mg,Fe,Al)(Si,Al)_2O_6$ mono.

Cross-references: *Clinopyroxenes; Meteoritic Minerals; Pyroxene Group.*

FAUJASITE A hydrated sodium calcium aluminosilicate of the ***zeolite*** group. First named by Damour in 1842 in honor of Faujas de Saint Fond.

Composition: $(Na_2,Ca)[Al_2Si_5O_{14}] \cdot 6.6H_2O$. SiO_2, 47.4%; Al_2O_3, 16.5%; CaO, 4.5%; Na_2O, 5.0%; H_2O, 26.6% by wt. Here, Na_2 : Ca=1 : 1.

Properties: Isometric $Fd3m$; a=24.65 Å, H=5; G=1.92; luster vitreous; colorless to white; transparent to opaque; habit in octahedrons; cleavage distinct on {111}; twinning on {111}. Optically, isotropic; refractive index, n=1.48. Isomorphous replacements of the type NaAl⇌Si may occur and Na is exchanged with varous ions; the most open of all the ***zeolites,*** with a minimum diameter of the largest channels in the aluminosilicate framework of 9 Å, it shows the highest molecular absorption of all the ***zeolites;*** normal and iso-paraffins, cyclopentane, cyclohexane, benzene, and toluene can permeate the channels; Na, K, Tl, Ag, NH_4, Mg, Ca, Sr, Ba, Co, Ni, and many organically substituted ammonium ions are exchangeable; decomposed by acid.

Occurrence: Rather rare; originally reported from Sasbach, Germany, in limburgite; also found near Eisenbuch in the Aar and St. Gotthard massifs, Switzerland, associated with other ***zeolites;*** recently found in the cement of melilite-nephelinite paragonite tuffs on Oahu, Hawaii, associated with *phillipsite, gismondine,* and *chabazite* which were formed by reaction of nephelinite glass with percolating groundwater.

Uses: Synthetic ***zeolite*** with the faujasite structure is used for a molecular sieve.

A. IIJIMA
M. UTADA

References

Am. Min. **49**, 697–704; **53**, 1293–1303; **54**, 149–155.
Helvetia Chim, Acta **39**, 519.
N.J. Min. Mh., 1958, 193–200; 1975, 433–443.

Cross-reference: *Zeolites.*

FAUSTITE (***Turquoise*** gr.) *Am. Min.* **38**, 964
$(Zn,Cu)Al_6(PO_4)_4(OH)_8 \cdot 5H_2O$ tric.

FAYALITE An iron silicate of the ***olivine*** group. Named by Gmelin in 1840 after Fayal Island in the Azores where it was believed to have occurred in a local volcanic rock, but it was more probably obtained from slag carried as ship's ballast.

Composition: $4[Fe_2SiO_4]$; forms a continuous isomorphous series with *forsterite*, Mg_2SiO_4, to give the ***olivine*** series but *fayalite* itself is restricted to 10% of the forsterite molecule; FeO, 66–70%; MgO, 0–6%; SiO_2, 29–30% by wt.

Properties: Orthorhombic, *Pbnm;* H=6.5; G=4.2–4.4; luster vitreous; color: greenish yellow or yellow-amber; habit: equant or platy; cleavage: moderate on {010}, weak on {100}; twinning on {100}; streak: white. Refractive indices α=1.814–1.827, β=1.840–1.869, γ=1.852–1.879; $2V_\alpha$=46°–54°; straight extinction; pale yellow in thin section. Insoluble in most acids, gelatinizes slightly in HCl; soluble in molten Na_2CO_3. Any relatively high values of Fe^{3+} are due to alteration and oxidation of Fe^{2+}. Mn may substitute appreciably for Fe^{2+} and there is a continuous series from *fayalite* to *tephroite*, Mn_2SiO_4.

Occurrence: Thermally metamorphosed or more commonly regionally metamorphosed iron-rich sediments; relatively common in *quartz*-bearing syenites and may be present in small amounts in granites and in some felsic and alkaline volcanic rocks; an important constituent of fayalite ferrogabbro. Often found associated with *hedenbergite, grunerite, almandine, eulite, arfvedsonite.*

R. A. HOWIE

References

AJS, ser. 5, **25**, 273.
Am. Min. **24**, 18; **35**, 1067.
DHZ, 1, 1.
J. Geol. Soc. Japan **47**, 228.

Cross-references: *Clinopyroxenes; Glass, Devitrification of Volcanic; Magnetic Minerals; Mineral Classification: Principles; Minerals, Uniaxial and Biaxial; Olivine Group; Orthopyroxenes; Thermometry, Geologic.*

FEDORITE mono. *Am. Min.* **52**, 561
$(Na,K)Ca(Si,Al)_4(O,OH)_{10} \cdot 1.5H_2O$

FEDOROVSKITE orth. *Am. Min.* **62**, 173
$Ca_2Mg_2(OH)_4[B_4O_7(OH)_2]$

FEITNECHTITE *Am. Min.* **50**, 1296
β–MnO(OH) hexa.

FELDSPAR GROUP–*Albite, andesine, anorthite, anorthoclase, banalsite, buddingtonite, bytownite, celsian, hyalophane, labradorite, microcline, oligoclase, orthoclase, paracelsian,* ***plagioclase***, *reedmergnerite*, and *slawsonite*

Cross-references: *Alkali Feldspars; Blacksand Minerals; Crystal Growth; Diagenetic Minerals; Feldspar Group; Pegmatite Minerals; Plagioclase Feldspars; Soil Mineralogy.*

FELDSPATH = Feldspar *Min. Mag.* **43**, 1053

FELDSPATHOIDS–see *Feldspathoid Group*

FELSOBANYITE *Am. Min.* **50**, 812
$Al_4(SO_4)(OH)_{10} \cdot 5H_2O$ orth.(?)

FELSPAR = ***Feldspar*** *Min. Mag.* **43**, 1053

FEMOLITE = Ferrian *molybdenite* (?)
Am. Min. **50**, 261

FENAKSITE *Am. Min.* **45**, 252
$FeNaKSi_4O_{10}$ tric.

FERBERITE An iron tungstate and end member of the ***wolframite*** series. Named by Liebe (1863) for the German, Rudolph Ferber.

Composition: $FeWO_4$; with Mn replacing Fe up to 20 mol% (Mn>20% is *wolframite*).

Properties: Monoclinic, *P2/c*; striated prismatic crystals in subparallel groups, also lamellar or massive granular; simple contact twins; one perfect cleavage {010}; H=4.5; G=7.5; submetallic to metallic luster; black with brownish black to black streak; weakly magnetic.

Occurrence: Originally found in *quartz* veins near Aquilas, Spain; later with *cryolite* at Ivigtut, Greenland. In the USA in mesothermal *quartz* veins in granite and gneiss in Boulder County, Colorado; also in Maricopa County, Arizona; Camas County, Idaho; and Colfax County, New Mexico.

Use: Ore of tungsten.

K. FRYE

Reference: *Dana*, II, 1064–1072.

Cross-references: *Magnetic Minerals; Wolframite Group.*

FERDISILICITE (artifact?)
$FeSi_2$ *Am. Min.* **54**, 1737
tetr. *Min. Abst.* **30**, 67

FERGUSONITE A niobate of yttrium-earth elements named by Haidinger in 1827 after the Scottish physician Robert Ferguson (1799–1865).

Composition: $8[YNbO_4]$, in which Ta may replace Nb toward *formanite.* Tetragonal. *β-fergusonite* is monoclinic and may have substantial rare-earth elements replacing Y.

Occurrence: In granite pegmatites, skarns, and carbonatites, especially those rich in the rare-earth elements, where it is associated with *calcite,* ***apatite,*** *zircon, biotite, magnetite,* and *monazite.* Originally from Kikertaursuk, Julianehaab district, Greenland, embedded in *quartz.*

J. A. GILBERT

References

Am. Min. **46,** 1516; **60,** 485; **62,** 397.
Can. Min. **6,** 72.
GSA Bull. **63,** 1235.

Cross-references: *Blacksand Minerals; Metamict State; Pegmatite Minerals.*

FERMORITE (***Apatite*** gr.) *Dana,* II, 904
$(Ca,Sr)_5(AsO_4,PO_4)_3(F,OH)$
hexa.

FERNANDINITE *Am. Min.* **44,** 322
$CaV_2^{4+}V_{10}^{5+}O_{30}\cdot 14H_2O$(?)

FEROXYHYTE *Am. Min.* **62,** 1057
δ'-FeO(OH) hexa. *Min. Abst.* **29,** 200

FERRAZITE *Dana,* II, 832
$(Pb,Ba)_3(PO_4)_2\cdot 8H_2O$(?)

FERRICOPIAPITE trig. *Dana,* II, 623
$Fe_5^{3+}(SO_4)_6O(OH)\cdot 20H_2O$

FERRIDRAVITE (***Tourmaline*** gr.)
$NaMg_3Fe_6^{3+}(BO_3)_3Si_6O_{18}(OH,F)_4$
Am. Min. **64,** 945
trig. *Min. Rec.* **11,** 111

FERRIERITE A hydrated sodium potassium magnesium aluminosilicate of the ***zeolite*** group. First named by Graham in 1918 in honor of the late W. F. Ferrier of the Canadian Geological Survey.

Composition: $(Na,K)_2Mg_2[Al_6Si_{30}O_{72}]\cdot 18H_2O$. SiO_2, 67.9%; Al_2O_3, 11.5%; MgO, 3.0%; Na_2O, 4.7%; H_2O, 12.9% by wt.

Properties: Orthorhombic, *Immm; a*=19.12; *b*=14.14, *c*=7.48 Å, H=3–3.5; G=2.14; colorless to white; spherical aggregates or in crystals tabular on (010). Optically, refractive indices, α=1.478, β=1.479, γ=1.482; birefringence = 0.004; $2V=50°$ (+); optic axial plane (010); $\alpha=x$, $\beta=y$, $\gamma=z$. Considerable MgO and high SiO_2 are unusual for a ***zeolite;*** rehydration good; molecular sieve properties more selective; nearly insoluble in HCl.

Occurrence: From the north shore of Kamloops Lake, British Columbia, enclosed in chalcedony in fractures in olivine basalt; in druses, in cavities, and crack fillings of igneous rocks; alteration product of silicic tuffs surrounding Kuroko type Cu-Pb-Zn exhalative mineral deposits of Nishi Aizu and Waga-Omono Districts, northeastern Japan, associated with *clinoptilolite* and(or) *mordenite.*

A. IIJIMA
M. UTADA

References

Acta Cryst. **21,** 983–990.
Am. Min. **40,** 1095–1099; **50,** 484–489; **61,** 60–66.
Mining Geol. (Tokyo) **20,** 295–304.
Trans. Roy. Soc. Canada, ser. 3, **12,** 185–190.

Cross-reference: *Zeolites.*

FERRIHYDRITE A hydrated iron oxide.

Composition: $5Fe_2O_3\cdot 9H_2O$.

Properties: Hexagonal.

Occurrence: In hot and cold springs as a result of bacterial action.

D. H. SPEIDEL

Reference: *Am. Min.* **60,** 485.

FERRIMOLYBDITE *Am. Min.* **48,** 14
$Fe_2(MoO_4)_3\cdot 8H_2O$(?)

FERRINATRITE *Dana,* II, 456
$Na_3Fe(SO_4)_3\cdot 3H_2O$ trig. *Min. Abst.* **28,** 396

FERRIPYROPHYLLITE *Min. Mag.* **31,** 355
$Fe_2^{3+}Si_4O_{10}(OH)_2$ mono.

FERRISICKLERITE *Dana,* II, 672
$Li(Fe^{3+},Mn^{2+})PO_4$ orth. *Min. Abst.* **28,** 395

Cross-reference: *Pegmatite Minerals.*

FERRISYMPLESITE (doubtful validity)
$Fe_3(AsO_4)_2(OH)_3\cdot 5H_2O$
amorph. *Dana,* II, 753

FERRITUNGSTITE iso. *Am. Min.* **42,** 83
$Ca_2Fe_2^{2+}Fe_2^{3+}(WO_4)_7\cdot 9H_2O$

FERROACTINOLITE–see ***Amphibole Group***

FERRONANTHOPHYLLITE–see ***Amphibole** Group*

FERROBARROISITE–see ***Amphibole Group***

FERROAXINITE (***Axinite*** gr.)
$Ca_2FeAl_2BSi_4O_{15}(OH)$ *Am. Min.* **65**, 1119
tric. DHZ, 1, 320

FERROBUSTAMITE *Am. Min.* **59**, 632
$Ca(Fe,Ca,Mn)Si_2O_6$ tric.

FERROCARPHOLITE orth. *Am. Min.* **36**, 736
$(Fe,Mg)Al_2Si_2O_6(OH)_4$ *Min. Abst.* **27**, 20

FERROCOLUMBITE *Dana,* I, 783
$FeNb_2O_6$ orth.

FERROECKERMANNITE (***Amphibole*** gr.)–see *Eckermannite*

FERROEDENITE (***Amphibole*** gr.)–see *Edenite*

FERROFERRITSCHERMAKITE (***Amphibole*** gr.)–see *Tschermakite*

FERROGEDRITE (***Amphibole*** gr.)–see *Gedrite*

FERROGLAUCOPHANE (***Amphibole*** gr.)–see *Glaucophane*

FERROHEXAHYDRITE *Am. Min.* **48**, 433
$FeSO_4 \cdot 6H_2O$ mono.

FERROHOLMQUISTITE (***Amphibole*** gr.)–see *Holmquistite*

FERROHORNBLENDE (***Amphibole*** gr.)–see *Hornblende*

FERROKAERSUTITE (***Amphibole*** gr)–see *Kaersutite*

FERROPARGASITE (***Amphibole*** gr.)–see *Pargasite*

FERROPUMPELLYITE *Am. Min.* **56**, 2158
$Ca_2FeAl_2(SiO_4)(Si_2O_7)(OH)_2 \cdot H_2O$
mono. DHZ, 1, 227

FERRORICHTERITE (***Amphibole*** gr.)–see *Richterite*

FERROSELITE (***Marcasite*** gr.)
$FeSe_2$ orth. *Am. Min.* **41**, 671

FERROSILITE (***Pyroxene*** gr)–see *Clinoferrosilite, Orthoferrosilite*

FERROTANTALITE *Dana,* I, 783
$FeTa_2O_6$ orth.

FERROTSCHERMAKITE (***Amphibole*** gr.)–see *Tschermakite*

FERROWINCHITE (***Amphibole*** gr.)–see *Winchite*

FERROWYLLIEITE mono. *Am. Min.* **65**, 810
$(Na,Ca,Mn)(Fe,Mn)(Fe,Mg)Al(PO_4)_3$
Min. Mag. **43**, 227

FERRUCCITE A sodium borofluoride named by Carobbi in 1933 after Ferruccio Zambonini, Italian mineralogist.

Composition: $4[NaBF_4]$. Orthorhombic.

Occurrence: Rare, found as minute crystals in admixed fumarolic sublimates on Vesuvius associated with *hieratite, avogadrite, sassolite,* and *malladrite.*

J. A. GILBERT

References

Am. Min. **19**, 555.
Dana, II, 98.
Min. Mag. **23**, 629.

FERSILICITE (artifact(?))
FeSi iso. *Am. Min.* **54**, 1737

FERSMANITE tric. *Am. Min.* **64**, 658
$(Ca,Na)_4(Ti,Nb)_2Si_2O_{11}(F,OH)_2$
Can. Min. **15**, 87

FERSMITE orth. *Am. Min.* **32**, 373;
$(Ca,Ce,Na)(Nb,Ta,Ti)_2(O,OH,F)_6$ **44**, 1

FERVANITE *Am. Min.* **44**, 322
$Fe_4(VO_4)_4 \cdot 5H_2O$ mono.

FIBROFERRITE *Am. Min.* **61**, 398
$Fe(SO_4)(OH) \cdot 5H_2O$ hexa.(?) *Dana,* II, 614

FIBROLITE = Fibrous, felted *sillimanite*

FIEDLERITE *Dana,* II, 67
$Pb_3Cl_4(OH)_2$ mono.

FILLOWITE trig. *Am. Min.* **50**, 1647
$(Na_{33}Ca_{12})(Mn_{102}Fe_{24})(PO_4)_{103}$

FINNEMANITE *Dana,* II, 1038
$Pb_5(AsO_3)_3Cl$ hexa.

FISCHESSERITE *Am. Min.* **57**, 1554
Ag_3AuSe_2 iso.

FIZELYITE *Am. Min.* **60**, 621
$Pb_5Ag_2Sb_8S_{18}$ orth. *Dana,* I, 450

FLAGSTAFFITE (*cis*-Terpinhydrate)
$C_{10}H_{18}(OH)_2(H_2O)$ orth. *Am. Min.* **50**, 2109

FLEISCHERITE hexa. *Am. Min.* **45**, 1313
$Pb_3Ge(SO_4)_2(OH)_6 \cdot 3H_2O$

FLETCHERITE *Am. Min.* **62**, 1057
$Cu(Ni,Co)_2S_4$ iso. *Econ. Geol.* **72**, 480

FLINKITE *Am. Min.* **52**, 1603
$Mn_3(AsO_4)(OH)_4$ orth.

FLINT – Nodular chalcedony *Dana,* III, 219

Cross-references: *Electron Microscopy (Transmission); Mineral Industries; Soil Mineralogy.*

FLORENCITE (***Crandallite*** gr.)
$CeAl_3(PO_4)_2(OH)_6$ *Can. Min.* **18**, 301
trig. *Min. Abst.* **27**, 179

FLUELLITE A hydrated aluminum fluophosphate hydroxide, originally thought to be an aluminum fluoride and named in allusion to that composition.

Composition: $Al_2(PO_4)F_2(OH) \cdot 7H_2O$.

Properties: Orthorhombic, *Fddd,* dipyramidal crystals; H=3; G=2.16; vitreous luster; transparent; colorless to white.

Occurrence: Found in pegmatites as minute crystals lining cavities, originally described from Stenna Gwyn, Cornwall, England, where it occurred as crystals on *quartz* associated with *fluorite, wavellite, arsenopyrite,* and *torbernite;* also from Oberpfalz, Bavaria, West Germany.

K. FRYE

Reference: *Am. Min.* **51**, 1579–1592.

FLUOBORITE A magnesium fluoborate named by Geijer in 1926 in allusion to its composition.

Composition: $B_3[Mg_9(F,OH)_9O_9]$.

Occurrence: Relatively rare, originally found at the Tallgruvan mine, Norberg, Sweden, in a contact metasomatic *magnetite* deposit associated with *ludwigite* and *chondrodite.* Also at Sterling Hill, New Jersey, as secondary hydrothermal veinlets associated with *willemite, fluorite, hydrozincite,* and *pyrochroite,* and cutting the *franklinite* ore.

J. A. GILBERT

References

Am. Min. **61**, 88.
GSA Bull. **69**, 1678.
Min. Mag. **21**, 564.

FLUOCERITE *Dana,* II, 48
$(Ce,La)F_3$ hexa. *Min. Abst.* **27**, 256

FLUORAPATITE A calcium fluophosphate of the ***apatite*** group named from the Greek *apate,* deceit, because the mineral was early confused with *tourmaline, beryl,* ***olivine,*** etc. *Fluorapatite* is the most common of the ***apatites.***

Composition: $2[Ca_5(PO_4)_3F]$. (See ***Apatite*** group for compositional variation.)

Properties: Hexagonal ($P6_3/m$); crystals short to long prismatic or tabular, also globular, massive, granular, earthy, or concretionary; oriented inclusions of *rutile* or *monazite;* zoned crystals with other ***apatite*** minerals; H=5; G=3.2; vitreous to subresinous luster; all colors; commonly fluorescent or phosphorescent.

Occurrence: Common accessory in igneous rocks, a pegmatite mineral (q.v.), a vein mineral (q.v.), in metasomatized calc-silicate and impure carbonate rocks, and a major constituent of phosphorites.

K. FRYE

References

DHZ, 5, 323–338.
McConnell, D., 1973. *Apatite.* New York: Springer-Verlag, 111p.

Cross-references: *Apatite Group; Green River Mineralogy; Minerals, Uniaxial and Biaxial; Soil Mineralogy.*

FLUORAPOPHYLLITE (***Apophyllite*** gr.)
$KCa_4Si_8O_{20}F \cdot 8H_2O$ tetr. *Am. Min.* **63**, 196

FLUORITE Calcium fluoride named by Agricola in 1529 as "fluores" from the Latin *fluere,* "to flow," because it melts easily and in allusion to its use as a flux.

Composition: $4[CaF_2]$; Ca, 51.3%; F, 48.7%, by wt.

Properties: Isometric, *Fm3m*; crystals often twinned, may also be massive, granular, or columnar; perfect octahedral cleavage {111}; H=4; G=3.18; vitreous luster; transparent to translucent; color varies widely but most commonly light green, yellow, bluish green, or purple.

Occurrence: Found in veins, as a gangue mineral with metallic ores, in dolomites and limestone, and as a minor accessory mineral in igneous rocks. Associated with many different minerals – e.g., *calcite, dolomite, gypsum, celestite, barite, quartz, galena, sphalerite, cassiterite, topaz,* ***tourmaline,*** and ***apatite.*** Well-known locations are Derbyshire, England; Frieberg, East Germany; Harz, West Germany; Kongsberg, Norway; Ivigtut, Greenland; and in Hardin and Pope Counties, Illinois.

Uses: As a flux in the manufacture of steel, in the manufacture of opalescent glass, in enamelling cooking utensils, for the preparation of hydrofluoric acid, and some varieties are used for coatings of lenses and prisms.

J. A. GILBERT

References

Am. Min. **45,** 884.
Dana, II, 29.
Min. Mag. **30,** 327.

Cross-references: *Crystal Growth; Crystal Habits; Crystallography: Morphological; Green River Mineralogy; Magnetic Minerals; Minerals, Uniaxial and Biaxial; Mohs Scale of Hardness; Museums, Mineralogical; Optical Mineralogy; Pegmatite Minerals; Phantom Crystals; Pleochroic Halos; Pseudomorphism; Refractive Index; Refractory Minerals; Rock-Forming Minerals; Staining Techniques; Thermoluminescence; Vein Minerals.*

FLUORSPAR = *Fluorite*

Cross-reference: *Mineral Industries.*

FOGGITE orth. *Am. Min.* **60,** 957
$CaAl(PO_4)(OH)_2 \cdot H_2O$ *Min. Abst.* **27,** 257

FORBESITE – Mixture of *annabergite* and *arsenolite* *Can. Min.* **14,** 414

FORMANITE A tantalate of yttrium-earth elements named by Berman and Frondel in 1944 after F. G. Forman, government geologist of Western Australia.

Composition: $8[YTaO_4]$; Ta replaced by Nb toward *fergusonite.* Tetragonal.

Occurrence: In granite pegmatites. It is abundant in Western Australia at Cooglegong in placers with *cassiterite, monazite, euxenite,* and *gadolinite.*

J. A. GILBERT

References

Am. Min. **39,** 667.
Dana, I, 757.

Cross-reference: *Pegmatite Minerals.*

FORNACITE *Am. Min.* **49,** 447
$(Pb,Cu)_3[(Cr,As)O_4]_2(OH)$ mono.

FORSTERITE A magnesium silicate of the ***olivine*** group named by Lévy in 1824 after J. R. Forster (1729–1798), a German naturalist who founded the Henland Cabinet. Magnesium-rich end member of the ***olivine*** series which extends to the iron-rich end member, *fayalite.*

Composition: Mg_2SiO_4

Properties: Orthorhombic, *Pbnm,* usually in granular masses of imperfect crystals and as rounded grains in igneous rocks; H=7, G=3.22; color: white to light green (lemon yellow), streak: white; vitreous luster, transparent to translucent; gelatinizes in HCl.

Occurrence: Typically in implanted crystals in a number of metamorphic rocks; particularly, as an early product in the thermal metamorphism of dolomitic limestones. Important localities include: Mt. Somma, Vesuvius; Snarum, Norway, and at Zlatoust, Urals, USSR.

Use: Magnesium-rich ***olivines,*** such as *forsterite,* have a very high melting point and this enables it to be used in the manufacture of refractory bricks.

G. C. MOERSCHELL

Reference: DHZ, 1, 1-33.

Cross-references: *Clinopyroxenes; Diamond; Mantle Mineralogy; Meteoritic Minerals; Minerals, Uniaxial and Biaxial; Olivine Group; Polysomatism; Refractory Minerals.*

FOSHAGITE *Am. Min.* **43,** 1
$Ca_4Si_3O_9(OH)_2$ mono.

FOSHALLASITE *Am. Min.* **23,** 667
$Ca_3Si_2O_7 \cdot 3H_2O$(?) mono.(?)

FOURMARIERITE A hydrated lead and uranium oxide of the ***curite*** group named by Buttgenbach in 1924 after P. Fourmarier, once Professor of Geology at the University of Liége.

Composition: $PbO \cdot 4UO_3 \cdot 5H_2O$.

Properties: Orthorhombic; color is red; highly radioactive.

Occurrence: Very rare; as an alteration product of *uraninite,* associated with *torbernite, kasolite,* and *curite* at Kasolo, Shaba, Zaire.

D. H. SPEIDEL

References

Am. Min. **45,** 1026.
Dana, I, 628.
Ann. Soc. Géol. Belgique **47,** 41.

Cross-reference: *Museums, Mineralogical.*

FOWLERITE = zincian *rhodonite*
DHZ, 2A, 588

FRAIPONTITE (***Kaolinite-serpentine*** gr.)
$(Zn,Al)_3(Si,Al)_2O_5(OH)_4$ *Am. Min.* **62,** 175
mono. *Min. Abst.* **28,** 444

FRANCEVILLITE orth. *Am. Min.* **43,** 180
$Ba(UO_2)_2(VO_4)_2 \cdot 5H_2O$ *Min. Abst.* **29,** 232

FRANCKEITE *Dana,* I, 448
$Pb_5Sn_3Sb_2S_{14}$ tric. *Min. Abst.* **29,** 423

FRANCOANELLITE trig. *Am. Min.* **61,** 1054
$H_6K_3Al_5(PO_4)_8 \cdot 13H_2O$

FRANCOLITE (***Apatite*** gr.) DHZ, 5, 323
$Ca_5(PO_4,CO_3)_3F$ hexa. *Dana,* II, 879

Cross-references: *Apatite Group; Invertebrate and Plant Mineralogy.*

FRANKDICKSONITE *Am. Min.* **59,** 885
BaF_2 iso.

FRANKLINITE A zinc iron oxide of the *magnetite* series of the ***spinel*** group, named by Berthier in 1819 for the locality, Franklin Furnace, New Jersey, and also in honor of Benjamin Franklin.

Composition: $ZnFe_2O_4$, with Zn replaced by Mn^{2+} and Fe^{3+} replaced by Mn^{3+}

Properties: Cubic, *Fd3m*; rounded octahedra, massive, or granular; G=5.3; H=6; black with metallic luster; weakly magnetic.

Occurrence: Associated with *calcite, zincite,* etc. at Franklin Furnace, New Jersey.

Use: An ore of zinc.

D. H. SPEIDEL

References

Am. Min. **50,** 1670.
Dana, I, 698.
GSA Bull. **69,** 775.

Cross-references: *Magnetic Minerals; Naming of Minerals.*

FRANZINITE (***Cancrinite*** gr.)
$(Na,Ca)_{34}(Si,Al)_{60}O_{120}(SO_4,CO_3,OH)_{14} \cdot 4.3H_2O$ *Am. Min.* **62,** 1259
hexa. *Min. Abst.* **29,** 418

FREBOLDITE *Am. Min.* **41,** 164; **44,** 907
CoSe hexa.

FREIBERGITE *Am. Min.* **60,** 489
$(Ag,Cu)_{12}(Sb,As)_4S_{13}$ iso.

Cross-reference: *Ore Microscopy.*

FREIESLEBENITE *Dana,* I, 416
$PbAgSbS_3$ mono.

FRESNOITE *Am. Min.* **50,** 314
$Ba_2TiSi_2O_8$ tetr.

FREUDENBERGITE *Am. Min.* **46,** 765
$Na_2(Ti,Fe)_8O_{16}$ mono.

FRIEDELITE A hydrous manganese and iron silicate named by Bertrand in 1876 after the French chemist and mineralogist Ch. Friedel.

Composition: $6[(Mn,Fe)_8Si_6O_{15}(OH,Cl)_{10}]$; MnO, 45–50%; FeO, 0–12%; SiO_2, 31–34%; H_2O, 9–10%; Cl, 2–3% by wt.

Properties: Trigonal; H=5; G=3.07–3.17; vitreous luster; rose-red to brown color; habit commonly tabular on (0001); cleavage moderate on (0001). Refractive indices ϵ=1.63, ω=1.66; almost colorless in thin section. Gelatinizes in HCl; soluble in molten Na_2CO_3; Fe^{2+} substitutes extensively for Mn; Ca and Zn may also substitute for Mn; As may substitute for Si.

Occurrence: Manganese-bearing skarn deposits as at Franklin, New Jersey, where it is often associated with *schallerite, bementite, leucophoenicite,* and *willemite.*

Uses: Not itself of economic importance, but with other Mn silicates may be used as an ore of manganese.

R. A. HOWIE

References

Am. Min. **13**, 341; **38**, 755–760.
Zap. Vses. Min. Obshch. 97, 342–348 (in Russian).

FRIEDRICHITE *Am. Min.* **64**, 654
$Pb_5Cu_5Bi_7S_{18}$ orth. *Can. Min.* **16**, 127

FRIGIDITE = *Tetrahedrite* + Ni sulfides
Min. Mag. **43**, 99

FRITZSCHEITE (***Autunite*** gr.)
$Mn(UO_2)_2(PO_4)_2 \cdot 10H_2O$
tetr. *USGS Bull.* **1064**, 195

FROHBERGITE (***Marcasite*** gr.)
$FeTe_2$ orth. *Am. Min.* **33**, 210

FROLOVITE *Am. Min.* **43**, 385
$Ca[B(OH)_4]_2$ tric. *Min. Abst.* **29**, 25

FRONDELITE (***Frondelite*** gr.)
$Mn^{2+}Fe_4^{3+}(PO_4)_3(OH)_5$ *Am. Min.* **34**, 541
orth. *Dana,* II, 867

Cross-references: *Naming of Minerals; Pegmatite Minerals.*

FROODITE *Am. Min.* **44**, 207
$PdBi_2$ mono. *Can. Min.* **6**, 200

FUCHSITE = Chromian *muscovite*

Cross-reference: *Mineral Classification: Principles.*

FUKALITE *Am. Min.* **63**, 793
$Ca_4Si_2O_6(OH,F)_2(CO_3)$
ortho. *Min. Mag.* **43**, 1060

FUKUCHILITE (***Pyrite*** gr.)
Cu_3FeS_8 iso. *Am. Min.* **55**, 1811

FÜLÖPPITE (***Plagionite*** gr.)
$Pb_3Sb_8S_{15}$ *Am. Min.* **59**, 1127
mono. *Min. Abst.* **27**, 22

FURONGITE tric. *Am. Min.* **63**, 425
$Al_2(UO_2)(PO_4)_2(OH)_2 \cdot 8H_2O$
Min. Abst. **31**, 85

G

GABRIELSONITE *Can. Min.* **18**, 191
$PbFe(AsO_4)(OH)$ orth. *Min. Rec.* **11**, 37

GADOLINITE A rare-earth element, iron, beryllium silicate of the ***datolite*** group, named for the Swedish chemist J. Gadolin (1760–1852).

Composition: $REE_2Fe^{2+}Be_2Si_2O_{10}$, named *gadolinite-(Y)* or *gadolinite-(Ce)* depending upon whether the small or the large rare-earth elements predominate.

Properties: Monoclinic, $P2_1/a$; prismatic crystals, commonly massive; H=6.5-7; G=4.0–4.5; vitreous to greasey luster; color: black, greenish black, or brown.

Occurrence: Found in pegmatites and black-sands. From Kopparberg and Ytterby, Sweden; Telemark and Vest-Agder, Norway; Baveno, Italy; and Llano County, Texas, USA.

K. FRYE

References: *Am. Min.* **59**, 700–708; **63**, 188–195.

Cross-references: *Blacksand Minerals; Metamict State; Museums, Mineralogical; Pegmatite Minerals.*

GAGARINITE *Am. Min.* **47**, 805
$NaCaY(F,Cl)_6$ hexa.

GAGEITE orth. *Am. Min.* **64**, 1056
$(Mn,Mg,Zn)_{40}Si_{15}O_{50}(OH)_{40}$

GAHNITE Zinc aluminate of the ***spinel*** group, named by von Moll in 1807 after J. G. Gahn, Swedish chemist.

Composition: $ZnAl_2O_4$, with Fe and Mn replacing Zn. Kreittonite and dysluite are Fe- and Mn-bearing varieties, respectively.

Properties: Cubic, *Fd3m*; H=7.5–8; G=4.62; *n*=1.805; dark bluish green and gray streak.

Occurrence: In granitic pegmatites, contact-altered limestones, or metasomatic replacement veins such as at Franklin, New Jersey, USA.

D. H. SPEIDEL

References

Dana, I, 689.
DHZ, 5, 56.

Cross-reference: *Pegmatite Minerals.*

GAIDONNAYITE *Can. Min.* **12**, 316
$Na_2ZrSi_3O_9 \cdot 2H_2O$ orth.

GAITITE *Can. Min.* **18**, 197
$H_2Ca_2Zn(AsO_4)_2(OH)_2$
tric. *Min. Mag.* **43**, 1060

GALAXITE Manganese aluminum oxide named by Ross and Kerr in 1932 after a plant, *Galax*, abundant in the region of Alleghany County, North Carolina. Manganese-bearing member of the ***spinel*** group.

Composition: $MnAl_2O_4$, always contains an appreciable amount of Fe^{2+}.

Properties: Isometric, *Fd3m*; H=7.5–8; G=4.04; color: mahogany-red to black; streak: red-brown.

Occurrence: Associated with manganese-rich vein deposits and *tephroite, rhodonite,* and *hausmannite*. Known in manganese vein deposits at Bald Knob, North Carolina, USA, and Ioi mine, Shiga Prefecture, Japan.

D. H. SPEIDEL

References

Dana, I, 689.
DHZ, 5, 56–67.

GALEITE *Am. Min.* **56**, 174
$Na_{15}(SO_4)_5F_4Cl$ trig. *Min. Mag.* **40**, 357

Cross-reference: *Polycrystal.*

GALENA Lead sulfide named by Pliny in 77 AD from the Latin, a name given to lead ore or the dross from melted lead.

Composition: 4[PbS]

Properties: Isometric, *Fm3m*; most common habit is the cube; perfect cubic {100} cleavage; H=2.5; G=7.58; bright metallic luster; color: lead-gray; oxidizes to *anglesite* and *cerussite*.

Occurrence: Widely distributed in many different types of deposits–in sedimentary rocks, in hydrothermal veins, in cryolitic pegmatites, and in areas of contact metamorphism, often associated with *sphalerite, pyrite, marcasite, chalcopyrite, cerussite, anglesite, dolomite, calcite, quartz, barite, fluorite,* and almost always with silver, in the form of *argentite* when found in veins. May also be found associated with limestone either as veins, open space fillings, authigenic mineral, or replacement deposits.

Use: Practically the only source of lead and an important ore of silver.

J. A. GILBERT

References

Dana, I, 200.
DHZ, 5, 180.

Cross-references: *Blowpipe Analysis; Crystallography: Morphological; Metallic Minerals; Mineral Classification: Principles; Mineral and Ore Deposits; Mohs Scale of Hardness; Ore Microscopy; Plastic Flow in Minerals; Rock-Forming Minerals; Staining Techniques; Vein Minerals.*

GALENITE = *Galena*

Cross-reference: *Naming of Minerals.*

GALENOBISMUTITE *Am. Min.* **62**, 346
$PbBi_2S_4$ orth. *Dana*, I, 471

Cross-reference: *Ore Microscopy.*

GALKHAITE *Am. Min.* **59**, 208
$HgAsS_2$ iso.

GALLITE *Am. Min.* **44**, 906
$CuGaS_2$ tetr.

GAMAGARITE mono. *Am. Min.* **28**, 329
$Ba_4(Fe,Mn)_2V_4O_{15}(OH)_2$

γ-MnO_2 = *Nsutite*

GANISTER = Quartzite

Cross-reference: *American Society for Testing and Materials.*

GANOMALITE A basic calcium lead silicate named by Nordenskiöld in 1876 from the Greek word for "luster."

Composition: $Ca_4Pb_6Si_6O_{21}(OH)_2$; may contain up to 2.3% MnO.

Properties: Hexagonal; usually occurs in granular masses; H=3; G=5.7; resinous to vitreous luster; colorless to gray; habit, prismatic; cleavage, perfect on {0001} (basal) and $\{10\bar{1}0\}$ (prismatic); refractive indices = 1.901–1.945; gelatinizes in HCl and HNO_3.

Occurrence: In *calcite* veins from Långban, Sweden; Jakobsberg, Nordmark (Sweden); and Franklin, New Jersey, USA. Associated with *tephroite,* native *lead, calcite,* and *jacobsite* at Långban; with *manganophyllite, calcite,* and *jacobsite* at Jakobsberg.

L. H. FUCHS

References

Dana, E. S., 1892. *System of Mineralogy,* 6th ed. New York: Wiley, 422.

Neumann, H.; Sverdrup, T.; and Saebo, P. C., 1957. X-ray powder patterns for mineral identification. III–Silicates. *Avhandl. Norske Videnskaps–Akad, Oslo I. Mat.–Naturv. Kl.,* no. 6, 18 p.

Nordenskiöld, A. E., 1876–77. New mineral from Långban, *Geol. För. Förh.,* **3**, 376.

Winchell, A. N., 1956. *Elements of Optical Mineralogy,* 4th ed. New York: Wiley, 478.

Zenzen, N., 1915. Mineralogical notes–The crystal system of ganomalite, *Geol. För. Förh.,* **37**, 294.

GANOPHYLLITE RRW, 229
$(Na,K)(Mn,Fe,Al)_5(Si,Al)_6O_{15}(OH)_5 \cdot 2H_2O$
mono. or tric. *Min. Abst.* **30**, 16

GARNET GROUP–*Almandine, andradite, calderite, goldmanite, grossular, hydrogrossular, hydrougrandite, kimzeyite, knorringite, majorite, pyrope, schorlomite, spessartine,* and *uvarovite.*

Cross-references: *Abrasive Materials; Blacksand Minerals; Crystal Growth; Crystallography: Morphological; Density Measurements; Diamond; Fire Clays; Garnet Group; Gemology; Jade; Lunar Minerals; Mantle Mineralogy; Minerals; Uniaxial and Biaxial; Optical Mineralogy; Order-Disorder; Orthopyroxenes; Placer Deposits; Refractive Index; Refractory Minerals; Rock-Forming Minerals; Soil Mineralogy; Synthetic Minerals; Thermometry, Geological.*

GARNIERITE–A ni-rich ***serpentine***

Cross-references: *Blowpipe analysis; Metallic Minerals.*

GARRELSITE *Am. Min.* **41**, 672; **59**, 632
$Ba_3NaSi_2B_7O_{16}(OH)_4$ mono.

Cross-reference: *Green River Mineralogy.*

GARRONITE A hydrated sodium calcium aluminosilicate of the ***zeolite*** group. First named by Walker in 1962 after the original locality, the Garron Plateau, Antrim, Northern Ireland.

Composition: $NaCa_{2.5}[Al_6Si_{10}O_{32}]\cdot 13.5H_2O$. Closely related to *phillipsite* and *gismondine;* differs from *phillipsite* in having a very low content of K_2O and from *gismondine* in containing less CaO, Al_2O_3, and H_2O; small amounts of K and lesser of Ba may enter.

Properties: Tetragonal, $I4_1/amd$; a=9.88; c=10.30 Å; G=2.13–2.17; habit in radiating aggregates; cleavage in two directions at about 90°, both parallel to the length of crystals. Optically, refractive indices, ϵ= 1.503±0.002, ω=1.510±0.002; birefringence = 0.001–0.004; uniaxial or biaxial.

Occurrence: In amygdules only in olivine basalt lavas in the Tertiary basalts of Antrim and eastern Ireland, occasionally associated with other relatively silica-poor ***zeolites*** such as *chabazite, levyne, thomsonite, phillipsite,* and *gismondine;* in veins in metamorphosed basalt and dolerite of greenschist facies, associated with *chabazite* and *quartz* in the area of the Tanzawa Mountains, Japan.

A. IIJIMA
M. UTADA

References

J. Jap. Assoc. Min. Petrol. Econ. Geol., **61**, 241–249.
Min. Mag. **33**, 173–186.

GASPEITE (***Calcite*** gr.) *Am. Min.* **51**, 677
$(Ni,Mg,Fe)CO_3$ trig.

GATUMBAITE mono. *Am. Min.* **63**, 794
$Ca_2Al_4(PO_4)_4(OH)_4\cdot 2H_2O$

GAUDEFROYITE hexa. *Am. Min.* **50**, 806
$Ca_4Mn^{3+}_{3-x}(BO_3)_3(CO_3)(O_{1-x}(OH)_x)_3$

GAYLUSSITE A hydrated sodium calcium carbonate named in 1826 after the French chemist L. J. Guy-Lussac.

Composition: $Na_2Ca(CO_3)_2\cdot 5H_2O$

Properties: Monoclinic, $I2/a$; elongate or flattened wedge-shaped crystals; perfect prismatic cleavage {110}; H=2.5–3; G=1.99; vitreous luster; color: white, yellowish, or grayish.

Occurrence: From soda lakes in arid regions.

K. FRYE

Reference: *Dana*, II, 234–235.

Cross-references: *Green River Mineralogy; Saline Minerals.*

GEARKSUTITE *Dana*, II, 119
$CaAl(OH)F_4\cdot H_2O$ mono.

GEDRITE A hydroxyl aluminosilicate of sodium, magnesium, ferrous iron, manganese, and aluminum, in the ***amphibole*** group, orthorhombic division. Named by Dufrénoy in 1836 for Gèdres, Hautes Pyrénées, France.

Composition: May vary among end members
Magnesio-gedrite $Mg_5Al_2(Si_6Al_2O_{22})(OH)_2$
Ferrogedrite $Fe^{2+}_5Al_2(Si_6Al_2O_{22})(OH)_2$
Sodium gedrite $Na(Mg,Fe)_6Al(Si_6Al_2O_{22})(OH)_2$
Mn may replace (Mg,Fe) and (F,Cl) may replace (OH)

Occurrence: The original *gedrite* is from Gèdres, France, in a crystalline schist containing microscopic crystals of chromian ***spinel,*** also observed at Hilsen, Kragerø, and Bamble, Norway, in the zone of *sillimanite* gneisses of the Bamble Formation associated with amphibolites and gabbroic rocks.

G. C. MOERSCHELL

References

Am. Min. **33**, 263; **63**, 1029.
DHZ, 2, 211–229.
Min. Mag. **26**, 257.

Cross-reference: *Minerals, Uniaxial and Biaxial.*

GEHLENITE A calcium aluminosilicate named by Fuchs in 1815 in honor of his colleague Gehlen. The solid-solution series *akermanite-gehlenite* constitutes the ***melilite*** group.

Composition: $Ca_2Al(AlSi)O_7$,MgSi replaces AlAl toward *akermanite.*

Occurrence: Common constituent of feldspathoidal rocks formed by the reaction of limestones and mafic magmas as at Scawt Hill, Antrim, Northern Ireland, in a contact metamorphic zone between dolerite and chalk and associated with *augite, aegirine, nepheline, labradorite,* and *perovskite.* Found at Monte Somma and Vesuvius, Italy, and Merapi, Java, in blocks of impure limestone in leucitite and basalt. Also found in granular aggregates in the contact zone between limestone and diorite in the Velardeña mining district, Durango, Mexico.

G. C. MOERSCHELL

References

Am. Min. **14**, 389.
DHZ, 1, 236–255.

Cross-references: *Meteoritic Minerals; Portland Cement Mineralogy.*

GEIKIELITE A magnesium titanate of the ***ilmenite*** group, named by Dick in 1892 after

Sir Archibald Geikie (1835–1924), a former Director of the Geological Survey of Great Britain.

Composition: $MgTiO_3$; often appreciable Fe^{2+} substitution for Mg toward *ilmenite.*

Properties: Trigonal, $R\bar{3}$; thick tabular {0001} habit; color: brownish-black; metallic luster; streak: purplish brown.

Occurrence: Found in gem gravel placers of the Rakwana and Balangoda district, Ceylon.

D. H. SPEIDEL

References

Am. Min. **44,** 879–882.
Dana, I, 535.
Min. Mag. **11,** 326.

GENKINITE *Am. Min.* **64,** 654
$(Pt,Pd)_4S_3$ tet. *Can. Min.* **15,** 389

GENTHELVITE A zinc beryllosilicate sulphide of the ***helvite*** group. Named by Glass, Jahns, and Stevens in 1944 in honor of F. A. Genth, who first described (in 1892) the species as a zinc-rich *danalite* from West Cheyenne Canyon, Colorado.

Composition: $2[Zn_4Be_3(SiO_4)_3S]$; with Zn replaced by Fe toward *danalite* and by Mn toward *helvite.* Isometric.

Occurrence: Crystals have been found in the miarolitic pegmatites of the Pikes Peak granite, El Paso County, Colorado, associated with *microcline* perthite, *quartz, phenakite, zircon,* and *siderite.* Also found in an *albite* vein cutting an albite-biotite granite of the Jos-Bokuru complex, Nigeria.

G. C. MOERSCHELL

References

AJS, ser. 3, **44,** 385.
Am. Min. **29,** 163.
DHZ, 4, 303–309.

GENTNERITE
$Cu_8Fe_3Cr_{11}S_{18}$

Cross-reference: *Meteoritic Minerals.*

GEOCRONITE *Am. Min.* **39,** 908; **61,** 963
$Pb_{28}(As,Sb)_{12}S_{46}$ mono.

GEORGEITE *Am. Min.* **64,** 1330
$Cu_5(CO_3)_3(OH)_4 \cdot 6H_2O$
amorph. *Min. Mag.* **43,** 97

GEORGIADESITE *Dana,* II, 791
$Pb_3(AsO_4)Cl_3$ mono.

GERASIMOVSKITE
$(Mn,Ca)_2(Nb,Ti)_5O_{12} \cdot 9H_2O(?)$
amorph. *Am. Min.* **43,** 1220

GERHARDTITE orth. *Dana,* II, 308
$Cu_2(NO_3)(OH)_3$

Cross-references: *Museum, Mineralogical; Nitrates.*

GERMANITE *Am. Min.* **38,** 794
$Cu_3(Ge,Fe)(S,As)_4$ iso.

GERSDORFFITE A nickel sulfide-arsenide named in 1843 after the von Gersdorffs, owners of the nickel mine at Schladming, Styria, Austria.

Composition: NiAsS with Ni replaced by Fe and Co and As by Sb.

Properties: Isometric, $P2_13$; crystals octahedra, cubo-octahedra, or pyritohedra, also lamellar or granular massive; cubic {100} cleavage; H=5.5; G=5.9; metallic luster, silver white to steel gray, but commonly tarnished.

Use: An ore of nickel.

Occurrence: In veins with other sulfides and nickel minerals.

K. FRYE

Reference: *Dana,* I, 298–300.

Cross-reference: *Ore Microscopy.*

GERSTLEYITE mono. *Am. Min.* **41,** 839
$(Na,Li)_4As_2Sb_8S_{17} \cdot 6H_2O$

GERSTMANNITE orth. *Am. Min.* **62,** 51
$(Mn,Mg)Mg(OH)_2(ZnSiO_4)$

GETCHELLITE *Am. Min.* **50,** 1817
$AsSbS_3$ mono.

GEVERSITE (***Pyrite*** gr.) *Am. Min.* **46,** 1518
$PtSb_2$ iso. *Min. Mag.* **32,** 833

GIANELLAITE *Am. Min.* **62,** 1057
$(NHg_2)_2(SO_4)$ iso.

GIANNETTITE tric. *Am. Min.* **34,** 770
$Na_3Ca_3Mn(Zr,Fe)TiSi_6O_{21}Cl(?)$

GIBBSITE An aluminum oxyhydroxide named by Torrey in 1822 after Colonel George Gibbs (1777–1834), original owner of the

Gibbs mineral collection at Yale College. Also called hydroargillite. One of the principal constituents of bauxite. Lowest-temperature stable hydrous aluminum oxide; alters at 150°–200°C to *boehmite*; stability increases with increasing water pressure.

Composition: $Al(OH)_3$

Properties: Monoclinic, $P2_1/n$; usually in lamellar-radiate pisolitic concretions and stalactitic aggregates; H=2.5–3.5; G=2.4; perfect basal {001} cleavage. Structure composed of two sheets of OH in hexagonal-close packing with 2/3 of octahedral sites between filled by Al (see *brucite*). Optically, α=1.56–1.58; β=1.56–1.58; γ=1.58–1.60; color: white or pastel; vitreous to pearly luster; transparent.

Occurrence: Constituent of bauxite, chiefly as a secondary mineral formed from the alteration of aluminous minerals; also as a low-temperature hydrothermal mineral in veins associated with aluminum-rich igneous rocks. Often associated with *boehmite, diaspore,* and *corundum.* Principal localities are Slatoust, Urals, USSR; Ouro Prêto and Pocos de Caldas, Minas Gerais, Brazil; and Richmond, Massachusetts.

Use: Primarily as a source for aluminum, but also in the manufacture of artificial abrasives, chemicals, and refactories.

D. H. SPEIDEL

References

AJS **264**(4), 289–309.
Dana, I, 663.
DHZ, 5, 91–101.
Soil Sci. Soc. Am. Proc. **29**(5), 531–534.

Cross-references: *Minerals, Uniaxial and Biaxial; Polysomatism; Rock-Forming Minerals; Soil Mineralogy.*

GIESSENITE *Am. Min.* **50**, 264
$Pb_9CuBi_6Sb_{1.5}S_{30}$(?) orth.

GILALITE *Am. Min.* **65**, 1065
$Cu_5Si_6O_{17}\cdot 7H_2O$ mono. *Min. Mag.* **43**, 639

GILLESPITE *Am. Min.* **59**, 1166; **60**, 938
$BaFe^{2+}Si_4O_{10}$ tetr.

GILPINITE – *Johannite*

GILSONITE
Hydrocarbon

Cross-reference: *Green River Mineralogy.*

GINIITE orth. *Min. Abst.* **31**, 356
$Fe^{2+}Fe^{3+}_4(H_2O)(OH)_2(PO_4)_4$
Min. Mag. **43**, 1061

GINORITE *Am. Min.* **42**, 56
$Ca_2B_{14}O_{23}\cdot 8H_2O$ mono.

GIORGIOSITE *Am. Min.* **55**, 1457
Hydrous Mg carbonate

GIRDITE mono. *Am. Min.* **65**, 809
$Pb_3H_2(TeO_3)(TeO)_6$ *Min. Mag.* **43**, 453

GISMONDINE A hydrated calcium alminosilicate of the ***zeolite*** group, first named by Leonhard in 1817 in honor of Professor C. G. Gismondi.

Composition: $Ca_4[Al_8Si_8O_{32}]\cdot 16H_2O$; SiO_2, 34.3%; Al_2O_3, 29.1%; CaO, 16.0%; H_2O, 20.6% by wt. The replacement $(K,Na)_2 \rightleftharpoons Ca$ may occur; K normally dominant over Na; gelatinized by HCl.

Properties: Monoclinic, $P2_1/c$; a=10.01 Å, b=10.61 Å, c=9.81 Å, β=92°20′; H=4.5; G=2.2; vitreous luster; colorless, white, bluish white, grayish or reddish; transparent to translucent; habit in spherulitic aggregates or in square octahedra. Refractive indices, α=1.515, β=1.54, γ=1.546; birefringence = 0.008–0.017; $2V$=15°–90° (–); optic axial plane ⊥ (010).

Occurrence: Rather rare; in cavities in leucite tephrite in Czechoslovakia and in leucitic lava in Italy; in cavities in basalt of Antrim, Ireland, characteristically associated with *chabazite, thomsonite,* and *phillipsite;* on **chlorite** in cavities of highly altered granite in Queensland; in cement of porous melilite-nephelinite palagonite tuffs of the Tantalus Craters on Oahu, Hawaii, associated with *phillipsite, chabazite,* and magnesian *calcite.*

A. IIJIMA
M. UTADA

References

Am. Min. **48**, 664–672; **54**, 149–155.
Hay, R. L., and Iijima A., 1967. Nature and origin of zeolitic palagonite tuffs on the Honolulu Series on Oahu, Hawaii, *Geol. Soc. Am., Prof. H. Williams Vol.*, 331–376.
J. Am. Chem. Soc. **78**, 5963.
Min. Mag. **33**, 187–201.

GITTINSITE *Can. Min.* **18**, 201
$CaZrSi_2O_7$ mono. *Min. Mag.* **43**, 1061

GLADITE *Am. Min.* **61**, 15
$PbCuBi_5S_9$ orth. *Dana*, I, 483

GLASERITE = *Aphthitalite*

GLAUBERITE A sodium calcium sulfate named by Brongniart in 1808, after the German chemist, J. R. Glauber (1604–1668).

Composition: $4[Na_2Ca(SO_4)_2]$

Properties: Monoclinic, *C2/c*; crystals usually thin tabular; H=2.5–3; G=2.75–2.85; vitreous luster; transparent to translucent; slightly salty taste; color: pale yellow or gray; streak: white; deliquesces slowly.

Occurrence: Widespread as a saline deposit formed as a precipitate from evaporating salt lakes, also under arid conditions as isolated crystals embedded in clastic sediments. Often found associated with *halite, thenardite, polyhalite,* and *anhydrite* in nitrate deposits of the arid regions.

Use: Sodium sulfate within the *glauberite* is important in paper and glass manufacture, metallurgical, and industrial purposes.

J. A. GILBERT

References

Am. Min. **52**, 1272.
Lindgren, W., 1933. *Mineral Deposits.* New York: McGraw-Hill, 323.

Cross-reference: *Saline Minerals.*

GLAUBER'S SALT = *Mirabilite*

GLAUCOCHROITE (***Olivine*** gr.)
$CaMnSiO_4$ orth. *Am. Min.* **36**, 918

GLAUCODOT *Dana*, I, 322
(Co,Fe)AsS orth.

Cross-reference: *Ore Microscopy.*

GLAUCOKERINITE *Dana*, II, 574
$(Cu,Zn)_{10}Al_4(SO_4)(OH)_{30} \cdot 2H_2O(?)$

GLAUCONITE A hydroxyl potassium, iron, aluminum phyllosilicate of the ***mica*** group named by Keferstein (1828) from the Greek *glaucos,* meaning "bluish green," but first described by von Humboldt (1823).

Composition:
$4[(K,Na)(Al,Fe^{3+},Mg)_2(Al,Si)_4O_{10}(OH)_2]$; substitutions of Ca, Na, Mg, Fe^{2+} common. Considered to be an intermediate member of the isomorphous series between aluminum-rich skolite, and aluminum-free *celadonite.*

Properties: Monoclinic, *C2/m*; H=2; G=2.4–3.0, variable from iron and moisture content primarily; luster is earthy, dull; color: dull green, ranging from light green to greenish black; habit: pelletal, flakey, vermiform, and coatings; cleavage: perfect basal but often unobservable because of fine flakey aggregates; streak: dull green. Optically (—); α=1.592–1.610, $\beta=\gamma$=1.614–1.641; $2V$:0°–20°; $\alpha \perp (001)$. Readily attacked by HCl; extensive substitution may occur; Mg and Fe^{2+} for Fe^{3+} and Al; Ca and Na for K, Al for Si; commonly, statistical formulae have fractional subscripts, as are common to ***illites***; water content is variable, apparently occurring with K in layers between flakes.

Occurrence: Most common as an authigenic mineral in sedimentary rocks of marine origin although some nonmarine ***illites*** have the composition of *glauconite. Celadonite,* although a similar mineral, occurs as an alteration product of volcanic rocks. Often in association with ***apatite*** (phosphate nodules), *pyrite,* various clay minerals, and *calcite.*

Use: Radioactive age dating of host sedimentary rocks.

A. J. EHLMANN

References

Am. Min. **43**, 481–497.
DHZ, 3, 35–41.
J. Sed. Petrol. **33**, 87–96.
Zbl. Geol. Paläontol. **1**, 974–1017.

Cross-references: *Authigenic Minerals; Clays, Clay Minerals; Minerals, Uniaxial and Biaxial; Phyllosilicates.*

GLAUCOPHANE A sodium magnesium hydroxyl aluminosilicate of the ***amphibole*** group named by Hausmann in 1845 from the Greek meaning "to appear bluish green." The name is for the iron-free end member of the *glaucophane-riebeckite* series. Intermediate members are *crossite.*

Composition: $Na_2Mg_3Al_2Si_8O_{22}(OH)_2$. Fe^{2+} replaces Mg and Fe^{3+} replaces Al toward *crossite.*

Properties: Monoclinic, *C2/m*; usually prismatic or aggregates of acicular crystals; H=6; G=3.1–3.3; pale bluish gray to lavender blue, deepening in color with increasing iron content; white to pale blue streak; vitreous luster; strongly pleochroic blue, to violet, to green, but less intense than *riebeckite.*

Occurrence: Relatively common, only as a constituent of low-grade metamorphic rocks, especially the glaucophane schists. High pressure conditions are necessary for the formation of the glaucophane facies which are rich in sodium. Some of the more important minerals in a *glaucophane* association are *epidote*, ***chlorite***, *lawsonite*, *zoisite*, *muscovite*, *jadeite*, and *almandine*. Glaucophane schists are widespread in the Franciscan series of the coastal ranges, California; in the Kanto Mountains, Japan; along eastern Corsica and at various locations in the Swiss Alps.

G. C. MOERSCHELL

References

Am. Min. **63,** 1036

DHZ, 2, 333–351.

Kostov, I., 1968. *Mineralogy.* London: Oliver and Boyd, 346–347.

Cross-references: *Amphibole Group; Clinopyroxenes; Minerals, Uniaxial and Biaxial.*

GLAUKOSPHAERITE *Can. Min.* **14,** 574
$(Cu,Ni)_2(CO_3)(OH)_2$ mono. *Min. Abst.* **29,** 24

GLIMMER = ***Mica***

GLIMMERTON = ***Illite***

GLUCINE *Am. Min.* **49,** 1152
$CaBe_4(PO_4)_2(OH)_4 \cdot \frac{1}{2}H_2O$

GLUSHINSKITE *Am. Min.* **66,** 439
β-$MgC_2O_4 \cdot 2H_2O$ mono. *Min. Mag.* **43,** 837

GMELINITE A hydrated sodium calcium aluminosilicate of the ***zeolite*** group named by Brooke in 1825 in honor of C. G. Gmelin (1792–1860).

Composition: $(Na_2,Ca)_4[Al_8Si_{16}O_{48}] \cdot 24H_2O$. Here Ca:$Na_2$=1:1.

Properties: Hexagonal, $P6_3/mmc$; a=13.72 Å, c=9.95 Å; H=4.5; G=2.1; vitreous luster; colorless, yellowish, greenish or reddish white, or flesh-red; transparent to translucent; rhombohedral, pyramidal, prismatic, or tabular habit; cleavage good on $\{10\bar{1}0\}$ and {0001}. Optically, refractive indices, ϵ= 1.474–1.480, ω=1.476–1.494; birefringence = 0.002–0.015; uniaxial (−). Closely related to *chabazite;* appreciable amounts of K may enter the (Na_2,Ca) group; Sr may replace Ca to a small extent; minimum diameter of the largest channel in the aluminosilicate framework is 6.4 Å; cation sieve properties very selective; Na is exchanged by K; decomposed by HCl.

Occurrence: Mainly in amygdules of basalt, associated with *chabazite, analcime, levyne, phillipsite, aragonite,* and *calcite.* Principal localities are Antrim, Northern Ireland; Bergen Hill, New Jersey, and Cape Blomidon, Nova Scotia.

A. IIJIMA
M. UTADA

References

DHZ, 4, 386–400.

Doklady Acad. Sci. USSR **26,** 659.

Min. Mag. **32,** 202–217.

N.J. Min. Mh. 1956, 250–259.

Smith, J. V., 1963. Structural classification of zeolites, *Mineral. Soc. Am. Spec. Paper 1,* 281–290.

Cross-reference: *Zeolites.*

GODLEVSKITE *Am. Min.* **55,** 317
$(Ni,Fe)_7S_6$ orth.

GOEDKENITE mono. *Am. Min.* **60,** 957
$(Sr,Ca)_2Al(PO_4)_2(OH)$ *Min. Abst.* **27,** 257

GOERGEYITE *Am. Min.* **39,** 403
$K_2Ca_5(SO_4)_6 \cdot H_2O$ mono.

Cross-reference: *Saline Minerals.*

GOETHITE An iron oxyhydroxide isostructural with *diaspore,* named by Lenz in 1806 after the poet and philosopher Goethe (1749–1832). The name was originally applied to its polymorph γ-FeO(OH), *lepidocrocite,* but was transferred to the more common α-phase because of long-established usage.

Composition: α-FeO(OH).

Properties: Orthorhombic, *Pbnm;* may be massive, reniform, foliated, or in radiating fibrous masses; H=5-5.5; G=4.37; adamantine to dull luster, may be silky in the fibrous variety; subtranslucent; color: yellowish brown to brown, blackish brown in crystals, and ochre-yellow when earthy; streak: brownish yellow.

Occurrence: Found as a weathering product at ordinary temperatures and pressures of other iron-bearing minerals such as *siderite, magnetite, pyrite,* etc., or as a direct precipitate in bogs and springs. A major constituent of limonite often in association with *lepidocrocite, hematite,* psilomelane, *manganite, quartz,* and *calcite.*

Uses: An economically important sedimentary ore of iron (bog iron ore) in the Lorraine basin, France, and in Knob Lake, Canada. The pigment yellow ochre.

J. A. GILBERT

References

Dana, I, 680.
DHZ, 5, 118-121.

Cross-references: *Gossan; Invertebrate and Plant Mineralogy; Magnetic Minerals; Minerals, Uniaxial and Biaxial; Naming of Minerals; Ore Microscopy; Rock-Forming Minerals; Soil Mineralogy; Syntaxy; Topotaxy.*

GOLD A native element, the name from Old English and common to many Teutonic dialects.

Composition: 4Au commonly alloyed with Ag (Ag>20%, *electrum*), Pd, Rh, Cu, or Bi.

Properties: Isometric, *Fm3m*; crystals octahedral, dodecahedral, or cubic, also parallel groups and twinned aggregates, also reticulated, dendritic, arborescent, filiform, or spongy, also massive, granular, or scaley; commonly twinned {111}; maleable and ductile; H=2.5; G=19.3; metallic luster; color: golden, if pure, silver white to orange red for impure varieties.

Use: A major ore of gold.

Occurrence: In hydrothermal veins with *quartz* and *pyrite*, in pegmatites in blacksands; and in placers.

K. FRYE

References

Collins, R. S., 1975. Gold, *Mineral Dossier 14.* London: Institute of Geological Sciences, 66p.
Jensen, M. L., and Bateman, A. M., 1979. *Economic Mineral Deposits,* 3rd ed. New York: Wiley, 271-293.

Cross-references: *Blacksand Minerals; Crystallography Morphological; Gossan; Metallic Minerals; Mineral Classification: Principles; Mineral Deposits: Classification; Mineral Industries; Museums, Mineralogical; Native Elements and Alloys; Ore Microscopy; Placer Deposits; Salt Economy; Skeletal Crystals.* Vol. IVA: *Gold: Element and Geochemistry.*

GOLDFIELDITE (***Tetrahedrite*** gr.)
$Cu_{12}(Sb,As)_4(Te,S)_{13}$ iso. *Am. Min.* **53**, 2105

GOLDICHITE *Am. Min.* **40**, 469
$KFe(SO_4)_2 \cdot 4H_2O$ mono.

GOLDMANITE
(***Garnet*** gr.) *Am. Min.* **49**, 644
$Ca_3V_2(SiO_4)_3$ iso.

GONNARDITE A hydrated sodium calcium aluminosilicate of the ***zeolite*** group named by Lacroix in 1896 in honor of M. Gonnard.

Composition: $Na_4Ca_2[Al_8Si_{12}O_{40}] \cdot 14H_2O$; SiO_2, 45.6%; Al_2O_3 25.8%; CaO, 4.6%; Na_2O, 10.3%; H_2O, 13.7% by wt.

Properties: Orthorhombic (pseudo-tetragonal), *Pbmn; a*=13.38 Å, *b*=13.38 Å, *c*= 6.66 Å; H=5; G=2.3; colorless, white, pink, or brown; transparent to translucent; habit in spherulites. Optically, refractive indices, α=1.497-1.506, γ=1.499-1.508; birefringence = 0.002; $2V$=50° (−), α=z. Intermediate member of the isomorphous series between *thomsonite* and *natrolite;* small amounts of K may substitute for Na; considerable replacements CaAl⇌NaSi and $Ca \rightleftharpoons Na_2$; molecular sieve properties good but not too selective due to adaptability of the aluminosilicate framework; gelatinized with HCl.

Occurrence: In cavities in basalt and leucite tephrite, associated with *thomsonite* and *phillipsite;* alteration product of *nepheline* and ***plagioclase;*** in cement of nephelinite palagonite tuffs on Oahu, Hawaii, in association with *phillipsite, chabazite, natrolite,* and *calcite* which were formed by reaction of nephelinite glass, *nepheline,* and ***melilite*** with percolating groundwater. Principle localities are Chaux de Bergonne, Pay-de-Dome, France, and Langesundsfjord, Norway.

A. IIJIMA
M. UTADA

References

Am. Min. **54**, 149-155.
Foster, M. D., 1965. Compositional relations among thomsonites, gonnardites, and natrolites, *U.S. Geol. Surv. Prof. Paper 504-E,* 10p.
Hay, R. L., and Iijima, A., 1967. Nature and origin of zeolitic palagonite tuffs of the Honolulu Series on Oahu, Hawaii, *Geol. Soc. Am., Prof. H. Williams Vol.*, 331-376.
Min. Mag. **31**, 265-271.

Cross-reference: *Zeolites.*

GONYERITE (***Chlorite*** gr.)
$(Mn,Mg)_5Fe^{3+}(Si_3Fe^{3+})O_{10}(OH)_8$
orth. *Am. Min.* **40**, 1090

GOONGARRITE = Mixture of *cosalite* and *galena;* pseudomorphous after *heyrovskyite*
Am. Min. **62**, 397

GOOSECREEKITE *Can. Min.* **18**, 323
$CaAl_2Si_6O_{16} \cdot 5H_2O$ mono.

GORCEIXITE (***Crandallite*** gr.)
$BaAl_3(PO_4)_2(OH)_5 \cdot H_2O$ trig. *Dana,* II, 975

GOSHENITE–Colorless variety of *beryl*

GOSARLITE *Dana,* II, 513
$ZnSO_4 \cdot 7H_2O$ orth.

GOTZENITE tric. *Am. Min.* **45**, 221
$(Ca,Na)_7(Ti,Al)_2Si_4O_{15}(F,OH)_3$
Min. Mag. **31**, 503

GOUDEYITE hexa. *Am. Min.* **63**, 704
$Cu_6Al(AsO_4)_3(OH)_6 \cdot 3H_2O$
Min. Mag. **43**, 1061

GOWERITE *Am. Min.* **57**, 381
$CaB_6O_{10}(OH)_4 \cdot 3H_2O$ mono.

GOYAZITE (***Crandallite*** gr.)
$SrAl_3(PO_4)_2(OH)_5 \cdot H_2O$ trig. *Dana,* II, 834

Cross-reference: *Pegmatite Minerals.*

GRAEMITE *Min. Abst.* **29**, 200
$CuTeO_3 \cdot H_2O$ orth. *Min. Rec.* **6**, 32

GRAFTONITE mono. *Am. Min.* **53**, 742
$(Fe,Mn,Ca)_3(PO_4)_2$ *Min. Abst.* **29**, 82

Cross-reference: *Meteoritic Minerals.*

GRANDIDIERITE *Min. Mag.* **43**, 651
$(Mg,Fe)Al_3O_2(BO_3)(SiO_4)$
orth. Winchell, II, 497

GRANDITE–*Grossular-andradite* series of the ***garnet*** gr.

GRANTSITE *Am. Min.* **48**, 1511
$Na_4CaV^{4+}_2V^{5+}_{10}O_{32} \cdot 8H_2O$ mono.

GRAPHITE Carbon, native nonmetal, named by Werner in 1789 from the Greek meaning "to write," referring to its use as a crayon. Polymorphous with *chaoite, diamond,* and *lonsdaleite.*

Composition: 4[C]; often impure with admixed clays and iron oxides.

Properties: Hexagonal, $P6_3/mmc$; usually in foliated or scaly masses, may be radiated or granular; H=1-2; G=2.2; metallic luster; greasy feel; folia flexible but not elastic; sectile; color: black to steel gray; streak: black to dark steel-gray shining.

Occurrence: Most commonly found in metamorphic rocks containing carbonaceous material and occasionally as an original constituent of igneous rocks. Important sources include the Alibert mine, Botogolsk, USSR; Southern, Western and Sabaragamuwa Provinces, Sri Lanka; Ch'ungch'ong-pukto Province, Korea; Tamatave to Marovintsy, Malagasy Republic; Ticonderoga, New York, and Sterling Hill, New Jersey.

Use: To coat foundry facings; in steelmaking; in lubricants; in batteries and dry cells; in crucibles for melting nonferrous metals; in lead pencils; as "brushes" in electrical equipment.

J. A. GILBERT

References

Jensen, M. L., and Bateman, A. M., 1979. *Economic Mineral Deposits,* 3rd ed. New York: Wiley, 529–531.

Lamey, C. A., 1966. *Metallic and Industrial Mineral Deposits.* New York: McGraw-Hill, 485–493.

Cross-references: *Abrasive Materials; Allotropy; Diamond; Lunar Minerals; Mantle Mineralogy; Meteoritic Minerals; Mineral Classification: Principles; Mineral Industries; Native Elements and Alloys; Ore Microscopy; Pigments and Fillers; Polymorphism; Refractory Minerals; Synthetic Minerals.*

GRATONITE *Dana,* I, 397
$Pb_9As_4S_{15}$ trig. *Min. Abst.* **27**, 278

Cross-reference: *Pegmatite Minerals.*

GRAYITE *Am. Min.* **47**, 419
$(Th,Pb,Ca)PO_4 \cdot H_2O$ pseudo hexa.

GRAY TIN = *Tetrahedrite*

GREENALITE (***Kaolinite-Serpentine*** gr.)
$(Fe^{2+},Fe^{3+})_{2-3}Si_2O_5(OH)_4$ *Am. Min.* **65**, 11
mono. *Can. Min.* **18**, 208

Cross-references: *Authigenic Minerals; Clays, Clay Minerals.*

GREENOCKITE Cadmium sulfide named by Jameson in 1840 after Lord Greenock (later the Earl Cathcart).

Composition: 2[CdS]. Often Zn substitutes for Cd and a complete solid solution series exists between *greenockite* and *wurtzite.* Dimorphous with *hawleyite.*

Properties: Hexagonal, $C6_3/mc$; usually pulverulent and as powdery incrustations; H=3–

3.5; G=4.9; brittle; adamantine to resinous, earthy luster; color: yellow to orange; streak: orange-yellow to brick-red; translucent.

Occurrence: The most common mineral containing cadmium, but is found in few localities and in small amounts, usually as an earthy coating on zinc ores, particularly *sphalerite* and *smithsonite*. Occasionally crystals are found in amygdaloidal cavities in mafic igneous rocks. Found with zinc ores at Pribram, Czechoslovakia; Pierrefitte, France; and in the USA as yellow coatings at Joplin, Missouri; Marion County, Arkansas; and Franklin, New Jersey.

Use: Ceramic glazes; low-melting point alloys; control rods in nuclear reactors; yellow pigment (cadmium yellow) in enamels; nickel-cadmium batteries, and with selenium in ruby glass.

J. A. GILBERT

References

Dana, I, 228.

Vanders, I., and Kerr, P. F., 1967. *Mineral Recognition.* New York: Wiley, 178.

GREIGITE (***Linnaeite*** gr.)
$FeFe_2S_4$ iso. *Am. Min.* **53**, 2087; *Min. Mag.* **43**, 733

GRIMALDIITE
$CrO(OH)$ trig. *Am. Min.* **62**, 593; *Min. Abst.* **29**, 337

GRIMSELITE hexa. *Am. Min.* **58**, 139
$K_3Na(UO_2)(CO_3)_3 \cdot H_2O$

GRIPHITE iso. *Am. Min.* **60**, 1333; *Min. Abst.* **30**, 296
$(Mn,Na,Ca,Li,Fe)_{24}Ca_4Fe_3^{2+}Al_8(PO_4)_{24}(F,OH)_8$

Cross-reference: *Pegmatite Minerals.*

GROSSULAR (Grossularite) A calcium aluminum silicate of the ***garnet*** group, so named because of resemblance to the gooseberry, *R. grossularia.*

Composition: $8[Ca_3Al_2Si_3O_{12}]$ with Al replaced by Fe^{3+} toward *andradite* and by Cr toward *uvarovite.*

Properties: Isometric, *Ia3d;* crystals dodecahedra or trapezohedra, simple or mutually modified, in irregular grains, or massive; H=6.5–7; G=3.5; vitreous luster; colorless, white, pink, green, yellow, or brown.

Occurrence: Contact and regionally metamorphosed impure calcareous rocks, also in skarns.

K. FRYE

References

DHZ, 1, 77–96.

Phillips, W. R., and Griffen, D. T., 1981. *Optical Mineralogy.* San Francisco: Freeman, 117–118.

Cross-references: *Garnet Group; Meteoritic Minerals; Minerals, Uniaxial and Biaxial.*

GROUTITE A manganexe oxyhydroxide, named by Gruner in 1945 after Professor F. F. Grout, late of the University of Minnesota.

Composition: $4[MnO(OH)]$.

Properties: Orthorhombic, *Pbnm;* wedge-shaped crystals; H=3.5–4; G=4.14; color: jet black; streak: dark brown; adamantine luster; strong pleochroism. Polymorphic with *manganite* and *feitknechtite.* Structure similar to that of *diaspore* and *ramsdellite.*

Occurrence: Originally from the Cuyuna Range, Minnesota, associated with *manganite, quartz, hematite,* and *goethite* in vugs within the iron ores. Also from Talcville, near Gouverneur, New York, in vugs in the *talc* associated with *calcite* and "hexagonite", lilac variety of *tremolite.*

D. H. SPEIDEL

References: *Am. Min.* **30**, 169; **32**, 654–569; **44**, 877–878.

GROVESITE (***Chlorite*** gr.)
$(Mn,Mg,Al)_6(Si,Al)_4O_{10}(OH)_8$ tric. *Am. Min.* **59**, 1155; **65**, 335

GRUENLINGITE *Dana,* I, 164
$Bi_4TeS_3(?)$ trig.

GRUNERITE An iron magnesium hydroxyl silicate of the ***amphibole*** group. Named by Kenngott in 1853 after E. L. Grüner, who first analyzed it. It is the iron-rich end-member of the *cummingtonite-grunerite* series.

Composition: $2[Fe_7^{2+}Si_8O_{22}(OH)_2]$. Monoclinic.

Occurrence: In contact or regionally metamorphosed iron-rich siliceous sediments, *grunerite* is associated with *almandine, fayalite,* and *hedenbergite.* Such schists are found in the *staurolite* zones of iron formations in northern Michigan and in the Mesabi formations around Duluth, Minnesota.

G. C. MOERSCHELL

References

AJS **277**, 735.
Am. Min. **63**, 1030.
DHZ, 2, 234–248.

Cross-references: *Amphibole Group; Orthopyroxenes.*

GUANAJUATITE *Dana*, I, 278
Bi_2Se_3 orth.

GUANGLINITE = *Isomertieite(?)*
Am. Min. **65**, 408

GUDMUNDITE *Dana*, I, 325
FeSbS mono. *Min. Abst.* **28**, 79

GUERINITE mono. *Am. Min.* **50**, 812
$Ca_5H_2(AsO_4)_4 \cdot 9H_2O$

GUETTARDITE *Am. Min.* **53**, 1425
$Pb(Sb,As)_2S_4$ mono. *Can. Min.* **18**, 13

GUILDITE mono. *Am. Min.* **55**, 502; **63**, 478
$CuFe^{3+}(SO_4)_2(OH) \cdot 4H_2O$

GUILLEMINITE orth. *Am. Min.* **50**, 2103
$Ba(UO_2)_3(SeO_3)_2(OH)_4 \cdot 3H_2O$

GUMMITE–Mixture of secondary uranium oxides *Am. Min.* **41**, 539

GUNNINGITE *Am. Min.* **47**, 1218
$(Zn,Mn)SO_4 \cdot H_2O$ mono. *Min. Abst.* **29**, 514

GUSTAVITE *Am. Min.* **56**, 633
$PbAgBi_3S_6$ orth. *Min. Abst.* **29**, 140

GUTSEVICHITE *Am. Min.* **46**, 1200
$(Al,Fe)_3(PO_4,VO_4)_2(OH)_3 \cdot 8H_2O(?)$

GUYANAITE *Am. Min.* **62**, 593
CrO(OH) orth. *Min. Abst.* **30**, 419

GYPSUM A hydrated calcium sulfate named by Theophrastus in 315 BC or earlier from the Greek name for the mineral, but more especially for the calcined mineral, or "plaster."

Composition: $8[CaSO_4 \cdot 2H_2O]$

Properties: Monoclinic, *A*2/*a*; twins common; often massive, foliated, fibrous, granular (fibrous variety with silky luster is satin spar; fine-grained, massive is alabaster; variety with broad colorless and transparent cleavage folia is selenite); H=2; G=2.32; vitreous luster; transparent to translucent; color: white, gray, or colorless with shades of yellow, brown, or red from impurities; streak: white.

Occurrence: Widespread, found as beds in sedimentary rocks often interstratified with limestone, shale, and *halite*, as an authigenic mineral and in saline lakes and salt pans. May be found in volcanic regions in native *sulfur* deposits and as a gangue mineral in metallic veins. Usually in an association with several of the following: *halite, anhydrite, dolomite, calcite, pyrite, quartz,* or *sulfur.* Well-known localities include the Paris basin, France; Girgenti, Sicily; Chihuahua, Mexico; Wayne County, Utah; Tully and Syracuse, New York; and White Sands, Alamogordo, New Mexico.

Uses: In the production of plaster of paris; as a fertilizer; in the production of Portland cement; a flux in glass manufacture and for ornamental purposes.

J. A. GILBERT

References

DHZ, 5, 202–218.
Jensen, M. L., and Bateman, A. M., 1979. *Economic Mineral Deposits,* 3rd ed. New York: Wiley, 202–204, 513–515.
Lamey, C. A., 1966. *Metallic and Industrial Mineral Deposits.* New York: McGraw-Hill, 516–519.

Cross-references: *Anhydrite and Gypsum; Authigenic Minerals; Blowpipe Analysis; Cave Minerals; Crystal Habits; Crystallography: Morphological; Diagenetic Minerals; Fire Clays; Green River Mineralogy; Human and Vertebrate Mineralogy; Minerals, Uniaxial and Biaxial; Mohs Scale of Hardness; Optical Orientation; Portland Cement Mineralogy; Pseudomorphism; Rock-Forming Minerals; Soil Mineralogy; Staining Techniques.*

GYROLITE *Am. Min.* **46**, 913
$Ca_2Si_3O_7(OH)_2 \cdot H_2O$
hexa. *Min. Abst.* **29**, 473

H

HAAPALAITE hexa. *Am. Min.* **58**, 1111
$(Fe,Ni)_4(Mg,Fe)_3S_4(OH)_6$

HACKMANITE—S-rich *sodalite*

HAFNON tetr. *Am. Min.* **61**, 175
$(Hf,Zr)SiO_4$

Cross-reference: *Pegmatite Minerals.*

HAGENDORFITE mono. *Am. Min.* **40**, 553
$(Na,Ca)(Fe,Mn)_2(PO_4)_2$ *GSA Bull.* **67**, 1694

HAGGITE *Am. Min.* **45**, 1144; **65**, 210
$V_2O_2(OH)_3$ mono.

HAIDINGERITE *Dana*, II, 708
$CaHAsO_4 \cdot H_2O$ orth.

HAIWEEITE mono. *Am. Min.* **44**, 839
$Ca(UO_2)_2Si_6O_{15} \cdot 5H_2O$

HAKITE (***Tetrahedrite*** gr.) *Am. Min.* **57**, 1553
$(Cu,Hg)_{12}Sb_4(S,Se)_{13}$ iso.

HALITE Sodium chloride named by Glocker in 1847 from the Greek word meaning "salt" or "sea" because of its recovery through the evaporation of sea water, also known as rock salt.

Composition: 4[NaCl]

Properties: Isometric, *Fm3m*; found in crystals or granular masses; cubic cleavage {100}; H=2.5; G=2.16; diathermanous; salty taste; transparent to translucent; color: white but various shades of yellow, red, blue, or purple result from impurities; colors flame a deep sodium yellow.

Occurrence: *Halite* results as a precipitant of sea water or an evaporite in salt lakes and often occurs interstratified with and covered by beds of sedimentary rocks including shale, limestone, dolomite, and gypsum or anhydrite. Also, commonly as an authigenic mineral. Often associated with *sylvite, calcite,* ***clay***, and native *sulfur*. Deformation of thick stratified salt beds, usually at considerable depth, may result in the extrusion, through the overlying sedimentary burden, of massive, circular plug-like salt bodies known as salt domes. Principal deposits include Iletsk near Orenburg, USSR; Hall near Innsbruck, Austria; Stassfurt, E. Germany; Salt Range, Punjab, India; and in the USA in central and western New York; the Michigan basin; Borax Lake, California; the Great Salt Lake Region, Utah; and along the coast of the Gulf of Mexico.

Uses: Source of sodium and chlorine, in fertilizers, stock feeds, weed killers, preparation of food, as a preservative, (see *Salt Economy*).

J. A. GILBERT

References

DHZ, 5, 357–361.

Jensen, M. L., and Bateman, A. M., 1979. *Economic Mineral Deposits,* 3rd ed. New York: Wiley, 204–205, 553–557.

Cross-references: *Authigenic Minerals; Blowpipe Analysis; Crystal Growth; Crystal Habits; Crystallography: Morphological; Green River Mineralogy; Human and Vertebrate Mineralogy; Magnetic Minerals; Mineral Classification: Principles; Mineral and Ore Deposits; Minerals, Uniaxial and Biaxial; Mohs Scale of Hardness; Phantom Crystals; Plastic Flow in Minerals; Rock-Forming Minerals; Saline Minerals; Salt Economy; Skeletal Crystals.*

HALLIMONDITE *Am. Min.* **50**, 1143
$Pb_2(UO_2)(AsO_4)_2$ tric.

HALLOYSITE Hydrous aluminosilicate of the ***kaolinite-serpentine*** group, named by Berthier in 1826 in honor of Omalius d'Halloy who had first noted the mineral. The distinctive tubular morphology can be readily recognized in electron photomicrographs. This morphology has been attributed to differences in *a* and *b* dimensions of alumina, silica, and water sheets with curvature resulting from adjustments in the sheets for better fit.

Composition: $2[Al_4Si_4(OH)_8O_{10}]$.

Properties: Monoclinic; ultramicroscopic tubular crystals; white, grayish, greenish, yellowish, bluish, or reddish shades of white; H=1–2; G=2.0–2.2; translucent to opaque;

pearly or waxy to dull luster; conchoidal fracture in massive pieces; massive; approaches transparency in water; colorless in thin section; n=1.54; decomposed by acids; yields water on heating.

Occurrence: In veins and beds of ore as a secondary product, suggesting hydrothermal origin. Derived from the decomposition of the aluminous minerals, *kaolinite, allophane,* and *alunite,* with which it is associated.

E. BOOY

References

Clay Mineral. Bull. **2**, 294.
Dana, E. S., (Ford, W. E., ed.), 1949. *A Textbook of Mineralogy.* New York: Wiley, 384p.
DHZ, 3, 191–192.
Grim, R. E., 1968. *Clay Mineralogy.* New York: McGraw-Hill, 596p.
Kerr, P. F., ed., 1951. *Reference Clay Minerals, American Petroleum Institute Research Project 49.* New York: Columbia University, 701p.

Cross-references; *Clays, Clay Minerals; Electron Microscopy (Transmission); Soil Mineralogy; Staining Techniques.*

HALOTRICHITE A hydrated ferrous iron and aluminum sulfate of the ***halotrichite*** group named in 1839 from the Latin *halotrichum,* hair salts, based on the older German *Haarsalz,* in allusion to its fibrous habit.

Composition: $4[Fe^{2+}Al_2(SO_4)_4 \cdot 22H_2O]$ with Fe^{2+} replaced by Mg (Mg>Fe, *pickeringite*) and by Mn^{2+} toward *apjohnite.*

Properties: Monoclinic, $P2/m$; acicular or hair-like crystals, tufted or matted; H=1.5; G=1.9; vitreous luster; colorless, white, yellowish, or greenish.

Occurrence: Weathering product of *pyrite*-bearing rocks, especially in sheltered places such as mines, also in arid regions, and around hot springs and fumaroles.

K. FRYE

References

Dana, II, 523–527.
Min. Mag. **40**, 599.

HALOTRICHITE GROUP–*Apiohnite, bilinite, dietrichite, halotrichite, pickeringite,* and *redingtonite*

HALURGITE *Am. Min.* **47**, 1217
$Mg_2B_8O_{10}(OH)_8 \cdot H_2O$ mono.

HAMBERGITE A beryllium hydroxyl borate named in 1890 for the Swedish mineralogist Axel Hamberg.

Composition: $8[Be_2(BO_3)(OH)]$ with (OH) replaced by F.

Properties: Orthorhombic, *Pbca;* equant or striated prismatic crystals; one perfect {010} and one good {100} cleavage; H=7.5; G=2.36; vitreous luster; colorless, grayish, or yellowish.

Occurrence: Alkali pegmatites and placers.

K. FRYE

References

Am. Min. **50**, 85–95.
Dana, II, 370–372.

HAMMARITE *Am. Min.* **61**, 15
$Pb_2Cu_2Bi_4S_9$ orth. *Can. Min.* **14**, 536

HANCOCKITE (***Epidote*** gr.)
$(Pb,Ca,Sr)_2(Al,Fe)_3Si_3O_{12}(OH)$
mono. DHZ, 1, 194

HANKSITE hexa. *Am. Min.* **58**, 799
$Na_{22}K(SO_4)_9(CO_3)_2Cl$ *Dana,* II, 628

HANNAYITE tric. *Am. Min.* **48**, 635
$(NH_4)_2Mg_3H_4(PO_4)_4 \cdot 8H_2O$
Min. Abst. **28**, 395

Cross-reference: *Human and Vertebrate Mineralogy.*

HARADAITE *Am. Min.* **56**, 1123; **60**, 340
$SrVSi_2O_7$ orth.

HARDYSTONITE (***Melilite*** gr.) RRW, 259
$Ca_2ZnSi_2O_7$ tetr.

HARKERITE trig. *Am. Min.* **62**, 263
$Ca_{24}Mg_8[AlSi_4(O,OH)_{16}]_2(BO_3)_8(CO_3)_8(H_2O,Cl)$

HARMOTOME A hydrated barium aluminosilicate of the ***zeolite*** group. First named by Haüy in 1801 from the Greek word meaning "a cutting joint," in allusion to the morphology of the twinned crystals.

Composition: $Ba_2[Al_4Si_{12}O_{32}] \cdot 12H_2O$; SiO_2, 49.8%; Al_2O_3, 14.1%; BaO, 21.2%; H_2O, 14.9% by wt.

Properties: Monoclinic, $P2_1/m$ or $P2_1$; a= 9.87 Å, b=14.14 Å, c=8.72 Å, β=124°50′; H=4.5; G=2.41–2.47; vitreous luster; colorless, white, gray, pink, or yellow; subtransparent to translucent; habit in interpenetrant complex twins or in radiated aggregates; cleavage good on {010}; twinning on {001}, {021}, {110} (interpenetrant). Refractive indices, α=1.503–1.508, β=1.505–1.509, γ= 1.508–1.514; birefringence = 0.005–0.008; $2V$=80° (+); optic axial plane ⊥ (010); γ=

y, $\alpha:x$=63°–67°. Considerable replacement Si⇌Al with a corresponding introduction of alkalies into the Ba position; K and sometimes Na are the dominant alkali metals; Ca is generally only in trace amounts; minimum diameter of the largest channel in the aluminosilicate framework is 3.2 Å; molecular sieve properties very selective for H_2, N_2, CO_2, and NH_3; decomposed by HCl.

Occurrence: In veins and vugs in igneous rocks; in thin bands traversing manganese ores, associated with barium ***feldspar;*** in hydrothermal veins, together with *sphalerite, galena, barite,* and *calcite,* in schist and gneiss; in fissures in altered tuffs associated with *calcite, laumontite,* and *pyrite;* alteration product of Pliocene tuffs deposited in saline lake in Arizona. Important localities include Strontian, Argyllshire, Scotland; Andreasburg, West Germany; Kongsberg, Norway; and Rabbit Mountain near Port Arthur, Ontario.

A. IIJIMA
M. UTADA

References

Acta Cryst. **14,** 1153–1163.
J. Chem. Soc. 1959, 1521.
Min. Mag. **30,** 136–138.
Sheppard, R. A., and Gude, A. J., 1971. Sodic harmotome in lacustrine Pliocene tuff near Wilkieup, Mohana County, Arizona, *USGS Prof. Paper 750-D,* 50–55.

Cross-references: *Green River Mineralogy; Zeolites.*

HARSTIGITE orth. *Am. Min.* **53,** 1418
$MnCa_6(Be_2OOH)_2(Si_3O_{10})_2$

HASTINGSITE An iron hydroxyl aluminosilicate of calcium and sodium of the ***amphibole*** group. Named by Adams and Harrington in 1896 from the original locality, Dungannon, Hastings County, Ontario.

Composition: $2[NaCa_2Fe^{2+}_4Fe^{3+}(Al_2Si_6)O_{22}(OH)_2]$ with substitution of Mg for Fe^{2+} toward *magnesio-hastingsite.*

Occurrence: In a ***pyroxene***-free nepheline syenite at Dungannon, Canada; and intergrown with *aegirine* in a nepheline syenite from Umptek, Kola Peninsula, USSR.

G. C. MOERSCHELL

References

AJS, ser. 4, **1,** 210–214.
Am. Min. **13,** 287–296; **63,** 1032.
DHZ, 2, 264–314.

Cross-reference: *Amphibole Group.*

HASTITE (***Marcasite*** gr.) *Am. Min.* **41,** 164
$CoSe_2$ orth.

HATCHETTOLITE = *Uranpyrochlore*
Am. Min. **62,** 406

HATCHITE tric. *Am. Min.* **56,** 361
$(Pb,Tl)_2AgAs_2S_5$

HATRURITE *Am. Min.* **63,** 425
$Ca_3O(SiO_4)$ tric.(?) *Min. Abst.* **29,** 481

HAUCHECORNITE *Am. Min.* **66,** 436
Ni_9BiSbS_8 tetr. *Min. Mag.* **43,** 873

HAUERITE (***Pyrite*** gr.) *Dana,* I, 293
MnS_2 iso. *Min. Abst.* **30,** 258

HAUCKITE hexa. *Am. Min.* **65,** 192
$(Mg,Mn)_{24}Zn_{18}Fe^{3+}_3(SO_4)_4(CO_3)_2(OH)_{81}$ (?)
Min. Mag. **43,** 1060

HAUSMANNITE A manganese oxide named by Haidinger in 1827 in honor of J. F. L. Hausmann (1782–1859), Professor of Mineralogy at the University of Göttingen.

Composition: $Mn^{2+}_2Mn^{4+}O_4$

Properties: Tetragonal, $I4_1/amd$; often pseudo-octahedral with bipyramids, or massive; distorted spinel structure, degree of distortion decreases on heating and above 1160°C it has cubic (isometric) spinel structure (see *Spinel Group*); H=5.5; G=4.84; color: brownish black; streak: chestnut brown; submetallic luster; brittle.

Occurrence: Widespread, usually in high-temperature hydrothermal veins, also in contact metamorphic deposits, and in manganese ores as an alteration product formed by circulating meteoric waters. Often found in association with *braunite, magnetite, barite,* psilomelane, *pyrolusite, hematite,* and some manganese silicates. Notable localities include Ilmenau and Ilfeld, East Germany; Jacobsberg and Långban, Sweden; Miguel Burnier, Minas Gerais, Brazil; Kacharwahi, Nagpur district, India, and the Batesville district, Arkansas, USA.

D. H. SPEIDEL

References

Dana, I, 712.
Kostov, I., 1968. *Mineralogy.* Edinburgh: Oliver and Boyd, p. 236.

Cross-references: *Magnetic Minerals; Ore Microscopy.*

HAUYNE A sodium and calcium aluminosilicate with sulfate, of the ***sodalite*** group and related to *lazurite*. Named by Bruun-Weergard in 1807 in honor of the French mineralogist Abbé Haüy (1734–1822).

Composition: $[(Na,Ca)_{4-8}Al_6Si_6O_{24}(SO_4,S)_{1-2}]$

Properties: Isometric, $P\bar{4}3n$; dodecahedral and octahedral crystals with polysynthetic, contact, and penetration twins, but usually as rounded grains in masses with fused surfaces; distinct dodecahedral cleavage; uneven fracture; brittle; H=5.5–6; G=2.4–2.5; vitreous to greasy luster; color: bright sky-blue, greenish blue, yellow to white; streak: bluish to colorless; translucent; gelatanized by HCl.

Occurrence: Common in igneous rocks, such as phonolites, tephrite, haüynophyre, nepheline syenites, and other related silica-undersaturated types. Usually in association with *nepheline, leucite,* or other feldspathoids. Abundant in the lavas of Vesuvius, Campagna, Rome, and the Alban Hills, Italy; in ankaratrites at Jebel Tourguejid, Morocco; and in theralite of the Crazy Mountains, Montana, USA.

G. C. MOERSCHELL

Reference: DHZ, 4, 289–302.

Cross-reference: *Gemology.*

HAWLEYITE *Am. Min.* **40,** 55
CdS iso.

HAXONITE *Am. Min.* **59,** 209
$(Fe,Ni)_{23}C_6$ iso.

Cross-reference: *Meteoritic Minerals.*

HAYCOCKITE *Am. Min.* **57,** 689
$Cu_4Fe_5S_8$ orth.

HEAVY SPAR = *Barite*

HEAZLEWOODITE *Am. Min.* **62,** 341
Ni_3S_2 trig.

Cross-references: *Geologic Thermometry; Meteoritic Minerals.*

HECTORITE (***Smectite*** gr.) DHZ, 3, 226
$Na_{0.33}(Mg,Li)_3Si_4O_{10}(F,OH)_2$ mono.

Cross-references: *Clays, Clay Minerals; Staining Techniques; Soil Mineralogy.*

HEDENBERGITE Calcium ferrous iron silicate, a monoclinic member of the ***pyroxene*** group. Named by Berzelius in 1819 after M. A. Ludwig Hedenberg, the Swedish chemist who discovered it.

Composition: $4[CaFe^{2+}Si_2O_6]$; Ca, 20–22; FeO, 20–23; MgO, 0–4; SiO_2, 47–49% by wt.; forms a continuous series toward *diopside,* $CaMgSi_2O_6$, with substitution of some Mg for Fe^{2+}. Although the end-member composition is $CaFe^{2+}Si_2O_6$, in natural ***pyroxenes*** there is a complete series of compositions from $CaFe^{2+}Si_2O_6$ to $CaMgSi_2O_6$ and a separate series (*ferrohedenbergite*) some way toward $FeSiO_3$; some Mn may substitute for Fe^{2+}.

Properties: Monoclinic, $C2/c$; habit stumpy prismatic; H=6; G=3.50–3.56; luster vitreous; color: brownish green, dark green, or black; streak: white; cleavage, good on {110}, partings on {100} and {010}, (110):(1$\bar{1}$0), ≈ 87°; simple and multiple twinning on {100} and {001}. Refractive indices α=717–1.732, β=1.723–1.736, γ=1.741–1.755; $2V_\gamma$ 52°–64°; $\gamma:c$ 47°–48°; pale green, yellow-green, or brownish green in thin section. Insoluble in HCl; soluble in HF or in molten Na_2CO_3.

Occurrence: Thermally and regionally metamorphosed iron-rich sediments; eulysite; limestone skarns; *quartz*-bearing syenites and some *fayalite* granites and granophyres. Often in association with *grunerite, fayalite,* and ***garnet.*** *Ferrohedenbergite* is a major constituent of some *fayalite* ferrogabbros.

R. A. HOWIE

References

Am. Min. **34,** 621.
DHZ, 2A, 198–293.
Doklady Akad. Sci. USSR **105,** 814.
Geol. Mag. **92,** 367.

Cross-references: *Clinopyroxenes; Pyroxene Group.*

HEDLEYITE *Am. Min.* **48,** 435
Bi_7Te_3 trig. *Min. Abst.* **27,** 336

HEDYPHANE (***Apatite*** gr.) *Dana,* II, 900
$(Ca,Pb)_5(AsO_4)_3Cl$ hexa.

HEIDEITE *Am. Min.* **59,** 465
$(Fe,Cr)_{1+x}(Ti,Fe)_2S_4$ mono.

Cross-reference: *Meteoritic Minerals.*

HEIDORNITE mono. *Am. Min.* **42,** 120
$Na_2Ca_3B_5O_8(SO_4)_2Cl(OH)_2$

HEINRICHITE (*Autunite* gr.)
$Ba(UO_2)_2(AsO_4)_2 \cdot 10\text{–}12H_2O$
tetr. *Am. Min.* **43**, 1134

HELIOPHYLLITE *Dana*, II, 1037
$Pb_6As_2O_7Cl_4$(?) orth.

HELIOTROPE–red spotted variety of chalcedony *Dana*, III, 219

HELLANDITE mono. *Am. Min.* **62**, 89
$Ca_{5.5}(Y,RE)_5AlFe(OH)_4[Si_8B_8O_{40}(OH)_4]$
Can. Min. **11**, 760

HELLYERITE *Am. Min.* **44**, 533
$NiCO_3 \cdot 6H_2O$

HELMUTWINKLERITE *Am. Min.* **65**, 1067
$PbZn_2(AsO_4)_2 \cdot 2H_2O$
tric. *Min. Mag.* **43**, 1061

HELVITE A maganese beryllosilicate with sulfide named by Werner in 1816 in allusion to its yellow color from the Greek, *helios*, "the sun." It is the manganese end member of the ***helvite*** group.

Composition: $2[Mn_4Be_3Si_3O_{12}S]$; with Mn replaced by Fe toward *danalite* and by Zn toward *genthelvite*.

Properties: Isometric, $P\bar{4}3n$; tetrahedral crystals common, also spherical masses; brittle with uneven fracture; H=6, G=3.2–3.4; vitreous to resinous luster; honey-yellow, brown, red-brown color with pale brown to colorless streak; translucent; decomposed by HCl with the evolution of H_2S.

Occurrence: Typically in contact zones of metasomatic rocks and in granites and pegmatites. Commonly associated with *petalite*, *spodumene*, *spessartine*, or *corundum* as in the Ural Mountains, USSR, and at Modium, Norway. Most frequently formed, however, in large amounts at contact deposits and skarns. This paragenesis is exemplified by the Butte district, Montana; Iron Mountain, New Mexico; Hörtekollen, near Oslo, Norway; and Yagaisawa Mine, Nagano, Japan.

G. C. MOERSCHELL

References

DHZ, 4, 303–309.
Thrush, P. W., 1968. *A Dictionary of Mining, Mineral, and Related Terms.* Washington, D.C.: U.S. Bur. Mines, p. 538.

HELVITE Group–*Danalite, genthelvite,* and *helvite.*

HEMAFIBRITE *Dana*, II, 919
$Mn_3(AsO_4)(OH)_3 \cdot H_2O$ orth.

HEMATITE Iron oxide named by Theophrastus in 325 BC from Greek *haema*, blood. "Red ochre" is the earthy variety.

Composition: $\alpha\text{-}Fe_2O_3$; forms solid-solution series with *ilmenite* ($FeTiO_3$), *geikielite* ($MgTiO_3$), and *pyrophanite* ($MnTiO_3$); may contain up to about 10% Al_2O_3 by wt.

Properties: Trigonal, $R\bar{3}c$; H=5–5; G=5.26; color: reddish brown to black with a characteristic cherry or indian red streak; corundum structure ($\gamma\text{-}Fe_2O_3$, *maghemite*, has the cubic spinel structure); no cleavage; massive variety can be botryoidal and is called kidney ore; martite is dodecahedral or octahedral *hematite* pseudomorphous after *magnetite* or *pyrite* crystals. Optically, ω=3.15–3.22, ϵ=2.87–2.94. Color can be dark gray to metallic black with red streak, then called specular *hematite*, specularite, or iron glance. Dissociates to Fe_3O_4 in air at 1390°C.

Occurrence: Red coloring in sediments; Precambrian metamorphosed banded iron ores throughout world; an oxidation and weathering product of iron-containing minerals in its association, such as *magnetite*, limonite, *siderite*. Rare in igneous rocks.

Uses: The most important iorn ore; a red pigment; and as a polishing powder called "rouge."

D. H. SPEIDEL

References

DHZ, 5, 21–27.
J. Petrol. **1**(2), 178–217.

Cross-references: *Abrasive Materials; Authigenic Minerals; Blacksand Minerals; Blowpipe Analysis; Crystal Habits; Crystallography: Morphological; Diagenetic Minerals; Fire Clays; Gossan; Lunar Minerals; Magnetic Minerals; Metallic Minerals; Mineral and Ore Deposits; Ore Microscopy; Pegmatite Minerals; Pigments and Fillers; Plagioclase Feldspars; Rock-Forming Minerals; Saline Minerals; Soil Mineralogy; Spinel Group; Thermometry, Geologic; Topotaxy.* Vol. IVA: *Iron: Element and Geochemistry; Iron: Economic Deposits.*

HEMATOLITE trig. *Dana*, II, 777
$(Mn,Mg,Al)_{15}(OH)_{23}(AsO_4)_2(AsO_3)$
Am. Min. **63**, 150

HEMATOPHANITE *Am. Min.* **56**, 625
$Pb_4Fe_3O_8Cl$ tetr. *Min. Mag.* **39**, 49

HEMIHEDRITE tric. *Am. Min.* **55**, 1088
$Pb_{10}Zn(CrO_4)_6(SiO_4)_2F_2$

HEMIMORPHITE A hydrated zinc hydroxyl silicate named by Kenngott in 1853 from the hemimorphic nature of the crystals. Originally known as calamine, a name also used for the zinc carbonate, *smithsonite.*

Composition: $2[Zn_4(Si_2O_7)(OH)_2 \cdot H_2O]$

Properties: Orthorhombic, *Imm*2; striated tabular or prismatic crystals, commonly singly terminated, also stalactitic, botryoidal, massive, or granular; H=4.5–5; G=3.4–3.5; vitreous luster; transparent to translucent; strongly pyroelectric and piezoelectric; color: white, yellow, or brown; streak: white.

Occurrence: Secondary mineral found in the oxidized portions of zinc deposits, and associated with *smithsonite, sphalerite, cerussite, galena,* and *anglesite.*

Use: Ore of zinc.

J. A. GILBERT

References

Min. Abst. **30,** 347.
Vanders, I., and Kerr, P. F., 1967. *Mineral Recognition.* New York: Wiley, 295–6.

Cross-references: *Gossan; Topotaxy.*

HEMUSITE *Am. Min.* **55**, 1847
Cu_6SnMoS_8 iso.

HENDERSONITE *Am. Min.* **47**, 1252
$Ca_2V^{4+}V_8^{5+}O_{24} \cdot 8H_2O$ orth.

HENDRICKSITE (***Mica*** gr.)
$K(Zn,Mn)_3(Si_3Al)O_{10}(OH)_2$
mono. *Am. Min.* **51**, 1107

HENRITERMIERITE *Am. Min.* **54**, 1739
$Ca_3(Mn,Al)_2(SiO_4)_2(OH)_4$
tetr.

HERCYNITE An iron aluminum oxide of the ***spinel*** group, named by Zippe in 1847 from the Latin name for the Bohemian Forest where it was first found.

Composition: $FeAl_2O_4$, often with some Mg,Fe^{3+}, or both; forms series with *chromite, gahnite,* and *spinel.*

Properties: Isometric, *Fd*3*m*; twinning in the spinel law; H=7.5–8; G=4.40; *n*=1.835; color: dark green to black; streak: dark green; picotite is used for the variety which contains considerable amounts of chromium.

Occurrence: Common in metamorphosed argillaceous sediments; metamorphosed ferruginous bauxites and emery deposits; some ultramafic rocks; evolved from *magnetite* in some mafic rocks. In association with *sillimanite, corundum,* and ***garnet*** near Schenkenzell, West Germany; at Le Prese, Switzerland; and at Peekskill, New York.

D. H. SPEIDEL

References

Dana, I, 689.
DHZ, 5, 56–65.

Cross-references: *Meteoritic Minerals; Refractory Minerals.*

HERDERITE A calcium beryllium fluophosphate named in 1828 after S. A. W. von Herder, mining official of Freiberg, Germany.

Composition: $4[CaBe(PO_4)F]$ with F replaced by (OH) (OH>F, *hydroxylherderite*).

Properties: Monoclinic, $P2_1/a$ (pseudo-orthorhombic); crystals stout prismatic, also botryoidal aggregates; H=5–5.5; G=3; vitreous luster; colorless, yellowish, greenish.

Occurrence: Late-stage pegmatite mineral.

K. FRYE

Reference: *Am. Min.* **63,** 913–917.

Cross-reference: *Pegmatite Minerals.*

HERSCHELITE–Sodiun variety of *chabazite*(?)

HERZENBERGITE *Dana,* I, 259
SnS orth. *Min. Abst.* **28,** 279

HESSITE A silver telluride named in 1843 for G. H. Hess of St. Petersburg (Leningrad).

Composition: $4[Ag_2Te]$ with Ag replaced by Au toward *petzite.*

Properties: Monoclinic, $P2_1/c$; massive, compact, or fine grained; H=2–3; G=8.24–8.45; metallic luster; color: lead to steel gray.

Occurrence: In hydrothermal veins with other tellurides, *gold, galena,* and *tellurium.*

K. FRYE

Reference: *Dana,* I, 184–186.

Cross-reference: *Ore Microscopy.*

HETAEROLITE A zinc and manganese oxide member of the ***hausmannite*** group. Named by Moore in 1877 from the Greek meaning "companion" in allusion to the original association with *chalcophanite.*

Composition: $ZnMn_2O_4$

Properties: Tetragonal, $I4_1/amd$ hausmannite structure; fibrous with {001} cleavage; H=6, G=5.2; color: dark brown to black; streak: dark brown.

Occurrence: In zinc and manganese ore deposits, at Sterling Hill, New Jersey, as crystal coatings with *franklinite,* and massive with *chalcophanite.*

D. H. SPEIDEL

Reference: *Dana,* I, 715.

HETEROGENITE-3R — *Am. Min.* **48**, 216
CoO(OH) trig. — *Min. Mag.* **33**, 253

HETEROGENITE-2H — *Am. Min.* **59**, 381
CoO(OH) hexa.

HETEROMORPHITE (***Plagionite*** gr.)
$Pb_7Sb_8S_{19}$ — *Am. Min.* **59**, 1127
mono. — *Min. Abst.* **27**, 23

HETEROSITE — *Am. Min.* **55**, 135
$FePO_4$ orth. — *Dana,* II, 675

Cross-reference: *Pegmatite Minerals.*

HEULANDITE A hydrated calcium sodium aluminosilicate of ***zeolite*** group. First named by Brooke in 1822 in honor of H. Heuland, an English mineral collector and naturalist.

Composition: $(Ca,Na_2)_4[Al_8Si_{28}O_{72}]\cdot 24H_2O$. Remarkable variation in the Si:Al ratio with a corresponding variation in the Ca:Na_2 ratio; sometimes K may substitute for and predominate over Na.

Properties: Triclinic (pseudo-monoclinic), $C2/m$; a=17.85 Å, b=17.84 Å, c=7.46 Å, β=91°26′; H=3.5–4; G=2.1–2.2; luster vitreous to strongly pearly; colorless, white, gray, yellow, pink, red, or brown; transparent to subtranslucent; habit in tablets parallel to (010); cleavage perfect on {010}. Refractive indices, α=1.476–1.505, β=1.477–1.503, γ= 1.479–1.512; birefringence = 0.002–0.006; $2V$=34° (+), but variable; optic axial plane usually ⊥ (010); dispersion $r>v$. *Clinoptilolite* is richer in silica and alkali and has different characteristics; *clinoptilolite* remains stable up to 700°C on heating while *heulandite* transforms into "heulandite B" at 230°C and becomes amorphous at 350°C; dehydration stepwise, losing 17.2% H_2O at maximum; rehydration poor; molecular sieve properties more selective; *heulandite* gelatinizes with HCl, but *clinoptilolite* is not attacked by HCl.

Occurrence: Typical *heulandite* occurs in amygdaloidal cavities in basalt and andesite closely associated with *stilbite* and other ***zeolites;*** in skarn druses associated with ***garnet*** and ***axinite;*** *clinoptilolite* occurs regionally as a diagenetic alteration product of silicic vitric tuffs accumulated in fresh and marine waters, often associated with silica minerals, *mordenite, montmorillonite,* and *celadonite;* in bentonitic clays; reaction product of silicic vitric ash with alkaline saline lake or interstitial water, associated with *phillipsite, chabazite, erionite,* and opal; authigenic mineral in limestone and chalk or in cement of sandstone. Best specimens are from Berufjord region, Iceland; Faeroe Islands; Bombay, India; and Peter's Point, Nova Scotia.

Uses: *Clinoptilolite* is used, like bentonite, for improvement of soils of paddyfields, for absorber, and for paper manufacture as soft clay.

A. IIJIMA
M. UTADA

References

Am. Min. **45**, 351–367; **52**, 273–275; **57**, 1448–1462, 1463–1493.

Pure & Appl. Chem. **52**, 2115–2130.

Sheppard, R. A., and Gude, A. J., 1968. Distribution and genesis of authigenic silicate minerals in tuffs of Pleistocene Lake Tecopa, Inyo County, California, *U.S. Geol. Surv. Prof. Pap. 597,* 1–38.

Univ. Calif. Pub. Geol. Sci. **42**, 199–262.

Cross-references: *Authigenic Minerals; Minerals, Uniaxial and Biaxial; Zeolites.*

HEWETTITE — *Am. Min.* **44**, 322
$CaV_6O_{16}\cdot 9H_2O$ orth.

HEXAHYDRITE — *Dana,* II, 494
$MgSO_4\cdot 6H_2O$ mono. — *Min. Abst.* **30**, 70

Cross-references: *Cave Minerals; Human and Vertebrate Mineralogy; Saline Minerals.*

HEXAHYDROBORITE — *Am. Min.* **62**, 1259; **63**, 1283
$CaB_2O_4\cdot 6H_2O$ mono.

HEXATESTIBIO-PANICKELITE
$(Ni,Pd)_2SbTe$ hexa. — *Am. Min.* **61**, 182

HEYITE — *Am. Min.* **59**, 382
$Pb_5Fe_2(VO_4)_2O_4$ mono. — *Min. Mag.* **39**, 65

HEYROVSKYITE — *Am. Min.* **62**, 397
$Pb_{10}AgBi_5S_{18}$ orth.

HIBONITE hexa. *Am. Min.* **42**, 119
$(Ca,RE)(Al,Fe,Ti,Si,Mg)_{12}O_{19}$
Min. Mag. **43**, 995

Cross-reference: *Meteoritic Minerals.*

HIBSCHITE = *Hydrogrossular*
Am. Min. **29**, 247

HIDALGOITE (***Beudantite*** gr.)
$PbAl_3(SO_4)(AsO_4)(OH)_6$
trig. *Min. Rec.* **2**, 212

HIDDENITE–emerald-green *spodumene*

HIERATITE *Dana,* II, 103
K_2SiF_6 iso.

HILAIRITE trig. *Can. Min.* **12**, 237
$Na_2ZrSi_3O_9 \cdot 3H_2O$ *Min. Abst.* **27**, 81

HILGARDITE *Am. Min.* **64**, 187
$Ca_2(B_5O_9)Cl \cdot H_2O$ mono. *Min. Abst.* **30**, 219

HILLEBRANDITE RRW, 274
$Ca_2SiO_3(OH)_2$ mono.

HINSDALITE (***Beudantite*** gr.)
$(Pb,Sr)Al_3(PO_4)(SO_4)(OH)_6$
trig. *Dana,* II, 1004

HIORTDAHLITE tric. *Can. Min.* **12**, 241
$(Ca,Na)_3ZrSi_2O_7(O,OH,F)_2$

HISINGERITE *Am. Min.* **46**, 1412
$Fe_2^{3+}Si_2O_5(OH)_4 \cdot 2H_2O$
mono.(?) *Min. Abst.* **27**, 128

HJELMITE = *Yttromicrolite*
Am. Min. **64**, 890

Cross-reference: *Blacksand Minerals.*

HOCARTITE *Am. Min.* **54**, 573
Ag_2FeSnS_4 tetr.

HODGKINSONITE *Am. Min.* **49**, 415
$MnZn_2SiO_4(OH)_2$ mono.

HODRUSHITE *Am. Min.* **56**, 633
$Pb_2Cu_8Bi_{10}S_{22}$ mono. *Min. Mag.* **37**, 641

HOEGBOMITE = *HÖGBOMITE*

HOERNESITE *Am. Min.* **52**, 1588
$Mg_3(AsO_4)_2 \cdot 8H_2O$ mono.

HÖGBOMITE An oxide of magnesium, aluminum, ferric iron and titanium named for Professor A. G. Högbom of the University of Upsala.

Composition: $Mg(Al,Fe,Ti)_4O_7$.

Properties: Hexagonal; black in mass with metallic luster.

Occurrence: Found in iron ore in Peekskill, New York, and Swedish Lapland.

D. H. SPEIDEL

Reference: *Am. Min.* **49**, 445.

HOHMANNITE tric. *Min. Mag.* **42**, 144
$Fe_2(H_2O)_4(SO_4)_2 \cdot 4H_2O$

HOLDENITE orth. *Am. Min.* **62**, 513
$Mn_6(OH)_8Zn_3(AsO_4)_2(SiO_4)$
Dana, II, 775

HOLLANDITE An oxide of manganese and barium named for T. H. Holland, Director of the Geological Survey of India.

Composition: $MnBaMn_6O_{14}$· Forms a series with *cryptomelane* ($MnK_2Mn_6O_{14}$).

Properties: Monoclinic, pseudotetragonal; prisms with striated faces; silver gray color. Coronadite is a mixture of *hollandite* and a lead oxide.

Occurrence: In *quartz* veins in central India.

D. H. SPEIDEL

References

Dana, I, 743.
Science **212**, 1024.

Cross-reference: *Ore Microscopy.*

HOLLINGWORTHITE *Am. Min.* **50**, 1068
$(Rh,Pt,Pd)(As,S)_2$ iso.

HOLMQUISTITE (***Amphibole*** gr.)
$Li_2Mg_3Al_2(Si_4O_{11})_2(OH)_2$
orth. *Am. Min.* **63**, 1030
DHZ, 2, 230

Cross-reference: *Pegmatite Minerals.*

HOLTEDAHLITE *Am. Min.* **65**, 809
$Mg_2(PO_4)(OH)$ hexa.

HOLTITE orth. *Am. Min.* **57**, 1556
$(Al,Sb,Ta)_7(B,Si)_4O_{18}$(?)
Min. Mag. **38**, 21

HOMILITE Winchell, II, 356
$Ca_2(Fe,Mg)B_2Si_2O_{10}$ mono.

HONESSITE *Am. Min.* **44**, 995
Sulfate of Fe and Ni

HONGQUIITE iso. *Am. Min.* **61**, 184
TiO

HOPEITE *Am. Min.* **61**, 987
$Zn_3(PO_4)_2 \cdot 4H_2O$ orth. *Min. Abst.* **28**, 261

Cross-reference: *Human and Vertebrate Mineralogy.*

HORNBLENDE A calcium magnesium iron and hydroxyl aluminosilicate of the ***amphibole*** group, named by Werner in 1789 from an old German word for any dark prismatic mineral found in ores but having no recoverable metal.

Composition: $2[(Ca,Na)_{2-3}(Mg,Fe^{2+},Fe^{3+},Al)_5(Si,Al)_8O_{22}(OH,F)_2]$ with Ca usually ≈2 and Si from 6 to 7 per formula, but this varies with much solid solution. Common *hornblende* ranges in composition among the end-members; *actinolite,* $Ca_2(Mg,Fe^{2+})_5Si_8O_{22}(OH)_2$; *edenite,* $NaCa_2Mg_5Si_7AlO_{22}(OH)_2$; *hastingsite,* $NaCa_2Si_6Al_2O_{22}(OH)_2$; *tschermakite,* $Ca_2Mg_3(Al,Fe)_2Si_6Al_2O_{22}(OH)_2$; *pargasite,* $NaCa_2Mg_4AlSi_6Al_2O_{22}(OH)_2$; and the analogous "ferro-molecules."

Properties: Monoclinic, *C2/m*; prismatic crystals with a hexagonal cross section, but may be fibrous, granular or columnar; H=5–6, G=3.02–3.45, increasing with increasing iron content; vitreous luster; translucent; color: dark green to black; streak: white to gray.

Occurrence: Widespread as a rock-forming mineral in both igneous and medium-grade metamorphic rocks, frequently as an alteration of ***pyroxene.*** When it is the major constituent, the rock is termed amphibolite. Common in syenites, diorites, and pegmatites, because the structure favors crystallization under pressure. The wide range in igneous paragenesis helps account for the large variation in the composition of *hornblende.* Associated with its hydrothermal alteration products, including *calcite,* ***chlorite,*** and *epidote.*

J. A. GILBERT

References

Am. Min. **63**, 1032.
DHZ, 2, 263–314.
Kostov, I., 1968. *Mineralogy.* Edinburgh: Oliver and Boyd, 345–346.

Cross-references: *Amphibole Group; Crystal Growth; Crystallography: Morphological; Mantle Mineralogy; Minerals, Uniaxial and Biaxial; Optical Mineralogy; Plastic Flow in Minerals; Rock-Forming Minerals; Soil Mineralogy.*

HORN SILVER = *Chlorargyrite*

HORSFORDITE *Dana,* I, 173
Cu_5Sb

HORTONOLITE = Magnesian *fayalite*

Cross-reference: *Mineral Classification: Principles.*

HOSHIITE = Nickeloan *magnesite*
Am. Min. **50**, 2100

HOWIEITE tric. *Am. Min.* **59**, 86
$Na(Fe,Mn)_{10}(Fe,Al)_2Si_{12}O_{31}(OH)_{13}$

HOWLITE A calcium hydroxyl silico-borate named by Dana in 1868 in honor of Henry How, Canadian mineralogist.

Composition: $Ca_2B_5SiO_9(OH)_5$.

Properties: Monoclinic, $P2_1/c$; procelaneous nodules, also earthy; H=3.5; G=2.58; white.

Occurrence: In large nodules and veins in southern California, USA, deserts and in small nodules in *anhydrite* or *gypsum* in Hauts Co., Nova Scotia, Canada.

K. FRYE

Reference: *Am. Min.* **55**, 716–728.

HSIANGHUALITE *Am. Min.* **44**, 1327; **46**, 244
$Ca_3Li_2Be_3(SiO_4)_3F_2$ iso.

HUANGHOITE *Am. Min.* **48**, 1179
$BaCe(CO_3)_2F$ hexa.

HUEBNERITE Manganese tungstate of the ***wolframite*** group, named after Adolph Hübner, a metallurgist at Freiberg, Saxony, GDR.

Composition: $2[MnWO_4]$ with Mn replaced by Fe toward *wolframite.*

Properties: Monoclinic, *P2/c*; crystals prismatic, commonly striated, often in parallel or radiating groups; one perfect cleavage {010}; H=4; G=7.12; submetallic to resinous luster; color: yellowish or reddish brown; streak: yellow, reddish brown, or greenish gray.

Use: A minor ore of tungsten.

Occurrence: A vein mineral in some cases associated with granitic intrusions, also in placers.

K. FRYE

Reference: *Dana*, II, 1064–1072.

Cross-references: *Magnetic Minerals; Wolframite Group.*

HUEMULITE *Am. Min.* **51**, 1
$Na_4MgV_{10}O_{28} \cdot 24H_2O$ tric.

HÜGELITE mono. *Am. Min.* **47**, 418
$Pb_2(UO_2)_3(AsO_4)_2(OH)_4 \cdot 3H_2O$

HÜHNERKOBELITE = *Aluaudite*
Min. Mag. **43**, 227

HULSITE *Am. Min.* **50**, 249; **61**, 116
$(Fe^{2+},Mg)_2(Fe^{3+},Sn)BO_5$ mono.

HUMBERSTONITE trig. *Am. Min.* **55**, 1518
$K_3Na_7Mg_2(SO_4)_6(NO_3)_2 \cdot 6H_2O$

HUMBOLDTINE *Dana*, II, 1102
$FeC_2O_4 \cdot 2H_2O$ mono.(?)

HUMITE (***Humite*** gr.) *Am. Min.* **54**, 391
$(Mg,Fe)_7(SiO_4)_3(F,OH)_2$ orth. DHZ, 1, 47

Cross-references: *Humite Group; Minerals, Uniaxial and Biaxial; Rock-Forming Minerals.*

HUMITE GROUP–*Alleghanyite, chondrodite, clinchumite, humite, manganhumite, norbergite,* and *sonolite.*

HUMMERITE *Am. Min.* **36**, 326
$KMgV_5O_{14} \cdot 8H_2O$ tric.

HUNGCHAOITE *Am. Min.* **62**, 1135; **64**, 369
$Mg(H_2O)_5B_4O_5(OH)_4 \cdot 2H_2O$ tric.

HUNTITE A calcium magnesium carbonate named in 1953 in honor of W. F. Hunt, University of Michigan, USA.

Composition: $CaMg_3(CO_3)_4$.

Properties: Trigonal, $R32$; chalky masses; soft; G=2.7; earthy luster, white.

Occurrence: A cave mineral and a vug and vein filler in magnesium-rich rocks, also a weathering product.

K. FRYE

References

Am. Min. **38**, 4–24.
DHA, 5, 302–304.

Cross-references: *Cave Minerals; Rock-Forming Minerals.*

HUREAULITE A hydrated manganese phosphate named by Alluaud in 1825 from the locality where he first found it.

Composition: $Mn_5(H_2O)_4(PO_4)_2[PO_3(OH)]_2$, with some Fe^{2+} replacing Mn. Monoclinic.

Occurrence: In cavities of *triphylite* or its alteration product, *heterosite;* with *vivianite, rockbridgeite,* and *cacoxenite* at the pegmatite quarry at Huréaux, St. Sylvestre, France. Earlier found in the United States, at Branchville, Connecticut, associated with *fairfieldite, dickinsonite,* and *reddingite* as a hydrothermal alteration of *lithiophilite,* but now recognized from many pegmatites throughout the world where the ***triphylite*** group occurs.

G. C. MOERSCHELL

References

Am. Min. **49**, 398; **58**, 302.
Bull. Soc. fr. **99**, 261–273.

Cross-reference: *Pegmatite Minerals.*

HURLBUTITE *Am. Min.* **37**, 931
$CaBe_2(PO_4)_2$ orth.

Cross-reference: *Pegmatite Minerals.*

HUTCHINSONITE *Dana*, I, 468
$(Pb,Tl)_2As_5S_9$ orth.

HUTTONITE An exceedingly rare form of thorium silicate named by Pabst in 1950 in honor of its discoverer, Professor C. O. Hutton of Stanford University, USA.

Composition: $4[ThSiO_4]$; ThO_2, 76.6%; SiO_2, 19.7%; Fe_2O_3, 1.2%; Ce_2O_3, 2.6% by wt.

Properties: Monoclinic; colorless; mean refractive index = 1.900; G=7.1; isostructural with *monazite.* It is probably the high temperature polymorph of $ThSiO_4$, the common mineral form of which is *thorite.*

Occurrence: As minute grains in beach sands of Westland, South Island, New Zealand.

A. PABST

Reference: *Am. Min.* **36**, 60–69.

HYACINTH–Red *zircon*

HYALITE–Colorless opal

HYALOPHANE A potassium barium aluminosilicate of the *orthoclase-celsian* series of the ***alkali feldspar*** group, named in 1855 by Sartorius von Walterhausen for its glassy appearance.

Composition: $4[(K,Ba)Al(Al,Si)_3O_8]$ with 15–30 mol% Ba and small amounts of (Na,Ca) replacing (K,Ba).

Properties: Monoclinic or triclinic; prismatic crystals or massive; prismatic cleavage {001}, {010}; H=6–6.5; G=2.6–2.8; vitreous luster; colorless, white; yellow, or gray.

Occurrence: Characteristically associated with manganese ore deposits; but igneous at Broken Hill, N.S.W., Australia, in equal amounts with ***plagioclase*** (*An* 70); in veins in gneiss with *phlogopite* and *calcite* at Slyudyanka, Siberia, USSR; and with pegmatitic ***apatite*** at Nisikkatch Lake, Saskatchewan, Canada.

K. FRYE

References

DHZ, 4, 166–178.
Smith, J. V., 1974. *Feldspar Minerals,* vol. 1. New York and Heidelberg: Springer-Verlag, 353–355, 455–457.

HYALOTEKITE mono. Winchell, II, 401
$(Pb,Ca,Ba)_4BSi_6O_{17}(F,OH)$

HYDRARGILLITE = *Gibbsite*
Min. Mag. **43**, 1053

HYDROBASALUMINITE *Dana,* II, 586
$Al_4(SO_4)(OH)_{10}\cdot 36H_2O$ *Min. Mag.* **43**, 931

HYDROBORACITE *Dana,* II, 353
$CaMgB_6O_{11}\cdot 6H_2O$ mono. *Min. Rec.* **9**, 379

Cross-reference: *Saline Minerals.*

HYDROCALUMITE *Dana,* I, 667
$Ca_2Al(OH)_7\cdot 3H_2O$ mono.

HYDROCERUSSITE *Dana,* II, 270
$Pb_3(CO_3)_2(OH)_2$ hexa.

HYDROCHLOR-BORITE
$Ca_2[B_3O_3(OH)_4\cdot OB(OH)_3]Cl\cdot 7H_2O$
mono. *Am. Min.* **62**, 147; **63**, 814

HYDRODRESSERITE tric. *Am. Min.* **64**, 654
$BaAl_2(CO_3)_2(OH)_4\cdot 3H_2O$ *Can. Min.* **15**, 399

HYDROGARNET–A ***garnet*** having SiO_4 partially replaced by $(OH)_4$.

HYDROGLAUBERITE *Am. Min.* **55**, 321
$Na_{10}Ca_3(SO_4)_8\cdot 6H_2O$ orth.(?)

HYDROGROSSULAR (***Garnet*** gr.)
$Ca_3Al_2(SiO_4)_{3-x}(OH)_{4x}$ iso. DHZ, 1, 77

HYDROHALITE *Dana,* II, 15
$NaCl\cdot 2H_2O$ mono. *Min. Abst.* **27**, 101

Cross-reference: *Saline Minerals.*

HYDROHETAEROLITE *Am. Min.* **41**, 268
$Zn_2Mn_4O_8\cdot H_2O$(?) tetr.(?)

HYDROMAGNESITE A hydrated magnesium hydroxylcarbonate named in 1835 for its composition.

Composition: $Mg_5(CO_3)_4(OH)_2\cdot 4H_2O$.

Properties: Monoclinic, $P2_1/c$; small tufted crystals or chalky crusts; twinning {100} gives pseudo-orthorhombic symmetry; one perfect cleavage {010}; H=3.5; G=2.1–2.3; crystals vitreous and crusts earthy luster; white.

Occurrence: In low-temperature hydrothermal veinlets and crusts in serpentinite and other magnesium-rich rocks associated with other carbonate minerals.

K. FRYE

References

Dana, II, 271–274.
Min. Rec. **4**, 18–20.

Cross-references: *Authigenic Minerals; Cave Minerals.*

HYDROMICA (***Mica*** gr.) mono. DHZ, 3, 217
$(K,Na,H_3O)Al_2(AlSi_3)O_{10}[(OH)_2,H_2O]$

Cross-reference: *Soil Mineralogy.*

HYDROMOLYSITE *Am. Min.* **44**, 908; **51**, 1551
$FeCl_3\cdot 6H_2O$

HYDROMUSCOVITE = *Illite*

HYDRONIUM JAROSITE (***Alunite*** gr.)
$(H_3O)^+Fe_3(SO_4)_2(OH)_6$
trig. *Am. Min.* **50**, 1595

HYDROPHILITE = Unknown hydrate of $CaCl_2$ *Min. Mag.* **43**, 682

HYDROROMARCHITE *Am. Min.* **58**, 552
$Sn_3O_2(OH)_2$ tetr. *Can. Min.* **10**, 916

HYDROSCARBROITE — Am. Min. **45**, 910
$Al_{14}(CO_3)_3(OH)_{36} \cdot nH_2O$
tric. — Min. Mag. **32**, 353

HYDROTALCITE A hydrated magnesium aluminum hydroxide carbonate named by Hochstetter in 1842 from its resemblance to talc and its high water content.

Conposition: $Mg_6Al_2(OH)_{16}CO_3 \cdot 4H_2O$.

Properties: Trigonal; basal plates with basal cleavage; H=2; G=2.05; color: white with a brownish tint; streak: white; greasy feel; pearly to waxy luster; polymorphic with *manasseite*.

Occurrence: An alteration product of *spinel* closely associated with *manasseite* in a serpentine found in Snarum and Nordmark, Norway. In the USA as an alteration product of *spinel* at Somerville, Amith, Rossie, and Oxbow, New York.

D. H. SPEIDEL

References

Dana, I, 653.
Kostov, I., 1968. *Mineralogy*. Edinburgh: Oliver and Boyd, 526.

Cross-reference: *Minerals, Uniaxial and Biaxial.*

HYDROTUNGSTITE — Am. Min. **48**, 935
$H_2WO_4 \cdot H_2O$ mono.

HYDROUGRANDITE (***Garnet*** gr.)
$(Ca,Mg,Fe)_3(Fe,Al)_2(SiO_4)_{3-x}(OH)_{4x}$
iso. — Am. Min. **50**, 2100

HYDROXYAPATITE (Hydroxylapatite)
(***Apatite*** gr.) — *Dana*, II, 878
$Ca_5(PO_4)_3(OH)$

Cross-references: *Apatite Group; Cave Minerals; Human and Vertebrate Mineralogy; Minerals, Uniaxial and Biaxial; Pegmatite Minerals; Soil Mineralogy.*

HYDROXYAPOPHYLLITE
(***Apophyllite*** gr.) — Am. Min. **63**, 196
$KCa_4Si_8O_{20}(OH) \cdot 8H_2O$
tetr. — Min. Abst. **29**, 341

HYDROXYLBASTNAESITE Am. Min. **50**, 805
$(Ce,La)(CO_3)(OH,F)$ hexa.

HYDROXYLELLESTADITE
(***Apatite*** gr.) — Am. Min. **56**, 1507
$Ca_{10}(SiO_4)_3(SO_4)_3(OH,Cl,F)_2$ hexa.

HYDROXYLHERDERITE A calcium beryllium hydroxylphosphate of the *herderite* series, named after the German mining official S. A. W. von Herder and the predominance of hydroxyl over fluoride ion.

Composition: $4[CaBe(PO_4)(OH)]$ with (OH) replaced by F (F>OH, *herderite*)

Properties: Monoclinic, $P2_1/c$; stout prismatic crystals, also botryoidal; commonly twinned (t.p. {001} {100}); one cleavage {110}; H=5.5–5.5; G=2.95–3.01; yellowish to greenish white.

Occurrence: Late-stage pegmatite mineral (q.v.).

K. FRYE

Reference: *Am. Min.* **63**, 913–917.

HYDROZINCITE A zinc hydroxyl carbonate named by Kenngott in 1853 in allusion to its composition. Zinc prospectors use it as an indicator of zinc deposits; hence, its synonym, "zinc bloom."

Composition: $2[Zn_5(CO_3)_2(OH)_6]$.

Properties: Monoclinic, $C2/m$; minute lathlike crystals, or more commonly, as dense, dull white masses or crusts; earthy to chalky luster; H=2–2.5, G=3.5–3.8; dull white to gray color and streak; blue luminescence under UV light; effervescence in HCl.

Occurrence: Secondary mineral in the weathered zone of many zinc deposits, as crusts on other zinc minerals or on limonite. Usually in association with *sphalerite, aurichalcite, hemimorphite,* and *smithsonite.* In abundance at Santander, Spain; Carinthia (Austria); Goodsprings, Nevada, and Socorro, New Mexico.

G. C. MOERSCHELL

References

Kostov, I., 1968. *Mineralogy*. Edinburgh: Oliver and Boyd, 542–543.
Vanders, I., and Kerr, P. F., 1977. *Mineral Recognition.* New York: Wiley, 223.

Cross-reference: *Gossan.*

HYPERCINNABAR — Am. Min. **63**, 1151
$Hg_{1-x}S$ hexa. — Min. Mag. **43**, 1061

HYPERSTHENE A magnesium and ferrous silicate intermediate in the *enstatite-ferrosilite* series of the ***pyroxene*** group, named from the Greek *huper,* over, and *sthenos,* strength, since it is harder than *hornblende* with which it was originally confused.

Composition: $16[(Mg,Fe^{2+})SiO_3]$ with Fe^{2+} from 39–50 mol%. Up to 10 mol% total of

Al, Ca, Mn, Fe^{3+}, Ti, Cr, and Ni may be present.

Properties: Orthorhombic, *Pbca*; crystals rare, commonly foliated massive; distinct cleavages at near right angles {110}; H=5–6; G=3.4–3.5; pearly luster; color: brown, brownish green, grayish, or greenish black.

Occurrence: ***Orthopyroxene*** of this composition occurs mainly in mafic plutonic rocks, e.g., norite, some contaminated by assimilation of aluminous host rock. Also in medium to high grade metamorphic rocks, both regional (amphibolites and granulites) and contact (hornfels). Found in certain meteorites.

K. FRYE

Reference: DHZ, 2A, 20–161.

Cross-references: *Mineral Classification: Principles; Minerals, Uniaxial and Biaxial; Orthopyroxenes; Pyroxene Group; Skeletal Crystals; Soil Mineralogy.*

I

IANTHINITE A hydrated uranium oxide named by Schoep in 1926 from the Greek meaning "violet colored" in allusion to its color. The uranium carbonate *wyartite* is often mistaken for *ianthinite*.

Composition: $(UO_2)_6O_2(OH)_8 \cdot 6H_2O$.

Properties: Orthorhombic; small rectangular plates, thick tabular, or prismatic; perfect {001} cleavage; H=2–3; G=5.15; submetallic luster; color: violet black that alters to yellow; streak: brown violet; transparent.

Occurrence: In cavities of *uraninite* as an alteration product associated with *schoepite* and *becquerelite* at Shinkolobwe-Kasolo, Shaba, Zaire.

D. H. SPEIDEL

References

Am. Min. **12,** 355; **40,** 943; **44,** 1103.
Dana, I, 633.

ICE Dihydrogen oxide named from the common English term for the substance. Unlike most minerals, it is molecular, built of polar water molecules linked together so that each has four closest neighbors located at the apices of a nearly regular tetrahedron. The ordered arrangement of water molecules characteristic of *ice* persists in part in the liquid state, up to 4°C above the melting point. At this temperature, the structure becomes unstable and collapses into a more dense yet more random arrangement.

Composition: $4[H_2O]$.

Properties: Hexagonal; occurs most abundantly in massive and granular form; H=1.5, G=0.917; vitreous luster; conchoidal fracture; colorless to white in thin layers, greenish blue in thick layers; colorless streak; melting point 0°C.

Occurrence: *Ice* forms on the surface of lakes and ponds at low temperature. Is precipitated as snow, frost, hail, etc., from water vapor in the atmosphere. Appears as massive sheets known as glaciers as a result of the recrystallization of snow; forms permanent fields of snow at definite altitudes depending on the latitude; and as permanent *ice* (permafrost) under Arctic tundras.

Uses: Many cold-weather forms of recreation (skiing, skating, etc.) depend on it. In areas without electricity, it is placed in insulated boxes as a heat absorber.

J. A. GILBERT

Reference: *Dana,* I, 494.

Cross-references: *Crystal Growth; Crystal Habits; Crystallography: History; Diagenetic Minerals; Minerals, Uniaxial and Biaxial; Plastic Flow in Minerals; Refractive Index; Skeletal Crystals.*

ICELAND SPAR–Transparent variety of *calcite*

IDAITE A copper iron sulfide, named for the Ida mine at Khan, Namibia.

Composition: Cu_3FeS_4.

Properties: Hexagonal(?); lamellar masses; H=2.5; G=4.2; metallic luster; color: copper red to brown.

Occurrence: A decomposition product of *bornite* or *chalcopyrite*.

K. FRYE

References

Am. Min. **60,** 1013–1018.
Min. Mag. **43,** 193–200.

IDDINGSITE–Mixture of silicates formed by the alteration of ***olivine*** *Am. Min.* **46,** 92

IDOCRASE = VESUVIANITE

IDRIALITE A natural hydrocarbon named for the type locality, Idria, Yugoslavia.

Composition: $4[C_{22}H_{14}]$.

Properties: Orthorhombic; small flakes or tabular crystals; $H<2$; G=1.2; vitreous to adamantine luster; greenish yellow to light brown; bluish white fluorescence.

Occurrence: Associated with mercury ore.

K. FRYE

References

Am. Min. **55**, 1073; **61**, 1055.
Chem. Geol. **16**, 245.

IIMORIITE *Am. Min.* **58**, 140
$Y_5(SiO_4)_3(OH)_3$ tric.

IKAITE *Am. Min.* **49**, 439
$CaCO_3 \cdot 6H_2O$

IKUNOLITE *Am. Min.* **45**, 477
$Bi_4(S,Se)_3$ trig.

ILESITE mono. *Dana,* II, 486
$(Mn,Zn,Fe)SO_4 \cdot 4H_2O$

ILIMAUSSITE hexa. *Am. Min.* **54**, 992
$Ba_2Na_4CeFeNb_2Si_8O_{28} \cdot 5H_2O$

ILLITE (Hyromuscovite) A group of hydrous potassium aluminosilicates of the ***mica*** group. The name was proposed by Grim to denote the clay-grade ***mica*** minerals, many of which had been studied in Illinois.

Composition: $(K,H_3O)(Al,Mg,Fe)_2(Al,Si)_4O_{10}[(OH)_2,H_2O]$.

Properties: Monoclinic; white or other pale colors; G=2.642–2.688; vitreous luster; perfect basal {001} cleavage. Colorless in thin section; optically (–); $2V_\alpha$ usually less than 10°; α=1.54–1.57, β=1.57–1.61, γ=1.57–1.61, δ=0.03; translucent; impure samples may show some pleochroism. Readily attached by acids; includes both dioctahedral (like *muscovite*) and trioctahedral (like *biotite*) minerals. ***Illites*** have more water and less potassium than *muscovite*.

Occurrence: The most common mineral in shales and mudstones. Found in marine environments, soils, and in zones of hydrothermal alteration as an authigenic mineral.

E. BOOY

References

DHZ, 3, 213–225.
Grim, R. E., 1968. *Clay Mineralogy.* New York: McGraw-Hill, 596p.
Kerr, P. F., ed., 1951. *Reference Clay Minerals, American Petroleum Institute Research Project 49.* New York: Columbia University, 701p.

Cross-references: *Authigenic Minerals; Clays, Clay Minerals; Diagentic Minerals; Fire Clays; Green River Mineralogy; Phyllosilicates; Soil Mineralogy; Staining Techniques.*

ILMAJOKITE mono.(?) *Am. Min.* **58**, 139
$(Na,RE,Ba)_{10}Ti_5Si_{14}O_{22}(OH)_{44} \cdot nH_2O$

ILMENITE An iron titanium oxide with variable amounts of magnesium and manganese, named by Kupffer in 1827 for Ilmen Mts., USSR.

Composition: $FeTiO_3$. Forms solid-solution series with *geikielite* ($MgTiO_3$), *pyrophanite* ($MnTiO_3$), and *hematite* (Fe_2O_3). Stability is strongly related to oxygen content.

Properties: Trigonal, $R\bar{3}$; usually massive, crystals tabular; ordered corundum structure with Fe^{2+} and Ti^{4+} occupying alternate layers; H=5–6; G=4.70–4.78; opaque, n = 2.7; color: iron black; metallic to submetallic luster; streak: black to brownish red; may be magnetic.

Occurrence: Common accessory mineral in igneous rocks (gabbro and diorite); metamorphic rocks (orthogneisses, granulites; ubiquitous mineral in detrital sediments, concentrated in some heavy blacksands. Usually in association with *magnetite, rutile, zircon,* and *monazite.* Important sources include: Allard Lake, Quebec; Blaafjeldite, Sogndal, Norway; Elizabethtown, New York, and beach deposits along east cost of Florida near St. Augustine, USA; the west coast of South Island, New Zealand; and comprises 15–20% of lunar oxides.

Use: A source of titanium.

D. H. SPEIDEL

References

DHZ, 5, 28–33.
Econ. Geol. **61**, 798.
J. Petrology **5**, 310.

Cross-references: *Blacksand Minerals; Diamond; Lunar Minerals; Magnetic Minerals; Mantle Mineralogy; Metallic Minerals; Mineral Industries; Mineral and Ore Deposits; Minerals, Uniaxial and Biaxial; Ore Microscopy; Pigments and Fillers; Rock-Forming Minerals; Spinel Group; Thermometry, Geologic.*

ILMENORUTILE = Niobian *rutile*

ILSEMANNITE *Dana,* I, 603
$Mo_3O_8 \cdot nH_2O(?)$

ILVAITE orth. Winchell, II, 511
$CaFe_2^{2+}Fe^{3+}O(Si_2O_7)(OH)$

IMANDRITE orth. *Am. Min.* **65**, 810
$Na_{12}Ca_3Fe_2^{3+}Si_{12}O_{36}$ *Min. Abst.* **31**, 496

IMGREITE *Am. Min.* **49**, 1151
NiTe(?) hexa.

IMHOFITE — Am. Min. **54**, 1498
$Tl_{5.6}As_{16}S_{5.3}$ mono. — Z. Krist. **144**, 323

INCAITE — Am. Min. **60**, 486
$Pb_4FeSn_4Sb_2S_{15}$ mono.

INDERBORITE mono. — Dana, II, 355
$CaMg[B_3O_3(OH)_5]_2 \cdot 6H_2O$

INDERITE mono. — Am. Min. **41**, 927
$MgB_3O_3(OH)_5 \cdot 5H_2O$ — Min. Abst. **28**, 260

INDIALITE — Am. Min. **40**, 787
$Mg_2Al_4Si_5O_{18}$ hexa. — Can. Min. **15**, 43

Cross-reference: *Pegmatite Minerals.*

INDICOLITE = Blue *elbaite*

INDIGIRITE — Am. Min. **57**, 326
$Mg_2Al_2(CO_3)_4(OH)_2 \cdot 15H_2O$

INDITE (***Linnaeite*** gr.) — Am. Min. **49**, 439
$FeIn_2S_4$ iso.

INDIUM — Am. Min. **52**, 299
In

Cross-references: *Lunar Minerals; Native Elements and Alloys.*

INESITE tric. — Am. Min. **53**, 1614; **63**, 563
$Ca_2Mn_7Si_{10}O_{28}(OH)_2 \cdot 5H_2O$

INNELITE tric. — Am. Min. **47**, 805
$Na_2(Ba,K)_4(Ca,Mg,Fe)Ti_3Si_4O_{18}(OH,F)_{1.5}(SO_4)$

INSIZWAITE (***Pyrite*** gr.) — Am. Min. **58**, 805
$Pt(Bi,Sb)_2$ iso. — Min. Mag. **38**, 794

INYOITE — Am. Min. **38**, 912
$Ca_2B_6O_{11} \cdot 13H_2O$ mono.

Cross-reference: *Saline Minerals.*

IODARGYRITE An iodide of silver named for its composition.

Composition: AgI.

Properties: Hexagonal, *P*6_3*mc*; prismatic or tabular crystals; H=1.5; G=5.7; sectile; resinous to adamantine luster except pearly on cleavages; colorless, yellow, or greenish yellow, the colorless samples becoming yellowish on exposure to light; streak: yellow, usually transparent.

Occurrence: Found in the oxidized zone of silver-ore deposits and associated with *chlorargyrite, bromargyrite,* native *silver,* and *calcite* and other carbonates.

Use: An ore of silver.

K. FRYE

References

Am. Min. **49**, 224.
Dana, II, 22.

Cross-reference: *Gossan.*

IODOBROMITE = Iodian *bromargyrite*

IODYRITE = *Iodargyrite* — Am. Min. **49**, 224

IOLITE = *Cordierite*

IOWAITE hexa. — Am. Min. **54**, 296
$Mg_4Fe^{3+}(OH)_8OCl \cdot 2\text{–}4H_2O$

IRANITE — Am. Min. **48**, 1417
$PbCrO_4 \cdot H_2O$ tric.

IRAQUITE — Am. Min. **61**, 1054
$KCa_4REE_3Si_{16}O_{40}$ tetr. — Min. Mag. **40**, 441

IRARSITE — Am. Min. **52**, 1580
(Ir,Ru,Rh,Pt)AsS iso.

IRHTEMITE — Am. Min. **59**, 209
$Ca_4MgH_2(AsO_4)_4 \cdot 4H_2O$ mono.

IRIDARSENITE — Am. Min. **61**, 177
$IrAs_2$ mono. — Can. Min. **12**, 280

IRIDIUM
Ir iso.

Cross-reference: *Native Elements.*

IRIDOSMINE — Dana, I, 111
(Os,Ir) hexa.

Cross-references: *Blacksand Minerals; Native Elements and Alloys.*

IRIDOSMIUM = *Iridosmine*

IRIGINITE — Am. Min. **45**, 257
$(UO_2)Mo_2O_7 \cdot 3H_2O$ mono.

IRON Native *iron* with usually some nickel, named from the common English term for the material.

Composition: α-Fe, with about 2% Ni.

Properties: Isometric, *Im*3*m*; usually found in blebs and large masses, and in meteorites (as *kamacite*) in plates and lamellar masses; H=4.5, G=7.3–7.9; fracture hackly; poor cubic cleavage; malleable; metallic luster; opaque; color: steel-gray to black; strongly magnetic; soluble in HCl.

Occurrence: Very rare, always found alloyed with some nickel and frequently with small amounts of cobalt, copper, manganese, sulfur, and carbon. Terrestrial *iron* is regarded as a primary magmatic constituent or reduction product of iron compounds. Most important locality is on Disko Island off the west coast of Greenland, where it is in a basalt as large masses and small embedded grains. Meteoritic low-nickel iron is found as the natural alloy, *kamacite*, with up to 5.5% nickel.

Use: Too rare to be of economic importance, except at the Greenland occurrence where several-ton masses are mined as ore.

J. A. GILBERT

References

Dana, I, 114–116.

Kostov, I., 1968. *Mineralogy.* Edinburgh: Oliver and Boyd, 90–91.

Cross-references: *Allotropy; Magnetic Minerals; Metallurgy; Native Elements and Alloys.* Vol. IVA: *Iron: Element and Geochemistry.*

IRON CORDIERITE = *Sekaninaite*

IRON GLANCE = Specular *hematite*

IRON PYRITE = *Pyrite*

ISHIKAWAITE orth. RRW, 305
(U,Fe,Y,Ca)(Nb,Ta)O_4(?)

ISOCLASITE *Dana,* II, 933
$Ca_2(PO_4)(OH)\cdot 2H_2O$ mono.(?)

ISOFERROPLATINUM *Am. Min.* **61**, 338
Pt_3Fe iso. *Can. Min.* **13**, 117

ISOKITE *Am. Min.* **41**, 167
$CaMg(PO_4)F$ mono. *Min. Mag.* **30**, 681

ISOMERTIEITE *Am. Min.* **59**, 1330
$(Pb,Cu)_5(Sb,As)_2$ iso. *Min. Mag.* **39**, 528

ISOSTANNITE = *Kesterite* *Can. Min.* **13**, 309

ITOITE *Am. Min.* **45**, 1313
$Pb_3Ge(SO_4)_2O_2(OH)_2$ orth.

IWAKIITE *Am. Min.* **65**, 406
$Mn^{2+}(Fe^{3+},Mn^{3+})_2O_4$ tetr. *Min. Mag.* **43**, 1061

IXIOLITE An iron niobate named after Ixion, mythical character related to Tantalus.

Composition: $FeNbO_4$, with Ta, Sn, and Mn substituting.

Properties: Orthorhombic, *Pnab*; medium-temperature polymorph stable between 1085°C and 1380°C; *c* axis is 1/3 that of *columbite.*

D. H. SPEIDEL

Reference: *Am. Min.* **48**, 961–979.

J

JACOBSITE Manganese iron oxide named for the original locality, Jakobsberg, Sweden. It is a member of the *magnetite* series of the ***spinel*** group.

Composition: $MnFe_2O_4$, Isometric.

Occurrence: Commonly associated with manganese ores in Sweden, India, and Japan.

D. H. SPEIDEL

References

Dana, I, 698.
DHZ, 5, 75.

Cross-references: *Ore Microscopy; Spinel Group.*

JADE–Microcrystalline gem variety of *jadeite* or nephrite (*actinolite*)

Cross-references: *Gemology; Jade; Placer Deposits.*

JADEITE A sodium aluminosilicate of the ***pyroxene*** group, named by Damour in 1863, in order to distinguish it from jade of the ***amphibole*** group.

Composition: $4[NaAlSi_2O_6]$; Na_2O, 15.4%; Al_2O_3, 25.2%; SiO_2, 59.4% by wt. Some Al replaced by Fe^{3+}.

Properties: Monoclinic, *C2/c*; usually granular or fibrous masses; H=6.5–7; G=3.3–3.5; vitreous luster, pearly on cleavage faces; color: usually green, but sometimes white, brown; streak: white.

Occurrence: *Jadeite* is formed as a result of high-pressure, low-temperature metamorphism of ***plagioclase.*** It is always associated with *albite* and commonly with *quartz,* ***feldspathoids,*** *lawsonite,* ***chlorite,*** *serpentine, zoisite, titanite,* ***garnet,*** *stilpnomelane,* and ***mica.*** Localities include: Tawmaw, Burma; Kotaki, Japan; and San Benito County, California, USA.

Use: An ornamental stone called jade. The term jade also includes nephrite (*actinolite*).

J. A. GILBERT

References

Am. Min. **51,** 956–975.
C. R. Acad. Sci., **56,** 861.
DHZ, 2, 278–279.
Vanders, I., and Kerr, P. F., 1967. *Mineral Recognition.* New York: Wiley, 278–279.

Cross-references: *Alkali Feldspars; Clinopyroxenes; Gemology; Jade; Minerals, Uniaxial and Biaxial; Pyroxene Group.*

JAGOITE *Am. Min.* **43**, 387
$Pb_3FeSi_3O_{10}(OH,Cl)$ trig.

JAGOWERITE *Am. Min.* **61**, 175
$BaAl_2(PO_4)_2(OH)_2$ tric. *Can. Min.* **12**, 135

JAHNSITE mono.
$CaMn(Mg,Fe)_2Fe_2^{3+}(PO_4)_4(OH)_2 \cdot 8H_2O$
Am. Min. **59**, 964; **62**, 692

JALPAITE *Am. Min.* **53**, 1530, 1778
Ag_3CuS_2 tetr. *Min. Abst.* **29**, 80

JAMBORITE hexa. *Am. Min.* **58**, 835
$(Ni^{2+},Ni^{3+},Fe)(OH)_2(OH,S,H_2O)$(?)

JAMESONITE A lead iron antimony sulfide named by Haidinger in 1825 after the Scottish mineralogist Robert Jameson (1774–1854).

Composition: $2[Pb_4FeSb_6S_{14}]$; Pb, 40.16%; Fe, 2.71%; Sb, 35.39%; S, 27.74% by wt.

Properties: Monoclinic, $P2_1/a$; usually in matted fibrous or feather-like crystalline masses; good cleavage {001} across length; H=2.5–3, G=5.65; metallic luster; opaque; color and streak lead-gray to black, sometimes tarnished iridescent surface.

Occurrence: Associated with *galena, sphalerite, tetrahedrite, quartz, stibnite, pyrite,* and *siderite* in low- and medium-temperature veins. Many of the principal localities include mining districts such as: Příbram, Czechoslovakia; Felsőbánya, Rumania; the Harz mountains, East and West Germany; Cornwall, England; and in Nevada, USA.

Use: Minor ore of lead, nined in the Czechoslovakian, German and English occurrences.

J. A. GILBERT

References

Dana, I, 451.

Vanders, L. and Kerr, P. F., 1967. *Mineral Recognition.* New York: Wiley, 176.

JANGGUNITE orth. *Am. Min.* **63**, 794
$Mn^{4+}_{4.8}(Mn^{2+},Fe^{3+})_{1.2}O_8(OH)_6$
Min. Mag. **41**, 519

JAROSITE A potassium iron hydroxyl sulfate named by Breithaupt in 1852 from the original locality in the Jaroso ravine in the Sierra Almagrera, Spain. It is an end-member in the ***alunite*** group.

Composition: $[KFe_3(SO_4)_2(OH)_6]$; K_2O, 9.41%; Fe_2O_3, 47.83%; SO_3, 31.971%; H_2O, 10.79% by wt. A very small amount of Al is usually present in substitution for Fe^{3+}, but there are only few occurrences which indicate the possibility of a complete natural series between *alunite* and *jarosite.* Sodium commonly substitutes for potassium, and a complete series extends towards *natrojarosite.*

Properties: Trigonal, $R\bar{3}m$; usually as a coating, massive, fibrous, concretionary, or earthy; H=2.5–3.5; G=2.35; subadamantine to vitreous luster on faces, resinous on cleavages; translucent; pyroelectric; color: amber yellow (ocherous) to dark brown; streak: pale yellow.

Occurrence: *Jarosite* is a widespread secondary mineral, commonly as crusts and coatings on ferruginous ores and in cracks of adjoining rocks and as a constituent of limonitic gossans, frequently confused with earthy limonite. Usually in association with limonite, *hematite, barite, pyrite, quartz,* and coal. Principal localities include: Chuquicamata, Chile; Czechoslovakia; Laurium, Greece; Bisbee, Arizona; Custer County, Idaho, Lincoln, Nevada, and Tintic, Utah, USA.

J. A. GILBERT

References

Am. Min. 47, 112–126.
Dana, II, 560.
Mason, B., and Berry, L. G., 1968. *Elements of Mineralogy.* San Francisco: Freeman, 373–374.

Cross-references: *Alunite Group; Gossan; Soil Mineralogy.*

JARGON = Colorless gem *zircon*

JARLITE *Dana,* II, 118
$NaSr_3Al_3(F,OH)_{16}$ mono.

JASPER–Red chalcedony

Cross-reference: *Gemology.*

JASPILITE–Interbedded jasper and iron oxides

JEFFERSONITE–Mn and Zn rich ***pyroxene*** *Am. Min.* **51**, 1406

JENNITE *Am. Min.* **51**, 56; **62**, 356
$9CaO \cdot 6SiO_2 \cdot 11H_2O$ tric.

JEREMEJEVITE *Am. Min.* **61**, 88
$Al_6(OH)_3(BO_3)_5$ hexa. *Min. Mag.* **20**, 452

Cross-reference: *Pegmatite Minerals.*

JET–variety of coal *Min. Abst.* **28**, 158

Cross-reference: *Gemology.*

JEZEKITE = *MORINITE* *Am. Min.* **47**, 398

JIMBOITE *Am. Min.* **48**, 1416
$Mn_3B_2O_6$ orth.

JIMTHOMPSONITE orth. *Am. Min.* **63**, 1000
$(Mg,Fe)_{10}Si_{12}O_{32}(OH)_4$ *Min. Abst.* **29**, 341

Cross-reference: *Biopyribole.*

JIXIANITE *Am. Min.* **64**, 1330
$Pb(W,Fe)_2(O,OH)_7$ iso. *Min. Mag.* **43**, 1062

JOAQUINITE mono. *Am. Min.* **60**, 435,872
$Ba_2NaCe_2Fe(Ti,Nb)_2Si_8O_{26}(OH,F)_2$

JOESMITHITE mono. *Am. Min.* **54**, 577
$PbCa_2(Mg,Fe^{2+})_4Fe^{3+}(Si,Be)_6O_{12}(OH)_4(O,OH)_8$ *Min. Mag.* **36**, 387

JOHACHIDOLITE *Am. Min.* **62**, 327
$CaAlB_3O_7$ orth. *Nature* **240**, 63

JOHNSOMERVILLEITE *Am. Min.* **66**, 437
$Na_{10}Ca_6Mg_{18}(Fe,Mn)_{25}(PO_4)_{36}$ trig. *Min. Mag.* **43**, 833

JOHANNITE A hydrated sulfate of uranium and copper named by Haidinger in 1830 in honor of Archduke Johann of Austria.

Composition: $Cu(UO_2)_2(SO_4)_2(OH)_2 \cdot 6H_2O$.

Properties: Triclinic, $P\bar{1}$; prismatic or thick tabular crystals; H=2–2.5; G=3.3; vitreous emerald to apple green with a slightly paler streak; transparent to translucent; bitter taste.

Occurrence: Found at Joachimsthal and Johanngeorgenstadt, Czechoslovakia; Cornwall, England; Mounana, Gabon; and Gilpin County, Colorado. USA.

K. FRYE

References

Dana, II, 606.
USGS Bull., **1064,** 130–135.

JOHANNSENITE A calcium manganese silicate of the ***pyroxene*** group, named by Schaller in 1938 in honor of A. Johannsen, University of Chicago. The manganese end member of the *diopside* series of the ***clinopyroxenes.***

Composition: $4[CaMnSi_2O_6]$, with Fe^{2+} substitution for Mn.

Properties: Monoclinic, *C2/c*; commonly in columnar and radiating aggregates of fibrous prisms; vitreous luster; H=6, G=3.44–3.55; color: clove-brown to grayish-green; streak: colorless; surface alteration to black manganese oxide helps distinguish from other ***pyroxenes.***

Occurrence: It is generally associated with *bustamite* and *rhodonite* in metasomatized limestones and with some manganese ores. In copper, lead, and zinc ores, *johannsenite* is found with *chalcopyrite, galena, sphalerite,* and *pyrite.* From Fukui and Okayama Prefectures, Japan; Pueblo and Hidalgo, Mexico; Vanadium, New Mexico; Franklin, New Jersey, USA.

G. C. MOERSCHELL

References

Am. Min. **23,** 575.
DHZ, 2A, 415–422.

Cross-references: *Clinopyroxenes; Pyroxene Group.*

JOHNBAUMITE (***Apatite*** gr). *Am. Min.* **65,** 1143
$Ca_5(AsO_4)_3(OH)$ hexa.

JOKOKUITE *Am. Min.* **64,** 655
$MnSO_4 \cdot 5H_2O$ tric. *Min. Mag.* **43,** 1062

JOLIOTITE *Min. Abst.* **28,** 208
$UO_2CO_3 \cdot 1.5\text{–}2H_2O$ orth.

JONESITE orth. *Min. Abst.* **29,** 482
$Ba_4(K,Na)_2Ti_4Al_2Si_{10}O_{36} \cdot 6H_2O$
Min. Rec. **8,** 453

JORDANITE mono. *Am. Min.* **61,** 963
$(Pb,Tl)_{14}(As,Sb)_6S_{23}$ *Dana,* I, 398

JORDISITE *Dana,* I, 331
MoS_2 amorph.

JOSEITE *Dana,* I, 166
$Bi_3Te(Se,S)$ trig. *Min. Abst.* **29,** 477,480

JOSEPHINITE = *Nickel-iron* and *andradite*
AJS **276,** 241

JOURAVSKITE hexa. *Am. Min.* **50,** 2102
$Ca_3Mn^{4+}(SO_4)(CO_3)(OH)_6 \cdot 12H_2O$

JULGOLDITE mono. *Am. Min.* **56,** 2157
$Ca_2Fe^{2+}(Fe^{3+},Al)_2(SiO_4)(Si_2O_7)(OH)_2 \cdot H_2O$
Can. Min. **12,** 219

JULIENITE *Dana,* II, 1106
$Na_2Co(SCN)_4 \cdot 8H_2O$(?) tetr.

JUNGITE orth. *Am. Min.* **65,** 1067
$Ca_2(H_2O)_6Zn_4Fe_8^{3+}(OH)_9 \cdot 10H_2O$
Min. Abst. **31,** 496

JUNITOITE *Am. Min.* **61,** 1255
$CaZn_2Si_2O_7 \cdot H_2O$ orth.

JUNOITE *Am. Min.* **60,** 548
$Pb_3Cu_2Bi_8(S,Se)_{16}$ mono. *Can. Min.* **18,** 353

JURBANITE *Am. Min.* **61,** 1
$AlSO_4OH \cdot 5H_2O$ mono. *Min. Abst.* **27,** 258

K

KAERSUTITE (*Amphibole* gr.)
$NaCa_2Mg_4TiSi_6Al_2(O+OH)_{24}$
mono. *Am. Min.* **63**, 1032
DHZ, 2, 231

Cross-references: *Amphibole Group; Meteoritic Minerals; Minerals, Uniaxial and Biaxial.*

KAFEHYDROCYANITE (Artifact?)
$K_4Fe(CN)_6 \cdot 3H_2O$ *Am. Min.* **59**, 209

KAHLERITE (*Autunite* gr.)
$Fe(UO_2)_2(AsO_4)_2 \cdot nH_2O$
tetr. *Am. Min.* **39**, 1038

KAINITE *Dana*, II, 594
$MgSO_4 \cdot KCl \cdot 3H_2O$ mono. *Min. Abst.* **27**, 307

Cross-reference: *Saline Minerals.*

KAINOSITE orth. *Am. Min.* **47**, 328
$Ca_2(REE)_2Si_4O_{12}(CO_3) \cdot H_2O$

KALIBORITE mono. *Am. Min.* **50**, 1079
$HKMg_2B_{12}O_{16}(OH)_{10} \cdot 4H_2O$

KALICINITE *Dana*, II, 136
$KHCO_3$ mono.

KALINITE *Dana*, II, 471
$KAl(SO_4)_2 \cdot 11H_2O$ mono.(?)

KALIOPHILITE DHZ, 4, 232
$KAlSiO_4$ hexa.

KALIPYROCHLORE (*Pyrochlore* gr.)
$(K,Na)_2(Nb,Ta,Ti)_2O_6(O,OH,F)$
iso. *Am. Min.* **62**, 405; **63**, 528

KALISTRONTITE *Am. Min.* **48**, 708
$K_2Sr(SO_4)_2$ trig.

KALSILITE A potassium aluminosilicate, named by Bannister in 1942 in reference to the chemical formula.

Composition: $2[KAlSiO_4]$, usually with 2–5% Na_2O.

Properties: Hexagonal, $P6_322$; usually white granular masses; poor cleavage; H=6; G=2.6–2.63; colorless, white, or gray; colorless streak; low birefringence; distinguished from *nepheline* by X-ray diffraction.

Occurrence: It is an important constituent in the groundmass of some K-rich, silica-poor lavas; also as *nepheline-kalsilite* phenocrysts in less K-rich lavas, but unknown in plutonic rocks. *Kalsilite* is present in the groundmass of the Katunga, Uganda, and San Venanzo, Italy, lavas and in the phenocrysts of Mt. Nyiragongo, Zaire, lavas associated with *nepheline* and *leucite*. Sometimes formed on the fine clay refractory linings when alkali vapor attacks the blast furnace bricks.

G. C. MOERSCHELL

References

Am. Min. **63**, 1225–1240.
DHZ, 4, 231–270.
Min. Mag. **26**, 218.

KAMACITE *Dana*, I, 115
α-(Fe,Ni) iso.

Cross-references: *Lunar Minerals; Metallurgy; Meteoritic Minerals; Native Elements and Alloys; Rock-Forming Minerals; Widmanstätten Structures.*

KÄMMERERITE = Chromian *clinochlore*

KANDITE = ***Kaolinite*** group

KANEMITE orth. *Am. Min.* **59**, 210; **62**, 763
$NaH(Si_2O_4)(OH)_2 \cdot 2H_2O$

KANKITE *Am. Min.* **62**, 594
$FeAsO_4 \cdot 3\frac{1}{2}H_2O$ mono. *Min. Abst.* **28**, 208

KANOITE (*Pyroxene* gr.) *Am. Min.* **63**, 598
$(Mn,Mg)_2Si_2O_6$ mono. *Min. Abst.* **30**, 423

KANONAITE *Am. Min.* **64**, 655
$Mn^{3+}AlSiO_5$ orth. *Min. Abst.* **29**, 482

KAOLIN = *Kaolinite*

KAOLINITE A hydrous aluminosilicate, the

most important of the ***kaolinite-serpentine*** group of minerals. The name was derived by Johnson in 1867 from the rock term "kaolin" (from the Chinese "kauling" or "high ridge" near Jauchau Fu where it was originally found).

Composition: $2[Al_2Si_2O_5(OH)_4]$; Al_2O_3, 41.2%; SiO_2, 48.0%; H_2O, 10.8% by wt.

Properties: Monoclinic; off-white when impurities are present, otherwise white; sometimes in thin hexagonal plates with pearly luster, usually in compact friable masses with pearly to earthy luster; perfect basal {001} cleavage; H=2–2.5; G=2.60–2.63. Colorless in thin section; translucent; occasionally slightly pleochroic; optically (-); $2V$=24°–50°; α= 1.553–1.565, β=1.559–1.569, γ=1.560–1.570, $\delta \approx$0.006; attacked by HCl; polymorphous with *dickite* and *nacrite*.

Occurrence: Formed by the hydrothermal or weathering alteration of other aluminosilicates, especially the ***feldspars*** and ***feldspathoids*** in granites and pegmatites under acid conditions. A *kaolinite* association usually includes: *quartz*, iron oxides, *pyrite, siderite, muscovite,* or other clay minerals. Large deposits are at Cornwall and Devon, England; Yrieix, near Limoges, France; Zinnwald, Czechoslovakia; Jauchau Fu, China; and Georgia, USA.

Uses: As a filler in paper (10–15% clay in newsprint and 25–35% clay in glossy magazine and book pages), rubber, paints, etc.; a ceramic raw material ("china clay"); and a whitening agent in textiles and paper.

E. BOOY

References

Dana, E. S. (W. E. Ford, ed.), 1949. *A Textbook of Mineralogy*. New York: Wiley, 384p.

DHZ, 3, 194–212.

Grim, R. E., 1968. *Clay Mineralogy*. New York: McGraw-Hill, 596p.

Kerr, P. F., ed., 1951. *Reference Clay Minerals, American Petroleum Institute Research Project 49*. New York: Columbia University Press, 701p.

Cross-references: *Authigenic Minerals; Blowpipe Analysis; Clays, Clay Minerals; Diagenetic Minerals; Electron Microscopy (Transmission); Fire Clays; Green River Mineralogy; Human and Vertebrate Mineralogy; Pegmatite Minerals; Phyllosilicates; Pigments and Fillers; Soil Mineralogy; Staining Techniques; Synthetic Minerals.*

KAOLINITE-SERPENTINE GROUP—*Amesite, antigorite, baumite, berthierine, brindleyite, caryopilite, clinochrysotile, cronstedtite, dickite, endellite, fraipontite, greenalite, halloysite, kaolinite, kellyite, lizardite, nacrite, nepouite, orthoantigorite, orthochrysotile, parachrysotile,* and *pecoraite.*

KARELIANITE *Am. Min.* **48**, 33
V_2O_3 trig.

KARIBIBITE *Am. Min.* **59**, 382
$Fe_2^{3+}As_4^{3+}O_8(OH)$ orth.

KARNASURTITE hexa.(?) *Am. Min.* **45**, 1133
$(Ce,La,Th)(Ti,Nb)(Al,Fe)(Si,P)_2O_7(OH)_4 \cdot 3H_2O$(?)

KARPATITE (Coronene) *Am. Min.* **54**, 329
$C_{24}H_{12}$ mono. *Chem. Geol.* **16**, 245

KARPINSKITE mono.(?) *Am. Min.* **42**, 584
$(Mg,Ni)_2Si_2O_5(OH)_2$(?)

KARPINSKYITE—mixture of *leifite* and Zn-bearing clay *Am. Min.* **57**, 1006

KASOLITE A hydrated lead uranosilicate named for the locality Kasolo, Katanga, Zaire.

Composition: $Pb(UO_2)SiO_4 \cdot H_2O$.

Properties: Monoclinic, $P2_1/a$; minute, cleavable prismatic crystals; H=4–5; G=6.0; resinous to greasy luster; color: yellow to brown.

Occurrence: Oxidation product of *uraninite.*

K. FRYE

Reference: *USGS Bull.* **1064**, 315–319.

KASSITE *Am. Min.* **52**, 559
$CaTi_2O_4(OH)_2$ orth.

KATOPHORITE (Catophorite) A hydroxyl aluminosilicate of iron, magnesium, calcium, sodium, and potassium. A member of the ***amphibole*** group named by Brögger in 1894 from the Greek meaning "a carrying down," in allusion to its volcanic origin.

Composition: $NaCaNa(Mg,Fe^{2+})_4(Al,Fe^{3+})(Si_7AlO_{22})(OH)_2$; monoclinic.

Occurrence: It is found in mafic alkali rocks, such as theralite and shonkinite, often associated with *arfvedsonite, aegirine,* and *aenigmatite.* Important localities are the Sande Cauldron, Oslo, Norway; the laccolith of the Shields River, Montana; Lake Malawi (Nyasa), Tanzania; and Lake Naivasha, Kenya.

J. A. GILBERT

References

Am. Min. **63**, 1023–1052.

DHZ, 2, 259–263.

KATOPTRITE *Am. Min.* **51**, 1494; **62**, 396
$(Mn_5^{2+}Sb_2^{5+})(Mn_8^{2+}Al_4Si_2)O_{28}$
mono.

KAWAZULITE *Am. Min.* **57**, 1312
$BiTe_2Se$ trig.

KAZAKOVITE *Am. Min.* **60**, 161
$Na_6H_2TiSi_6O_{18}$ trig.

KEATITE (tetr. polymorph of SiO_2)
Dana, III, 307

KECKITE mono. *Am. Min.* **64**, 1330
$Ca(Mn,Zn)_2Fe_3^{3+}(PO_4)_4(OH)_3 \cdot 2H_2O$
Min. Mag. **43**, 1062

KEGELITE *Am. Min.* **62**, 175
$Pb_{12}(Zn,Fe)_2Al_4(Si_{11}S_4)O_{54}$
Naturw. **62**, 137

KEHOEITE *Am. Min.* **49**, 1500
$Zn_{11}Ca_5H_{96}Al_{32}P_{32}O_{192} \cdot 32H_2O$(?)
Min. Mag. **33**, 799

KEITHCONNITE *Can. Min.* **17**, 589
$Pd_{3-x}Te$ trig. *Min. Rec.* **11**, 111

KELDYSHITE *Am. Min.* **55**, 1072
$NaZrSi_2O_6(OH)$ tric. *Min. Abst.* **30**, 217

KELLYITE (***Kaolinite-Serpentine*** gr.)
$(Mn,Mg)_2Al(SiAl)O_5(OH)_4$
hexa. *Am. Min.* **59**, 1153

KEMMLITZITE
(***Beudantite*** gr.) *Am. Min.* **55**, 320
$SrAl_3(AsO_4)(SO_4)(OH)_6$ trig.

KEMPITE *Dana*, I, 73
$Mn_2Cl(OH)_3$ orth.

KENNEDYITE *Am. Min.* **46**, 766
$Fe_2MgTi_3O_{10}$ orth. *Min. Mag.* **32**, 676

KENTROLITE
$Pb_2Mn_2^{3+}Si_2O_9$ orth. *Am. Min.* **52**, 1085

KENYAITE mono. *Am. Min.* **53**, 2061
$Na_2Si_{22}O_{41}(OH)_8 \cdot 6H_2O$ *Science* **157**, 1177

KERMESITE *Dana*, I, 279
Sb_2S_2O mono.

KERNITE A hydrated sodium borate named by Schaller (1927) after Kern County, California, in which it occurs in the Kramer borate district.

Composition: $4[Na_2B_4O_7 \cdot 4H_2O]$; Na_2O, 22.7%; B_2O_3, 51.0%; H_2O, 26.3% by wt.

Properties: Monoclinic, $P2_1/n$; usually in coarse cleavable aggregates; long splintery cleavage; H=2.5; G=1.91; vitreous to satiny luster; colorless to white; white streak; on exposure to air may alter to chalky white *tincalconite.*

Occurrence: At Boron, California in the Mojave Desert. It is thought to have been formed from *borax* by recrystallization brought about due to increased temperature and pressure. Associated with *borax, ulexite, tincalconite*, and *kramerite* as a large mass, and as large crystalline veins in the clayshales of Kern County.

Uses: A major source of *borax* in boron production in the USA.

J. A. GILBERT

References

Dana, II, 335.
Vanders, I., and Kerr, P. F., 1967. *Mineral Recognition*. New York: Wiley, 226.

Cross-reference: *Saline Minerals.*

KEROLITE–Variety of *talc* *Min. Mag.* **41**, 443

KERSTENITE *Am. Min.* **39**, 850
$PbSeO_4$ orth.

KESTERITE *Am. Min.* **44**, 1329
$Cu_2(Zn,Fe)SnS_4$ tetr. *Can. Min.* **16**, 131

KETTNERITE *Am. Min.* **43**, 385
$CaBi(CO_3)OF$ tetr.

KEYITE *Am. Min.* **62**, 1259
$(Cu,Zn,Cd)_3(AsO_4)_2$ mono. *Min. Rec.* **8**, 87

K-*FELDSPAR* = *Microcline, orthoclase*, or *sanidine*

KHADEMITE=*Rostite* *Min. Abst.* **30**, 424

KHIBINSKITE *Am. Min.* **59**, 1140; **60**, 340
$K_2ZrSi_2O_7$ mono.

KHINITE *Am. Min.* **63**, 1016
$Cu_3PbTeO_4(OH)_6$ orth. *Min. Mag.* **43**, 1062

KHLOPINITE = *Samarskite* *Am. Min.* **57**, 329

KHUNIITE = *Iranite* *Am. Min.* **61**, 186

KIDWELLITE *Min. Abst.* **29**, 200
$NaFe_9^{3+}(OH)_{10}(PO_4)_6 \cdot 5H_2O$ mono. *Min. Mag.* **42**, 137

KIESERITE A hydrated magnesium sulfate named by Reichardt in 1861 after Mr. Kieser, a president of the Academy of Jena.

Composition: $4[MgSO_4 \cdot H_2O]$; SO_3, 58%; MgO, 29%; H_2O, 13%; by wt.

Properties: Monoclinic, *C2/c*; usually compact massive or coarse to fine granular; perfect {110} and {111} or {100} cleavage; friable; H=3–3.5, G=2.57; vitreous luster; color: white, grayish to yellowish; translucent; slightly soluble.

Occurrence: Very abundant in salt deposits of Stassfurt, E. Germany; Hannover, W. Germany; Kalusz, Galicia, Spain; and May Mines, Pendschab, India, in association with *halite, carnallite, gypsum, anhydrite, glauberite, epsomite*, and other evaporites.

G. C. MOERSCHELL

References

Dana, II, 477.

Kostov, I., 1968. *Mineralogy*. Edinburgh: Oliver and Boyd, 494–496.

Cross-reference: *Saline Minerals.*

KILCHOANITE Am. Min. **46**, 1203
$Ca_3Si_2O_7$ orth. *Nature* **189**, 743

KILLALAITE Am. Min. **59**, 1331
$Ca_3Si_2O_7 \cdot ½H_2O$ mono. *Min. Mag.* **39**, 544

KIMZEYITE (***Garnet*** gr.) *Am. Min.* **46**, 533
$Ca_3(Zr,Ti)_2(Al,Si)_3O_{12}$

KINGITE tric. *Am. Min.* **55**, 515
$Al_3(PO_4)_2(OH,F)_3 \cdot 9H_2O$ *Min. Mag.* **31**, 351

KINGSMOUNTITE *Can. Min.* **17**, 579
$Ca_4FeAl_4(PO_4)_6(OH)_4 \cdot 12H_2O$
mono. *Min. Abst.* **31**, 496

KINOITE *Am. Min.* **55**, 709; **62**, 1032
$Ca_2Cu_2Si_3O_{10} \cdot 2H_2O$ mono.

KINOSHITALITE (***Mica*** gr.)
$(Ba,K)(Mg,Mn,Al)_3(Al_2Si_2)O_{10}(OH)_2$
mono. *Min. Abst.* **29**, 84

KIRSCHSTEINITE (***Olivine*** gr.)
$CaFeSiO_4$ *Am. Min.* **43**, 790
orth. *Min. Mag.* **31**, 698

KITKAITE *Am. Min.* **50**, 581
NiTeSe trig.

KLADNOITE *Am. Min.* **31**, 605
$C_6H_4(CO)_2NH$ mono.

KLAPROTHITE = *Wittichenite*

KLEBELSBERGITE *Am. Min.* **65**, 499, 931
$Sb_4O_4(OH)_2SO_4$ orth.

KLEBERITE *Am. Min.* **64**, 655
$FeTi_6O_{13} \cdot 4H_2O$(?) hexa. *Min. Mag.* **43**, 1062

KLEEMANITE *Am. Min.* **64**, 1331
$ZnAl_2(PO_4)_2(OH)_2 \cdot 3H_2O$
mono. *Min. Mag.* **43**, 93

KLEINITE hexa. *Am. Min.* **63**, 322
$Hg_2N(Cl,SO_4) \cdot nH_2O$ *Dana*, II, 87

KLOCKMANNITE *Dana*, I, 251
CuSe hexa.

Cross-reference: *Ore Microscopy.*

KMAITE = *Celadonite* *Am. Min.* **47**, 808

KNEBELITE = Manganoan *fayalite* (***Olivine*** gr.)

KNIPOVICHITE = Chromian ***alumohydrocalcite*** *Min. Rec.* **6**, 180

KNOPITE = Cerian *perovskite*

KNORRINGITE (***Garnet*** gr.)
$Mg_3Cr_2(SiO_4)_3$ iso. *Am. Min.* **53**, 1833

Cross-reference: *Mantle Mineralogy.*

KOASHVITE orth. *Am. Min.* **60**, 487
$Na_6(Ca,Mn)(Ti,Fe)Si_6O_{18} \cdot H_2O$

KOBEITE amorph. *Am. Min.* **42**, 342
$(Y,U)(Ti,Nb)_2(O,OH)_6$(?) *Min. Abst.* **29**, 79

KOBELLITE *Am. Min.* **54**, 573
$Pb_5Bi_8S_{17}$ mono. *Can. Min.* **9**, 371

Cross-reference: *Ore Microscopy.*

KOECHLINITE *Dana*, II, 1092
Bi_2MoO_6 orth.

KOENENITE hexa. *Dana*, II, 86
$Na_4Mg_9Al_4Cl_{12}(OH)_{22}$

KOETTIGITE mono. *Am. Min.* **64**, 376
$Zn_3(AsO_4)_2 \cdot 8H_2O$ *Can. Min.* **14**, 437

KOGARKOITE *Am. Min.* **58**, 116
$Na_3(SO_4)F$ mono. *Min. Mag.* **43**, 753

KOKTAITE *Dana*, II, 444
$(NH_4)_2Ca(SO_4)_2 \cdot H_2O$ mono.

KOLBECKITE *Am. Min.* **45**, 257
$ScPO_4 \cdot 2H_2O$ mono.

KOLICITE *Am. Min.* **64**, 708
$Mn_7Zn_4(AsO_4)_2(SiO_4)_2(OH)_6$
orth. *Min. Abst.* **31**, 230

KOLOVRATITE *Dana*, II, 1048
hydrous vanadate of Ni and Zn

KOLSKITE–Mixture of *lizardite* and *sepiolite* *Am. Min.* **59**, 212

KOLWEZITE *Am. Min.* **65**, 1067
$(Cu,Co)_2(CO_3)(OH)_2$
tric. *Min. Mag.* **43**, 1062

KOLYMITE *Am. Min.* **66**, 218
Cu_7Hg_6 iso. *Min. Mag.* **43**, 1063

KOMAROVITE orth. *Am. Min.* **57**, 1315
$(Ca,Mn)Nb_2Si_2O_9(O,F) \cdot 3½H_2O$

KONINCKITE *Dana*, II, 763
$Fe^{3+}PO_4 \cdot 3H_2O$(?) orth.(?)

KORITNIGITE *Am. Min.* **65**, 206
$Zn(AsO_3)(OH) \cdot H_2O$ tric.

KORNELITE *Dana*, II, 530
$Fe_2^{3+}(SO_4)_3 \cdot 7H_2O$ mono.

KORNERUPINE A magnesium aluminosilicate, named by Lorenzen in 1884 after the Danish geologist, Kornerup.

Composition:
$Mg_3Al_6[Si_2O_7(Al,Si)_2SiO_{10}]O_4(OH)$ with Si replaced by B.

Properties: Orthorhombic; fibrous to columnar aggregates, similar to *sillimanite*; perfect prismatic cleavage; H=6.5; G=3.27; vitreous luster; transparent to translucent; colorless, white to brown; the iron-rich gem variety prismatine is a sea-green color.

Occurrence: At Fiskernäs, Greenland, with *sapphirine*, ***mica***, *gedrite* and *cordierite*. At Waldheim, E. Germany, interlayered with *albite* in a granulite. Large clear crystals are near Betroka, Malagasy Republic.

Use: A minor gemstone.

G. C. MOERSCHELL

References

Am. Min. 37, 531–541.
Min. Abst. 30, 135, 347.
Science 159, 524–526.

KORZHINSKITE *Am. Min.* **49**, 441
$CaB_2O_4 \cdot H_2O$

KOSMOCHLOR = *Ureyite* *Am. Min.* **53**, 511

KOSTOVITE *Am. Min.* **51**, 29
$CuAuTe_4$ mono.

KOTOITE *Dana*, II, 328
$Mg_3B_2O_6$ orth.

KÖTTIGITE = *KOETTIGITE*

KOTULSKITE *Am. Min.* **48**, 1181
Pd(Te,Bi) hexa. *Min. Mag.* **35**, 815

KOUTEKITE *Am. Min.* **46**, 467
Cu_5As_2 hexa.

KOZULITE (***Amphibole*** gr.)
$Na_3Mn_4(Fe,Al)Si_8O_{22}(OH,F)_2$
mono. *Am. Min.* **55**, 1815; **63**, 1036

KRAISSLITE hexa. *Am. Min.* **63**, 938
$Mn_{24}Zn_4(AsO_4)_4(SiO_4)_8(OH)_{12}$
Min. Abst. **30**, 170

KRAMERITE = *Probertite*

KRATOCHVILITE *Am. Min.* **23**, 667
$C_{13}H_{10}$ orth. *Min. Mag.* **25**, 635

KRAUSITE *Am. Min.* **50**, 1929
$KFe^{3+}(SO_4)_2 \cdot H_2O$ mono.

KRAUSKOPFITE *Am. Min.* **50**, 314
$BaSi_2O_5 \cdot 3H_2O$ mono.

KRAUTITE mono. *Am. Min.* **64**, 1248
$Mn(H_2O)(AsO_3OH)$ *Bull. Soc. fr.* **98**, 78

KREITTONITE = Iron-bearing *gahnite*

KREMERSITE *Dana*, II, 101
$(NH_4,K)_2FeCl_5 \cdot H_2O$ orth.

KRENNERITE A gold telluride named by vom Rath in 1877 in honor of the Hungarian mineralogist, Joseph A. Krenner.

Composition: $AuTe_2$, with Ag substituting for Au.

Properties: Orthorhombic, *Pma*2; in short prismatic crystals; H=2–3; G=8.6; metallic luster; color: silver-white to light brass yellow.

Occurrence: Found in Transylvania, Rumania, associated with *quartz* and *pyrite*. Also found at Kalgoorlie and Mulgabbie,

Western Australia; Montbray Township, Quebec, Canada; and Cripple Creek, Colorado, USA.

K. FRYE

References

Am. Min. **35**, 959–984.
Dana, I, 333.

KREUZBERGITE = *FLUELLITE* *Am. Min.* **51**, 1579

KRIBERGITE *Dana*, II, 1011
$Al_5(PO_4)_3(SO_4)(OH)_4 \cdot 2H_2O$(?)

KRINOVITE *Am. Min.* **54**, 578
$NaMg_2CrSi_3O_{10}$ tric. *Science* **161**, 786

Cross-reference: *Meteoritic Minerals.*

KROEHNKITE *Dana*, II, 444
$Na_2Cu(SO_4)_2 \cdot 2H_2O$ mono.

KRUPKAITE *Am. Min.* **60**, 300, 737; **61**, 15
$PbCuBi_3S_6$ orth.

KRUTAITE (***Pyrite*** gr.)
$CuSe_2$ iso. *Am. Min.* **59**, 210

KRUTOVITE *Am. Min.* **62**, 173
$Ni_{1-x}As_2$ iso.

KRYZHANOVSKITE orth. *Am. Min.* **56**, 1
$Fe_3^{3+}(OH)_3(PO_4)_2$ *Min. Mag.* **43**, 789

Cross-reference: *Pegmatite Minerals.*

KTENASITE *Am. Min.* **64**, 446
$(Cu,Zn)_3(SO_4)(OH)_4 \cdot 2H_2O$
mono. *Min. Mag.* **41**, 65

KULANITE tric. *Am. Min.* **62**, 174
$Ba(Fe,Mn,Mg,Ca)_2(Al,Fe)_2(PO_4)_3(OH)_3$
Can. Min. **14**, 127

KULKEITE (*Talc-clinochlore* interlayer)
$Mg_8Al(Si_7Al)O_{20}(OH)_{10}$
mono. *Am. Min.* **66**, 218

KULLERUDITE (***Marcasite*** gr.)
$NiSe_2$ orth. *Am. Min.* **50**, 520

KUNZITE—Lilac gem variety of *spodumene*

Cross-reference: *Pegmatite Minerals.*

KUPLETSKITE (***Astrophyllite*** gr.)
$(K,Na)_3(Mn^{2+},Fe)_7(Ti,Nb)_2Si_8O_{24}(O,OH)_7$
tric. *Am. Min.* **42**, 118

KURAMITE *Am. Min.* **65**, 1067
Cu_3SnS_4 tetr. *Min. Abst.* **31**, 496

KURANAKHITE *Am. Min.* **61**, 339
$PbMnTeO_6$ orth.

KURCHATOVITE *Am. Min.* **51**, 1817
$Ca(Mg,Mn)B_2O_5$ orth. *Min. Abst.* **29**, 25

KURGANTAITE *Am. Min.* **40**, 941
$(Sr,Ca)_2B_4O_8 \cdot H_2O$(?)

KURNAKOVITE *Dana*, II, 360
$Mg_2B_6O_{11} \cdot 15H_2O$ tric. *Min. Abst.* **28**, 260

KURUMSAKITE *Am. Min.* **42**, 583
$(Zn,Ni,Ca)_8Al_8V_2Si_5O_{35} \cdot 27H_2O$(?)

KUSUITE tetr. *Am. Min.* **62**, 1058
$(Ce^{3+},Pb^{2+},Pb^{4+})VO_4$ *Bull. Soc. fr.* **100**, 39

KUTINAITE *Am. Min.* **55**, 1083
Cu_2AgAs iso.

KUTNAHORITE = *KUTNOHORITE*

KUTNOHORITE A calcium manganese carbonate named by Bukowsky in 1901 from quarries in Kutná Hora, Czechoslovakia. *Kutnohorite* is an ordered-phase member of the ***dolomite*** group with composition intermediate between *calcite* and *rhodochrosite.*

Composition: 2 $[Ca(Mn,Mg,Ca,Fe)(CO_3)_2]$. Composition is variable, but a typical range is: CaO, 24–28%; MgO, 2–6%; MnO, 23–28%; FeO, 1–5%; CO_2, 41–42% by wt.

Properties: Trigonal; physically similar to other members of the ***dolomite*** group, e.g., *rhodochrosite*; G=3–3.1; and occurs in rare, granular to coarse cleavable masses; color: white to pale rose.

Occurrence: Found in carbonate veinlets. Has been reported from Czechoslovakia and Franklin, New Jersey, USA.

L. WILSON

References

Am. Min. **40**, 748–760; **52**, 1751–1761.
Dana, II, 217.

KYANITE An aluminum silicate named by Werner in 1789 from a Greek word meaning "blue."

Composition: $4[Al_2SiO_5]$.

Properties: Triclinic, $P\bar{1}$; usually in long, flexible tabular crystals or bladed aggregates; H=5 (along length parallel to *c*), 7 (perpendicular to length, parallel to *b*); G=3.55–3.66; vitreous to pearly luster; color: usually patchy blue, may be gray, green, or white; streak: white.

Occurrence: In medium-grade metamorphic rocks, e.g., gneiss and mica schist, as an accessory mineral, and in veins and pegmatites cutting these rocks. Commonly associated with ***garnet***, *staurolite*, *corundum*, and *quartz*. Economically important localities include Lapsa Buru, India; Burnsville, North Carolina, and the Chocolate Mts., California, USA.

Use: Manufacture of highly refractory porcelains.

J. A. GILBERT

References

DHZ, 1, 137–144.

Science **151**(3715), 1222–1225.

Varley, E. R., 1965. *Sillimanite*. London: Her Majesty's Stat. Off., 165p.

Cross-references: *Blacksand Minerals; Diamond; Mantle Mineralogy; Mineral and Ore Deposits; Minerals, Uniaxial and Biaxial; Refractory Minerals; Rock-Forming Minerals; Soil Mineralogy; Synthetic Minerals.*

KYANOPHYLLITE = *Paragonite* + *muscovite* *Am. Min.* **58**, 807

L

LABRADORITE A calcium sodium aluminosilicate of the ***feldspar*** group, named by Werner in 1780 for a locality of the Labrador coast.

Composition: $(Ca,Na)[(Al,Si)AlSi_2O_8]$; with some K replacing (Ca,Na). In the ***plagioclase feldspars,*** *labradorite* is assigned the range *Ab* 30–50, *An* 70–50 mol%.

Properties: Triclinic, $P\bar{1}$, commonly in cleavable masses or in lath-shaped grains; twinning according to the albite, Carlsbad, or pericline laws; prismatic {001} {010} cleavage; H=2.7; G=6; vitreous to pearly luster; transparent to translucent; color: white to gray to black; may show irridescence (labradorescence).

Occurrence: A common rock-forming mineral as phenocrysts and ground mass in basalts; essential constituent in gabbros, diabases, and many anorthosites; in metamorphic rocks in some hornfels contact rocks and regionally metamorphosed siliceous marbles; and as detrital grains in some arkoses.

K. FRYE

References*

DHZ, 4, 94–165.

*See *Feldspar Group* and *Plagioclase Feldspars* for additional references.

Cross-references: *Feldspar Group; Gemology; Minerals, Uniaxial and Biaxial; Plagioclase Feldspars.*

LABUNTSOVITE mono. *Am. Min.* **41**, 163
$(K,Ba,Na)(Ti,Nb)(Si,Al)_2(O,OH)_7 \cdot H_2O$

Cross-reference: *Green River Mineralogy.*

LACROIXITE *Am. Min.* **57**, 1914
$NaAl(PO_4)(F,OH)$ mono.

Cross-reference: *Pegmatite Minerals.*

LAFFITTITE *Am. Min.* **60**, 945
$AgHgAsS_3$ mono.

LAIHUNITE *Am. Min.* **62**, 1058
$Fe^{3+}Fe^{2+}_{0.6}SiO_4$ orth.

LAITAKARITE *Am. Min.* **47**, 806
$Bi_4(Se,S)_3$ trig. *Can. Min.* **18**, 353

LAMPROBOLITE = *Oxyhornblende* *Am. Min.* **63**, 1051

LAMPROPHYLLITE mono. RRW, 339
$Na_2(Sr,Ba)_2Ti_3(SiO_4)_4(OH,F)_2$

LANARKITE *Dana,* II, 550
$Pb_2(SO_4)O$ mono.

LANDAUITE (***Crichtonite*** gr.)
$Na[MnZn_2(Ti,Fe)_6Ti_{12}]O_{38}$
Can. Min. **16**, 63
ps. trig. *Am. Min.* **63**, 28

LANDESITE orth. *Am. Min.* **49**, 1122
$Mn^{2+}_2Fe^{3+}(H_2O)(OH)_2(PO_4)_2$
Min. Mag. **43**, 789

LANDSBERGITE = *Moschellandsbergite*

LANGBANITE trig. *Am. Min.* **55**, 1496
$(Mn^{2+},Ca)_4(Mn^{3+},Fe^{3+})_9SbSi_2O_{24}$

LANGBEINITE *Dana,* II, 434
$K_2Mg_2(SO_4)_3$ iso. *Min. Abst.* **30**, 353

Cross-reference: *Saline Minerals.*

LANGISITE *Am. Min.* **57**, 1910
(Co,Ni)As hexa. *Can. Min.* **9**, 597

LANGITE orth. *Dana,* II, 583
$Cu_4(SO_4)(OH)_6 \cdot 2H_2O$ *Min. Abst.* **30**, 295

LANSFORDITE *Dana,* II, 228
$MgCO_3 \cdot 5H_2O$ mono.

LANTHANITE orth. *Am. Min.* **62**, 142
$(REE)_2(CO_3)_3 \cdot 8H_2O$ *Dana,* II, 241

LAPIS LAZULI–*Lazurite*-bearing rock
Can. Min. **18**, 59

Cross-references: *Gemology; Pigments and Fillers.*

LAPLANDITE *Am. Min.* **60**, 487
$Na_4CeTiPSi_7O_{22} \cdot 5H_2O$ orth.

LAPPARENTITE = *Rostite* or *tamarugite* *Min. Abst.* **30**, 424

LARDERELLITE *Am. Min.* **45**, 1087
$(NH_4)B_5O_6(OH)_4$ mono.

LARNITE A calcium silicate named by Tilley in 1929 for Larne County, Ireland, its type locality.

Composition: $4[\beta\text{-}Ca_2SiO_4]$.

Properties: Monoclinic, $P2_1/n$; granular aggregates; brittle; vitreous luster; color: gray; streak: white to off-white; H=5.5–6; birefringence 0.01.

Occurrence: In a limestone contact zone at Scawt Hill, County Larne, Antrim, Ireland, associated with *spurrite, melilite, merwinite,* and *spinel.* Formed by the reaction between the calcium carbonate of the Tertiary chalk and the silica of internal flint nodules.

G. C. MOERSCHELL

References

Am. Min. **51,** 1766–1774.
Min. Mag. **22,** 77–86.

Cross-references: *Portland Cement Mineralogy; Rock-Forming Minerals.*

LAROSITE *Am. Min.* **59**, 382
$(Cu,Ag)_{21}(Pb,Bi)_2S_{13}$ orth. *Can. Min.* **11**, 886

LARSENITE *Am. Min.* **51**, 269
$PbZnSiO_4$ orth.

LATIUMITE mono. *Am. Min.* **58**, 466
$K(Ca,Na)_3(Al,Si)_5O_{11}(SO_4,CO_3,OH)$
Min. Mag. **30**, 39

LATRAPPITE (***Perovskite*** gr.)
$(Ca,Na)(Nb,Ti,Fe)O_3$ *Am. Min.* **50**, 265
orth. *Can. Min.* **8**, 121

LAUBMANNITE orth. *Am. Min.* **55**, 135
$Fe_3^{2+}Fe_6^{3+}(OH)_{12}(PO_4)_4$

LAUEITE tric. *Am. Min.* **54**, 1312; **55**, 135
$Mn^{2+}Fe_2^{3+}(OH)_2(PO_4)_2 \cdot 8H_2O$

Cross-reference: *Pegmatite Minerals.*

LAUMONTITE A hydrated calcium aluminosilicate of the ***zeolite*** group, first named by Haüy in 1808 in honor of Gillet Laumont, who first observed it in 1785 in the lead mines at Huelgoet, Brittany, France.

Composition: $Ca_4[Al_8Si_{16}O_{48}] \cdot 16H_2O$.

Properties: Monoclinic, *A*2/*m*; *a*=7.6 Å, *b*=14.8 Å, *c*=13.1 Å, γ=112°; H=3–3.5; G=2.2–2.3; luster vitreous, but pearly on cleavage surfaces; colorless, white, yellow, red, or brown; transparent to translucent, becoming opaque and pulverulent on exposure; cleavage good on {010} and {110}; twinning on {100}. Refractive indices, α=1.502–1.514, β=1.512–1.522, γ=1.514–1.525; birefringence = 0.011–0.012; 2*V*=26–47 (–); optic axial plane (010); β=*y*, $\gamma\Lambda z$=8°–33°; dispersion $r<v$, strong. The replacements of the types NaSi⇌CaAl and Ca⇌$(Na,K)_2$ may occur; loses its water to form *leonhardite,* $Ca(Al_2Si_4O_{12}) \cdot 3½H_2O$, in air or on gentle heating; dehydration stepwise, losing 14% H_2O at maximum; rehydration good; molecular sieve properties selective.

Occurrence: Commonly in veins and cavities in igneous rocks such as granite, diorite, diabase, quartz-porphyry, basalt, and andesite; probably in hydrothermal veins in joints in slates, quartzite, thermally metamorphosed andesite, and in crevices in amphibolite; as a large-scale replacement product of vitric tuffs and as a degradation product of ***plagioclase*** under incipiently metamorphic conditions; also as laumontitic bedded deposits in the thick series of graywackes and tuffs in New Zealand formed by burial metamorphism; as an authigenic mineral in cement of sandstones.

A. IIJIMA
M. UTADA

References

Am. Min. **37,** 812–830; **40,** 923–925.
N.J. Min. Mh. **2/3,** 33–42; **7,** 298–310.

Cross-references: *Authigenic Minerals, Zeolites.*

LAUNAYITE *Am. Min.* **53**, 1423
$Pb_{22}Sb_{26}S_{61}$ mono. *Can. Min.* **9**, 191

LAURIONITE *Dana,* II, 62
PbCl(OH) orth.

LAURITE (***Pyrite*** gr.) *Am. Min.* **54**, 1330
RuS_2 iso. *Can. Min.* **17**, 469

LAUSENITE *Dana,* II, 530
$Fe_2^{3+}(SO_4)_3 \cdot 6H_2O$ mono.

LAUTARITE *Dana,* II, 312
$Ca(IO_3)_2$ mono. *Min. Abst.* **30**, 115

Cross-reference: *Saline Minerals.*

LAUTITE *Dana,* I, 327
CuAsS orth.

LAVENDULAN *Am. Min.* **42,** 123
$NaCaCu_5(AsO_4)_4Cl \cdot 5H_2O$ orth.

LAVENITE A sodium calcium zirconium silicate fluoride named by Brögger in 1885 after Låven Island in Langesundfjörd, Norway, its type locality.

Composition: $NaCaMnZrSi_2O_8F$.

Properties: Monoclinic, $P2_1a$; prismatic and tabular crystals, also embedded grains; brittle; H=6; G=3.53; vitreous luster; color: light yellow to colorless, dark yellow to dark brown; translucent; strongly pleochroic; α= wine-yellow, β=yellowish-green, γ=deep red-brown; strong double refraction.

Occurrence: Associated with *eudialyte* and *catapleiite,* in nepheline syenite at its type locality; near Minas Gerais and Sao Paulo, Brazil, and on one of the Los islands, French Guyana.

G. C. MOERSCHELL

Reference: Kostov, I., 1968. *Mineralogy,* Edinburgh: Oliver and Boyd, 302.

Cross-reference: *Rock-Forming Minerals.*

LAVROVITE = Chromian *diopside*
Am. Min. **65,** 814

LAWRENCITE *Dana,* II, 40
(Fe,Ni)Cl_2 trig.

Cross-references: *Electron Probe Microanalysis; Lunar Minerals; Meteorite Minerals.*

LAWSONBAUERITE *Am. Min.* **64,** 949
$(Mn,Mg)_5Zn_2(SO)_4(OH)_{12} \cdot 4H_2O$
mono. *Min. Abst.* **31,** 230

LAWSONITE A hydrous calcium aluminosilicate named by Ransome in 1895 after Andrew C. Lawson.

Composition: $4[CaAl_2Si_2O_7(OH)_2 \cdot H_2O]$, by wt.

Properties: Orthorhombic, *Ccmm;* frequent polysynthetic twinning; H=8; G=3.09; vitreous to greasy luster; translucent; pale bluish to colorless; streak: white.

Occurrence: As grains in metamorphic rocks, particularly glaucophane schists, and as veins in these schists associated with and replaced by *pumpellyite.* Principal localities include the San Francisco Bay area and Healdsburg, California; Mendoke and Rumbia Mountains, South East Celebes; New Caledonia; Corsica; and the Kanto Mts., Japan.

J. A. GILBERT

References

Am. Min. **63,** 311.
DHZ, 1, 221–226.
Min. Mag. **39,** 121.

Cross-references: *Minerals, Uniaxial and Biaxial; Rock-Forming Minerals.*

LAZAREVICITE = *Arsenosulvanite*
Am. Min. **46,** 465

LAZULITE A magnesium aluminum hydroxyl phosphate named by Klaproth in 1795 from the older German name, *lazurstein,* meaning blue stone.

Composition: $2[MgAl_2(PO_4)_2(OH_2]$; a complete series exists between *lazulite* and *scorzalie,* $2[FeAl_2(PO_4)_2(OH)_2]$.

Properties: Monoclinic, $P2_1/c$; usually massive, granular to compact; H=5.5–6; G=3.0–3.1; vitreous luster; translucent; color: azure blue; streak: white.

Occurrence: A rare mineral, usually found in granite pegmatites, quartzites and in *quartz* veins in metamorphic rocks; often in association with *kyanite, andalusite, corundum,* and *rutile.* Good crystals are from Zermatt, Valais, Switzerland; Salzburg, Austria; Dattas, Minas Gerais, Brazil; and Graves Mt., Georgia.

Use: A minor gemstone.

J. A. GILBERT

References

Dana, II, 908.
Vanders, I., and Kerr, P. F., 1967. *Mineral Recognition.* New York: Wiley, 246–247.

Cross-references: *Pegmatite Minerals.*

LAZURITE A sodium calcium aluminosilicate with sulfide sulfate of the ***sodalite*** group, named by Brögger in 1890 because of its color resemblance to *azurite,* from an obsolete synonym for *azurite.*

Composition: $(Na,Ca)_8(Al,Si)_{12}O_{24}(S,SO_4)$.

Properties: Isometric, $P\bar{4}3m$; commonly compact; H=5–5.5; G=2.4–2.45; vitreous luster; translucent; color: deep blue to greenish blue.

Occurrence: A rare mineral which usually occurs as a product of contact metamorphism in crystalline limestone. Lapis lazuli is usually a mixture of *lazurite* with *calcite,* ***pyroxene,*** and other silicates.

Uses: A minor gem stone, lapis lazuli, for vases and ornamental furniture. Also employed in the manufacture of mosaics and paints.

J. A. GILBERT

References

Can. Min. **18**, 59–70.
Vanders, I., and Kerr, P. F., 1967. *Mineral Recognition.* New York: Wiley, 266.

Cross-references: *Gemology; Green River Mineralogy; Pigments and Fillers.*

LEAD *Dana,* I, 102
Pb iso *Min. Abst.* **29**, 116

Cross-reference: *Native Elements and Alloys.*

LEAD GLANCE = *Galena*

LEADHILLITE A lead hydroxylsulfate carbonate named in 1832 for the locality Leadhills, Lanarkshire, Scotland.

Composition: $8[Pb_4(SO_4)(CO_3)_2(OH)_2]$.

Properties: Monoclinic, $P2_1/a$; pseudohexagonal tabular crystals; one perfect cleavage; H=2.5–3; G=2.5; resinous to pearly luster; colorless, white, bluish, greenish, yellowish; some samples fluoresce yellow; polymorphous with *susannite.*

Occurrence: Found in the oxidized zone of lead-ore deposits with other secondary lead minerals.

K. FRYE

References

Can. Min. **10**, 141.
Dana, II, 295–298.

LECHATELIERITE This name is given to naturally fused gray siliceous glass. Lechatelierite was named by Lacroix in 1915 after the French chemist, Henry Le Chatelier, under the impression that it was a mineral. Actually it is a minor rock type, varying in composition according to the original sand type. Characteristically it contains 90–99.5% silica and has a refractive index between 1.458 and 1.462.

K. FRYE

Reference: *Dana,* III, 325.

Cross-reference: *Mineral Classification: Principles.*

LECONTITE *Am. Min.* **48**, 180
$(NH_4,K)Na(SO_4)\cdot 2H_2O$ orth.

LEGRANDITE *Am. Min.* **61**, 95
$Zn_2(AsO_4)(OH)\cdot H_2O$ mono.

LEHIITE *Dana,* II, 942
$(Na,K)_2Ca_5Al_8(PO_4)_8(OH)_{12}\cdot 6H_2O$

LEIFITE *Am. Min.* **57**, 1006
$Na_2(Si,Al,Be)_7(O,OH,F)_{14}$ trig.

LEIGHTONITE *Dana,* II, 461
$K_2Ca_2Cu(SO_4)_4\cdot 4H_2O$ tric.

LEITEITE *Am. Min.* **62**, 1259
$ZnAs_2O_4$ mono. *Min. Rec.* **8**, 95

LEMOYNITE mono. *Am. Min.* **57**, 1913
$(Na,Ca)_3Zr_2Si_8O_{22}\cdot 8H_2O$ *Can. Min.* **9**, 585

LENGENBACHITE *Dana,* I, 398
$Pb_6(Ag,Cu)_2As_4S_{13}$ mono.

LENOBLITE *Am. Min.* **56**, 635
$V_2O_4\cdot 2H_2O$

LEONHARDITE = partially dehydrated *Laumontite*

LEONITE mono. *Dana,* II, 450
$K_2Mg(SO_4)_2\cdot 4H_2O$

Cross-reference: *Saline Minerals.*

LEPIDOCROCITE An iron oxyhydroxide named from the Greek words meaning "scale" and "thread."

Composition: γ-FeO(OH).

Properties: Orthorhombic, *Amam;* yellow to orange red, but brownish to red in mass; strongly pleochroic; alters on heating to *maghemite,* γ=Fe_2O_3; polymorphous with *akaganeite* and *goethite,* isostructural with *boehmite.*

Occurrence: Occurs as a weathering product of iron-bearing minerals and is the pigment in brown ochre.

D. H. SPEIDEL

Reference: *Dana,* I, 642–645.

Cross-references: *Invertebrate and Plant Mineralogy; Magnetic Minerals; Minerals, Uniaxial and Biaxial; Ore Microscopy; Rock-Forming Minerals; Soil Mineralogy.*

LEPIDOLITE A fluohydroxyl aluminosilicate with potassium and lithium, belonging to the ***mica*** group and named by Klaproth in 1792 from a Greek word meaning "scale."

Composition: $K(Li,Al)_2(Si,Al)_4O_{10}(F,OH)_2$.

Properties: Monoclinic, trigonal, and orthombic; commonly in granular and scaley aggregates; H=2.5–4; G=2.8–2.9; pearly luster; color: shades of pink and purple.

Occurrence: Found in Li-rich granite pegmatites associated with other Li-bearing minerals–*amblygonite, spodumene,* and *zinnwaldite.*

Use: A source of lithium.

K. FRYE

References

Am. Min. **63,** 203, 332.
DHZ, 3, 85–91.

Cross-references: *Clinopyroxenes; Minerals, Uniaxial and Biaxial; Pegmatite Minerals; Phyllosilicates; Soil Mineralogy.*

LEPIDOMELANE = Iron-rich *biotite*

Cross-reference: *Mineral Classification Principles.*

LERMONTOVITE *Am. Min.* **43,** 379
$(U,Ca,Ce)_3(PO_4)_4 \cdot 6H_2O(?)$

LESSERITE = *Inderite* *Am. Min.* **45,** 732

LETOVICITE *Dana,* II, 397
$(NH_4)_3H(SO_4)_2$ mono.

LEUCITE A potassium aluminosilicate of the ***feldspathoid*** group, named by Werner in 1791 from the Greek word, *leukos,* meaning white.

Composition: $16[KAlSi_2O_6]$.

Properties: Tetragonal, $I4_1/a$, but cubic, $Ia3d$, above 625°C; pseudocubic crystals or disseminated grains; H=5.5–6; G=2.45–2.5; vitreous to dull luster; translucent; color: white to gray; may exsolve to a mixture of potassium ***feldspar*** and *nepheline,* called pseudoleucite.

Occurrence: It is characteristic of potassium-rich, silica-poor lavas of Tertiary to recent age notably at Nyiragongo volcano, Zaire; Ruwenzori, Uganda; West Kimberly, Australia; and in the USA in the Leucite Hills, Wyoming, and the Bearpaw Mts., Montana.

K. FRYE

Reference: DHZ, 4, 276–288.

Cross-references: *Alkali Feldspars; Feldspathoid Group; Glass, Devitrification of Volcanic; Minerals, Uniaxial and Biaxial; Refractory Minerals.*

LEUCOPHANITE
$(Na,Ca)_2BeSi_2(O,OH,F)_7$ pseudo. tetr.

LEUCOPHOENICITE *Am. Min.* **55,** 1146
$Mn_7(SiO_4)_3(OH)_2$ mono.

LEUCOPHOSPHITE *Am. Min.* **57,** 397
$KFe_2^{3+}(PO_4)_2(OH) \cdot 2H_2O$ mono.

Cross-reference: *Pegmatite Minerals.*

LEUCOSPHENITE *Am. Min.* **57,** 1801
$BaNa_4Ti_2B_2Si_{10}O_{30}$ mono.

Cross-reference: *Green River Mineralogy.*

LEUCOXENE–Fine-grained alteration products of *ilmenite*

Cross-references: *Authigenic Minerals; Metallic Minerals.*

LEVYNE A hydrated calcium aluminosilicate of ***zeolite*** group, named by Brewster in 1825 in honor of A. Lévy.

Composition: $Ca_3[Al_6Si_{12}O_{36}] \cdot 18H_2O$.

Properties: Trigonal, $R\bar{3}m$; a=13.3 Å, c=22.5 Å; H=4.5; G=2.1; luster vitreous; colorless, white, grayish, greenish, reddish, or yellowish; transparent to translucent; habit in thin tabular crystals or in sheaf-like aggregates; twinning interpenetrant; refractive indices; ϵ=1.491–1.500, ω=1.496–1.505; birefringence = 0.002–0.006; uniaxial (–). Closely related to *chabazite;* there is appreciable replacement $Si \rightleftharpoons Al$ and correspondingly $Ca \rightleftharpoons Na_2$ or K_2; minimum diameter of the largest channel in the aluminosilicate framework is 2.7 Å; molecular sieve properties very selective.

Occurrence: Uncommon; in amygdules in basalts in Iceland, the Faeroe Islands, and Northern Ireland, and in alkali basalt on Iki Island, western Japan, associated with *erionite.*

A. IIJIMA
M. UTADA

References

Smith, J. V., 1963. Structural classification of zeolites, *Mineral. Soc. Am. Spec. Paper 1,* 281–290.
Trans. Farady Soc. **55,** 1915.
Tsch. Miner. Petrol., Mitt. **22,** 117–129.

LEWISITE (***Stibiconite*** gr.) *Dana,* II, 1021
$(Ca,Fe,Na)_2(Sb,Ti)_2O_7$ iso.

LEWISTONITE = *Fluorapatite*
Min. Mag. **42,** 282

LIANDRATITE *Am. Min.* **63,** 941
$U^{6+}(Nb,Ta)_2O_8$ hexa. *Min. Mag.* **43,** 1063

LIBERITE *Am. Min.* **50,** 519
Li_2BeSiO_4 orth.

LIBETHENITE A copper hydroxylphosphate of the ***olivenite*** group named in 1823 for the locality Libethen (Lúbietová), Czechoslovakia.

Composition: $4[Cu_2(PO_4)(OH)]$.

Properties: Orthorhombic (*Pnnm*); short prismatic, bladed, or pseudo-octahedral crystals in druses and sprays; H=4; G=3.9; vitreous to greasy luster; color: light to very dark green.

Occurrence: Found in the oxidized zone of copper ore deposits with secondary copper carbonates at the type locality and near Kitwe, Zambia.

K. FRYE

References

Can. Min. **16**, 153–157.
Min. Rec. **9**, 341–346.

LIDDICOATITE (***Tourmaline*** gr.)
$Ca(Li,Al)_3Al_6B_3Si_6O_{27}(O,OH)_3(OH,F)$
trig. *Am. Min.* **62**, 1121
Min. Abst. **29**, 342

LIEBENBERGITE (***Olivine*** gr.)
$(Ni,Mg)_2SiO_4$ orth. *Am. Min.* **58**, 733

LIEBIGITE *Min. Mag.* **43**, 665
$Ca_2UO_2(CO_3)_3 \cdot 10H_2O$
orth. *USGS Bull.* **1064**, 108

LIKASITE *Am. Min.* **40**, 942; **63**, 599
$Cu_{12}(NO_3)_4(PO_4)_2(OH)_{14}$(?) orth.

LILLIANITE *Am. Min.* **54**, 579
$Pb_3Bi_2S_6$ orth. *Min. Abst.* **29**, 140

LIME *Dana*, I, 503
CaO iso.

Cross-references: *Portland Cement Mineralogy; Refractory Minerals.*

LIMONITE A mixture of cryptocrystalline *goethite* and(or) *lepidocrocite* with adsorbed water and with other minerals present in varying amounts, usually small.

Composition: $FeO(OH) \cdot nH_2O$.

Occurrence: An alteration product of iron-bearing minerals found in all types of rocks.

D. H. SPEIDEL

Reference

Phillips, W. R., and Griffen, D. T., 1981. *Optical Mineralogy: The Nonopaque Minerals.* San Francisco: Freeman, p. 48.

Cross-references: *Authigenic Minerals; Blowpipe Analysis; Fire Clays; Gossan; Metallic Minerals; Minerals, Uniaxial and Biaxial; Pigments and Fillers; Pseudomorphism; Rock-Forming Minerals; Staining Techniques.*

LINARITE A lead copper hydroxyl sulfate named by Glocker in 1839 from the type locality of Linares, Spain.

Composition: $2[PbCu(SO_4)(OH)_2]$; SO_3, 20.0%; PbO, 55.7%; CuO, 19.8%; H_2O, 4.5% by wt.

Properties: Monoclinic, $P2_1/m$; small elongated prismatic or tabular crystals, usually in crusts or indistinct crystal aggregates; vitreous to subadamantine luster; one perfect and one imperfect cleavage; brittle, conchoidal fracture; color: deep azure-blue; streak: pale blue; translucent; distinguishable from *azurite* by lack of effervescence in HCl; alters to *antlerite* and *cerussite*.

Occurrence: Widespread secondary mineral, in small amounts in the oxidized zone of lead, copper, and silver deposits, often in association with its alteration products, *aurichalcite, anglesite, chrysocolla, hemimorphite,* and other minerals of the secondary sulfide-enrichment zone. Found in Cumberland, England; Tsumeb, Namibia; Broken Hill, New South Wales, Australia; Mammoth Mine, Tiger, Arizona; and Inyo, California.

G. C. MOERSCHELL

References

Dana, II, 553–555.
Vanders, I., and Kerr, P. F., 1967. *Mineral Recognition.* New York: Wiley, 236.

LINDACKERITE *Am. Min.* **42**, 124
$H_2Cu_5(AsO_4)_4 \cdot 8\text{–}9H_2O$ mono.

LINDGRENITE *Dana*, II, 1094
$Cu_3(MoO_4)_2(OH)_2$ mono.

LINDSTRÖMITE *Am. Min.* **61**, 15
$Pb_3Cu_3Bi_7S_{15}$ orth. *Can. Min.* **15**, 527

LINNAEITE Cobalt sulfide named by Haidinger in 1845 in honor of the Swedish botanist, Linnaeus (1707–1778). The ***linnaeite*** group is structurally analogous to the ***spinel*** group.

Composition: $8[Co_3S_4]$; S, 42.1%; Co, 57.9% by wt. Co is replaced by Ni, Fe, and Cu in varying proportions.

Properties: Isometric, *Fd3m*; octahedral

crystals, also massive granular to compact; imperfect cubic cleavage; brittle; uneven to subconchoidal fracture; H=5.5; G=4.8–5.0; metallic luster; color: pale steel-gray, tarnishing to copper-red; streak: blackish gray; nearly indistinguishable from *cobaltite* and *skutterudite* without chemical tests.

Occurrence: Associated with *chalcopyrite, pyrrhotite, pyrite, millerite,* and other related sulfides in veins with *quartz* and *siderite* gangue. Large crystals are present at Katanga, Zaire; Sigegn district, Westphalia, West Germany; and Mineral Hill, Maryland.

Use: A source of cobalt and nickel.

G. C. MOERSCHELL

References

Kostov, I., 1968. *Mineralogy.* Edinburgh: Oliver and Boyd, 123–124.
Min. Mag. **43,** 733–739.

LINNAEITE GROUP–*Bornhardtite, carrollite, daubreelite, fletcherite, greigite, indite, linnaeite, polydymite, siegenite, trüstedtite, tyrrellite,* and *violarite.*

LIOTTITE (**Cancrinite** gr.) hexa. *Am. Min.* **62,** 321
$(Ca,Na,K)_{24}(Si,Al)_{36}O_{72}(SO_4,CO_3,Cl,OH)_{12}\cdot 2H_2O$ *Min. Abst.* **29,** 84

LIPSCOMBITE *Am. Min.* **38,** 612; **55,** 135
$(Fe^{2+},Mn)Fe_2^{3+}(OH)_2(PO_4)_2$ tetr.

LIROCONITE *Am. Min.* **36,** 484
$Cu_2Al(AsO_4)(OH)_4\cdot 4H_2O$ mono.

LISKEARDITE *Dana,* II, 924
$(Al,Fe)_3(AsO_4)(OH)_6\cdot 5H_2O$ orth.

LITHARGE A low-temperature lead oxide dimorphous with *massicot* from which it is derived and with which it is commonly mixed.

Composition: PbO.

Properties: Tetragonal, *P4/nmm*(?); basal tablets with {110} cleavage; H=2; G=9.13; color: yellowish red.

Occurrence: Fort Tejon, California, USA.

D. H. SPEIDEL

Reference: *Dana,* I, 514–515.

LITHIDIONITE (Litidionite) *Am. Min.* **60,** 471
$CuNaKSi_4O_{10}$ tric.

LITHIOPHILITE A lithium manganese phosphate of the ***triphylite*** group, so named for its association with other lithium-bearing minerals.

Composition: $4[LiMnPO_4]$ with Mn^{2+} replaced by Fe^{2+} (Fe>Mn, *triphylite*).

Properties: Orthorhombic, *Pmnb*; massive, cleavable to compact; H=4–5; G=3.34–3.50; vitreous to resinous luster; color: pink, yellow, or brown.

Occurrence: In granite pegmatites with other lithium and phosphate minerals.

Use: An ore of lithium.

K. FRYE

Reference

Phillips, W. R., and Griffen, D. T., 1981. *Optical Mineralogy: The Nonopaque Minerals.* San Francisco: Freeman, 79–81.

Cross-reference: *Pegmatite Minerals.*

LITHIOPHORITE *Am. Min.* **52,** 1545
$(Al,Li)Mn^{4+}O_2(OH)_2$ mono. *Can. Min.* **18,** 529

LITHIOPHOSPHATE *Am. Min.* **54,** 1467
Li_3PO_4 orth. *Min. Abst.* **30,** 354

Cross-reference: *Pegmatite Minerals.*

LIVEINGITE *Dana,* I, 462
$Pb_9As_{13}S_{28}$ mono.

LIVINGSTONITE *Min. Abst.* **27,** 133; **28,** 267
$HgSb_4S_8$ mono.

LIZARDITE (***Kaolinite-serpentine*** gr.)
$Mg_3Si_2O_5(OH)_4$ mono. *Can. Min.* **17**(4)
DHZ, 3, 170

Cross-reference: *Soil Mineralogy.*

LODESTONE = *Magnetite*

LOELLINGITE (*Löllingite*) Iron arsenide named by Haidinger in 1845 after the locality where it was discovered by Mohs in 1820, the Lölling-Hüttenberg district, Austria.

Composition: $2[FeAs_2]$; As, 72.8%; Fe, 27.2% by wt. Some Fe replaced by Co and As by S.

Properties: Orthorhombic, *Pnnm*; crystalline masses; good basal cleavage; brittle; uneven fracture; H=5–5.5; G=7.45; metallic luster; color: silver-white to steel-gray; streak: grayish black.

Occurrence: Associated with *siderite, bismuth, nickeline,* and *barite* in veins with *quartz.* Well known localities include: Ehrenfriedersdorf, East Germany, and Andreasberg,

West Germany; Fossam, Norway; Brush Creek, Gunnison County, Colorado, and Alexander County, North Carolina, USA.

Use: A source of arsenic.

G. C. MOERSCHELL

Reference: *Dana,* I, 303–307.

Cross-references: *Ore Microscopy; Pegmatite Minerals.*

LOEWEITE (Löwite) *Am. Min.* **55**, 378
$Na_{12}Mg_7(SO_4)_{13} \cdot 15H_2O$ trig.

Cross-reference: *Saline Minerals.*

LOKKAITE *Am. Min.* **56**, 1838
$(Y,Ca)_2(CO_3)_3 \cdot 2H_2O$ orth.

LOMONOSOVITE *Am. Min.* **35**, 1092
$Na_2Ti_2Si_2O_9 \cdot Na_3PO_4$
tric. *Min. Abst.* **27**, 131

LONSDALEITE *Nature* **214**, 587
C hexa. *Science* **155**, 995

Cross-references: *Allotropy; Diamond; Meteoritic Minerals; Native Elements and Alloys.*

LOPARITE = *Perovskite* rich in Nb and *REE*

LOPEZITE *Dana,* II, 645
$K_2Cr_2O_7$ tric.

LORANDITE *Dana,* I, 437
$TlAsS_2$ mono. *Min. Abst.* **27**, 113

LORANSKITE = *Euxenite*(?)

LORENZENITE RRW, 364
$Na_2Ti_2Si_2O_9$ orth.

LORETTOITE–An artifact *Am. Min.* **64**, 1303

LOSEYITE *Dana,* II, 244
$(Mn,Zn)_7(CO_3)_2(OH)_{10}$ mono.

LOUGHLINITE *Am. Min.* **45**, 270
$Na_2Mg_3Si_6O_{16} \cdot 8H_2O$ orth.

Cross-reference: *Green River Mineralogy.*

LOVDARITE orth. *Am. Min.* **59**, 874
$(Na,K,Ca)_4(Be,Al)_2Si_6O_{16} \cdot 4H_2O$

LOVERINGITE (***Crichtonite*** gr.)
$(Ca,REE)(Ti,Fe^{3+},Cr)_{21}O_{38}$ *Am. Min.* **63**, 28
trig. *Can. Min.* **17**, 635

LOVOZERITE *Am. Min.* **59**, 633
$(Na,Ca)_3(Zr,Ti)Si_6(O,OH)_{18}$ trig.

LÖWEITE = Loeweite

LUCINITE = *Variscite*

LUDLAMITE mono. *Am. Min.* **55**, 135
$(Fe^{2+},Mg,Mn)_3(PO_4)_2 \cdot 4H_2O$ *Dana,* II, 952

Cross-reference: *Pegmatite Minerals.*

LUDLOCKITE *Am. Min.* **57**, 1003
$(Fe,Pb)As_2O_6$ tric. *Min. Abst.* **29**, 342

LUDWIGITE *Dana,* II, 321
$Mg_2Fe^{3+}BO_5$ orth. *Am. Min.* **46**, 335

LUENEBURGITE (*Lüneburgite*)
$Mg_3B_2(PO_4)_2(OH)_6 \cdot 5H_2O$
mono. *Dana,* II, 385

LUESHITE (***Perovskite*** gr.)
$NaNbO_3$ mono. *Am. Min.* **46**, 1004

LUETHEITE mono. *Am. Min.* **62**, 1058
$Cu_2Al_2(AsO_4)_2(OH)_4 \cdot H_2O$ *Min. Mag.* **41**, 27

LUSAKITE = Cobaltoan *staurolite* *Am. Min.* **20**, 316

LUSUNGITE (***Crandallite*** gr.)
$(Sr,Pb)Fe_3(PO_4)_2(OH)_5 \cdot H_2O$
trig. *Am. Min.* **44**, 906

LUZONITE *Am. Min.* **42**, 766
Cu_3AsS_4 tetr.

M

MACDONALDITE — *Am. Min.* **50**, 314
$BaCa_4Si_{15}O_{35} \cdot 11H_2O$ orth.

MACEDONITE — *Am. Min.* **56**, 387
$PbTiO_3$ tetr.

MACFALLITE — *Am. Min.* **65**, 406
$Ca_2(Mn,Al)_3(SiO_4)(Si_2O_7)(OH)_3$
mono. *Min. Mag.* **43**, 325

MACHATSCHKIITE — *Am. Min.* **62**, 1260
$Ca_3(AsO_4)_2 \cdot 9H_2O$ trig. *Min. Abst.* **29**, 200

MACKAYITE — *Am. Min.* **55**, 1072
$FeTe_2O_5(OH)$ tetr. *Min. Abst.* **28**, 394

MACKINAWITE — *Am. Min.* **48**, 215
$(Fe,Ni)_9S_8$ tetr. *Min. Mag.* **42**, 516

Cross-references: *Lunar Minerals; Meteoritic Minerals; Soil Mineralogy.*

MACKINTOSHITE = *Thorogummite*
Am. Min. **38**, 1007

MADOCITE — *Am. Min.* **53**, 1421
$Pb_{17}(Sb,As)_{16}S_{41}$ orth. *Can. Min.* **9**, 191

MAGADIITE — *Am. Min.* **54**, 1034; **60**, 642
$NaSi_7O_{13}(OH)_3 \cdot 3H_2O$ mono.(?)

MAGBASITE — *Am. Min.* **51**, 530
$KBa(Al,Sc)(Mg,Fe)_6Si_6O_{20}F_2$

MAGHAGENDORFITE — *Am. Min.* **65**, 810
$NaMn(Mg,Fe)_3(PO_4)$ mono. *Min. Mag.* **43**, 227

MAGHEMITE Ferric oxide of the *magnetite* series of the ***spinel*** group, named by Wagner in 1927 from the first syllables of *magnetite* and *hematite*.

Composition: γ-Fe_2O_3.

Properties: Isometric, *Fd3m*; defect spinel structure, isostructural with γ-Al_2O_3; H=5; G=4.88; strongly magnetic; inverts to *hematite* on heating.

Occurrence: *Maghemite* results from the oxidation of *magnetite* or the dehydration of *lepidocrocite*.

D. H. SPEIDEL

Reference: *Dana,* I, 708–709.

Cross-references: *Magnetic Minerals; Ore Microscopy; Spinel Group; Soil Mineralogy.*

MAGNESIOAXINITE (***Axinite*** gr.)
$Ca_2MgAl_2(BO_3)(Si_4O_{12})(OH)$
tric. *Am. Min.* **61**, 503

MAGNESIOCARPHOLITE — *Am. Min.* **65**, 406
$MgAl_2Si_2O_6(OH)_4$ orth.

MAGNESIOCHROMITE (***Spinel*** gr.)
$MgCr_2O_4$ iso. *Dana,* I, 709

Cross-reference: *Refractory Minerals.*

MAGNESIOCOPIAPITE tric. *Am. Min.* **58**, 314
$MgFe^{3+}_4(SO_4)_6(OH)_2 \cdot 2OH_2O$ *Dana,* II, 623

MAGNESIOFERRITE Mangesium iron oxide member of the *magnetite* series of the ***spinel*** group, named by Dana in 1892 for its composition.

Composition: $MgFe_2O_4$.

Properties: Cubic, *Fd3m*; G=4.52; *n*=2.38; color: black to brownish black; slightly magnetic.

Occurrence: Pure end member is rare; found in volcanic regions.

D. H. SPEIDEL

References

Am. Min. **52,** 1139–1152.
Ark. Min. Geol. **5**, 1–10.

Cross-reference: *Refractory Minerals.*

MAGNESIOKATOPHORITE (***Amphibole*** gr.)
$Na_2CaMg_4(Fe^{3+},Al)AlSi_7O_{22}(OH,F)_2$
mono. DHZ, 2, 359

MAGNESIORIEBECKITE (***Amphibole*** gr.)
$Na_2(Mg,Fe^{2+},Fe^{3+})_5Si_8O_{22}(OH)_2$
mono. DHZ, 2, 333

MAGNESITE Magnesium carbonate of the

calcite group, named by Karsten in 1808 for its composition.

Composition: $2[MgCO_3]$; with up to total replacement of Mg by Fe toward *siderite* and partial replacement by Ni toward *gaspeite*.

Properties: Trigonal, $R\bar{3}c$; crystals rare, commonly massive, granular, compact, porcellanous, earthy, chalky, or fibrous; H=3.7–4.3; G=3.0; perfect rhombohedral $\{10\bar{1}1\}$ cleavage; vitreous luster; transparent to translucent; color: white, grayish, yellowish to brown.

Occurrence: Found commonly in veins derived from the alteration of magnesium-rich rocks (e.g., serpentinite or peridotite); in beds of crystalline masses in magnesium-rich schists; or of authigenic origin as a replacement of *calcite*-bearing rocks. Major localities include: Carinthia and Styria, Austria; Gömör, Czechoslovakia; Macedonia, Greece; Piedmont and Tuscany, Italy; Almeria and Santander, Spain; Katanga, Zaire; Sheng-king, Manchuria; and in the USA in the Coast Ranges and Sierra Nevadas, California; Paradise Range, Nevada; and in the counties of Dona Ana, Grant, and Eddy, New Mexico.

K. FRYE

Reference: *Dana,* II, 162–166.

Cross-references: *Authigenic Minerals; Blowpipe Analysis; Calcite Group; Cave Minerals; Fire Clays; Green River Mineralogy; Meteoritic Minerals; Minerals, Uniaxial and Biaxial; Refractory Minerals; Rock-Forming Minerals; Soil Mineralogy; Staining Techniques; Synthetic Minerals.*

MAGNESIUM CHLOROPHOENICITE
$(Mg,Mn)_5(AsO_4)(OH)_7$ mono. *Dana,* II, 780

MAGNESIUM ZIPPEITE
$Mg_2(UO_2)_6(SO_4)_3(OH)_{10}\cdot 16H_2O$
orth. *Can. Min.* **14**, 429
Min. Mag. **43**, 1063

MAGNETITE Ferrous ferric oxide of the ***spinel*** group. One of the most abundant and ubiquitous oxide minerals and igneous and metamorphic rocks. Named by Haidinger in 1845 from the locality, Magnesia, bordering on Macedonia.

Composition: Fe_3O_4; variations $(Fe,Mg,Ni,Zn,Mn)(Fe,Ti)_2O_4$.

Properties: Isometric, $Fd3m$; G=5.2; magnetic, octahedral crystals; properties vary with composition; n=2.42 (varies with composition); black with black streak; opaque; gray in reflected light; end member of the *magnetite* solid-solution series; see *magnesioferrite, franklinite, jacobsite, trevorite, cuprospinel* and *ulvöspinel;* stable form of iron oxide, in air, above 1388°C; composition dependent on oxygen content of environment; varies with rock type.

Occurrence: An accessory mineral in many igneous rocks, sometimes concentrated in segregations (Bushveld); skarn deposits associated with *andradite* and ***pyroxene;*** thermally metamorphosed sediments (Norway); common detrital mineral (see *Blacksand Minerals*).

Use: Ore of iron.

D. H. SPEIDEL

References

Am. Min. **52,** 1139–1152.
Econ. Geol. **61,** 798.
J. Petrology **5,** 310.

Cross-references: *Abrasive Materials; Blacksand Minerals; Crystal Growth; Crystallography: Morphological; Diamond; Fire Clays; Glass, Devitrification of Volcanic; Invertebrate and Plant Mineralogy; Magnetic Minerals; Mantle Mineralogy; Metallic Minerals; Metallurgy; Meteoritic Minerals; Mineral and Ore Deposits; Mohs Scale of Hardness; Ore Microscopy; Pigments and Fillers; Pseudomorphism; Refractory Minerals; Rock-Forming Minerals; Soil Mineralogy; Synthetic Minerals; Thermometry, Geologic.*

MAGNETOPLUMBITE A lead manganese iron oxide named by Aminoff in 1925 in reference to its lead content and magnetic properties.

Composition: $Pb(Mn,Fe)_6O_{10}$.

Properties: Hexagonal; strongly magnetic; black with dark brown streak.

Occurrence: Associated with manganoan *biotite* at Långban in Vermland, Sweden.

D. H. SPEIDEL

References: *Am. Min.* **36,** 512–514; **53,** 869–879.

MAGNIOBORITE = *Suanite* *Am. Min.* **48,** 915

MAGNIOPHILITE = *Beusite* *Am. Min.* **53,** 1799

MAGNIOTRIPLITE *Am. Min.* **37,** 359
$(Mg,Fe,Mn)_2(PO_4)F$ mono.

MAGNOCOLUMBITE *Am. Min.* **48,** 1182
$(Mg,Fe,Mn)(Nb,Ta)_2O_6$ orth.

MAGNOPHORITE = Titanian potassian *richterite* *Am. Min.* **63,** 1051

MAGNUSSONITE — *Am. Min.* **64**, 390
$Mn_5(AsO_3)_3(OH,Cl)$ iso. — *Min. Abst.* **27**, 258

MAJAKITE — *Am. Min.* **62**, 1260
PdNiAs hexa. — *Min. Abst.* **28**, 449

MAJORITE (***Garnet*** gr.) — *Am. Min.* **55**, 1815
$Mg_3(Fe,Al,Si)_2(SiO_4)_3$ iso. — *Science* **168**, 832

Cross-reference: *Meteoritic Minerals.*

MAKATITE — *Am. Min.* **55**, 358
$Na_2Si_4O_9 \cdot 5H_2O$ orth.

MÄKINENITE — *Am. Min.* **50**, 520
γ-NiSe trig.

MALACHITE A copper hydroxyl carbonate named by Pliny in 77 AD from the Greek word for mallows in reference to its color.

Composition: $4[Cu_2CO_3(OH)_2]$; CuO, 71.9%; CO_2, 19.9%; H_2O, 8.2% by wt.

Properties: Monoclinic, $P2_1/a$; commonly in radiating fibers forming stalactitic or botryoidal masses; H=3.5–4; G=4.05; silky, dull, or earthy; translucent; color: bright green; streak: pale green.

Occurrence: A supergene mineral found in the oxidized portions of copper deposits, especially those in limestone, rarely an authigenic mineral. Associated with *azurite, cuprite,* iron oxides, native *copper,* and the various sulfides of copper and iron.

Uses: Ore of copper and as an ornamental material.

K. FRYE

Reference: *Dana,* II, 252–256.

Cross-references: *Authigenic Minerals; Blowpipe Analysis; Crystal Habits; Gemology; Gossan; Metallic Minerals; Minerals, Uniaxial and Biaxial; Museums, Mineralogical; Pseudomorphism.*

MALACON–Metamict *zircon*

MALANITE — *Am. Min.* **61**, 185
$(Cu,Pt,Ir)S_2$ iso.

MALAYAITE — *Am. Min.* **62**, 801
$CaSnSiO_5$ mono. — *Min. Mag.* **35**, 622

MALDONITE — *Dana,* I, 95
Au_2Bi iso.

Cross-reference: *Native Elements and Alloys.*

MALLADRITE — *Dana,* II, 105
Na_2SiF_6 trig.

MALLARDITE — *Dana,* II, 507
$MnSO_4 \cdot 7H_2O$ mono.

MANANDONITE (***Chlorite*** gr.)
$LiAl_4BSi_3O_{10}(OH)_8$ mono. — RRW, 378

MANASSEITE A hydrated magnesium aluminum hydroxyl carbonate named by Frondel in 1941 for Professor Ernesto Manasse of the University of Florence.

Composition: $Mg_6Al_2(OH)_{16}(CO_3) \cdot 4H_2O$.

Properties: Hexagonal; basal plates with basal cleavage; white to bluish color; polymorphic with *hydrotalcite.*

Occurrence: With *serpentine* at the Kongsberg, Norway, and Amity, New York, USA.

D. H. SPEIDEL

Reference: *Dana,* I, 658–659.

MANDARINOITE — *Am. Min.* **65**, 206
$Fe_2^{3+}Se_3O_9 \cdot 4H_2O$ mono. — *Can. Min.* **16**, 605

MANGANAXINITE — *Am. Min.* **65**, 1119
$Ca_2MnAl_2BSi_4O_{15}(OH)$ tric. — DHZ, 1, 320

MANGANBABINGTONITE — *Am. Min.* **53**, 1064
$Ca_2(Mn,Fe^{2+})Fe^{3+}Si_5O_{14}(OH)$ tric.

MANGANBELYANKINITE
$(Mn,Ca)(Ti,Nb)_5O_{12} \cdot 9H_2O$
amorph. — *Am. Min.* **43**, 1220

MANGANBERZELIITE
$(Ca,Na)_3(Mn,Mg)_2(AsO_4)_3$ — *Dana,* II, 681
iso. — *Min. Abst.* **30**, 292

MANGANESE-HOERNESITE
$(Mn,Mg)_3(AsO_4)_2 \cdot 8H_2O$
mono. — *Am. Min.* **39**, 159

MANGANESE-SHADLUNITE
(***Pentlandite*** gr.) — *Am. Min.* **58**, 1114
$(Mn,Pb,Cd)(Fe,Cu)_8S_8$ iso.

MANGANHUMITE (***Humite*** gr.)
$(Mn,Mg)_7(OH)_2(SiO_4)_3$ — *Min. Abst.* **29**, 200
orth. — *Min. Mag.* **42**, 133

MANGANITE A manganese oxyhydroxide named by Haidinger in 1827 for its composition.

Composition: MnO(OH).

Properties: Monoclinic, $B2_1/d$; crystals with many faces; perfect {010} cleavage; steel gray to black with reddish brown streak. Structurally related to *boehmite*.

Occurrence: *Manganite* is a hydrothermal vein mineral associated with *pyrolusite* and *goethite*. Prominent localities include: the Harz mountains, Wesphalia, and Rhineland, West Germany; Ilmenau, East Germany; Ariège and Hautes-Pyrénées, France; Cumberland and Devonshire, England; Aberdeen, Scotland; and in the USA in the Lake Superior iron district, Michigan; Alameda County, California; Shenandoah County, Virginia; and Bartow County, Georgia.

D. H. SPEIDEL

Reference: *Dana,* I, 646–650.

Cross-reference: *Magnetic Minerals.*

MANGAN-NEPTUNITE *Can. Min.* **7**, 679
$Na_2KLi(Mn,Fe)Ti_2Si_8O_{24}$ mono.

MANGANOCHROMITE *Am. Min.* **63**, 1166
$(Mn,Fe)(Cr,V)_2O_4$ iso. *Min. Mag.* **43**, 1064

MANGANOCOLUMBITE
$(Mn,Fe^{2+})(Nb,Ta)_2O_6$ orth.

MANGANOLANGBEINITE *Dana,* II, 435
$K_2Mn_2(SO_4)_3$ iso.

MANGANOPHYLLITE = Manganoan *biotite*

MANGANOSITE A manganese oxide named by Blomstrand in 1874 for its composition.

Composition: MnO.

Properties: Isometric, *Fm3m*; isostructural with *halite*; H=5.6; G=5.0–5.4; emerald green to black with brown streak.

Occurrence: With other manganese minerals at Franklin, New Jersey, USA, and Långban, Sweden.

D. H. SPEIDEL

Reference: *Dana,* I, 501–502.

MANGANOSTIBITE *Am. Min.* **55**, 1489
$(Mn,Fe)_7Sb^{5+}As^{5+}O_{12}$ orth.

MANGANOTANTALITE *Dana,* I, 780
$MnTa_2O_6$ orth.

MANGANPYROSMALITE
$(Mn,Fe^{2+})_8Si_6O_{15}(OH,Cl)_{10}$
hexa. *Am. Min.* **38**, 755

MANJIROITE *Am. Min.* **53**, 2103
$(Na,K)Mn_8O_{16} \cdot nH_2O$ tetr.

MANSFIELDITE *Dana,* II, 763
$Al(AsO_4) \cdot 2H_2O$ orth.

MAPIMITE mono. *Min. Mag.* **43**, 1064
$Zn_2Fe_3^{3+}(AsO_4)_3(OH)_4 \cdot 10H_2O$

MARCASITE An iron sulfide of the ***marcasite*** group, named by Haidinger in 1845 from a word of Arabic or Moorish origin formerly used for a variety of minerals.

Composition: FeS_2; Fe, 46.6%; S, 53.4% by wt., generally quite pure.

Properties: Orthorhombic, *Pnnm;* crystals commonly twinned and in radiating or cockscomb shapes, also stalactitic with an inner core, radiating structure; and covered on the outside with irregular crystal faces, also globular or reniform, intergrown or overgrown on *pyrite*; H=6–6.5; G=4.9; metallic luster; pale bronze yellow to nearly white on fresh fracture.

Occurrence: Most commonly as an authigenic mineral or replacement in sedimentary rocks (e.g., limestone, clays, marls, or shales) formed at low temperatures from acid solutions; also as a supergene mineral associated with lead and zinc ores.

K. FRYE

References

Am. Min. **63,** 210.
Dana, I, 311–315.

Cross-references: *Authigenic Minerals; Gossan; Green River Mineralogy; Ore Microscopy; Polymorphism; Pseudomorphism.*

MARCASITE GROUP—*Ferroselite, frohbergite, hastite, kullerudite, marcasite,* and *mattagamite.*

MARGARITE A calcium hydroxyl aluminosilicate of the ***mica*** group, named from the Greek word meaning pearl in allusion to its luster.

Composition: $4[CaAl_2(Al_2Si_2O_{10})(OH)_2]$.

Properties: Monoclinic, *C2/c*; crystals uncommon, usually in foliated masses; H=3.5–4.5; G=3.0–3.1; brittle folia; pearly to vitreous luster; color: grayish pink, pale yellow, to pale green; one perfect cleavage {001}.

Occurrence: Associated with and probably and alteration product of *corundum,* it is found in the Ural mountains, USSR; in emery deposits in the Cyclades Islands, Greece; in Trentino-Alto Adige; Italy; and in the USA near Chester, Massachusetts; in

Chester and Delaware Counties, Pennsylvania; and in Clay County, North Carolina.

K. FRYE

References

Am. Min. **63**, 186.
DHZ, 3, 95–98.

Cross-references: *Minerals, Uniaxial and Biaxial; Phyllosilicates; Soil Mineralogy.*

MARGAROSANITE *Am. Min.* **49**, 781
$Pb(Ca,Mn)_2Si_3O_9$ tric.

MARIALITE (***Scapolite*** gr.)
$3NaAlSi_3O_8 \cdot NaCl$ tetr. DHZ, 4, 321

MARICITE *Am. Min.* **64**, 655
$NaFe(PO_4)$ orth. *Can. Min.* **15**, 396

MARIGNACITE = *Ceriopyrochlore*
Am. Min. **62**, 406

MARIPOSITE = Chromian siliceous *muscovite*

MARMATITE = Ferroan *sphalerite*

MAROKITE *Am. Min.* **49**, 817
$CaMn_2^{3+}O_4$ orth.

MARRITE *Am. Min.* **50**, 812
$PbAgAsS_3$ mono.

MARSHITE *Dana*, II, 20
CuI iso.

MARSTURITE *Am. Min.* **63**, 1187
$Mn_3CaNaHSi_5O_{15}$ tric. *Min. Mag.* **43**, 1064

MARTHOZITE orth. *Am. Min.* **55**, 533
$Cu(UO_2)_3(SeO_3)_3(OH)_2 \cdot 7H_2O$

MARTITE = *Hematite* pseudomorph of *magnetite*

Cross-reference: *Pseudomorphism.*

MASCAGNITE *Dana*, II, 398
$(NH_4)_2SO_4$ orth.

MASKELYNITE = Glass formed from a ***plagioclase***

Cross-reference: *Meteoritic Minerals.*

MASLOVITE (***Pyrite*** gr.) *Am. Min.* **65**, 406
PtBiTe iso.

MASSICOT A lead oxide, the name being derived from a French word for its composition.

Composition: PbO.

Properties: Orthorhombic, *Pcma*; H=2; G=9.3–9.6; massive; perfect {100} cleavage; yellow with light yellow streak; high-temperature polymorphous with *litharge*.

Occurrence: In the oxidized zone of lead-ore deposits such as at Leadville, Colorado, USA.

D. H. SPEIDEL

Reference: *Dana*, I, 516–517.

MASUTOMILITE (***Mica*** gr.)
$K_2(LiMnAl)_2(AlSi_3)_2O_{20}(F,OH)_4$
mono. *Am. Min.* **62**, 594
Min. Abst. **30**, 297

MASUYITE *Am. Min.* **45**, 1026
Hydrous oxide U orth. *Min. Mag.* **41**, 51

MATILDITE *Dana*, I, 429
$AgBiS_2$ hexa.

MATLOCKITE A lead chlorofluoride named after the original locality, Matlock, in Derbyshire, England.

Composition: 2[PbFCl].

Properties: Tetragonal, *P4/nmm*; tabular crystals or coarsely lamellar; basal cleavage; adamantine to pearly luster; color: yellow or greenish.

Occurrence: Alteration of *galena* or of lead-bearing slags.

K. FRYE

Reference: *Dana*, I, 59–60.

MATRAITE *Am. Min.* **45**, 1131
ZnS trig.

MATTAGAMITE (***Marcasite*** gr.)
$(Co,Fe)Te_2$ *Am. Min.* **59**, 382
orth. *Can. Min.* **12**, 55

MATTEUCCITE *Am. Min.* **39**, 848
$NaHSO_4 \cdot H_2O$ mono.

MAUCHERITE *Dana*, I, 192
$Ni_{11}As_8$ tetr.

Cross-reference: *Ore Microscopy.*

MAUFITE *Am. Min.* **15**, 275
$(Mg,Ni)Al_4Si_3O_{13} \cdot 4H_2O(?)$ *Min. Mag.* **22**, 624

MAWSONITE *Am. Min.* **50**, 900
$Cu_6Fe_2SnS_8$ tetr. *Can. Min.* **17**, 125

MAYENITE *Am. Min.* **50**, 2106
$Ca_{12}Al_{14}O_{33}$ iso.

MAZZITE (***Zeolite*** gr.)
$K_2CaMg_2(Al,Si)_{36}O_{72}\cdot 28H_2O$
hexa. *Am. Min.* **60**, 340
Cont. Min. Pet. **45**, 99

MBOZIITE = Potassian *taramite*
Am. Min. **63**, 1051

MCALLISTERITE *Am. Min.* **52**, 1776
$Mg_2B_{12}O_{20}\cdot 15H_2O$ trig.

MCCONNELLITE *Am. Min.* **62**, 593
$CuCrO_2$ trig. *GSA Abst.* (1967), 151

MCGILLITE *Can. Min.* **18**, 31
$Mn_8Si_6O_{15}(OH)_8Cl_2$ trig. *Min. Mag.* **43**, 1063

MCGOVERNITE trig. *Am. Min.* **63**, 150;
$(Mn,Mg)_{19}Zn_3(OH)_{21}(AsO_3)$ **65**, 957
$(AsO_4)_3(SiO_4)_3$(?)

MCGUINNESSITE *Min. Rec.* **12**, 143
$(Mg,Cu)_2CO_3(OH)_2$ mono.

MCKELVEYITE *Am. Min.* **56**, 2156
$NaCaBa_3(CO_3)_6\cdot 3H_2O$
pseudo. trig. *Can. Min.* **16**, 335

Cross-reference: *Green River Mineralogy.*

MCKINSTRYITE *Am. Min.* **52**, 1253
$(Ag,Cu)_2S$ orth. *Econ. Geol.* **61**, 1383

MEDMONTITE–Mixture of
chrysocolla and ***mica*** *Am. Min.* **54**, 994

MEERSCHAUM = *Sepiolite*

MEIONITE (***Scapolite*** gr.) *Can. Min.* **17**, 53
$3CaAl_2Si_2O_8\cdot CaCO_3$ tetr. DHZ, 4, 321

Cross-reference: *Vein Minerals.*

MEIXNERITE *Am. Min.* **61**, 176
$Mg_6Al_2(OH)_{18}\cdot 4H_2O$
trig. *Tsch. Min. Pet.* **22**, 79

MELANITE = Titanian *andradite*

MELANOCERITE hexa. RRW, 388
$(Ce,Ca)_5(Si,B)_3O_{12}(OH,F)\cdot nH_2O$(?)

MELANOPHLOGITE A low-temperature cubic polymorph of silica; it is yellow to brownish and on heating turns black, hence the name, from the Greek *melan,* black, and *phlogos,* to be burnt.

Composition: SiO_2 but always with (C+H+S) at least 6%; SiO_2, 88–94%; SO_3, 5–8%; H, 1–2%; C, 1–3% by wt.

Properties: Cubic; H=6.5–7; G=2.05; luster is vitreous; color: usually yellowish to brownish, also colorless or white; habit: regular cubes up to 2 mm or more commonly as rounded aggregates; regular color zoning is generally present parallel to the cube faces; refractive index = 1.467 on pigmented crystals, 1.425 after heating to break down organic matter; the organic matter is anisotropic causing the *melanophlogite* to exhibit a weak birefringence. Insoluble in acids except HF; stable to 800°C, but above 900°C it is converted to *cristobalite.*

Occurrence: As coatings on crystals as sulfur and other minerals in the sulfur deposits of Sicily, associated with *sulfur, celestite, calcite, opal, quartz,* and bitumen.

R. A. HOWIE

References

Am. Min. **48**, 854–867.
Bull. Soc. fr. **13**, 356–372.
Dana, III, 283–284.

MELANOSTIBITE *Am. Min.* **53**, 1104
$Mn(Sb,Fe)O_3$ trig.

MELANOTEKITE *Am. Min.* **52**, 1085
$Pb_2Fe_2^{3+}Si_2O_9$ orth.

MELANOVANADITE *Dana,* II, 1058
$Ca_2V_4^{4+}V_6^{5+}O_{25}\cdot nH_2O$ tric.

MELANTERITE *Can. Min.* **17**, 65
$FeSO_4\cdot 7H_2O$ mono. *Dana,* II, 499

Cross-reference: *Halotrichite Group.*

MELILITE An alkali magnesium iron aluminosilicate named by Delametherie in 1796 from the Greek for honey, in allusion to its color. It is the principal member of the ***melilite*** group which has the end members *gehlenite* and *åkermanite.*

Composition:
$2[(Ca,Na,K)_2(Mg,Fe^{2+},Fe^{3+},Al)(Si,Al)_2O_7]$

Properties: Tetragonal, $P\bar{4}2_1m$; usually in short square prisms or rectangular prisms; brittle, conchoidal fracture; H=5–6; G=2.95–3.05; vitreous to resinous luster; color: honey-yellow, brown, green-brown; streak: colorless; sometimes shows optical anomalies.

Occurrence: Characteristic of high-temperature mineral associations in skarns and basalts, also in furnace slags. Localities with *melilite*-rich rocks include Iron Hill, Gunnison County, Colorado, USA, and Scawt Hill, Antrim, Northern Ireland.

G. C. MOERSCHELL

References

Carnegie Inst. Wash. Yr. Book 76, 478–485.
DHZ, 1, 236–255.

Cross-references: *Lunar Minerals; Meteoritic Minerals.*

MELILITE GROUP–*Akermanite, gehlenite, hardystonite,* and *melilite.*

MELIPHANITE RRW, 390
$(Ca,Na)_2Be(Si,Al)_2(O,OH,F)_7$ tetr.

MELKOVITE *Am. Min.* **55**, 320
$CaFe^{3+}H_6(MoO_4)_4(PO_4)\cdot 6H_2O$

MELLITE tetr. *Dana,* II, 1104
$Al_2C_6(COO)_6\cdot 18H_2O$ *Min. Abst.* **27**, 257

MELNIKOVITE = *GREIGITE*
GREGITE *Am. Min.* **49**, 543

MELONITE *Min. Mag.* **43**, 775
$NiTe_2$ hexa.

MELONJOSEPHITE orth. *Am. Min.* **60**, 946;
$CaFe^{2+}Fe^{3+}(PO_4)_2(OH)$ **62**, 60

MENDIPITE *Min. Mag.* **43**, 901
$Pb_3Cl_2O_2$ orth.

MENDOZITE *Dana,* II, 469
$NaAl(SO_4)_2\cdot 11H_2O$ mono.

MENEGHINITE *Can. Min.* **16**, 393
$CuPb_8(Sb,Pb)_{12}S_{24}$ orth.

MERCALLITE *Dana,* II, 395
$KHSO_4$ orth. *Min. Abst.* **28**, 260

MERCURY *Dana,* I, 103
Hg liquid (above –39°C)

MERENSKYITE *Am. Min.* **52**, 926
$(Pd,Pt)(Te,Bi)_2$ hexa. *Min. Mag.* **35**, 815

MERLINOITE (***Zeolite*** gr.)
$(K_2,Ca,Ba,Na_2)_{4.5}[Al_9Si_{23}O_{64}]\cdot 23H_2O$
orth. *Am. Min.* **63**, 598
Min. Abst. **29**, 84

MERRIHUEITE (***Osumilite*** gr.)
$(K,Na)_2(Fe^{2+},Mg)_5Si_{12}O_{30}$
Am. Min. **50**, 2096
hexa. *Science* **149**, 972

Cross-reference: *Meteoritic Minerals.*

MERTIEITE (I and II) *Am. Min.* **58**, 1;
$Pd_5(Sb,As)_2$ mono. and trig. **61**, 1249

MERWINITE *Am. Min.* **57**, 1355;
$Ca_3Mg(SiO_4)_2$ mono. **59**, 1117

Cross-reference: *Rock-Forming Minerals.*

MESOLITE A hydrated sodium calcium aluminosilicate of ***zeolite*** group, first named by Fuchs and Gehlen in 1813 from the Greek for middle stone in allusion to its intermediate chemistry between *natrolite* and *scolecite.*

Composition: $Na_2Ca_2[Al_2Si_3O_{10}]_3\cdot 8H_2O$; SiO_2, 46.5%; Al_2O_3, 26.2%; CaO, 9.6%; Na_2O, 5.3%; H_2O, 12.4% by wt.

Properties: Monoclinic (pseudo-orthorhombic), *C*2 (pseudo *Fdd*2); a=3×18.9 Å, b=6.55 Å, c=18.48 Å, β=90°; H=5; G=2.26; luster vitreous, but silky in fibrous masses; colorless, white, gray, yellow, pink, or red; transparent to translucent; habit in invariably twinned prisms or in fine fibrous aggregates elongated along the *y* axis, often intergrown with *natrolite, scolecite,* or *thomsonite;* cleavage perfect on {101} and {10$\bar{1}$}; twinning always on {100}. Optically, refractive indices, β=1.504–1.508; birefringence = 0.001; 2V=80°(+); optic axial plane (010); β=y, α:z=8°; dispersion $r<v$, strong, Chemically, a solid solution of varying amounts of *natrolite* and *scolecite;* considerable replacement of Si by Al; minor substitution of the type $Ca\rightleftharpoons Na_2$; small amounts of K may substitute for Na; base exchanged by K, Li, Ag, NH_4, Tl, and Na; molecular sieve properties good but not too selective due to adaptability of the aluminosilicate framework; gelatinized by HCl.

Occurrences: In cavities in basalt and andesite in association with *calcite* and other ***zeolites*** such as *chabazite, stilbite, natrolite,* and *analcime;* in Iceland, in cavities of the Tertiary basalts as the lower *mesolite-scolecite* zone; in geodes, in phonolite in Sardinia; in hydrothermal veins, associated with manganese ores and barium minerals of Achinsk in Siberia. Also occurs on the Cyclopean Islands, Sicily; Scotland; Ireland; Faeroe Islands; Greenland; in India in the western Ghats; Richmond, Victoria, Australia. In the USA it occurs on Fritz Island in the Schuylkill River, Pennsylvania, and at Table Mountain, Colorado. Also in Nova Scotia from the region of the Bay of Fundy.

A. IIJIMA
M. UTADA

References

Foster, M. D., 1965. Studies of the zeolites of the natrolite group, *U.S. Geol. Surv. Prof. Paper 504-D*, 7p.
J. Geol. **68,** 515–528.
Min. Mag. **23,** 421–447.

Cross-reference: *Zeolites.*

MESSELITE tric. *Am. Min.* **40**, 828; **44**, 469
$Ca_2(Fe^{2+},Mn)(PO_4)_2 \cdot 2H_2O$

META-ALUMINITE *Am. Min.* **53**, 717
$Al_2(SO_4)(OH)_4 \cdot 5H_2O$ mono.

META-ALUNOGEN RRW, 395
$Al_4(SO_4)_6 \cdot 27H_2O$ mono.

META-ANKOLEITE (***Meta-autunite*** gr.)
$K_2(UO_2)_2(PO_4)_2 \cdot 6H_2O$
tetr. *Am. Min.* **52**, 560

META-AUTUNITE (***Meta-autunite*** gr.)
$Ca(UO_2)_2(PO_4)_2 \cdot 2\text{–}6H_2O$
Am. Min. **48**, 1389
tetr. *Dana,* II, 984

META-AUTUNITE GROUP—*Abernathyite, bassetite, meta-ankoleite, meta-autunite, meta-heinrichite, metakahlerite, metakirchheimerite, metalodevite, metanovacekite, metatorbernite, meta-uranocircite, meta-uranospinite, meta-zeunerite, sodium uranospinite,* and *uramphite.*

METABORITE *Am. Min.* **50**, 261
HBO_2 iso.

METACALCIOURANOITE *Am. Min.* **58**, 1111
$(Ca,Na,Ba)U_2O_7 \cdot 2H_2O$

METACALTSURANOITE = METACALCIOURANOITE

METACINNABAR *Am. Min.* **63**, 1143
HgS iso *Dana,* I, 215

METADELRIOITE *Am. Min.* **55**, 185
$CaSrV_2O_6(OH)_2$ tric.

METAHAIWEEITE mono.(?) *Am. Min.* **44**, 839
$Ca(UO_2)_2Si_6O_{15} \cdot nH_2O$

METAHEINRICHITE (***Meta-autunite*** gr.)
$Ba(UO_2)_2(AsO_4)_2 \cdot 8H_2O$
tetr. *Am. Min.* **43**, 1134

METAHEWETTITE *Am. Min.* **44**, 322
$CaV_6O_{16} \cdot 9H_2O$ orth. *Min. Mag.* **43**, 550

METAHOHMANNITE *Dana,* II, 608
$Fe_2(SO_4)_2(OH)_2 \cdot 3H_2O$

METAKAHLERITE (***Meta-autunite*** gr.)
$Fe^{2+}(UO_2)_2(AsO_4)_2 \cdot 8H_2O$
tetr. *Am. Min.* **45**, 254

METAKIRCHHEIMERITE (***Meta-autunite*** gr.)
$Co(UO_2)_2(AsO_4)_2 \cdot 8H_2O$
tetr.(?) *Am. Min.* **44**, 466

METALODEVITE (***Meta-autunite*** gr.)
$Zn(UO_2)_2(AsO_4)_2 \cdot 10H_2O$
tetr. *Am. Min.* **59**, 210

METANOVACEKITE (***Meta-autunite*** gr.)
$Mg(UO_2)_2(AsO_4)_2 \cdot 4\text{–}8H_2O$
Min. Abst. **27**, 80
tetr. RRW, 398

METAROSSITE *Dana,* II, 1054
$CaV_2O_6 \cdot 2H_2O$ tric.

METASCHODERITE *Am. Min.* **47**, 637
$Al_2(PO_4)(VO_4) \cdot 6H_2O$ mono.

METASCHOEPITE *Am. Min.* **45**, 1026
$UO_3 \cdot 1\text{–}2H_2O$ orth.

METASIDERONATRITE
$Na_4Fe_2^{3+}(SO_4)_4(OH)_2 \cdot 3H_2O$
orth. *Dana,* II, 603

METASTIBNITE *Am. Min.* **55**, 2104
Sb_2S_3 amorph.

METATORBERNITE *Metatorbernite I* and *II* are low temperature, partially dehydrated forms of *torbernite,* a hydrous phosphate of copper and uranium named after Swedish chemist Torbern Bergmann.

Composition: $Cu(UO_2)_2(PO_4)_2 \cdot nH_2O$ (n=8 or less).

Properties: Tetragonal, $P4/n$; tabular; G= 3.5–3.7; vitreous to pearly luster; micaceous cleavage on {001}; color: emerald to apple green; streak: pale green; usually occurs in tabular masses; brittle; refractive indices ω=1.59, ϵ=1.58; distinct dichroism; not fluorescent; translucent to transparent.

Occurrence: A secondary mineral, resulting from the dehydration of *torbernite,* found in granite pegmatites associated with copper sulfides and *uraninite.*

W. E. HILL

References

Am. Min. **49**, 1578–1621.
Berry, L. G., and Mason, B., 1959. *Mineralogy.* San Francisco: Freeman, 458–459.

Cross-reference: *Dispersion, Optical.*

METATYUYAMUNITE *Am. Min.* **41**, 187
$Ca(UO_2)_2(VO_4)_2 \cdot 3\text{-}5H_2O$
orth. *USGS Bull.* **1064**, 254

META-URANOCIRCITE (***Meta-autunite*** gr.)
$Ba(UO_2)_2(PO_4)_2 \cdot 8H_2O$ *Min. Abst.* **17**, 695
orth. *USGS Bull.* **1064**, 211

META-URANOPILITE *Am. Min.* **37**, 350
$(UO_2)_6(SO_4)(OH)_{10} \cdot 5H_2O$

META-URANOSPINITE (***Meta-autunite*** gr.)
$Ca(UO_2)_2(AsO_4)_2 \cdot 8H_2O$
tetr. *Am. Min.* **45**, 254

METAVANDENDRIESSCHEITE
$PbU_7O_{22} \cdot nH_2O$ *Am. Min.* **45**, 1026

METAVANURALITE tric. *Am. Min.* **56**, 637
$Al(UO_2)_2(VO_4)_2(OH) \cdot 8H_2O$

METAVARISCITE *Dana,* II, 767
$AlPO_4 \cdot 2H_2O$ mono.

METAVAUXITE mono. *Dana,* II, 971
$Fe^{2+}Al_2(PO_4)_2(OH)_2 \cdot 8H_2O$

METAVIVIANITE *Am. Min.* **59**, 896
$Fe_3^{2+}(PO_4)_2 \cdot 8H_2O$ tric.

METAVOLTINE hexa. *Min. Abst.* **28**, 396
$K_5[Fe_3^{3+}O(H_2O)_3(SO_4)_6] \cdot 7H_2O$
Min. Mag. **43**, 669

METAZELLERITE *Am. Min.* **51**, 1567
$Ca(UO_2)(CO_3)_2 \cdot 3H_2O$ orth.

METAZEUNERITE (***Meta-autunite*** gr.)
$Cu(UO_2)_2(AsO_4)_2 \cdot 8H_2O$ *Am. Min.* **42**, 222
tetr. *USGS Bull.* **1064**, 215

MEYERHOFFERITE *Am. Min.* **38**, 912
$Ca_2B_6O_{11} \cdot 7H_2O$ tric. *Dana,* II, 356

MEYMACITE *Am. Min.* **53**, 1065
$WO_3 \cdot 2H_2O$ amorph.

MIARGYRITE A silver antimony sulfide, the name (1829) indicating that it contains less (*meion*) silver (*argyros*) than *pyrargyrite* or *proustite.*

Composition: $8[AgSbS_2]$ with minor Cu substitution for Ag.

Properties: Monoclinic, *Aa*; thick tabular crystals with striated faces; one imperfect cleavage; H=2.5; G=5.2; metallic adamantine luster; iron black to steel gray; red streak.

Occurrence: In low-temperature hydrothermal veins with other silver sulfosalts, other sulfides, *calcite,* and *quartz.*

Use: An ore of silver.

K. FRYE

References

Am. Min. **60**, 621–633.
Dana, I, 424–427.

Cross-reference: *Ore Microscopy.*

MICA GROUP–*Anandite, annite, biotite, bityite, brammallite, celadonite, chernykhite, clintonite, ephesite, glauconite, hendricksite, hydromica, illite, kinoshitalite, lepidolite, margarite, masutomilite, montdorite, muscovite, paragonite, phlogopite, polylithionite, preiswerkite, roscoelite, siderophyllite, taeniolite, tarasovite, wonesite, zinnwaldite.*

Cross-references: *Crystallography: Morphological; Diagenetic Minerals; Pegmatite Minerals; Phyllosilicates; Refractory Minerals; Rock-Forming Minerals; Soil Mineralogy; Synthetic Minerals.*

MICHENERITE (***Pyrite*** gr.) *Am. Min.* **44**, 207
(Pd,Pt)BiTe iso *Can. Min.* **11**, 903

MICROCLINE Potassium aluminosilicate of the ***feldspar*** group. The name is derived from the Greek words *mikros,* little, and *klinein,* incline, since the angle between the cleavages deviates from 90° by as much as ½°.

Composition: $4[KAlSi_3O_8]$, always with small amounts of Na, Ca, Ba, Rb, Cs, or Fe.

Properties: Triclinic, $C\bar{1}$; crystals prismatic parallel to *z* or to *x* with monoclinic habit or pseudorhombohedra with triclinic (adularia) habit, also in cleavable masses, irregular grains, or as cores mantled by sodic ***plagioclase*** (rapakivi texture); prismatic cleavage with {001} perfect and {010} good; combined albite- and pericline-law twinning yields typical tartan pattern (authigenic *microcline* untwinned); H=6; G=2.56; commonly intergrown with *albite* (crypto- to macroperthite), less commonly with *quartz* (graphic granite); color: white or shades of yellow, red, or green; streak: white; polymorphic with *sanidine* and *orthoclase*.

Occurrence: A major rock-forming mineral in plutonic rocks and pegmatites, in gneisses and as porphyroblasts in other metamorphic rocks, in arkoses as a detrital mineral, and authigenic on detrital *microcline* cores.

K. FRYE

References

Can. Min. **17**, 515–525.
DHZ, 4, 1–93.
Ribbe, P. H., ed., 1975. *Feldspar Mineralogy.* Washington, D. C.: Mineral Soc. Am. Short Course notes, vol. 2.
Smith, J. V., 1974. *Feldspar Minerals,* vols. 1 and 2. New York: Springer-Verlag.

Cross-references: *Alkali Feldspars; Crystallography: Morphological; Feldspar Group; Gemology; Minerals, Uniaxial and Biaxial; Order–Disorder; Polymorphism; Soil Mineralogy; Rock-Forming Minerals.*

MICROLITE (***Pyrochlore*** gr.)
$(Na,Ca)_2Ta_2O_6(O,OH,F)$ *Am. Min.* **62**, 405
iso. *USGS Bull.* **1064**, 326

Cross-reference: *Pegmatite Minerals.*

MICROSOMMITE = DAVYNE(?)
Can. Min. **17**, 49

MIERSITE *Dana,* II, 19
(Ag,Cu)I iso.

MIHARAITE *Am. Min.* **65**, 784
$Cu_4FePbBiS_6$ orth.

MILARITE (***Osumilite*** gr.)
$K_2Ca_4Be_4Al_2Si_{24}O_{60}\cdot H_2O$
hexa. *Can. Min.* **18**, 41

MILLERITE A nickel sulfide named by Haidinger in 1845 after W. H. Miller who first studied the crystals.

Composition: 9[NiS]; Ni, 64.7%, S, 35.3% by wt.

Properties: Trigonal, $R3m$; characteristically in hair-like tufts and radiating groups of slender to capillary crystals; perfect rhombohedral cleavage $\{10\bar{1}1\}$ and $\{01\bar{1}2\}$; H=3.0–3.5; G=5.5; metallic luster; color: pale brass to bronze yellow commonly tarnished to irridescence; streak: greenish black.

Occurrence: Forms at low temperatures in rock cavities and carbonate veins as an alteration product of other Ni minerals or as inclusions in other minerals. Found in Jáchymov, Czechoslovakia; Dillenburg, Saarbrücken, and Siegen, GFR; Freiberg, GDR; Glamorgan, Wales, UK; Malartic, Quebec, and Sudbury, Ontario, Canada; and in the USA at Antwerp, New York; Lancaster County, Pennsylvania; Keokuk, Iowa; St. Louis, Missouri; and Milwaukee, Wisconsin.

K. FRYE

References

Can. Min. **12**, 248–257.
Dana, I, 239–241.

Cross-references: *Magnetic Minerals; Ore Microscopy; Pegmatite Minerals; Refractory Minerals; Thermometry, Geologic.*

MILLISITE tetr.(?) *Am. Min.* **64**, 626
$(Ca,Na)_2Al_6(PO_4)_4(OH,O)_9\cdot 3H_2O$

MILLOSEVICHITE *Am. Min.* **59**, 1140
$(Al,Fe)_2(SO_4)_3$

MIMETITE A lead chloroarsenate of the ***apatite*** group, named by Beudant in 1832 from the Greek word meaning imitator in reference to its resemblance to *pyromorphite.*

Composition: $Pb_5Cl(AsO_4)_3$; with (AsO_4) replaced by (PO_4), $(PO_4)>(AsO_4)$ is *pyromorphite.*

Properties: Monoclinic, $P2_1/b$ (pseudohexagonal); barrel-shaped crystals common, acicular crystals less common, also globular, reniform, or botryoidal; H=3.5–4; G=7.3; resinous to subadamantine luster; transparent to translucent; color: pale yellow, yellowish brown, or orange yellow; streak: white.

Occurrence: A secondary mineral found in the oxidized parts of lead deposits associated with *cerussite* and limonite; localities include Baden and Hesse, West Germany; Saxony, East Germany; Příbram, Czechoslovakia; Nerchinsk, USSR; Puy-de-Dome, France; Cornwall, Cumberland, and Leadhills, UK; Långban, Sweden; Tsumeb, Namibia; and in the USA in Pinal and Yavapai Counties, Arizona; Inyo County, California; Tintic, Utah; and Eureka, Nevada.

K. FRYE

Reference: *Am. Min.* **54**, 993.

MINASRAGRITE *Am. Min.* **58**, 531
$VO(SO_4)\cdot 5H_2O$ mono.

MINGUZZITE *Am. Min.* **41**, 370
$K_3Fe^{3+}(C_2O_4)_3\cdot 3H_2O$ mono.

MINIUM A lead oxide, the name was originally applied to *cinnabar* which could be adulterated with the "red lead."

Composition: Pb_3O_4.

Properties: Tetragonal, *P4/mbc*; found only as fine powder; color: vivid red; streak: orange yellow.

Occurrence: Found as an alteration product of *galena* or *cerussite* as at Leadville, Colorado, or Badenweiler, West Germany.

D. H. SPEIDEL

Reference: *Dana,* I, 517–519.

MINNESOTAITE *Am. Min.* **50**, 148; **65**, 15
$(Fe,Mg)_3Si_4O_{10}(OH)_2$ mono.

Cross-reference: *Soil Mineralogy.*

MINYULITE *Am. Min.* **62**, 256
$KAl_2F(H_2O)_4(PO_4)_2$ orth. *Can. Min.* **17**, 99

MIRABILITE A hydrated sodium sulfate named by Haidinger in 1845 alluding to German chemist J. R. Glauber's (1603–1668) expression of surprise at the formation of the salt.

Composition: $[Na_2SO_4 \cdot 10H_2O]$.

Properties: Monoclinic, $P2_1/a$; short prismatic to acicular crystals, massive, as efflorescent crusts, granular, stalactitic; perfect {100} cleavage; conchoidal fracture; H=1.5–2; G=1.49; colorless and transparent to white and opaque; white streak; cool taste, then weakly saline and bitter.

Occurrence: Typical evaporite deposit in saline lakes, playas, and hot springs, and as a soil efflorescence. Low solubility at low temperatures means that it is often deposited in the winter from saline bodies. Associated with *thenardite, gypsum,* and *halite* at the Karabergas Gulf, Caspian Sea, and the Great Salt Lake, Utah, USA.

Use: Mined as Glauber Salt.

G. C. MOERSCHELL

References

Dana, II, 439–442.

Kostov, I., 1968. *Mineralogy.* Edinburgh: Oliver and Boyd, 502–503.

Cross-references: *Cave Minerals, Saline Minerals.*

MISENITE *Can. Min.* **11**, 569
$K_2SO_4 \cdot 6KHSO_4$(?) mono.

MISERITE tric. *Can. Min.* **14**, 515
$KCa_5(Si_2O_7)(Si_6O_{15})(OH)F$
Min. Abst. **29**, 20

MISPICKEL = *ARSENOPYRITE*
Min. Mag. **43**, 1053

MITHRAIL–Species status dubious
Min. Abst. **28**, 118

MITRIDATITE mono. *Am. Min.* **64**, 169
$Ca_3Fe_4^{3+}(PO_4)_4(OH)_6 \cdot 3H_2O$
Min. Mag. **40**, 863

Cross-reference: *Pegmatite Minerals.*

MITSCHERLICHITE *Dana,* II, 100
$K_2CuCl_4 \cdot 2H_2O$ tetr.

MIXITE hexa. *Dana,* II, 943
$BiCu_6(AsO_4)_3(OH)_6 \cdot 3H_2O$

MIZZONITE–Intermediate composition in the ***scapolite*** gr. DHZ, 4, 322

MOCTEZUMITE *Am. Min.* **50**, 1158
$Pb(UO_2)(TeO_3)_2$ mono.

MOHRITE mono. *Am. Min.* **50**, 805
$(NH_4)_2Fe^{2+}(SO_4)_2 \cdot 6H_2O$

MOISSANITE (Carborundum)
α-SiC hexa. *Dana,* I, 123

Cross-references: *Lunar Minerals; Meteoritic Minerals; Native Elements and Alloys.*

MOLURANITE amorph. *Am. Min.* **45**, 258
$H_4U^{4+}(UO_2)_3(MoO_4)_7 \cdot 18H_2O$

MOLYBDENITE-2H Molybdenum sulfide named by Dana in 1837; the name *molybdenite* originally applied to the element now called molybdenum but derived from the Greek word for lead.

Composition: $2[MoS_2]$; Mo, 59.9%; S, 40.1% by wt.

Properties: Hexagonal, *C6/mmc*; commonly foliated, massive or in scales; perfect basal {0001} cleavage; H=1–1.5; G=4.62–4.73; lamellae flexible but not elastic; greasy feel; metallic luster; lead gray with greenish streak on glazed porcelain; polymorphic with *molybdenite-3R* and *jordisite.*

Occurrence: Found as an accessory mineral in some granites and associated aplites and pegmatites; in deep veins associated with *topaz, scheelite, wolframite,* and *fluorite,* and in contact metamorphic deposits; *Molybdenite* was the major ore of molybdenum at Climax, Colorado, USA.

K. FRYE

References

Am. Min. **64**, 758–775.
Dana, I, 328–331.

Cross-references: *Blowpipe Analysis; Gossan; Metallic Minerals; Mineral and Ore Deposits; Ore Microscopy.*

MOLYBDENITE-3R *Am. Min.* **64**, 758
MoS_2 trig. *Can. Min.* **7**, 524

MOLYBDITE *Am. Min.* **49**, 1497
MoO_3 orth.

MOLYBDOMENITE *Am. Min.* **50**, 812
$PbSeO_3$ mono. *Can. Min.* **8**, 149

MOLYBDOPHYLLITE RRW, 413
$Pb_2Mg_2Si_2O_7(OH)_2$ hexa.

MOLYSITE *Dana*, II, 47
$FeCl_3$ trig.

MONAZITE A phosphate of cerium and rare earths, named by Breithaupt in 1829 from the Greek meaning to be solitary, in allusion to its supposed scarcity.

Composition: (Ce,La,Y,Th)PO_4; usually the La earths and Ce are in the ratio approximately 1:1; ThO_2 content may be as high as 12%.

Properties: Monoclinic; generally of tabular habit; H=5–5.5; G=4.6–5.4; luster resinous to waxy, color: yellow to brown; cleavage poorly developed. Optically, biaxial positive, n_m=1.79–1.80, birefringence = 0.05–0.06, $2V$ small, not notably pleochroic; slowly decomposed by acids; isostructural with *huttonite* and *crocoite*.

Occurrence: Widely disseminated as an accessory in granitic rocks; in relatively large crystals in pegmatites; greatly concentrated in some detrital deposits and beach sands (see *Blacksand Minerals*). In pegmatites associated with *gadolinite, samarskite,* and *columbite;* in sands and detrital deposits with *zircon,* ***apatite,*** *titanite (sphene),* and *magnetite.*

Uses: An important source of thorium and of rare earths.

A. PABST

Reference: *Am. Min.* **44**, 510–532.

Cross-references: *Blacksand Minerals; Metallic Minerals; Mineral and Ore Deposits; Pegmatite Minerals; Pleochroic Halos; Refractory Minerals; Rock-Forming Minerals.*

MONCHEITE *Am. Min.* **48**, 1181
(Pt,Pd)(Te,Bi)$_2$ hexa.

MONETITE *Dana*, II, 660
$CaHPO_4$ tric. *Min. Abst.* **30**, 20

Cross-references: *Cave Minerals; Human and Vertebrate Mineralogy.*

MONHEIMITE = Composition between *siderite* and *smithsonite*

MONIMOLITE (***Roméite*** gr.) *Dana*, II, 1023
(Pb,Ca)$_2Sb_2O_7$ iso.

MONOHYDROCALCITE *Am. Min.* **49**, 1151
$CaCO_3 \cdot H_2O$ hexa.

Cross-reference: *Human and Vertebrate Mineralogy.*

MONSMEDITE *Am. Min.* **54**, 1496
$H_8K_2Tl_2^{3+}(SO_4)_8 \cdot 11H_2O$ iso.

MONTANITE *Dana*, II, 636
$Bi_2TeO_6 \cdot 2H_2O$ mono.(?)

MONTBRAYITE *Am. Min.* **57**, 146
Au_2Te_3 tric.

MONTEBRASITE A lithium aluminum hydroxylphosphate of the ***amblygonite*** series, named for the original locality of Montebras, Creuse, France.

Composition: 2[LiAl(PO_4)(OH)], with Li replaced by Na toward *natromontebrasite* and (OH) by F toward *amblygonite*.

Properties: Triclinic, $P\bar{1}$; equant crystals, rough if large, also cleavable masses, columnar, compact; one perfect, one good, and one distinct cleavage; H=5.5–6; G=3.0; vitreous to greasy luster; white to yellowish, pinkish, greenish, or bluish.

Occurrence: In granite pegmatites some giant crystals and masses of hundreds of tons mined near Keystone, South Dakota, USA, with other lithium-bearing minerals.

Use: An ore of lithium.

K. FRYE

References:

Am. Min. **63**, 1249; 58, 291.
Dana, II, 823–827.

MONTEPONITE *Dana*, I, 502
CdO iso.

MONTEREGIANITE
$(Na,K)_6(Y,Ca)_2Si_{16}O_{38} \cdot 10H_2O$
orth. *Am. Min.* **65**, 207
Can. Min. **16**, 561

MONTESITE = Plumboan *herzenbergite*
Am. Min. **60**, 163

MONTGOMERYITE mono. *Am. Min.* **59**, 843
$Ca_4Mg(H_2O)_{12}[Al_4(OH)_4(PO_4)_6]$ **61**, 12

Cross-reference: *Pegmatite Minerals.*

MONTICELLITE A magnesium calcium silicate of the ***olivine*** group, named for T. Monticelli, Italian mineralogist.

Composition: $4[MgCaSiO_4]$, with small amounts of Fe replacing Mg, Mn replacing Ca.

Properties: Orthorhombic, *Pnma;* small prismatic crystals or grains; H=5.5; G=3.1–3.3; colorless to gray.

Occurrence: From contact metamorphism of siliceous dolomitic marble, also in ultramafic rocks, localities include Vesuvius and near Monzoni, Italy; Shannon Tier, Tasmania; and in the USA at Keene, New York; Magnet Cove, Arkansas; and Crestmore, California.

K. FRYE

Reference: DHZ, 1, 41–46.

MONTMORILLONITE A hydrated aluminosilicate of calcium, sodium, magnesium, and iron; chief member of the ***smectite*** group; and name for the original locality, Montmorillon, France.

Composition:
$Na_{0.7}(Al_{3.3}Mg_{0.7})(Si_8O_{20})(OH)_4 \cdot nH_2O$

Properties: Monoclinic; fine-grained aggregates; H=1–2; G=2–3; perfect basal {001} cleavage; white, yellow, or greenish; stacking of layers disordered or interstratified with ***chlorites;*** layers expansive by water up-take; cations exchangeable.

Occurrence: Constituent of bentonite clays formed by alteration of mafic volcanic ash; weathering product of mafic igneous rocks; hydrothermal alteration product of ***plagioclases.***

Uses: Drilling muds and as fillers.

K. FRYE

Reference: DHZ, 3, 226–245.

Cross-references: *Authigenic Minerals; Clays, Clay Minerals; Diagenetic Minerals; Electron Microscopy (Transmission); Fire Clays; Green River Mineralogy; Phyllosilicates; Pigments and Fillers; Smectite Group; Soil Mineralogy; Staining Techniques.*

MONTROSEITE A vanadium oxyhydroxide first named by Weeks, Cisney, and Sherwood in 1953 from the locality Montrose County, Colorado, USA.

Composition: VO(OH); some Fe commonly replaces V.

Properties: Orthorhombic, *Pbnm*; G=4.0–4.2; submetallic luster, color: black; streak: black; habit: bladed microscopic crystals; cleavage good parallel to length of blade {010} and {110}; brittle; opaque. Distinguished from other black vanadium minerals by its X-ray powder pattern. Soluble in H_2SO_4; oxidizes to *paramontroseite* (VO_2).

Occurrence: As fine-grained mixtures disseminated in sandstone, believed to be a primary mineral. Widespread deposits found on the Colorado (USA) Plateau uranium-vanadium ores, and associated with pitchblende, *coffinite, pyrite, galena,* and other sulfides.

Uses: Ore of vanadium.

L. H. FUCHS

References: *Am. Min.* **38**, 1235; **40**, 861.

MONTROYDITE *Dana*, I, 511
HgO orth.

MOOIHOEKITE *Am. Min.* **57**, 689
$Cu_9Fe_9S_{16}$ tetr.

MOONSTONE—Bluish variety of alkali ***feldspar***

Cross-reference: *Gemology.*

MOOREITE mono. *Am. Min.* **54**, 973
$(Mn,Zn,Mg)_8(SO_4)(OH)_{14} \cdot 4H_2O$

MOORHOUSEITE *Am. Min.* **50**, 808
$(Co,Ni,Mn)SO_4 \cdot 6H_2O$ mono. *Can. Min.* **8**, 166

MORAESITE *Am. Min.* **38**, 1126
$Be_2(PO_4)(OH) \cdot 4H_2O$ mono.

MORDENITE A hydrated sodium potassium calcium aluminosilicate of the ***zeolite*** group, named by How in 1864 after the original locality near Morden, Nova Scotia.

Composition: $(Na_2,K_2,Ca)[Al_2Si_{10}O_{24}] \cdot 7H_2O$; SiO_2, 67.2%; Al_2O_3, 11.4%; CaO, 2.1%;

Na_2O, 3.5%; K_2O, 1.7%; H_2O, 14.1% by wt. Here $Na_2:K_2:Ca=3:1:2$.

Properties: Orthorhombic, *Cmcm* or *Cmc2;* a=18.13 Å; b=20.49 Å, c=7.52 Å; G=2.12–2.15; H=3–4; luster vitreous; colorless, white, yellow, red, or brown; transparent on the edge; habit acicular or fibrous elongated in z direction. Optically, refractive indices, α=1.472–1.483, β=1.475–1.485, γ=1.477–1.487; birefringence = 0.004–0.005; $2V$= 76°–90° (±); optic axial plane (100); $\alpha=z$, $\beta=x$, $\gamma=y$. Alkalis usually dominant over Ca, but Ca-rich *mordenite* is not uncommon; K generally less abundant than Na; Si:Al ratio is also variable; dehydration gradual, losing 16% H_2O at maximum; rehydration good; molecular sieve properties selective for NH_3, CO_2, SO_2, etc.; diameter of the largest channels is 6.6 Å, which is reduced to about 4 Å by stacking faults in the aluminosilicate framework; Li, Na, K, NH_4^+, Ca, and Ba salts give the base-exchange products; not attacked by HCl.

Occurrence: In veins and amygdules in rhyolite and andesite, sometimes associated with *quartz;* less commonly in vesicles of basalt and dolerite; alteration product of volcanic glass of silicic vitric tuffs, associated with *clinoptilolite, opal,* and *montmorillonite;* in bentonitic clays in the Tertiary tuffaceous sediments in Japan; in cement of sandstone and in chalks as authigenic minerals.

A. IIJIMA
M. UTADA

References

GCA **19**, 145–146.
Mineral J. (Japan), **2**, 196–199.
Z. Krist. **115**, 439–450.

Cross-references: *Green River Mineralogy; Minerals, Uniaxial and Biaxial; Zeolites.*

MORENOSITE *Dana*, II, 516
$NiSO_4 \cdot 7H_2O$ orth.

MORGANITE–Gem variety of pink *beryl*

Cross-references: *Gemology; Pegmatite Minerals.*

MORINITE *Am. Min.* **43**, 585
$Ca_2Na[Al_2F_4(OH)(H_2O)_2(PO_4)_2]$
mono. *Can. Min.* **17**, 93

Cross-references: *Apatite Group; Pegmatite Minerals.*

MORION = Smoky *quartz*

MOSANDRITE tric. RRW, 420
$(Ca,Na)_3(Ti,Zr)(F,O)_2(Si_2O_7)$

MOSCHELLANDSBERGITE *Dana*, I, 103
Ag_2Hg_3 iso.

Cross-reference: *Native Elements and Alloys.*

MOSESITE iso. *Am. Min.* **38**, 1225
$Hg_2N(SO_4,MoO_4,Cl) \cdot H_2O$

MOSSITE = Variety of *columbite*
Am. Min. **65**, 814

Cross-reference: *Blacksand Minerals.*

MOTTRAMITE A copper lead hydroxylvandate of the ***olivenite*** group, named in 1876 after the locality Mottram, St. Andrew, Cheshire, England.

Composition: $4[PbCu(VO_4)(OH)]$, with Cu replaced by Zn (Zn >Cu, *descloizite*).

Properties: Orthorhombic, *Pnam*; crystals varied and drusy; commonly massive or mammiliform; H=3–3.5; G=5.9–6.2; greasy luster; color: green or reddish brown; streak: yellowish to brownish red.

Occurrence: A secondary mineral in the oxidized zone of ore deposits, some pseudomorphous after *vanadinite*.

K. FRYE

References

Dana, II, 811–815.
Min. Mag. **31**, 289.

MOTUKOREAITE hexa. *Am. Min.* **63**, 598
$NaMg_{19}Al_{12}(CO_3)_{6.5}(SO_4)_4(OH)_{54} \cdot 28H_2O$
Min. Mag. **41**, 389

MOUNANAITE *Am. Min.* **54**, 1738
$PbFe_2^{3+}(VO_4)_2(OH)_2$ tric.

MOUNTAINITE mono. *Am. Min.* **43**, 624
$(Ca,Na_2,K_2)_2Si_4O_{10} \cdot 3H_2O$
Min. Mag. **31**, 611

MOURITE *Am. Min.* **56**, 163
$U^{4+}Mo_5^{6+}O_{12}(OH)_{10}$ mono.

MPOROROITE *Am. Min.* **58**, 1112
$(Al,Fe)_2W_2O_9 \cdot 6H_2O$ mono.

MROSEITE *Am. Min.* **61**, 339
$CaCO_3TeO_2$ orth. *Can. Min.* **13**, 286

MUIRITE tetr. *Am. Min.* **50**, 1500
$Ba_{10}Ca_2MnTiSi_{10}O_{30}(OH,Cl,F)_{10}$

MUKHINITE (***Epidote*** gr.)
$Ca_2(Al_2V)O(Si_2O_7)(SiO_4)(OH)$
mono. *Am. Min.* **55**, 321

MULLITE An aluminum silicate named after the first recorded natural occurrence in Mull, Scotland.

Composition: $3Al_2O_3 \cdot 2SiO_2$; synthetic *mullite* may be richer in Al_2O_3 with a formula near $2Al_2O_3 \cdot SiO_2$.

Properties: Orthorhombic, *Pbam;* long prismatic crystals or fibrous aggregates; cleavage, distinct on {010}; H=6–7; G=3.15–3.26; vitreous luster; color: white or pale pink; streak: white; refractive indices, α=1.640–1.670, β=1.642–1.675, γ=1.651–1.690; $2V_\gamma$= 45°–61°; pleochroism α=β colorless, γ pinkish. Difficult to distinguish from *sillimanite* except by X-ray cell dimensions or infrared absorption spectra. Insoluble in most acids; the theoretical composition has 60 mol% Al_2O_3, but natural and synthetic *mullites* have up to 63 mol% Al_2O_3 and synthetic *mullite* can contain 67 mol% Al_2O_3; Fe^{3+} and Ti may substitute for Al.

Occurrence: In pelitic xenoliths (buchites) in mafic igneous rocks; iron-*mullite* has been described from a thermally metamorphosed lateritic lithomarge. Associated with *spinel* and ***plagioclase.***

Uses: *Kyanite* (or the other Al_2SiO_5 polymorphs, *andalusite* and *sillimanite*) can be converted to a *mullite* and glass refractory product by calcining at high temperatures.

R. A. HOWIE

References

Am. Min. **38,** 725.
J. Am. Ceram. Soc. **43,** 69.
J. Wash. Acad. Sci. **14,** 183.

Cross-references: *Mineral Properties; Portland Cement Mineralogy; Refractory Minerals; Rock-Forming Minerals; Synthetic Minerals.*

MURATAITE iso. *Am. Min.* **59**, 172
$(Na,Y)_4(Zn,Fe)_3(Ti,Nb)_6O_{18}(F,OH)_4$

MURDOCHITE *Am. Min.* **40**, 905
$PbCu_6(O,Cl,Br)_8$ iso.

MURMANITE tric. *Am. Min.* **48**, 1413
$Na_2(Ti,Nb)_2Si_2O_9 \cdot nH_2O$

MUSCOVITE A hydroxyl potassium aluminosilicate of the ***mica*** group, named by Dana in 1850 from its popular name, muscovy-glass, after the Russian Province, Muscovy.

Composition: $2[K_2Al_4[Si_6Al_2O_{20}](OH,F)_4$; with K replaced by Na, Ba, and H_3O^+; octahedral Al by Li, V, Cr, Mn, Mg, and Fe^{2+}; and tetrahedral Al by Si.

Properties: Monoclinic, *C2/c*; most commonly foliated in sheets and scales, also large cleavable blocks or tabular crystals, may be cryptocrystalline or massive; perfect basal {001} cleavage; H=2.5–3; G=2.77–2.88; vitreous to pearly luster; colorless or light shades of yellow, brown, green, or red; transparent in thin sheets; electrically and thermally insulating.

Occurrence: Of wide-spread distribution as a rock-forming mineral, characteristic of silica-rich plutonic rocks and their pegmatites; the chief constituent of mica schists; an alteration product of ***feldspars*** (sericite) from weathering and hydrothermal action; as a detrital mineral in clastic sediments; and an authegenic mineral.

K. FRYE

References*

DHZ, 3, 11–30.
*See *Phyllosilicates* for additional references.

Cross-references: *Authigenic Minerals; Clays, Clay Minerals; Crystal Habits; Mica Group; Mineral Classification: Principles; Mineral and Ore Deposits; Minerals, Uniaxial and Biaxial; Pegmatite Minerals; Phyllosilicates; Pigments and Fillers; Pleochroic Halos; Rock-Forming Minerals; Soil Mineralogy.*

MUSKOXITE trig. *Am. Min.* **54**, 684
$Mg_7Fe_4^{3+}O_{13} \cdot 10H_2O$

MUTHMANNITE *Dana,* I, 260
(Ag,Au)Te

N

NACAPHITE Am. Min. **66**, 218
$Na_2Ca(PO_4)F$ orth. *Min. Mag.* **43**, 1064

NACRITE—Polymorphous with *kaolinite* DHZ, 3, 191

Cross-references: *Clays, Clay Minerals; Phyllosilicates; Soil Mineralogy; Staining Techniques.*

NADORITE *Dana*, II, 1039
$PbSbO_2Cl$ orth.

NAGELSCHMIDTITE *Am. Min.* **63**, 425
$Ca_{3-4}(Si,P)_2O_8$ *Min. Abst.* **29**, 481

NAGYAGITE *Dana*, I, 168
$Pb_5Au(Te,Sb)_4S_{5-8}$ orth.(?)

NAHCOLITE *Dana*, II, 134
$NaHCO_3$ mono.

Cross-references: *Green River Mineralogy; Saline Minerals.*

NAKAURIITE orth. *Am. Min.* **62**, 594
$(Mn,Ni,Cu)_8(SO_4)_4(CO_3)(OH)_6 \cdot 48H_2O$
Min. Abst. **31**, 497

NAMBULITE tric. *Am. Min.* **58**, 1112
$NaLiMn_8Si_{10}O_{28}(OH)_2$ *Min. Abst.* **30**, 289

NANLINGITE *Am. Min.* **62**, 1058
$CaMg_4(AsO_3)_2F_4$ trig.

NANTOKITE *Dana*, II, 18
CuCl iso.

NARSARSUKITE *Am. Min.* **47**, 539
$Na_4Ti_2O_2(Si_8O_{20})$ tetr.

NASINITE orth. *Am. Min.* **48**, 709
$Na_2[B_5O_8(OH)] \cdot 2H_2O$ *Min. Abst.* **28**, 260

NASLEDOVITE *Am. Min.* **44**, 1325
$PbMn_3Al_4(CO_3)_4(SO_4)O_5 \cdot 5H_2O$

NASONITE *Am. Min.* **56**, 1174
$Pb_6Ca_4Si_6O_{21}Cl_2$ hexa.

NATISITE *Am. Min.* **61**, 339
$Na_2TiOSiO_4$ tetr.

NATROALUNITE (***Alunite*** gr.) *Dana*, II, 555
$NaAl_3(SO_4)_2(OH)_6$ trig. *Min. Abst.* **28**, 79

NATROAPOPHYLLITE *Am. Min.* **66**, 410
$NaCa_4Si_8O_{20}F \cdot 8H_2O$ orth.

NATROCHALCITE *Dana*, II, 602
$NaCu_2(SO_4)_2(OH) \cdot H_2O$ mono.

NATROFAIRCHILDITE = NYEREREITE(?)
$Na_2Ca(CO_3)_2$ orth. *Am. Min.* **60**, 488

NATROJAROSITE (***Alunite*** gr.)
$NaFe^{3+}_3(SO_4)_2(OH)_6$ trig. *Dana*, II, 563

NATROLITE A hydrated sodium aluminosilicate of the ***zeolite*** group, named by Klaproth in 1803 from the Greek, "soda stone."

Composition: $Na_{16}[Al_{16}Si_{24}O_{80}] \cdot 16H_2O$; SiO_2, 47.4%; Al_2O_3, 26.8%; Na_2O, 16.3%; H_2O, 9.5% by wt.

Properties: Orthorhombic (pseudo-tetragonal), *Fdd2*; a=18.30Å, b=18.63Å, c=6.60Å; H=5; G=2.20–2.26; luster vitreous, but pearly in fibrous masses; colorless, white, grayish, or yellowish; transparent to translucent; habit in slender prisms, in fine needles, or in radiating clusters elongated in the z direction, cleavage very good on {110} and {$1\bar{1}0$}; twinning on {110}, {011} and {031}, rare; shows pyroelectricity and piezoelectricity. Optically, refractive indices, α=1.473–1.483, β=1.476–1.486, γ=1.485–1.496; birefringence = 0.012; $2V$=58°–64° (+); optic axial plane (010); $\alpha=x$, $\beta=y$, $\gamma=z$. Small amounts of K and Ca may substitute for Na; dehydration sudden, losing 12% H_2O at maximum; rehydration poor; Na is exchanged with Li, Ag, K, NH_4, and Tl; molecular sieve properties good but not too selective due to adaptability of the aluminosilicate framework; transition to metanatrolite at 400°–600°C; gelatinized by HCl.

Occurrence: Late ***zeolite*** to crystallize in cavities in basalt, associated with *analcime, chabazite, calcite*, etc.; alteration product of *nepheline* or *sodalite* in nepheline syenite and phonolite; on joint surfaces in gneiss; reaction product of nephelinite glass and *nepheline* with percolating groundwater, forming the cement of nephelinite palagonite tuffs on Oahu, Hawaii, in association with *phillipsite, chabazite, gonnardite*, and *calcite*.

A. IIJIMA
M. UTADA

References

Am. Min. **56**, 560–569.
Foster, M. D., 1965. Studies of the zeolites of the natrolite group, *U.S. Geol. Surv. Prof. Paper 504-D*, 7p.
Hay, R. L., and Iijima, A., 1967. Nature and origin of zeolitic palagonite tuffs of the Honolulu Series on Oahu, Hawaii, *Geol. Soc. Am. Memoir. 116*, 331–376.
Z. Krist. **113**, 430–444.

Cross-references: *Authigenic Minerals; Green River Mineralogy; Minerals, Uniaxial and Biaxial; Zeolites.*

NATROMONTEBRASITE (***Amblygonite*** gr.)
(Na,Li)Al(PO_4)(OH,F) tric. *Dana,* II, 823

NATRON *Dana*, II, 230
$Na_2CO_3 \cdot 10H_2O$ mono.

Cross-references: *Green River Mineralogy; Minerals, Uniaxial and Biaxial.*

NATRONIOBITE *Am. Min.* **47**, 1483
$NaNbO_3$ mono.(?)

NATROPHILITE *Am. Min.* **50**, 1096
$NaMnPO_4$ orth.

Cross-reference: *Pegmatite Minerals.*

NATROPHOSPHATE iso. *Am. Min.* **58**, 139
$Na_6H(PO_4)_2(F,OH) \cdot 17H_2O$

NATROSILITE mono. *Am. Min.* **61**, 339
$Na_2Si_2O_5$

NAUJAKASITE mono. *Am. Min.* **53**, 1780
$Na_6(Fe,Mn,Ca)(Al,Fe)_4Si_8O_{26}$
Min. Abst. **27**, 131

NAUMANNITE *Am. Min.* **35**, 337
Ag_2Se orth.

Cross-reference: *Ore Microscopy.*

NAVAJOITE *Am. Min.* **44**, 322
$V_2O_5 \cdot 3H_2O$ mono.

NEALITE *Min. Rec.* **11**, 209
$Pb_4Fe^{2+}(AsO_4)_2Cl_4$ tric.

NEIGHBORITE *Am. Min.* **46**, 378
$NaMgF_3$ orth.

Cross-reference: *Green River Mineralogy.*

NEKOITE *Am. Min.* **65**, 1270
$Ca_3Si_6O_{12}(OH)_6 \cdot 5H_2O$
tric. *Min. Mag.* **31**, 5

NEMALITE = Fibrous *brucite*

NENADKEVICHITE orth. *Am. Min.* **40**, 1154
$(Na,Ca,K)(Nb,Ti)Si_2O_6(O,OH) \cdot 2H_2O$

NENADKEVITE = Mixture *uraninite* + *boltwoodite* *Am. Min.* **62**, 1261

NEOTANTALITE = *Microlite*
Min. Mag. **43**, 1054

NEOTOCITE *Am. Min.* **46**, 1412
hydrous silicate of Mn and Fe

NEPHELINE A sodium potassium aluminosilicate of the ***feldspathoid*** group, the name being derived from the Greek *nephele*, a cloud, because it becomes cloudy upon immersion in strong acid.

Composition: $Na_3K[AlSiO_4]_4$; ratio 3Na:K varies little in plutonic specimens, widely in volcanic ones.

Properties: Hexagonal, $P6_3$; crystals rare, usually in grains or massive; distinct prismatic $\{10\bar{1}0\}$ and imperfect {0001} cleavage; H=5.5–6; G=2.55–2.66; vitreous to greasy luster; transparent to translucent; white, colorless, or yellowish.

Occurrence: As a primary phase in alkaline plutonic and volcanic rocks, a product of metasomatism, or a reaction product between magmas and calcium-rich sediments; associated with ***feldspars, olivines,*** *augite, diopside*, and sodium-rich ***pyroxenes*** and ***amphiboles.*** Distinctive localities include in the Ilmen Mts., USSR; in the Transylvanian Alps, Rumania; near Heidelberg, West Germany; in syenites in southern Norway, southern Greenland, and Dungannon, Ontario, Canada; and in the USA at Litchfield, Maine, and Magnet Cove, Arkansas.

K. FRYE

References*

DHZ, 4, 231–270.
*See *Feldspathoid Group* for additional references.

Cross-references: *Feldspathoid Group; Meteoritic Minerals; Refractory Minerals; Rock-Forming Minerals.*

NEPHELITE = *Nepheline*

NEPHRITE–Jade variety of *actinolite*

Cross-references: *Gemology; Jade.*

NEPOUITE (***Kaolinite-Serpentine*** gr.)
$Ni_3Si_2O_5(OH)_4$ mono. *Am. Min.* **60**, 863

NEPTUNITE (***Pyroxene*** gr.) *Am. Min.* **57**, 85
$Na_2KLi(Fe,Mn)_2Ti_2Si_8O_{24}$
mono. *Min. Abst.* **28**, 319

NESQUEHONITE *Dana*, II, 225
$Mg(HCO_3)(OH)\cdot 2H_2O$ mono.

Cross-reference: *Cave Minerals.*

NEWBERYITE *Dana*, II, 709
$MgHPO_4\cdot 3H_2O$ orth.

Cross-references: *Cave Minerals; Human and Vertebrate Minerals.*

NEYITE *Am. Min.* **55**, 1444
$Pb_7(Cu,Ag)_2Bi_6S_{17}$ mono. *Can. Min.* **10**, 90

NICCOLITE = *Nickeline* *Min. Mag.* **43**, 1053

NICHROMITE (***Spinel*** gr.) *Am. Min.* **65**, 811
$(Ni,Co,Fe^{2+})(Cr,Fe^{3+},Al)_2O_4$
iso. *Min. Mag.* **43**, 1064

NICKEL *Am. Min.* **53**, 348
Ni iso.

Cross-references: *Magnetic Minerals; Native Elements and Alloys.*

NICKELBISCHOFITE *Am. Min.* **65**, 207
$NiCl_2\cdot 6H_2O$ mono. *Can. Min.* **17**, 107

NICKELBLOEDITE *Am. Min.* **62**, 1059
$Na_2Ni(SO_4)_2\cdot 4H_2O$ mono. *Min. Mag.* **41**, 37

NICKEL-BLOOM = *Annabergite*

NICKEL-HEXAHYDRITE *Am. Min.* **51**, 529
$(Ni,Mg,Fe)(SO_4)\cdot 6H_2O$ mono.

NICKELINE Nickel arsenide named by Beudant in 1832. The name for the element is derived from *kupfernickel*, an older name for the mineral.

Composition: 2[NiAs]; Ni, 43.9%; As, 56.1% by wt.

Properties: Hexagonal, $P6_3/mmc$; usually massive, reniform with columnar structure; H=5–5.5; G=7.78; metallic luster; color: pale copper red with gray to blackish tarnish; streak: pale brownish black.

Occurrence: Commonly found in, or in ores derived from, norite; also in veins with cobalt and silver minerals; as a nickel ore in the Cobalt, Gowanda, and Sudbury districts, Ontario, Canada.

K. FRYE

Reference: *Dana*, I, 236–238.

Cross-references: *Ore Microscopy; Refractory Minerals.*

NICKEL-IRON = *Kamacite, taenite,* or *tetrataenite*

Cross-reference: *Metallurgy.*

NICKELITE = *Nickeline* *Min. Mag.* **43**, 1053

NICKEL-SKUTTERUDITE *Dana*, I, 342
$NiAs_{2-3}$ iso.

NICKEL-ZIPPEITE orth. *Can. Min.* **14**, 429
$Ni_2(UO_2)_6(SO_4)_3(OH)_{10}\cdot 16H_2O$
Min. Mag. **43**, 1064

NIFONTOVITE *Am. Min.* **47**, 172
$Ca_3B_6O_6(OH)_{12}\cdot 2H_2O$ mono.

NIGERITE trig. *Am. Min.* **52**, 864
$(Zn,Mg,Fe)(Sn,Zn)_2(Al,Fe)_{12}O_{22}(OH)_2$
Min. Abst. **29**, 478

NIGGLIITE *Am. Min.* **56**, 360
PtSn hexa.

NIMESITE = *Brindleyite* *Am. Min.* **63**, 484

NIMITE (***Chlorite*** gr.) *Am. Min.* **55**, 18
$(Ni,Mg,Fe,Al)_6(AlSi_3)O_{10}(OH)_8$
mono.

Cross-reference: *Minerals, Uniaxial and Biaxial.*

NINGYOITE orth. *Am. Min.* **44**, 633
$(U,Ca,Ce)_2(PO_4)_2\cdot 1\text{–}2H_2O$

NININGERITE *Am. Min.* **52**, 925
(Mg,Fe,Mn)S iso. *Science* **155**, 451

Cross-references: *Lunar Mineralogy; Meteoritic Minerals.*

NIOBO-AESCHYNITE orth. *Am. Min.* **47**, 417
$(Ce,Ca,Th)(Nb,Ti)_2(O,OH)_6$

NIOBOPHYLLITE (***Astrophyllite*** gr.)
$(K,Na)_3(Fe^{2+},Mn)_6(Nb,Ti)_2Si_8(O,OH,F)_{31}$
Am. Min. **50**, 263
tric. *Can. Min.* **8**, 40

NIOCALITE — *Am. Min.* **41**, 785
$Ca_4NbSi_2O_{10}(O,F)_2$ tric. — *Can. Min.* **6**, 264

NISBITE — *Am. Min.* **56**, 631
$NiSb_2$ orth. — *Can. Min.* **18**, 165

NISSONITE mono. — *Am. Min.* **52**, 927
$Cu_2Mg_2(PO_4)_2(OH)_2 \cdot 5H_2O$

NITER Potassium nitrate of the *aragenite* group, the name being derived from the Greek *nitron*, a general term for saline materials. Also known as saltpeter.

Composition: $4[KNO_3]$; generally quite pure.

Properties: Orthorhombic, *Pmcn*; generally as encrustations or delicate acicular crystals, also massive, granular, or earthy; pseudo-hexagonal groups result from twinning; H=2; G=2.11; vitreous luster; transparent; white; soluble in water; cooling salty taste.

Occurrence: Found as a crust on sheltered surfaces in arid regions and in soils with *nitratine, epsomite, nitrocalcite*, and *gypsum*; localities include the deserts of northern Chile; Cape Province, South Africa; Cochabamba, Bolivia; and in the USA in caves in the middlewest and Socorro Countv. New Mexico, and in soil in Death Valley, California.

K. FRYE

Reference: *Dana*, II, 303–305.

Cross-references: *Nitrates; Saline Minerals.*

NITRATINE (SODA NITER) Sodium nitrate, named by Haidinger in 1845.

Composition: $NaNO_3$.

Properties: Trigonal, $R\bar{3}c$; isostructural with *calcite*; usually massive, as incrustations, or in beds; perfect rhombohedral $\{10\bar{1}1\}$ cleavage; H=1.5–2; G=2.25; vitreous luster; deliquescent; transparent to translucent; color is white, colorless, reddish brown, gray, or yellow; soluble in water; cooling taste.

Occurrence: Found only in very arid regions often as a bed interstratified with sand, *halite, gypsum*, and other evaporites, principally in the deserts of the coast range of northern Chile.

J. A. GILBERT

Reference: *Dana*, II, 300–302.

Cross-references: *Minerals, Uniaxial and Biaxial; Nitrates; Saline Minerals.*

NITROBARITE — *Dana*, II, 305
$Ba(NO_3)_2$ iso.

Cross-reference: *Nitrates.*

NITROCALCITE — *Dana*, II, 306
$Ca(NO_3)_2 \cdot 4H_2O$ mono.

Cross-reference: *Nitrates.*

NITROGLAUBERITE = *Darapskite*
Am. Min. **55**, 776

NITROMAGNESITE — *Dana*, II, 307
$Mg(NO_3)_2 \cdot 6H_2O$ mono.

Cross-reference: *Nitrates.*

NOBLEITE — *Am. Min.* **46**, 560
$CaB_6O_{10} \cdot 4H_2O$ mono.

NOLANITE — *Am. Min.* **52**, 734
$Fe_3V_7O_{16}$ hexa.

NONTRONITE (CHLOROPAL) A hydrated layered aluminosilicate of sodium and ferric iron, belonging to the ***smectite*** group and named for the locality Nontron, Dordogne, France.

Composition: $Na_{0.33}Fe_2^{3+}(Al,Si)_4O_{10}(OH)_2 \cdot nH_2O$.

Properties: Monoclinic, *C2/m*; compact massive; H=2.5–4.5; G=1.7–1.8; earthy or opalescent luster; color: greenish yellow to green.

Use: A "swelling" clay used as a filler.

Occurrence: In veins and as an alteration product of volcanic glass.

K. FRYE

Reference: DHZ, 3, 226–245.

Cross-references: *Clays, Clay Minerals; Green River Mineralogy; Phyllosilicates; Pigments and Fillers; Smectite Group; Soil Mineralogy; Staining Techniques.*

NORBERGITE A magnesium fluohydroxyl-silicate of the ***humite*** group, and named for the locality Norberg, Sweden.

Composition: $[Mg_3(SiO_4)(F,OH)_2]$, with Mg replaced by Fe^{2+}.

Properties: Orthorhombic, *Pbnm*; H=6.5; G=3.1–3.2; color: tawny to chamois.

Occurrence: With other skarn minerals in carbonate rock metasomatized by felsic or alkaline plutonic rocks.

K. FRYE

Reference: DHZ, 1, 47–58.

Cross-references: *Humite Group; Polysomatism.*

NORDENSKIÖLDINE Dana, II, 332
$CaSnB_2O_6$ trig.

NORDITE orth. *Am. Min.* **55**, 1167
$(La,Ce)(Sr,Ca)Na_2(Na,Mn)(Zn,Mg)Si_6O_{17}$

NORDSTRANDITE *Min. Abst.* **29**, 478
$Al(OH)_3$ tric. *Nature* **196**, 265

Cross-reference: *Green River Mineralogy.*

NORSETHITE *Am. Min.* **46**, 420
$BaMg(CO_3)_2$ trig.

Cross-reference: *Green River Mineralogy.*

NORDSTRÖMITE *Am. Min.* **65**, 789
$Pb_3CuBi_7S_{10}Se_4$ mono. *Can. Min.* **18**, 343

NORTHUPITE *Dana*, II, 278
$Na_3Mg(CO_3)_2Cl$ iso.

Cross-references: *Green River Mineralogy; Saline Minerals.*

NOSEAN, NOSELITE
(***Sodalite*** gr.) DHZ, 4, 289
$Na_8(AlSiO_4)_6(SO_4)$ iso.

NOVACEKITE (***Autunite*** gr.)
$Mg(UO_2)_2(AsO_4)_2 \cdot 12H_2O$
Am. Min. **36**, 680
tetr. *USGS Bull.* **1064**, 177

NOVACULITE—White microgranular *quartz* *Dana*, III, 222

Cross-reference: *Electron Microscopy (Transmission).*

NOVAKITE *Am. Min.* **46**, 885
$(Cu,Ag)_4As_3$ tetr.

NOWACKIITE *Am. Min.* **54**, 1497
$Cu_6Zn_3As_4S_{12}$ trig.

NSUTITE *Am. Min.* **50**, 170
$Mn_x^{2+}Mn_{1-x}^{4+}O_{2-2x}(OH)_{2x}$
hexa. *Science* **212**, 1024

NUFFIELDITE *Am. Min.* **57**, 319
$Pb_2Cu(Pb,Bi)Bi_2S_7$ orth. *Can. Min.* **9**, 439

NUKUNDAMITE *Am. Min.* **66**, 398
$Cu_{3.4}Fe_{0.6}S_4$ trig. *Min. Mag.* **43**, 193

NYERERERITE *Am. Min.* **63**, 600
$Na_2Ca(CO_3)_2$ orth.

O

OBOYERITE tric. *Am. Min.* **65**, 809
$Pb_6H_6(Te^{4+}O_3)_3(Te^{6+}O_6)_2 \cdot 2H_2O$
Min. Mag. **43**, 453

OBRUCHEVITE = *Yttropyrochlore*
Am. Min. **62**, 407

OCHRE–Many brownish, yellowish, and reddish fine-grained oxides and hydroxides

OCTAHEDRITE = *Anatase*
Min. Mag. **43**, 1053

O'DANIELITE *Am. Min.* **66**, 218
$Na(Zn,Mg)_3H_2(AsO_4)_3$ mono.

OFFRETITE (***Zeolite*** gr.)
$(K,Ca)_3(Al_5Si_{13}O_{36}) \cdot 14H_2O$
hexa. *Am. Min.* **54**, 875
Nature **214**, 1005

Cross-reference: *Naming of Minerals.*

OJUELAITE mono. *Min. Mag.* **43**, 1065
$ZnFe_2^{3+}(AsO_4)_2(OH)_2 \cdot 4H_2O$

OKANOGANITE trig. *Am. Min.* **65**, 1138
$(Na,Ca)_3(Y,RE)_{12}Si_6B_2O_{27}F_{14}$

OKENITE *Min. Mag.* **43**, 677
$CaSi_2O_4(OH)_2 \cdot H_2O$ tric. Winchell, II, 358

OLDHAMITE *Dana*, I, 208
(Ca,Mn)S iso.

Cross-reference: *Meteoritic Minerals.*

OLGITE *Am. Min.* **66**, 438
$Na(Sr,Ba)PO_4$ trig. *Min. Mag.* **43**, 1065

OLIGISTE = *Hematite* *Min. Mag.* **43**, 1053

OLIGOCLASE A sodium calcium aluminosilicate of the ***feldspar*** group, named by Breithaupt in 1826 from the Greek *oligos*, little, and *klasis*, fracture on the notion that the cleavage was less perfect than that of *albite*.

Composition: $(Na,Ca)[Si,Al)AlSi_2O_8]$; with some K replacing (Na,Ca). In the ***plagioclase feldspars***, *oligoclase* occupies the composition range *Ab* 10–30, *An* 90–70 mol%.

Properties: Triclinic, $C\bar{1}$; commonly in cleavable masses or irregular grains; mostly twinned according to the albite, Carlsbad, or pericline laws; prismatic {001} {010} cleavage; H=6; G=2.65; vitreous to pearly luster; transparent to translucent; white to gray or slightly tinted; intergrowths with *albite* (peristerite) sometimes chatoyant, also with potash ***feldspar*** (perthite), or with bleby *quartz* (myrmekite).

Occurrence: A rock-forming mineral found as phenocrysts in felsic porphyries; an essential constituent of some granites, granodiorites, and syenites; in some pegmatites; in amphibolites; as nearly the entire mass of some anorthosites; and as detrital grains in some arkoses.

K. FRYE

References*

DHZ, 4, 94–165.
*See *Feldspar Group* and *Plagioclase Feldspars* for additional references.

Cross-references: *Feldspar Group; Minerals; Uniaxial and Biaxial; Plagioclase Feldspars.*

OLIGONITE = Manganoan *siderite*

OLIVENITE A copper hydroxylarsenate named in 1820 for its color.

Composition: $4[Cu_2(AsO_4)(OH)]$ with As replaced by P toward *libethenite* or Cu by Zn toward *adamite.*

Properties: Monoclinic, $P2_1/n$; prismatic in druses and reniform or globular shapes; H=3; G=3.9–4.4; adamantine to vitreous or silky luster; olive green.

Occurrence: In the oxidized zone of ore deposits with other secondary copper minerals.

K. FRYE

References

Am. Min. **36**, 484–503.
Dana, II, 859–861.

OLIVINE GROUP–*Fayalite, forsterite, liebenbergite,* and *tephroite.*

Cross-references: *Crystal Growth; Crystallography: Morphological; Density Measurements; Diamond; Electron Probe Microanalysis; Gemology; Glass, Devitrification of Volcanic; Lunar Minerals; Mantle Mineralogy; Meteoritic Minerals; Mineral Properties; Naming of Minerals; Orthopyroxenes; Portland Cement Mineralogy; Refractory Minerals; Rock-Forming Minerals; Soil Mineralogy; Spinel Group; Topotaxy.*

OLMSTEADITE orth. *Am. Min.* **61,** 5
$K_2 Fe_4^{2+}(Nb,Ta)_2^{5+}O_4(H_2O)_4(PO_4)_4$
Min. Abst. **27,** 258

OLSACHERITE *Am. Min.* **54,** 1519
$Pb_2(SeO_4)(SO_4)$ orth.

OLSHANSKYITE *Am. Min.* **54,** 1737
$Ca_3B_4(OH)_{18}$ mono.(?)

OLYMPITE *Am. Min.* **66,** 438
Na_2PO_4 ortho.

OMEIITE *Am. Min.* **64,** 464
$OsAs_2$ orth. *Min. Mag.* **43,** 1065

OMPHACITE (***Pyroxene*** gr.) DHZ, 2, 154
$(Ca,Na)(Mg,Fe^{2+},Fe^{3+},Al)Si_2O_6$
mono. *Min. Mag.* **44,** 37

Cross-references: *Clinopyroxenes; Diamond; Minerals, Uniaxial and Biaxial.*

ONOFRITE = Selenian *metacinnabar*

ONORATOITE *Am. Min.* **54,** 1219
$Sb_8O_{11}Cl_2$ tric. *Min. Mag.* **36,** 1037

ONYX A chalcedonic variety of *quartz* with slight impurities, giving color distributed in parallel bands rather than in curved or other patterns as in agate; the name is from the Greek, *onux,* fingernail.

Composition: SiO_2. SiO_2, 98–100%; Fe_2O_3, 0–2%.

Properties: Microcrystalline *quartz* with submicroscopic pores; G=2.60–2.64; H=6.5; vitreous luster; habit, nodular, filling cavities; onyx is strictly a variety of agate with parallel banding and with white bands alternating with deep brown or black bands; fracture conchoidal. Sardonyx is an onyx with red bands (sard). Optically, indices $\omega \cong 1.534$, $\epsilon \cong 1.540$, i.e., slightly lower than for ordinary *quartz.* Insoluble in acids other than HF; attached by alkaline solutions; soluble in molten Na_2CO_3.

Occurrence: Associated with agate and *quartz* in amygdaloidal cavities in altered mafic lavas; filling late-stage veins; as detrital cobbles from the weathering of amygdaloidal lavas.

Uses: In the carving of cameos, the figure being carved in relief in the white layer, with the dark layer as background; in intaglios, the figure being cut through the black layer to show the white material beneath. Mexican or Algerian onyx is a color-banded variety of decorative marble and is softer (H=3).

R. A. HOWIE

References

Dana, III, 214.
Herbert Smith, G. F., 1958. *Gemstones* (revised by F. C. Phillips). London. Methuen.

Cross-reference: *Gemology.*

OOSTERBOSCHITE *Am. Min.* **57,** 1553
$(Pd,Cu)_7Se_5$ orth.

OPAL *Dana,* III, 287
$SiO_2 \cdot nH_2O$ amorph.

Cross-references: *Authigenic Minerals; Diagenetic Minerals; Gemology; Human and Vertebrate Mineralogy; Invertebrate and Plant Mineralogy; Mineral Classification: Principles; Minerals, Uniaxial and Biaxial; Museums, Mineralogical; Opal; Soil Mineralogy.*

ORCELITE *Am. Min.* **45,** 753
$Ni_{5-x}As_2$ hexa.

ORDOÑEZITE *Am. Min.* **40,** 64
$ZnSb_2O_6$ tetr.

OREGONITE *Am. Min.* **45,** 1130
Ni_2FeAs_2 hexa.

Cross-reference: *Naming of Minerals.*

ORIENTITE *Am. Min.* **43,** 381
$Ca_2Mn_2[Si_3O_{10}](OH)_2 \cdot 2H_2O$
orth. *Min. Mag.* **43,** 325

ORYZITE = *Epistilbite* *Am. Min.* **57,** 592

ORPHEITE trig.(?) *Am. Min.* **61,** 176
$H_6Pb_{10}Al_{20}(PO_4)_{12}(SO_4)_5(OH)_{40} \cdot 11H_2O$

ORPIMENT Arsenic sulfide named by Wallerius in 1747 from a corrpution of the Latin *auripigmentum,* golden paint, in reference to its color and the supposition that it contained gold, *aurum.*

Composition: $4[As_2S_3]$; As, 60.9%; S, 39.1% by wt.

Properties: Monoclinic, $P2_1/n$; generally in foliated, columnar, or fibrous masses; one perfect cleavage {010}; cleavage plates flexible and inelastic; H=1.5–2; G=3.4–3.5; pearly luster on cleavages, but otherwise resinous; lemon yellow with paler streak.

Occurrence: A low-temperature alteration product of other arsenic-bearing minerals, especially *realgar* with which it is associated along with *stibnite,* native *arsenic, calcite, barite,* and *gypsum.* Notable localities include Morococha and Acabambilla, Peru; Bosnia, Yugoslavia; Macedonia, Greece; and in the USA in Tooele County, Utah, and Humbolt and Nye Counties, Nevada.

K. FRYE

Reference: *Dana,* I, 266–269.

Cross-reference: *Ore Microscopy.*

ORTHITE = *Allanite* *Min. Mag.* **43**, 1053

ORTHOANTIGORITE
(***Kaolinite-Serpentine*** gr.)
$Mg_3Si_2O_5(OH)_4$ orth. DHZ, 3, 170

ORTHOBRANNERITE *Am. Min.* **64**, 656
$U_2Ti_4O_{12}(OH)_2$ orth. *Min. Mag.* **43**, 1065

ORTHOCHAMOSITE (***Chlorite*** gr.)
$(Fe^{2+},Mg,Fe^{3+})_5Al(Si_3AlO_{10})(O,OH)_8$
orth. DHZ, 3, 164

ORTHOCHRYSOLITE
(***Kaolinite-Serpentine*** gr.)
$Mg_3Si_2O_5(OH)_4$ orth. DHZ, 3, 170
Can. Min. **17**, 679

ORTHOCLASE A potassium aluminosilicate of the ***feldspar*** group, named by Breithaupt in 1823 from the Greek *ortho,* upright, and *klasis,* fracture, in reference to the mineral's 90° prismatic cleavage.

Composition: $KAlSi_3O_8$; with K replaced by Na, Ba, or Rb and Al by Fe^{3+}.

Properties: Monoclinic, $C2/m$; crystals prismatic elongated parallel to *c* or to *a* or tabular parallel to (010), also massive, coarsely cleavable to granular; prismatic cleavage {001} {010}; commonly twinned according to the Carlsbad law, contact or interpenetrant; H=5; G=2.56; vitreous luster; colorless, white, pale yellow, reddish, or gray; commonly perthitically intergrown with *albite;* polymorphous with *microcline* and *sanidine.*

Occurrence: Common rock-forming mineral and essential constituent of alkali and felsic plutonic and some volcanic rocks, as cores of *plagioclase*-mantled crystals in rapakivi texture; as metasomatically produced porphyroblasts in schists and gneisses around granites; as a medium- to high-grade metamorphic minerals; as authigenic overgrowths on *orthoclase* or *microcline* cores in sediments; and as detrital grains in arkoses.

K. FRYE

References*

DHZ, 4, 1–93.

*See *Alkali Feldspars* and *Feldspar Group* for additional references.

Cross-references: *Alkali Feldspars; Authigenic Minerals; Blowpipe Analysis; Crystallography: Morphological; Feldspar Group; Fire Clays; Glass, Devitrification of Volcanic; Green River Mineralogy; Jade; Minerals, Uniaxial and Biaxial; Mohs Scale of Hardness; Polymorphism; Pseudomorphism; Rock-Forming Minerals; Soil Mineralogy; Thermometry, Geologic; Twinning; Vein Minerals.*

ORTHOERICSSONITE *Am. Min.* **56**, 2157
$BaMn_2(Fe^{3+}O)(Si_2O_7)(OH)$
orth. *Lithos* **4**, 137

ORTHOFERROSILITE (***Pyroxene*** gr.)
$(Fe^{2+},Mg)_2Si_2O_6$ orth. DHZ, 2A, 20

Cross-references: *Mantle Mineralogy; Mineral Classification-Principles; Orthopyroxenes.*

ORTHOPINAKIOLITE *Am. Min.* **46**, 768
$(Mg,Mn^{2+})_2Mn^{3+}(BO_3)O_2$
orth. *Can. Min.* **16**, 475

ORTHOSE = *ORTHOCLASE*
Min. Mag. **43**, 1053

OSARIZAWAITE *Am. Min.* **65**, 1287
$PbCuAl_2(SO_4)_2(OH)_6$
trig. *Min. Abst.* **28**, 324

OSARSITE *Am. Min.* **57**, 1029
(Os,Ru)AsS mono.

OSBORNITE *Dana,* I, 124
TiN iso.

Cross-references: *Meteoritic Minerals; Native Elements and Alloys.*

OSMIRIDIUM *Am. Min.* **49**, 818
(Ir,Os) iso. *Min. Mag.* **33**, 712

Cross-references: *Iridosmine; Native Elements and Alloys.*

OSMIUM *Am. Min.* **49**, 818
(Os,Ir) hexa. *Min. Mag.* **33**, 712

Cross-reference: *Native Elements and Alloys.*

OSUMILITE, OSUMILITE-Mg Hydrated aluminosilicates of potassium, sodium, magnesium, and iron, named for the locality, Osumi, a former province in Japan, where first identified.

Composition: $2[(K,Na)(Fe^{2+},Mg)_2(Al,Fe)_3(Si,Al)_{12}O_{30} \cdot H_2O]$. If $Mg > Fe^{2+}$, the mineral is *osumilite-(Mg)*.

Properties: Hexagonal, *P6/mcc*; short prismatic or tabular crystals; G=2.64; vitreous luster; color: dark blue to black.

Occurrence: In vugs in felsic volcanic rocks associated with high-temperature minerals.

K. FRYE

References

Am. Min. **41,** 104; **54,** 101; **59,** 383; **63,** 490.
Min. Mag. 39, 189.

OSUMILITE GROUP–*Brannockite, darapiosite, eifelite, emeleusite, merrihueite, milarite, osumilite, osumilite-(Mg), roedderite, sogdianite, sugilite,* and *yagiite.*

OTAVITE (***Calcite*** gr.) *Dana,* II, 181
$CdCO_3$ trig.

OTTEMANNITE *Am. Min.* **51,** 1551
Sn_2S_3 orth.

OTTRELITE mono. and tric. *Min. Abst.* **30,** 285
$(Mn,Fe,Mg)Al_4Si_2O_{10}(OH)_4$

OTWAYITE orth. *Am. Min.* **62,** 999
$(Ni,Mg)_2(OH)_2(CO_3) \cdot H_2O$
Min. Abst. **29,** 201

OURAYITE *Am. Min.* **64,** 244
$Ag_{25}Pb_{30}Bi_{41}S_{104}$ orth.

OVERITE mono. *Am. Min.* **59,** 48; **62,** 692
$CaMgAl(PO_4)_2(OH) \cdot 4H_2O$

OWYHEEITE *Am. Min.* **34,** 401
$Ag_2Pb_5Sb_6S_{15}$ orth.

OXAMMITE *Dana,* II, 1103
$(NH_4)_2C_2O_4 \cdot H_2O$ orth.

OXYAPATITE–see *Apatite group*

OXYHORNBLENDE = HORNBLENDE
with $(OH+F+Cl) < 1.0$ *Am. Min.* **63,** 1026

P

PABSTITE *Am. Min.* **50**, 1164
$Ba(Sn,Ti)Si_3O_9$ hexa.

PACHNOLITE *Dana*, II, 114
$NaCaAlF_6 \cdot H_2O$ mono.

PAIGEITE = *VONSENITE*

PAINITE *Am. Min.* **61**, 88
$CaAl_9ZrO_{15}(BO_3)$ hexa. *Min. Mag.* **42**, 518

PALAGONITE–Devitrified basaltic glass

PALERMOITE orth. *Am. Min.* **50**, 777;
$(Li,Na)_2(Sr,Ca)Al_4(PO_4)_4(OH)_4$ **60**, 460

Cross-reference: *Pegmatite Minerals.*

PALLADIUM *Dana*, I, 109
Pd iso.

Cross-references: *Blackcard Minerals; Native Elements and Alloys.*

PALLADOARSENIDE *Am. Min.* **60**, 162
Pd_2As mono.

PALLADOBISMUTHARSENIDE
$Pd_2As_{0.8}Bi_{0.2}$ *Can. Min.* **14**, 410
orth. *Min. Abst.* **29**, 84

PALLADSEITE *Am. Min.* **62**, 1059
$Pd_{17}Se_{15}$ iso. *Min. Mag.* **41**, 123

PALLITE = Ferrian *millisite*
Am. Min. **45**, 256

PALMIERITE *Dana*, II, 403
$(K,Na)_2Pb(SO_4)_2$ trig.

PALYGORSKITE This term is used both for the group of fibrous clay minerals in the *attapulgite-sepiolite* series, and for the specific mineral specimens that are similar to *sepiolite* except for some replacement of magnesium by aluminum.

Composition:
$(OH_2)_4(OH)_2(Mg,Al)_5(Si,Al)_8O_{20} \cdot 4H_2O$

Properties: Monoclinic and orthorhombic; G=2.20–2.36; similar in many properties to *sepiolite*. Optically mean value n=1.50–1.555; biaxial (-); $2V$ small to large; γ-α= 0.025–0.035; z is parallel to elongation.

Occurrence: A significant component of desert soils.

E. BOOY

References

Am. Min. **54**, 198–205; **60**, 328–330.

Grim, R. E., 1968. *Clay Mineralogy*. New York: McGraw–Hill, 596p.

Kerr, P. F. ed. 1951. *Reference clay minerals, American Petroleum Institute Research Project 49*. New York: Columbia University, 701p.

Ovcharenko, F. D. et al., 1963. *Kolloidnaya Khimiya Palygorskita [The Colloid Chemistry of Palygorskite]*. Kiev: Izd. Akad. Nauk Ukr. S.S.R., 120p.

Cross-references: *Clays, Clay Minerals; Diagenetic Minerals; Soil Mineralogy; Staining Techniques.*

PANABASE = *TETRAHEDRITE*
Min. Mag. **43**, 1053

PANDAITE = *Bariopyrochlore*
Am. Min. **62**, 407

PANDERMITE = *PRICEITE* *Dana*, II, 341

PANETHITE mono. *Am. Min.* **53**, 509
$(Na,Ca,K)_2(Mg,Fe,Mn)_2(PO_4)_2$
GCA **31**, 1711

Cross-reference: *Meteoritic Minerals.*

PAOLOVITE *Am. Min.* **59**, 1331
Pd_2Sn orth.

PAPAGOITE *Am. Min.* **45**, 599
$CaCuAlSi_2O_6(OH)_3$ mono.

PARA-ALUMOHYDROCALCITE
$CaAl_2(CO_3)_2(OH)_4 \cdot 6H_2O$ *Am. Min.* **63**, 794

PARABUTLERITE *Am. Min.* **61**, 398
$Fe^{3+}(SO_4)(OH) \cdot 2H_2O$
orth. *Dana,* II, 610

PARACELSIAN (**Feldspar** gr.) DHZ, 4, 167
$BaAl_2Si_2O_8$ mono. *Min. Abst.* **28**, 139

PARACHRYSOTILE (***Kaolinite-Serpentine*** gr.)
$Mg_3Si_2O_5(OH)_4$ orth. — *Can. Min.* **17**, 693; DHZ, 3, 173

PARACOQUIMBITE — *Am. Min.* **56**, 1567
$Fe_2^{3+}(SO_4)_3 \cdot 9H_2O$ trig.

Cross-reference: *Syntaxy.*

PARACOSTIBITE — *Am. Min.* **56**, 631
CoSbS orth. — *Can. Min.* **18**, 165

PARADAMITE — *Am. Min.* **65**, 353
$Zn_2(AsO_4)(OH)$ trig.

PARADOCRASITE — *Am. Min.* **56**, 1127
$Sb_2(Sb,As)_2$ mono.

PARAGONITE A sodium hydroxyl aluminosilicate of the ***mica*** group, named from the Greek *paragon*, misleading, because it was originally mistaken for *talc*.

Composition: $NaAl_2(AlSi_3O_{10})(OH)_2$; with partial replacement of Na by K.

Properties: Monoclinic, *C2/c* or *Cc*; massive, fine-grained, scaley aggregates; H=2.5; G=2.85; perfect basal {001} cleavage; colorless to pale yellow; commonly mistaken for *muscovite*.

Occurrence: In phyllites, schists, and gneisses; in *quartz* veins; and in fine-grained sediments.

K. FRYE

References

Am. Min. 65, 1277.
DHZ, 3, 31–34.
Min. Mag. 26, 307.

Cross-references: *Minerals, Uniaxial and Biaxial; Phyllosilicates; Soil Mineralogy.*

PARAGUANAJUATITE — *Am. Min.* **34**, 619
$Bi_2(Se,S)_3$ trig.

PARAHILGARDITE — *Dana*, II, 383
$Ca_2B_5ClO_8(OH)_2$ tric.

PARAHOPEITE — *Dana*, II, 733
$Zn_3(PO_4)_2 \cdot 4H_2O$ tric.

PARAJAMESONITE — *Am. Min.* **34**, 133
$Pb_4FeSb_6S_{14}$ orth.

PARAKELDYSHITE — *Am. Min.* **64**, 656
$Na_2ZrSi_2O_7$ tric. — *Min. Mag.* **43**, 1065

PARAKHINITE — *Am. Min.* **63**, 1016
$Cu_3PbTeO_4(OH)_6$ hexa. — *Min. Abst.* **30**, 170

PARALAURIONITE — *Dana*, II, 64
PbCl(OH) mono.

PARALSTONITE — *Am. Min.* **66**, 219
$(Ba,Sr)Ca(CO_3)_2$ trig. — *Min. Abst.* **31**, 356

PARAMELACONITE — *Dana*, I, 510
$Cu_2^{+}Cu_2^{2+}O_3$ tetr. — *Am. Min.* **63**, 180

PARAMONTROSEITE — *Am. Min.* **40**, 861
VO_2 orth.

PARANATROLITE — *Can. Min.* **18**, 85
$Na_2Al_2Si_3O_{10} \cdot 3H_2O$ pseudo orth. — *Min. Mag.* **43**, 1065

PARAPIERROTITE — *Am. Min.* **61**, 504
$TlSb_5S_8$ mono.

PARARAMMELSBERGITE — *Dana*, I, 310
$NiAs_2$ orth.

Cross-reference: *Ore Microscopy.*

PARAREALGAR — *Can. Min.* **18**, 525
AsS mono.

PARASCHACHNERITE — *Am. Min.* **58**, 347
Ag_3Hg_2 orth.

PARASCHOEPITE — *Am. Min.* **45**, 1026
$UO_3 \cdot 2H_2O$(?) orth.

PARASPURRITE — *Am. Min.* **62**, 1003
$Ca_5(SiO_4)_2CO_3$ mono. — *Min. Abst.* **29**, 201

PARASYMPLESITE — *Am. Min.* **40**, 368
$Fe_3^{3+}(AsO_4)_2 \cdot 8H_2O$ mono.

PARATACAMITE — *Min. Abst.* **27**, 24
$Cu_2(OH)_3Cl$ trig. — *Min. Mag.* **43**, 547

PARATELLURITE — *Am. Min.* **45**, 1272
TeO_2 tetr.

PARAVAUXITE — *Am. Min.* **47**, 1
$Fe_2^{2+}(H_2O)_4[Al_4(OH)_4(H_2O)_4(PO_4)_4] \cdot 4H_2O$ tric. — *Dana*, II, 972

PARAVEATCHITE = *P-veatchite*

PARAWOLLASTONITE — DHZ, 2, 167
$CaSiO_3$ mono.

PARGASITE A hydroxylaluminosilicate of sodium, calcium, and magnesium. A member of the ***amphibole*** group, named for the locality Pargas, Finland.

Composition: $NaCa_2Mg_4AlSi_6Al_2O_{22}(OH)_2$,

with Mg replaced by Fe^{2+} toward *ferropargasite*, tetrahedral Al by Si toward *hornblende*, and octahedral Al by Fe^{3+}.

Properties: Monoclinic, *C2/m*; prismatic crystals, commonly twinned; prismatic cleavage at 56°; H=5–6; G=3.1; vitreous luster; color: brown, bluish green, or gray black.

Occurrence: Mainly metamorphosed impure dolomitic limestones, commonly resulting from metasomatism and from regionally metamorphosed skarns.

K. FRYE

References

Am. Min. **63**, 1023–1052.
DHZ, 2, 264–314.

Cross-references: *Amphibole Group; Mantle Mineralogy; Minerals, Uniaxial and Biaxial.*

PARISITE A calcium fluocarbonate with cerium-group rare-earth elements, named in 1845 after J. J. Paris, proprietor of the mine where it was discovered (Muzo, Columbia).

Composition: $6[(Ce,La)_2Ca(CO_3)_3F_2]$.

Properties: Trigonal, *R*3; slender pyramidal crystals; one distinct cleavage or parting; H=4.5; G=4.36; Luster vitreous to resinous or pearly; brownish yellow.

Occurrence: In veins, pockets, and pegmatites.

K. FRYE

Reference: *Dana*, II, 282–285.

Cross-reference: *Syntaxy.*

PARKERITE *Am. Min.* **58**, 435
$Ni_3(Bi,Pb)_2S_2$ mono.

PARNAUITE orth. *Am. Min.* **63**, 704
$Cu_9(AsO_4)_2(SO_4)(OH)_{10}\cdot 7H_2O$
Min. Mag. **43**, 1065

PARSETTENSITE Phillips and Griffen, 444
$KMn_{10}Si_{12}O_{30}(OH)_{12}$ hexa.

PARSONSITE *Am. Min.* **35**, 245
$Pb_2(UO_2)(PO_4)_2\cdot 2H_2O$
tric. *USGS Bull.* **1064**, 233

PARTHEITE *Am. Min.* **65**, 1068
$CaAl_2Si_2O_8\cdot 2H_2O$ mono. *Min. Mag.* **43**, 1065

PARTZITE iso. *Am. Min.* **39**, 407
$Cu_2Sb_2(O,OH)_7(?)$ *Min. Mag.* **30**, 100

PARWELITE *Am. Min.* **55**, 323
$(Mn,Mg)_5Sb(As,Si)_2O_{12}$ mono.

PASCOITE *Dana*, II, 1055
$Ca_3V_{10}O_{28}\cdot 17H_2O$ mono.

PATERNOITE = *KALIBORITE*
Am. Min. **50**, 1079

PATRONITE *Dana*, I, 347
$VS_4(?)$ mono.

PAULINGITE A hydrated calcium barium sodium potassium aluminosilicate of the ***zeolite*** group named by Kamb and Oke (1960) in honor of Prof. L. C. Pauling.

Composition:
$(K_2,Na_2,Ca,Ba)_{76}[Al_{152}Si_{525}O_{1354}]\cdot 700H_2O$.

Properties: Isometric, *Im*3*m*, *a*=35.10Å; H=5; luster vitreous; colorless; transparent; habit in rhombic dodecahedra; no cleavage. Optically isotropic; refractive index, *n*=1.473.

Occurrence: In vesicles in basalt boulders from the Columbia River near Wenatchee, Washington, USA, associated with *erionite, heulandite, phillipsite, calcite*, and *pyrite*.

A. IIJIMA
M. UTADA

Reference: *Am. Min.* **45**, 79–91.

Cross-reference: *Zeolites.*

PAULMOOREITE *Am. Min.* **64**, 352
$Pb_2As_2O_5$ mono. *Min. Mag.* **43**, 1065

PAVONITE *Am. Min.* **39**, 409
$(Ag,Cu)(Bi,Pb)_3S_5$ mono. *Can. Min.* **13**, 408

PAXITE *Am. Min.* **47**, 1484
Cu_2As_3 orth.

PEACOCK ORE = *Bornite*

PEARCEITE *Dana*, I, 353
$Ag_{16}As_2S_{11}$ mono.

Cross-reference: *Ore Microscopy.*

PEARL—Biogenic *calcite* secretion

Cross-reference: *Gemology.*

PEARL SPAR = *Dolomite*

PECORAITE (***Kaolinite-Serpentine*** gr.)
$Ni_3Si_2O_5(OH)_4$ *Am. Min.* **54**, 1740
mono. *Science* **165**, 59

PECTOLITE A sodium calcium hydroxyl silicate named by Kobell in 1828 from the Greek *pektos*, congealed, referring to its translucent appearance.

Composition: $NaCa_2Si_3O_8(OH)$.

Properties: Triclinic, $P\bar{1}$ (rarely monoclinic); usually in crystal aggregates often with a fibrous radiating appearance, may be massive; H=5; G=2.7–2.8; vitreous to silky luster; colorless, white, or gray.

Occurrence: A secondary mineral found in cavities in basalt associated with ***zeolites***, *calcite*, and *prehnite*; also lime-rich metamorphic rocks.

J. A. GILBERT

References

Am. Min. **63**, 274.
DHZ, 2A, 564–574.

Cross-references: *Minerals, Uniaxial and Biaxial; Pyroxenoids; Rock-Forming Minerals.*

PEKOITE — *Am. Min.* **61**, 15
$CuPbBi_{11}S_{18}$ orth. *Can. Min.* **14**, 322

PELLYITE orth. *Am. Min.* **61**, 67
$Ba_2Ca(Fe^{2+},Mg)_2Si_6O_{17}$ *Can. Min.* **11**, 444

PENDLETONITE = *Karpatite*
Am. Min. **54**, 329

PENFIELDITE — *Dana*, II, 66
$Pb_2Cl_3(OH)$ hexa.

PENIKISITE tric. *Am. Min.* **64**, 656
$Ba(Mg,Fe)_2Al_2(PO_4)_2(OH)_3$
Can. Min. **15**, 393

PENKVILKSITE — *Am. Min.* **60**, 340
$Na_4Ti_2Si_8O_{22}\cdot 5H_2O$
mono. or orth.

PENNANTITE (***Chlorite*** gr.)
$(Mn,Al)_6(Si,Al)_4O_{10}(OH)_8$
mono. DHZ, 3, 146

PENNINE, PENNINITE—Variety of *clinochlore*

PENROSEITE (***Pyrite*** gr.)
$(Ni,Co,Cu)Se_2$ iso. *Am. Min.* **35**, 337

PENTAGONITE — *Am. Min.* **58**, 405
$Ca(VO)Si_4O_{10}\cdot 4H_2O$ orth.

PENTAHYDRITE — *Dana*, II, 492
$MgSO_4\cdot 5H_2O$ tric.

Cross-references: *Saline Minerals.*

PENTAHYDROBORITE — *Am. Min.* **47**, 1482
$CaB_2O(OH)_6\cdot 2H_2O$ tric.

PENTLANDITE An iron nickel sulfide named by Dufrenoy in 1856 after J. B. Pentland, who first noted the mineral.

Composition: $(Fe,Ni)_9S_8$; Fe:Ni generally close to 1:1; may contain small amounts of Co.

Properties: Isometric, *Fm3m*; granular aggregates or massive; H=3.5–4; G=4.6–5.0; brittle; metallic luster; opaque; color: bronze yellow; streak: bronze brown.

Occurrence: The principal nickel ore at Sudbury, Ontario, Canada, also from the Bushveld district, South Africa; Oppland and Nordland, Norway; southeast Alaska; and Clarke County, Nevada, USA. Associated with *pyrrhotite* in mafic igneous rocks.

K. FRYE

References: *Can. Min.* 7, 1–207; 17, 275–285.

Cross-references: *Diamond; Lunar Minerals; Metallic Minerals; Meteoritic Minerals; Mineral and Ore Deposits; Ore Microscopy; Refractory Minerals; Thermometry, Geologic.*

PERCYLITE (doubtful species)
$PbCuCl_2(OH)_2$ iso. *Min. Rec.* **5**, 280

PERETAITE mono. *Am. Min.* **65**, 936
$CaSb_4O_4(OH)_2(SO_4)_2\cdot 2H_2O$

PERHAMITE hexa. *Min. Abst.* **29**, 84
$3CaO\cdot 3.5Al_2O_3\cdot 3SiO_2\cdot 18H_2O$
Min. Mag. **41**, 437

PERICLASE Magnesium oxide named from the Greek *periklasis*, to break around, in reference to its cubic cleavage.

Composition: MgO; with Mg replaced by Fe, Zn, Mn, and Ni.

Properties: Isometric, *Fm3m*; halite structure; octahedral or cubic crystals, also irregular grains; cubic {100} cleavage; G=3.56–3.68; H=2.5; n=1.735; colorless to yellowish or green.

Occurrence: Results from metamorphism of dolomites and magnesian limestones (appears before *monticellite* but after *wollastonite*); alters to *brucite*, ***serpentine***, or *hydromagnesite*.

D. H. SPEIDEL

Reference: DHZ, 5, 1–4.

Cross-references: *Mantle Mineralogy; Mineral Classification: Principles; Mineral Properties; Minerals, Uniaxial and Biaxial; Refractory Minerals; Rock-Forming Minerals; Synthetic Minerals.*

PERIDOT–Green gem variety of *forsterite*

Cross-references: *Gemology; Jade; Museums, Mineralogical.*

PERISTERITE–***plagioclase feldspar*** intergrowth

PERITE Am. Min. **46**, 765
$PbBiO_2Cl$ orth. *Min. Mag.* **40**, 537

PERLITE–Pearly volcanic glass

PERLOFFITE mono. *Am. Min.* **62**, 1059
$Ba(Mn,Fe)_2Fe_2^{3+}(OH)_3(PO_4)_3$
Min. Rec. **8**, 112

PERMINGEATITE *Am. Min.* **57**, 1554
Cu_3SbSe_4 tetr.

PEROVSKITE A calcium titanate named after Count Perovski of St. Petersburg.

Composition: $CaTiO_3$; rare earths substitute for Ca, and Nb and Ta for Ti; knopite, $CeTiO_3$; dysanalyte, $(Ca,Na)(Ti,Nb)O_3$, with up to 18% Nb; *loparite* $(Ce,Na)(Ti,Nb)O_3$. Zn and Fe^{2+} also substitute. Synthetic $BaTiO_3$ is important in semiconductors.

Properties: Orthorhombic, *Pnma* (pseudocubic); Ti in octahedral coordination on corners of a distorted cube, Ca in 12-fold coordination in the body center; forms cubic crystals or octahedral crystals with poor cubic cleavage; G=3.98–4.26; H=5.5; colorless to dark brown; *n*=2.30–2.38; high relief, weak birefringence associated with penetration twins.

Occurrence: Accessory mineral in mafic and alkaline rocks, associated with *melilite, leucite*, and *nepheline*; also found in contact metamorphased limestones (Ce, Nb varieties) associated with *larnite, wollastonite*, and *spinel.*

D. H. SPEIDEL

References

DHZ, 5, 48–55.
Econ. Geol. **60**(4), 672–792.
Sov. Geol. 8(11), 132–136 (in Russian).

Cross-references: *Lunar Minerals; Mantle Mineralogy; Meteoritic Minerals; Refractory Minerals; Rock-Forming Minerals.*

PERRIERITE
$(Ca,REE,Th)_4(Mg,Fe)_2(Ti,Fe^{3+})_3Si_4O_{22}$
mono. *Am. Min.* **51**, 1394; **63**, 499

PERRYITE *Am. Min.* **56**, 1123
$(Ni,Fe)_5(Si,P)_2$ *Min. Mag.* **37**, 905

Cross-reference: *Meteoritic Minerals.*

PERTHITE–***Alkali feldspar*** intergrowth

Cross-references: *Alkali Feldspars; Pegmatite Minerals.*

PETALITE A lithium aluminosilicate named by d'Andrada in 1840 from the Greek *petalon*, leaf, in reference to its cleavage.

Composition: $LiAlSi_4O_{10}$; with small amounts of Na, K, or Ca replacing Li.

Properties: Monoclinic, *Pa*; commonly massive, foliated; H=6–6.5; G=2.4; vitreous luster; transparent to translucent; colorless, gray, white, pink, or green.

Occurrence: Found in granite pegmatites with other lithium minerals as in the Transbaikal region, USSR; Tammela and Somero, Finland; Varuträsk, Sweden; Karibib, Namibia; Londonderry, Western Australia; and Nagatare, Japan.

K. FRYE

References

Acta Cryst. **14**, 399.
DHZ, 4, 271–275.

Cross-reference: *Pegmatite Minerals.*

PETARASITE mono. *Can. Min.* **18**, 497
$Na_5Zr_2Si_6O_{18}(Cl,OH)\cdot 2H_2O$

PETROVICITE *Am. Min.* **62**, 594
$Cu_3HgPbBiSe_5$ orth. *Bull. Soc. fr.* **99**, 310

PETSCHECKITE *Am. Min.* **63**, 941
$U^{4+}Fe^{2+}(Nb,Ta)_2O_8$ hexa. *Min. Mag.* **43**, 1065

PETZITE A gold silver telluride named in 1845 for W. Petz who described the mineral.

Composition: Ag_3AuTe_2; Ag, 41.71%; Au, 25.42%; Te, 32.87% by wt.

Properties: Isometric, $I4_132$; massive or granular to compact; cubic cleavage; H=2.5–3; G=8.7–9.0; metallic luster; color: steel gray to iron black.

Occurrence: In vein deposits with other tellurides.

K. FRYE

References

Am. Min. **44**, 693–701.
Dana, I, 186–187.

PHARMACOLITE *Dana*, II, 706

$CaH(AsO_4)\cdot 2H_2O$ mono.

PHARMACOSIDERITE A hydrated basic ferric arsenate named by Hausmann in 1813 from the Greek for poison and iron.

Composition: $KFe_4(AsO_4)_3(OH)_4\cdot 6\text{-}7H_2O$; P may substitute for As.

Properties: Isometric, $P\bar{4}3m$; G=2.8–2.9; H=2.5; adamantine to greasy luster; color: olive-green, yellow, brown, red, or emerald green; habit, usually in cubes, also tetrahedral; imperfect to good cubic {001} cleavage; uneven fracture; n=1.68–1.70; transparent to translucent.

Occurrence: Found as an oxidation product of *arsenopyrite* and other arsenic minerals with various arsenate minerals. Also as a hydrothermal deposit. Found in several localities in Germany, also in France, Czechoslovakia, England, Algiers, and USA.

L. H. FUCHS

Reference: *Dana*, II, 955.

PHENAKITE A beryllium silicate named by Nordenskiöld in 1833 from the Greek meaning a deceiver, referring to its having been mistaken for *quartz*.

Composition: $Be_2(SiO_4)$.

Properties: Trigonal, $R\bar{3}$; disk-like to stubby prismatic crystals; H=7.5–8; G=2.97–3.00; vitreous luster; transparent to translucent; colorless to white.

Occurrence: In granite pegmatites associated with *topaz, chrysoberyl, beryl*, and ***apatite*** as at Sverdlovsk, USSR; Kragerø, Norway; Minas Gerais, Brazil; and in the USA near Colorado Springs and in Chaffee County, Colorado; Oxford County, Maine; Carroll County, New Hampshire; and Amelia County, Virginia.

J. A. GILBERT

References: *Am. Min.* **24**, 791; **63**, 664.

Cross-references: *Blacksand Minerals; Pegmatite Minerals.*

PHENGITE–Variety of *muscovite* with high Si content

Cross-reference: *Phyllosilicates.*

PHILLIPSITE A hydrated calcium sodium potassium aluminosilicate of the ***zeolite*** group named by Lévy in 1825 in honor of W. Phillips.

Composition: $(\frac{1}{2}Ca,Na,K)_6[Al_6Si_{10}O_{32}]\cdot 12H_2O$; here ½Ca:Na:K=1:1:1.

Properties: Monoclinic, $P2_1/m$; C^2_{2h}; a=10.02Å, b=14.28Å, c=8.64Å, β=125°40′; H=4–4.5; G=2.2; luster, vitreous; colorless, white, gray, pink, yellow, or green; transparent to opaque; habit in penetration twins, in spherules, or prismatic; cleavage good on {010} and {100}; twinning on {001}, {021}, {110} (interpenetrant). Optically, refractive indices, α=1.483–1.504, β=1.484–1.509, γ=1.486–1.514; birefringence = 0.003–0.010 (high alkali and silica varieties having much lower refractive indices and birefringence, sometimes nearly isotropic); $2V$=60°–80° (+); optic axial plane ⊥ (010); $\alpha=y$, $\beta: x$= 46°–65°; dispersion $r<v$. Considerable substitution of Al for Si may occur with a corresponding increase in the (½Ca,Na,K) group, the possible substitution being (½Ca,Na,K) Al⇌Si; K is usually dominant over Na; in some *phillipsite* Ca<(Na+K) while in others Ca>(Na+K); Ba and lesser Sr may enter the (½Ca,Na,K) group; minimum diameter of the largest channel in the aluminosilicate framework is 3.2 Å; molecular sieve properties very selective for small molecules, such as NH_3 and CO_2; gelatinized by HCl.

Occurrence: Commonly in cavities or amygdules in basalt, often associated with *chabazite* and other ***zeolites***; in red clay of pelagic sediments; reaction product of basaltic glass with percolating groundwater in palagonite tuffs, associated with *chabazite, analcime*, or other ***zeolites***; alteration product of silicic vitric ash deposited in alkaline saline lakes, associated with *chabazite, clinoptilolite, erionite*, or *analcime*; in alkaline soils; from wallwork of Roman baths with emerging hot springs of maximum 70°C, together with *chabazite*; the Ba-bearing *wellsite* occurs in fissures in igneous rock and from a *corundum* mine as a primary mineral in the groundmass of alkali basalt in New Zealand and southeastern California, USA.

Uses: Synthetic ***zeolite*** with the phillipsite structure is utilized as a molecular seive.

A. IIJIMA
M. UTADA

References

Acta Cryst. **15**, 644–651.

Am. Min. **49**, 656–682, 1366–1387; **57**, 1125–1145.

Edinborgh J. Sci. 7, 140–143.

Cross-references: *Authigenic Minerals; Minerals, Uniaxial and Biaxial; Zeolites.*

PHLOGOPITE A potassium magnesium aluminosilicate of the ***mica*** group, named by Breithaupt in 1841 from the Greek *phlogos*, firelike, in reference to its common reddish tinge.

Composition: $KMg_3(AlSi_3O_{10})(OH,F)_2$; with K replaced by Na, Rb, Cs, and Ba; Mg by Fe (*biotite*), Mn, and Al; and Si by Al.

Properties: Monoclinic, *Cm* and *C*2/*c*, and trigonal, $P3_112$; foliated masses; H=2–2.5; G=2.86; vitreous to pearly luster; perfect basal {001} cleavage; transparent to translucent; color: yellowish or reddish brown, green, or white.

Occurrence: In contact metamorphosed or metasomatized marble as in Val Malenco, Italy; Ardgour, Argyllshire, UK, Pyrenees Mountains, France; Carlingford, Eire; and Iron Hill, Colorado, USA. Also in kimberlites as at Kimberley, South Africa, and Bachelor Lake, Quebec, Canada.

K. FRYE

Reference: DHZ, 3, 42–54.

Cross-references: *Diamond; Mantle Mineralogy; Minerals, Uniaxial and Biaxial; Phyllosilicates; Soil Mineralogy.*

PHOENICOCHROITE *Am. Min.* **55**, 784
$Pb_2(CrO_4)O$ mono.

PHOSGENITE A lead chlorocarbonate, so named in 1841 because phosgene (carbonyl chloride) also contains carbon, oxygen, and chlorine.

Composition: $4[Pb_2(CO_3)Cl_2]$.

Properties: Tetragonal, *P*4/*mbm*; prismatic crystals, also massive or granular; three distinct right-angle cleavages; H=2–3; G=6.1; adamantine luster; color: white, gray, or yellow.

Occurrence: A secondary mineral formed by weathering of lead minerals, also by the action of seawater on lead-bearing slags and artifacts.

K. FRYE

Reference: *Dana*, II, 256–259.

PHOSINAITE orth. *Am. Min.* **60**, 488
$H_2Na_3(Ca,Ce)(SiO_4)(PO_4)$

PHOSPHAMMITE Hey, 228
$(NH_4)_2HPO_4$ *Min. Mag.* **39**, 346

Cross-references: *Cave Minerals; Human and Vertebrate Mineralogy.*

PHOSPHOFERRITE *Am. Min.* **55**, 135
$(Fe^{2+},Mn)_3(PO_4)_2 \cdot 3H_2O$
orth. *Min. Mag.* **43**, 789

Cross-reference: *Pegmatite Minerals.*

PHOSPHOPHYLLITE *Am. Min.* **62**, 812
$Zn_2(Fe^{2+},Mn)(PO_4)_2 \cdot 4H_2O$
mono. *Dana*, II, 738

Cross-reference: *Pegmatite Minerals.*

PHOSPHORROESSLERITE *Dana*, II, 713
$MgHPO_4 \cdot 7H_2O$ mono.

Cross-reference: *Human and Vertebrate Mineralogy.*

PHOSPHOSIDERITE A hydrated iron phosphate of the ***metavariscite*** group and named in 1890 for its composition.

Composition: $4[Fe^{3+}PO_4 \cdot 2H_2O]$; with Fe^{3+} replaced by Al toward *metavariscite.*

Properties: Monoclinic, $P2_1/n$; crystals tabular or prismatic, commonly twinned, also as incrustations; one good cleavage; H=3.5–4; G=2.76; vitreous red.

Occurrence: In iron ores and in pegmatites.

K. FRYE

References

Am. Min. **55**, 135.
Dana, II, 769–771.

PHOSPHURANYLITE
$Ca(UO_2)_4(PO_4)_2(OH)_4 \cdot 7H_2O$
orth. *USGS Bull.* **1064**, 222

PHURCALITE orth. *Am. Min.* **63**, 1283
$Ca_2(UO_2)_3(PO_4)_2(OH)_4 \cdot 4H_2O$
Min. Abst. **30**, 72

PIANLINITE *Am. Min.* **65**, 1068
$Al_2Si_2O_6(OH)_2$

PICKERINGITE (MAGNESIA ALUM) A hydrated magnesium aluminum sulfate of the ***halotrichite*** group, named in 1844 for John Pickering.

Composition: $4[MgAl_2(SO_4)_4 \cdot 22H_2O]$; with Mg replaced by Fe^{2+} (Fe>Mg, *halotrichite*) or Mn^{2+} toward *apjohnite.*

Properties: Monoclinic, *P*2; fibrous masses; H=1.5; G=1.8; vitreous luster; white or colorless; astringent taste.

Occurrence: Product of weathering of pyritic aluminous rocks especially in sheltered places, such as mines, and in arid regions.

K. FRYE

Reference: *Dana*, II, 523–527.

PICOTITE = Chromian *spinel*

PICOTPAULITE *Am. Min.* **57**, 1909
$TlFe_2S_3$ orth.

PICROCHROMITE = *Magnesiochromite*

PICROLITE = Asbestiform *antigorite*

PICROMERITE *Dana*, II, 453
$K_2Mg(SO_4)_2 \cdot 6H_2O$ mono.

Cross-reference: *Saline Minerals.*

PICROPHARMACOLITE
$Ca_4Mg(H_2O)_7(AsO_3OH)_2(AsO_4)_2 \cdot 4H_2O$
tric. *Am. Min.* **61**, 326; **66**, 385

PIEDMONTITE = *Piemontite*

PIEMONTITE A calcium aluminum iron manganese silicate of the ***epidote*** group, named by Kenngott in 1853 after the locality, Piemonte, Italy.

Composition:
$Ca_2(Mn^{3+},Al,Fe)_3OH(Si_2O_7)(SiO_4)O$

Properties: Monoclinic, $P2_1/m$; prismatic crystals, massive; basal {001} cleavage; uneven fracture; H=6.5; G=3.45, vitreous to pearly luster; color: reddish brown to black, columbine red; streak: reddish; strongly pleochroic, α, yellow, β, amethyst, γ, red.

Occurrence: Usually in crystalline schists and low-grade metamorphic rocks. Manganese-rich varieties are associated with metasomatic manganese ore deposits, best examples at type locality of St. Marcel, Aosta Valley, Piemonte, Italy.

G. C. MOERSCHELL

References

Am. Min. **54**, 710–717.
DHZ, 1, 194–210.

PIERROTITE *Am. Min.* **57**, 1909
$Tl_2(Sb,As)_{10}S_{17}$ orth.

PIGEONITE A calcium, magnesium, ferrous silicate of the ***pyroxene*** group, named by A. N. Winchell for its occurrence at Pigeon Point, Minnesota, USA. The composition is intermediate between *clinoenstatite* and *diopside*.

Composition:
$4[(Mg,Fe^{2+},Ca)(Mg,Fe^{2+})Si_2O_6]$.

Properties: Monoclinic, $P2_1/c$; short prismatic crystals, microphenocrysts, or coronas around olivine or an ***orthopyroxene***; prismatic cleavage {110} at near right angles; H=6; G=3.3–3.4; color: greenish borwn, brown, or black.

Occurrence: Quickly chilled andesite and dacite lavas or small gabbroic intrusions, in the latter commonly inverted to ***orthopyroxene***, also in the Moore Co. (North Carolina, USA) meteorite.

K. FRYE

References

Am. Min. **55**, 1195–1209.
DHZ, 2A, 162–196.

Cross-references: *Clinopyroxenes; Meteoritic Minerals; Minerals, Uniaxial and Biaxial; Orthopyroxenes; Pyroxene Group; Skeletal Crystals.*

PIMELITE (***Smectite*** gr.)
$(Ni,Mg)_3Si_4O_{10}(OH)_2 \cdot 4H_2O$
mono. *Am. Min.* **64**, 615

PINAKIOLITE mono. *Am. Min.* **59**, 985
$(Mg,Mn^{2+})_2Mn^{3+}BO_5$ *Can. Min.* **16**, 475

PINCHITE *Am. Min.* **61**, 340
$Hg_5O_4Cl_2$ orth. *Can. Min.* **12**, 417

PINNOITE *Dana*, II, 334
$MgB_2O_4 \cdot 3H_2O$ tetr.

PINTADOITE *Dana*, II, 1053
$Ca_2V_2O_7 \cdot 9H_2O$

PIRSSONITE *Dana*, II, 232
$Na_2Ca(CO_3)_2 \cdot 2H_2O$ orth.

Cross-references: *Green River Mineralogy; Saline Minerals.*

PISANITE = Cuproan *melanterite*

PISTACITE = *Epidote*

PITCHBLENDE = Massive *uraninite*

Cross-references: *Mineral Classification: Principles.*

PITCHSTONE—Dark resinous volcanic glass

PITTICITE Hydrous ferric arsenate-sulfate
Dana, II, 1014

PLAGIOCLASE–The series *albite, oligoclase, andesine, labradorite, bytownite*, and *anorthite* of the ***feldspar*** group

Cross-references: *Crystallography: Morphological; Feldspar Group; Meteoretic Minerals; Orthopyroxenes; Plagioclase Feldspars; Refractive Index; Refractory Minerals; Rock-Forming Minerals; Soil Mineralogy; Thermoluminescence.*

PLAGIONITE (***Plagionite*** gr.)
$Pb_5Sb_8S_{17}$ *Am. Min.* **59**, 1127
mono. *Min. Mag.* **37**, 442

PLAGIONITE GROUP–*Fülöppite, plagionite, heteromorphite*, and *semseyite.*

PLANCHÉITE orth. *Am. Min.* **62**, 491
$Cu_8Si_8O_{22}(OH)_4 \cdot H_2O$ *Science* **154**, 506

PLASMA–Green variety of chalcedony
Dana, III, 218

PLATARSITE *Am. Min.* **64**, 657
$Pt(As,S)_2$ iso. *Can. Min.* **15**, 385

PLATINIRIDIUM *Dana*, I, 110
(Ir,Pt) iso.

Cross-reference: *Native Elements and Alloys.*

PLATINUM The native metal named by d'Ulloa in 1748 from the Spanish *platina*, diminutive of *plata*, silver.

Composition: 4[Pt], with variable amounts of Pd, Ir, Fe, and Ni.

Properties: Isometric, *Fm*3*m*; commonly in small scales or grains; H=4–4.5; G=21.45 (pure), native alloys 14–19; malleable and ductile; color: steel to dark gray; metallic luster; insoluble except in hot aqua regia; magnetism dependent upon Fe content.

Occurrence: In ultramafic rocks, especially dunites, in *quartz* veins, and in blacksands and placers as in the Pinto River, Colombia; Tura River, USSR; Ivalo River, Finland; Rhine River, West Germany; Takaka and Gorge Rivers, New Zealand; with *gold* sands in Quebec and British Columbia, Canada; and in placers in the Sierra Nevada, California, USA.

K. FRYE

References

Can. Min. **15**, 59–69.
Dana, I, 106–109.

Cross-references: *Blacksand Minerals; Metallic Minerals; Mineral and Ore Deposits; Native Elements and Alloys; Ore Microscopy; Placer Deposits.* Vol. IVA: *Platinum: Element and Geochemistry.*

PLATTNERITE A lead oxide named after Prof. K. F. Plattner of Freiberg.

Composition: PbO_2.

Properties: Tetragonal, *P*4/*mnm*; rutile structure; commonly massive; color: iron black; streak: brown.

Occurrence: Found in lead mines in Idaho and Scotland.

D. H. SPEIDEL

Reference: *Dana*, I, 581–583.

PLATYNITE *Dana*, I, 474
$PbBi_2(Se,S)_3$ trig.

PLAYFAIRITE *Am. Min.* **53**, 1424
$Pb_{16}Sb_{18}S_{43}$ mono. *Can. Min.* **9**, 191

PLEONASTE = Ferroan *spinel*

PLEONECTITE = *Hedyphane*
Am. Min. **58**, 562

PLESSITE–Intergrowth of *kamacite* and *taenite*

Cross-reference: *Widmanstätten Structures.*

PLOMBIERITE orth. *Am. Min.* **39**, 1038
$Ca_5H_2Si_6O_{18} \cdot 6H_2O$(?)

PLUMALSITE *Am. Min.* **53**, 349
$Pb_4Al_2Si_7O_{21}$(?) orth.

PLUMBAGO = *Graphite*

PLUMBOBETAFITE (***Pyrochlore*** gr.)
$Pb_2Ti_2O_6(OH)$ iso. *Am. Min.* **62**, 405

PLUMBOFERRITE *Dana*, I, 726
$PbFe_4O_7$ trig.

PLUMBOGUMMITE (***Crandallite*** gr.)
$PbAl_3(PO_4)(PO_3OH)(OH)_6$
trig. *Dana*, II, 831

PLUMBOJAROSITE (***Alunite*** gr.)
$PbFe^{3+}_6(SO_4)_4(OH)_{12}$
trig. *Dana*, II, 568

PLUMBOMICROLITE (***Pyrochlore*** gr.)
$Pb_2(Ta,Nb)_2O_6(OH)$ iso. *Am. Min.* **62**, 405

PLUMBOPALLADINITE *Am. Min.* **56**, 1121
Pd_3Pb_2 hexa.

PLUMBOPYROCHLORE (***Pyrochlore*** gr.)
$(Pb,Y,U,Ca)_{2-x}Nb_2O_6(OH)$
Am. Min. **55**, 1068;
iso. **62**, 405

POITEVINITE *Am. Min.* **50**, 263
$(Cu,Fe,Zn)SO_4 \cdot H_2O$ mono. *Can. Min.* **8**, 109

POLARITE *Am. Min.* **55**, 1810
Pd(Pb,Bi) orth.

POLHEMUSITE tetr. *Am. Min.* **63**, 1153
(Zn,Hg)S *Min. Mag.* **43**, 1066

POLIANITE = *Pyrolusite*

POLKOVICITE *Am. Min.* **66**, 437
$(Fe,Pb)_3(Ge,Fe)S_4$ iso.

POLLUCITE iso. *Am. Min.* **52**, 1515
$(Cs,Na)_2Al_2Si_4O_{12} \cdot H_2O$

POLYBASITE A silver sulfosalt named by Rose in 1829 from the Greek *poly*, many, and *basis*, bases, in reference to the many metallic bases present.

Composition: $(Ag,Cu)_{16}Sb_2S_{11}$.

Properties: Monoclinic, *C2/m*; pseudohexagonal tabular crystals; also massive; imperfect basal {001} cleavage; metallic luster; color: steel gray to iron black; streak: black.

Occurrence: In low-temperature veins with other Ag and Pb sulfosalts, various sulfides, carbonates, and *quartz* as at Joachimstal, Pribram, and Schemnitz, Czechoslovakia; Freiberg, East Germany; the silver districts of Chile, Peru, and Mexico; at Cobalt and Red Lake, Ontario, Canada; and in the USA from the silver veins in Arizona, Colorado, Idaho, Montana, and Nevada.

K. FRYE

Reference: *Dana*, I, 351-353.

POLYCRASE orth. *Can. Min.* **14**, 111
$(Y,Ca,Ce,U,Th)(Ti,Nb,Ta)_2O_6$

POLYDYMITE (***Linnaeite*** gr.)
$NiNi_2S_4$ iso. *Min. Mag.* **43**, 733

Cross-reference: *Thermometry, Geologic.*

POLYHALITE A hydrated potassium calcium magnesium sulfate named by Stromeyer in 1818 from the Greek *poly*, many, and *hals*, salt, in reference to its many ions.

Composition: $K_2Ca_2Mg(SO_4)_4 \cdot 2H_2O$.

Properties: Triclinic; commonly in granular, fibrous, or foliated masses; polysynthetically twinned; H=3.5; G=2.78; resinous luster; translucent; color: white to gray except made pink or red by fine *hematite* inclusions; bitter taste.

Occurrence: As grains or beds in the salt deposits of Upper Austria; Stassfurt, East Germany; Lorraine, France; northwestern Iran; western Kazakhstan, USSR; and Texas-New Mexico potash region, USA. Also from volcanic action at Vesuvius, Italy.

K. FRYE

Reference: *Dana*, II, 458-461.

Cross-reference: *Saline Minerals.*

POLYLITHIONITE (***Mica*** gr.)
$KLi_2AlSi_4O_{10}(F,OH)_2$ mono. DHZ, 3, 87

Cross-reference: *Phyllosilicates.*

POLYMIGNITE orth. *Dana*, I, 764
$(REE,Ca,Zr)(Ta,Nb,Ti)O_4$

PORTLANDITE Calcium hydroxide named by Tilley (1933) for its occurrence in Portland cement.

Composition: $Ca(OH)_2$.

Properties: Trigonal, $C\bar{3}m$; brucite structure; pearly luster; alters to *calcite.*

Occurrence: Found at Scawt Hill, Ireland.

D. H. SPEIDEL

Reference: *Dana*, I, 641-642.

Cross-reference: *Portland Cement Mineralogy.*

POSNJAKITE *Am. Min.* **52**, 1582
$Cu_4(SO_4)(OH)_6 \cdot H_2O$ mono.

POTARITE *Am. Min.* **45**, 1093
PdHg tetr.

POTASH ALUM *Dana*, II, 472
$KAl(SO_4)_2 \cdot 12H_2O$ iso.

POTASH ***FELDSPAR*** = Potassium ***Feldspar:*** *microcline, orthoclase*, and *sanidine*

Cross-reference: *Staining Techniques.*

POUBAITE *Am. Min.* **63**, 1283
$PbBi_2(Se,Te,S)_4$ trig. *Min. Abst.* **29**, 342

POUGHITE orth. *Am. Min.* **53**, 1075
$Fe_2^{3+}(TeO_3)_2(SO_4)\cdot 3H_2O$

POWELLITE A calcium molybdate of the ***scheelite*** group, named in 1891 for American geologist and explorer John Wesley Powell.

Composition: $4[CaMoO_4]$ with Mo replaced by W toward *scheelite.*

Properties: Tetragonal, $I4_1/a$; crystals bipyramids or tablets, also massive, foliated, or incrusting; H=3.5, G=4.2; luster commonly greasy or pearly; color: yellow, brown, blue, or black.

Occurrence: A secondary mineral often formed from the oxidation of *molybdenite.*

K. FRYE

Reference: *Dana*, II, 1079–1081.

PRASE–Green variety of chalcedony

PREHNITE A calcium aluminosilicate hydroxide, first named by Werner (1790) in honor of Col. von Prehn who first found it at the Cape of Good Hope.

Composition: $Ca_2[Al_2Si_3O_{10}](OH)_2$; SiO_2, 43.6%; Al_2O_3, 24.8%; CaO, 27.2%; H_2O, 4.4% by wt. Appreciable amounts of Fe substitute for Al; a small replacement of Si by Al; dehydration stepwise, losing 9.9% H_2O at maximum; rehydration poor; slowly gelatinized by HCl.

Properties: Orthorhombic, *P2cm; a*=4.61Å, *b*=5.47Å, *c*=18.48Å; H=6–6.5; G=2.90–2.95; luster vitreous to weakly pearly; color: pale green, yellow, gray, or white; subtransparent to translucent; habit in tabular groups, in barrel-shaped aggregates, or in reniform globular masses; cleavage good on {001}, weak on {110}; twinning fine lamellar. Optically, refractive indices, α=1.611–1.632, β=1.615–1.642, γ=1.632–1.665; birefringence = 0.022–0.035; $2V$=65°–69° (+); optic axial plane (010); $\alpha=x$, $\beta=y$, $\gamma=z$; dispersion usually $r>v$; sometimes shows "hour-glass" structure.

Occurrence: Commonly as a secondary or hydrothermal mineral in veins, cavities, and amygdules in mafic volcanic rocks, frequently associated with ***zeolites***; in veins in granite, monzonite, and diorite, pseudomorphous after *laumontite,* ***axinite***, and *clinozoisite*; in contact-metamorphosed limestone and marl; regionally occurs in the zone characterized by *pumpellyite* and(or) *prehnite* of low-grade metamorphosed graywackes and mafic tuffaceous rocks.

A. IIJIMA
M. UTADA

References

Coombs, D. S., 1960. Lower grade mineral facies in New Zealand, *21st Internat. Geol. Cong. Rept.* Norden, pt. 13, 339–351.

GSA Bull. **70**, 879–920.

J. Geol. Soc. Japan, **70**, 180–192.

Cross-references: *Minerals, Uniaxial and Biaxial; Naming of Minerals; Rock-Forming Minerals; Vein Minerals.*

PREISWERKITE mono. *Am. Min.* **65**, 1134
$NaMg_2Al[(OH)_2(Al_2Si_2O_{10})]$

PREOBRAZHENSKITE *Am. Min.* **55**, 1071
$Mg_3B_{11}O_{15}(OH)_9$ orth.

PRICEITE *Dana*, II, 341
$Ca_4B_{10}O_{19}\cdot 7H_2O$ tric.(?)

Cross-reference: *Saline Minerals.*

PRIDERITE *Am. Min.* **36**, 793
$(K,Ba)(Ti,Fe)_8O_{16}$ tetr. *Min. Mag.* **29**, 496

PRIORITE = *Aeschynite-(Y)*
Am. Min. **51**, 152

Cross-reference: *Blacksand Minerals.*

PROBERTITE *Am. Min.* **44**, 712
$NaCaB_5O_9\cdot 5H_2O$ mono.

PROCHLORITE = *Ripidolite*

PROSOPITE *Dana*, II, 121
$CaAl_2(F,OH)_8$ mono. *Min. Abst.* **28**, 157

PROSPERITE mono. *Am. Min.* **65**, 208
$HCaZn_2(AsO_4)_2(OH)$ *Can. Min.* **17**, 87

PROUDITE mono. *Am. Min.* **61**, 839
$CuPb_{7.5}Bi_{9.3}(S,Se)_{22}$

PROUSTITE A silver sulfosalt named by Beudant in 1832 after the French chemist, J. L. Proust (1755–1826).

Composition: Ag_3AsS_3; Ag, 65.4%; As, 15.2%; S, 19.4% by wt.

Properties: Trigonal, $R\bar{3}c$; prismatic or rhombohedral crystals, commonly massive, compact, or in disseminated grains; rhombohedral $\{10\bar{1}1\}$ cleavage; H=2–2.5; G=5.57;

brittle; adamantine luster; translucent; color: ruby red; streak: red.

Occurrence: A low-temperature or secondary-enrichment vein mineral associated with *pyrargyrite* and other silver minerals. Localities include Joachimsthal, Czechoslovakia, Saxony, East Germany; Chanarcillo, Chile; and in the USA in the silver districts of the western mountain states.

K. FRYE

Reference: *Dana*, I, 366–369.

Cross-reference: *Ore Microscopy.*

PRZHEVALSKITE *Am. Min.* **43**, 381
$Pb(UO_2)_2(PO_4)_2 \cdot 4H_2O$ orth.

Cross-reference: *Naming of Minerals.*

PSEUDOAUTUNITE tetr.(?) *Am. Min.* **50**, 1505
$(H_3O)_4Ca_2(UO_2)_2(PO_4)_4 \cdot 5H_2O$(?)

PSEUDOBOLEITE tetr. *Min. Mag.* **44**, 101
$Pb_5Cu_4Cl_{10}(OH)_8 \cdot 2H_2O$

PSEUDOBROOKITE An iron titanium oxide once thought to be similar to *brookite.*

Composition: Fe_2TiO_5.

Properties: Orthorhombic, *Cmcm*; {100} tablets with {010} cleavage; color: dark brown to black in mass.

Occurrence: Found in cavities in igneous rocks as at Vesuvius, Italy, and Crater Lake, California, USA.

D. H. SPEIDEL

Reference: *Dana*, I, 736–738.

Cross-reference: *Magnetic Minerals.*

PSEUDOCOTUNNITE *Dana*, II, 96
K_2PbCl_4(?) orth.(?)

PSEUDOLAUEITE
$Mn^{2+}Fe_2^{3+}(PO_4)_2(OH)_2 \cdot 8H_2O$
mono. *Am. Min.* **54**, 1312

Cross-reference: *Pegmatite Minerals.*

PSEUDOMALACHITE
$Cu_5(PO_4)_2(OH)_4 \cdot H_2O$
mono. *Am. Min.* **66**, 176

PSEUDOWALLASTONITE = *Cyclowollastonite* *Am. Min.* **58**, 560

Cross-reference: *Portland Cement Mineralogy.*

PSILOMELANE = Wad, massive oxides of Mn not otherwise identified
Science **212**, 1024

Cross-references: *Authigenic Minerals; Blowpipe Analysis; Metallic Minerals; Ore Microscopy.*

PTILOLITE = *Mordenite* *Am. Min.* **43**, 1224

PUCHERITE *Am. Min.* **37**, 423
$BiVO_4$ orth.

PUMPELLYITE mono. *Am. Min.* **61**, 176
$Ca_2MgAl_2(SiO_4)(Si_2O_7)(OH)_2 \cdot H_2O$
DHZ, 1, 227

Cross-reference: *Rock-Forming Minerals.*

PURPURITE *Dana*, II, 675
$Mn^{3+}PO_4$ orth.

Cross-reference: *Pegmatite Minerals.*

PUTORANITE *Min. Mag.* **43**, 1066
$Cu_{16-18}(Fe,Ni)_{18-19}S_{32}$ iso.

P-VEATCHITE mono. *Am. Min.* **45**, 1221
$Sr_2B_{11}O_{16}(OH)_5 \cdot H_2O$

PYRALSPITE–*PYrope-ALmandine-SPessartine* subgroup of the ***garnet*** group (q.v.)

PYRARGYRITE A silver sulfosalt named by Glocker in 1831 from the Greek *pyr*, fire, and *argyros*, silver, for its color and composition.

Composition: Ag_3SbS_3; Ag, 59.7%; Sb, 22.5%; S, 17.8% by wt.

Properties: Trigonal, *R3c*; prismatic or rhombohedral crystals, commonly twinned, also massive compact; rhombohedral $\{10\bar{1}1\}$ cleavage; H=2.5; G=5.85; adamantine luster; translucent; color: deep red with a purplish streak.

Occurrence: A late-primary or secondary-enrichment vein mineral more common than *proustite* with which it is associated. Localities include Saxony, East Germany; Colquechaca, Bolivia; Atacama, Chile; Guanajuato, Mexico; Cobalt, Ontario; and in the Comstock Lode, Nevada, USA.

K. FRYE

Reference: *Dana*, I, 362–366.

Cross-reference: *Ore Microscopy.*

PYRIBOLE = Pyroxenes + amphiboles
Am. Min. **63**, 239

PYRITE An iron sulfide named by Dioscorides in A.D. 50, from the Greek *pyr*, fire, referring to the sparks it gives off when struck with steel.

Composition: FeS_2. Fe, 46.6%; S, 53.4% by wt; may contain Ni or Co. A complete solid solution may exist between *pyrite* and *bravoite*, $(Ni,Fe)S_2$.

Properties: Isometric, $Pa3$ (also triclinic pseudocubic); most common forms are the cube, the faces of which are usually striated, the pyritohedron, and the octahedron, may be intergrown with or overgrown on *marcasite*; H=6–6.5; G=5.02; brittle; luster: metallic, splendent; color: pale brass-yellow; easily altered to oxides of iron, usually limonite.

Occurrence: It is the most common and widespread of the sulfides; found as direct magmatic segregation and as an accessory mineral in igneous rock, in contact metamorphic, and in vein deposits. It is a common mineral in sedimentary rocks, being both primary and secondary in origin.

Uses: A source of sulfur and also mined for the gold or copper associated with it. Rocks that contain *pyrite* are unsuitable for structural purposes because the ready oxidation of the *pyrite* in them would serve both to disintegrate the rock and to stain it with iron oxide.

J. A. GILBERT

References

Am. Min. **62**, 1168–1172.
Dana, I, 282–290.

Cross-references: *Authigenic Minerals; Beudantite Group; Blowpipe Analysis; Crystal Habits; Crystallography: Morphological; Fire Clays; Gossan; Green River Mineralogy; Halotrichite Group; Magnetic Minerals; Meteoritic Minerals; Mineral Deposits: Classification; Mineral Industries; Ore Microscopy; Pegmatite Minerals; Pigments and Fillers; Polymorphism; Pseudomorphism; Rock-Forming Minerals; Soil Mineralogy; Staining Techniques; Thermometry, Geologic; Vein Minerals.*

PYRITE GROUP–*Aurostibite, bravoite, cattierite, erlichmanite, fukuchilite, geversite, hauerite, insizwaite, krutaite, laurite, malanite, maslovite, michenerite, penroseite, pyrite, sperrylite, testibiopalladite, trogtalite, vaesite*, and *villamaninite.*

PYROAURITE A hydrous magnesium iron hydroxyl carbonate.

Composition: $Mg_6Fe_2^{3+}(OH)_{16}CO_3 \cdot 4H_2O$.

Properties: Rhombohedral, $R\bar{3}m$; basal plates with basal {0001} cleavage; color: brown to white; polymorphic with *sjögrenite*; isostructural with *hydrotalcite* and *stichtite*.

Occurrence: Found in veins at Långban, Sweden.

D. H. SPEIDEL

Reference: *Am. Min.* **54**, 296–299.

PYROBELONITE *Can. Min.* **10**, 117
$PbMn(VO_4)(OH)$ orth. *Min. Mag.* **41**, 85

PYROCHLORE (***Pyrochlore*** gr.)
$(Na,Ca)_2(Nb,Ta,Ti)_2O_6(O,OH,F)$
iso. *Am. Min.* **62**, 403

PYROCHLORE GROUP–***Pyrochlore*** subgroup: *Pyrochlore, kalipyrochlore, bariopyrochlore, yttropyrochlore, ceriopyrochlore, plumbopyrochlore, uranpyrochlore;* ***Microlite*** subgroup: *Microlite, stannomicrolite, bariomicrolite, plumbomicrolite, bismutomicrolite, uranmicrolite;* ***Betafite*** subgroup: *yttrobetafite, plumbobetafite, betafite.*

PYROCHROITE A manganese hydroxide.

Composition: $Mn(OH)_2$.

Properties: Trigonal, $C\bar{3}m$; brucite structure; colorless to yellow or brown.

Occurrence: A hydrothermal mineral found at Franklin, New Jersey, USA.

D. H. SPEIDEL

Reference: *Dana*, I, 639–641.

PYROLUSITE A manganese oxide.

Composition: MnO_2.

Properties: Tetragonal, $P4/mnm$; rutile structure; usually massive or powdery; metallic luster; color: steel-gray; streak: black.

Occurrence: An important ore of manganese in Westphalia, West Germany; Hillsborough, Canada; and Batesville, Arkansas, USA.

D. H. SPEIDEL

Reference: *Dana*, I, 562–566.

Cross-references: *Magnetic Minerals; Metallic Minerals; Mineral and Ore Deposits; Ore Microscopy; Skeletal Crystals; Soil Mineralogy.*

PYROMORPHITE A lead chlorophosphate of the ***apatite*** group, named by Hausmann in 1813 from the Greek *pyr*, fire, and *morph*, form, because a molten globule assumes a crystalline shape upon cooling.

Composition: $Pb_5Cl(PO_4)$, with Ca and Fe replacing Pb and AsO_4 replacing PO_4 ($AsO_4 > PO_4$ is *mimetite*).

Properties: Hexagonal, $C6_3/m$; prismatic, barrel-shaped, or skeletal crystals, also globular, reniform, or bottryoidal; H=3.5–4; G=6–7; resinous to subadamantine luster; transparent to translucent; color: yellow, orange, green, brownish, or reddish.

Occurrence: A secondary mineral in the oxidized portions of lead-ore veins and associated with *cerussite, smithsonite, hemimorphite, anglesite, pyrite, vanadinite, wulfenite,* and *mottramite.* Localities include Corrèze, France; the Rhineland, West Germany; Saxony, East Germany; Pribram, Czechoslovakia; Cornwall and Cumberland, England; Beresovsk, USSR; New South Wales, Australia; mines in Namibia; Zacatecas and Durango, Mexico; the Steele mining district, British Columbia; and in the USA in the Coeur d'Alene district, Idaho, and Gila County, Arizona.

K. FRYE

References

Dana, II, 889–895.
Rocks Min. **51,** 130–133.

Cross-references: *Apatite Group; Museums, Mineralogical; Skeletal Crystals.*

PYROPE A magnesium aluminum silicate of the ***garnet*** group named "fire-like" for its color and appearance.

Composition: $8[Mg_3Al_2Si_3O_{12}]$ with Mg replaced by Fe^{2+} toward *almandine* and by Mn toward *spessartine* and with Al replaced by Cr^{3+} toward *knorringite.*

Properties: Isometric *Ia*3*d*; crystals dodecahedra or trapezohedra, singly or in combination, also granular or massive; H=6.5–7.5; G=3.5; vitreous luster; color: deep red to black.

Occurrence: An indicator of high pressure formation in ultramafic rocks and their derivatives and in high-grade metamorphic rocks, also placers (q.v.).

Uses: Gemstone and abrasive material.

K. FRYE.

Reference: DHZ, 1, 77–112.

Cross-references: *Abrasive Materials; Clinopyroxenes; Garnet Group; Gemology; Mantle Mineralogy; Minerals, Uniaxial and Biaxial.*

PYROPHANITE A manganese titanium oxide.

Composition: $MnTiO_3$; end member of the ilmenite-type structure solid-solution series.

Properties: Trigonal, $R\bar{3}$; color: deep blood-red; streak: brownish yellow.

Occurrence: Found in ore deposits in Sweden.

D. H. SPEIDEL

Reference: *Dana,* I, 535–541.

PYROPHYLLITE A hydroxyl aluminosilicate which bears a marked physical resemblance to *talc,* named from the Greek *pyr,* fire, and *phyllon,* leaf. It is slightly harder than *talc* and is sometimes grouped with *talc* in commercial sales.

Composition: $Al_4(Si_8O_{20})(OH)_4$; commonly found rather pure when separated from admixed minerals.

Properties: Monoclinic, $C2/c$; foliated, radiated lamellar, granular to compact, the compact variety sometimes slaty; H=1–2; G=2.65–2.9; perfect basal {001} cleavage; pearly luster on folia, dull to glistening luster when compact; color: white, yellow, pale blue, grayish, brownish to apple green; greasy feel. Optically (–), $2V$=53°–62°; α=1.534–1.556, β=1.586–1.589, γ=1.596–1.601, $\delta \cong 0.050$; pleochroism, absorption greater for vibration directions in (001) plane; colorless in thin section.

Occurrence: In some schistose rocks; with *quartz* and ***micas*** in hydrothermal veins.

Uses: In pencils; thermal and electrical insulation; in lubricants; fillers for paper and rubber; as a carrier for insecticides.

E. BOOY

References

Am. Min. **63,** 96–108.
Dana, E. S. (Ford, W. E., ed.), 1949. *A Textbook of Mineralogy.* New York: Wiley, 683p.
DHZ, 3, 115–120.
Grim, R. E., 1968. *Clay Mineralogy.* New York: McGraw-Hill, 596p.
Kerr, P. F. ed., 1951. *Reference Clay Minerals, American Petroleum Institute Research Project 49.* New York: Columbia University, 701p.

Cross-references: *Clays, Clay Minerals; Mineral Industries; Minerals, Uniaxial and Biaxial; Phyllosilicates; Pigments and Fillers; Polysomatism; Pseudomorphism; Rock-Forming Minerals; Soil Mineralogy; Staining Techniques.*

PYROSMALITE trig. Winchell, II, 359
$(Fe^{2+},Mn)_8Si_6O_{15}(OH,Cl)_{10}$

PYROSTILPNITE Dana, I, 369
Ag_3SbS_3 mono.

PYROXENE GROUP–*Acmite, aegirine, aegirine-augite, augite,* bronzite, *clinoenstatite, clinoferrosilite, clinohypersthene, diopside, enstatite, fassaite, hedenbergite, hypersthene, jadeite, johannsenite, kanoite,* omphacite, *orthoferrosilite, pigeonite, spodumene,* and *ureyite.*

Cross-references: *Amphibole Group; Clinopyroxenes; Crystal Growth; Orthopyroxenes; Pyroxene Group; Refractory Minerals; Rock-Forming Minerals; Soil Mineralogy.*

PYROXFERROITE *Am. Min.* **55**, 2137
$(Fe^{2+},Ca)SiO_3$ tric. DHZ, 2A, 600

Cross-reference: *Lunar Minerals.*

PYROXMANGITE DHZ, 2A, 600
$MnSiO_3$ tric.

Cross-reference: *Rock-Forming Minerals.*

PYRRHOARSENITE = *Berzeliite*
Am. Min. **58**, 562

PYRRHOTITE (PYRRHOTINE) An iron sulfide named "pyrrotin" by Breithaupt in 1835 from the Greek *pyrros,* reddish; changed by Dana in 1869.

Composition: FeS; but nearly always deficient in Fe; small amounts of Fe replaced by Ni, Co, Mn, and Cu, although most Ni results from oriented intergrowths with *pentlandite. Troilite* is stroichiometric FeS of meteritic origin.

Properties: Monoclinic and hexagonal; tabular crystals rare, generally massive granular or lamellar; H=3.5–4.5; G=4.58–4.65; metallic luster; color is bronze yellow with dark brown, sometimes irridescent tarnish; variably magnetic.

Occurrence: In mafic igneous rocks associated with *pentlandite, chalcopyrite,* and other sulfides; also found in contact metamorphic deposits; in high-temperature veins; in granite pegmatites; and occasional sediments. Localities include Kisbanya, Rumania; Trepča, Yugoslavia; the Bushveld, South Africa; Mt. Isa, Queensland, Australia; Minas Gerais, Brazil; Llallagua, Bolivia; St. Stephen, New Brunswick, and the Sudbury district, Ontario, Canada; and Cumberland County, Maine, USA.

K. FRYE

References

Dana, I, 231–235.
DHZ, 5, 144-157.

Cross-references: *Diamond; Green River Mineralogy; Magnetic Minerals; Mantle Mineralogy; Metallurgy; Meteoritic Minerals; Ore Microscopy; Pegmatite Minerals; Rock-Forming Minerals; Thermoluninescence; Thermometry, Geologic.*

Q

QUARTZ A silicon oxide, the name being derived from the German *Querklufterz*, cross-vein ore.

Composition: SiO_2.

Properties: Trigonal, $C3_2 2$ (right handed) or $C3_1 2$ (left handed); prismatic crystals with combined positive and negative rhombohedral terminations most common, also massive or granular; conchoidal fracture; H=7; G=2.65; colorless, white, yellow, pink, blue, brown, or black; transparent. Optically uniaxial (+), ω=1.544; ϵ=1.553. Inverts to high *quartz* at 573°C; polymorphic with *tridymite, cristobalite, coesite,* and *stishovite.*

Occurrence: *Quartz* is one of the most widespread minerals, found in felsic volcanic and plutonic rocks, in metamorphic rocks, as detrital grains in sedimentary rocks, as authigenic overgrowths, in mineral veins, in meteorites, and in lunar rocks.

C. G. I. FRIEDLAENDER

References*

Dana, III, 334p.
DHZ, 4, 179–230.
*See *Quartz* for additional references.

Cross-references: *Abrasive Materials; Alkali Feldspars; Authigenic Minerals; Blacksand Minerals; Collecting Minerals; Crystal Growth; Crystal Habits; Crystallography: History; Crystallography: Morphological; Diagenetic Minerals; Electron Microscopy (Scanning); Electron Microscopy (Transmission); Etch Pits; Fire Clays; Gemology; Gossan; Green River Mineralogy; Human and Vertebrate Mineralogy; Invertebrate and Plant Mineralogy; Jade; Lunar Minerals; Magnetic Minerals; Mantle Mineralogy; Meteoritic Minerals; Mineral Classification: Principles; Mineral Properties; Minerals, Uniaxial and Biaxial; Mohs Scale of Hardness; Museums, Mineralogical; Optical Mineralogy; Ore Microscopy; Pegmatite Minerals; Phantom Crystals; Piezoelectricity; Plastic Flow in Minerals; Pleochroic Halos; Polymorphism; Portland Cement Mineralogy; Pseudomorphism; Quartz; Refractory Minerals; Rock-Forming Minerals; Sands, Glass and Building; Skeletal Crystals; Soil Mineralogy; Staining Techniques; Synthetic Minerals; Thermometry, Geologic; Twinning; Vein Minerals.* Vol. IVA: *Geochemistry of Sedimentary Silica.*

QUEITITE mono. *Am. Min.* **65**, 407
$Pb_4Zn_2(SiO_4)(Si_2O_7)(SO_4)$
Min. Mag. **43**, 1066

QUENSELITE A lead manganese oxyhydroxide.

Composition: $PbMnO_2(OH)$.

Properties: Monoclinic, *P*2; crystals tabular parallel to {010}; basal {001} cleavage; pitch black with brown-gray streak.

Occurrence: Found at Långban, Sweden.

D. H. SPEIDEL

Reference: *Dana,* I, 729–730.

QUENSTEDTITE *Dana,* II, 535
$Fe_2^{3+}(SO_4)_3 \cdot 10H_2O$ tric.

QUETZALCOATLITE *Am. Min.* **59**, 874
$Zn_8Cu_4(TeO_3)_3(OH)_{18}$
hexa. *Min. Mag.* **39**, 261

QUICKSILVER = *Mercury*

R

RABBITTITE mono. *Am. Min.* **40**, 201
$Ca_3Mg_3(UO_2)_2(CO_3)_6(OH)_4 \cdot 18H_2O$

RAGUINITE *Am. Min.* **54**, 1495, 1741
$TlFeS_2$ orth.

RAITE orth. *Am. Min.* **58**, 1113
$Na_4Mn_3Si_8(O,OH)_{24} \cdot 9H_2O(?)$

RAJITE *Am. Min.* **64**, 1331
$CuTe_2^{4+}O_5$ mono. *Min. Mag.* **43**, 91

RALSTONITE iso. *Am. Min.* **50**, 1851
$Na_xMg_xAl_{2-x}(F,OH)_6 \cdot H_2O$

RAMDOHRITE *Am. Min.* **39**, 161; **60**, 621
$PbAgSb_3S_6$ orth.

RAMEAUITE *Am. Min.* **58**, 805
$K_2CaU_6O_{20} \cdot 9H_2O$ mono. *Min. Mag.* **38**, 781

RAMMELSBERGITE A nickel arsenide named by Dana in 1854 in honor of the German mineral chemist, K. F. Rammelsberg (1813–1899).

Composition: $2[NiAs_2]$; Ni, 28.15%; As, 71.85% by wt.

Properties: Orthorhombic, *Pnnm;* crystals rare, usually massive, granular to prismatic or fibrous in structure; uneven fracture; H=5.5–6, G=7.1; metallic luster; color: tin-white tinged red; streak: grayish black; alters readily to *annabergite*.

Occurrence: Found in vein deposits with other nickel and cobalt minerals. Often in association with *loellingite, smaltite, nickeline, arsenopyrite, cobaltite,* and *safflorite*. Well-known locations are Schneeberg, East Germany; Lölling-Hüttenberg; Austria; Valais, Switzerland, and the Eldorado mine, Great Bear Lake, Northwest Territories, Canada.

Use: A minor source of nickel and arsenic.

G. C. MOERSCHELL

Reference: *Dana,* I, 309–310.

Cross-reference: *Ore Microscopy.*

RAMSDELLITE *Am. Min.* **47**, 47
MnO_2 orth. *Science* **212**, 1024

RANCIEITE hexa. *Am. Min.* **54**, 1741; **63**, 762
$(Ca,Mn^{2+})Mn_4^{4+}O_9 \cdot 3H_2O$

RANKAMAITE orth. *Am. Min.* **55**, 1814
$(Na,K,Pb,Li)_3(Ta,Nb,Al)_{11}(O,OH)_{30}$

RANKINITE *Min. Abst.* **30**, 216, 285
$Ca_3Si_2O_7$ mono. RRW, 509

Cross-references: *Portland Cement Mineralogy; Rock-Forming Minerals.*

RANQUILITE = *Haiweeite* *Am. Min.* **45**, 1078

RANSOMITE *Am. Min.* **55**, 729
$CuFe_2^{3+}(SO_4)_4 \cdot 6H_2O$ mono.

RANUNCULITE mono. *Min. Mag.* **43**, 321
$AlH(UO_2)(PO_4)(OH)_3 \cdot 4H_2O$
Min. Rec. **11**, 112

RASPITE *Dana,* 11, 1089
$PbWO_4$ mono.

RASVUMITE *Am. Min.* **64**, 776; **65**, 477
KFe_2S_3 orth.

RATHITE *Dana,* I, 455
$(Pb,Tl)_3As_5S_{10}$ mono.

RATHITE-II = *Liveingite* *Am. Min.* **54**, 1498

RAUENTHALITE mono. *Am. Min.* **50**, 805
$Ca_3(AsO_4)_2 \cdot 10H_2O$ or tric.

RAUVITE *USGS Bull.* **1064**, 263
$Ca(UO_2)_2V_{10}O_{28} \cdot 16H_2O$

REALGAR An arsenic sulfide named by Wallerius in 1747 from the Arabic *Raj al ghar,* powder of the mine.

Composition: AsS; As, 70.0%; S, 30.0% by wt.

Properties: Monoclinic, $P2_1/n$; prismatic crystals short and vertically striated, commonly coarse to fine granular, or as an incrustation; H=1.5–2; G=3.56; resinous to greasy luster; transparent when fresh; red to orange color and streak.

Occurrence: Found in lead-, silver-, and gold-ore veins associated with *orpiment, stibnite,* and other arsenic minerals; as a volcanic sublimate; and as a hot-spring deposit. Localities include the Binnatal, Switzerland; Transylvania, Rumania; Matra, Corsica; Kern County, California, and Humbolt County, Nevada, USA.

Uses: Formerly used to give brillian white to fireworks and as a pigment.

K. FRYE

Reference: *Dana,* I, 255–258.

Cross-references: *Blowpipe Analysis; Ore Microscopy.*

RECTORITE–Interstratified *beidellite-**mica***

Cross-reference: *Clays, Clay Minerals.*

REDDINGITE *Min. Mag.* **43**, 789
$Mn_3^{2+}(H_2O)_3(PO_4)_2$ orth.

REDINGTONITE (***Halotrichite*** gr.)
$(Fe,Mg,Ni)(Cr,Al)_2(SO_4)_4 \cdot 22H_2O$
mono. *Dana,* II, 529

REDLEDGEITE tetr. *Am. Min.* **46**, 1201
$Mg_4Cr_6Ti_{23}Si_2O_{61}(OH)_4$(?)

REEDMERGNERITE (***Feldspar*** gr.)
$NaBSi_3O_8$ *Am. Min.* **50**, 1827
tric. *Min. Mag.* **43**, 905

Cross-references: *Alkali Feldspars; Green River Mineralogy.*

REEVESITE trig. *Am. Min.* **52**, 1190
$Ni_2Fe_6^{3+}(CO_3)(OH)_{16} \cdot 4H_2O$

REFIKITE *Am. Min.* **50**, 2110
$C_{20}H_{32}O_2$ orth.

REINERITE *Am. Min.* **62**, 1129
$Zn_3(AsO_3)_2$ orth.

RENARDITE orth. *USGS Bull.* **1064**, 227
$Pb(UO_2)_4(PO_4)_2(OH)_4 \cdot 7H_2O$

RENIERITE tetr. *Am. Min.* **38**, 794
$Cu_3(Fe,Ge,Zn)(S,As)_4$ *Econ. Geol.* **52**, 612

RETGERSITE *Dana,* II, 497
$NiSO_4 \cdot 6H_2O$ tetr.

RETZIAN *Am. Min.* **52**, 1603
$Mn_2Y(AsO_4)(OH)_4$ orth.

REYERITE trig. *Am. Min.* **58**, 517
$(Na,K)_4Ca_{14}(Si,Al)_{24}O_{60}(OH)_5 \cdot 5H_2O$

REZBANYITE *Dana,* I, 470
$Pb_3Cu_2Bi_{10}S_{19}$ orth.(?)

RHABDOPHANE *Am. Min.* **65**, 1065
(*REE*)$PO_4 \cdot H_2O$ hexa. *Dana,* II, 774

RHENIUM hexa. *Am. Min.* **63**, 1283
Re *Min. Mag.* **43**, 1066

RHINESTONE–*Quartz* or other material cut to imitate *Diamond*

Cross-reference: *Gemology*

RHODESITE orth. *Am. Min.* **54**, 251
$(Ca,Na_2,K_2)_8Si_{16}O_{40} \cdot 11H_2O$
Min. Mag. **31**, 607

RHODIUM *Am. Min.* **61**, 340
(Rh,Pt) iso. *Can. Min.* **12**, 399

RHODIZITE *Am. Min.* **51**, 533
$CsAl_4Be_4B_{11}O_{25}(OH)_4$ iso.

Cross-reference: *Pegmatite Minerals.*

RHODOCHROSITE Manganese carbonate of the ***calcite*** group, named by Hausmann in 1813 from the Greek *rhodochros,* rose colored, in reference to its typical color.

Composition: $MnCO_3$, with Mn replaced by Fe, Ca, Mg, Zn, Co, and Cd.

Properties: Trigonal, $R\bar{3}c$; rhombohedral crystals uncommon, generally massive, botryoidal, or incrusting; perfect rhombohedral $\{10\bar{1}1\}$ cleavage; H=3.5–4; G=3.7; vitreous luster; transparent to translucent; color: pink, rose-red, or brown; streak: white.

Occurrence: Found in hydrothermal veins with other carbonates, as a secondary mineral in residual manganese deposits, and in pegmatites. Localities include Freiberg, East Germany; Westphalia, Hesse, and Rhineland, West Germany; Kapnik, Rumania; Calamarca provine, Argentina; and in the USA, in Butte, Montana, and in veins at many Colorado localities.

K. FRYE

References

Dana, II, 171–175.
DHZ, 5, 263–271.

Cross-references: *Authigenic Minerals; Calcite Group; Magnetic Minerals; Minerals, Uniaxial and Biaxial; Pegmatite Minerals; Rock-Forming Minerals; Staining Techniques.*

RHODONITE A manganese silicate named by Jasche in 1819 from the Greek *rhodon*, rose, in reference to its color.

Composition: $MnSiO_3$; always containing some Ca, Fe, Mg, or Zn.

Properties: Triclinic, $C\bar{1}$; crystals rare, commonly fine-grained massive; prismatic {110} {1$\bar{1}$0} cleavage at 92.5°; H=5.5–6.5; G=3.6–3.8; vitreous luster; transparent to translucent; color: rose pink to brownish red.

Occurrence: Found in metasomatic manganese ore bodies; in pegmatites that assimilated these ores; and as contact metamorphic product of *rhodochrosite*. Localities include Långban, Sweden; Devonshire and Cornwall, UK; Forno, Italy; Postmasburg, South Africa; Bhandara district, India; Honshu, Japan; Broken Hill, New South Wales, Australia; and Franklin, New Jersey, and Balmat, New York, USA.

K. FRYE

References

Am. Min. **63**, 1137–1142.
DHZ, 2A, 586–599.
Min. Abst. **28**, 26.

Cross-reference: *Clinopyroxenes; Gemology; Magnetic Minerals; Rock-Forming Minerals; Topotaxy.*

RHODOSTANNITE *Am. Min.* **54**, 1218
$Cu_2FeSn_3S_8$ hexa. *Min. Mag.* **36**, 1045

RHODUSITE = *Magnesioriebeckite*
Am. Min. **63**, 1051

RHOMBOCLASE orth. *Can. Min.* **17**, 63
$(H_3O)^+Fe^{3+}(SO_4)_2 \cdot 3H_2O$ *Dana*, II, 436

RHÖNITE tric. *Am. Min.* **55**, 864
$Ca_2(Mg,Ti,Al,Fe)_6(Si,Al)_6O_{20}$

Cross-reference: *Meteoritic Minerals.*

RICHELLITE amorph. *Am. Min.* **55**, 135
$Ca_3Fe^{3+}_{10}(OH,F)_{12}(PO_4)_8 \cdot nH_2O$
Dana, II, 956

RICHETITE *USGS Bull.* **1064**, 91
oxide of Pb and U mono.(?)

RICHTERITE A sodium calcium hydroxyl silicate of magnesium, iron, manganese, and aluminum, member of the ***amphibole*** group, and named by Breithaupt in 1865 after Professor J. L. Richter, mineral chemist.

Composition:
$Na_2Ca(Mg,Fe,Mn,Al)_5(Si_8O_{22})(OH,F)_2$.

Properties: Monoclinic, $C2/m$; elongate crystals, fibrous, or asbestiform; prismatic [110] cleavage; H=5–6; G=2.97–3.45; color: brown, yellow, brownish red, or pale to dark green; streak: colorless, pale green, or yellow; variably pleochroic.

Occurrence: Found in thermally metamorphosed limestones, in skarns, and as a hydrothermal product in alkaline igneous rocks, the manganese-rich varieties being associated with manganese-ore deposits. Localities include Långban, Sweden; Chikla, India; Sauka, Burma; West Kimberley, Western Australia; and in the USA at Iron Hill, Colorado; Libby, Montana; and in the Leucite Hills, Wyoming.

K. FRYE

References

Am. Min. **60**, 367–374; **63**, 1023–1052.
DHZ, 2, 352–359.

Cross-references: *Amphibole Group; Isomorphism; Meteoritic Minerals.*

RICKARDITE *Dana*, I, 198
Cu_7Te_5 orth. *Min. Abst.* **27**, 227

RIEBECKITE A hydroxyl sodium iron silicate of the ***amphibole*** group, named by Sauer in 1888 in honor of Dr. E. Riebeck.

Composition: $2[Na_2Fe^{2+}_3Fe^{3+}_2Si_8O_{22}(OH)_2]$; Mg replaces Fe^{2+} to *magnesioriebectite* and F replaces OH.

Properties: Monoclinic, $C2/m$; embedded longitudinally striated prismatic crystals; good prismatic {110} cleavage at 56°; H=5; G=3.02–3.42; vitreous luster; color: dark blue to black; very strongly pleochroic.

Occurrence: Usually in an asbestiform variety known as "crocidolite" (blue asbestos) in metamorphic and igneous rocks; especially in soda-rich rhyolites, granites and pegmatites. Important localities include northern Nigeria; South Africa; Hammersley Ranges, Western Australia, and Lusaka, Zambia.

Use: When crocidolite is replaced by *quartz*, it is used as a gemstone known as "tiger-eye".

G. C. MOERSCHELL

References

Am. Min. **63**, 1023–1052.
DHZ, 2, 333–351.

Cross-references: *Amphibole Group; Asbestos; Clinopyroxenes; Gemology; Minerals, Uniaxial and Biaxial.*

RIJKEBOERITE = *Bariomicrolite*
Am. Min. **62**, 407

RINGWOODITE (***Spinel*** gr.)
$(Mg,Fe)_2SiO_4$ *Am. Min.* **54**, 1219
iso. *Nature* **221**, 943

Cross-reference: *Meteoritic Minerals.*

RINKITE, RINKOLITE = *Mosandrite*
Am. Min. **43**, 795

RINNEITE *Dana*, II, 107
$K_3NaFe^{2+}Cl_6$ trig.

Cross-reference: *Saline Minerals.*

RIPIDOLITE (***Chlorite*** gr.)
$(Mg,Fe,Al)_6(Si,Al)_4O_{10}(OH)_8$ DHZ, 3, 137

RIVADAVITE *Am. Min.* **52**, 326
$Na_6MgB_{24}O_{40}\cdot 22H_2O$ mono.

RIVERSIDEITE orth. *Am. Min.* **39**, 1038
$Ca_5Si_6O_{16}(OH)_2\cdot 2H_2O$
Min. Mag. **30**, 155

ROBERTSITE mono. *Am. Min.* **59**, 48
$Ca_3Mn_4^{3+}(PO_4)_4(OH)_6\cdot 3H_2O$

ROBINSONITE *Am. Min.* **60**, 621
$Pb_7Sb_{11}S_{25}$ tric. *Min. Abst.* **29**, 140

ROCKBRIDGEITE (***Frondelite*** gr.)
$(Fe^{2+},Mn)Fe_4^{3+}(OH)_5(PO_4)_3$
Am. Min. **34**, 513;
orth. **55**, 135

ROCK CRYSTAL = *Quartz*

ROCK SALT = *Halite*

ROCK WOOL = Asbestos

RODALQUILARITE *Am. Min.* **53**, 2104
$H_3Fe_2^{3+}(TeO_3)_4Cl$ tric.

ROEBLINGITE mono. *Am. Min.* **51**, 504
$Pb_2Ca_7Si_6O_{14}(OH)_{10}(SO_4)_2$

ROEDDERITE (***Osumilite*** gr.)
$(Na,K)_2(Mg,Fe)_5Si_{12}O_{30}$
hexa. *Am. Min.* **51**, 949

Cross-reference: *Meteoritic Minerals.*

ROEMERITE A hydrated ferric and ferrous sulfate named in 1858 after the German geologist Friedrich Adolph Roemer.

Composition: $Fe^{2+}Fe_2^{3+}(SO_4)_4\cdot 14H^2O$.

Properties: Triclinic, $P\bar{1}$; tabular, granular, or massive; one perfect cleavage; H=3–3.5; G=2.2; color: yellow to rust brown; astringent saline taste.

Occurrence: With other iron sulfates from the oxidation of *pyrite*.

K. FRYE

References

Am. Min. **55**, 78–89.
Dana, II, 520–522.

ROENTGENITE *Am. Min.* **38**, 868, 932
$Ca_2(REE)_3(CO_3)_5F_3$ trig.

Cross-reference: *Syntaxy.*

ROEPPERITE = zincian *tephroite*

ROESSLERITE *Dana*, II, 712
$MgHAsO_4\cdot 7H_2O$ mono.

ROGERSITE = *Weinschenkite*

ROGGIANITE tetr. *Am. Min.* **55**, 322
$NaCa_6Al_9Si_{13}O_{46}\cdot 20H_2O$

ROHAITE *Am. Min.* **65**, 208
$TlCu_5SbS_2$ tetr. *Min. Abst.* **30**, 421

ROKÜHNITE *Am. Min.* **66**, 219
$Fe^{2+}Cl_2\cdot 2H_2O$ mono. *Min. Mag.* **43**, 1066

ROMANÈCHITE *Science* **212**, 1024
$BaMn^{2+}Mn_8^{4+}O_{16}(OH)_4$ mono.

ROMARCHITE *Am. Min.* **57**, 1555
SnO tetr. *Can. Min.* **10**, 916

ROMEITE (***Stibiconite*** gr.)
$(Ca,Fe,Mn,Na)_2(Sb,Ti)_2O_6(O,OH,F)$
iso. *Dana*, II, 1020

RÖMERITE = ROEMERITE

RÖNTGENITE = ROENTGENITE

ROOSEVELITE *Dana*, II, 697
$BiAsO_4$ mono.

ROQUESITE *Am. Min.* **48**, 1178
$CuInS_2$ tetr. *Can. Min.* **18**, 361

ROSASITE A copper zinc hydroxylcarbonate named in 1908 for the Rosas mine, Sukis, Sardinia, Italy.

Composition: $4[(Cu,Zn)_2(CO_3)(OH_2]$.

Properties: Monoclinic; green to blue spherules; H≅4.5; G=4.0–4.2.

Occurrence: Secondary mineral in oxidized zone of zinc-copper-lead ore deposits.

K. FRYE

Reference: *Dana,* II, 251–252.

ROSCHERITE A hydrated basic phosphate of Be, Ca, Mn, and Fe, named by Slavik in 1914 for Walter Roscher.

Composition:
$(Ca,Mn,Fe)_3Be_3(PO_4)_3 \cdot 2H_2O$.

Properties: Monoclinic, *C2/c*, or triclinic, $C\bar{1}$; G=2.93–2.94; H=4.5; color: dark brown to olive green; prismatic crystals parallel to [001] with 8- or 6-sided cross sections, largest crystals about ≅1 mm long; good {001} and distinct {010} cleavage; refractive index, n=1.64–1.68, transparent.

Occurrence: In drusy cavities in granite in Saxony, and from pegmatites in North Carolina and Maine, USA, and Minas Gerais, Brazil. Associated with *morinite, lacroixite, childrenite,* ***apatite, tourmaline,*** *beryl, frondelite, faheyite, variscite,* and *quartz.*

L. H. FUCHS

References

Am. Min. **43,** 824–838.
Tsch. Min. Pet. **24,** 169–178 (*Am. Min.* **63**, 427).

ROSCOELITE (***Mica*** gr.) mono. DHZ, 3, 14
$K(V,Al,Mg)_3(AlSi_3)O_{10}(OH)_2$

Cross-reference: *Metallic Minerals.*

ROSELITE mono. and tric. *Can. Min.* **15**, 36
$Ca_2(Co,Mg)(AsO_4)_2 \cdot 2H_2O$ *Dana,* II, 723

ROSEMARYITE mono. *Am. Min.* **65**, 811
$(Na,Ca,Mn)(Mn,Fe^{2+})(Fe,Mg)Al(PO_4)_3$
Min. Mag. **43**, 227

ROSENBUSCHITE tric. RRW, 526
$(Ca,Na)_3(Zr,Ti)Si_2O_8F$ Winchell, II, 518

Cross-reference: *Rock-Forming Minerals.*

ROSENHAHNITE *Am. Min.* **62**, 503
$Ca_3Si_3O_8(OH)_2$ tric. *Min. Abst.* **29**, 473

ROSE *QUARTZ* = Pink to red *quartz* *Dana,* III, 186

ROSICKYITE RRW, 527
γ-S mono.

Cross-references: *Allotropy; Native Elements and Alloys.*

ROSSITE *Dana,* II, 1053
$CaV_2O_6 \cdot 4H_2O$ tric.

ROSTITE *Am. Min.* **64**, 1331
$Al(SO_4)(OH) \cdot 5H_2O$ orth. *Min. Mag.* **43,** 1066

RÖSSLERITE = ROESSLERITE

ROUBAULTITE tric. *Am. Min.* **57**, 1912
$Cu_2(UO_2)_3(OH)_{10} \cdot 5H_2O$
Min. Abst. **22**, 229

ROUTHIERITE *Am. Min.* **60**, 947
$TlHgAsS_3$ tetr.

ROWEITE *Am. Min.* **59**, 60
$Ca_2Mn_2B_4O_7(OH)_6$ orth. *Min. Abst.* **27**, 338

ROWLANDITE amorph. *Am. Min.* **63**, 754
$(Y,RE,Fe,Mn)_3(SiO_4)_2(F,OH)$
Min. Abst. **27**, 67

ROZENITE *Can. Min.* **7**, 751
$Fe^{2+}SO_4 \cdot 4H_2O$ mono. *Min. Abst.* **15**, 364

RUBELLITE = Pink *elbaite*

Cross-references: *Gemology, Pegmatite Minerals.*

RUBY–Red gem variety of *corundum*

Cross-references: *Crystal Growth; Gemology; Mineral and Ore Deposits; Placer Deposits; Refractory Minerals; Synthetic Minerals.*

RUBY SILVER *Proustite, pyrargyrite or cuprite*

RUCKLIDGEITE *Am. Min.* **63**, 599
$(Bi,Pb)_3Te_4$ trig. *Min. Abst.* **28**, 449

RUIZITE mono. *Min. Abst.* **29**, 84
$CaMn^{3+}(SiO_3)_2(OH) \cdot 2H_2O$
Min. Mag. **41**, 429

RUSAKOVITE *Am. Min.* **45**, 1315
$(Fe,Al)_5(VO_4,PO_4)_2(OH)_9 \cdot 3H_2O$

RUSSELLITE *Dana,* I, 604
$nBi_2O_3 \cdot yWO_3$ tetr.

RUSTENBURGITE — *Am. Min.* **61**, 340
$(Pt,Pd)_3Sn$ iso. — *Can. Min.* **13**, 146

RUSTUMITE — *Am. Min.* **50**, 2104
$Ca_4Si_2O_7(OH)_2$ orth. — *Min. Mag.* **34**, 1

RUTHENARSENITE — *Am. Min.* **61**, 177
(Ru,Ni)As orth. — *Can. Min.* **12**, 280

RUTHENIUM — *Am. Min.* **61**, 177
Ru hexa.(?)

RUTHERFORDINE — *Am. Min.* **41**, 127
$(UO_2)CO_3$ orth. *USGS Bull.* **1064**, 104, 844

RUTILE A widespread accessory mineral in metamorphic and sedimentary rocks named from Latin *rutilus*, red.

Composition: TiO_2, with ferrous and ferric iron, Nb and Ta as impurities; possibly a solid solution among *rutile, tapiolite,* and *mossite* called *strüverite*; melts at 1825°C; is the high temperature polymorph of *brookite* and *anastase.*

Properties: Tetragonal, $P4_2/mnm$; characteristic structure with Ti in 6-fold coordination; H=6–6.5; G=4.23–5.5; color: reddish brown but may be yellow, blue, violet, or black; streak: pale brown; optically ω= 2.605–2.613; ϵ=2.899–2.901; very strong dispersion and high relief are characteristic.

Occurrence: Widely distributed, common in metamorphic rocks such as amphibolites and eclogites as well as contact metamorphics. Found in granitic pegmatites and associated veins.

Uses: Source of Ti, white pigment, and a gem stone.

D. H. SPEIDEL

References

DHZ, 5, 34-39.
J. Am. Chem. Soc., 48(8), 391–398.

Cross-references: *Authigenic Minerals; Black-sand Minerals; Blowpipe Analysis; Crystal Growth; Crystal Habits; Crystallography: Morphological; Diamond; Fire Clays; Gemology; Lunar Minerals; Mantle Mineralogy; Meteoritic Minerals; Mineral Industries; Mineral and Ore Deposits; Minerals, Uniaxial and Biaxial; Ore Microscopy; Pegmatite Minerals; Pigments and Fillers; Polymorphism; Pseudomorphism; Refractory Minerals; Rock-Forming Minerals; Soil Mineralogy; Synthetic Minerals; Vein Minerals.*

RUTILE GROUP—*Cassiterite, plattnerite, pyrolusite, rutile*, and *stishovite.*

RYNERSONITE — *Am. Min.* **63**, 709
$Ca(Ta,Nb)_2O_6$ orth. — *Min. Mag.* **43**, 1066

S

SAAMITE = Strontian *apatite*
Am. Min. **51**, 1306

SABATIERITE *Am. Min.* **64**, 1331
Cu_6TlSe_4 orth.

SABINAITE *Can. Min.* **18**, 25
$Na_9Zr_4Ti_2O_9(CO_3)_8$
mono. *Min. Mag.* **43**, 1067

SABUGALITE
(***Autunite*** gr.) *Am. Min.* **36**, 671
$HAl(UO_2)_4$ tetr. *USGS Bull.* **1064**, 196
$(PO_4)_4 \cdot 16H_2O$

SAFFLORITE Cobalt arsenide, named by Breithaupt in 1835 from "bastard saffron" in allusion to its use as a pigment.

Composition: $4[CoAs_2]$; considerable Fe present, with small amounts of Ni, Bi, Cu, and S.

Properties: Orthorhombic, *Pnnm;* crystals short to long prismatic, usually massive with a radial fibrous structure; good {100} cleavage; brittle, uneven fracture; metallic luster; H=4.5-5, G=7.4; color: tin-white, tarnishes to dark gray; streak: grayish black; strongly anisotropic.

Occurrence: Commonly in hydrothermal veins with *rammelsbergite, nickeline, smaltite, loellingite,* and other Co-Ni minerals. Found at Schneeberg, East Germany; Wittichen, Baden, West Germany; Nordmarken, Sweden; and Great Bear Lake, Canada.

G. C. MOERSCHELL

References

Am. Min. **53,** 1856-1881.
Dana, I, 307-309.

SAHAMALITE *Am. Min.* **38**, 741
$(Mg,Fe)Ce_2(CO_3)_4$ mono.

SAHLINITE *Dana,* II, 775
$Pb_{14}(AsO_4)_2O_9Cl_4$ mono.

SAINFELDITE *Am. Min.* **50**, 806
$H_2Ca_5(AsO_4)_4 \cdot 4H_2O$ mono.

SAKHAITE iso. *Am. Min.* **62**, 263
$Ca_{12}Mg_4(CO_3)_4(BO_3)_8(H_2O)$
Min. Abst. **27**, 133

SAKHAROVAITE *Am. Min.* **41**, 814;
$(Pb,Fe)(Bi,Sb)_2S_4$ orth. **45**, 1134

SAKURAIITE tetr. *Am. Min.* **53**, 1421
$(Cu,Fe,Zn)_3(In,Sn)S_4$

SALAMMONIAC *Dana,* II, 15
NH_4Cl iso.

SALÉEITE (***Autunite*** gr.)
$Mg(UO_2)_2(PO_4)_2 \cdot 8H_2O$
tetr. *USGS Bull.* **1064**, 177

SALESITE *Acta Cryst.* **15**, 1105
$Cu(IO_3)(OH)$ orth. *Am. Min.* **63**, 172

SALITE–Iron-rich *diopside* DHZ, 2A, 3

SALMONSITE = *Hureaulite* + *Jahnsite*
Am. Min. **64**, 466

SALT–See *halite*

SALTPETER = *Niter*

SAMARSKITE A rare-earth, uranium, iron, calcium niobotantalate of the ***euxenite*** group, named by Rose in 1847 for the Russian Col. von Samarski.

Composition: $(Y,Er,Ce,U,Fe,Ca)(Nb,Ta)_2O_6$.

Properties: Orthorhombic; elongated crystals; G=5.7; H=5-6; commonly metamict; color and streak: black to brown.

Occurrence: Found in granite pegmatite in Minsk, USSR, and Nuevo, California, USA.

D. H. SPEIDEL

Reference: *Dana,* I, 797-800.

Cross-references: *Blacksand Minerals; Pegmatite Minerals.*

SAMIRESITE = plumbian *uranpyrochlore*
Am. Min. **62**, 407

SAMPLEITE orth. *Min. Mag.* **42**, 369
$NaCaCu_5(PO_4)_4Cl \cdot 5H_2O$

SAMSONITE *Dana*, I, 393
$Ag_4MnSb_2S_6$ mono.

SAMUELSONITE (***Apatite*** gr.)
$(Ca,Ba)Fe_2^{2+}Mn_2^{2+}Ca_8Al_2(OH)_2(PO_4)_{10}$
Am. Min. **62**, 229
mono. *Min. Abst.* **27**, 257

SANBORNITE *Am. Min.* **43**, 517
$BaSi_2O_5$ orth.

SANDERITE *Am. Min.* **37**, 1072
$MgSO_4 \cdot 2H_2O$

SANIDINE A potassium sodium aluminosilicate of the ***feldspar*** group, the name derived from the Greek words *sanis,* tablet, and *-idos,* appearance, for its typical tabular habit.

Composition: $(K,Na)AlSi_3O_8$, with small amounts of Ba and Ca replacing (K,Na).

Properties: Monoclinic, $C2/m$; crystals tabular or square prisms; Carlsbad twinning common; H-6; G=2.57–2.58; vitreous luster; transparent to translucent; perfect prismatic cleavage {001}, {010}.

Occurrence: As phenocrysts in volcanic rocks as at Vogelsberg, Hesse, and Eifel, Rhineland, West Germany; Mt. Cimino, Italy; Végardó, Czechoslovakia; Zvečan, Yugoslavia; Taijii Kii, Japan; and in the USA at Kokomo, Colorado, and in the Mitchell Mesa rhyolite, Texas.

K. FRYE

References*

DHZ, 4, 1–93.
*For additional references see *Alkali Feldspars* and *Feldspar Group.*

Cross-references: *Alkali Feldspars; Feldspar Group; Mantle Mineralogy; Meteoritic Minerals; Minerals, Uniaxial and Biaxial; Order-Disorder; Plagioclase Feldspars; Polymorphism; Rock-Forming Minerals; Soil Mineralogy.*

SANJUANITE mono.(?) *Am. Min.* **53**, 1
$Al_2(PO_4)(SO_4)(OH) \cdot 9H_2O$

SANMARTINITE (***Wolframite*** gr.)
$(Zn,Fe)WO_4$ *Dana*, II, 1072
mono. *Min. Abst.* **29**, 478

SANTACLARAITE
$CaMn_4[Si_5O_{14}(OH)](OH) \cdot H_2O$
tric. *Am. Min.* **66**, 154

SANTAFEITE
$(Sr,Na)Mn_2^{3+}(OH)_2(VO_4)_2 \cdot 4H_2O$
orth. *Am. Min.* **43**, 677

SANTANAITE *Am. Min.* **58**, 966
$9PbO \cdot 2PbO_2 \cdot CrO_3$ hexa.

SANTITE *Am. Min.* **56**, 636
$KB_5O_8 \cdot 4H_2O$ orth.

SAPONITE (***Smectite*** gr.)
$(½Ca,Na)_{0.33}(Mg,Fe)_3(Si,Al)_4O_{10}(OH)_2 \cdot 4H_2O$
mono. DHZ, 3, 226

Cross-references: *Clay, Clay Minerals; Soil Mineralogy.*

SAPPHIRE—Blue gem variety of *corundum*

Cross-references: *Crystal Growth; Gemology; Mineral and Ore Deposits; Piezoelectricity; Placer Deposits; Synthetic Minerals.*

SAPPHIRINE *Am. Min.* **65**, 821
$(Mg,Al)_8(Al,Si)_6O_{20}$
mono. and tric. *Can. Min.* **18**, 373

Cross-reference: *Rock-Forming Minerals.*

SARABAUITE *Am. Min.* **63**, 715
$CaSb_{10}O_{10}S_6$ mono.

SARCOLITE tetr. *Am. Min.* **64**, 245
$Na_2Ca_6[Al_4Si_6O_{23}](OH,H_2O)_2[(Si,P)O_4]_{0.5}(CO_3,Cl)_{0.5}$

SARCOPSIDE *Am. Min.* **57**, 24
$(Fe^{2+},Mn)_3(PO_4)_2$ mono.

Cross-references: *Meteoritic Minerals; Pegmatite Minerals.*

SARD—Brown chalcedony *Dana*, III, 206

Cross-reference: *Gemology.*

SARDONYX—Banded sard

Cross-reference: *Gemology.*

SARKINITE *Min. Mag.* **43**, 681
$Mn_2^{2+}(AsO_4)(OH)$ mono.

SARMIENTITE mono. *Am. Min.* **53**, 2077
$Fe_2^{3+}(AsO_4)(SO_4)(OH) \cdot 5H_2O$

SARTORITE *Dana*, I, 478
$PbAs_2S_4$ mono.

SARYARKITE tetr. *Am. Min.* **49**, 1775
$(Ca,Y,Th)_2Al_4(SiO_4,PO_4)_4(OH)_6 \cdot 9H_2O$

SASSOLITE Boric acid named for the Tuscan locality, Sasso, Italy.

Composition: $B(OH)_3$.

Properties: Triclinic, $P\bar{1}$; related to gibbsite structure; crystals 6-sided basal plates; white to grayish.

Occurrence: Found about Tuscan volcanoes.

D. H. SPEIDEL

Reference: *Dana,* I, 662–663.

Cross-reference: *Saline Minerals.*

SATIMOLITE orth. *Am. Min.* **55**, 1069
$KNa_2Al_4B_6O_{15}Cl_3 \cdot 13H_2O$

SATIN SPAR–Fibrous, pearly *gypsum*

SATPAEVITE orth.(?) *Am. Min.* **44**, 1325
$Al_{12}V_2^{4+}V_6^{5+}O_{37} \cdot 30H_2O$(?)

SATTERLYITE *Am. Min.* **64**, 657
$(Fe,Mg)PO_4(OH)$ trig. *Can. Min.* **16**, 411

SAUCONITE (***Smectite*** gr.)
$Na_{0.33}Zn_3(Si,Al)_4O_{10}(OH)_2 \cdot 4H_2O$
Am. Min. **31**, 141
mono. DHZ, 3, 226

Cross-reference: *Soil Mineralogy.*

SAUSSURITE = *Albite* + an ***epidote***

SAZHINITE *Am. Min.* **60**, 162
$Na_3CeSi_6O_{15} \cdot 6H_2O$ orth.

SBORGITE *Am. Min.* **43**, 378
$NaB_5O_8 \cdot 5H_2O$ mono.

SCACCHITE *Dana,* II, 40
$MnCl_2$ trig.

SCAPOLITE GROUP–*Marialite,* dipyre, mizzonite, and *meionite.*

References

Am. Min. **64,** 1188
DHZ, 4, 321.

Cross-references: *Plagioclase Feldspars; Rock-Forming Minerals; Vein Minerals.*

SCARBROITE *Am. Min.* **43**, 384
$Al_5(CO_3)(OH)_{13} \cdot 5H_2O$
tric. *Min. Mag.* **43**, 615

SCAWTITE mono. *Am. Min.* **40**, 505
$Ca_7Si_6O_{18}(CO_3) \cdot 2H_2O$

SCHACHNERITE *Am. Min.* **58**, 347
$Ag_{1.1}Hg_{0.9}$ hexa.

SCHAFARZIKITE *Am. Min.* **37**, 136
$FeSb_2O_4$ tetr. *Min. Abst.* **27**, 300

SCHAIRERITE *Am. Min.* **56**, 174
$Na_{21}(SO_4)_7F_6Cl$ trig.

Cross-references: *Polycrystal; Saline Minerals.*

SCHALLERITE RRW, 542
$(Mn,Fe^{2+})_8Si_6As(O,OH,Cl)_{26}$

SCHAURTEITE hexa. *Am. Min.* **53**, 507
$Ca_3Ge(SO_4)_2(OH)_6 \cdot 3H_2O$

SCHEELITE Calcium tungstate named "scheelerz" by Karsten in 1800 after K. W. Scheele, the Swedish chemist who proved in 1781 the presence of tungsten oxide in the mineral; the name modified to its present form by Leonhard in 1821.

Composition: $CaWO_4$, with W replaced by Mo toward *powellite,* and by small amounts of Nb and Ta; Ca replaced by *REE.*

Properties: Tetragonal, $I4_1/a$; crystals commonly octahedral, also massive or columnar; prismatic {101} cleavage; H=4.5–5; G=5.9–6.1; adamantine to vitreous luster; translucent; colorless, white, brownish, greenish, or yellow; streak: white; fluoresces bright blue.

Occurrence: Found in limestone metamorphosed at the contact with granite, in *quartz* veins near granites, and in granite pegmatites, commonly associated with *wolfranite.* Localities include Saxony, East Germany; Zinnwald, Czechoslovakia; Piemont, Italy; Andalucia, Spain; Cornwall and Cumberland, UK; Huancaya, Peru; and in the USA in Inyo and Kern Counties, California; Cochise County, Arizona; Beaver County, Utah; in the Humbolt range, Nevada; and Fairfield, Connecticut.

Use: An ore of tungsten.

K. FRYE

Reference: *Dana,* II, 1074–1079.

Cross-references: *Blowpipe Analysis; Crystal Growth; Metallic Minerals; Mohs Scale of Hardness; Naming of Minerals; Ore Microscopy; Staining Techniques; Wolframite Group.*

SCHERTELITE *Am. Min.* **48**, 635
$(NH_4)_2MgH_2(PO_4)_2 \cdot 4H_2O$ orth.

SCHETELIGITE = *Betafite*(?)
Am. Min. **62**, 407

SCHIEFFELINITE *Am. Min.* **66**, 219
$Pb(Te,S)O_4 \cdot H_2O$ orth. *Min. Mag.* **43**, 771

SCHIRMERITE *Am. Min.* **64**, 244
$Ag_3Pb_{3-6}Bi_{9-7}S_{18}$ orth. *Can. Min.* **11**, 952

SCHIZOLITE = Manganoan *pectolite*
DHZ, 2A, 565

SCHLOSS-MACHERITE trig. *Am. Min.* **65**, 1068
$(H_3O,Ca)Al_3(SO_4,AsO_4)_2(OH)_6$ *Min. Abst.* **31**, 497

SCHMIEDERITE *Am. Min.* **49**, 1498
$(Pb,Cu)_2SeO_4(OH)_2$(?) mono.(?)

SCHMITTERITE *Am. Min.* **56**, 411
$(UO_2)TeO_3$ orth. *Min. Abst.* **28**, 324

SCHNEIDERHÖHNITE *Am. Min.* **59**, 1139
$Fe_8^{2+}As_{10}^{3+}O_{23}$ tric.

SCHODERITE *Am. Min.* **64**, 713
$Al_2(PO_4)(VO_4)\cdot 8H_2O$ mono.

SCHOENFLIESITE *Am. Min.* **57**, 1557
$MgSn(OH)_6$ iso.

SCHOEPITE *Am. Min.* **45**, 1026
$UO_2(OH)_2\cdot H_2O$ orth. *Can. Min.* **7**, 331

SCHOLZITE *Am. Min.* **60**, 1019
$CaZn_2(PO_4)_2\cdot 2H_2O$ orth.

SCHOONERITE orth. *Am. Min.* **62**, 246
$ZnMn^{2+}Fe_2^{2+}Fe^{3+}(OH)_2(H_2O)_7(PO_4)_3\cdot 2H_2O$
Min. Abst. **28**, 449

SCHORL An iron-rich, aluminum borosilicate of the ***tourmaline*** group. The term *schorl* has a complex derivation and has been used in the past for several other mineral species.

Composition:
$3[NaFe_3^{2+}Al_6(BO_3)_3Si_6O_{18}(OH,F)_4]$.

Properties: Trigonal, $R3m$; prismatic crystals with triangular or hexagonal cross section, also massive, columnar, and in parallel and radiating groups; cleavage poorly developed; H=7–7.5; G=3.10–3.25; luster vitreous to resinous; color: black; strongly pyroelectric and piezoelectric; refractive indices, ω= 1.655–1.675, ϵ=1.640–1.655; unit-cell dimensions, a=15.93–16.03 Å, c=7.12–7.19 Å.

Occurrence: Usually found in granites, granite pegmatites, and related reaction zones (greisen). Also a common accessory mineral in schists and gneisses. Localities include the pegmatite districts of New England, Southern California, and South Dakota, USA. Excellent crystals occur at Pierrepont, St. Lawrence Co., New York and in the Ramona district, San Diego Co., California.

P. E. ROSENBERG
F. F. FOIT

References

Battey, M. H., 1972. *Mineralogy for Students.* Edinburgh: Oliver and Boyd, 255–256.
Can. Min. **13**, 173–177.
Sinkankas, J., 1966. *Mineralogy: A First Course,* 2nd ed. Princeton, N.J.: Van Nostrand, 510–514.

Cross-references: *Minerals, Uniaxial and Biaxial; Pegmatite Minerals; Tourmaline Group.*

SCHORLOMITE—Black titanian *andradite garnet*

SCHREIBERSITE *Dana,* I, 124
$(Fe,Ni)_3P$ tetr.

Cross-references: *Meteoritic Minerals; Native Elements and Alloys.*

SCHREYERITE *Am. Min.* **62**, 395; **63**, 1182
$V_2Ti_3O_9$ mono.

SCHROECKINGERITE A hydrated sulfate-carbonate fluoride of uranium, sodium, and calcium, named in 1875 for J. von Schroeckinger, who discovered it at Joachimsthal, Bohemia, Czechoslovakia.

Composition:
$NaCa_3(UO_2)(CO_3)_3(SO_4)F\cdot 10H_2O$.

Properties: Orthorhombic; clusters of scaley crystals with one perfect cleavage; H=2.5; G=2.5; greenish yellow color and fluorescence.

Occurrence: An alteration product of *uraninite* associated with *gypsum.*

K. FRYE

References

Am. Min. **44**, 1020–1025.
USGS Bull. **1064**, 121.

SCHUBNELITE *Am. Min.* **57**, 1556
$Fe^{3+}VO_4\cdot H_2O$ tric.

SCHUCHARDTITE = Interlayered Ni-rich ***vermiculites*** and ***chlorites*** *Am. Min.* **64**, 1334

SCHUETTEITE *Am. Min.* **44**, 1026
$Hg_3(SO_4)O_2$ hexa.

SCHUILINGITE mono. *Am. Min.* **43**, 796
$Pb_3Ca_6Cu_2(CO_3)_8(OH)_6\cdot 6H_2O$

SCHULTENITE *Dana,* II, 661
$PbHAsO_4$ mono.

SCHWARTZEMBERGITE
$Pb_6(IO_3)_2Cl_4O_2(OH)_2$ orth. *Dana,* II, 317

SCHWATZITE, SCHWAZITE =
Mercurian *tetrahedrite*

SCOLECITE Hydrated calcium aluminosilicate of the ***zeolite*** group named by Gehlen and Fuchs (1813) from the Greek *skolex,* worm, alluding to the curl on a borax bead.

Composition: $Ca_8[Al_{16}Si_{24}O_{80}] \cdot 24H_2O$. Small amounts of Na and K may substitute for Ca; minor replacement of Al by Si may occur; Ca is exchanged with K, Li, NH_4, Ag, and Tl; molecular sieve properties good but not too selective due to adaptability of the aluminosilicate framework; transition to metascolecite at about 255°C; gelantinized by HCl.

Properties: Monoclinic (pseudo-tetragonal), *Cc;* a=18.52 Å, b=18.99 Å, c=6.55 Å, β= 90°39′; H=5; G=2.25–2.29; luster, vitreous or silky when fibrous; colorless, white, gray, yellow, or pink; transparent to subtranslucent; habit in prisms or radiating fibrous masses elongated in z direction; cleavage very good on {110} and {1$\bar{1}$0}; twinning common on {100}. Optically, refractive indices, α= 1.507–1.513, β=1.516–1.520, γ=1.517–1.521; birefringence = 0.007; $2V$=36°–56° (−); optic axial plane ⊥(010); $\gamma=y$, $\alpha\Lambda z$=18°; dispersion $r<v$, strong.

Occurrence: As *thomsonite-scolecite* in gabbro and *scolecite* believed to be of magmatic origin; in cavities, in basalt from Mull as the latest product of the sequence: *albite*→***chlorite***→*epidote*→*prehnite*→*scolecite*→(*heulandite*); in Iceland, in vesicles of the Tertiary basalts as the lower *scolecite-mesolite* zone; in marl, in contact with phonolite; as a hydrothermal vein in crevices in biotite gneiss near a pegmatite; and as a hydrothermal vein in a fracture in diorite, Antarctic Peninsula.

A. IIJIMA
M. UTADA

References

Antarctic J. **11,** 258–259.
J. Geol. **68,** 515–528.
Min. Mag. **24,** 227–253.

SCORODITE A hydrated iron arsenate named by Breithaupt in 1817 from the Greek *skorodion,* garlic-like, in reference to its odor when heated.

Composition: $FeAsO_4 \cdot 2H_2O$, with Fe replaced by Al (Al>Fe is *mansfieldite*), also partial replacement of AsO_4 by PO_4.

Properties: Orthorhombic, *Pcab;* rare crystals pyramidal, tabular, or prismatic, also encrusting, earthy, or massive; H=3.5–4; G=3.3 (pure); vitreous to adamantine luster; colorless, pale green, or brown.

Occurrence: Found in gossans from the oxidation of other arsenic minerals; in the oxidized portions of metallic veins; and around some hot springs. Localities include Lölling, Austria; Creuse, France; Laurium, Greece; Cornwall, UK; Tsumeb, Southwest Africa; and Minas Gerais, Brazil.

K. FRYE

Reference: *Dana,* II, 763–767.

Cross-reference: *Naming of Minerals.*

SCORZALITE mono. *Am. Min.* **35,** 1
$(Fe,Mg)Al_2(PO_4)_2(OH)_2$

Cross-reference: *Pegmatite Minerals.*

SEAMANITE *Am. Min.* **56,** 1527
$Mn_3(OH)_2(PO_4)B(OH)_4$ orth.

SEARLESITE *Am. Min.* **35,** 1014;
$NaBSi_2O_5(OH)_2$ mono. **61,** 123

Cross-reference: *Green River Mineralogy.*

SEDERHOLMITE *Am. Min.* **50,** 519
β-NiSe hexa.

SEDOVITE *Am. Min.* **51,** 530
$U(MoO_4)_2$ orth.

SEELIGERITE *Am. Min.* **57,** 327
$Pb_3Cl_3(IO_3)O$ orth.

SEGELERITE orth. *Am. Min.* **59,** 48;
$CaMgFe^{3+}(PO_4)_2(OH) \cdot 4H_2O$ **62,** 692

SEIDOZERITE mono. *Am. Min.* **44,** 467
$(Na,Ca)_2(Zr,Ti,Mn)_2Si_2O_7(O,F)_2$

SEINÄJOKITE *Am. Min.* **62,** 1059
$FeSb_2$ orth. *Min. Abst.* **28,** 327

SEKANINAITE *Am. Min.* **50,** 264;
$(Fe,Mg)_2Al_4Si_5O_{18}$ orth. **62,** 395

SELENITE = Transparent *gypsum* crystals

SELENIUM *Dana,* I, 136
Se trig.

Cross-reference: *Native Elements and Alloys.*

SELENJOSEITE = *LAITAKARITE*
Can. Min. 7, 677

SELENOLITE = *DOWNEYITE*
Am. Min. **62**, 316

SELEN-TELLURIUM *Dana,* I, 137
(Se,Te) trig.

SELIGMANNITE *Dana,* I, 411
$PbCuAsS_3$ orth.

SELLAITE *Dana,* II, 37
MgF_2 tetr.

SEMENOVITE orth. *Am. Min.* **64**, 202
$(Fe,Mn,Zn,Ti)RE_2Na_{0-2}(Ca,Na)_8(Si,Be)_{20}(O,OH,F)_{48}$

SEMSEYITE *Am. Min.* **59**, 1127
$Pb_9Sb_8S_{21}$ mono. *Dana,* I, 466

SENAITE (***Crichtonite*** gr.)
$Pb(Fe,Mn)_7Ti_{14}O_{38}$ *Am. Min.* **61**, 1203;
trig. **63**, 28

SENARMONTITE Antimony trioxide named by Dana in 1851 after Professor Henri de Sénarmont who first described the species.

Composition: Sb_2O_3.

Properties: Cubic, *Fd3m*; octahedral crystals; high-temperature polymorph (460°C) of *valentinite* with arsenolite structure; colorless.

Occurrence: An oxidation product of *stibnite.*

D. H. SPEIDEL

Reference: *Dana,* I, 544–545.

SENEGALITE *Am. Min.* **64**, 1243
$Al_2(OH)_3(H_2O)(PO_4)$ orth.

SENGIERITE *Am. Min.* **66**, 220
$Cu(UO_2)_2(VO_4)_2(OH)_2 \cdot 6H_2O$
mono. *USGS Bull.* **1064**, 258

SEPIOLITE A clay-like fibrous magnesiosilicate which, in dry masses, floats on water. Named by Glocker in 1847 from the Greek word for cuttlefish, whose bone is light and porous. Well known as "meerschaum," from the German word for sea froth, used in pipemaking.

Composition: $(H_2O)_4(OH)_2Mg_8Si_{12}O_{30} \cdot 8H_2O$; Mg end member of the series with *palygorskite.*

Properties: Monoclinic; smooth, earthy, or clay-like texture; H=2–2.5; G=2; color: white or faintly gray, yellow, red, or bluish-green shades of white. Optically (−); colorless to gray in thin section; $2V$=0°–60°; α=1.490–1.520, β=1.505–1.530, δ=0.009–0.015.

Occurrence: A secondary mineral with ***serpentine;*** with *opal;* in stratified deposits associated with *magnesite.*

Uses: In pipes and formerly as a substitute for soap (Morocco).

E. BOOY

References

Grim, R. E., 1968. *Clay Mineralogy.* New York: McGraw-Hill, 596p.

Kerr, P. F. ed., 1951. *Reference Clay Minerals, American Petroleum Institute Research Project 49.* New York: Columbia University, 701p.

Cross-references: *Clays, Clay Minerals; Diagentic Minerals; Green River Mineralogy; Soil Mineralogy.*

SEPTECHLORITE = Serpentine

SERANDITE *Am. Min.* **63**, 274
$Na_2(Mn,Ca)_4Si_6O_{17} \cdot H_2O$
tric. DHZ, 2A, 565

SERENDIBITE DHZ, 2A, 659
$Ca_2(Mg,Al)_6O_2[(Si,Al,B)_6O_{18}]$ tric.

SERGEEVITE *Min. Mag.* **43**, 1067
$Ca_2Mg_{11}(CO_3)_{13} \cdot 10H_2O$ trig.(?)

SERICITE-Fine-grained ***mica*** group minerals, commonly ***muscovite*** but also *paragonite* or ***illite***

Cross-references: *Pigments and Fillers; Plagioclase Feldspars.*

SERPENTINE GROUP—Magnesian members of the ***kaolinite-serpentine*** group, *antigorite, clinochrysotile, orthochrysotile,* and *lizardite.* Serpentine is also used as a name of a rock composed predominantly of these minerals.

Reference: *Can. Min.* **17**(4)

Cross-references: *Asbestos; Orthopyroxenes; Phyllosilicates; Rock-Forming Minerals.*

SERPIERITE mono. *Am. Min.* **54**, 328
$Ca(Cu,Zn)_4(SO_4)_2(OH)_6 \cdot 3H_2O$
Min. Abst. **30**, 295

SEYBERTITE = *CLINTONITE*
Am. Min. **52**, 1122

SHADLUNITE (***Pentlandite*** gr.)
$(Fe,Cu)_8(Pb,Cd)S_8$ iso. *Am. Min.* **58**, 1114

SHANDITE Am. Min. **35**, 425
$Ni_3Pb_2S_2$ trig. *Min. Abst.* **30**, 19

SHARPITE *Am. Min.* **24**, 658
$(UO_2)CO_3 \cdot H_2O$(?)
orth.(?) *USGS Bull.* **1064**, 106

SHATTUCKITE *Am. Min.* **62**, 491
$Cu_5(SiO_3)_4(OH)_2$ orth. *Science* **154**, 506

SHCHERBAKOVITE
$(K,Na,Ba)_3(Ti,Nb)_2Si_4O_{14}$
mono. *Am. Min.* **40**, 788

SHCHERBINAITE *Am. Min.* **58**, 560
V_2O_5 orth. *Econ. Geol.* **58**, 1186

SHERIDANITE (***Chlorite*** gr.)
$(Mg,Al)_6(Si,Al)_4O_{10}(OH)_8$ DHZ, 3, 137
Min. Mag. **30**, 277

SHERWOODITE tetr. *Am. Min.* **43**, 749;
$Ca_{4.5}(AlV^{5+}_{12}V^{4+}_2O_{40} \cdot 28H_2O$ **63**, 863

SHORTITE *Dana*, II, 222
$Na_2Ca_2(CO_3)_3$ orth.

Cross-references: *Green River Mineralogy; Saline Minerals.*

SIBERSKITE *Am. Min.* **48**, 433
$CaHBO_3$ mono.(?)

SICKLERITE *Dana*, II, 672
$Li(Mn^{2+},Fe^{3+})PO_4$ orth.

SIDERAZOT *Dana*, I, 126
Fe_5N_2 hexa.

Cross-reference: *Native Elements and Alloys.*

SIDERITE Iron carbonate of the ***calcite*** group and named from the Greek *sideros*, iron. The name "spherosiderite" was originally given to the globular variety by Hausmann in 1813 and subsequently shortened to its present spelling by Haidinger in 1845.

Composition: $FeCO_3$, with Fe replaced by Mn toward *rhodochrosite* or by Mg toward *magnesite;* also some Ca, Co, or Zn, rarely pure.

Properties: Trigonal, $R\bar{3}c$; rhombohedral crystals, commonly with curved faces, also earthy, globular, concretionary, compact, or massive granular; perfect rhombohedral $\{10\bar{1}1\}$ cleavage; H=3.5–4; G=3.96 (pure); vitreous luster; transparent to translucent; color: light to dark brown.

Occurrence: Found as bedded deposits with clay as near Radom, Poland; as black-band ores as in Durham, Northumberland, Somerset, and Yorkshire, UK; and as blackband ore and clay ironstone in the coal measures of Pennsylvania, Illinois, Indiana, Ohio, Kentucky, and West Virginia, USA. Also found in hydrothermal veins, in cavities in mafic igneous rocks, in pegmatites, and as limestone replacement.

K. FRYE

References

Dana, II, 166–171.
DHZ, 5, 272–277.

Cross-references: *Authigenic Minerals; Blowpipe Analysis; Calcite Group; Crystallography: Morphological; Fire Clays; Green River Mineralogy; Magnetic Minerals; Mineral and Ore Deposits; Minerals, Uniaxial and Biaxial; Pegmatite Minerals; Rock-Forming Minerals; Soil Mineralogy; Staining Techniques; Vein Minerals.*

SIDEROMELANE = Basaltic glass

SIDERONATRITE orth. *Can. Min.* **17**, 63
$Na_2Fe^{3+}(SO_4)_2)(OH) \cdot 3H_2O$

SIDEROPHYLLITE (***Mica*** gr.)
$K(Fe^{2+}_2Al)(Al_2Si_2)O_{10}(F,OH)_2$
mono. DHZ, 3, 57

SIDEROTIL *Am. Min.* **49**, 820
$Fe^{2+}SO_4 \cdot 5H_2O$ tric. *Can. Min.* **7**, 751

SIDORENKITE *Am. Min.* **64**, 1332
$Na_3Mn(PO_4)(CO_3)$ mono. *Min. Mag.* **43**, 1067

SIEGENITE (***Linnaeite*** gr.)
$(Ni,Co)_3S_4$ iso. *Min. Mag.* **43**, 733

Cross-reference: *Ore Microscopy.*

SIGLOITE tric. *Am. Min.* **47**, 1
$(Fe^{2+},Fe^{3+})Al_2(PO_4)_2(O,OH)_2 \cdot 8H_2O$

SILHYDRITE *Am. Min.* **57**, 1053
$3SiO_2 \cdot H_2O$ orth.

SILLÉNITE *Am. Min.* **28**, 521
Bi_2O_3 iso. *Dana*, I, 601

SILLIMANITE An aluminum silicate named after Benjamin Silliman, professor of chemistry at Yale University and founder of the *American Journal of Science.*

Composition: Al_2SiO_5, with little Fe^{3+} replacing Al.

Properties: Orthorhombic, *Pbnm*; prismatic, commonly in long, slender crystals, parallel groups, or compact fibrous aggregates; one perfect {010} cleavage; H=6.5–7.5; G=3.23–3.27; vitreous luster; transparent to translucent; colorless, white, bluish, greenish, or brown.

Occurrence: Found in gneiss or schists of the granulite facies associated with *corundum* or *quartz* and in sands and gravels derived therefrom.

K. FRYE

References

DHZ, 1, 121–128.
Science, **151**(3715), 1222–1225.
Varley, E. R., 1965. *Sillimanite.* London: Her Majesty's Stat. Off., 165p.

Cross-references: *Blacksands; Mineral and Ore Deposits; Minerals, Uniaxial and Biaxial; Optical Mineralogy; Refractory Minerals; Rock-Forming Minerals; Soil Mineralogy.*

SILVER A native metal, the name of which is common to Teutonic languages but of unknown derivation.

Composition: Ag; commonly alloyed with Au or Hg.

Properties: Isometric, *Fm3m*; rarely hexagonal, *P6₃mc*; commonly elongated, reticulated; arborescent, wiry, massive, or in scales; H=2.5–3; G=10.5 (pure Ag); malleable and ductile; metallic luster; hackly fracture; color: silver-white; streak: commonly tarnished gray to black.

Occurrence: Found in the oxidized zone of hydrothermal deposits with other silver minerals, sulfides, and ***zeolites*** as at Kongsberg, Norway; with nickel, cobalt, and silver arsenides and sulfides as at Cobalt, Ontario; with *uraninite* and nickel-cobalt minerals as at Joachimstal, Czechoslovakia; and in blacksands and placers. Other localities include Saxony, East Germany; Baden, West Germany; NE USSR; Tsumeb, Namibia; Broken Hill, New South Wales, Australia; the Northwest Territories and British Columbia, Canada; the states of Chihuahua, Durango, Sonora, and Zacatecas, Mexico; and in the USA at Bisbee, Arizona; in the Cascades, Colorado; the Silver City district, Idaho; the Keeweenaw Peninsula, Michigan; and at Butte, Montana.

K. FRYE

References

Dana, I, 96–99.
Min. Abst. **31,** 497–498.

Cross-references: *Blacksand Minerals; Crystal Habits; Gossan; Mineral Classification: Principles; Mineral and Ore Deposits; Museums, Mineralogical; Native Elements and Alloys; Ore Microscopy; Placers; Pseudomorphism; Refractive Index; Skeletal Crystals.* Vol. IVA: *Silver: Element and Geochemistry; Silver: Economic Deposits.*

SIMONELLITE *Am. Min.* **55,** 1818
C_1H_2 orth.

SIMPLOTITE *Am. Min.* **43,** 16
$CaV_4O_9 \cdot 5H_2O$ mono.

SIMPSONITE *Dana,* I, 771
$Al_4Ta_3O_{13}(F,OH)$ trig.

Cross-reference: *Pegmatite Minerals.*

SINCOSITE *Dana,* II, 1057
$CaV_2^{4+}(PO_4)_2(OH)_4 \cdot 3H_2O$ tetr.

SINHALITE *Am. Min.* **37,** 700, 1072
$MgAlBO_4$ orth. *Min. Mag.* **29,** 841

SINJARITE *Min. Abst.* **31,** 356
$CaCl_2 \cdot 2H_2O$ *Min. Mag.* **43,** 643

SINNERITE *Am. Min.* **57,** 824; **60,** 998
$Cu_6As_4S_9$ tric.

SINOITE *Am. Min.* **50,** 521
Si_2N_2O orth. *Science,* **146,** 256

Cross-reference: *Meteoritic Minerals.*

SISERSKITE = *Iridosmine*

SISMONDINE = Mg-rich *chloritoid*

SJÖGRENITE hexa. *Dana,* I, 658
$Mg_6Fe_2^{3+}(OH)_{16}(CO_3) \cdot 4H_2O$

SKINNERITE *Am. Min.* **59,** 889
Cu_3SbS_3 mono.

SKLODOWSKITE *USGS Bull.* **1064,** 300
$Mg(UO_2)_2Si_2O_7 \cdot 6H_2O$ mono.

SKUTTERUDITE Cobalt arsenide named by Haidinger in 1845 for the locality Skutterud, Norway.

Composition: $CoAs_{2-3}$, with Co replaced by Ni (Ni>Co is *nickel-skutterudite*), also by Fe.

Properties: Cubic, *I3m*; crystals cubes or octahedra, some skeletal; cubic and octahedral cleavage; H=5.5–6; G=6.5; brittle, metallic luster; tin white to silver gray with iridescent tarnish; black streak.

Occurrence: Found with other Co and Ni arsenides as at Skutterud, Norway; Schneeberg, East Germany; the Turtmannthal, Switzerland; Cornwall and Lancashire, UK; Bou-Azzer, Morocco; Cobalt, Ontario, Canada; and in Gunnison County, Colorado, USA.

K. FRYE

References

Dana, I, 342–346.
Min. Rec. **9**, 69–73.

Cross-references: *Mineral and Ore Deposits; Ore Microscopy.*

SLAVIKITE *Dana,* II, 621
$Na(SO_4)[Mg(H_2O)_6]_2[Fe_5(OH)_6(H_2O)_6(SO_4)_6]\cdot 12H_2O$

SLAVYANSKITE = *Tunisite*
Am. Min. **65**, 1070

SLAWSONITE *Am. Min.* **62**, 31
$SrAl_2Si_2O_8$ mono. *Min. Abst.* **28**, 327

SMALTITE—As-deficient *skutterudite*

Cross-references: *Blowpipe Analysis; Refractory Minerals.*

SMECTITE GROUP—*Beidellite, hectorite, montmorillonite, nontronite,* and *saponite.*

Cross-references: *Clays, Clay Minerals; Phyllosilicates; Soil Mineralogy.*

SMITHITE *Dana,* I, 430
$AgAsS_2$ mono.

SMITHSONITE Zinc carbonate of the ***calcite*** group, named by Beudant in 1832 after James Smithson, founder of the Smithsonian Institution of Washington, D.C.

Composition: $ZnCO_3$; rarely pure, Zn may be replaced extensively by Fe, less so by Co, Cu, Mn^{2+}, Ca, Cd, Mg, or Pb.

Properties: Trigonal, $R\bar{3}c$; rhombohedral crystals rare, commonly botryoidal, reniform, stalactitic, encrusting, or honeycomb masses (dry-bone ore); rhombohedral $\{10\bar{1}1\}$ cleavage; H=4–4.5; G=4.00–4.45; vitreous to pearly luster; commonly brown but may be green, blue, gray, yellow, or colorless.

Occurrence: Found altered from *sphalerite* in the oxidized zone of limestone replacement. Localities include Sardinia; Laurium, Greece; Tsumeb, Namibia; Broken Hill, New South Wales, Australia; and in the USA in Marion County, Arkansas; Inyo County, California; and Socorro County, New Mexico.

K. FRYE

Reference: *Dana,* II, 176–181.

Cross-references: *Calcite Group; Gossan; Metallic Minerals; Staining Techniques.*

SMOKEY *QUARTZ*—Light to very dark brown *quartz* *Dana,* III, 182

SMOLIANINOVITE orth. *Am. Min.* **59**, 1141
$(Co,Ni,Ca,Mg)_2(Fe,Al)(AsO_4)_2(OH)\cdot 5H_2O$

SMYTHITE *Am. Min.* **57**, 1571
Fe_3S_4 trig. *Min. Abst.* **27**, 77

SNOW—See *ice*

SOAPSTONE = *Talc*

SOBOLEVSKITE *Am. Min.* **61**, 1054
PdBi hexa. *Min. Abst.* **27**, 258

SOBOTKITE = Al-rich *saponite*
Am. Min. **61**, 177

SODA = *Natron*

SODA ALUM (Sodium alum) *Dana,* II, 474
$NaAl(SO_4)_2\cdot 12H_2O$ iso.

SODA *FELDSPAR* = *Albite*

SODALITE A sodium aluminosilicate chloride named by Thomson in 1811, in reference to its sodium content.

Composition: $2[Na_4Al_3Si_3O_{12}Cl]$.

Properties: Isometric, $P\bar{4}3m$; but commonly massive or in grains; H=5.5–6; G=2.27–2.33; vitreous luster; transparent to translucent; colorless, blue, white, gray, green or pale pink.

Occurrence: A rare rock-forming mineral found in silica-poor igneous rocks in association with *nepheline* or *leucite*. Found in the Ilmen Mts., USSR; Transylvania of Rumania; at Mt. Somma, Vesuvius; Southern Norway; Western Greenland; in the USA, Maine and Massachusetts; in Canada in the nepheline-syenites of Quebec; Bancroft, Ontario; and at Ice River, British Columbia.

J. A. GILBERT

References

Can. Min. **17**, 39–46.
DHZ, 4, 289–302.

Cross-references: *Meteoritic Minerals; Minerals, Uniaxial and Biaxial; Pegmatite Minerals; Rock-Forming Minerals.*

SODALITE GROUP–*Hauyne, lazurite, nosean,* and *sodalite*

SODA NITER = *Nitratine* *Min. Mag.* **43**, 1053

SODDYITE orth. *Am. Min.* **37**, 839
$(UO_2)_{15}(OH)_{20}(Si_6O_{17}) \cdot 8H_2O$
USGS Bull. **1064**, 312

SODIUM AUTUNITE (***Autunite*** gr.)
$Na_2(UO_2)_2(PO_4)_2 \cdot 8H_2O$
tetr. *Am. Min.* **43**, 383

SODIUM BETPAKDALITE
$(Na,Ca)_3Fe_2^{3+}(As_2O_4)(MoO_4)_6 \cdot 15H_2O$
mono. *Am. Min.* **57**, 1312

SODIUM BOLTWOODITE
$(H_3O)(Na,K)(UO_2)(SiO_4) \cdot 2.5H_2O$
orth. *Am. Min.* **61**, 1054

SODIUM ***FELDSPAR*** = *Albite*

SODIUM URANOSPINITE
(***Meta-autunite*** gr.) *Am. Min.* **43**, 383
$(Na_2,Ca)(UO_2)_2(AsO_4)_2 \cdot 5H_2O$ tetr.

SODIUM ZIPPEITE orth. *Can. Min.* **14**, 429
$Na_4(UO_2)_6(SO_4)_3(OH)_{10} \cdot 4H_2O$
Min. Mag. **43**, 1067

SOGDIANITE (***Osumilite*** gr.)
$(K,Na)_2Li_2(Li,Fe,Al)_2ZrSi_{12}O_{30}$
hexa. *Am. Min.* **54**, 1221

SÖHNGEITE *Am. Min.* **51**, 1815
$Ga(OH)_3$ iso.

SOLONGOITE *Am. Min.* **60**, 162
$Ca_4B_6O_8Cl(OH)_9$ mono.

SONOLITE (***Humite*** gr.)
$Mn_9(SiO_4)_4(OH,F)_2$
mono. *Am. Min.* **48**, 1413

SONORAITE *Am. Min.* **53**, 1828
$Fe^{3+}Te^{4+}O_3(OH) \cdot H_2O$ mono.

SORBYITE *Am. Min.* **53**, 1425
$Pb_{17}(Sb,As)_{22}S_{50}$ mono. *Can. Min.* **9**, 191

SORENSENITE mono. *Am. Min.* **51**, 1547;
$Na_4SnBe_2Si_6O_{16}(OH)_4$ **52**, 928

SOUČEKITE *Am. Min.* **65**, 209
$PbCuBi(S,Se)_3$ orth. *Min. Mag.* **43**, 1067

SOUZALITE mono.
$(Mg,Fe)_3(Al,Fe)_4(OH)_6(PO_4)_4 \cdot 2H_2O$
Am. Min. **34**, 83; **55**, 135

Cross-reference: *Pegmatite Minerals.*

SPADAITE RRW, 571
$MgSiO_2(OH)_2 \cdot H_2O(?)$

SPANGOLITE trig. *Dana,* II, 576
$Cu_6Al(SO_4)(OH)_{12}Cl \cdot 3H_2O$

SPECULARITE = Specular *hematite*

SPENCERITE mono. *Dana,* II, 931
$Zn_4(PO_4)_2(OH)_2 \cdot 3H_2O$

SPENCITE = *Tritomite-(Y)*

SPERRYLITE A platinum arsenide of the ***pyrite*** group, named in 1889 after its discoverer, Canadian chemist Francis L. Sperry.

Composition: $4[PtAs_2]$.

Properties: Isometric, *Pa*3; minute cubes sometimes modified by octahedral, pyritohedral, or diploidal faces; H=6–7; G=10.6; color: tin white; streak: black.

Occurrence: With heavy-metal ores in the Sudbury District, Ontario, Canada, and the Bushveld District, South Africa, also in *gold*-bearing placers.

K. FRYE

Reference: *Can. Min.* **17**, 117–123.

SPESSARTINE (SPESSARTITE) A manganese aluminum silicate of the ***garnet*** group named for the locality in the Spessart Mts., NW Bavaria, West Germany.

Composition: $8[Mn_3Al_2Si_3O_{12}]$ with Fe^{2+} replacing Mn^{2+} toward *almandine,* and Mg toward *pyrope.*

Properties: Isometric, *Ia3d*; crystals dodecahedra or trapezhedra, also massive or granular; H=6.5–7.5; G=4.2; adamantine luster; color: red, green, yellowish, or black.

Occurrence: From manganese-rich skarn deposits, granite pegmatites, and metamorphosed graywacke.

Use: Gemstone.

K. FRYE

Reference: DHZ, 1, 99–101.

Cross-references: *Gemology; Minerals, Uniaxial and Biaxial; Orthopyroxenes; Pegmatite Minerals.*

SPHAEROCOBALTITE = *Spherocobaltite*
Min. Mag. **43**, 1053

SPHALERITE Zinc sulfide named from the Greek meaning treacherous. It is also called "zinc blende" because, although it often resembles *galena,* it yields no lead; the word in German means blind or deceiving.

Composition: ZnS. Almost always contains Fe, less commonly Mn or Cd. The higher the pressure of formation, the less the amount of iron tolerated in the 300–600°C range. Thus, *sphalerite* may be used in geologic barometry (see *Barometry, Geologic*).

Properties: Isometric, $F\bar{4}3m$; crystals are frequently highly complex, usually malformed, or in rounded aggregates, often show polysynthetic twinning; usually found in cleavable masses, coarse to fine granular; perfect dodecahedral {110} cleavage, dimorphic with *wurtzite;* H=3.5–4; G=3.9–4.1; nonmetallic and resinous to submetallic luster, also adamantine; transparent to translucent; color: white when pure, green when nearly so, may also be yellow, brown, black, darkening with increased Fe present.

Occurrence: Widely distributed, but is chiefly in veins and irregular replacement in limestone, also in veins in igneous rocks and in contact metamorphic deposits; associated with *galena, pyrite, marcasite, chalcopyrite, pyrrhotite, smithsonite, calcite,* and *dolomite.*

Use: The most important ore of zinc.

J. A. GILBERT

References

Am. Min. **63**, 250–257.
Dana, **I**, 210–215.

Cross-references: *Authigenic Minerals; Barometry, Geologic; Blowpipe Analysis; Crystal Habits; Crystallography: Morphological; Isomorphism; Lunar Minerals; Metallic Minerals; Meteoritic Minerals; Mineral and Ore Deposits; Minerals, Uniaxial and Biaxial; Naming of Minerals; Optical Mineralogy; Ore Microscopy; Pegmatite Minerals; Refractive Index; Rock-Forming Minerals; Staining Techniques; Thermometry, Geologic; Vein Minerals.*

SPHENE = TITANITE

SPHEROCOBALTITE (***Calcite*** gr.)
$CoCO_3$ trig. *Dana,* II, 175

SPINEL A magnesium aluminum oxide named by Agricola in 1546. It is a compositional end member of the ***spinel*** group, or, more generally, with aluminum as the dominant trivalent cation.

Composition: $MgAl_2O_4$; variations: $(Mg,Fe,Zn,Mn)(Al,Fe,Cr)_2O_4$. $MgAl_2O_4$ is the most common mineral in the ***spinel*** series with *hercynite* (Fe), *gahnite* (Zn), *galaxite* (Mn). Pleonaste or ceylonite has Mg: Fe from 3 : 1.

Properties: Isometric, $Fd3m$; H=7.5–8; G=3.6; octahedral crystals. Optically, n=1.719; color varies with composition, usually clear to light greens and browns.

Occurrence: High-temperature metamorphic rocks, Al-rich xenoliths, contact metamorphosed limestones; Cr-rich varieties found in serpentines, talc-schists, and ultramafic igneous rocks. Alters to *talc,* ***mica, serpentine,*** *corundum, manasseite.*

Use: Gem mineral.

D. H. SPEIDEL

References

DHZ, 5, 56–67.
Econ. Geol. **61**, 795–796.

Cross-references: *Crystal Growth; Gemology; Lunar Minerals; Mantle Mineralogy; Meteoritic Minerals; Mineral Properties; Minerals, Uniaxial and Biaxial; Orthopyroxenes; Refractory Minerals; Rock-Forming Minerals; Spinel Group.*

SPINEL GROUP—*Brunoqeierite, chromite, cochromite, coulsonite, cuprospinel, franklinite, gahnite, galaxite, hercynite, jacobsite, magnesiochromite, magnesioferrite, magnetite, manganochromite, nichromite, spinel, trevorite,* and *ulvöspinel.*

SPIONKOPITE *Can. Min.* **18**, 511
$Cu_{39}S_{28}$ hexa.

SPIROFFITE *Am. Min.* **49**, 444
$(Mn,Zn)_2Te_3O_8$ mono.

SPODIOSITE (species doubtful) *Dana,* II, 848
$Ca_2(PO_4)F$ orth.

SPODUMENE Lithium aluminum silicate of the ***pyroxene*** group, named by d'Andrada in 1800 from the Greek *spodumenos,* reduced to ashes, for its common grayish-white color.

Composition: $LiAl(Si_2O_6)$; with little Al replaced by Fe^{3+}

Properties: Monoclinic, $C2/c$; crystals vertically striated and up to 14 m long; may also be massive; H=6.5–7; G=3.03–3.20; vitreous luster; good prismatic {110} cleavage; transparent to translucent; colorless, pink, yellow, white, green, gray (transparent lavender gem variety is "kunzite," transparent green gem variety is "hiddenite").

Occurrence: Found sporadically in granite pegmatites as at Varuträsk and on Kluntarna Island, Sweden; Graz, Austria; Kolar gold field, India; Katumba, Zaire; and in the USA in the *spodumene* pegmatite district, North Carolina, and the Black Hills, South Dakota.

Use: Source of lithium and as a gemstone.

J. A. GILBERT

References

Am. Min. **60**, 919–923.
DHZ, 2A, 527–544.

Cross-references: *Clinopyroxenes; Jade; Mineral and Ore Deposits; Minerals, Uniaxial and Biaxial; Pegmatite Minerals; Pyroxene Group; Refractory Minerals; Staining Techniques.*

SPURRITE *Min. Abst.* **28**, 71
$Ca_5(SiO_4)_2(CO_3)$ mono. Winchell, II, 516

Cross-reference: *Rock-Forming Minerals.*

STANFIELDITE *Am. Min.* **53**, 508
$Ca_4(Mg,Fe,Mn)_5(PO_4)_6$
mono. *Science* **158**, 910

Cross-reference: *Meteoritic Minerals.*

STANNITE Copper, iron, tin sulfide, named by Beudant in 1832 from the Latin name for tin.

Composition: $Cu_2(Fe,Zn)SnS_4$.

Properties: Tetragonal, $I\bar{4}2m$; crystals rare, commonly massive; H=4; G=4.4; metallic luster; color: steel-gray to iron-black; streak: blackish.

Occurrence: Found in tin-bearing veins associated with *cassiterite, chalcopyrite, wolframite, pyrite,* and *quartz.* It is a rare mineral found in Bohemia, Czechoslovakia; in various places in Cornwall, UK; and in the tin ores of Bolivia.

Use: A minor ore of tin.

J. A. GILBERT

Reference: *Can. Min.* **17**, 125–135.

Cross-reference: *Ore Microscopy.*

STANNOIDITE *Am. Min.* **54**, 1495
$Cu_8(Fe,Zn)_3Sn_2S_{12}$ orth. *Can. Min.* **17**, 132

STANNOMICROLITE (***Pyrochlore*** gr.)
$Sn_2Ta_2O_7$ *Am. Min.* **53**, 2103;
iso **62**, 405

STANNOPALLADINITE *Am. Min.* **56**, 360
$(Pd,Cu)_3Sn_2(?)$ hexa.

STARINGITE tetr. *Am. Min.* **55**, 1446
$(Fe,Mn)_x(Nb,Ta)_{2x}(Sn,Ti)_{6-3x}O_{12}$
Min. Mag. **37**, 447

STARKEYITE *Am. Min.* **41**, 662
$MgSO_4 \cdot 4H_2O$ mono.

Cross-reference: *Green River Mineralogy.*

STAUROLITE An iron aluminum hydroxyl silicate named by Delamétherie in 1792 from the Greek *stauros,* cross, and *lithos,* stone, referring to its cruciform twins.

Composition: $Fe_2Al_9O(OH)(SiO_4)_4$; with Fe^{3+} replaced by Mg or Zn and Al by Fe^{2+}.

Properties: Monoclinic, $C2/m$; cruciform twins very common, also pseudohexagonal prisms; H=7.5; G=3.74–3.83; resinous to vitreous luster on fresh surface; translucent; color: red brown, yellow, to dark brown.

Occurrence: Found as a common accessory mineral in schists of medium-grade regional metamorphism, also associated with ***garnet,*** *kyanite,* and ***tourmaline.***

Use: Transparent specimens from Brazil are occassionally used as a gem, cruciform twins as "fairy crosses."

J. A. GILBERT

References

Am. Min. **53**, 1139–1155.
DHZ, 1, 151–160.

Cross-references: *Crystallography: Morphological; Minerals, Uniaxial and Biaxial; Rock-Forming Minerals; Soil Mineralogy.*

STEATITE = *Talc*

STEENSTRUPINE hexa. RRW, 577
$(Ce,La,Na,Mn)_6(Si,P)_6O_{18}(OH)$

STEIGERITE *Am. Min.* **44**, 322
$AlVO_4 \cdot 3H_2O$ mono.

STELLERITE (***Zeolite*** gr.)
$Ca(Al_2Si_7)O_{18} \cdot 7H_2O$ *Am. Min.* **53**, 511
orth. *Min. Abst.* **30**, 66

STENHUGGARITE *Am. Min.* **56**, 636
$CaFe^{3+}Sb^{3+}As_2^{3+}O_7$ tetr. *Min. Abst.* **30**, 18

STENONITE mono. *Am. Min.* **48**, 1178
$(Sr,Ba,Na)_2Al(CO_3)F_5$

STEPANOVITE trig. *Am. Min.* **49**, 442
$NaMgFe^{3+}(C_2O_4)_3 \cdot 8\text{–}9H_2O$

STEPHANITE A silver antimony sulfide named by Haidinger in 1845 after Archduke Stephan, formerly mining director of Austria.

Composition: Ag_5SbS_4; Ag, 68.3%; Sb, 15.4%; S, 16.3% by wt.

Properties: Orthorhombic, *Cmc*2; short prismatic to tabular crystals, may be massive, compact, or disseminated; H=2.0–2.5; G= 6.2–6.3; metallic luster; brittle; iron black color and streak.

Occurrence: One of the last vein minerals to form in many silver deposits, usually in small amounts. Associated with other silver sulfosalts, *argentite, silver, tetrahedrite,* and common sulfides.

Use: An ore of silver.

J. A. GILBERT

Reference: *Dana,* I, 358–361.

STERCORITE *Dana,* II, 698
$H(NH_4)Na(PO_4) \cdot 4H_2O$ tric.

STERLINGHILLITE *Am. Min.* **66**, 182
$Mn_3(AsO_4)_2 \cdot 4H_2O$(?)

STERNBERGITE *Dana,* I, 246
$AgFe_2S_3$ orth.

STERRYITE *Am. Min.* **53**, 1423
$Pb_{12}(Sb,As)_{10}S_{27}$ orth. *Can. Min.* **9**, 191

STETEFELDTITE (***Stibiconite*** gr.)
$Ag_2Sb_2(O,OH)_7$(?) *Am. Min.* **39**, 408
iso. *Min. Mag.* **30**, 100

STEVENSITE *Am. Min.* **44**, 343
$Mg_3Si_4O_{10}(OH)_2$ mono. *Min. Abst.* **28**, 16

Cross-reference: *Green River Mineralogy.*

STEWARTITE tric. *Am. Min.* **59**, 1272
$Mn^{2+}Fe_2^{3+}(OH)_2(PO_4)_2 \cdot 8H_2O$

Cross-reference: *Pegmatite Minerals.*

STIBICONITE An antimony oxide hydroxide.

Composition: $Sb_3O_6(OH)$.

Properties: Isometric, *Fd*3*m*; massive to powdery; pale yellow to white.

Occurrence: Associated with antimony ore deposits.

D. H. SPEIDEL

Reference: *Am. Min.* **37**, 982–999.

STIBICONITE GROUP–*Bindheimite, lewisite, partzite, romeite, stetefeldtite,* and *stibiconite.*

STIBIOBETAFITE iso *Can. Min.* **17**, 583
$(Ca,Sb)_2(Ti,Nb,Ta)_2(O,OH)_7$
Min. Mag. **43**, 1067

STIBIOCOLUMBITE *Dana,* I, 767
$SbNbO_4$ orth.

STIBIOPALLADINITE *Am. Min.* **58**, 1;
Pd_7Sb_3 hexa. **61**, 1249

STIBIOTANTALITE An antimony tantalum oxide named for its composition.

Composition: $SbTaO_4$; with Ta replaced by Nb (Nb>Sb is *stibiocolumbite*).

Properties: Orthorhombic, *Pbn*2; prismatic crystals with {010} cleavage; H=5; G=5.7 (Nb) to 7.5 (Ta); color: light to dark brown; streak: yellow to brown.

Occurrence: Found in pegmatites and placers.

D. H. SPEIDEL

References

Dana, I, 767–769.
GSA Bull. **68**, 1744

STIBIVANITE *Can. Min.* **18**, 329
Sb_2VO_5 mono.

STIBNITE Antimony sulfide named by Beudant in 1832 from the Greek *stibi,* antimony.

Composition: Sb_2S_3; Sb, 71.7%; S, 28.3% by wt.

Properties: Orthorhombic, *Pbnm;* stout to slender prismatic crystals; vertically striated, sometimes bent or curved, often in radiating groups or in bladed forms with prominent {010} cleavage, also massive, coarse to fine granular; H=2; G=4.63; metallic luster; lead gray to steel gray color and streak; black tarnish.

Occurrence: A low-temperature hydrothermal mineral in veins and around hot springs, usually in association with *quartz,* other antimony minerals, *galena, cinnabar, sphalerite, barite, realgar,* and *orpiment.*

Uses: Chief ore of antimony.

J. A. GILBERT

Reference

Jensen, M. L., and Bateman, A. M., 1979. *Economic Mineral Deposits,* 3rd ed. New York: Wiley, 443–445.

Cross-references: *Blowpipe Analysis; Metallic Minerals; Mineral and Ore Deposits; Ore Microscopy.*

STICHTITE A magnesium chromium hydrated hydroxyl carbonate named by Pettard in 1910 for Robert Sticht.

Composition: $Mg_6Cr_2(OH)_{16}CO_3 \cdot 4H_2O$.

Properties: Trigonal; basal plates with basal cleavage; polymorphic with *barbertonite;* color: lilac to pink.

Occurrence: Found in Tasmania and Mt. Keith, Western Australia.

D. H. SPEIDEL

Reference: *Dana,* I, 655–656.

STILBITE A hydrated calcium sodium potassium aluminosilicate of the ***zeolite*** group, named by Brooke in 1822 from the Greek *stilbein,* gleam.

Composition: $Na_2Ca_4[Al_{10}Si_{26}O_{72}] \cdot 34H_2O$. Considerable replacements Na(K) Si⇌CaAl may occur; small amounts of Sr and Ba may enter; dehydration stepwise, losing 19% H_2O at maximum; rehydration poor; a small amount of base exchange with $FeSO_4$, but not with $FeCl_3$, $AlCl_3$, and $ZnCl_2$; molecular sieve properties more selective; absorbs ammonia; decomposed by HCL.

Properties: Monoclinic, $C2/m$; a=13.63Å, b=18.17Å, c=11.31Å; β=129°10′; H=3.5–4; G=2.1–2.2; luster vitreous; colorless, white, yellow, brown, or red; transparent to translucent; habit tabular, or in sheaf-like aggregates; cleavage very good on {010}; twinning common on {001}, cruciform, interpenetrant. Optically: refractive indices, α=1.484–1.500, β=1.492–1.507, γ=1.494–1.513; birefringence = 0.010; $2V$=30°–49° (−); optic axial plane (010); $\beta = y$, $\alpha \wedge Z$=5°; dispersion $r < v$.

Occurrence: In cavities in basalt and many other volcanic and plutonic rocks, often associated with *chabazite* and *heulandite;* alteration product of ***plagioclase*** in gabbro; in crevices in metamorphic rocks with other minerals of hydrothermal origin, e.g., *quartz, epidote, prehnite,* and adularia; along joints in gneiss associated with *natrolite, heulandite,* and *laumontite;* as a hydrothermal mineral in pyroclastic rocks and conglomerate in the Tertiary Green Tuff formations in central Japan, replacing andesitic glass and ***plagioclase.***

A. IIJIMA
M. UTADA

References

Acta Cryst. **B27,** 833–841.
Arkiv. Mineral. Geol. **1,** 519–526.
Mem. Coll. Sci., Univ. Kyoto, ser. B, **31,** 199–214.
N.J. Min. Mh. 1956, 1–9.

Cross-reference: *Zeolites.*

STILLEITE — *Am. Min.* **42,** 584
ZnSe — iso.

STILLWATERITE — *Am. Min.* **62,** 1060
Pd_8As_3 — hexa. — *Can. Min.* **13,** 321

STILLWELLITE — *Am. Min.* **41,** 370
$(REE,Ca)BSiO_5$ — trig. — *Nature* **176,** 509

STILPNOMELANE — DHZ, 3, 103
$K_{0.6}(Fe,Mg)_6(Si_8Al)(O,OH)_{27} \cdot 2\text{–}4H_2O$
mono. and tric. — *Min. Mag.* **42,** 361

Cross-references: *Minerals, Uniaxial and Biaxial; Rock-Forming Minerals.*

STISHOVITE
(***Rutile*** gr.) — *Am. Min.* **47,** 807
SiO_2 — tetr. — *Dana,* III, 317

Cross-references: *Coesite and Stishovite; Mantle Mineralogy; Polymorphism.*

STISTAITE — *Am. Min.* **56,** 358
SnSb — iso.

STOIBERITE — *Am. Min.* **64,** 941
$Cu_5V_2O_{10}$ — mono. — *Min. Abst.* **31,** 230

STOKESITE — *Min. Mag.* **33,** 615
$CaSnSi_3O_9 \cdot 2H_2O$ — orth.

STOLZITE A lead tungstate of the ***wulfenite*** group, named in 1845 after Dr. Stolz of Teplitz (Teplice), Bohemia, Czechoslovakia.

Composition: $4[PbWO_4]$.

Properties: Tetragonal, $I4_1/a$; crystals commonly bypyramidal; H=2.5–3; G=7.9–8.3; resinous to subadamantine luster; color: red, brown, yellow. Dimorphous with *raspite.*

Occurrence: Secondary mineral in oxidized zone of tungsten deposits.

K. FRYE

Reference: *Dana*, II, 1087–1089.

STOTTITE *Am. Min.* **43**, 1006
$Fe^{2+}Ge(OH)_6$ tetr.

STRANSKIITE *Am. Min.* **63**, 213
$(Zn_2Cu)_3(AsO_4)_2$ tric.

STRASHIMIRITE *Am. Min.* **54**, 1221
$Cu_8(AsO_4)_4(OH)_4 \cdot 5H_2O$ mono.

STRÄTLINGITE *Am. Min.* **62**, 395
$2CaO \cdot Al_2O_3 \cdot SiO_2 \circ 8H_2O$ trig.

STRELKINITE orth. *Am. Min.* **60**, 488
$Na_2(UO_2)_2(VO_4)_2 \cdot 6H_2O$

STRENGITE *Am. Min.* **55**, 135
$Fe^{3+}PO_4 \cdot 2H_2O$ orth.

Cross-reference: *Pegmatite Minerals.*

STRINGHAMITE *Am. Min.* **61**, 189
$CuCaSiO_4 \cdot 2H_2O$ mono. *Min. Abst.* **27**, 339

STROMEYERITE *Dana*, I, 190
AgCuS orth.

Cross-reference: *Ore Microscopy.*

STRONTIANITE A strontium carbonate of the ***aragonite*** group, named in 1791 for the town of Strontian, Argyll, Scotland.

Composition: $4[SrCO_3]$ with Sr replaced by Ca and Ba.

Properties: Orthorhombic, *Pmcn;* crystals prismatic, commonly multiply twinned, also massive, fibrous, or granular; nearly perfect prismatic cleavage; H=3.5; G=3.7; vitreous luster; colorless, white, pale green, gray, or yellowish.

Occurrence: Hydrothermal vein and cavity filling with other carbonate minerals, in limestones and marls, and rarely as an authigenic mineral.

Use: An ore of strontium.

K. FRYE

References

Am. Min. **61**, 1001–1004.
DHZ, 5, 316–318.

Cross-references: *Authigenic Minerals; Blowpipe Analysis; Green River Mineralogy; Minerals, Uniaxial and Biaxial; Refractory Minerals; Rock-Forming Minerals; Staining Techniques.*

STRONT…
$(Sr,Ca)_5$(…

STRONTIOB…
$Sr[B_8O_{11}(OH)$…

STRONTIODRE…
$(Sr,Ca)Al_2(CO_3)_2$(…
orth.

STRONTIOGINORITE … **55**, 1911
$(Sr,Ca)_2B_{14}O_{23} \cdot 8H_2O$ …

STRONTIOHILGARDITE = Strontian *hilgardite*

STRONTIUM-APATITE = *Strontiapatite*

STRUNZITE A hydrated manganese and iron hydroxyl phosphate named in 1958 for Dr. Hugo Strunz of Berlin.

Composition: $4[Mn^{2+}Fe_2^{3+}(PO_4)_2(OH)_2 \cdot 6H_2O]$.

Properties: Triclinic; tufts or hair-like or lath-like crystals; G=2.5; yellow.

Occurrence: Weathering product of phosphate minerals in pegmatites and phosphate rock.

K. FRYE

References

Am. Min. **43**, 793; **55**, 135.
Can. Min. **15**, 405.

STRÜVERITE = Tantalian *rutile*

STRUVITE *Dana*, II, 715
$(NH_4)MgPO_4 \cdot 6H_2O$ orth.

Cross-references: *Cave Minerals; Human and Vertebrate Mineralogy; Museums, Mineralogical.*

STUDTITE *Am. Min.* **59**, 166
$UO_4 \cdot 4H_2O$ mono.

STUETZITE *Am. Min.* **49**, 325
$Ag_{5-x}Te_3$ hexa.

STUMPFLITE *Am. Min.* **59**, 211
Pt(Sb,Bi) hexa.

STURTITE
(species doubtful) *Am. Min.* **65**, 210
hydrous Mn silicate amorph. RRW, 589

STÜTZITE = *Stuetzite*

Am. Min. **48**, 915
mono.

...ITE—Variety of amber

SUDBURYITE *Am. Min.* **61**, 178
(Pd,Ni)Sb hexa. *Can. Min.* **12**, 275

SUDOITE (***Chlorite*** gr.)
$(Al,Mg,Fe)_{4-5}(Si,Al)_4O_{10}(OH)_8$
mono. *Am. Min.* **52**, 673

Cross-reference: *Soil Mineralogy.*

SUGILITE (***Osumilite*** gr.)
$(K,Na)(Na,H_2O)_2(Fe,Na)_2(Li,Al,Fe)_3Si_{12}O_{30}$
Can. Min. **18**, 37
hexa. *Min. Mag.* **43**, 947

SUKULAITE = *Stannomicrolite*
Am. Min. **62**, 407

Cross-reference:
Cross-reference: *Pegmatite Minerals.*

SULFOBORITE *Dana*, II, 387
$Mg_3B_2(SO_4)(OH)_{10}$ orth.

SULFOHALITE *Dana*, II, 548
$Na_6(SO_4)_2FCl$ iso.

Cross-reference: *Saline Minerals.*

SULFUR A native element, the name being derived from the Anglo-French *sulf(e)re* (12th century) and in similar form in many western European languages. Also known as brimstone.

Composition: 128[S] (alpha), with very minor Se and Te.

Properties: Orthorhombic, *Fddd*; crystals are bipyramids or disphenoids, tabular, massive; H=1.5–2.5; G=2.07; resinous luster; color: yellow. *Beta sulfur* and *rosickyite* are monoclinic and occur with *alpha sulfur* at some fumaroles.

Occurrence: From volcanic fumaroles, by sulfur-reducing bacteria, in Tertiary sedimentary rocks associated with gypsum and limestone, in salt domes, associated with bituminous deposits.

Use: Manufacture of sulfuric and sulfurous acid.

K. FRYE

Reference: *Dana*, I, 140–146.

Cross-references: *Allotropy; Authigenic Minerals; Color in Minerals; Crystal Growth; Crystallography, Morphological; Green River Mineralogy; Meteoritic Minerals; Mineral Classification: Principles; Mineral Industries; Mineral and Ore Deposits; Native Elements and Alloys; Thermometry, Geologic.* Vol. IVA: *Sulfur: Element and Geochemistry.*

SULVANITE *Am. Min.* **51**, 890
Cu_3VS_4 iso.

SUN STONE—Aventurescent *oligoclase*

Cross-reference: *Plagioclase Feldspars.*

SUNDIUSITE *Am. Min.* **65**, 506
$Pb_{10}(SO_4)Cl_2O_8$ mono.

SUNGULITE = Mixture of *lizardite* + *sepiolite* *Am. Min.* **59**, 212

SUOLUNITE *Am. Min.* **53**, 349
$Ca_2Si_2O_5(OH)_2 \cdot H_2O$ orth.

SURINAMITE mono. *Am. Min.* **61**, 193
$(Al,Mg,Fe)_3(Si,Al)_2(O,OH)_8$
Min. Abst. **27**, 339

SURITE mono. *Am. Min.* **63**, 1175
$Pb(Pb,Ca)_{1.17}(CO_3)_2$ *Min. Mag.* **43**, 1068
$(Al,Fe,Mg)_2(Si,Al)_4O_{10}(OH)_2$

SURSASSITE *Am. Min.* **49**, 168
$Mn_5Al_4Si_5O_{21} \cdot 3H_2O$(?) mono.

SUSANNITE trig. *Am. Min.* **55**, 1449
$Pb_4(SO_4)(CO_3)_2(OH)_2$ *Can. Min.* **10**, 141

SUSSEXITE *Dana*, II, 375
$Mn_2(OH)[B_2O_4(OH)]$
mono. *Min. Abst.* **30**, 68

SVABITE (***Apatite*** gr.) *Dana*, II, 899
$Ca_5(AsO_4)_3F$ hexa.

SVANBERGITE (***Beudantite*** gr.)
$SrAl_3(PO_4)(SO_4)(OH)_6$ *Dana*, II, 1005
trig. *Min. Abst.* **30**, 422

SVETLOZARITE (***Zeolite*** gr.)
$(Ca,K_2,Na_2)Al_2(Si,Al)_{12}O_{28} \cdot 6H_2O$
Am. Min. **62**, 1060
orth. *Min. Abst.* **28**, 327

SWARTZITE *USGS Bull.* **1064**, 117
$CaMg(UO_2)(CO_3)_3 \cdot 12H_2O$ mono.

SWEDENBORGITE *Dana*, II, 1027
$NaBe_4SbO_7$ hexa.

SWINEFORDITE (***Smectite*** gr.)
$(Li,½Ca)_{.7}(Al,Li,Mg)_4(Si,Al)_8O_{20}(OH,F)_4 \cdot nH_2O$ *Am. Min.* **60**, 540
mono. *Min. Abst.* **27**, 82

SWITZERITE *Am. Min.* **52**, 1595
$(Mn,Fe)_3(PO_4)_2 \cdot 4H_2O$ mono.

SYLVANITE A gold silver telluride of the ***krennerite*** group, named in 1835 after Transylvania (Rumania). (Sylvanium was a name proposed for tellurium.)

Composition: $2[AgAuTe_4]$.

Properties: Monoclinic, *P2/c*; crystals prismatic or tabular, commonly skeletal and twinned to resemble writing, also bladed, columnar, or granular; one perfect cleavage {010}; H=1.5; G=8.1; brilliant metallic luster; color and streak steel gray, silver white, or yellowish.

Occurrence: In *quartz* veins with sulfides and other tellurides.

Use: An ore of gold and silver.

K. FRYE

Reference: *Dana,* I, 338–341.

Cross-reference: *Ore Microscopy.*

SYLVINITE–Mixture of *halite* and *sylvite*

SYLVITE Potassium chloride, the name derived from *Sal digestivus Sylvii* (1847).

Composition: 4[KCl].

Properties: Isometric, *Fm3m*; crystals cubic, also granular crystalline; cubic cleavage; H=2; G=1.99; color: white or off-white; bitter salty taste.

Occurrence: In evaporite deposits, also from fumaroles.

K. FRYE

Reference: *Dana,* II, 7–9.

Cross-references: *Blowpipe Analysis; Crystal Growth; Electron Microscopy (Transmission) Saline Minerals.*

SYMPLESITE *Dana,* II, 752
$Fe^{2+}_3(AsO_4)_2 \cdot 8H_2O$ tric.

SYNADELPHITE orth. *Am. Min.* **55**, 2023
$Mn^{2+}_9(AsO_3)(AsO_4)_2(OH)_9 \cdot 2H_2O$

SYNCHYSITE *Am. Min.* **60**, 351; **64**, 658
(*REE*)$Ca(CO_3)_2F$ hexa.

Cross-references: *Pegmatite Minerlas; Syntaxy.*

SYNCHYSITE-(Y) *Am. Min.* **47**, 337
(Y,*REE*)$Ca(CO_3)_2F$ hexa. *Can. Min.* **16**, 361

SYNGENITE *Dana,* II, 442
$K_2Ca(SO_4)_2 \cdot H_2O$ mono. *Min. Abst.* **30**, 220

Cross-reference: *Saline Minerals.*

SZAIBELYITE *Am. Min.* **60**, 273
$Mg_2(OH)[B_2O_4(OH)]$ mono.

Cross-reference: *Saline Minerals.*

SZMIKITE *Dana,* II, 481
$MnSO_4 \cdot H_2O$ mono.

SZOMOLNOKITE *Dana,* II, 479
$Fe^{2+}SO_4 \cdot H_2O$ mono.

Cross-reference: *Green River Mineralogy.*

T

TAAFFEITE *Am. Min.* **37**, 360
$(Mg,Be)_2Al_4O_8$ trig. or hexa. *Min. Mag.* **43**, 575

TACHARANITE *Am. Min.* **61**, 1055
$Ca_{12}Al_2Si_{18}O_{69}H_{36}$ mono. *Min. Mag.* **40**, 113

TACHYHYDRITE *Dana*, II, 95
$CaMg_2Cl_6 \cdot 12H_2O$ trig.

Cross-reference: *Saline Minerals.*

TACHYLITE = Basaltic glass

TADZHIKITE
$Ca_3(Ce,Y)_2(Ti,Al,Fe)B_4Si_4O_{22}$
mono. *Am. Min.* **56**, 1838

TAENIOLITE
(***Mica*** gr.) *Min. Abst.* **30**, 218
$KLiMg_2Si_4O_{10}F_2$ mono. RRW, 660

TAENITE *Am. Min.* **51**, 37
γ-(Fe,Ni) iso.

Cross-references: *Lunar Minerals; Metallurgy; Meteoritic Minerals; Native Elements and Alloys; Rock-Forming Minerals; Widmanstätten Structures.*

TAKANELITE *Am. Min.* **56**, 1487
$(Mn^{2+},Ca)Mn_4^{4+}O_9 \cdot H_2O$ hexa.

TAKEUCHIITE *Am. Min.* **65**, 1130
$(Mg,Mn,Fe)_3BO_5$ orth.

TAKOVITE *Am. Min.* **57**, 1559; 62, 458
$Ni_6Al_2(OH)_{16}(CO_3,OH) \cdot 4H_2O$
trig.

TALC A magnesium hydroxyl silicate which is the major constituent of soapstone or steatite. The name "talc" is probably of Arabic derivation.

Composition: $Mg_6(Si_8O_{20})(OH)_4$.

Properties: Monoclinic and triclinic; in tabular crystals, foliated, massive, coarse or fine granular massive, fibrous pseudomorphs after *enstatite* or *tremolite*, compact; perfect {001} cleavage (basal cleavage); sectile; greasy feel; H=1; G=2.58–2.83; color: white, colorless, light to dark green, brown; streak: white to light green (in dark green specimens); pearly luster on cleavage surfaces. Optically (–); $2V$=0°–30°; colorless in thin section; α=1.539–1.550, β=1.589–1.594, γ=1.589–1.600, $\delta \cong 0.05$.

Occurrence: Found as a product of hydrothermal alteration of ultramafic rocks; formed by low grade thermal metamorphism of siliceous dolomites. Associated with *dolomite, chrysotile, actinolite,* ***tourmaline***, *magnetite,* ***pyroxene***, and ***amphibole***.

Uses: Thermal and electrical insulation; ceramic raw material; filler for paper, paints, and rubber; in soaps and cosmetics; as a lubricant; as soapstone for tabletops, washtubs, etc.

E. BOOY

References

Am. Min. **53**, 751–769.
DHZ, 3, 121–130.

Cross-references: *Clays, Clay Minerals; Green River Mineralogy; Minerals, Uniaxial and Biaxial; Mohs Scale of Hardness; Phyllosilicates; Pigments and Fillers; Plastic Flow in Minerals; Polysomatism: Refractory Minerals; Soil Mineralogy.*

TALMESSITE *Am. Min.* **50**, 813
$Ca_2Mg(AsO_4)_2 \cdot 2H_2O$ tric. *Min. Rec.* **9**, 69

TALNAKHITE *Am. Min.* **56**, 2159
$Cu_9(Fe,Ni)_8S_{16}$ iso. *Econ. Geol.* **66**, 673

TAMARUGITE mono. *Am. Min.* **54**, 19
$NaAl(SO_4)_2 \cdot 6H_2O$ *Min. Mag.* **40**, 642

TANCOITE *Can. Min.* **18**, 185
$HNa_2LiAl(PO_4)_2(OH)$
orth. *Min. Mag.* **43**, 1068

TANEYAMALITE tric. *Min. Mag.* **44**, 51
$(Na,Ca)(Mn^{2+},Mg,Fe^{3+},Al)_{12}Si_{12}(O,OH)_{44}$

TANGEITE = *CALCIOVOLBORTHITE*

TANTALAESCHYNITE-(Y)
(Y,*REE*,Ca)$(Ta,Ti,Nb)_2O_6$ *Am. Min.* **59**, 1331
orth. *Min. Mag.* **39**, 571

TANTALITE *Dana*, I, 780
$(Fe^{2+},Mn)(Ta,Nb)_2O_6$
orth. *Min. Abst.* **27**, 334

Cross-references: *Blacksand Minerals; Metallic Minerals; Ore Microscopy; Pegmatite Minerals.*

TANTALUM = TANTALUM CARBIDE
Am. Min. **47**, 786

TANTEUXENITE orth. *Am. Min.* **45**, 756
$(Y,Ce,Ca)(Ta,Nb,Ti)_2(O,OH)_6$

TANZANITE–A blue gem variety of *zoisite*
Am. Min. **54**, 702

Cross-reference: *Gemology.*

TAPIOLITE An iron manganese tantalate niobate named for an ancient Finnish diety, *Tapio.*

Composition: $(Fe,Mn)(Ta,Nb)_2O_6$; Nb>Ta is *mossite.*

Properties: Tetragonal, *P4/mnm*; prismatic crystals; G=7.3–7.8; H=6–6.5; color: black; streak: brown.

Occurrence: Found in pegmatites and placers in Berg, Norway, associated with *columbite.*

D. H. SPEIDEL

Reference: *Dana*, I, 775–778.

Cross-references: *Blacksands; Pegmatite Minerals; Placers.*

TARAMELLITE orth. *Am. Min.* **65**, 123
$Ba_4(Fe^{3+},Ti,Fe^{2+},Mg,V)_4B_2Si_6O_{29}Cl$

TARAMITE (***amphibole*** gr.)
$NaCaNa(Fe^{2+},Mg)_3(Fe^{3+},Al)_2(Si_6Al_2)O_{22}(OH,F)_2$ *Am. Min.* **63**, 1036
mono. DHZ, 2, 360

TARANAKITE trig. *Am. Min.* **44**, 138; **60**, 331
$KAl_3(PO_4)_3(OH)\cdot 9H_2O$

Cross-reference: *Cave Minerals.*

TARAPACAITE *Dana*, II, 644
K_2CrO_4 orth.

TARASOVITE–Interlayered ***mica-clay***
Am. Min. **56**, 1123

TARBUTTITE *Am. Min.* **51**, 1218
$Zn_2(PO_4)(OH)$ tric.

TARNOWITZITE, TARNOWSKITE = Plumboan *aragonite* *Am. Min.* **65**, 1069

TATARSKITE orth.(?) *Am. Min.* **49**, 1151
$Ca_6Mg_2(SO_4)_2(CO_3)_2Cl_4(OH)_4\cdot 7H_2O$

TAVISTOCKITE = *Apatite*
Am. Min. **54**, 1742

TAVORITE *Am. Min.* **40**, 952; **55**, 135
$LiFe^{3+}(OH)(PO_4)$ tric.

TAWMAWITE = Cr-rich *epidote*

TAYLORITE *Am. Min.* **36**, 590
$(K,NH_4)_2SO_4$ orth.

TAZHERANITE *Am. Min.* **55**, 318
$(Zr,Ca,Ti)O_2$ iso. *Min. Abst.* **30**, 68

TEALLITE *Dana*, I, 439
$PbSnS_2$ orth.

TEEPLEITE *Am. Min.* **44**, 875
$Na_2BO_2Cl\cdot 2H_2O$ tetr.

TEINEITE *Am. Min.* **46**, 466
$CuTeO_3\cdot 2H_2O$ orth. *Min. Abst.* **29**, 140

TELARGPALITE iso. *Am. Min.* **60**, 489
Palladium-silver telluride

TELLURANTIMONY *Am. Min.* **59**, 382
Sb_2Te_3 trig. *Can. Min.* **12**, 55

TELLURITE Tellurium oxide named for its composition.

Composition: TeO_2.

Properties: Orthorhombic, *Pcab*; brookite-type structure; colorless to yellow.

Occurrence: Found with native Te in Boulder, Colorado.

D. H. SPEIDEL

Reference: *Dana*, I, 593–595.

TELLURIUM A native element named after *tellus*, the earth.

Composition: 3[Te], with minor Se.

Properties: Trigonal; crystals minute prisms; also granular or massive; prismatic cleavage; H=2–2.5; G=6.2; a tin-white metal with gray streak.

Occurrence: In hydrothermal *quartz* veins with *gold*, sulfides, and tellurides.

K. FRYE

Reference: *Dana*, I, 138–139.

Cross-references: *Native Elements and Alloys.* Vol. IVA: *Tellurium: Element and Geochemistry.*

TELLUROBISMUTHITE Dana, I, 160
Bi_2Te_3 trig.

Cross-reference: *Ore Microscopy.*

TELLUROHAUCHECORNITE
Ni_9BiTeS_8 *Am. Min.* **66**, 436
tetr. *Min. Mag.* **43**, 877

TELLUROPALLADINITE *Can. Min.* **17**, 589
Pd_9Te_4 mono.

TEMAGAMITE *Am. Min.* **60**, 947
Pd_3HgTe_3 orth. *Can. Min.* **12**, 193

TENGERITE *Dana*, II, 275
$CaY_3(CO_3)_4(OH)_3 \cdot 3H_2O$

TENNANTITE *Can. Min.* **18**, 173
$(Cu,Fe)_{12}As_4S_{13}$ iso. *Min. Abst.* **29**, 197

TENORITE Copper oxide named after Italian botanist M. Tenore.

Composition: CuO.

Properties: Monoclinic, *C2/c*; lath-shaped crystals; H=3.5; G=6.4; color and streak black to iron gray.

Occurrence: Found in oxide zones of copper deposits.

D. H. SPEIDEL

Reference: *Dana*, I, 507–510.

Cross-references: *Magnetic Minerals; Ore Microscopy.*

TEPHROITE A manganese silicate of the ***olivine*** group named from the Greek *tephros* for its ash color.

Composition: $4[Mn_2SiO_4]$ with Mn^{2+} replaced by Fe^{2+} with complete crystal solution to *fayalite*, or partly by Zn.

Properties: Orthorhombic, *Pbnm*; granular; H=6; G=4.1; color: flesh red to ash gray.

Occurrence: Mainly in iron-manganese ore deposits and their skarns, also metamorphosed manganese-rich sedimentary rocks.

K. FRYE

References

Am. Min. **65**, 1263.
DHZ, 1, 34–40.

Cross-references: *Magnetic Minerals; Minerals, Uniaxial and Biaxial; Olivine Group.*

TERLINGUAITE *Dana*, II, 52
Hg_2ClO mono.

TERNOVSKITE =
Magnesio-riebeckite *Am. Min.* **63**, 1052

TERRUGGITE mono. *Am. Min.* **53**, 1815
$Ca_4MgAs_2B_{12}O_{22}(OH)_{12} \cdot 12H_2O$
Min. Abst. **29**, 412

TERTSCHITE *Am. Min.* **39**, 849
$Ca_4B_{10}O_{19} \cdot 20H_2O$ mono.(?) *Min. Abst.* **29**, 412

TESCHEMACHERITE *Dana*, II, 137
$(NH_4)HCO_3$ orth.

TESTIBIOPALLADITE *Am. Min.* **61**, 182
Pd(Sb,Bi)Te iso.

TETRADYMITE *Am. Min.* **60**, 994
$Bi_{14}Te_{13}S_8$ trig. *Dana*, I, 161

TETRAFERROPLATINUM *Am. Min.* **61**, 341
PtFe tetr. *Can. Min.* **13**, 117

TETRAHEDRITE A copper iron antimony sulfide named in 1845 in allusion to the shape of its crystals.

Composition: $(Cu,Ag)_{10}(Fe,Zn)_2(Sb,As)_4S_{13}$; As>Sb is *tennantite*.

Properties: Isometric; $I\bar{4}3m$; crystals tetrahedra, some twinned or modified, also massive, granular, or compact; H=3–4.5; G=4.6–5.1; metallic luster; color: flint gray to iron black.

Occurrence: In mineral veins, but also contact metamorphic deposits with sulfides of copper, lead, silver, etc.

Use: An ore of copper and the other metals contained.

K. FRYE

References

Can. Min. **17**, 619–633.
Min. Abst. **29**, 197; **30**, 22.

Cross-references: *Metallic Minerals; Naming of Minerals; Ore Microscopy.*

TETRAKALSILITE *Am. Min.* **64**, 658
$(K,Na)AlSiO_4$ hexa. DHZ, 4, 232

TETRANATROLITE *Can. Min.* **18**, 77
$Na_2Al_2Si_3O_{10} \cdot 2H_2O$ tetr.

TETRATAENITE *Am. Min.* **65**, 624
FeNi tetr.

TETRAWICKMANITE *Am. Min.* **58**, 966
$MnSn(OH)_6$ tetra. *Min. Rec.* **4**, 24

TEXASITE — *Am. Min.* **62**, 1006
$Pr_2O_2(SO_4)$ — orth. — *Min. Rec.* **9**, 251

THADEUITE — orth. — *Am. Min.* **64**, 359
$CaMg(Mg,Fe,Mn)_2(PO_4)_2(OH,F)_2$
Min. Mag. **43**, 1068

THALCUSITE — *Am. Min.* **62**, 396
$Tl(Cu,Fe)_2S_2$ — tetr. — *Min. Abst.* **28**, 81

THALENITE — *Am. Min.* **58**, 545
$RE_2Si_2O_7$ — mono.

Cross-reference: *Pegmatite Minerals.*

THALFENSITE — iso. — *Min. Mag.* **43**, 1068
$Tl_6(Fe,Ni,Cu)_{25}S_{26}Cl$

THAUMASITE — hexa. — *Min. Abst.* **28**, 333
$Ca_3Si(CO_3)(SO_4)(OH)_6 \cdot 12H_2O$
Winchell, II, 179

THENARDITE A sodium sulfate named by Casaseca in 1826 after the French chemist, Louis J. Thenard.

Composition: Na_2SO_4; small amounts of impurities reported are due to admixture.

Properties: Orthorhombic, *Fddd* H=2.5–3; G=2.67; luster is vitreous but somewhat resinous; colorless when pure, also gray, yellow and red; habit, dipyramidal and tabular. Commonly forms crystals several inches in size, also crusted masses; cleavage, perfect on {010}, fair on {101}, poor on {100}. Forms cruciform twins and butterfly twins similar to *gypsum*. Fracture is uneven. Taste is faintly salty. Optically, refractive index, n=1.464–1.485; translucent to transparent.

Occurrence: Deposited in layers, sometimes several feet thick, from lakes and playas in arid regions. Crusted deposits on desert soils, around fumaroles, and on recent lavas. Associated with *epsomite, gypsum, natron, halite*, and borates.

Uses: Source of sodium sulfate, deposits in Arizona contain about 90% sodium sulfate and are easily mined.

L. H. FUCHS

Reference: *Dana*, II, 404–407.

Cross-references: *Cave Minerals; Saline Minerals.*

THERMONATRITE — *Dana*, II, 224
$Na_2CO_3 \cdot H_2O$ — orth. — *Min. Abst.* **27**, 23

Cross-references: *Green River Mineralogy; Saline Minerals.*

THOMSENOLITE — *Dana*, II, 116
$NaCaAlF_6 \cdot H_2O$ — mono. — *Min. Abst.* **29**, 340

THOMSONITE A hydrated sodium calcium aluminosilicate of the ***zeolite*** group, named by Brooke in 1820 in honor of Dr. T. Thomson.

Composition: $Na_4Ca_8[Al_{20}Si_{20}O_{80}] \cdot 24H_2O$. An isomorphous series exists between *thomsonite* and *natrolite* with *gonnardite* as the intermediate member; considerable replacement $CaAl \rightleftharpoons NaSi$ and $Ca \rightleftharpoons Na_2$; small amounts of Sr may replace Ca; dehydration stepwise, losing 15% H_2O at maximum; rehydration good; Na is exchanged with Ag, Tl, and K, though base-exchange reactions are extremely sluggish; molecular sieve properties good but not too selective due to adaptability of the aluminosilicate framework; transition to metathomsonite around 270°–300°C; gelatinized by HCl.

Properties: Orthohombic (pseudo-tetragonal), *Pnn*2; a=13.03 Å, b=13.13 Å, c=13.23 Å; H=5–5.5; G=2.10–2.39; luster vitreous, more or less pearly; colorless, white, pink or brown; transparent to translucent; habit in prisms, in plates, or in radiating or columnar aggregates; cleavage perfect on {010}, good on {100}; twinning on {110}. Optically, refractive indices α=1.497–1.530, β=1.513–1.533, γ=1.518–1.544; birefringence = 0.006–0.015; $2V$=42°–75° (+); optic axial plane (001); $\alpha=x$, $\beta=z$, $\gamma=y$; dispersion $r>v$.

Occurrence: In amygdules and crevices in basalt; as the upper *chabazite-thomsonite* zone in vesicles of the Tertiary basalts in Iceland; alteration product of *anorthite* in peridotite; in marl xenolith in teschenite sill; replacing wood and cementing the nephelinite palagonite tuffs on Oahu, Hawaii, associated with *phillipsite, chabazite, gonnardite, natrolite*, and *calcite*.

A. IIJIMA
M. UTADA

References

Foster, M. D., 1965. Compositional relations among thomsonites, gonnardites, and natrolites, *U.S. Geol. Surv. Prof. Paper 504-E*, 10p.
Min. Mag. **23**, 51–125; **29**, 952–954.
N.J. Min. Mh. 1975, 396–411.

Cross-references: *Pegmatite Minerals; Zeolites.*

THORBASTNAESITE hexa. *Am. Min.* **50**, 1505
$Th(Ca,RE)(CO_3)_2F_2 \cdot 3H_2O$

THOREAULITE — *Am. Min.* **59**, 1036
$SnTa_2O_6$ — mono. — *Min. Abst.* **27**, 300

THORIANITE Thorium oxide named for its composition.

Composition: ThO_2; with Th replaced by U and Ce.

Properties: Isometric; *Fm*3*m*; fluorite structure; H=6.5; G=9.7; color: black; streak: gray; commonly metamict when UO_2 exceeds 15 wt.%.

Occurrence: Found in placers and pegmatites in Sri Lanka; Madagascar; Siberia; and Pennsylvania, USA.

D. H. SPEIDEL

References

Am. Min. **49**, 1469–1471.
USGS Bull. **1064**, 47–53.

Cross-references: *Blacksands; Human and Vertebrate Mineralogy; Ore Microscopy.*

THORITE A widespread silicate of thorium named by Berzelius in 1829.

Composition: $ThSiO_4$; uranium may substitute for thorium up to 50%, yielding the variety uranothorite. Some substitution of Ce for Th and P for Si has also been found.

Properties: Tetragonal, *I*4/*amd*; invariably metamict; color: usually brown to black, rarely green; conchoidal fracture, H=5, G= 4.4–4.8; optical properties obscure due to metamictization but mean index may be about 1.68.

Occurrence: A widespread minor accessory in granitic rocks, whence it is concentrated with other heavy detrital minerals. Such *thorite* occurs in minute crystals resembling *zircon* but darker, optically isotropic, cloudy, with dull edges and less luster. Also found in some pegmatites in larger crystals, and in detrital deposits with *zircon*, ***apatite***, and *monazite*. Isostructural with *zircon*. *Huttonite* is a dimorphous form of $ThSiO_4$.

A. PABST

Reference: *Am. Min.* **38**, 1007–1018.

Cross-references: *Blacksands; Metamict State; Mineral Classification: Principles; Pegmatite Minerals; Staining Techniques.*

THOROGUMMITE tetr. *Am. Min.* **38**, 1007
$Th(SiO_4)_{1-x}(OH)_{4x}$ *USGS Bull.* **1064**, 280

THOROSTEENSTRUPINE
$(Ca,Th,Mn)_3Si_4O_{11}F \cdot 6H_2O$
amorph. *Am. Min.* **48**, 433

THORTVEITITE Hey, 104
$(Sc,Y)Si_2O_7$ mono.

Cross-reference: *Pegmatite Minerals.*

THORUTITE *Am. Min.* **48**, 1419
$(Th,U,Ca)Ti_2(O,OH)_6$ mono.

THREADGOLDITE
$Al(UO_2)_2(PO_4)_2(OH) \cdot 8H_2O$
mono. *Am. Min.* **65**, 209
Min. Abst. **31**, 87

THULITE—Intense pink *zoisite*

THURINGITE (***Chlorite*** gr.)
$(Fe^{2+},Fe^{3+},Mg,Al)_6(Si,Al)_4O_{10}(O,OH)_8$
DHZ, 3, 137
Min. Mag. **30**, 277

TIEMANNITE A mercury selenide of the ***sphalerite*** group named in 1855 for Tiemann, who discovered the mineral.

Composition: 4[HgSe] with minor Cd and S; either Hg or Se may be in excess.

Properties: Cubic, *F*$\bar{4}$3*m*; commonly massive; H=2.5; G=8.2; metallic luster; color: steel or blackish lead gray.

Occurrence: Harz Mts. of Clausthal, W. Germany, and Piute Co., Colorado, USA.

K. FRYE

References: *Am. Min.* **35**, 358–360; **65**, 797–799.

TIENSHANITE *Am. Min.* **53**, 1426
$BaNa_2MnTiB_2Si_6O_{20}$ hexa. *Min. Abst.* **30**, 217

TIGER-EYE—Asbestos partly or completely replaced by *quartz* *Dana*, III, 236

TIKHONENKOVITE *Am. Min.* **49**, 1774
$SrAlF_4(OH) \cdot H_2O$ mono.

TILASITE *Min. Rec.* **9**, 385
$CaMg(AsO_4)F$ mono.

TILLEYITE *Min. Mag.* **28**, 151
$Ca_5Si_2O_7(CO_3)_2$ mono.

Cross-reference: *Rock-Forming Minerals.*

TIN *Dana*, I, 126
Sn tetr. and iso.

Cross-references: *Allotropy; Gossan; Lunar Minerals; Native Elements and Alloys; Placer Deposits.*

TINAKSITE tric. *Am. Min.* **50**, 2098
$K_2NaCa_2TiSi_7O_{19}(OH)$

TINCAL = *Borax*

TINCALONITE — *Am. Min.* **58**, 523
$Na_2B_4O_5(OH)_4 \cdot 3H_2O$ hexa. — *Min. Rec.* **6**, 74

TIN STONE = *Cassiterite*

TINTICITE — *Am. Min.* **55**, 135
$Fe^{3+}_6(OH)_6(PO_4)_4 \cdot 7H_2O$ — *Dana*, II, 970

TINTINAITE — *Am. Min.* **54**, 573
$Pb_5(Sb,Bi)_8S_{17}$ orth. — *Can. Min.* **9**, 371

TINZENITE (***Axinite*** gr.)
$(Ca,Mn,Fe)_3Al_2BSi_4O_{15}(OH)$
Am. Min. **53**, 1407;
tric. **65**, 1119

TIRAGALLOITE — *Am. Min.* **65**, 947
$Mn_4[AsSi_3O_{12}(OH)]$ mono.

TIRODITE (***Amphibole*** gr.) *Am. Min.* **49**, 963;
$Mn_2Mg_5Si_8O_{22}(OH)_2$ mono. **63**, 1030

TISINALITE trig. *Am. Min.* **66**, 219
$Na_3H_3(Mn,Ca,Fe)TiSi_6(O,OH)_{18} \cdot 2H_2O$
Min. Mag. **43**, 1068

TITANAUGITE = Titanian *augite*

TITANITE (*SPHENE*) A calcium titanosilicate named by Klaproth in 1795 for the titanium content. *Sphene* was named by Haüy in 1801 from the Greek *sphenos*, wedge, in reference to the characteristic cross section of the crystals.

Composition: $CaTi(SiO_4)$ (O,OH,F), with Ca replaced by Na, *REE*, Mn, Sr, and Ba; Ti by Al, Fe, Mg, Nb, Ta, V, and Cr.

Properties: Monoclinic, *C2/c*; commonly in wedge-shaped crystals, also lamellar; one {110} cleavage; H=5; G=3.45–3.55; resinous to adamantine luster; transparent to translucent; colorless, brown, yellow, green, or black.

Occurrence: A common accessory mineral in syenites, diorites, and granites; also in some metamorphic rocks such as gneisses, chlorite schists, and marble; associated with ***chlorites, pyroxenes***, iron ores, ***scapolites, apatites***, *zircon, quartz*, ***feldspars***, and ***amphiboles***.

Use: An ore of titanium.

J. A. GILBERT

References

Am. Min. **61**, 238–247, 878–888.
DHZ, 1, 69–76.

Cross-references: *Blacksand Minerals; Crystallography: Morphological; Lunar Minerals; Minerals, Uniaxial and Biaxial; Pleochroic Halos; Refractory Minerals; Rock-Forming Minerals; Soil Mineralogy; Vein Minerals.*

TLALOCITE mono.(?) *Am. Min.* **61**, 504
$Cu_{10}Zn_6(TeO_3)(TeO_4)_2Cl(OH)_{25} \cdot 27H_2O$
Min. Mag. **40**, 221

TLAPALLITE mono. *Am. Min.* **64**, 465
$H_6(Ca,Pb)_2(Cu,Zn)_3(SO_4)(TeO_3)_4(TeO_6)$
Min. Mag. **42**, 183

TOBERMORITE orth. *Am. Min.* **39**, 1038
$Ca_5Si_6O_{17} \cdot 5H_2O$ — *Min. Mag.* **31**, 361

Cross-reference: *Portland Cement Mineralogy.*

TOCHILINITE mono. *Am. Min.* **57**, 1552
$6Fe_{0.9}S \cdot 5(Mg,Fe)(OH)_2$

TOCORNALITE — *Am. Min.* **58**, 384
iodide of Ag and Hg

TODOROKITE — *Am. Min.* **63**, 827
Manganese oxides — *Science* **212**, 1024

TOERNEBOHMITE — RRW, 623
$(REE)_3Si_2O_8(OH)$ hexa.

TOMBARTHITE mono. *Am. Min.* **54**, 327
$Y_4(Si,H_4)_4O_{12-x}(OH)_{4+2x}$

Cross-reference: *Naming of Minerals.*

TOMICHITE mono. *Am. Min.* **65**, 811
$(V,Fe)_4Ti_3AsO_{13}(OH)$ — *Min. Mag.* **43**, 469

TOPAZ An aluminosilicate with fluoride and hydroxyl, named for *Topazios*, the ancient Greek name for an island in the Red Sea said to abound with the gem. Originally the term referred to gem ***olivine*** (peridot), whereas chrysolite, a modern member of the ***olivine*** group, referred to modern *topaz*.

Composition: $4[Al_2SiO_4(F,OH)]$.

Properties: Triclinic, pseudo-orthorhombic, prismatic crystals, also columnar, or coarsely or finely granular; one perfect cleavage; H=8; G=3.4–3.6; vitreous luster; color: white, yellow, grayish, greenish, bluish, or reddish.

Occurrence: In cavities in granites, granite pegmatites, or rhyolites, or in surrounding metamorphic rocks; associated with *fluorite, cassiterite,* ***tourmaline****, beryl*, and *quartz*.

Use: Gemstone.

K. FRYE

References

Am. Min. **56**, 24–30.
DHZ, 1, 145–150.
Min. Mag. **43**, 741–742.

Cross-references: *Abrasive Materials; Blacksand Minerals; Crystallography: Morphological; Gemology; Jade; Minerals, Uniaxial and Biaxial; Mohs Scale of Hardness; Pegmatite Minerals; Rock-Forming Minerals; Soil Mineralogy.*

TOPAZOLITE–Yellow *andradite*

TORBERNITE A hydrated copper uranium phosphate of the ***autunite*** group, named for the Swedish chemist Torbern Bergmann (1735–1784).

Composition: $Cu(UO_2)_2(PO_4)_2 \cdot 8\text{–}12H_2O$. P_2O_5, 14.1%; UO_3, 56.5%; CuO, 7.9%; H_2O, 21.4% by wt; As substitutes for P.

Properties: Tetragonal, *I4/mmm*; G=3.22; H=2.5; vitreous to pearly luster; color: emerald green; streak: pale green; tabular thin crystals, scaly aggregates, crystals usually square in outline, rarely pyramidal; brittle; micaceous cleavage {001}, distinct {100}; lamina brittle; optically (-), ω=1.59, ϵ=1.58; translucent to transparent; not fluorescent. Upon heating, loses water and reverts to *metatorbernite I* below 100°C; undergoes another change at 130°C to *metatorbernite II.*

Occurrence: A secondary mineral derived from the oxidation of *uraninite*. Found in granite pegmatite and in copper deposits. Associated with *autunite*, copper sulfides, *uraninite*, and other uranium minerals.

W. E. HILL

References

Min. Rec. **6**, 237–249.
USGS Bull. **1064**, 170–177.

TÖRNEBOHMITE = TOERNEBOHMITE

TORREYITE mono. *Am. Min.* **64**, 952
$(Mg,Mn)Zn_2(SO_4)(OH)_{12} \cdot 4H_2O$

TOURMALINE GROUP–*Buergerite, dravite, ferridravite, elbaite, liddicoatite, schorl*, and *uvite*.

Cross-references: *Authigenic Minerals; Crystal Growth; Fire Clays; Gemology; Pegmatite Minerals; Phantom Crystals; Piezoelectricity; Placer Deposits; Pleochroic Halos; Rock-Forming Minerals; Soil Mineralogy; Tourmaline Group.*

TRANQUILLITYITE amorph *Am. Min.* **58**, 140
$Fe_8(Zr,Y)_2Ti_3Si_3O_{24}$ *Min. Abst.* **30**, 155

Cross-reference: *Lunar Minerals.*

TRASKITE hexa. *Am. Min.* **50**, 1500
$Ba_9Fe_2^{2+}Ti_2(Si_{12}O_{36})(OH,Cl,F)_6 \cdot 6H_2O$

TREASURITE *Am. Min.* **64**, 244
$Ag_7Pb_6Bi_{15}S_{32}$ mono. *Min. Abst.* **29**, 140

TRECHMANNITE *Dana*, I, 432
$AgAsS_2$ trig.

TREMOLITE A calcium magnesium hydroxyl silicate of the ***amphibole*** group. The name is from the locality, Val Tremola, south of the St. Gotthard.

Composition: $2[Ca_2Mg_5Si_8O_{22}(OH,F)_2]$, with Mg replaced by Fe^{2+} to *actinolite*, and Mg and Si by Al toward *hornblende*.

Properties: Monoclinic, *C2/m*; long-bladed or short crystals, fibrous, or massive; H=5–6; G=2.9–3.2; perfect prismatic {110} cleavage at 56°; color: white to pale green with increasing iron content.

Occurrence: Product of low-grade, regional metamorphism of carbonate, mafic, and ultramafic rocks. Associated with *talc*, ***chlorite****, epidote, albite, lawsonite*, or *glaucophane*.

Uses: The variety nephrite is the gemstone jade. Byssolite is asbestiform *tremolite*.

K. FRYE

References

Am. Min. **53**, 751–769; **63**, 1023–1052.
DHZ, 2, 249–262.

Cross-references: *Amphibole Group; Asbestos; Gemology; Human and Vertebrate Mineralogy; Isomorphism; Jade; Minerals, Uniaxial and Biaxial.*

TREVORITE (***Spinel*** gr.) *Dana*, I, 698
$NiFe_2O_4$ iso.

TRIDYMITE A high-temperature silica polymorph. The name is from the trillings produced by twinning, which gives wedge-shaped crystals of three or more individuals.

Composition: SiO_2, with Al_2O_3, 0–3%; K_2O, 0–1%; and Na_2O, 0–1%.

Properties: Orthorhombic (pseudo-hexagonal). (The high-temperature β phase is hexagonal.) H=7; G=2.26; luster is vitreous;

colorless or white; streak: white; there is a poor prismatic cleavage; twinning is common on {110}, giving wedge-shaped crystals. Optically, refractive indices α=1.469–1.479, β=1.470–1.480, γ=1.473–1.483; $2V\gamma$=40°–90°.

Occurrence: Felsic volcanic rocks such as rhyolite, obsidian, trachyte, andesite, and dacite; in some rhyolitic tuffs, where it forms after the tuffs are deposited; occasionally in highly metamorphosed arkose adjacent to a mafic igneous intrusion. Associated with *sanidine, augite, fayalite*, and *cristobalite*.

Uses: Natural *tridymite* does not occur in economic amounts but this phase is an important constituent of many refractories.

R. A. HOWIE

References

AJS, ser. 4, **36**, 331.
Dana, III, 259–272.

Cross-references: *Lunar Minerals; Meteoritic Minerals; Mineral Properties; Polymorphism; Portland Cement Mineralogy; Rock-Forming Minerals.*

TRIGONITE *Dana*, II, 1032
$Pb_3MnH(AsO_3)_3$ mono.

TRIKALSILITE *Am. Min.* **42**, 286
$(K,Na)AlSiO_4$ hexa. DHZ, 4, 239

TRIMERITE *Am. Min.* **50**, 1170
$CaMn_2(BeSiO_4)_3$ mono. *Min. Abst.* **29**, 263

TRIPHANE = *SPODUMENE*
Min. Mag. **43**, 1053

TRIPHYLITE A lithium ferrous manganous phosphate, named in allusion to its three cations. The name also refers to the series and the group of which it is a member.

Composition: $4[LiFe^{2+}(PO_4)]$ with Fe^{2+} replaced by Mn^{2+} (Mn>Fe, *lithiophilite*).

Properties: Orthorhombic, *Pmnb*; massive, but single crystals attain 4 m in greatest dimension; one perfect and one imperfect cleavage; H=4–5; G=3.3–3.5; vitreous to subresinous luster; color: bluish to greenish gray or brown.

Occurrence: In granite pegmatites associated with other lithium and phosphate minerals.

K. FRYE

References

Am. Min. **55**, 135.
Dana, II, 665–669.

Cross-reference: *Pegmatite Minerals.*

TRIPLITE A fluophosphate of iron, manganese, calcium, and magnesium, the name referring to its three cleavage directions.

Composition:
$8[(Mn^{2+},Fe^{2+},Mg,Ca)_2(PO_4)(F,OH)]$.

Properties: Monoclinic, *I2/m*; massive; three cleavages; H=5–5.5; G=3.5–3.9; vitreous to resinous luster; color: various shades of brown.

Occurrence: Phosphate-rich pegmatites.

K. FRYE

References

Am. Min. **36**, 256–271.
Dana, II, 849–852.

Cross-reference: *Pegmatite Minerals.*

TRIPLOIDITE mono. *Min. Mag.* **43**, 179
$(Mn,Fe^{2+})_2(PO_4)(OH)$

Cross-reference: *Pegmatite Minerals.*

TRIPOLI—Diatomaceous earth

Cross-reference: *Abrasive Materials.*

TRIPPKEITE *Dana*, II, 1034
$CuAs_2^{3+}O_4$ tetr. *Min. Abst.* **27**, 300

TRIPUHYITE *Dana*, II, 1024
$FeSb_2O_6$ tetr.

TRITOMITE trig. *Am. Min.* **47**, 9
$(RE,Y,Th)_5(Si,B)_3(O,OH,F)_{13}$(?)

TRITOMITE-(Y) trig.(?) *Am. Min.* **47**, 9
$(Y,Ca,RE,Fe)_5(Si,B,Al)_3(O,OH,F)_{13}$(?)

TROEGERITE *Min. Abst.* **27**, 80
$(UO_2)_3(AsO_4)_2 \cdot 12H_2O$(?)
tetr.(?) *USGS Bull.* **1064**, 187

TROGTALITE (***Pyrite*** gr.)
$CoSe_2$ iso. *Am. Min.* **41**, 164

TROILITE *Dana*, I, 233
FeS hexa.

Cross-references: *Lunar Minerals; Magnetic Minerals; Meteoritic Minerals.*

TROLLEITE *Am. Min.* **61**, 409; **64**, 1175
$Al_4(PO_4)_3(OH)_3$ mono.

TRONA A hydrated sodium carbonate, the name being a shortened form of the Arabic *natrun*.

Composition: $Na_3(CO_3)(HCO_3) \cdot 2H_2O$.

Properties: Monoclinic, *I2/a*; fibrous or massive; one {100} perfect cleavage; H=2.5–3; G=2.14; vitreous luster; colorless or whitish; alkaline taste.

Occurrence: Saline lake deposits and soils in arid regions, associated with *natron, thermonatrite, halite, glauberite, thenardite, mirabilite*, and *gypsum*. Localities include the Nile Valley, Egypt; Sudan; Tanzania; Kenya; Iran; Mongolia; and in the USA in soda lakes and oil shale in Wyoming; saline lakes and playas in Nevada; and with borates at Searles Lake, California.

Use: Mined for the carbonate and bicarbonate of soda.

K. FRYE

References

Am. Min. **44**, 274–281.
Dana, II, 138–140.

Cross-references: *Green River Mineralogy; Saline Minerals.*

TRUSCOTTITE trig. *Am. Min.* **53**, 511
$(Ca,Mn)_2Si_4O_9(OH)_2$ *Min. Mag.* **30**, 450

TRÜSTEDTITE (***Linnaeite*** gr.)
$NiNi_2Se_4$ iso. *Am. Min.* **50**, 519

TSCHEFFKINITE = *Chevkinite*

TSCHERMAKITE (***Amphibole*** gr.)
$Ca_2(Mg,Fe^{2+})_3(Al,Fe^{3+})_2(Al_2Si_6)O_{22}(OH,F)_2$
Am. Min. **63**, 1032
mono. DHZ, 2, 263

TSCHERMIGITE = Ammonium *alum*

Cross-reference: *Crystal Growth.*

TSILAISITE–Synthetic Mn *tourmaline*

TSUMCORITE mono. *Am. Min.* **57**, 1558
$PbZnFe^{2+}(AsO_4)_2 \cdot H_2O$

TSUMEBITE mono. *Dana*, II, 918
$Pb_2Cu(PO_4)(SO_4)(OH)$

TSUMOITE *Am. Min.* **63**, 1162
BiTe trig. *Min. Mag.* **43**, 1068

TUCEKITE *Min. Abst.* **29**, 482
$Ni_9Sb_2S_8$ tetr. *Min. Mag.* **43**, 873

TUGARINOVITE *Am. Min.* **66**, 438
MoO_2 mono.

TUGTUPITE *Am. Min.* **48**, 1178
$Na_4AlBeSi_4O_{12}Cl$ tetr. *Min. Mag.* **42**, 251

TUHUALITE orth. *Am. Min.* **62**, 418
$(Na,K)_2Fe_2^{2+}Fe_2^{3+}Si_{12}O_{30} \cdot H_2O$
Science **166**, 1399

TULAMEENITE *Am. Min.* **59**, 383
Pt_2FeCu tetr. *Can. Min.* **12**, 21

TUNDRITE tric. *Am. Min.* **59**, 633
$Na_3(REE)_4(Ti,Nb)_2(SiO_4)_2(CO_3)_3O_4(OH) \cdot 2H_2O$

TUNDRITE-(Nd) tric. *Am. Min.* **53**, 1780
$Na_2(Nd,La)_4(Ti,Nb)_2(SiO_4)_2(CO_3)_3O_4(OH) \cdot 2H_2O$

TUNELLITE *Am. Min.* **47**, 416; **49**, 1549
$SrB_6O_9(OH)_2 \cdot 3H_2O$ mono.

TUNGSTENITE *Dana*, I, 331
WS_2 *Min. Abst.* **28**, 279
hexa. and trig.

TUNGSTITE A tungsten oxyhydroxide named by Dana in 1868 for its composition.

Composition: $WO_2(OH)_2$.

Properties: Orthorhombic (?); microscopic scales; color: golden yellow to green; pearly luster.

Occurrence: An oxidation product of tungsten minerals as found in Bolivia and in Salida, Colorado.

D. H. SPEIDEL

Reference: *Dana*, I, 605–606.

TUNGUSITE *Am. Min.* **52**, 927
$Ca_4Fe_2^{2+}Si_6O_{15}(OH)_6$

TUNISITE tetr. *Am. Min.* **54**, 1
$NaHCa_2Al_4(CO_3)_4(OH)_{10}$

TURANITE orth.(?) *Dana*, II, 818
$Cu_5(VO_4)_2(OH)_4$(?)

TURGITE, TURJITE = *Hematite* with adsorbed water

TURQUOISE A hydrated hydroxylphosphate of copper and aluminum, the name meaning Turkish, probably because the mineral was brought through Turkey to Europe.

Composition: $CuAl_6(PO_4)_4(OH)_8 \cdot 5H_2O$ with Fe^{3+} replacing Al (Fe^{3+}>Al is *chalcosiderite*).

Properties: Triclinic, $P\bar{1}$; massive, dense, or cryptocrystalline to finely granular; H=5–6; G=2.6–2.8; waxy luster; color: blue to green.

Occurrence: Secondary mineral formed in arid regions by surface waters acting on aluminous rock.

Use: Gemstone.

K. FRYE

References

Am. Min. **50**, 283.
Dana, II, 947–951.

Cross-references: *Gemology; Minerals, Uniaxial and Biaxial.*

TUSCANITE mono. *Am. Min.* **62**, 1110
$(K,H_2O)_2Ca_6(Si,Al)_{10}O_{22}(SO_4,CO_3OH,O_4H_4)_2$ *Min. Abst.* **29**, 342

TVALCHRELIDZEITE *Am. Min.* **62**, 174
$Hg_{12}(Sb,As)_8S_{15}$ mono. *Min. Abst.* **28**, 81

TVEITITE mono. *Am. Min.* **62**, 1060
$Ca_{1-x}(Y,REE)_xF_{2+x}$ *Min. Abst.* **29**, 201

TWINNITE *Am. Min.* **53**, 1424
$Pb(Sb,As)_2S_4$ orth. *Can. Min.* **9**, 191

TYCHITE iso. *Am. Min.* **54**, 302
$Na_6Mg_2(CO_3)_4(SO_4)$

Cross-references: *Green River Mineralogy; Saline Minerals.*

TYRETSKITE *Am. Min.* **53**, 2084
$Ca_2B_5O_8(OH)_3$ tric. *Min. Abst.* **17**, 500

TYROLITE orth. *Am. Min.* **42**, 123
$Cu_5Ca(AsO_4)_2(CO_3)(OH)_4 \cdot 6H_2O$

TYRRELLITE (***Linnaeite*** gr.)
$(Cu,Co,Ni)_3Se_4$ iso. *Am. Min.* **37**, 542

TYSONITE = *FLUOCERITE*

TYUYAMUNITE This mineral is the calcium analog of *carnotite*, named from Tyuya Muyum in southeastern Turkistan, USSR.

Composition: $Ca(UO_2)_2(VO_4)_2 \cdot 5\text{–}8H_2O$; water varies with moisture of atmosphere.

Properties: Orthorhombic, *Pnan*; H=2; G=3.7–4.3; earthy to waxy luster; color: canary yellow to greenish yellow; streak: yellow; yellow green fluorescence; habit commonly massive to cryptocrystalline, also small crystals as scales or laths flattened on the (001) and elongated on the *a* axis; perfect micaceous cleavage on {001}, distinct on {100} and {010}. Optically (-), $2V=36°$, $\alpha=1.670$, $\beta=1.870$, $\gamma=1.895$.

Occurrence: A secondary mineral of uranium and vanadium found in limestones, sandstones, or concentrated with vegetable matter in Russia, and in the USA in the Colorado Plateau and adjacent areas of Utah, New Mexico, and Arizona. Associated with *carnotite* and other uranium and vanadium minerals.

Use: An ore of uranium.

W. E. HILL

Reference: *USGS Bull.* **1064**, 248–253.

Cross-reference: *Naming of Minerals.*

U

UGRANDITE—Calcium ***garnets:*** *Uvarovite + grossular + andradite*

UHLIGITE iso. *Dana,* I, 735
$Ca_3(Ti,Al,Zr)_9O_{20}$(?)

UKLONSKOVITE *Am. Min.* **50**, 520
$NaMg(SO_4)(OH)\cdot 2H_2O$ mono.

ULEXITE A hydrated sodium, calcium borate named in 1850 after the German chemist, G. L. Ulex, who determined its composition.

Composition: $2[NaCaB_5O_6(OH)_6\cdot 5H_2O]$.

Properties: Triclinic, $P\bar{1}$; nodules, lenses, bundles of fibers, and encrustations; one perfect and one good cleavage; H=2.5; G=1.96; silky to satiny luster, white, tasteless.

Occurrence: Salt playas and desiccated saline lakes associated with *borax, halite, glauberite*, and other saline minerals.

K. FRYE

References: *Am. Min.* **44**, 712–719; **63**, 160–171.

Cross-reference: *Saline Minerals.*

ULLMANNITE A nickel sulfide-antimonide of the ***cobaltite*** group, named in 1843 after J. C. Ullmann, discoverer of the mineral.

Composition: 4[NiSbS] with Ni replaced by Co and Fe and Sb by As and Bi.

Properties: Triclinic, $P1$; pseudocubic; crystals cubes or less commonly octahedra and pyritohedra; cubic cleavage; H=5–5.5; G=6.65; metallic luster; color: steel gray to silver white.

Occurrence: Found in veins with other nickel minerals.

K. FRYE

References

Am. Min. **62**, 369–373.
Can. Min. **18**, 165–171.

Cross-reference: *Ore Microscopy.*

ULTRAMARINE = Synthetic *lazurite*

ULVITE = *Ulvöspinel*

ULVÖSPINEL An iron titanium oxide, the titanium end member of the ***magnetite*** series of the ***spinel*** group.

Composition: Fe_2TiO_4; exact composition dependent on oxygen content.

Properties: Isometric, $Fd3m$; common as exsolution-type blebs within *magnetite*; darker than *magnetite* in reflected light.

Occurrence: Found in mafic igneous rocks and metamorphosed equivalents, e.g., the Skaergaard, and as minor oxide in lunar rocks.

D. H. SPEIDEL

Reference: *J. Petrology* **5**, 310.

Cross-references: *Magnetic Minerals; Spinel Group.*

UMANGITE *Am. Min.* **35**, 354
Cu_3Se_2 tetr. *Min. Abst.* **29**, 198

Cross-reference: *Ore Microscopy.*

UMBOZERITE amorph. *Am. Min.* **60**, 341
$Na_3Sr_2ThSi_8(O,OH)_{24}$

UMOHOITE *Am. Min.* **44**, 1248
$(UO_2)MoO_4\cdot 4H_2O$ orth. or mono.
Min. Abst. **28**, 447

UNGEMACHITE hexa. *Dana*, II, 596
$Na_8K_3Fe^{3+}(SO_4)_6(OH)_2\cdot 10H_2O$

UPALITE orth. *Am. Min.* **65**, 208
$Al(UO_2)_3(PO_4)_2(OH)_3$

URALBORITE *Am. Min.* **47**, 1482
$CaB_2O_4\cdot 2H_2O$ mono.

URALITE—Fibrous ***amphibole*** pseudomorphous after ***pyroxene***

URALOLITE mono. *Am. Min.* **49**, 1776
$CaBe_3(PO_4)_2(OH)_2\cdot 4H_2O$ *Min. Rec.* **9**, 99

URAMPHITE (***Meta-autunite*** gr.)
$(NH_4)(UO_2)(PO_4)\cdot 3H_2O$
orth.(?) *Am. Min.* **44**, 464

URANINITE A uranium oxide named for its composition; called "pitchblende" when metamict.

Composition: UO_2; with Th, Pb, He, Ar commonly present due to radioactive decay.

Properties: Cubic, *Fm*3*m*; fluorite structure; H=5.6; G=10.9; usually opaque with submetallic luster; streak: brownish black or green.

Occurrence: Found in pegmatite and veins in Zaire and at Great Bear Lake, Canada.

D. H. SPEIDEL

Reference: *Dana*, I, 611–620.

Cross-references: *Autunite Group; Blacksand Minerals; Magnetic Minerals; Metallic Minerals; Mineral and Ore Deposits; Pegmatite Minerals.*

URANITE–Uranyl phosphates and arsenates of the ***autunite*** and ***meta-autunite*** groups

URANMICROLITE (***Pyrochlore*** gr.)
$U_2(Ta,Nb)_2O_6(OH,F)$ iso. *Am. Min.* **62**, 405

URANOCIRCITE (***Autunite*** gr.)
$Ba(UO_2)_2(PO_4)_2\cdot 12H_2O$
tetr. *USGS Bull.* **1064**, 211

URANOPHANE A hydrated silicate of calcium and hexavalent uranium, named by M. Websky in 1853 (from Uran, uranium, and Greek, to appear).

Composition: $Ca(UO_2)_2(SiO_3)_2(OH)_2\cdot 5H_2O$; Mg, Ba, Pb, Th, and Y may substitute for Ca.

Properties: Monoclinic, $P2_1$; H=2.5; G=3.7–3.9; luster is vitreous, pearly on good cleavage surfaces, massive material is earthy to waxy; color: various shades of yellow; habit: minute needles elongated parallel to the *b*-axis; cleavage: one perfect {100}, and another poor parallel to the elongation of the needles; usually occurs as tufted aggregates or as crusts. Crystals are weakly fluorescent in ultraviolet light but the massive material is commonly not fluorescent.

Occurrence: Deposited as coatings in crevices of rocks by the action of meteoric waters on primary uranium minerals. *Uranophane* is a common secondary mineral and occurs in practically all uranium localities associated with *β-uranophane* (same chemical composition as *uranophane* but differs in crystal structure), *kasolite, meta-autunite, phosphuranylite*, and various hydrated uranium oxides.

Use: An ore of uranium.

L. H. FUCHS

References

Am. Min. **42**, 594.
Heinrich, E. W., 1958. *Mineralogy and Geology of Radioactive Raw Materials.* New York: McGraw-Hill, 116.
USGS Bull. **1064**, 294.

URANOPILITE
$(UO_2)_6(SO_4)(OH)_{10}\cdot 12H_2O$
mono.(?) *USGS Bull.* **1064**, 135

URANOSPATHITE *Min. Mag.* **42**, 117
$(HAl)_{0.5}(UO_2)_2(PO_4)_2\cdot 20H_2O$
tetr. *USGS Bull.* **1064**, 194

URANOSPHAERITE *Am. Min.* **42**, 905
$Bi_2U_2O_9\cdot 3H_2O$ mono. *USGS Bull.* **1064**, 98

URANOSPINITE (***Autunite*** gr.)
$Ca(UO_2)_2(AsO_4)_2\cdot 10H_2O$
tetr. *USGS Bull.* **1064**, 183

URANOTHORIANITE *Min. Abst.* **28**, 117
$(Th,U)O_2$ iso. *Min. Mag.* **23**, 638

URANOTHORITE = Uranian *thorite*
Am. Min. **36**, 557–562

URANOTILE = *Uranophane*

URANPYROCHLORE (***Pyrochlore*** gr.)
$(U,Ca,Ce)_2(NbTa)_2(OH,F)_6$
iso. *Am. Min.* **62**, 405

UREA *Am. Min.* **59**, 874
$CO(NH_2)_2$ tetr.

Cross-reference: *Cave Minerals.*

UREYITE (***Pyroxene*** gr.)
$NaCrSi_2O_6$ *Am. Min.* **50**, 1096
mono. *Science* **149**, 742

Cross-references: *Clinopyroxenes; Mantle Mineralogy; Meteoritic Minerals.*

URVANTSEVITE — *Am. Min.* **62**, 1260
$Pd(Bi,Pb)_2$ hexa.

USOVITE — *Am. Min.* **52**, 1582; **60**, 739
$Ba_2MgAl_2F_{12}$ mono.

USSINGITE — *Am. Min.* **59**, 335
$Na_2AlSi_3O_8(OH)$ tric.

USTARASITE — *Am. Min.* **41**, 814
$Pb(Bi,Sb)_6S_{10}$

UVANITE — *USGS Bull.* **1064**, 261
$U_2V_6O_{21} \cdot 15H_2O(?)$ orth.

UVAROVITE A calcium chromium silicate of the ***garnet*** group, named after S. S. Uvarov, President of the St. Petersburg (Leningrad) Academy.

Composition: $8[Ca_3Cr_2Si_3O_{12}]$ with Cr replaced by Al toward *grossular* and by Fe^{3+} toward *andradite.*

Properties: Isometric, *Ia3d*; crystals are dodecahedra and trapezohedra, simple or in combination; H=7.5; G=3.4–3.5; vitreous luster; color: emerald green.

Occurrence: In serpentinite associated with *chromite*, in metamorphosed carbonates, and in skarns.

K. FRYE

Reference: DHZ, 1, 77–112.

Cross-references: *Garnet Group; Minerals, Uniaxial and Biaxial; Naming of Minerals.*

UVITE (***Tourmaline*** gr.) — *Min. Abst.* **28**, 414
$CaMg_3(MgAl_5)(BO_3)_3Si_6O_{18}(OH)_4$
trig. — *Min. Rec.* **8**, 100

UYTENBOGAARDTITE — *Am. Min.* **65**, 209
Ag_3AuS_2 tetr. — *Can. Min.* **16**, 651

UZBEKITE = *Volborthite* — *Am. Min.* **50**, 2111

V

VAESITE (*Pyrite* gr.)
NiS$_2$ *Am. Min.* **30**, 483
iso. *Min. Rec.* **9**, 111

Cross-reference: *Geologic Thermometry.*

VALENTINITE Antimony oxide named for Basil Valentine, 15th-century alchemist.

Composition: Sb_2O_3.

Properties: Orthorhombic, *Pccn*; prismatic {110} cleavage; H=2.5–3; G=5.76; colorless; low-temperature polymorph of *senarmontite.*

Occurrence: An oxidation product of antimony ores.

D. H. SPEIDEL

Reference: *Dana*, I, 547–550.

VALLERIITE *Am. Min.* **62**, 1030
4(Fe,Cu)S·3(Mg,Al)$(OH)_2$
trig. *Min. Abst.* **29**, 198

Cross-reference: *Meteoritic Minerals.*

VANADINITE A lead chloro-vanadate of the *pyromorphite* series of the ***apatite*** group, named in 1838 for its composition.

Composition: $2[Pb_5(VO_4)_3Cl]$; with P and As replacing V and Ca replacing Pb.

Properties: Hexagonal, $P6_3/m$; prismatic crystals; H=2.75–3; G=6.9; subresinous to subadamantine luster; color: red, brown, or yellow.

Occurrence: In oxidized zone of lead-ore deposits with *pyromorphite, mimetite*, and other lead-bearing minerals as at Mibladen, Morocco, and the Old Yuma Mine, Arizona, USA.

K. FRYE

References

Dana, II, 895–898.
Min. Abst. **29**, 514–515.

Cross-references: *Apatite Group; Skeletal Crystals.*

VANALITE mono. *Am. Min.* **57**, 597
$NaAl_8V_{10}O_{38}\cdot 30H_2O$

VANDENBRANDEITE *Am. Min.* **36**, 394
$CuUO_2(OH)_4$ tric. *USGS Bull.* **1064**, 100

VANDENDRIESSCHEITE *Am. Min.* **45**, 1026
$PbU_7O_{22}\cdot 12H_2O$ orth.

VANOXITE *Dana*, I, 601
$V_4^{4+}V_2^{5+}O_{13}\cdot 8H_2O$(?)

VANTHOFFITE *Dana*, II, 430
$Na_6Mg(SO_4)_4$ mono.

Cross-reference: *Saline Minerals.*

VANURALITE mono. *Am. Min.* **56**, 639
$Al(UO_2)_2(VO_4)_2(OH)\cdot 11H_2O$

VANURANYLITE orth.(?) *Am. Min.* **51**, 1548
$(H_3O,Ba,Ca,K)_{1.6}(UO_2)_2(VO_4)_2\cdot 4H_2O$(?)

VARISCITE A hydrated aluminum phosphate of the ***variscite*** group, named in 1837 after Variscia, former name for a district of Saxony, GDR, where the mineral was first found.

Composition: $8[AlPO_4\cdot 2H_2O]$, with complete substitution of Al by Fe^{3+} toward *strengite.*

Properties: Orthorhombic, *Pcab*; fine-grained masses, nodules, veinlets, or crusts; one good cleavage; H=4.5; G=2.6; vitreous to waxy luster; color: blue, green, or white; polymorphic with *metavariscite.*

Occurrence: Formed in cavities by phosphate meteoric water acting on aluminous rocks, with *wavellite, crandallite, metavariscite,* ***apatite***, chalcedony, and limonite.

K. FRYE

References

Am. Min. **57**, 36–44.
Dana, II, 756–761.

VARLAMOFFITE = *Cassiterite*
Hey, 47

VARULITE (***Alluaudite*** gr.)
(Na,Ca)Mn(Mn,Fe)$_2$$(PO_4)_3$ *Dana,* II, 669
mono. *Min. Mag.* **43**, 227

VASHEGYITE amorph. *Dana*, II, 999
$Al_4(PO_4)_3(OH)_3 \cdot 13H_2O(?)$

VATERITE (***Calcite*** gr.)
$CaCO_3$ *Am. Min.* **45**, 1316
hexa. *Min. Mag.* **32**, 535

Cross-references: *Human and Vertebrate Mineralogy; Invertebrate and Plant Mineralogy; Syntaxy.*

VAUQUELINITE mono. *Dana*, II, 650
$Pb_2Cu(CrO_4)(PO_4)(OH)$

VAUXITE
$Fe_2^{2+}(H_2O)_4[Al_4(H_2O)_4(PO_4)_4] \cdot 4H_2O$
tric. *Am. Min.* **53**, 1025; **59**, 843

VÄYRYNENITE *Am. Min.* **41**, 371
$MnBe(PO_4)(OH,F)$ mono.

VEATCHITE *Am. Min.* **64**, 362
$Sr_2B_{11}O_{16}(OH)_5 \cdot H_2O$
tric. or mono. *Min. Mag.* **43**, 1069

VEENITE *Am. Min.* **53**, 1422
$Pb_2(Sb,As)_2S_5$ orth. *Can. Min.* **9**, 7

VELIKITE *Am. Min.* **62**, 1260
$(Cu,Hg)_5Sn_2S_8$ tetr. *Min. Abst.* **29**, 140

VERDELITE = Green *elbaite*

VERMICULITE Hydrous aluminosilicates (clay minerals) first named by Webb in 1824 from the Latin to grow worms. They are similar to the ***smectites*** but notable for exfoliation when rapidly heated.

Composition:
$(Mg,Ca)_{0.7}(Mg,Fe^{3+},Al)_6(Al,Si)_8O_{20}(OH)_4 \cdot 8H_2O$

Properties: Monoclinic, *Cc*; H=1.5; G=2.3; colorless, yellow, green, brown; perfect {001} cleavage (basal cleavage); pearly or bronze-like luster; rapid heating causes exfoliation into long worm-like threads, whence the name; streak white. Optically (-); $2V\alpha = 0°–8°$; colorless in thin section; α=1.525–1.564, β=1.545–1.583, γ=1.545–1.585, δ= 0.02–0.03; pleochroism: α paler shades than β and γ.

Occurrence: Alteration product of *biotite* by weathering or hydrothermal attack; in regions of contact between felsic intrusives and mafic or ultramafic rocks; in soils. Associated with *corundum*, ***apatite***, ***serpentine***, ***chlorite***, and *talc* in regions of felsic-mafic rock contact.

Uses: When expanded, as lightweight aggregate in concretes and plasters; as a packing material; in agriculture; for thermal and acoustic insulation; as a filler or extender in paper, plastics, and paints.

E. BOOY

References

DHZ, 3, 246–257.
Grim, R. E., 1968. *Clay Mineralogy*. New York: McGraw-Hill, 596p.
Kerr, P. F. ed. 1951. *Reference Clay Minerals, American Petroleum Institute Research Project 49*. New York: Columbia University, 701p.

Cross-references: *Clays, Clay Minerals; Human and Vertebrate Mineralogy; Mineral Industries; Minerals, Uniaxial and Biaxial; Pigments and Fillers; Soil Mineralogy.*

VERPLANCKITE hexa. *Am. Min.* **50**, 1500
$Ba_2(Mn,Fe,Ti)Si_2O_6(O,OH,Cl,F)_2 \cdot 3H_2O$

VERSILIAITE *Am. Min.* **64**, 1230
$Fe_2^{2+}Fe_4^{3+}Sb_6O_{16}S$ orth. *Min. Abst.* **31**, 356

VERTUMNITE mono. *Am. Min.* **62**, 1061
$Ca_4Al_4Si_4O_6(OH)_{24} \cdot 3H_2O$

VESIGNIEITE *Am. Min.* **40**, 942
$BaCu_3(VO_4)_2(OH)_2$ mono. *Min. Abst.* **29**, 85

VESUVIANITE (IDOCRASE) A hydrous calcium magnesium iron aluminosilicate named for Mt. Vesuvius.

Composition:
$4Ca_{10}[(Mg,Fe)_2Al_4(SiO_4)_5(Si_2O_7)_2(OH)_4]$.

Properties: Tetragonal, *P4/nnc*; prismatic crystals with vertical striations, often in vertically striated columnar aggregates, and may be massive or granular; H=6–7; G=3.33–3.43; vitreous to resinous luster; subtransparent to translucent; color: green or brown, may be yellow, blue, or red.

Occurrence: Most commonly in crystalline limestone as a result of contact metamorphism; in association with ***garnet***, *wollastonite*, *diopside*, and ***tourmaline***. Well-formed crystals are from Eden Mills, Vermont; Olmsteadville, New York; Magnet Cove, Arkansas; and Crestmore, California, in the USA; and Ala, Italy.

Uses: A minor gem stone.

J. A. GILBERT

References

Am. Min. **53**, 880; **64**, 367.
DHZ, 1, 113–120.

VESZELYITE mono. *Dana*, II, 916
$(Cu,Zn)_3(PO_4)(OH)_3 \cdot 2H_2O$
Min. Abst. **28**, 207

VIGEZZITE orth. *Am. Min.* **65**, 811
$(Ca,Ce)(Nb,Ta,Ti)_2O_6$ *Min. Mag.* **43**, 459

VIKINGITE *Am. Min.* **64**, 243
$Ag_5Pb_8Bi_{13}S_{30}$ mono.

VILLAMANINITE (***Pyrite*** gr.)
$(Cu,Ni,Co,Fe)S_2$ iso. *Min. Mag.* **41**, 545

VILLIAUMITE *Am. Min.* **55**, 126
NaF iso. *Dana*, II, 10

VIMSITE *Am. Min.* **54**, 1219
$CaB_2O_2(OH)_4$ mono. *Min. Abst.* **28**, 260

VINCENTITE *Am. Min.* **59**, 1332
$(Pd,Pt)_3(As,Sb,Te)$ *Min. Mag.* **39**, 525

VINOGRADOVITE mono. *Am. Min.* **42**, 308
$(Na,Ca,K)_4Ti_4AlSi_6O_{23}(OH)\cdot 2H_2O$

Cross-reference: *Green River Mineralogy.*

VIOLAN–A purple ***pyroxene***
Am. Min. **65**, 813

VIOLARITE (***Linnaeite*** gr.)
Ni_2FeS_4 *Can. Min.* **17**, 309
iso. *Min. Mag.* **43**, 733

VIRGILITE *Am. Min.* **63**, 461
$LiAlSi_2O_6$ hexa.

VIRIDINE = Green manganese-rich *andalusite*

VISÉITE iso. DHZ, 4, 342, 408
$5CaO\cdot 6Al_2O_3\cdot 3SiO_2\cdot 3.5P_2O_5\cdot 1.5F\cdot 36H_2O$
Min. Mag. **41**, 437

VISHNEVITE–Sodium sulfate-rich *cancrinite* *Can. Min.* **17**, 49

VITUSITE *Am. Min.* **65**, 812
$Na_3(REE)_3(PO_4)_2$ orth. *Min. Mag.* **43**, 1069

VIVIANITE A hydrated ferrous iron phosphate, named in 1817 after the English mineralogist J. G. Vivian.

Composition: $2[Fe_3^{2+}(PO_4)_2\cdot 8H_2O]$; but usually partly oxidized to Fe^{3+}.

Properties: Monoclinic, $C2/m$; crystals prismatic, sometimes flattened, equant, or tabular; globular or tabular masses or concretions; one perfect cleavage; H=1.5–2; G=2.68; vitreous or earthy luster; colorless and transparent when fresh, but becoming increasingly darker blue because of oxidation upon exposure to air; polymorphic with *metavivianite.*

Occurrence: Secondary mineral of ore deposits, a weathering product of pegmatite phosphates, in clays associated with bone and other organic remains, and in anaerobic lake sediments.

K. FRYE

References

Am. Min. **55**, 135.
Dana, II, 742–746.

Cross-references: *Human and Vertebrate Mineralogy; Magnetic Minerals; Pegmatite Minerals.*

VLADIMIRITE *Am. Min.* **56**, 639
$Ca_5H_2(AsO_4)_4\cdot 5H_2O$ mono.

VLASOVITE *Am. Min.* **46**, 1202
$Na_2ZrSi_4O_{11}$ mono. and tric.

VOELCKERITE = *Oxyapatite*

VOGLITE tric. *USGS Bull.* **1064**, 126
$Ca_2Cu(UO_2)(CO_3)_4\cdot 6H_2O$

VOLBORTHITE *Am. Min.* **45**, 1307; **59**, 372
$Cu_3(VO_4)_2\cdot 3H_2O$ mono.

VOLKOVSKITE *Am. Min.* **51**, 1550
$(Ca,Sr)B_6O_{10}\cdot 3H_2O$ mono.

VOLTAITE iso. *Dana*, II, 464
$K_2Fe_5^{2+}Fe_4^{3+}(SO_4)_{12}\cdot 18H_2O$

VOLYNSKITE *Am. Min.* **51**, 531
$AgBiTe_2$ orth.

VONSENITE *Am. Min.* **50**, 249
$Fe_2^{2+}Fe^{3+}BO_5$ orth.

VRBAITE *Am. Min.* **53**, 351
$Tl_4Hg_3Sb_2As_8S_{20}$ orth.

VUAGNATITE *Am. Min.* **61**, 825
$CaAl(OH)SiO_4$ orth. *Min. Abst.* **30**, 413

VULCANITE *Am. Min.* **46**, 258
CuTe orth.

VUONNEMITE tric. *Am. Min.* **59**, 875
$Na_4TiNb_2Si_4O_{17}\cdot 2Na_3PO_4$

VYSOTSKITE *Am. Min.* **63**, 832
$(Pd,Ni,Pt)S$ tetr. *Can. Min.* **17**, 453

W

WAD = Psilomelane, an earthy mixture of manganeses oxides and hydroxides

WADEITE Min. Abst. **29**, 20
$K_2Zr(Si_3O_9)$ hexa. RWW, 659

WAGNERITE Min. Abst. **29**, 340
$(Mg,Fe)_2(PO_4)F$
mono. *USGS Prof. Pap.* **955**, 1

Cross-reference: *Soil Mineralogy.*

WAIRAKITE A hydrated calcium aluminosilicate related to the ***zeolite*** group, named by Steiner in 1955 after the original locality, the Wairakei geothermal area in New Zealand.

Composition: $Ca_8[Al_{16}Si_{32}O_{96}]\cdot 16H_2O$; SiO_2, 55.3%; Al_2O_3, 23.5%; CaO, 12.9%; H_2O, 8.3% by wt. *Wairakite* is the Ca analogue of *analcime*; there may be a complete solid-solution series between the two.

Properties: Monoclinic (pseudo-cubic), *I2/a*; a=13.69 Å, b=13.64 Å, c=13.56 Å, β=90.5°; H=5.5–5; G=2.26; luster vitreous; colorless to white; transparent to translucent; twinning on {110}, with {100}. Refractive indices, α=1.498, γ=1.502; birefringence = 0.004; $2V$=70°–105° (+) $\alpha=y$, $\beta=x$, $\gamma=z$; weak dispersion $r>v$. Its optical properties are variable even within a single grain.

Occurrence: In tuffaceous sandstones and breccias, vitric tuffs and ignimbrite altered by alkaline hydrothermal fluids associated with geothermal steam at Wairakei, New Zealand, at a depth having a temperature range of 200°–250°C and a hydrostatic pressure of 55–265 atm; in low-grade metamorphosed tuffs and tuff breccias in Japan; alteration product of effusive rocks in geothermal areas of Japan.

A. IIJIMA
M. UTADA

References

Am. Min. **64**, 993–1001.
J. Japan. Assoc. Min. Petrol. Econ. Geol. **55**, 254–261; **56**, 30–40.
Min. Mag. **30**, 691–701.

Cross-reference: *Minerals, Uniaxial and Biaxial.*

WAIRAUITE Am. Min. **50**, 521
CoFe iso Min. Mag. **33**, 942

WAKABAYASHILITE Am. Min. **57**, 1311
$Sb_2As_{20}S_{36}$ mono. Can. Min. **13**, 418

WAKEFIELDITE Am. Min. **56**, 395
YVO_4 tetr.

WALLISITE Am. Min. **54**, 1497
$PbTl(Cu,Ag)As_2S_5$ tric.

WALPURGITE tric.
$(BiO)_4(UO_2)(AsO_4)_2\cdot 3H_2O$
USGS Bull. **1064**, 239

WALSTROMITE Am. Min. **53**, 9
$BaCa_2Si_3O_9$ tric.

WARDITE Am. Min. **37**, 849
$NaAl_3(PO_4)_2(OH)_4\cdot 2H_2O$ tetr.

Cross-reference: *Pegmatite Minerals.*

WARDSMITHITE Am. Min. **55**, 349
$Ca_5MgB_{24}O_{42}\cdot 30H_2O$ hexa.

WARIKAHNITE tric. Am. Min. **65**, 408
$Zn_3(AsO_4)_2\cdot 2H_2O$ Min. Mag. **43**, 1069

WARWICKITE orth. Am. Min. **59**, 1004
$(Mg,Ti,Fe^{3+},Al)_2(BO_3)O$ Min. Abst. **30**, 116

WATER = Melted *ice*

WATTEVILLITE Dana, II, 452
$Na_2Ca(SO_4)_2\cdot 4H_2O$ orth. or mono.

WAVELLITE A hydrated aluminum hydroxylphosphate named in 1805 after the English physician William Wavell.

Composition: $4[Al_3(PO_4)_2(OH)_3\cdot 5H_2O]$.

Properties: Orthorhombic, *Pcmn*; globular crusts or stalactitic with radial fibers; three cleavages; H=3.5–4; G=2.3; pearly or resinous luster; colorless, yellow, green, brown, or blue.

Occurrence: In low-grade aluminous phosphatic rocks or in veins.

K. FRYE

Reference: *Am. Min.* **51**, 422–428.

Cross-reference: *Soil Mineralogy.*

WAYLANDITE (***Crandalite*** gr.)
$(Bi,Ca)Al_3(PO_4,SiO_4)_2(OH)_6$
trig. *Am. Min.* **48**, 216

WEBERITE *Dana*, II, 127
Na_2MgAlF_7 orth.

WEDDELLITE *Dana*, II, 1101
$CaC_2O_4 \cdot 2H_2O$ tetr.

Cross-references: *Cave Minerals; Human and Vertebrate Mineralogy; Invertebrate and Plant Mineralogy.*

WEEKSITE *Am. Min.* **45**, 39
$K_2(UO_2)_2Si_6O_{15} \cdot 4H_2O$ orth.

WEGSCHEIDERITE *Am. Min.* **48**, 400
$Na_5(CO_3)(HCO_3)_3$ tric.

Cross-reference: *Green River Mineralogy.*

WEHRLITE *Min. Abst.* **30**, 421
Bi_4Te_3 trig.

WEIBULLITE *Am. Min.* **65**, 789
$Pb_5Bi_8Se_6S_{11}$ orth. *Can. Min.* **18**, 1

WEILBERITE (***Beudantite*** gr.)
$BaAl_3(AsO_4)(SO_4)(OH)_6$(?)
trig. *Am. Min.* **52**, 1588

WEILITE *Am. Min.* **49**, 816
$CaHAsO_4$ tric.

WEINSCHENKITE = *Churchite*
Am. Min. **39**, 851

WEISSBERGITE *Am. Min.* **63**, 720
$TlSbS_2$ tric. *Min. Abst.* **30**, 171

WEISSITE *Am. Min.* **34**, 357
Cu_5Te_3 ps. cubic

WELINITE hexa. *Am. Min.* **53**, 1064
$Mn^{4+}Mn_3^{2+}SiO_7$

WELLSITE (***Zeolite*** gr.)
$(Ba,Ca,K_2)(Al_2Si_3)O_{10} \cdot 3H_2O$
mono. DHZ, 4, 387, 399

WELOGANITE *Am. Min.* **54**, 576
$SrZrNa_2(CO_3)_6 \cdot 3H_2O$
ps. trig. *Can. Min.* **13**, 209

WELSHITE (***Aenigmatite*** gr.)
$Ca_2Mg_4Fe^{3+}Sb^{5+}O_2[Si_4Be_2O_{18}]$
Min. Abst. **29**, 201
tric. *Min. Mag.* **42**, 129

WENKITE hexa. *Am. Min.* **48**, 213
$Ba_4Ca_6(Si,Al)_{20}O_{39}(SO_4)_3(OH)_2 \cdot nH_2O$
Min. Abst. **29**, 263

WERMLANDITE hexa. *Am. Min.* **57**, 327
$Ca_2Mg_{14}(Al,Fe)_4(CO_3)(OH)_{42} \cdot 29H_2O$
Lithos **4**, 213

WERNERITE = Intermediate ***scapolite*** composition

Cross-reference: *Naming of Minerals.*

WESTERVELDITE *Am. Min.* **57**, 354
FeAs ortho. *Min. Abst.* **29**, 198

WESTGRENITE = *Bismutomicrolite*
Am. Min. **62**, 408

Cross-reference: *Pegmatite Minerals.*

WHERRYITE A lead copper chlorocarbonate sulfate named in 1950 for the American mineralogist Edgar T. Wherry.

Composition: $Pb_4Cu(CO_3)(SO_4)_2(Cl,OH)_2O$.

Properties: Monoclinic; massive; G=6.45; light green.

Occurrence: A secondary mineral in Pinal Co., Arizona, USA.

K. FRYE

References: *Am. Min.* **35**, 93–98; **55**, 505–508.

WHEWELLITE A hydrated calcium oxalate named in 1852 in honor of the English sicentist and philosopher William Whewell.

Composition: $CaC_2O_4 \cdot H_2O$.

Properties: Monoclinic, $P2_1/n$; equant or short prismatic crystals; one good and one imperfect cleavage; H=2.5–3; G=2.2; vitreous luster; colorless or slightly yellowish or brownish.

Occurrence: Generally associated with coal seams, rarely in veins, also as calculi in human urinary tract.

K. FRYE

References

Am. Min. **53**, 455–463.
Dana, II, 1099–1101.

Cross-references: *Human and Vertebrate Mineralogy; Meteoritic Minerals; Minerals, Uniaxial and Biaxial.*

WHITEITE mono. *Am. Min.* **64**, 465
$Ca(Fe,Mn)Mg_2Al_2(OH)_2(H_2O)_8(PO_4)_4$
Min. Mag. **42**, 309

WHITEITE-(Mn) mono. *Am. Min.* **64**, 465
$Mn(Fe,Mn)Mg_2Al_2(OH)_2(H_2O)_8(PO_4)_4$
Min. Mag. **42**, 309

WHITLOCKITE *Dana*, II, 684
$Ca_9(Mg,Fe)H(PO_4)_7$ trig.

Cross-references: *Cave Minerals; Human and Vertebrate Mineralogy; Lunar Minerals; Metoeritic Minerals; Pegmatite Minerals.*

WHITMOREITE mono. *Am. Min.* **59**, 900
$Fe^{2+}Fe_2^{3+}(PO_4)_2(OH)_2 \cdot 4H_2O$
Can. Min. **17**, 100

WICKENBURGITE hexa. *Am. Min.* **53**, 1433
$Pb_3CaAl_2Si_{10}O_{24}(OH)_6$

WICKMANITE *Am. Min.* **56**, 1488
$MnSn(OH)_6$ iso. *Can. Min.* **15**, 437

WIGHTMANITE mono. *Am. Min.* **59**, 985
$Mg_5(BO_3)O(OH)_5 \cdot 2H_2O$

WIIKITE–Mixture of ***pyrochlores***
Am. Min. **62**, 408

WILKEITE (***Apatite*** gr.)
$Ca_5(SiO_4,PO_4,SO_4)_3(O,OH,F)$
hexa. *Dana*, II, 905

WILKMANITE *Am. Min.* **50**, 519
Ni_3Se_4 mono.

WILLEMITE A zinc silicate of the ***phenakite*** group, named in honor of William I of the Netherlands.

Composition: $18[Zn_2SiO_4]$, with Zn replaced by Mn and Fe.

Properties: Trigonal, $R\bar{3}$; prismatic hexagonal crystals with basal and prismatic cleavage, also massive, as disseminated grains, or fibrous; H=5.5; G=3.9–4.1; vitreous to resinous luster; color: white, yellow, green, red, gray, or brown.

Occurrence: Found with other zinc minerals in zinc-ore deposits.

Use: An ore of zinc.

K. FRYE

References

Min. Mag. **41**, 71–75.
Winchell, II, 497.

Cross-references: *Spinel Group; Wolframite Group.*

WILLEMSEITE *Am. Min.* **55**, 31
$(Ni,Mg)_3Si_4O_{10}(OH)_2$ mono.

WILLYAMITE *Am. Min.* **56**, 361
(Co,Ni)SbS ps. cubic *Min. Abst.* **29**, 198

WINCHITE (***amphibole*** gr.)
$CaNa(Mg,Fe^{2+})_4(Al,Fe^{3+})Si_8O_{22}(OH)_2$
Am. Min. **63**, 1034
mono. *Can. Min.* **18**, 101

WINSTANLEYITE *Am. Min.* **65**, 809
$TiTe_3O_8$ iso *Min. Mag.* **43**, 456

WISERITE *Am. Min.* **45**, 258
$Mn_4B_2O_5(OH,Cl)_4$ tetr.

WITHERITE Barium carbonate with the aragonite structure, named in 1790 for the English physician and mineralogist William Withering.

Composition: $4[BaCO_3]$, with Ba replaced by Sr and Ca.

Properties: Orthorhombic, *Pmcn;* crystals pseudohexagonal pyramids and bipyramids, also globular, botryoidal, columnar, granular, or fibrous; one distinct and one imperfect cleavage; H=3–3.5; G=4.3; vitreous to resinous luster; colorless, milky, or slightly off white.

Occurrence: Low-temperature hydrothermal veins with *barite* and *galena,* an authigenic mineral.

Use: An ore of barium.

K. FRYE

References

Am. Min. **56**, 578–767.
DHZ, 5, 319–322.

Cross-references: *Authigenic Minerals; Green River Mineralogy; Minerals, Uniaxial and Biaxial; Rock-Forming Minerals; Staining Techniques.*

WITTICHENITE *Econ. Geol.* **42**, 147
Cu_3BiS_3 orth. *Min. Mag.* **43**, 109

Cross-reference: *Ore Microscopy.*

WITTITE *Am. Min.* **65**, 789
(Pb,Bi)(S,Se) mono. *Dana*, I, 451

WODGINITE A manganese iron tantalate named after the locality Wodgina, Australia.

Composition: $Mn^{2+}_{2-2x}Fe^{3+}_{3x}Ta_{4-x}O_{12}$, with $0.08 < x < 0.29$.

Properties: Monoclinic, structurally similar to *columbite-tantalite* and *wolframite*.

Occurrence: Found in Ta-rich granitic pegmatites at Wodgina, Australia, and Bernic Lake, Manitoba.

D. H. SPEIDEL

References

Am. Min. **48**, 1417.
Can. Min. **7**, 390–402.

WOEHLERITE A zironium niobium sorosilicate with sodium and calcium.

Composition: $NaCa_2(Zr,Nb)Si_2O_7(O,OH,F)_2$.

Properties: Monoclinic, $P2_1/m$; prismatic or tabular crystals; one cleavage; H-5.5–6; G=3.4; color: yellow to brown.

Occurrence: Found in syenite in the Langesundfiord, Norway; and in Carroll County, New Hampshire, USA.

K. FRYE

Reference

Roberts, W. L., et al., 1974. *Encyclopedia of Minerals.* New York: Van Nostrand Reinhold, 673.

WOLFEITE *Am. Min.* **55**, 135
$(Fe^{2+},Mn)_2(OH)(PO_4)$
mono. *Min. Mag.* **43**, 505

Cross-reference: *Pegmatite Minerals.*

WOLFRAMITE An iron manganese tungstate of the ***wolframite*** group, intermediate in composition between *huebnerite* and *ferberite*. The name is of ancient and uncertain origin.

Composition: $2[(Fe^{2+},Mn)WO_4]$.

Properties: Monoclinic, $P2/c$; prismatic or tabular crystals, lamellar or massively granular; one perfect cleavage; H=4–4.5; G=7.2–7.6; submetallic to metallic-adamantine luster; color: dark grayish or brownish to black.

Occurrence: In *quartz* veins, often associated with tin minerals and commonly related to late-stage granite intrusives, in blacksands.

Use: Major ore of tungsten.

K. FRYE

Reference: *Dana*, II, 1064–1072.

Cross-references: *Blacksand Minerals; Blowpipe Analysis; Metallic Minerals; Mineral and Ore Deposits; Minerals, Uniaxial and Biaxial; Ore Microscopy; Wolframite Group.*

WOLFRAMOIXIOLITE *Am. Min.* **55**, 318
$(Nb,W,Ta,Fe,Mn)_3O_6$ mono.

WOLLASTONITE A calcium silicate named after the British mineralogist and chemist W. H. Wollaston. It is a single chain silicate differing from the pyroxene structure in repeat length.

Composition: $6[CaSiO_3]$; with Ca slightly replaced by Fe and Mn.

Properties: Triclinic; tabular or short prismatic crystals, massive, fibrous, or compact; prismatic cleavage; H=4.5–5; G=2.8–2.9; vitreous to pearly luster; color: white, grayish, yellowish, reddish, or brownish white; polymorphic with *cyclowollastonite* and *parawollastonite*.

Occurrence: Contact metamorphosed limestone or in limestone xenoliths in volcanic rocks.

K. FRYE

References

Am. Min. **58**, 560; **63**, 274.
DHZ, 2A, 547–563.

Cross-references: *Meteoritic Minerals; Polysomatism; Pyroxenoids; Rock-Forming Minerals; Topotaxy.*

WÖLSENDORFITE *Am. Min.* **42**, 919
$(Pb,Ca)U_2O_7 \cdot 2H_2O$ orth. *Min. Mag.* **41**, 51

WONESITE (***Mica*** gr.)
$(Na,K)_{1-1.5}(Mg,Fe,Al)_6(Si_{7-6.5}Al_{1-1.5})$
Am. Min. **66**, 100

WOODHOUSEITE (***Beudantite*** gr.)
$CaAl_3(PO_4)(SO_4)(OH)_6$ *Dana*, II, 1006
trig. *Min. Abst.* **27**, 255

WOODRUFFITE mono. *Am. Min.* **38**, 761
$(Zn,Mn^{2+})Mn^{4+}_3O_7 \cdot 1\text{–}2H_2O$

WOOD OPAL—Opal with woody texture

WOOD TIN—Variety of *cassiterite*
Am. Min. **62**, 100

WOODWARDITE *Am. Min.* **62**, 599
$Cu_4Al_2(SO_4)(OH)_{12} \cdot 2H_2O(?)$
Min. Mag. **40**, 644

WROEWOLFEITE *Am. Min.* **61**, 179
$Cu_4(SO_4)(OH)_6 \cdot 2H_2O$
mono. *Min. Mag.* **40**, 893

WULFENITE A lead molybdate of the *scheelite* group, named for the Austrian mineralogist Franz Xaver Wülfen.

Composition: $4[PbMoO_4]$, with some substitution of W for Mo and Ca for Pb.

Properties: Tetragonal; crystals commonly square tabular, less commonly pseudo-octahedral, prismatic, or cuboidal; prismatic cleavage; H=2.75–3; G=6.5–7.0; resinous to adamantine luster; color: yellow, grayish white, green, brown, yellow brown.

Occurrence: A secondary mineral found in the oxidized zone of lead-molybdenum deposits in association with *pyromorphite, vanadinite, mimetite, cerussite,* wad, limonite, and *calcite.*

K. FRYE

Reference: *Dana,* II, 1081–1086.

Cross-references: *Gossan; Museums, Mineralogical.*

WURTZITE A zinc sulfide named for the French chemist Adolphe Wurtz.

Composition: $2[ZnS]$ with up to 40% Fe.

Properties: Hexagonal, $P6_3mc$; pyramidal or short prismatic crystals, tabular, fibrous, or banded crusts; prismatic cleavage; H=3.5–4; G=3.98; resinous luster; brownish black; polymorphic with *sphalerite* and *matraite.*

Occurrence: Commonly intergrown with *sphalerite* and other sulfides.

K. FRYE

Reference: *Dana,* I, 227–228.

Cross-references: *Green River Mineralogy; Skeletal Crystals.*

WÜSTITE *Dana,* I, 503
FeO iso.

Cross-references: *Magnetic Minerals; Mantle Mineralogy.*

WYARTITE orth. *Am. Min.* **45**, 200
$Ca_3U^{4+}(UO_2)_6(CO_3)_2(OH)_{18} \cdot 3\text{–}5H_2O$

WYLLIEITE *Am. Min.* **59**, 280
$Na_2Fe_2^{2+}Al(PO_4)_3$ mono. *Min. Rec.* **4**, 131

Cross-reference: *Pegmatite Minerals.*

X

XANTHIOSITE *Am. Min.* **50**, 2108
$Ni_3(AsO_4)_2$ mono. *Min. Mag.* **35**, 72

XANTHOCONITE *Dana*, I, 371
Ag_3AsS_3 mono.

XANTHOPHYLLITE = *Clintonite*
Am. Min. **52**, 1122

XANTHOSIDERITE = *Goethite*

XANTHOXENITE
$Ca_4Fe_2^{3+}(OH)_2(H_2O)_3(PO_4)_4$
mono. or tric. *Am. Min.* **64**, 466

Cross-reference: *Pegmatite Minerals.*

XENOTIME Yttrium phosphate, named by Beudent in 1832 from the Greek words for "vain" and "honor" in allusion to the fact that the yttrium in it had been mistaken for a new element.

Composition: YPO_4, with yttrium sometimes substituted by rare earth elements, especially erbium; may contain uranium or thorium.

Properties: Tetragonal; crystals mostly prismatic, resembling *zircon* in habit; H=4–5; G=4.25 (for pure YPO_4), vitreous, pale colored, commonly yellowish; rarely metamict; uniaxial (+), ω=1.721; ϵ=1.816; isostructural with *zircon* and *thorite.*

Occurrence: Widespread as a minor accessory in granitic rocks, also in larger crystals in granite pegmatites; a common detrital mineral. Associated with *monazite, zircon, rutile*, and *magnetite*.

A. PABST

References: *Am. Min.* **32**, 141–145; **34**, 830–834.

Cross-references: *Blacksand Minerals; Pegmatite Minerals; Pleochroic Halos.*

XIANGJIANGITE Am. Min. **64**, 466
$(Fe,Al)(UO_2)_4(PO_4)_2(SO_4)_2(OH)\cdot 22H_2O$
pseudo. tetr. *Min. Abst.* **29**, 483

XINGZHONGITE iso.(?) *Am. Min.* **61**, 185
(Ir,Cu,Rh)S

XOCOMECATLITE *Am. Min.* **61**, 504
$Cu_3(TeO_4)(OH)_4$ orth. *Min. Mag.* **40**, 221

XONOTLITE A calcium hydroxylsilicate named for the location Tetela de Xonotla, Mexico.

Composition: $Ca_6Si_6O_{17}(OH)_2$.

Properties: Monoclinic, $P2/a$; massive, in matted fibers, or in needles; one cleavage; H=6.5; G=2.7; color: white, gray, or pink.

Occurrence: In serpentinite or contact zones as in the Coast Ranges, California, USA; and Boso peninsula, Japan.

K. FRYE

References

Am. Min. **38**, 860–862.
Min. Abst. **29**, 191.

Y

YAGIITE (***Osumilite*** gr.)
$(Na,K)_3 Mg_4 (Al,Mg)_6 (Si,Al)_{24} O_{60}$
hexa. *Am. Min.* **54**, 14

Cross-reference: *Meteoritic Minerals.*

YAROSLAVITE *Am. Min.* **51**, 1546, 1820
$Ca_3Al_2F_{10}(OH)_2 \cdot H_2O$ orth.

YARROWITE *Can. Min.* **18**, 511
Cu_9S_8 hexa.

YAVAPAIITE *Am. Min.* **44**, 1105
$KFe^{3+}(SO_4)_2$ mono.

YEATMANITE tric. *Am. Min.* **62**, 396; **65**, 196
$(Mn_5^{2+}Sb_2^{5+})(Mn_2^{2+}Zn_8 Si_4)O_{28}$

YEDLINITE *Am. Min.* **59**, 1157
$Pb_6CrCl_6(O,OH)_8$ trig.

YFTISITE orth. *Am. Min.* **62**, 396
$(Y,REE)_4(F,OH)_6(Ti,Sn)O(SiO_4)_2$(?)

YIXUNITE = Indum-bearing *platinum*
Am. Min. **65**, 408

YODERITE mono. *Am. Min.* **45**, 753
$(Mg,Al)_8 Si_4 (O,OH)_{20}$ *Min. Mag.* **32**, 782

YOFORTIERITE mono.(?) *Am. Min.* **61**, 341
$Mn_5 Si_8 O_{20} (OH)_2 (H_2 O)_4 \cdot 4\text{-}5H_2 O$
Can. Min. **13**, 68

YOSHIMURAITE
$(Ba,Sr)_2 TiMn_2 (SiO_4)_2 (PO_4,SO_4)(OH,Cl)$
tric. *Am. Min.* **45**, 479; **46**, 1515

YTTRIALITE–Variety of
thalenite(?) *Am. Min.* **58**, 545

YTTROBETAFITE (***Pyrochlore*** gr.)
$(Y,REE)_2 (Ti,Nb,Ta)_2 O_6 (OH)$
iso. *Am. Min.* **62**, 405

YTTROCERITE = Cerian *fluorite*

YTTROCOLUMBITE *Am. Min.* **25**, 155
$(Y,U,Fe)(Nb,Ta)O_4$ orth.

YTTROCRASITE (***Euxenite*** gr.)
$(Y,Th,Ca,U,REE)(Ti,Fe^{3+},W)_2 (O,OH)_6$
orth. *Am. Min.* **62**, 1009

YTTROFLUORITE = Rare-earth *fluorite*

YTTROMICROLITE (***Pyrochlore*** gr.)
$(Y,Ca)(Ta,Nb)_2 O_6 (OH)$ *Am. Min.* **64**, 890
iso. *Min. Abst.* **31**, 231

YTTROPYROCHLORE (***Pyrochlore*** gr.)
$(Y,Na,Ca,U)(Nb,Ta,Ti)_2 (O,OH)_7$
iso. *Am. Min.* **62**, 405

YTTROTANTALITE mono.(?) *Dana*, I, 763
$(Y,Fe,U,Ca)(Ta,Nb)O_4$

YTTROTUNGSTITE *Am. Min.* **36**, 641
$YW_2O_6(OH)_3$ mono(?)

YUGAWARALITE A hydrated calcium aluminosilicate of the ***zeolite*** group, named by Sakurai and Hayashi in 1952 after the original locality near Yugawara Hot Spring in central Japan.

Composition: $Ca_2 [Al_4Si_{12}O_{32}] \cdot 8H_2O$.

Properties: Monoclinic, *Pc*; a=6.73 Å, b=13.95 Å, c=10.03 Å, β=111°30'; H=4.5; G=2.2; luster vitreous; colorless to white; transparent to translucent; habit flat; cleavage poor on {010}. Refractive indices, α=1.495, β=1.497, γ=1.504; birefringence = 0.009; $2V$=56°–89° (+); optic axial plane ⊥ (010). A very small amount of Na may substitute for Ca; dehydration stepwise; not attacked by HCl.

Occurrence: Originally reported from Yugawara and afterward from several other places in Japan, in network veins in the altered Miocene andesite tuff, associated with *chabazite, laumontite, wairakite*, and *calcite*; in low-grade metamorphic tuffs in the Tanzawa Mountain-land in central Japan; from Heinabersjokull, southeastern Iceland.

A. IIJIMA
M. UTADA

References

Acta Cryst. **B25**, 1183–1190.
Am. Min. **50**, 484–489.
J. Japan. Assoc. Min. Petrol. Econ. Geol. **56**, 107–111.
Sakurai, K., and Hayashi, A., 1952. Yugawaralite, a new zeolite, *Sci. Rep. Yokohama National Univ.*, sec. II, no. 1, 69–77.
Z. Krist. **130**, 88–99.

YUKONITE
$Ca_3Fe_7^{3+}(AsO_4)_6(OH)_9 \cdot 18H_2O$(?)
amorph. *Dana*, II, 953

YUKSPORITE *Am. Min.* **62**, 1262
$(Na,K)(Ca,Sr,Ba)(Ti,Al,Fe)_{0.6}Si_2O_4(F,Cl)_{0.5} \cdot H_2O$

Z

ZAHERITE *Am. Min.* **62**, 1125
$Al_{12}(SO_4)_5(OH)_{26} \cdot 20H_2O$
Min. Abst. **29**, 343

ZAIRITE (***Cranadallite*** gr.)
$Bi(Fe,Al)_3(PO_4)_2(OH)_6$ *Am. Min.* **62**, 174
trig. *Min. Abst.* **27**, 339

ZAPATALITE tetr. *Am. Min.* **57**, 1911
$Cu_3Al_4(PO_4)_3(OH)_9 \cdot 4H_2O$ *Min. Mag.* **38**, 541

ZARATITE A hydrated nickel hydroxyl-carbonate named after the Spaniard Sr. Zarate.

Composition: $Ni_3(CO_3)(OH)_4 \cdot 4H_2O$.

Properties: Isometric; H=3; G=2.6; emerald green; massive or mammilliform crusts.

Occurrence: A secondary mineral in mafic or ultramafic rocks associated with other carbonates and hydroxides.

K. FRYE

Reference: *Dana*, II, 245–246.

ZAVARITSKITE *Am. Min.* **48**, 210
BiOF tetr.

ZEKTZERITE *Am. Min.* **62**, 416; **63**, 304
$LiNa(Zr,Ti,Hf)Si_6O_{15}$ orth.

ZELLERITE *Am. Min.* **51**, 1567
$Ca(UO_2)(CO_3)_2 \cdot 5H_2O$ orth.

ZEMANNITE hexa. *Am. Min.* **55**, 1448
$(Zn,Fe)_2(TeO_3)_3(Na_xH_{2-x}) \cdot nH_2O$
Can. Min. **10**, 139

ZEOLITE GROUP–*Amicite, barrerite, bikitaite, brewsterite, chabazite, clinoptilolite, cowlesite, dachiardite, edingtonite, epistilbite, erionite, faujasite, ferrierite, garronite, gismondine, gmelinite, gonnardite, harmotome, herschellite, heulandite, laumontite, levyne, mazzite, mertinoite, mesolite, mordenite, natrolite, offretite, paranatrolite, paulingite, phillipsite, scolecite, stellerite, stilbite, svetlozarite, thomsonite, wairakite, wellsite*, and *yugawaralite*.

Cross-references: *Authigenic Minerals; Collecting Minerals; Diagenetic Minerals; Mineral Classification: Principles; Museums, Mineralogical; Naming of Minerals; Soil Mineralogy; Synthetic Minerals; Vein Minerals; Zeolite Group.*

ZEOPHYLLITE *Min. Mag.* **31**, 726
$Ca_4Si_3O_7(OH,F)_6$ tric.

ZEUNERITE (***Autunite*** gr.)
$Cu(UO_2)_2(AsO_4)_2 \cdot 10\text{–}16H_2O$
tetr. *Am. Min.* **42**, 905

ZHEMCHUZHNIKOVITE
$NaMg(Al,Fe)(C_2O_4)_3 \cdot 8H_2O$
trig. *Am. Min.* **49**, 442

ZIESITE *Am. Min.* **65**, 1146
$\beta\text{-}Cu_2V_2O_7$ mono.

ZINALSITE = *FRAIPONTITE*(?)
$Zn_7Al_4(SiO_4)_6(OH)_2 \cdot 9H_2O$(?)
mono.(?) *Am. Min.* **44**, 208

ZINC *Can. Min.* **6**, 692
Zn hexa. *Min. Abst.* **18**, 200

Cross-reference: *Native Elements and Alloys.*

ZINCALUMINITE *Dana*, II, 579
$Zn_6Al_6(SO_4)_2(OH)_{26} \cdot 5H_2O$ hexa.(?)

ZINC BLENDE = *Sphalerite*

ZINCITE Zinc oxide named for its composition.

Composition: ZnO.

Properties: Hexagonal, *C6mc;* würtzite structure; usually foliated massive; H=4; G=5.7; color: deep red; streak: orange yellow.

Occurrence: Found at Franklin, New Jersey, USA, associated with *franklinite, willemite,* and *smithsonite.*

D. H. SPEIDEL

Reference: *Dana,* I, 504–506.

Cross-references: *Crystal Growth; Ore Microscopy.*

ZINCKENITE = *Zinkenite*

ZINC-MELANTERITE *Dana,* II, 508
$(Zn,Cu,Fe)SO_4 \cdot 7H_2O$ mono.

ZINCOBOTRYOGEN
$(Zn,Mg,Mn)Fe^{3+}(SO_4)_2(OH) \cdot 7H_2O$
mono. *Am. Min.* **49**, 1776

ZINCOCOPIAPITE tric. *Am. Min.* **49**, 1777
$ZnFe_4^{3+}(SO_4)_6(OH)_2 \cdot 18H_2O$

ZINCROSASITE *Am. Min.* **44**, 1323
$(Zn,Cu)_2(CO_3)(OH)_2$ mono.

ZINCSILITE mono.(?) *Am. Min.* **46**, 241
$Zn_3Si_4O_{10}(OH)_2 \cdot 4H_2O$

ZINC-ZIPPEITE orth. *Can. Min.* **14**, 429
$Zn_2(UO_2)_6(SO_4)_3(OH)_{10} \cdot 16H_2O$
Min. Mag. **43**, 1069

ZINKENITE *Dana,* I, 476
$PbSb_4S_7$ hexa. *Min. Abst.* **29**, 401

Cross-reference: *Ore Microscopy.*

ZINNWALDITE A lithium iron ***mica***-group mineral named for the locality Zinnwald, Germany.

Composition:
$K_2(LiFe^{2+}Al)_2(AlSi_3)_2O_{20}(F,OH)_4$.

Properties: Monoclinic; basal cleavage; H= 2.5–4; G=2.90–3.02; color: pale violet, yellowish or grayish brown.

Occurrence: In granites and pegmatites with *cassiterite, topaz, lepidolite, spodumene, beryl,* ***tourmaline,*** and *monazite.*

K. FRYE

Reference: DHZ, 3, 92–94.

Cross-references: *Pegmatite Minerals; Phyllosilicates; Soil Mineralogy.*

ZIPPEITE *Can. Min.* **14**, 429
$K_4(UO_2)_6(SO_4)_3$ *USGS Bull.* **1064**, 141
$(OH)_{10} \cdot 16H_2O$

ZIRCON Zirconium silicate, named by A. G. Werner in 1783.

Composition: $ZrSiO_4$; Hf and Th substitute for Zr to some extent.

Properties: Tetragonal, $I4_1/amd$; H-7.5; G= 4.68–4.70; adamantine luster; colorless, also pale yellowish, yellowish green; etc.; mostly in small prismatic crystals, some highly modified; cleavage not conspicuous; uniaxial (+) high birefringence, ω=1.923–1.960. Many *zircons* are metamict, are optically isotropic, and have much lower density, ca. 4.0. Names used for gem varieties include hyacinth for reddish translucent gems and jargon for colorless or smoky gem *zircons.*

Occurrence: One of the most widespread accessory minerals in granitic and intermediate igneous rocks, also in large crystals in pegmatites. Commonly concentrated as a placer mineral in sands with other heavy, resistant minerals. Associated with ***apatite,*** *titanite,* and *magnetite.*

A. PABST

References

Acta. Cryst. 8, 291–300.
AJS **253,** 433–461; **254,** 521–554.
Sedimentology, **10,** 209–216.

Cross-references: *Blacksand Minerals; Color in Minerals; Crystal Habits; Crystallography: Morphological; Gemology; Lunar Minerals; Metamict State; Meteoritic Minerals; Mineral Classification: Principles; Mineral and Ore Deposits; Minerals, Uniaxial and Biaxial; Placer Deposits; Pleochroic Halos, Soil Mineralogy.*

ZIRCONOLITE = *ZIRKELITE*
Am. Min. **60**, 341

ZIRCOPHYLLITE (***Astrophyllite*** gr.)
$(K,Na,Ca)_3(Mn,Fe^{3+})_7(Zr,Nb)_2Si_8O_{27}(OH,F)_4$
tric. *Am. Min.* **58,** 967

ZIRCOSULFATE *Am. Min.* **51**, 529
$Zr(SO_4)_2 \cdot 4H_2O$ orth.

ZIRKELITE mono. *Am. Min.* **60**, 341;
(Ca,Th,*REE*)$Zr(Ti,Nb,Fe)_2O_7$ **62**, 408

ZIRKLERITE trig. *Dana,* II, 87
$(Fe,Mg)_9Al_4Cl_{18}(OH)_{12} \cdot 14H_2O$(?)

ZIRKOPHYLLITE = *Zircophyllite*

ZIRSINALITE trig. *Am. Min.* **60**, 489
$Na_6(Ca,Mn,Fe)ZrSi_6O_{18}$

ZOISITE A calcium aluminum hydroxyl-silicate of the ***epidote*** group, named after Baron von Zois from whom Werner received his first specimen.

Composition: $Ca_2Al_3(Si_2O_7)(SiO_4)O(OH)$; CaO, 25–26%; Al_2O_3, 32–35%; SiO_2, 37–

39%; H_2O, 1–2% by wt; small amounts of Mn may substitute for Ca giving the pink thulite variety; Fe^{3+} may substitute for Al.

Properties: Orthorhombic, *Pnmc*; H=6; G=3.2–3.4; vitreous luster; color: gray or greenish brown (manganese-bearing variety thulite is pink); habit often acicular with elongation along *b* axis; cleavage; perfect on {100}, imperfect on {001}; white streak. Optically, refractive indices, α=1.685–1.705, β=1.688–1.710, γ=1.697–1.725; $2v_\alpha$=0°–60°; straight extinction; may show anomalous blue interference colors; the optic axial plane is parallel to (100) in α-zoisites and parallel to (010) in β-zoisites, which are often the iron-bearing variety; thulite has pleochroism, α is pale pink, β is nearly colorless; and γ is pale yellow.

Occurrence: Medium-grade regional metamorphic rocks, particularly in rocks of pelitic calcareous sandstone composition and in calcareous shales; less commonly in lower-grade thermally metamorphosed impure limestones; regionally metamorphosed mafic igneous rocks; occasionally as a primary mineral in hornblende sclogite; in hydrothermal alteration (saussuritization) of calcic ***plagioclase***. Associated with *calcite*, sericite, *albite*, *prehnite*, ***amphibole***, etc.

Use: Rare occurrences have been reported of clear transparent blue crystals of gem quality.

R. A. HOWIE

References

Am. Min. **27**, 519, 638.
Geol. Mag. **86**, 43.

Cross-references: *Epidote Group; Gemology; Vein Minerals.*

ZORITE orth. *Am. Min.* **58**, 1113
$Na_2Ti(Si,Al)_3O_9 \cdot nH_2O$

ZUNYITE A complex aluminosilicate named for the Zuni mine, San Juan Co., Colorado, USA.

Composition: $Al_{13}Si_5O_{20}(OH,F)_{18}Cl$.

Properties: Isometric, $F\bar{4}3m$; minute transparent tetrahedra; H-7; G-2.88.

Occurrence: Found in aluminous shales in Postmasburg, South Africa; in the USA, San Juan Co., and Ouray Co., Colorado.

K. FRYE

Reference: *Am. Min.* **37**, 960–965.

ZUSSMANITE trig. *Am. Min.* **61**, 470
$KFe_{13}Si_{17}AlO_{42}(OH)_{14}$ *Min. Mag.* **43**, 605

ZVYAGINTSEVITE *Am. Min.* **52**, 299, 1587
$(Pd,Pt,Au)_3(Pb,Sn)$ iso. *Can. Min.* **8**, 541

ZWIESELITE *Dana*, II, 849
$(Fe^{2+},Mn)_2(PO_4)F$ mono.

Cross-reference: *Pegmatite Minerals.*

ZYKAITE orth. *Am. Min.* **63**, 1284
$Fe_4^{3+}(AsO_4)_3(SO_4)(OH) \cdot 15H_2O$
Min. Abst. **29**, 483

AUTHOR CITATION INDEX

Authors cited in the Mineral Glossary are not included in this index.

SUBJECT INDEX

Entries in capital letters represent article titles; boldface numbers indicate page(s) of main articles. Valid mineral species are set in italic type; series or group names are set in boldface italic type; varietal names, synonyms, and discredited names are set in roman type.